Second Edition

Signals and Systems Simplified

for Anna University

ECE Course

Other CBS Titles by *A Nagoor Kani*

- A Simplified Text in Electrical Machine Design, 3/e
- Advanced Control Theory, 3/e
- Circuit Analysis, 2/e (*Anna University/ECE*)
- Circuit Theory, 2/e (*Anna University/EEE*)
- Control Systems, 5/e
- Control Systems (*Anna University/EEE*)
- Control Systems Engineering (*Anna University/ECE*)
- Control Systems Engineering, 2/e
- Design of Electrical Apparatus (*Anna University/EEE*)
- Microprocessors and Microcontrollers, 2/e (*Anna University/ECE*)
- Microprocessors and Microcontrollers (*Anna University/EEE*)
- Power System Analysis
- 8085 Microprocessor and its Applications, 4/e
- 8086 Microprocessor and its Applications, 3/e
- Digital Electronics (*in Press*)
- Digital Logic Circuits (*in Press*)
- Digital Electronics (*Anna University/ECE*) (*in Press*)
- Digital Logic Circuits (*Anna University/EEE*) (*in Press*)
- Digital Principles and System Design (*in Press*)
- Digital Principles and System Design (*Anna University/CSE/IT*) (*in Press*)
- Digital Signal Processing, 3/e (*in Press*)
- Digital Signal Processing (*Anna University/EEE*) (*in Press*)
- Discrete Time Signal Processing (*Anna University/ECE*) (*in Press*)
- Electric Circuits, 2/e (*in Press*)
- Microprocessors and Microcontrollers, 3/e (*in Press*)

Second Edition

Signals and Systems Simplified

for Anna University

ECE Course

A Nagoor Kani

Founder, RBA Educational Group
Chennai

CBSPD

CBS Publishers & Distributors Pvt Ltd

New Delhi • Bengaluru • Chennai • Kochi • Kolkata • Lucknow • Mumbai
Hyderabad • Jharkhand • Nagpur • Patna • Pune • Uttarakhand

Second Edition

Signals and Systems Simplified

for Anna University
ECE Course

ISBN: 978-93-5466-101-3

Second Edition: 2022
Repirnt: 2023, **2025**
First Edition: 2018

Published by **Satish Kumar Jain** and produced by **Varun Jain** for

CBS Publishers and Distributors Pvt Ltd

4819/XI Prahlad Street, 24 Ansari Road, Daryaganj, New Delhi 110 002, India.
Ph: 011-23289259, 23266838   Website: www.cbspd.com
                                              e-mail: delhi@cbspd.com;

Corporate Office: 204 FIE, Industrial Area, Patparganj, Delhi 110 092, India
Ph: 011-4934 4934   Fax: 011-4934 4935   e-mail: publishing@cbspd.com;
                                              publicity@cbspd.com

Branches

- **Bengaluru:** Seema House 2975, 17th Cross, K.R. Road, Banasankari 2nd Stage, Bengaluru 560 070, Karnataka, India
  Ph: +91-80-26771678/79   Fax: +91-80-26771680   e-mail: bangalore@cbspd.com
- **Chennai:** 7, Subbaraya Street, Shenoy Nagar, Chennai 600 030, Tamil Nadu, India.
  Ph: +91-44-26680620, 26681266   Fax: +91-44-42032115   e-mail: chennai@cbspd.com
- **Kochi:** 42/1325, 36, Power House Road, Opposite KSEB, Kochi-682 018, Kerala, India
  Ph: +91-484-4059061-67   Fax: +91-484-4059065   e-mail: kochi@cbspd.com
- **Kolkata:** 147, Hind Ceramics Compound, 1st Floor, Nilgunj Road, Belghoria, Kolkata-700 056, West Bengal, India
  Ph: +91-33-25633055-56   e-mail: kolkata@cbspd.com
- **Lucknow:** Basement, Khushnuma Complex, 7-Meerabai Marg (Behind Jawahar Bhawan), Lucknow-226 001, Uttar Pradesh, India
  Ph: +91-522-4000032   e-mail:tiwari.lucknow@cbspd.com
- **Mumbai:** PWD Shed, Gala No. 25/26, Ramchandra Bhatt Marg, Next to JJ Hospital Gate No. 2,
  Opp Union Bank of India, Noorbaug, Mumbai-400009, Maharashtra, India
  Ph: +91-22-66661880, 66661889   e-mail: mumbai@cbspd.com

Representatives

• **Hyderabad**	0-9885175004	• **Jharkhand**	0-9811541605	• **Nagpur**	0-8692091830
• **Patna**	0-9334159340	• **Pune**	0-9664372571	• **Uttarakhand**	0-9716462459

Printed at: Mudrak, Noida, UP, India.

to

Er T A Benazir

PREFACE

The main objective of this book is to explore the basic concepts of signals and systems in a simple and easy-to-understand manner.

This text on signals and systems has been crafted and designed to meet students' requirements. Considering the highly mathematical nature of this subject, more emphasis has been given on the problem-solving methodology. Considerable effort has been made to elucidate mathematical derivations in a step-by-step manner. Exercise problems with varied difficulty levels are given in the text to help students get an intuitive grasp on the subject.

This book with its lucid writing style and germane pedagogical features will prove to be a master text for engineering students and practitioners.

Salient Features

- Proof of important concepts and theorems are clearly highlighted by shaded boxes
- Wherever required, problems are solved in multiple methods
- Additional explanations for solutions and proofs are provided in separate boxes
- Different types of fonts are used for text, proof and solved problems for better clarity
- Keywords are highlighted by bold, italic fonts
- Easy, concise and accurate study material
- Extremely precise edition where concepts are reinforced by pedagogy
- Demonstration of multiple techniques in problem solving, additional explanations and proofs highlighted
- Ample figures and examples to enhance students' understanding
- Practice through MCQ's

Pedagogy

- Solved Numerical Examples: $44 + 46 + 35 + 38 + 59 = 222$ (excluding a, b, c, d,...)
- Short-Answer Questions: $27 + 32 + 18 + 20 + 35 = 132$
- Figures: $58 + 33 + 18 + 13 + 22 = 144$ (numbered figures excluding figures in problems and QA)
- Exercise Numerical Problems: $21 + 21 + 22 + 12 + 20 = 96$ (exercise problems excluding a, b, c, d,...)
- Review Questions: $40 + 35 + 18 + 15 + 32 = 140$
- MCQs: $35 + 48 + 13 + 18 + 18 = 132$
- Fill in the blanks and True/False: $55 + 46 + 21 + 20 + 31 = 173$

Organization

In this book, the concepts of signals and systems are organized in five chapters. Each chapter provides the foundations and practical implications of their own topic with large number of solved numerical examples and illustrative figures for better understanding. The important concepts are summarized at the end of each chapter which can help in quick reference. Another significant aspect of this book is that, it contains MATLAB based computer exercises for each chapter with complete explanation, which will be of great assistance to both instructor and student.

Chapter 1 starts with a general introduction about various types of signals, systems and their importance in real life. Basic definitions of signals, their mathematical representation, significance of their frequency domain analysis and usage of MATLAB in this course are presented in brief manner.

Chapter 1 explores various standard continuous time and discrete time signals, classifications of continuous time and discrete time signals and possible mathematical operations on signals such as amplitude and time scaling, folding, time shifting, addition, multiplication, etc. The concept of generation of discrete time signals is also presented in this chapter. The classification and properties of continuous time and discrete time systems are also presented in chapter 1 with appropriate examples.

Chapter 2 deals with analysis of continuous time signals using Fourier series, Fourier transform and Laplace transform. Chapter 2 starts with Fourier analysis of continuous time signals which forms the basics for frequency domain analysis. Fourier series in both trigonometric and exponential forms, Fourier coefficients of various signals with symmetry, properties of Fourier series, frequency spectrum using Fourier series and Gibbs phenomenon have been discussed. Then the Chapter 2 explains the development of Fourier transform from Fourier series, Fourier transform of some standard signals, various properties of Fourier transform and frequency spectrum via Fourier transform. The proof of properties of Fourier transform of continuous time signal are also presented with clear steps.

Also, Chapter 2 discusses about analysis of continuous time signals using Laplace transform. The properties of Laplace transform and proof with clear steps are presented. The rational functions of 's' and their representation in terms of poles and zeros, region of convergence of Laplace transform and its properties are presented in a crisp and clear manner. Further, chapter 2 discusses about the inverse Laplace transform using partial fraction method and convolution theorem. The relation between Fourier transform and Laplace transform of continuous time signals is also discussed in chapter 2.

Chapter 3 deals with analysis of Linear Time Invariant (LTI) continuous time systems in time domain, frequency domain and s-domain. The differential equation governing the LTI continuous time system in time domain and their direct solutions in time domain are discussed with examples. The time domain convolution operation that can be used to find the response of LTI system from its impulse response is explained with clear numerical examples. Another important thing in chapter 3 is that the graphical convolution operation of continuous time signals is discussed by clearly separating the shift index and time index that will aid in clear understanding.

Also, the Chapter 3 deals with detailed analysis of continuous time systems in s-domain using Laplace transform. The transfer function in s-domain, impulse response, response for specific inputs, convolution and deconvolution operations using Laplace transform are presented with appropriate numerical examples. The stability of the LTI systems in s-domain via Laplace transform is dealt clearly.

In addition, the Chapter 3 deals with analysis of continuous time systems in frequency domain using Fourier transform. The transfer function of continuous time system in frequency domain, impulse response, response for specific inputs and convolution using Fourier transform are presented with appropriate numerical examples. The computation of frequency response of continuous time LTI systems using Fourier transform is also explained.

The standard realization structures for the continuous time systems characterized by differential equations are also presented in chapter 3.

Chapter 4 deals with analysis of discrete time signals using discrete time fourier transform (DTFT) and $\mathcal{Z}$-transform. Chapter 4 starts with Fourier transform analysis of discrete time signals which forms the basics for frequency domain analysis. The frequency spectrum, various properties of Fourier transform and Fourier transform of some standard signals are presented. The proof of properties of DTFT are presented with clear steps. The concept of sampling and aliasing in frequency spectrum are also discussed.

Also, Chapter 4 discusses about analysis of discrete time signals using $\mathcal{Z}$-transform. The properties of $\mathcal{Z}$-transform and proof with clear steps are presented. The rational functions of 'z' and their representation in terms of poles and zeros, region of convergence of $\mathcal{Z}$-transform and its properties are presented in detail with appropriate examples. Further, chapter 4 discusses about the various methods of inverse $\mathcal{Z}$-transform. The relation between Fourier transform and $\mathcal{Z}$-transform of discrete time signals is also discussed in chapter 4.

Chapter 5 deals with analysis of discrete time systems in time domain, frequency domain and z-domain. The difference equation governing the LTI discrete time system in time domain and their direct solutions in time domain are discussed with examples. The time domain discrete convolution operation that can be used to find the response of LTI discrete time system from its impulse response is explained with clear numerical examples. The graphical convolution operation of discrete time signals is illustrated with figures for each step that will aid in clear understanding.

Also, Chapter 5 deals with analysis of discrete time systems in frequency domain using Fourier transform. The transfer function of discrete time system in frequency domain, impulse response, response for specific inputs and convolution using Fourier transform are presented with appropriate numerical examples The computation of frequency response of discrete time LTI systems using Fourier transform is also explained with examples.

In addition, Chapter 5 deals with detailed analysis of discrete time systems in z-domain using $\mathcal{Z}$-transform. The transfer function in z-domain, impulse response, response for specific inputs, convolution and deconvolution operations using $\mathcal{Z}$-transform are presented with appropriate numerical examples. The stability of the LTI systems in z-domain via $\mathcal{Z}$-transform is dealt clearly.

Also, Chapter 5 focuses on structures for realization of discrete time systems with special attention to IIR and FIR systems.

A Nagoor Kani

ACKNOWLEDGEMENTS

I express my heartfelt thanks to my wife Ms C Gnanaparanjothi Nagoor Kani and my sons N Bharath Raj alias Chandrakani Allaudeen and N Vikram Raj for the support, encouragement and cooperation they have extended to me throughout my career.

I thank Ms T A Benazir for the affection and care on my day-to-day activities.

It's my pleasure to acknowledge the contributions of our technical editors, Ms. E R Suhasini, Ms M A Aswathy, Ms C Mohana Priya, Ms Devi Asokan and Ms L Sharmila for editing, proof-reading and type-setting of the manuscript and preparing the layout of the book.

I thank all my office staff for their support and help in carrying my office work.

My sincere thanks to all reviewers for their valuable suggestions and comments which helps me to explore the subject to greater depth.

I am also grateful to Mr Satish K Jain, CMD, CBS Publishers & Distributors, for his keen interest in publishing this work in CBS banner. My sincere thanks to all team members of CBS Publishers & Distributors, for their concern and care in publishing this work.

Finally, a special note of appreciation is due to my sisters, brothers, relatives, friends, students and the entire teaching community for their overwhelming support and encouragement to my writing.

A Nagoor Kani

CONTENTS

Contents

CHAPTER 4: ANALYSIS OF DISCRETE TIME SIGNALS   4.1–4.104

CHAPTER 5: LINEAR TIME INVARIANT DISCRETE TIME SYSTEMS   **5.1–5.170**

LIST OF SYMBOLS AND ABBREVIATIONS

Symbols

a_o, a_n, b_n	Fourier coefficients of trignometric form of Fourier series of $x(t)$
B	Bandwidth in Hz
c_n	Fourier coefficients of exponential form of Fourier series of $x(t)$
E	Energy of a signal
f	Frequency of discrete time signal in Hz/sample
F	Frequency of continuous time signal in Hz
F_o	Fundamental frequency of continuous time signal in Hz
F_m	Maximum frequency of continuous time signal
F_s	Sampling frequency of continuous time signal in Hz
$\mathcal{H}$	System operator
j	Complex operator, $\sqrt{-1}$
L	Inductance
$n\Omega_o$	Harmonic angular frequency, where $n = 1,2,3,...$
P	Power of a signal
p	Pole
R	Resistor
s	Complex frequency ($s = \sigma + j\Omega$)
t	Time in seconds
T	Time period in seconds
W	Phase factor or twiddle factor
z	Complex variable ($z = u + jv$)
z	Unit advance operator or zero
z^{-1}	Unit delay operator

Ω	Angular frequency of continuous time signal in rad/sec
Ω_o	Fundamental angular frequency
Ω_{max}	Maximum angular frequency in rad/sec
ω	Angular frequency of discrete time signal
ω_k	Sampling frequency point
σ	Neper frequency (Real part of s)
$*$	Convolution operator
$\oint$	Integration operator
$\dfrac{d}{dt}$	Differentiation operator

Standard/Input/Output Signals

h(n)	Impulse response of discrete time system
h(t)	Impulse response of continuous time system
sgn(t)	Signum signal
sinc (t)	Sinc signal
u(n)	Discrete time unit step signal
u(t)	Continuous time unit step signal
x(n)	Discrete time signal
x(n)	Input of discrete time system
$x_o(n)$	Odd part of discrete time signal x(n)
$x_e(n)$	Even part of discrete time signal x(n)
x(n–m)	Delayed or linearly shifted x(n) by m units
x(t)	Continuous time signal or Input of continuous time system
$x_o(t)$	Odd part of continuous time signal x(t)

$x_e(t)$	Even part of continuous time signal $x(t)$
$x(t{-}m)$	Delayed or linearly shifted $x(t)$ by m units
$y(n)$	Output/Response of discrete time system
$y_{zs}(n)$	Zero state response of discrete time system
$y_{zi}(n)$	Zero input response of discrete time time system
$y(t)$	Output /Response of continuous time system
$y_{zs}(t)$	Zero state response of continuous time system
$y_{zi}(t)$	Zero input response of continuous time system
$\delta(t)$	Continuous time impulse signal
$\delta(n)$	Discrete time impulse signal
$\Pi(t)$	Unit pulse signal

Transform Operators and Functions

$\mathcal{F}$	Fourier transform
$\mathcal{F}^{-1}$	Inverse Fourier transform
$H(s)$	Laplace transform of $h(t)$

$\mathcal{L}$	Laplace transform
$\mathcal{L}^{-1}$	Inverse Laplace transform
$X(e^{j\omega})$	Discrete time fourier transform of $x(n)$
$X_r(e^{j\omega})$	Real part of $X(e^{j\omega})$
$X_i(e^{j\omega})$	Imaginary part of $X(e^{j\omega})$
$X(j\Omega)$	Fourier transform of $x(t)$
$X(s)$	Laplace transform of $x(t)$
$X(z)$	$\mathcal{Z}$-transform of $x(n)$
$\mathcal{Z}$	$\mathcal{Z}$-transform
$\mathcal{Z}^{-1}$	Inverse $\mathcal{Z}$-transform

Abbreviations

BIBO	**Bounded Input Bounded Output**
CT	**Continuous Time**
CTFS	**Continuous Time Fourier Series**
CTFT	**Continuous Time Fourier Transform**
DT	**Discrete Time**
DTFT	**Discrete Time Fourier Transform**
FIR	**Finite Impulse Response**
IIR	**Infinite Impulse Response**
LHP	**Left Half Plane**
LTI	**Linear Time Invariant**
RHP	**Right Half Plane**
ROC	**Region of Convergence**

CHAPTER 1

Classification of Signals and Systems

1.1 Introduction to Signals and Systems

1.1.1 Signals

Any physical phenomenon that conveys or carries some information can be called a signal. The music, speech, motion pictures, still photos, heart beat, etc., are examples of signals that we normally encounter in day to day life.

Usually, the information carried by a signal will be a function of an independent variable. The independent variable can be time, spatial coordinates, intensity of colours, pressure, temperature, etc. The most popular independent variable in signals is time and it is represented by the letter "t".

The value of a signal at any specified value of the independent variable is called its *amplitude*. The sketch or plot of the amplitude of a signal as a function of independent variable is called its *waveform*.

Mathematically, any signal can be represented as a function of one or more independent variables. Therefore, *a signal is defined as any physical quantity that varies with one or more independent variables*.

Table 1.1: Examples of Signals

Basis for Classification	Type	Definition	Example
Number of sources	One-channel signals	Signals that are generated by a single source or sensor	i) Record of room temperature. ii) Audio output of monospeaker.
	Multi-channel signals	Signals that are generated by multiple sources or speaker	i) Record of ECG at eight different places in a human body. ii) Audio output of two stereo speakers.
Number of dimensions	One-dimensional signals	Signal which is a function of single independent variable	i) Music, Speech and heart beat which are function of single independent variable, time. ii) $x_1(t) = 0.7t$.
	Multi-dimensional signals	Signal which is a function of two or more independent variables	i) Photograph is 2D signal. ii) Motion picture of a black and while TV is a 3D signal.

Table 1.1: Continued...

Basis for Classification	Type	Definition	Example
Whether the dependent variable is continuous or discrete	Analog or continuous signal	Signal which is defined continuously for any value of independent variable is called analog signal. When independent variable of analog signal is time it is called continuous time signal.	Most of the signals encountered in science and engineering are analog.
	Discrete signal	Signal which is defined for discrete intervals of indepedent variable is called discrete signal. When the indepedent variable of discrete signal is time it is called discrete time signal.	Sampled version of analog signal.

1.1.2 Systems

Any process that exhibits cause and effect relation can be called a ***system***. A system will have an input signal and an output signal. The output signal will be a processed version of the input signal. A system is either interconnection of hardware devices or software / algorithm.

A system is denoted by letter $\mathcal{H}$. The diagrammatic representation of a system is shown in Fig 1.1.

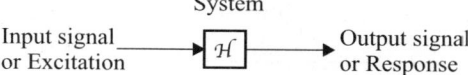

Fig 1.1: *Representation of a system.*

The operation performed by a system on input signal to produce output signal can be expressed as,

$$\text{Output} = \mathcal{H}\{\text{Input}\}$$

where $\mathcal{H}$ denotes the system operation (also called ***system operator***).

The systems can be classified in many ways.

Depending on type of energy used to operate the systems, the systems can be classified into Electrical systems, Mechanical systems, Thermal systems, Hydraulic systems, etc.

Depending on the type of input and output signals, the systems can be classified into Continuous time systems and Discrete time systems.

1.1.3 Frequency Domain Analysis of Continuous Time Signals and Systems

Physically, we realize any signal or system in time domain. In time domain, the continuous time systems are governed by differential equations. The analysis of continuous time signals and systems in time domain involves solution of differential equations. The solution of differential equations are difficult due to assumption of a solution and then solving the constants using initial conditions.

In order to simplify the task of analysis, the signals can be transformed to some other domain, where the analysis is easier. One such transform exists for continuous time signals is Laplace transform. The ***Laplace transform***, will transform a function of time "t" into a function of complex frequency "s" where $s = \sigma + j\Omega$. Therefore, Laplace transform of a continuous time signal will transform the time domain signal into s-domain signal.

On taking Laplace transform of the differential equation governing the system, it becomes algebraic equation in "s" and the solution of algebraic equation will give the response of the system as a function of "s" and it is called s-domain response. The inverse Laplace transform of the s-domain response, will give the time domain response of the continuous time system. Also, the stability analysis of the continuous time systems are much easier in s-domain.

Another important characteristics of any signal is frequency, and for most of the applications the frequency content of the signal is an important criteria. The frequency contents of a signal can be studied by taking Fourier transform of a signal. The Fourier transform of a signal is a particular class of Laplace transform in which $s = j\Omega$, where "Ω" is real frequency.

The ***Fourier transform***, will transform a function of time "t" into a function of real frequency "Ω". Therefore, Fourier transform of a continuous time signal will transform the time domain signal into frequency domain signal. From the Fourier transform of a continuous time signal, the frequency spectrum of the signal can be obtained which is used to study the frequency content of a signal. The frequency range of some of the signals are listed in Table 1.2 and 1.3.

Table 1.2: Frequency Range of Some Biological and Seismic Signals

Type of Signal	Frequency Range (Hz)
Electroretinogram	0 to 20
Electronystagmogram	0 to 20
Pneumogram	0 to 40
Electrocardiogram (ECG)	0 to 100
Electroencephalogram (EEG)	0 to 100
Electromyogram	10 to 200
Sphygmomanogram	0 to 200
Speech	100 to 4000
Wind noise	100 to 1000
Seismic exploration signals	10 to 100
Earthquake and nuclear explosion signals	0.01 to 10
Seismic noise	0.1 to 1

Table 1.3: Frequency Range of Some Electromagnetic Signals

Type of Signal	Wavelength (m)	Frequency Range (Hz)
Radio broadcast	10^4 to 10^2	3×10^4 to 3×10^6
Shortwave radio signals	10^2 to 10^{-2}	3×10^6 to 3×10^{10}
Radar / Space communications	1 to 10^{-2}	3×10^8 to 3×10^{10}
Common-carrier microwave	1 to 10^{-2}	3×10^8 to 3×10^{10}
Infrared	10^{-3} to 10^{-6}	3×10^{11} to 3×10^{14}
Visible light	3.9×10^{-7} to 8.1×10^{-7}	3.7×10^{14} to 7.7×10^{14}
Ultraviolet	10^{-7} to 10^{-8}	3×10^{15} to 3×10^{16}
Gamma rays and x-rays	10^{-9} to 10^{-10}	3×10^{17} to 3×10^{18}

1.1.4 Frequency Domain Analysis of Discrete Time Signals and Systems

Mostly, the discrete time systems are designed for analysis of discrete time signals. Physically, the discrete time systems are also realized in time domain. In time domain, the discrete time systems are governed by difference equations. The analysis of discrete time signals and systems in time domain involves solution of difference equations. The solution of difference equations are difficult due to assumption of a solution and then solving the constants using initial conditions.

In order to simplify the task of analysis, the discrete time signals can be transformed to some other domain, where the analysis may be easier. One such transform exists for discrete time signals is $\mathcal{Z}$-transform. The $\mathcal{Z}$-transform, will transform a function of discrete time "n" into a function of complex variable "z" where, $z = re^{j\omega}$. Therefore, $\mathcal{Z}$-*transform* of a discrete time signal will transform the time domain signal into z-domain signal.

On taking $\mathcal{Z}$-transform of the difference equation governing the discrete time system, it becomes algebraic equation in "z" and the solution of algebraic equation will give the response of the system as a function of "z" and it is called z-domain response. The inverse $\mathcal{Z}$-transform of the z-domain response, will give the time domain response of the discrete time system. Also, the stability analysis of the discrete systems are much easier in z-domain.

The frequency contents of a discrete time signal can be studied by taking Fourier transform of the discrete time signal. The Fourier transform of discrete time signal is a particular class of $\mathcal{Z}$-transform in which $z = e^{j\omega}$, where "ω" is the frequency of the discrete time signals.

The Fourier transform, will transform a function of discrete time "n" into a function of frequency "ω". Therefore, Fourier transform of a discrete time signal will transform the discrete time signal into frequency domain signal. From the Fourier transform of the discrete time signal, the frequency spectrum of the discrete time signal can be obtained which is used to study the frequency content of the discrete time signal.

1.1.5 Importance of Signals and Systems

Every part of the universe is a system which generates or processes some type of signal [of course the universe itself is a system and said to be controlled by signals (or commands) issued by God].

The signals and systems play a vital role in almost every field of Science and Engineering. Some of the applications of signals and systems in various field of Science and Engineering are listed here.

1. Biomedical

❖ ECG is used to predict heart diseases.

❖ EEG is used to study normal and abnormal behaviour of the brain.

❖ EMG is used to study the condition of muscles.

❖ X-ray images are used to predict the bone fractures and tuberclosis.

❖ Ultrasonic scan images of kidney and gall bladder are used to predict stones.

❖. Ultrasonic scan images of foetus are used to predict abnormalities in a baby.

❖ MRI scan is used to study minute inner details of any part of the human body.

2. Speech Processing

❖ Speech compression and decompression to reduce memory requirement of storage systems.

❖ Speech compression and decompression for effective use of transmission channels.

❖ Speech recognization for voice operated systems and voice based security systems.

❖ Speech recognization for conversion of voice to text.

❖ Speech synthesis for various voice based warnings or annoucements.

3. Audio and Video Equipments

❖ The analysis of audio signals will be useful to design systems for special effects in audio systems like stereo, woofer, karoke, equalizer, attenuator, etc.

❖ Music synthesis and composing using music keyboards.

❖ Audio and video compression for storage in DVDs.

4. Communication

❖ The spectrum analysis of modulated signals helps to identify the information bearing frequency component that can be used for transmission.

❖ The analysis of signals received from radars are used to detect flying objects and their velocity.

❖ Generation and detection of DTMF signals in telephones.

❖ Echo and noise cancellation in transmission channels.

5. Power Electronics

❖ The spectrum analysis of the output of converters and inverters will reveal the harmonics present in the output, which in turn helps to design suitable filter to eliminate the harmonics.

❖ The analysis of switching currents and voltages in power devices will help to reduce losses.

6. Image Processing

❖ Image compression and decompression to reduce memory requirement of storage systems.

❖ Image compression and decompression for effective use of transmission channels.

❖ Image recognization for security systems.

❖ Filtering operations on images to extract the features or hidden information.

7. Geology

❖ The seismic signals are used to determine the magnitude of earthquake and volconic eruptions.

❖ The seismic signals are also used to predict nuclear explosions.

❖ The seismic noise are also used to predict the movement of earth layers (tectonic plates).

8. Astronomy

❖ The analysis of light received from a star is used to determine the condition of the star.

❖ The analysis of images of various celestial bodies gives vital information about them.

1.1.6 Use of MATLAB in Signals and Systems

The *MATLAB* (**MAT**rix **LAB**oratory) is a software developed by The MathWork Inc, USA, which can run on any windows platform in a PC (Personal Computer). This software has number of tools for the study of various engineering subjects. It includes a tool for signal processing also. Using this tool a wide variety of studies can be made on signals and systems. Some of the analysis that is relevant to this particular text book are given below:

❖ Sketch or plot of signals as a function of independent variable.

❖ Spectrum analysis of signals.

❖ Solution of LTI systems.

❖ Perform convolution and deconvolution operations on signals.

❖ Perform various transforms on signals like Laplace transform, Fourier transform, Z-transform, Fast Fourier Transform (FFT), etc.

❖ Determination of state model from transfer function and viceversa.

❖ Stability analysis of signals and systems in various domains.

1.2 Continuous Time Signals

In a signal with time as independent variable, if the signal is defined continuously for any value of the independent variable time "t", then the signal is called *continuous time signal*. The continuous time signal is denoted as "x(t)".

The continuous time signal is defined for every instant of the independent variable time and so the magnitude (or the value) of continuous time signal is continuous in the specified range of time. Here both the magnitude of the signal and the independent variable are continuous.

1.2.1 Standard Continuous Time Signals

(AU May'13, 10 Marks)

1. Impulse Signal

(AU Dec'13, 2 Marks)

The impulse signal is a signal with infinite magnitude and zero duration, but with an area of A. Mathematically, impulse signal is defined as,

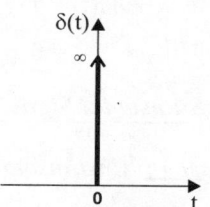

$$\text{Impulse Signal, } \delta(t) = \infty \, ; \; t = 0 \qquad \text{and} \qquad \int_{-\infty}^{+\infty} \delta(t)\, dt = A$$

$$= 0 \, ; \; t \neq 0$$

The unit impulse signal is a signal with infinite magnitude and zero duration, but with unit area. Mathematically, unit impulse signal is defined as,

Fig 1.2: Impulse signal (or Unit Impulse signal).

$$\text{Unit Impulse Signal, } \delta(t) = \infty \, ; \; t = 0 \quad \text{and} \quad \int_{-\infty}^{+\infty} \delta(t)\, dt = 1$$

$$= 0 \, ; \; t \neq 0$$

2. Step Signal

The step signal is defined as,

$$x(t) = A \, ; \, t \geq 0$$
$$= 0 \, ; \, t < 0$$

The unit step signal is defined as,

$$u(t) = 1 \, ; \, t \geq 0$$
$$= 0 \, ; \, t < 0$$

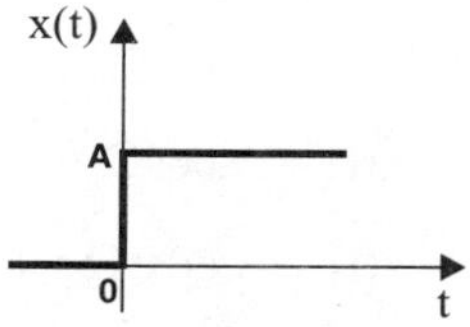

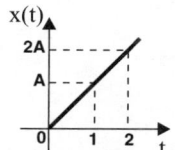

Fig 1.3: Step signal.

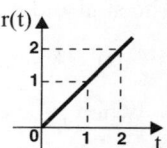

Fig 1.4: Unit step signal.

3. Ramp Signal

The ramp signal is defined as,

$$x(t) = At \, ; \; t \geq 0$$
$$= 0 \; ; \; t < 0$$

The unit ramp signal is defined as,

$$r(t) = t \; ; \; t \geq 0$$
$$= 0 \; ; \; t < 0$$

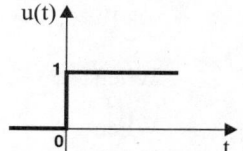

Fig 1.5: Ramp signal.

Fig 1.6: Unit ramp signal.

4. Parabolic Signal

The parabolic signal is defined as,

$$x(t) = \frac{At^2}{2} \quad ; \; \text{for } t \geq 0$$
$$= 0 \quad ; \qquad t < 0$$

The unit parabolic signal is defined as,

$$x(t) = \frac{t^2}{2} \quad ; \; \text{for } t \geq 0$$
$$= 0 \quad ; \qquad t < 0$$

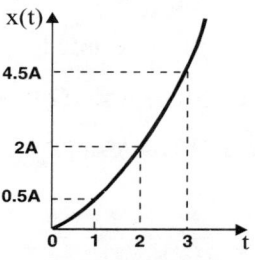

Fig 1.7: Parabolic signal.

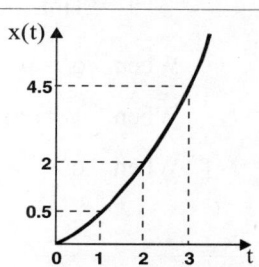

Fig 1.8: Unit parabolic signal.

5. Unit Pulse Signal

The unit pulse signal is defined as,

$$\Pi(t) = 1 \quad ; \quad -0.5 < t < 0.5$$
$$= 0 \quad ; \quad t < -0.5 \text{ and } t > 0.5$$

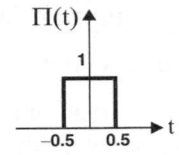

Fig 1.9: Unit pulse signal.

6. Sinusoidal Signal

Case i : Cosinusoidal Signal

The cosinusoidal signal is defined as,

$$x(t) = A \cos(\Omega_0 t + \phi)$$

where,

$$\Omega_0 = 2\pi F_0 = \frac{2\pi}{T} = \text{Angular frequency in } rad/sec$$

F_0 = Frequency in cycles/sec or *Hz*

T = Time period in *sec*

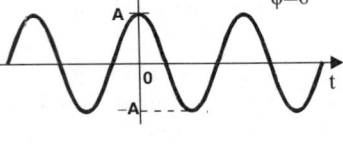

When $\phi = 0,$ $x(t) = A \cos\Omega_0 t$

When $\phi = $ Positive, $x(t) = A \cos(\Omega_0 t + \phi)$

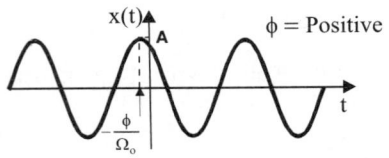

When $\phi = $ Negative, $x(t) = A \cos(\Omega_0 t - \phi)$

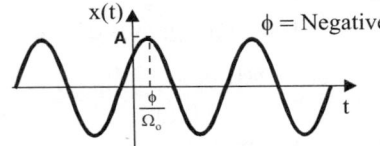

Fig 1.10: Cosinusoidal signal.

Case ii : Sinusoidal Signal

The sinusoidal signal is defined as,

$$x(t) = A \sin(\Omega_0 t + \phi)$$

where,

$$\Omega_0 = 2\pi F_0 = \frac{2\pi}{T} = \text{Angular frequency in } rad/sec$$

F_0 = Frequency in cycles/sec or *Hz*

T = Time period in *sec*

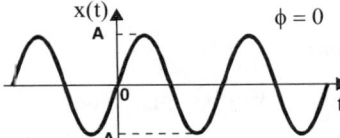

When $\phi = 0,$ $x(t) = A \sin\Omega_0 t$

When $\phi = $ Positive, $x(t) = A \sin(\Omega_0 t + \phi)$

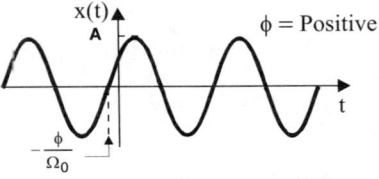

When $\phi = $ Negative, $x(t) = A \sin(\Omega_0 t - \phi)$

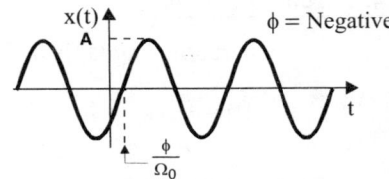

Fig 1.11: Sinusoidal signal.

7. Exponential Signal   (AU Dec'13, 6 Marks)

Case i : Real exponential signal

The real exponential signal is defined as,

$$x(t) = A\,e^{bt} \quad \text{where, A and b are real}$$

Here, when b is positive, the signal x(t) will be an exponentially rising signal; and when b is negative the signal x(t) will be an exponentially decaying signal.

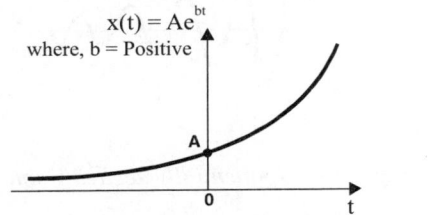 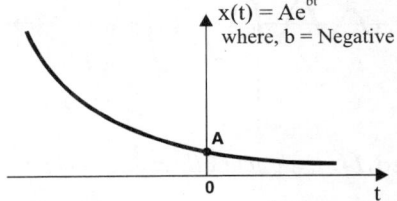

Fig 1.12: *Real exponential signal.*

Case ii : Complex exponential signal

The complex exponential signal is defined as,

$$x(t) = Ae^{j\Omega_0 t}$$

$$\text{where,} \quad \Omega_0 = 2\pi F_0 = \frac{2\pi}{T} = \text{Angular frequency in } rad/sec$$

$$F_0 = \text{Frequency in cycles/sec or } Hz$$

$$T = \text{Time period in } sec$$

The complex exponential signal can be represented in a complex plane by a rotating vector, which rotates with a constant angular velocity of Ω_0 *rad/sec.*

The complex exponential signal can be resolved into real and imaginary parts as shown below:

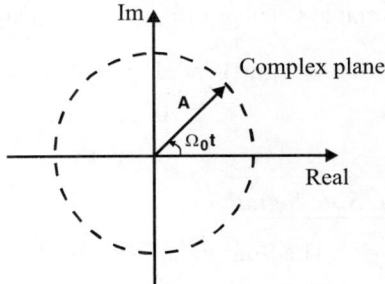

$$x(t) = Ae^{j\Omega_0 t} = A(\cos\Omega_0 t + j\sin\Omega_0 t)$$

$$= A\cos\Omega_0 t + jA\sin\Omega_0 t$$

$$\therefore \ A\cos\Omega_0 t = \text{Real part of } x(t)$$

$$A\sin\Omega_0 t = \text{Imaginary part of } x(t)$$

Fig 1.13: *Complex exponential signal.*

From the above equation, we can say that a complex exponential signal is the vector sum of two sinusoidal signals of the form $\cos\Omega_0 t$ and $\sin\Omega_0 t$.

8. Exponentially Rising/Decaying Sinusoidal Signal

The exponential rising/decaying sinusoidal signal is defined as,

$$x(t) = Ae^{bt}\sin\Omega_0 t$$

$$\text{where,} \quad \Omega_0 = 2\pi F_0 = \frac{2\pi}{T} = \text{Angular frequency in } rad/sec$$

$$F_0 = \text{Frequency in cycles/sec or } Hz$$

$$T = \text{Time period in } sec$$

Here, A and b are real constants. When b is positive, the signal x(t) will be an exponentially rising sinusoidal signal; and when b is negative, the signal x(t) will be an exponentially decaying sinusoidal signal.

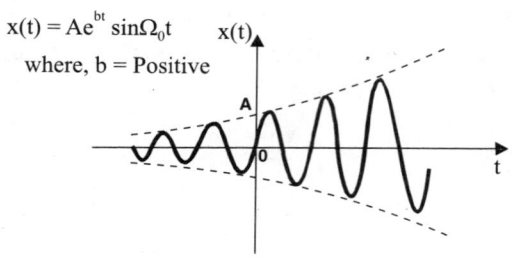

$x(t) = Ae^{bt} \sin\Omega_0 t$

where, b = Positive

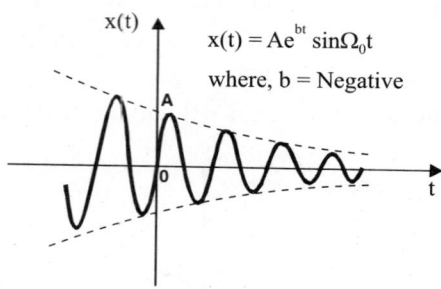

$x(t) = Ae^{bt} \sin\Omega_0 t$

where, b = Negative

Fig 1.14: Exponentially rising sinusoid.   *Fig 1.15: Exponentially decaying sinusoid.*

9. Triangular Pulse Signal

The Triangular pulse signal is defined as

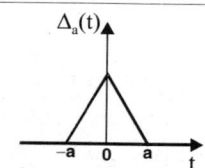

$$\Delta_a(t) = 1 - \frac{|t|}{a} \quad ; \quad |t| \le a$$
$$= 0 \quad\quad ; \quad |t| > a$$

Fig 1.16: Triangular pulse signal.

10. Signum Signal

The Signum signal is defined as the sign of the independent variable t. Therefore, the Signum signal is expressed as,

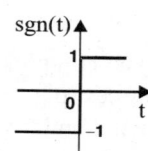

$$sgn(t) = 1 \quad ; \quad t \ge 0$$
$$= 0 \quad ; \quad t = 0$$
$$= -1 \quad ; \quad t \le 0$$

Fig 1.17: Signum signal.

11. Sinc Signal

The Sinc signal is defined as,

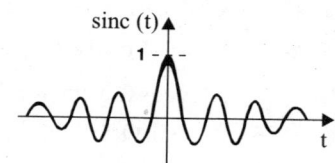

$$sinc(t) = \frac{\sin t}{t} \quad ; \quad -\infty < t < \infty$$

Fig 1.18: Sinc signal.

12. Gaussian Signal

The Gaussian signal is defined as,

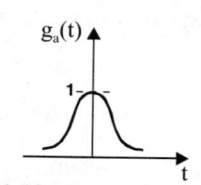

$$g_a(t) = e^{-a^2 t^2} \quad ; \quad -\infty < t < \infty$$

Fig 1.19: Gaussian signal.

1.2.2 Properties of Impulse Signal

Property - 1: $\displaystyle\int_{-\infty}^{+\infty}\delta(t)\,dt = 1$

Proof:

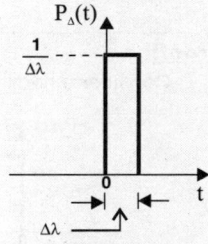

Consider a narrow pulse signal, $P_\Delta(t)$ of width $\Delta\lambda$ and height $1/\Delta\lambda$ as shown in Fig 1.20.

Now the pulse signal is defined as,

$$P_\Delta(t) = \frac{1}{\Delta\lambda} \quad ; \; 0 \le t \le \Delta\lambda$$
$$= 0 \quad ; \; t > \Delta\lambda$$

Now the impulse signal can be represented as,

$$\delta(t) = \underset{\Delta\lambda\to 0}{Lt}\; P_\Delta(t)$$

On integrating the above equation we get,

$$\int_{-\infty}^{+\infty}\delta(t)\,dt = \int_{-\infty}^{+\infty}\underset{\Delta\lambda\to 0}{Lt}\;P_\Delta(t)\,dt = \underset{\Delta\lambda\to 0}{Lt}\int_{0}^{\Delta\lambda}\frac{1}{\Delta\lambda}\,dt = \underset{\Delta\lambda\to 0}{Lt}\frac{1}{\Delta\lambda}\int_{0}^{\Delta\lambda}dt$$

$$= \underset{\Delta\lambda\to 0}{Lt}\frac{1}{\Delta\lambda}[t]_{0}^{\Delta\lambda} = \underset{\Delta\lambda\to 0}{Lt}\frac{1}{\Delta\lambda}[\Delta\lambda - 0] = \underset{\Delta\lambda\to 0}{Lt}\; 1 = 1$$

Fig 1.20.

Property - 2: $\displaystyle\int_{-\infty}^{+\infty}x(t)\,\delta(t)\,dt = x(0)$

Proof:

$$\int_{-\infty}^{+\infty}x(t)\,\delta(t)\,dt = \int_{-\infty}^{+\infty}x(0)\,\delta(t)\,dt$$

$$= x(0)\int_{-\infty}^{+\infty}\delta(t)\,dt = x(0)\times 1 = x(0)$$

> Since $\delta(t)$ is nonzero only at $t=0$, $x(t)$ is replaced by $x(0)$.

> Since $x(0)$ is constant, it is taken outside integration.

> Using property-1

Property - 3: $\displaystyle\int_{-\infty}^{+\infty}x(t)\,\delta(t - t_0)\,dt = x(t_0)$

Proof:

$$\int_{-\infty}^{+\infty}x(t)\,\delta(t - t_0)\,dt = \int_{-\infty}^{+\infty}x(t_0)\,\delta(t - t_0)\,dt$$

$$= x(t_0)\int_{-\infty}^{+\infty}\delta(t - t_0)\,dt = x(t_0)\times 1 = x(t_0)$$

> Since $\delta(t_0)$ is nonzero only at $t=t_0$, $x(t)$ is replaced by $x(t_0)$.

> Since $x(t_0)$ is constant, it is taken outside integration.

> Using property-1

Property - 4: $\displaystyle\int_{-\infty}^{+\infty}x(\lambda)\,\delta(t - \lambda)\,d\lambda = x(t)$

Proof:

Consider the property-3 of impulse signal, $\displaystyle\int_{-\infty}^{+\infty}x(t)\,\delta(t - t_0)\,dt = x(t_0)$

On substituting $t = \lambda$ in the above equation we get, $\displaystyle\int_{-\infty}^{+\infty}x(\lambda)\,\delta(\lambda - t_0)\,d\lambda = x(t_0)$

On substituting $t_0 = t$ in the above equation we get, $\displaystyle\int_{-\infty}^{+\infty} x(\lambda)\,\delta(\lambda - t)\,d\lambda = x(t)$

Since impulse signal is even, $\delta(\lambda-t) = \delta(t-\lambda)$, Therefore the above equation is written as shown below:

$$\int_{-\infty}^{+\infty} x(\lambda)\,\delta(t - \lambda)\,d\lambda = x(t)$$

Property - 5: $\delta(at) = \dfrac{1}{|a|}\,\delta(t)$

Proof:

Consider a narrow pulse signal, $P_\Delta(t)$ of width $\Delta\lambda$ and height $1/\Delta\lambda$ as shown in Fig 1.21(a).

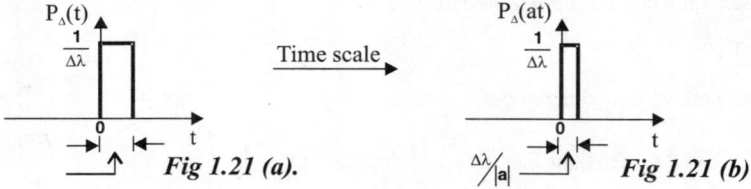

Fig 1.21 (a).   *Fig 1.21 (b).*

Fig 1.21: *A pulse signal and its time scaled version.*

Now the signal, $P_\Delta(at)$ will be time scaled version of signal $P_\Delta(t)$ as shown in Fig 1.21(b). Now the pulse signal and time scaled pulse signal can be mathematically defined as,

$$P_\Delta(t) = \frac{1}{\Delta\lambda} \quad ; 0 \le t \le \Delta\lambda \qquad\qquad P_\Delta(at) = \frac{1}{\Delta\lambda} \quad ; 0 \le t \le \frac{\Delta\lambda}{|a|}$$
$$\qquad = 0 \quad ; t > \Delta\lambda \qquad\qquad\qquad\qquad = 0 \quad ; t > \frac{\Delta\lambda}{|a|}$$

$$\therefore \int_{-\infty}^{+\infty} P_\Delta(t)\,dt = \int_0^{\Delta\lambda} \frac{1}{\Delta\lambda}\,dt \qquad\qquad\qquad\qquad\qquad\qquad\qquad\qquad .....(1.1)$$

$$\int_{-\infty}^{+\infty} P_\Delta(at)\,dt = \int_0^{\Delta\lambda/|a|} \frac{1}{\Delta\lambda}\,dt = \frac{1}{|a|}\int_0^{\Delta\lambda} \frac{1}{\Delta\lambda}\,dt = \frac{1}{|a|}\int_{-\infty}^{+\infty} P_\Delta(t)\,dt \quad \boxed{\text{Using equation (1.1)}} \;\;.....(1.2)$$

Now the impulse signal and time scaled impulse signal can be represented as,

$$\delta(t) = \underset{\Delta\lambda \to 0}{\mathrm{Lt}}\; P_\Delta(t) \qquad \text{and} \qquad \delta(at) = \underset{\Delta\lambda \to 0}{\mathrm{Lt}}\; P_\Delta(at)$$

On integrating the time scaled impulse signal we get,

$$\boxed{\text{Using equation (1.2)}}$$

$$\int_{-\infty}^{+\infty} \delta(at)\,dt = \int_{-\infty}^{+\infty} \underset{\Delta\lambda \to 0}{\mathrm{Lt}}\; P_\Delta(at)\,dt = \underset{\Delta\lambda \to 0}{\mathrm{Lt}} \int_{-\infty}^{+\infty} \underset{\Delta\lambda \to 0}{\mathrm{Lt}}\; P_\Delta(at)\,dt = \underset{\Delta\lambda \to 0}{\mathrm{Lt}} \frac{1}{|a|}\int_{-\infty}^{+\infty} P_\Delta(t)\,dt$$

$$= \frac{1}{|a|}\int_{-\infty}^{+\infty} \underset{\Delta\lambda \to 0}{\mathrm{Lt}}\; P_\Delta(t)\,dt = \frac{1}{|a|}\int_{-\infty}^{+\infty} \delta(t)\,dt \qquad \boxed{\begin{array}{l}\text{Using definition of}\\ \text{impulse signal}\end{array}}$$

$$\therefore \int_{-\infty}^{+\infty} \delta(at)\,dt = \frac{1}{|a|}\int_{-\infty}^{+\infty} \delta(t)\,dt$$

On differentiating the above equation we get,

$$\delta(at) = \frac{1}{|a|}\,\delta(t)$$

1.2.3 Representation of a Continuous Time Signal as Integral of Impulses

Let x(t) be a continuous time signal as shown in Fig 1.22.

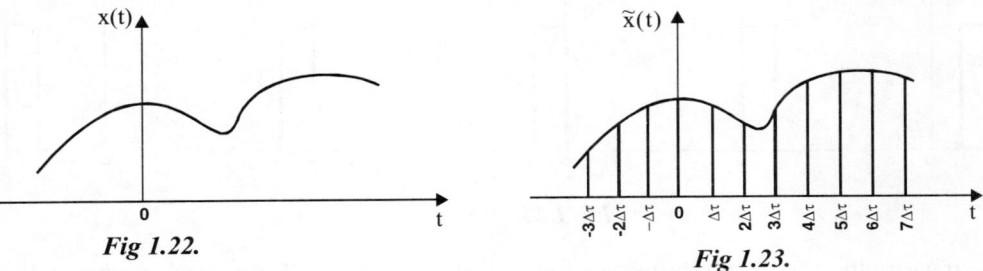

Fig 1.22.   *Fig 1.23.*

Let us divide x(t) as narrow pulses of width $\Delta\tau$ as shown in Fig 1.23.

Now the signal x(t) can be expressed as,

$$x(t) = \underset{\Delta\tau \to 0}{Lt} \, \tilde{x}(t)$$

Each narrow pulse of Fig 1.23 can be interpretted as shown below:

$$\vdots$$

$$
\begin{aligned}
x(-2\,\triangle\tau) &= \tilde{x}(t) \; ; \quad \text{for} -2\,\triangle\tau < t < -\triangle\tau \\
x(-\triangle\tau) &= \tilde{x}(t) \; ; \quad \text{for} -\triangle\tau < t < 0 \\
x(0) &= \tilde{x}(t) \; ; \quad \text{for } 0 < t < \triangle\tau \\
x(\triangle\tau) &= \tilde{x}(t) \; ; \quad \text{for } \triangle\tau < t < 2\,\triangle\tau \\
x(2\,\triangle\tau) &= \tilde{x}(t) \; ; \quad \text{for } 2\,\triangle\tau < t < 3\,\triangle\tau
\end{aligned}
$$

$$\vdots$$

$$\therefore \; x(t) = \underset{\Delta\tau \to 0}{Lt} \, \tilde{x}(t)$$

$$= \underset{\Delta\tau \to 0}{Lt} \, \left[.....x(-2\,\triangle\tau) + x(-\triangle\tau) + x(0) + x(\triangle\tau) + x(2\,\triangle\tau) + ... \right] \qquad .....(1.3)$$

Consider the pulse signal of width $\Delta\tau$ and height $1/\Delta\tau$ as shown in Fig 1.24. This pulse signal can be expressed as,

$$
\begin{aligned}
P_\triangle(t) &= \frac{1}{\Delta\tau} \; ; \quad 0 \le t \le \Delta\tau \\
&= 0 \quad ; \quad \text{otherwise}
\end{aligned}
$$

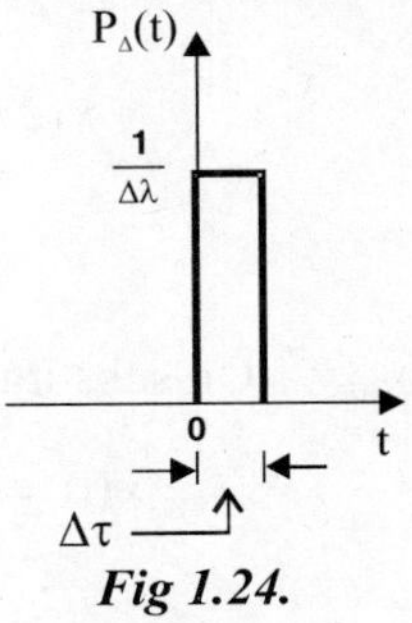

Now, $P_\triangle(t) \times \Delta\tau = A$ pulse of unit amplitude.

$\therefore$ On multiplying $P_\triangle(t) \times \Delta\tau$ with the signal, $\tilde{x}(t)$ the signal $\tilde{x}(0)$ is selected.

$$\therefore \; \tilde{x}(0) = \tilde{x}(t) \, P_\triangle(t) \, \triangle\tau$$

Consider the shifted version of the pulse signal of Fig 1.24, as shown in Fig 1.25.

Fig 1.24.

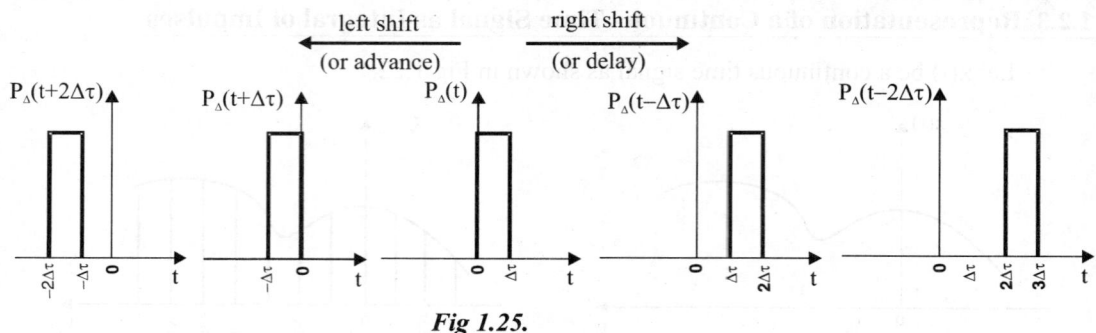

Fig 1.25.

If we multiply $\tilde{x}(t)$ with shifted pulse signals shown in Fig 1.25, then each product will select one pulse of the signal $\tilde{x}(t)$ as shown below:

$$\vdots$$

$$x(-2\,\triangle\,\tau) = \tilde{x}(t)\, P_\triangle\,(t+2\,\triangle\,\tau)\,\triangle\,\tau$$
$$x(-\triangle\tau) = \tilde{x}(t)\, P_\triangle\,(t+\triangle\tau)\,\triangle\,\tau$$
$$x(0) = \tilde{x}(t)\, P_\triangle\,(t)\,\triangle\,\tau$$
$$x(\triangle\,\tau) = \tilde{x}(t)\, P_\triangle\,(t-2\,\triangle\,\tau)\,\triangle\,\tau$$
$$x(2\,\triangle\,\tau) = \tilde{x}(t)\, P_\triangle\,(t-2\,\triangle\,\tau)\,\triangle\,\tau$$

$$\vdots$$

In the above equation, $\tilde{x}(t)$ can be replaced by respective selected pulses itself as shown below:

$$\vdots$$

$$\therefore\ x(-2\,\triangle\,\tau) = \tilde{x}(-2\,\triangle\,\tau)\, P_\triangle\,(t+2\,\triangle\,\tau)\,\triangle\,\tau$$
$$x(-\triangle\tau) = \tilde{x}(-\triangle\tau)\, P_\triangle\,(t+\triangle\tau)\,\triangle\,\tau$$
$$x(0) = \tilde{x}(0)\, P_\triangle\,(t)\,\triangle\,\tau$$
$$x(\triangle\,\tau) = \tilde{x}(\triangle\,\tau)\, P_\triangle\,(t-\triangle\tau)\,\triangle\,\tau$$
$$x(2\,\triangle\,\tau) = \tilde{x}(2\,\triangle\,\tau)\, P_\triangle\,(t-2\,\triangle\,\tau)\,\triangle\,\tau$$

$$\vdots$$

On substituting the above equations in equation (1.3) we get,

$$\tilde{x}(t) = \operatorname*{Lt}_{\triangle\tau\to 0}\left[\begin{array}{l}.....\tilde{x}(-2\,\triangle\,\tau)\, P_\triangle(t+2\,\triangle\,\tau)\,\triangle\,\tau+\tilde{x}(-\triangle\tau)\, P_\triangle(t+\triangle\tau)\,\triangle\,\tau+\tilde{x}(0)P_\triangle(t)\,\triangle\,\tau\\ +\,\tilde{x}(\triangle\,\tau)\, P_\triangle(t-\triangle\tau)\,\triangle\,\tau+\tilde{x}(2\,\triangle\,\tau)\, P_\triangle(t-2\,\triangle\,\tau)\,\triangle\,\tau+.....................\end{array}\right]$$

$$= \operatorname*{Lt}_{\triangle\tau\to 0}\sum_{n=-\infty}^{+\infty}\tilde{x}(n\,\triangle\,\tau)\, P_\triangle(t-n\,\triangle\,\tau)\,\triangle\,\tau$$

On applying limit $\Delta\tau \to 0$ the signal $\tilde{x}(n\,\Delta\,\tau)$ becomes continuous, the signal $P_\Delta(t - n\Delta\tau)$ becomes an impulse and so the summation becomes integration.

Hence the above equation can be expressed as,

$$x(t) = \int\limits_{-\infty}^{+\infty} x(\tau)\,\delta(t - \tau)\,d\tau \qquad\qquad ....(1.4)$$

The equation (1.4) is used to represent any continuous time signal $x(t)$ as an integral of impulses.

1.2.4 Mathematical Operations on Continuous Time Signals

The mathematical operations that can be performed on continuous time signals are,

1. Amplitude scaling
2. Time scaling
3. Folding (or Time folding)
4. Shifting (or Time shifting): Right shift (or advance) and left shift (or delay)
5. Addition
6. Multiplication
7. Differentiation and Integration

1. Amplitude Scaling of Continuous Time Signals

The *amplitude scaling* is performed by multiplying the amplitude of the signal by a constant.

Let $x(t)$ be a continuous time signal. Now $Ax(t)$ is the amplitude scaled version of $x(t)$, where A is a constant.

When $|A| > 1$, then $Ax(t)$ is the amplitude magnified version of $x(t)$ and when $|A| < 1$, then $Ax(t)$ is the amplitude attenuated version of $x(t)$.

Example: 1

Let, $x(t) = at + be^{-ct}$

Let $x_1(t)$ and $x_2(t)$ be the amplitude scaled versions of $x(t)$, scaled by constants 4 and 0.25 respectively.

Now, $x_1(t) = 4x(t) = 4(at + be^{-ct}) = 4at + 4be^{-ct}$

$x_2(t) = 0.25x(t) = 0.25(at + be^{-ct}) = 0.25at + 0.25be^{-ct}$

Example: 2

A continuous time signal and its amplitude scaled version are shown in Fig 1.26.

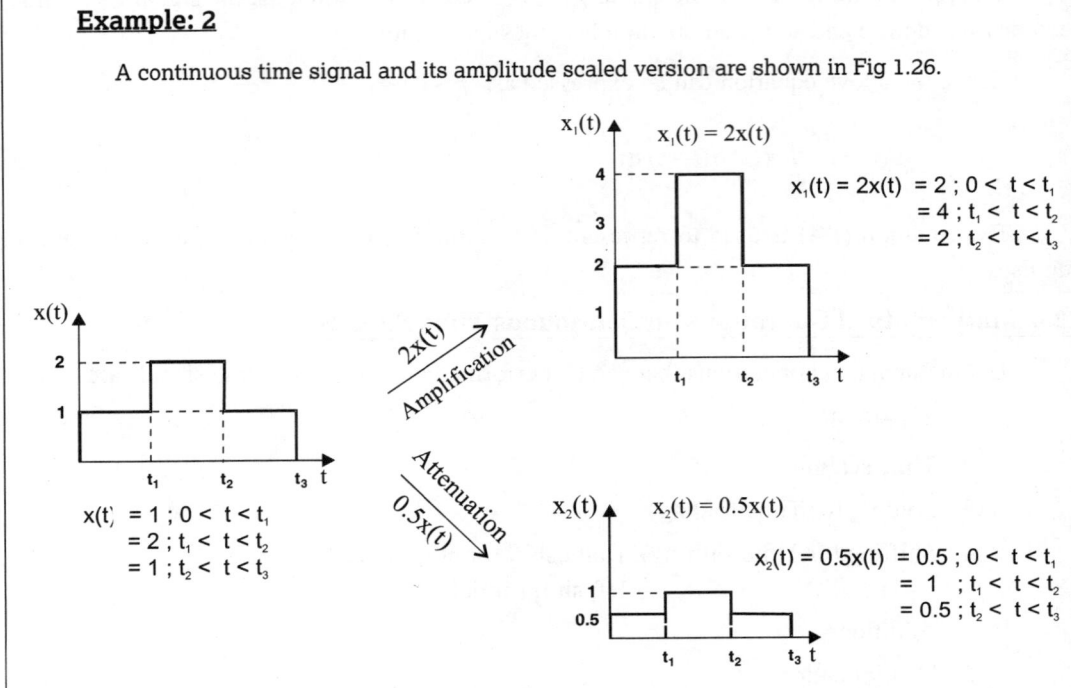

Fig 1.26: A continuous time signal and its amplitude scaled version.

2. Time Scaling of Continuous Time Signals

The **time scaling** is performed by multiplying the variable time by a constant.

If $x(t)$ is a continuous time signal, then $x(At)$ is the time scaled version of $x(t)$, where A is a constant.

When $|A| > 1$, then $x(At)$ is the time compressed version of $x(t)$ and when $|A| < 1$, then $x(At)$ is the time expanded version of $x(t)$.

Example: 1

Let, $x(t) = at + be^{-ct}$

Let $x_1(t)$ and $x_2(t)$ be the time scaled versions of $x(t)$, scaled by constants 4 and 0.25 respectively.

Now, $x_1(t) = x(4t) = a \times 4t + be^{-c \times 4t} = 4at + be^{-4ct}$

$x_2(t) = x(0.25t) = a \times 0.25t + be^{-c \times 0.25t} = 0.25at + be^{-0.25ct}$

Example: 2

A continuous time signal and its time scaled version are shown in Fig 1.27.

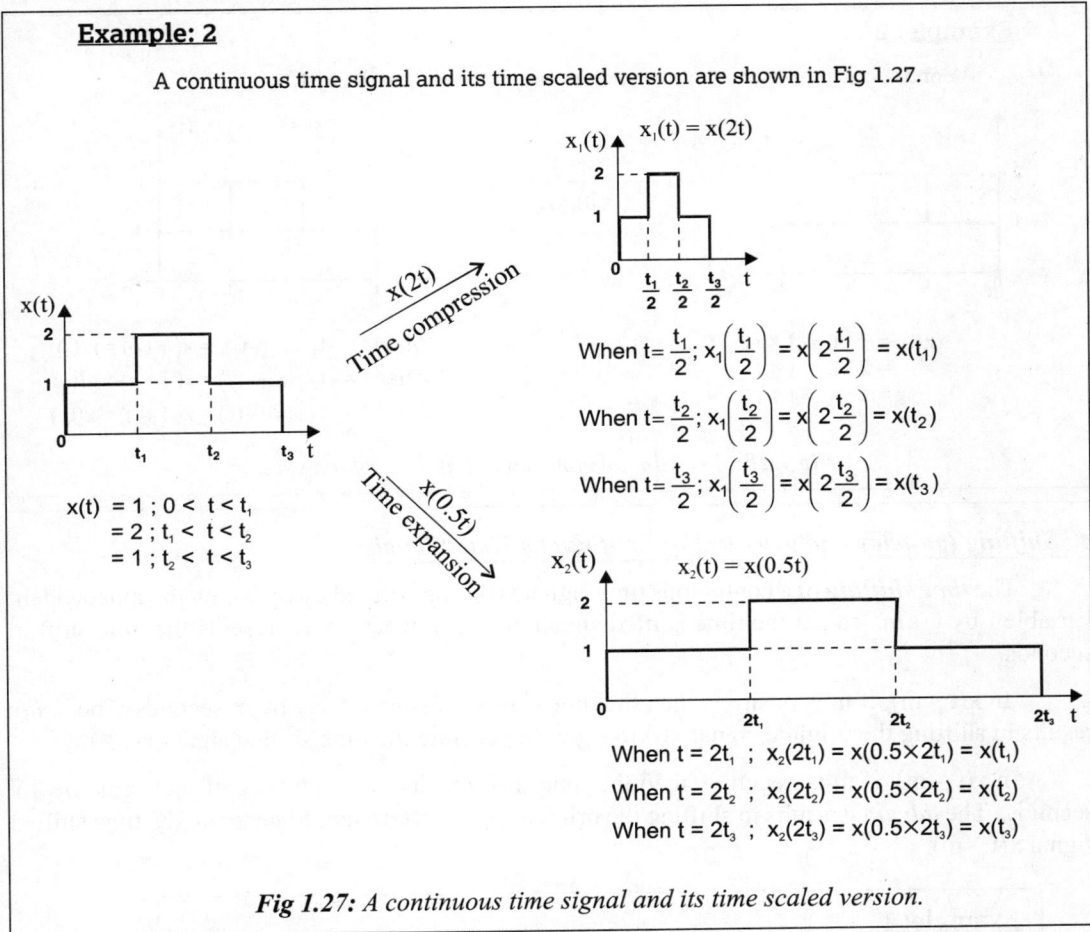

$$x(t) = 1 \; ; \; 0 < t < t_1$$
$$= 2 \; ; \; t_1 < t < t_2$$
$$= 1 \; ; \; t_2 < t < t_3$$

When $t = \dfrac{t_1}{2} \; ; \; x_1\!\left(\dfrac{t_1}{2}\right) = x\!\left(2\dfrac{t_1}{2}\right) = x(t_1)$

When $t = \dfrac{t_2}{2} \; ; \; x_1\!\left(\dfrac{t_2}{2}\right) = x\!\left(2\dfrac{t_2}{2}\right) = x(t_2)$

When $t = \dfrac{t_3}{2} \; ; \; x_1\!\left(\dfrac{t_3}{2}\right) = x\!\left(2\dfrac{t_3}{2}\right) = x(t_3)$

When $t = 2t_1 \; ; \; x_2(2t_1) = x(0.5 \times 2t_1) = x(t_1)$

When $t = 2t_2 \; ; \; x_2(2t_2) = x(0.5 \times 2t_2) = x(t_2)$

When $t = 2t_3 \; ; \; x_2(2t_3) = x(0.5 \times 2t_3) = x(t_3)$

Fig 1.27: *A continuous time signal and its time scaled version.*

3. Folding (or Reflection or Transpose) of Continuous Time Signals

The *folding* of a continuous time signal x(t) is performed by changing the sign of time base t in the signal x(t).

The folding operation produces a signal x(−t) which is a mirror image of the original signal x(t) with respect to the time origin t = 0.

Example: 1

Let, $x(t) = at + be^{-ct}$

Let $x_1(t)$ be folded version of x(t).

Now, $x_1(t) = x(-t) = a(-t) + be^{-c(-t)} = -at + be^{ct}$

Example: 2

A continuous time signal and its folded version are shown in Fig 1.28.

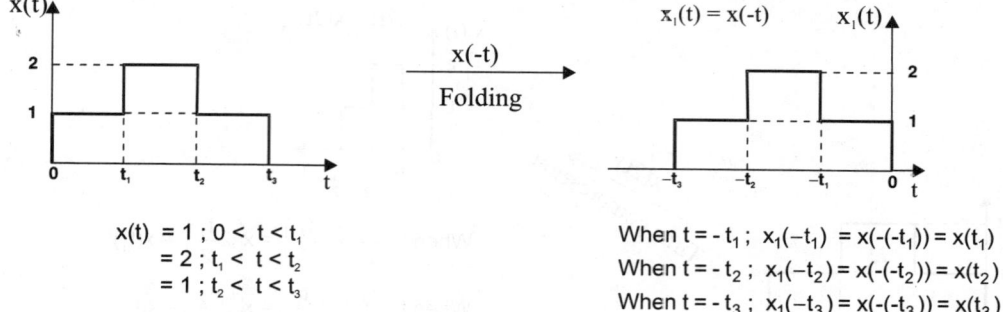

$$x(t) = 1 \; ; \; 0 < t < t_1$$
$$= 2 \; ; \; t_1 < t < t_2$$
$$= 1 \; ; \; t_2 < t < t_3$$

When $t = -t_1$; $x_1(-t_1) = x(-(-t_1)) = x(t_1)$
When $t = -t_2$; $x_1(-t_2) = x(-(-t_2)) = x(t_2)$
When $t = -t_3$; $x_1(-t_3) = x(-(-t_3)) = x(t_3)$

Fig 1.28: *A continuous time signal and its folded version.*

4. Shifting (or delay and advance) of Continuous Time Signals

The *time shifting* of a continuous time signal x(t) is performed by replacing the independent variable t by t − m, to get the time shifted signal x(t − m), where m represents the time shift in seconds.

In x(t − m), if m is positive, then the time shift results in a delay by m seconds. The *delay* results in shifting the original signal x(t) to right, to generate the time shifted signal x(t − m).

In x(t − m), if m is negative, then the time shift results in an advance of the signal by |m| seconds. The *advance* results in shifting the original signal x(t) to left, to generate the time shifted signal x(t − m).

Example: 1

Let, $x(t) = at + be^{-ct}$

Let $x_1(t)$ and $x_2(t)$ be time shifted version of x(t), shifted by m units of time.

Let $x_1(t)$ be delayed version of x(t) and $x_2(t)$ be advanced version of x(t).

Now, $x_1(t) = a(t - m) + be^{-c(t-m)}$

$x_2(t) = a(t + m) + be^{-c(t+m)}$

Note : *The shifting of folded signal will be opposite to that of signal without folding.*

Example: 1. *x(t+m) is left shifted or advanced version of x(t)*
whereas x(−t+m) is right shifted or delayed version of x(−t).

2. *x(t−m) is right shifted or delayed version of x(t)*
whereas x(−t−m) is left shifted or advanced version of x(−t).

Example: 2

A signal and its shifted version are shown in Fig 1.29.

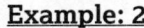

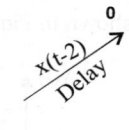

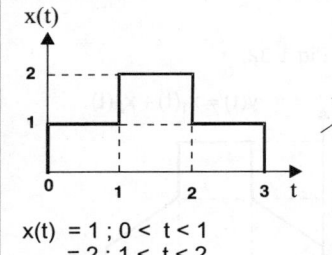

$x(t) = 1 ; 0 < t < 1$
$\quad = 2 ; 1 < t < 2$
$\quad = 1 ; 2 < t < 3$

When $t = 2$; $x_1(2) = x(2-2) = x(0) = 1$
When $t = 3$; $x_1(3) = x(3-2) = x(1) = 1$
When $t = 4$; $x_1(4) = x(4-2) = x(2) = 2$
When $t = 5$; $x_1(5) = x(5-2) = x(3) = 1$

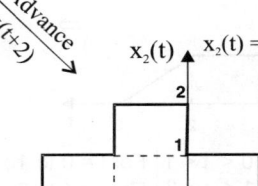

When $t = -2$; $x_1(-2) = x(-2+2) = x(0) = 1$
When $t = -1$; $x_1(-1) = x(-1+2) = x(1) = 1$
When $t = 0$; $x_1(0) = x(0+2) = x(2) = 2$
When $t = 1$; $x_1(1) = x(1+2) = x(3) = 1$

Fig 1.29: *A continuous time signal and its shifted version.*

Delayed Unit Impulse Signal

The unit impulse signal is defined as,

$$\delta(t) = \infty \;\; ; t = 0 \quad \text{and} \quad \int_{-\infty}^{\infty} \delta(t) \; dt = 1$$

$$= 0 \;\; ; t \neq 0$$

The unit impulse signal delayed by m units of time is denoted as $\delta(t - m)$, and it is defined as,

$$\delta(t-m) = \infty \;\; ; t = m \quad \text{and} \quad \int_{-\infty}^{\infty} \delta(t-m) \, dt = 1$$

$$= 0 \;\; ; t \neq m$$

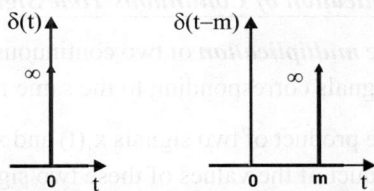

Fig 1.30a: *Impulse.* **Fig 1.30b:** *Delayed impulse.*
Fig 1.30: *Impulse and delayed impulse signal.*

Delayed Unit Step Signal

The unit step signal is defined as,

$$u(t) = 1 \; ; \text{ for } t \geq 0$$

$$= 0 \; ; \text{ for } t < 0$$

The unit step signal delayed by m units of time is denoted as $u(t - m)$, and it is defined as,

$$u(t - m) = 1 \; ; t \geq m$$

$$= 0 \; ; t < m$$

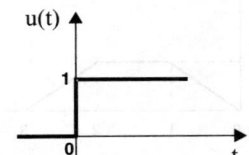 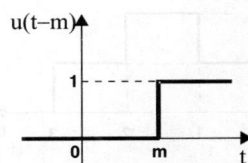

Fig 1.31a: *Unit step signal.* **Fig 1.31b:** *Delayed unit step signal.*
Fig 1.31: *Unit step and delayed unit step signal.*

5. Addition of Continuous Time Signals

The **addition** of two continuous time signals is performed by adding the value of the two signals corresponding to the same instant of time.

The sum of two signals $x_1(t)$ and $x_2(t)$ is a signal $y(t)$, whose value at any instant is equal to the sum of the value of these two signals at that instant.

i.e., $y(t) = x_1(t) + x_2(t)$

Example:

Graphical addition of two continuous time signals is shown in Fig 1.32.

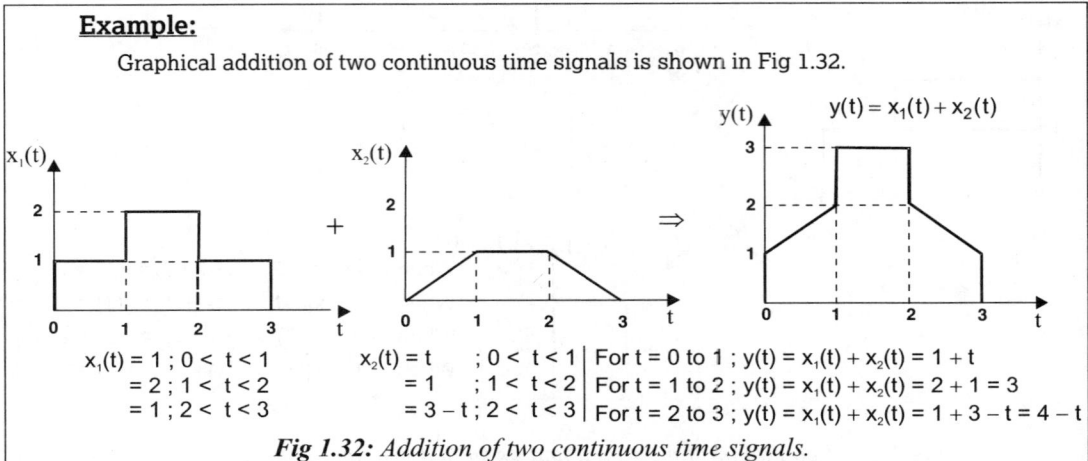

Fig 1.32: *Addition of two continuous time signals.*

6. Multiplication of Continuous Time Signals

The **multiplication** of two continuous time signals is performed by multiplying the value of the two signals corresponding to the same instant of time.

The product of two signals $x_1(t)$ and $x_2(t)$ is a signal $y(t)$, whose value at any instant is equal to the product of the values of these two signals at that instant.

i.e., $y(t) = x_1(t) \times x_2(t)$

Example:

Graphical multiplication of two continuous time signals is shown in Fig 1.33.

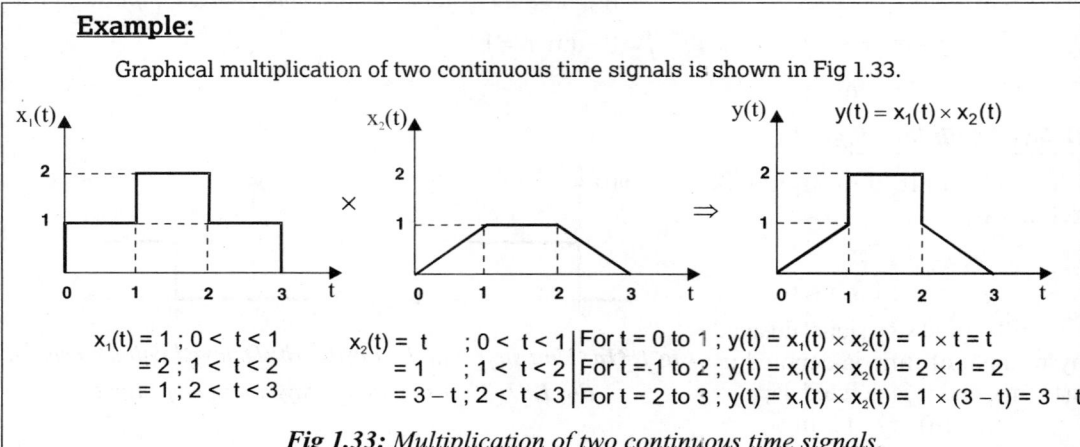

Fig 1.33: *Multiplication of two continuous time signals.*

7. Differentiation and Integration of Continuous Time Signals

Differentiation is a mathematical operation used to estimate the rate of change of a continuous time signal at any instant of time.

Differentiation is denoted by the operator $\dfrac{d}{dt}$.

Therefore, the differentiation of a continuous time signal x(t) is denoted by $\dfrac{d}{dt}$ x(t) $\left(\text{or } \dfrac{dx(t)}{dt} \right)$.

The differentiation of a continuous time signal x(t) is defined as,

$$\frac{dx(t)}{dt} = \underset{\Delta t \to 0}{Lt} \frac{x(t) - x(t - \Delta t)}{\Delta t}$$

Integration, is the inverse process of differentiation. More appropriately, Integration is the process of identifying the signal from its differentiation.

The integration is denoted by the operaor $\int$ .... dt

Therefore, the integration of a continuous time signal x(t) is denoted by $\int$ x(t) dt .

Differentiation and integration of standard continuous time signals are listed in Table 1.4.

Table 1.4: Differentiation and Integration of Standard Continuous Time Signals

Signal, x(t)	Differentiation of x(t), $\dfrac{dx(t)}{dt}$	Integration of x(t), $\int$ x(t) dt
$\delta(t)$	—	1
u(t)	$\delta(t)$	t
t	u(t)	$\dfrac{t^2}{2}$
t^2	2t	$\dfrac{t^3}{3}$
sin t	cos t	$-\cos t$
cos t	$-\sin t$	sin t
e^{-at}	$-ae^{-at}$	$\dfrac{e^{-at}}{-a}$
e^{at}	ae^{at}	$\dfrac{e^{at}}{a}$
$\sin \Omega_0 t$	$\Omega_0 \cos \Omega_0 t$	$\dfrac{-\cos \Omega_0 t}{\Omega_0}$
$\cos \Omega_0 t$	$-\Omega_0 \sin \Omega_0 t$	$\dfrac{\sin \Omega_0 t}{\Omega_0}$

The standard signals such as impulse, step, ramp and parabolic signals are related through integration and differentiation as shown below: **(AU Dec'15, 2 Marks & Jun'13, 10 Marks)**

$$\delta(t) \xrightarrow{\text{integration}} u(t) \xrightarrow{\text{integration}} t \xrightarrow{\text{integration}} \frac{t^2}{2}$$
(Impulse)   (Unit step)   (Unit ramp)   (Unit parabolic)

$$\delta(t) \xleftarrow{\text{differentiation}} u(t) \xleftarrow{\text{differentiation}} t \xleftarrow{\text{differentiation}} \frac{t^2}{2}$$
(Impulse)   (Unit step)   (Unit ramp)   (Unit parabolic)

Note: $\left. \dfrac{d}{dt} u(t) \right|_{t=0} = \left. \dfrac{\Delta u}{\Delta t} \right|_{t=0} = \left. \dfrac{u(0^+) - u(0^-)}{\Delta t} \right|_{t=0} = \left. \dfrac{1-0}{0} \right|_{t=0} = \infty \big|_{t=0} = Impulse$

Table 1.5: Mathematical operations on ramp signal, $r(t) = t$; $t \geq 0$

Signal	Waveform	Slope (m)
$r(t)$ (Ramp)		$m = \dfrac{r(t_2) - r(t_1)}{t_2 - t_1} = \dfrac{2-1}{2-1} = 1$
$r_1(t) = 2r(t)$ (Amplitude Scaled Ramp)		$m = \dfrac{r_1(t_2) - r_1(t_1)}{t_2 - t_1} = \dfrac{6-4}{3-2} = 2$
$r_2(t) = r(-t)$ (Folded Ramp)		$m = \dfrac{r_2(t_2) - r_2(t_1)}{t_2 - t_1} = \dfrac{2-1}{-2+1} = -1$
$r_3(t) = r(t-2)$ (Shifted Ramp)		$m = \dfrac{r_3(t_2) - r_3(t_1)}{t_2 - t_1} = \dfrac{3-2}{5-4} = 1$

Table 1.5: Continued...

Signal	Waveform	Slope (m)
$r_4(t) = r(2t)$ (Time Scaled Ramp - Compression)		$m = \dfrac{r_4(t_2) - r_4(t_1)}{t_2 - t_1} = \dfrac{4-2}{2-1} = 2$
$r_5(t) = r\left(\dfrac{t}{2}\right)$ (Time Scaled Ramp - Expansion)		$m = \dfrac{r_5(t_2) - r_5(t_1)}{t_2 - t_1} = \dfrac{3-2}{6-4} = \dfrac{1}{2}$

1.2.5 Order of Mathematical Operations on Signals with Multiple Operations

When four mathematical operations like folding, shifting, time scaling and amplitude scaling have to be performed on a signal then convert the signal to the form $Ax(-at+b)$. The order of performing operation should be as follows,

     First operation   :   Folding

     Second operation :   Shifting

     Third operatioon   :   Time Scaling

     Fourth operation   :   Amplitude Scaling

> ***Note:*** *The priority of amplitude scaling will not affect the result. Therefore, amplitude scaling can be performed prior to any operation, but preferrably performed at the end.*

Example 1.1

Consider the ramp signal, $x(t) = t$; $t \geq 0$. Plot and find the slope of the signal $2x\left(\dfrac{-4t+8}{2}\right)$.

Solution:

Given that, $x(t) = t$; $t \geq 0$.

In order to plot $2x\left(\dfrac{-4t+8}{2}\right)$, first convert $2x\left(\dfrac{-4t+8}{2}\right)$ into the form $Ax(-at+b)$.

$\therefore \; 2x\left(\dfrac{-4t+8}{2}\right) = 2x\left(\dfrac{-4}{2}t + \dfrac{8}{2}\right) = 2x(-2t+4)$

Now, the order of Mathematical operations to be performed are given below:

     i)   Folding $\rightarrow$ $x(-t)$

     ii)   Shifting $\rightarrow$ $x(-t+4)$

     iii)   Time Scaling $\rightarrow$ $x(-2t+4)$

     iv)   Amplitude Scaling $\rightarrow$ $2x(-2t+4)$

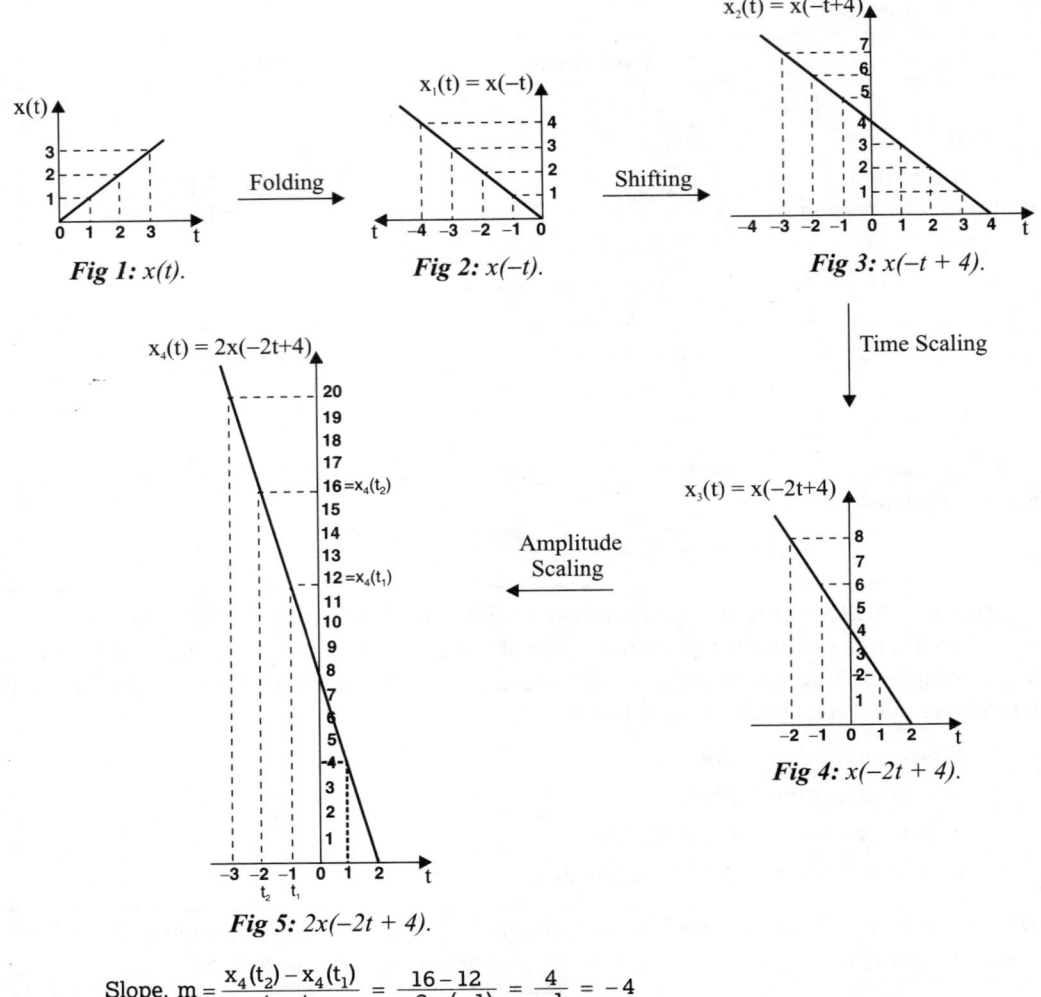

Fig 1: x(t).   Fig 2: x(−t).   Fig 3: x(−t + 4).

Fig 5: 2x(−2t + 4).

Fig 4: x(−2t + 4).

$$\text{Slope, } m = \frac{x_4(t_2) - x_4(t_1)}{t_2 - t_1} = \frac{16 - 12}{-2 - (-1)} = \frac{4}{-1} = -4$$

Example 1.2

A continuous time signal is defined as,

$$x(t) = t \;\; ; \;\; 0 \leq t \leq 3$$
$$= 0 \;\; ; \;\; t > 3$$

Sketch the waveform of x(−t) and x(2 − t).

Solution:

The signal x(−t) is the folded version of x(t).

The signal x(2 − t) = x(−t + 2) is obtained by doing folding operation first and then time shifting operation next.

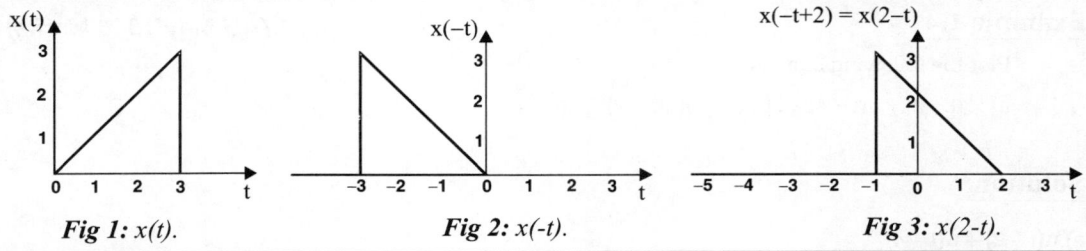

Fig 1: x(t).   *Fig 2: x(-t).*   *Fig 3: x(2-t).*

Example 1.3

(AU Jun'14, 8 Marks)

A continous time signal is defined as,

$$x(t) = \frac{1}{6}(t+2) \; ; \quad -2 \leq t \leq 4$$
$$= 0 \qquad ; \quad \text{otherwise}$$

Setch the waveforms:

a) $x(t)$ **b)** $x(t+1)$ **c)** $x(2t)$ **d)** $x\left(\frac{t}{2}\right)$

Solution

Given that, $x(t) = \frac{1}{6}(t+2) \; ; \; -2 \leq t \leq 4$

When $t = -2$; $x(-2) = \frac{1}{6}(-2+2) = 0$ | When $t = 1$; $x(1) = \frac{1}{6}(1+2) = \frac{3}{6}$

When $t = -1$; $x(-1) = \frac{1}{6}(-1+2) = \frac{1}{6}$ | When $t = 2$; $x(2) = \frac{1}{6}(2+2) = \frac{4}{6}$

When $t = 0$; $x(0) = \frac{1}{6}(0+2) = \frac{2}{6}$ | When $t = 3$; $x(3) = \frac{1}{6}(3+2) = \frac{5}{6}$

                                                When $t = 4$; $x(4) = \frac{1}{6}(4+2) = \frac{6}{6}$

a)

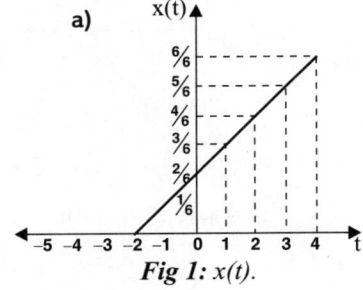

Fig 1: x(t).

Time shift (shift right) by one unit of time →

b)

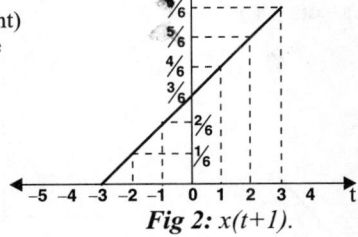

Fig 2: x(t+1).

Time scaling (compression) by 2 ↓

Time scaling (expansion) by $\frac{1}{2}$ ↘

c)

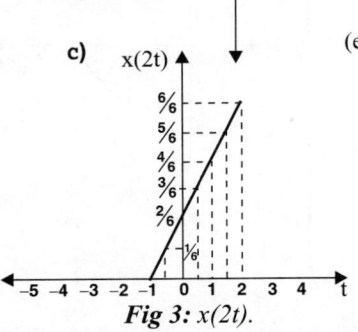

Fig 3: x(2t).

d)

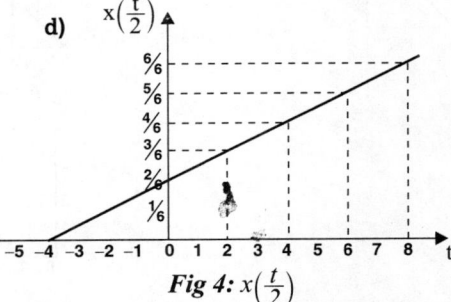

Fig 4: $x\left(\frac{t}{2}\right)$

Example 1.4

Plot the following signals

a) u(t − 2) + u(t − 4)   **b)** (t − 4)(u(t − 2) − u(t − 4))

Solution:

a) u(t − 2) + u(t − 4)

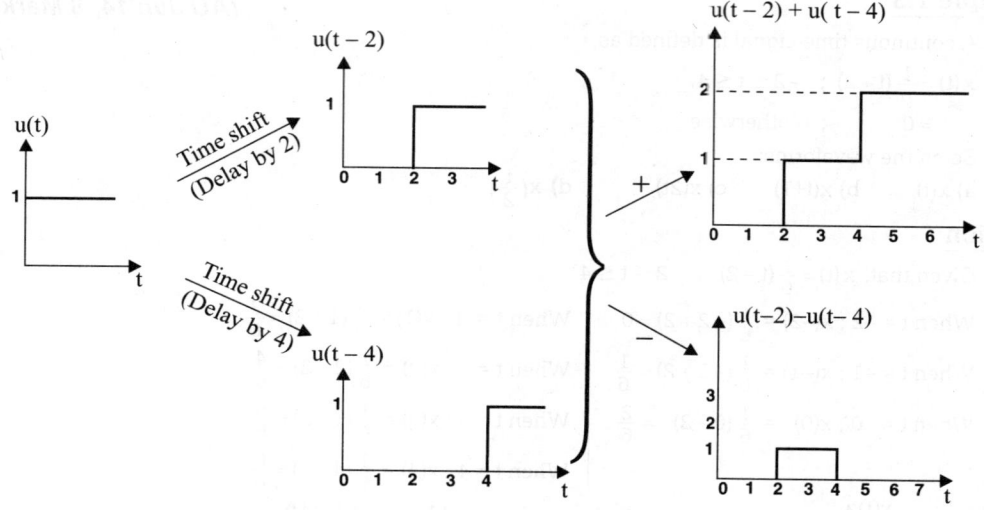

b) (t − 4)(u(t − 2) − u(t − 4))

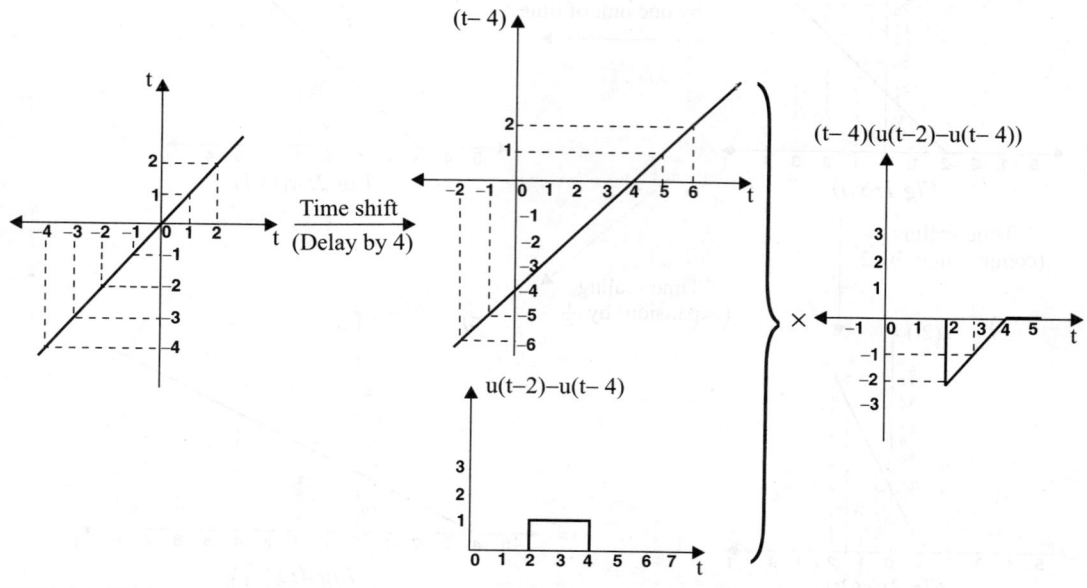

Example 1.5

Plot the following signals

　　a)　$x(t) = r(t)$　　　　b) $x(t) = r(-t+2)$　　　　c)　$x(t) = -2r(t)$

Solution:

a) $x(t) = r(t)$　　　　　　　　　　　　　　　　**b)** $x(t) = r(-t+2)$

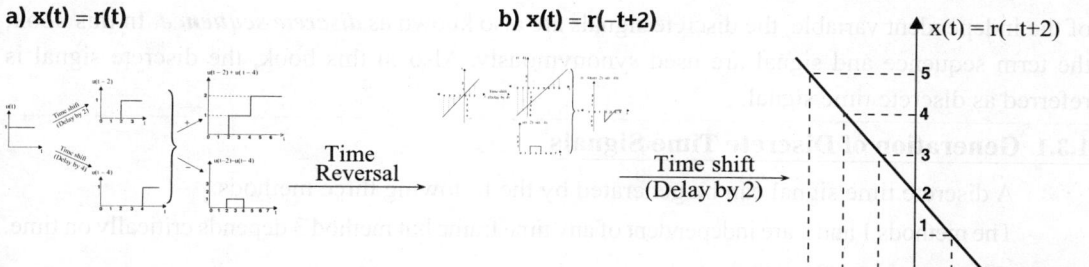

Fig 3: $x(t) = r(-t+2)$.

c) $x(t) = -2r(t)$

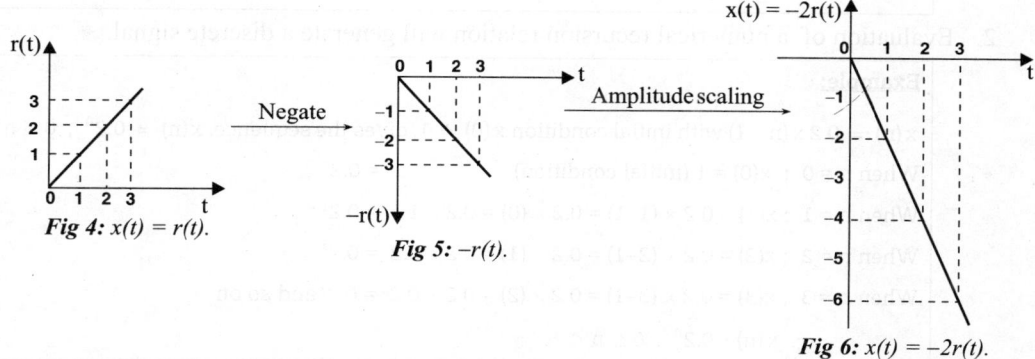

Fig 4: $x(t) = r(t)$.　　　Fig 5: $-r(t)$.　　　Fig 6: $x(t) = -2r(t)$.

1.3 Discrete Time Signals

The *discrete signal* is a function of a discrete independent variable. The independent variable is divided into uniform intervals and each interval is represented by an integer. The letter "n" is used to denote the independent variable. The discrete or digital signal is denoted by x(n).

The discrete signal is defined for every integer value of the independent variable "n". The magnitude (or value) of discrete signal can take any discrete value in the specified range. Here both the value of the signal and the independent variable are discrete. The discrete signal can be represented by a one-dimensional array as shown in the following example.

> ### Example:
>
> 　　$x(n) = \{ 2, 4, -1, 3, 3, 4 \}$
>
> Here the discrete signal x(n) is defined for,　n = 0, 1, 2, 3, 4, 5
>
> 　∴　$x(0) = 2$;　　$x(1) = 4$;　　$x(2) = -1$;　　$x(3) = 3$;　　$x(4) = 3$;　　$x(5) = 4$

When the independent variable is time t, the discrete signal is called *discrete time signal*. In discrete time signal, the time is divided uniformly using the relation $t = nT$, where T is the sampling time period. The sampling time period is the inverse of sampling frequency. The discrete time signal is denoted by $x(n)$ or $x(nT)$.

Since the discrete signals have a sequence of numbers (or values) defined for integer values of the independent variable, the discrete signals are also known as *discrete sequence*. In this book, the term sequence and signal are used synonymously. Also in this book, the discrete signal is referred as discrete time signal.

1.3.1 Generation of Discrete Time Signals

A discrete time signal can be generated by the following three methods.

The methods 1 and 2 are independent of any time frame but method 3 depends critically on time.

1. Generate a set of numbers and arrange them as a sequence.

> **Example:**
> The numbers 0, 2, 4, ..., 2N form a sequence of even numbers and can be expressed as,
> $$x(n) = 2n \quad ; \quad 0 \le n \le N$$

2. Evaluation of a numerical recursion relation will generate a discrete signal.

> **Example:**
> $x(n) = 0.2\,x(n-1)$ with initial condition $x(0) = 1$, gives the sequence, $x(n) = 0.2^n$; $0 \le n < \infty$
>
> When $n = 0$; $x(0) = 1$ (initial condition) $= 0.2^0$
>
> When $n = 1$; $x(1) = 0.2 \times (1-1) = 0.2 \times (0) = 0.2 \times 1$ $= 0.2^1$
>
> When $n = 2$; $x(2) = 0.2 \times (2-1) = 0.2 \times (1) = 0.2 \times 0.2 = 0.2^2$
>
> When $n = 3$; $x(3) = 0.2 \times (3-1) = 0.2 \times (2) = 0.2 \times 0.2^2 = 0.2^3$ and so on
>
> $$\therefore \; x(n) = 0.2^n \; ; \; 0 \le n < \infty$$

3. A third method is by uniformly sampling a continuous time signal and using the amplitudes of the samples to form a sequence.

Let, $x(t)$ = Continuous time signal

Now, Discrete signal, $x(nT) = x(t)\big|_{t=nT} \; ; \; -\infty < n < \infty$

where, T is the sampling interval.

1.3.2 Digital Signal

The *digital signal* is same as discrete signal except that the magnitude of the signal is quantized. The magnitude of the signal can take one of the values in a set of quantized values. Here quantization is necessary to represent the signal in binary codes.

The generation of a discrete time signal by sampling a continuous time signal and then quantizing the samples in order to convert the signal to digital signal is shown in the following example.

Let, x(t) = Continuous time signal

T = Sampling time

A typical continuous time signal and the sampling of this continuous time signal at uniform interval are shown in Fig 1.34a and Fig 1.34b respectively. The samples of the continuous time signal as a function of sampling time instants are shown in Fig 1.34c. In Fig 1.34c, 1T, 2T, 3T, ... etc., represents sampling time instants and the value of the samples are functions of this sampling time instants.

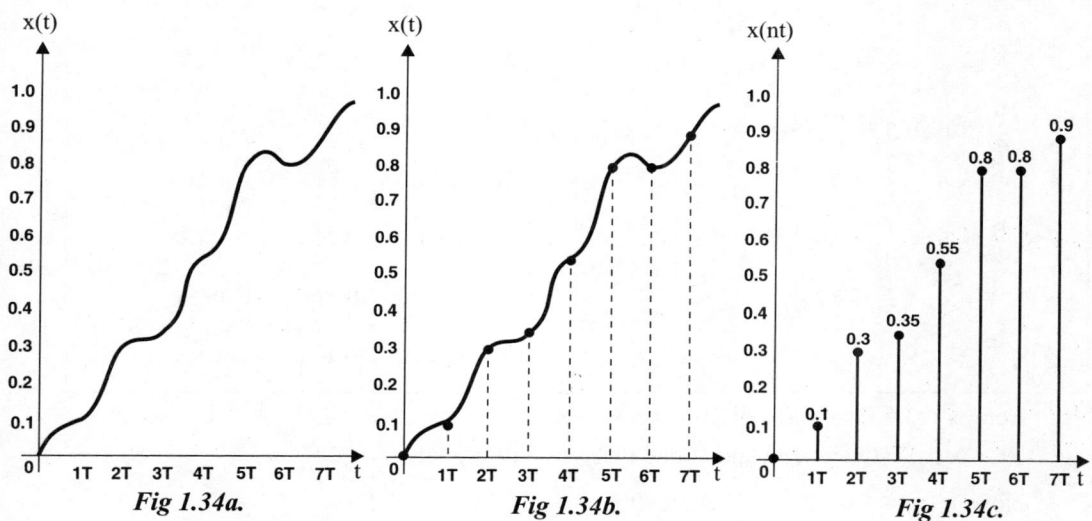

Fig 1.34a.   Fig 1.34b.   Fig 1.34c.

Fig 1.34: Sampling a continuous time signal to generate discrete time signal.

When t = 0 ; x(t) = 0

When t = 1T ; x(t) = 0.1

When t = 2T ; x(t) = 0.3

When t = 3T ; x(t) = 0.35

When t = 4T ; x(t) = 0.55

When t = 5T ; x(t) = 0.8

When t = 6T ; x(t) = 0.8

When t = 7T ; x(t) = 0.9

In general, the sampling time instants can be represented as, "nT", where "n" is an integer. When we drop the sampling time "T" , then the samples are functions of the integer variable "n" alone. Therefore, the samples of the continuous time signal will be a discrete time signal, denoted as x(n), which is a function of an integer variable "n" as shown below:

x(n) = { 0, 0.1, 0.3, 0.35, 0.55, 0.8, 0.8, 0.9 }

Here the discrete signal x(n) is defined for, n = 0, 1, 2, 3, 4, 5, 6, 7

∴ x(0) = 0 ;   x(1) = 0.1 ;   x(2) = 0.3 ;   x(3) = 0.35 ;

x(4) = 0.55 ;   x(5) = 0.8 ;   x(6) = 0.8 ;   x(7) = 0.9

The values of the samples lies in the range of 0 to 1.

Let us choose 3-bit binary to represent the values of the samples in binary code. Now, the possible binary codes are $2^3 = 8$, and so the range can be divided into eight quantization levels, and each sample is assigned, one of the quantization level as shown in the following table.

Table 1.6: Quantization of Analog Value 0 to 1 using 3 Bit Binary

Quantization Level (R = Range = 1)	Binary Code	Range Represented by Quantization Level for Quantization by Truncation		
$0 \times \dfrac{R}{2^3} = 0 \times \dfrac{1}{8} = 0$	000	$0.000 \leq x(n) < 0.125$	$\Rightarrow$	0.000
$1 \times \dfrac{R}{2^3} = 1 \times \dfrac{1}{8} = 0.125$	001	$0.125 \leq x(n) < 0.250$	$\Rightarrow$	0.125
$2 \times \dfrac{R}{2^3} = 2 \times \dfrac{1}{8} = 0.25$	010	$0.250 \leq x(n) < 0.375$	$\Rightarrow$	0.250
$3 \times \dfrac{R}{2^3} = 3 \times \dfrac{1}{8} = 0.375$	011	$0.375 \leq x(n) < 0.500$	$\Rightarrow$	0.375
$4 \times \dfrac{R}{2^3} = 4 \times \dfrac{1}{8} = 0.5$	100	$0.500 \leq x(n) < 0.625$	$\Rightarrow$	0.500
$5 \times \dfrac{R}{2^3} = 5 \times \dfrac{1}{8} = 0.625$	101	$0.625 \leq x(n) < 0.75$	$\Rightarrow$	0.625
$6 \times \dfrac{R}{2^3} = 6 \times \dfrac{1}{8} = 0.75$	110	$0.750 \leq x(n) < 0.875$	$\Rightarrow$	0.750
$7 \times \dfrac{R}{2^3} = 7 \times \dfrac{1}{8} = 0.875$	111	$0.875 \leq x(n) < 0.100$	$\Rightarrow$	0.875

Let, $x_q(n)$ = Quantized discrete time signal.

$x_c(n)$ = Quantized and coded discrete time signal.

Now, $x_q(n)$ = { 0, 0, 0.25, 0.25, 0.5, 0.75, 0.75, 0.875 }

$x_c(n)$ = { 000, 000, 010, 010, 100, 110, 110, 111}

The quantized and coded discrete time signal $x_c(n)$ is called digital signal.

1.3.3 Representation of Discrete Time Signals

The discrete time signal can be represented by the following methods.

1. Functional Representation

In functional representation, the signal is represented as a mathematical equation, as shown in the following example.

$$
\begin{aligned}
x(n) &= -0.5 \ ; \ n = -2 \\
&= \ \ \ 1.0 \ ; \ n = -1 \\
&= -1.0 \ ; \ n = \ \ 0 \\
&= \ \ \ 0.6 \ ; \ n = \ \ 1 \\
&= \ \ \ 1.2 \ ; \ n = \ \ 2 \\
&= \ \ \ 1.5 \ ; \ n = \ \ 3 \\
&= \ \ \ 0 \ \ \ ; \ \text{other } n
\end{aligned}
$$

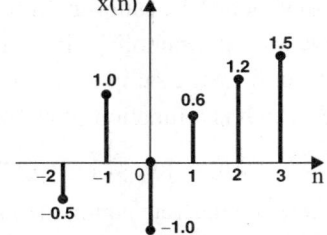

Fig 1.35: Graphical representation of a discrete time signal.

2. Graphical Representation

In graphical representation, the signal is represented in a two-dimensional plane. The independent variable is represented in the horizontal axis and the value of the signal is represented in the vertical axis as shown in Fig 1.35.

3. Tabular Representation

In tabular representation, two rows of a table are used to represent a discrete time signal. In the first row, the independent variable "n" is tabulated and in the second row the value of the signal for each value of "n" are tabulated as shown in the following table.

n	....	−2	−1	0	1	2	3	....
x(n)	....	−0.5	1.0	−1.0	0.6	1.2	1.5	....

4. Sequence Representation

In sequence representation, the discrete time signal is represented as a one-dimensional array as shown in the following examples:

An infinite duration discrete time signal with the time origin, n = 0, indicated by the symbol ↑ is represented as,

$$x(n) = \{ \ ..... -0.5, \underset{\uparrow}{1.0}, -1.0, 0.6, 1.2, 1.5, ... \}$$

An infinite duration discrete time signal that satisfies the condition x(n) = 0 for n < 0 is represented as,

$$x(n) = \{ \underset{\uparrow}{-1.0}, 0.6, 1.2, 1.5, ... \} \quad \text{or} \quad x(n) = \{-1.0, 0.6, 1.2, 1.5, ... \}$$

An infinite duration discrete time signal that satisfies the condition x(n) = 0 for n > 0 is represented as,

$$x(n) = \{ \ ....-0.5, 1.0, \underset{\uparrow}{-1.0}\}$$

A finite duration discrete time signal with the time origin, n = 0, indicated by the symbol ↑ is represented as,

$$x(n) = \{ -0.5, \underset{\uparrow}{1.0}, -1.0, 0.6, 1.2, 1.5 \}$$

A finite duration discrete time signal that satisfies the condition x(n) = 0 for n < 0 is represented as,

$$x(n) = \{ \underset{\uparrow}{-1.0}, -0.6, 1.2, 1.5 \} \quad \text{or} \quad x(n) = \{ -1.0, 0.6, 1.2, 1.5\}$$

A finite duration discrete time signal that satisfies the condition x(n) = 0 for n > 0 is represented as,

$$x(n) = \{ -0.5, 1.0, \underset{\uparrow}{-1.0}\}$$

1.3.4 Standard Discrete Time Signals   *(AU Jun'11, 2 Marks & Dec'13, 2 Marks)*

1. Discrete Impulse Signal or Unit Sample Sequence

Impulse signal, $\delta(n) = 1 \ ; \ n = 0$
$$= 0 \ ; \ n \neq 0$$

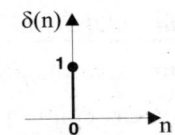

Fig 1.36: *Discrete impulse signal.*

2. Unit Step Signal

Unit step signal, $u(n) = 1 \ ; \ n \geq 0$
$$= 0 \ ; \ n < 0$$

(AU May'11, 2 Marks)

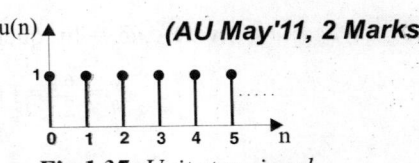

Fig 1.37: *Unit step signal.*

3. Ramp Signal

Ramp signal, $u_r(n) = n$; $n \geq 0$
$$= 0 \; ; \; n < 0$$

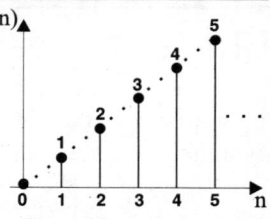

Fig 1.38: Ramp signal.

4. Exponential Signal

Exponential signal, $g(n) = a^n$; $n \geq 0$
$$= 0 \; ; \; n < 0$$

(AU Dec'12, 4 Marks)

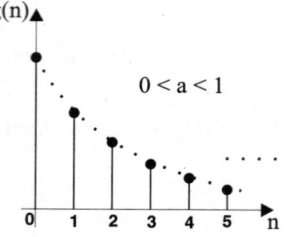

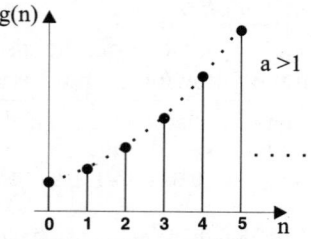

Fig 1.39a: Decreasing exponential signal.   *Fig 1.39b: Increasing exponential signal.*

Fig 1.39: Exponential signal.

5. Discrete Time Signum Signal

$\text{sgn}(n) = 1$; $n > 0$
$$= 0 \; ; \; n = 0$$
$$= -1 \; ; \; n < 0$$

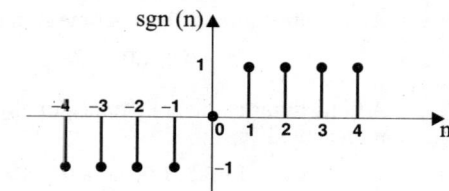

Fig 1.40: Discrete time signum signal.

6. Discrete Rectangular Signal

$\text{rect}(n) = 1$; $-N < n < N$
$$= 0 \; ; \; n < -N \text{ and } n > N$$

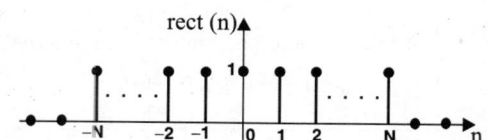

Fig 1.41: Discrete rectangular signal.

7. Discrete Time Sinusoidal Signal

The discrete time sinusoidal signal may be expressed as,

$$x(n) = A\cos(\omega_0 n + \theta) \; ; \text{ for n in the range } -\infty < n < +\infty$$

$$x(n) = A\sin(\omega_0 n + \theta) \; ; \text{ for n in the range } -\infty < n < +\infty$$

where, ω_0 = Frequency in *radians/sample* ; θ = Phase in *radians*

$$f_0 = \frac{\omega_0}{2\pi} = \text{Frequency in } cycles/sample$$

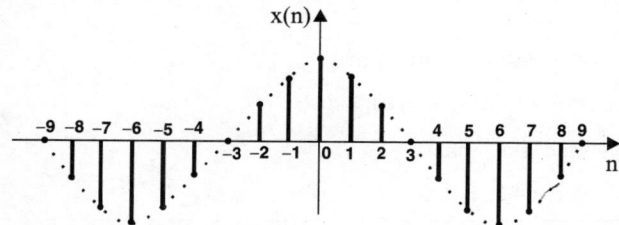

Fig 1.42a: *Discrete time sinusoidal signal represented by equation x(n) = A cos(ω₀n).*

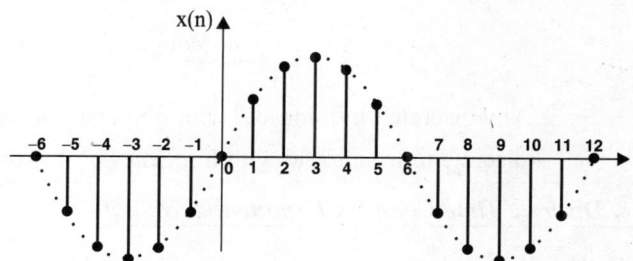

Fig 1.42b: *Discrete time sinusoidal signal represented by the equation x(n) = A sin(ω₀n).*

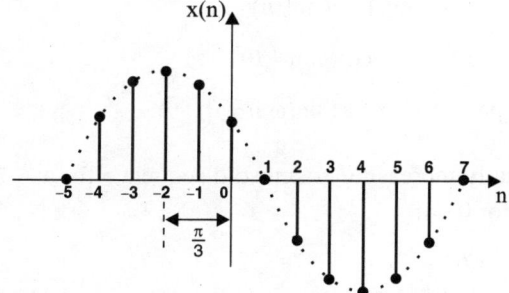

Fig 1.42c: *Discrete time sinusoidal signal represented by equation,*

$$x(n) = A\cos\left(\frac{\pi}{6}n + \frac{\pi}{3}\right) \; ; \omega_0 = \frac{\pi}{6} \;, \theta = \frac{\pi}{3}$$

Fig 1.42: *Discrete time sinusoidal signals.*

Properties of Discrete Time Sinusoid

1. A discrete time sinusoid is periodic only if its frequency f_0 is a rational number, (i.e., ratio of two integers).

2. Discrete time sinusoids whose frequencies are separated by integer multiples of 2π are identical.

$$\therefore \; x(n) = A \cos[(\omega_0 + 2\pi k) n + \theta], \quad \text{for} \;\; k = 0,1,2\ldots\ldots\ldots\text{are identical in the interval}$$
$$-\pi \le \omega_0 \le \pi \;\; \text{and so they are indistinguishable.}$$

Proof:

$$\cos[(\omega_0 + 2\pi k)\,n + \theta] = \cos(\omega_0 n + 2\pi nk + \theta) = \cos[(\omega_0 n + \theta) + 2\pi nk]$$

$$= \cos(\omega_0 n + \theta)\cos 2\pi nk - \sin(\omega_0 n + \theta)\sin 2\pi nk$$

Since n and k are integers, $\cos 2\pi nk = 1$ and $\sin 2\pi nk = 0$

$$\therefore \cos[(\omega_0 + 2\pi k)\,n + \theta] = \cos(\omega_0 n + \theta), \quad \text{for } k = 0, 1, 2, 3, \ldots..$$

Conclusion

1. The sequences of any two sinusoids with frequencies in the range, $-\pi \le \omega_o \le \pi$ (or $-1/2 \le f_0 \le 1/2$), are distinct.

$$[-\pi \le \omega \le \pi \xrightarrow{\text{divide by } 2\pi} -1/2 \le f \le 1/2]$$

2. Any discrete time sinusoid with frequency $\omega_o > |\pi|$ (or $f_0 > |1/2|$) will be identical to another discrete time sinusoid with frequency $\omega_o < |\pi|$ (or $f_0 < |1/2|$).

8. Discrete Time Complex Exponential Signal   (AU Dec'12, 4 Marks)

The discrete time complex exponential signal is defined as,

$$x(n) = a^n\,[\cos(\omega_0 n + \theta) + j\,\sin(\omega_0 n + \theta)]$$

$$= a^n \cos(\omega_0 n + \theta) + j\,a^n \sin(\omega_0 n + \theta) = x_r(n) + j\,x_i(n)$$

$$\text{where,} \quad x_r(n) = \text{Real part of } x(n) = a^n \cos(\omega_0 n + \theta)$$

$$x_i(n) = \text{Imaginary part of } x(n) = a^n \sin(\omega_0 n + \theta)$$

The real part of $x(n)$ will give an exponentially increasing cosinusoid sequence for $a > 1$ and exponentially decreasing cosinusoid sequence for $0 < a < 1$.

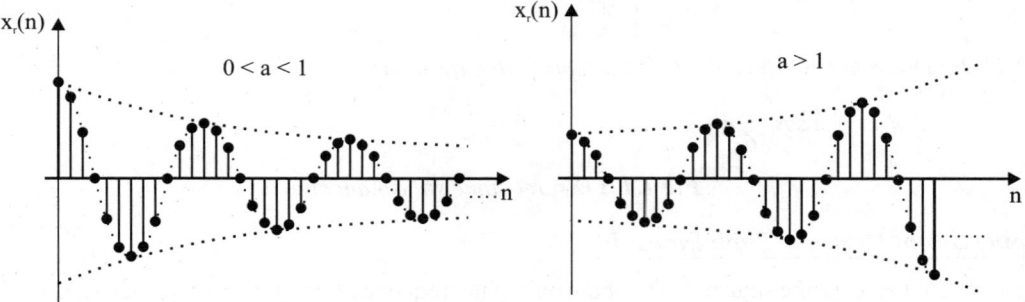

Fig 1.43a: The discrete time sequence represented by the equation, $x_r(n) = a^n \cos\omega_0 n$ for $0 < a < 1$.

Fig 1.43b: The discrete time sequence represented by the equation, $x_r(n) = a^n \cos\omega_0 n$ for $a > 1$.

Fig 1.43: *Real part of complex exponential signal.*

The imaginary part of x(n) will give rise to an exponentially increasing sinusoid sequence for a > 1 and exponentially decreasing sinusoid sequence for 0 < a < 1.

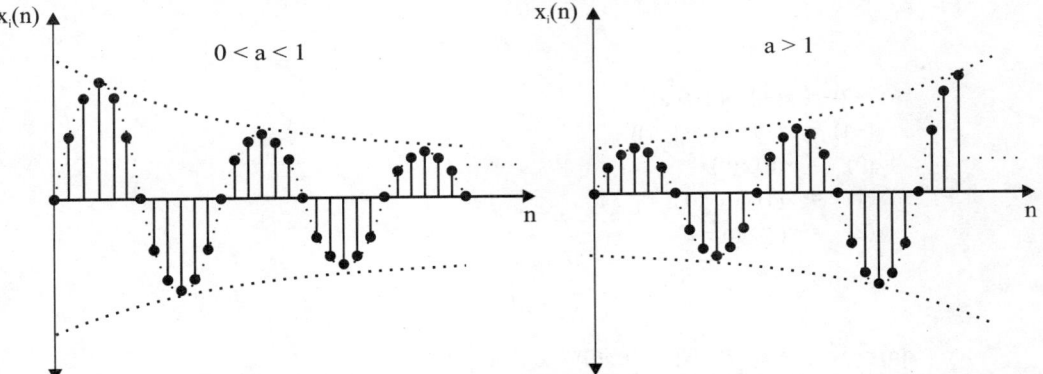

Fig 1.44a: *The discrete time sequence represented by the equation,* $x_i(n) = a^n \sin \omega_0 n$ *for 0 < a < 1.*

Fig 1.44b: *The discrete time sequence represented by the equation,* $x_i(n) = a^n \sin \omega_0 n$ *for a > 1.*

Fig 1.44: *Imaginary part of complex exponential signal.*

1.3.5 Representation of a Discrete Time Signal as Summation of Impulses

A discrete time signal can be expressed as summation of impulses as shown below:

$$x(n) = \sum_{m=-\infty}^{+\infty} x(m)\,\delta(n-m)$$

where,   x(n) = Discrete time signal defined for all n

         $\delta(n)$ = Unit impulse signal

         $\delta(n-m)$ = Delayed impulse signal

Proof:

Let x(n) be a discrete time signal defined for all n.

We know that, $\delta(n) = 1$; at n = 0

                   $= 0$; when $n \neq 0$

and, $\delta(n-m) = 1$; at n = m

                      $= 0$; when $n \neq m$

If we multiply the signal x(n) with the delayed impulse $\delta(n-m)$ then the product is nonzero only at n = m and zero for all other values of n. Also at n = m, the value of product signal is m[th] sample x(m) of the signal x(n).

     $\therefore$ x(n) $\delta(n-m)$ = x(m)

Each multiplication of the signal x(n) by an unit impulse at some delay m, in essence picks out the single value x(m) of the signal x(n) at n = m, where the unit impulse is nonzero.

For example,   x(n) $\delta(n-(-2))$ =   x(-2)

                x(n) $\delta(n-(-1))$ =   x(-1)

                x(n) $\delta(n)$ =   x(0)

                x(n) $\delta(n-1)$ =   x(1)

                x(n) $\delta(n-2)$ =   x(2)

From the above products we can say that each sample of x(n) can be expressed as a product of the sample and delayed impulse, as shown below:

$$\vdots$$

$$
\begin{aligned}
x(-2) &= x(-2)\,\delta(n-(-2)) \\
x(-1) &= x(-1)\,\delta(n-(-1)) \\
x(0) &= x(0)\,\delta(n) \\
x(1) &= x(1)\,\delta(n-1) \\
x(2) &= x(2)\,\delta(n-2)
\end{aligned}
$$

$$\vdots$$

$$
\begin{aligned}
\therefore\quad x(n) &= \cdots\cdots + x(-2) + x(-1) + x(0) + x(1) + x(2) + \cdots\cdots\cdots \\
&= \cdots\cdots + x(-2)\,\delta(n-(-2)) + x(-1)\,\delta(n-(-1)) + x(0)\,\delta(n) + x(1)\,\delta(n-1) \\
&\quad + x(2)\,\delta(n-2) + \cdots\cdots\cdots \\
&= \sum_{m=-\infty}^{+\infty} x(m)\,\delta(n-m) \qquad\qquad\qquad\qquad \dots\dots(1.5)
\end{aligned}
$$

In equation (1.5) each product $x(m)\,\delta(n-m)$ is an impulse and the summation of impulses gives the sequence x(n).

1.3.6 Mathematical Operations on Discrete Time Signals

The mathematical operations that can be performed on discrete time signals are,

1. Amplitude scaling
2. Time scaling
3. Folding (or Time folding)
4. Shifting (or Time shifting) : Right shift (or advance) and
    left shift (or delay)
5. Addition
6. Multiplication

1. *Amplitude Scaling (or Scalar Multiplication) of Discrete Time Signals*

Amplitude scaling of a discrete time signal by a constant A is accomplished by multiplying the value of every signal sample by the constant A.

Example:

Let $x_1(n)$ be amplitude scaled signal of x(n), then $x_1(n) = A\,x(n)$

Let, $x(n) = 10$; $n = 0$ and $A = 0.2$, When $n = 0$; $x_1(0) = A\,x(0) = 0.2 \times 10 = 2.0$

$\qquad\qquad = 16$; $n = 1$ $\qquad\qquad$ When $n = 1$; $x_1(1) = A\,x(1) = 0.2 \times 16 = 3.2$

$\qquad\qquad = 20$; $n = 2$ $\qquad\qquad$ When $n = 2$; $x_1(2) = A\,x(2) = 0.2 \times 20 = 4.0$

2. *Time Scaling (or Downsampling and Upsampling) of Discrete Time Signals*

There are two ways of time scaling a discrete time signal. They are downsampling and upsampling.

In a signal x(n), if n is replaced by Dn, where D is an integer, then it is called ***downsampling***.

In a signal x(n), if n is replaced by $\frac{n}{I}$, where I is an integer, then it is called ***upsampling***.

> **Note:** *In upsampling, when $\frac{n}{I}$ is not an integer, the signal value is zero.*

Example:

If $x(n) = b^n$; $n \geq 0$; $0 < b < 1$, then

$x_1(n) = x(2n)$ is a down sampled version of x(n)

$x_2(n) = x\left(\frac{n}{2}\right)$ is an up sampled version on x(n).

When $n = 0$; $x_1(0) = x(0) = b^0$

When $n = 1$; $x_1(1) = x(2) = b^2$

When $n = 2$; $x_1(2) = x(4) = b^4$ and so on

When $n = 0$; $x_2(0) = x\left(\frac{0}{2}\right) = x(0) = b^0$

When $n = 1$; $x_2(1) = x\left(\frac{1}{2}\right) = 0$

When $n = 2$; $x_2(2) = x\left(\frac{2}{2}\right) = x(1) = b^1$

When $n = 3$; $x_2(3) = x\left(\frac{3}{2}\right) = 0$

When $n = 4$; $x_2(4) = x\left(\frac{4}{2}\right) = x(2) = b^2$ and so on

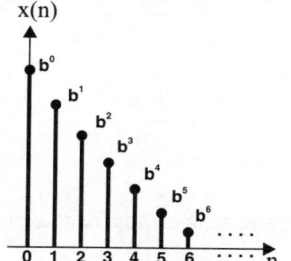 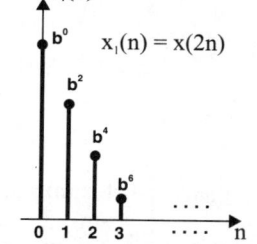 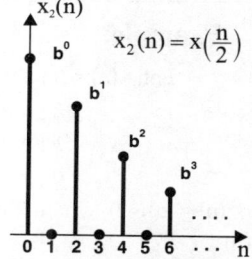

Fig 1.45a: A Discrete time signal x(n).   *Fig 1.45b:* Down sampled signal of x(n).   *Fig 1.45c:* Up sampled signal of x(n).

Fig 1.45: *A discrete time signal and its time scaled version.*

3. Folding (or Reflection or Transpose) of Discrete Time Signals

The folding of a discrete time signal x(n) is performed by changing the sign of the time base n in x(n). The folding operation produces a signal x(–n) which is a mirror image of the signal x(n) with respect to time origin n = 0.

Example:

If $x(n) = 0.8n$; $-2 \leq n \leq 2$. Now the folded signal, $x_1(n) = x(-n) = -0.8n$; $-2 \leq n \leq 2$.

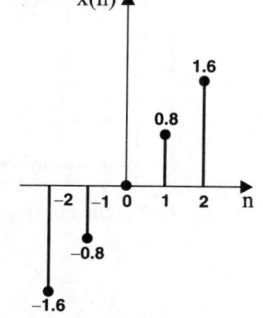 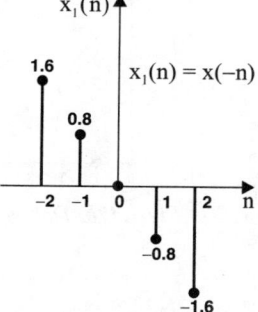

Fig 1.46a: *A discrete time signal x(n).*   **Fig 1.46b:** *Folded signal of x(n).*

Fig 1.46: *A discrete time signal and its folded version.*

4. Shifting (or delay and advance) of Discrete Time Signals

A signal x(n) may be shifted in time by replacing the independent variable n by n – m, where m is an integer. [i.e, x(n–m) is shifted version of x(n)].

In x(n–m), if m is a positive integer, the time shift results in a delay by m units of time. The *delay* results in shifting each sample of x(n) to the right.

If m is a negative integer, the time shift results in an advance of the signal by |m| units in time. The *advance* results in shifting each sample of x(n) to the left.

Note: *The shifting of folded signal will be opposite to that of signal without folding.*

      Example: *1. x(n+m) is left shifted or advanced version of x(n)*

               whereas x(–n+m) is right shifted or delayed version of x(–n).

         2. x(n–m) is right shifted or delayed version of x(n)

               whereas x(–n–m) is left shifted or advanced version of x(–n).

Example:

Let, x(n) = 3 ; n = 2

           = 2 ; n = 3

           = 1 ; n = 4

           = 0 ; for other n

Let, $x_1(n) = x(n–2)$, where $x_1(n)$ is delayed signal of x(n)

    When n = 4 ; $x_1(4) = x(4–2) = x(2) = 3$

    When n = 5 ; $x_1(5) = x(5–2) = x(3) = 2$

    When n = 6 ; $x_1(6) = x_1 6–2) = x(4) = 1$

The sample x(2) is available at n = 2 in the original sequence x(n), but the same sample is available at n = 4 in $x_1(n)$. Similarly every sample of x(n) is delayed by two sampling time.

Let, $x_2(n) = x(n+2)$, where $x_2(n)$ is an advanced signal of x(n)

    When n = 0 ; $x_2(0) = x(0+2) = x(2) = 3$

    When n = 1 ; $x_2(1) = x(1+2) = x(3) = 2$

    When n = 2 ; $x_2(2) = x(2+2) = x(4) = 1$

The sample x(2) is available at n = 2 in the original sequence x(n), but the same sample is available at n = 0 in $x_2(n)$. Similarly every sample of x(n) is advanced by two sampling time. Hence the signal $x_2(n)$ is an advanced version of x(n).

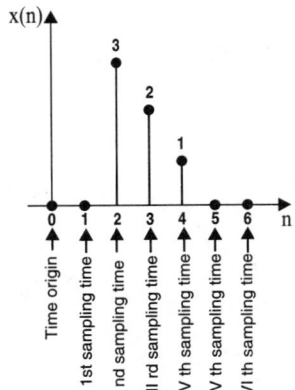

Fig 1.47a: *A discrete time signal x(n).*

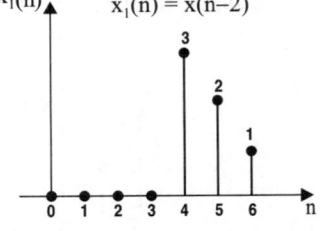

Fig 1.47b: *Delayed signal of x(n).*

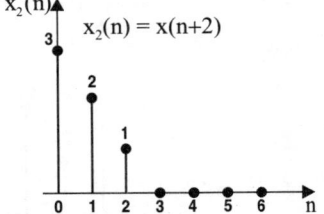

Fig 1.47c: *Advanced signal of x(n).*

Fig 1.47: *A discrete time signal and its shifted version.*

Delayed Discrete Unit Impulse Signal

The unit impulse signal is defined as,

$$\delta(n) = 1 \; ; \; \text{for } n = 0$$
$$= 0 \; ; \; \text{for } n \ne 0$$

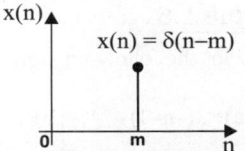

Fig 1.48: Delayed unit impulse.

The unit impulse signal delayed by m units of time is denoted as $\delta(n - m)$.

Now, $\delta(n - m) = 1 \; ; \; n = m$
$$= 0 \; ; \; n \ne m$$

Delayed Discrete Unit Step Signal

The unit step signal is defined as,

$$u(n) = 1 \; ; \; \text{for } n \ge 0$$
$$= 0 \; ; \; \text{for } n < 0$$

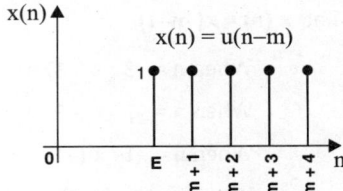

Fig 1.49: Delayed unit step signal.

The unit step signal delayed by m units of time is denoted as $u(n - m)$.

Now, $u(n - m) = 1 \; ; \; n \ge m$
$$= 0 \; ; \; n < m$$

5. Addition of Discrete Time Signals

The **addition** of two discrete time signals is performed on a sample-by-sample basis.

The sum of two signals $x_1(n)$ and $x_2(n)$ is a signal $x_3(n)$, whose value at any instant is equal to the sum of the samples of these two signals at that instant.

i.e., $x_3(n) = x_1(n) + x_2(n) \; ; \; -\infty < n < \infty.$

> **Example:**
>
> Let, $x_1(n) = \{2, 2, -1\}$ and $x_2(n) = \{-1, 1, 2\}$
>
> When n = 0 ; $x_3(0) = x_1(0) + x_2(0) = 2 + (-1) = 1$
>
> When n = 1 ; $x_3(1) = x_1(1) + x_2(1) = 2 + 1 \quad = 3$
>
> When n = 2 ; $x_3(2) = x_1(2) + x_2(2) = -1 + 2 \quad = 1$
>
> $\therefore \; x_3(n) = x_1(n) + x_2(n) = \{1, 3, 1\}$

6. Multiplication of Discrete Time Signals

The **multiplication** of two discrete time signals is performed on a sample-by-sample basis. The product of two signals $x_1(n)$ and $x_2(n)$ is a signal $x_3(n)$, whose value at any instant is equal to the product of the samples of these two signals at that instant. The product is also called **modulation**.

> **Example:**
>
> Let, $x_1(n) = \{ 2, 2, -1 \}$ and $x_2(n) = \{ -1, 1, 2 \}$
>
> When n = 0 ; $x_3(0) = x_1(0) \times x_2(0) = 2 \times (-1) = -2$
>
> When n = 1 ; $x_3(1) = x_1(1) \times x_2(1) = 2 \times 1 \quad = 2$
>
> When n = 2 ; $x_3(2) = x_1(2) \times x_2(2) = -1 \times 2 \quad = -2$
>
> $\therefore \; x_3(n) = x_1(n) \times x_2(n) = \{-2, 2, -2\}$

Example 1.6

Plot the following signals, if $x(n) = \{1, 4, 3, -1, 2\}$
                                              ↑

a) $x(-n-1)$ **b)** $x\left(-\frac{n}{2}\right)$ **c)** $x(-2n+1)$ **d)** $x\left(-\frac{n}{2}+2\right)$

Solution:

Given that, $x(n) = \{1, 4, 3, -1, 2\}$
                      ↑

$\therefore x(-2) = 1, x(-1) = 4, x(0) = 3, x(1) = -1, x(2) = 2$

a) Let, $x_1(n) = x(-n-1)$

When $n = -3$; $x_1(-3) = x(-(-3)-1) = x(2)$ $= 2$

When $n = -2$; $x_1(-2) = x(-(-2)-1) = x(1)$ $= -1$

When $n = -1$; $x_1(-1) = x(-(-1)-1) = x(0)$ $= 3$

When $n = 0$; $x_1(0) = x(-0-1)$ $= x(-1) = 4$

When $n = 1$; $x_1(1) = x(-1-1)$ $= x(-2) = 1$

For $n < -3$ and $n > 1$, $x_1(n)$ will be zero.

$\therefore x_1(n) = x(-n-1) = \{2, -1, 3, 4, 1\}$
                                      ↑

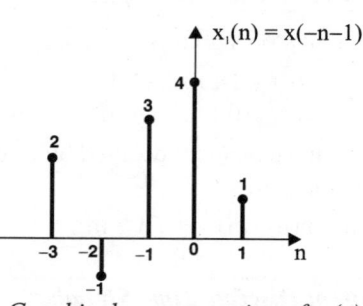

Fig 1: Graphical representation of $x_1(n) = x(-n-1)$.

b) Let, $x_2(n) = x\left(-\frac{n}{2}\right)$

When $n = -4$; $x_2(-4) = x\left(-\left(\frac{-4}{2}\right)\right) = x(2) = 2$ When $n = 1$; $x_2(1) = x\left(-\frac{1}{2}\right) = 0$

When $n = -3$; $x_2(-3) = x\left(-\left(\frac{-3}{2}\right)\right) = 0$ When $n = 2$; $x_2(2) = x\left(-\frac{2}{2}\right) = x(-1) = 4$

When $n = -2$; $x_2(-2) = x\left(-\left(\frac{-2}{2}\right)\right) = x(1) = -1$ When $n = 3$; $x_2(3) = x\left(-\frac{3}{2}\right) = 0$

When $n = -1$; $x_2(-1) = x\left(-\left(\frac{-1}{2}\right)\right) = x\left(\frac{1}{2}\right) = 0$ When $n = 4$; $x_2(4) = x\left(-\frac{4}{2}\right) = x(-2) = 1$

When $n = 0$; $x_2(0) = x\left(\frac{0}{2}\right) = x(0) = 3$

For $n < -4$ and $n > 4$, the $x_2(n)$ will be zero.

$\therefore x_2(n) = x\left(-\frac{n}{2}\right) = \{2, 0, -1, 0, 3, 0, 4, 0, 1\}$
                                                    ↑

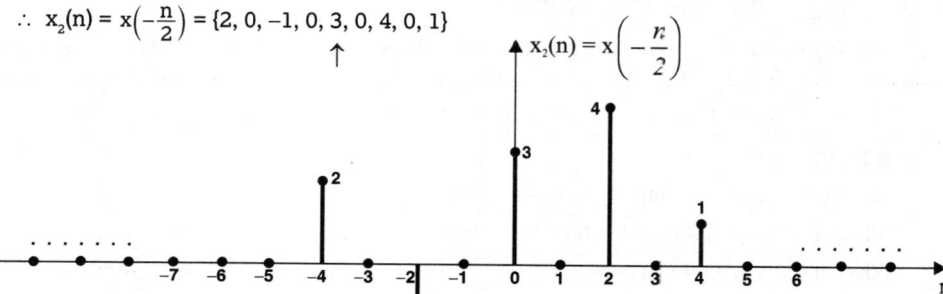

Fig 2: Graphical representation of $x_2(n) = x\left(-\frac{n}{2}\right)$.

c) Let, $x_3(n) = x(-2n+1)$

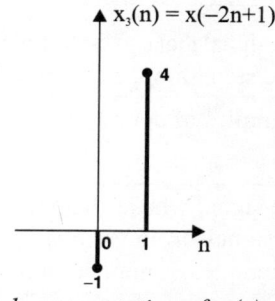

When $n = -1$; $x_3(-1) = x(-2(-1) + 1) = x(3) = 0$

When $n = 0$; $x_3(0) = x(-2(0) + 1) = x(1) = -1$

When $n = 1$; $x_3(1) = x(-2(1) + 1) = x(-1) = 4$

When $n = 2$; $x_3(2) = x(-2(2) + 1) = x(-3) = 0$

For $n < -1$ and $n > 1$, the $x_3(n)$ will be zero.

$\therefore x_3(n) = x(-2n + 1) = \{-1, 4\}$
$\qquad\qquad\qquad\qquad\quad\uparrow$

Fig 3: Graphical representation of $x_3(n)=x(-2n+1)$.

---

d) Let, $x_4(n) = x\left(-\dfrac{n}{2}+2\right)$

When $n = 0$; $x_4(0) = x\left(\dfrac{0}{2}+2\right) = x(2) = 2$

When $n = 1$; $x_4(1) = x\left(-\dfrac{1}{2}+2\right) = x\left(\dfrac{3}{2}\right) = 0$

When $n = 2$; $x_4(2) = x\left(-\dfrac{2}{2}+2\right) = x(1) = -1$

When $n = 3$; $x_4(3) = x\left(-\dfrac{3}{2}+2\right) = x\left(\dfrac{1}{2}\right) = 0$

When $n = 4$; $x_4(4) = x\left(-\dfrac{4}{2}+2\right) = x(0) = 3$

When $n = 5$; $x_4(5) = x\left(-\dfrac{5}{2}+2\right) = x\left(-\dfrac{1}{2}\right) = 0$

When $n = 6$; $x_4(6) = x\left(-\dfrac{6}{2}+2\right) = x(-1) = 4$

When $n = 7$; $x_4(7) = x\left(-\dfrac{7}{2}+2\right) = x\left(-\dfrac{3}{2}\right) = 0$

When $n = 8$; $x_4(8) = x\left(-\dfrac{8}{2}+2\right) = x(-2) = 1$

For $n < 0$ and $n > 8$, the $x_4(n)$ will be zero.

$\therefore x_4(n) = x\left(-\dfrac{n}{2}+2\right) = \{2, 0, -1, 0, 3, 0, 4, 0, 1\}$
$\qquad\qquad\qquad\qquad\qquad\qquad\quad\uparrow$

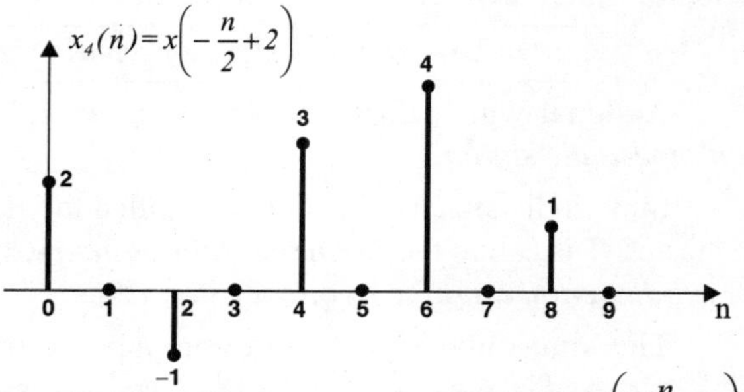

Fig 4: Graphical representation of $x_4(n) = x\left(-\dfrac{n}{2}+2\right)$.

1.4 Classification of Continuous Time Signals *(AU May'15, 10 Marks & Dec'12, 8 Marks)*

The continuous time signals are classified depending on their characteristics. Some ways of classifying continuous time signals are,

1. Deterministic and Nondeterministic (Random) signals
2. Periodic and Nonperiodic signals
3. Symmetric and Antisymmetric signals (Even and Odd signals)
4. Energy and Power signals
5. Causal and Noncausal signals

1.4.1 Deterministic and Nondeterministic (Random) Signals   *(AU May'11, 2 Marks)*
(AU Jun'13, 2 Marks)

The signal that can be completely specified by a mathematical equation is called a *deterministic signal*. The step, ramp, exponential and sinusoidal signals are examples of deterministic signals.

> Examples of deterministic signals: $x_1(t) = At$
> $$x_2(t) = X_m \sin\Omega_0 t$$

The signal whose characteristics are random in nature is called a *nondeterministic signal.* The noise signals from various sources like electronic amplifiers, oscillators, radio receivers, etc., are best examples of nondeterministic signals.

1.4.2 Periodic and Nonperiodic Signals

A periodic signal will have a definite pattern that repeats again and again over a certain period of time. Therefore, the signal which satisfies the condition,

> $x(t + T) = x(t)$ is called a *periodic signal.*

A signal which does not satisfy the condition, $x(t + T) = x(t)$ is called an *aperiodic or nonperiodic signal*.

In periodic signals, the term T is called the *fundamental time period* of the signal. Hence, inverse of T is called the *fundamental frequency,* F_0 in *cycles/sec* or *Hz*, and $2\pi F_0 = \Omega_0$ is called the *fundamental angular frequency* in *rad/sec*.

The sinusoidal signals and complex exponential signals are always periodic with a periodicity of T, where, $T = \dfrac{1}{F_0} = \dfrac{2\pi}{\Omega_0}$. The proof of this concept is given below:

Proof:

a) Cosinusoidal signal

Let, $x(t) = A\cos\Omega_0 t$

$\therefore x(t + T) = A\cos\Omega_0(t+T) = A\cos(\Omega_0 t + \Omega_0 T)$

$\qquad = A\cos\left(\Omega_0 t + \dfrac{2\pi}{T}\, T\right)$ $\boxed{\Omega_0 = 2\pi F_0 = \dfrac{2\pi}{T}}$

$\qquad = A\cos(\Omega_0 t + 2\pi) = A\cos\Omega_0 t = x(t)$ $\boxed{\cos(\theta + 2\pi) = \cos\theta}$

b) Sinusoidal signal

Let, $x(t) = A\sin\Omega_0 t$

$\therefore x(t + T) = A\sin\Omega_0(t+T) = A\sin(\Omega_0 t + \Omega_0 T)$

$\qquad = A\sin\left(\Omega_0 t + \dfrac{2\pi}{T}\, T\right)$ $\boxed{\Omega_0 = 2\pi F_0 = \dfrac{2\pi}{T}}$

$\qquad = A\sin(\Omega_0 t + 2\pi) = A\sin\Omega_0 t = x(t)$ $\boxed{\sin(\theta + 2\pi) = \sin\theta}$

c) Complex exponential signal

Let, $x(t) = A\sin\Omega_0 t$

$\therefore x(t + T) \doteq A e^{j\Omega_0(t+T)} = A e^{j\Omega_0 t} e^{j\Omega_0 T} = A e^{j\Omega_0 t} e^{j\frac{2\pi}{T}T} = A e^{j\Omega_0 t} e^{j2\pi}$ $\boxed{e^{j\theta} = \cos\theta + j\sin\theta}$

$\qquad = A e^{j\Omega_0 t}(\cos 2\pi + j\sin 2\pi) = A e^{j\Omega_0 t}(1 + j\,0) = x(t)$ $\boxed{\cos 2\pi = 1, \quad \sin 2\pi = 0}$

Periodicity of Sum of Periodic Signals

When a continuous time signal is a sum of two periodic signals with fundamental time periods T_1 and T_2, then the continuous time signal will be periodic, if the ratio of T_1 and T_2 (i.e., T_1/T_2) is a rational number. Now the periodicity of the continuous time signal will be the LCM (Least Common Multiple) of T_1 and T_2.

Note: *1. The ratio of two integers is called a rational number.*

Example of rational number: $\dfrac{5}{2}, \dfrac{7}{9}, \dfrac{8}{11}$.

Example of non-rational number: $\dfrac{\sqrt{2}}{5}, \dfrac{7}{2\pi}, \dfrac{4}{\sqrt{7}}$.

2. When T_1/T_2 is a rational number, then F_{01}/F_{02} and Ω_{01}/Ω_{02} are also rational numbers.

To Find LCM

Consider two numbers T_1, T_2.

Case i) If T_1 and T_2 are not divisible by common factor then LCM is given by product of T_1 and T_2. (i.e., LCM = $T_1 T_2$).

Example : $T_1 = 3$, $T_2 = 4$; LCM = $3 \times 4 = 12$.

Case ii) If T_1 and T_2 are divisible by common divisor then determine the greatest common divisor(GCD). Now the LCM is given by product of $T_1 T_2$ divided by GCD.

Example : $T_1 = 4$, $T_2 = 18$. Now, GCD = 2.

$$\boxed{\text{LCM} = \dfrac{T_1 T_2}{\text{GCD}}}$$

$$\text{LCM} = \dfrac{T_1 T_2}{\text{GCD}} = \dfrac{4 \times 18}{2} = 36$$

Example 1.7

Verify whether the following continuous time signals are periodic. If periodic, find the fundamental period.

a) $x(t) = 2 \cos \dfrac{t}{4}$ **b)** $x(t) = e^{\alpha t}$; $\alpha > 1$ **c)** $x(t) = e^{\frac{-j2\pi t}{7}}$

d) $x(t) = 3 \cos\left(5t + \dfrac{\pi}{6}\right)$ **e)** $x(t) = \cos^2\left(2t - \dfrac{\pi}{4}\right)$

Solution:

a) Given that, $x(t) = 2 \cos \dfrac{t}{4}$

The given signal is a cosinusoidal signal, which is always periodic.

On comparing x(t) with the standard form "$A \cos 2\pi F_0 t$" we get,

$$2\pi F_0 = \dfrac{1}{4} \quad \Rightarrow \quad F_0 = \dfrac{1}{8\pi}$$

Period, $T = \dfrac{1}{F_0} = 8\pi$

$\therefore$ x(t) is periodic with period, $T = 8\pi$

---

b) Given that, x(t) = $e^{\alpha t}$; $\alpha > 1$

$$\therefore \; x(t + T) = e^{\alpha(t + T)}$$

$$= e^{\alpha t}\, e^{\alpha T}$$

For any value of α, $e^{\alpha T} \ne 1$ and so $x(t + T) \ne x(t)$

Since $x(t + T) \ne x(t)$, the signal $x(t)$ is nonperiodic.

---

c) Given that, x(t) = $e^{\frac{-j2\pi t}{7}}$

The given signal is a complex exponential signal, which is always periodic.

On comparing $x(t)$ with the standard form "$A\,e^{-j2\pi F_0 t}$", we get,

$$2\pi F_0 = \frac{2\pi}{7} \quad \Rightarrow \quad F_0 = \frac{1}{7}$$

Peroid, $T = \dfrac{1}{F_0} = 7$

$\therefore$ $x(t)$ is periodic with period, $T = 7$

---

d) Given that, x(t) = $3\cos\left(5t + \dfrac{\pi}{6}\right)$

The given signal is a cosinusoidal signal, which is always periodic.

$$\therefore \;\; x(t + T) = 3\cos\left(5(t + T) + \frac{\pi}{6}\right) = 3\cos\left(5t + 5T + \frac{\pi}{6}\right) = 3\cos\left(\left(5t + \frac{\pi}{6}\right) + 5T\right)$$

Let $5T = 2\pi$, $\quad \therefore T = \dfrac{2\pi}{5}$

$$\therefore \;\; x(t + T) = 3\cos\left(\left(5t + \frac{\pi}{6}\right) + 5 \times \frac{2\pi}{5}\right) = 3\cos\left(\left(5t + \frac{\pi}{6}\right) + 2\pi\right)$$

$$= 3\cos\left(5t + \frac{\pi}{6}\right) = x(t)$$

For integer values of M, $\cos(\theta + 2\pi M) = \cos\theta$

Since $x(t + T) = x(t)$, the signal $x(t)$ is periodic with period, $T = \dfrac{2\pi}{5}$

---

e) Given that, x(t) = $\cos^2\left(2t - \dfrac{\pi}{3}\right)$

$$x(t) = \cos^2\left(2t - \frac{\pi}{3}\right) = \frac{1 + \cos 2\left(2t - \frac{\pi}{3}\right)}{2} = \frac{1 + \cos\left(4t - \frac{2\pi}{3}\right)}{2}$$

$\cos^2\theta = \dfrac{1 + \cos 2\theta}{2}$

$$\therefore \;\; x(t + T) = \frac{1 + \cos\left(4(t + T) - \frac{2\pi}{3}\right)}{2} = \frac{1 + \cos\left(4t + 4T - \frac{2\pi}{3}\right)}{2}$$

$$= \frac{1 + \cos\left(4t - \frac{2\pi}{3} + 4T\right)}{2}$$

Let $4T = 2\pi$, $\quad \therefore T = \dfrac{2\pi}{4} = \dfrac{\pi}{2}$

$$\therefore x(t + T) = \frac{1 + \cos\left(4t - \frac{2\pi}{3} + 4 \times \frac{\pi}{2}\right)}{2} = \frac{1 + \cos\left(\left(4t - \frac{2\pi}{3}\right) + 2\pi\right)}{2}$$

$$= \frac{1 + \cos\left(4t - \frac{2\pi}{3}\right)}{2} = \frac{1 + \cos 2\left(2t - \frac{\pi}{3}\right)}{2} = \cos^2\left(2t - \frac{\pi}{3}\right) = x(t)$$

Since $x(t + T) = x(t)$, the signal $x(t)$ is periodic with period, $T = \dfrac{\pi}{2}$

For integer values of M, $\cos(\theta + 2\pi M) = \cos\theta$

Example 1.8

Determine the periodicity of the following continuous time signals.

a) $x(t) = 2 \cos \dfrac{2\pi t}{3} + 3 \cos \dfrac{2\pi t}{7}$ b) $x(t) = 2 \cos 3t + 3 \sin 7t$ c) $x(t) = 5 \cos 4\pi t + 3 \sin 8\pi t$

d) $x(t) = \cos 2t + \sin \dfrac{t}{5}$ e) $x(t) = 4 \cos\left(3\pi t + \dfrac{\pi}{4}\right) + 2 \cos(4\pi t)$ f) $x(t) = 3 \cos t + 4 \cos \dfrac{t}{2}$

Solution:

a) Given that, $x(t) = 2 \cos \dfrac{2\pi t}{3} + 3 \cos \dfrac{2\pi t}{7}$

Let, $x_1(t) = 2 \cos \dfrac{2\pi t}{3}$

Let T_1 be the periodicity of $x_1(t)$. On comparing $x_1(t)$ with the standard form "A cos $2\pi F_{01}t$", we get,

$$2\pi F_{01} = \frac{2\pi}{3} \quad \Rightarrow \quad F_{01} = \frac{1}{3} \quad ; \quad \therefore \text{ Period, } T_1 = \frac{1}{F_{01}} = 3$$

Let, $x_2(t) = 3 \cos \dfrac{2\pi t}{7}$

Let T_2 be the periodicity of $x_2(t)$. On comparing $x_2(t)$ with the standard form "A sin $2\pi F_{02}t$", we get,

$$2\pi F_{02} = \frac{2\pi}{7} \quad \Rightarrow \quad F_{02} = \frac{1}{7} \quad ; \quad \therefore \text{ Period, } T_2 = \frac{1}{F_{02}} = 7$$

Now, $\dfrac{T_1}{T_2} = \dfrac{3}{7}$

Since $x_1(t)$ and $x_2(t)$ are periodic, and the ratio of T_1 and T_2 is a rational number, the signal $x(t)$ is also periodic. Let T be the periodicity of $x(t)$. Now the periodicity of $x(t)$ is the LCM (Least Common Multiple) of T_1 and T_2, i.e., LCM of 3 and 7, which is 21.

$$\therefore \text{ Period, } T = 21$$

Proof: $x(t + T) = 2 \cos \dfrac{2\pi(t+T)}{3} + 3 \cos \dfrac{2\pi(t+T)}{7} = 2 \cos\left(\dfrac{2\pi t}{3} + \dfrac{2\pi T}{3}\right) + 3 \cos\left(\dfrac{2\pi t}{7} + \dfrac{2\pi T}{7}\right)$

$$= 2 \cos\left(\frac{2\pi t}{3} + \frac{2\pi \times 21}{3}\right) + 3 \cos\left(\frac{2\pi t}{7} + \frac{2\pi \times 21}{7}\right) \qquad \boxed{\text{Put, } T = 21}$$

$$= 2 \cos\left(\frac{2\pi t}{3} + 14\pi\right) + 3 \cos\left(\frac{2\pi t}{7} + 6\pi\right)$$

$$= 2 \cos \frac{2\pi t}{3} + 3 \cos \frac{2\pi t}{7} = x(t) \qquad \boxed{\begin{array}{l}\text{For integer values of M,}\\ \cos(\theta + 2\pi M) = \cos\theta \end{array}}$$

---

b) Given that, $x(t) = 2 \cos 3t + 3 \sin 7t$

Let, $x_1(t) = 2 \cos 3t$

Let T_1 be the periodicity of $x_1(t)$. On comparing $x_1(t)$ with the standard form "A cos $2\pi F_{01}t$", we get,

$$2\pi F_{01} = 3 \quad \Rightarrow \quad F_{01} = \frac{3}{2\pi} \quad ; \quad \therefore \text{ Period, } T_1 = \frac{1}{F_{01}} = \frac{2\pi}{3}$$

Let, $x_2(t) = 3 \sin 7t$

Let T_2 be the periodicity of $x_2(t)$. On comparing $x_2(t)$ with the standard form "A sin $2\pi F_{02}t$", we get,

$$2\pi F_{01} = 7 \quad \Rightarrow \quad F_{02} = \frac{7}{2\pi} \quad ; \quad \therefore \text{ Period, } T_2 = \frac{1}{F_{02}} = \frac{2\pi}{7}$$

Now, $\dfrac{T_1}{T_2} = T_1 \times \dfrac{1}{T_2} = \dfrac{2\pi}{3} \times \dfrac{7}{2\pi} = \dfrac{7}{3}$

Since $x_1(t)$ and $x_2(t)$ are periodic and the ratio of T_1 and T_2 is a rational number, the signal $x(t)$ is also periodic. Let T be the periodicity of $x(t)$. Now the periodicity of $x(t)$ is the LCM (Least Common Multiple) of T_1 and T_2, which is calculated as shown below:

$$T_1 = \frac{2\pi}{3} = \frac{2\pi}{3} \times \frac{21}{2\pi} = 7$$

$$T_2 = \frac{2\pi}{7} = \frac{2\pi}{7} \times \frac{21}{2\pi} = 3$$

> **Note :** *To find LCM, first convert T_1 and T_2 to integers by multiplying by a common number. Find LCM of integer values of T_1 and T_2. Then divide this LCM by the common number.*

Now LCM of 7 and 3 is 21.

$$\therefore \text{Period, } T = 21 \div \frac{21}{2\pi} = 21 \times \frac{2\pi}{21} = 2\pi$$

Proof : $x(t + T) = 2 \cos 3(t + T) + 3 \sin 7(t + T)$

$$= 2 \cos(3t + 3T) + 3 \sin(7t + 7T)$$

$$= 2 \cos(3t + 3 \times 2\pi) + 3 \sin(7t + 7 \times 2\pi)$$ $\boxed{\text{Put, } T = 2\pi}$

$$= 2 \cos(3t + 6\pi) + 3 \sin(7t + 14\pi)$$

> For integer values of M,
> $\cos(\theta + 2\pi M) = \cos\theta$
> $\sin(\theta + 2\pi M) = \sin\theta$

$$= 2 \cos 3t + 3 \sin 7t = x(t)$$

---

c) Given that, $x(t) = 5 \cos 4\pi t + 3 \sin 8\pi t$ *(AU Dec'13, 5 Marks)*

Let, $x_1(t) = 5 \cos 4\pi t$

Let T_1 be the periodicity of $x_1(t)$. On comparing $x_1(t)$ with the standard form "A cos $2\pi F_{01}t$", we get,

$$2\pi F_{01} = 4\pi \qquad \Rightarrow \qquad F_{01} = 2 \; ; \qquad \qquad \therefore \text{Period, } T_1 = \frac{1}{F_{01}} = \frac{1}{2}$$

Let, $x_2(t) = 3 \sin 8\pi t$

Let T_2 be the periodicity of $x_2(t)$. On comparing $x_2(t)$ with the standard form "A sin $2\pi F_{02}t$", we get,

$$2\pi F_{02} = 8\pi \qquad \Rightarrow \qquad F_{02} = 4 \; ; \qquad \qquad \therefore \text{Period, } T_2 = \frac{1}{F_{02}} = \frac{1}{4}$$

Now, $\dfrac{T_1}{T_2} = T_1 \times \dfrac{1}{T_2} = \dfrac{1}{2} \times \dfrac{4}{1} = 2$

Since $x_1(t)$ and $x_2(t)$ are periodic and the ratio of T_1 and T_2 is a rational number, the signal $x(t)$ is also periodic. Let T be the periodicity of $x(t)$. Now, the periodicity of $x(t)$ is the LCM (Least Common Multiple) of T_1 and T_2, which is calculated as shown below:

$$T_1 = \frac{1}{2} = \frac{1}{2} \times 4 = 2$$

$$T_2 = \frac{1}{4} = \frac{1}{4} \times 4 = 1$$

> **Note :** *To find LCM, first convert T_1 and T_2 to integers by multiplying by a common number. Find LCM of integer values of T_1 and T_2. Then divide this LCM by the common number.*

Now LCM of 2 and 1 is 2.

$$\therefore \text{Period, } T = 2 \div 4 = 2 \times \frac{1}{4} = \frac{1}{2}$$

Proof: $x(t + T) = 5 \cos 4\pi(t + T) + 3 \sin 8\pi(t + T)$

$$= 5 \cos(4\pi t + 4\pi T) + 3 \sin(8\pi t + 8\pi T)$$

$$= 5 \cos\left(4\pi t + 4\pi \times \frac{1}{2}\right) + 3 \sin\left(8\pi t + 8\pi \times \frac{1}{2}\right)$$ $\boxed{\text{Put, } T = \dfrac{1}{2}}$

$$= 5 \cos(4\pi t + 2\pi) + 3 \sin(8\pi t + 4\pi)$$

> For integer values of M,
> $\cos(\theta + 2\pi M) = \cos\theta$
> $\sin(\theta + 2\pi M) = \sin\theta$

$$= 5 \cos 4\pi t + 3 \sin 8\pi t = x(t)$$

d) Given that, $x(t) = \cos 2t + \sin \dfrac{t}{5}$ (*AU Dec'14, 8 Marks*)

Let, $x_1(t) = \cos 2t$

Let T_1 be the periodicity of $x_1(t)$. On comparing $x_1(t)$ with the standard form "$A \cos 2\pi F_{01} t$", we get,

$$2\pi F_{01} = 2 \quad \Rightarrow \quad F_{01} = \frac{2}{2\pi} \quad ; \quad \therefore \text{ Period, } T_1 = \frac{1}{F_{01}} = \frac{2\pi}{2} = \pi$$

Let, $x_2(t) = \sin \dfrac{t}{5}$

Let T_2 be the periodicity of $x_2(t)$. On comparing $x_2(t)$ with the standard form "$A \sin 2\pi F_{02} t$", we get,

$$2\pi F_{02} = \frac{1}{5} \quad \Rightarrow \quad F_{02} = \frac{1}{5 \times 2\pi} = \frac{1}{10\pi} \quad ; \quad \therefore \text{ Period, } T_2 = \frac{1}{F_{02}} = 10\pi$$

Now, $\dfrac{T_1}{T_2} = T_1 \times \dfrac{1}{T_2} = \pi \times \dfrac{1}{10\pi} = \dfrac{1}{10}$

Since $x_1(t)$ and $x_2(t)$ are periodic, and the ratio of T_1 and T_2 is a rational number, the signal $x(t)$ is also periodic. Let T be the periodicity of $x(t)$. Now the periodicity of $x(t)$ is the LCM (Least Common Multiple) of T_1 and T_2, which is calculated as shown below:

$$T_1 = \pi = \pi \times \frac{1}{\pi} = 1$$

$$T_2 = 10\pi = 10\pi \times \frac{1}{\pi} = 10$$

> **Note:** *To find LCM, first convert T_1 and T_2 to integers by multiplying by a common number. Find LCM of integer values of T_1 and T_2. Then divide this LCM by the common number.*

Now LCM of 1 and 10 is 10.

$$\therefore \text{ Period, } T = 10 \div \frac{1}{\pi} = 10 \times \pi = 10\pi$$

Proof: $\quad x(t+T) = \cos 2(t+T) + \sin \left(\dfrac{t+T}{5}\right)$ $\boxed{\text{Put, } T = 10\pi}$

$$= \cos (2t + 10\pi) + \sin \left(\frac{t + 10\pi}{5}\right)$$

$$= \cos (2t + 10\pi) + \sin \left(\frac{t}{5} + 2\pi\right) = \cos 2t + \sin \frac{t}{5}$$

> For integer values of M,
> $\cos(\theta + 2\pi M) = \cos\theta$
> $\sin (\theta + 2\pi M) = \sin\theta$

---

e) Given that, $x(t) = 4 \cos \left(3\pi t + \dfrac{\pi}{4}\right) + 2 \cos (4\pi t)$ (*AU May'15, 6 Marks*)

Let, $x_1(t) = 4 \cos \left(3\pi t + \dfrac{\pi}{4}\right)$

Let T_1 be the periodicity of $x_1(t)$. On comparing $x_1(t)$ with the standard form "$A \cos 2\pi F_{01} t$" we get,

$$2\pi F_{01} = 3\pi \quad \Rightarrow \quad F_{01} = \frac{3}{2} \quad ; \quad \therefore \text{ Period, } T_1 = \frac{1}{F_{01}} = \frac{2}{3}$$

Let, $x_2(t) = 2 \cos(4\pi t)$

Let T_2 be the periodicity of $x_2(t)$. On comparing $x_2(t)$ with the standard form "$A \cos 2\pi F_{02} t$", we get,

$$2\pi F_{02} = 4\pi \quad \Rightarrow \quad F_{02} = 2 \quad ; \quad \therefore \text{ Period, } T_2 = \frac{1}{F_{02}} = \frac{1}{2}$$

Now, $\dfrac{T_1}{T_2} = T_1 \times \dfrac{1}{T_2} = \dfrac{2}{3} \times \dfrac{2}{1} = \dfrac{4}{3}$

Since $x_1(t)$ and $x_2(t)$ are periodic, and the ratio of T_1 and T_2 is a rational number, the signal $x(t)$ is also periodic. Let T be the periodicity of $x(t)$. Now, the periodicity of $x(t)$ is the LCM (Least Common Multiple) of T_1 and T_2, which is calculated as shown below:

$$T_1 = \frac{2}{3} = \frac{2}{3} \times 6 = 4$$

$$T_2 = \frac{1}{2} = \frac{1}{2} \times 6 = 3$$

> **Note:** *To find LCM, first convert T_1 and T_2 to integers by multiplying by a common number. Find LCM of integer values of T_1 and T_2. Then divide this LCM by the common number.*

Now LCM of 4 and 3 is 12.

$\therefore$ Period, $T = 12 \div 6 = 12 \times \dfrac{1}{6} = 2$

Proof : $x(t+T) = 4 \cos\left(3\pi(t+T)+\dfrac{\pi}{4}\right) + 2 \cos\left(4\pi(t+T)\right)$

$\qquad\qquad = 4 \cos\left(3\pi t + \dfrac{\pi}{4} + 3\pi T\right) + 2 \cos\left(4\pi t + 4\pi T\right)$

$\qquad\qquad = 4 \cos\left(3\pi t + \dfrac{\pi}{4} + 3\pi \times 2\right) + 2 \cos\left(4\pi t + 4\pi \times 2\right)$ $\qquad$ $\boxed{\text{Put, } T = 2}$

$\qquad\qquad = 4 \cos\left(3\pi t + \dfrac{\pi}{4} + 6\pi\right) + 2 \cos\left(4\pi t + 8\pi\right)$ $\qquad$ $\boxed{\begin{array}{l}\text{For integer values of M,}\\ \cos(\theta + 2\pi M) = \cos\theta\end{array}}$

$\qquad\qquad = 4 \cos\left(3\pi t + \dfrac{\pi}{4}\right) + 2 \cos\left(4\pi t\right) = x(t)$

f) Given that, $x(t) = 3 \cos t + 4 \cos\dfrac{t}{2}$ **(AU Dec'12, 4 Marks)**

Let, $x_1(t) = 3 \cos t$

Let T_1 be the periodicity of $x_1(t)$. On comparing $x_1(t)$ with the standard form "$A\cos 2\pi F_{01}t$", we get,

$2\pi F_{01} = 1 \qquad \Rightarrow \qquad F_{01} = \dfrac{1}{2\pi} \qquad ; \qquad \therefore \text{ Period, } T_1 = \dfrac{1}{F_{01}} = 2\pi$

Let, $x_2(t) = 4 \cos \dfrac{t}{2}$

Let T_2 be the periodicity of $x_2(t)$. On comparing $x_2(t)$ with the standard form "$A \cos 2\pi F_{02}t$", we get,

$2\pi F_{02} = \dfrac{1}{2} \qquad \Rightarrow \qquad F_{02} = \dfrac{1}{4\pi} \qquad ; \qquad \therefore \text{ Period, } T_2 = \dfrac{1}{F_{02}} = 4\pi$

Now, $\dfrac{T_1}{T_2} = T_1 \times \dfrac{1}{T_2} = 2\pi \times \dfrac{1}{4\pi} = \dfrac{1}{2}$

Since $x_1(t)$ and $x_2(t)$ are periodic, and the ratio of T_1 and T_2 is a rational number, the signal $x(t)$ is also periodic. Let T be the periodicity of $x(t)$. Now the periodicity of $x(t)$ is the LCM (Least Common Multiple) of T_1 and T_2, which is calculated as shown below:

$T_1 = 2\pi \times \dfrac{1}{\pi} = 2$

$T_2 = 4\pi \times \dfrac{1}{\pi} = 4$

$\boxed{\begin{array}{l}\textit{Note:} \ \textit{To find LCM, first convert } T_1 \textit{ and } T_2 \textit{ to integers}\\ \textit{by multiplying by a common number. Find LCM}\\ \textit{of integer values of } T_1 \textit{ and } T_2. \textit{ Then divide this}\\ \textit{LCM by the common number.}\end{array}}$

Now the LCM of 2 and 4 is 4

$\therefore$ Period, $T = 4 \div \dfrac{1}{\pi} = 4 \times \dfrac{\pi}{1} = 4\pi$

Proof: $x(t+T) = 3 \cos(t+T) + 4\cos\left(\dfrac{t+T}{2}\right) = 3 \cos(t+4\pi) + 4\cos\left(\dfrac{t+4\pi}{2}\right)$ $\quad$ $\boxed{\text{Put, } T = 4\pi}$

$\qquad\qquad = 3 \cos(t+4\pi) + 4\cos\left(\dfrac{t}{2}+2\pi\right)$ $\qquad$ $\boxed{\begin{array}{l}\text{For integer values of M,}\\ \cos(\theta + 2\pi M) = \cos\theta\end{array}}$

$\qquad\qquad = 3 \cos t + 4 \cos\dfrac{t}{2} = x(t)$

1.4.3 Symmetric (Even) and Antisymmetric (Odd) Signals

The signals may exhibit symmetry or antisymmetry with respect to $t = 0$.

When a signal exhibits symmetry with respect to $t = 0$, then it is called an *even signal*. Therefore, the even signal satisfies the condition, **x(−t) = x(t).**

When a signal exhibits antisymmetry with respect to $t = 0$, then it is called an *odd signal*. Therefore, the odd signal satisfies the condition, **x(−t) = −x(t).**

Since $\cos(-\theta) = \cos\theta$, the cosinusoidal signals are even signals and since $\sin(-\theta) = -\sin\theta$, the sinusoidal signals are odd signals.

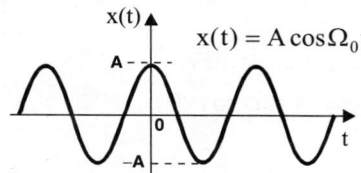
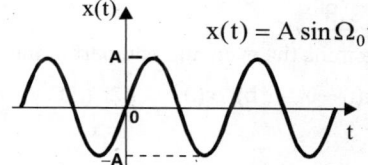

Fig 1.50a: *Symmetric (or even) signal.*   **Fig 1.50b:** *Antisymmetric (or odd) signal.*

Fig 1.50: *Symmetric and antisymmetric continuous time signals.*

The properties of signals with symmetry are given below without proof.

1. When a signal is even, then its odd part will be zero.
2. When a signal is odd, then its even part will be zero.
3. The product of two odd signals will be an even signal.
4. The product of two even signals will be an even signal.
5. The product of an even and odd signal will be an odd signal.

Neither Even Nor Odd Signals

A continuous time signal $x(t)$ which is neither even nor odd can be expressed as a sum of even and odd signal.

Let, $x(t) = x_e(t) + x_o(t)$

where, $x_e(t) = $ Even part of $x(t)$ and $x_o(t) = $ Odd part of $x(t)$

Now, it can be proved that,

$$x_e(t) = \frac{1}{2}\left[x(t) + x(-t)\right]$$

$$x_o(t) = \frac{1}{2}\left[x(t) - x(-t)\right]$$

Proof:

Let, $x(t) = x_e(t) + x_o(t)$ .....(1.6)

On replacing t by $-t$ in equation (1.6) we get,

$x(-t) = x_e(-t) + x_o(-t)$ .....(1.7)

Since $x_e(t)$ is even, $x_e(-t) = x_e(t)$

Since $x_o(t)$ is odd, $x_o(-t) = -x_o(t)$

Hence the equation (1.7) can be written as,

$x(-t) = x_e(t) - x_o(t)$ .....(1.8)

On adding equations (1.6) & (1.8) we get,

$x(t) + x(-t) = 2\,x_e(t)$

$\therefore\; x_e(t) = \frac{1}{2}[x(t) + x(-t)]$

On subtracting equation (1.8) from equation (1.6) we get,

$x(t) - x(-t) = 2\,x_o(t)$

$\therefore\; x_o(t) = \frac{1}{2}[x(t) - x(-t)]$

Example 1.9

Determine the even and odd part of the following continuous time signals.

a) $x(t) = e^t$ **b)** $x(t) = 3 + 2t + 5t^2$ **c)** $x(t) = \sin 2t + \cos t + \sin t \cos 2t$

Solution:

a) Given that, $x(t) = e^t$

$$\therefore x(-t) = e^{-t}$$

Even part, $x_e(t) = \frac{1}{2}[x(t) + x(-t)] = \frac{1}{2}[e^t + e^{-t}]$

Odd part, $x_o(t) = \frac{1}{2}[x(t) - x(-t)] = \frac{1}{2}[e^t - e^{-t}]$

b) Given that, $x(t) = 3 + 2t + 5t^2$

$$\therefore x(-t) = 3 + 2(-t) + 5(-t)^2$$

$$= 3 - 2t + 5t^2$$

Even part, $x_e(t) = \frac{1}{2}[x(t) + x(-t)] = \frac{1}{2}[3 + 2t + 5t^2 + 3 - 2t + 5t^2]$

$$= \frac{1}{2}[6 + 10t^2] = 3 + 5t^2$$

Odd part, $x_o(t) = \frac{1}{2}[x(t) - x(-t)] = \frac{1}{2}[3 + 2t + 5t^2 - 3 + 2t - 5t^2]$

$$= \frac{1}{2}[4t] = 2t$$

c) Given that, $x(t) = \sin 2t + \cos t + \sin t \cos 2t$

$$\therefore x(-t) = \sin 2(-t) + \cos(-t) + \sin(-t) \cos 2(-t)$$

| $\cos(-\theta) = \cos\theta$ |
| $\sin(-\theta) = -\sin\theta$ |

$$= -\sin 2t + \cos t - \sin t \cos 2t$$

Even part, $x_e(t) = \frac{1}{2}[x(t) + x(-t)] = \frac{1}{2}[\sin 2t + \cos t + \sin t \cos 2t - \sin 2t + \cos t - \sin t \cos 2t]$

$$= \frac{1}{2}[2 \cos t] = \cos t$$

Odd part, $x_o(t) = \frac{1}{2}[x(t) - x(-t)] = \frac{1}{2}[\sin 2t + \cos t + \sin t \cos 2t + \sin 2t - \cos t + \sin t \cos 2t]$

$$= \frac{1}{2}[2 \sin 2t + 2 \sin t \cos 2t] = \sin 2t + \sin t \cos 2t$$

Example 1.10

Sketch the even and odd parts of the following signals.

a)

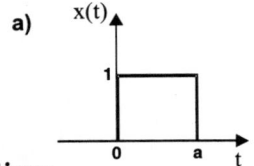

b)

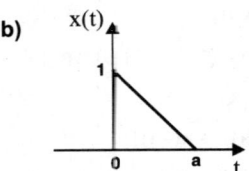

Solution:

a) The even part of the signal is given by, $x_e(t) = \frac{1}{2}[x(t) + x(-t)] = \frac{1}{2}x(t) + \frac{1}{2}x(-t)$ .....(1)

The odd part of the signal is given by, $x_o(t) = \frac{1}{2}[x(t) - x(-t)] = \frac{1}{2}x(t) - \frac{1}{2}x(-t)$ .....(2)

From equations (1) and (2), it is observed that the even and odd parts of the signal can be obtained from the folded and scaled versions of the signal. Hence the given signal is folded, scaled and then graphically added and subtracted to get the even and odd parts as shown below:

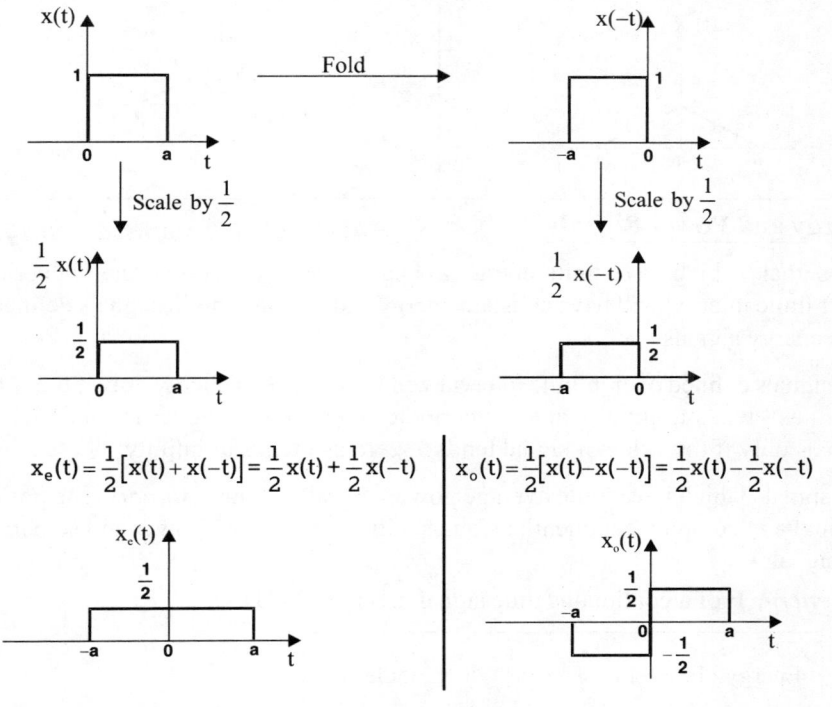

$$x_e(t) = \frac{1}{2}[x(t) + x(-t)] = \frac{1}{2}x(t) + \frac{1}{2}x(-t) \qquad x_o(t) = \frac{1}{2}[x(t) - x(-t)] = \frac{1}{2}x(t) - \frac{1}{2}x(-t)$$

b) The even part of the signal is given by, $x_e(t) = \frac{1}{2}[x(t) + x(-t)] = \frac{1}{2}x(t) + \frac{1}{2}x(-t)$ .....(1)

The odd part of the signal is given by, $x_o(t) = \frac{1}{2}[x(t) - x(-t)] = \frac{1}{2}x(t) - \frac{1}{2}x(-t)$ .....(2)

From equations (1) and (2), it is observed that the even and odd parts of the signal can be obtained from the folded and scaled versions of the signal. Hence the given signal is folded, scaled and then graphically added and subtracted to get the even and odd parts as shown below:

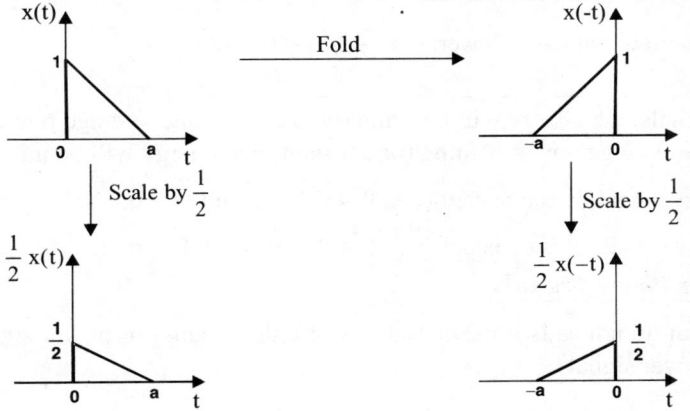

$$x_e(t) = \frac{1}{2}\big[x(t) + x(-t)\big] = \frac{1}{2}x(t) + \frac{1}{2}x(-t) \qquad \qquad x_0(t) = \frac{1}{2}\big[x(t) - x(-t)\big] = \frac{1}{2}x(t) - \frac{1}{2}x(-t)$$

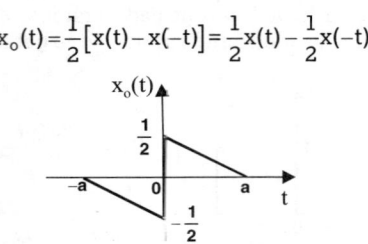

1.4.4 Energy and Power Signals   *(AU May'15, 2 Marks & Jun'13, 4 Marks)*

The signals which have finite energy are called *energy signals*. The nonperiodic signals defined over finite interval will have constant energy and so nonperiodic signals defined over finite interval are energy signals.

The signals defined over infinite interval will have constant energy only if the value of signal tends to zero as t tends to infinity and nonperiodic signals defined over infinite intervals will be energy signals only if the value of signal tends to zero as t tends to infinity.

The signals which have finite average power are called *power signals*. The periodic signals like sinusoidal and complex exponential signals will have constant power and so periodic signals are power signals.

The *energy* E of a continuous time signal x(t) is defined as,

$$\text{Energy,} \quad E = \underset{T \to \infty}{\text{Lt}} \int_{-T}^{T} |x(t)|^2 \, dt \text{ in joules}$$

The average *power* of a continuous time signal x(t) is defined as,

$$\text{Power,} \quad P = \underset{T \to \infty}{\text{Lt}} \frac{1}{2T} \int_{-T}^{T} |x(t)|^2 \, dt \text{ in watts}$$

For periodic signals, the average power over one period will be same as average power over an infinite interval.

$$\therefore \text{ For periodic signals, Power, } P = \frac{1}{T} \int_{0}^{T} |x(t)|^2 \, dt$$

For energy signals, the energy will be finite (or constant) and average power will be zero. For power signals the average power is finite (or constant) and energy will be infinite.

i.e., For energy signal, E is constant (i.e., $0 < E < \infty$) and $P = 0$.

For power signal, P is constant (i.e., $0 < P < \infty$) and $E = \infty$.

Neither Energy Nor Power Signals

The signals that do not satisfy the conditions of either energy or power signals are called neither energy nor power signals.

Proof:

The energy of a signal x(t) is defined as ,

$$E = \underset{T \to \infty}{Lt} \int_{-T}^{T} |x(t)|^2 \, dt \qquad \qquad \qquad \qquad .....(1.9)$$

The power of a signal is defined as ,

$$P = \underset{T \to \infty}{Lt} \frac{1}{2T} \int_{-T}^{T} |x(t)|^2 \, dt = \underset{T \to \infty}{Lt} \frac{1}{2T} \underset{T \to \infty}{Lt} \int_{-T}^{T} |x(t)|^2 \, dt \qquad \qquad .....(1.10)$$

Using equation (1.9), the equation (1.10) can be written as,

$$P = \underset{T \to \infty}{Lt} \frac{1}{2T} \times E \qquad \qquad \qquad \qquad .....(1.11)$$

In equation (1.11), When E = constant,

$$P = E \times \underset{T \to \infty}{Lt} \frac{1}{2T}$$

$$= E \times \frac{1}{2 \times \infty} = E \times 0 = 0$$

From the above analysis, we can say that when a signal has finite energy the power will be zero. Also, from the above analysis we can say that the power is finite only when energy is infinite.

Example 1.11

Determine the power and energy for the following continuous time signals.

a) $x(t) = e^{-2t} u(t)$ **b)** $x(t) = e^{j\left(2t + \frac{\pi}{4}\right)}$ **c)** $x(t) = 3 \cos 5\Omega_0 t$ **d)** $x(t) = 2t^2 u(t)$

e) $x(t) = rect\left(\dfrac{t}{T_0}\right)$ **f)** $x(t) = \cos^2(\Omega_0 t)$

Solution:

a) Given that, x(t) = e⁻²ᵗ u(t)   *(AU Dec'12, 2 Marks)*

Here, $x(t) = e^{-2t} u(t)$; for all t

$\therefore x(t) = e^{-2t}$; for t ≥ 0

$$\therefore \int_{-T}^{T} |x(t)|^2 \, dt = \int_{0}^{T} \left(|e^{-2t}|\right)^2 dt = \int_{0}^{T} (e^{-2t})^2 \, dt = \int_{0}^{T} e^{-4t} \, dt = \left[\frac{e^{-4t}}{-4}\right]_{0}^{T} = \left[\frac{e^{-4T}}{-4} - \frac{e^0}{-4}\right]$$

$$\therefore \int_{-T}^{T} |x(t)|^2 \, dt = \left[\frac{1}{4} - \frac{e^{-4T}}{4}\right] \qquad \qquad .....(1)$$

Energy, $E = \underset{T \to \infty}{Lt} \int_{-T}^{T} |x(t)|^2 \, dt = \underset{T \to \infty}{Lt} \left[\frac{1}{4} - \frac{e^{-4T}}{4}\right]$ $\boxed{\text{Using equation (1)}}$

$$= \frac{1}{4} - \frac{e^{-\infty}}{4} = \frac{1}{4} - \frac{0}{4} = \frac{1}{4} \text{ joules}$$

Power, $P = \underset{T \to \infty}{Lt} \frac{1}{2T} \int_{-T}^{T} |x(t)|^2 \, dt = \underset{T \to \infty}{Lt} \frac{1}{2T} \left[\frac{1}{4} - \frac{e^{-4T}}{4}\right]$ $\boxed{\text{Using equation (1)}}$

$$= \frac{1}{\infty} \left[\frac{1}{4} - \frac{e^{-\infty}}{4}\right] = 0 \times \left[\frac{1}{4} - 0\right] = 0$$

Since energy is constant and power is zero, the given signal is an energy signal.

b) Given that, $x(t) = e^{j\left(2t + \frac{\pi}{4}\right)}$

Here, $x(t) = e^{j\left(2t + \frac{\pi}{4}\right)} = 1\angle\left(2t + \frac{\pi}{4}\right)$

$\therefore |x(t)| = 1$

$$\int_{-T}^{T} |x(t)|^2 \, dt = \int_{-T}^{T} 1 \times dt = [t]_{-T}^{T} = T + T = 2T \qquad \text{.....(1)}$$

Energy, $E = \underset{T \to \infty}{\text{Lt}} \int_{-T}^{T} |x(t)|^2 \, dt = \underset{T \to \infty}{\text{Lt}} \, 2T = \infty$ ┃Using equation (1)┃

Power, $P = \underset{T \to \infty}{\text{Lt}} \dfrac{1}{2T} \int_{-T}^{T} |x(t)|^2 \, dt = \underset{T \to \infty}{\text{Lt}} \dfrac{1}{2T} \times 2T = 1 \, watt$ ┃Using equation (1)┃

Since power is constant and energy is infinite, the given signal is a power signal.

---

c) Given that, $x(t) = 3 \cos 5\Omega_0 t$

$$\therefore \int_{-T}^{T} |x(t)|^2 \, dt = \int_{-T}^{T} (|3 \cos 5\Omega_0 t|)^2 \, dt = \int_{-T}^{T} |(3 \cos 5\Omega_0 t)^2| \, dt = \int_{-T}^{T} (3 \cos 5\Omega_0 t)^2 \, dt$$

$$= \int_{-T}^{T} 9 \cos^2 5\Omega_0 t \, dt = 9 \int_{-T}^{T} \left(\frac{1 + \cos 10\Omega_0 t}{2}\right) dt \qquad \boxed{\cos^2 \theta = \frac{1 + \cos 2\theta}{2}}$$

$$= \frac{9}{2} \int_{-T}^{T} (1 + \cos 10\Omega_0 t) \, dt = \frac{9}{2}\left[t + \frac{\sin 10\Omega_0 t}{10\Omega_0}\right]_{-T}^{T}$$

$$= \frac{9}{2}\left[T + \frac{\sin 10\Omega_0 T}{10\Omega_0} - \left(-T + \frac{\sin 10\Omega_0(-T)}{10\Omega_0}\right)\right]$$

$$= \frac{9}{2}\left[2T + 2\frac{\sin 10\Omega_0 T}{10\Omega_0}\right] = \frac{9}{2}\left[2T + 2\frac{\sin 10\frac{2\pi}{T}T}{10\frac{2\pi}{T}}\right] \qquad \boxed{\sin(-\theta) = -\sin\theta} \\ \boxed{\Omega_0 = 2\pi F_0 = \frac{2\pi}{T}}$$

$$= \frac{9}{2}\left[2T + \frac{T}{10\pi}\sin 20\pi\right] = \frac{9}{2}\left[2T + \frac{T}{10\pi} \times 0\right] = 9T \qquad \text{.....(1)} \qquad \boxed{\begin{array}{l}\text{For integer M,}\\ \sin \pi M = 0\end{array}}$$

Energy, $E = \underset{T \to \infty}{\text{Lt}} \int_{-T}^{T} |x(t)|^2 \, dt = \underset{T \to \infty}{\text{Lt}} \, 9T = \infty$ ┃Using equation (1)┃

Power, $P = \underset{T \to \infty}{\text{Lt}} \dfrac{1}{2T} \int_{-T}^{T} |x(t)|^2 \, dt = \underset{T \to \infty}{\text{Lt}} \dfrac{1}{2T} \times 9T = \underset{T \to \infty}{\text{Lt}} \dfrac{9}{2} = \dfrac{9}{2} = 4.5 \, watts$ ┃Using equation (1)┃

Since energy is infinite and power is constant, the given signal is a power signal.

---

d) Given that, $x(t) = 2t^2 \, u(t)$

Here, $x(t) = 2t^2 \, u(t)$; for all t

$x(t) = 2t^2$; for $t \geq 0$

$$\therefore \int_{-T}^{T} |x(t)|^2 \, dt = \int_{0}^{T} |2t^2|^2 \, dt = \int_{0}^{T} 4t^4 \, dt = 4\left[\frac{t^5}{5}\right]_{0}^{T} = \frac{4}{5}T^5 \qquad \text{.....(1)}$$

Energy, $E = \underset{T \to \infty}{\text{Lt}} \int_{-T}^{T} |x(t)|^2 \, dt = \underset{T \to \infty}{\text{Lt}} \left[\frac{4}{5}T^5\right] = \infty$ ┃Using equation (1)┃

$$\text{Power, P} = \underset{T \to \infty}{Lt}\ \frac{1}{2T} \int\limits_{-T}^{T} |x(t)|^2\, dt = \underset{T \to \infty}{Lt}\ \frac{1}{2T} \left[\frac{4}{5}\, T^5\right]$$

Using equation (1)

$$= \frac{2}{5} \underset{T \to \infty}{Lt}\ [T^4] = \infty$$

Since both energy and power are infinte, the given signal is neither energy nor power signal.

e) Given that, $x(t) = \text{rect}\left(\frac{t}{T_0}\right)$ *(AU Jun'13, 4 Marks)*

$$\text{rect}(t) = 1\ ,\ -\frac{1}{2} < t < \frac{1}{2}\ \ ;\ \ \ \text{rect}\left(\frac{t}{T_0}\right) = 1\ ;\ \ \frac{-1}{2} < \frac{t}{T_0} < \frac{1}{2}\ \text{(or)}\ \frac{-T_0}{2} < t < \frac{T_0}{2}$$
$$= 0\,,\ \text{otherwise}$$

$$\therefore \int\limits_{-T}^{T} |x(t)|^2 dt = \int\limits_{-\frac{T_0}{2}}^{\frac{T_0}{2}} 1^2\, dt = \left[t\right]_{-\frac{T_0}{2}}^{\frac{T_0}{2}} = \frac{T_0}{2} - \left(-\frac{T_0}{2}\right) = T_0 \qquad\qquad .....(1)$$

$$\text{Energy, E} = \underset{T \to \infty}{Lt} \int\limits_{-T}^{T} |x(t)|^2 dt = \underset{T \to \infty}{Lt}\ T_0 = T_0\ \text{joules}$$

Using equation (1)

$$\text{Power, P} = \underset{T \to \infty}{Lt}\ \frac{1}{2T} \int\limits_{-T}^{T} |x(t)|^2 dt = \underset{T \to \infty}{Lt}\ \frac{1}{2T} \times T_0 = 0$$

Using equation (1)

Since energy is constant and power is zero, the given signal is an energy signal.

f) Given that, $x(t) = \cos^2(\Omega_0 t)$ *(AU Jun'13, 4 Marks)*

$$\therefore \int\limits_{-T}^{T} |x(t)|^2\, dt = \int\limits_{-T}^{T} \left(|\cos^2 \Omega_0 t|\right)^2\, dt = \int\limits_{-T}^{T} (\cos^2 \Omega_0 t)^2\, dt$$

$$\cos^2\theta = \frac{1 + \cos 2\theta}{2}$$

$$= \int\limits_{-T}^{T} \left(\frac{1 + \cos 2\Omega_0 t}{2}\right)^2\, dt$$

$$= \int\limits_{-T}^{+T} \frac{1}{4}(1 + \cos 2\Omega_0 t)^2 = \frac{1}{4} \int\limits_{-T}^{+T} (1 + 2\cos 2\Omega_0 t + \cos^2 2\Omega_0 t)\, dt$$

$$= \frac{1}{4} \int\limits_{-T}^{+T} \left(1 + 2\cos 2\Omega_0 t + \frac{1 + \cos 4\Omega_0 t}{2}\right) dt = \frac{1}{4} \int\limits_{-T}^{+T} \frac{2 + 4\cos 2\Omega_0 t + 1 + \cos 4\Omega_0 t}{2}\, dt$$

$$= \frac{1}{8} \int\limits_{-T}^{+T} (3 + 4\cos 2\Omega_0 t + \cos 4\Omega_0 t)\, dt = \frac{1}{8}\left[3t + \frac{4\sin 2\Omega_0 t}{2\Omega_0} + \frac{\sin 4\Omega_0 t}{4\Omega_0}\right]_{-T}^{+T}$$

$$= \frac{1}{8}\left[3T - (-3T) + \frac{4\sin 2\Omega_0 T}{2\Omega_0} - \frac{4\sin 2\Omega_0(-T)}{2\Omega_0} + \frac{\sin 4\Omega_0 T}{4\Omega_0} - \frac{\sin 4\Omega_0(-T)}{4\Omega_0}\right]$$

$$= \frac{1}{8}\left[6T + \frac{8\sin 2\Omega_0 T}{2\Omega_0} + \frac{2\sin 4\Omega_0 T}{4\Omega_0}\right] = \frac{6}{8}T + \frac{\sin 2\Omega_0 T}{2\Omega_0} + \frac{\sin 4\Omega_0 T}{16\Omega_0}$$

$$= \frac{3}{4}T + \frac{\sin 2 \times \frac{2\pi}{T} \times T}{2 \times \frac{2\pi}{T}} + \frac{\sin 4 \times \frac{2\pi}{T} \times T}{16 \times \frac{2\pi}{T}}$$

$$\Omega_0 = 2\pi F_0 = \frac{2\pi}{T}$$

For integer M,
$\sin \pi M = 0$

$$= \frac{3}{4}T + \frac{\sin 4\pi}{\frac{4\pi}{T}} + \frac{\sin 8\pi}{\frac{32\pi}{T}} = \frac{3}{4}T \qquad\qquad .....(1)$$

Energy, $E = \underset{T \to \infty}{Lt} \int_{-T}^{+T} |x(t)|^2 \, dt = \underset{T \to \infty}{Lt} \frac{3}{4} T = \infty$ \qquad $\boxed{\text{Using equation (1)}}$

Power, $P = \underset{T \to \infty}{Lt} \frac{1}{2T} \int_{-T}^{+T} |x(t)|^2 \, dt = \underset{T \to \infty}{Lt} \frac{1}{2T} \times \frac{3}{4} T = \underset{T \to \infty}{Lt} \frac{3}{8} = \frac{3}{8} \, watt$ \qquad $\boxed{\text{Using equation (1)}}$

Since energy is infinite and power is constant, the given signal is a power signal.

1.4.5 Causal, Noncausal and Anticausal Signals

A signal is said to be *causal*, if it is defined for $t \geq 0$.

Therefore if x(t) is causal, then x(t) = 0, for t < 0.

A signal is said to be *noncausal*, if it is defined for either $t \leq 0$, or for both $t \leq 0$ and $t > 0$.

Therefore if x(t) is noncausal, then $x(t) \neq 0$, for t < 0.

When a noncausal signal is defined only for $t \leq 0$, it is called *anticausal signal*.

Examples of Causal and Noncausal Signals

Step signal,   x(t) = A   ; $t \geq 0$

Unit step signal,   x(t) = u(t) = 1   ; $t \geq 0$

Exponential signal, $x(t) = A \, e^{bt} \, u(t)$

Complex exponential signal, $x(t) = Ae^{j\Omega_0 t} \, u(t)$

⎫ Causal signals

Unit step signal,   x(t) = u(-t) = 1 ; t < 0

Exponential signal, $x(t) = Ae^{bt} \, u(-t)$

⎱ Anticausal signals

Exponential signal, $x(t) = A \, e^{bt}$; for all t

Complex exponential signal, $x(t) = Ae^{j\Omega_0 t}$; for all t

⎬ Noncausal signals

Note: On multiplying a noncausal signal by u(t), it becomes causal.

1.5 Classification of Discrete Time Signals   *(AU Dec'11, 16 Marks & May'15, 10 Marks)* *(AU Dec'12, 8 Marks)*

The discrete time signals are classified depending on their characteristics. Some ways of classifying discrete time signals are,

     1. Deterministic and Nondeterministic (Random) Signals

     2. Periodic and Nonperiodic Signals

     3. Symmetric (Even) and Antisymmetric (Odd) Signals

     4. Energy and Power Signals

     5. Causal and Noncausal Signals

1.5.1 Deterministic and Nondeterministic (Random) Signals

The signals that can be completely specified by mathematical equations are called *deterministic signals*. The step, ramp, exponential and sinusoidal signals are examples of deterministic signals.

The signals whose characteristics are random in nature are called *nondeterministic signals.* The noise signals from various sources are best examples of nondeterministic signals.

1.5.2 Periodic and Nonperiodic Signals

When a discrete time signal x(n), satisfies the condition **x(n + N) = x(n)** for integer values of N, then the discrete time signal x(n) is called *periodic signal*. Here N is the number of samples of a period.

> i.e, if, **x(n + N) = x(n)**, for all n, then x(n) is periodic.

The smallest value of N for which the above equation is true is called *fundamental period*. If there is no value of N that satisfies the above equation, then x(n) is called *aperiodic* or *nonperiodic* signal.

When N is the fundamental period, the periodic signals will also satisfy the condition x(n + kN) = x(n), where k is an integer. The periodic signals are power signals. The discrete time sinusoidal and complex exponential signals are periodic signals when their fundamental frequency, f_0 is a rational number.

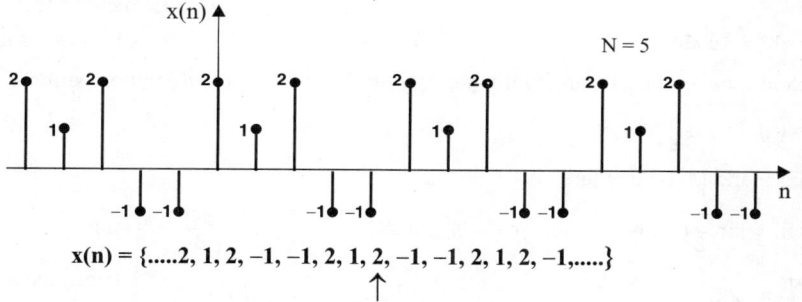

$$x(n) = \{.....2, 1, 2, -1, -1, 2, 1, 2, -1, -1, 2, 1, 2, -1,.....\}$$

Fig 1.51: *Periodic discrete time signal.*

Periodicity of Sum of Periodic Signal

When a discrete time signal is a sum or product of two periodic signals with fundamental periods N_1 and N_2, then the discrete time signal will be periodic with period given by LCM of N_1 and N_2.

Example 1.12

Determine whether following signals are periodic or not. If periodic find the fundamental period.

a) $x(n) = \cos\left(\frac{5\pi}{9}n + 1\right)$ **b)** $x(n) = \sin\left(\frac{n}{9} - \pi\right)$ **c)** $x(n) = \sin\frac{\pi}{8}n^2$ **d)** $x(n) = e^{\frac{j7\pi n}{4}}$

e) $x(n) = 2\cos\frac{5\pi}{3}n + 3e^{j\frac{3\pi n}{4}}$ **f)** $x(n) = 3 + \cos\frac{\pi}{2}n + \cos 2n$ **g)** $x(n) = \sin\left(\frac{3\pi}{7}n + \frac{\pi}{4}\right) + \cos\frac{\pi}{3}n$

h) $x(n) = e^{\frac{j3\pi}{5}\left(n + \frac{1}{2}\right)}$

Solution

a) Given that, x(n) $= \cos\left(\frac{5\pi}{9}n + 1\right)$ **(AU Jun'11, 8 Marks & Jun'12, 2 Marks)**

Let N and M be two integers.

Now, $x(n + N) = \cos\left(\frac{5\pi}{9}(n + N) + 1\right) = \cos\left(\frac{5\pi n}{9} + 1 + \frac{5\pi}{9}N\right)$

Let, $\frac{5\pi}{9}N = M \times 2\pi$

> Since, $\cos(\theta + 2\pi M) = \cos\theta$, for periodicity $\frac{5\pi}{9}N$ should be integral multiple of 2π.

$$\therefore N = M \times 2\pi \times \frac{9}{5\pi} = \frac{18M}{5}$$

Here N is an integer if, M = 5, 10, 15, 20, ...

| For integer values of M, $\cos(\theta + 2\pi M) = \cos\theta$ |

Let,   M = 5; $\therefore$ N = 18

When N = 18 ; $x(n+N) = \cos\left(\frac{5\pi n}{9} + 1 + \frac{5\pi}{9} \times 18\right) = \cos\left(\frac{5\pi n}{9} + 1 + 10\pi\right) = \cos\left(\frac{5\pi n}{9} + 1\right) = x(n)$

Hence x(n) is periodic with fundamental period of 18 samples.

b) Given that, $x(n) = \sin\left(\frac{n}{9} - \pi\right)$

Let N and M be two integers.

Now, $x(n+N) = \sin\left(\frac{n+N}{9} - \pi\right) = \sin\left(\frac{n}{9} + \frac{N}{9} - \pi\right) = \sin\left(\frac{n}{9} - \pi + \frac{N}{9}\right)$

Let, $\frac{N}{9} = M \times 2\pi$

$$\therefore N = 18\,\pi M$$

| Since, $\sin(\theta + 2\pi M) = \sin\theta$, for periodicity $\frac{N}{9}$ should be equal to integral multiple of 2π. |

Here N cannot be an integer for any integer value of M and so x(n) will not be periodic.

c) Given that, $x(n) = \sin\frac{\pi}{8}n^2$

Let N, M, M_1 and M_2 be integers.

$$\therefore x(n+N) = \sin\frac{\pi}{8}(n+N)^2 = \sin\frac{\pi}{8}(n^2 + N^2 + 2nN) = \sin\left(\frac{\pi}{8}n^2 + \frac{\pi N^2}{8} + \frac{\pi N}{4}n\right)$$

Let, $\frac{\pi N^2}{8} = 2\pi M_1$ Let, $\frac{\pi N}{4} = 2\pi M_2$

$$\therefore N = 4\sqrt{M_1}$$ $$\therefore N = 8M_2$$

| Since, $\sin(\theta + 2\pi M) = \sin\theta$, for periodicity, both $\frac{\pi N^2}{8}$ $\frac{\pi N}{4}$ should be integral multiple of 2π. |

Now, N is integer for $M_1 = 1^2, 2^2, 3^2, 4^2 \ldots$ | Now, N is integer for $M_2 = 1, 2, 3, 4 \ldots$

When $M_1 = 2^2$ and $M_2 = 1$, we get a common value for N as, N = 8.

When N = 8 ; $x(n+N) = \sin\left(\frac{\pi}{8}n^2 + \frac{\pi 8^2}{8} + \frac{\pi 8}{4}n\right)$

$$= \sin\left(\left(\frac{\pi}{8}n^2 + 2\pi n\right) + 4 \times 2\pi\right) = \sin\left(\frac{\pi}{8}n^2 + 2\pi n\right)$$

| For integer values of M, $\sin(\theta + 2\pi M) = \sin\theta$ |

$$= \sin\frac{\pi}{8}n^2 = x(n)$$

$\therefore$ x(n) is periodic with fundamental period, N = 8 samples.

d) Given that, $x(n) = e^{\frac{j7\pi n}{4}}$

Let N and M be two integers.

Now, $x(n+N) = e^{\frac{j7\pi(n+N)}{4}} = e^{\frac{j7\pi n}{4}} e^{\frac{j7\pi N}{4}}$

Let, $\frac{7\pi N}{4} = M \times 2\pi$,

| Since, $e^{j2\pi M} = 1$, for periodicity $\frac{7\pi N}{4}$ should be integral multiple of 2π. |

$$\therefore N = M \times 2\pi \times \frac{4}{7\pi} = \frac{8M}{7}$$

Here, N is integer, when M = 7, 14, 21, ...

When M = 7 ; N = 8

$\therefore$ x(n) is periodic with fundamental period of 8 samples.

e) **Given that, $x(n) = 2\cos\dfrac{5\pi n}{3} + 3e^{j\frac{3\pi n}{4}}$**

Let, $x(n) = x_1(n) + x_2(n)$

where, $x_1(n) = 2\cos\dfrac{5\pi n}{3}$

$\qquad x_2(n) = 3e^{j\frac{3\pi n}{4}}$

Let N_1, N_2, M_1 and M_2 be integers.

Now, $x_1(n + N_1) = 2\cos\dfrac{5\pi(n+N_1)}{3}$

$\qquad\qquad\qquad = 2\cos\left(\dfrac{5\pi n}{3} + \dfrac{5\pi N_1}{3}\right)$

Since, $\sin(\theta + 2\pi M_1) = \sin\theta$, for periodicity $\dfrac{5\pi N_1}{3}$ should be integral multiple of 2π.

Let, $\dfrac{5\pi N_1}{3} = 2\pi M_1 \quad\Rightarrow\quad N_1 = \dfrac{6}{5}M_1$

Let, $M_1 = 5$; $\therefore$ $N_1 = 6$

$\therefore$ $x_1(n+N_1) = 2\cos\left(\dfrac{5\pi n}{3} + \dfrac{5\pi}{3}\times 6\right)$

$\qquad\qquad\qquad = 2\cos\left(\dfrac{5\pi n}{3} + 5\times 2\pi\right)$

$\qquad\qquad\qquad = 2\cos\dfrac{5\pi n}{3} = x_1(n)$

$\therefore$ $x_1(n)$ is periodic with fundamental period, $N_1 = 6$ samples.

Now, $x_2(n + N_2) = 3e^{j\frac{3\pi(n+N_2)}{4}}$

$\qquad\qquad\qquad = 3e^{j\left(\frac{3\pi n}{4} + \frac{3\pi N_2}{4}\right)}$

Since, $e^{j(\theta + 2\pi M_2)} = e^{j\theta}$, for periodicity, $\dfrac{3\pi N_2}{4}$ should be integral multiple of 2π.

Let, $\dfrac{3\pi N_2}{4} = 2\pi M_2 \quad\Rightarrow\quad N_2 = \dfrac{8}{3}M_2$

Let, $M_2 = 3$; $\therefore$ $N_2 = 8$

$\therefore$ $x_2(n+N_2) = 3e^{j\left(\frac{3\pi n}{4} + \frac{3\pi\times 8}{4}\right)}$

$\qquad\qquad\qquad = 3e^{j\left(\frac{3\pi n}{4} + 3\times 2\pi\right)}$

$\qquad\qquad\qquad = 3e^{j\frac{3\pi n}{4}} = x_2(n)$

$\therefore$ $x_2(n)$ is periodic with fundamental period, $N_2 = 8$ samples.

From the above analysis we can say that $x(n)$ is sum of two periodic signals.

Therefore, $x(n)$ is periodic with period N, where N is LCM of N_1 and N_2.

The LCM of 6 and 8 is 24.

$\therefore$ $N = 24$

The GCD of 6 and 8 is 2. $\therefore$ $\text{LCM} = \dfrac{N_1 N_2}{\text{GCD}} = \dfrac{6\times 8}{2} = 24$

$\therefore$ $x(n)$ is periodic with fundamental period, $N = 24$.

---

f) **Given that, $x(n) = 3 + \cos\dfrac{\pi}{2}n + \cos 2n$**   ***(AU Dec'14, 8 Marks)***

Let, $x(n) = 3 + x_1(n) + x_2(n)$

where, $x_1(n) = \cos\dfrac{\pi}{2}n$

$\qquad x_2(n) = \cos 2n$

Let N_1, N_2, M, M_1 and M_2 be integers.

Since, $\cos(\theta + 2\pi M) = \cos\theta$, for periodicity, both $\dfrac{\pi N_1}{2}$ and $2N_2$ should be integral multiple of 2π.

Now, $x_1(n + N_1) = \cos\dfrac{\pi}{2}(n+N_1) = \cos\left(\dfrac{\pi n}{2} + \dfrac{\pi N_1}{2}\right)$

Let, $\dfrac{\pi N_1}{2} = 2\pi M_1 \Rightarrow N_1 = 4M_1$

Now, $M_1 = 1$; $\therefore$ $N_1 = 4$

$\therefore$ $x_1(n)$ is periodic with period $N = 4$

Now, $x_2(n+N_2) = \cos 2(n+N_2) = \cos(2n + 2N_2)$

Let, $2N_2 = 2\pi M_2 \Rightarrow N_2 = \pi M_2$

Here N_2 cannot be an integer for any integer value of M_2

$\therefore$ $x_2(n)$ is nonperiodic.

From the above analysis we can say that $x(n)$ is sum of periodic and nonperiodic signals.

Therefore, $x(n)$ will be nonperiodic.

g) Given that, $x(n) = \sin\left(\frac{3\pi}{7}n + \frac{\pi}{4}\right) + \cos\frac{\pi}{3}n$ *(AU Jun'14, 8 Marks)*

Let, $x(n) = x_1(n) + x_2(n)$

where, $x_1(n) = \sin\left(\frac{3\pi n}{7} + \frac{\pi}{4}\right)$

$x_2(n) = \cos\left(\frac{\pi}{3}n\right)$

Let N_1, N_2, M_1 and M_2 be integers.

Now, $x_1(n+N_1) = \sin\left(\frac{3\pi}{7}(n+N_1) + \frac{\pi}{4}\right)$ Now, $x_2(n+N_2) = \cos\left(\frac{\pi}{3}(n+N_2)\right)$

$= \sin\left(\frac{3\pi}{7}n + \frac{3\pi}{7}N_1 + \frac{\pi}{4}\right)$ $= \cos\left(\frac{\pi}{3}n + \frac{\pi}{3}N_2\right)$

Since, $\sin(\theta+2\pi M_1) = \sin\theta$, for periodicity, $\frac{2\pi N_1}{7}$ should be integral multiple of 2π.	Since, $\cos(\theta+2\pi M_2) = \cos\theta$, for periodicity, $\frac{\pi}{3}N_2$ should be integral multiple of 2π.

Let, $\frac{3\pi N_1}{7} = 2\pi M_1 \Rightarrow N_1 = \frac{14}{3}M_1$ Let, $\frac{\pi N_2}{3} = 2\pi M_2 \Rightarrow N_2 = 6M_2$

Let, $M_1 = 3$; $\therefore N_1 = 14$ Let, $M_2 = 1$; $\therefore N_2 = 6$

$\therefore x_1(n+N_1) = \sin\left(\frac{3\pi}{7}n + \frac{3\pi}{7}\times14 + \frac{\pi}{4}\right)$ $\therefore x_2(n+N_2) = \cos\left(\frac{\pi}{3}n + \frac{\pi}{3}\times6\right)$

$= \sin\left(\frac{3\pi}{7}n + \frac{\pi}{4}\right) = x_1(n)$ $= \cos\left(\frac{\pi}{3}n\right) = x_2(n)$

For integer M, $\cos(\theta+2\pi M) = \cos\theta$ $\sin(\theta+2\pi M) = \sin\theta$

$\therefore$ $x_1(n)$ is periodic with fundamental period, $\therefore$ $x_2(n)$ is periodic with fundamental period,

$N_1 = 14$ samples. $N_2 = 6$ samples.

From the above analysis we can say that x(n) is sum of two periodic signals.

Therefore, x(n) is periodic with period N, where N is LCM of N_1 and N_2.

The LCM of 14 and 6 is 42.

$\therefore$ N = 42

The GCD of 14 and 6 is 2. $\therefore$ $LCM = \frac{N_1 N_2}{GCD} = \frac{14\times6}{2} = 42$

$\therefore$ x(n) is periodic with fundamental period, N = 42.

h) Given that, $x(n) = e^{\frac{j3\pi}{5}\left(n+\frac{1}{2}\right)}$ *(AU Jun'11, 8 Marks)*

Let N and M be two integers.

Now, $x(n+N) = e^{\frac{j3\pi}{5}\left(n+N+\frac{1}{2}\right)} = e^{\frac{j3\pi}{5}\left(n+\frac{1}{2}\right)} e^{\frac{j3\pi}{5}N}$

Let, $\frac{3\pi}{5}N = M\times2\pi, \Rightarrow N = \frac{10M}{3}$

Since, $e^{j2\pi M} = 1$, for periodicity, $\frac{3\pi N}{5}$ should be integral multiple of 2π.

Here, N is integer, when M $= 3, 6, 9, \ldots$

When M $= 3$; N $= 10$

$\therefore$ x(n) is periodic with fundamental period of 10 samples.

Example 1.13   *(AU Jun'13, 6 Marks)*

Determine whether the given signal is periodic or not. If periodic find the fundamental period.

$x(n) = \cos\left(\frac{n\pi}{2}\right) - \sin\left(\frac{n\pi}{8}\right) + 3\cos\left(\frac{n\pi}{4} + \frac{\pi}{3}\right)$

Solution:

Given that, $x(n) = \cos\left(\frac{n\pi}{2}\right) - \sin\left(\frac{n\pi}{8}\right) + 3\cos\left(\frac{n\pi}{4} + \frac{\pi}{3}\right)$

Let, $x(n) = x_1(n) + x_2(n) + x_3(n)$

where, $x_1(n) = \cos\left(\frac{n\pi}{2}\right)$; $x_2(n) = \sin\left(\frac{\pi n}{8}\right)$; $x_3(n) = 3\cos\left(\frac{\pi n}{4} + \frac{\pi}{3}\right)$

Let N_1, N_2, N_3, M_1, M_2 and M_3 be integers.

Now, $x_1(n+N_1) = \cos\left(\frac{\pi(n+N_1)}{2}\right) = \cos\left(\frac{n\pi}{2} + \frac{\pi N_1}{2}\right)$

Since, $\cos(\theta + 2\pi M_1) = \cos\theta$, for periodicity, $\frac{\pi N_1}{2}$ should be integral multiple of 2π.

Let, $\frac{\pi N_1}{2} = 2\pi M_1 \Rightarrow N_1 = 4M_1$

Let, $M_1 = 1$; $\therefore N_1 = 4$

$\therefore x_1(n + N_1) = \cos\left(\frac{n\pi}{2} + \frac{\pi}{2} \times 4\right) = \cos\left(\frac{n\pi}{2} + 2\pi\right) = \cos\left(\frac{n\pi}{2}\right) = x_1(n)$

$\therefore x_1(n)$ is periodic with fundamental period, $N_1 = 4$.

Now, $x_2(n+N_2) = \sin\left(\frac{\pi(n+N_2)}{8}\right) = \sin\left(\frac{n\pi}{8} + \frac{\pi N_2}{8}\right)$

Since, $\sin(\theta + 2\pi M_2) = \sin\theta$, for periodicity, $\frac{\pi N_2}{8}$ should be integral multiple of 2π.

Let, $\frac{\pi N_2}{8} = 2\pi M_2 \Rightarrow N_2 = 16 M_2$

Let, $M_2 = 1$; $\therefore N_2 = 16$

$\therefore x_2(n + N_2) = \sin\left(\frac{\pi n}{8} + \frac{\pi}{8} \times 16\right) = \sin\left(\frac{\pi n}{8} + 2\pi\right) = \sin\left(\frac{\pi n}{8}\right) = x_2(n)$

$\therefore x_2(n)$ is periodic with fundamental period, $N_2 = 16$.

Now, $x_3(n+N_3) = 3\cos\left(\frac{\pi(n+N_3)}{4} + \frac{\pi}{3}\right) = 3\cos\left(\frac{n\pi}{4} + \frac{\pi N_3}{4} + \frac{\pi}{3}\right)$

Since, $\cos(\theta + 2\pi M_3) = \cos\theta$, for periodicity, $\frac{\pi N_3}{4}$ should be integral multiple of 2π.

Let, $\frac{\pi N_3}{4} = 2\pi M_3 \Rightarrow N_3 = 8M_3$

Let, $M_3 = 1$; $\therefore N_3 = 8$

$\therefore x_3(n + N_3) = 3\cos\left(\frac{n\pi}{4} + \frac{\pi}{4} \times 8 + \frac{\pi}{3}\right) = 3\cos\left(\frac{n\pi}{4} + 2\pi + \frac{\pi}{3}\right) = 3\cos\left(\frac{n\pi}{4} + \frac{\pi}{3}\right) = x_3(n)$

$\therefore x_3(n)$ is periodic with fundamental period, $N_3 = 8$.

From the above analysis we can say that $x(n)$ is sum of three periodic signals. Therefore $x(n)$ is periodic with period N, where N is LCM of N_1, N_2 and N_3.

The LCM of 4,16 and 8 is 16.

$\therefore N = 16$

$\therefore x(n)$ is periodic with fundamental period, N = 16.

GCD of 4 and 16 is 4 (GCD_1). $\therefore LCM_1 = \frac{4 \times 16}{GCD_1} = \frac{4 \times 16}{4} = 16$ GCD of 16 and 8 is 8 (GCD_2). $\therefore LCM_2 = \frac{16 \times 8}{GCD_2} = \frac{16 \times 8}{8} = 16$ LCM_2 is LCM of 4, 16 and 8.

1.5.3 Symmetric (Even) and Antisymmetric (Odd) Signals

The discrete time signals may exhibit symmetry or antisymmetry with respect to $n = 0$. When a discrete time signal exhibits symmetry with respect to $n = 0$ then it is called an ***even signal.*** Therefore, the even signal satisfies the condition,

$$x(-n) = x(n)$$

When a discrete time signal exhibits antisymmetry with respect to $n = 0$, then it is called an ***odd signal.*** Therefore the odd signal satisfies the condition,

$$x(-n) = -x(n)$$

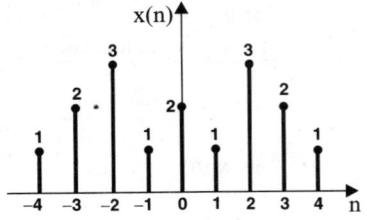

$$x(n) = \{1, 2, 3, 1, 2, 1, 3, 2, 1\}$$
$$\uparrow$$

Fig 1.52a: Symmetric (or even) signal.

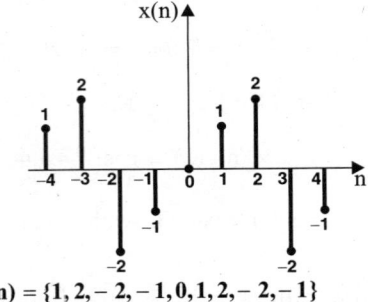

$$x(n) = \{1, 2, -2, -1, 0, 1, 2, -2, -1\}$$
$$\uparrow$$

Fig 1.52b: Antisymmetric (or odd) signal.

***Fig 1.52:** Symmetric and antisymmetric discrete time signal.*

Neither Even Nor Odd Signals

A discrete time signal $x(n)$ which is neither even nor odd can be expressed as a sum of even and odd signal.

Let, $x(n) = x_e(n) + x_o(n)$

where, $x_e(n) =$ Even part of $x(n)$

   $x_o(n) =$ Odd part of $x(n)$

Note: *If $x(n)$ is even then its odd part will be zero. If $x(n)$ is odd then its even part will be zero.*

Now, it can be proved that,

Even part, $x_e(n) = \frac{1}{2}[x(n) + x(-n)]$

Odd part, $x_o(n) = \frac{1}{2}[x(n) - x(-n)]$

Proof:

Let, $x(n) = x_e(n) + x_o(n)$ .....(1.12)

On replacing n by $-n$ in equation (1.12) we get,

$x(-n) = x_e(-n) + x_o(-n)$ .....(1.13)

Since $x_e(n)$ is even, $x_e(-n) = x_e(n)$

Since $x_o(n)$ is odd, $x_o(-n) = -x_o(n)$

Hence the equation (1.13) can be written as,

$$x(-n) = x_e(n) - x_o(n) \qquad \qquad .....(1.14)$$

On adding equation (1.12) and (1.14) we get,

$$x(n) + x(-n) = 2\,x_e(n)$$

$$\therefore\ x_e(n) = \frac{1}{2}[x(n) + x(-n)]$$

On subtracting equation (1.14) from equation (1.12) we get,

$$x(n) - x(-n) = 2\,x_o(n)$$

$$\therefore\ x_o(n) = \frac{1}{2}[x(n) - x(-n)]$$

Example 1.14

Determine the even and odd parts of the following discrete time signals.

a) $x(n) = 3^n$ **b)** $x(n) = 3\,e^{j\frac{\pi}{5}n}$ **c)** $x(n) = \{2, -2, 6, -2\}$

Solution:

a) Given that, $x(n) = 3^n$

$$\therefore x(-n) = 3^{-n}$$

Even part, $x_e(n) = \frac{1}{2}[x(n) + x(-n)] = \frac{1}{2}[3^n + 3^{-n}]$

Odd part, $x_o(n) = \frac{1}{2}[x(n) - x(-n)] = \frac{1}{2}[3^n - 3^{-n}]$

b) Given that, $x(n) = 3\,e^{j\frac{\pi}{5}n}$

$$x(n) = 3\,e^{j\frac{\pi}{5}n} = 3\cos\frac{\pi}{5}n + j3\sin\frac{\pi}{5}n$$

$$\therefore x(-n) = 3\,e^{j\frac{\pi}{5}n} = 3\cos\frac{\pi}{5}n - j3\sin\frac{\pi}{5}n$$

Even part, $x_e(n) = \frac{1}{2}[x(n) + x(-n)]$

$$= \frac{1}{2}\left[3\cos\frac{\pi}{5}n + j3\sin\frac{\pi}{5}n + 3\cos\frac{\pi}{5}n - j3\sin\frac{\pi}{5}n\right] = \frac{1}{2}\left[6\cos\frac{\pi}{5}n\right] = 3\cos\frac{\pi}{5}n$$

Odd part, $x_o(n) = \frac{1}{2}[x(n) - x(-n)]$

$$= \frac{1}{2}\left[3\cos\frac{\pi}{5}n + j3\sin\frac{\pi}{5}n - 3\cos\frac{\pi}{5}n + j3\sin\frac{\pi}{5}n\right] = \frac{1}{2}\left[j6\sin\frac{\pi}{5}n\right] = j3\sin\frac{\pi}{5}n$$

c) Given that, $x(n) = \{2, -2, 6, -2\}$
$$\uparrow$$

Given that, $x(n) = \{2, -2, 6, -2\}$, $\therefore\ x(0) = 2$; $x(1) = -2$; $x(2) = 6$; $x(3) = -2$
$$\uparrow$$

$x(-n) = \{-2, 6, -2, 2\}$, $\therefore\ x(-3) = -2$; $x(-2) = 6$; $x(-1) = -2$; $x(0) = 2$
$$\uparrow$$

Even part, $x_e(n) = \frac{1}{2}[x(n)+x(-n)]$	Odd part, $x_o(n) = \frac{1}{2}[x(n)-x(-n)]$
At $n=-3$; $x(n)+x(-n) = 0+(-2) = -2$	At $n=-3$; $x(n)-x(-n) = 0-(-2) = 2$
At $n=-2$; $x(n)+x(-n) = 0+6 = 6$	At $n=-2$; $x(n)-x(-n) = 0-6 = -6$
At $n=-1$; $x(n)+x(-n) = 0+(-2) = -2$	At $n=-1$; $x(n)-x(-n) = 0-(-2) = 2$
At $n=0$; $x(n)+x(-n) = 2+2 = 4$	At $n=0$; $x(n)-x(-n) = 2-2 = 0$
At $n=1$; $x(n)+x(-n) = -2+0 = -2$	At $n=1$; $x(n)-x(-n) = -2-0 = -2$
At $n=2$; $x(n)+x(-n) = 6+0 = 6$	At $n=2$; $x(n)-x(-n) = 6-0 = 6$
At $n=3$; $x(n)+x(-n) = -2+0 = -2$	At $n=3$; $x(n)-x(-n) = -2-0 = -2$
$\therefore\ x(n)+x(-n) = \{-2,6,-2,4,-2,6,-2\}$	$\therefore\ x(n)-x(-n) = \{2,-6,2,0,-2,6,-2\}$

$$\therefore\ x_e(n) = \frac{1}{2}[x(n)+x(-n)] \qquad\qquad \therefore\ x_o(n) = \frac{1}{2}[x(n)-x(-n)]$$

$$= \{-1,3,-1,2,-1,3,-1\} \qquad\qquad = \{1,-3,1,0,-1,3,-1\}$$

1.5.4 Energy and Power Signals

The *energy* E of a discrete time signal x(n) is defined as,

$$\boxed{\text{Energy, } E = \sum_{n=-\infty}^{\infty} |x(n)|^2} \qquad\qquad .....(1.15)$$

The energy of a signal may be finite or infinite, and can be applied to complex valued and real valued signals.

If energy E of a discrete time signal is finite and nonzero, then the discrete time signal is called an *energy signal.* The exponential signals are examples of energy signals.

The average *power* of a discrete time signal x(n) is defined as,

$$\boxed{\text{Power, } P = \lim_{N \to \infty} \frac{1}{2N+1} \sum_{n=-N}^{N} |x(n)|^2} \qquad\qquad .....(1.16)$$

If power P of a discrete time signal is finite and nonzero, then the discrete time signal is called a *power signal*. The periodic signals are examples of power signals.

For energy signals, the energy will be finite and average power will be zero. For power signals the average power is finite and energy will be infinite.

$$\therefore \quad \text{For energy signal, } 0 < E < \infty \text{ and } P = 0$$

$$\text{For power signal, } 0 < P < \infty \text{ and } E = \infty$$

Neither Energy Nor Power Signals

The signals that do not satisfy the condition of either energy or power signals are called neither energy nor power signals.

Example 1.15

Determine whether the following signals are energy or power signals.

a) $x(n) = \left(\frac{1}{4}\right)^n u(n)$ **b)** $x(n) = \sin\left(\frac{\pi}{3}n\right)$ **c)** $x(n) = u(n)$ **d)** $x(n) = 2n^2 u(n)$

Solution:

a) Given that, x(n) = $\left(\frac{1}{4}\right)^n u(n)$ *(AU Dec'13, 5 Marks & Jun'13, 4 Marks)*

Here, $x(n) = \left(\frac{1}{4}\right)^n u(n)$ for all n.

$$\therefore x(n) = \left(\frac{1}{4}\right)^n = 0.25^n \; ; \; n \geq 0$$

Energy, $E = \sum_{n=-\infty}^{\infty} |x(n)|^2 = \sum_{n=0}^{\infty} |(0.25)^n|^2 = \sum_{n=0}^{\infty} (0.25^{\,2})^n$

> Infinite geometric series sum formula.
> $$\sum_{n=0}^{\infty} C^n = \frac{1}{1-C}$$

$$= \sum_{n=0}^{\infty} (0.0625)^n = \frac{1}{1-0.0625}$$

> Using infinite geometric series sum formula.

$$= 1.067 \text{ joules}$$

Power, $P = \underset{N\to\infty}{Lt} \frac{1}{2N+1} \sum_{n=-N}^{N} |x(n)|^2 = \underset{N\to\infty}{Lt} \frac{1}{2N+1} \sum_{n=0}^{N} |(0.25)^n|^2$

$$= \underset{N\to\infty}{Lt} \frac{1}{2N+1} \sum_{n=0}^{N} |(0.25)^2|^n = \underset{N\to\infty}{Lt} \frac{1}{2N+1} \sum_{n=0}^{N} (0.0625)^n$$

> Using infinite geometric series sum formula.

$$= \underset{N\to\infty}{Lt} \frac{1}{2N+1} \frac{(0.0625)^{N+1}-1}{0.0625-1}$$

> Finite geometric series sum formula.
> $$\sum_{n=0}^{N} C^n = \frac{C^{N+1}-1}{C-1}$$

$$= \frac{1}{\infty} \times \frac{0.0625^{\infty}-1}{0.0625-1}$$

$$= 0$$

Here E is finite and P is zero and so x(n) is an energy signal.

---

b) Given that, x(n) = $\sin\left(\frac{\pi}{3}n\right)$

Energy, $E = \sum_{n=-\infty}^{\infty}|x(n)|^2 = \sum_{n=-\infty}^{\infty} \sin^2\left(\frac{\pi}{3}n\right) = \sum_{n=-\infty}^{\infty} \frac{1-\cos\frac{2\pi}{3}n}{2}$

> $\sin^2\theta = \dfrac{1-\cos 2\theta}{2}$

$$= \frac{1}{2}\left(\sum_{n=-\infty}^{\infty}\left(1-\cos\frac{2\pi}{3}n\right)\right) = \frac{1}{2}\left(\sum_{n=-\infty}^{\infty}1^n - \sum_{n=-\infty}^{\infty}\cos\frac{2\pi}{3}n\right) = \frac{1}{2}(\infty-0) = \infty$$

> **Note:** Sum of infinite 1's is infinity. Sum of samples of one period of cosinusoidal signal is zero.

Power, $P = \underset{N\to\infty}{Lt} \frac{1}{2N+1} \sum_{n=-N}^{N} |x(n)|^2 = \underset{N\to\infty}{Lt} \frac{1}{2N+1} \sum_{n=-N}^{N} \sin^2\frac{\pi n}{3}$

$$\therefore P = \underset{N\to\infty}{Lt} \frac{1}{2N+1} \sum_{n=-N}^{N} \frac{\left(1-\cos\frac{2\pi}{3}n\right)}{2}$$

$$= \underset{N\to\infty}{Lt} \frac{1}{2N+1} \frac{1}{2}\left[\sum_{n=-N}^{N} 1^n - \sum_{n=-N}^{N} \cos\frac{2\pi}{3}n\right]$$

$$= \underset{N\to\infty}{Lt} \frac{1}{2N+1} \frac{1}{2}\left[\underbrace{1+1+.....+1}_{N-\text{terms}}+1+\underbrace{1+.....+1+1}_{N-\text{terms}}-0\right]$$

$$\therefore P = \underset{N \to \infty}{Lt} \frac{1}{2N+1} \frac{1}{2} [2N+1] = \underset{N \to \infty}{Lt} \frac{1}{2} = \frac{1}{2} \, watts$$

Since P is finite and E is infinite, x(n) is a power signal.

> **Note:** The term $\cos\frac{2\pi}{3}n$ is periodic with periodicity of 3 samples. Samples of $\cos\frac{2\pi}{3}n$ for two periods are given below. It can be observed that sum of samples of a period is zero.
>
> When $n = 0$; $\cos\frac{2\pi}{3}n = 1$,   When $n = 1$; $\cos\frac{2\pi}{3}n = -0.5$,   When $n = 2$; $\cos\frac{2\pi}{3}n = -0.5$
>
> When $n = 3$; $\cos\frac{2\pi}{3}n = 1$,   When $n = 4$; $\cos\frac{2\pi}{3}n = -0.5$,   When $n = 5$; $\cos\frac{2\pi}{3}n = -0.5$

c) Given that, x(n) = u(n)

$$E = \sum_{n=-\infty}^{\infty} \left| x(n) \right|^2 = \sum_{n=0}^{\infty} (u(n))^2$$

$$= \sum_{n=0}^{\infty} u(n) = 1 + 1 + 1 \ldots \ldots \ldots \infty = \infty$$

$$P = \underset{N \to \infty}{Lt} \frac{1}{2N+1} \sum_{n=-N}^{N} \left| x(n) \right|^2 = \underset{N \to \infty}{Lt} \frac{1}{2N+1} \sum_{n=0}^{N} u(n) = \underset{N \to \infty}{Lt} \frac{1}{2N+1} \left(\underbrace{\frac{1 + 1 + 1 + \ldots \ldots \ldots + 1}{N+1 \text{ terms}}} \right)$$

$$= \underset{N \to \infty}{Lt} \frac{1}{2N+1} (N+1) = \underset{N \to \infty}{Lt} \frac{N\left(1 + \frac{1}{N}\right)}{N\left(2 + \frac{1}{N}\right)} = \frac{1 + \frac{1}{\infty}}{2 + \frac{1}{\infty}} = \frac{1+0}{2+0} = \frac{1}{2} \, watts$$

Since P is finite and E is infinite, x(n) is a power signal.

d) Given that, x(n) = 2n² u(n)

Here, $x(n) = 2n^2 u(n)$; for all n.

$$\therefore x(n) = 2n^2 \text{ ; for } n \geq 0$$

$$\text{Energy, } E = \sum_{n=-\infty}^{\infty} \left| x(n) \right|^2 = \sum_{n=0}^{\infty} ((2)^n)^2 = \sum_{n=0}^{\infty} 4n^4$$

$$= 4 \times 0^4 + 4 \times 1^4 + 4 \times 2^4 + 4 \times 3^4 + 4 \times 4^4 \ldots \ldots \infty = \infty$$

$$\text{Power, } P = \underset{N \to \infty}{Lt} \frac{1}{2N+1} \sum_{n=-N}^{N} \left| x(n) \right|^2 = \underset{N \to \infty}{Lt} \frac{1}{2N+1} \sum_{n=0}^{N} (2n^2)^2$$

$$= \underset{N \to \infty}{Lt} \frac{1}{2N+1} \sum_{n=0}^{N} 4n^4 = \underset{N \to \infty}{Lt} \frac{1}{2N+1} (4 \times 0^4 + 4 \times 1^4 + 4 \times 2^4 + \ldots \ldots + 4 \times N^4)$$

$$= \frac{1}{\infty + 1} (4 \times 0^4 + 4 \times 1^4 + 4 \times 2^4 + \ldots \ldots + \infty) = 0 \times \infty = 0$$

Here, the conditions for energy or power signal are not satisfied , therfore the given signal is neither energy nor power signal.

1.5.5 Causal, Noncausal and Anticausal Signals

A discrete time signal is said to be *causal*, if it is defined for n ≥ 0. Therefore if x(n) is causal, then x(n) = 0 for n < 0.

A discrete time signal is said to be *noncausal*, if it is defined for either n ≤ 0, or for both n ≤ 0 and n > 0. Therefore if x(n) is noncausal, then x(n) ≠ 0 for n < 0. A noncausal signal can be converted to causal signal by multiplying the noncausal signal by a unit step signal, u(n).

When a noncausal discrete time signal is defined only for n ≤ 0, it is called an *anticausal signal*.

Examples of Causal and Noncausal Signals

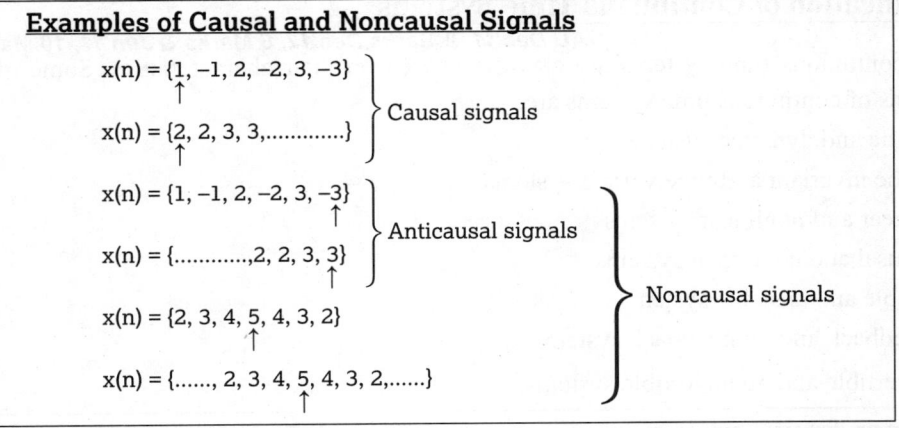

1.6 Continuous Time Systems

A system which can process continuous time signal is called **continuous time system**, and so the input and output signals of a continuous time system are continuous time signals.

A continuous time system is denoted by letter $\mathcal{H}$. The input of continuous time system is denoted as x(t) and the output of continuous time system is denoted as y(t). The diagrammatic representation of a continuous time system is shown in Fig 1.53.

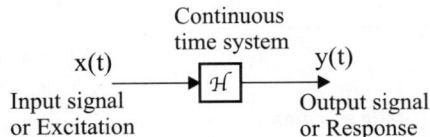

Fig 1.53 : *Representation of continuous time system.*

The operation performed by a continuous time system on input to produce output or response can be expressed as,

Response, $\quad y(t) = \mathcal{H}\{x(t)\}$

where, $\mathcal{H}$ denotes the system operation (also called system operator).

When a continuous time system satisfies the properties of linearity and time invariance then it is called **LTI (Linear Time Invariant) continuous time system.**   **(AU Dec'13, 2 Marks)**

Most of the practical systems that we encounter in science and engineering are LTI systems.

The input-output relation of an LTI continuous time system is represented by constant coefficient differential equation shown below(equation (1.17)).

$$a_o \frac{d^N}{dt^N} y(t) + a_1 \frac{d^{N-1}}{dt^{N-1}} y(t) + a_2 \frac{d^{N-2}}{dt^{N-2}} y(t) + \dots\dots + a_{N-1} \frac{d}{dt} y(t) + a_N y(t) = b_o \frac{d^M}{dt^M} x(t)$$

$$+ b_1 \frac{d^{M-1}}{dt^{M-1}} x(t) + b_2 \frac{d^{M-2}}{dt^{M-2}} x(t) + \dots\dots + b_{M-1} \frac{d}{dt} x(t) + b_M x(t) \qquad \dots\dots(1.17)$$

where, N = Order of the system, $M \le N$, and $a_0 = 1$.

The solution of the above differential equation is the response y(t) of the continuous time system, for the input x(t).

1.7 Classification of Continuous Time Systems

(AU Dec'11, 2 Marks, Dec'12, 6 Marks & Jun'11, 10 Marks)

The continuous time systems are classified based on their characteristics. Some of the classifications of continuous time systems are,

1. Static and dynamic systems

2. Time invariant and time variant systems

3. Linear and nonlinear systems

4. Causal and noncausal systems

5. Stable and unstable systems

6. Feedback and nonfeedback systems

7. Invertible and noninvertible systems.

Note: *Static or dynamic nature, time invariance, linearity, causality, stability and invertibility are also called basic properties of continuous time systems.*

1.7.1 Static and Dynamic Systems

A continuous time system is called *static* or *memoryless* if its output at any instant of time t depends on present input signal but not on the past or future input. A continuous time system is called *dynamic* or *memory* system if its output at any instant t depends on past and future inputs in addition to present input.

Example:

$y(t) = a\, x(t)$

$y(t) = t\, x(t) + 6\, x^3(t)$ } Static systems

$y(t) = t\, x(t) + 3\, x(t^2)$

$y(t) = x(t) + 3\, x(t-2)$ } Dynamic systems

1.7.2 Time Invariant and Time Variant Systems

A system is said to be *time invariant* if its input-output characteristics does not change with time. In a system, if the input-output characteristics change with time then the system is called *time variant* system.

Definition: A relaxed system $\mathcal{H}$ is *time invariant* or *shift invariant* if and only if

$$x(t) \xrightarrow{\ \mathcal{H}\ } y(t) \text{ implies that, } x(t-m) \xrightarrow{\ \mathcal{H}\ } y(t-m)$$

for every input signal x(t) and every time shift m.

Alternative Definition for Time Invariance

A system $\mathcal{H}$ is *time invariant* if the response to a shifted (or delayed) version of the input is identical to a shifted (or delayed) version of the response based on the unshifted (or undelayed) input.

The diagrammatic explanation of the above definition of time invariance is shown in Fig 1.54.

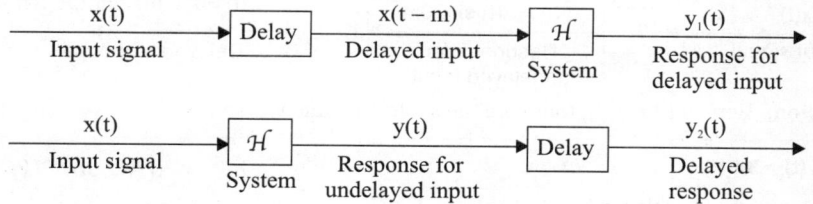

If $y_1(t) = y_2(t)$ then the system is time invariant

Fig 1.54: Diagrammatic explanation of time invariance.

Procedure To Test For Time Invariance

1. Delay the input signal by m units of time and determine the response of the system for this delayed input signal. Let this response be $y_1(t)$.

2. Delay the response of the system for unshifted input by m units of time. Let this delayed response be $y_2(t)$.

3. If $y_1(t) = y_2(t)$ then the system is time invariant.

4. If $y_1(t) \neq y_2(t)$ then the system is time variant.

Example 1.16

State whether the following systems are time invariant or not.

a) $y(t) = 2t\, x(t)$ **b)** $y(t) = x(t)\sin 20\pi t$ **c)** $y(t) = 3x(t^2)$ **d)** $y(t) = x(-t)$

Solution:

a) Given that, $y(t) = 2t\, x(t)$

Test 1: Response for delayed input

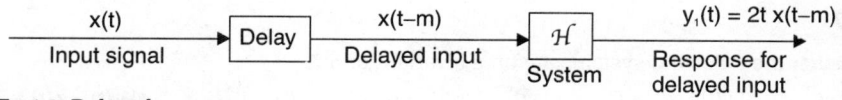

Test 2: Delayed response

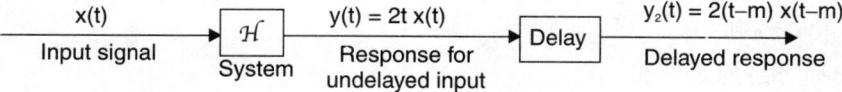

Conclusion: Here, $y_1(t) \neq y_2(t)$, therefore the system is time variant.

---

b) Given that, $y(t) = x(t)\sin 20\pi t$

Test 1: Response for delayed input

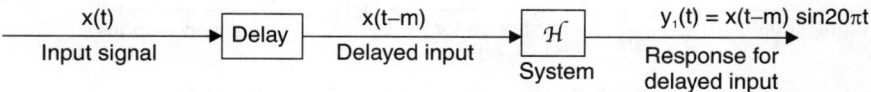

Test 2: Delayed response

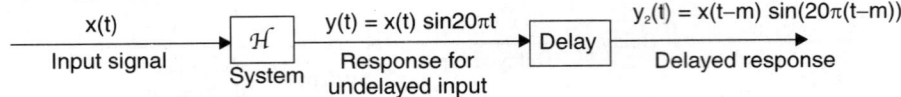

Conclusion: Here, $y_1(t) \neq y_2(t)$, therefore the system is time variant.

---

c) Given that, y(t) = 3x(t²)   ***(AU Jun'12, 2 Marks)***

Test 1: Response for delayed input

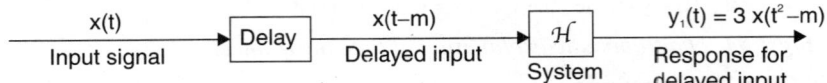

Test 2: Delayed response

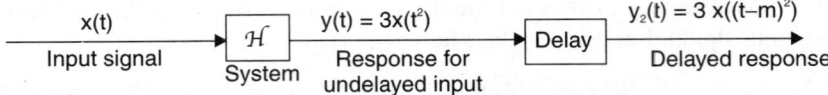

Conclusion: Here, $y_1(t) \neq y_2(t)$, therefore the system is time variant.

---

d) Given that, y(t) = x(−t)

Test 1: Response for delayed input

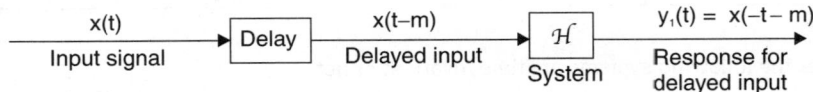

Test 2: Delayed response

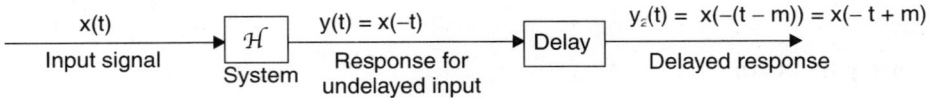

Conclusion: Here, $y_1(t) \neq y_2(t)$, therefore the system is time variant.

Example 1.17

State whether the following systems are time invariant or not.

a) $y(t) = 2\,e^{x(t)}$ **b)** $y(t) = x(t) + C$ **c)** $y(t) = 3x^2(t)$ **d)** $y(t) = x(t) + \dfrac{dx(t)}{dt}$ **e)** $y(t) = x(t) + \int x(t)\,dt$

Solution:

a) Given that, y(t) = 2 e^{x(t)}

Test 1: Response for delayed input

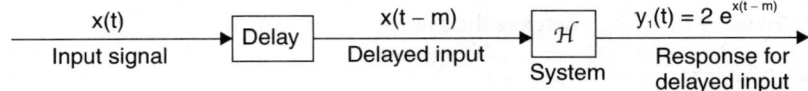

Test 2: Delayed response

Conclusion: Here, $y_1(t) = y_2(t)$, therefore the system is time invariant.

b) Given that, y(t) = x(t) + C

    <u>Test 1:</u> Response for delayed input

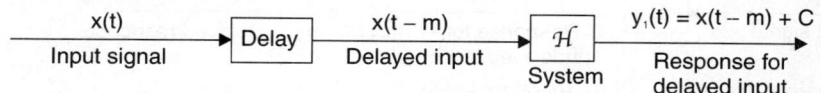

    <u>Test 2:</u> Delayed response

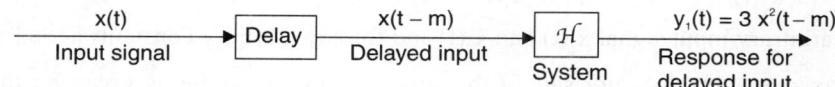

    <u>Conclusion:</u> Here, $y_1(t) = y_2(t)$, therefore the system is time invariant.

---

c) Given that, y(t) = 3x²(t)

    <u>Test 1:</u> Response for delayed input

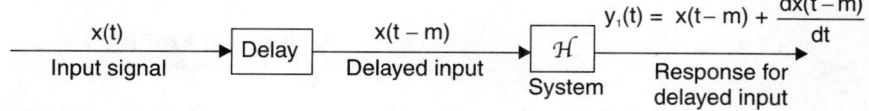

    <u>Test 2:</u> Delayed response

    <u>Conclusion:</u> Here, $y_1(t) = y_2(t)$, therefore the system is time invariant.

---

d) Given that, $y(t) = x(t) + \dfrac{dx(t)}{dt}$

    <u>Test 1:</u> Response for delayed input

    <u>Test 2:</u> Delayed response

    <u>Conclusion:</u> Here, $y_1(t) = y_2(t)$, therefore the system is time invariant.

---

e) Given that, $y(t) = x(t) + \displaystyle\int x(t)\, dt$

    <u>Test 1:</u> Response for delayed input

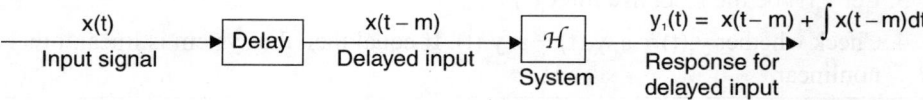

Test 2: Delayed response

$$x(t) \rightarrow \boxed{\mathcal{H}} \rightarrow y(t) = x(t) + \int x(t)dt \rightarrow \boxed{\text{Delay}} \rightarrow y_2(t) = x(t-m) + \int x(t-m)dt$$

Input signal | System | Response for undelayed input | | Delayed response

Conclusion: Here, $y_1(t) = y_2(t)$, therefore the system is time invariant.

1.7.3 Linear and Nonlinear Systems

A *linear system* is the one that satisfies the superposition principle.

The *principle of superposition* requires that the response of a system to a weighted sum of the signals is equal to the corresponding weighted sum of the responses to each of the individual input signals.

Definition: A relaxed system $\mathcal{H}$ is *linear* if

$$\mathcal{H}\{a_1 x_1(t) + a_2 x_2(t)\} = a_1 \mathcal{H}\{x_1(t)\} + a_2 \mathcal{H}\{x_2(t)\}$$

for any arbitrary input signal $x_1(t)$ and $x_2(t)$ and for any arbitrary constants a_1 and a_2.

If a relaxed system does not satisfy the superposition principle as given by the above definition, the system is *nonlinear*. The diagrammatic explanation of linearity is shown in Fig. 1.55

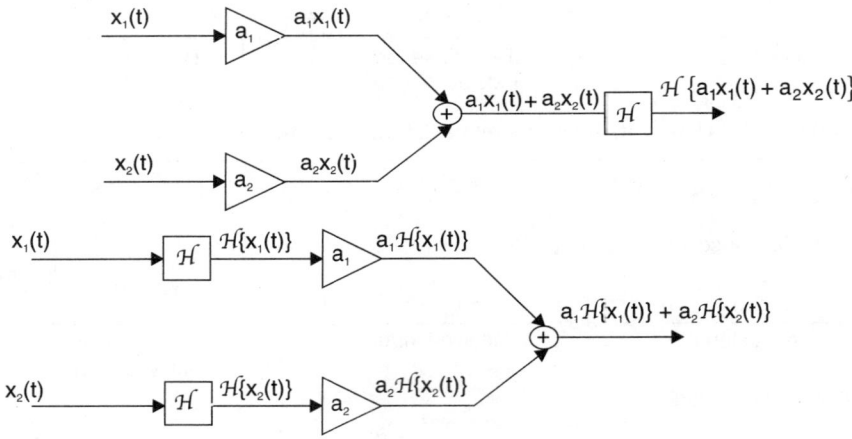

The system, $\mathcal{H}$ is linear if and only if, $\mathcal{H}\{a_1 x_1(t) + a_2 x_2(t)\} = a_1 \mathcal{H}\{x_1(t)\} + a_2 \mathcal{H}\{x_2(t)\}$

Fig 1.55: Diagrammatic explanation of linearity.

Procedure To Test For Linearity

1. Let $x_1(t)$ and $x_2(t)$ be two inputs to the system $\mathcal{H}$, and $y_1(t)$ and $y_2(t)$ be the corresponding responses.

2. Consider a signal, $x_3(t) = a_1 x_1(t) + a_2 x_2(t)$ which is a weighed sum of $x_1(t)$ and $x_2(t)$.

3. Let $y_3(t)$ be the response for $x_3(t)$.

4. Check whether $y_3(t) = a_1 y_1(t) + a_2 y_2(t)$. If equal then the system is linear, otherwise it is nonlinear.

Example 1.18

Test the following systems for linearity.

a) $y(t) = t\, x(t)$, **b)** $y(t) = x(t^2)$, **c)** $y(t) = x^2(t)$, **d)** $y(t) = A\, x(t) + B$, **e)** $y(t) = e^{x(t)}$.

Solution:

a) Given that, $y(t) = t\, x(t)$

Let $\mathcal{H}$ be the system operating on $x(t)$ to produce, $y(t) = \mathcal{H}\{x(t)\} = t\, x(t)$.

Consider two signals $x_1(t)$ and $x_2(t)$.

Let $y_1(t)$ and $y_2(t)$ be the response of the system $\mathcal{H}$ for inputs $x_1(t)$ and $x_2(t)$ respectively.

$\therefore\ y_1(t) = \mathcal{H}\{x_1(t)\} = t\, x_1(t)$ .....(1)

$y_2(t) = \mathcal{H}\{x_2(t)\} = t\, x_2(t)$ .....(2)

Let $x_3(t) = a_1 x_1(t) + a_2 x_2(t)$. | A linear combination of inputs $x_1(t)$ and $x_2(t)$ |

Let $y_3(t)$ be the response of the system $\mathcal{H}$ for input $x_3(t)$.

$\therefore\ y_3(t) = \mathcal{H}\{\, x_3(t)\} = t x_3(t)$

$= t[a_1 x_1(t) + a_2 x_2(t)] = a_1 t\, x_1(t) + a_2 t\, x_2(t)$

$= a_1 y_1(t) + a_2 y_2(t))$ | Using equations (1) and (2) |

Since, $y_3(t) = a_1 y_1(t) + a_2 y_2(t)$, the given system is linear.

b) Given that, $y(t) = x(t^2)$ *(AU Jun'12, 2 Marks)*

Let $\mathcal{H}$ be the system operating on $x(t)$ to produce, $y(t) = \mathcal{H}\{x(t)\} = x(t^2)$.

Consider two signals $x_1(t)$ and $x_2(t)$.

Let $y_1(t)$ and $y_2(t)$ be the response of the system $\mathcal{H}$ for inputs $x_1(t)$ and $x_2(t)$ respectively.

$\therefore\ y_1(t) = \mathcal{H}\{x_1(t)\} = x_1(t^2)$ ....(1)

$y_2(t) = \mathcal{H}\{x_2(t)\} = x_2(t^2)$ .....(2)

Let $x_3(t) = a_1 x_1(t) + a_2 x_2(t)$. | A linear combination of inputs $x_1(t)$ and $x_2(t)$ |

Let $y_3(t)$ be the response of the system $\mathcal{H}$ for input $x_3(t)$.

$\therefore\ y_3(t) = \mathcal{H}\{x_3(t)\} = x_3(t^2)$

$= a_1 x_1(t^2) + a_2 x_2(t^2)$

$= a_1 y_1(t) + a_2 y_2(t)$ | Using equations (1) and (2) |

Since, $y_3(t) = a_1 y_1(t) + a_2 y_2(t)$, the given system is linear.

c) Given that, $y(t) = x^2(t)$

Let $\mathcal{H}$ be the system operating on $x(t)$ to produce, $y(t) = \mathcal{H}\{x(t)\} = x^2(t)$.

Consider two signals $x_1(t)$ and $x_2(t)$.

Let $y_1(t)$ and $y_2(t)$ be the response of the system $\mathcal{H}$ for inputs $x_1(t)$ and $x_2(t)$ respectively.

$\therefore\ y_1(t) = \mathcal{H}\{x_1(t)\} = x_1^2(t)$ .....(1)

$y_2(t) = \mathcal{H}\{x_2(t)\} = x_2^2(t)$ .....(2)

Let $x_3(t) = a_1 x_1(t) + a_2 x_2(t)$. | A linear combination of inputs $x_1(t)$ and $x_2(t)$ |

Let $y_3(t)$ be the response of the system $\mathcal{H}$ for input $x_3(t)$.

$$\therefore y_3(t) = \mathcal{H}\{x_3(t)\} = x_3^2(t) = [a_1 x_1(t) + a_2 x_2(t)]^2$$

$$= a_1^2 x_1^2(t) + a_2^2 x_2^2(t) + 2 a_1 a_2 x_1(t) x_1(t)$$

$$= a_1^2 y_1(t) + a_2^2 y_2(t) + 2 a_1 a_2 x_1(t) x_1(t) \qquad \boxed{\text{Using equations (1) and (2)}}$$

Here, $y_3(t) \ne a_1 y_1(t) + a_2 y_2(t)$. Hence the given system is nonlinear.

d) Given that, $y(t) = A x(t) + B$

Let $\mathcal{H}$ be the system operating on $x(t)$ to produce, $y(t) = \mathcal{H}\{x(t)\} = A x(t) + B$.

Consider two signals $x_1(t)$ and $x_2(t)$.

Let $y_1(t)$ and $y_2(t)$ be the response of the system $\mathcal{H}$ for inputs $x_1(t)$ and $x_2(t)$ respectively.

$$\therefore y_1(t) = \mathcal{H}\{x_1(t)\} = A x_1(t) + B \qquad\qquad .....(1)$$

$$y_2(t) = \mathcal{H}\{x_2(t)\} = A x_2(t) + B \qquad\qquad ....(2)$$

Let $x_3(t) = a_1 x_1(t) + a_2 x_2(t)$. $\qquad \boxed{\text{A linear combination of inputs } x_1(t) \text{ and } x_2(t)}$

Let $y_3(t)$ be the response of the system $\mathcal{H}$ for input $x_3(t)$.

$$\therefore y_3(t) = \mathcal{H}\{x_3(t)\} = A x_3(t) + B$$

$$= A [a_1 x_1(t) + a_2 x_2(t)] + B$$

$$= a_1 A x_1(t) + a_2 A x_2(t) + B$$

$$= a_1 [y_1(t) - B] + a_2 [y_2(t) - B] + B \qquad \boxed{\text{Using equations (1) and (2)}}$$

Here, $y_3(t) \ne a_1 y_1(t) + a_2 y_2(t)$. Hence the given system is nonlinear.

e) Given that, $y(t) = e^{x(t)}$ *(AU Jun'14, 8 Marks)*

Let $\mathcal{H}$ be the system operating on $x(t)$ to produce, $y(t) = \mathcal{H}\{x(t)\} = e^{x(t)}$.

Consider two signals $x_1(t)$ and $x_2(t)$.

Let $y_1(t)$ and $y_2(t)$ be the response of the system $\mathcal{H}$ for inputs $x_1(t)$ and $x_2(t)$ respectively.

$$\therefore y_1(t) = \mathcal{H}\{x_1(t)\} = e^{x_1(t)} \qquad\qquad .....(1)$$

$$y_2(t) = \mathcal{H}\{x_2(t)\} = e^{x_2(t)} \qquad\qquad .....(2)$$

Let $x_3(t) = a_1 x_1(t) + a_2 x_2(t)$. $\qquad \boxed{\text{A linear combination of inputs } x_1(t) \text{ and } x_2(t)}$

Let $y_3(t)$ be the response of the system $\mathcal{H}$ for input $x_3(t)$.

$$\therefore y_3(t) = \mathcal{H}\{x_3(t)\} = e^{x_3(t)}$$

$$= e^{[a_1 x_1(t) + a_2 x_2(t)]} = e^{a_1 x_1(t)} e^{a_2 x_2(t)}$$

$$= [e^{x_1(t)}]^{a_1} [e^{x_2(t)}]^{a_2} = [y_1(t)]^{a_1} [y_2(t)]^{a_2} \qquad \boxed{\text{Using equations (1) and (2)}}$$

Here, $y_3(t) \ne a_1 y_1(t) + a_2 y_2(t)$. Hence the given system is nonlinear.

Example 1.19

Test the following systems for linearity.

a) $y(t) = 4 x(t) + 2 \dfrac{dx(t)}{dt}$ $\qquad\qquad$ **b)** $\dfrac{d^2 y(t)}{dt^2} + 2 \dfrac{dy(t)}{dt} + 3 y(t) = x(t)$

Solution:

a) Given that, $y(t) = 4x(t) + 2\dfrac{dx(t)}{dt}$

Let $\mathcal{H}$ be the system operating on $x(t)$ to produce, $y(t) = \mathcal{H}\{x(t)\} = 4x(t) + 2\dfrac{dx(t)}{dt}$

Consider two signals $x_1(t)$ and $x_2(t)$.

Let $y_1(t)$ and $y_2(t)$ be the response of the system $\mathcal{H}$ for inputs $x_1(t)$ and $x_2(t)$ respectively.

$$\therefore y_1(t) = \mathcal{H}\{x_1(t)\} = 4x_1(t) + 2\frac{dx_1(t)}{dt} \qquad\qquad .....(1)$$

$$y_2(t) = \mathcal{H}\{x_2(t)\} = 4x_2(t) + 2\frac{dx_2(t)}{dt} \qquad\qquad .....(2)$$

Let $x_3(t) = a_1 x_1(t) + a_2 x_2(t)$. | A linear combination of inputs $x_1(t)$ and $x_2(t)$ |

Let $y_3(t)$ be the response of the system $\mathcal{H}$ for input $x_3(t)$.

$$\therefore y_3(t) = \mathcal{H}\{x_3(t)\}$$

$$= 4x_3(t) + 2\frac{dx_3(t)}{dt}$$

$$= 4[a_1 x_1(t) + a_2 x_2(t)] + 2\frac{d}{dt}[a_1 x_1(t) + a_2 x_2(t)]$$

$$= 4a_1 x_1(t) + 4a_2 x_2(t) + 2a_1 \frac{dx_1(t)}{dt} + 2a_2 \frac{dx_2(t)}{dt}$$

$$= a_1\left[4x_1(t) + 2\frac{dx_1(t)}{dt}\right] + a_2\left[4x_2(t) + 2\frac{dx_2(t)}{dt}\right] \qquad | \text{Using equations (1) and (2)} |$$

$$= a_1 y_1(t) + a_2 y_2(t)$$

Since, $y_3(t) = a_1 y_1(t) + a_2 y_2(t)$, the given system is linear.

---

b) Given that, $\dfrac{d^2 y(t)}{dt^2} + 2\dfrac{dy(t)}{dt} + 3y(t) = x(t)$

Let $\mathcal{H}$ be the system operating on $x(t)$ to produce, $y(t)$.

Consider two signals $x_1(t)$ and $x_2(t)$.

Let $y_1(t)$ and $y_2(t)$ be the response of the system $\mathcal{H}$ for inputs $x_1(t)$ and $x_2(t)$ respectively.

When the input is $x_1(t)$, the response is $y_1(t)$. Hence the system equation for the input $x_1(t)$ can be written as,

$$\frac{d^2 y_1(t)}{dt^2} + 2\frac{dy_1(t)}{dt} + 3y_1(t) = x_1(t) \qquad\qquad .....(1)$$

When the input is $x_2(t)$, the response is $y_2(t)$. Hence the system equation for the input $x_2(t)$ can be written as,

$$\frac{d^2 y_2(t)}{dt^2} + 2\frac{dy_2(t)}{dt} + 3y_2(t) = x_2(t) \qquad\qquad .....(2)$$

Let, $x_3(t) = a_1 x_1(t) + a_2 x_2(t)$. | A linear combination of inputs $x_1(t)$ and $x_2(t)$ |

Let, $y_3(t)$ be the response of the system $\mathcal{H}$ for input $x_3(t)$.

When the input is $x_3(t)$, the response is $y_3(t)$. Hence the system equation for the input $x_3(t)$ is given by,

$$\frac{d^2 y_3(t)}{dt^2} + 2\frac{dy_3(t)}{dt} + 3y_3(t) = x_3(t) \qquad\qquad .....(3)$$

Let us multiply equation (1) by a_1.

$$\therefore a_1 \frac{d^2 y_1(t)}{dt^2} + 2a_1 \frac{dy_1(t)}{dt} + 3a_1 y_1(t) = a_1 x_1(t)$$

.....(4)

Let us multiply equation (2) by a_2.

$$\therefore a_2 \frac{d^2 y_2(t)}{dt^2} + 2a_2 \frac{dy_2(t)}{dt} + 3a_2 y_2(t) = a_2 x_2(t)$$

.....(5)

On adding equation (4) and (5) we get,

$$a_1 \frac{d^2 y_1(t)}{dt^2} + 2a_1 \frac{dy_1(t)}{dt} + 3a_1 y_1(t) + a_2 \frac{d^2 y_2(t)}{dt^2} + 2a_2 \frac{dy_2(t)}{dt} + 3a_2 y_1(t) = a_1 x_1(t) + a_2 x_2(t)$$

$$\frac{d^2}{dt^2}[a_1 y_1(t) + a_2 y_2(t)] + 2\frac{d}{dt}[a_1 y_1(t) + a_2 y_2(t)] + 3[a_1 y_1(t) + a_2 y_2(t)] = a_1 x_1(t) + a_2 x_2(t) \quad .....(6)$$

On comparing equations (3) and (6) we can say that,

if, $x_3(t) = a_1 x_1(t) + a_2 x_2(t)$, then $y_3(t) = a_1 y_1(t) + a_2 y_2(t)$

Hence the sytem is linear.

1.7.4 Causal and Noncausal Systems

(AU Dec'11 & Jun'14, 2 Marks)

Definition: A system is said to be *causal* if the output of the system at any time t depends only on the present input, past inputs and past outputs but does not depend on the future inputs and outputs.

If the system output at any time t depends on future inputs or outputs then the system is called a *noncausal* system.

The causality refers to a system that is realizable in real time. It can be shown that an LTI system is causal if and only if the impulse response is zero for t < 0, (i.e., h(t) = 0 for t < 0).

Example 1.20

Test the casuality of the following systems.

a) $y(t) = x(t) - x(t - 1)$ **b)** $y(t) = x(t) + 2 x(3 - t)$ **c)** $y(t) = t\, x(t)$

d) $y(t) = x(t) + \int_0^t x(\lambda)\, d\lambda$ **e)** $y(t) = x(t) + \int_0^{3t} x(\lambda)\, d\lambda$ **f)** $y(t) = 2x(t) + \frac{dx(t)}{dt}$

Solution:

a) Given that, $y(t) = x(t) - x(t - 1)$

When t = 0, y(0) = x(0) − x(−1) ⇒ The response at t = 0, i.e., y(0) depends on the present input x(0) and past input x(−1).

When t = 1, y(1) = x(1) − x(0) ⇒ The response at t = 1, i.e., y(1) depends on the present input x(1) and past input x(0).

From the above analysis we can say that for any value of t, the system output depends on present and past inputs. Hence the system is causal.

b) Given that, $y(t) = x(t) + 2 x(3 - t)$

When t = − 1, y(− 1) = x(− 1) + 2 x(4) ⇒ The response at t = − 1, i.e., y(− 1) depends on the present input x(− 1) and future input x(4).

When t = 0,   y(0) = x(0) + 2 x(3) ⇒ The response at t = 0, i.e., y(0) depends on the present input x(0) and future input x(3).

When t = 1,   y(1)   = x(1) + 2 x(2) $\Rightarrow$ The response at t = 1, i.e., y(1) depends on the present input x(1) and future input x(2).

When t = 2,   y(2)   = x(2) + 2 x(1) $\Rightarrow$ The response at t = 2, i.e., y(2) depends on the present input x(2) and past input x(1).

From the above analysis we can say that for t< 2, the system output depends on present and future inputs. Hence the system is noncausal.

c) Given that, $y(t) = t\, x(t)$

When t = 0, y(0) = 0 × x(0) $\Rightarrow$ The response at t = 0, i.e., y(0) depends on the present input x(0).

When t = 1, y(1) = 1 × x(1) $\Rightarrow$ The response at t = 1, i.e., y(1) depends on the present input x(1).

When t = 2, y(2) = 2 × x(2) $\Rightarrow$ The response at t = 2, i.e., y(2) depends on the present input x(2).

From the above analysis we can say that the response for any value of t depends on the present input. Hence the system is causal.

d) Given that, $y(t) = x(t) + \int_{0}^{t} x(\lambda)\, d\lambda$

$$y(t) = x(t) + \int_{0}^{t} x(\lambda)\, d\lambda = x(t) + [z(\lambda)]_{0}^{t} = x(t) + z(t) - z(0), \qquad \text{where } z(\lambda) = \int x(\lambda)\, d\lambda$$

When t = 0,   y(0) = x(0) + z(0) − z(0) $\Rightarrow$ The response at t = 0, i.e., y(0) depends on present input.

When t = 1,   y(1) = x(1) + z(1) − z(0) $\Rightarrow$ The response at t = 1, i.e., y(1) depends on present and past input.

When t = 2,   y(2) = x(2) + z(1) − z(0) $\Rightarrow$ The response at t = 2, i.e., y(2) depends on present and past input.

From the above analysis we can say that the response for any value of t depends on the present and past input. Hence the system is causal.

e) Given that, $y(t) = x(t) + \int_{0}^{3t} x(\lambda)\, d\lambda$

$$y(t) = x(t) + \int_{0}^{3t} x(\lambda)\, d\lambda = x(t) + [z(\lambda)]_{0}^{3t} = x(t) + z(3t) - z(0), \qquad \text{where, } z(\lambda) = \int x(\lambda)\, d\lambda$$

When t = 0,   y(0) = x(0)+z(0)− z(0) $\Rightarrow$ The response at t = 0, i.e., y(0) depends on present input.

When t = 1,   y(1) = x(1) + z(3) − z(0) $\Rightarrow$ The response at t =1, i.e., y(1) depends on present, past and future inputs.

When t = 2,   y(2) = x(2) + z(6) − z(0) $\Rightarrow$ The response at t = 2, i.e., y(2) depends on present, past and future inputs.

From the above analysis we can say that the response for t > 0 depends on the present, past and future inputs. Hence the system is noncausal.

f) Given that, $y(t) = 2x(t) + \dfrac{dx(t)}{dt}$

$$y(t) = 2x(t) + \frac{dx(t)}{dt}$$

$$\therefore y(t) = 2x(t) + \underset{\Delta t \to 0}{Lt} \frac{x(t) - x(t - \Delta t)}{\Delta t}$$

Using definition of differentiation,
Refer Section(1.2.4, Page.No: 1.21).

In the above equation, for any value of t, the x(t) is present input and x(t–Δt) is the past input. Therefore we can say that the response for any value of t depends on present and past input. Hence the system is causal.

Example 1.21

Test the causality of the following systems.

a) y(t) = x(t) + 3 x(t + 4)   **b)** y(t) = x(t²)   **c)** y(t) = x(2t)   **d)** y(t) = x(–t)

Solution:

a) Given that, y(t) = x(t) + 3 x(t + 4)

When t = 0, y(0) = x(0) + 3 x(4)   ⇒   The response at t = 0, i.e., y(0) depends on the present input x(0) and future input x(4).

When t = 1, y(1) = x(1) + 3 x(5)   ⇒   The response at t = 1, i.e., y(1) depends on the present input x(1) and future input x(5).

From the above analysis we can say that the response for any value of t depends on present and future inputs. Hence the system is noncausal.

b) Given that, y(t) = x(t²)

When t = –1 ; y(–1) = x(1)   ⇒   The response at t = –1, depends on the future input x(1).

When t = 0 ; y(0) = x(0)   ⇒   The response at t = 0, depends on the present input x(0).

When t = 1 ; y(1) = x(1)   ⇒   The response at t = 1, depends on the present input x(1).

When t = 2 ; y(2) = x(4)   ⇒   The response at t = 2, depends on the future input x(4).

From the above analysis we can say that the response for any value of t (except t = 0 and t = 1) depends on future input. Hence the system is noncausal.

c) Given that, y(t) = x(2t)

When t = –1 ; y(–1) = x(–2)   ⇒   The response at t = –1, depends on the past input x(–2).

When t = 0 ; y(0) = x(0)   ⇒   The response at t = 0, depends on the present input x(0).

When t = 1 ; y(1) = x(2)   ⇒   The response at t = 1, depends on the future input x(2).

From the above analysis we can say that the response of the system for t > 0, depends on future input. Hence the system is noncausal.

d) Given that, y(t) = x(–t)

When t = –2 ; y(–2) = x(2)   ⇒   The response at t = –2, depends on the future input x(2).

When t = –1 ; y(–1) = x(1)   ⇒   The response at t = –1, depends on the future input x(1).

When t = 0 ; y(0) = x(0)   ⇒   The response at t = 0, depends on the present input x(0).

When t = 1 ; y(1) = x(–1)   ⇒   The response at t = 1, depends on the past input x(–1).

From the above analysis we can say that the response of the system for t < 0 depends on future input. Hence the system is noncausal.

1.7.5 Stable and Unstable Systems

An arbitrary relaxed system is said to be **BIBO stable** (Bounded Input-Bounded Output stable) if and only if every bounded input produces a bounded output.

A system is said to be **unstable system** if any bounded input produces unbounded output. (The term unbounded output refers to infinite output).

Let x(t) be the input of continuous time system and y(t) be the response or output for x(t).

The term **bounded input** refers to finite value of the input signal x(t) for any value of t. Hence if input x(t) is bounded then there exists a constant M_x such that $|x(t)| \leq M_x$ and $M_x < \infty$, for all t.

> Examples of bounded input signal are step signal, decaying exponential signal and impulse signal. Examples of unbounded input signal are ramp signal and increasing exponential signal.

The term **bounded output** refers to finite and predictable output for any value of t. Hence if output y(t) is bounded then there exists a constant M_y such that $|y(t)| \leq M_y$ and $M_y < \infty$, for all t.

In general, the test for stability of the system is performed by applying specific input. On applying a bounded input to a system if the output is bounded then the system is said to be BIBO stable.

Condition for Stability of an LTI System

For an LTI (Linear Time Invariant) system, the condition for BIBO stability can be transformed to a condition on impulse response, h(t). For BIBO stability of an LTI continuous time system, the integral of impulse response should be finite.

$$\therefore \quad \int_{-\infty}^{+\infty} \left| h(t) \right| dt < \infty, \text{ for stability of an LTI system} \qquad \qquad .....(1.18)$$

Proof:

Let, x(t) = Input of LTI system.

y(t) = Response of LTI system for the input x(t).

Now, by convolution formula,

$$y(t) = x(t) * h(t) = h(t) * x(t) = \int_{-\infty}^{+\infty} h(\tau) x(t - \tau) d\tau$$

> Convolution satisfy commutative property.

$$\therefore |y(t)| = \left| \int_{-\infty}^{+\infty} h(\tau) x(t - \tau) d\tau \right|$$

> Taking absolute value on both sides.

$$= \int_{-\infty}^{+\infty} \left| h(\tau) x(t - \tau) \right| d\tau$$

> For linear system the order integration and absolute value can be interchanged.

$$= \int_{-\infty}^{+\infty} \left| h(\tau) \right| \left| x(t - \tau) \right| d\tau$$

> For linear system the order of multiplication and absolute value can be interchanged.

$$|y(t)| = \int_{-\infty}^{+\infty} h(\tau) M_x \, d\tau$$

$$= M_x \int_{-\infty}^{+\infty} |h(\tau)| \, d\tau$$

$$= M_x \int_{-\infty}^{+\infty} |h(t)| \, dt \qquad \qquad \dots(1.19)$$

If input is bounded, then $|x(t-\tau)| =$ Constant $= M_x$

M_x is independant of integration index τ.

Change index τ to t.

In the above equation, if

$$\int_{-\infty}^{+\infty} |h(t)| \, dt < \infty \qquad \qquad \dots(1.20)$$

then the response y(t) is bounded.

Example 1.22

Test the stability of the following systems.

a) $y(t) = \cos[x(t)]$ **b)** $y(t) = x(-t-2)$ **c)** $y(t) = t\,x(t)$ **d)** $y(t) = e^{x(t)}$

Solution:

a) Given that, $y(t) = \cos[x(t)]$

The given system is a nonlinear system, and so the test for stability should be performed for specific inputs.

The value of $\cos\theta$ lies between -1 to $+1$ for any value of θ. Therefore the output y(t) is bounded for any value of input x(t). Hence the given system is stable.

b) Given that, $y(t) = x(-t-2)$

The given system is a time variant system, and so the test for stability should be performed for specific inputs.

The operations performed by the system on the input signal are folding and shifting. A bounded input signal will remain bounded even after folding and shifting. Therefore in the given system, the output will be bounded as long as input is bounded. Hence the given system is BIBO stable.

c) Given that, $y(t) = t\,x(t)$

The given system is a time variant system, and so the test for stability should be performed for specific inputs.

Case i : Let x(t) tends to ∞ or constant, as **t** tends to infinity. In this case, y(t) = t x(t) will be infinity as **t** tends to inifnity and so the system is unstable.

Case ii : Let x(t) tends to 0, as **t** tends to infinity. In this case y(t) = t x(t) will be zero as **t** tends to infinity and so the system is stable.

d) Given that, $y(t) = e^{x(t)}$ **(AU Jun'14, 8 Marks)**

The given system is a time invariant system, and so the stability can be determined from impulse response.

Let the input x(t) be an impulse input $\delta(t)$.

Now, the response y(t) will be impulse response h(t).

$\therefore$ Impulse response, $h(t) = e^{\delta(t)}$.

$$\therefore \int_{-\infty}^{+\infty} |h(t)| \, dt = \int_{-\infty}^{+\infty} e^{\delta(t)} dt = e^{\delta(t)} \, dt \big|_{t=0} = e^1 = \text{Constant}$$

Here, the integral of impulse response is constatnt and so the system is stable.

Example 1.23

Test the stability of the LTI systems, whose impulse responses are given below:

a) $h(t) = e^{-5|t|}$ **b)** $h(t) = e^{4t} u(t)$ **c)** $h(t) = e^{-4t} u(t)$

d) $h(t) = t \, e^{-3t} u(t)$ **e)** $h(t) = t \, cost \, u(t)$ **f)** $h(t) = e^{-t} \sin t \, u(t)$

Solution:

a) Given that, $h(t) = e^{-5|t|}$

For stability, $\int_{-\infty}^{+\infty} |h(t)| \, dt < \infty$

$$\therefore \int_{-\infty}^{+\infty} |h(t)| \, dt = \int_{-\infty}^{+\infty} \left| e^{-5|t|} \right| dt = \int_{-\infty}^{+\infty} e^{-5|t|} \, dt$$

$$= \int_{-\infty}^{0} e^{5t} \, dt + \int_{0}^{+\infty} e^{-5t} \, dt = \left[\frac{e^{5t}}{5} \right]_{-\infty}^{0} + \left[\frac{e^{-5t}}{-5} \right]_{0}^{\infty}$$

$$= \frac{e^0}{5} - \frac{e^{-\infty}}{5} + \frac{e^{-\infty}}{-5} - \frac{e^{-0}}{-5} = \frac{1}{5} - 0 + 0 + \frac{1}{5} = \frac{2}{5}$$

Here, $\int_{-\infty}^{+\infty} |h(t)| \, dt = \frac{2}{5} = \text{Constant}$. Hence the system is stable.

b) Given that, $h(t) = e^{4t} u(t)$

For stability, $\int_{-\infty}^{+\infty} |h(t)| \, dt < \infty$

$$\therefore \int_{-\infty}^{+\infty} |h(t)| \, dt = \int_{-\infty}^{+\infty} \left| e^{4t} u(t) \right| dt = \int_{-\infty}^{+\infty} e^{4t} u(t) \, dt$$

$$= \int_{0}^{+\infty} e^{4t} \, dt = \left[\frac{e^{4t}}{4} \right]_{0}^{\infty} = \frac{e^{\infty}}{4} - \frac{e^0}{4} = \infty - \frac{1}{4} = \infty$$

Here, $\int_{-\infty}^{+\infty} |h(t)| \, dt = \infty$. Hence the system is unstable.

c) Given that, $h(t) = e^{-4t} u(t)$

For stability, $\int_{-\infty}^{+\infty} |h(t)| \, dt < \infty$

$$\therefore \int_{-\infty}^{+\infty} |h(t)| \, dt = \int_{-\infty}^{+\infty} \left| e^{-4t} u(t) \right| dt = \int_{-\infty}^{+\infty} e^{-4t} u(t) \, dt$$

$$= \int_{0}^{+\infty} e^{-4t} \, dt = \left[\frac{e^{-4t}}{-4} \right]_{0}^{\infty} = \frac{e^{-\infty}}{-4} - \frac{e^0}{-4} = 0 + \frac{1}{4} = \frac{1}{4}$$

Here, $\int_{-\infty}^{+\infty} |h(t)| \, dt = \frac{1}{4} = \text{Constant}$. Hence the system is stable.

d) Given that, h(t) = t e^{-3t} u(t)

For stability, $\displaystyle\int_{-\infty}^{+\infty} |h(t)| \, dt < \infty$

$\displaystyle \therefore \int_{-\infty}^{+\infty} |h(t)| \, dt = \int_{-\infty}^{+\infty} |t e^{-3t} u(t)| \, dt = \int_{0}^{+\infty} t e^{-3t} \, dt$

$\qquad\qquad\qquad = \left[t \dfrac{e^{-3t}}{-3} - \int \dfrac{e^{-3t}}{-3} \times 1 \, dt \right]_{0}^{\infty} = \left[-\dfrac{te^{-3t}}{3} - \dfrac{e^{-3t}}{9} \right]_{0}^{\infty}$

$\qquad\qquad\qquad = -\dfrac{\infty \times e^{-\infty}}{3} - \dfrac{e^{-\infty}}{9} + \dfrac{0 \times e^{0}}{3} + \dfrac{e^{0}}{9}$

$\qquad\qquad\qquad = -\dfrac{\infty \times 0}{3} - 0 + 0 + \dfrac{1}{9} = \dfrac{1}{9}$

$\int u \, dv = u\,v - \int v \, du$
$u = t \quad \Rightarrow \quad du = 1$
$dv = e^{-3t} \Rightarrow \quad v = \dfrac{e^{-3t}}{-3}$

Since, $\displaystyle\int_{-\infty}^{+\infty} |h(t)| \, dt = \dfrac{1}{9} = $ Constant, the system is stable.

---

e) Given that, h(t) = t cost u(t)

For stability, $\displaystyle\int_{-\infty}^{+\infty} |h(t)| \, dt < \infty$

$\displaystyle \therefore \int_{-\infty}^{+\infty} |h(t)| \, dt = \int_{-\infty}^{+\infty} |t \cos t \, u(t)| \, dt = \int_{0}^{+\infty} t \cos t \, dt$

$\qquad\qquad\qquad = \left[t \sin t - \int \sin t \times 1 \, dt \right]_{0}^{\infty} = [t \sin t + \cos t]_{0}^{\infty}$

$\int u \, dv = u\,v - \int v \, du$
$u = t \quad \Rightarrow \quad du = 1$
$dv = \cos t \Rightarrow \quad v = \sin t$

$\qquad\qquad\qquad = \infty \times \sin \infty + \cos \infty - 0 \times \sin 0 - \cos 0 = \infty + \cos \infty - 0 - 1 = \infty$

Since, $\displaystyle\int_{-\infty}^{+\infty} |h(t)| \, dt = \infty$, the system is unstable.

---

f) Given that, h(t) = e^{-t} sint u(t)

For stability, $\displaystyle\int_{-\infty}^{+\infty} |h(t)| \, dt < \infty$

$\int u \, dv = u\,v - \int v \, du$
$u = e^{-t} \quad \Rightarrow \quad du = -e^{-t}$
$dv = \sin t \Rightarrow \quad v = -\cos t$

.....(1)

$\displaystyle \therefore \int_{-\infty}^{+\infty} |h(t)| \, dt = \int_{-\infty}^{+\infty} |e^{-t} \sin t \, u(t)| \, dt = \int_{0}^{+\infty} e^{-t} \sin t \, dt$

$\displaystyle \int_{0}^{\infty} e^{-t} \sin t \, dt = [e^{-t}(-\cos t)]_{0}^{\infty} - \int_{0}^{+\infty} (-\cos t)(-e^{-t}) \, dt$

$\qquad\qquad\qquad = [-e^{-t} \cos t]_{0}^{\infty} - \int_{0}^{\infty} e^{-t} \cos t \, dt$

$\int u \, dv = u\,v - \int v \, du$
$u = e^{-t} \quad \Rightarrow \quad du = -e^{-t}$
$dv = \cos t \Rightarrow \quad v = \sin t$

.....(2)

$\qquad\qquad\qquad = [-e^{-t} \cos t]_{0}^{\infty} - \left[[e^{-t} \sin t]_{0}^{\infty} - \int_{0}^{\infty} \sin t\,(-e^{-t}) dt \right]$

$\qquad\qquad\qquad = [-e^{-t} \cos t]_{0}^{\infty} - [e^{-t} \sin t]_{0}^{\infty} - \int_{0}^{\infty} e^{-t} \sin t \, dt$

From equation (2) we can write,

$\displaystyle 2 \int_{0}^{\infty} e^{-t} \sin t \, dt = [-e^{-t} \cos t]_{0}^{\infty} - [e^{-t} \sin t]_{0}^{\infty}$

$$\therefore \int_0^\infty e^{-t} \sin t \ dt = \frac{1}{2}[-e^{-t}\cos t]_0^\infty - \frac{1}{2}[e^{-t}\sin t]_0^\infty \qquad \dots(3)$$

Using equation (3), the equation (1) can be written as,

$$\int_{-\infty}^{+\infty} |h(t)| \ dt = \int_0^\infty e^{-t}\sin t\, dt = \frac{1}{2}[-e^{-t}\cos t]_0^\infty - \frac{1}{2}[e^{-t}\sin t]_0^\infty$$

$$= \frac{1}{2}[-e^{-\infty}\cos\infty + e^0 \cos 0] - \frac{1}{2}[e^{-\infty}\sin\infty - e^0 \sin 0]$$

$$= \frac{1}{2}[-0\times\cos\infty + 1] - \frac{1}{2}[0\times\sin\infty - 0]$$

$$= \frac{1}{2}[0+1] - \frac{1}{2}[0+0] = \frac{1}{2}$$

Since, $\int_{-\infty}^{+\infty} |h(t)| \ dt = \frac{1}{2}$ = Constant, the system is stable.

Example 1.24

Determine the range of values of "a" and "b" for the stability of LTI system with impulse response.
h(t) = e^{at} u(t) + e^{-bt} u(t)

Solution:

Given that, **h(t) = e^{at} u(t) + e^{-bt} u(t).**

$$\int_{-\infty}^{+\infty} |h(t)| \, dt = \int_{-\infty}^{+\infty} |e^{at}u(t) + e^{-bt}u(t)| \ dt = \left| \int_{-\infty}^{+\infty} (e^{at}u(t) + e^{-bt}u(t))\, dt \right|$$

$$= \left| \int_0^\infty e^{at}\,dt + \int_0^\infty e^{-bt}\,dt \right| = \left| \left[\frac{e^{at}}{a}\right]_0^\infty + \left[\frac{e^{-bt}}{-b}\right]_0^\infty \right|$$

$$= \left| \frac{e^{a\times\infty}}{a} - \frac{e^{a\times 0}}{a} + \frac{e^{-b\times\infty}}{-b} - \frac{e^{-b\times 0}}{-b} \right| = \left| \frac{e^{a\times\infty}}{a} - \frac{1}{a} + \frac{e^{-b\times\infty}}{-b} + \frac{1}{b} \right|$$

In the above equation if "a" is negative and "b" is positive then it converges to finite value.

Therefore, when "a" is negative and "b" is positive.

$$\int_0^\infty |h(t)| \ dt = \left| \frac{0}{a} - \frac{1}{a} + \frac{0}{-b} + \frac{1}{b} \right|$$

$$= \left| -\frac{1}{a} + \frac{1}{b} \right| = \left| \frac{a-b}{ab} \right| = \text{Constant}$$

Here the integral of impulse response is a constant when "a" is negative and "b" is positive.

Therefore the range of values of "a" and "b" for stability of LTI system are, **a < 0 and b > 0.**

1.7.6 Feedback and Nonfeedback Systems

The system in which the output y(t) at any time t depends on past output, past input and present input is called a *feedback system.* The integration and differentiation of a signal at any time depends on past value and so the equations governing feedback systems will have terms involving differentiations and integrations of output and input.

The equations governing feedback systems will be in the form,

$$a_0 \frac{d^N}{dt^N} y(t) + a_1 \frac{d^{N-1}}{dt^{N-1}} y(t) + a_2 \frac{d^{N-2}}{dt^{N-2}} y(t) + ... + a_{N-1} \frac{d}{dt} y(t) + a_N y(t) = b_0 \frac{d^M}{dt^M} x(t)$$

$$+ b_1 \frac{d^{M-1}}{dt^{M-1}} x(t) + b_2 \frac{d^{M-2}}{dt^{M-2}} x(t) + ... + b_{M-1} \frac{d}{dt} x(t) + b_M x(t)$$

The system in which the output depends only on the present and past input is called a **nonfeedback system.** The equations governing nonfeedback systems will not have terms involving differentiations and integrations of input.

The equations governing nonfeedback systems will be in the form,

$$y(t) = b_0 \frac{d^M}{dt^M} x(t) + b_1 \frac{d^{M-1}}{dt^{M-1}} x(t) + b_2 \frac{d^{M-2}}{dt^{M-2}} x(t) + ... + b_{M-1} \frac{d}{dt} x(t) + b_M x(t)$$

1.7.7 Invertible and Noninvertible Systems

A system is said to be **invertible** if distinct inputs produce distinct outputs. Therefore, for an invertible system an inverse system exist.

A system is said to be **noninvertible** if distinct inputs produce same output or any input produces zero output. Therefore, for a noninvertible system we cannot determine an inverse system.

Table 1.7: Examples of Invertible Systems

System	Inverse System
$y(t) = A\, x(t)$	$x(t) = \frac{1}{A} y(t)$
$y(t) = x(t+T_0)$	$x(t) = y(t-T_0)$
$y(t) = \int x(t)\, dt$	$x(t) = \frac{d}{dt} y(t)$

Table 1.8: Examples of Noninvertible Systems

System	Reasons for Nonexistence of Inverse System
$y(t) = \delta(t)\, \delta(t - T_0)$	$\delta(t) = 1\;;\,t = 0$ $= 0\;;\,t \neq 0$ $\delta(t - T_0) = 1\;;\,t = T_0$ $= 0\;;\,t \neq T_0$ $\therefore\; y(t) = 0$
$y(t) = A\sin(t + 2\pi k)$ where, k is integer	Here, $y(t)$ will produce same value for any integer value of k
$y(t) = x^2(t)$	Here, $x(t) = \pm\sqrt{y(t)}$ and so we cannot predict the sign of $x(t)$.

Example 1.25

(AU Dec'15, 16 Marks)

State whether the system $y(t) = t\, x(-t)$ is causal, stable, linear and time invariant.

Solution:

Given that, $y(t) = t\, x(-t)$

i) Casuality

When $t = -1$, $y(-1) = -1 \times x(1)$ $\Rightarrow$ The response at $t = -1$, depends on the future input $x(1)$.

When $t = 0$, $y(0) = 0 \times x(0)$ $\Rightarrow$ The response at $t = 0$, depends on the present input $x(0)$.

When $t = 1$, $y(1) = 1 \times x(-1)$ $\Rightarrow$ The response at $t = 1$, depends on the past input $x(-1)$.

From the above analysis we can say that the response of the system for $t < 0$ depends on future input. Hence the system is noncausal.

ii) Linearity

Let $\mathcal{H}$ be the system operating on $x(t)$ to produce, $y(t) = \mathcal{H}\{x(t)\} = t\, x(-t)$.

Consider two signals $x_1(t)$ and $x_2(t)$.

Let $y_1(t)$ and $y_2(t)$ be the response of the system $\mathcal{H}$ for inputs $x_1(t)$ and $x_2(t)$ respectively.

$\therefore\quad y_1(t) = \mathcal{H}\{x_1(t)\} = t\, x_1(-t)$ .....(1)

$\quad y_2(t) = \mathcal{H}\{x_2(t)\} = t\, x_2(-t)$ .....(2)

Let, $x_3(t) = a_1 x_1(t) + a_2 x_2(t)$.   A linear combination of inputs $x_1(t)$ and $x_2(t)$

Let, $y_3(t)$ be the response of the system $\mathcal{H}$ for input $x_3(t)$.

$\therefore\quad y_3(t) = \mathcal{H}\{x_3(t)\} = t\, x_3(-t)$

$\qquad = t[a_1 x_1(-t) + a_2 x_2(-t)] = a_1 t\, x_1(-t) + a_2 t\, x_2(-t)$ Using equations (1) and (2)

$\qquad = a_1 y_1(t) + a_2 y_2(t))$

Since, $y_3(t) = a_1 y_1(t) + a_2 y_2(t)$, the given system is linear.

iii) Time invariance

Test 1: Response for delayed input

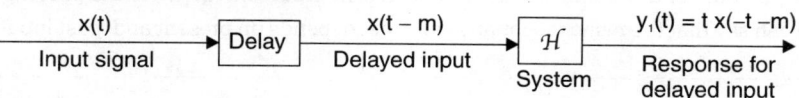

Test 2: Delayed response

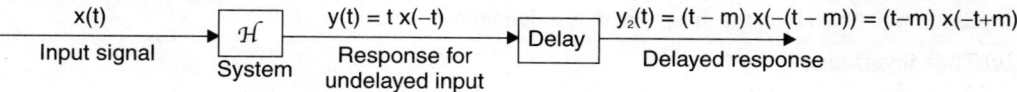

Conclusion: Here, $y_1(t) \neq y_2(t)$, therefore the system is time variant.

iv) Stability

The given system is a time variant system, and so the test for stability should be performed for specific inputs.

Case i : Let x(t) tends to ∞ or constant, as **t** tends to infinity. In this case, y(t) = t x(–t) will be infinity as t tends to inifnity and so the system is unstable.

Case ii : Let x(t) tends to 0, as **t** tends to infinity. In this case y(t) = t x(–t) will be zero as t tends to infinity and so the system is stable.

Example 1.26   *(AU Dec'14, 16 Marks & Jun'12, 4 Marks)*

State whether the system $y(t) = \dfrac{dx(t)}{dt}$ is linear, causal, dynamic, time invariant and stable.

Solution:

Given that, $y(t) = \dfrac{dx(t)}{dt}$

i) Linearity

Let, $\mathcal{H}$ be the system operating on x(t) to produce, $y(t) = \mathcal{H}\{x(t)\} = \dfrac{dx(t)}{dt}$

Consider two signals, $x_1(t)$ and $x_2(t)$.

Let, $y_1(t)$ and $y_2(t)$ be the response of the system $\mathcal{H}$ for inputs $x_1(t)$ and $x_2(t)$ respectively.

$$\therefore \quad y_1(t) = \mathcal{H}\{x_1(t)\} = \frac{dx_1(t)}{dt} \qquad\qquad\qquad .....(1)$$

$$y_2(t) = \mathcal{H}\{x_2(t)\} = \frac{dx_2(t)}{dt} \qquad\qquad\qquad .....(2)$$

Let, $x_3(t) = a_1 x_1(t) + a_2 x_2(t)$. | A linear combination of inputs $x_1(t)$ and $x_2(t)$ |

Let $y_3(t)$ be the response of the system $\mathcal{H}$ for input $x_3(t)$.

$$\therefore \quad y_3(t) = \mathcal{H}\{x_3(t)\} = \frac{dx_3(t)}{dt}$$

$$= \frac{d}{dt}[a_1 x_1(t) + a_2 x_2(t)] = a_1 \frac{dx_1(t)}{dt} + a_2 \frac{dx_2(t)}{dt} = a_1 y_1(t) + a_2 y_2(t) \quad \boxed{\text{Using equations (1) and (2)}}$$

Here, $y_3(t) = a_1 y_1(t) + a_2 y_2(t)$. Hence the given system is linear.

ii) Causality

$$y(t) = \frac{dx(t)}{dt}$$

$$= \underset{\Delta t \to 0}{\text{Lt}} \; \frac{x(t) - x(t - \Delta t)}{\Delta t} \quad \text{(Using definition of differentiation, Refer Section 1.2.4, (Page.No.1.21))}$$

In the above equation, for any value of t, the x(t) is present input and x(t–Δt) is the past input.

Therefore we can say that the response for any value of t depends on present and past input. Hence the system is causal.

iii) Dynamic

The output of the system depends on differentiation of the input which in turn depends on past and present input. Therefore, the given sytem is a dynamic system.

iv) Time invariance

Test 1: Response for delayed input

Test 2: Delayed response

Conclusion: Here, $y_1(t) = y_2(t)$, therefore the system is time invariant.

v) Stability

The given system is LTI system and so the stability can be determined from impulse response.

Let the input $x(t)$ be an impulse input $\delta(t)$.

Now, response $y(t)$ will be impulse response $h(t)$.

∴ Impulse response, $h(t) = \dfrac{d}{dt}\delta(t)$

$$\therefore \int_{-\infty}^{+\infty}\left|h(t)\right|\,dt = \int_{-\infty}^{+\infty}\left|\dfrac{d}{dt}\delta(t)\right|dt = \delta(t)$$

We know that, $\delta(t) = \infty$, at $t = 0$ and $\int \delta(t)\,dt = 1$ and so the system is stable.

Example 1.27

State whether the following systems are linear, casual, time variant and dynamic.

a) $y(t) = x(t-3) + (3-t)$ **b)** $\dfrac{d^3y(t)}{dt^3} + 4\dfrac{d^2y(t)}{dt^2} + 5\dfrac{dy(t)}{dt} + 2y^2(t) = x(t)$

Solution:

a) Given that, $y(t) = x(t-3) + (3-t)$ **(AU Jun'12, 4 Marks)**

i) Linearity

Let, $\mathcal{H}$ be the system operating on $x(t)$ to produce, $y(t) = \mathcal{H}\{x(t)\} = x(t-3) + (3-t)$

Consider two signals, $x_1(t)$ and $x_2(t)$.

Let, $y_1(t)$ and $y_2(t)$ be the response of the system $\mathcal{H}$ for inputs $x_1(t)$ and $x_2(t)$ respectively.

$$\therefore\ y_1(t) = \mathcal{H}\{x_1(t)\} = x_1(t-3) + (3-t) \qquad\qquad\qquad \dots\text{(1)}$$

$$y_2(t) = \mathcal{H}\{x_2(t)\} = x_2(t-3) + (3-t) \qquad\qquad\qquad \dots\text{(2)}$$

Let, $x_3(t) = a_1 x_1(t) + a_2 x_2(t)$. $\boxed{\text{A linear combination of inputs } x_1(t) \text{ and } x_2(t)}$

Let $y_3(t)$ be the response of the system $\mathcal{H}$ for input $x_3(t)$.

$$\therefore\ y_3(t) = \mathcal{H}\{x_3(t)\} = x_3(t-3) + (3-t)$$

$$= a_1 x_1(t-3) + a_2 x_2(t-3) + (3-t)$$

$$= a_1\left[y_1(t) - (3-t)\right] + a_2\left[y_2(t) - (3-t)\right] + (3-t) \qquad \boxed{\text{Using equations (1) and (2)}}$$

Here, $y_3(t) \ne a_1 y_1(t) + a_2 y_2(t)$. Hence the given system is nonlinear.

ii) Causality

When t = 0, y(0) = x(−3) + 3   ⇒   The response at t = 0, i.e., y(0) depends on the past input x(−3).

When t = 1, y(1) = x(−2) + 2   ⇒   The response at t = 1, i.e., y(1) depends on the past input x(−2).

From the above analysis we can say that the response for any value of t depends on the past input. Hence the system is causal.

iii) Dynamic

For any values of t, y(t) depends on past input x(t−3) and so system require memory for storage. Therfore, the given system is dynamic system.

iv) Time invariance

Test 1: Response for delayed input

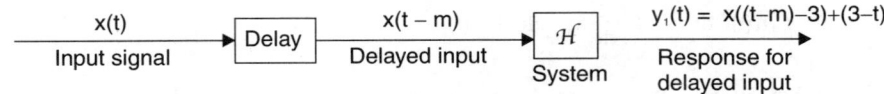

$$x(t) \xrightarrow{\text{Input signal}} \boxed{\text{Delay}} \xrightarrow[\text{Delayed input}]{x(t-m)} \boxed{\mathcal{H}}_{\text{System}} \xrightarrow{y_1(t) = x((t-m)-3)+(3-t)} \text{Response for delayed input}$$

Test 2: Delayed response

$$x(t) \xrightarrow{\text{Input signal}} \boxed{\mathcal{H}}_{\text{System}} \xrightarrow[\substack{\text{Response for}\\\text{undelayed input}}]{y(t) = x(t-3)+(3-t)} \boxed{\text{Delay}} \xrightarrow{y_2(t) = x((t-m)-3)+(3-(t-m))} \text{Delayed response}$$

Conclusion: Here, $y_1(t) \neq y_2(t)$, therefore the system is time variant.

b) Given that, $\dfrac{d^3y(t)}{dt^3} + 4\dfrac{d^2y(t)}{dt^2} + 5\dfrac{dy(t)}{dt} + 2y^2(t) = x(t)$ **(AU Dec'11, 8 Marks)**

i) Linearity

Given that, $\dfrac{d^3y(t)}{dt^3} + 4\dfrac{d^2y(t)}{dt^2} + 5\dfrac{dy(t)}{dt} + 2y^2(t) = x(t)$

$$\therefore \ y(t) = \sqrt{\frac{1}{2}\left[x(t) - \frac{d^3y(t)}{dt^3} - 4\frac{d^2y(t)}{dt^2} - 5\frac{dy(t)}{dt}\right]}$$

The response y(t) involves square root operation which is nonlinear operation and so system response is nonlinear.

ii) Causality

The response of the system depends on differentiation of the input which inturn depends on past and present inputs. Hence the system is causal system.

iii) Dynamic

The output or response of the system depends on differentiation of the input which in turn depends on past and present inputs. Therefore, the given system is dyanamic system.

iv) Time invariance

Test 1: Response for delayed input

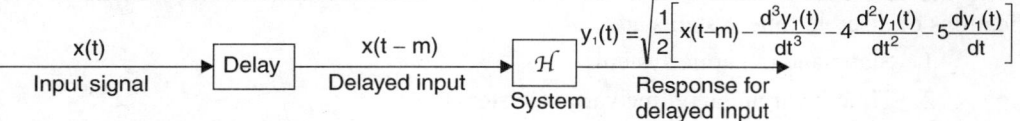

Test 2: Delayed response

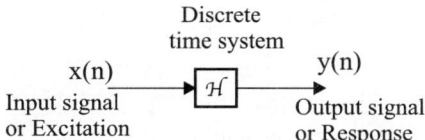

Conclusion: Here, $y_1(t) = y_2(t)$, therefore the system is time invariant.

1.8 Discrete Time Systems

A system which can process discrete time signal is called ***discrete time system***, and so the input and output signals of a discrete time system are discrete time signals.

A discrete time system is denoted by the letter $\mathcal{H}$. The input of discrete time system is denoted as "x(n)" and the output of discrete time system is denoted as "y(n)". The diagrammatic representation of a discrete time system is shown in Fig 1.56.

$$\underset{\substack{\text{Input signal} \\ \text{or Excitation}}}{x(n)} \longrightarrow \boxed{\mathcal{H}} \longrightarrow \underset{\substack{\text{Output signal} \\ \text{or Response}}}{y(n)}$$

Discrete time system

Fig 1.56: *Representation of discrete time system.*

The operation performed by a discrete time system on input to produce output or response can be expressed as,

$$\text{Response,} \quad y(n) = \mathcal{H}\{x(n)\} \qquad\qquad .....(1.21)$$

where, $\mathcal{H}$ denotes the system operation (also called system operator).

When a discrete time system satisfies the properties of linearity and time invariance then it is called ***LTI (Linear Time Invariant) discrete time system*** .

The input-output relation of an LTI discrete time system is represented by constant coefficient difference equation shown below:

$$y(n) = -\sum_{m=1}^{N} a_m\, y(n-m) + \sum_{m=0}^{M} b_m\, x(n-m) \qquad\qquad .....(1.22)$$

where, N = Order of the system, and $M \leq N$.

The solution of the above difference equation is the response y(n) of the discrete time system, for the input x(n).

1.9 Classification of Discrete Time Systems   *(AU Jun'11, 10 Marks)*

The discrete time systems are classified based on their characteristics. Some of the classifications of discrete time systems are,

1. Static and dynamic systems
2. Time invariant and time variant systems
3. Linear and nonlinear systems
4. Causal and noncausal systems
5. Stable and unstable systems
6. FIR and IIR systems
7. Recursive and nonrecursive systems
8. Invertible and Noninvertible systems.

> *Note: Static or dynamic nature, time invariance, linearity, causality, stability and invertibility are also called basic properties of discrete time systems.*

1.9.1 Static and Dynamic Systems

A discrete time system is called *static* or *memoryless* system if its output at any instant n depends on present input sample but not on the past or future samples of the input. A discrete time system is called *dynamic or memory system* if its output at any instant n depends on past and future inputs in addition to present input.

Example:

$$y(n) = a\, x(n)$$
$$y(n) = n\, x(n) + 6\, x^3(n) \qquad \Big\} \text{ Static systems}$$

$$y(n) = x(n) + 3\, x(n-1)$$
$$y(n) = \sum_{m=0}^{N} x(n-m) \qquad \Big\} \text{ Finite memory is required}$$
$$y(n) = \sum_{m=0}^{\infty} x(n-m) \qquad \Big\} \text{ Inifinte memory is required} \qquad \Bigg\} \text{ Dynamic systems}$$

1.9.2 Time Invariant and Time Variant Systems

A system is said to be *time invariant* if its input-output characteristics do not change with time. In a system if the input-output characteristics change with time then the system is called *time variant* system.

Definition : A relaxed system $\mathcal{H}$ is *time invariant* or *shift invariant* if and only if

$$\mathcal{H}\{x(n)\} = y(n) \text{ implies that, } \mathcal{H}\{x(n-m)\} = y(n-m)$$

for every input signal x(n) and every time shift m.

Alternative Definition for Time Invariance

A system $\mathcal{H}$ is *time invariant* if the response to a shifted (or delayed) version of the input is identical to a shifted (or delayed) version of the response based on the unshifted (or undelayed) input.

∴ In a time invariant system, $\mathcal{H}\{x(n-m)\} = z^{-m}\,\mathcal{H}\{x(n)\}$

In a time variant system, $\mathcal{H}\{x(n-m)\} \neq z^{-m}\,\mathcal{H}\{x(n)\}$

The operator z^{-m} represents a signal delay of m samples.

The diagrammatic explanation of the above definition of time invariance is shown in Fig 1.57.

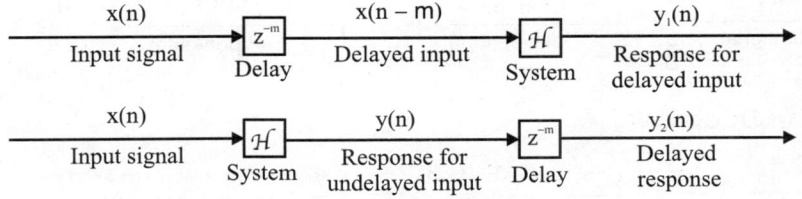

If, $y_1(n) = y_2(n)$, then the system is time invariant

Fig 1.57: *Diagrammatic explanation of time invariance.*

Procedure to Test for Time Invariance

1. Delay the input signal by m units of time and determine the response of the system for this delayed input signal. Let this response be $y_1(n)$.

2. Delay the response of the system for undelayed input by m units of time. Let this delayed response be $y_2(n)$.

3. If $y_1(n) = y_2(n)$, then the system is time invariant.

4. If $y_1(n) \neq y_2(n)$, then the system is time variant.

Example 1.28

Test the following systems for time invariance.

a) $y(n) = 2n\, x(n)$ **b)** $y(n) = x(n) - b\, x(n-1)$

Solution:

a) Given that, $y(n) = 2n\, x(n)$

Test 1: Response for delayed input

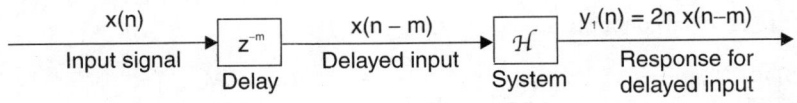

Test 2: Delayed response

$$
\begin{array}{c}
\text{x(n)} \\
\text{Input signal}
\end{array}
\longrightarrow
\boxed{\mathcal{H}} \underset{\text{System}}{}
\begin{array}{c}
\text{y(n) = 2n x(n)} \\
\text{Response for} \\
\text{undelayed input}
\end{array}
\longrightarrow
\boxed{z^{-m}} \underset{\text{Delay}}{}
\begin{array}{c}
\text{y}_2\text{(n) = 2(n--m) x(n--m)} \\
\text{Delayed response}
\end{array}
$$

Conclusion: Here, $y_1(n) \neq y_2(n)$, therefore the system is time variant.

b) Given that, $y(n) = x(n) - b\,x(n-1)$

    <u>**Test 1:** Response for delayed input</u>

$x(n)$				
Input signal	z^{-m} Delay	$x(n-m)$ Delayed input	$\mathcal{H}$ System	$y_1(n) = x(n-m) - b\,x(n-m-1)$ Response for delayed input

    <u>**Test 2:** Delayed response</u>

$x(n)$				
Input signal	$\mathcal{H}$ System	$y(n) = x(n) - b\,x(n-1)$ Response for undelayed input	z^{-m} Delay	$y_2(n) = x(n-m) - b\,x(n-m-1)$ Delayed response

    <u>**Conclusion:**</u> Here, $y_1(n) = y_2(n)$, therefore the system is time invariant.

Example 1.29

    Test the following systems for time invariance.

    a) $y(n) = x(n) + B$ **b)** $y(n) = n\,x^3(n)$ **c)** $y(n) = b^{x(n)}$ **d)** $y(n) = \displaystyle\sum_{k=0}^{M} b_k\,x(n-k) - \sum_{k=1}^{N} a_k\,y(n-k)$

Solution:

a) Given that, $y(n) = x(n) + B$

    <u>**Test 1:** Response for delayed input</u>

$x(n)$				
Input signal	z^{-m} Delay	$x(n-m)$ Delayed input	$\mathcal{H}$ System	$y_1(n) = x(n-m) + B$ Response for delayed input

    <u>**Test 2:** Delayed response</u>

$x(n)$				
Input signal	$\mathcal{H}$ System	$y(n) = x(n) + B$ Response for undelayed input	z^{-m} Delay	$y_2(n) = x(n-m) + B$ Delayed response

    <u>**Conclusion:**</u> Here, $y_1(n) = y_2(n)$, therefore the system is time invariant.

b) Given that, $y(n) = n\,x^3(n)$

    <u>**Test 1:** Response for delayed input</u>

$x(n)$				
Input signal	z^{-m} Delay	$x(n-m)$ Delayed input	$\mathcal{H}$ System	$y_1(n) = n\,x^3(n-m)$ Response for delayed input

    <u>**Test 2:** Delayed response</u>

$x(n)$				
Input signal	$\mathcal{H}$ System	$y(n) = n\,x^3(n)$ Response for undelayed input	z^{-m} Delay	$y_2(n) = (n-m)\,x^3(n-m)$ Delayed response

    <u>**Conclusion:**</u> Here, $y_1(n) \neq y_2(n)$, therefore the system is time variant.

c) Given that, $y(n) = b^{x(n)}$

<u>Test 1:</u> Response for delayed input

$$
\begin{array}{ccccc}
\underset{\text{Input signal}}{x(n)} \longrightarrow & \boxed{z^{-m}} & \underset{\substack{\text{Delayed input}}}{x(n-m)} \longrightarrow & \boxed{\mathcal{H}} & \underset{\substack{\text{Response for}\\ \text{delayed input}}}{y_1(n) = b^{x(n-m)}} \longrightarrow \\
& \text{Delay} & & \text{System} &
\end{array}
$$

<u>Test 2:</u> Delayed response

$$
\begin{array}{ccccc}
\underset{\text{Input signal}}{x(n)} \longrightarrow & \boxed{\mathcal{H}} & \underset{\substack{\text{Response for}\\ \text{undelayed input}}}{y(n) = b^{x(n)}} \longrightarrow & \boxed{z^{-m}} & \underset{\substack{\text{Delayed response}}}{y_2(n) = b^{x(n-m)}} \longrightarrow \\
& \text{System} & & \text{Delay} &
\end{array}
$$

<u>Conclusion:</u> Here, $y_1(n) = y_2(n)$, therefore the system is time invariant.

d) Given that, $y(n) = \displaystyle\sum_{k=0}^{M} b_k\, x(n-k) - \sum_{k=1}^{N} a_k\, y(n-k)$

<u>Test 1:</u> Response for delayed input

$$
\begin{array}{ccccc}
\underset{\text{Input signal}}{x(n)} \longrightarrow & \boxed{z^{-m}} & \underset{\substack{\text{Delayed input}}}{x(n-m)} \longrightarrow & \boxed{\mathcal{H}} & \underset{\substack{\text{Response for}\\ \text{delayed input}}}{y_1(n)} \longrightarrow \\
& \text{Delay} & & \text{System} &
\end{array}
$$

Response for delayed input,

$$
y_1(n) = \mathcal{H}\{x(n-m)\} = \sum_{k=0}^{M} b_k\, x(n-m-k) - \sum_{k=1}^{N} a_k\, y(n-m-k)
$$

<u>Test 2:</u> Delayed response

$$
\begin{array}{ccccc}
\underset{\text{Input signal}}{x(n)} \longrightarrow & \boxed{\mathcal{H}} & \underset{\substack{\text{Response for}\\ \text{undelayed input}}}{y(n)} \longrightarrow & \boxed{z^{-m}} & \underset{\substack{\text{Delayed response}}}{y_2(n)} \longrightarrow \\
& \text{System} & & \text{Delay} &
\end{array}
$$

Response for undelayed input $= \mathcal{H}\{x(n)\} = \displaystyle\sum_{k=0}^{M} b_k\, x(n-k) - \sum_{k=1}^{N} a_k\, y(n-k)$

Delayed response, $y_2(n) = z^{-m}\, \mathcal{H}\{x(n)\}$

$$
= z^{-m} \left[\sum_{k=0}^{M} b_k\, x(n-k) - \sum_{k=1}^{N} a_k\, y(n-k) \right]
$$

$$
= \sum_{k=0}^{M} b_k\, x(n-m-k) - \sum_{k=1}^{N} a_k\, y(n-m-k)
$$

<u>Conclusion:</u> Here, $y_1(n) = y_2(n)$, therefore the system is time invariant.

1.9.3 Linear and Nonlinear Systems

A *linear system* is one that satisfies the superposition principle. The ***principle of superposition*** requires that the response of the system to a weighted sum of the signals is equal to the corresponding weighted sum of the responses of the system to each of the individual input signals.

__Definition:__   A relaxed system $\mathcal{H}$ is *linear* if

$$\mathcal{H}\{a_1 x_1(n) + a_2 x_2(n)\} = a_1 \mathcal{H}\{x_1(n)\} + a_2 \mathcal{H}\{x_2(n)\} \qquad .....(1.23)$$

for any arbitrary input sequences $x_1(n)$ and $x_2(n)$ and for any arbitrary constants a_1 and a_2.

If a relaxed system does not satisfy the superposition principle as given by the above definition, then the system is *nonlinear*.The diagrammatic explanation of linearity is shown in Fig 1.58.

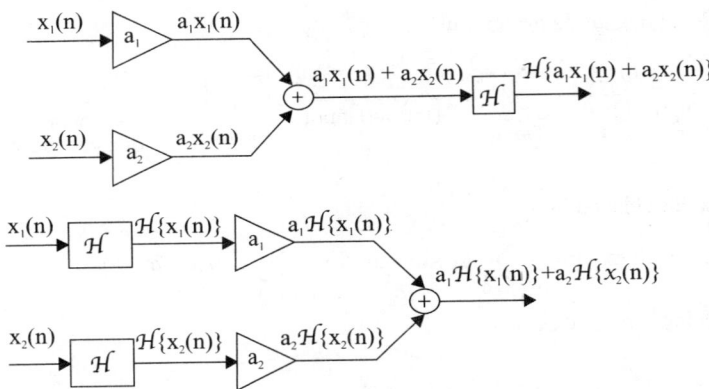

The system, $\mathcal{H}$ is linear if and only if, $\mathcal{H}\{a_1 x_1(n)+a_2 x_2(n)\} = a_1 \mathcal{H}\{x_1(n)\} + a_2 \mathcal{H}\{x_2(n)\}$

Fig 1.58: *Diagrammatic explanation of linearity.*

Procedure To Test For Linearity

1. Let $x_1(n)$ and $x_2(n)$ be two inputs to system $\mathcal{H}$, and $y_1(n)$ and $y_2(n)$ be corresponding responses.

2. Consider a signal, $x_3(n) = a_1 x_1(n) + a_2 x_2(n)$ which is a weighed sum of $x_1(n)$ and $x_2(n)$.

3. Let $y_3(n)$ be the response for $x_3(n)$.

4. Check whether $y_3(n) = a_1 y_1(n) + a_2 y_2(n)$. If they are equal then the system is linear, otherwise it is nonlinear.

Example 1.30

Test the following systems for linearity.

a) $y(n) = n\, x(n)$ 　　　**b)** $y(n) = x(n^2)$ 　　**c)** $y(n) = B\, x(n) + C$

Solution:

a) Given that, $y(n) = n\, x(n)$

Let $\mathcal{H}$ be the system represented by the equation, $y(n) = nx(n)$.

The system $\mathcal{H}$ operates on $x(n)$ to produce, $y(n)$.

$$\xrightarrow{\;x(n)\;}\boxed{\mathcal{H}}\xrightarrow{\;\;\;\;}\; y(n) = \mathcal{H}\{x(n)\} = n\, x(n)$$

Consider two signals, $x_1(n)$ and $x_2(n)$.

Let, $y_1(n)$ and $y_2(n)$ be the response of the system $\mathcal{H}$ for inputs $x_1(n)$ and $x_2(n)$ respectively.

$$\xrightarrow{\;x_1(n)\;}\boxed{\mathcal{H}}\xrightarrow{\;\;\;\;}\; y_1(n) = \mathcal{H}\{x_1(n)\} = n\, x_1(n)$$

$$\xrightarrow{\;x_2(n)\;}\boxed{\mathcal{H}}\xrightarrow{\;\;\;\;}\; y_2(n) = \mathcal{H}\{x_2(n)\} = n\, x_2(n)$$

$$\therefore\; a_1 y_1(n) + a_2 y_2(n) = a_1\, n\, x_1(n) + a_2\, n\, x_2(n) \qquad\qquad\qquad .....(1)$$

Consider a linear combination of inputs, $a_1 x_1(n) + a_2 x_2(n) = x_3(n)$.

Let, $y_3(n)$ be the response for $x_3(n)$.

$$\xrightarrow{\;x_3(n)\;}\boxed{\mathcal{H}}\xrightarrow{\;\;\;\;}\; y_3(n) = \mathcal{H}\{x_3(n)\}$$

$$\therefore\; y_3(n) = \mathcal{H}\{x_3(n)\} = n\, x_3(n) = n[a_1 x_1(n) + a_2 x_2(n)] = a_1\, n\, x_1(n) + a_2\, n\, x_2(n) \qquad .....(2)$$

The condition to be satisfied for linearity is, $y_3(n) = a_1 y_1(n) + a_2 y_2(n)$.

From equations (1) and (2) we can say that, $y_3(n) = a_1 y_1(n) + a_2 y_2(n)$. Hence the system is linear.

b) Given that, $y(n) = x(n^2)$

Let, $\mathcal{H}$ be the system represented by the equation, $y(n) = x(n^2)$.

The system $\mathcal{H}$ operates on $x(n)$ to produce, $y(n)$.

$$\xrightarrow{\;x(n)\;}\boxed{\mathcal{H}}\xrightarrow{\;\;\;\;}\; y(n) = \mathcal{H}\{x(n)\} = x(n^2)$$

Consider two signals, $x_1(n)$ and $x_2(n)$.

Let, $y_1(n)$ and $y_2(n)$ be the response of the system $\mathcal{H}$ for inputs $x_1(n)$ and $x_2(n)$ respectively.

$$\xrightarrow{\;x_1(n)\;}\boxed{\mathcal{H}}\xrightarrow{\;\;\;\;}\; y_1(n) = \mathcal{H}\{x_1(n)\} = x_1(n^2)$$

$$x_2(n) \longrightarrow \boxed{\mathcal{H}} \longrightarrow y_2(n) = \mathcal{H}\{x_2(n)\} = x_2(n^2)$$

$$\therefore a_1 y_1(n) + a_2 y_2(n) = a_1 x_1(n^2) + a_2 x_2(n^2) \qquad \qquad .....(1)$$

Consider a linear combination of inputs, $a_1 x_1(n) + a_2 x_2(n) = x_3(n)$.

Let, $y_3(n)$ be the response for $x_3(n)$.

$$x_3(n) \longrightarrow \boxed{\mathcal{H}} \longrightarrow y_3(n) = \mathcal{H}\{x_3(n)\}$$

$$\therefore y_3(n) = \mathcal{H}\{x_3(n)\} = x_3(n^2) = a_1 x_1(n^2) + a_2 x_2(n^2) \qquad \qquad .....(2)$$

The condition to be satisfied for linearity is, $y_3(n) = a_1 y_1(n) + a_2 y_2(n)$.

From equations (1) and (2) we can say that, $y_3(n) = a_1 y_1(n) + a_2 y_2(n)$. Hence the system is linear.

c) Given that, $y(n) = B\,x(n) + C$

Let, $\mathcal{H}$ be the system represented by the equation, $y(n) = B\,x(n) + C$.

The system $\mathcal{H}$ operates on $x(n)$ to produce, $y(n)$.

$$x(n) \longrightarrow \boxed{\mathcal{H}} \longrightarrow y(n) = \mathcal{H}\{x(n)\} = B\,x(n) + C$$

Consider two signals, $x_1(n)$ and $x_2(n)$.

Let, $y_1(n)$ and $y_2(n)$ be the response of the system $\mathcal{H}$ for inputs $x_1(n)$ and $x_2(n)$ respectively.

$$x_1(n) \longrightarrow \boxed{\mathcal{H}} \longrightarrow y_1(n) = \mathcal{H}\{x_1(n)\} = B\,x_1(n) + C$$

$$x_2(n) \longrightarrow \boxed{\mathcal{H}} \longrightarrow y_2(n) = \mathcal{H}\{x_2(n)\} = B\,x_2(n) + C$$

$$\therefore a_1 y_1(n) + a_2 y_2(n) = a_1[B\,x_1(n) + C] + a_2[B\,x_2(n) + C]$$
$$= B\,a_1 x_1(n) + C\,a_1 + B\,a_2 x_2(n) + C\,a_2 \qquad \qquad .....(1)$$

Consider a linear combination of inputs, $a_1 x_1(n) + a_2 x_2(n) = x_3(n)$.

Let, $y_3(n)$ be the response for $x_3(n)$.

$$x_3(n) \longrightarrow \boxed{\mathcal{H}} \longrightarrow y_3(n) = \mathcal{H}\{x_3(n)\}$$

$$\therefore y_3(n) = \mathcal{H}\{x_3(n)\} = B\,x_3(n) + C = B[a_1 x_1(n) + a_2 x_2(n)] + C = Ba_1 x_1(n) + B\,a_2 x_2(n) + C \quad .....(2)$$

The condition to be satisfied for linearity is, $y_3(n) = a_1 y_1(n) + a_2 y_2(n)$.

From equations (1) and (2) we can say that, $y_3(n) \neq a_1 y_1(n) + a_2 y_2(n)$. Hence the system is nonlinear.

Example 1.31

Test the following systems for linearity.

a) $y(n) = e^{x(n)}$ **b)** $y(n) = b^{x(n)}$ **c)** $y(n) = n\, x^2(n)$

Solution:

a) Given that, $y(n) = e^{x(n)}$

Let, $\mathcal{H}$ be the system represented by the equation, $y(n) = e^{x(n)}$.

The system $\mathcal{H}$ operates on $x(n)$ to produce, $y(n)$.

$$\xrightarrow{\;x(n)\;} \boxed{\;\mathcal{H}\;} \xrightarrow{} y(n) = \mathcal{H}\{x(n)\} = e^{x(n)}$$

Consider two signals, $x_1(n)$ and $x_2(n)$.

Let, $y_1(n)$ and $y_2(n)$ be the response of the system $\mathcal{H}$ for inputs $x_1(n)$ and $x_2(n)$ respectively.

$$\xrightarrow{\;x_1(n)\;} \boxed{\;\mathcal{H}\;} \xrightarrow{} y_1(n) = \mathcal{H}\{x_1(n)\} = e^{x_1(n)}$$

$$\xrightarrow{\;x_2(n)\;} \boxed{\;\mathcal{H}\;} \xrightarrow{} y_2(n) = \mathcal{H}\{x_2(n)\} = e^{x_2(n)}$$

$$\therefore\ a_1 y_1(n) + a_2 y_2(n) = a_1 e^{x_1(n)} + a_2 e^{x_2(n)} \qquad\qquad .....(1)$$

Consider a linear combination of inputs, $a_1 x_1(n) + a_2 x_2(n) = x_3(n)$.

Let, $y_3(n)$ be the response for $x_3(n)$.

$$\xrightarrow{\;x_3(n)\;} \boxed{\;\mathcal{H}\;} \xrightarrow{} y_3(n) = \mathcal{H}\{x_3(n)\}$$

$$\therefore\ y_3(n) = \mathcal{H}\{x_3(n)\} = e^{x_3(n)} = e^{\{a_1 x_1(n)\ +\ a_2 x_2(n)\}} = e^{a_1 x_1(n)}\, e^{a_2 x_2(n)} \qquad .....(2)$$

The condition to be satisfied for linearity is, $y_3(n) = a_1 y_1(n) + a_2 y_2(n)$.

From equations (1) and (2) we can say that, $y_3(n) \neq a_1 y_1(n) + a_2 y_2(n)$. Hence the system is nonlinear.

---

b) Given that, $y(n) = b^{x(n)}$

Let, $\mathcal{H}$ be the system represented by the equation, $y(n) = b^{x(n)}$.

The system $\mathcal{H}$ operates on $x(n)$ to produce, $y(n)$.

$$\xrightarrow{\;x(n)\;} \boxed{\;\mathcal{H}\;} \xrightarrow{} y(n) = \mathcal{H}\{x(n)\} = b^{x(n)}$$

Consider two signals, $x_1(n)$ and $x_2(n)$.

Let, $y_1(n)$ and $y_2(n)$ be the response of the system $\mathcal{H}$ for inputs $x_1(n)$ and $x_2(n)$ respectively.

$$\xrightarrow{\;x_1(n)\;} \boxed{\;\mathcal{H}\;} \xrightarrow{} y_1(n) = \mathcal{H}\{x_1(n)\} = b^{x_1(n)}$$

$$\xrightarrow{\;x_2(n)\;} \boxed{\;\mathcal{H}\;} \xrightarrow{} y_2(n) = \mathcal{H}\{x_2(n)\} = b^{x_2(n)}$$

$$\therefore\ a_1 y_1(n) + a_2 y_2(n) = a_1 b^{x_1(n)} + a_2 b^{x_2(n)} \qquad\qquad .....(1)$$

Consider a linear combination of inputs, $a_1 x_1(n) + a_2 x_2(n) = x_3(n)$.

Let, $y_3(n)$ be the response for $x_3(n)$.

$$x_3(n) \longrightarrow \boxed{\mathcal{H}} \longrightarrow y_3(n) = \mathcal{H}\{x_3(n)\}$$

$$\therefore \ y_3(n) \ = \ \mathcal{H}\{x_3(n)\} = b^{x_3(n)} = b^{[a_1 x_1(n) + a_2 x_2(n)]} = b^{a_1 x_1(n)} \, b^{a_2 x_2(n)} \qquad \qquad .....(2)$$

The condition to be satisfied for linearity is, $y_3(n) = a_1\, y_1(n) + a_2\, y_2(n)$.

From equations (1) and (2) we can say that, $y_3(n) \ne a_1\, y_1(n) + a_2\, y_2(n)$. Hence the system is nonlinear.

c) Given that, $y(n) = n\, x^2(n)$

Let, $\mathcal{H}$ be the system represented by the equation, $y(n) = n\, x^2(n)$.

The system $\mathcal{H}$ operates on $x(n)$ to produce, $y(n)$.

$$x(n) \longrightarrow \boxed{\mathcal{H}} \longrightarrow y(n) = \mathcal{H}\{x(n)\} = n\, x^2(n)$$

Consider two signals, $x_1(n)$ and $x_2(n)$.

Let, $y_1(n)$ and $y_2(n)$ be the response of the system $\mathcal{H}$ for inputs $x_1(n)$ and $x_2(n)$ respectively.

$$x_1(n) \longrightarrow \boxed{\mathcal{H}} \longrightarrow y_1(n) = \mathcal{H}\{x_1(n)\} = n\, x_1^2(n)$$

$$x_2(n) \longrightarrow \boxed{\mathcal{H}} \longrightarrow y_2(n) = \mathcal{H}\{x_2(n)\} = n\, x_2^2(n)$$

$$\therefore \ a_1\, y_1(n) + a_2\, y_2(n) = a_1\, n\, x_1^2(n) + a_2\, n\, x_2^2(n) \qquad \qquad .....(1)$$

Consider a linear combination of inputs, $a_1\, x_1(n) + a_2\, x_2(n) = x_3(n)$.

Let, $y_3(n)$ be the response for $x_3(n)$.

$$x_3(n) \longrightarrow \boxed{\mathcal{H}} \longrightarrow y_3(n) = \mathcal{H}\{x_3(n)\}$$

$$\therefore \ y_3(n) = \mathcal{H}\{x_3(n)\} = n\, x_3^2(n) \ = n[a_1\, x_1(n) + a_2\, x_2(n)]^2$$

$$= n\, a_1^2\, x_1^2(n) + n\, a_2^2\, x_2^2(n) + 2\, n\, a_1\, a_2\, x_1(n)\, x_2(n) \qquad \qquad .....(2)$$

The condition to be satisfied for linearity is, $y_3(n) = a_1\, y_1(n) + a_2\, y_2(n)$.

From equations (1) and (2) we can say that, $y_3(n) \ne a_1\, y_1(n) + a_2\, y_2(n)$. Hence the system is nonlinear.

Example 1.32

Test the following systems for linearity.

a) $y(n) = x(n) - 2x(n-1)$ b) $y(n) = \displaystyle\sum_{m=0}^{M} b_m\, x(n-m) - \sum_{m=1}^{N} c_m\, y(n-m)$

Solution:

a) Given that, $y(n) = x(n) - 2x(n-1)$

Let, $\mathcal{H}$ be the system represented by the equation, $y(n) = x(n) - 2\, x(n-1)$.

The system $\mathcal{H}$ operates on $x(n)$ to produce, $y(n)$.

$$x(n) \longrightarrow \boxed{\mathcal{H}} \longrightarrow y(n) = \mathcal{H}\{x(n)\} = x(n) - 2x(n-1)$$

Consider two signals, $x_1(n)$ and $x_2(n)$.

Let, $y_1(n)$ and $y_2(n)$ be the response of the system $\mathcal{H}$ for inputs $x_1(n)$ and $x_2(n)$ respectively.

$$\xrightarrow{\quad x_1(n) \quad} \boxed{\mathcal{H}} \longrightarrow y_1(n) = \mathcal{H}\{x_1(n)\} = x_1(n) - 2x_1(n-1)$$

$$\xrightarrow{\quad x_2(n) \quad} \boxed{\mathcal{H}} \longrightarrow y_2(n) = \mathcal{H}\{x_2(n)\} = x_2(n) - 2x_2(n-1)$$

$$\therefore a_1 y_1(n) + a_2 y_2(n) = a_1 x_1(n) - a_1 2 x_1(n-1) + a_2 x_2(n) - a_2 2 x_2(n-1) \qquad \dots(1)$$

Consider a linear combination of inputs, $a_1 x_1(n) + a_2 x_2(n) = x_3(n)$.

Let, $y_3(n)$ be the response for $x_3(n)$.

$$\xrightarrow{\quad x_3(n) \quad} \boxed{\mathcal{H}} \longrightarrow y_3(n) = \mathcal{H}\{x_3(n)\}$$

$$\therefore y_3(n) = \mathcal{H}\{x_3(n)\} = x_3(n) - 2 x_3(n-1) = a_1 x_1(n) + a_2 x_2(n) - 2[a_1 x_1(n-1) + a_2 x_2(n-1)]$$

$$= a_1 x_1(n) - a_1 2 x_1(n-1) + a_2 x_2(n) - a_2 2 x_2(n-1) \qquad \dots(2)$$

The condition to be satisfied for linearity is, $y_3(n) = a_1 y_1(n) + a_2 y_2(n)$.

From equations (1) and (2) we can say that, $y_3(n) = a_1 y_1(n) + a_2 y_2(n)$. Hence the system is linear.

---

b) Given that, $y(n) = \sum\limits_{m=0}^{M} b_m x(n-m) - \sum\limits_{m=1}^{N} c_m y(n-m)$

Let, $\mathcal{H}$ be the system represented by the equation,

$$y(n) = \sum_{m=0}^{M} b_m x(n-m) - \sum_{m=1}^{N} c_m y(n-m)$$

Here, $y(n)$ is the response of the system $\mathcal{H}$ fo the input $x(n)$.

$$\therefore \mathcal{H}\{x(n)\} = y(n) = \sum_{m=0}^{M} b_m x(n-m) - \sum_{m=1}^{N} c_m y(n-m)$$

Consider two signals, $x_1(n)$ and $x_2(n)$.

Let, $y_1(n)$ and $y_2(n)$ be the response of the system $\mathcal{H}$ for inputs $x_1(n)$ and $x_2(n)$ respectively.

$$\therefore y_1(n) = \mathcal{H}\{x_1(n)\} = \sum_{m=0}^{M} b_m x_1(n-m) - \sum_{m=1}^{N} c_m y_1(n-m)$$

$$y_2(n) = \mathcal{H}\{x_2(n)\} = \sum_{m=0}^{M} b_m x_2(n-m) - \sum_{m=1}^{N} c_m y_2(n-m)$$

$$\therefore a_1 y_1(n) + a_2 y_2(n) = a_1\left(\sum_{m=0}^{M} b_m x_1(n-m) - \sum_{m=1}^{N} c_m y_1(n-m)\right)$$

$$+ a_2\left(\sum_{m=0}^{M} b_m x_2(n-m) - \sum_{m=1}^{N} c_m y_2(n-m)\right) \qquad \dots(1)$$

Consider a linear combination of inputs, $a_1 x_1(n) + a_2 x_2(n) = x_3(n)$.

Let, $y_3(n)$ be the response for the input $x_3(n)$.

$$\therefore \ y_3(n) = \mathcal{H}\{x_3(n)\} = \mathcal{H}\{a_1 x_1(n) + a_2 x_2(n)\}$$

$$= \sum_{m=0}^{M} b_m \left(a_1 x_1(n-m) + a_2 x_2(n-m)\right) - \sum_{m=1}^{N} c_m\, y_3(n-m)$$

$$= a_1 \sum_{m=0}^{M} b_m\, x_1(n-m) + a_2 \sum_{m=0}^{M} b_m\, x_2(n-m) - \sum_{m=1}^{N} c_m\, y_3(n-m) \quad .....(2)$$

By time invariant property,

If $y_3(n) = \mathcal{H}\{a_1 x_1(n) + a_2 x_2(n)\}$ then $y_3(n-m) = \mathcal{H}\{a_1 x_1(n-m) + a_2 x_2(n-m)\}$

If $y_2(n) = \mathcal{H}\{x_2(n)\}$ then $y_2(n-m) = \mathcal{H}\{x_2(n-m)\}$

If $y_1(n) = \mathcal{H}\{x_1(n)\}$ then $y_1(n-m) = \mathcal{H}\{x_1(n-m)\}$

$$\therefore \ y_3(n-m) = \mathcal{H}\{a_1 x_1(n-m) + a_2 x_2(n-m)\} = a_1\, \mathcal{H}\{x_1(n-m)\} + a_2\, \mathcal{H}\{x_2(n-m)\}$$

$$= a_1\, y_1(n-m) + a_2\, y_2(n-m) \quad .....(3)$$

Using equation (3), the equation (2) can be written as,

$$y_3(n) = a_1 \sum_{m=0}^{M} b_m\, x_1(n-m) + a_2 \sum_{m=0}^{M} b_m\, x_2(n-m) - \sum_{m=1}^{N} c_m \left[a_1 y_1(n-m) + a_2 y_2(n-m)\right]$$

$$= a_1 \sum_{m=0}^{M} b_m\, x_1(n-m) + a_2 \sum_{m=0}^{M} b_m\, x_2(n-m) - a_1 \sum_{m=1}^{N} c_m\, y_1(n-m) - a_2 \sum_{m=1}^{N} c_m\, y_2(n-m)$$

$$= a_1 \left(\sum_{m=0}^{M} b_m\, x_1(n-m) - \sum_{m=1}^{N} c_m\, y_1(n-m) \right) + a_2 \left(\sum_{m=0}^{M} b_m\, x_2(n-m) - \sum_{m=1}^{N} c_m\, y_2(n-m) \right)$$

$$.....(4)$$

The condition to be satisfied for linearity is, $y_3(n) = a_1\, y_1(n) + a_2\, y_2(n)$.

From equations (1) and (4) we can say that the condition for linearity is satisfied. Therefore the system is linear.

1.9.4 Causal and Noncausal Systems   *(AU May'15, 2 Marks)*

A system is said to be *causal* if the output of the system at any time n depends only on the present input, past inputs and past outputs but does not depend on the future inputs and outputs.

If the system output at any time n depends on future inputs or outputs then the system is called **noncausal** system.

The causality refers to a system that is realizable in real time. It can be shown that an LTI system is causal if and only if the impulse response is zero for n < 0, (i.e., h(n) = 0 for n < 0).

Let, x(n) = Present input   and   y(n) = Present output

$$\therefore x(n-1),\, x(n-2),\, ...,\ \text{are past inputs}$$

$$y(n-1),\, y(n-2),\, ...,\ \text{are past outputs}$$

In mathematical terms the output of a causal system satisfies the equation of the form,

$$y(n) = F\,[x(n),\, x(n-1),\, x(n-2),\, ...\,,\, y(n-1),\, y(n-2)\, ...]$$

where, F[·] is some arbitrary function.

Example 1.33

Test the causality of the following systems.

a) $y(n) = x(n) - x(n - 2)$ **b)** $y(n) = \sum\limits_{k=-\infty}^{n} x(k)$ **c)** $y(n) = b\, x(n)$ **d)** $y(n) = n\, x(n)$

Solution:

a) Given that, $y(n) = x(n) - x(n - 2)$

     When $n = 0$, $y(0) = x(0) - x(-2)$ $\Rightarrow$ The response at $n = 0$, i.e., $y(0)$ depends on the present input $x(0)$ and past input $x(-2)$

     When $n = 1$, $y(1) = x(1) - x(-1)$ $\Rightarrow$ The response at $n = 1$, i.e., $y(1)$ depends on the present input $x(1)$ and past input $x(-1)$.

From the above analysis we can say that for any value of n, the system output depends on present and past inputs. Hence the system is causal.

---

b) Given that, $y(n) = \sum\limits_{k=-\infty}^{n} x(k)$

     When $n = 0$, $y(0) = \sum\limits_{k=-\infty}^{0} x(k)$

         $= \ldots x(-2) + x(-1) + x(0)$ $\Rightarrow$ The response at $n = 0$, i.e., $y(0)$ depends on the present input $x(0)$ and past inputs $x(-1)$, $x(-2)$,.....

     When $n = 1$, $y(1) = \sum\limits_{k=-\infty}^{1} x(k)$

         $= \ldots x(-2) + x(-1) + x(0) + x(1)$ $\Rightarrow$ The response at $n = 1$, i.e., $y(1)$ depends on the present input $x(1)$ and past inputs $x(0)$, $x(-1)$, $x(-2)$,.....

From the above analysis we can say that for any value of n, the system output depends on present and past inputs. Hence the system is causal.

---

c) Given that, $y(n) = b\, x(n)$

     When $n = 0$, $y(0) = b\, x(0)$ $\Rightarrow$ The response at $n = 0$, i.e., $y(0)$ depends on the present input $x(0)$.

     When $n = 1$, $y(1) = b\, x(1)$ $\Rightarrow$ The response at $n = 1$, i.e., $y(1)$ depends on the present input $x(1)$.

From the above analysis we can say that the response for any value of n depends on the present input. Hence the system is causal.

---

d) Given that, $y(n) = n\, x(n)$

     When $n = 0$, $y(0) = 0 \times x(0)$ $\Rightarrow$ The response at $n = 0$, i.e., $y(0)$ depends on the present input $x(0)$.

     When $n = 1$, $y(1) = 1 \times x(1)$ $\Rightarrow$ The response at $n = 1$, i.e., $y(1)$ depends on the present input $x(1)$.

     When $n = 2$, $y(2) = 2 \times x(2)$ $\Rightarrow$ The response at $n = 2$, i.e., $y(2)$ depends on the present input $x(2)$.

From the above analysis we can say that the response for any value of n depends on the present input. Hence the system is causal.

Example 1.34

Test the causality of the following systems.

a) $y(n) = x(n) + 2\,x(n + 3)$ **b)** $y(n) = x(n^2)$ **c)** $y(n) = x(3n)$ **d)** $y(n) = x(-n)$

Solution:

a) Given that, $y(n) = x(n) + 2\,x(n + 3)$

When n = 0, $y(0) = x(0) + 2\,x(3)$ $\Rightarrow$ The response at n = 0, i.e., y(0) depends on the present input x(0) and future input x(3).

When n = 1, $y(1) = x(1) + 2\,x(4)$ $\Rightarrow$ The response at n = 1, i.e., y(1) depends on the present input x(1) and future input x(4).

From the above analysis we can say that the response for any value of n depends on present and future inputs. Hence the system is noncausal.

b) Given that, $y(n) = x(n^2)$

When n = –1 ; $y(-1) = x(1)$ $\Rightarrow$ The response at n = –1, depends on the future input x(1).

When n = 0 ; $y(0) = x(0)$ $\Rightarrow$ The response at n = 0, depends on the present input x(0).

When n = 1 ; $y(1) = x(1)$ $\Rightarrow$ The response at n = 1, depends on the present input x(1).

When n = 2 ; $y(2) = x(4)$ $\Rightarrow$ The response at n = 2, depends on the future input x(4).

From the above analysis we can say that the response for any value of n (except n = 0 and n = 1) depends on future inputs. Hence the system is noncausal.

c) Given that, $y(n) = x(3n)$ *(AU Dec'12, 2 Marks)*

When n = –1 ; $y(-1) = x(-3)$ $\Rightarrow$ The response at n = –1, depends on the past input x(–3).

When n = 0 ; $y(0) = x(0)$ $\Rightarrow$ The response at n = 0, depends on the present input x(0).

When n = 1 ; $y(1) = x(3)$ $\Rightarrow$ The response at n = 1, depends on the future input x(3).

From the above analysis we can say that the response of the system for n > 0, depends on future inputs. Hence the system is noncausal.

d) Given that, $y(n) = x(-n)$ *(AU Jun'13, 2 Marks)*

When n = –2 ; $y(-2) = x(2)$ $\Rightarrow$ The response at n = –2, depends on the future input x(2).

When n = –1 ; $y(-1) = x(1)$ $\Rightarrow$ The response at n = –1, depends on the future input x(1).

When n = 0 ; $y(0) = x(0)$ $\Rightarrow$ The response at n = 0, depends on the present input x(0).

When n = 1 ; $y(1) = x(-1)$ $\Rightarrow$ The response at n = 1, depends on the past input x(–1).

From the above analysis we can say that the response of the system for n < 0 depends on future inputs. Hence the system is noncausal.

1.9.5 Stable and Unstable Systems

An arbitrary relaxed system is said to be ***BIBO stable system*** (Bounded Input-Bounded Output stable) if and only if every bounded input produces a bounded output.

A system is said to be ***unstable system*** if any bounded input produces unbounded output. (The term unbounded output refers to infinite output).

Let x(n) be the input of discrete time system and y(n) be the response or output for x(n). The term ***bounded input*** refers to finite value of the input signal x(n) for any value of n. Hence if input x(n) is bounded then there exits a constant M_x such that $|x(n)| \le M_x$ and $M_x < \infty$, for all n.

Examples of bounded input signal are step signal, decaying exponential signal and impulse signal. Examples of unbounded input signal are ramp signal and increasing exponential signal.

The term ***bounded output*** refers to finite and predictable output for any value of n. Hence if output y(n) is bounded then there exists a constant M_y such that $|y(n)| \le M_y$ and $M_y < \infty$, for all n.

In general, the test for stability of the system is performed by applying specific input. On applying a bounded input to a system if the output is bounded then the system is said to be BIBO stable. For LTI (Linear Time Invariant) systems the condition for BIBO stability can be transformed to a condition on impulse response as shown below:

Condition for Stability of LTI System   *(AU May'15, 2 Marks)*

The condition for stability of an LTI system is,

$$\sum_{-\infty}^{+\infty} |h(n)| < \infty \qquad\qquad .....(1.24)$$

i.e., an LTI system is ***stable*** if the impulse response is absolutely summable.

Proof:

Let, x(n) = Input to LTI system.

y(n) = Response of LTI system for the input x(n).

Now, by convolution sum formula,

$$y(n) = x(n) * h(n) = h(n) * x(n) = \sum_{m=-\infty}^{+\infty} h(m)\,x(n-m)$$

| Convolution satisfy commutative property. |

$$\therefore\ |y(n)| = \left| \sum_{m=-\infty}^{+\infty} h(m)x(n-m) \right|$$

| Taking absolute value on both sides. |

$$= \sum_{m=-\infty}^{+\infty} |h(m)x(n-m)|$$

| For linear system the order summation and absolute value can be interchanged. |

$$= \sum_{m=-\infty}^{+\infty} |h(m)|\,|x(n-m)|$$

| For linear system the order of multiplication and absolute value can be interchanged. |

$$= \sum_{m=-\infty}^{+\infty} |h(m)|\,M_x$$

| If input is bounded, then $|x(n-m)| =$ Constant $= M_x$ |

$$= M_x \sum_{m=-\infty}^{+\infty} |h(m)|$$

| M_x is independant of summation index m. |

$$= M_x \sum_{n=-\infty}^{+\infty} |h(n)|$$

| Change index m to n. |

In the above equation, if

$$\sum_{n=-\infty}^{+\infty} |h(n)| < \infty \qquad\qquad .....(1.25)$$

then the response y(n) is bounded.

Example 1.35

Test the stability of the following systems.

a) $y(n) = \cos[x(n)]$ **b)** $y(n) = x(-n-3)$ **c)** $y(n) = n\,x(n)$ **d)** $y(n) = x(n-1)$

Solution:

a) Given that, $y(n) = \cos[x(n)]$

The given system is a nonlinear system, and so the test for stability should be performed for specific inputs.

The value of $\cos\theta$ lies between -1 to $+1$ for any value of θ. Therefore the output $y(n)$ is bounded for any value of input $x(n)$. Hence the given system is stable.

b) Given that, $y(n) = x(-n-3)$

The given system is a time variant system, and so the test for stability should be performed for specific inputs.

The operations performed by the system on the input signal are folding and shifting. A bounded input signal will remain bounded even after folding and shifting. Therefore in the given system, the output will be bounded as long as input is bounded. Hence the given system is BIBO stable.

c) Given that, $y(n) = n\,x(n)$

The given system is a time variant system, and so the test for stability should be performed for specific inputs.

Case i: If $x(n)$ tends to infinity or constant, as "n" tends to infinity, then $y(n) = n\,x(n)$ will be infinite as "n" tends to infinity. So the system is unstable.

Case ii: If $x(n)$ tends to zero as "n" tends to infinity, then $y(n) = n\,x(n)$ will be zero as "n" tends to infinity. So the system is stable.

d) Given that, $y(n) = x(n-1)$ ***(AU Jun'14, 8 Marks)***

The given system is an LTI system, and so the stability can be determined from impulse response.

Let the input $x(n)$ be an impulse input $x(n)$.

Now, the response $y(n)$ will be impulse response $h(n)$.

$\therefore$ Impulse response, $h(n) = \delta(n-1)$.

$$\therefore \sum_{n=0}^{\infty} |h(n)| = \sum_{n=0}^{\infty} |\delta(n-1)| = \delta(n-1)\big|_{n=1} = 1$$

Since, $\displaystyle\sum_{n=0}^{\infty} |h(n)| < \infty$, the system is stable.

Example 1.36   ***(AU Jun'12, 16 Marks)***

Test the stability of the system $y(n) = y^2(n-1) + x(n)$, when the input signal $x(n) = 2\delta(n)$ is applied to the system and the system is initially relaxed.

Solution:

Given that, $y(n) = y^2(n-1) + x(n)$ and $x(n) = 2\delta(n)$

We know that, $\delta(n) = 1$; $n = 0$

$\qquad\qquad\qquad = 0$; $n \neq 0$

$\therefore x(n) = 2$; $n = 0$

$\qquad\quad = 0$; $n \neq 0$

Since the given system is time variant, the test for stability should be performed for specific inputs.

$\therefore$ When $n = 0$; $y(0) = y^2(0-1) + x(0) = 0 + 2 = 2$

When $n = 1$; $y(1) = y^2(1-1) + x(1) = y^2(0) + 0 = (2)^2 + 0 = 2^2 = 2^{2^1}$

When n = 2 ; $y(2) = y^2(2{-}1) + x(2) = y^2(1) + 0 = ((2)^2)^2 + 0 = 2^4 = 2^{2^2}$

When n = 3 ; $y(3) = y^2(3{-}1) + x(3) = y^2(2) + 0 = (2^4)^2 + 0 = 2^8 = 2^{2^3}$

$\therefore \ y(n) = 2^{2^n}$

From the above analysis we can infer that as "n" tends to infinity, y(n) tends to infinite and so system is unstable.

Example 1.37

Determine the range of values of "p" and "q" for the stability of LTI system with impulse response,

$h(n) = p^n$; $n < 0$

$\quad\ \ = q^n$; $n \geq 0$

Solution:

The condition to be satisfied for the stability of the system is, $\displaystyle\sum_{n=-\infty}^{\infty} |h(n)| < \infty$

Given that, $h(n) = p^n$; $n < 0$

$\qquad\qquad\quad = q^n$; $n \geq 0$

$$\sum_{n=-\infty}^{\infty} |h(n)| = \sum_{n=-\infty}^{-1} |p^n| + \sum_{n=0}^{\infty} |q^n| = \sum_{n=1}^{\infty} |p^{-n}| + \sum_{n=0}^{\infty} |q^n|$$

$$= \sum_{n=1}^{\infty} \left|\frac{1}{p^n}\right| + \sum_{n=0}^{\infty} |q^n| = \sum_{n=1}^{\infty} \frac{1}{|p|^n} + \sum_{n=0}^{\infty} |q|^n \qquad \boxed{\text{n is always positive.}}$$

$$= \sum_{n=0}^{\infty} \left(\frac{1}{|p|}\right)^n - 1 + \sum_{n=0}^{\infty} |q|^n \qquad \boxed{|p|^0 = 1}$$

The summation of infinite terms in the above equation converges if, $0 < \dfrac{1}{|p|} < 1$ and $0 < |q| < 1$. Hence by using infinite geometric series formula,

$$\sum_{n=-\infty}^{\infty} |h(n)| = \frac{1}{1 - \dfrac{1}{|p|}} - 1 + \frac{1}{1 - |q|}$$

$$\qquad\qquad\quad = \text{Constant}$$

> Infinite geometric series sum formula
>
> $$\sum_{n=0}^{\infty} C^n = \frac{1}{1-C}$$
>
> if $0 < |C| < 1$

Therefore, the system is stable if |p| > 1 and |q| < 1.

Example 1.38

Test the stability of LTI systems, whose impulse responses are,

a) $h(n) = 0.2^n \, u(n)$ b) $h(n) = 0.3^n \, u(n) + 2^n \, u(n)$

c) $h(n) = 4^n \, u(-n)$ d) $h(n) = 0.2^n \, u(-n) + 3^n \, u(-n)$

Solution:

a) h(n) = 0.2ⁿ u(n)

$$\therefore \sum_{n=-\infty}^{+\infty} |h(n)| = \sum_{n=-\infty}^{+\infty} |0.2^n \, u(n)| = \sum_{n=0}^{\infty} 0.2^n$$

$$= \frac{1}{1 - 0.2} = 1.25$$

Since, $\displaystyle\sum_{n=-\infty}^{+\infty} |h(n)| < \infty$, system is stable.

> Infinite geometric series sum formula
>
> $$\sum_{n=0}^{\infty} C^n = \frac{1}{1-C}$$
>
> if $0 < |C| < 1$

b) h(n) = 0.3ⁿ u(n) + 2ⁿ u(n)

$$\therefore \sum_{n=-\infty}^{+\infty} |h(n)| = \sum_{n=-\infty}^{+\infty} |0.3^n u(n) + 2^n u(n)|$$

$$= \sum_{n=0}^{\infty} 0.3^n + \sum_{n=0}^{\infty} 2^n u(n) = \frac{1}{1-0.3} + \infty = \infty$$

Since, $\sum_{n=-\infty}^{+\infty} |h(n)| = \infty$, system is unstable.

$$\boxed{\begin{array}{l} \sum_{n=0}^{\infty} C^n = \infty \\ \text{if } C > 1 \end{array}}$$

---

c) h(n) = 4ⁿ u(−n)

$$\therefore \sum_{n=-\infty}^{+\infty} |h(n)| = \sum_{n=-\infty}^{+\infty} |4^n u(-n)| = \sum_{n=-\infty}^{0} 4^n = \sum_{n=0}^{+\infty} 4^{-n}$$

$$= \sum_{n=0}^{\infty} \frac{1}{4^n} = \sum_{n=0}^{\infty} \left(\frac{1}{4}\right)^n = \sum_{n=0}^{\infty} 0.25^n = \frac{1}{1-0.25} = 1.3333$$

Since, $\sum_{n=-\infty}^{+\infty} |h(n)| < \infty$, system is stable.

$$\boxed{\begin{array}{l} \text{Infinite geometric} \\ \text{series sum formula} \\[4pt] \sum_{n=0}^{\infty} C^n = \frac{1}{1-C} \\ \text{if } 0 < |C| < 1 \end{array}}$$

---

d) h(n) = 0.2ⁿ u(−n) + 3ⁿ u(−n)

$$\therefore \sum_{n=-\infty}^{+\infty} |h(n)| = \sum_{n=-\infty}^{+\infty} |0.2^n u(-n) + 3^n u(-n)|$$

$$= \sum_{n=-\infty}^{0} 0.2^n + \sum_{n=-\infty}^{0} 3^n = \sum_{n=0}^{+\infty} 0.2^{-n} + \sum_{n=0}^{+\infty} 3^{-n}$$

$$= \sum_{n=0}^{\infty} \frac{1}{0.2^n} + \sum_{n=0}^{\infty} \frac{1}{3^n} = \sum_{n=0}^{\infty} \left(\frac{1}{0.2}\right)^n + \sum_{n=0}^{\infty} \left(\frac{1}{3}\right)^n$$

$$= \sum_{n=0}^{\infty} 5^n + \sum_{n=0}^{\infty} 0.333^n = \infty + \frac{1}{1-0.333} = \infty$$

Since, $\sum_{n=-\infty}^{+\infty} |h(n)| = \infty$, system is unstable.

1.9.6 FIR and IIR Systems

In ***FIR system*** (**F**inite duration **I**mpulse **R**esponse system), the impulse response consists of finite number of samples. The convolution formula for FIR system is given by,

$$y(n) = \sum_{m=0}^{N-1} h(m) x(n-m) \qquad\qquad .....(1.26)$$

where, $h(n) = 0$; for $n < 0$ and $n \geq N$

From equation (1.26) it can be concluded that the impulse response selects only N samples of the input signal. In effect, the system acts as a window that views only the most recent N input signal samples in forming the ouput. Thus a FIR system requires memory of length N. In general, a FIR system is described by the difference equation,

$$y(n) = \sum_{m=0}^{N-1} b_m x(n-m) \qquad\qquad .....(1.27)$$

where, $b_m = h(m)$; for $m = 0$ to $N-1$

In IIR system (**I**nfinite duration **I**mpulse **R**esponse system), the impulse response has infinite number of samples. The convolution formula for IIR systems is given by,

$$y(n) = \sum_{m=0}^{\infty} h(m)\, x(n-m) \qquad \qquad .....(1.28)$$

Since this weighted sum involves the present and all the past input sample, we can say that the IIR system requires infinite memory. In general, an IIR system is described by the difference equation,

$$y(n) = -\sum_{m=1}^{N} a_m\, y(n-m) + \sum_{m=0}^{M} b_m\, x(n-m)$$

1.9.7 Recursive and Nonrecursive Systems

(AU Dec'15 & Jun'11, 2 Marks)
(AU May'15, 4 Marks)

A system whose output $y(n)$ at time n depends on any number of past output values as well as present and past inputs is called a ***recursive system.*** The past outputs are $y(n-1)$, $y(n-2)$, $y(n-3)$, etc.,.

Hence for recursive system, the output $y(n)$ is given by,

$$y(n) = F\ [y(n-1),\, y(n-2),...y(n-N),\, x(n),\, x(n-1),...x(n-M)]$$

A system whose output does not depend on past output but depends only on the present and past input is called a ***nonrecursive system.***

Hence for nonrecursive system, the output $y(n)$ is given by,

$$y(n) = F\ [x(n),\, x(n-1)\,,......,\, x(n-M)]$$

In a recursive system, in order to compute $y(n_0)$, we need to compute all the previous values $y(0), y(1)\,,........,\, y(n_0-1)$ before calculating $y(n_0)$. Hence the output samples of a recursive system has to be computed in order [i.e., $y(0), y(1), y(2), ...$]. The IIR systems are recursive systems.

In nonrecursive system, $y(n_0)$ can be computed immediately without having $y(n_0-1)$, $y(n_0-2).....$ Hence the output samples of nonrecursive system can be computed in any order [i.e. $y(50), y(5), y(2), y(100),....$]. The FIR systems are nonrecursive systems.

1.9.8 Invertible and Noninvertible Systems

A system is said to be ***invertible*** if distinct inputs produce distinct outputs. Therefore, for an invertible system an inverse system exist.

A system is said to be ***noninvertible*** if distinct inputs produce same output or any input produces zero output. Therefore, for a noninvertible system we cannot determine an inverse system.

Table 1.9: Examples of Invertible Systems

System	Inverse System
$y(n) = A\, x(n)$	$x(n) = \dfrac{1}{A}\, y(n)$
$y(n) = x(n+m)$	$x(n) = y(n-m)$

Table 1.10: Examples of Noninvertible Systems

System	Reason for Nonexistence of Inverse System
$y(n) = \delta(n)\,\delta(n-m)$	$\delta(n) = 1 \; ; \; n = 0$ $= 0 \; ; \; n \neq 0$ $\delta(n-m) = 1 \; ; \; n = m$ $= 0 \; ; \; n \neq m$ $\therefore \; y(n) = 0$
$y(n) = A\sin(n + 2\pi k)$ where, k is integer	Here, $y(n)$ will produce same value for any integer value of k
$y(n) = x^2(n)$	Here, $x(n) = \pm y\sqrt{n}$ and so we cannot predict the sign of $x(n)$.

Example 1.39
                                                                       (AU Dec'14, 8 Marks)

State whether the system is causal, linear, time invariant and stable.

$y(n) = x(n)-x(n-1)$

Solution:

Given that, $y(n) = x(n)-x(n-1)$

i) Casuality

When n = 0, $y(0) = x(0) - x(-1)$ $\Rightarrow$ The response at n = 0, i.e., $y(0)$ depends on the present input $x(0)$ and past input $x(-1)$.

When n = 1, $y(1) = x(1) - x(0)$ $\Rightarrow$ The response at n = 1, i.e., $y(1)$ depends on the present input $x(1)$ and past input $x(0)$.

From the above analysis we can say that for any value of n, the system output depends on present and past inputs. Hence the system is causal.

ii) Linearity

Let, $\mathcal{H}$ be the system represented by the equation, $y(n) = x(n) - x(n-1)$.

The system $\mathcal{H}$ operates on $x(n)$ to produce, $y(n)$.

$$x(n) \longrightarrow \boxed{\mathcal{H}} \longrightarrow y(n) = \mathcal{H}\{x(n)\} = x(n) - x(n-1)$$

Consider two signals, $x_1(n)$ and $x_2(n)$.

Let, $y_1(n)$ and $y_2(n)$ be the response of the system $\mathcal{H}$ for inputs $x_1(n)$ and $x_2(n)$ respectively.

$$x_1(n) \longrightarrow \boxed{\mathcal{H}} \longrightarrow y_1(n) = \mathcal{H}\{x_1(n)\} = x_1(n) - x_1(n-1)$$

$$x_2(n) \longrightarrow \boxed{\mathcal{H}} \longrightarrow y_2(n) = \mathcal{H}\{x_2(n)\} = x_2(n) - x_2(n-1)$$

$$\therefore \; a_1 y_1(n) + a_2 y_2(n) = a_1 x_1(n) - a_1 x_1(n-1) + a_2 x_2(n) - a_2 x_2(n-1) \qquad .....(1)$$

Consider a linear combination of inputs, $a_1 x_1(n) + a_2 x_2(n) = x_3(n)$.

Let, $y_3(n)$ be the response for $x_3(n)$.

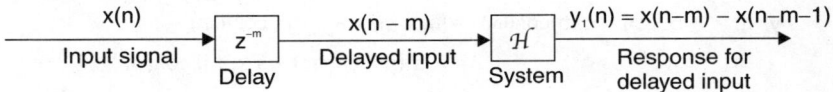

$$\therefore y_3(n) = \mathcal{H}\{x_3(n)\} = x_3(n) - x_3(n-1) = a_1 x_1(n) + a_2 x_2(n) - [a_1 x_1(n-1) + a_2 x_2(n-1)]$$

$$= a_1 x_1(n) - a_1 x_1(n-1) + a_2 x_2(n) - a_2 x_2(n-1) \qquad .....(2)$$

The condition to be satisfied for linearity is, $y_3(n) = a_1 y_1(n) + a_2 y_2(n)$.

From equations (1) and (2) we can say that, $y_3(n) = a_1 y_1(n) + a_2 y_2(n)$. Hence the system is linear.

iii) Time invariance

<u>Test 1:</u> Response for delayed input

```
    x(n)      ┌──────┐  x(n – m)    ┌──────┐   y₁(n) = x(n–m) – x(n–m–1)
  ─────────▶  │ z⁻ᵐ  │ ──────────▶ │  H   │ ─────────────────────────▶
 Input signal └──────┘ Delayed input└──────┘   Response for
               Delay                 System    delayed input
```

<u>Test 2:</u> Delayed response

```
    x(n)      ┌──────┐ y(n) = x(n) – x(n –1) ┌──────┐ y₂(n) = x(n–m) – x(n–m–1)
  ─────────▶  │  H   │ ────────────────────▶ │ z⁻ᵐ  │ ──────────────────────────▶
 Input signal └──────┘ Response for          └──────┘  Delayed response
               System  undelayed input        Delay
```

<u>Conclusion:</u> Here, $y_1(n) = y_2(n)$, therefore the system is time invariant.

iv) Stability

The given system is an LTI system, and so the stability can be determined from impulse response.

Given that, $y(n) = x(n) - x(n-1)$

Let, the input $x(n)$ be an impulse input $\delta(n)$.

Now, response $y(n)$ will be impulse response $h(n)$.

$\therefore$ Impulse response, $h(n) = \delta(n) - \delta(n-1)$

$$\therefore \sum_{n=0}^{\infty} |h(n)| = \sum_{n=0}^{\infty} |\delta(n) - \delta(n-1)| = \delta(n)|_{n=0} - \delta(n-1)|_{n=1} = 1 - 1 = 0$$

Since, $\displaystyle\sum_{n=0}^{\infty} |h(n)| < \infty$, the system is stable.

Example 1.40   *(AU Dec'13, 16 Marks)*

State whether the system $y(n) = x(n) \cos(\omega n)$ is stable, linear, causal, time invariant and memoryless.

Solution:

i) Linearity

Let, $\mathcal{H}$ be the system represented by the equation, $y(n) = x(n) \cos(\omega n)$.

The system $\mathcal{H}$ operates on $x(n)$ to produce, $y(n)$.

```
    x(n)      ┌──────┐
  ─────────▶  │  H   │ ──────▶  y(n) = H{x(n)} = x(n)cos(ωn)
              └──────┘
```

Consider two signals, $x_1(n)$ and $x_2(n)$.

Let, $y_1(n)$ and $y_2(n)$ be the response of the system $\mathcal{H}$ for inputs $x_1(n)$ and $x_2(n)$ respectively.

$$x_1(n) \rightarrow \boxed{\mathcal{H}} \rightarrow y_1(n) = \mathcal{H}\{x_1(n)\} = x_1(n)\cos(\omega n)$$

$$x_2(n) \rightarrow \boxed{\mathcal{H}} \rightarrow y_2(n) = \mathcal{H}\{x_2(n)\} = x_2(n)\cos(\omega n)$$

$$\therefore a_1 y_1(n) + a_2 y_2(n) = a_1 x_1(n)\cos(\omega n) + a_2 x_2 \cos(\omega n) \qquad .....(1)$$

Consider a linear combination of inputs, $a_1 x_1(n) + a_2 x_2(n) = x_3(n)$.

Let, $y_3(n)$ be the response for $x_3(n)$.

$$x_3(n) \rightarrow \boxed{\mathcal{H}} \rightarrow y_3(n) = \mathcal{H}\{x_3(n)\}$$

$$\therefore y_3(n) = \mathcal{H}\{x_3(n)\} = x_3(n)\cos(\omega n) = [a_1 x_1(n) + a_2 x_2(n)]\cos(\omega n)$$

$$= a_1 x_1(n)\cos(\omega n) + a_2 x_2(n)\cos(\omega n) \qquad .....(2)$$

The condition to be satisfied for linearity is, $y_3(n) = a_1 y_1(n) + a_2 y_2(n)$.

From equations (1) and (2) we can say that, $y_3(n) = a_1 y_1(n) + a_2 y_2(n)$. Hence the system is linear

ii) Stability

The value of $\cos \theta$ lies between -1 to $+1$ for any value of θ. Therefore the output $y(n)$ is bounded as long as $x(n)$ is bounded.

iii) Causality

When $n = 0$, $y(0) = x(0) \times \cos(0)$ $\qquad \Rightarrow \qquad$ The response at $n = 0$, i.e., $y(0)$ depends on the present input $x(0)$.

When $n = 1$, $y(1) = x(1) \times \cos(\omega)$ $\qquad \Rightarrow \qquad$ The response at $n = 1$, i.e., $y(1)$ depends on the present input $x(1)$.

From the above analysis we can say that the response for any value of n depends on the present input. Hence the system is causal.

iv) Memoryless

For any values of n, $y(n)$ depends only on present input therefore it is a static (memoryless) system.

v) Time invariance

Test 1: Response for delayed input

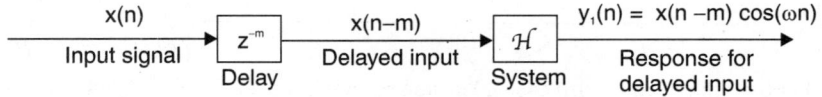

Test 2: Delayed response

$$x(n) \rightarrow \boxed{\mathcal{H}} \xrightarrow{y(n) = x(n)\cos(\omega n)} \boxed{z^{-m}} \rightarrow y_2(n) = x(n-m)\cos(\omega(n-m))$$

Input signal — System — Response for undelayed input — Delay — Delayed response

Conclusion: Here, $y_1(n) \neq y_2(n)$, therefore the system is time variant.

Example 1.41

State whether the following systems are linear, causal, time variant, and dynamic.

a) $y(n) = nx(n) + bx^2(n)$ **b)** $x(n-1)$ **c)** $y(n) = x(n) + \dfrac{1}{x(n-1)}$

Solution:

a) Given that, $y(n) = nx(n) + bx^2(n)$ **(AU Jun'12, 8 Marks)**

i) Linearity

Let, $\mathcal{H}$ be the system represented by the equation, $y(n) = nx(n) + bx^2(n)$

The system $\mathcal{H}$ operates on $x(n)$ to produce, $y(n)$.

$$x(n) \longrightarrow \boxed{\mathcal{H}} \longrightarrow y(n) = \mathcal{H}\{x(n)\} = nx(n) + bx^2(n)$$

Consider two signals, $x_1(n)$ and $x_2(n)$.

Let, $y_1(n)$ and $y_2(n)$ be the response of the system $\mathcal{H}$ for inputs $x_1(n)$ and $x_2(n)$ respectively.

$$x_1(n) \longrightarrow \boxed{\mathcal{H}} \longrightarrow y_1(n) = \mathcal{H}\{x_1(n)\} = nx_1(n) + bx_1^2(n)$$

$$x_2(n) \longrightarrow \boxed{\mathcal{H}} \longrightarrow y_2(n) = \mathcal{H}\{x_2(n)\} = nx_2(n) + bx_2^2(n)$$

$$\therefore a_1 y_1(n) + a_2 y_2(n) = a_1 nx_1(n) + a_1 bx_1^2(n) + a_2 nx_2(n) + a_2 bx_2^2(n) \qquad .....(1)$$

Consider a linear combination of inputs, $a_1 x_1(n) + a_2 x_2(n) = x_3(n)$.

Let, $y_3(n)$ be the response for $x_3(n)$.

$$x_3(n) \longrightarrow \boxed{\mathcal{H}} \longrightarrow y_3(n) = \mathcal{H}\{x_3(n)\}$$

$$\therefore y_3(n) = \mathcal{H}\{x_3(n)\} = nx_3(n) + bx_3^2(n)$$

$$= n[a_1 x_1(n) + a_2 x_2(n)] + b[a_1 x_1(n) + a_2 x_2(n)]^2$$

$$= n[a_1 x_1(n) + a_2 x_2(n)] + b[a_1^2 x_1^2(n) + a_2^2 x_2^2(n) + 2 a_1 a_2 x_1(n) x_2(n)]. \qquad .....(2)$$

The condition to be satisfied for linearity is, $y_3(n) = a_1 y_1(n) + a_2 y_2(n)$.

From equations (1) and (2) we can say that, $y_3(n) \neq a_1 y_1(n) + a_2 y_2(n)$. Hence the system is nonlinear.

ii) Causality

When $n = 0$, $y(0) = 0 \times x(0) + b\, x^2(0)$ $\Rightarrow$ The response at $n = 0$, i.e., $y(0)$ depends on the present input $x(0)$.

When $n = 1$, $y(1) = 1 \times x(1) + b\, x^2(1)$ $\Rightarrow$ The response at $n = 1$, i.e., $y(1)$ depends on the present input $x(1)$.

When $n = 2$, $y(2) = 2 \times x(2) + b\, x^2(2)$ $\Rightarrow$ The response at $n = 2$, i.e., $y(2)$ depends on the present input $x(2)$.

From the above analysis we can say that the response for any value of n depends on the present input. Hence the system is causal.

iii) Dynamic

For any values of n, $y(n)$ depends only on present input. Hence the system does not require memory and so the given system is static system.

iv) Time invariance

Test 1: Response for delayed input

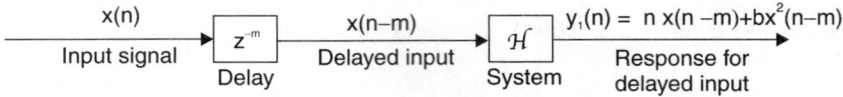

Test 2: Delayed response

$$x(n) \xrightarrow{\text{Input signal}} \boxed{\mathcal{H}} \xrightarrow[\text{Response for}]{y(n) = nx(n)+bx^2(n)} \boxed{z^{-m}} \xrightarrow[\text{Delayed response}]{y_2(n) = (n–m)x(n–m)+bx^2(n–m)}$$

System   undelayed input   Delay

Conclusion: Here, $y_1(n) \neq y_2(n)$, therefore the system is time variant.

---

b) Given that, y(n) = x(n–1)   **(AU Jun'12, 16 Marks & Jun'14, 8 Marks)**

i) Linearity

Let, $\mathcal{H}$ be the system represented by the equation, $y(n) = x(n–1)$

The system $\mathcal{H}$ operates on $x(n)$ to produce, $y(n)$.

$$x(n) \xrightarrow{} \boxed{\mathcal{H}} \xrightarrow{} y(n) = \mathcal{H}\{x(n)\} = x(n–1)$$

Consider two signals, $x_1(n)$ and $x_2(n)$.

Let, $y_1(n)$ and $y_2(n)$ be the response of the system $\mathcal{H}$ for inputs $x_1(n)$ and $x_2(n)$ respectively.

$$x_1(n) \xrightarrow{} \boxed{\mathcal{H}} \xrightarrow{} y_1(n) = \mathcal{H}\{x_1(n)\} = x_1(n–1)$$

$$x_2(n) \xrightarrow{} \boxed{\mathcal{H}} \xrightarrow{} y_2(n) = \mathcal{H}\{x_2(n)\} = x_2(n–1)$$

$$\therefore a_1 y_1(n) + a_2 y_2(n) = a_1 x_1(n–1) + a_2 x_2(n–1) \qquad \text{.....(1)}$$

Consider a linear combination of inputs, $a_1 x_1(n) + a_2 x_2(n) = x_3(n)$.

Let, $y_3(n)$ be the response for $x_3(n)$.

$$x_3(n) \xrightarrow{} \boxed{\mathcal{H}} \xrightarrow{} y_3(n) = \mathcal{H}\{x_3(n)\}$$

$$\therefore y_3(n) = \mathcal{H}\{x_3(n)\} = x_3(n–1) = a_1 x_1(n–1) + a_2 x_2(n–1). \qquad \text{.....(2)}$$

The condition to be satisfied for linearity is, $y_3(n) = a_1 y_1(n) + a_2 y_2(n)$.

From equations (1) and (2) we can say that, $y_3(n) = a_1 y_1(n) + a_2 y_2(n)$. Hence the system is linear.

---

ii) Causality

| When $n = 0$, $y(0) = x(–1)$ | $\Rightarrow$ | The response at $n = 0$, i.e., $y(0)$ depends on the past input $x(–1)$. |
| When $n = 1$, $y(1) = \dot{x}(0)$ | $\Rightarrow$ | The response at $n = 1$, i.e., $y(1)$ depends on the past input $x(0)$. |

From the above analysis we can say that the response for any value of n depends on the past input. Hence the system is causal.

---

iii) Dynamic

For any values of n, $y(n)$ depends on past input. Hence the system requires memory and so the given system is dynamic system.

iv) Time invariance

Test 1: Response for delayed input

Test 2: Delayed response

Conclusion: Here, $y_1(n) = y_2(n)$, therefore the system is time invariant.

c) Given that, $y(n) = x(n) + \dfrac{1}{x(n-1)}$ **(AU Dec'11, 8 Marks)**

i) Linearity

Let, H be the system represented by the equation, $y(n) = x(n) + \dfrac{1}{x(n-1)}$
The system H operates on $x(n)$ to produce, $y(n)$.

$$x(n) \rightarrow \boxed{H} \rightarrow y(n) = H\{x(n)\} = x(n) + \frac{1}{x(n-1)}$$

Consider two signals, $x_1(n)$ and $x_2(n)$.

Let, $y_1(n)$ and $y_2(n)$ be the response of the system H for inputs $x_1(n)$ and $x_2(n)$ respectively.

$$x_1(n) \rightarrow \boxed{H} \rightarrow y_1(n) = H\{x_1(n)\} = x_1(n) + \frac{1}{x_1(n-1)}$$

$$x_2(n) \rightarrow \boxed{H} \rightarrow y_2(n) = H\{x_2(n)\} = x_2(n) + \frac{1}{x_2(n-1)}$$

$$\therefore \; a_1 y_1(n) + a_2 y_2(n) = a_1\left(x_1(n) + \frac{1}{x_1(n-1)}\right) + a_2\left(x_2(n) + \frac{1}{x_2(n-1)}\right) \qquad .....(1)$$

Consider a linear combination of inputs, $a_1 x_1(n) + a_2 x_2(n) = x_3(n)$.
Let, $y_3(n)$ be the response for $x_3(n)$.

$$x_3(n) \rightarrow \boxed{H} \rightarrow y_3(n) = H\{x_3(n)\}$$

$$\therefore \; y_3(n) = H\{x_3(n)\} = x_3(n) + \frac{1}{x_3(n-1)} = a_1 x_1(n) + a_2 x_2(n) + \frac{1}{a_1 x_1(n-1) + a_2 x_2(n-1)} \qquad .....(2)$$

The condition to be satisfied for linearity is, $y_3(n) = a_1 y_1(n) + a_2 y_2(n)$.

From equations (1) and (2) we can say that, $y_3(n) \neq a_1 y_1(n) + a_2 y_2(n)$. Hence the system is nonlinear

ii) Causality

When $n = 0$, $y(0) = x(0) + \dfrac{1}{x(-1)}$	$\Rightarrow$	The response at $n = 0$, i.e., $y(0)$ depends on the present input $x(0)$ and past input $x(-1)$.
When $n = 1$, $y(1) = x(1) + \dfrac{1}{x(0)}$	$\Rightarrow$	The response at $n = 1$, i.e., $y(1)$ depends on the present input $x(1)$ and past input $x(0)$.
When $n = 2$, $y(2) = x(2) + \dfrac{1}{x(1)}$	$\Rightarrow$	The response at $n = 2$, i.e., $y(2)$ depends on the present input $x(2)$ and past input $x(1)$.

From the above analysis we can say that the response for any value of n depends on the present and past input. Hence the system is causal.

iii) Dynamic

For any values of n, y(n) depends on present and past inputs. Therefore, it is a dynamic system.

iv) Time invariance

Test 1: Response for delayed input

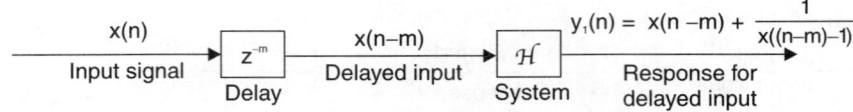

Test 2: Delayed response

Conclusion: Here, $y_1(n) = y_2(n)$, therefore the system is time invariant.

Example 1.42

(AU May'15, 10 Marks)

State whether the system is linear, causal, time variant and stable.

$y(n) = \log x(n)$

Solution:

Given that, $y(n) = \log x(n)$

i) Linearity

Let $\mathcal{H}$ be the system represented by the equation, $y(n) = \log x(n)$.

The system $\mathcal{H}$ operates on $x(n)$ to produce, $y(n)$.

$$x(n) \longrightarrow \boxed{\mathcal{H}} \longrightarrow y(n) = \mathcal{H}\{x(n)\} = \log x(n)$$

Consider two signals $x_1(n)$ and $x_2(n)$.

Let $y_1(n)$ and $y_2(n)$ be the response of the system $\mathcal{H}$ for inputs $x_1(n)$ and $x_2(n)$ respectively.

$$x_1(n) \longrightarrow \boxed{\mathcal{H}} \longrightarrow y_1(n) = \mathcal{H}\{x_1(n)\} = \log x_1(n)$$

$$x_2(n) \longrightarrow \boxed{\mathcal{H}} \longrightarrow y_2(n) = \mathcal{H}\{x_2(n)\} = \log x_2(n)$$

$\therefore a_1 \, y_1(n) + a_2 \, y_2(n) = a_1 \log x(n) + a_2 \log x(n)$ .....(1)

Consider a linear combination of inputs, $a_1 \, x_1(n) + a_2 \, x_2(n) = x_3(n)$.

Let $y_3(n)$ be the response for $x_3(n)$.

$$x_3(n) \longrightarrow \boxed{\mathcal{H}} \longrightarrow y_3(n) = \mathcal{H}\{x_3(n)\}$$

$\therefore y_3(n) = \mathcal{H}\{x_3(n)\} = \log x_3(n) = \log [a_1 \, x_1(n) + a_2 \, x_2(n)]$ .....(2)

The condition to be satisfied for linearity is, $y_3(n) = a_1 \, y_1(n) + a_2 \, y_2(n)$.

From equation (1) and (2), we can say that, $y_3(n) \neq a_1 \, y_1(n) + a_2 \, y_2(n)$. Hence the system is nonlinear.

ii) Causality

When n = 0, y(0) = log x(0) $\Rightarrow$ The response at n = 0, i.e., y(0) depends on the present input x(0).

When n = 1, y(1) = log x(1) $\Rightarrow$ The response at n = 1, i.e., y(1) depends on the present input x(1).

When n = 2, y(2) = log x(2) $\Rightarrow$ The response at n = 2, i.e., y(2) depends on the present input x(2).

From the above analysis we can say that the response for any value of n depends on the present input. Hence the system is causal.

iii) Time invariance

Test 1: Respose for delayed input.

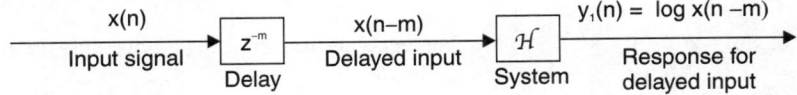

Test 2: Delayed response

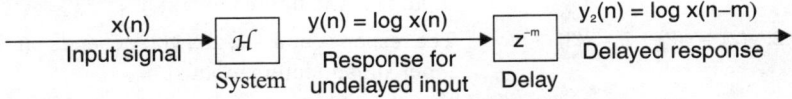

Conclusion: Here, $y_1(n) = y_2(n)$, therefore the system is time invariant.

iii) Stability

The given system is nonlinear system, and so the test for stsbility should be performed for specific inputs.

Case i: If x(n) tends to zero or infinity, as "n" tends to infinity then y(n) = log x(n) will be infinite as "n" tends to infinity. So the system is unstable.

Case ii: If x(n) tends to constant as "n" tends to infinity, then y(n) = log x(n) will be finite as "n" tends to infinity. So the system is stable.

Example 1.43
(AU May'11, 8 Marks)

State whether the given system y(n) = x(n) + nx(n+1) is linear, causal, time variant and stable.

Solution:

Given that, y(n) = x(n) + nx(n+1)

i) Linearity

Let $\mathcal{H}$ be the system represented by the equation, y(n) = x(n) + nx(n+1)

The system $\mathcal{H}$ operates on x(n) to produce, y(n).

$$x(n) \longrightarrow \boxed{\mathcal{H}} \longrightarrow y(n) = \mathcal{H}\{x(n)\} = x(n) + nx(n+1)$$

Consider two signals $x_1(n)$ and $x_2(n)$.

Let $y_1(n)$ and $y_2(n)$ be the response of the system $\mathcal{H}$ for inputs $x_1(n)$ and $x_2(n)$ respectively.

$$x_1(n) \longrightarrow \boxed{\mathcal{H}} \longrightarrow y_1(n) = \mathcal{H}\{x_1(n)\} = x_1(n) + nx_1(n+1)$$

$$x_2(n) \longrightarrow \boxed{\mathcal{H}} \longrightarrow y_2(n) = \mathcal{H}\{x_2(n)\} = x_2(n) + nx_2(n+1)$$

$\therefore a_1 y_1(n) + a_2 y_2(n) = a_1 x_1(n) + a_1 n x_1(n+1) + a_2 x_2(n) + a_2 n x_2(n+1)$ .....(1)

Consider a linear combination of inputs, $a_1 x_1(n) + a_2 x_2(n) = x_3(n)$.

Let $y_3(n)$ be the response for $x_3(n)$.

$$x_3(n) \longrightarrow \boxed{\mathcal{H}} \longrightarrow y_3(n) = \mathcal{H}\{x_3(n)\}$$

$\therefore y_3(n) = \mathcal{H}\{x_3(n)\} = x_3(n) + n\, x_3(n+1) = a_1 x_1(n) + a_2 x_2(n) + n[a_1 x_1(n+1) + a_2 x_2(n+1)]$

$$= a_1 x_1(n) + a_1 n x_1(n+1) + a_2 x_2(n) + a_2 n x_2(n+1) \qquad \text{.....(2)}$$

The condition to be satisfied for linearity is, $y_3(n) = a_1 y_1(n) + a_2 y_2(n)$.

From equations (1) and (2), we can say that, $y_3(n) = a_1 y_1(n) + a_2 y_2(n)$. Hence the system is linear.

ii) Causality

When $n = 0$, $y(0) = x(0) + 0 \times x(1)$ $\Rightarrow$ The response at $n = 0$, i.e., $y(0)$ depends on the present input $x(0)$.

When $n = 1$, $y(1) = x(1) + 1 \times x(2)$ $\Rightarrow$ The response at $n = 1$, i.e., $y(1)$ depends on the present input $x(1)$ and future input $x(2)$.

When $n = 2$, $y(2) = x(2) + 2 \times x(3)$ $\Rightarrow$ The response at $n = 2$, i.e., $y(2)$ depends on the present input $x(2)$ and future input $x(3)$.

From the above analysis we can say that the response for any value of n (except n = 0) depends on the present and future inputs. Hence the system is noncausal.

iii) Time invariance

Test 1: Respose for delayed input.

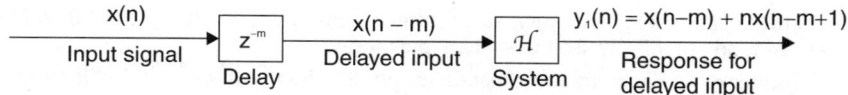

$$x(n) \xrightarrow{\text{Input signal}} \boxed{z^{-m}} \xrightarrow{\substack{x(n-m)\\ \text{Delayed input}}} \boxed{\mathcal{H}} \xrightarrow{y_1(n) = x(n-m) + nx(n-m+1)}$$

Input signal   Delay   Delayed input   System   Response for delayed input

Test 2: Delayed response

$$x(n) \xrightarrow{\text{Input signal}} \boxed{\mathcal{H}} \xrightarrow{y(n) = x(n) + nx(n+1)} \boxed{z^{-m}} \xrightarrow{y_2(n) = x(n-m) + (n-m)x(n-m+1)}$$

Input signal   System   Response for undelayed input   Delay   Delayed response

Conclusion: Here $y_1(n) \neq y_2(n)$, therefore the system is time variant.

iii) Stability

The given system is time variant system, and so the test for stability should be performed for specific inputs.

Case i: If $x(n)$ tends to constant or infinity, as "n" tends to infinity then $y(n) = x(n) + nx(n+1)$ will be infinite as "n" tends to infinity. So the system is unstable.

Case ii: If $x(n)$ tends to zero as "n" tends to infinity, then $y(n) = x(n) + nx(n+1)$ will be finite as "n" tends to infinity. So the system is stable.

Example 1.44
                                                                      (AU Dec'12, 10 Marks)

Verify whether the following systems are linear, time invariant, stable and invertible.

a) $y(n) = x^2(n)$ **b)** $y(n) = x(-n)$

Solution:

a) Given that, $y(n) = x^2(n)$

i) Linearity

Let, $\mathcal{H}$ be the system represented by the equation, $y(n) = x^2(n)$.

The system $\mathcal{H}$ operates on $x(n)$ to produce, $y(n)$.

$$x(n) \longrightarrow \boxed{\mathcal{H}} \longrightarrow y(n) = \mathcal{H}\{x(n)\} = x^2(n)$$

Consider two signals, $x_1(n)$ and $x_2(n)$.

Let, $y_1(n)$ and $y_2(n)$ be the response of the system $\mathcal{H}$ for inputs $x_1(n)$ and $x_2(n)$ respectively.

$$x_1(n) \longrightarrow \boxed{\mathcal{H}} \longrightarrow y_1(n) = \mathcal{H}\{x_1(n)\} = x_1^2(n)$$

$$x_2(n) \longrightarrow \boxed{\mathcal{H}} \longrightarrow y_2(n) = \mathcal{H}\{x_2(n)\} = x_2^2(n)$$

$$\therefore a_1 y_1(n) + a_2 y_2(n) = a_1 x_1^2(n) + a_2 x_2^2(n) \qquad .....(1)$$

Consider a linear combination of inputs, $a_1 x_1(n) + a_2 x_2(n) = x_3(n)$.

Let, $y_3(n)$ be the response for $x_3(n)$.

$$x_3(n) \longrightarrow \boxed{\mathcal{H}} \longrightarrow y_3(n) = \mathcal{H}\{x_3(n)\}$$

$$\therefore y_3(n) = \mathcal{H}\{x_3(n)\} = x_3^2(n) = [a_1 x_1(n) + a_2 x_2(n)]^2$$

$$= a_1^2 x_1^2(n) + a_2^2 x_2^2(n) + 2 a_1 a_2 x_1(n) x_2(n) \qquad .....(2)$$

The condition to be satisfied for linearity is, $y_3(n) = a_1 y_1(n) + a_2 y_2(n)$.

From equations (1) and (2) we can say that, $y_3(n) \neq a_1 y_1(n) + a_2 y_2(n)$. Hence the system is nonlinear.

ii) Time invariant

Test 1: Response for delayed input

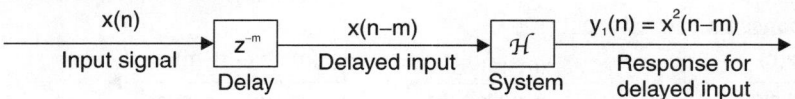

$$x(n) \atop \text{Input signal} \longrightarrow \boxed{z^{-m} \atop \text{Delay}} \xrightarrow{x(n-m) \atop \text{Delayed input}} \boxed{\mathcal{H} \atop \text{System}} \longrightarrow y_1(n) = x^2(n-m) \atop \text{Response for} \atop \text{delayed input}$$

Test 2: Delayed response

$$x(n) \atop \text{Input signal} \longrightarrow \boxed{\mathcal{H} \atop \text{System}} \xrightarrow{y(n) = x^2(n) \atop \text{Response for} \atop \text{undelayed input}} \boxed{z^{-m} \atop \text{Delay}} \longrightarrow y_2(n) = x^2(n-m) \atop \text{Delayed response}$$

Conclusion: Here, $y_1(n) = y_2(n)$, therfore the system is time invariant

ii) Stability

The given system is nonlinear system, and so the test for stability should be performed for specific inputs.

Case i: If $x(n)$ tends to infinity or constant, as "n" tends to infinity, then $y(n) = x^2(n)$ will be infinite as "n" tends to infinity. So the system is unstable.

Case ii: If $x(n)$ tends to zero as "n" tends to infinity, then $y(n) = x^2(n)$ will be zero as "n" tends to infinity. So the system is stable.

iii) Invertibile

Given that, $y(n) = x^2(n)$

$$\therefore x(n) = \pm\sqrt{y(n)}$$

Here, for any value of n, the x(n) calculated using above equation cannot predict the sign of x(n) and so inverse system does not exist. Therefore the system is non invertible.

b) Given that, $y(n) = x(-n)$

i) Linearity

Let, $\mathcal{H}$ be the system represented by the equation, $y(n) = x(-n)$.

The system $\mathcal{H}$ operates on x(n) to produce, y(n).

$$x(n) \longrightarrow \boxed{\mathcal{H}} \longrightarrow y(n) = \mathcal{H}\{x(n)\} = x(-n)$$

Consider two signals, $x_1(n)$ and $x_2(n)$.

Let, $y_1(n)$ and $y_2(n)$ be the response of the system $\mathcal{H}$ for inputs $x_1(n)$ and $x_2(n)$ respectively.

$$x_1(n) \longrightarrow \boxed{\mathcal{H}} \longrightarrow y_1(n) = \mathcal{H}\{x_1(n)\} = x_1(-n)$$

$$x_2(n) \longrightarrow \boxed{\mathcal{H}} \longrightarrow y_2(n) = \mathcal{H}\{x_2(n)\} = x_2(-n)$$

$$\therefore a_1 y_1(n) + a_2 y_2(n) = a_1 x_1(-n) + a_2 x_2(-n) \qquad \qquad .....(1)$$

Consider a linear combination of inputs, $a_1 x_1(n) + a_2 x_2(n) = x_3(n)$.

Let, $y_3(n)$ be the response for $x_3(n)$.

$$x_3(n) \longrightarrow \boxed{\mathcal{H}} \longrightarrow y_3(n) = \mathcal{H}\{x_3(n)\}$$

$$\therefore y_3(n) = \mathcal{H}\{x_3(n)\} = x_3(-n) = a_1 x_1(-n) + a_2 x_2(-n) \qquad \qquad .....(2)$$

The condition to be satisfied for linearity is, $y_3(n) = a_1 y_1(n) + a_2 y_2(n)$.

From equations (1) and (2) we can say that, $y_3(n) = a_1 y_1(n) + a_2 y_2(n)$. Hence the system is linear.

ii) Time invariance:

Test 1: Response for delayed input

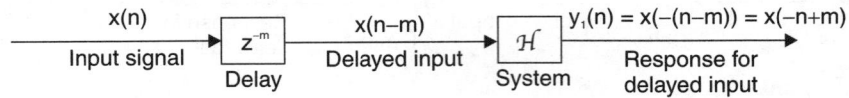

Test 2: Delayed response

$$x(n) \longrightarrow \underset{\text{System}}{\boxed{\mathcal{H}}} \xrightarrow{\underset{\text{Response for undelayed input}}{y(n) = x(-n)}} \underset{\text{Delay}}{\boxed{z^{-m}}} \xrightarrow{\underset{\text{Delayed response}}{y_2(n) = x(-n-m)}}$$

Conclusion : Here, $y_1(n) \neq y_2(n)$, therefore the system is time variant.

iii) Stability

The given system is time variant system, and so the test for stability should be performed for specific inputs

The operation performed by the system on the input signal is folding. A bounded input signal will remain bounded even after folding. Therfore in the given system, the ouput will remain bounded as long as input is bounded. Hence the system is BIBO stable.

iv) Invertible:

Given that, $y(n) = x(-n)$

When $n = -2$; $y(-2) = x(2)$ $\Rightarrow$ $x(2) = y(-2)$

When $n = -1$; $y(-1) = x(1)$ $\Rightarrow$ $x(1) = y(-1)$

When $n = 0$; $y(0) = x(0)$ $\Rightarrow$ $x(0) = y(0)$

When $n = 1$; $y(1) = x(-1)$ $\Rightarrow$ $x(-1) = y(1)$

When $n = 2$; $y(2) = x(-2)$ $\Rightarrow$ $x(-2) = y(2)$ and so on.

From the above analysis we can say that the inverse system is defined by the equation,

$x(n) = y(-n)$.

1.10 Summary of Important Concepts

1. An analog signal is a continuous function of an independent variable.

2. When the independent variable of an analog signal is time 't' then the analog signal is called continuous time (CT) signal.

3. In a continuous time signal the magnitude and the independent variable are continuous.

4. Ideally, a continuous time impulse signal is a signal with infinite magnitude and zero duration.

5. Practically, a continuous time impulse signal is a signal with large magnitude and short duration.

6. The discrete signal is a function of a discrete independent variable.

7. In a discrete time signal, the value of discrete time signal and the independent variable time are discrete.

8. The digital signal is same as discrete signal except that the magnitude of the signal is quantized.

9. The sampling is the process of conversion of continuous time signal into discrete time signal.

10. The time interval between successive samples is called sampling time or sampling period.

11. The inverse of sampling period is called sampling frequency.

12. The signals that can be completely specified by mathematical equations are called deterministic signals.

13. The signals whose characteristics are random in nature are called nondeterministic signals.

14. The continuous time sinusoidal and complex exponential signals are always periodic.

15. The sum of two periodic signals is also periodic if the ratio of their fundamental periods is a rational number.

16. A discrete time signal $x(n)$ is periodic with periodicity of N samples if $x(n + N) = x(n)$.

17. A discrete time sinusoid is periodic only if its frequency is a rational number.

18. Discrete time sinusoids whose frequencies are separated by an integer multiple of 2π are identical.

19. When a continuous time signal exhibits symmetry with respect to $t = 0$, then it is called an even signal.

20. When a continuous time signal exhibits antisymmetry with respect to $t = 0$, then it is called an odd signal.

21. A signal which is neither even nor odd can be expressed as a sum of even and odd signals.

22. When a discrete time signal exhibits symmetry with respect to $n = 0$ then it is called an even signal.

23. When a discrete time signal exhibits antisymmetry with respect to $n = 0$, then it is called an odd signal.

24. For energy signals, the energy will be finite and average power will be zero.

25. For power signals the average power is finite and energy will be infinite.

26. Periodic signals are power signals and nonperiodic signals are energy signals.

27. A continuous time signal is said to be causal if it is defined only for $t \geq 0$.

28. A continuous time signal is said to be noncausal if it is defined for either $t \leq 0$ or all t.

29. A continuous time signal is said to be anticausal if it is defined for $t \leq 0$.

30. A discrete time signal is said to be causal, if it is defined for $n \geq 0$.

31. A discrete time signal is said to be noncausal, if it is defined for either $n \leq 0$ or all n.

32. A discrete time signal is said to be anticausal, if it is defined for $n \leq 0$.

33. A continuous time system is a physical device that operates on continuous time signal.

34. A discrete time system is a device or algorithm that operates on a discrete time signal.

35. When a system satisfies the properties of linearity and time invariance, it is called an LTI system.

36. When the input to a continuous time system is unit impulse $\delta(t)$, then the output is called impulse response, h(t).

37. When the input to a discrete time system is unit impulse $\delta(n)$, the output is called impulse response, h(n).

38 . The response of a static system depends on present input whereas, the response of a dynamic system depends on present, past and future inputs.

39. The dynamic system requires memory whereas the static system does not require memory.

40. A system is said to be time invariant if its input-output characteristics do not change with time.

41. A system is said to be time variant if its input-output characteristics change with time.

42. In time invariant systems, if a delay is introduced either at input or at output, the response remains same.

43. A linear system is one that satisfies the superposition principle.

44. In linear systems, the response for weighted sum of inputs is equal to similar weighted sum of individual responses.

45. A system is said to be causal if the output does not depends on future inputs/outputs.

46. When a system output depends on future inputs/outputs, it is called a noncausal system.

47. System is said to be BIBO stable if and only if every bounded input produces a bounded output.

48. A system is said to be unstable if a bounded input produces an unbounded output.

49. When the output of a continuous time system depends on past outputs then the system is called feedback system.

50. When the output of a discrete time system at any time n depends on past outputs, it is called a recursive system.

1.11 Short-answer Questions

Q1.1 *Prove that sinc (0) = 1.*

Proof:

$$\text{sinc}(0) = \underset{t \to 0}{\text{Lt}} \ \text{sinc}(t) = \underset{t \to 0}{\text{Lt}} \ \frac{\sin t}{t} = 1 \qquad \text{(Using L' Hospital's rule)}$$

Q1.2 *Consider the complex valued exponential signal $x(t) = Ae^{\alpha t + j\Omega t}$, a > 0. Evaluate the real and imaginary components of x(t) for the following cases.*

    i) α *real*, $\alpha = \alpha_1$ *ii)* α *imaginary*, $\alpha = j\Omega_1$ *iii)* α *complex*, $\alpha = \alpha_1 + j\Omega_1$

Solution:

case i: $\alpha = \alpha_1$

$$x(t) = Ae^{\alpha t + j\Omega t} = Ae^{\alpha_1 t + j\Omega t} = Ae^{\alpha_1 t} e^{j\Omega t} = Ae^{\alpha_1 t}(\cos \Omega t + j \sin \Omega t)$$

$\therefore$ Real part of x(t) = $Ae^{\alpha_1 t} \cos \Omega t$;   Imaginary part of x(t) = $Ae^{\alpha_1 t} \sin \Omega t$

case ii: $\alpha = j\Omega_1$

$$x(t) = Ae^{\alpha t + j\Omega t} = Ae^{j\Omega_1 t + j\Omega t} = Ae^{j(\Omega_1 + \Omega)t} = A(\cos(\Omega_1 + \Omega)t + j \sin(\Omega_1 + \Omega)t)$$

$\therefore$ Real part of x(t) = $A \cos(\Omega_1 + \Omega)t$;   Imaginary part of x(t) = $A \sin(\Omega_1 + \Omega)t$

case iii: $\alpha = \alpha_1 + j\Omega_1$

$$x(t) = Ae^{\alpha t + j\Omega t} = Ae^{(\alpha_1 + j\Omega_1)t + j\Omega t} = Ae^{\alpha_1 t + j(\Omega_1 + \Omega)t} = Ae^{\alpha_1 t} e^{j(\Omega_1 + \Omega)t}$$

$$= Ae^{\alpha_1 t}[\cos(\Omega_1 + \Omega)t + j\sin(\Omega_1 + \Omega)t]$$

$\therefore$ Real part of x(t) = $A e^{\alpha_1 t} \cos(\Omega_1 + \Omega)t$;   Imaginary part of x(t) = $A e^{\alpha_1 t} \sin(\Omega_1 + \Omega)t$

Q1.3 *Determine whether the signal, $x(t) = 3\cos\sqrt{2}t + 7\cos 5\pi t$ is periodic.*

Solution:

Given that $x(t) = 3\cos\sqrt{2}t + 7\cos 5\pi t$

Let, $x_1(t) = 3\cos\sqrt{2}t$. Let T_1 be period of $x_1(t)$. On comparing $x_1(t)$ with standard form "A $\cos 2\pi F_{01} t$:" we get,

$$2\pi F_{01} = \sqrt{2} \quad \Rightarrow \quad F_{01} = \frac{\sqrt{2}}{2\pi}, \qquad \therefore T_1 = \frac{1}{F_{01}} = \frac{2\pi}{\sqrt{2}}$$

Let, $x_2(t) = 7\cos 5\pi t$. Let T_2 be period of $x_2(t)$. On comparing $x_2(t)$ with standard form "A $\cos 2\pi F_{02} t$" we get,

$$2\pi F_{02} = 5\pi \quad \Rightarrow \quad F_{02} = \frac{5}{2}, \qquad \therefore T_2 = \frac{1}{F_{02}} = \frac{2}{5}$$

Now, $\dfrac{T_1}{T_2} = T_1 \times \dfrac{1}{T_2} = \dfrac{2\pi}{\sqrt{2}} \times \dfrac{5}{2} = \dfrac{5\pi}{\sqrt{2}}$. Here T_1/T_2 is not a rational number and so x(t) is nonperiodic.

Q1.4 *Determine the period of the signal $x(t) = 0.1 e^{-j\frac{2\pi}{3}t} + 0.3 \sin \pi t$.*

Solution:

Let, $x_1(t) = 0.1 e^{-j\frac{2\pi}{3}t}$. Let T_1 be period of $x_1(t)$. On comparing $x_1(t)$ with standard form "A $e^{-j2\pi F_{01}t}$" we get,

$$2\pi F_{01} = \frac{2\pi}{3} \quad \Rightarrow \quad F_{01} = \frac{1}{3} \qquad \therefore T_1 = \frac{1}{F_{01}} = 3$$

Let, $x_2(t) = 0.3 \sin\pi t$. Let T_2 be period of $x_2(t)$. On comparing $x_2(t)$ with standard form "A $\sin 2\pi F_{02}t$" we get,

$$2\pi F_{02} = \pi \quad \Rightarrow \quad F_{02} = \frac{1}{2} \qquad \therefore T_2 = \frac{1}{F_{02}} = 2$$

Here, $\dfrac{T_1}{T_2} = \dfrac{3}{2}$ = Rational number , $\therefore$ x(t) is periodic.

Let T be the period of x(t), which is given by LCM of T_1 and T_2.   T = LCM of 2 and 3 = 2 × 3 = 6.

Q1.5 *Determine whether the signal, $x(n) = \sin 3n$ is periodic.*   **(AU Jun'13, 2 Marks)**

Solution:

Given that, $x(n) = \sin 3n$

Let N and M be two integers.

Now, $x(n+N) = \sin 3(n+N) = \sin(3n+3N)$ .....(1)

For integer values of M, $\sin(\theta+2\pi M) = \sin \theta$.

Therefore, for x(n) to be periodic, the term 3N in equation (1) should be integral multiple of 2π.

Let, $3N = 2\pi M$

$$\therefore N = 2\pi M \times \frac{1}{3} = \frac{2\pi M}{3}$$

Here N cannot be an integer for any integer value of M and so x(n) will not be periodic.

Q1.6   Determine the even and odd parts of a continuous time complex exponential signal.

Solution:

Complex exponential signal, $x(t) = Ae^{j\Omega_0 t} = A(\cos\Omega_0 t + j\sin\Omega_0 t) = A\cos\Omega_0 t + jA\sin\Omega_0 t$

Now, $x(-t) = A\cos\Omega_0(-t) + jA\sin\Omega_0(-t) = A\cos\Omega_0 t - jA\sin\Omega_0 t$

Even part, $x_e(t) = \dfrac{1}{2}[x(t) + x(-t)] = A\cos\Omega_0 t$

Odd part, $x_o(t) = \dfrac{1}{2}[x(t) - x(-t)] = jA\sin\Omega_0 t$

Q1.7   Sketch the even parts and odd parts of a continuous time unit step signal.

Solution:

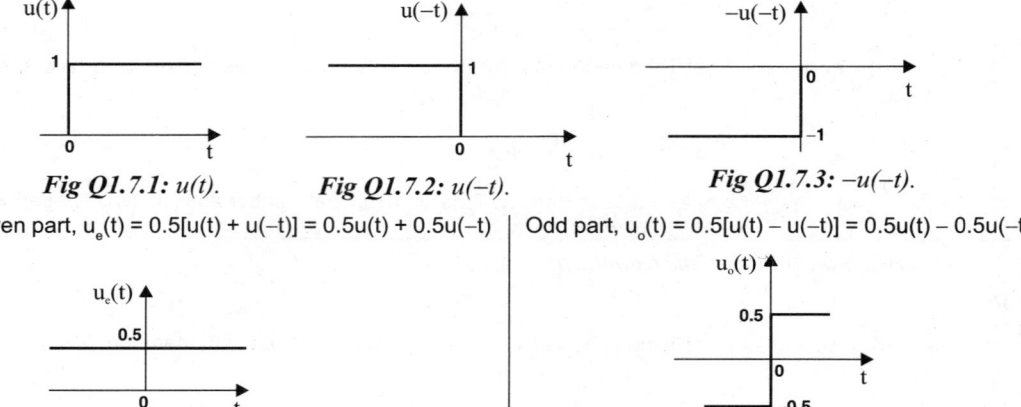

Fig Q1.7.1: *u(t).*   **Fig Q1.7.2:** *u(-t).*   **Fig Q1.7.3:** *-u(-t).*

Even part, $u_e(t) = 0.5[u(t) + u(-t)] = 0.5u(t) + 0.5u(-t)$ | Odd part, $u_o(t) = 0.5[u(t) - u(-t)] = 0.5u(t) - 0.5u(-t)$

Fig Q1.7.4: $u_e(t)$. | **Fig Q1.7.5:** $u_o(t)$.

Q1.8   Determine the energy and power of a continuous time unit step signal.

Solution:

Unit step signal, $u(t) = 1 ; t \geq 0$

$\qquad\qquad\qquad\quad = 0 ; t < 0$

$\therefore \displaystyle\int_{-T}^{T} |x(t)|^2 \, dt = \int_{-T}^{T} |u(t)|^2 \, dt = \int_{0}^{T} 1^2 \, dt = \int_{0}^{T} dt = [t]_0^T = T - 0 = T \quad .....(1)$

Energy, $E = \displaystyle\underset{T\to\infty}{\text{Lt}} \int_{-T}^{T} |x(t)|^2 \, dt = \underset{T\to\infty}{\text{Lt}} \; T = \infty$ $\boxed{\text{Using equation (1)}}$

Power, $P = \displaystyle\underset{T\to\infty}{\text{Lt}} \frac{1}{2T} \int_{-T}^{T} |x(t)|^2 \, dt = \underset{T\to\infty}{\text{Lt}} \frac{1}{2T} \times T = \frac{1}{2}$ watts $\boxed{\text{Using equation (1)}}$

Q1.9   Compare energy and power signals.

Energy Signal	Power Signal
1. Energy of the signal is constant.	1. Energy of the signal is infinite.
2. Power of the signal is zero.	2. Power of the signal is constant.
3. Nonperiodic signals.	3. Periodic signals.

Q1.10 *Test whether the system, y(n) = x(2n) is static.*   **(AU Dec'12, 2 Marks)**

Solution:

Given that, $y(n) = x(2n)$

When, $n = -1$; $y(-1) = x(-2)$ $\Rightarrow$ The response at n =-1, depends on the past input x(-2)

When, $n = 0$; $y(0) = x(0)$ $\Rightarrow$ The response at n = 0, depends on the present input x(0)

When, $n = 1$; $y(1) = x(2)$ $\Rightarrow$ The response at n = 1, depends on the future input x(2).

From the above analysis we can say that the response of the system except for n = 0, depends on past and future input. Hence the system requires memory and so it is dynamic system.

Q1.11 *Differentiate between continuous time causal and noncausal signals.*

The causal signals are defined only for t ≥ 0, whereas the noncausal signals are defined for either t ≤ 0 or for all t (i.e, for both t ≤ 0 and t > 0).

Q1.12 *Sketch the signal, x(t) = 2 u(t) + t u(t) – (t – 1) u(t – 1) – 3 u(t – 2)*

Solution:

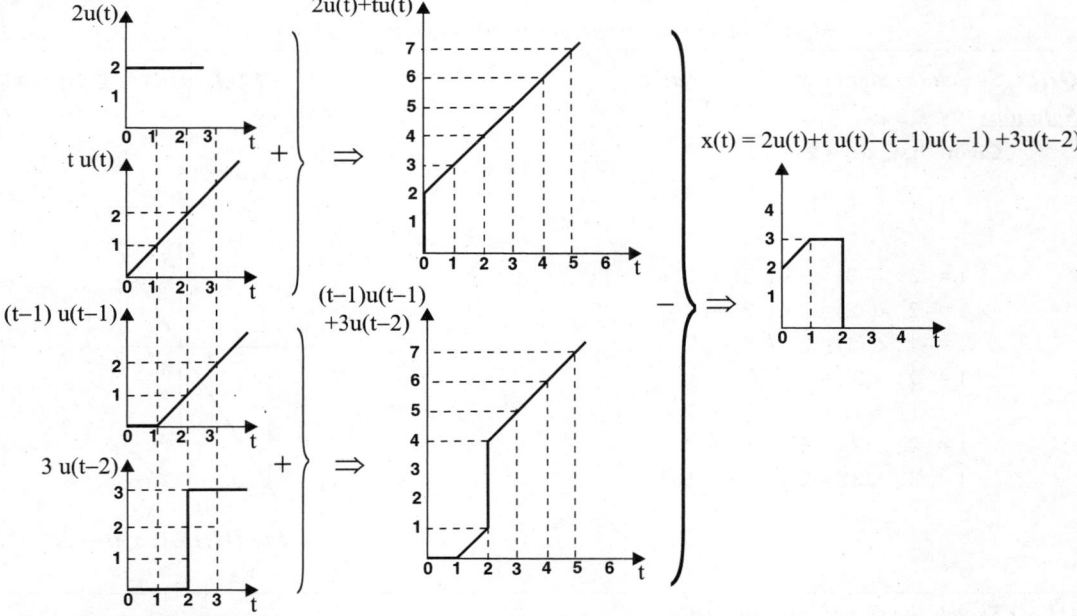

Fig Q1.12.1: *Various stages in construction of signal x(t).*

Q1.13 *Sketch the signal, x(t) = Π(2t+3) when* $x(t) = \begin{cases} 1, & -0.5 < t < 0.5 \\ 0, & eleswhere \end{cases}$ **(AU Dec'11, 2 Marks)**

Solution:

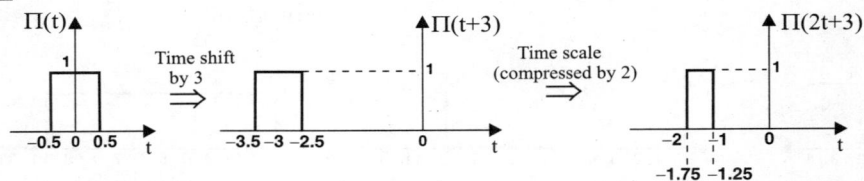

Fig Q1.13.1: Π(t).   **Fig Q1.13.2:** Π(t + 3).   **Fig Q1.13.3:** Π(2t + 3).

Q1.14 *Sketch the signal,* $x(t) = \Pi\left(\frac{t-1}{2}\right) + \Pi(t-1)$.   **(AU Dec'14, 2 Marks)**

Solution:

Given that, $x(t) = \Pi\left(\frac{t-1}{2}\right) + \Pi(t-1) = \Pi\left(\frac{t}{2} - \frac{1}{2}\right) + \Pi(t-1)$

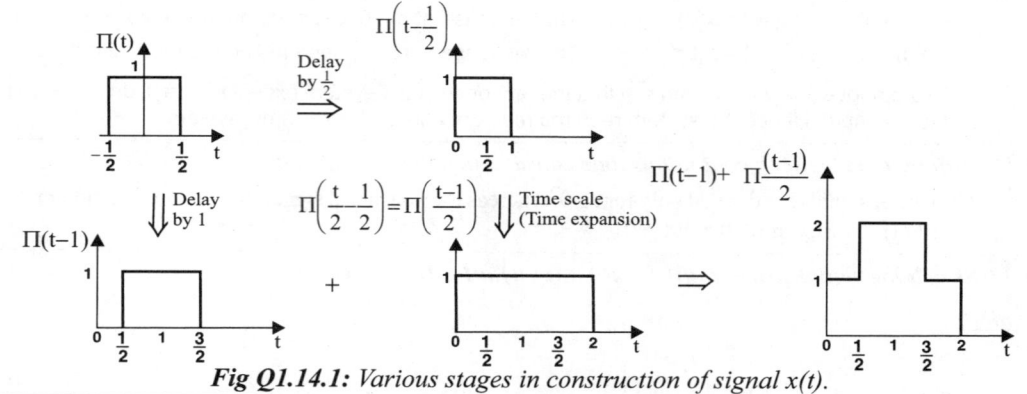

Fig Q1.14.1: *Various stages in construction of signal x(t).*

Q1.15 *Sketch the signal,* $x(t) = 2t$ *for all t.*   **(AU Jun'14, 2 Marks)**

Solution:

Given that, x(t) = 2t

.
.
.

t = –3 ; x(–3) = 2 × (–3) = –6
t = –2 ; x(–2) = 2 × (–2) = –4
t = –1 ; x(–1) = 2 × (–1) = –2
t = 0 ; x(0) = 2 × 0 = 0
t = 1 ; x(1) = 2 × 1 = 2
t = 2 ; x(2) = 2 × 2 = 4
t = 3 ; x(3) = 2 × 3 = 6
.
.
.

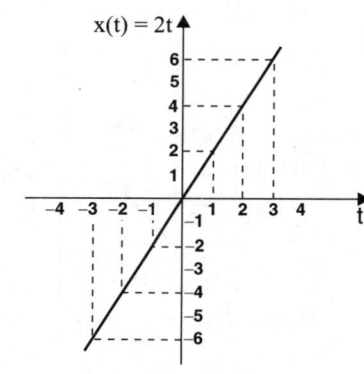

Fig Q1.15.1: *x(t) = 2t.*

Q1.16 *Sketch the signal,* u(t)–u(t–10).   **(AU Dec'14, 2 Marks)**

Solution:

Given that,   u(t)   = 1 ; t ≥ 0

u(t–10) = 1 ; t ≥ 10

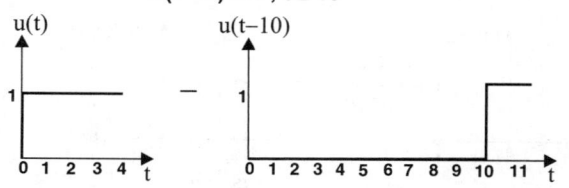

Fig Q1.16.1: *u(t).*   **Fig Q1.16.2:** *Delayed unit step signal.*

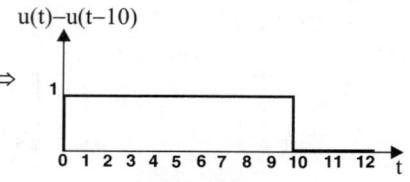

Fig Q1.16.3: *u(t)–u(t–10).*

Q1.17 *Sketch the signal, x(n) = 2n–3*   **(AU Jun'14, 2 Marks)**

Solution:

Given that, x(n) = 2n – 3

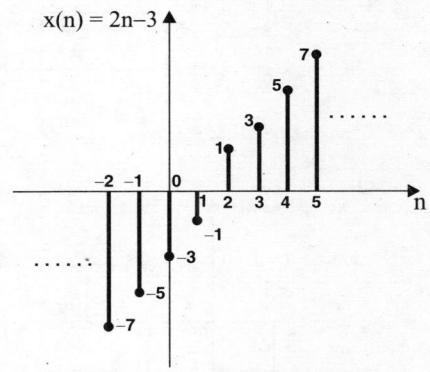

$$x(n) = 2n-3$$

Fig Q1.17: x(n) = 2n–3.

n = –2 ; x(–2) = 2 × (–2) – 3 = –7

n = –1 ; x(–1) = 2 × (–1) – 3 = –5

n = 0 ; x(0) = 2 × 0 – 3 = –3

n = 1 ; x(1) = 2 × 1 – 3 = –1

n = 2 ; x(2) = 2 × 2 – 3 = 1

n = 3 ; x(3) = 2 × 3 – 3 = 3

n = 4 ; x(4) = 2 × 4 – 3 = 5

n = 5 ; x(5) = 2 × 5 – 3 = 7 and so on.

Q1.18 *Sketch the signal, $x(n) = \left(\frac{1}{2}\right)^n u(n-1)$*   **(AU Dec'14, 2 Marks)**

Solution:

Given that,   u(n–1) = 1 ; n ≥ 0

                = 0 ; n ≥ 0

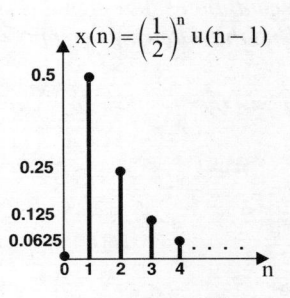

$$x(n) = \left(\frac{1}{2}\right)^n u(n-1)$$

Fig Q1.18.1: $\left(\frac{1}{2}\right)^n u(n-1)$.

$\therefore x(n) = \left(\frac{1}{2}\right)^n u(n-1) = \left(\frac{1}{2}\right)^n$; n ≥ 1

n = 1 ; $\left(\frac{1}{2}\right)^n = \left(\frac{1}{2}\right)^1 = 0.5$

n = 2 ; $\left(\frac{1}{2}\right)^n = \left(\frac{1}{2}\right)^2 = 0.25$

n = 3 ; $\left(\frac{1}{2}\right)^n = \left(\frac{1}{2}\right)^3 = 0.125$

n = 4 ; $\left(\frac{1}{2}\right)^n = \left(\frac{1}{2}\right)^4 = 0.0625$

Q1.19 *A continuous time signal is shown in fig Q 1.19.*
Sketch the following versions of the signal.

a) x (t – 3)   *b) – 2x (t)*   *c) x (t – 3) – 2x (t)*   *d) $\dfrac{dx(t)}{dt}$*

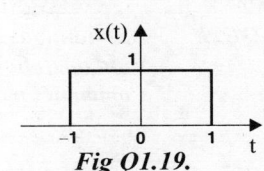

Fig Q1.19.

Solution

a)

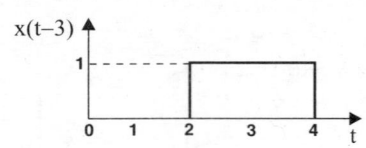

Fig Q1.19.1 : *x(t – 3).*

b)

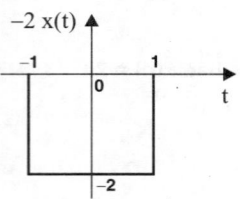

Fig Q1.19.2 : *– 2x(t)).*

c)

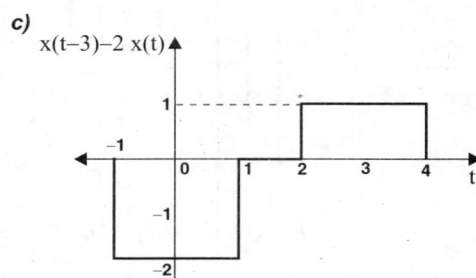

Fig Q1.19.3 : *x(t – 3) - 2x(t).*

d) $x(t) = u(t+1) - u(t-1)$

$\therefore \dfrac{d}{dt} x(t) = \dfrac{d}{dt} u(t+1) - \dfrac{d}{dt} u(t-1)$

$= \delta(t+1) - \delta(t-1)$

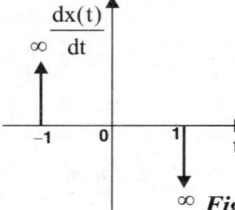

$\dfrac{d}{dt} u(t) = \delta(t)$
Using time invariant property
$\dfrac{d}{dt} u(t \pm k) = \delta(t \pm k)$

Fig Q1.19.4 : $\dfrac{dx(t)}{dt}.$

Q1.20 *A continuous time signal is shown in Fig Q 1.20.*
Find the following versions of the signal.

a) *x(– t)*   b) *– x(t)*

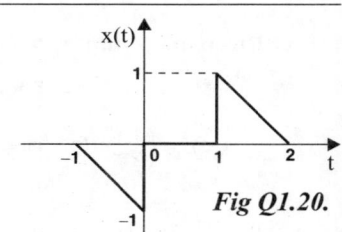

Fig Q1.20.

Solution:

a)

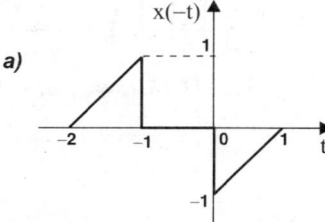

Fig Q1.20.1 : *x(−t).*

b)

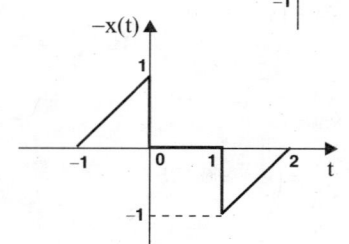

Fig Q1.20.2 : *−x(t).*

Q1.21   *A continuous time signal is shown in Fig Q 1.21 .*
Find the following versions of the signal.
Comment on the result.

a) *x(t–2)*   b) *x(–t)*   c) *x(–t +2)*

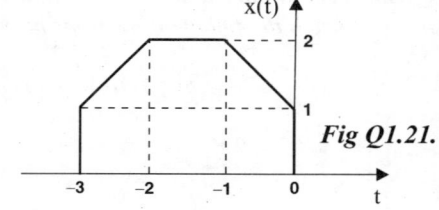

Fig Q1.21.

Solution:

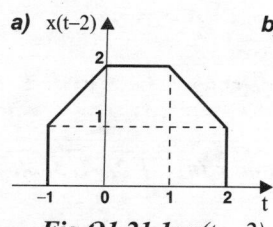

a) x(t–2)

Fig Q1.21.1: x(t – 2).

b) x(–t)

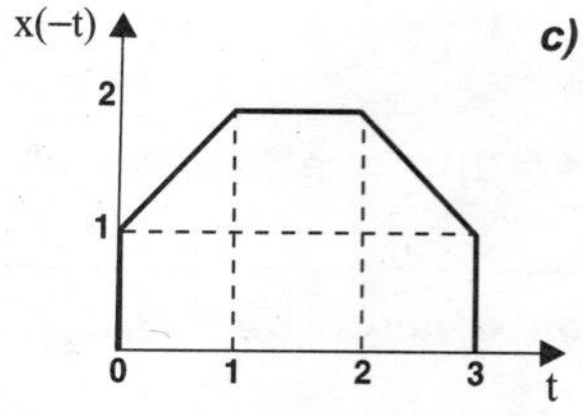

Fig Q1.21.2: x(– t).

c) x(–t+2)

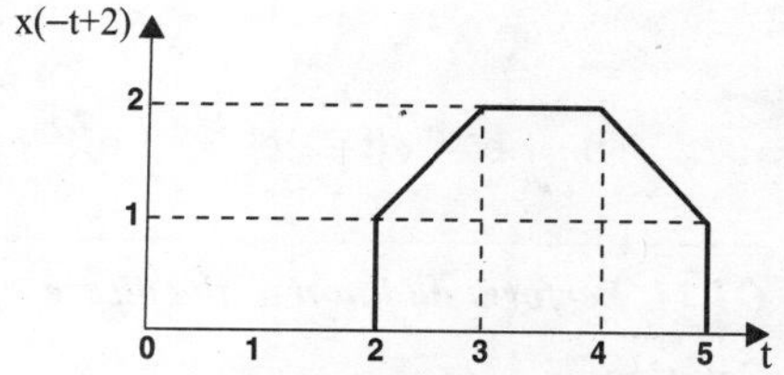

Fig Q1.21.3: x(– t + 2).

Comment : The signal x(–t+2) is obtained by doing folding operation first ie., x(–t) and then time shifting is done x(–t+2).

Q1.22 *A continuous time signal is shown in Fig Q 1.22.*

Find the following versions of the signal.
Comment on the result.

a) x(t +2)   b) x(–t)   c) x(–t – 2)

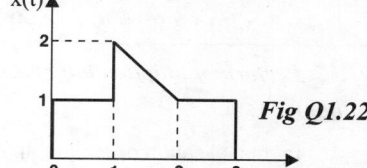

x(t)

Fig Q1.22.

Solution:

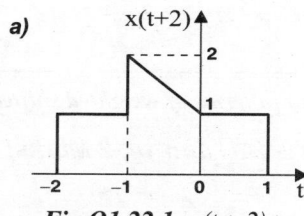

a) x(t+2)

Fig Q1.22.1: x(t + 2).

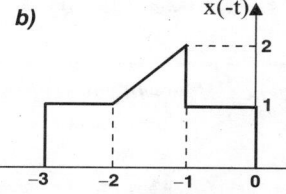

b) x(-t)

Fig Q1.22.2: x(– t).

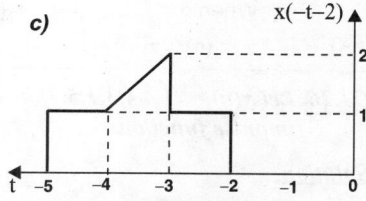

c) x(–t–2)

Fig Q1.22.3: x(– t + 2).

Comment : The signal x(–t–2) is obtained by doing folding operation first ie., x(–t) and then time shifting is done x(–t–2).

Q1.23 *Evaluate the following integrals,*

i) $\int_{-\infty}^{\infty} \delta(t+3)\, e^{-t}\, dt$

ii) $\int_{-\infty}^{\infty} \left[\delta(t)\cos t + \delta(t-1)\sin t\right] dt$

iii) $\int_{0}^{5} \delta(t)\sin 2\pi t\, dt$

iv) $\int_{-\infty}^{\infty} e^{-2t}\,\delta(t+2)\, dt$

(AU Dec'15, 2 Marks)

Solution:

i) $\int_{-\infty}^{\infty} \delta(t+3)\, e^{-t}\, dt = \int_{-\infty}^{\infty} e^{-t}\,\delta(t-(-3))\, dt = e^{-t}\big|_{t=-3} = e^{3} = 20.0855$

> Using property of impulse
> $\int x(t)\delta(t-t_0)\, dt = x(t_0)$

ii) $\int_{-\infty}^{\infty} \left[\delta(t)\cos t + \delta(t-1)\sin t\right] dt = \int_{-\infty}^{\infty} \cos t\,\delta(t)\, dt + \int_{-\infty}^{\infty} \sin t\,\delta(t-1)\, dt = \cos t\big|_{t=0} + \sin t\big|_{t=1}$

$= \cos 0 + \sin 1 = 1 + 0.8415 = 1.8415$

> Using property of impulse
> $\int x(t)\delta(t)\, dt = x(0)$

iii) $\int_{0}^{5} \delta(t) \sin 2\pi t \ dt = \sin 2\pi t \big|_{t=0} = \sin 0 = 0$

iv) $\int_{-\infty}^{\infty} e^{-2t} \delta(t+2)dt = \int_{-\infty}^{\infty} e^{-2t} \delta(t-(-2))dt = e^{-2t}\big|_{t=-2} = e^{4} = 54.5982$

Q1.24 Perform addition of the discrete time signals, $x_1(n) = \{2, 2, 1, 2\}$ and $x_2(n) = \{-2, -1, 3, 2\}$.

Solution:

In addition operation, the samples corresponding to same value of n are added.

When n = 0, $x_1(0) + x_2(0) = 2 + (-2) = 0$ | When n = 2, $x_1(2) + x_2(2) = 1 + 3 = 4$

When n = 1, $x_1(1) + x_2(1) = 2 + (-1) = 1$ | When n = 3, $x_1(3) + x_2(3) = 2 + 2 = 4$

$\therefore x_1(n) + x_2(n) = \{0, 1, 4, 4\}$

Q1.25 Perform multiplication of discrete time signals, $x_1(n) = \{2, 2, 1, 2\}$ and $x_2(n) = \{-2, -1, 3, 2\}$.

Solution:

In multiplication operation, the samples corresponding to same value of n are multiplied.

When n = 0, $x_1(0) \times x_2(0) = 2 \times (-2) = -4$ | When n = 2, $x_1(2) \times x_2(2) = 1 \times 3 = 3$

When n = 1, $x_1(1) \times x_2(1) = 2 \times (-1) = -2$ | When n = 3, $x_1(3) \times x_2(3) = 2 \times 2 = 4$

$\therefore x_1(n) \times x_2(n) = \{-4, -2, 3, 4\}$

Q1.26 Let $x(n) = \{1, -4, 3, 1, 5, 2\}$ be a sequence. How will you represent $x(n)$ in terms of weighted shifted impulse functions.

(AU Jun'14, 2 Marks)

Solution:

Given that $x(n) = \{1, -4, 3, 1, 5, 2\}$

$\therefore x(0) = 1, x(1) = -4, x(2) = 3, x(3) = 1, x(4) = 5, x(5) = 2$

The shifted impulse, $\delta(n-k) = 1$; for $n = k$

$= 0$; for $n \neq k$

Therefore, if we multiply $x(n)$ by $\delta(n-k)$ then it selects only k^{th} sample of $x(n)$. If we multiply $x(n)$ by $\delta(n-k)$ for all possible shifts and sum up the product then we get the signal $x(n)$.

$\therefore$ The weighted shifted impulse function is,

$x(n) = \delta(n) - 4\delta(n-1) + 3\delta(n-2) + \delta(n-3) + 5\delta(n-4) + 2\delta(n-5)$

Q1.27 Express the discrete time signal $x(n)$ as a summation of impulses.

If we multiply a signal $x(n)$ by a delayed unit impulse $\delta(n - m)$, then the product is $x(m)$, where $x(m)$ is the signal sample at $n = m$ [because $\delta(n - m)$ is 1 only at $n = m$ and zero for other values of n]. Therefore, if we repeat this multiplication over all possible delays in the range $-\infty < m < \infty$ and sum all the product sequences, then the result will be a sequence that is equal to the sequence $x(n)$.

$\therefore x(n) = \dots x(-2) \delta(n + 2) + x(-1) \delta(n + 1) + x(0) \delta(n) + x(1) \delta(n - 1) + x(2) \delta(n - 2) + \dots$

$$= \sum_{m=-\infty}^{\infty} x(m) \delta(n - m)$$

1.12 MATLAB Programs

Program 1.1

Write a MATLAB program to generate standard signals like unit impulse, unit step, unit ramp, parabolic, sinusoidal, triangular pulse, signum, sinc and Gaussian signals.

```
%****************** program to plot some standard signals
tmin=-5; dt=0.1; tmax=5;
t=tmin:dt:tmax;                       %set a time vector

%****************** unit impulse signal
x1=1;
x2=0;
x=x1.*(t==0)+x2.*(t~=0);              %generate unit impulse signal
subplot(3,3,1);plot(t,x);             %plot the generated unit impulse signal
xlabel('t');ylabel('x(t)');title('unit impulse signal');

%****************** unit step signal
x1=1;
x2=0;
x=x1.*(t>=0)+x2.*(t<0);               %generate unit step signal
subplot(3,3,2);plot(t,x);             %plot the generated unit step signal
xlabel('t');ylabel('x(t)');title('unit step signal');

%****************** unit ramp signal
x1=t;
x2=0;
x=x1.*(t>=0)+x2.*(t<0);               %generate unit ramp signal
subplot(3,3,3);plot(t,x);             %plot the generated unit ramp signal
xlabel('t');ylabel('x(t)');title('unit ramp signal');

%****************** parabolic signal
A=0.4;
x1=(A*(t.^2))/2;
x2=0;
x=x1.*(t>=0)+x2.*(t<0);               %generate parabolic signal
subplot(3,3,4);plot(t,x);             %plot the generated parabolic signal
xlabel('t');ylabel('x(t)');title('parabolic signal');

%****************** sinusoidal signal
T=2;                                  %declare time period
F=1/T;                                %compute frequency
x=sin(2*pi*F*t);                      %generate sinusoidal signal
subplot(3,3,5);plot(t,x);             %plot the generated sinusoidal signal
xlabel('t');ylabel('x(t)');title('sinusoidal signal');
```

```
%******************** triangular pulse signal
a=2;
x1=1-abs(t)/a;
x2=0;
x=x1.*(abs(t)<=a)+x2.*(abs(t)>a); %generate triangular pulse signal
subplot(3,3,6);plot(t,x);        %plot the triangular pulse signal
xlabel('t');ylabel('x(t)');title('triangular pulse signal');

%******************** signum signal
x1=1;
x2=0;
x3=-1;
x=x1.*(t>0)+x2.*(t==0)+x3.*(t<0);          %generate signum signal
subplot(3,3,7);plot(t,x);                  %plot the generated signum signal
xlabel('t');ylabel('x(t)');title('signum signal');
```

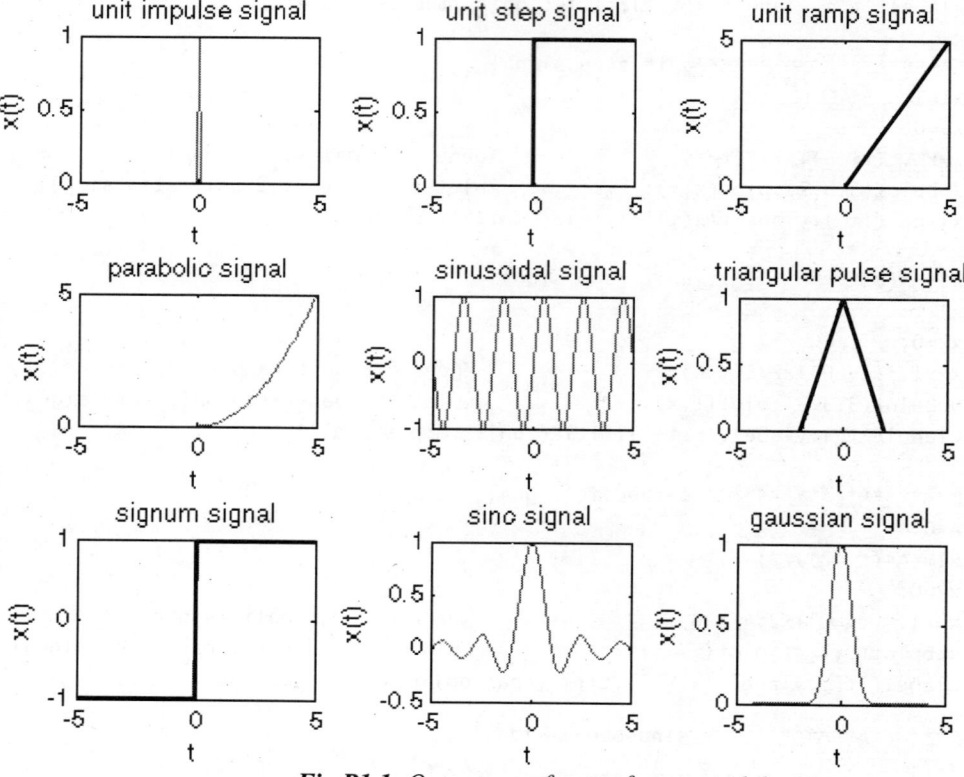

Fig P1.1: *Output waveforms of program 1.1.*

```
%******************** sinc Pulse
x=sinc(t);                              %generate sinc Pulse
subplot(3,3,8);plot(t,x);               %plot the generated sinc Pulse
xlabel('t');ylabel('x(t)');title('sinc signal');

%******************** gaussian signal
a=2;
x=exp(-a.*(t.^2));                      %generate gaussian signal
subplot(3,3,9);plot(t,x);               %plot the generated gaussian signal
xlabel('t');ylabel('x(t)');title('gaussian signal');
```

OUTPUT

The output waveforms of program 1.1 are shown in fig P1.1.

Program 1.2

Write a MATLAB program to find the even and odd parts of the signal $x(t)=e^{2t}$.

```
%To find the even and odd parts of the signal, x(t)=exp(2*t)

tmin=-3; tmax=3; dt=.1;
t=tmin:dt:tmax;           %set a time vector

x1=exp(2*t);             %generate the given signal
x2=exp(-2*t);            %generate the time folded signal

if(x2==x1)
    disp('"The given signal is even signal"');
else if (x2==(-x1))
    disp('"The given signal is odd signal"');
else
    disp('"The given signal is neither even nor odd signal"');
end
end

xe=(x1+x2)/2;            %compute even part
xo=(x1-x2)/2;            %compute odd part

ymin=min([min(x1), min(x2), min(xe), min(xo)]);
ymax=max([max(x1), max(x2), max(xe), max(xo)]);

subplot(2,2,1);plot(t,x1);axis([tmin tmax ymin ymax]);
xlabel('t');ylabel('x1(t)');title('signal x(t)');

subplot(2,2,2);plot(t,x2);axis([tmin tmax ymin ymax]);
xlabel('t');ylabel('x2(t)');title('signal x(-t)');

subplot(2,2,3);plot(t,xe);axis([tmin tmax ymin ymax]);
xlabel('t');ylabel('xe(t)');title('even part of x(t)');

subplot(2,2,4);plot(t,xo);axis([tmin tmax ymin ymax]);
xlabel('t');ylabel('xo(t)');title('odd part of x(t)');
```

OUTPUT

"The given signal is neither even nor odd signal"
The input and output waveforms of the program 1.2 are shown in fig P1.2.

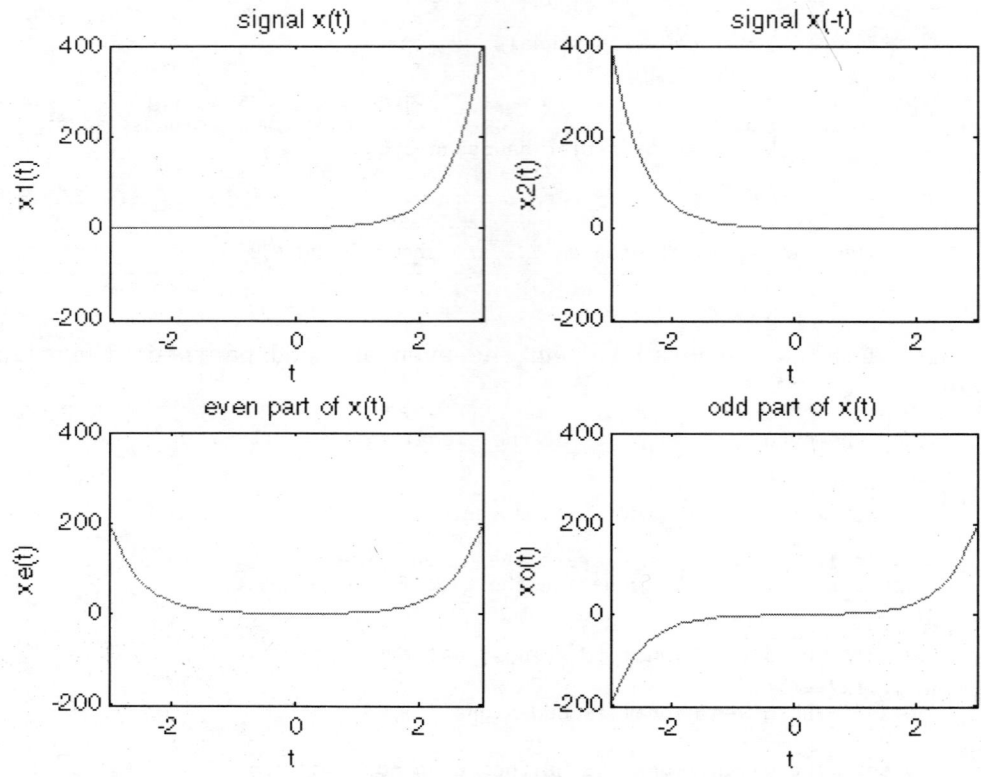

Fig P1.2: *Input and Output waveforms of program 1.2.*

Program 1.3

Write a MATLAB program to find the energy and power of the signal $x(t)=10\sin(10\pi t)$.

```
%Program to compute the signal energy and power of the signal
%x(t)=10sin(10*pi*t)

tmax=10; dt=.01; T0=10;
t=-tmax:dt:tmax;          %set up a time vector
x=10*sin(10*pi*t);        %generate the given signal x(t)
xsq = x.^2 ;              %compute the square of the given signal x(t)
E = trapz (t, xsq) ;      %use trapezoidal-rule numerical integration to
                          %find energy
P = E /(2*T0);            %divide the energy by period to compute power
disp(['  Energy, E  =',num2str(E),' Joules']);
disp(['  Power, P =',num2str(P),' Watts']);
```
OUTPUT

```
Energy, E  = 1000 Joules
Power,  P  = 50 Watts
```

Program 1.4

Write a MATLAB program to perform Amplitude scaling, Time scaling, and Time shift on the signal x(t) = 1+t; for t = 0 to 2.

Program to declare the given signal as function y(t)

```
% declare the given signal as function y(t)
function x = y(t)
x=(1.0 + t).*(t>=0 & t<=2);
```

> Note: The above program should be stored as a separate file in the
>   current working directory

Program to perform amplitude & time scaling and time shift on y(t)

```
%To perform Amplitude scaling, Time scaling and Time shift
%on the signal x(t)=1.0+t; for t= 0 to 2
%include y.m file in work directory which declare the given signal
%as function y(t)

tmin=-3; tmax=5; dt=0.2;
t=tmin:dt:tmax;     %set a time vector

y0 =y(t);              %assign the given signal as y0
y1 =1.5*y(t);          %compute the amplified version of x(t)
y2 =0.5*y(t);          %compute the attenuated version of x(t)
y3=y(2*t);             %compute the time compressed version of x(t)
y4=y(0.5*t);           %compute the time expanded version of x(t)
y5=y(t-2);             %compute the delayed version of x(t)
y6=y(t+2);             %compute the advanced version of x(t)

%compute the min and max value for y-axis
ymin=min([min(y0), min(y1), min(y2), min(y3), min(y4), min(y5),min(y6)]);
ymax=max([max(y0), max(y1), max(y2), max(y3), max(y4), max(y5),max(y6)]);

%plot the given signal and amplitude scaled signal
subplot(3,3,1);plot(t,y0);axis([tmin tmax ymin ymax]);
xlabel('t');ylabel('x(t)');title('Signal x(t)');
subplot(3,3,2);plot(t,y1);axis([tmin tmax ymin ymax]);
xlabel('t');ylabel('x1(t)');title('Amplified signal 1.5x(t)');
subplot(3,3,3);plot(t,y2);axis([tmin tmax ymin ymax]);
xlabel('t');ylabel('x2(t)');title('Attenuated signal 0.5x(t)');

%plot the given signal and time scaled signal
subplot(3,3,4);plot(t,y0);axis([tmin tmax ymin ymax]);
xlabel('t');ylabel('x(t)');title('Signal x(t)');
subplot(3,3,5);plot(t,y3);axis([tmin tmax ymin ymax]);
xlabel('t');ylabel('x3(t)');title('Time comp. signal x(2t)');
subplot(3,3,6);plot(t,y4);axis([tmin tmax ymin ymax]);
xlabel('t');ylabel('x4(t)');title('Time expan. signal x(0.5t)');

%plot the given signal and time shifted signal
subplot(3,3,7);plot(t,y0);axis([tmin tmax ymin ymax]);
xlabel('t');ylabel('x(t)');title('Signal x(t)');
subplot(3,3,8);plot(t,y5);axis([tmin tmax ymin ymax]);
xlabel('t');ylabel('x5(t)');title('Delayed signal x(t-2)');
subplot(3,3,9);plot(t,y6);axis([tmin tmax ymin ymax]);
xlabel('t');ylabel('x6(t)');title('Advanced signal x(t+2)');
```

OUTPUT

The input and output waveforms of program 1.4 are shown in fig P1.4.

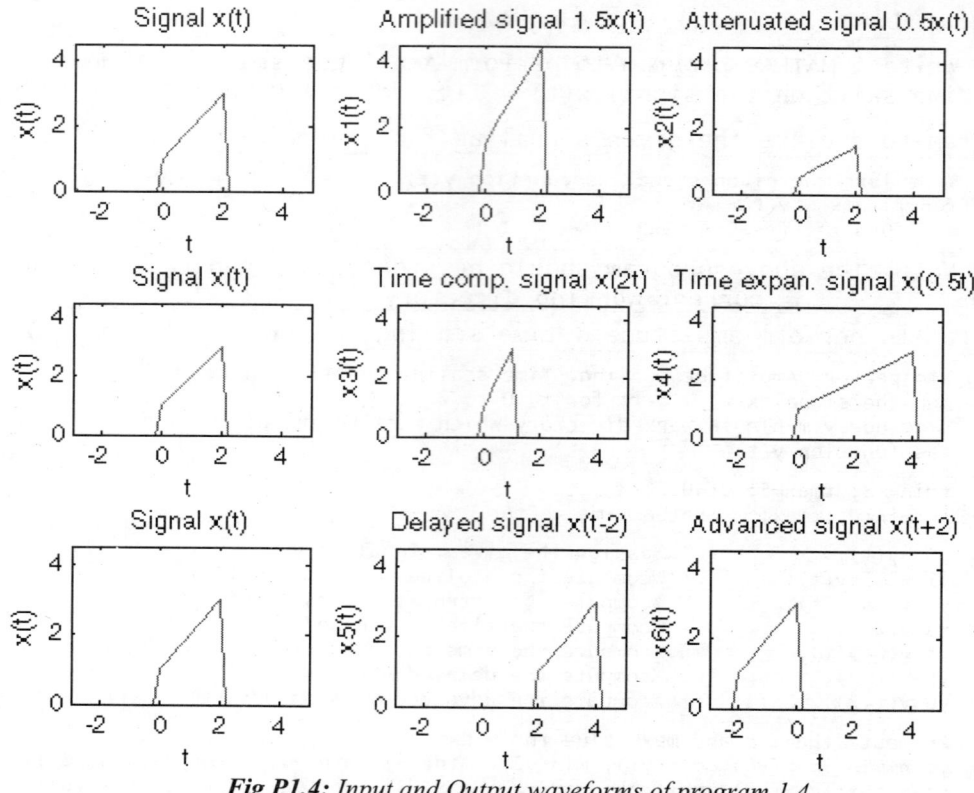

Fig P1.4: *Input and Output waveforms of program 1.4.*

Program 1.5

Write a MATLAB program to perform addition and multiplication on the following two signals.

xa(t)=1; 0<t<1   xb(t)=t;   0<t<1
=2; 1<t<2   =1;   1<t<2
=1; 2<t<3   =3-t; 2<t<3

```
%To perform addition and multiplication of the following two signals
%1) xa(t)=1;0<t<1        2) xb(t)=t;0<t<1
%       =2;1<t<2              =1;1<t<2
%       =1;2<t<3              =3-t;2<t<3

tmin=-1; tmax=5; dt=0.1;
t=tmin:dt:tmax;          %Set a time vector

x1=1;
x2=2;
x3=3-t;
xa=x1.*(t>0&t<1)+x2.*(t>=1&t<=2)+x1.*(t>2&t<3);
xb=t.*(t>0&t<1)+x1.*(t>=1&t<=2)+x3.*(t>2&t<3);
xadd=xa+xb;         %Add the two signals
xmul=xa.*xb;        %Multiply two signals

xmin=min([min(xa), min(xb), min(xadd), min(xmul)]);
xmax=max([max(xa), max(xb), max(xadd), max(xmul)]);
```

```
subplot(2,3,1);plot(t,xa);axis([tmin tmax xmin xmax]);
xlabel('t');ylabel('xa(t)');title('Signal xa(t)');
subplot(2,3,2);plot(t,xb);axis([tmin tmax xmin xmax]);
xlabel('t');ylabel('xb(t)');title('Signal xb(t)');
subplot(2,3,3);plot(t,xadd);axis([tmin tmax xmin xmax]);
xlabel('t');ylabel('xadd(t)');title('Sum of xa(t) and xb(t)');
subplot(2,3,4);plot(t,xa);axis([tmin tmax xmin xmax]);
xlabel('t');ylabel('xa(t)');title('Signal xa(t)');

subplot(2,3,5);plot(t,xb);axis([tmin tmax xmin xmax]);
xlabel('t');ylabel('xb(t)');title('Signal xb(t)');
subplot(2,3,6);plot(t,xmul);axis([tmin tmax xmin xmax]);
xlabel('t');ylabel('xmul(t)');title('Product of xa(t) and xb(t)');
```

OUTPUT

The input and output waveforms of program 1.5 are shown in Fig P1.5.

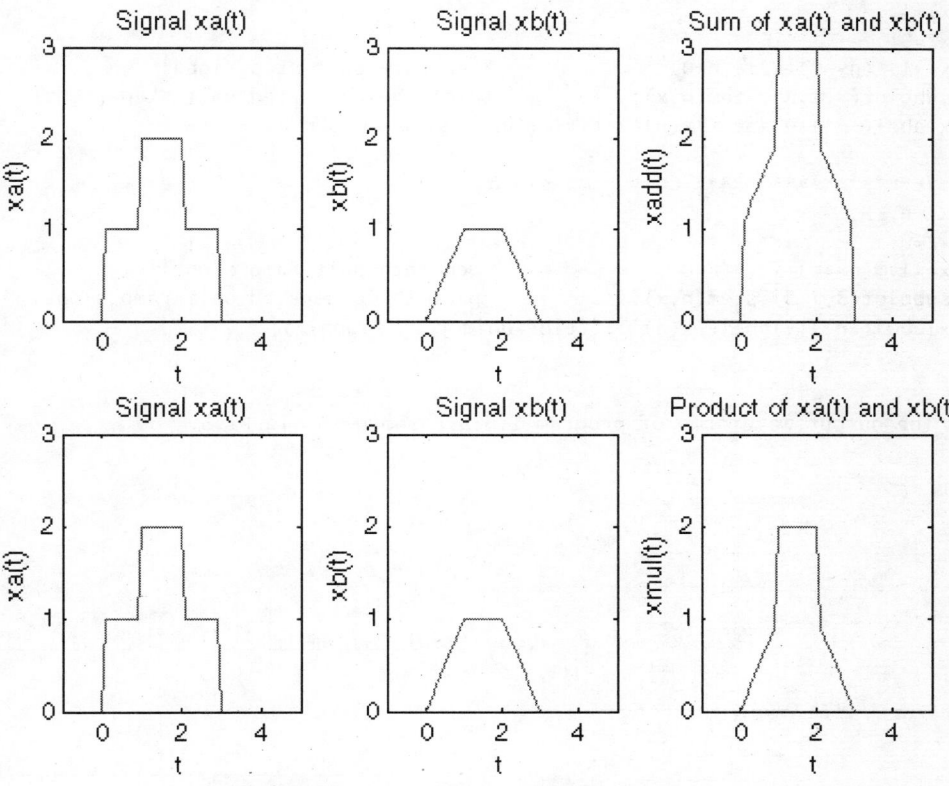

Fig P1.5: Input and Output waveforms of program 1.5.

Program 1.6

Write a MATLAB program to generate the standard discrete time signals unit impulse, unit step and unit ramp signals.

```
%******************* program to plot some standard signals
n=-20 : 1 : 20;                           %specify the range of n

%****************** unit impulse signal
x1=1;
x2=0;
x=x1.*(n==0)+x2.*(n~=0);                  %generate unit impulse signal
subplot(3,1,1);stem(n,x);                 %plot the generated unit impulse signal
xlabel('n');ylabel('x(n)');title('unit impulse signal');

%****************** unit step signal
x1=1;
x2=0;
x=x1.*(n>=0)+x2.*(n<0);                   %generate unit step signal
subplot(3,1,2);stem(n,x);                 %plot the generated unit step signal
xlabel('n');ylabel('x(n)');title('unit step signal');

%****************** unit ramp signal
x1=n;
x2=0;
x=x1.*(n>=0)+x2.*(n<0);                   %generate unit ramp signal
subplot(3,1,3);stem(n,x);                 %plot the generated unit ramp signal
xlabel('n');ylabel('x(n)');title('unit ramp signal');
```

OUTPUT

The output waveforms of program 1.6 are shown in Fig P1.6.

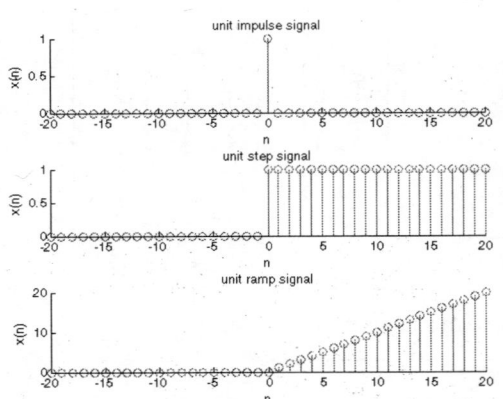

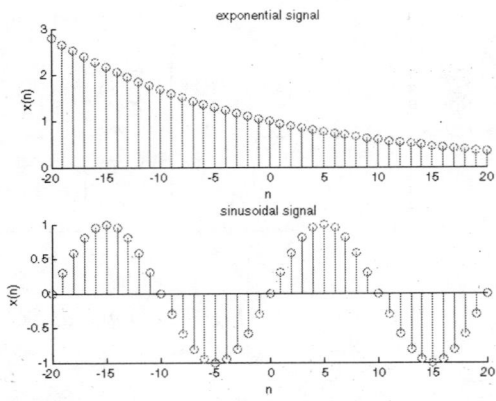

Fig P1.6: *Output waveforms of program 1.6.*   ***Fig P1.7:*** *Output waveforms of program 1.7.*

Program 1.7

Write a MATLAB program to generate the standard discrete time signals exponential and sinusoidal signals.

```
%****************** program to plot some standard signals

n=-20 : 1 : 20;                      %specify the range of n

%****************** exponential signal
A=0.95;
x=A.^n;                              %generate exponential signal
subplot(2,1,1);stem(n,x);            %plot the generated exponential signal
xlabel('n');ylabel('x(n)');title('exponential signal');

%****************** sinusoidal signal
N=20;                                %declare periodicity
f=1/20;                              %compute frequency
x=sin(2*pi*f*n);                     %generate sinusoidal signal
subplot(2,1,2);stem(n,x);            %plot the generated sinusoidal signal
xlabel('n');ylabel('x(n)');title('sinusoidal signal');
```

OUTPUT

The output waveforms of program 1.7 are shown in Fig P1.7.

Program 1.8

Write a MATLAB program to find the even and odd parts of the signal $x(n)=0.8^n$.

```
%To find the even and odd parts of the signal, x(n)= 0.8^n

n= -5 :1 :5;            %specify the range of n
A=0.8;
x1=A.^n;                %generate the given signal
x2=A.^(-n);             %generate the folded signal
if(x2==x1)
    disp('"The given signal is even signal"');
else if (x2==(-x1))
    disp('"The given signal is odd signal"');
else
    disp('"The given signal is neither even nor odd signal"');
end
end

xe=(x1+x2)/2;           %compute even part
xo=(x1-x2)/2;           %compute odd part

subplot(2,2,1);stem(n,x1);
xlabel('n');ylabel('x1(n)');title('signal x(n)');
```

```
subplot(2,2,2);stem(n,x2);
xlabel('n');ylabel('x2(n)');title('signal x(-n)');

subplot(2,2,3);stem(n,xe);
xlabel('n');ylabel('xe(n)');title('even part of x(n)');

subplot(2,2,4);stem(n,xo);
xlabel('n');ylabel('xo(n)');title('odd part of x(n)');
```

OUTPUT

"The given signal is neither even nor odd signal"
The output waveforms of program 1.8 are shown in Fig P1.8.

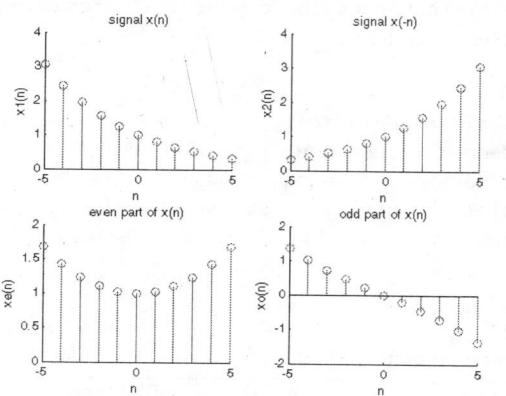

Fig P1.8: *Output waveforms of program 1.8.*   **Fig P1.9:** *Output waveforms of program 1.9.*

Program 1.9

Write a MATLAB program to perform amplitude scaling and time shift on the signal x(n) = 1+n; for n = 0 to 2.

Program to declare the given signal as function y(n)

```
% declare the given signal as function y(n)

function x = y(n)
x=(1.0 + n).*(n>=0 & n<=2);
```

Program to perform amplitude scaling and time shift on y(n)

```
%To Perform Amplitude scaling and Time shift on signal x(n)=1+n; for n= 0 to 2
%include y.m file in current work directory which declare given signal as
%function y(n)

n=-5:1:5;              %specify range of n

y0 =y(n);              %assign the given signal as y0

y1 =1.5*y(n);          %compute the amplified version of x(n)

y2 =0.5*y(n);          %compute the attenuated version of x(n)

y3 =y(n-2);            %compute the delayed version of x(n)

y4 =y(n+2);            %compute the advanced version of x(n)
```

```
%plot the given signal and amplitude scaled signal
subplot(2,3,1);stem(n,y0);
xlabel('n');ylabel('x(n)');title('Signal x(n)');
subplot(2,3,2);stem(n,y1);
xlabel('n');ylabel('x1(n)');title('Amplified signal 1.5x(n)');
subplot(2,3,3);stem(n,y2);
xlabel('n');ylabel('x2(n)');title('Attenuated signal 0.5x(n)');

%plot the given signal and time shifted signal
subplot(2,3,4);stem(n,y0);
xlabel('n');ylabel('x(n)');title('Signal x(n)');
subplot(2,3,5);stem(n,y3);
xlabel('n');ylabel('x3(n)');title('Delayed signal x(n-2)');
subplot(2,3,6);stem(n,y4);
xlabel('n');ylabel('x4(n)');title('Advanced signal x(n+2)');
```

OUTPUT

The input and output waveforms of program 1.9 are shown in Fig P1.9.

1.13 Exercises

I. Fill in the blanks with appropriate words.

1. The ———— signal is continuous function of an independent variable

2. The ———— signal can be represented by a rotating vector in a complex plane.

3. The sum of two periodic signals will also be periodic if the ratio of their fundamental periods is a ———— number.

4. When a periodic signal is a sum of two or more periodic signals then the period of the signal is given by ———— of the period of its components.

5. When a signal exhibits ———— with respect to t = 0 it is called even signal.

6. When a signal exhibits antisymmetry with respect to t = 0 it is called ———— signal.

7. The periodic signals will have constant ————.

8. The nonperiodic signals will have constant ————.

9. The ———— signals are defined for t ≥ 0.

10. When the sign of t is changed in x(t) the resultant signal is called ———— signal.

11. The ———— is the reverse process of differentiation.

12. A signal with large magnitude and short duration is called ————.

13 The response of a system for impulse input is called ————.

14. The systems that do not require memory are called ———— systems.

15. A system is said to be ———— if its input-output characteristics do not change with time.

16. A ———— system is one that satisfies the superposition principle.

17. A ———— system is one in which the response does not depend on future inputs/outputs.

18. A signal x(n) may be shifted in time by m units by replacing the independent variable n by _____.

19. The _____ of a signal x(n) is performed by changing the sign of the time base n.

20. If the average power of a signal is finite then it is called _____.

21. The smallest value of N for which x(n + N) = x(n) is true is called _____.

22. In a discrete time signal x(n), if x(n) = x(−n) then it is called _____ signal.

23. In a discrete time signal x(n), if x(−n) = −x(n) then it is called _____ signal.

24. A discrete time system is _____ if it obeys the principle of superposition.

25. A discrete time system is _____ if its input-output relationship do not change with time.

Answers

1. analog	11. integration	21. fundamental period
2. complex exponential	12. impulse	22. symmetric
3. rational	13. impulse response	23. antisymmetric
4. LCM (Least Common Multiple)	14. static	24. linear
5. symmetry	15. time invariant	25. time invariant
6. odd	16. linear	
7. power	17. causal	
8. energy	18. n-m	
9. causal	19. folding	
10. folded	20. power signal	

II. State whether the following statements are True/False.

1. In an analog signal, both the magnitude of the signal and the independent variable are continuous.

2. A complex exponential signal is a three dimensional signal.

3. The sinc signal is a periodic signal.

4. The continuous time sinusoidal and complex exponential signals are always periodic.

5. The product of two odd signals will be an odd signal.

6. When the energy of a signal is finite then the power will be infinite.

7. When the power of a signal is finite then the energy will be zero.

8. The continuous time causal signals are defined for all t.

9. The time shift of a signal x(t) is obtained by replacing t by t ± m .

10. The rate of change of a signal is called differentiation.

11. The continuous time impulse signal has infinite magnitude but constant area.

12. A system which satisfies the property of linearity and time invariance is called an LTI system.

13. The systems that require memory are called dynamic systems.
14. In a time invariant system, if a delay is introduced either at input or output, then the overall response will remain same.
15. A nonlinear system is one which satisfies the principle of superposition.
16. The response of a noncausal system depends only on past inputs.
17. The discrete signals are continuous function of an independent variable.
18. In digital signal the magnitudes of the signal are unquantized.
19. A discrete time signal x(n) is defined for noninteger values of n.
20. A discrete time impulse signal has a nonzero sample only for one value of n.
21. When we multiply a discrete time signal by unit step signal, the signal is converted to one-sided signal.
22. Shifting a signal to left is called delay and shifting to right is called advance.
23. Any discrete time signal can be expressed as a summation of impulses.
24. Periodic signals are power signals.
25. The output of a system for impulse input is called impulse response.
26. A system can be realized in real time only if it is noncausal and stable.
27. Dynamic systems does not require memory but static systems require memory.
28. A system is time invariant if the response to a shifted version of the input is identical to a shifted version of the response based on the unshifted input.
29. An LTI system is unstable if the impulse response is absolutely summable.
30. A system whose output depends only on the present and past input is called a recursive system.

Answers

1. True	8. False	15. False	22. False	29. False
2. True	9. True	16. False	23. True	30. False
3. False	10. True	17. False	24. True	
4. True	11. True	18. False	25. True	
5. False	12. True	19. False	26. False	
6. False	13. True	20. True	27. False	
7. False	14. True	21. True	28. True	

III. Choose the right answer for the following questions.

1. *The unit impulse is defined as,*

   a) $\delta(t) = \infty\,;\, t = 0$

   b) $\delta(t) = \infty;\, t = 0$
   $= 0\,;\, t \neq 0$

   c) $\delta(t) = \infty\,;\, t = 0$
   and $\int_{-\infty}^{+\infty} \delta(t)\, dt = A$

   d) $\delta(t) = \infty\,;\, t = 0$
   $= 0\,;\, t \neq 0$
   and $\int_{-\infty}^{+\infty} \delta(t)\, dt = 1$

2. *Which of the following is a periodic signal?*

   a) $x(t) = A\, u(t)$ b) $x(t) = A\, e^{-jbt}$ c) $x(t) = A\, e^{bt}$ d) $x(t) = A\, t$

3. Which of the following is a nonperiodic signal?

a) $x(t) = A e^{-j\sqrt{bt}}$ **b)** $x(t) = A e^{-j\frac{\pi t}{b}}$ **c)** $x(t) = A e^{bt}$ **d)** $x(t) = A e^{-jb\pi t}$

4. Which of the following statements is false?

i) the product of two odd signals is an odd signal.
ii) the product of two even signals is an even signal.
iii) the product of even and odd signals is an even signal.
iv) the product of even and odd signals is an odd signal.

a) i) and iii)   **b)** i) only   **c)** iii) only   **d)** iv) only

5. Which of the following is a purely even signal?

a) $x(t) = e^{at}$ **b)** $x(t) = \cos\Omega_0 t$ **c)** $x(t) = e^{j\Omega_0 t}$ **d)** $x(t) = u(t)$

6. Which of the following is a purely odd signal?

a) $x(t) = e^{jbt}$ **b)** $x(t) = Ae^{bt}$ **c)** $x(t) = A \sin n\Omega_0 t$ **d)** $x(t) = t^2$

7. Which of the following is an energy signal?

a) $x(t) = Ae^{j\Omega_0 t}$ **b)** $x(t) = A \sin \Omega_0 t$ **c)** $x(t) = B \cos \Omega_0 t$ **d)** $x(t) = e^{-at}u(t)$

8. Which of the following statements are true?

i) the periodic signals are power signals.   ii) the nonperiodic signals are energy signals.
iii) for energy signals, the power is zero.   iv) for power signals, the energy is zero.

a) i only   **b)** i) and ii) only   **c)** i), ii) and iii) only   **d)** all of the above

9. Which of the following signal is a causal signal?

a) $x(t) = A$ **b)** $x(t) = t$ **c)** $x(t) = u(t)$ **d)** $x(t) = e^{-at}$

10. Which of the following represents the pulse signal shown in fig 10.?

a) u (t + 0.5) – u (t – 0.5)
b) u (t – 0.5) + u (t + 0.5)
c) u (t – 0.5) – u (t + 0.5)
d) u (t + 0.5) + u (t – 0.5)

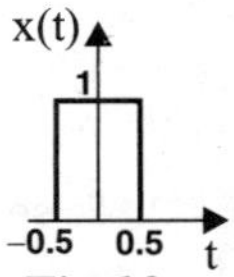

Fig 10.

11. The differentiation of a unit step signal is,

a) ramp signal   **b)** impulse signal   **c)** exponential signal   **d)** parabolic signal

12. The integral of a unit impulse signal is,

a) infinity   **b)** zero   **c)** one   **d)** constant

13. If x(t) is a continuous time signal and δ(t) is a unit impulse signal then, $\int_{-\infty}^{+\infty} x(t)\delta(t - t_0)\, dt$ is equal to,

a) $x(t)$ **b)** $\delta(t)$ **c)** $x(t_0)$ **d)** $\delta(t_0)$

14. If x(t) is a continuous time signal and δ(t) is unit impulse signal then,

$\int_{-\infty}^{+\infty} x(\tau)\delta(\tau - t)\, d\tau$ is equal to,

a) $x(t)$ **b)** $\delta(t)$ **c)** $x(t)\delta(t)$ **d)** $\delta(t - \tau)$

15. Which of the following is not a static system?

   a) $y(t) = x(t^2)$ **b)** $y(t) = x^2(t)$ **c)** $y(t) = t\, x(t)$ **d)** $y(t) = e^{x(t)}$

16. Which of the following is a time invariant system?

   a) $y(t) = t\, x(t)$ **b)** $y(t) = e^{x(t)}$ **c)** $y(t) = x(-t)$ **d)** $y(t) = x(t^2)$

17. Which of the following is a linear system?

   a) $y(t) = x(t^2)$ **b)** $y(t) = x^2(t)$ **c)** $y(t) = x(t)+C$ **d)** $y(t) = e^{x(t)}$

18. Which of the following is a nonlinear system?

   a) $y(t) = t\, x(t)$ **b)** $y(t) = \int x(t)\, dt$ **c)** $y(t) = \dfrac{dx(t)}{dt}$ **d)** $y(t) = \sqrt{x(t)}$

19. Which of the following is a causal system?

   a) $y(t) = x(t^2)$ **b)** $y(t) = x^2(t)$ **c)** $y(t) = x(-t)$ **d)** $y(t) = x(2t)$

20. Which of the following is a noncausal system?

   a) $y(t) = \dfrac{dx(t)}{dt}$ **b)** $y(t) = \int_{0}^{t} x(\lambda)\, d\lambda$ **c)** $y(t) = e^{x(t)}$ **d)** $y(t) = x(t^2)$

21. Which of the following statements are true ?

   i)   an LTI system is always stable.

   ii) an LTI system is stable only if the integral of its impulse response is finite.

   iii) in a system, if the input is bounded then the output is always bounded.

   iv) in a system, even if the input is unbounded the output can be bounded

   a) ii) only   **b)** ii) and iv) only   **c)** iii) only   **d)** i) and iv) only

22. Which of the following is a stable system?

   a) $y(t) = t\, x(t)$ **b)** $y(t) = t^2\, x(t)$ **c)** $y(t) = e^{t}\, x(t)$ **d)** $y(t) = e^{-t}\, x(t)$

23. $x(n) = \dfrac{x(n-1)}{4}$ *with initial condition x(0) = –1, gives the sequence,*

   a) $x(n) = \left(\dfrac{1}{4}\right)^{n}$ **b)** $x(n) = -\left(\dfrac{1}{4}\right)^{n}$ **c)** $x(n) = \left(\dfrac{1}{4}\right)^{-n}$ **d)** $x(n) = \left(\dfrac{-1}{4}\right)^{-n}$

24. The process of conversion of continuous time signal into discrete time signal is known as,

   a) aliasing   **b)** sampling   **c)** convolution   **d)** None

25. Which of the following signal is the example for deterministic signal?

   a) step   **b)** ramp   **c)** exponential   **d)** all of the above

26. For energy signals, the energy will be finite and the average power will be,

   a) infinite   **b)** finite   **c)** zero   **d)** not defined

27. In a signal x(n), if 'n' is replaced by $\frac{n}{I}$, then it is called,

   a) upsampling   **b)** folded version   **c)** downsampling   **d)** shifted version

28. *The unit step signal u(n) delayed by 3 units of time is denoted as,*

    a) $u(n+3) = 1$; $n \geq 3$
    $= 0$; $n < 3$

    b) $u(3-n) = 1$; $n \geq 3$
    $= 0$; $n < 3$

    c) $u(n-3) = 1$; $n \geq 3$
    $= 0$; $n < 3$

    d) $u(3n) = 1$; $n > 3$
    $= 0$; $n < 3$

29. *The discrete time system, y(n) = x(n–3) – 4x(n–10) is a,*

    a) dynamic system   **b)** memoryless system   **c)** time varying system **d)** none

30. *An LTI discrete time system is causal if and only if,*

    a) $h(n) \neq 0$ for $n < 0$ **b)** $h(n) = 0$ for $n < 0$ **c)** $h(n) \neq \infty$ for $n < 0$ **d)** $h(n) \neq 0$ for $n > 0$

31. *Which of the following system is causal?*

    a) $h(n) = n\left(\frac{1}{2}\right)^n u(n+1)$

    b) $y(n) = x^2(n) - x(n+1)$

    c) $y(n) = x(-n) + x(2n-1)$

    d) $h(n) = n\left(\frac{1}{2}\right)^n u(n)$

32. *An LTI system is stable, if the impulse response is,*

    a) $\sum_{n=-\infty}^{\infty} |h(n)| = 0$ **b)** $\sum_{n=-\infty}^{\infty} |h(n)| < \infty$ **c)** $\sum_{n=-\infty}^{\infty} |h(n)| \neq 0$ **d)** either a or b

33. *The system y(n) = sin[x(n)] is,*

    a) stable   **b)** BIBO stable   **c)** unstable   **d)** none

34. *For a system, y(n) = nx(n), the inverse system will be,*

    a) $y\left(\frac{1}{n}\right)$ **b)** $\frac{1}{n} y(n)$ **c)** $ny(n)$ **d)** $n^{-1}y(n)$

35. *For a system y(n) = x(n–3) the impulse response of the system and the inverse system will be* _____ *and* _____ *respectively.*

    a) $h(n) = \delta(n+3)$, $x(n) = y(n-3)$

    b) $h(n) = \delta(3n)$, $x(n) = y\left(\frac{n}{3}\right)$

    c) $h(n) = \delta(n-3)$, $x(n) = y(n+3)$

    d) $h(n) = \delta(n+3)$, $x(n) = y(3n)$

Answers

1. d	6. c	11. b	16. b	21. b	26. c	31. d
2. b	7. d	12. c	17. a	22. d	27. a	32. d
3. c	8. c	13. c	18. d	23. b	28. c	33. a
4. a	9. c	14. a	19. b	24. b	29. a	34. b
5. b	10. a	15. a	20. d	25. d	30. b	35. c

IV. Answer the following questions.

1. Define and sketch continuous time impulse and unit impulse signal.
2. Define and sketch continuous time step and unit step signal.
3. Define and sketch continuous time ramp signal.
4. Define and sketch continuous time sinc signal.
5. Define and sketch continuous time signum signal.

6. What are the various ways of classifying the continuous time signals?

7. Define deterministic and nondeterministic continuous time signals. Give examples.

8. Define periodic and nonperiodic continuous time signals. Give examples.

9. Define even and odd continuous time signals. Give examples.

10. Prove that the even part of a continuous time signal is given by $x_e(t) = \frac{1}{2}[x(t) + x(-t)]$ and the odd part of a signal is given by $x_0(t) = \frac{1}{2}[x(t) - x(-t)]$.

11. Define energy of a continuous time signal. Give examples.

12. Define power of a continuous time signal. Give examples.

13. Prove that power of a continuous time energy signal is zero.

14. Prove that energy of a continuous time power signal is infinite.

15. Define causal and noncausal continuous time signals. Give examples.

16. List the properties of continuous time impulse signal.

17. Write any two properties of continuous time impulse signal and prove.

18. Show that any continuous time signal x(t) can be expressed as an integral of impulses.

19. Define static and dynamic continuous time systems. Give examples.

20. Explain the time invariant property of a continuous time system.

21. Define linear continuous time systems.

22. Define causal and noncausal continuous time systems. Give examples.

23. Define stable and unstable continuous time systems.

24. Define discrete and digital signal.

25. Explain briefly, the various methods of representing discrete time signal with examples.

26. Define the discrete time impulse and unit step signal.

27. Show that any discrete time signal x(n) can be expressed as a summation of impulses.

28. How will you classify the discrete time signals?

29. What are energy and power signals?

30. When a discrete time signal is called periodic?

31. What is discrete time system?

32. Write the difference equation governing the Nth order LTI discrete time system.

33. List the various methods of classifying discrete time systems.

34. Define time invariant discrete time system.

35. What is linear and nonlinear discrete time systems?

36. What is the importance of causality?

37. What is BIBO stability of discrete time system ? What is the condition to be satisfied for stability of discrete time LTI system ?

38. What are FIR and IIR discrete time systems?

39. What are recursive and nonrecursive discrete time systems? Give examples.

40. What is inverse system? What is its importance?

V. Solve the following problems.

E1.1 Verify whether the following signals are periodic. If periodic find the fundamental period.

    a) $x(t) = 4 \sin 7t$ **b)** $x(t) = 2e^{0.7t}$ **c)** $x(t) = 3e^{-j0.1\pi t}$ **d)** $x(t) = 9 \sin\left(6t + \frac{\pi}{3}\right)$ **e)** $x(t) = \sin^2\left(3t - \frac{\pi}{5}\right)$

E1.2 Determine the periodicity of the following continuous time signals.

    a) $x(t) = 4 \sin \frac{2\pi t}{7} + 5 \sin \frac{2\pi t}{9}$ **b)** $x(t) = 0.5e^{-j\frac{2\pi t}{5}} + 0.7 \cos 8\pi t$ **c)** $x(t) = 6 \sin 5t + 3 \cos 4\pi t$

E1.3 Determine the even part and odd part of the following continuous time signals.

    a) $x(t) = e^{j2t}$ **b)** $x(t) = 4 + e^{3t}$ **c)** $x(t) = t + 3t^2 + \cos^2 t$ **d)** $x(t) = \sin^2 t + e^{j5\pi t}$

E1.4 Determine the energy and power of the following continuous time signals.

    a) $x(t) = 0.9\, e^{-3t}\, u(t)$ **b)** $x(t) = 3\, e^{-j0.5\pi t}$ **c)** $x(t) = 1.2 \sin 7\Omega_0 t$ **d)** $x(t) = t\, u(t)$

E1.5 Sketch the even and odd parts of the signals.

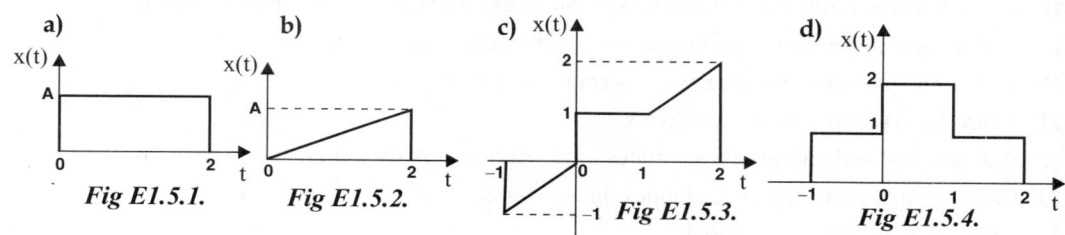

    a)   **b)**   **c)**   **d)**

     Fig E1.5.1.   *Fig E1.5.2.*   *Fig E1.5.3.*   *Fig E1.5.4.*

E1.6 A continuous time signal is shown in Fig E1.6
    Determine the following versions of the signal.

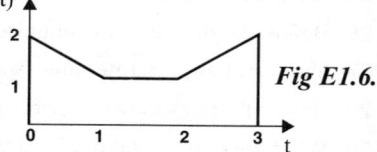

    a) $x(-t)$ **b)** $-x(t)$ **c)** $x(t + 3)$ *Fig E1.6.*

    d) $x(-t + 3)$ **e)** $x(t - 3)$ **f)** $x(-t - 3)$

E1.7 Evaluate the following integrals.

    a) $\displaystyle\int_{-\infty}^{+\infty} e^{-0.5t}\, \delta(t-3)\, dt$ **b)** $\displaystyle\int_{-\infty}^{+\infty} e^{j0.1t}\, \delta(t-10)\, dt$ **c)** $\displaystyle\int_{-\infty}^{+\infty} t^3\, \delta(t+1)\, dt$

E1.8 Verify whether the following systems are time invariant or time variant.

    a) $y(t) = e^t\, x(t)$ **b)** $y(t) = \cos t\, x(t)$ **c)** $y(t) = x(-t + 2)$

    d) $y(t) = 3\, x(t) + 5$ **e)** $y(t) = 3\, \dfrac{dx(t)}{dt} + 7\, x(t)$ **f)** $y(t) = e^{x(t)} + \displaystyle\int x(t)\, dt$

E1.9 Verify whether the following systems are linear or nonlinear.

    a) $y(t) = t^2 x(t)$ **b)** $y(t) = \sqrt{x(t)}$ **c)** $y(t) = e^t\, x(t)$

    d) $y(t) = \displaystyle\int x(t)\, dt$ **e)** $y(t) = t + x(t)$ **f)** $y(t) = \cos t\, x(t)$

E1.10 Determine the linearity of the LTI systems governed by the following differential equations.

a) $\dfrac{d^2 y(t)}{dt^2} + 0.3 \dfrac{dy(t)}{dt} + 0.5y(t) = 2\,x(t)$ b) $y(t) = 4\dfrac{d^2 x(t)}{dt^2} + 2\dfrac{dx(t)}{dt} + 5\,x(t)$

E1.11 Verify whether the following systems are causal or noncausal.

a) $y(t) = e^t\,x(t)$ b) $y(t) = (t-2)\,u(t+2)$ c) $y(t) = (t+2)\,x(t)$

d) $y(t) = e^{x(t)} + \dfrac{dx(t)}{dt}$ e) $y(t) = \cos t\,x(t)$ f) $y(t) = \sin t\,x(t+2)$

E1.12 Verify whether the following systems are stable or unstable.

a) $y(t) = e^t\,x(t)$ b) $y(t) = e^{-t}\,x(t)$ c) $y(t) = t^2\,x(t)$

d) $y(t) = \sin t\,x(t)$ e) $y(t) = e^{-t}\sin t\,x(t)$ f) $y(t) = x(t+2)$

E1.13 Verify the stability of LTI systems whose impulse response are given below:

a) $h(t) = t^2\,u(t)$ b) $h(t) = e^{-7t}\,u(t)$ c) $h(t) = e^{5t}\,u(t)$

d) $h(t) = t \sin t\,u(t)$ e) $h(t) = (A + Be^{-Ct})\,u(t)$ f) $h(t) = 2e^{-3t}\cos t\,u(t)$

E1.14 Determine whether the following signals are periodic or not. If periodic, find the fundamental period.

a) $x(n) = \sin\left(\dfrac{5\pi}{8}n + 6\right)$ b) $x(n) = \sin\left(\dfrac{7n}{3} + \pi\right)$ c) $x(n) = \cos\left(\dfrac{4\pi n}{12}\right)$

d) $x(n) = \cos\left(\dfrac{\pi}{32}n^2\right)$ e) $x(n) = e^{j9n}$ f) $x(n) = 4\sin\dfrac{3\pi n}{2} + 5\cos\dfrac{3\pi n}{4}$

E1.15 Determine the even and odd parts of the signals.

a) $x(n) = \dfrac{1}{a^{2n}}$ b) $x(n) = 8e^{-j\frac{\pi}{6}n}$ c) $x(n) = \{6, 4, 2, 2\}$
                                                                                             ↑

E1.16 Determine whether the following signals are energy or power signals.

a) $x(n) = \left(\dfrac{5}{9}\right)^n u(n)$ b) $x(n) = \cos\left(\dfrac{3\pi}{4}n\right)$ c) $x(n) = u(2n)$ d) $x(n) = 2\,u(3-n)$

E1.17 Test the following systems for time invariance.

a) $y(n) = x(n+1) + x(n+2)$ b) $y(n) = na^{x(n)}$

c) $y(n) = x^2(n+2) + C$ d) $y(n) = (n-1)\,x^2(n) + C$

E1.18 Test the following systems for linearity.

a) $y(n) = x^2(n) + x^3(n-1)$ b) $y(n) = bx(n+2) + ne^{x(n)}$ c) $y(n) = a\sqrt{x(n)} + bx(n)$

d) $y(n) = \sqrt{x(n)} + \dfrac{1}{\sqrt{x(n)}}$ e) $y(n) = \displaystyle\sum_{m=-1}^{N} b_m\,x(n+m) + \sum_{m=0}^{M} c_m\,y(n+m)$

E1.19 **Test the causality of the following systems.**

    a) $y(n) = x(n) - x(-n - 2) + x(n - 1)$ **b)** $y(n) = a\,x(2n) + x(n^2)$

    c) $y(n) = \displaystyle\sum_{m=-1}^{n} x(m) + \sum_{m=-\infty}^{n} x(2m)$ **d)** $y(n) = (0.3)^n u(n+2)$ **e)** $y(n) = \displaystyle\sum_{k=-4}^{4} x(n-k)$

E1.20 **Test the stability of the following discrete time systems.**

    a) $y(n) = x^2(n) + x(n+1)$ **b)** $y(n) = nx(n-1)$ **c)** $h(n) = (0.4)^n\,u(n+3)$

    d) $h(n) = 8^n\,u(4-n)$ **e)** $y(n) = x(n-3)$

E1.21 **Determine the range of values of 'a' and 'b' for the stability of an LTI system with impulse response,**

$$h(n) = (-4a)^n \;\;;\; n \ge 0$$
$$= \;\; 2b^{-n} \;\;;\; n < 0$$

Answers

E1.1   **a)** periodic, $T = \dfrac{2\pi}{7}$ **b)** nonperiodic   **c)** periodic, $T = 20$

          d) periodic, $T = \dfrac{\pi}{3}$ **e)** periodic, $T = \dfrac{\pi}{3}$

E1.2   **a)** periodic, $T = 63$ **b)** periodic, $T = 5$ **c)** nonperiodic

E1.3   **a)** $x_e(t) = \cos 2t$; $x_o(t) = j\sin 2t$ **b)** $x_e(t) = 4 + \dfrac{1}{2}(e^{3t} + e^{-3t})$; $x_o(t) = \dfrac{1}{2}(e^{3t} - e^{-3t})$

          c) $x_e(t) = 3t^2 + \cos^2 t$; $x_o(t) = t$ **d)** $x_e(t) = \sin^2 t + \cos 5\pi t$; $x_o(t) = j\sin 5\pi t$

E1.4   **a)** $E = 0.135$ *joules,* $P = 0$, Energy Signal   **b)** $E = \infty$, $P = 9$ *watts,* Power Signal

          c) $E = \infty$, $P = 0.72$ *watts,* Power Signal   **d)** $E = \infty$, $P = \infty$, Neither Energy nor Power Signal

E1.5   **a)**

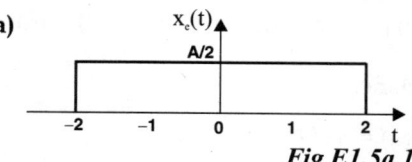

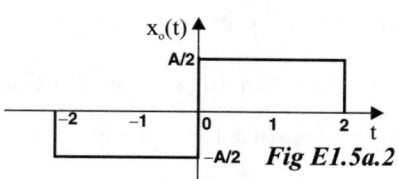

Fig E1.5a.1   *Fig E1.5a.2*

    b)

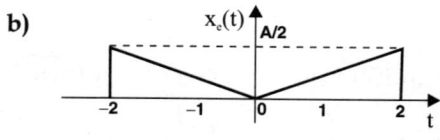

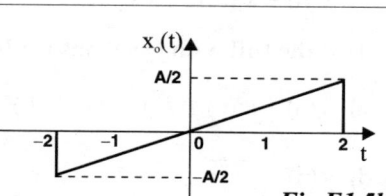

Fig E1.5b.1   *Fig E1.5b.2*

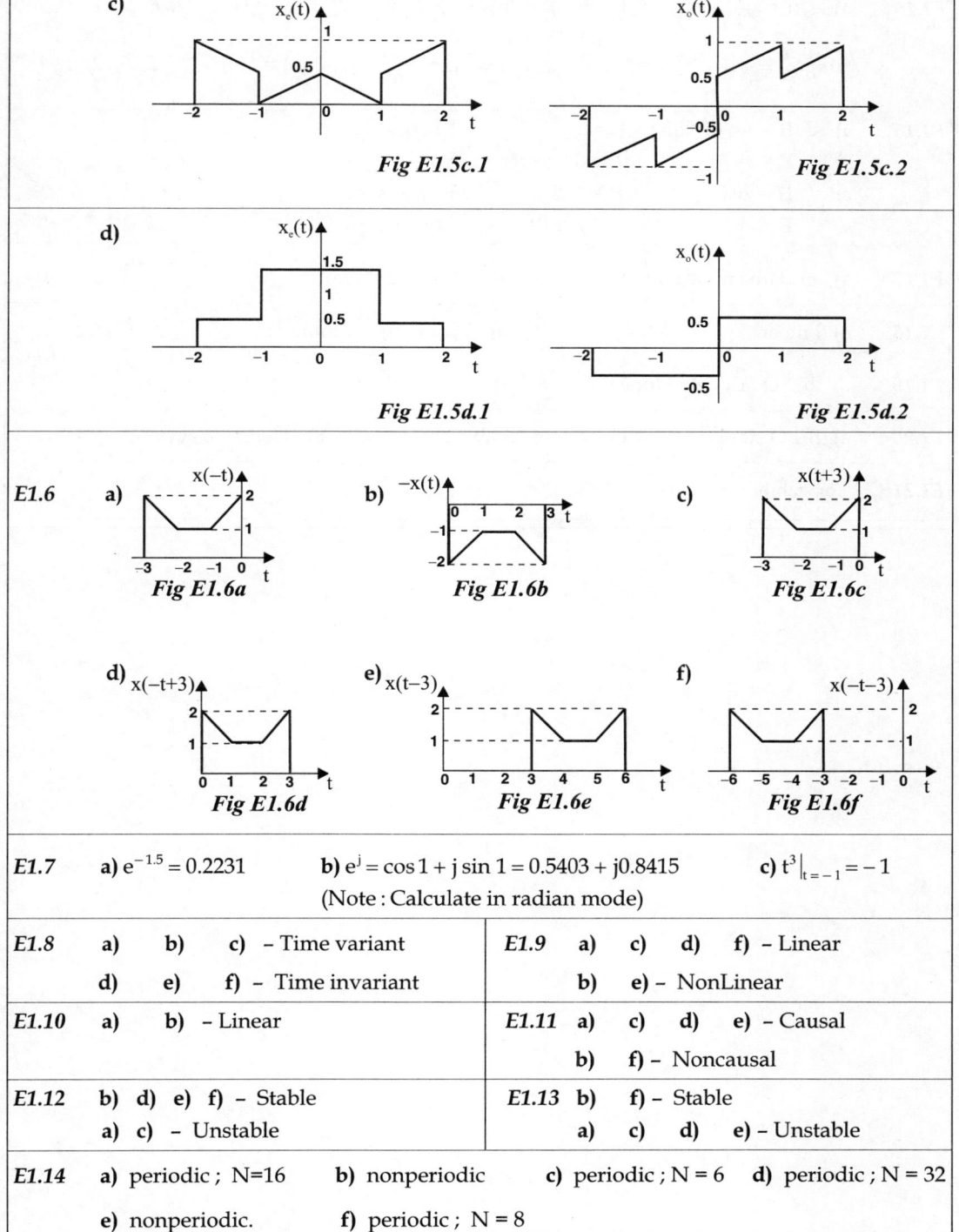

c)

Fig E1.5c.1

Fig E1.5c.2

d)

Fig E1.5d.1

Fig E1.5d.2

E1.6   a)   Fig E1.6a

b)   Fig E1.6b

c)   Fig E1.6c

d)   Fig E1.6d

e)   Fig E1.6e

f)   Fig E1.6f

E1.7	a) $e^{-1.5} = 0.2231$	b) $e^{j} = \cos 1 + j\sin 1 = 0.5403 + j0.8415$	c) $t^{3}\big|_{t=-1} = -1$

(Note : Calculate in radian mode)

E1.8	a) b) c) – Time variant	E1.9	a) c) d) f) – Linear
	d) e) f) – Time invariant		b) e) – NonLinear
E1.10	a) b) – Linear	E1.11	a) c) d) e) – Causal
			b) f) – Noncausal
E1.12	b) d) e) f) – Stable	E1.13	b) f) – Stable
	a) c) – Unstable		a) c) d) e) – Unstable

E1.14	a) periodic ; N=16 b) nonperiodic c) periodic ; N = 6 d) periodic ; N = 32
	e) nonperiodic. f) periodic ; N = 8

E1.15	a) $x_e(n) = \frac{1}{2}[a^{-2n} + a^{2n}]$	b) $x_e(n) = 8\cos\frac{\pi}{6}n$	c) $x_e(n) = \{1,\ 1,\ 2,\ \overset{\uparrow}{6},\ 2,\ 1,\ 1\}$
	$x_o(n) = \frac{1}{2}[a^{-2n} - a^{2n}]$	$x_o(n) = -j8\sin\frac{\pi}{6}n$	$x_o(n) = \{-1,\ -1,\ -2,\ \overset{\uparrow}{0},\ 2,\ 1,\ 1\}$

E1.16	a)	$E = 1.435\ joules$; $P = 0$	; Energy signal.
	b)	$E = \infty$; $P = 0.5\ watts$;	Power signal.
	c)	$E = \infty$; $P = 0.25\ watts$;	Power signal.
	d)	$E = \infty$; $P = 2\ watts$;	Power signal.

E1.17	a) c) Time invariant	b) d) Time variant

E1.18	e) Linear	a) b) c) d) Nonlinear

E1.19	a) b) c) d) e) Noncausal

E1.20	a) BIBO stable	c) d) e) Stable system	b) Unstable system

E1.21	For stability, $0 < |a| < \frac{1}{4}$ and $0 < |b| < \frac{1}{2}$

CHAPTER 2

Analysis of Continuous Time Signals

2.1 Fourier Series Analysis of Continuous Time Signals

The French mathematician Jean Baptiste Joseph Fourier (J.B.J. Fourier) has shown that any periodic nonsinusoidal signal can be expressed as a linear weighted sum of harmonically related sinusoidal signals. This leads to a method called *Fourier series* in which a periodic signal is represented as a function of frequency.

The Fourier representation of periodic signals has been extended to nonperiodic signals by letting the fundamental period T tend to infinity, and this Fourier method of representing nonperiodic signals as a function of frequency is called *Fourier transform*. The Fourier represention of signals is also known as frequency domain representation. In general, the Fourier series representation can be obtained only for periodic signals, but the Fourier transform technique can be applied to both periodic and nonperiodic signals to obtain the frequency domain representation of the signals.

The Fourier representation of signals can be used to perform frequency domain analysis of signals, in which we can study the various frequency components present in the signal and the magnitude and phase of various frequency components. The graphical plots of magnitude and phase as a function of frequency are also drawn. The plot of magnitude versus frequency is called *magnitude spectrum* and the plot of phase versus frequency is called *phase spectrum*. In general, these plots are called *frequency spectrum.*

2.1.1 Trigonometric Form of Fourier Series   (AU Dec'13, 2 Marks)

The *trigonometric form of Fourier series* of a periodic signal, x(t), with period T is defined as,

$$\boxed{x(t) = \frac{1}{2}a_0 + \sum_{n=1}^{\infty} a_n \cos n\Omega_0 t + \sum_{n=1}^{\infty} b_n \sin n\Omega_0 t}$$

.....(2.1)

$$\therefore \ x(t) = \frac{1}{2}a_0 + a_1 \cos \Omega_0 t + a_2 \cos 2\Omega_0 t + a_3 \cos 3\Omega_0 t + ...$$
$$+ b_1 \sin \Omega_0 t + b_2 \sin 2\Omega_0 t + b_3 \sin 3\Omega_0 t + ...$$

where, $\Omega_0 = 2\pi F_0 = \dfrac{2\pi}{T}$ = Fundamental frequency in *rad/sec*

$ F_0$ = Fundamental frequency in *cycles/sec* or Hz

$ n$ = Harmonic order

$ n\Omega_0$ = Harmonic frequencies

$ a_0, a_n, b_n$ = Fourier coefficients of trigonometric form of Fourier series

> **Note:** 1. Here $a_0/2$ is the value of constant component of the signal x(t).
> 2. The Fourier coefficient a_n and b_n are maximum amplitudes of nth harmonic components.

The Fourier coefficients can be evaluated using the following formulae.

$$a_0 = \frac{2}{T} \int_{-T/2}^{+T/2} x(t)\, dt \qquad \text{(or)} \qquad a_0 = \frac{2}{T} \int_{0}^{T} x(t)\, dt \qquad .....(2.2)$$

$$a_n = \frac{2}{T} \int_{-T/2}^{+T/2} x(t) \cos n\Omega_0 t \, dt \quad \text{(or)} \quad a_n = \frac{2}{T} \int_{0}^{T} x(t) \cos n\Omega_0 t \, dt \qquad .....(2.3)$$

$$b_n = \frac{2}{T} \int_{-T/2}^{+T/2} x(t) \sin n\Omega_0 t \, dt \quad \text{(or)} \quad b_n = \frac{2}{T} \int_{0}^{T} x(t) \sin n\Omega_0 t \, dt \qquad .....(2.4)$$

In the above formulae, the limits of integration are either $-T/2$ to $+T/2$ or 0 to T. In general, the limit of integration is one period of the signal and so the limits can be from t_0 to $t_0 + T$, where t_0 is any time instant. The proof of equations for a_0, a_n, b_n are left as exercise to readers.

2.1.2 Conditions for Existence of Fourier Series
(AU Dec'15, Dec'14, Jun'14, 2 Marks)
(AU Dec'13, 2 Marks & Dec'12, 4 Marks)

The Fourier series exists only if the following Dirichlet's conditions are satisfied.

1. The signal $x(t)$ is well defined and single valued, except possibly at a finite number of points.

2. The signal $x(t)$ must possess only a finite number of discontinuities in the period T.

3. The signal must have a finite number of positive and negative maxima in the period T.

Note: *1. The value of signal x(t) at $t = t_0$ is $x(t_0)$ if $t = t_0$ is a point of continuity.*

         2. The value of signal x(t) at $t = t_0$ is $\dfrac{x(t_0^+) + x(t_0^-)}{2}$ if $t = t_0$ is a point of discontinuity.

Table 2.1: Some of the Waveforms that Violate Dirichlet's Conditions

Waveform	Comment
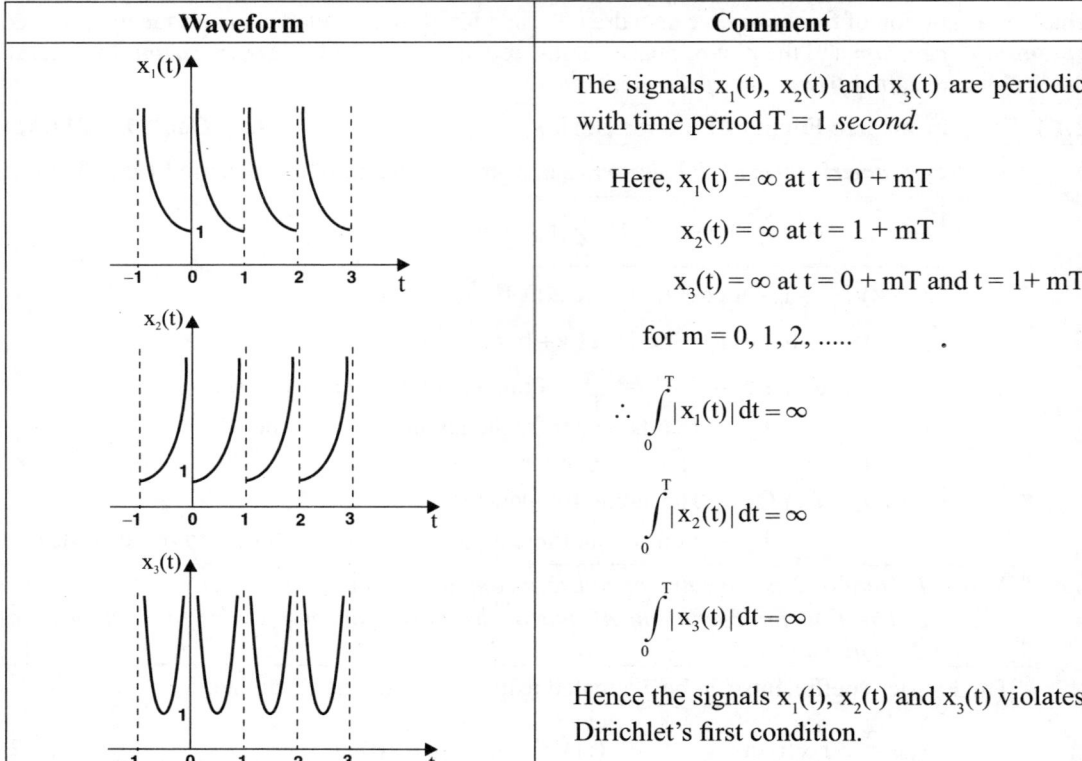	The signals $x_1(t)$, $x_2(t)$ and $x_3(t)$ are periodic with time period $T = 1$ *second*. Here, $x_1(t) = \infty$ at $t = 0 + mT$ $x_2(t) = \infty$ at $t = 1 + mT$ $x_3(t) = \infty$ at $t = 0 + mT$ and $t = 1 + mT$ for $m = 0, 1, 2, ...$ $\therefore \int_{0}^{T} |x_1(t)| \, dt = \infty$ $\int_{0}^{T} |x_2(t)| \, dt = \infty$ $\int_{0}^{T} |x_3(t)| \, dt = \infty$ Hence the signals $x_1(t)$, $x_2(t)$ and $x_3(t)$ violates Dirichlet's first condition.

Table 2.1 : Continued...

Waveform	Comment
$x_4(t)$ waveform	The signal $x_4(t)$ is periodic with time period $T = 2$ *seconds.*
	The signal $x_4(t)$ has infinite number of discontinuties in the period $T = 2$ *seconds.*
	Hence the signal $x_4(t)$ violates the second Dirichlet's condition.
$x_5(t)$ waveform	The signal $x_5(t)$ is periodic with time period $T = 2$ *seconds.*
	The signal $x_5(t)$ has infinite number of positive maxima and negative maxima in the period $T = 2$ *seconds.*
	Hence the signal $x_5(t)$ violates the third Dirichlet's condition.

2.1.3 Exponential Form of Fourier Series

The *exponential form of Fourier series* of a periodic signal $x(t)$ with period T is defined as,

$$x(t) = \sum_{n=-\infty}^{+\infty} c_n e^{jn\Omega_0 t} \qquad .....(2.5)$$

where, $\Omega_0 = 2\pi F_0 = \dfrac{2\pi}{T} =$ Fundamental frequency in *rad/sec*

$F_0 =$ Fundamental frequency in *cycles/sec* or Hz

$n =$ Harmonic frequencies

$\pm n\Omega_0 =$ Harmonic frequencies

$c_n =$ Fourier coefficients of exponential form of Fourier series.

The *Fourier coefficient c_n* can be evaluated using the following equation.

$$c_n = \frac{1}{T} \int_{-T/2}^{+T/2} x(t)\, e^{-jn\Omega_0 t}\, dt \qquad (or) \qquad c_n = \frac{1}{T} \int_{0}^{T} x(t)\, e^{-jn\Omega_0 t}\, dt \qquad .....(2.6)$$

In equation (2.6), the limits of integration are either **−T/2 to +T/2 or 0 to T.** In general, the limit of integration is one period of the signal and so the limits can be from $\mathbf{t_0}$ **to** $\mathbf{t_0 + T}$, where t_0 is any time instant. The proof of equation of c_n is left as exercise to readers.

2.1.4 Relation Between Fourier Coefficients of Trigonometric and Exponential Form

The relation between Fourier coefficients of trigonometric form and exponential form are given below.

$$c_0 = \frac{a_0}{2} \qquad\qquad\qquad\qquad\qquad\qquad\qquad\qquad\qquad\qquad .....(2.7)$$

$$c_n = \frac{1}{2}(a_n - jb_n) \qquad \text{for n} = 1,2,3,4,..... \qquad\qquad .....(2.8)$$

$$c_{-n} = \frac{1}{2}(a_n + jb_n) \quad \text{for} -n = -1,-2,-3,-4,..... \qquad .....(2.9)$$

$$\therefore |c_n| = \frac{1}{2}\sqrt{a_n^2 + b_n^2} \quad \text{for all values of n, except when n} = 0. \qquad .....(2.10)$$

Proof:

Consider the trigonometric form of Fourier series of x(t), [equation (2.1)].

$$x(t) = \frac{a_0}{2} + \sum_{n=1}^{\infty} a_n \cos n\Omega_0 t + \sum_{n=1}^{\infty} b_n \sin n\Omega_0 t$$

$$= \frac{a_0}{2} + \sum_{n=1}^{\infty} a_n \left[\frac{e^{jn\Omega_0 t} + e^{-jn\Omega_0 t}}{2} \right] + \sum_{n=1}^{\infty} b_n \left[\frac{e^{jn\Omega_0 t} - e^{-jn\Omega_0 t}}{2j} \right]$$

$$= \frac{a_0}{2} + \sum_{n=1}^{\infty} \left[\frac{a_n}{2} e^{jn\Omega_0 t} + \frac{a_n}{2} e^{-jn\Omega_0 t} - j\frac{b_n}{2} e^{jn\Omega_0 t} + j\frac{b_n}{2} e^{-jn\Omega_0 t} \right]$$

$$= \frac{a_0}{2} + \sum_{n=1}^{\infty} \left[\frac{a_n}{2} - j\frac{b_n}{2} \right] e^{jn\Omega_0 t} + \sum_{n=1}^{\infty} \left[\frac{a_n}{2} + j\frac{b_n}{2} \right] e^{-jn\Omega_0 t}$$

$$\boxed{\cos\theta = \frac{e^{j\theta} + e^{-j\theta}}{2}}$$
$$\boxed{\sin\theta = \frac{e^{j\theta} - e^{-j\theta}}{2j}}$$

$$\boxed{\frac{1}{j} = \frac{j}{j^2} = -j}$$

Let, $c_0 = \frac{a_0}{2}$; $c_n = \frac{a_n - jb_n}{2}$; $c_n^* = \frac{a_n + jb_n}{2}$

$$\therefore x(t) = c_0 + \sum_{n=1}^{\infty} c_n e^{jn\Omega_0 t} + \sum_{n=1}^{\infty} c_n^* e^{-jn\Omega_0 t}$$

$$= c_0 + \sum_{n=1}^{\infty} c_n e^{jn\Omega_0 t} + \sum_{n=-\infty}^{-1} c_{-n} e^{jn\Omega_0 t}$$

$$\boxed{c_{-n} = c_n^*}$$

$$= \sum_{n=-\infty}^{-1} c_{-n} e^{jn\Omega_0 t} + c_0 + \sum_{n=1}^{\infty} c_n e^{jn\Omega_0 t}$$

$$= \sum_{n=-\infty}^{+\infty} c_n e^{jn\Omega_0 t}$$

$$\boxed{\begin{array}{l} e^{jn\Omega_0 t} = 1 \\ \text{for n} = 0 \end{array}}$$

$$\therefore c_0 = \frac{a_0}{2}$$

$$c_n = \frac{1}{2}(a_n - jb_n) \qquad \text{for n} = 1,2,3,.....\infty$$

$$c_{-n} = \frac{1}{2}(a_n + jb_n) \qquad \text{for} -n = -1,-2,-3.......,-\infty$$

2.1.5 Properties of Fourier Series

The properties of exponential form of Fourier series coefficients are listed in Table 2.2. The proof of these properties are left as exercise to the readers.

Table 2.2: Properties of Exponential Form of Fourier Series Coefficients

Note : c_n and d_n are exponential form of Fourier series coefficients of x(t) and y(t) respectively.

Property	Continuous Time Periodic Signal	Fourier Series Coefficient
Linearity	$A\,x(t) + B\,y(t)$	$A\,c_n + B\,d_n$
Time shifting	$x(t - t_0)$	$c_n e^{-jn\Omega_0 t_0}$
Frequency shifting	$e^{-jm\Omega_0 t}\,x(t)$	c_{n-m}
Conjugation	$x^*(t)$	c^*_{-n}
Time reversal	$x(-t)$	c_{-n}
Time scaling	$x(\alpha t)\;;\;\alpha > 0$ [x(t) is periodic with period T/α]	c_n (No change in Fourier coefficient)
Multiplication	$x(t)\,y(t)$	$\displaystyle\sum_{m=-\infty}^{+\infty} c_m\,d_{n-m}$
Differentiation	$\dfrac{d}{dt}x(t)$	$jn\Omega_0 c_n$
Integration	$\displaystyle\int_{-\infty}^{t} x(t)\,dt$ (Finite valued and periodic only if $a_0 = 0$)	$\dfrac{1}{jn\Omega_0}\,c_n$
Periodic convolution	$\displaystyle\int_{T} x(\tau)\,y(t-\tau)\,d\tau$	$T\,c_n\,d_n$
Symmetry of real signals	x(t) is real	$c_n = c^*_{-n}$ $\lvert c_n\rvert = \lvert c_{-n}\rvert\;;\;\angle c_n = -\angle c_{-n}$ $\mathrm{Re}\{c_n\} = \mathrm{Re}\{-c_n\}$ $\mathrm{Im}\{c_n\} = -\mathrm{Im}\{c_{-n}\}$
Real and even	x(t) is real and even	c_n are real and even
Real and odd	x(t) is real and odd	c_n are imaginary and odd
Parseval's relation	Average power, P of x(t) is defined as $P = \dfrac{1}{T}\displaystyle\int_{T} \lvert x(t)\rvert^2\,dt$	The average power, P in terms of Fourier series coefficients is $P = \displaystyle\sum_{n=-\infty}^{+\infty}\lvert c_n\rvert^2$
	$\dfrac{1}{T}\displaystyle\int_{T} \lvert x(t)\rvert^2\,dt = \displaystyle\sum_{n=-\infty}^{+\infty}\lvert c_n\rvert^2$	

Note: 1. *The term $\lvert c_n\rvert^2$ respresents the power in n^{th} harmonic component of x(t). The total average power in a periodic signal is equal to the sum of power in all of its harmonics.*

2. *The term $\lvert c_n\rvert^2$ for n = 0, 1, 2, ... is the distribution of power as a function of frequency and so it is called **power density spectrum** or **power spectral density** of the periodic signal.*

2.2 Fourier Coefficients of Signals with Symmetry

2.2.1 Even Symmetry

A signal, x(t) is called *even signal*, if the signal satisfies the condition *x(–t) = x(t).*

The waveform of an even periodic signal exhibits symmetry with respect to t = 0 (i.e. with respect to vertical axis) and so the symmetry of a waveform with respect to t = 0 or vertical axis is called *even symmetry.*

In order to determine the even symmetry of a waveform, fold the waveform with respect to vertical axis. After folding, if the waveshape remains same then it is said to have even symmetry.

For even signals the Fourier coefficient a_0 is optional, a_n exists and b_n are zero. The Fourier coefficient a_0 is zero if the average value of one period is equal to zero. For an even signal the Fourier coefficients are given by,

$$a_0 = \frac{4}{T} \int_0^{T/2} x(t)\, dt \qquad \text{(or)} \qquad a_0 = \frac{4}{T} \int_{-T/4}^{+T/4} x(t)\, dt$$

$$a_n = \frac{4}{T} \int_0^{T/2} x(t) \cos n\Omega_0 t\, dt \qquad \text{(or)} \qquad a_n = \frac{4}{T} \int_{-T/4}^{+T/4} x(t) \cos n\Omega_0 t\, dt \; ; \quad b_n = 0$$

Proof:

Consider the equation for a_0, [equation (2.2)].

$$a_0 = \frac{2}{T} \int_{-T/2}^{T/2} x(t)\, dt = \frac{2}{T} \int_{-T/2}^{0} x(t)\, dt + \frac{2}{T} \int_{0}^{T/2} x(t)\, dt \qquad \boxed{\text{Dividing the integral into two parts.}}$$

$$= \frac{2}{T} \int_{T/2}^{0} x(-\tau)(-d\tau) + \frac{2}{T} \int_{0}^{T/2} x(t)\, dt \qquad \boxed{\begin{array}{l}\text{Change of integral index,}\\ \text{Let, } t = -\tau \; ; \; \therefore \; dt = -d\tau \\ \text{When } t = 0, \; \tau = -t = 0 \\ \text{When } t = -T/2, \; \tau = -t = -(-T/2) = T/2\end{array}}$$

$$= -\frac{2}{T} \int_{0}^{T/2} x(-\tau)(-d\tau) + \frac{2}{T} \int_{0}^{T/2} x(t)\, dt$$

$$= \frac{2}{T} \int_{0}^{T/2} x(-\tau)\, d\tau + \frac{2}{T} \int_{0}^{T/2} x(t)\, dt = \frac{2}{T} \int_{0}^{T/2} x(-t)\, dt + \frac{2}{T} \int_{0}^{T/2} x(t)\, dt \qquad \boxed{\text{Since } \tau \text{ is dummy variable, let } \tau = t.}$$

$$= \frac{2}{T} \int_{0}^{T/2} x(t)\, dt + \frac{2}{T} \int_{0}^{T/2} x(t)\, dt = \frac{4}{T} \int_{0}^{T/2} x(t)\, dt \qquad \boxed{\text{Since } x(t) \text{ is even, } x(-t) = x(t).}$$

Consider the equation for a_n, [equation (2.3)].

$$\boxed{\begin{array}{l}\text{Dividing the integral}\\ \text{into two parts.}\end{array}}$$

$$a_n = \frac{2}{T} \int_{-T/2}^{T/2} x(t) \cos n\Omega_0\, dt = \frac{2}{T} \int_{-T/2}^{0} x(t) \cos n\Omega_0\, dt + \frac{2}{T} \int_{0}^{T/2} x(t) \cos n\Omega_0\, dt$$

$$= \frac{2}{T} \int_{T/2}^{0} x(-\tau) \cos[n\Omega_0(-\tau)](-d\tau) + \frac{2}{T} \int_{0}^{T/2} x(t) \cos n\Omega_0 t\, dt \qquad \boxed{\begin{array}{l}\text{Change of integral index,}\\ \text{Let, } t = -\tau ; \; \therefore \; dt = -d\tau \\ \text{When } t = 0, \; \tau = -t = 0 \\ \text{When } t = -T/2, \; \tau = -t = -(-T/2) = T/2\end{array}}$$

$$= -\frac{2}{T} \int_{0}^{T/2} x(-\tau) \cos[n\Omega_0(-\tau)](-d\tau) + \frac{2}{T} \int_{0}^{T/2} x(t) \cos n\Omega_0 t\, dt$$

$$\therefore\ a_n = \frac{2}{T}\int_0^{T/2} x(-\tau)\cos n\Omega_0\tau\,d\tau + \frac{2}{T}\int_0^{T/2} x(t)\cos n\Omega_0 t\,dt$$

$$\boxed{\cos(-\theta) = \cos\theta}$$
$$\boxed{\text{Since } \tau \text{ is dummy variable, let } \tau = t.}$$

$$= \frac{2}{T}\int_0^{T/2} x(-t)\cos n\Omega_0 t\,dt + \frac{2}{T}\int_0^{T/2} x(t)\cos n\Omega_0 t\,dt$$

$$\boxed{\text{Since } x(t) \text{ is even, } x(-t) = x(t).}$$

$$= \frac{2}{T}\int_0^{T/2} x(t)\cos n\Omega_0 t\,dt + \frac{2}{T}\int_0^{T/2} x(t)\cos n\Omega_0 t\,dt = \frac{4}{T}\int_0^{T/2} x(t)\cos n\Omega_0 t\,dt$$

Consider the equation for b_n, [equation (2.4)].

$$b_n = \frac{2}{T}\int_{-T/2}^{T/2} x(t)\sin n\Omega_0 t\,dt = \frac{2}{T}\int_{-T/2}^{0} x(t)\sin n\Omega_0 t\,dt + \frac{2}{T}\int_0^{T/2} x(t)\sin n\Omega_0\,dt$$

$$\boxed{\begin{array}{l}\text{Dividing the integral}\\ \text{into two parts.}\end{array}}$$

$$= \frac{2}{T}\int_{T/2}^{0} x(-\tau)\sin[n\Omega_0(-\tau)](-d\tau) + \frac{2}{T}\int_0^{T/2} x(t)\sin n\Omega_0 t\,dt$$

$$\boxed{\begin{array}{l}\text{Change of integral index,}\\ \text{Let, } t=-\tau;\ \therefore\ dt=-d\tau\\ \text{When } t=0,\ \tau=-t=0\\ \text{When } t=-T/2,\ \tau=-t=-(-T/2)=T/2\end{array}}$$

$$= -\frac{2}{T}\int_0^{T/2} x(-\tau)\sin[n\Omega_0(-\tau)](-d\tau) + \frac{2}{T}\int_0^{T/2} x(t)\sin n\Omega_0 t\,dt$$

$$= \frac{2}{T}\int_0^{T/2} x(-\tau)\sin[n\Omega_0(-\tau)]\,d\tau + \frac{2}{T}\int_0^{T/2} x(t)\sin n\Omega_0 t\,dt$$

$$= -\frac{2}{T}\int_0^{T/2} x(-t)\sin n\Omega_0 t\,dt + \frac{2}{T}\int_0^{T/2} x(t)\sin n\Omega_0 t\,dt$$

$$\boxed{\sin(-\theta) = -\sin\theta}$$
$$\boxed{\text{Since } \tau \text{ is dummy variable, let } \tau = t.}$$

$$= -\frac{2}{T}\int_0^{T/2} x(t)\sin n\Omega_0 t\,dt + \frac{2}{T}\int_0^{T/2} x(t)\sin n\Omega_0 t\,dt = 0$$

$$\boxed{\text{Since } x(t) \text{ is even, } x(-t) = x(t).}$$

2.2.2 Odd Symmetry

A signal, $x(t)$ is called *odd signal* if it satisfies the condition *$x(-t) = -x(t)$.*

The waveform of an odd periodic signal will exhibit antisymmetry with respect to $t = 0$ (i.e. with respect to vertical axis) and so the antisymmetry of a waveform with respect to $t = 0$ or vertical axis is called *odd symmetry.*

In order to determine the odd symmetry of a waveform, invert either the right side or the left side of the waveform with respect to horizontal axis and then fold the waveform with respect to vertical axis. After inverting one half and folding, if the waveshape remains same then it is said to have odd symmetry.

For odd signals a_0 and a_n are zero and b_n exists. For odd signal the Fourier coefficients are given by,

$$a_0 = 0 \quad ; \quad a_n = 0$$

$$b_n = \frac{4}{T}\int_0^{T/2} x(t)\sin n\Omega_0 t\,dt \quad \text{(or)} \quad b_n = \frac{4}{T}\int_{-T/4}^{+T/4} x(t)\sin n\Omega_0 t\,dt$$

Proof:

Consider the equation for a_0, (equation (2.2)).

$$a_0 = \frac{2}{T} \int_{-T/2}^{T/2} x(t)\, dt = \frac{2}{T} \int_{-T/2}^{0} x(t)\, dt + \frac{2}{T} \int_{0}^{T/2} x(t)\, dt$$

> Dividing the integral into two parts.

> Change of integral index,
> Let, $t = -\tau;\ \therefore\ dt = -d\tau$
> When $t = 0,\ \tau = -t = 0$
> When $t = -T/2,\ \tau = -t = -(-T/2) = T/2$

$$= \frac{2}{T} \int_{T/2}^{0} x(-\tau)(-d\tau) + \frac{2}{T} \int_{0}^{T/2} x(t)\, dt$$

$$= -\frac{2}{T} \int_{0}^{T/2} x(-\tau)(-d\tau) + \frac{2}{T} \int_{0}^{T/2} x(t)\, dt$$

$$= \frac{2}{T} \int_{0}^{T/2} x(-\tau)\, d\tau + \frac{2}{T} \int_{0}^{T/2} x(t)\, d = \frac{2}{T} \int_{0}^{T/2} x(-t)\, dt + \frac{2}{T} \int_{0}^{T/2} x(t)\, dt$$

> Since τ is dummy variable, let $\tau = t$.

$$= -\frac{2}{T} \int_{0}^{T/2} x(t)\, dt + \frac{2}{T} \int_{0}^{T/2} x(t)\, dt = 0$$

> Since $x(t)$ is odd, $x(-t) = -x(t)$.

Consider the equation for a_n, [equation (2.3)].

$$a_n = \frac{2}{T} \int_{-T/2}^{T/2} x(t) \cos n\Omega_0 t\, dt = \frac{2}{T} \int_{-T/2}^{0} x(t) \cos n\Omega_0 t\, dt + \frac{2}{T} \int_{0}^{T/2} x(t) \cos n\Omega_0 t\, dt$$

> Dividing the integral into two parts.

> Change of integral index,
> Let, $t = -\tau;\ \therefore\ dt = -d\tau$
> When $t = 0,\ \tau = -t = 0$
> When $t = -T/2,\ \tau = -t = -(-T/2) = T/2$

$$= \frac{2}{T} \int_{T/2}^{0} x(-\tau) \cos n\Omega_0 (-\tau)(-d\tau) + \frac{2}{T} \int_{0}^{T/2} x(t) \cos n\Omega_0 t\, dt$$

$$= -\frac{2}{T} \int_{0}^{T/2} x(-\tau) \cos [n\Omega_0(-\tau)](-d\tau) + \frac{2}{T} \int_{0}^{T/2} x(t) \cos n\Omega_0 t\, dt$$

$$= \frac{2}{T} \int_{0}^{T/2} x(-\tau) \cos n\Omega_0 \tau\, d\tau + \frac{2}{T} \int_{0}^{T/2} x(t) \cos n\Omega_0 t\, dt$$

> $\cos(-\theta) = \cos\theta$

$$= \frac{2}{T} \int_{0}^{T/2} x(-t) \cos n\Omega_0 t\, dt + \frac{2}{T} \int_{0}^{T/2} x(t) \cos n\Omega_0 t\, dt$$

> Since τ is dummy variable, let $\tau = t$.

> Since $x(t)$ is odd, $x(-t) = -x(t)$.

$$= -\frac{2}{T} \int_{0}^{T/2} x(t) \cos n\Omega_0 t\, dt + \frac{2}{T} \int_{0}^{T/2} x(t) \cos n\Omega_0 t\, dt = 0$$

Consider the equation for b_n, [equation (2.4)].

> Dividing the integral into two parts.

$$b_n = \frac{2}{T} \int_{-T/2}^{T/2} x(t) \sin n\Omega_0 t\, dt = \frac{2}{T} \int_{-T/2}^{0} x(t) \sin n\Omega_0 t\, dt + \frac{2}{T} \int_{0}^{T/2} x(t) \sin n\Omega_0 t\, dt$$

> Change of integral index,
> Let, $t = -\tau;\ \therefore\ dt = -d\tau$
> When $t = 0,\ \tau = -t = 0$
> When $t = -T/2,\ \tau = -t = -(-T/2) = T/2$

$$= \frac{2}{T} \int_{T/2}^{0} x(-\tau) \sin [n\Omega_0(-\tau)](-d\tau) + \frac{2}{T} \int_{0}^{T/2} x(t) \sin n\Omega_0 t\, dt$$

$$\therefore b_n = -\frac{2}{T}\int_0^{T/2} x(-\tau)\sin[n\Omega_0(-\tau)](-d\tau) + \frac{2}{T}\int_0^{T/2} x(t)\sin n\Omega_0 t\, dt$$

$$= -\frac{2}{T}\int_0^{T/2} x(-\tau)\sin n\Omega_0\tau\, d\tau + \frac{2}{T}\int_0^{T/2} x(t)\sin n\Omega_0 t\, dt \qquad \boxed{\sin(-\theta) = -\sin\theta}$$

$$= -\frac{2}{T}\int_0^{T/2} x(-t)\sin n\Omega_0 t\, dt + \frac{2}{T}\int_0^{T/2} x(t)\sin n\Omega_0 t\, dt \qquad \boxed{\text{Since } \tau \text{ is dummy variable, let } \tau = t.}$$

$$\boxed{\text{Since } x(t) \text{ is odd, } x(-t) = -x(t)}$$

$$= \frac{2}{T}\int_0^{T/2} x(t)\sin n\Omega_0 t\, d\tau + \frac{2}{T}\int_0^{T/2} x(t)\sin n\Omega_0 t\, dt = \frac{4}{T}\int_0^{T/2} x(t)\sin n\Omega_0 t\, dt$$

2.2.3 Half Wave Symmetry (or Alternation Symmetry)

The periodic waveform in which each period/cycle consists of two equal and opposite half period/cycle are called alternating waveforms, because this type of waveform will have alternate positive and negative half cycles. Such waveforms are said to have *half wave symmetry* or *alternation symmetry.*

The waveforms with half wave symmetry will satisfy the condition,

$$\boxed{x\left(t \pm \frac{T}{2}\right) = -x(t)}$$

When a waveform has half wave symmetry, the Fourier series will consist of odd harmonic terms alone. Certain waveform will exhibit half wave symmetry after subtraction of the dc component $(a_0/2)$.

2.2.4 Quarter Wave Symmetry

A waveform with half wave symmetry if in addition has even/odd symmetry then it is said to have *quarter wave symmetry*. In a waveform with quarter wave symmetry, each quarter period will have identical shape, but may have opposite sign. The existence of the type of Fourier coefficients for waveform with quarter wave symmetry is shown below:

$$x\left(t \pm \frac{T}{2}\right) = -x(t)$$

x(t) has half wave symmetry.
Fourier series has odd harmonic terms.

$x(-t) = x(t)$ and $x\left(t \pm \frac{T}{2}\right) = -x(t)$	$x(-t) = -x(t)$ and $x\left(t \pm \frac{T}{2}\right) = -x(t)$
x(t) has even and half wave symmetries. [i.e. x(t) has quarter wave symmetry]. Fourier series will have odd harmonics of cosine terms.	x(t) has even and half wave symmetrices. [i.e. x(t) has quarter wave symmetry]. Fourier series will have odd harmonics of sine terms.

Table 2.3 : Waveform with symmetry

Waveform	Type of Symmetry	Fourier Series	Comments
	Even, half wave and quarter wave symmetry.	$x(t) = \dfrac{4A}{\pi}\left[\dfrac{\cos\Omega_0 t}{1} - \dfrac{\cos 3\Omega_0 t}{3} + \dfrac{\cos 5\Omega_0 t}{5} - \dfrac{\cos 7\Omega_0 t}{7} + \dfrac{\cos 9\Omega_0 t}{9} - ......\right]$ $\boxed{\Omega_0 = \dfrac{2\pi}{T}}$ (Refer Example 2.1 for derivation of Fourier series)	$a_0 = 0$ $b_n = 0$ $a_n = 0$; for even n $a_n \neq 0$; for odd n Fourier series consists of odd harmonics of cosine terms alone.
	Even symmetry. If component $(a_0/2)$ is subtracted from the waveform, it will have half wave and quarter wave symmetry.	$x(t) = \dfrac{A}{2} + \dfrac{2A}{\pi}\left[\dfrac{\cos\Omega_0 t}{1} - \dfrac{\cos 3\Omega_0 t}{3} + \dfrac{\cos 5\Omega_0 t}{5} - \dfrac{\cos 7\Omega_0 t}{7} + \dfrac{\cos 9\Omega_0 t}{9} - ......\right]$ $\boxed{\Omega_0 = \dfrac{2\pi}{T}}$ (Refer Example 2.3 for derivation of Fourier series)	$a_0 \neq 0$ $b_n = 0$ $a_n = 0$; for even n $a_n \neq 0$; for odd n Fourier series has a dc component and consists of odd harmonics of cosine terms.
	Even symmetry.	$x(t) = \dfrac{2A}{\pi} + \dfrac{4A}{\pi}\left[\dfrac{\cos 2\Omega_0 t}{2^2-1} - \dfrac{\cos 4\Omega_0 t}{4^2-1} + \dfrac{\cos 6\Omega_0 t}{6^2-1} - \dfrac{\cos 8\Omega_0 t}{8^2-1} + ...\right]$ $\boxed{\Omega_0 = \dfrac{2\pi}{T}}$	$a_0 \neq 0$ $b_n = 0$ $a_n \neq 0$; for even n $a_n = 0$; for odd n Fourier series has a dc component and consists of even harmonics of cosine terms.

Table 2.3 : Continued.....

Waveform	Type of Symmetry	Fourier Series	Comments
	Even symmetry.	$$x(t) = \frac{2A}{\pi} - \frac{4A}{\pi}\left[\frac{\cos 2\Omega_0 t}{2^2-1} + \frac{\cos 4\Omega_0 t}{4^2-1} + \frac{\cos 6\Omega_0 t}{6^2-1} + \frac{\cos 8\Omega_0 t}{8^2-1} + \ldots\ldots\right]$$ $$\boxed{\Omega_0 = \frac{2\pi}{T}}$$ (Refer Example 2.4 for derivation of Fourier series.)	$a_0 \neq 0$ $b_n = 0$ $a_n = 0$; for even n $a_n \neq 0$; for odd n Fourier series has dc component and consists of even harmonics of cosine terms.
	Even symmetry. If dc component $(a_0/2)$ is subtracted from the waveform, it will have half wave and quarter wave symmetry.	$$x(t) = \frac{A}{2} - \frac{4A}{\pi^2}\left[\frac{\cos \Omega_0 t}{1^2} + \frac{\cos 3\Omega_0 t}{3^2} + \frac{\cos 5\Omega_0 t}{5^2} + \frac{\cos 7\Omega_0 t}{7^2} + \ldots\ldots\right]$$ $$\boxed{\Omega_0 = \frac{2\pi}{T}}$$ (Refer Example 2.2 for derivation of Fourier series.)	$a_0 \neq 0$ $b_n = 0$ $a_n = 0$; for even n $a_n \neq 0$; for odd n Fourier series has dc component and consists of odd harmonics of cosine terms.
	Even symmetry. If dc component $(a_0/2)$ is subtracted from the waveform, it will have half wave and quarter wave symmetry.	$$x(t) = \frac{A}{2} + \frac{4A}{\pi^2}\left[\frac{\cos \Omega_0 t}{1^2} + \frac{\cos 3\Omega_0 t}{3^2} + \frac{\cos 5\Omega_0 t}{5^2} + \frac{\cos 7\Omega_0 t}{7^2} + \ldots\ldots\right]$$ $$\boxed{\Omega_0 = \frac{2\pi}{T}}$$ (Refer Example 2.11 for derivation of Fourier series.)	$a_0 \neq 0$ $b_n = 0$ $a_n = 0$; for even n $a_n \neq 0$; for odd n Fourier series has dc component and consists of odd harmonics of cosine terms.

Table 2.3 : Continued.....

Waveform	Type of Symmetry	Fourier Series	Comments
	Odd, half wave and quarter wave symmetry.	$$x(t) = \frac{4A}{\pi}\left[\sin\Omega_0 t + \frac{\sin 3\Omega_0 t}{3} + \frac{\sin 5\Omega_0 t}{5} + \frac{\sin 7\Omega_0 t}{7} + \dots\right]$$ $\boxed{\Omega_0 = \frac{2\pi}{T}}$ (Refer Example 2.5 for derivation of Fourier series.)	$a_0 = 0$ $a_n = 0$ $b_n = 0$; for even n $b_n \neq 0$; for odd n Fourier series consists of odd harmonics of sine terms.
	Odd, half wave and quarter wave symmetry.	$$x(t) = \frac{8A}{\pi^2}\left[\frac{\sin\Omega_0 t}{1} - \frac{\sin 3\Omega_0 t}{3^2} + \frac{\sin 5\Omega_0 t}{5^2} - \frac{\sin 7\Omega_0 t}{7^2} + \dots\right]$$ $\boxed{\Omega_0 = \frac{2\pi}{T}}$ (Refer Example 2.6 for derivation of Fourier series.)	$a_0 = 0$ $a_n = 0$ $b_n = 0$; for even n $b_n \neq 0$; for odd n Fourier series consists of odd harmonics of sine terms.
	Odd symmetry.	$$x(t) = \frac{2A}{\pi}\left[\frac{\sin\Omega_0 t}{1} - \frac{\sin 2\Omega_0 t}{2} + \frac{\sin 3\Omega_0 t}{3} - \frac{\sin 4\Omega_0 t}{4} + \frac{\sin 5\Omega_0 t}{5} + \dots\right]$$ $\boxed{\Omega_0 = \frac{2\pi}{T}}$ (Refer Example 2.7 for derivation of Fourier series.)	$a_0 = 0$ $a_n = 0$ $b_n \neq 0$; for all n Fourier series consists of both even and odd harmonics of sine terms.

Table 2.3 : Continued.....

Waveform	Type of Symmetry	Fourier Series	Comments
	Neither even nor odd. If dc component ($a_0/2$) is subtracted from the waveform, it becoome odd signal.	$x(t) = \dfrac{A}{2} - \dfrac{A}{\pi}\left[\dfrac{\sin\Omega_0 t}{1} + \dfrac{\sin 2\Omega_0 t}{2} + \dfrac{\sin 3\Omega_0 t}{3} + \dfrac{\sin 4\Omega_0 t}{4} + \dfrac{\sin 5\Omega_0 t}{5} + \ldots \right]$ $\boxed{\Omega_0 = \dfrac{2\pi}{T}}$ (Refer Example 2.8 for derivation of Fourier series.)	$a_0 \neq 0$ $a_n = 0$ $b_n \neq 0$; for all n Fourier series has a dc component and consists of all harmonics of sine terms.
	Half wave symmetry.	$x(t) = -\dfrac{4A}{\pi^2}\left(\cos\Omega_0 t + \dfrac{\cos 3\Omega_0 t}{3^2} + \dfrac{\cos 5\Omega_0 t}{5^2} + \ldots\right)$ $\quad + \dfrac{2A}{\pi}\left(\sin\Omega_0 t + \dfrac{\sin 3\Omega_0 t}{3} + \dfrac{\sin 5\Omega_0 t}{5} + \ldots\right)$ $\boxed{\Omega_0 = \dfrac{2\pi}{T}}$ (Refer Example 2.10 for derivation of Fourier series.)	$a_0 = 0$ $a_n = 0$; for even n $a_n \neq 0$; for odd n $b_n = 0$; for even n $b_n \neq 0$; for odd n Fourier series consists of odd harmonics of cosine and sine terms.
	Half wave symmetry.	$x(t) = \dfrac{4A}{\pi^2}\left(\cos\Omega_0 t + \dfrac{\cos 3\Omega_0 t}{3^2} + \dfrac{\cos 5\Omega_0 t}{5^2} + \ldots\right)$ $\quad - \dfrac{2A}{\pi}\left(\sin\Omega_0 t + \dfrac{\sin 3\Omega_0 t}{3} + \dfrac{\sin 5\Omega_0 t}{5} + \ldots\right)$ $\boxed{\Omega_0 = \dfrac{2\pi}{T}}$	$a_0 = 0$; for even n $a_n \neq 0$; for odd n $b_n = 0$; for even n $b_n \neq 0$; for odd n Fourier series consists of odd harmonics of cosine and sine terms.

2.3 Gibbs Phenomenon

The exponential form of Fourier series of a continuous time periodic signal x(t) is given by,

$$x(t) = \sum_{n=-\infty}^{+\infty} c_n \, e^{jn\Omega_0 t}$$

The above equation is frequency domain representation of the signal x(t) as a sum of infinite series with each term in the series representing a harmonic frequency component. When the signal x(t) is reconstructed or synthesised with only N number of terms of the infinite series, the reconstructed signal exhibits oscillations (or overshoot or ripples), especially in signals with discontinuities.

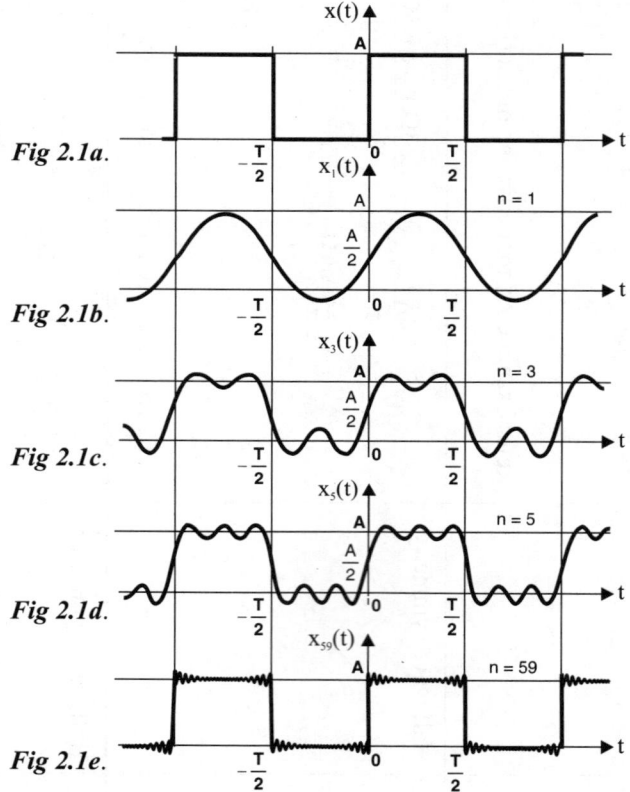

Fig 2.1a.

Fig 2.1b.

Fig 2.1c.

Fig 2.1d.

Fig 2.1e.

Fig 2.1: *Successively closer approximation to a square pulse signal.*

Consider a periodic square pulse signal shown in Fig 2.1a. The reconstructed signal using N-terms of Fourier series are shown in Fig 2.1b, c, d and e. (Refer MATLAB program 2.5 in section 2.14). It can be observed that the reconstructed signal exhibits oscillations and the oscillations are compressed towards points of discontinuities with increasing value of N.

Also it can be observed that, at the points of discontinuity, the Fourier series converges to average value of the signal on either side of discontinuity. This phenomenon was named after a famous mathematician, Josiah Gibbs, as *Gibbs phenomenon* and the oscillations are called *Gibbs oscillations* (Josiah Gibbs is a famous mathematician who first provided mathematical explanation for this phenomenon).

Also in reconstruction of signal with finite terms of Fourier series it can be observed that the peak overshoot of the oscillations remains constant irrespective of the value of N, but with increasing N the peak overshoot shifts towards the point of discontinuity.

2.4 Spectrum of Continuous Time Signals from Fourier Series

Let x(t) be a periodic continuous time signal. Now, exponential form of Fourier series of x(t) is,

$$x(t) = \sum_{n=-\infty}^{+\infty} c_n e^{jn\Omega_0 t}$$

where, c_n is the Fourier coefficient of nth harmonic component.

The Fourier coefficient, c_n is a complex quantity and so it can be expressed in the polar form as shown below:

$$c_n = |c_n| \angle c_n$$

where, $|c_n|$ = Magnitude of c_n ; $\angle c_n$ = Phase of c_n

The term, $|c_n|$ represents the magnitude of nth harmonic component and the term $\angle c_n$ represents the phase of the nth harmonic component.

The plot of harmonic magnitude / phase of a signal versus "n" (or harmonic frequency $n\Omega_0$) is called *Frequency spectrum (or Line spectrum).* The plot of harmonic magnitude versus "n" (or $n\Omega_0$) is called *magnitude (line) spectrum* and the plot of harmonic phase versus "n" (or $n\Omega_0$) is called *phase (line) spectrum.*

Consider the periodic ramp signal shown in Fig 2.2. The Fourier coefficient c_n for this ramp signal is given by,

$$c_0 = \frac{A}{2} , \qquad c_n = \frac{jA}{2n\pi}$$

(Please refer example 2.12 for derivation of c_n)

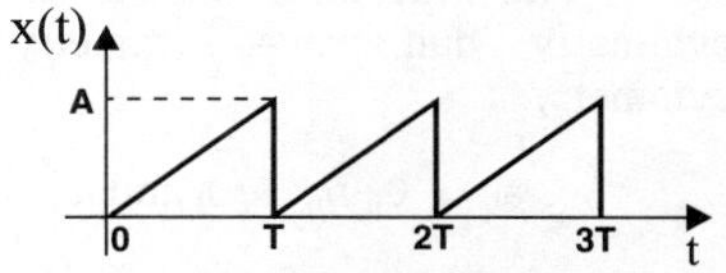

Let, $A = 20$, $\therefore\ c_n = \dfrac{j20}{2n\pi} = \dfrac{j10}{n\pi}$

Fig 2.2: Periodic ramp signal.

.
.
.

When $n = -3$, $c_{-3} = -j\dfrac{10}{3\pi} = -j1.061 = 1.061 \angle -90° = 1.061 \angle -\pi/2$

When $n = -2$, $c_{-2} = -j\dfrac{10}{2\pi} = -j1.592 = 1.592 \angle -90° = 1.592 \angle -\pi/2$

When $n = -1$, $c_{-1} = -j\dfrac{10}{\pi} = -j3.183 = 3.183 \angle -90° = 3.183 \angle -\pi/2$

When $n = 0$, $c_0 = \dfrac{20}{2} = 10 = 10 \angle 0$

When n = 1, c_1 = $j\dfrac{10}{\pi}$ = j3.183   = 3.183 $\angle +90°$ = 3.183 $\angle \pi/2$

When n = 2, c_2 = $j\dfrac{10}{2\pi}$ = j1.592   = 1.592 $\angle +90°$ = 1.592 $\angle \pi/2$

When n = 3, c_3 = $j\dfrac{10}{3\pi}$ = j1.061   = 1.061 $\angle +90°$ = 1.061 $\angle \pi/2$

⋮

Using the above calculated values the magnitude and phase spectrums are sketched as shown in Fig 2.3 and Fig 2.4 respectively.

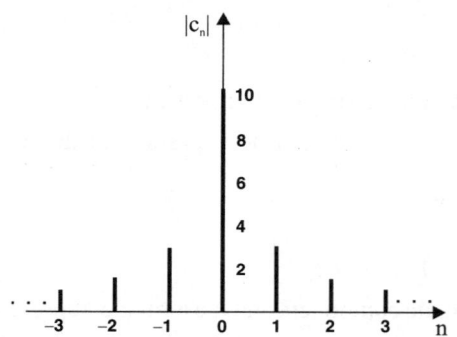

Fig 2.3: *Magnitude spectrum of periodic ramp.*

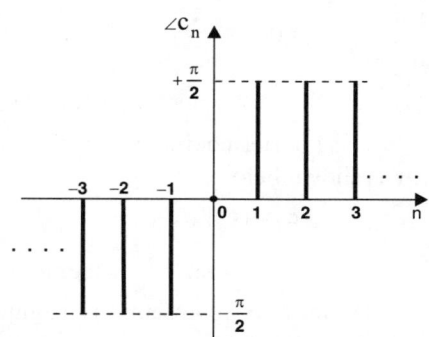

Fig 2.4: *Phase spectrum of periodic ramp.*

Example 2.1

Determine the trigonometric form of Fourier series of the waveform shown in Fig 2.1.1.

Solution:

The waveform shown in Fig 2.1.1 has even symmetry, half wave symmetry and quarter wave symmetry.

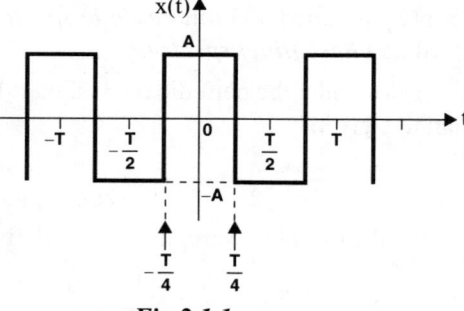

Fig 2.1.1.

$$\therefore a_0 = 0, \ b_n = 0 \ \text{and} \ a_n = \frac{4}{T} \int_{-T/4}^{T/4} x(t) \cos n\Omega_0 t \, dt$$

The mathematical equation of the square wave is,

$$x(t) = A \quad ; \quad \text{for} \quad t = 0 \ \text{to} \ t = -\frac{T}{4} \ \text{to} \ \frac{T}{4}$$

> **Note:** Here x(t) is governed by single mathematical equation in the range $-\dfrac{T}{4}$ to $+\dfrac{T}{4}$. And so the calculations will be simple, if the integral limit is $-\dfrac{T}{4}$ to $+\dfrac{T}{4}$

Evaluation of a_n

$$a_n = \frac{4}{T} \int_{-T/4}^{T/4} x(t) \cos n\Omega_0 t \, dt = \frac{4}{T} \int_{-T/4}^{T/4} A \cos n\Omega_0 t \, dt$$

$$= \frac{4A}{T} \left[\frac{\sin n\Omega_0 t}{n\Omega_0} \right]_{-T/4}^{T/4} = \frac{4A}{T} \left[\frac{\sin n\frac{2\pi}{T} t}{n\frac{2\pi}{T}} \right]_{-T/4}^{T/4}$$

$$\boxed{\Omega_0 = \frac{2\pi}{T}}$$

$$= \frac{4A}{T} \left[\frac{\sin\left(n\frac{2\pi}{T}\frac{T}{4}\right)}{n\frac{2\pi}{T}} - \frac{\sin\left(n\frac{2\pi}{T}\left(-\frac{T}{4}\right)\right)}{n\frac{2\pi}{T}} \right]$$

$$\boxed{\sin(-\theta) = -\sin\theta}$$

$$= \frac{4A}{T}\left[\frac{T}{2n\pi}\sin\frac{n\pi}{2} + \frac{T}{2n\pi}\sin\frac{n\pi}{2}\right] = \frac{4A}{T}\left[\frac{T}{n\pi}\sin\frac{n\pi}{2}\right] = \frac{4A}{n\pi}\sin\frac{n\pi}{2}$$

For even values of n, $\sin\frac{n\pi}{2} = 0$

For odd values of n, $\sin\frac{n\pi}{2} = \pm 1$

$\therefore a_n = 0$; for even values of n

$a_n = \frac{4A}{n\pi}\sin\frac{n\pi}{2}$; for odd values of n

$\therefore a_1 = \frac{4A}{1\times\pi}\sin\frac{\pi}{2} = +\frac{4A}{\pi}$

$a_3 = \frac{4A}{3\times\pi}\sin\frac{3\pi}{2} = -\frac{4A}{3\pi}$

$a_5 = \frac{4A}{5\times\pi}\sin\frac{5\pi}{2} = +\frac{4A}{5\pi}$

$a_7 = \frac{4A}{7\times\pi}\sin\frac{7\pi}{2} = -\frac{4A}{7\pi}$ and so on.

Fourier Series

The trigonometric form of Fourier series of x(t) is,

$$x(t) = \frac{a_0}{2} + \sum_{n=1}^{\infty} a_n \cos n\Omega_0 t + \sum_{n=1}^{\infty} b_n \sin n\Omega_0 t$$

Here, $a_0 = 0$, $b_n = 0$ and a_n exists only for odd values of n.

$$\therefore x(t) = \sum_{n=odd} a_n \cos n\Omega_0 t$$

$$= a_1 \cos\Omega_0 t + a_3 \cos 3\Omega_0 t + a_5 \cos 5\Omega_0 t + a_7 \cos 7\Omega_0 t + \dots\dots\dots$$

$$= \frac{4A}{\pi}\cos\Omega_0 t - \frac{4A}{3\pi}\cos 3\Omega_0 t + \frac{4A}{5\pi}\cos 5\Omega_0 t - \frac{4A}{7\pi}\cos 7\Omega_0 t + \dots\dots\dots$$

$$= \frac{4A}{\pi}\left[\cos\Omega_0 t - \frac{\cos 3\Omega_0 t}{3} + \frac{\cos 5\Omega_0 t}{5} - \frac{\cos 7\Omega_0 t}{7} + \dots\dots\dots\right]$$

Example 2.2

Find the Fourier series of the waveform shown in Fig 2.2.1.

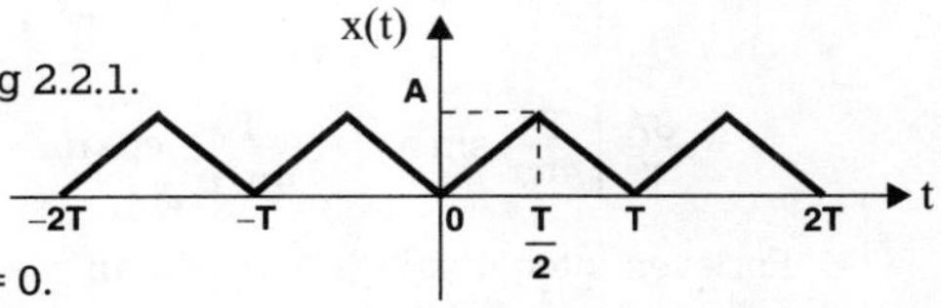

Solution:

The given waveform has even symmetry and so $b_n = 0$.

$$a_0 = \frac{4}{T}\int\limits_{0}^{T/2} x(t)\, dt \quad ; \quad a_n = \frac{4}{T}\int\limits_{0}^{T/2} x(t)\,\cos n\Omega_0\, dt \quad ; \quad b_n = 0$$

Fig 2.2.1

Note: *Here x(t) is governed by single mathematical equation in the range 0 to $\frac{T}{2}$. And so the calculations will be simple, if the integral limit is 0 to $\frac{T}{2}$.*

To Find Mathematical Equation for x(t)

Consider the equation of straight line, $\dfrac{y - y_1}{y_1 - y_2} = \dfrac{x - x_1}{x_1 - x_2}$

Here, $y = x(t), \quad x = t$.

$\therefore$ The equation of straight line can be written as, $\dfrac{x(t) - x(t_1)}{x(t_1) - x(t_2)} = \dfrac{t - t_1}{t_1 - t_2}$ .....(1)

Consider points P and Q, as shown in Fig 1.

Coordinates of point-P $= [t_1, x(t_1)] = [0, 0]$

Coordinates of point-Q $= [t_2, x(t_2)] = \left[\dfrac{T}{2}, A\right]$

On substituting the coordinates of points P and Q in equation (1) we get,

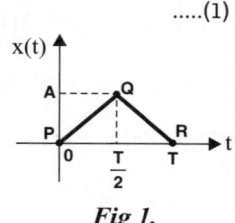

Fig 1.

$$\dfrac{x(t) - 0}{0 - A} = \dfrac{t - 0}{0 - \dfrac{T}{2}} \Rightarrow \dfrac{x(t)}{-A} = \dfrac{-2t}{T} \Rightarrow x(t) = \dfrac{2A}{T}t$$

$$\therefore \quad x(t) = \dfrac{2A}{T}t \quad ; \quad \text{for } t = 0 \text{ to } \dfrac{T}{2}$$

Evaluation of a_0

$$a_0 = \dfrac{4}{T}\int_0^{T/2} x(t)\,dt = \dfrac{4}{T}\int_0^{T/2} \dfrac{2A}{T}t\,dt = \dfrac{8A}{T^2}\int_0^{T/2} t\,dt$$

$$= \dfrac{8A}{T^2}\left[\dfrac{t^2}{2}\right]_0^{T/2} = \dfrac{8A}{T^2}\left[\dfrac{(T/2)^2}{2} - 0\right] = \dfrac{8A}{T^2}\left[\dfrac{T^2}{8} - 0\right] = A$$

Evaluation a_n

$$a_n = \dfrac{4}{T}\int_0^{T/2} x(t)\cos n\Omega_0 t\,dt = \dfrac{4}{T}\int_0^{T/2} \dfrac{2A}{T}t\cos n\Omega_0 t\,dt = \dfrac{8A}{T^2}\int_0^{T/2} t\cos n\Omega_0 t\,dt$$

$$= \dfrac{8A}{T^2}\left[t\dfrac{\sin n\Omega_0 t}{n\Omega_0} - \int\left(\dfrac{\sin n\Omega_0 t}{n\Omega_0}\right)\times 1\,dt\right]_0^{T/2}$$

$\int u\,dv = uv - \int v\,du$	
$u = t$	$\Rightarrow du = 1$
$dv = \cos n\Omega_0 t \Rightarrow v = \dfrac{\sin n\Omega_0 t}{n\Omega_0}$	

$$= \dfrac{8A}{T^2}\left[\dfrac{t\sin n\Omega_0 t}{n\Omega_0} - \left(\dfrac{-\cos n\Omega_0 t}{n^2\Omega_0^2}\right)\right]_0^{T/2} = \dfrac{8A}{T^2}\left[\dfrac{t\sin n\dfrac{2\pi}{T}t}{n\dfrac{2\pi}{T}} + \dfrac{\cos n\dfrac{2\pi}{T}t}{n^2\dfrac{4\pi^2}{T^2}}\right]_0^{T/2}$$

$\boxed{\Omega_0 = \dfrac{2\pi}{T}}$

$$= \dfrac{8A}{T^2}\left[\dfrac{T}{2}\dfrac{\sin n\dfrac{2\pi}{T}\dfrac{T}{2}}{n\dfrac{2\pi}{T}} + \dfrac{\cos n\dfrac{2\pi}{T}\dfrac{T}{2}}{n^2\dfrac{4\pi^2}{T^2}} - \dfrac{0\times\sin 0}{n\dfrac{2\pi}{T}} - \dfrac{\cos 0}{n^2\dfrac{4\pi^2}{T^2}}\right]$$

$\boxed{\sin 0 = 0}$
$\boxed{\cos 0 = 1}$

$$= \dfrac{8A}{T^2}\left[\dfrac{T^2}{4n\pi}\sin n\pi + \dfrac{T^2}{4n^2\pi^2}\cos n\pi - \dfrac{T^2}{4n^2\pi^2}\right] = \dfrac{2A}{n^2\pi^2}[\cos n\pi - 1]$$

$\boxed{\begin{array}{l}\sin n\pi = 0 \\ \text{for integer n}\end{array}}$

For even integer values of n, $\cos n\pi = +1$

For odd integer values of n, $\cos n\pi = -1$

$$\therefore \quad a_n = 0 \quad ; \quad \text{for even values of n, and}$$

$$a_n = \dfrac{2A}{n^2\pi^2}[\cos n\pi - 1] = \dfrac{2A}{n^2\pi^2}[-1 - 1] = -\dfrac{4A}{n^2\pi^2} \quad ; \quad \text{for odd values of n.}$$

$$\therefore \quad a_1 = -\dfrac{4A}{1^2\pi^2} \quad ; \quad a_3 = -\dfrac{4A}{3^2\pi^2} \quad ; \quad a_5 = -\dfrac{4A}{5^2\pi^2} \quad ; \quad \text{and so on.}$$

Fourier Series

The trigonometric form of Fourier series of x(t) is,

$$x(t) = \frac{a_0}{2} + \sum_{n=1}^{\infty} a_n \cos n\Omega_0 t + \sum_{n=1}^{\infty} b_n \sin n\Omega_0 t$$

Here, $b_n = 0$, and a_n exists only for odd values of n.

$$\therefore \; x(t) = \frac{a_0}{2} + \sum_{n=odd} a_n \cos n\Omega_0 t$$

$$= \frac{a_0}{2} + a_1 \cos \Omega_0 t + a_3 \cos 3\Omega_0 t + a_5 \cos 5\Omega_0 t + \dots\dots\dots\dots\dots$$

$$= \frac{A}{2} - \frac{4A}{1^2 \pi^2} \cos \Omega_0 t - \frac{4A}{3^2 \pi^2} \cos 3\Omega_0 t - \frac{4A}{5^2 \pi^2} \cos 5\Omega_0 t - \dots\dots\dots$$

$$= \frac{A}{2} - \frac{4A}{\pi^2} \left[\cos \Omega_0 t + \frac{\cos 3\Omega_0 t}{3^2} + \frac{\cos 5\Omega_0 t}{5^2} + \dots\dots\dots\dots\dots\dots \right]$$

Example 2.3 (AU Dec'14, 16 Marks)

Determine the trigonometric form of Fourier series of the waveform shown in Fig 2.3.1.

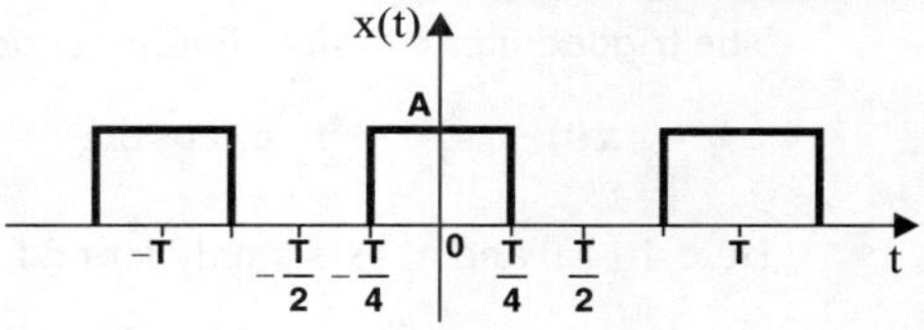

Solution:

Fig 2.3.1.

The waveform of Fig 2.3.1 has even symmetry.

$$\therefore \; b_n = 0, \; a_0 = \frac{4}{T} \int_0^{T/2} x(t)\, dt \;\; ; \;\; a_n = \frac{4}{T} \int_0^{T/2} x(t) \cos n\Omega_0 t\, dt$$

The mathematical equation of the given periodic rectangular pulse is,

$$x(t) = A \;\; ; \;\; \text{for } t = 0 \text{ to } \frac{T}{4}$$

$$= 0 \;\; ; \;\; \text{for } t = \frac{T}{4} \text{ to } \frac{T}{2}$$

Evaluation of a_0

$$a_0 = \frac{4}{T} \int_0^{T/2} x(t)\, dt \; = \; \frac{4}{T} \int_0^{T/4} A\, dt \; = \; \frac{4}{T} [At]_0^{T/4}$$

$$= \frac{4}{T} \left[A\frac{T}{4} - 0 \right] = A$$

Evaluation of a_n

$$a_n = \frac{4}{T} \int_0^{T/2} x(t) \cos n\Omega_0 t\, dt = \frac{4}{T} \int_0^{T/4} A \cos n\Omega_0 t\, dt = \frac{4A}{T} \left[\frac{\sin n\Omega_0 t}{n\Omega_0} \right]_0^{T/4}$$

$$= \frac{4A}{T} \left[\frac{\sin n\frac{2\pi}{T} t}{n\frac{2\pi}{T}} \right]_0^{T/4} = \frac{4A}{T} \left[\frac{\sin n\frac{2\pi}{T} \frac{T}{4}}{n\frac{2\pi}{T}} - \frac{\sin 0}{n\frac{2\pi}{T}} \right]$$

$$\boxed{\begin{array}{l} \Omega_0 = \frac{2\pi}{T} \\ \hline \sin 0 = 0 \end{array}}$$

$$= \frac{4A}{T} \times \frac{T}{2n\pi} \sin \frac{n\pi}{2} = \frac{2A}{n\pi} \sin \frac{n\pi}{2}$$

For even values of n, $\sin\frac{n\pi}{2} = 0$

For odd values of n, $\sin\frac{n\pi}{2} = \pm 1$

$\therefore \ a_n = 0 \ ; \ $ for even values of n, and

$\qquad a_n = \frac{2A}{n\pi}\sin\frac{n\pi}{2} \ ; \ $ for odd values of n.

$\therefore \ a_1 = \frac{2A}{1\times\pi}\ \sin\frac{\pi}{2} = +\frac{2A}{\pi}$

$\qquad a_3 = \frac{2A}{3\times\pi}\sin\frac{3\pi}{2} = -\frac{2A}{3\pi}$

$\qquad a_5 = \frac{2A}{5\times\pi}\ \sin\frac{5\pi}{2} = +\frac{2A}{5\pi}$

$\qquad a_7 = \frac{2A}{7\times\pi}\sin\frac{7\pi}{2} = -\frac{2A}{7\pi} \ $ and so on.

Fourier Series

The trigonomteric form of Fourier series of x(t) is,

$$x(t) = \frac{a_0}{2} + \sum_{n=1}^{\infty} a_n \cos n\Omega_0 t + \sum_{n=1}^{\infty} b_n \sin n\Omega_0 t$$

Here, $b_n = 0$ and a_n exists only for odd values of n.

$\therefore \ x(t) = \frac{a_0}{2} + \sum_{n=odd} a_n \cos n\Omega_0 t$

$\qquad\quad = \frac{a_0}{2} + a_1\cos\Omega_0 t + a_3\cos 3\Omega_0 t + a_5\cos 5\Omega_0 t + a_7 \cos 7\Omega_0 t + \dots\dots\dots$

$\therefore \ x(t) = \frac{A}{2} + \frac{2A}{\pi}\cos\Omega_0 t - \frac{2A}{3\pi}\cos 3\Omega_0 t + \frac{2A}{5\pi}\cos 5\Omega_0 t - \frac{2A}{7\pi}\cos 7\Omega_0 t + \dots\dots$

$\qquad\quad = \frac{A}{2} + \frac{2A}{\pi}\left[\cos\Omega_0 t - \frac{\cos 3\Omega_0 t}{3} + \frac{\cos 5\Omega_0 t}{5} - \frac{\cos 7\Omega_0 t}{7} + \dots\dots\dots\right]$

Example 2.4   *(AU Dec'11, 16 Marks)*

Determine the trigonometric form of Fourier series of the full wave rectified sine wave shown in Fig 2.4.1.

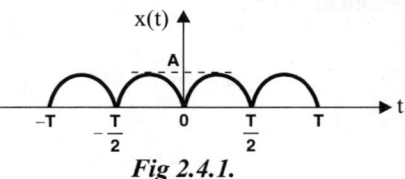

Fig 2.4.1.

Solution:

The waveform shown in Fig 2.4.1 is the output of full wave rectifier and it has even symmetry.

$$\therefore \ b_n = 0, \ a_0 = \frac{4}{T}\int_0^{T/2} x(t)\,dt \ ; \ a_n = \frac{4}{T}\int_0^{T/2} x(t)\cos n\Omega_0 t\,dt$$

The mathematical equation of full wave rectified output is,

$$x(t) = A\sin\Omega_0 t \ ; \ \text{for } t = 0 \text{ to } \frac{T}{2} \ \text{ and } \ \Omega_0 = \frac{2\pi}{T}$$

$$a_0 = \frac{4}{T}\int_0^{T/2} x(t)\,dt = \frac{4}{T}\int_0^{T/2} A\sin\Omega_0 t\,dt = \frac{4A}{T}\left[-\frac{\cos\Omega_0 t}{\Omega_0}\right]_0^{T/2}$$

$$\therefore a_0 = \frac{4A}{T}\left[-\frac{\cos\frac{2\pi}{T}t}{\frac{2\pi}{T}}\right]_0^{T/2} = \frac{4A}{T}\left[-\frac{\cos\frac{2\pi}{T}\frac{T}{2}}{\frac{2\pi}{T}} + \frac{\cos 0}{\frac{2\pi}{T}}\right]$$

$\boxed{\Omega_0 = \frac{2\pi}{T}}$

$$= \frac{2A}{\pi}\left[-\cos\pi + \cos 0\right] = \frac{2A}{\pi}\left[1+1\right] = \frac{4A}{\pi}$$

$\boxed{\cos\pi = -1 \mid \cos 0 = 1}$

Evaluation of a_n

$$a_n = \frac{4}{T}\int_0^{T/2} x(t)\cos n\Omega_0 t\, dt = \frac{4}{T}\int_0^{T/2} A\sin\Omega_0 t\,\cos n\Omega_0 t\, dt$$

$\boxed{2\sin A\cos B = \sin(A+B) + \sin(A-B)}$

$$= \frac{4A}{T}\int_0^{T/2}\frac{\sin(\Omega_0 t + n\Omega_0 t) + \sin(\Omega_0 t - n\Omega_0 t)}{2}\, dt$$

$$= \frac{2A}{T}\int_0^{T/2}\sin(1+n)\Omega_0 t\, dt + \frac{2A}{T}\int_0^{T/2}\sin(1-n)\Omega_0 t\, dt$$

$$= \frac{2A}{T}\left[\frac{-\cos(1+n)\Omega_0 t}{(1+n)\Omega_0}\right]_0^{T/2} + \frac{2A}{T}\left[\frac{-\cos(1-n)\Omega_0 t}{(1-n)\Omega_0}\right]_0^{T/2}$$

$$= \frac{2A}{T}\left[\frac{-\cos(1+n)\frac{2\pi}{T}t}{(1+n)\frac{2\pi}{T}}\right]_0^{T/2} + \frac{2A}{T}\left[\frac{-\cos(1-n)\frac{2\pi}{T}t}{(1-n)\frac{2\pi}{T}}\right]_0^{T/2}$$

$\boxed{\Omega_0 = \frac{2\pi}{T}}$

$$= \frac{2A}{T}\left[\frac{-\cos(1+n)\frac{2\pi}{T}\frac{T}{2}}{(1+n)\frac{2\pi}{T}} + \frac{\cos 0}{(1+n)\frac{2\pi}{T}}\right] + \frac{2A}{T}\left[\frac{-\cos(1-n)\frac{2\pi}{T}\frac{T}{2}}{(1-n)\frac{2\pi}{T}} + \frac{\cos 0}{(1-n)\frac{2\pi}{T}}\right]$$

$\boxed{\cos 0 = 1}$

$$= -\frac{A\cos(1+n)\pi}{(1+n)\pi} + \frac{A}{(1+n)\pi} - \frac{A\cos(1-n)\pi}{(1-n)\pi} + \frac{A}{(1-n)\pi} \qquad\qquad .....(1)$$

The equation (1) for a_n can be evaluated for all values of n except n = 1. For n = 1, a_n has to be estimated seperately as shown below:

$$a_1 = \frac{4}{T}\int_0^{T/2} x(t)\cos\Omega_0 t\, dt = \frac{4}{T}\int_0^{T/2} A\sin\Omega_0 t\cos\Omega_0 t\, dt$$

$$= \frac{4}{T}\int_0^{T/2} A\frac{\sin 2\Omega_0 t}{2}\, dt = \frac{2A}{T}\int_0^{T/2}\sin 2\Omega_0 t\, dt$$

$\boxed{\sin 2\theta = 2\sin\theta\cos\theta}$

$$= \frac{2A}{T}\left[\frac{-\cos 2\Omega_0 t}{2\Omega_0}\right]_0^{T/2} = \frac{2A}{T}\left[\frac{-\cos\left(2\times\frac{2\pi}{T}\times\frac{T}{2}\right)}{2\times\frac{2\pi}{T}} + \frac{\cos 0}{2\times\frac{2\pi}{T}}\right]$$

$\boxed{\Omega_0 = \frac{2\pi}{T}}$

$$= \frac{2A}{T}\left[-\frac{T}{4\pi}\cos 2\pi + \frac{T}{4\pi}\right] = \frac{2A}{T}\left[-\frac{T}{4\pi} + \frac{T}{4\pi}\right] = 0$$

$\boxed{\cos 2\pi = \cos 0 = 1}$

For values of n > 1, the a_n are calculated using equation (1) as shown below:

$$a_n = -\frac{A\cos(1+n)\pi}{(1+n)\pi} + \frac{A}{(1+n)\pi} - \frac{A\cos(1-n)\pi}{(1-n)\pi} + \frac{A}{(1-n)\pi}$$

When n is even integer, $(1+n)$ and $(1-n)$ will be odd, $\therefore \cos(1+n)\pi = -1$; $\cos(1-n)\pi = -1$

When n is odd integer, $(1+n)$ and $(1-n)$ will be even, $\therefore \cos(1+n)\pi = 1$; $\cos(1-n)\pi = 1$

$$\therefore a_n = 0 \quad ; \quad \text{for odd values of } n$$

$$a_n = \frac{A}{(1+n)\pi} + \frac{A}{(1+n)\pi} + \frac{A}{(1-n)\pi} + \frac{A}{(1-n)\pi} \quad ; \quad \text{for even values of } n$$

$$= \frac{2A}{(1+n)\pi} + \frac{2A}{(1-n)\pi} = \frac{2A(1-n)+2A(1+n)}{(1+n)(1-n)\pi} = \frac{4A}{(1-n^2)\pi} = -\frac{4A}{(n^2-1)\pi}$$

$$\therefore a_2 = -\frac{4A}{(2^2-1)\pi} = -\frac{4A}{3\pi}$$

$$a_4 = -\frac{4A}{(4^2-1)\pi} = -\frac{4A}{15\pi}$$

$$a_6 = -\frac{4A}{(6^2-1)\pi} = -\frac{4A}{35\pi}$$

$$a_8 = -\frac{4A}{(8^2-1)\pi} = -\frac{4A}{63\pi} \quad \text{and so on}$$

Fourier Series

The trigonometric form of Fourier series of x(t) is,

$$x(t) = \frac{a_0}{2} + \sum_{n=1}^{\infty} a_n \cos n\Omega_0 t + \sum_{n=1}^{\infty} b_n \sin n\Omega_0 t$$

Here, $b_n = 0$, and a_n exists only for even values of n.

$$\therefore x(t) = \frac{a_0}{2} + \sum_{n=\text{even}} a_n \cos n\Omega_0 t$$

$$= \frac{a_0}{2} + a_2 \cos 2\Omega_0 t + a_4 \cos 4\Omega_0 t + a_6 \cos 6\Omega_0 t + a_8 \cos 8\Omega_0 t + \dots\dots\dots$$

$$= \frac{2A}{\pi} - \frac{4A}{3\pi}\cos 2\Omega_0 t - \frac{4A}{15\pi}\cos 4\Omega_0 t - \frac{4A}{35\pi}\cos 6\Omega_0 t - \frac{4A}{63\pi}\cos 8\Omega_0 t - \dots$$

$$= \frac{2A}{\pi} - \frac{4A}{\pi}\left[\frac{\cos 2\Omega_0 t}{2^2-1} + \frac{\cos 4\Omega_0 t}{4^2-1} + \frac{\cos 6\Omega_0 t}{6^2-1} + \frac{\cos 8\Omega_0 t}{8^2-1} + \dots\dots\dots\right]$$

Example 2.5   *(AU Dec'13, 10 Marks)*

Determine the Fourier series of the square wave shown in Fig 2.5.1.

Solution:

The given waveform has odd symmetry, half-wave symmetry and quarter wave symmetry.

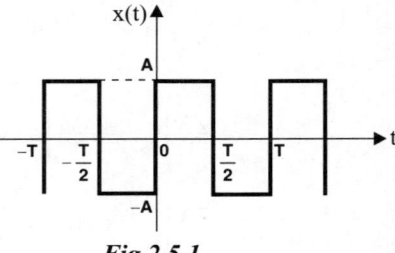

Fig 2.5.1.

$$\therefore a_0 = 0, \quad a_n = 0, \quad b_n = \frac{4}{T}\int_0^{T/2} x(t)\sin n\Omega_0 t \, dt$$

The mathematical equation of the given waveform is,

$$x(t) = A \quad ; \quad \text{for } t = 0 \text{ to } \frac{T}{2}$$

$$= -A \quad ; \quad \text{for } t = \frac{T}{2} \text{ to } T$$

Evaluation of b_n

$$b_n = \frac{4}{T} \int_0^{T/2} x(t) \sin n\Omega_0 t\, dt = \frac{4}{T} \int_0^{T/2} A \sin n\Omega_0 t\, dt = \frac{4A}{T} \left[\frac{-\cos n\Omega_0 t}{n\Omega_0} \right]_0^{T/2}$$

$$\boxed{\Omega_0 = \frac{2\pi}{T}}$$
$$\boxed{\cos 0 = 1}$$

$$= \frac{4A}{T} \left[\frac{-\cos n\frac{2\pi}{T}t}{n\frac{2\pi}{T}} \right]_0^{T/2} = \frac{4A}{T} \left[\frac{-\cos n\frac{2\pi}{T}\frac{T}{2}}{n\frac{2\pi}{T}} + \frac{\cos 0}{n\frac{2\pi}{T}} \right] = \frac{4A}{T} \left[-\frac{T}{2n\pi} \cos n\pi + \frac{T}{2n\pi} \right]$$

$\cos n\pi = -1,$ for n = odd

$\cos n\pi = +1,$ for n = even

$\therefore b_n = 0$; for even values of n

$\qquad = \frac{4A}{T} \left[\frac{T}{2n\pi} + \frac{T}{2n\pi} \right] = \frac{4A}{n\pi}$; for odd values of n

$\therefore b_1 = \frac{4A}{\pi}$; $b_3 = \frac{4A}{3\pi}$; $b_5 = \frac{4A}{5\pi}$ and so on.

Fourier Series

The trigonometric form of Fourier series of $x(t)$ is,

$$x(t) = \frac{a_0}{2} + \sum_{n=1}^{\infty} a_n \cos n\Omega_0 t + \sum_{n=1}^{\infty} b_n \sin n\Omega_0 t$$

Here, $a_0 = 0$, $a_n = 0$ and b_n exists only for odd values of n.

$$\therefore x(t) = \sum_{n=odd} b_n \sin n\Omega_0 t$$

$$= b_1 \sin \Omega_0 t + b_3 \sin 3\Omega_0 t + b_5 \sin 5\Omega_0 t + \text{..........................}$$

$$= \frac{4A}{\pi} \sin \Omega_0 t + \frac{4A}{3\pi} \sin 3\Omega_0 t + \frac{4A}{5\pi} \sin 5\Omega_0 t + \text{.....................}$$

$$= \frac{4A}{\pi} \left[\sin \Omega_0 t + \frac{\sin 3\Omega_0 t}{3} + \frac{\sin 5\Omega_0 t}{5} + \text{..........................} \right]$$

Example 2.6

Determine the trigonometric form of Fourier series of the signal shown in Fig 2.6.1.

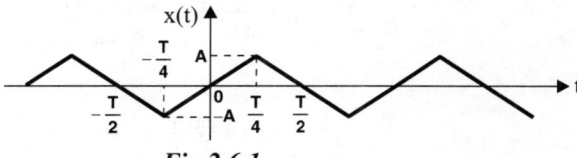

Solution:

The given signal has odd symmetry, half-wave symmetry and quarter wave symmetry, and so

Fig 2.6.1.

$a_0 = 0$, $a_n = 0$,

$$b_n = \frac{4}{T} \int_0^{T/2} x(t) \sin n\Omega_0 t\, dt \quad \text{(or)} \quad b_n = \frac{4}{T} \int_{-T/4}^{+T/4} x(t) \sin n\Omega_0 t\, dt$$

Note: *Here $x(t)$ is governed by single mathematical equation in the range $-\frac{T}{4}$ to $+\frac{T}{4}$. And so the calculations will be simple, if the integral limit is $-\frac{T}{4}$ to $+\frac{T}{4}$*

To Find Mathematical Equation for x(t)

Consider the equation of straight line, $\dfrac{y-y_1}{y_1-y_2} = \dfrac{x-x_1}{x_1-x_2}$

Here, $y = x(t)$, $x = t$.

$\therefore$ The equation of straight line can be written as, $\dfrac{x(t)-x(t_1)}{x(t_1)-x(t_2)} = \dfrac{t-t_1}{t_1-t_2}$ .....(1)

Consider points P and Q, as shown in Fig 1.

Coordinates of point-P = $[t_1, x(t_1)] = \left[-\dfrac{T}{4}, -A\right]$

Coordinates of point-Q = $[t_2, x(t_2)] = \left[\dfrac{T}{4}, A\right]$

On substituting the coordinates of points P and Q in equation (1) we get,

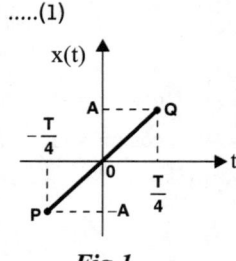

Fig 1.

$$\dfrac{x(t)-(-A)}{-A-A} = \dfrac{t-\left(-\dfrac{T}{4}\right)}{-\dfrac{T}{4}-\dfrac{T}{4}} \quad\Rightarrow\quad \dfrac{x(t)+A}{-2A} = \dfrac{t+\dfrac{T}{4}}{-\dfrac{T}{2}}$$

$$\Downarrow$$

$$-\dfrac{x(t)}{2A}-\dfrac{1}{2} = -\dfrac{2t}{T}-\dfrac{1}{2} \quad\Rightarrow\quad -\dfrac{x(t)}{2A} = -\dfrac{2t}{T} \quad\Rightarrow\quad x(t) = \dfrac{4A}{T}t$$

$\therefore x(t) = \dfrac{4A}{T}t$; for $t = -\dfrac{T}{4}$ to $+\dfrac{T}{4}$

Evaluation of b_n

$$b_n = \dfrac{4}{T}\int_{-T/4}^{+T/4}x(t)\sin n\Omega_0 t\,dt = \dfrac{4}{T}\int_{-T/4}^{+T/4}\dfrac{4A}{T}t\sin n\Omega_0 t\,dt = \dfrac{16A}{T^2}\int_{-T/4}^{+T/4}t\sin n\Omega_0 t\,dt$$

$$= \dfrac{16A}{T^2}\left[t\left(\dfrac{-\cos n\Omega_0 t}{n\Omega_0}\right) - \int\left(\dfrac{-\cos n\Omega_0 t}{n\Omega_0}\right)\times 1\,dt\right]_{-T/4}^{+T/4}$$

$$\boxed{\begin{array}{l}\int u\,dv = u\,v - \int v\,du \\ u = t \qquad\qquad \Rightarrow du = 1 \\ dv = \sin n\Omega_0 t \Rightarrow v = \dfrac{-\cos n\Omega_0 t}{n\Omega_0}\end{array}}$$

$$= \dfrac{16A}{T^2}\left[-t\dfrac{\cos n\Omega_0 t}{n\Omega_0} + \dfrac{\sin n\Omega_0 t}{n^2\Omega_0^2}\right]_{-T/4}^{T/4} = \dfrac{16A}{T^2}\left[-t\dfrac{\cos n\dfrac{2\pi}{T}t}{n\dfrac{2\pi}{T}} + \dfrac{\sin n\dfrac{2\pi}{T}t}{n^2\dfrac{4\pi^2}{T^2}}\right]_{-T/4}^{T/4}$$

$$\boxed{\Omega_0 = \dfrac{2\pi}{T}}$$

$$= \dfrac{16A}{T^2}\left[-\dfrac{T}{4}\dfrac{\cos n\dfrac{2\pi}{T}\dfrac{T}{4}}{n\dfrac{2\pi}{T}} + \dfrac{\sin n\dfrac{2\pi}{T}\dfrac{T}{4}}{n^2\dfrac{4\pi^2}{T^2}} + \dfrac{T}{4}\dfrac{\cos n\dfrac{2\pi}{T}\left(-\dfrac{T}{4}\right)}{n\dfrac{2\pi}{T}} - \dfrac{\sin n\dfrac{2\pi}{T}\left(-\dfrac{T}{4}\right)}{n^2\dfrac{4\pi^2}{T^2}}\right]$$

$$= \dfrac{16A}{T^2}\left[-\dfrac{T^2}{8n\pi}\cos\dfrac{n\pi}{2} + \dfrac{T^2}{4n^2\pi^2}\sin\dfrac{n\pi}{2} + \dfrac{T^2}{8n\pi}\cos\dfrac{n\pi}{2} + \dfrac{T^2}{4n^2\pi^2}\sin\dfrac{n\pi}{2}\right]$$

$$= -\dfrac{2A}{n\pi}\cos\dfrac{n\pi}{2} + \dfrac{4A}{n^2\pi^2}\sin\dfrac{n\pi}{2} + \dfrac{2A}{n\pi}\cos\dfrac{n\pi}{2} + \dfrac{4A}{n^2\pi^2}\sin\dfrac{n\pi}{2}$$

$$\boxed{\begin{array}{l}\cos(-\theta) = \cos\theta \\ \sin(-\theta) = -\sin\theta\end{array}}$$

$$= \dfrac{8A}{n^2\pi^2}\sin\dfrac{n\pi}{2}$$

For odd integer values of n, $\sin\frac{n\pi}{2} = \pm 1$

For even integer values of n, $\sin\frac{n\pi}{2} = 0$

$$\therefore \quad b_n = 0 \qquad ; \text{ for even values of n}$$

$$= \frac{8A}{n^2\pi^2}\sin\frac{n\pi}{2} \; ; \text{ for odd values of n}$$

$$\therefore \; b_1 = \frac{8A}{1^2\pi^2}\sin\frac{\pi}{2} = +\frac{8A}{\pi^2}$$

$$b_3 = \frac{8A}{3^2\pi^2}\sin\frac{3\pi}{2} = -\frac{8A}{3^2\pi^2}$$

$$b_5 = \frac{8A}{5^2\pi^2}\sin\frac{5\pi}{2} = +\frac{8A}{5^2\pi^2}$$

$$b_7 = \frac{8A}{7^2\pi^2}\sin\frac{7\pi}{2} = -\frac{8A}{7^2\pi^2} \text{ and so on.}$$

Fourier Series

The trigonometric form of Fourier series of x(t) is given by,

$$x(t) = \frac{a_0}{2} + \sum_{n=1}^{\infty} a_n\cos n\Omega_0 t + \sum_{n=1}^{\infty} b_n\sin n\Omega_0 t$$

Here, $a_0 = 0$, $a_n = 0$ and b_n exists only for odd values of n.

$$\therefore \; x(t) = \sum_{n=odd} b_n\sin n\Omega_0 t$$

$$= b_1\sin\Omega_0 t + b_3\sin 3\Omega_0 t + b_5\sin 5\Omega_0 t + b_7\sin 7\Omega_0 t$$

$$= \frac{8A}{\pi^2}\sin\Omega_0 t - \frac{8A}{3^2\pi^2}\sin 3\Omega_0 t + \frac{8A}{5^2\pi^2}\sin 5\Omega_0 t - \frac{8A}{7^2\pi^2}\sin 7\Omega_0 t + \ldots\ldots\ldots$$

$$= \frac{8A}{\pi^2}\left[\sin\Omega_0 t - \frac{\sin 3\Omega_0 t}{3^2} + \frac{\sin 5\Omega_0 t}{5^2} - \frac{\sin 7\Omega_0 t}{7^2} + \ldots\ldots\ldots\ldots\ldots\ldots\ldots\right]$$

Example 2.7

Determine the trigonometric form of Fourier series for the signal shown in Fig 2.7.1.

Solution:

The given signal has odd symmetry and so $a_0 = 0$, $a_n = 0$,

$$b_n = \frac{4}{T}\int_0^{T/2} x(t)\sin n\Omega_0 t \, dt$$

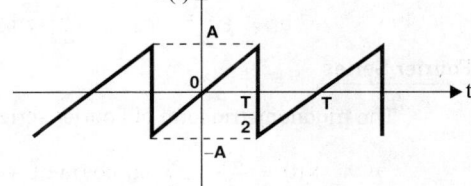

Fig 2.7.1.

To Find Mathematical Equation for x(t)

Consider the equation of straight line, $\dfrac{y - y_1}{y_1 - y_2} = \dfrac{x - x_1}{x_1 - x_2}$

Here, $y = x(t)$, $x = t$.

$\therefore$ The equation of straight line can be written as, $\dfrac{x(t) - x(t_1)}{x(t_1) - x(t_2)} = \dfrac{t - t_1}{t_1 - t_2}$ .....(1)

Consider points P and Q, as shown in Fig 1.

Coordinates of point-P = $[t_1, x(t_1)] = [0, 0]$

Coordinates of point-Q = $[t_2, x(t_2)] = \left[\dfrac{T}{2}, A\right]$

On substituting the coordinates of points P and Q in equation (1) we get,

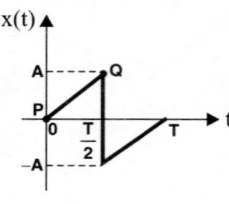

$$\frac{x(t) - 0}{0 - A} = \frac{t - 0}{0 - \dfrac{T}{2}} \Rightarrow \frac{x(t)}{-A} = \frac{t}{-\dfrac{T}{2}} \Rightarrow x(t) = \frac{2At}{T}$$

$$\therefore x(t) = \frac{2At}{T} \quad ; \quad \text{for } t = 0 \text{ to } \frac{T}{2}$$

Fig 1.

Evaluation of b_n

$$b_n = \frac{4}{T}\int_0^{T/2} x(t)\sin n\Omega_0 t\, dt = \frac{4}{T}\int_0^{T/2}\frac{2At}{T}\sin n\Omega_0 t\, dt = \frac{8A}{T^2}\int_0^{T/2} t\sin n\Omega_0 t\, dt$$

$$= \frac{8A}{T^2}\left[t\left(\frac{-\cos n\Omega_0 t}{n\Omega_0}\right) - \int\left(\frac{-\cos n\Omega_0 t}{n\Omega_0}\right)\times 1 dt\right]_0^{T/2}$$

$$\begin{array}{|l|}
\hline
\int u\, dv = u\, v - \int v\, du \\
\hline
u = t \qquad\qquad \Rightarrow du = 1 \\
dv = \sin n\Omega_0 t \Rightarrow v = \dfrac{-\cos n\Omega_0 t}{n\Omega_0} \\
\hline
\end{array}$$

$$= \frac{8A}{T^2}\left[-t\,\frac{\cos n\Omega_0 t}{n\Omega_0} + \frac{\sin n\Omega_0 t}{n^2\Omega_0^2}\right]_0^{T/2} = \frac{8A}{T^2}\left[-t\,\frac{\cos n\dfrac{2\pi}{T}t}{n\dfrac{2\pi}{T}} + \frac{\sin n\dfrac{2\pi}{T}t}{n^2\dfrac{4\pi^2}{T^2}}\right]_0^{T/2}$$

$$\boxed{\Omega_0 = \frac{2\pi}{T}}$$

$$= \frac{8A}{T^2}\left[-\frac{\dfrac{T}{2}\cos n\dfrac{2\pi}{T}\dfrac{T}{2}}{n\dfrac{2\pi}{T}} + \frac{\sin n\dfrac{2\pi}{T}\dfrac{T}{2}}{n^2\dfrac{4\pi^2}{T^2}} + \frac{0\times\cos 0}{n\dfrac{2\pi}{T}} - \frac{\sin 0}{n^2\dfrac{4\pi^2}{T^2}}\right]$$

$$\boxed{\begin{array}{l}\sin 0 = 0\\ \cos 0 = 1\end{array}}$$

$$= \frac{8A}{T^2}\left[-\frac{T^2}{4n\pi}\cos n\pi + \frac{T^2}{4n^2\pi^2}\sin n\pi\right] = -\frac{2A}{n\pi}\cos n\pi$$

$$\boxed{\begin{array}{l}\sin n\pi = 0\\ \text{for integer } n\end{array}}$$

For even integer values of n, $\cos n\pi = +1$

For odd integer values of n, $\cos n\pi = -1$

$$\therefore b_n = -\frac{2A}{n\pi} \quad \text{for } n = \text{even}$$

$$= +\frac{2A}{n\pi} \quad \text{for } n = \text{odd}$$

$$\therefore b_1 = +\frac{2A}{\pi} \; ; \; b_2 = -\frac{2A}{2\pi} \; ; \; b_3 = +\frac{2A}{3\pi} \; ; \; b_4 = -\frac{2A}{4\pi} \; ; \; b_5 = +\frac{2A}{5\pi} \text{ and so on.}$$

Fourier Series

The trigonometric form of Fourier series of x(t) is,

$$x(t) = \frac{a_0}{2} + \sum_{n=1}^{\infty} a_n\cos n\Omega_0 t + \sum_{n=1}^{\infty} b_n\sin n\Omega_0 t$$

Here, $a_0 = 0$, $a_n = 0$

$$\therefore x(t) = \sum_{n=1}^{\infty} b_n\sin n\Omega_0 t$$

$$= b_1\sin\Omega_0 t + b_2\sin 2\Omega_0 t + b_3\sin 3\Omega_0 t + b_4\sin 4\Omega_0 t + b_5\sin 5\Omega_0 t + \dots\dots$$

$$\therefore \; x(t) = \frac{2A}{\pi} \sin \Omega_0 t - \frac{2A}{2\pi} \sin 2\Omega_0 t + \frac{2A}{3\pi} \sin 3\Omega_0 t - \frac{2A}{4\pi} \sin 4\Omega_0 t + \frac{2A}{5\pi} \sin 5\Omega_0 t - .....$$

$$= \frac{2A}{\pi} \left[\frac{\sin \Omega_0 t}{1} - \frac{\sin 2\Omega_0 t}{2} + \frac{\sin 3\Omega_0 t}{3} - \frac{\sin 4\Omega_0 t}{4} + \frac{\sin 5\Omega_0 t}{5} - .... \right]$$

Example 2.8   *(AU May'15, 10 Marks)*

Determine the trigonometric form of the Fourier series of the ramp signal shown in Fig 2.8.1

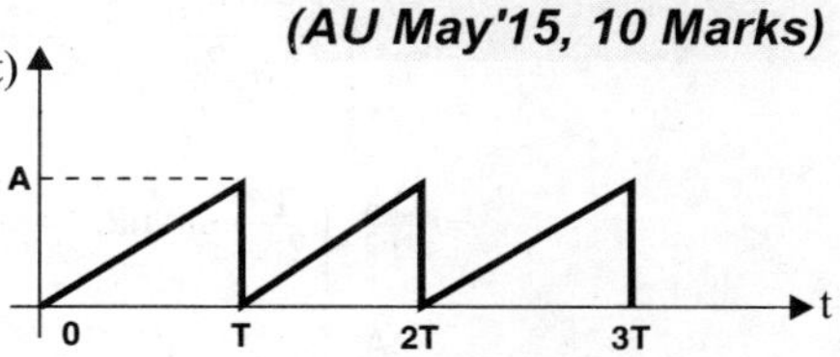

Solution:

The given signal is neither even nor odd.

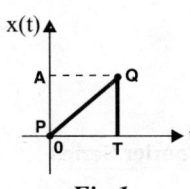

Fig 2.8.1.

$$\therefore \; a_0 = \frac{2}{T} \int_0^T x(t)\, dt \quad ; \quad a_n = \frac{2}{T} \int_0^T x(t) \cos n\Omega_0 t\, dt \quad ; \quad b_n = \frac{2}{T} \int_0^T x(t) \sin n\Omega_0 t\, dt$$

> **Note :** *It can be shown that after subtracting $a_0/2$ from the signal, it becomes odd. Hence a_n will be equal to zero.*

To Find Mathematical Equation for x(t)

Consider the equation of straight line, $\dfrac{y - y_1}{y_1 - y_2} = \dfrac{x - x_1}{x_1 - x_2}$

Here, $y = x(t), \quad x = t.$

$\therefore$ The equation of straight line can be written as, $\dfrac{x(t) - x(t_1)}{x(t_1) - x(t_2)} = \dfrac{t - t_1}{t_1 - t_2}$

.....(1)

Consider points P and Q, as shown in Fig 1.

Coordinates of point-P = $[t_1, x(t_1)] = [0, 0]$

Coordinates of point-Q = $[t_2, x(t_2)] = [T, A]$

On substituting the coordinates of points P and Q in equation (1) we get,

$$\frac{x(t) - 0}{0 - A} = \frac{t - 0}{0 - T} \;\Rightarrow\; \frac{x(t)}{-A} = \frac{t}{-T} \;\Rightarrow\; x(t) = \frac{At}{T}$$

$$\therefore \; x(t) = \frac{At}{T} \quad ; \quad \text{for } t = 0 \text{ to } T$$

Fig 1.

Evaluation of a_0

$$a_0 = \frac{2}{T} \int_0^T x(t)\, dt = \frac{2}{T} \int_0^T \frac{A}{T} t\, dt = \frac{2A}{T^2} \int_0^T t\, dt$$

$$= \frac{2A}{T^2} \left[\frac{t^2}{2} \right]_0^T = \frac{2A}{T^2} \left[\frac{T^2}{2} - 0 \right] = A$$

Evaluation of a_n

$$a_n = \frac{2}{T} \int_0^T x(t) \cos n\Omega_0 t\, dt = \frac{2}{T} \int_0^T \frac{At}{T} \cos n\Omega_0 t\, dt = \frac{2A}{T^2} \int_0^T t \cos n\Omega_0 t\, dt$$

$$= \frac{2A}{T^2} \left[t \left(\frac{\sin n\Omega_0 t}{n\Omega_0} \right) - \int \left(\frac{\sin n\Omega_0 t}{n\Omega_0} \right) \times 1 \; dt \right]_0^T$$

$\int u\, dv = u\, v - \int v\, du$	
$u = t$	$\Rightarrow du = 1$
$dv = \cos n\Omega_0 t \Rightarrow$	$v = \dfrac{\sin n\Omega_0 t}{n\Omega_0}$

$$\therefore a_n = \frac{2A}{T^2}\left[t\frac{\sin n\Omega_0 t}{n\Omega_0} - \left(\frac{-\cos n\Omega_0 t}{n^2\Omega_0^2}\right)\right]_0^T = \frac{2A}{T^2}\left[\frac{t\sin n\frac{2\pi}{T}t}{n\frac{2\pi}{T}} + \frac{\cos n\frac{2\pi}{T}t}{n^2\frac{4\pi^2}{T^2}}\right]_0^T$$

$$\boxed{\Omega_0 = \frac{2\pi}{T}}$$

$$= \frac{2A}{T^2}\left[\frac{T\sin n\frac{2\pi}{T}T}{n\frac{2\pi}{T}} + \frac{\cos n\frac{2\pi}{T}T}{n^2\frac{4\pi^2}{T^2}} - \frac{0\times\sin 0}{n\frac{2\pi}{T}} - \frac{\cos 0}{n^2\frac{4\pi^2}{T^2}}\right]$$

$$\boxed{\begin{array}{l}\sin 0 = 0\\ \cos 0 = 1\end{array}}$$

$$= \frac{2A}{T^2}\left[\frac{T^2}{2n\pi}\sin n2\pi + \frac{T^2}{4n^2\pi^2}\cos n2\pi - 0 - \frac{T^2}{4n^2\pi^2}\right]$$

$$\boxed{\begin{array}{l}\sin n2\pi = 0\\ \cos n2\pi = 1\\ \text{for integer } n\end{array}}$$

$$= \frac{2A}{T^2}\left[0 + \frac{T^2}{4n^2\pi^2} - \frac{T^2}{4n^2\pi^2}\right] = 0$$

Evaluation of b_n

$$b_n = \frac{2}{T}\int_0^T x(t)\sin n\Omega_0 t\,dt = \frac{2}{T}\int_0^T \frac{At}{T}\sin n\Omega_0 t\,dt = \frac{2A}{T^2}\int_0^T t\sin n\Omega_0 t\,dt$$

$$\boxed{\begin{array}{l}\int u\,dv = uv - \int v\,du\\[4pt] u = t \qquad\qquad \Rightarrow du = 1\\[4pt] dv = \sin n\Omega_0 t \Rightarrow v = \frac{-\cos n\Omega_0 t}{n\Omega_0}\end{array}}$$

$$= \frac{2A}{T^2}\left[t\left(\frac{-\cos n\Omega_0 t}{n\Omega_0}\right) - \int\left(\frac{-\cos n\Omega_0 t}{n\Omega_0}\right)\times 1\,dt\right]_0^T$$

$$= \frac{2A}{T^2}\left[-t\frac{\cos n\Omega_0 t}{n\Omega_0} + \frac{\sin n\Omega_0 t}{n^2\Omega_0^2}\right]_0^T = \frac{2A}{T^2}\left[-t\frac{\cos n\frac{2\pi}{T}t}{n\frac{2\pi}{T}} + \frac{\sin n\frac{2\pi}{T}t}{n^2\frac{4\pi^2}{T^2}}\right]_0^T$$

$$= \frac{2A}{T^2}\left[-\frac{T\cos n\frac{2\pi}{T}T}{n\frac{2\pi}{T}} + \frac{\sin n\frac{2\pi}{T}T}{n^2\frac{4\pi^2}{T^2}} + \frac{0\times\cos 0}{n\frac{2\pi}{T}} - \frac{\sin 0}{n^2\frac{4\pi^2}{T^2}}\right]$$

$$\boxed{\Omega_0 = \frac{2\pi}{T}}$$

$$\boxed{\begin{array}{l}\sin 0 = 0\\ \cos 0 = 1\end{array}}$$

$$= \frac{2A}{T^2}\left[-\frac{T^2}{2n\pi}\cos n2\pi + \frac{T^2}{4n^2\pi^2}\sin n2\pi\right] = -\frac{A}{n\pi}$$

$$\therefore b_1 = -\frac{A}{\pi} \;;\; b_2 = -\frac{A}{2\pi} \;;\; b_3 = -\frac{A}{3\pi}\; b_4 = -\frac{A}{4\pi}\;\text{ and so on.}$$

$$\boxed{\begin{array}{l}\sin n2\pi = 0\\ \cos n2\pi = 1\\ \text{for integer } n\end{array}}$$

Fourier Series

The trigonometric form of Fourier series of $x(t)$ is given by,

$$x(t) = \frac{a_0}{2} + \sum_{n=1}^{\infty} a_n\cos n\Omega_0 t + \sum_{n=1}^{\infty} b_n\sin n\Omega_0 t$$

Here, $a_n = 0$

$$\therefore x(t) = \frac{a_0}{2} + \sum_{n=1}^{\infty} b_n\sin n\Omega_0 t$$

$$= \frac{a_0}{2} + b_1\sin\Omega_0 t + b_2\sin 2\Omega_0 t + b_3\sin 3\Omega_0 t + b_4\sin 4\Omega_0 t + \ldots\ldots\ldots\ldots$$

$$= \frac{A}{2} - \frac{A}{\pi}\sin\Omega_0 t - \frac{A}{2\pi}\sin 2\Omega_0 t - \frac{A}{3\pi}\sin 3\Omega_0 t - \frac{A}{4\pi}\sin 4\Omega_0 t - \ldots\ldots\ldots$$

$$= \frac{A}{2} - \frac{A}{\pi}\left[\frac{\sin\Omega_0 t}{1} + \frac{\sin 2\Omega_0 t}{2} + \frac{\sin 3\Omega_0 t}{3} + \frac{\sin 4\Omega_0 t}{4} + \ldots\ldots\ldots\ldots\right]$$

Example 2.9   *(AU Jun'14, 16 Marks)*   *(AU Jun'13, 8 Marks & Dec'12, Jun'11 12 Marks)*

Determine the Fourier series representation of the halfwave rectifier output shown in Fig 2.9.1.

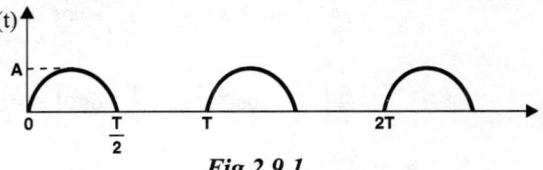

Solution:

The signal shown in Fig 2.9.1 is neither even nor odd.

Fig 2.9.1.

$$\therefore \ a_0 = \frac{2}{T}\int_0^T x(t)\,dt \ ; \ a_n = \frac{2}{T}\int_0^T x(t)\cos n\Omega_0 t\,dt \ ; \ b_n = \frac{2}{T}\int_0^T x(t)\sin n\Omega_0 t\,dt$$

The mathematical equation representing halfwave rectified output is,

$$x(t) = A\sin\Omega_0 t \quad ; \ \text{for } t = 0 \text{ to } \frac{T}{2}$$

$$= 0 \qquad\quad ; \ \text{for } t = \frac{T}{2} \text{ to } T$$

Evaluation of a_0

$$a_0 = \frac{2}{T}\int_0^T x(t)\,dt = \frac{2}{T}\int_0^{T/2} A\sin\Omega_0 t\,dt = \frac{2A}{T}\left[\frac{-\cos\Omega_0 t}{\Omega_0}\right]_0^{T/2}$$

$$= \frac{2A}{T}\left[\frac{-\cos\frac{2\pi}{T}t}{\frac{2\pi}{T}}\right]_0^{T/2} = \frac{2A}{T}\left[\frac{-\cos\frac{2\pi}{T}\frac{T}{2}}{\frac{2\pi}{T}} + \frac{\cos 0}{\frac{2\pi}{T}}\right] \qquad \boxed{\Omega_0 = \frac{2\pi}{T}}$$

$$\boxed{\cos 0 = 1}$$
$$\boxed{\cos \pi = -1}$$

$$= \frac{2A}{T}\left[-\frac{T}{2\pi}\cos\pi + \frac{T}{2\pi}\right] = \frac{2A}{T}\left[-\frac{T}{2\pi}\times(-1) + \frac{T}{2\pi}\right]$$

$$= \frac{2A}{T}\times\frac{T}{\pi} = \frac{2A}{\pi}$$

Evaluation of a_n

$$a_n = \frac{2}{T}\int_0^T x(t)\cos n\Omega_0 t\,dt = \frac{2}{T}\int_0^{T/2} A\sin\Omega_0 t\cos n\Omega_0 t\,dt \qquad \boxed{2\sin A\cos B = \sin(A+B) + \sin(A-B)}$$

$$= \frac{2A}{T}\int_0^{T/2}\frac{\sin(\Omega_0 t + n\Omega_0 t) + \sin(\Omega_0 t - n\Omega_0 t)}{2}\,dt = \frac{A}{T}\int_0^{T/2}\left[\sin(1+n)\Omega_0 t + \sin(1-n)\Omega_0 t\right]dt$$

$$= \frac{A}{T}\left[-\frac{\cos(1+n)\Omega_0 t}{(1+n)\Omega_0} - \frac{\cos(1-n)\Omega_0 t}{(1-n)\Omega_0}\right]_0^{T/2} = \frac{A}{T}\left[\frac{-\cos(1+n)\frac{2\pi}{T}t}{(1+n)\frac{2\pi}{T}} - \frac{\cos(1-n)\frac{2\pi}{T}t}{(1-n)\frac{2\pi}{T}}\right]_0^{T/2}$$

$$= \frac{A}{T}\left[\frac{-\cos(1+n)\frac{2\pi}{T}\frac{T}{2}}{(1+n)\frac{2\pi}{T}} - \frac{\cos(1-n)\frac{2\pi}{T}\frac{T}{2}}{(1-n)\frac{2\pi}{T}} + \frac{\cos 0}{(1+n)\frac{2\pi}{T}} + \frac{\cos 0}{(1-n)\frac{2\pi}{T}}\right] \qquad \boxed{\Omega_0 = \frac{2\pi}{T}}$$

$$\boxed{\cos 0 = 1}$$

$$= -\frac{A\cos(1+n)\pi}{(1+n)2\pi} - \frac{A\cos(1-n)\pi}{(1-n)2\pi} + \frac{A}{(1+n)2\pi} + \frac{A}{(1-n)2\pi}$$

The above expression for a_n can be evaluated for all values of n except for $n = 1$. For $n = 1$, a_n has to be evaluated separately as shown below.

$$a_1 = \frac{2}{T}\int_0^T x(t)\cos\Omega_0 t\,dt = \frac{2}{T}\int_0^T A\sin\Omega_0 t\cos\Omega_0 t\,dt \qquad \boxed{\sin 2\theta = 2\sin\theta\cos\theta}$$

$$\therefore \; a_1 = \frac{2A}{T}\int_0^T \frac{\sin 2\Omega_0 t}{2}\,dt \; = \; \frac{A}{T}\left[\frac{-\cos 2\Omega_0 t}{2\Omega_0}\right]_0^T \; = \; \frac{A}{T}\left[\frac{-\cos\frac{4\pi}{T}t}{\frac{4\pi}{T}}\right]_0^T$$

$\boxed{\Omega_0 = \dfrac{2\pi}{T}}$

$$= \; \frac{A}{T}\left[-\frac{T}{4\pi}\cos\frac{4\pi T}{T} + \frac{T}{4\pi}\cos 0\right] = \frac{A}{T}\left[-\frac{T}{4\pi} + \frac{T}{4\pi}\right] = 0$$

$\boxed{\cos 4\pi = 1}$
$\boxed{\cos 0 \; = 1}$

$$\therefore \; a_1 \; = \; 0$$

$$a_n \; = \; -\frac{A\cos(1+n)\pi}{(1+n)2\pi} - \frac{A\cos(1-n)\pi}{(1-n)2\pi} + \frac{A}{(1+n)2\pi} + \frac{A}{(1-n)2\pi} \; ; \quad \text{for all n except } n = 1$$

When n is even, the terms $(n + 1)$ and $(n - 1)$ are odd, $\qquad \therefore \cos(1+n)\pi = -1, \;\; \cos(1-n)\pi = -1$

When n is odd, the terms $(n + 1)$ and $(n - 1)$ are even, $\qquad \therefore \cos(1+n)\pi = \; 1, \;\; \cos(1-n)\pi = \; 1$

$$\therefore \; a_n = 0 \; ; \quad \text{for odd values of n}$$

$$a_n \; = \; \frac{A}{(1+n)2\pi} + \frac{A}{(1-n)2\pi} + \frac{A}{(1+n)2\pi} + \frac{A}{(1-n)2\pi} \; ; \quad \text{for even values of n}$$

$$= \; \frac{A}{(1+n)\pi} + \frac{A}{(1-n)\pi} = \frac{A(1-n) + A(1+n)}{(1+n)(1-n)\pi} = \frac{2A}{(1-n^2)\pi} = -\frac{2A}{(n^2-1)\pi}$$

$$\therefore \; a_2 = -\frac{2A}{(2^2-1)\pi} = -\frac{2A}{3\pi}$$

$$a_4 = -\frac{2A}{(4^2-1)\pi} = -\frac{2A}{15\pi}$$

$$a_6 = -\frac{2A}{(6^2-1)\pi} = -\frac{2A}{35\pi}$$

$$a_8 = -\frac{2A}{(8^2-1)\pi} = -\frac{2A}{63\pi} \; \text{and so on.}$$

Evaluation of b_n

$$b_n \; = \; \frac{2}{T}\int_0^T x(t)\sin n\Omega_0 t\,dt \; = \; \frac{2}{T}\int_0^{T/2} A\sin\Omega_0 t\sin n\Omega_0 t\,dt$$

$\boxed{2\sin A \sin B = \cos(A-B) - \cos(A+B)}$

$$= \; \frac{2A}{T}\int_0^{T/2}\frac{\cos(\Omega_0 t - n\Omega_0 t) - \cos(\Omega_0 t + n\Omega_0 t)}{2}\,dt \; = \; \frac{A}{T}\int_0^{T/2}[\cos(1-n)\Omega_0 t - \cos(1+n)\Omega_0 t]\,dt$$

$$= \; \frac{A}{T}\left[\frac{\sin(1-n)\Omega_0 t}{(1-n)\Omega_0} - \frac{\sin(1+n)\Omega_0 t}{(1+n)\Omega_0}\right]_0^{T/2} \; = \; \frac{A}{T}\left[\frac{\sin(1-n)\frac{2\pi}{T}t}{(1-n)\frac{2\pi}{T}} - \frac{\sin(1+n)\frac{2\pi}{T}t}{(1+n)\frac{2\pi}{T}}\right]_0^{T/2}$$

$\boxed{\Omega_0 = \dfrac{2\pi}{T}}$

$$= \; \frac{A}{T}\left[\frac{\sin(1-n)\frac{2\pi}{T}\frac{T}{2}}{(1-n)\frac{2\pi}{T}} - \frac{\sin(1+n)\frac{2\pi}{T}\frac{T}{2}}{(1+n)\frac{2\pi}{T}} - \frac{\sin 0}{(1-n)\frac{2\pi}{T}} + \frac{\sin 0}{(1+n)\frac{2\pi}{T}}\right]$$

$\boxed{\sin 0 = 0}$

$$= \; \frac{A\sin(1-n)\pi}{(1-n)2\pi} - \frac{A\sin(1+n)\pi}{(1+n)2\pi}$$

The above expression for b_n can be evaluated for all values of n except for $n = 1$. For $n = 1$, b_n has to be evaluated separately as shown below:

$$b_1 \; = \; \frac{2}{T}\int_0^T x(t)\sin\Omega_0 t\,dt \; = \; \frac{2}{T}\int_0^{T/2} A\sin\Omega_0 t\sin\Omega_0 t\,dt \; = \; \frac{2A}{T}\int_0^{T/2}\sin^2\Omega_0 t\,dt$$

$\boxed{\sin^2\theta = \dfrac{1-\cos 2\theta}{2}}$

$$b_1 = \frac{2A}{T} \int_0^{T/2} \frac{1 - \cos 2\Omega_0 t}{2} \, dt = \frac{A}{T} \int_0^{T/2} (1 - \cos 2\Omega_0 t) \, dt = \frac{A}{T} \left[t - \frac{\sin 2\Omega_0 t}{2\Omega_0} \right]_0^{T/2}$$

$$= \frac{A}{T} \left[t - \frac{\sin \frac{4\pi}{T} t}{\frac{4\pi}{T}} \right]_0^{T/2} = \frac{A}{T} \left[\frac{T}{2} - \frac{\sin \frac{4\pi}{T} \frac{T}{2}}{\frac{4\pi}{T}} - 0 + \frac{\sin 0}{\frac{4\pi}{T}} \right]$$

$$\boxed{\Omega_0 = \frac{2\pi}{T}}$$
$$\boxed{\sin 0 = 0}$$

$$= \frac{A}{2} - \frac{A}{4\pi} \sin 2\pi = \frac{A}{2}$$

$$\boxed{\sin 2\pi = 0}$$

$$\therefore \; b_1 = \frac{A}{2}$$

$$b_n = \frac{A \sin(1-n)\pi}{(1-n)2\pi} - \frac{A \sin(1+n)\pi}{(1+n)2\pi} \; ; \; \text{for all values of n except n = 1.}$$

For integer values of n, except when n = 1, $\sin(1 - n)\pi = 0$ and $\sin(1 + n)\pi = 0$.

$$\therefore \; b_n = 0 \text{ for all values of n except n = 1.}$$

Fourier Series

The trigonometric form of Fourier series of x(t) is given by,

$$x(t) = \frac{a_0}{2} + \sum_{n=1}^{\infty} a_n \cos n\Omega_0 t + \sum_{n=1}^{\infty} b_n \sin n\Omega_0 t$$

Here, a_n exists only for even values of n and $b_n = 0$ for all values of n except when n = 1.

$$\therefore \; x(t) = \frac{a_0}{2} + \sum_{n = \text{even}} a_n \cos n\Omega_0 t + b_1 \sin \Omega_0 t$$

$$= \frac{a_0}{2} + a_2 \cos 2\Omega_0 t + a_4 \cos 4\Omega_0 t + a_6 \cos 6\Omega_0 t + a_8 \cos 8\Omega_0 t + \ldots + b_1 \sin \Omega_0 t$$

$$= \frac{A}{\pi} - \frac{2A}{3\pi} \cos 2\Omega_0 t - \frac{2A}{15\pi} \cos 4\Omega_0 t - \frac{2A}{35\pi} \cos 6\Omega_0 t - \frac{2A}{63\pi} \cos 8\Omega_0 t - \ldots + \frac{A}{2} \sin \Omega_0 t$$

$$= \frac{A}{\pi} + \frac{2A}{\pi} \left[\frac{\pi}{4} \sin \Omega_0 t - \frac{\cos 2\Omega_0 t}{2^2 - 1} - \frac{\cos 4\Omega_0 t}{4^2 - 1} - \frac{\cos 6\Omega_0 t}{6^2 - 1} - \frac{\cos 8\Omega_0 t}{8^2 - 1} - \ldots \right]$$

Example 2.10

Find the Fourier series of the signal shown in Fig 2.10.1.

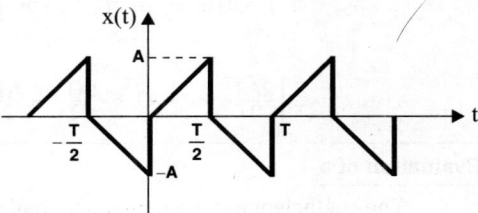

Fig 2.10.1.

Solution:

The given signal has halfwave symmetry and so a_0 will be zero. The Fourier coefficients a_n and b_n will exist only for odd integer values of n.

$$\therefore \; a_0 = 0 \; ; \; a_n = \frac{2}{T} \int_0^T x(t) \cos n\Omega_0 t \, dt \; ; \; b_n = \frac{2}{T} \int_0^T x(t) \sin n\Omega_0 t \, dt$$

To Find Mathematical Equation for x(t)

Consider the equation of straight line, $\dfrac{y - y_1}{y_1 - y_2} = \dfrac{x - x_1}{x_1 - x_2}$

Here, $y = x(t)$, $x = t$.

∴ The equation of straight line can be written as, $\dfrac{x(t) - x(t_1)}{x(t_1) - x(t_2)} = \dfrac{t - t_1}{t_1 - t_2}$ .....(1)

Consider points P and Q, as shown in Fig 1.

Coordinates of point-P = $[t_1, x(t_1)] = [0, 0]$

Coordinates of point-Q = $[t_2, x(t_2)] = \left[\dfrac{T}{2}, A\right]$

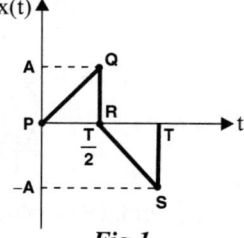

Fig 1.

On substituting the coordinates of points P and Q in equation (1) we get,

$$\dfrac{x(t) - 0}{0 - A} = \dfrac{t - 0}{0 - \dfrac{T}{2}} \;\Rightarrow\; \dfrac{x(t)}{-A} = \dfrac{t}{-\dfrac{T}{2}} \;\Rightarrow\; x(t) = \dfrac{2At}{T}$$

Consider points R and S, as shown in Fig 1.

Coordinates of point-R = $[t_3, x(t_3)] = \left[\dfrac{T}{2}, 0\right]$

Coordinates of point-S = $[t_4, x(t_4)] = [T, -A]$

On substituting the coordinates of points R and S in equation (1) we get,

$$\dfrac{x(t) - 0}{0 - (-A)} = \dfrac{t - \dfrac{T}{2}}{\dfrac{T}{2} - T} \;\Rightarrow\; \dfrac{x(t)}{A} = \dfrac{t - \dfrac{T}{2}}{-\dfrac{T}{2}} \;\Rightarrow\; \dfrac{x(t)}{A} = -\dfrac{2t}{T} + 1 \;\Rightarrow\; x(t) = A - \dfrac{2At}{T}$$

Now the mathematical equation of the waveform is given by,

$$x(t) = \dfrac{2At}{T} \quad ; \text{ for } t = 0 \text{ to } \dfrac{T}{2}$$

$$= A - \dfrac{2At}{T} \; ; \text{ for } t = \dfrac{T}{2} \text{ to } T$$

Evaluation of a_0

The given signal has half wave symmetry and so $a_0 = 0$.

Proof:

$$a_0 = \dfrac{2}{T} \int_0^T x(t)\, dt = \dfrac{2}{T}\left[\int_0^{T/2} \dfrac{2At}{T}\, dt + \int_{T/2}^T \left(A - \dfrac{2At}{T}\right) dt\right] = \dfrac{2}{T}\left[\left[\dfrac{2At^2}{2T}\right]_0^{T/2} + \left[At - \dfrac{2At^2}{2T}\right]_{T/2}^T\right]$$

$$= \dfrac{2}{T}\left[\dfrac{2AT^2}{8T} - 0 + AT - \dfrac{2AT^2}{2T} - \dfrac{AT}{2} + \dfrac{2AT^2}{8T}\right] = \dfrac{2}{T}\left[\dfrac{AT}{4} + AT - AT - \dfrac{AT}{2} + \dfrac{AT}{4}\right] = 0$$

Evaluation of a_n

The coefficient a_n for a signal with half-wave symmetry is,

$$a_n = \dfrac{2}{T}\int_0^T x(t)\cos n\Omega_0 t\; dt$$

$$= \dfrac{2}{T}\int_0^{T/2} \dfrac{2At}{T}\cos n\Omega_0 t\; dt + \dfrac{2}{T}\int_{T/2}^T \left(A - \dfrac{2At}{T}\right)\cos n\Omega_0 t\; dt$$

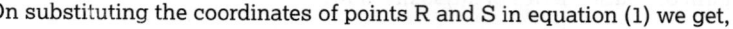

$$\therefore a_n = \frac{4A}{T^2} \int_0^{T/2} t \cos n\Omega_0 t \ dt + \frac{2A}{T} \int_{T/2}^{T} \cos n\Omega_0 t \ dt - \frac{4A}{T^2} \int_{T/2}^{T} t \cos n\Omega_0 t \ dt$$

$$= \frac{4A}{T^2}\left[t \times \left(\frac{\sin n\Omega_0 t}{n\Omega_0}\right) - \int \left(\frac{\sin n\Omega_0 t}{n\Omega_0}\right) \times 1 \, dt \right]_0^{T/2} + \frac{2A}{T}\left[\frac{\sin n\Omega_0 t}{n\Omega_0} \right]_{T/2}^{T}$$

$$- \frac{4A}{T^2}\left[t \times \left(\frac{\sin n\Omega_0 t}{n\Omega_0}\right) - \int \left(\frac{\sin n\Omega_0 t}{n\Omega_0}\right) \times 1 \, dt \right]_{T/2}^{T}$$

$\int u \, dv = u \, v - \int v \, du$
$u = t \qquad\qquad \Rightarrow du = 1$
$dv = \cos n\Omega_0 t \Rightarrow \quad v = \dfrac{\sin n\Omega_0 t}{n\Omega_0}$

$$= \frac{4A}{T^2}\left[\frac{t\sin n\Omega_0 t}{n\Omega_0} - \left(\frac{-\cos n\Omega_0 t}{n^2 \Omega_0{}^2}\right) \right]_0^{T/2} + \frac{2A}{T}\left[\frac{\sin n\Omega_0 t}{n\Omega_0} \right]_{T/2}^{T}$$

$$- \frac{4A}{T^2}\left[\frac{t\sin n\Omega_0 t}{n\Omega_0} - \left(\frac{-\cos n\Omega_0 t}{n^2 \Omega_0{}^2}\right) \right]_{T/2}^{T}$$

$$= \frac{4A}{T^2}\left[\frac{\frac{T}{2}\sin n\frac{2\pi}{T}\frac{T}{2}}{n\frac{2\pi}{T}} + \frac{\cos n\frac{2\pi}{T}\frac{T}{2}}{n^2\frac{4\pi^2}{T^2}} - \frac{0 \times \sin 0}{n\frac{2\pi}{T}} - \frac{\cos 0}{n^2\frac{4\pi^2}{T^2}} \right]$$

$$\boxed{\Omega_0 = \frac{2\pi}{T}}$$

$$+ \frac{2A}{T}\left[\frac{\sin n\frac{2\pi}{T}T}{n\frac{2\pi}{T}} + \frac{\sin n\frac{2\pi}{T}\frac{T}{2}}{n\frac{2\pi}{T}} \right]$$

$$\boxed{\begin{array}{l} \sin 0 = 0 \\ \cos 0 = 1 \end{array}}$$

$$- \frac{4A}{T^2}\left[\frac{T\sin n\frac{2\pi}{T}T}{n\frac{2\pi}{T}} + \frac{\cos n\frac{2\pi}{T}T}{n^2\frac{4\pi^2}{T^2}} - \frac{\frac{T}{2}\times\sin \frac{n2\pi}{T}\frac{T}{2}}{n\frac{2\pi}{T}} - \frac{\cos \frac{n2\pi}{T}\frac{T}{2}}{n^2\frac{4\pi^2}{T^2}} \right]$$

$$= \frac{A}{n\pi}\sin n\pi + \frac{A}{n^2\pi^2}\cos n\pi - \frac{A}{n^2\pi^2} + \frac{A}{n\pi}\sin n2\pi - \frac{A}{n\pi}\sin n\pi$$

$$- \frac{2A}{n\pi}\sin n2\pi - \frac{A}{n^2\pi^2}\cos n2\pi + \frac{A}{n\pi}\sin n\pi + \frac{A}{n^2\pi^2}\cos n\pi$$

$$\boxed{\begin{array}{l} \text{For integer n} \\ \sin n\pi = 0 \\ \sin n2\pi = 0 \\ \cos n2\pi = 1 \end{array}}$$

$$= 0 + \frac{A}{n^2\pi^2}\cos n\pi - \frac{A}{n^2\pi^2} + 0 - 0 - 0 - \frac{A}{n^2\pi^2} + 0 + \frac{A}{n^2\pi^2}\cos n\pi$$

$$= \frac{2A}{n^2\pi^2}\cos n\pi - \frac{2A}{n^2\pi^2} = \frac{2A}{n^2\pi^2}(\cos n\pi - 1)$$

When n is even integer, $\cos n\pi = +1$

When n is odd integer, $\cos n\pi = -1$

$$\therefore a_n = \frac{2A}{n^2\pi^2}(\cos n\pi - 1) = \frac{2A}{n^2\pi^2}(1-1) = 0 \qquad ; \text{ for even integer values of n}$$

$$= \frac{2A}{n^2\pi^2}(\cos n\pi - 1) = \frac{2A}{n^2\pi^2}(-1-1) = -\frac{4A}{\pi^2 n^2} \quad ; \text{ for odd integer values of n}$$

$$\therefore a_1 = -\frac{4A}{1^2\pi^2} \quad ; \quad a_3 = -\frac{4A}{3^2\pi^2} \quad ; \quad a_5 = -\frac{4A}{5^2\pi^2} \text{ and so on.}$$

Evaluation of b_n

The coefficient b_n for a signal with half wave symmetry is,

$$b_n = \frac{2}{T} \int_0^T x(t) \sin n\Omega_0 t \ dt$$

$$= \frac{2}{T} \int_0^{T/2} \frac{2At}{T} \sin n\Omega_0 t \ dt + \frac{2}{T} \int_{T/2}^T \left(A - \frac{2At}{T}\right) \sin n\Omega_0 t \ dt$$

$$= \frac{4A}{T^2} \int_0^{T/2} t \sin n\Omega_0 t \ dt + \frac{2A}{T} \int_{T/2}^T \sin n\Omega_0 t \ dt - \frac{4A}{T^2} \int_{T/2}^T t \sin n\Omega_0 t \ dt$$

$$= \frac{4A}{T^2} \left[t \times \left(\frac{-\cos n\Omega_0 t}{n\Omega_0}\right) - \int \left(\frac{-\cos n\Omega_0 t}{n\Omega_0}\right) \times 1 \ dt \right]_0^{T/2} + \frac{2A}{T} \left[\frac{-\cos n\Omega_0 t}{n\Omega_0}\right]_{T/2}^T$$

$$- \frac{4A}{T^2} \left[t \times \left(\frac{-\cos n\Omega_0 t}{n\Omega_0}\right) - \int \left(\frac{-\cos n\Omega_0 t}{n\Omega_0}\right) \times 1 \ dt \right]_{T/2}^T \qquad \boxed{\begin{array}{l} \int u \ dv = u\,v - \int v \ du \\ \hline u = t \qquad\qquad \Rightarrow du = 1 \\ dv = \sin n\Omega_0 t \Rightarrow \ v = \dfrac{-\cos n\Omega_0 t}{n\Omega_0} \end{array}}$$

$$= \frac{4A}{T^2} \left[\frac{-t\cos n\Omega_0 t}{n\Omega_0} - \left(\frac{-\sin n\Omega_0 t}{n^2 \Omega_0{}^2}\right) \right]_0^{T/2} + \frac{2A}{T} \left[\frac{-\cos n\Omega_0 t}{n\Omega_0}\right]_{T/2}^T$$

$$- \frac{4A}{T^2} \left[\frac{-t\cos n\Omega_0 t}{n\Omega_0} - \left(\frac{-\sin n\Omega_0 t}{n^2 \Omega_0{}^2}\right) \right]_{T/2}^T$$

$$= \frac{4A}{T^2} \left[\frac{-\dfrac{T}{2}\cos n\dfrac{2\pi}{T}\dfrac{T}{2}}{n\dfrac{2\pi}{T}} + \frac{\sin n\dfrac{2\pi}{T}\dfrac{T}{2}}{n^2 \dfrac{4\pi^2}{T^2}} + \frac{0 \times \cos 0}{n\dfrac{2\pi}{T}} - \frac{\sin 0}{n^2 \dfrac{4\pi^2}{T^2}} \right] \qquad \boxed{\begin{array}{l} \Omega_0 = \dfrac{2\pi}{T} \\ \hline \sin 0 = 0 \end{array}}$$

$$+ \frac{2A}{T} \left[-\frac{\cos n\dfrac{2\pi}{T}T}{n\dfrac{2\pi}{T}} + \frac{\cos n\dfrac{2\pi}{T}\dfrac{T}{2}}{n\dfrac{2\pi}{T}} \right]$$

$$- \frac{4A}{T^2} \left[\frac{-T\cos n\dfrac{2\pi}{T}T}{n\dfrac{2\pi}{T}} + \frac{\sin n\dfrac{2\pi}{T}T}{n^2 \dfrac{4\pi^2}{T^2}} + \frac{\dfrac{T}{2} \times \cos n\dfrac{2\pi}{T}\dfrac{T}{2}}{n\dfrac{2\pi}{T}} - \frac{\sin n\dfrac{2\pi}{T}\dfrac{T}{2}}{n^2 \dfrac{4\pi^2}{T^2}} \right]$$

$$= -\frac{A}{n\pi} \cos n\pi + \frac{A}{n^2\pi^2} \sin n\pi - \frac{A}{n\pi} \cos n2\pi + \frac{A}{n\pi} \cos n\pi + \frac{2A}{n\pi} \cos n2\pi$$

$$- \frac{A}{n^2\pi^2} \sin n2\pi - \frac{A}{n\pi} \cos n\pi + \frac{A}{n^2\pi^2} \sin n\pi$$

$$= -\frac{A}{n\pi} \cos n\pi + 0 - \frac{A}{n\pi} + \frac{A}{n\pi} \cancel{\cos n\pi} + \frac{2A}{n\pi} - 0 - \frac{A}{n\pi} \cancel{\cos n\pi} + 0 \qquad \boxed{\begin{array}{l} \text{For integer } n \\ \sin n\pi = 0 \\ \sin n2\pi = 0 \\ \cos n2\pi = 1 \end{array}}$$

$$= \frac{A}{n\pi} - \frac{A}{n\pi} \cos n\pi = \frac{A}{n\pi} (1 - \cos n\pi)$$

When n is even integer, $\cos n\pi = +1$

When n is odd integer, $\cos n\pi = -1$

$$\therefore \ b_n = \frac{A}{n\pi}(1 - \cos n\pi) = \frac{A}{n\pi}(1 - 1) = 0 \quad ; \text{ for even integer values of n.}$$

$$= \frac{A}{n\pi}(1 - \cos n\pi) = \frac{A}{n\pi}(1 + 1) = \frac{2A}{n\pi} \quad ; \text{ for odd integer values of n.}$$

$$\therefore \ b_1 = \frac{2A}{\pi} \ ; \ b_3 = \frac{2A}{3\pi} \ ; \ b_5 = \frac{2A}{5\pi} \text{ and so on.}$$

Fourier Series of x(t)

The Fourier series of x(t) is

$$x(t) = \frac{a_0}{2} + \sum_{n=1}^{\infty} a_n \cos n\Omega_0 t + \sum_{n=1}^{\infty} b_n \sin n\Omega_0 t$$

Here $a_0 = 0$ and the Fourier coefficients a_n and b_n exist only for odd values of n.

$$\therefore \ x(t) = \sum_{n=\text{odd}} a_n \cos n\Omega_0 t + \sum_{n=\text{odd}} b_n \sin n\Omega_0 t$$

$$= a_1 \cos \Omega_0 t + a_3 \cos 3\Omega_0 t + a_5 \cos 5\Omega_0 t + \ldots\ldots\ldots\ldots\ldots\ldots\ldots\ldots$$

$$+ b_1 \sin \Omega_0 t + b_3 \sin 3\Omega_0 t + b_5 \sin 5\Omega_0 t + \ldots\ldots\ldots\ldots$$

$$\therefore \ x(t) = -\frac{4A}{1^2\pi^2} \cos \Omega_0 t - \frac{4A}{3^2\pi^2} \cos 3\Omega_0 t - \frac{4A}{5^2\pi^2} \cos 5\Omega_0 t - \ldots\ldots\ldots\ldots\ldots$$

$$+ \frac{2A}{\pi} \sin \Omega_0 t + \frac{2A}{3\pi} \sin 3\Omega_0 t + \frac{2A}{5\pi} \sin 5\Omega_0 t + \ldots\ldots\ldots\ldots$$

$$= -\frac{4A}{\pi^2} \left(\frac{\cos \Omega_0 t}{1^2} + \frac{\cos 3\Omega_0 t}{3^2} + \frac{\cos 5\Omega_0 t}{5^2} + \ldots\ldots\ldots\ldots\ldots\ldots\ldots \right)$$

$$+ \frac{2A}{\pi} \left(\frac{\sin \Omega_0 t}{1} + \frac{\sin 3\Omega_0 t}{3} + \frac{\sin 5\Omega_0 t}{5} + \ldots\ldots\ldots\ldots \right)$$

Example 2.11

Determine the exponential form of the Fourier series representation of the signal shown in Fig 2.11.1. Hence determine the trigonometric form of Fourier series.

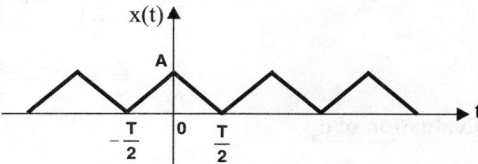

Solution:

To Find Mathematical Equation for x(t)

Fig 2.11.1.

Consider the equation of straight line, $\dfrac{y - y_1}{y_1 - y_2} = \dfrac{x - x_1}{x_1 - x_2}$

Here, $y = x(t)$, $x = t$.

$\therefore$ The equation of straight line can be written as, $\dfrac{x(t) - x(t_1)}{x(t_1) - x(t_2)} = \dfrac{t - t_1}{t_1 - t_2}$ .....(1)

Consider points P, Q and R as shown in Fig 1.

Coordinates of point-P $= [t_1, x(t_1)] = \left[\dfrac{-T}{2}, 0 \right]$

Coordinates of point-Q $= [t_2, x(t_2)] = [0, A]$

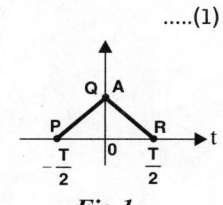

Fig 1.

Coordinates of point-R = $[t_3, x(t_3)] = \left[\dfrac{T}{2}, 0\right]$

On substituting the coordinates of points P and Q in equation (1) we get,

$$\dfrac{x(t) - 0}{0 - A} = \dfrac{t + \dfrac{T}{2}}{-\dfrac{T}{2} - 0} \quad \Rightarrow \quad \dfrac{x(t)}{-A} = \dfrac{-2t}{T} - 1 \quad \Rightarrow \quad x(t) = A + \dfrac{2At}{T}$$

On substituting the coordinates of points Q and R in equation (1) we get,

$$\dfrac{x(t) - A}{A - 0} = \dfrac{t - 0}{0 - \dfrac{T}{2}} \quad \Rightarrow \quad \dfrac{x(t)}{A} - 1 = \dfrac{-2t}{T} \quad \Rightarrow \quad x(t) = A - \dfrac{2At}{T}$$

$$\therefore \ x(t) = A + \dfrac{2At}{T} \ ; \ \text{for } t = -\dfrac{T}{2} \text{ to } 0$$

$$= A - \dfrac{2At}{T} \ ; \ \text{for } t = 0 \text{ to } \dfrac{T}{2}$$

Evaluation of c_0

$$c_n = \dfrac{1}{T} \int_{-T/2}^{+T/2} x(t)\, e^{-jn\Omega_0 t} \, dt$$

When $n = 0$, $c_0 = \dfrac{1}{T} \displaystyle\int_{-T/2}^{+T/2} x(t)\, e^0 \, dt = \dfrac{1}{T} \displaystyle\int_{-T/2}^{+T/2} x(t)\, dt$

$$= \dfrac{1}{T} \int_{-T/2}^{0} \left(A + \dfrac{2At}{T}\right) dt + \dfrac{1}{T} \int_{0}^{+T/2} \left(A - \dfrac{2At}{T}\right) dt$$

$$= \dfrac{A}{T} \int_{-T/2}^{0} dt + \dfrac{2A}{T^2} \int_{-T/2}^{0} t\, dt + \dfrac{A}{T} \int_{0}^{T/2} dt - \dfrac{2A}{T^2} \int_{0}^{T/2} t\, dt$$

$$= \dfrac{A}{T} \left[t\right]_{-T/2}^{0} + \dfrac{2A}{T^2} \left[\dfrac{t^2}{2}\right]_{-T/2}^{0} + \dfrac{A}{T} \left[t\right]_{0}^{T/2} - \dfrac{2A}{T^2} \left[\dfrac{t^2}{2}\right]_{0}^{T/2}$$

$$= \dfrac{A}{T} \left[0 + \dfrac{T}{2}\right] + \dfrac{2A}{T^2} \left[0 - \dfrac{T^2}{8}\right] + \dfrac{A}{T} \left[\dfrac{T}{2} - 0\right] - \dfrac{2A}{T^2} \left[\dfrac{T^2}{8} - 0\right]$$

$$= \dfrac{A}{2} - \dfrac{A}{4} + \dfrac{A}{2} - \dfrac{A}{4} = \dfrac{2A}{2} - \dfrac{2A}{4} = A - \dfrac{A}{2} = \dfrac{A}{2}$$

Evaluation of c_n

$$c_n = \dfrac{1}{T} \int_{-T/2}^{+T/2} x(t)\, e^{-jn\Omega_0 t} \, dt$$

$$= \dfrac{1}{T} \int_{-T/2}^{0} \left(A + \dfrac{2At}{T}\right) e^{-jn\Omega_0 t} \, dt + \dfrac{1}{T} \int_{0}^{T/2} \left(A - \dfrac{2At}{T}\right) e^{-jn\Omega_0 t} \, dt$$

$$= \dfrac{A}{T} \int_{-T/2}^{0} e^{-jn\Omega_0 t} \, dt + \dfrac{2A}{T^2} \int_{-T/2}^{0} t\, e^{-jn\Omega_0 t} \, dt + \dfrac{A}{T} \int_{0}^{T/2} e^{-jn\Omega_0 t} \, dt - \dfrac{2A}{T^2} \int_{0}^{T/2} t\, e^{-jn\Omega_0 t} \, dt$$

$$= \dfrac{A}{T} \left[\dfrac{e^{-jn\Omega_0 t}}{-jn\Omega_0}\right]_{-T/2}^{0} + \dfrac{2A}{T^2} \left[t\, \dfrac{e^{-jn\Omega_0 t}}{-jn\Omega_0} - \int \dfrac{e^{-jn\Omega_0 t}}{-jn\Omega_0} \times 1 \, dt\right]_{-T/2}^{0}$$

$$+ \dfrac{A}{T} \left[\dfrac{e^{-jn\Omega_0 t}}{-jn\Omega_0}\right]_{0}^{T/2} - \dfrac{2A}{T^2} \left[t\, \dfrac{e^{-jn\Omega_0 t}}{-jn\Omega_0} - \int \dfrac{e^{-jn\Omega_0 t}}{-jn\Omega_0} \times 1 \, dt\right]_{0}^{T/2}$$

$\int u\, dv = u\, v - \int v\, du$
$u = t \qquad\qquad \Rightarrow du = 1$
$dv = e^{-jn\Omega_0 t} \Rightarrow \ v = \dfrac{e^{-jn\Omega_0 t}}{-jn\Omega_0}$

$$\therefore c_n = \frac{A}{T}\left[\frac{e^{-jn\Omega_0 t}}{-jn\Omega_0}\right]_{-T/2}^{0} + \frac{2A}{T^2}\left[t\,\frac{e^{-jn\Omega_0 t}}{-jn\Omega_0} - \frac{e^{-jn\Omega_0 t}}{(-jn\Omega_0)^2}\right]_{-T/2}^{0} + \frac{A}{T}\left[\frac{e^{-jn\Omega_0 t}}{-jn\Omega_0}\right]_{0}^{T/2}$$

$$- \frac{2A}{T^2}\left[\frac{te^{-jn\Omega_0 t}}{-jn\Omega_0} - \frac{e^{-jn\Omega_0 t}}{(-jn\Omega_0)^2}\right]_{0}^{T/2} \qquad \boxed{\Omega_0 = \frac{2\pi}{T}}$$

$$= \frac{A}{T}\left[\frac{e^{0}}{-jn\frac{2\pi}{T}} - \frac{e^{-jn\frac{2\pi}{T}\left(-\frac{T}{2}\right)}}{-jn\frac{2\pi}{T}}\right] + \frac{2A}{T^2}\left[\frac{0\times e^{0}}{-jn\frac{2\pi}{T}} - \frac{e^{0}}{-n^2\frac{4\pi^2}{T^2}} + \frac{T}{2}\frac{e^{-jn\frac{2\pi}{T}\left(-\frac{T}{2}\right)}}{-jn\frac{2\pi}{T}} + \frac{e^{-jn\frac{2\pi}{T}\left(-\frac{T}{2}\right)}}{-n^2\frac{4\pi^2}{T^2}}\right]$$

$$+ \frac{A}{T}\left[\frac{e^{-jn\frac{2\pi}{T}\frac{T}{2}}}{-jn\frac{2\pi}{T}} - \frac{e^{0}}{-jn\frac{2\pi}{T}}\right] - \frac{2A}{T^2}\left[\frac{T}{2}\frac{e^{-jn\frac{2\pi}{T}\frac{T}{2}}}{-jn\frac{2\pi}{T}} - \frac{e^{-jn\frac{2\pi}{T}\frac{T}{2}}}{-n^2\frac{4\pi^2}{T^2}} - \frac{0\times e^{0}}{-jn\frac{2\pi}{T}} + \frac{e^{0}}{-n^2\frac{4\pi^2}{T^2}}\right]$$

$$= -\frac{\cancel{A}}{\cancel{j}2n\pi} + \frac{Ae^{jn\pi}}{\cancel{j}2n\pi} - 0 + \frac{A}{2n^2\pi^2} - \frac{Ae^{jn\pi}}{\cancel{j}2n\pi} - \frac{Ae^{jn\pi}}{2n^2\pi^2} - \frac{Ae^{-jn\pi}}{\cancel{j}2n\pi}$$

$$+ \frac{\cancel{A}}{\cancel{j}2n\pi} + \frac{Ae^{-jn\pi}}{\cancel{j}2n\pi} - \frac{Ae^{-jn\pi}}{2n^2\pi^2} - 0 + \frac{A}{2n^2\pi^2}$$

$$= \frac{A}{n^2\pi^2} - \frac{Ae^{jn\pi}}{2n^2\pi^2} - \frac{Ae^{-jn\pi}}{2n^2\pi^2}$$

We know that,

$$e^{\pm jn\pi} = \cos n\pi \pm j\sin n\pi$$

$$= +1 \pm j0 = 1 \quad ; \text{ for even } n.$$

$$= -1 \pm j0 = -1 \; ; \text{ for odd } n.$$

$\therefore$ When n is even,

$$c_n = \frac{A}{n^2\pi^2} - \frac{A}{2n^2\pi^2} - \frac{A}{2n^2\pi^2} = \frac{A}{n^2\pi^2} - \frac{A}{n^2\pi^2} = 0$$

$\therefore$ When n is odd,

$$c_n = \frac{A}{n^2\pi^2} + \frac{A}{2n^2\pi^2} + \frac{A}{2n^2\pi^2} = \frac{A}{n^2\pi^2} + \frac{A}{n^2\pi^2} = \frac{2A}{n^2\pi^2}$$

$$\therefore c_{-1} = \frac{2A}{(-1)^2\pi^2} = \frac{2A}{1^2\pi^2} \qquad\qquad c_1 = \frac{2A}{1^2\pi^2}$$

$$c_{-3} = \frac{2A}{(-3)^2\pi^2} = \frac{2A}{3^2\pi^2} \qquad\qquad c_3 = \frac{2A}{3^2\pi^2}$$

$$c_{-5} = \frac{2A}{(-5)^2\pi^2} = \frac{2A}{5^2\pi^2} \qquad\qquad c_5 = \frac{2A}{5^2\pi^2}$$

$$\text{and so on} \qquad\qquad\qquad\qquad \text{and so on}$$

Exponential Form of Fourier Series

The exponential form of Fourier series is,

$$x(t) = \sum_{n=-\infty}^{+\infty} c_n\, e^{jn\Omega t} = \sum_{n=-\infty}^{-1} c_n\, e^{jn\Omega t} + c_0 + \sum_{n=1}^{\infty} c_n\, e^{jn\Omega t}$$

Here c_n exist only for odd values of n.

$$\therefore\ x(t) = \sum_{\substack{n=\text{negative} \\ \text{odd integer}}} c_n\, e^{jn\Omega t} + c_0 + \sum_{\substack{n=\text{Positive} \\ \text{odd integer}}} c_n\, e^{jn\Omega t}$$

$$= \ldots\ldots + c_{-5}\, e^{-j5\Omega t} + c_{-3}\, e^{-j3\Omega t} + c_{-1}\, e^{-j1\Omega t} + c_0 + c_1\, e^{j\Omega t} + c_3\, e^{j3\Omega t} + c_5\, e^{j5\Omega t} + \ldots\ldots$$

$$x(t) = \ldots\ldots + \frac{2A}{5^2\pi^2}\, e^{-j5\Omega t} + \frac{2A}{3^2\pi^2}\, e^{-j3\Omega t} + \frac{2A}{1^2\pi^2}\, e^{-j\Omega t} + \frac{A}{2} + \frac{2A}{1^2\pi^2}\, e^{j\Omega t}$$

$$+ \frac{2A}{3^2\pi^2}\, e^{j3\Omega t} + \frac{2A}{5^2\pi^2}\, e^{j5\Omega t} + \ldots\ldots\ldots\ldots\ldots\ldots\ldots\ldots\ldots$$

$$= \frac{2A}{\pi^2}\left(\ldots\ldots + \frac{1}{5^2}\, e^{-j5\Omega t} + \frac{1}{3^2}\, e^{-j3\Omega t} + \frac{1}{1^2}\, e^{-j\Omega t} \right) + \frac{A}{2}$$

$$+ \frac{2A}{\pi^2}\left(\frac{1}{1^2}\, e^{j\Omega t} + \frac{1}{3^2}\, e^{j3\Omega t} + \frac{1}{5^2}\, e^{j5\Omega t} + \ldots\ldots\ldots\ldots\ldots\ldots \right)$$

Trigonometric Form of Fourier Series

The trigonometric form of Fourier series can be obtained as shown below.

$$x(t) = \frac{A}{2} + \frac{2A}{\pi^2}\left[\frac{1}{1^2}(e^{j\Omega t} + e^{-j\Omega t}) + \frac{1}{3^2}(e^{j3\Omega t} + e^{-j3\Omega t}) + \frac{1}{5^2}(e^{j5\Omega t} + e^{-j5\Omega t}) + \ldots\ldots \right]$$

$$= \frac{A}{2} + \frac{2A}{\pi^2}\left[\frac{1}{1^2}\, 2\cos\Omega_0 t + \frac{1}{3^2}\, 2\cos 3\Omega_0 t + \frac{1}{5^2}\, 2\cos 5\Omega_0 t + \ldots\ldots \right]$$

$$= \frac{A}{2} + \frac{4A}{\pi^2}\left[\frac{\cos\Omega_0 t}{1^2} + \frac{\cos 3\Omega_0 t}{3^2} + \frac{\cos 5\Omega_0 t}{5^2} + \ldots\ldots\ldots\ldots\ldots \right] \qquad \boxed{\cos\theta = \frac{e^{j\theta} + e^{-j\theta}}{2}}$$

Example 2.12

Determine the exponential form of the Fourier series representation of the signals shown in Fig 2.12.1. Hence determine the trigonometric form of Fourier series.

Solution:

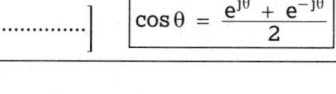

Fig 2.12.1.

To Find Mathematical Equation for x(t)

Consider the equation of straight line, $\dfrac{y - y_1}{y_1 - y_2} = \dfrac{x - x_1}{x_1 - x_2}$

Here, $y = x(t)$, $x = t$.

$\therefore$ The equation of straight line, $\dfrac{x(t) - x(t_1)}{x(t_1) - x(t_2)} = \dfrac{t - t_1}{t_1 - t_2}$ .....(1)

Consider points P and Q, as shown in Fig 1.

Coordinates of point-P = $[t_1, x(t_1)] = [0, 0]$

Coordinates of point-Q = $[t_2, x(t_2)] = [T, A]$

On substituting the coordinates of points P and Q in equation (1) we get,

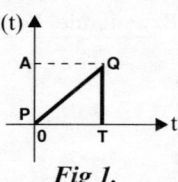

Fig 1.

$$\frac{x(t) - 0}{0 - A} = \frac{t - 0}{0 - T} \quad \Rightarrow \quad \frac{x(t)}{-A} = \frac{t}{-T} \quad \Rightarrow \quad x(t) = \frac{At}{T}$$

$$\therefore \; x(t) = \frac{At}{T} \; ; \; \text{for } t = 0 \text{ to } T$$

Evaluation of c_0

$$c_n = \frac{1}{T} \int_0^T X(t) \; e^{-jn\Omega_0 t} \; dt$$

$$\text{When } n = 0, \; c_0 = \frac{1}{T} \int_0^T x(t) \; e^0 \; dt = \frac{1}{T} \int_0^T x(t) \; dt$$

$$= \frac{1}{T} \int_0^T \frac{At}{T} \; dt = \frac{A}{T^2} \int_0^T t \; dt = \frac{A}{T^2} \left[\frac{t^2}{2} \right]_0^T$$

$$= \frac{A}{T^2} \left[\frac{T^2}{2} - 0 \right] = \frac{A}{2}$$

Evaluation of c_n

$$c_n = \frac{1}{T} \int_0^T x(t) \; e^{-jn\Omega_0 t} \; dt = \frac{1}{T} \int_0^T \frac{At}{T} \; e^{-jn\Omega_0 t} \; dt = \frac{A}{T^2} \int_0^T t \, e^{-jn\Omega_0 t} \; dt$$

$$= \frac{A}{T^2} \left[t \, \frac{e^{-jn\Omega_0 t}}{-jn\Omega_0} - \int \frac{e^{-jn\Omega_0 t}}{-jn\Omega_0} \times 1 \, dt \right]_0^T$$

$\int u \, dv = u \, v - \int v \, du$
$u = t \qquad \Rightarrow du = 1$
$dv = e^{-jn\Omega_0 t} \Rightarrow \; v = \dfrac{e^{-jn\Omega_0 t}}{-jn\Omega_0}$

$$= \frac{A}{T^2} \left[\frac{te^{-jn\Omega_0 t}}{-jn\Omega_0} - \frac{e^{-jn\Omega_0 t}}{(-jn\Omega_0)^2} \right]_0^T$$

$$= \frac{A}{T^2} \left[\frac{te^{-jn\frac{2\pi}{T}t}}{-jn\frac{2\pi}{T}} + \frac{e^{-jn\frac{2\pi}{T}t}}{n^2 \frac{4\pi^2}{T^2}} \right]_0^T = \frac{A}{T^2} \left[\frac{Te^{-jn\frac{2\pi}{T}T}}{-jn\frac{2\pi}{T}} + \frac{e^{-jn\frac{2\pi}{T}T}}{n^2 \frac{4\pi^2}{T^2}} - 0 - \frac{e^0}{n^2 \frac{4\pi^2}{T^2}} \right] \qquad \boxed{\Omega_0 = \frac{2\pi}{T}}$$

$$= -\frac{A}{jn2\pi} e^{-jn2\pi} + \frac{A}{n^2 4\pi^2} e^{-jn2\pi} - \frac{A}{n^2 4\pi^2}$$

$e^{-jn2\pi} = \cos n2\pi - j\sin n2\pi$
$= 1 - j0 = 1; \text{ for integer } n$

$$= -\frac{A}{jn2\pi} + \frac{A}{n^2 4\pi^2} - \frac{A}{n^2 4\pi^2} = -\frac{A}{jn2\pi}$$

$$\therefore \; c_{-1} = \frac{A}{j2\pi} \qquad\qquad c_1 = -\frac{A}{j2\pi}$$

$$c_{-2} = \frac{A}{j4\pi} \qquad\qquad c_2 = -\frac{A}{j4\pi}$$

$$c_{-3} = \frac{A}{j6\pi} \qquad\qquad c_3 = -\frac{A}{j6\pi}$$

and so on.   and so on.

Exponential Form of Fourier Series

The exponential form of Fourier series is,

$$x(t) = \sum_{n=-\infty}^{+\infty} c_n e^{jn\Omega_0 t} = \sum_{n=-\infty}^{-1} c_n e^{jn\Omega_0 t} + c_0 + \sum_{n=1}^{\infty} c_n e^{jn\Omega_0 t}$$

$$= .....c_{-3} e^{-j3\Omega_0 t} + c_{-2} e^{-j2\Omega_0 t} + c_{-1} e^{-j1\Omega_0 t} + c_0 + c_1 e^{j1\Omega_0 t} + c_2 e^{j2\Omega_0 t} + c_3 e^{j3\Omega_0 t} + ...$$

$$= ..... + \frac{A}{j6\pi} e^{-j3\Omega_0 t} + \frac{A}{j4\pi} e^{-j2\Omega_0 t} + \frac{A}{j2\pi} e^{-j\Omega_0 t} + \frac{A}{2} - \frac{A}{j2\pi} e^{j\Omega_0 t}$$

$$- \frac{A}{j4\pi} e^{j2\Omega_0 t} - \frac{A}{j6\pi} e^{j3\Omega_0 t} .....$$

$$= \frac{A}{j2\pi} \left[..... + \frac{e^{-j3\Omega_0 t}}{3} + \frac{e^{-j2\Omega_0 t}}{2} + \frac{e^{-j\Omega_0 t}}{1} \right] + \frac{A}{2} - \frac{A}{j2\pi} \left[\frac{e^{j\Omega_0 t}}{1} + \frac{e^{j2\Omega_0 t}}{2} + \frac{e^{j3\Omega_0 t}}{3} + ... \right]$$

Trigonometric Form of Fourier Series

The trigonometric form of Fourier series can be obtained as shown below:

$$x(t) = \frac{A}{2} - \frac{A}{\pi} \left[\frac{1}{1} \left(\frac{e^{j\Omega_0 t} - e^{-j\Omega_0 t}}{2j} \right) + \frac{1}{2} \left(\frac{e^{j2\Omega_0 t} - e^{-j2\Omega_0 t}}{2j} \right) + \frac{1}{3} \left(\frac{e^{j3\Omega_0 t} - e^{-j3\Omega_0 t}}{2j} \right) + ... \right]$$

$$= \frac{A}{2} - \frac{A}{\pi} \left[\frac{\sin\Omega_0 t}{1} + \frac{\sin 2\Omega_0 t}{2} + \frac{\sin 3\Omega_0 t}{3} + ... \right] \qquad \boxed{\sin\theta = \frac{e^{j\theta} - e^{-j\theta}}{2j}}$$

2.5 Fourier Transform

2.5.1 Development of Fourier Transform From Fourier Series

The exponential form of Fourier series representation of a periodic signal is given by:

$$x(t) = \sum_{n=-\infty}^{+\infty} c_n e^{jn\Omega_0 t} \qquad .....(2.11)$$

$$\text{where, } c_n = \frac{1}{T} \int_{-T/2}^{T/2} x(t) e^{-jn\Omega_0 t} dt \qquad .....(2.12)$$

In the Fourier representation using equation (2.11), the c_n for various values of n are the spectral components of the signal x(t), located at intervals of fundamental frequency Ω_0. Therefore, the frequency spectrum is discrete in nature.

The Fourier representation of a signal using equation (2.11) is applicable for periodic signals. For Fourier representation of nonperiodic signals, let us consider that the fundamental period tends to infinity. When the fundamental period tends to infinity, the fundamental frequency Ω_0 tends to zero or becomes very small. Since fundamental frequency Ω_0 is very small, the spectral components will lie very close to each other and so the frequency spectrum becomes continuous.

In order to obtain the Fourier representation of a nonperiodic signal let us consider that the fundamental frequency Ω_0 is very small.

Let, $\Omega_0 \to \Delta\Omega$

On replacing Ω_0 by $\Delta\Omega$ in equation (2.11) we get,

$$x(t) = \sum_{n=-\infty}^{+\infty} c_n e^{jn\Delta\Omega t}$$

On substituting for c_n in the above equation from equation (2.12) (by taking τ as dummy variable for integration) we get,

$$x(t) = \sum_{n=-\infty}^{+\infty} \left[\frac{1}{T} \int_{-T/2}^{T/2} x(\tau)\, e^{-jn\Delta\Omega\tau}\, d\tau \right] e^{jn\Delta\Omega t} \quad\quad .....(2.13)$$

We know that, $\Omega_0 = 2\pi F_0 = \dfrac{2\pi}{T}$; $\therefore \dfrac{1}{T} = \dfrac{\Omega_0}{2\pi}$

Since $\Omega_0 \to \Delta\Omega$, $\dfrac{1}{T} = \dfrac{\Delta\Omega}{2\pi}$ $\quad\quad .....(2.14)$

On substituting for $\dfrac{1}{T}$ from equation (2.14) in equation (2.13) we get,

$$x(t) = \sum_{n=-\infty}^{+\infty} \left[\frac{\Delta\Omega}{2\pi} \int_{-T/2}^{T/2} x(\tau)\, e^{-jn\Delta\Omega\tau} d\tau \right] e^{jn\Delta\Omega t}$$

$$= \frac{1}{2\pi} \sum_{n=-\infty}^{+\infty} \left[\int_{-T/2}^{T/2} x(\tau)\, e^{-jn\Delta\Omega\tau} d\tau \right] e^{jn\Delta\Omega t} \, \Delta\Omega$$

For nonperiodic signals, the fundamental period T tends to infinity. On letting limit T tends to infinity in the above equation we get,

$$x(t) = \underset{T\to\infty}{\text{Lt}} \frac{1}{2\pi} \sum_{n=-\infty}^{+\infty} \left[\int_{-T/2}^{T/2} x(\tau)\, e^{-jn\Delta\Omega\tau}\, d\tau \right] e^{jn\Delta\Omega t} \, \Delta\Omega$$

When $T \to \infty$; $\sum \to \int$; $\Delta\Omega \to d\Omega$

$$\therefore\ x(t) = \frac{1}{2\pi} \int_{-\infty}^{+\infty} \left[\int_{-\infty}^{+\infty} x(\tau)\, e^{-jn\Omega\tau}\, d\tau \right] e^{jn\Omega t}\, d\Omega$$

$$= \frac{1}{2\pi} \int_{-\infty}^{+\infty} X(j\Omega)\, e^{jn\Omega t}\, d\Omega \quad\quad .....(2.15)$$

where, $X(j\Omega) = \displaystyle\int_{-\infty}^{+\infty} x(\tau)\ e^{-jn\Omega\tau}\, d\tau$

> Since τ is a dummy variable, change τ to t.

$$\therefore\ X(j\Omega) = \int_{-\infty}^{+\infty} x(t)\, e^{-jn\Omega t}\, dt \quad\quad ....(2.16)$$

The equation (2.16) is Fourier transform of x(t) and equation (2.15) is inverse Fourier transform of x(t). **(AU Dec'12, 2 Marks)**

Since the equation (2.16) extracts the frequency components of the signal, transformation using equation (2.16) is also called ***analysis*** of the signal x(t). Since the equation (2.15) combines the frequency components of the signal, the inverse transformation using equation (2.15) is also called ***synthesis*** of the signal x(t).

2.5.2 Definition of Fourier Transform

Let, $x(t)$ = Continuous time signal

$X(j\Omega)$ = Fourier transform of $x(t)$

The Fourier transform of continuous time signal, $x(t)$ is defined as,

$$X(j\Omega) = \int_{-\infty}^{+\infty} x(t) e^{-j\Omega t} \, dt$$

Also, $X(j\Omega)$ is denoted as $\mathcal{F}\{x(t)\}$ where "$\mathcal{F}$" is the symbol used to denote the Fourier transform operation.

$$\therefore \; \mathcal{F}\{x(t)\} = X(j\Omega) = \int_{-\infty}^{+\infty} x(t) e^{-j\Omega t} \, dt \qquad .....(2.17)$$

Note: *Sometimes the Fourier transform is expressed as a function of cyclic frequency F, rather than radian frequency Ω. The Fourier transform as a function of cyclic frequency F, is defined as,*

$$X(jF) = \int_{-\infty}^{+\infty} x(t) e^{-j2\pi F t} \, dt$$

2.5.3 Conditions for Existence of Fourier Transform

The Fourier transform of $x(t)$ exists if it satisfies the following Dirichlet's condition.

1. The $x(t)$ should be absolutely integrable.

i.e., $\displaystyle \int_{-\infty}^{+\infty} x(t) \, dt < \infty$

2. The $x(t)$ should have a finite number of maxima and minima within any finite interval.

3. The $x(t)$ can have a finite number of discontinuities within any interval.

2.5.4 Definition of Inverse Fourier Transform

The *inverse Fourier transform* of $X(j\Omega)$ is defined as,

$$x(t) = \mathcal{F}^{-1}\{X(j\Omega)\} = \frac{1}{2\pi} \int_{-\infty}^{+\infty} X(j\Omega) e^{j\Omega t} \, d\Omega \qquad .....(2.18)$$

The signals $x(t)$ and $X(j\Omega)$ are called *Fourier transform pair* and can be expressed as shown below:

$$x(t) \; \underset{\mathcal{F}^{-1}}{\overset{\mathcal{F}}{\rightleftharpoons}} \; X(j\Omega)$$

Note: *When Fourier transform is expressed as a function of cyclic frequency F, the inverse Fourier transform is defined as,*

$$x(t) = \mathcal{F}^{-1}\{X(jF)\} = \int_{-\infty}^{+\infty} X(jF) e^{j2\pi F t} \, dF$$

2.5.5 Frequency Spectrum Using Fourier Transform

The $X(j\Omega)$ is a complex function of Ω. Hence it can be expressed as a sum of real part and imaginary part as shown below:

$$\therefore \ X(j\Omega) = X_r(j\Omega) + jX_i(j\Omega)$$

where, $X_r(j\Omega)$ = Real part of $X(j\Omega)$

$X_i(j\Omega)$ = Imaginary part of $X(j\Omega)$

The magnitude of $X(j\Omega)$ is called ***Magnitude spectrum.***

$$\therefore \ \text{Magnitude spectrum,} \left| X(j\Omega) \right| = \sqrt{X_r^2(j\Omega) + X_i^2(j\Omega)} \qquad .....(2.19)$$

$$(or)$$

$$\text{Magnitude spectrum,} \left| X(j\Omega) \right| = \sqrt{X(j\Omega)\,X^*(j\Omega)} \qquad .....(2.20)$$

where, $X^*(j\Omega)$ = Conjugate of $X(j\Omega)$

The phase of $X(j\Omega)$ is called ***Phase spectrum***.

$$\therefore \ \text{Phase spectrum,} \ \angle X(j\Omega) = \tan^{-1} \frac{X_i(j\Omega)}{X_r(j\Omega)} \qquad .....(2.21)$$

The magnitude spectrum will always have even symmetry and phase spectrum will have odd symmetry. The magnitude and phase spectrum together called ***frequency spectrum***.

2.5.6 Comparison of Fourier Series and Fourier Transform

(AU Jun'11, 4 Marks)

Table 2.4: Fourier Series Vs Fourier Transform

Fourier Series	Fourier Transform
1. Defined only for periodic signals	1. Defined for both periodic and nonperiodic signals
2. The spectrum is discrete	2. The spectrum is continuous
3. Magnitude spectrum and phase spectrum are plotted by taking "Magnitude/phase" of a signal versus harmonic order"n"	3. Magnitude spectrum and phase spectrum are plotted by taking "Magnitude/phase" of a signal versus frequency"Ω"
4. Parseval's relation of Fourier series is used to calculate power spectral density of a periodic signal x(t).	4. Parseval's relation of Fourier transform is used to calculate energy spectral density of the signal x(t).

2.5.7 Properties of Fourier Transform

1. Linearity   ***(AU Dec'15, 16 Marks)***

Let, $\mathcal{F}\{x_1(t)\} = X_1(j\Omega) \ ; \ \mathcal{F}\{x_2(t)\} = X_2(j\Omega)$

The linearity property of Fourier transform says that,

$$\mathcal{F}\{a_1\,x_1(t) + a_2\,x_2(t)\} = a_1\,X_1(j\Omega) + a_2\,X_2(j\Omega)$$

Proof:

By definition of Fourier transform,

$$X_1(j\Omega) = \int_{-\infty}^{+\infty} x_1(t)\,e^{-j\Omega t}\,dt \quad \text{and} \quad X_2(j\Omega) = \int_{-\infty}^{+\infty} x_2(t)\,e^{-j\Omega t}\,dt \qquad .....(2.22)$$

Consider the linear combination $a_1 x_1(t) + a_2 x_2(t)$. On taking Fourier transform of this signal we get,

$$\mathcal{F}\{a_1x_1(t) + a_2x_2(t)\} = \int_{-\infty}^{+\infty} [a_1x_1(t) + a_2x_2(t)]e^{-j\Omega t} \, dt$$

$$= \int_{-\infty}^{+\infty} a_1x_1(t) \, e^{-j\Omega t} \, dt + \int_{-\infty}^{+\infty} a_2x_2(t)e^{-j\Omega t} \, dt \qquad \boxed{\text{Using equation (2.22).}}$$

$$= a_1 \int_{-\infty}^{+\infty} x_1(t) \, e^{-j\Omega t} \, dt + a_2 \int_{-\infty}^{+\infty} x_2(t) \, e^{-j\Omega t} \, dt = a_1X_1(j\Omega) + a_2X_2(j\Omega)$$

2. Time Shifting

(AU Jun'12, 8 Marks)

The time shifting property of Fourier transform says that,

If $\mathcal{F}\{x(t)\} = X(j\Omega)$ then

$$\mathcal{F}\{x(t-t_0\} = e^{-j\Omega t_0} X(j\Omega)$$

Proof:

By definition of Fourier transform,

$$\mathcal{F}\{x(t)\} = X(j\Omega) = \int_{-\infty}^{+\infty} x(t) \, e^{-j\Omega t} \, dt \qquad .....(2.23)$$

$$\therefore \mathcal{F}\{x(t-t_0)\} = \int_{-\infty}^{+\infty} x(t-t_0) \, e^{-j\Omega t} \, dt = \int_{-\infty}^{+\infty} x(\tau) \, e^{-j\Omega(\tau+t_0)} \, d\tau \qquad \boxed{\begin{array}{l}\text{Let, } t - t_0 = \tau \\ \therefore t = \tau + t_0 \\ \text{On differentiating} \\ dt = d\tau\end{array}}$$

$$= \int_{-\infty}^{+\infty} x(\tau) \, e^{-j\Omega\tau} \times e^{-j\Omega t_0} \, d\tau = e^{-j\Omega t_0} \int_{-\infty}^{+\infty} x(\tau) \, e^{-j\Omega\tau} \, d\tau \qquad \boxed{\begin{array}{l}\text{Since } \tau \text{ is a dummy} \\ \text{variable for integration} \\ \text{we can change } \tau \text{ to t.}\end{array}}$$

$$= e^{-j\Omega t_0} \int_{-\infty}^{+\infty} x(t) \, e^{-j\Omega t} \, dt = e^{-j\Omega t_0} X(j\Omega) \qquad \boxed{\text{Using equation (2.23)}}$$

3. Time Scaling

The time scaling property of Fourier transform says that,

If $\mathcal{F}\{x(t)\} = X(j\Omega)$ then

$$\mathcal{F}\{x(at)\} = \frac{1}{|a|} X\left(\frac{j\Omega}{a}\right)$$

Proof:

By definition of Fourier transform,

$$\mathcal{F}\{x(t)\} = X(j\Omega) = \int_{-\infty}^{+\infty} x(t) \, e^{-j\Omega t} \, dt$$

$$\therefore \mathcal{F}\{x(at)\} = \int_{-\infty}^{+\infty} x(at)e^{-j\Omega t} \, dt = \int_{-\infty}^{+\infty} x(\tau)e^{-j\Omega(\frac{\tau}{a})} \frac{d\tau}{a} \qquad \boxed{\text{Put, } at = \tau \; ; \; \therefore t = \frac{\tau}{a} \; ; \; dt = \frac{d\tau}{a}}$$

$$= \frac{1}{a} \int_{-\infty}^{+\infty} x(\tau) \, e^{-j(\frac{\Omega}{a})\tau} \, d\tau = \frac{1}{a} X\left(\frac{j\Omega}{a}\right)$$

The above transform is applicable for positive values of "a". If "a" happens to be negative then it can be proved that,

$$\mathcal{F}\{x(at)\} = -\frac{1}{a} X\left(\frac{j\Omega}{a}\right)$$

Hence in general,

$$\mathcal{F}\{x(at)\} = \frac{1}{|a|} X\left(\frac{j\Omega}{a}\right) \quad \text{for both positive and negative values of "a"}$$

The term $\int_{-\infty}^{+\infty} x(\tau) \, e^{-j(\frac{\Omega}{a})\tau} \, d\tau$ is similar to the form of Fourier transform except that Ω is replaced by $\left(\frac{\Omega}{a}\right)$.

$$\therefore \int_{-\infty}^{+\infty} x(\tau) \, e^{-j(\frac{\Omega}{a})\tau} \, d\tau = X\left(\frac{j\Omega}{a}\right)$$

4. Time Reversal

The time reversal property of Fourier transform says that,

If $\mathcal{F}\{x(t)\} = X(j\Omega)$ then

$\mathcal{F}\{x(-t)\} = X(-j\Omega)$

Proof:

From time scaling property we know that,

$$\mathcal{F}\{x(at)\} = \frac{1}{|a|}\, X\left(\frac{j\Omega}{a}\right)$$

Let, a = –1.

$\therefore \mathcal{F}\{x(-t)\} = X(-j\Omega)$

5. Conjugation

The conjugation property of Fourier transform says that,

If $\mathcal{F}\{x(t)\} = X(j\Omega)$ then

$\mathcal{F}\{x^*(t)\} = X^*(-j\Omega)$

Proof:

By definition of Fourier transform,

$$\mathcal{F}\{x(t)\} = X(j\Omega) = \int_{-\infty}^{+\infty} x(t)\, e^{-j\Omega t}\, dt$$

$$\therefore \mathcal{F}\{x^*(t)\} = \int_{-\infty}^{+\infty} x^*(t)\, e^{-j\Omega t}\, dt$$

$$= \left[\int_{-\infty}^{+\infty} x(t)\, e^{j\Omega t}\, dt\right]^* = \left[\int_{-\infty}^{+\infty} x(t)\, e^{-j(-\Omega)t}\, dt\right]^*$$

$$[X(-j\Omega)]^* = X^*(-j\Omega)$$

The term $\int_{-\infty}^{+\infty} x(t)\, e^{-j(-\Omega)t}\, dt$

is similar to the form of Fourier transform except that Ω is replaced by $-\Omega$

$\therefore \int_{-\infty}^{+\infty} x(t)\, e^{-j(-\Omega)t}\, dt = X(-j\Omega)$

6. Frequency Shifting

(AU Jun'12, 8 Marks)

The frequency shifting property of Fourier transform says that,

If $\mathcal{F}\{x(t)\} = X(j\Omega)$ then

$\mathcal{F}\{e^{j\Omega_0 t}\, x(t)\} = X(j(\Omega - \Omega_0))$

Proof:

By definition of Fourier transform,

$$\mathcal{F}\{x(t)\} = X(j\Omega) = \int_{-\infty}^{+\infty} x(t)\, e^{-j\Omega t}\, dt$$

$$\therefore \mathcal{F}\{e^{j\Omega_0 t}\, x(t)\} = \int_{-\infty}^{+\infty} [e^{j\Omega_0 t}\, x(t)]\, e^{-j\Omega t}\, dt$$

$$= \int_{-\infty}^{+\infty} x(t)\, e^{j\Omega_0 t}\, e^{-j\Omega t}\, dt$$

$$= \int_{-\infty}^{+\infty} x(t)\, e^{-j(\Omega-\Omega_0)t}\, dt = X(j(\Omega - \Omega_0))$$

The term, $\int_{-\infty}^{+\infty} x(t)\, e^{-j(\Omega-\Omega_0)t}\, dt$ is similar

to the form of Fourier transform except that Ω is replaced by $\Omega - \Omega_0$

$\therefore \int_{-\infty}^{+\infty} x(t)\, e^{-j(\Omega-\Omega_0)t}\, dt = X(j(\Omega - \Omega_0))$

7. Time Differentiation

The differentiation property of Fourier transform says that,

If $\mathcal{F}\{x(t)\} = X(j\Omega)$ then

$$\mathcal{F}\left\{\frac{d}{dt}x(t)\right\} = j\Omega X(j\Omega)$$

Proof:

By definition of inverse Fourier transform,

$$x(t) = \frac{1}{2\pi}\int\limits_{-\infty}^{+\infty} X(j\Omega)e^{j\Omega t}\,d\Omega \qquad\qquad .....(2.24)$$

On differentiating the above equation we get,

$$\frac{dx(t)}{dt} = \frac{d}{dt}\frac{1}{2\pi}\int\limits_{-\infty}^{+\infty} X(j\Omega)e^{j\Omega t}\,d\Omega$$

> Interchanging the order of integration and differentiation

$$= \frac{1}{2\pi}\int\limits_{-\infty}^{+\infty} X(j\Omega)\frac{d}{dt}e^{j\Omega t}\,d\Omega = \frac{1}{2\pi}\int\limits_{-\infty}^{+\infty} X(j\Omega)\,j\Omega\,e^{j\Omega t}\,d\Omega$$

$$= \frac{1}{2\pi}\int\limits_{-\infty}^{+\infty}[j\Omega X(j\Omega)]e^{j\Omega t}\,d\Omega$$

> The term $\dfrac{1}{2\pi}\int\limits_{-\infty}^{+\infty}[j\Omega X(j\Omega)]e^{j\Omega t}\,d\Omega$ is inverse Fourier transform of $j\Omega X(j\Omega)$.

$$= \mathcal{F}^{-1}\{j\Omega X(j\Omega)\}$$

On taking Fourier transform of the above equation we get,

$$\mathcal{F}\left\{\frac{dx(t)}{dt}\right\} = j\Omega X(j\Omega)$$

8. Time Integration

The integration property of Fourier transform says that,

If $\mathcal{F}\{x(t)\} = X(j\Omega)$ and $X(0) = 0$ then

$$\mathcal{F}\left\{\int\limits_{-\infty}^{t} x(\tau)\,d\tau\right\} = \frac{1}{j\Omega}X(j\Omega)$$

> Note : If $X(0) \neq 0$, then
> $$\mathcal{F}\left\{\int\limits_{-\infty}^{t} x(\tau)\,d\tau\right\} = \frac{1}{j\Omega}X(j\Omega) + \pi X(0)\,\delta(\Omega)$$

Proof:

Consider a continuous time signal x(t). Let $X(j\Omega)$ be Fourier transform of x(t). Since integration and differentiation are inverse operations, x(t) can be expressed as shown below:

$$\frac{d}{dt}\left[\int\limits_{-\infty}^{t} x(\tau)\,d\tau\right] = x(t)$$

On taking Fourier transform of the above equation we get,

$$\mathcal{F}\left\{\frac{d}{dt}\left[\int\limits_{-\infty}^{t} x(\tau)\,d\tau\right]\right\} = \mathcal{F}\{x(t)\}$$

> Using time differentiation property of Fourier transform

$$j\Omega\,\mathcal{F}\left\{\int\limits_{-\infty}^{t} x(\tau)\,d\tau\right\} = \mathcal{F}\{x(t)\}$$

> $\mathcal{F}\{x(t)\} = X(j\Omega)$

$$\therefore \mathcal{F}\left\{\int\limits_{-\infty}^{t} x(\tau)\,d\tau\right\} = \frac{1}{j\Omega}X(j\Omega)$$

9. Frequency Differentiation

The frequency differentiation property of Fourier transform says that,

If $\mathcal{F}\{x(t)\} = X(j\Omega)$, then

$$\mathcal{F}\{t\,x(t)\} = j\frac{d}{d\Omega}X(j\Omega)$$

Proof :

By definition of Fourier transform,

$$X(j\Omega) = \mathcal{F}\{x(t)\} = \int\limits_{-\infty}^{+\infty} x(t)\,e^{-j\Omega t}\,dt$$

On differentiating the above equation with respect to Ω we get,

$$\frac{d}{d\Omega}X(j\Omega) = \frac{d}{d\Omega}\left(\int\limits_{-\infty}^{+\infty} x(t)\,e^{-j\Omega t}\,dt\right)$$

$$= \int\limits_{-\infty}^{+\infty} x(t)\left(\frac{d}{d\Omega}e^{-j\Omega t}\right)dt$$

Interchanging the order of integration and differentiation

$$= \int\limits_{-\infty}^{+\infty} x(t)\,(-jte^{-j\Omega t})\,dt = \frac{1}{j}\int\limits_{-\infty}^{+\infty} (t\,x(t))\,e^{-j\Omega t}\,dt$$

$$-j = -j \times \frac{j}{j} = \frac{1}{j}$$

$$= \frac{1}{j}\,\mathcal{F}\{t\,x(t)\}$$

Using definition of Fourier transform

$$\therefore\ \mathcal{F}\{t\,x(t)\} = j\frac{d}{d\Omega}X(j\Omega)$$

10. Convolution Theorem

The convolution theorem of Fourier transform says that, Fourier transform of convolution of two signals is given by the product of the Fourier transform of the individual signals.

i.e., if $\mathcal{F}\{x_1(t)\} = X_1(j\Omega)$ and $\mathcal{F}\{x_2(t)\} = X_2(j\Omega)$ then,

$$\mathcal{F}\{x_1(t) * x_2(t)\} = X_1(j\Omega)\,X_2(j\Omega) \qquad\qquad .....(2.25)$$

The equation (2.25) is also known as convolution property of Fourier transform.

By definition of convolution of continuous time signals,

$$x_1(t) * x_2(t) = \int\limits_{\tau=-\infty}^{\tau=+\infty} x_1(\tau)\,x_2(t-\tau)\,d\tau \qquad\qquad .....(2.26)$$

where τ is a dummy variable used for integration

Proof:

Let $x_1(t)$ and $x_2(t)$ be two time domain signals. Now, by definition of Fourier transform,

$$X_1(j\Omega) = \mathcal{F}\{x_1(t)\} = \int\limits_{-\infty}^{+\infty} x_1(t)\,e^{-j\Omega t}\,dt \qquad\qquad .....(2.27)$$

$$X_2(j\Omega) = \mathcal{F}\{x_2(t)\} = \int\limits_{-\infty}^{+\infty} x_2(t)\,e^{-j\Omega t}\,dt \qquad\qquad .....(2.28)$$

Using definition of Fourier transform we can write,

$$\mathcal{F}\{x_1(t) * x_2(t)\} = \int_{t=-\infty}^{t=+\infty} \left[x_1(t) * x_2(t)\right] e^{-j\Omega t}\ dt$$

$$= \int_{t=-\infty}^{t=+\infty} \left[\int_{\tau=-\infty}^{\tau=+\infty} x_1(\tau)\, x_2(t-\tau)\, d\tau\right] e^{-j\Omega t}\ dt \qquad \boxed{\text{Using equation (2.26)}} \quad .....(2.29)$$

Let, $e^{-j\Omega t} = e^{j\Omega \tau} \times e^{-j\Omega t} \times e^{-j\Omega t} = e^{-j\Omega \tau} \times e^{-j\Omega(t-\tau)} = e^{-j\Omega \tau} \times e^{-j\Omega M}$ \qquad .....(2.30)

    where, $M = t - \tau$ and so, $dM = dt$ \qquad .....(2.31)

Using equations (2.30) and (2.31), the equation (2.29) can be written as,

$$\mathcal{F}\{x_1(t) * x_2(t)\} = \int_{M=-\infty}^{M=+\infty} \int_{\tau=-\infty}^{\tau=+\infty} x_1(\tau)\, x_2(M)\, e^{-j\Omega \tau}\, e^{-j\Omega M}\ d\tau\ dM$$

$$= \int_{\tau=-\infty}^{\tau=+\infty} x_1(\tau)\, e^{-j\Omega \tau}\ d\tau \times \int_{M=-\infty}^{M=+\infty} x_2(M)\, e^{-j\Omega M}\ dM \qquad .....(2.32)$$

In equation (2.32), τ and M are dummy variables used for integration, and so they can be changed to t. Therefore equation (2.32) can be written as,

$$\mathcal{F}\{x_1(t) * x_2(t)\} = \int_{t=-\infty}^{t=+\infty} x_1(t)\, e^{-j\Omega t}\ dt \times \int_{t=-\infty}^{t=+\infty} x_2(t)\, e^{-j\Omega t}\ dt$$

$\boxed{\text{Using equations (2.27) and (2.28).}}$

$$= X_1(j\Omega)\, X_2(j\Omega)$$

11. *Frequency Convolution*

Let, $\mathcal{F}\{x_1(t)\} = X_1(j\Omega)$; $\mathcal{F}\{x_2(t)\} = X_2(j\Omega)$

The frequency convolution property of Fourier transform says that,

$$F\{x_1(t)x_2(t)\} = \frac{1}{2\pi} \int_{\lambda=-\infty}^{\lambda=+\infty} X_1(j\lambda)\, X_2(j(\Omega-\lambda))\ d\lambda$$

Proof:

By definition of Fourier transform,

$$\mathcal{F}\{x(t)\} = X(j\Omega) = \int_{-\infty}^{+\infty} x(t)\, e^{-j\Omega t}\ dt$$

$$\therefore\ \mathcal{F}\{x_1(t)x_2(t)\} = \int_{t=-\infty}^{t=+\infty} x_1(t)\, x_2(t)\, e^{-j\Omega t}\ dt \qquad .....(2.33)$$

By the definition of inverse Fourier transform we get,

$\boxed{\text{Here } \Omega \text{ is the variable used for integration. Let us change } \Omega \text{ to } \lambda.}$

$$x_1(t) = \mathcal{F}^{-1}\{X_1(j\Omega)\} = \frac{1}{2\pi} \int_{\Omega=-\infty}^{\Omega=+\infty} X_1(j\Omega)\, e^{j\Omega t}\ d\Omega = \frac{1}{2\pi} \int_{\lambda=-\infty}^{\lambda=+\infty} X_1(j\lambda)\, e^{j\lambda t}\ d\lambda \qquad .....(2.34)$$

On substituting for $x_1(t)$ from equation (2.34) in equation (2.33) we get,

$$\mathcal{F}\{x_1(t)x_2(t)\} = \int_{t=-\infty}^{t=+\infty} \left[\frac{1}{2\pi} \int_{\lambda=-\infty}^{\lambda=+\infty} X_1(j\lambda)\, e^{j\lambda t}\ d\lambda\right] x_2(t)\, e^{-j\Omega t}\ dt$$

$$\therefore \ \mathcal{F}\{x_1(t)\,x_2(t)\} = \frac{1}{2\pi} \int\limits_{\lambda=-\infty}^{\lambda=+\infty} X_1(j\lambda) \left[\int\limits_{t=-\infty}^{t=+\infty} x_2(t)\,e^{-j\Omega t}\,e^{j\lambda t}\ dt \right] d\lambda$$

> Interchanging the order of integration

$$= \frac{1}{2\pi} \int\limits_{\lambda=-\infty}^{\lambda=+\infty} X_1(j\lambda) \left[\int\limits_{t=-\infty}^{t=+\infty} x_2(t)\,e^{-j(\Omega-\lambda)t}\ dt \right] d\lambda$$

> The term, $\displaystyle\int\limits_{t=-\infty}^{t=+\infty} x_2(t)\,e^{-j(\Omega-\lambda)t}\ dt$
> is similar to the form of Fourier transform except that Ω is replaced by $\Omega-\lambda$.
> $$\therefore \int\limits_{t=-\infty}^{t=+\infty} x_2(t)\,e^{-j(\Omega-\lambda)t}\ dt = X_2(j(\Omega-\lambda))$$

$$= \frac{1}{2\pi} \int\limits_{\lambda=-\infty}^{\lambda=+\infty} X_1(j\lambda)\,X_2(j(\Omega-\lambda))\ d\lambda$$

12. Parseval's Relation   *(AU Jun'13, 6 Marks)*

The Parseval's relation says that,

If $\mathcal{F}\{x(t)\} = X(j\Omega)$ then

$$\int\limits_{-\infty}^{+\infty} |x(t)|^2\ dt = \frac{1}{2\pi} \int\limits_{-\infty}^{+\infty} |X(j\Omega)|^2\ d\Omega$$

> **Note:** *The term* $|X(j\Omega)|^2$ *represents the distribution of energy as function of* Ω *and so it is called* **energy density spectrum** *or* **energy spectral density** *of the signal x(t).*

Proof:

Let $x(t)$ be a continuous time signal and $x^*(t)$ be conjugate of $x(t)$.

Now, $|x(t)|^2 = x(t)\,x^*(t)$

On integrating the above equation with respect to t we get,

$$\int\limits_{t=-\infty}^{t=+\infty} |x(t)|^2\ dt = \int\limits_{t=-\infty}^{t=+\infty} x(t)\,x^*(t)\ dt \qquad\qquad .....(2.35)$$

By definition of inverse Fourier transform, we can write,

$$x(t) = \mathcal{F}^{-1}\{X(j\Omega)\} = \frac{1}{2\pi} \int\limits_{\Omega=-\infty}^{\Omega=+\infty} X(j\Omega)\,e^{j\Omega t}\ d\Omega$$

On taking conjugate of the above equation we get,

$$x^*(t) = \frac{1}{2\pi} \int\limits_{\Omega=-\infty}^{\Omega=+\infty} X^*(j\Omega)\,e^{-j\Omega t}\ d\Omega \qquad\qquad .....(2.36)$$

Using equation (2.36) the equation (2.35) can be written as,

$$\int\limits_{t=-\infty}^{t=+\infty} |x(t)|^2\ dt = \int\limits_{t=-\infty}^{t=+\infty} x(t) \left[\frac{1}{2\pi} \int\limits_{\Omega=-\alpha}^{\Omega=+\infty} X^*(j\Omega)\,e^{-j\Omega t}\ d\Omega \right] dt$$

$$= \frac{1}{2\pi} \int\limits_{\Omega=-\infty}^{\Omega=+\infty} X^*(j\Omega) \left[\int\limits_{t=-\infty}^{t=+\infty} x(t)\,e^{-j\Omega t}\ dt \right] d\Omega$$

> Interchanging the order of integration

$$= \frac{1}{2\pi} \int\limits_{\Omega=-\infty}^{\Omega=+\infty} X^*(j\Omega)\,X(j\Omega)\,d\Omega$$

> Using definition of Fourier transform

$$= \frac{1}{2\pi} \int\limits_{\Omega=-\infty}^{\Omega=+\infty} |X(j\Omega)|^2\,d\Omega$$

> $X(j\Omega)\,X^*(j\Omega) = |X(j\Omega)|^2$

13. Duality

If $\mathcal{F}\{x_1(t)\} = X_1(j\Omega)$ and $\mathcal{F}\{x_2(t)\} = X_2(j\Omega)$

and if, $x_2(t) \equiv X_1(j\Omega)$, [i.e., $x_2(t)$ and $X_1(j\Omega)$ are similar functions]

then, $X_2(j\Omega) \equiv 2\pi x_1(-j\Omega)$, [i.e., $X_2(j\Omega)$ are $2\pi x_2(-j\Omega)$ are similar functions]

Alternatively duality property is expressed as shown below:

If, $x_2(t) \Leftrightarrow X_1(j\Omega)$

then, $X_2(j\Omega) \Leftrightarrow 2\pi x_1(-j\Omega)$

Proof:

Let, $\mathcal{F}\{x_1(t)\} = X_1(j\Omega)$ and $\mathcal{F}\{x_2(t)\} = X_2(j\Omega)$

Let, $x_2(t)$ and $X_1(j\Omega)$ are similar in form,

$$\therefore\ x_2(t) = X_1(j\Omega)\big|_{j\Omega = t} \quad\quad\quad\quad .....(2.37)$$

By definition of inverse Fourier transform,

$$x_1(t) = \frac{1}{2\pi}\int_{-\infty}^{+\infty} X_1(j\Omega)\, e^{j\Omega t}\, d\Omega$$

$$\therefore\ x_1(-t) = \frac{1}{2\pi}\int_{-\infty}^{+\infty} X_1(j\Omega)\, e^{-j\Omega t}\, d\Omega \qquad \boxed{\text{Replacing t by }-t}$$

$$\therefore\ x_1(-t)\big|_{t=j\Omega} = \frac{1}{2\pi}\int_{-\infty}^{+\infty}\left(X_1(j\Omega)\big|_{j\Omega=t}\right)e^{-j\Omega t}\, d\Omega \qquad \boxed{\text{Interchanging j}\Omega\text{ and t}}$$

$$\therefore\ x_1(-j\Omega) = \frac{1}{2\pi}\int_{-\infty}^{+\infty} x_2(t)\, e^{-j\Omega t}\, d\Omega \qquad \boxed{\text{Using equation (2.37)}}$$

$$\therefore\ \int_{-\infty}^{+\infty} x_2(t)\, e^{-j\Omega t}\, d\Omega = 2\pi x_1(-j\Omega)$$

$$\boxed{\begin{array}{l}\text{Using definition of}\\\text{Fourier transform}\end{array}}$$

$$\therefore\ X_2(j\Omega) = 2\pi X_1(-j\Omega)$$

Note: For even function $x_1(-j\Omega) = x_1(j\Omega)$.
$$\therefore X_2(j\Omega) = 2\pi x_1(j\Omega)$$

14. Area Under a Time Domain Signal

Area under $x(t) = \int\limits_{-\infty}^{+\infty} x(t) \ dt$

If $x(t)$ and $X(j\Omega)$ are Fourier transform pair,

then, $\int\limits_{-\infty}^{+\infty} x(t) \ dt = X(0)$ where, $X(0) = \underset{j\Omega \to 0}{Lt} \ X(j\Omega)$

Proof:

By definition of Fourier transform,

$X(j\Omega) = \int\limits_{-\infty}^{+\infty} x(t) \ e^{-j\Omega t} \ dt$

$\therefore \ X(0) = \underset{j\Omega \to 0}{Lt} \ X(j\Omega) = \underset{j\Omega \to 0}{Lt} \int\limits_{-\infty}^{+\infty} x(t) \ e^{-j\Omega t} \ dt$

$= \int\limits_{-\infty}^{+\infty} x(t) \ e^0 \ dt = \int\limits_{-\infty}^{+\infty} x(t) \ dt$

$\therefore \ \int\limits_{-\infty}^{+\infty} x(t) \ dt = X(0)$

---

15. Area Under a Frequency Domain Signal

Area under $X(j\Omega) = \int\limits_{-\infty}^{+\infty} X(j\Omega) \ d\Omega$

If $x(t)$ and $X(j\Omega)$ are Fourier transform pair

then, $\int\limits_{-\infty}^{+\infty} X(j\Omega) \ d\Omega = 2\pi \, x(0)$ where, $x(0) = \underset{t \to 0}{Lt} \ x(t)$

Proof:

By definition of inverse Fourier transform,

$x(t) = \dfrac{1}{2\pi} \int\limits_{-\infty}^{+\infty} X(j\Omega) \ e^{j\Omega t} \ d\Omega$

$\therefore \ x(0) = \underset{t \to 0}{Lt} \ x(t) = \underset{t \to 0}{Lt} \dfrac{1}{2\pi} \int\limits_{-\infty}^{+\infty} X(j\Omega) \ e^{j\Omega t} \ d\Omega$

$= \dfrac{1}{2\pi} \int\limits_{-\infty}^{+\infty} X(j\Omega) \ e^0 \ d\Omega = \dfrac{1}{2\pi} \int\limits_{-\infty}^{+\infty} X(j\Omega) \ d\Omega$

$\therefore \ \int\limits_{-\infty}^{+\infty} X(j\Omega) \ d\Omega = 2\pi \, x(0)$

Table 2.5 : Summary of Properties of Fourier Transform

Let, $\mathcal{F}\{x(t)\} = X(j\Omega)$; $\mathcal{F}\{x_1(t)\} = X_1(j\Omega)$; $\mathcal{F}\{x_2(t)\} = X_2(j\Omega)$

Property	Time Domain Signal	Frequency Domain Signal
Linearity	$a_1 x_1(t) + a_2 x_2(t)$	$a_1 X_1(j\Omega) + a_2 X_2(j\Omega)$
Time shifting	$x(t - t_0)$	$e^{-j\Omega t_0} X(j\Omega)$
Time scaling	$x(at)$	$\frac{1}{|a|} X\left(\frac{j\Omega}{a}\right)$
Time reversal	$x(-t)$	$X(-j\Omega)$
Conjugation	$x^*(t)$	$X^*(-j\Omega)$
Frequency shifting	$e^{j\Omega_0 t} x(t)$	$X(j(\Omega - \Omega_0))$
Time differentiation	$\dfrac{d}{dt} x(t)$	$j\Omega X(j\Omega)$
Time integration	$\displaystyle\int_{-\infty}^{t} x(\tau)\, d\tau$	$\dfrac{X(j\Omega)}{j\Omega} + \pi X(0)\, \delta(\Omega)$
Frequency differentiation	$t\, x(t)$	$j\dfrac{d}{d\Omega} X(j\Omega)$
Time convolution	$x_1(t) * x_2(t) = \displaystyle\int_{-\infty}^{+\infty} x_1(\tau) x_2(t - \tau)\, d\tau$	$X_1(j\Omega)\, X_2(j\Omega)$
Frequency convolution (or Multiplication)	$x_1(t)\, x_2(t)$	$\dfrac{1}{2\pi} \displaystyle\int_{\lambda = -\infty}^{\lambda = +\infty} X_1(j\lambda)\, X_2(j(\Omega - \lambda))\, d\lambda$
Symmetry of real signals	$x(t)$ is real	$X(j\Omega) = X^*(j\Omega)$ $|X(j\Omega)| = |X(-j\Omega)|$; $\angle X(j\Omega) = -\angle X(-j\Omega)$ $\mathrm{Re}\{X(j\Omega)\} = \mathrm{Re}\{X(-j\Omega)\}$ $\mathrm{Im}\{X(j\Omega)\} = -\mathrm{Im}\{X(-j\Omega)\}$
Real and even	$x(t)$ is real and even	$X(j\Omega)$ are real and even
Real and odd	$x(t)$ is real and odd	$X(j\Omega)$ are imaginary and odd
Duality	If $x_2(t) \equiv X_1(j\Omega)$ [i.e. $x_2(t)$ and $X_1(j\Omega)$ are similar functions] then $X_2(j\Omega) \equiv 2\pi x_1(-j\Omega)$ [i.e. $X_2(j\Omega)$ and $2\pi x_1(-j\Omega)$ are similar functions]	
Area under a frequency domain signal	$\displaystyle\int_{-\infty}^{+\infty} X(j\Omega)\, d\Omega = 2\pi x(0)$	
Area under a time domain signal	$\displaystyle\int_{-\infty}^{+\infty} x(t)\, dt = X(0)$	
Parseval's relation	Energy in time domain is, $E = \displaystyle\int_{-\infty}^{+\infty} |x(t)|^2\, dt$	Energy in frequency domain is, $E = \dfrac{1}{2\pi} \displaystyle\int_{-\infty}^{+\infty} |X(j\Omega)|^2\, d\Omega$
	$\displaystyle\int_{-\alpha}^{+\infty} |x(t)|^2\, dt = \dfrac{1}{2\pi} \displaystyle\int_{-\infty}^{+\infty} |X(j\Omega)|^2\, d\Omega$	

2.6 Fourier Transform of Some Important Signals

2.6.1 Fourier Transform of Unit Impulse Signal

(AU May'15, 2 Marks)

The impulse signal is defined as,

$$x(t) = \delta(t) = \infty \quad ; \quad t = 0 \quad \text{and} \quad \int_{-\infty}^{+\infty} \delta(t) \, dt = 1$$

$$= 0 \quad ; \quad t \neq 0$$

By definition of Fourier transform,

$$X(j\Omega) = \mathcal{F}\{x(t)\} = \int_{-\infty}^{+\infty} x(t) \, e^{-j\Omega t} \, dt = \int_{-\infty}^{+\infty} \delta(t) \, e^{-j\Omega t} \, dt \qquad \boxed{\delta(t) \text{ exists only for } t = 0.}$$

$$= 1 \times e^{-j\Omega t}\big|_{t=0} = 1 \times e^{0} = 1$$

$$\boxed{\therefore \quad \mathcal{F}\{\delta(t)\} = 1}$$

The plot of impulse signal and its magnitude spectrum are shown in Fig 2.5 and Fig 2.6 respectively.

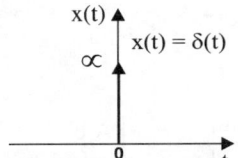

Fig 2.5: Impulse signal.

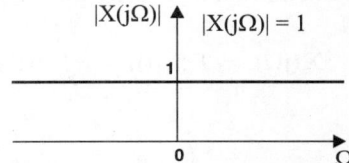

Fig 2.6: Magnitude spectrum of impulse signal.

2.6.2 Fourier Transform of Single Sided Exponential Signal

The single sided exponential signal is defined as,

$$x(t) = A\,e^{-at}\,u(t) = A\,e^{-at} \quad ; \quad \text{for } t \geq 0$$

By definition of Fourier transform,

$$X(j\Omega) = \mathcal{F}\{x(t)\} = \int_{-\infty}^{+\infty} x(t)\,e^{-j\Omega t}\,dt = \int_{0}^{+\infty} A\,e^{-at}\,e^{-j\Omega t}\,dt$$

$$= \int_{0}^{+\infty} A\,e^{-(a+j\Omega)t}\,dt = \left[\frac{A\,e^{-(a+j\Omega)t}}{-(a+j\Omega)}\right]_{0}^{+\infty} \qquad \boxed{e^{-\infty} = 0}$$

$$= \left[\frac{A\,e^{-\infty}}{-(a+j\Omega)} - \frac{A\,e^{0}}{-(a+j\Omega)}\right] = \frac{A}{a+j\Omega}$$

$$\boxed{\therefore \quad \mathcal{F}\{A\,e^{-at}\,u(t)\} = \frac{A}{a+j\Omega}} \qquad\qquad \text{.....(2.38)}$$

The plot of exponential signal and its magnitude spectrum are shown in Fig 2.7 and Fig 2.8 respectively.

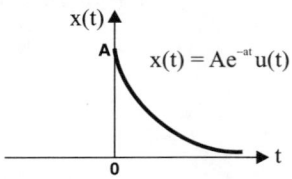

Fig 2.7: Exponential signal.

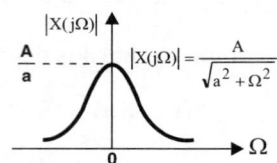

Fig 2.8: Magnitude spectrum of exponential signal.

2.6.3 Fourier Transform of Double Sided Exponential Signal

The double sided exponential signal is defined as,　　　　　**(AU Dec'13, 6 Marks)**

$$x(t) = A\,e^{-a|t|} \quad ; \quad \text{for all } t$$

$$\therefore\ x(t) = A\,e^{+at} \quad ; \quad \text{for } t = -\infty \text{ to } 0$$

$$= A\,e^{-at} \quad ; \quad \text{for } t = 0 \text{ to } +\infty$$

By definition of Fourier transform,

$$X(j\Omega) = \mathcal{F}\{x(t)\} = \int_{-\infty}^{+\infty} x(t)\,e^{-j\Omega t}\,dt = \int_{-\infty}^{0} A\,e^{at}\,e^{-j\Omega t}\,dt + \int_{0}^{+\infty} A\,e^{-at}\,e^{-j\Omega t}\,dt$$

$$= \int_{-\infty}^{0} A\,e^{(a-j\Omega)t}\,dt + \int_{0}^{\infty} A\,e^{-(a+j\Omega)t}\,dt = \left[\frac{A\,e^{(a-j\Omega)t}}{a-j\Omega}\right]_{-\infty}^{0} + \left[\frac{A\,e^{-(a+j\Omega)t}}{-(a+j\Omega)}\right]_{0}^{\infty}$$

$$= \frac{Ae^{0}}{a-j\Omega} - \frac{Ae^{-\infty}}{a-j\Omega} + \frac{Ae^{-\infty}}{-(a+j\Omega)} - \frac{Ae^{0}}{-(a+j\Omega)} = \frac{A}{a-j\Omega} + \frac{A}{a+j\Omega}$$

$$= \frac{A(a+j\Omega) + A(a-j\Omega)}{(a-j\Omega)(a+j\Omega)} = \frac{2aA}{a^2+\Omega^2}$$

$$\boxed{e^{-\infty} = 0}$$
$$\boxed{(a+b)(a-b) = a^2 - b^2 \mid j^2 = -1}$$

$$\therefore\ \mathcal{F}\{A\,e^{-a|t|}\} = \frac{2aA}{a^2+\Omega^2} \qquad\qquad ....(2.39)$$

The plot of double sided exponential signal and its magnitude spectrum are shown in Fig 2.9 and Fig 2.10 respectively.

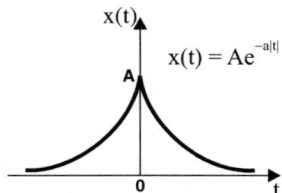

Fig 2.9: Double sided exponential signal.

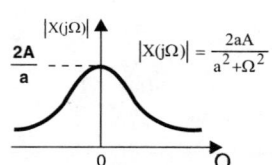

Fig 2.10: Magnitude spectrum of double sided exponential signal.

2.6.4 Fourier Transform of a Constant

(AU Jun'13, 2 Marks)

Let, $x(t) = A$, where A is a constant.

Let, $A = \underset{a \to 0}{Lt} \, A \, e^{-a|t|}$

$\therefore \; x(t) = \underset{a \to 0}{Lt} \, A \, e^{-a|t|}$

> If definition of Fourier transform is directly applied, the constant will not satisfy the condition,
>
> $$\int_{-\infty}^{+\infty} |x(t)| \, dt < \infty$$

On taking Fourier transform of the above equation we get,

> Hence the constant can be viewed as a double sided exponential with limit "a" tends to 0.

$X(j\Omega) = \mathcal{F}\{x(t)\} = \mathcal{F}\left\{ \underset{a \to 0}{Lt} \, Ae^{-a|t|} \right\}$

$= \underset{a \to 0}{Lt} \, \mathcal{F}\{Ae^{-a|t|}\}$

$= \underset{a \to 0}{Lt} \, \dfrac{2aA}{\Omega^2 + a^2}$

> Using equation (2.39)

The above equation is 0 for all values of Ω except at $\Omega = 0$.

At $\Omega = 0$, the above equation represents an impulse of magnitude "k".

$\therefore \; X(j\Omega) = k \, \delta(\Omega) \; ; \quad \Omega = 0$

$\qquad\qquad = 0 \qquad ; \quad \Omega \neq 0$

The magnitude "k" can be evaluated as shown below:

$k = \displaystyle\int_{-\infty}^{+\infty} \dfrac{2aA}{\Omega^2 + a^2} \, d\Omega = 2aA \int_{-\infty}^{+\infty} \dfrac{1}{\Omega^2 + a^2} \, d\Omega$

> $\displaystyle\int \dfrac{dx}{x^2 + a^2} = \dfrac{1}{a} \tan^{-1} \dfrac{x}{a}$

$= 2aA \left[\dfrac{1}{a} \tan^{-1}\left(\dfrac{\Omega}{a}\right) \right]_{-\infty}^{+\infty}$

$= 2aA \left[\dfrac{1}{a} \tan^{-1}(+\infty) - \dfrac{1}{a} \tan^{-1}(-\infty) \right]$

$= 2aA \left[\dfrac{1}{a} \dfrac{\pi}{2} - \dfrac{1}{a}\left(-\dfrac{\pi}{2}\right) \right]$

$= 2aA \left(\dfrac{\pi}{a} \right) = 2\pi A$

$\boxed{\therefore \; \mathcal{F}\{A\} = 2\pi A \, \delta(\Omega)}$ .....(2.40)

The plot of constant and its magnitude spectrum are shown in Fig 2.11 and Fig 2.12 respectively.

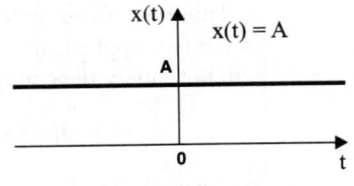

Fig 2.11: Constant.

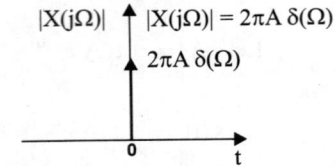

Fig 2.12: Magnitude spectrum of constant.

2.6.5 Fourier Transform of Signum Function

The signum function is defined as,

$$x(t) = \text{sgn}(t) = 1 \quad ; \quad t > 0$$
$$= -1 \quad ; \quad t < 0$$

The signum function can be expressed as a sum of two one sided exponential signal and taking limit "a" tends to 0 as shown below:

$$\therefore \ \text{sgn}(t) = \underset{a \to 0}{\text{Lt}} \left[e^{-at} u(t) - e^{at} u(-t) \right]$$

$$\therefore \ x(t) = \text{sgn}(t) = \underset{a \to 0}{\text{Lt}} \left[e^{-at} u(t) - e^{at} u(-t) \right]$$

By definition of Fourier transform,

$$X(j\Omega) = \mathcal{F}\{x(t)\} = \int_{-\infty}^{+\infty} x(t) e^{-j\Omega t} \ dt = \int_{-\infty}^{+\infty} \underset{a \to 0}{\text{Lt}} \left[e^{-at} u(t) - e^{at} u(-t) \right] e^{-j\Omega t} \ dt$$

$$= \underset{a \to 0}{\text{Lt}} \left[\int_{0}^{+\infty} e^{-at} e^{-j\Omega t} \ dt - \int_{-\infty}^{0} e^{at} e^{-j\Omega t} \ dt \right]$$

$$= \underset{a \to 0}{\text{Lt}} \left[\int_{0}^{+\infty} e^{-(a+j\Omega)t} \ dt - \int_{-\infty}^{0} e^{+(a-j\Omega)t} \ dt \right]$$

$$= \underset{a \to 0}{\text{Lt}} \left[\left[\frac{e^{-(a+j\Omega)t}}{-(a+j\Omega)} \right]_{0}^{\infty} - \left[\frac{e^{(a-j\Omega)t}}{(a-j\Omega)} \right]_{-\infty}^{0} \right] \qquad \boxed{e^0 = 1; \ e^{-\infty} = 0.}$$

$$= \underset{a \to 0}{\text{Lt}} \left[\frac{e^{-\infty}}{-(a+j\Omega)} - \frac{e^0}{-(a+j\Omega)} - \frac{e^0}{a-j\Omega} + \frac{e^{-\infty}}{a-j\Omega} \right]$$

$$= \underset{a \to 0}{\text{Lt}} \left[\frac{1}{a+j\Omega} - \frac{1}{a-j\Omega} \right] = \frac{1}{j\Omega} + \frac{1}{j\Omega} = \frac{2}{j\Omega}$$

$$\boxed{\therefore \ \mathcal{F}\{\text{sgn}(t)\} = \frac{2}{j\Omega}} \qquad\qquad \qquad .....(2.41)$$

The plot of signum function and its magnitude spectrum are shown in Fig 2.13 and Fig 2.14 respectively.

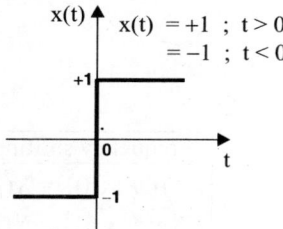

$x(t) = +1$; $t > 0$
$= -1$; $t < 0$

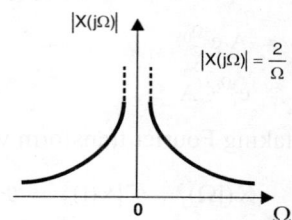

$|X(j\Omega)| = \dfrac{2}{\Omega}$

Fig 2.13: *Signum function.*   **Fig 2.14:** *Magnitude spectrum of signum function.*

2.6.6 Fourier Transform of Unit Step Signal

The unit step signal is defined as,

$$u(t) = 1 \quad ; \quad t \geq 0$$
$$= 0 \quad ; \quad t < 0$$

If can be proved that, $\text{sgn}(t) = 2u(t) - 1 \quad \Rightarrow \quad u(t) = \dfrac{1}{2}[1 + \text{sgn}(t)]$

$$\therefore \ x(t) = u(t) = \dfrac{1}{2}[1 + \text{sgn}(t)]$$

On taking Fourier transform of the above equation we get,

$$X(j\Omega) = \mathcal{F}\{x(t)\} = \mathcal{F}\left\{\dfrac{1}{2}[1 + \text{sgn}(t)]\right\}$$

$$= \mathcal{F}\left\{\dfrac{1}{2}\right\} + \mathcal{F}\left\{\dfrac{1}{2}\ \text{sgn}(t)\right\} = \dfrac{1}{2}\mathcal{F}\{1\} + \dfrac{1}{2}\mathcal{F}\{\text{sgn}(t)\}$$

$$= \dfrac{1}{2}[2\pi\delta(\Omega)] + \dfrac{1}{2}\left[\dfrac{2}{j\Omega}\right] = \pi\,\delta(\Omega) + \dfrac{1}{j\Omega}$$

Using equations (2.40) and (2.41).

$$\boxed{\therefore \ \ \mathcal{F}\{u(t)\} = \pi\,\delta(\Omega) + \dfrac{1}{j\Omega}}$$

.....(2.42)

The plot of unit step signal and its magnitude spectrum are shown in Fig 2.15 and Fig 2.16 respectively.

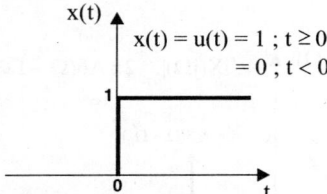

$x(t) = u(t) = 1$; $t \geq 0$
$= 0$; $t < 0$

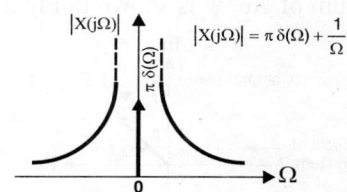

$|X(j\Omega)| = \pi\,\delta(\Omega) + \dfrac{1}{\Omega}$

Fig 2.15: *Unit step signal.*   **Fig 2.16:** *Magnitude spectrum of unit step signal.*

2.6.7 Fourier Transform of Complex Exponential Signal

The complex exponential signal is defined as,

$$x(t) = A\, e^{j\Omega_0 t}$$

$$= e^{j\Omega_0 t}\, A$$

On taking Fourier transform we get,

Frequency shifting property
If $\mathcal{F}\{x(t)\} = X(j\Omega)$ then,
$\mathcal{F}\{e^{j\Omega_0 t}\, x(t)\} = X(j(\Omega-\Omega_0))$

$$X(j\Omega) = \mathcal{F}\{x(t)\} = \mathcal{F}\{e^{j\Omega_0 t}\, A\}$$

$$= \mathcal{F}\{A\}\big|_{\Omega = \Omega - \Omega_0}$$

Using frequency shifting property

$$= 2\pi A\, \delta(\Omega)\big|_{\Omega = \Omega - \Omega_0}$$

Using equation (2.40)

$$= 2\pi A\, \delta(\Omega - \Omega_0)$$

$$\therefore\quad \mathcal{F}\{A\, e^{j\Omega_0 t}\} = 2\pi A\, \delta(\Omega - \Omega_0) \qquad .....(2.43)$$

Similarly, $\quad \boxed{\mathcal{F}\{e^{-j\Omega_0 t}\, A\} = 2\pi A\, \delta(\Omega + \Omega_0)} \qquad .....(2.44)$

The signal $Ae^{-j\Omega_0 t}$ can be represented by a rotating vector of magnitude, "A", in clockwise direction in a complex plane with an angular speed of $\Omega_0 t$ as shown in Fig 2.17 . The magnitude spectrum of $A\, e^{-j\Omega_0 t}$ is shown in Fig 2.18.

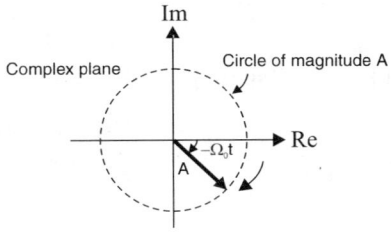

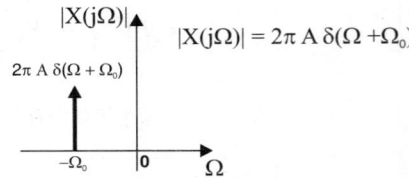

Fig 2.17: Complex exponential signal.   *Fig 2.18: Magnitude spectrum of $Ae^{-j\Omega_0 t}$.*

The signal $Ae^{j\Omega_0 t}$ can be represented by a rotating vector of magnitude "A", in anticlockwise direction in a complex plane with an angular speed of $\Omega_0 t$ as shown in Fig 2.19. The magnitude spectrum of $Ae^{j\Omega_0 t}$ is shown in Fig 2.20.

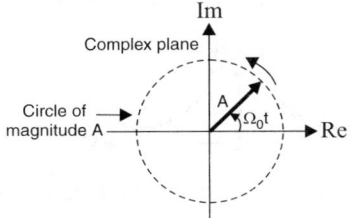

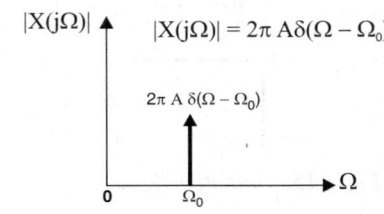

Fig 2.19: Complex exponential signal.   *Fig 2.20: Magnitude spectrum of $Ae^{+j\Omega_0 t}$.*

2.6.8 Fourier Transform of Sinusoidal Signal

The sinusoidal signal is defined as,

$$x(t) = A \sin\Omega_0 t = \frac{A}{2j}\left(e^{j\Omega_0 t} - e^{-j\Omega_0 t}\right)$$

$$\boxed{\sin\theta = \frac{e^{j\theta} - e^{-j\theta}}{2j}}$$

On taking Fourier transform we get,

$$X(j\Omega) = \mathcal{F}\{x(t)\} = \mathcal{F}\left\{\frac{A}{2j}\left(e^{j\Omega_0 t} - e^{-j\Omega_0 t}\right)\right\} = \frac{A}{2j}\left[\mathcal{F}\{e^{j\Omega_0 t}\} - \mathcal{F}\{e^{-j\Omega_0 t}\}\right] \quad \boxed{\begin{array}{l}\text{Using equations}\\ (2.43) \text{ and } (2.44)\end{array}}$$

$$= \frac{A}{2j}\left[2\pi\,\delta(\Omega - \Omega_0) - 2\pi\,\delta(\Omega + \Omega_0)\right] = \frac{A\pi}{j}\left[\delta(\Omega - \Omega_0) - \delta(\Omega + \Omega_0)\right]$$

$$\boxed{\therefore\ \mathcal{F}\{A\sin\Omega_0 t\} = \frac{A\pi}{j}\left[\delta(\Omega - \Omega_0) - \delta(\Omega + \Omega_0)\right]} \qquad \qquad \dots\dots(2.45)$$

The plot of sinusoidal signal and its spectrum are shown in Fig 2.21 and Fig 2.22.

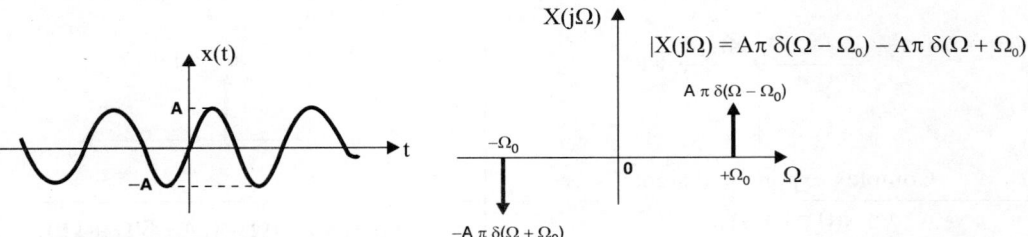

Fig 2.21: *Sinusoidal signal.*

Fig 2.22: *Spectrum of sinusoidal signal.*

2.6.9 Fourier Transform of Cosinusoidal Signal

$$\boxed{\cos\theta = \frac{e^{j\theta} + e^{-j\theta}}{2}}$$

The cosinusoidal signal is defined as,

$$x(t) = A\cos\Omega_0 t = \frac{A}{2}\left[e^{j\Omega_0 t} + e^{-j\Omega_0 t}\right]$$

On taking Fourier transform we get,

$$\mathcal{F}\{x(t)\} = \mathcal{F}\left\{\frac{A}{2}\left(e^{j\Omega_0 t} + e^{-j\Omega_0 t}\right)\right\} = \frac{A}{2}\left[\mathcal{F}\{e^{j\Omega_0 t}\} + \mathcal{F}\{e^{-j\Omega_0 t}\}\right] \quad \boxed{\begin{array}{l}\text{Using equations}\\ (2.43) \text{ and } (2.44)\end{array}}$$

$$= \frac{A}{2}\left[2\pi\,\delta(\Omega - \Omega_0) + 2\pi\,\delta(\Omega + \Omega_0)\right] = A\pi\left[\delta(\Omega - \Omega_0) + \delta(\Omega + \Omega_0)\right]$$

$$\boxed{\therefore\ \mathcal{F}\{A\cos\Omega_0 t\} = A\pi\left[\delta(\Omega - \Omega_0) + \delta(\Omega + \Omega_0)\right]} \qquad \qquad \dots\dots(2.46)$$

The plot of cosinusoidal signal and its magnitude spectrum are shown in Fig 2.23 and Fig 2.24.

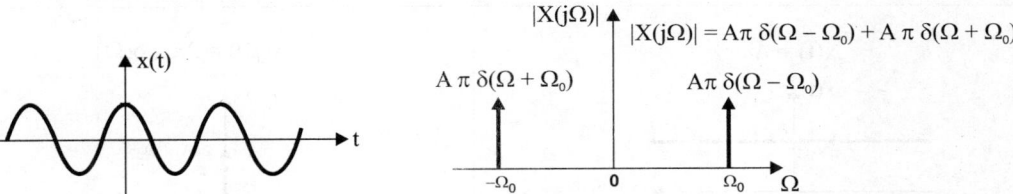

Fig 2.23: *Cosinusoidal signal.*

Fig 2.24: *Magnitude spectrum of cosinusoidal signal.*

Table 2.6: Fourier Transform of Standard Signals and their Magnitude Spectrum

x(t)	$X(j\Omega)$ and Magnitude Spectrum
$x(t) = \delta(t-t_0)$ Shifted impulse signal	$X(j\Omega) = e^{-j\Omega t_0}$
$x(t) = e^{j\Omega_0 t}$ Complex exponential signal	$X(j\Omega) = 2\pi\delta(\Omega - \Omega_0)$
$x(t) = \cos\Omega_0 t$ Cosinusoidal signal	$X(j\Omega) = \pi[\delta(\Omega - \Omega_0) + \delta(\Omega + \Omega_0)]$
$x(t) = \sin\Omega_0 t$ Sinusoidal signal	$X(j\Omega) = \dfrac{\pi}{j}[\delta(\Omega - \Omega_0) - \delta(\Omega + \Omega_0)]$
$x(t) = A$ Constant	$x(j\Omega) = 2\pi A\,\delta(\Omega)$

Table 2.6: Continued...

x(t)	X(jΩ) and Magnitude Spectrum
$x(t) = sgn(t) = \dfrac{t}{\lvert t \rvert} = 1 \quad ; t > 0$ $\qquad\qquad\qquad = -1 \quad ; t < 0$ Signum signal	$X(j\Omega) = \dfrac{2}{j\Omega}$
$x(t) = u(t) = 1 \quad ; \quad t \geq 0$ $\qquad\qquad = 0 \quad ; \quad t < 0$ unit step signal	$X(j\Omega) = \pi\,\delta(\Omega) + \dfrac{1}{j\Omega}$
$x(t) = e^{-at}u(t)$ Decaying exponential signal	$X(j\Omega) = \dfrac{1}{a + j\Omega}$
$x(t) = te^{-at}u(t)$ Product of ramp and decaying exponential signal	$X(j\Omega) = \dfrac{1}{(a + j\Omega)^2}$

Table 2.6: Continued...

x(t)	X(jΩ) and Magnitude Spectrum
$x(t)=e^{-a|t|}$ 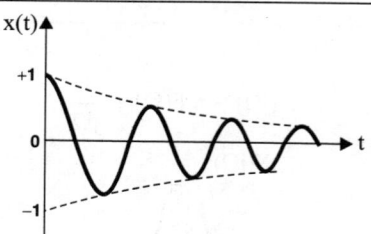 Double Exponential	$X(j\Omega)=\dfrac{2a}{a^2+\Omega^2}$ 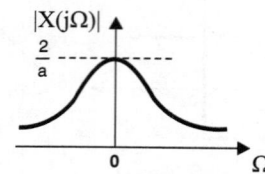
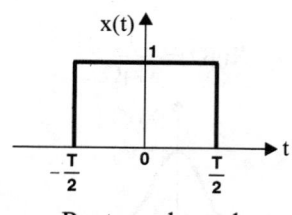 Exponentially decaying cosinusoidal signal	$X(j\Omega)=\dfrac{a+j\Omega}{(a+j\Omega)^2+\Omega_0^2}$ 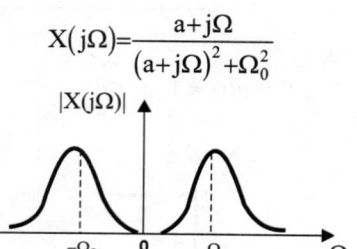
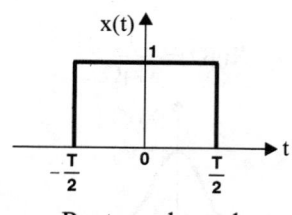 Rectangular pulse	$X(j\Omega)=T\dfrac{\sin\frac{\Omega T}{2}}{\frac{\Omega T}{2}}=T\mathrm{sinc}\left(\dfrac{\Omega T}{2}\right)$ **(AU May'15, 2 Marks)** 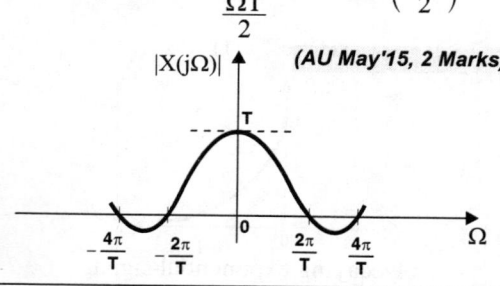
$x(t)=1+\dfrac{t}{T}\ ;\ t=-T\ \text{to}\ 0$ $\quad=1-\dfrac{t}{T}\ ;\ t=0\ \text{to}\ T$ Triangular pulse	$X(j\Omega)=T\left(\dfrac{\sin\frac{\Omega T}{2}}{\frac{\Omega T}{2}}\right)^2$ 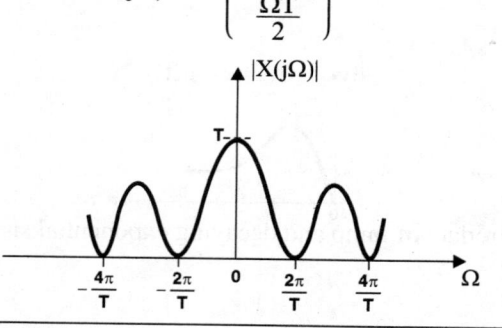

Table 2.6: Continued...

$x(t)$	$X(j\Omega)$ and Magnitude Spectrum
$x(t) = \dfrac{\Omega_0}{2\pi} \, \text{sinc}\!\left(\dfrac{\Omega_0}{2} t\right) = \dfrac{1}{\pi} \dfrac{\sin\!\left(\dfrac{\Omega_0}{2} t\right)}{t}$ Sinc pulse	$X(j\Omega) = \left[u\!\left(\Omega + \dfrac{\Omega_0}{2}\right) - u\!\left(\Omega - \dfrac{\Omega_0}{2}\right) \right]$ $\boxed{\Omega_0 = \dfrac{2\pi}{T}}$
$x(t) = e^{-a^2 t^2}$ Gaussian pulse	$X(j\Omega) = \dfrac{\sqrt{\pi}}{a} e^{-\left(\frac{\Omega}{2a}\right)^2}$
$x(t) = \displaystyle\sum_{n=-\infty}^{+\infty} \delta(t - nT)$ Impulse train	$X(j\Omega) = \dfrac{2\pi}{T} \displaystyle\sum_{n=-\infty}^{+\infty} \delta\!\left(\Omega - \dfrac{2\pi n}{T}\right)$

From Table 2.6 the following observations are made.

1. The Fourier transform of a Gaussian pulse will be another Gaussian pulse.

2. The Fourier transform of an impulse train will be another impulse train.

3. The Fourier transform of a rectangular pulse will be a sinc pulse and viceversa.

4. The Fourier transform of a triangular pulse will be a squared sinc pulse.

5. The Fourier transform of a constant will be an impulse and vice-versa.

Table 2.7: Standard Fourier Transform Pairs

$x(t)$	$X(j\Omega)$
$\delta(t)$	1
$\delta(t-t_0)$	$e^{-j\Omega t_0}$
A where, A is constant	$2\pi A\delta(\Omega)$
$u(t)$	$\pi\delta(\Omega) + \dfrac{1}{j\Omega}$
$sgn(t)$	$\dfrac{2}{j\Omega}$
$t\,u(t)$	$\dfrac{1}{(j\Omega)^2}$
$\dfrac{t^{m-1}}{(m-1)!}\,u(t)$ where, $m = 1,2,3, ...$	$\dfrac{1}{(j\Omega)^m}$
$t^m\,u(t)$ where, $m = 1,2,3, ...$	$\dfrac{m!}{(j\Omega)^{m+1}}$
$e^{-at}u(t)$	$\dfrac{1}{j\Omega + a}$
$t\,e^{-at}u(t)$	$\dfrac{1}{(j\Omega + a)^2}$
$Ae^{-a|t|}$	$\dfrac{2Aa}{a^2 + \Omega^2}$
$Ae^{j\Omega_0 t}$	$2\pi A\,\delta(\Omega - \Omega_0)$
$\sin\Omega_0 t$	$\dfrac{\pi}{j}[\delta(\Omega - \Omega_0) - \delta(\Omega + \Omega_0)]$
$\cos\Omega_0 t$	$\pi[\delta(\Omega - \Omega_0) + \delta(\Omega + \Omega_0)]$

2.6.10 Solved Problems in Fourier Transform

Example 2.13

Determine the Fourier transform of following continuous time domain signals.

a) $x(t) = 1 - t^2$; for $|t| < 1$ **b)** $x(t) = e^{-at} \cos \Omega_0 t \; u(t)$ **c)** $x(t) = t\, e^{-at} u(t)$

       $= 0$; for $|t| > 1$

Solution:

a) Given that, $x(t) = 1 - t^2$; for $|t| < 1$

$$\therefore x(t) = 1 - t^2 ; \quad \text{for } t = -1 \text{ to } + 1$$

By definition of Fourier transform,

$$X(j\Omega) = \mathcal{F}\{x(t)\} = \int_{-\infty}^{+\infty} x(t)\, e^{-j\Omega t} = \int_{-1}^{+1} (1 - t^2)\, e^{-j\Omega t}\; dt = \int_{-1}^{+1} e^{-j\Omega t}\; dt - \int_{-1}^{\div 1} t^2\, e^{-j\Omega t}\; dt$$

$$= \left[\frac{e^{-j\Omega t}}{-j\Omega}\right]_{-1}^{+1} - \left[t^2 \frac{e^{-j\Omega t}}{-j\Omega} - \int \frac{e^{-j\Omega t}}{-j\Omega} \times 2t\; dt\right]_{-1}^{1}$$

$$\boxed{\begin{array}{l} \int u\, dv = u\, v - \int v\, du \\ \hline u = t^2 \quad\quad \Rightarrow du = 2t \\ dv = e^{-j\Omega t} \Rightarrow v = \dfrac{e^{-j\Omega t}}{-j\Omega} \end{array}}$$

$$= \left[\frac{e^{-j\Omega t}}{-j\Omega}\right]_{-1}^{1} - \left[-\frac{t^2 e^{-j\Omega t}}{j\Omega} + \frac{2}{j\Omega} \int t\, e^{-j\Omega t}\; dt\right]_{-1}^{1}$$

$$= \left[\frac{e^{-j\Omega t}}{-j\Omega}\right]_{-1}^{1} - \left[-\frac{t^2 e^{-j\Omega t}}{j\Omega} + \frac{2}{j\Omega}\left(t\frac{e^{-j\Omega t}}{-j\Omega} - \int \frac{e^{-j\Omega t}}{-j\Omega} \times 1\, dt\right)\right]_{-1}^{1}$$

$$\boxed{\begin{array}{l} \int u\, dv = u\, v - \int v\, du \\ \hline u = t \quad\quad \Rightarrow du = 1 \\ dv = e^{-j\Omega t} \Rightarrow v = \dfrac{e^{-j\Omega t}}{-j\Omega} \end{array}}$$

$$= \left[\frac{e^{-j\Omega t}}{-j\Omega}\right]_{-1}^{1} - \left[-\frac{t^2 e^{-j\Omega t}}{j\Omega} + \frac{2}{(j\Omega)^2}\left(-t e^{-j\Omega t} + \int e^{-j\Omega t}\; dt\right)\right]_{-1}^{1}$$

$$= \left[\frac{e^{-j\Omega t}}{-j\Omega}\right]_{-1}^{1} - \left[-\frac{t^2 e^{-j\Omega t}}{j\Omega} - \frac{2}{\Omega^2}\left(-t e^{-j\Omega t} + \frac{e^{-j\Omega t}}{-j\Omega}\right)\right]_{-1}^{1}$$

$$= \left[\frac{e^{-j\Omega t}}{-j\Omega}\right]_{-1}^{1} - \left[-\frac{t^2 e^{-j\Omega t}}{j\Omega} + \frac{2t\, e^{-j\Omega t}}{\Omega^2} + \frac{2 e^{-j\Omega t}}{j\Omega^3}\right]_{-1}^{1}$$

$$= -\frac{e^{-j\Omega}}{j\Omega} + \frac{e^{j\Omega}}{j\Omega} - \left[-\frac{e^{-j\Omega}}{j\Omega} + \frac{2 e^{-j\Omega}}{\Omega^2} + \frac{2 e^{-j\Omega}}{j\Omega^3} + \frac{e^{j\Omega}}{j\Omega} + \frac{2 e^{j\Omega}}{\Omega^2} - \frac{2 e^{j\Omega}}{j\Omega^3}\right]$$

$$= -\frac{e^{-j\Omega}}{j\Omega} + \frac{e^{j\Omega}}{j\Omega} + \frac{e^{-j\Omega}}{j\Omega} - \frac{2 e^{-j\Omega}}{\Omega^2} - \frac{2 e^{-j\Omega}}{j\Omega^3} - \frac{e^{j\Omega}}{j\Omega} - \frac{2 e^{j\Omega}}{\Omega^2} + \frac{2 e^{j\Omega}}{j\Omega^3}$$

$$= -\frac{2}{\Omega^2}\left(e^{j\Omega} + e^{-j\Omega}\right) + \frac{2}{j\Omega^3}\left(e^{j\Omega} - e^{-j\Omega}\right)$$

$$\boxed{\sin\theta = \dfrac{e^{j\theta} - e^{-j\theta}}{2j} \quad\quad \cos\theta = \dfrac{e^{j\theta} + e^{-j\theta}}{2}}$$

$$= -\frac{2}{\Omega^2}\, 2\cos\Omega + \frac{2}{j\Omega^3}\, 2j\sin\Omega$$

$$= -\frac{4\cos\Omega}{\Omega^2} + \frac{4\sin\Omega}{\Omega^3}$$

$$= \frac{4}{\Omega^2}\left(\frac{\sin\Omega}{\Omega} - \cos\Omega\right)$$

b) Given that, $x(t) = e^{-at} \cos \Omega_0 t \, u(t)$

$$x(t) = e^{-at} \cos \Omega_0 t \, u(t) = e^{-at} \cos \Omega_0 t \; ; \quad \text{for} \quad t \geq 0$$

$$\boxed{\cos \theta = \dfrac{e^{j\theta} + e^{-j\theta}}{2}}$$

By definition of Fourier transform,

$$X(j\Omega) = \mathcal{F}\{x(t)\} = \int\limits_{-\infty}^{+\infty} x(t) \, e^{-j\Omega t} \, dt = \int\limits_{0}^{\infty} e^{-at} \cos \Omega_0 t \, e^{-j\Omega t} \, dt$$

$$= \int\limits_{0}^{\infty} e^{-at} \left(\frac{e^{j\Omega_0 t} + e^{-j\Omega_0 t}}{2} \right) e^{-j\Omega t} \, dt$$

$$= \frac{1}{2} \int\limits_{0}^{\infty} e^{-at} \, e^{j\Omega_0 t} \, e^{-j\Omega t} \, dt + \frac{1}{2} \int\limits_{0}^{\infty} e^{-at} \, e^{-j\Omega_0 t} \, e^{-j\Omega t} \, dt$$

$$= \frac{1}{2} \int\limits_{0}^{\infty} e^{-(a - j\Omega_0 + j\Omega)t} \, dt + \frac{1}{2} \int\limits_{0}^{\infty} e^{-(a + j\Omega_0 + j\Omega)t} \, dt$$

$$= \frac{1}{2} \left[\frac{e^{-(a - j\Omega_0 + j\Omega)t}}{-(a - j\Omega_0 + j\Omega)} \right]_{0}^{\infty} + \frac{1}{2} \left[\frac{e^{-(a + j\Omega_0 + j\Omega)t}}{-(a + j\Omega_0 + j\Omega)} \right]_{0}^{\infty}$$

$$= \frac{1}{2} \left[\frac{e^{-\infty}}{-(a - j\Omega_0 + j\Omega)} - \frac{e^{0}}{-(a - j\Omega_0 + j\Omega)} \right] + \frac{1}{2} \left[\frac{e^{-\infty}}{-(a + j\Omega_0 + j\Omega)} - \frac{e^{0}}{-(a + j\Omega_0 + j\Omega)} \right]$$

$$= \frac{1}{2} \left[0 + \frac{1}{a - j\Omega_0 + j\Omega} \right] + \frac{1}{2} \left[0 + \frac{1}{a + j\Omega_0 + j\Omega} \right]$$

$$\boxed{e^{-\infty} = 0 \; ; \; e^{0} = 1}$$

$$= \frac{1}{2} \left[\frac{1}{(a + j\Omega) - j\Omega_0} + \frac{1}{(a + j\Omega) + j\Omega_0} \right]$$

$$= \frac{1}{2} \left[\frac{(a + j\Omega) + j\Omega_0 + (a + j\Omega) - j\Omega_0}{[(a + j\Omega) - j\Omega_0][(a + j\Omega) + j\Omega_0]} \right]$$

$$\boxed{(a + b)(a - b) = a^2 - b^2 \mid j^2 = -1}$$

$$= \frac{1}{2} \frac{2(a + j\Omega)}{(a + j\Omega)^2 + \Omega_0^2} = \frac{a + j\Omega}{(a + j\Omega)^2 + \Omega_0^2}$$

c) Given that, $x(t) = t \, e^{-at} \, u(t)$ **(AU May'15, 6 Marks)**

$$x(t) = t \, e^{-at} u(t) = t \, e^{-at} \; ; \quad \text{for} \quad t \geq 0$$

By definition of Fourier transform,

$$X(j\Omega) = \mathcal{F}\{x(t)\} = \int\limits_{-\infty}^{+\infty} x(t) \, e^{-j\Omega t} \, dt = \int\limits_{0}^{\infty} t \, e^{-at} e^{-j\Omega t} \, dt$$

$$\boxed{\begin{array}{l} \displaystyle\int u \, dv = u \, v - \int v \, du \\[2mm] u = t \qquad \Rightarrow du = 1 \\[2mm] dv = e^{-(a + j\Omega)t} \Rightarrow v = \dfrac{e^{-(a + j\Omega)t}}{-(a + j\Omega)} \end{array}}$$

$$= \int\limits_{0}^{\infty} t \, e^{-(a + j\Omega)t} \, dt$$

$$= \left[\frac{t \, e^{-(a + j\Omega)t}}{-(a + j\Omega)} - \int \frac{e^{-(a + j\Omega)t}}{-(a + j\Omega)} \times 1 \, dt \right]_{0}^{\infty}$$

$$\therefore \mathcal{F}\{x(t)\} = \left[\frac{-t\,e^{-(a+j\Omega)t}}{a+j\Omega} - \frac{e^{-(a+j\Omega)t}}{(-(a+j\Omega))^2} \right]_0^\infty$$

$$= -\frac{\infty \times e^{-\infty}}{a+j\Omega} - \frac{e^{-\infty}}{(a+j\Omega)^2} + \frac{0 \times e^0}{a+j\Omega} + \frac{e^0}{(a+j\Omega)^2} \qquad \boxed{e^{-\infty}=0}$$

$$= \frac{1}{(a+j\Omega)^2}$$

Example 2.14

(AU May'15, 6 Marks)

If Fourier transform of x(t) is X(jΩ) then using time shifting property show that, Fourier transform of x(t + T) + x(t − T) is equal to 2X(jΩ) cos Ωt.

Solution:

Given that, x(t) = X(jΩ)

The time shifting property of Fourier transform says that,

$$\mathcal{F}\{x(t-t_0)\} = e^{-j\Omega t_0}\, X(j\Omega)$$

$$\therefore \mathcal{F}\{x(t+T)\} = e^{j\Omega T}\, X(j\Omega)$$

$$\mathcal{F}\{x(t-T)\} = e^{-j\Omega T}\, X(j\Omega)$$

Now, $\mathcal{F}\{x(t+T)+x(t-T)\} = \mathcal{F}\{x(t+T)\} + \mathcal{F}\{x(t-T)\} = e^{j\Omega T}\,X(j\Omega) + e^{-j\Omega T}\,X(j\Omega)$

$$\qquad\qquad = X(j\Omega)\,[e^{j\Omega T} + e^{-j\Omega T}] = X(j\Omega)[2\cos\Omega T] \qquad \boxed{\cos\theta = \frac{e^{j\theta}+e^{-j\theta}}{2}}$$

$$\therefore \mathcal{F}\{x(t+T)+x(t-T)\} = 2\,X(j\Omega)\cos\Omega T$$

Example 2.15

Determine the Fourier transform of the rectangular pulse shown in Fig 2.15.1.

Solution:

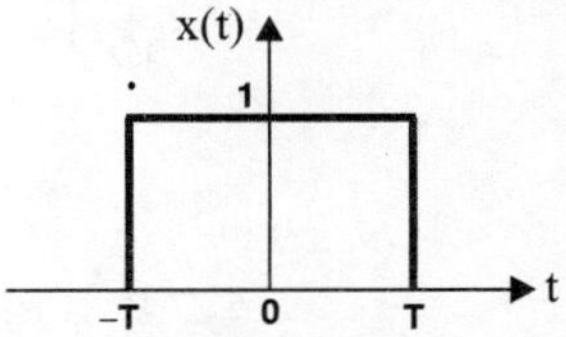

The mathematical equation of the rectangular pulse is,

$$x(t) = 1 \qquad ; \qquad \text{for } t = -T \text{ to } +T$$

By definition of Fourier transform,

Fig 2.15.1.

$$X(j\Omega) = \mathcal{F}\{x(t)\} = \int_{-\infty}^{+\infty} x(t)\, e^{-j\Omega t}\, dt = \int_{-T}^{+T} 1 \times e^{-j\Omega t}\, dt = \left[\frac{e^{-j\Omega t}}{-j\Omega} \right]_{-T}^{+T}$$

$$\boxed{\sin\theta = \frac{e^{j\theta}-e^{-j\theta}}{2j}}$$

$$= \frac{e^{-j\Omega T}}{-j\Omega} - \frac{e^{j\Omega T}}{-j\Omega} = \frac{1}{j\Omega}(e^{j\Omega T} - e^{-j\Omega T}) = \frac{1}{j\Omega}\, 2j\sin\Omega T$$

$$\boxed{\frac{\sin\theta}{\theta} = \text{sinc }\theta}$$

$$= 2\,\frac{\sin\Omega T}{\Omega} = 2T\,\frac{\sin\Omega T}{\Omega T}$$

$$= 2T \text{ sinc }\Omega T$$

Example 2.16

Determine the Fourier transform of the triangular pulse shown in Fig 2.16.1.

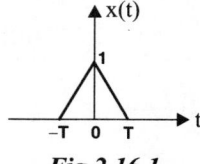

Fig 2.16.1.

Solution:

The mathematical equation of triangular pulse is,

$$x(t) = 1 + \frac{t}{T} \quad ; \quad \text{for } t = -T \text{ to } 0$$

$$= 1 - \frac{t}{T} \quad ; \quad \text{for } t = 0 \text{ to } T$$

> (Please refer example 2.11 for the mathematical equation of triangular pulse).

By definition of Fourier transform,

$$X(j\Omega) = \mathcal{F}\{x(t)\} = \int_{-\infty}^{+\infty} x(t)\,e^{-j\Omega t}\,dt = \int_{-T}^{0}\left(1 + \frac{t}{T}\right)e^{-j\Omega t}\,dt + \int_{0}^{T}\left(1 - \frac{t}{T}\right)e^{-j\Omega t}\,dt$$

$$\boxed{\begin{aligned} \int u\,dv &= u\,v - \int v\,du \\ u &= t \quad \Rightarrow \quad du = 1 \\ dv &= e^{-j\Omega t} \Rightarrow v = \frac{e^{-j\Omega t}}{-j\Omega} \end{aligned}}$$

$$= \int_{-T}^{0} e^{-j\Omega t}\,dt + \frac{1}{T}\int_{-T}^{0} t\,e^{-j\Omega t}\,dt + \int_{0}^{T} e^{-j\Omega t}\,dt - \frac{1}{T}\int_{0}^{T} t\,e^{-j\Omega t}\,dt$$

$$= \left[\frac{e^{-j\Omega t}}{-j\Omega}\right]_{-T}^{0} + \frac{1}{T}\left[t\frac{e^{-j\Omega t}}{-j\Omega} - \int\frac{e^{-j\Omega t}}{-j\Omega}\times 1\,dt\right]_{-T}^{0} + \left[\frac{e^{-j\Omega t}}{-j\Omega}\right]_{0}^{T} - \frac{1}{T}\left[t\frac{e^{-j\Omega t}}{-j\Omega} - \int\frac{e^{-j\Omega t}}{-j\Omega}\times 1\,dt\right]_{0}^{T}$$

$$= -\frac{1}{j\Omega}\left[e^{-j\Omega t}\right]_{-T}^{0} - \frac{1}{j\Omega T}\left[t\,e^{-j\Omega t} - \int e^{-j\Omega t}\,dt\right]_{-T}^{0} - \frac{1}{j\Omega}\left[e^{-j\Omega t}\right]_{0}^{T} + \frac{1}{j\Omega T}\left[t\,e^{-j\Omega t} - \int e^{-j\Omega t}\,dt\right]_{0}^{T}$$

$$= -\frac{1}{j\Omega}\left[e^{-j\Omega t}\right]_{-T}^{0} - \frac{1}{j\Omega T}\left[t\,e^{-j\Omega t} - \frac{e^{-j\Omega t}}{-j\Omega}\right]_{-T}^{0} - \frac{1}{j\Omega}\left[e^{-j\Omega t}\right]_{0}^{T} + \frac{1}{j\Omega T}\left[t\,e^{-j\Omega t} - \frac{e^{-j\Omega t}}{-j\Omega}\right]_{0}^{T}$$

$$= -\frac{1}{j\Omega}\left[e^{0} - e^{j\Omega T}\right] - \frac{1}{j\Omega T}\left[0 - \frac{e^{0}}{-j\Omega} + T\,e^{j\Omega T} + \frac{e^{j\Omega T}}{-j\Omega}\right] - \frac{1}{j\Omega}\left[e^{-j\Omega T} - e^{0}\right]$$

$$+ \frac{1}{j\Omega T}\left[T\,e^{-j\Omega T} - \frac{e^{-j\Omega T}}{-j\Omega} - 0 + \frac{e^{0}}{-j\Omega}\right]$$

$$= -\frac{1}{j\Omega} + \frac{e^{j\Omega T}}{j\Omega} - 0 + \frac{1}{T\Omega^2} - \frac{e^{j\Omega T}}{j\Omega} - \frac{e^{j\Omega T}}{T\Omega^2} - \frac{e^{-j\Omega T}}{j\Omega} + \frac{1}{j\Omega} + \frac{e^{-j\Omega T}}{j\Omega} - \frac{e^{-j\Omega T}}{T\Omega^2} - 0 + \frac{1}{T\Omega^2}$$

$$= \frac{2}{T\Omega^2} - \frac{1}{T\Omega^2}\left(e^{j\Omega T} + e^{-j\Omega T}\right) = \frac{2}{T\Omega^2} - \frac{1}{T\Omega^2}\,2\cos\Omega T$$

$$\boxed{\cos\theta = \frac{e^{j\theta} + e^{-j\theta}}{2}}$$

$$= \frac{2}{T\Omega^2}\,(1 - \cos\Omega T)$$

Alternatively the above result can be expressed as shown below:

$$X(j\Omega) = \frac{2}{T\Omega^2}(1 - \cos\Omega T) = \frac{2}{T\Omega^2}\left(1 - \cos 2\left(\frac{\Omega T}{2}\right)\right)$$

$$\therefore \; X(j\Omega) = \frac{2}{T\Omega^2}\left(2\sin^2\frac{\Omega T}{2}\right) = T\frac{4}{T^2\Omega^2}\sin^2\frac{\Omega T}{2} = T\frac{\sin^2\left(\frac{\Omega T}{2}\right)}{\left(\frac{\Omega T}{2}\right)^2}$$

$$\boxed{\sin^2\theta = \frac{1-\cos 2\theta}{2}}$$

$$= T\left[\frac{\sin\frac{\Omega T}{2}}{\frac{\Omega T}{2}}\right]^2 = T\left(\text{sinc }\frac{\Omega T}{2}\right)^2$$

$$\boxed{\frac{\sin\theta}{\theta} = \text{sinc }\theta}$$

Example 2.17

Determine the inverse Fourier transform of the function

$X(j\Omega) = 1$, $\quad -W < \Omega < W$

$\quad\quad = 0$, $\quad |\Omega| > W$

Solution:

Given that, $X(j\Omega) = 1$, $\quad -W < \Omega < W$

$\quad\quad\quad\quad\quad\quad\;\; = 0$, $\quad |\Omega| > W$

By definition of Inverse Fourier transform,

$$x(t) = \mathcal{F}^{-1}\{X(j\Omega)\} = \frac{1}{2\pi}\int_{-\infty}^{+\infty} X(j\Omega)\,e^{j\Omega t}\,d\Omega$$

$$= \frac{1}{2\pi}\int_{-W}^{W} 1\times e^{j\Omega t}\,d\Omega$$

$$= \frac{1}{2\pi}\left[\frac{e^{j\Omega t}}{jt}\right]_{-W}^{W} = \frac{1}{2\pi}\left[\frac{e^{jWt}-e^{-jWt}}{jt}\right]$$

$$= \frac{1}{\pi t}\left[\frac{e^{jWt}-e^{-jWt}}{2j}\right] = \frac{1}{\pi t}\sin Wt = \frac{W}{\pi}\frac{\sin Wt}{Wt}$$

$$\boxed{\sin\theta = \frac{e^{j\theta}-e^{-j\theta}}{2j}}$$

$$\boxed{\frac{\sin\theta}{\theta} = \text{sinc }\theta}$$

$$= \frac{W}{\pi}\text{ sinc } Wt$$

Example 2.18

Determine the inverse Fourier transform of the following functions, using partial fraction expansion technique.

a) $X(j\Omega) = \dfrac{3(j\Omega)+14}{(j\Omega)^2 + 7(j\Omega)+12}$

b) $X(j\Omega) = \dfrac{j\Omega+7}{(j\Omega+3)^2}$

Solution:

a) Given that, $X(j\Omega) = \dfrac{3(j\Omega)+14}{(j\Omega)^2 + 7(j\Omega)+12} = \dfrac{3(j\Omega)+14}{(j\Omega+3)(j\Omega+4)}$

By partial fraction expansion technique we can write,

$$X(j\Omega) = \frac{3(j\Omega)+14}{(j\Omega+3)(j\Omega+4)} = \frac{k_1}{j\Omega+3} + \frac{k_2}{j\Omega+4}$$

$$k_1 = \frac{3(j\Omega) + 14}{(j\Omega + 3)(j\Omega + 4)} \times (j\Omega + 3)\Big|_{j\Omega = -3} = \frac{3(-3) + 14}{-3 + 4} = 5$$

$$k_2 = \frac{3(j\Omega) + 14}{(j\Omega + 3)(j\Omega + 4)} \times (j\Omega + 4)\Big|_{j\Omega = -4} = \frac{3(-4) + 14}{-4 + 3} = -2$$

$$\therefore X(j\Omega) = \frac{5}{j\Omega + 3} - \frac{2}{j\Omega + 4}$$

$$\therefore x(t) = \mathcal{F}^{-1}\{X(j\Omega)\} = \mathcal{F}^{-1}\left\{\frac{5}{j\Omega + 3} - \frac{2}{j\Omega + 4}\right\}$$

$$= 5 \times \mathcal{F}^{-1}\left\{\frac{1}{j\Omega + 3}\right\} - 2 \times \mathcal{F}^{-1}\left\{\frac{1}{j\Omega + 4}\right\}$$

$$\boxed{\mathcal{F}\{e^{-at}\,u(t)\} = \frac{1}{j\Omega + a}}$$

$$= 5\,e^{-3t}\,u(t) - 2\,e^{-4t}\,u(t)$$

b) Given that, $X(j\Omega) = \dfrac{j\Omega + 7}{(j\Omega + 3)^2}$

By partial fraction expansion technique $X(j\Omega)$ can be written as,

$$\therefore X(j\Omega) = \frac{j\Omega + 7}{(j\Omega + 3)^2} = \frac{k_1}{(j\Omega + 3)^2} + \frac{k_2}{j\Omega + 3}$$

$$k_1 = \frac{j\Omega + 7}{(j\Omega + 3)^2} \times (j\Omega + 3)^2\Big|_{j\Omega = -3} = -3 + 7 = 4$$

$$k_2 = \frac{d}{d(j\Omega)}\left[\frac{j\Omega + 7}{(j\Omega + 3)^2} \times (j\Omega + 3)^2\right]\Big|_{j\Omega = -3} = \frac{d}{d(j\Omega)}[j\Omega + 7]\Big|_{j\Omega = -3} = 1$$

$$\therefore X(j\Omega) = \frac{4}{(j\Omega + 3)^2} + \frac{1}{j\Omega + 3}$$

$$\therefore x(t) = \mathcal{F}^{-1}\{X(j\Omega)\} = \mathcal{F}^{-1}\left\{\frac{4}{(j\Omega + 3)^2} + \frac{1}{j\Omega + 3}\right\}$$

$$\boxed{\mathcal{F}\{t\,e^{-at}\,u(t)\} = \frac{1}{(j\Omega + a)^2}}$$

$$= 4 \times \mathcal{F}^{-1}\left\{\frac{1}{(j\Omega + 3)^2}\right\} + \mathcal{F}^{-1}\left\{\frac{1}{j\Omega + 3}\right\}$$

$$\boxed{\mathcal{F}\{e^{-at}\,u(t)\} = \frac{1}{j\Omega + a}}$$

$$= 4t\,e^{-3t}\,u(t) + e^{-3t}\,u(t) = (4t + 1)\,e^{-3t}\,u(t)$$

Example 2.19   *(AU Jun' 11, 8 Marks)*

Determine the convolution of $x_1(t) = e^{-2t}\,u(t)$ and $x_2(t) = e^{-6t}\,u(t)$, using Fourier transform.

Solution:

Let, $X_1(j\Omega) = \mathcal{F}\{x_1(t)\}$

$\qquad X_2(j\Omega) = \mathcal{F}\{x_2(t)\}$

By convolution property of Fourier transform,

$$\mathcal{F}\{x_1(t) * x_2(t)\} = X_1(j\Omega)\,X_2(j\Omega)$$

Let, $X(j\Omega) = X_1(j\Omega)\,X_2(j\Omega)$

$$\therefore X(j\Omega) = \mathcal{F}\{e^{-2t}u(t)\} \times \mathcal{F}\{e^{-6t}u(t)\}$$

$$= \frac{1}{j\Omega + 2} \times \frac{1}{j\Omega + 6}$$

By partial fraction expansion technique $X(j\Omega)$ can be expressed as,

$$X(j\Omega) = \frac{1}{(j\Omega + 2)(j\Omega + 6)} = \frac{k_1}{j\Omega + 2} + \frac{k_2}{j\Omega + 6}$$

$$k_1 = \frac{1}{(j\Omega + 2)(j\Omega + 6)} \times (j\Omega + 2)\bigg|_{j\Omega = -2} = \frac{1}{-2 + 6} = \frac{1}{4} = 0.25$$

$$k_2 = \frac{1}{(j\Omega + 2)(j\Omega + 6)} \times (j\Omega + 6)\bigg|_{j\Omega = -6} = \frac{1}{-6 + 2} = -\frac{1}{4} = -0.25$$

$$\therefore X(j\Omega) = \frac{0.25}{j\Omega + 2} - \frac{0.25}{j\Omega + 6}$$

$$x(t) = \mathcal{F}^{-1}\{X(j\Omega)\} = \mathcal{F}^{-1}\left\{\frac{0.25}{j\Omega + 2} - \frac{0.25}{j\Omega + 6}\right\} = 0.25 \times \mathcal{F}^{-1}\left\{\frac{1}{j\Omega + 2}\right\} - 0.25 \times \mathcal{F}^{-1}\left\{\frac{1}{j\Omega + 6}\right\}$$

$$= 0.25\, e^{-2t}u(t) - 0.25\, e^{-6t}u(t)$$

$$= 0.25(e^{-2t} - e^{-6t})u(t)$$

$$\boxed{\mathcal{F}\{e^{-at}u(t)\} = \frac{1}{j\Omega + a}}$$

Example 2.20

The impulse response of an LTI system is, $h(t) = 2\,e^{-3t}u(t)$.

Find the response of the system for the input, $x(t) = 2e^{-5t}u(t)$, using Fourier transform.

Solution:

Given that, $x(t) = 2\,e^{-5t}u(t)$.

$$\boxed{\mathcal{F}\{e^{-at}u(t)\} = \frac{1}{j\Omega + a}}$$

$$\therefore X(j\Omega) = \mathcal{F}\{x(t)\} = \mathcal{F}\{2\,e^{-5t}u(t)\} = \frac{2}{j\Omega + 5} \qquad\qquad .....(1)$$

Given that, $h(t) = 2\,e^{-3t}u(t)$.

$$\therefore H(j\Omega) = \mathcal{F}\{h(t)\} = \mathcal{F}\{2\,e^{-3t}u(t)\} = \frac{2}{j\Omega + 3} \qquad\qquad .....(2)$$

For LTI system, the response,

$$y(t) = = x(t) * h(t)$$

On taking Fourier transform of above equation we get,

$$\mathcal{F}\{y(t)\} = \mathcal{F}\{x(t) * h(t)\}$$

Let, $\mathcal{F}\{y(t)\} = Y(j\Omega)$.

$$\therefore Y(j\Omega) = \mathcal{F}\{x(t) * h(t)\}$$

$$\boxed{\text{Using convolution property of Fourier transform}}$$

$$= X(j\Omega)\, H(j\Omega)$$

$$\boxed{\text{Using equations (1) and (2)}}$$

$$= \frac{2}{j\Omega + 5} \times \frac{2}{j\Omega + 3} = \frac{4}{(j\Omega + 5)(j\Omega + 3)}$$

By partial fraction expansion technique, the above equation can be written as,

$$Y(j\Omega) = \frac{4}{(j\Omega + 5)(j\Omega + 3)} = \frac{k_1}{j\Omega + 5} + \frac{k_2}{j\Omega + 3}$$

$$k_1 = \frac{4}{\cancel{(j\Omega + 5)}(j\Omega + 3)} \times \cancel{(j\Omega + 5)}\Big|_{j\Omega = -5} = \frac{4}{-5 + 3} = -2$$

$$k_2 = \frac{4}{(j\Omega + 5)\cancel{(j\Omega + 3)}} \times \cancel{(j\Omega + 3)}\Big|_{j\Omega = -3} = \frac{4}{-3 + 5} = 2$$

$$\therefore \ Y(j\Omega) = -\frac{2}{j\Omega + 5} + \frac{2}{j\Omega + 3}$$

On taking inverse Fourier transform of $Y(j\Omega)$ we get $y(t)$.

$$y(t) = \mathcal{F}^{-1}\{Y(j\Omega)\} = \mathcal{F}^{-1}\left\{-\frac{2}{j\Omega + 5} + \frac{2}{j\Omega + 3}\right\} = -2\times\mathcal{F}^{-1}\left\{-\frac{1}{j\Omega + 5}\right\} + 2\times\mathcal{F}^{-1}\left\{\frac{1}{j\Omega + 3}\right\}$$

$$= -2\,e^{-5t}\,u(t) + 2\,e^{-3t}\,u(t) = 2(e^{-3t} - e^{-5t})\,u(t)$$

Example 2.21

Determine the Fourier transform of the periodic pulse function shown in Fig 2.21.1.

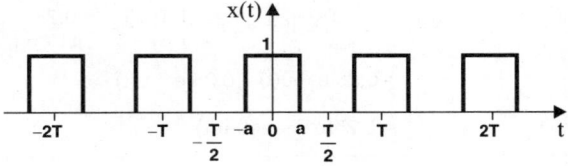

Fig 2.21.1.

Solution:

The mathematical equation for one period of the periodic pulse function is,

$$x(t) = 1 \ ; \ t = -a \text{ to } +a$$

$$= 0 \ ; \ t = -\frac{T}{2} \text{ to } -a \text{ and } t = a \text{ to } \frac{T}{2}$$

The Fourier coefficient c_n is given by,

$$c_n = \frac{1}{T}\int_{-T/2}^{+T/2} x(t)\,e^{-jn\Omega_0 t}\,dt = \frac{1}{T}\int_{-a}^{+a} e^{-jn\Omega_0 t}\,dt = \frac{1}{T}\left[\frac{e^{-jn\Omega_0 t}}{-jn\Omega_0}\right]_{-a}^{+a} \qquad \boxed{\Omega_0 = \frac{2\pi}{T}}$$

$$= \frac{1}{T}\left[\frac{e^{-jn\Omega_0 a}}{-jn\Omega_0} - \frac{e^{jn\Omega_0 a}}{-jn\Omega_0}\right] = \frac{1}{T}\frac{2}{n\Omega_0}\left[\frac{e^{jn\Omega_0 a} - e^{-jn\Omega_0 a}}{2j}\right] \qquad \boxed{\sin\theta = \frac{e^{j\theta} - e^{-j\theta}}{2j}}$$

$$= \frac{1}{T}\frac{2}{n}\frac{T}{2\pi}\sin(n\Omega_0 a) = \frac{1}{n\pi}\sin(an\Omega_0) \qquad \qquad ....(1)$$

The exponential Fourier series representation of the periodic pulse function is,

$$x(t) = \sum_{n=-\infty}^{+\infty} c_n\,e^{jn\Omega_0 t}$$

On taking Fourier transform of the above equation we get,

$$X(j\Omega) = \mathcal{F}\{x(t)\} = \mathcal{F}\left\{\sum_{n=-\infty}^{+\infty} c_n\,e^{jn\Omega_0 t}\right\}$$

$$\boxed{\begin{array}{l} c_n \text{ is independent} \\ \text{of time, t.} \end{array}}$$

$$= \sum_{n=-\infty}^{+\infty} c_n\,\mathcal{F}\{e^{jn\Omega_0 t}\}$$

$$\therefore\ X(j\Omega) = \sum_{n=-\infty}^{+\infty} c_n\, 2\pi\delta(\Omega - n\Omega_0)$$

$$\boxed{\mathcal{F}\{e^{jn\Omega_0 t}\} = 2\pi\,\delta(\Omega - n\Omega_0)}$$

Substituting for c_n from equation (1).

$$= \sum_{n=-\infty}^{+\infty} \frac{1}{n\pi} \sin(an\Omega_0)\ 2\pi\,\delta(\Omega - n\Omega_0) = \sum_{n=-\infty}^{+\infty} \frac{2\sin(an\Omega_0)}{n}\,\delta(\Omega - n\Omega_0)$$

$$\boxed{\frac{\sin\theta}{\theta} = \operatorname{sinc}\theta}$$

$$= \sum_{n=-\infty}^{+\infty} 2a\Omega_0\left(\frac{\sin an\Omega_0}{an\Omega_0}\right)\delta(\Omega - n\Omega_0) = \sum_{n=-\infty}^{+\infty} 2a\Omega_0\operatorname{sinc}(an\Omega_0)\,\delta(\Omega - n\Omega_0)$$

Example 2.22

Determine the Fourier transform of the periodic impulse function shown in Fig 2.22.1.

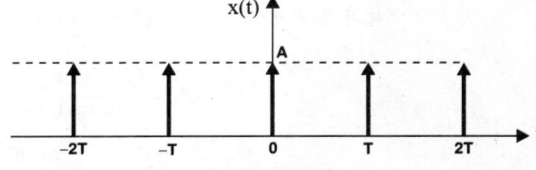

Solution:

The mathematical equation for one period of the periodic impulse function is,

$$x(t) = A\,\delta(t)\ ;\ \text{for}\ t = -\frac{T}{2}\ \text{to}\ +\frac{T}{2}$$

Fig 2.22.1.

$$\boxed{\Omega_0 = \frac{2\pi}{T}}$$

The Fourier coefficient c_n is given by,

$$c_n = \frac{1}{T}\int_{-T/2}^{+T/2} x(t)\,e^{-jn\Omega_0 t}\,dt = \frac{1}{T}\int_{-T/2}^{+T/2} A\,\delta(t)\,e^{-jn\Omega_0 t}\,dt = \frac{A}{T}e^{-jn\Omega_0 t}\Big|_{t=0} = \frac{A}{T} \qquad \ldots.(1)$$

The Exponential Fourier series representation of the periodic impulse train is,

$$x(t) = \sum_{n=-\infty}^{+\infty} c_n e^{jn\Omega_0 t}$$

On taking Fourier transform of the above equation we get,

$$X(j\Omega) = \mathcal{F}\{x(t)\} = \mathcal{F}\left\{\sum_{n=-\infty}^{+\infty} c_n\, e^{jn\Omega_0 t}\right\}$$

$$= \sum_{n=-\infty}^{+\infty} c_n\, \mathcal{F}\{e^{jn\Omega_0 t}\}$$

c_n is independent of time, t.

$$= \sum_{n=-\infty}^{+\infty} c_n\, 2\pi\,\delta(\Omega - n\Omega_0)$$

$$\boxed{\mathcal{F}\{e^{jn\Omega_0 t}\} = 2\pi\,\delta(\Omega - n\Omega_0)}$$

On substituting for c_n from equation (1).

$$= \sum_{n=-\infty}^{+\infty} \frac{A}{T}\, 2\pi\,\delta(\Omega - n\Omega_0) = \sum_{n=-\infty}^{+\infty} A\Omega_0\,\delta(\Omega - n\Omega_0)$$

$$\boxed{\Omega_0 = \frac{2\pi}{T}}$$

The magnitude spectrum of $X(j\Omega)$ is shown in Fig 1, which is also a periodic impulse function of Ω.

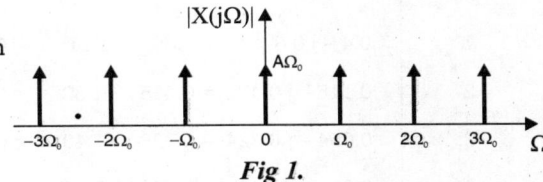

Fig 1.

Example 2.23   *(AU Dec'11, 16 Marks)*

Find the Fourier transform and sketch the magnitude and phase spectrum for the signal, $x(t) = e^{-at} u(t)$.

Solution:

Given that, $x(t) = e^{-at} u(t) = e^{-at}$; for $t \geq 0$

By definition of Fourier transform,

$$X(j\Omega) = \mathcal{F}\{x(t)\} = \int_{-\infty}^{\infty} x(t)\, e^{-j\Omega t}\, dt = \int_{0}^{\infty} e^{-at}\, e^{-j\Omega t}\, dt$$

$$= \int_{0}^{\infty} e^{-(a+j\Omega)t}\, dt = \left[\frac{e^{-(a+j\Omega)t}}{-(a+j\Omega)}\right]_{0}^{\infty} = \frac{e^{-\infty}}{-(a+j\Omega)} - \frac{e^{0}}{-(a+j\Omega)} = \frac{1}{a+j\Omega} \qquad \boxed{e^{-\infty} = 0}$$

$$= \frac{1}{a+j\Omega} \times \frac{a-j\Omega}{a-j\Omega} = \frac{a-j\Omega}{(a+j\Omega)(a-j\Omega)} = \frac{a-j\Omega}{a^2 - (j\Omega)^2} \qquad \boxed{(a+b)(a-b) = a^2 - b^2}$$

$$= \frac{a-j\Omega}{a^2 + \Omega^2} = \frac{a}{a^2 + \Omega^2} - j\frac{\Omega}{a^2 + \Omega^2}$$

The $X(j\Omega)$ is calculated for a = 0.5 and a = 1.0 and tabluted in Table 1 and Table 2 respectively. Using the values listed in Table 1 and Table 2, the magnitude and phase spectrum are sketched as shown in Fig 1 and Fig 2 respectively.

> **Note:** *The function X(jΩ) is calculated using complex mode of calculator, the magnitude and phase are calculated using rectangular to polar conversion technique.*

Table 1: Frequency Spectrum for a = 0.5

Ω	X(jΩ)			|X(jΩ)|	∠ X(jΩ) in rad
− 8	0.0080 + j 0.12	= 0.12 ∠1.506	= 0.12 ∠0.48π	0.12	0.48π
− 6	0.014 + j 0.167	= 0.166 ∠1.4866	= 0.166 ∠0.473π	0.166	0.473π
− 4	0.03 + j 0.246	= 0.248 ∠1.45	= 0.248 ∠0.46π	0.248	0.46π
− 3	0.054 + j 0.324	= 0.328 ∠1.40	= 0.328 ∠0.45π	0.328	0.45π
− 2	0.118 + j 0.47	= 0.485 ∠1.325	= 0.485 ∠0.422π	0.485	0.422 π
− 1	0.4 + j 0.8	= 0.89 ∠1.11	= 0.89 ∠0.353π	0.89	0.353π
0	2+ j 0	= 2 ∠0	= 2 ∠0	2.0	0
1	0.4 − j 0.8	= 0.89 ∠−1.11	= 0.89 ∠−0.353π	0.89	−0.353π
2	0.118 − j 0.47	= 0.485 ∠−1.325	= 0.485 ∠−0.422π	0.485	−0.422π
3	0.054 − j 0.324	= 0.328 ∠−1.40	= 0.282 ∠−0.45 π	0.282	−0.45π
4	0.03 − j 0.246	= 0.248 ∠−1.45	= 0.248 ∠−0.46 π	0.248	−0.46π
6	0.014 − j 0.167	= 0.166 ∠−1.4866	= 0.078 ∠− 0.473π	0.166	−0.473π
8	0.0080 − j 0.12	= 0.12 ∠−1.506	= 0.12 ∠−0.48 π	0.12	−0.48π

Table 2: Frequency Spectrum for a = 1

Ω	X(jΩ)			|X(jΩ)|	∠X(jΩ) in rad
− 8	0.015 + j 0.123	= 0.124 ∠1.45	= 0.124 ∠0.461π	0.124	0.461π
− 6	0.03 + j 0.162	= 0.165 ∠1.39	= 0.165 ∠0.442 π	0.165	0.442 π
− 4	0.059+ j 0.24	= 0.25 ∠1.33	= 0.25 ∠0.423π	0.25	0.423π
− 3	0.1 + j 0.3	= 0.316 ∠1.25	= 0.316 ∠0.398π	0.316	0.398π
− 2	0.2 + j 0.4	= 0.45 ∠1.11	= 0.45 ∠0.353π	0.45	0.353π
− 1	0.5 + j 0.5	= 0.707 ∠0.785	= 0.707 ∠0.25π	0.707	0.25π
0	1+ j 0 =1 ∠0	= 1 ∠0	= 1 ∠0	1.0	0
1	0.5 − j 0.5	= 0.707 ∠−0.785	= 0.707 ∠−0.25π	0.707	−0.25 π
2	0.2 − j 0.4	= 0.45 ∠−1.11	= 0.45 ∠−0.353π	0.45	−0.353 π
3	0.1 − j 0.3	= 0.316 ∠−1.25	= 0.316 ∠−0.398π	0.316	−0.398π
4	0.059 − j 0.24	= 0.25 ∠−1.33	= 0.25 ∠−0.423π	0.25	−0.423π
6	0.03 − j 0.162	= 0.165 ∠−1.39	= 0.165 ∠−0.442π	0.165	−0.442π
8	0.015 − j 0.123	= 0.124 ∠−1.45	= 0.124 ∠−0.461π	0.124	−0.461π

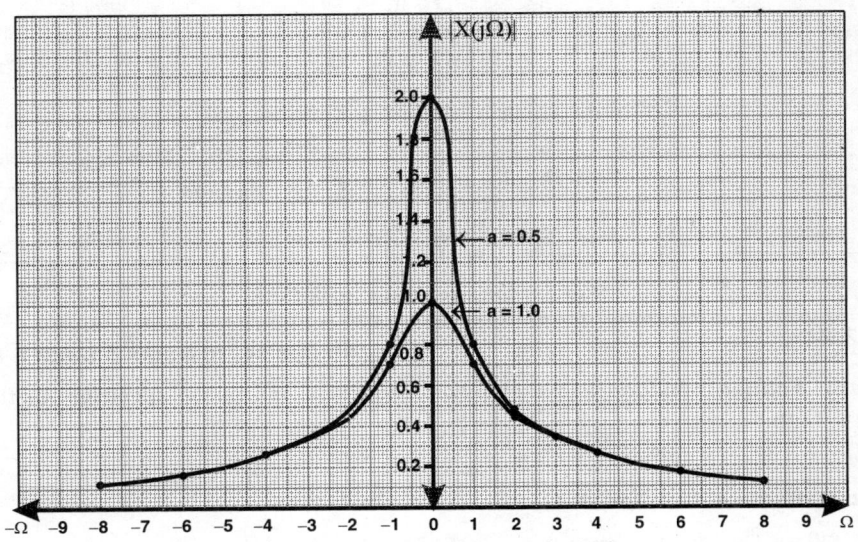

Fig 1 : Magnitude spectrum of X(jΩ).

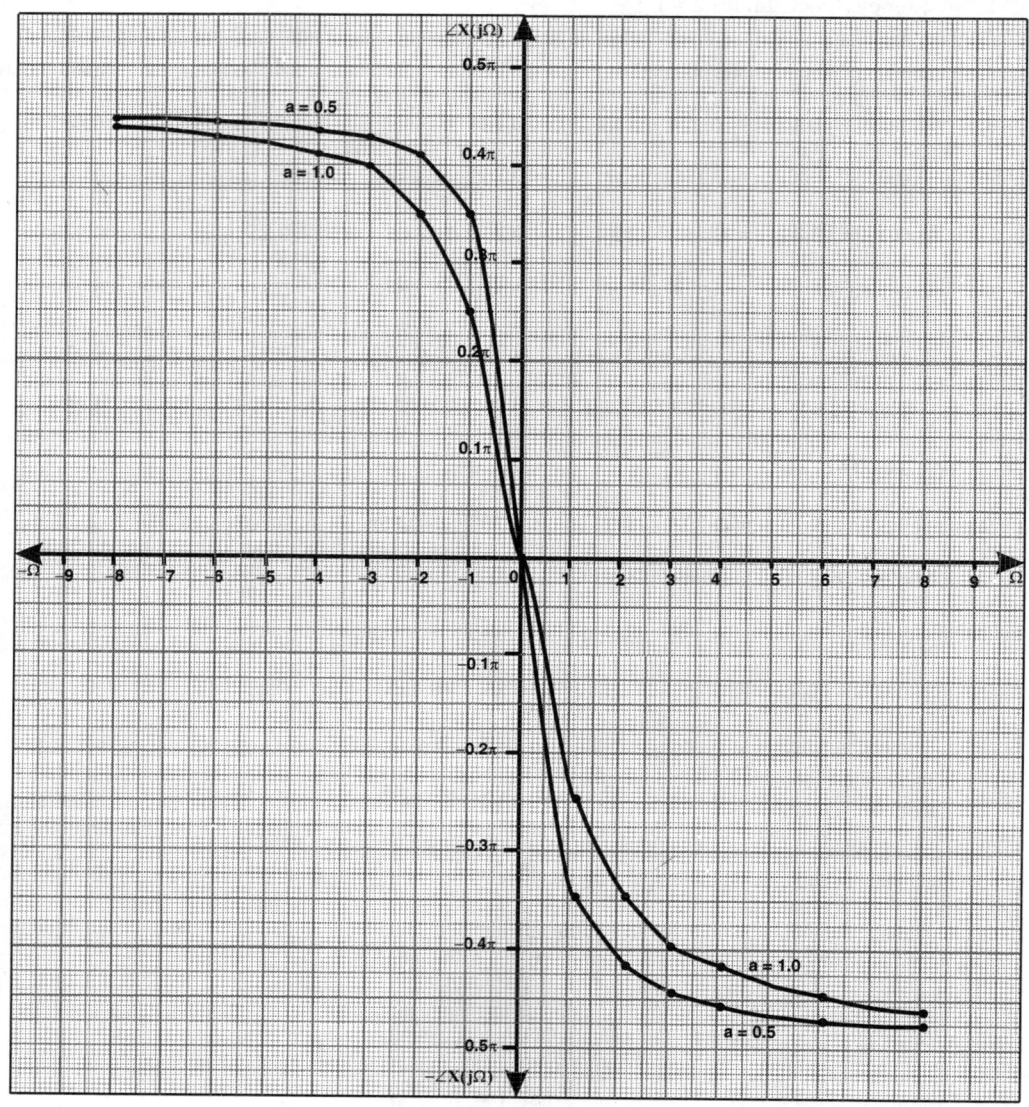

Fig 2: *Phase spectrum of X(jΩ)*

Example 2.24

(AU Dec'14, 16 Marks & Dec'12, 8 Marks)

Find the Fourier transform and sketch the magnitude and phase spectrum for the signal, $x(t) = e^{-2|t|} u(t)$.

Solution:

Given that, $x(t) = e^{-2|t|}$; **for all** t

$$\therefore \; x(t) = e^{+2t} \quad ; \text{ for } t = -\infty \text{ to } 0$$
$$= e^{-2t} \quad ; \text{ for } t = 0 \text{ to } +\infty$$

By definition of Fourier transform,

$$X(j\Omega) = \mathcal{F}\{x(t)\} = \int_{-\infty}^{+\infty} x(t)\, e^{-j\Omega t}\, dt = \int_{-\infty}^{0} e^{2t}\, e^{-j\Omega t}\, dt + \int_{0}^{+\infty} e^{-2t}\, e^{-j\Omega t}\, dt$$

$$= \int_{-\infty}^{0} e^{(2-j\Omega)t}\, dt + \int_{0}^{\infty} e^{-(2+j\Omega)t}\, dt = \left[\frac{e^{(2-j\Omega)t}}{(2-j\Omega)}\right]_{-\infty}^{0} + \left[\frac{e^{-(2+j\Omega)t}}{-(2+j\Omega)}\right]_{0}^{\infty}$$

$$= \frac{e^{0}}{(2-j\Omega)} - \frac{e^{-\infty}}{(2-j\Omega)} + \frac{e^{-\infty}}{-(2+j\Omega)} - \frac{e^{0}}{-(2+j\Omega)} = \frac{1}{2-j\Omega} + \frac{1}{2+j\Omega} \qquad \boxed{e^{-\infty}=0}$$

$$= \frac{(2+j\Omega)+(2-j\Omega)}{(2-j\Omega)(2+j\Omega)} = \frac{4}{2^{2}-(j\Omega)^{2}} = \frac{4}{2^{2}+\Omega^{2}} \qquad \boxed{(a+b)(a-b)=a^{2}-b^{2}}$$

The $|X(j\Omega)|$ and $\angle X(j\Omega)$ are calculated and tabluated in Table.1.

> **Note :** *Here $X(j\Omega)$ is purely real and so $\angle X(j\Omega) = 0$ for all values of Ω.*

Table 1: Magnitude and Phase of $X(j\Omega)$ for various values Ω

Ω	−5	−4	−3	−2	−1	0	1	2	3	4	5
$|X(j\Omega)|$	0.138	0.2	0.308	0.5	0.8	1	0.8	0.5	0.308	0.2	0.138
$\angle X(j\Omega)$	0	0	0	0	0	0	0	0	0	0	0

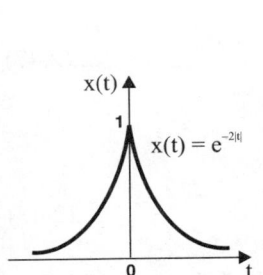

Fig 1: *Exponential signal.*

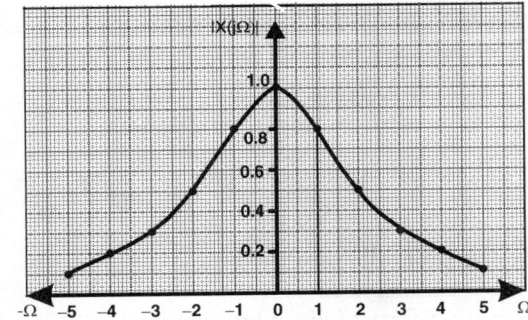

Fig 2: *Magnitude spectrum.*

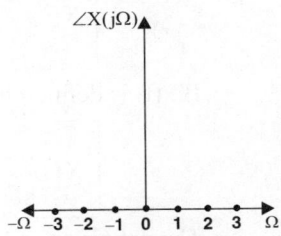

Fig 3: *Phase spectrum.*

2.7 Laplace Transform

The Laplace transform is used to transform a time domain signal to complex frequency domain. The complex frequency domain is also known as Laplace domain or s-domain. This transformation was first proposed by Laplace (in the year 1780) and later adopted for various engineering applications for solving differential equations. Hence, this transformation is called *Laplace transform.*

In signals and systems, the Laplace transform is used to transform a time domain system to s-domain. In time domain the equations governing a system will be in the form of differential equations. While transforming the system to s-domain, the differential equations are transformed to simple algebraic equations and so the analysis of systems will be much easier in s-domain.

2.7.1 s-Plane (or Complex Frequency Plane)

The complex frequency is defined as,

Complex frequency, $\sigma = \sigma + j\Omega$

where, σ = Real part of s

Ω = Imaginary part of s

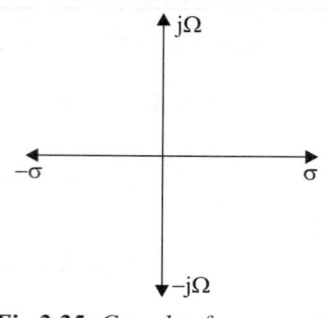

The σ and Ω can take values from $-\infty$ to $+\infty$. A two dimentional complex plane with values of σ on horizontal axis and values Ω on vertical axis as shown in Fig 2.25 is called complex frequency plane or s-plane.

Fig 2.25: Complex frequency plane or s-plane.

The s-domain is used to represent various critical frequencies (poles and zeros) of a s-domain function and to study the path taken by these critical frequencies when some parameters of the function are varied. This study will be useful for designing systems for a desired response.

2.7.2 Definition of Laplace Transform

Let, x(t) = Continuous time signal defined for all t

X(s) = Laplace transform of x(t)

Symbolically the *Laplace transform* of x(t) is denoted as,

$$X(s) = \mathcal{L}\{x(t)\}$$

Mathematically, the *Laplace transform* of x(t) is defined as,

$$X(s) = \mathcal{L}\{x(t)\} = \int_{-\infty}^{+\infty} x(t)\,e^{-st}\,dt \qquad \qquad .....(2.47)$$

If x(t) is defined for $t \geq 0$, (i.e., if x(t) is causal) then,

$$X(s) = \mathcal{L}\{x(t)\} = \int_{0}^{+\infty} x(t)\,e^{-st}\,dt \qquad \qquad .....(2.48)$$

The definition of Laplace transform as given by equation (2.47) is called *Two sided Laplace transform* or *Bilateral Laplace transform* and the definition of Laplace transform as given by equation (2.48) is called *One sided Laplace transform* or *Unilateral Laplace transform*.

The signal x(t) and X(s) are called *Laplace transform pair* and can be expressed as,

$$x(t) \underset{\mathcal{L}^{-1}}{\overset{\mathcal{L}}{\rightleftharpoons}} X(s)$$

2.7.3 Poles and Zeros of Rational Function of s

Let X(s) be Laplace transform of x(t). When X(s) is expressed as a ratio of two polynomials in s, then the s-domain signal X(s) is called a *rational function* of s.

The zeros and poles are two critical complex frequencies at which a rational function of s takes two extreme values, such as zero and infinity respectively.

Let $X(s)$ is expressed as a ratio of two polynomials in s as shown in equation (2.49).

$$X(s) = \frac{P(s)}{Q(s)}$$

$$= \frac{b_0 s^M + b_1 s^{M-1} + b_2 s^{M-2} + \..... + b_{M-1} s + b_M}{a_0 s^N + a_1 s^{N-1} + a_2 s^{N-2} + \..... + a_{N-1} s + a_N} \qquad .....(2.49)$$

where, $P(s)$ = Numerator polynomial of $X(s)$

$Q(s)$ = Denominator polynomial of $X(s)$

In equation (2.49) let us scale the coefficients of numerator polynomial by b_0 and the coefficients of denominator polynomial by a_0, and the equation (2.49) can be expressed in factorized form as shown in equation (2.50).

$$X(s) = \frac{b_0 \left(s^M + \frac{b_1}{b_0} s^{M-1} + \frac{b_2}{b_0} s^{M-2} + \..... + \frac{b_{M-1}}{b_0} s + \frac{b_M}{b_0} \right)}{a_0 \left(s^N + \frac{a_1}{a_0} s^{N-1} + \frac{a_2}{a_0} s^{N-2} + \..... + \frac{a_{N-1}}{a_0} s + \frac{a_N}{a_0} \right)}$$

$$= G \frac{(s - z_1)(s - z_2)\.....(s - z_M)}{(s - p_1)(s - p_2)\.....(s - p_N)} \qquad .....(2.50)$$

where, $G = \dfrac{b_0}{a_0}$ = Scaling factor

$z_1, z_2, \....., z_M$ = Roots of numerator polynomial, $P(s)$

$p_1, p_2, \....., p_N$ = Roots of denominator polynomial, $Q(s)$

In equation (2.50) if the value of s is equal to any one of the root of numerator polynomial then the signal $X(s)$ will become zero.

Therefore the roots of numerator polynomial $z_1, z_2, \....., z_M$ are called *zeros* of $X(s)$. Since s is complex frequency, the *zeros* can be defined as values of complex frequencies at which the signal $X(s)$ becomes zero.

In equation (2.50), if the value of s is equal to any one of the roots of the denominator polynomial then the signal $X(s)$ will become infinite. Therefore the roots of denominator polynomial $p_1, p_2, \....., p_N$ are called *poles* of $X(s)$. Since **s** is complex frequency, the *poles* can be defined as values of complex frequencies at which the signal $X(s)$ become infinite. *Since the signal X(s) attains infinte value at poles, the ROC of X(s) does not include poles.*

In a realizable system, *the number of zeros will be less than or equal to number of poles.* Also for every zero, we can associate one pole. (When number of finite zeros are less than poles, the missing zeros are assumed to exist at infinity).

Let z_i be the zero associated with the pole p_i. If we evaluate $|X(s)|$ for various values of s, then $|X(s)|$ will be zero for $s = z_i$ and infnite for $s = p_i$. Hence the plot of $|X(s)|$ in a three dimensional plane will look like a pole (or pillar like structure) and so the point $s = p_i$ is called a pole.

Representation of Poles and Zeros in s-Plane

We know that, Complex frequency, $s = \sigma + j\Omega$

where, σ = Real part of s, Ω = Imaginary part of s

Hence the s-plane is a complex plane, with σ on real axis and Ω on imaginary axis as shown in Fig 2.26. In the s-plane, the zeros are marked by small circle " $\circ$ " and the poles are marked by letter "**X**".

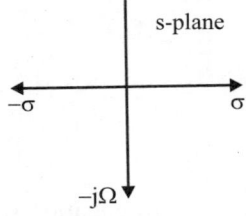

For example consider the rational function of s shown below:

$$X(s) = \frac{(s+2)(s+5)}{s(s^2+6s+13)}$$

The roots of quadratic $s^2 + 6s + 13 = 0$ are,

$$s_1, s_2 = \frac{-6 \pm \sqrt{36 - 4 \times 13}}{2} = \frac{-6 \pm j4}{2} = -3 + j2,\ -3 - j2$$

Fig 2.26: s-plane.

$$\therefore\ s^2 + 6s + 13 = (s - s_1)(s - s_2) = (s+3-j2)(s+3+j2)$$

$$\therefore\ X(s) = \frac{(s+2)(s+5)}{s(s^2+6s+13)} = \frac{(s+2)(s+5)}{s(s+3-j2)(s+3+j2)} \qquad \text{.....(2.51)}$$

The zeros of the above function are,

$$z_1 = -2$$
$$z_2 = -5$$

The poles of the above function are,

$$p_1 = 0$$
$$p_2 = -3 + j2$$
$$p_3 = -3 - j2$$

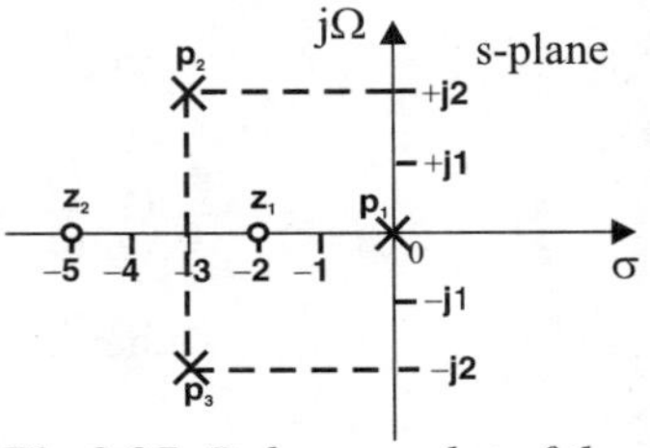

Fig 2.27: Pole-zero plot of the function of equation (2.51).

The pole-zero plot of rational function of s of equation (2.51) is shown in Fig 2.27.

2.8 Region of Convergence   *(AU Dec'12, 2 Marks)*

The Laplace transform of a signal is given by $\int\limits_{-\infty}^{+\infty} x(t)\,e^{-st}\ dt$. The values of s for which the integral $\int\limits_{-\infty}^{+\infty} x(t)\,e^{-st}\ dt$ converges is called *Region Of Convergence (ROC)*.

2.8.1 ROC of Continuous Time Signal

    Case i: Right sided (causal) signal

    Case ii: Left sided (anticausal) signal

    Case iii: Two sided signal.

Case i: Right sided (causal) signal

Let, $x(t) = e^{-at}u(t)$, where $a > 0$

$$= e^{-at} \text{ for } t \geq 0$$

Now, the Laplace transform of x(t) is given by,

$$X(s) = \mathcal{L}\{x(t)\} = \int\limits_{-\infty}^{+\infty} x(t)\,e^{-st}\ dt$$

$$= \int\limits_{0}^{+\infty} e^{-at}\,e^{-st}\ dt = \int\limits_{0}^{+\infty} e^{-(s+a)t}\ dt = \left[\frac{e^{-(s+a)t}}{-(s+a)} \right]_0^\infty$$

$$\therefore\ X(s) = \frac{e^{-(\sigma + j\Omega + a)\infty}}{-(s+a)} - \frac{e^{0}}{-(s+a)} \qquad \boxed{s = \sigma + j\Omega}$$

$$= -\frac{e^{-(\sigma + a)\times\infty}e^{-j\Omega\times\infty}}{(s+a)} + \frac{1}{s+a}$$

$$= -\frac{e^{-k\times\infty}e^{-j\Omega\times\infty}}{s+a} + \frac{1}{s+a}$$

where, $k = \sigma + a = \sigma -(-a)$

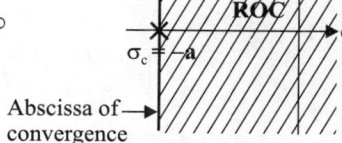

When $\sigma > -a$, $k = \sigma -(-a) = $ Positive, $\therefore e^{-k\infty} = e^{-\infty} = 0$

When $\sigma < -a$, $k = \sigma -(-a) = $ Negative, $\therefore e^{-k\infty} = e^{+\infty} = \infty$

Hence we can say that, $X(s)$ converges, when $\sigma > -a$, and does not converge for $\sigma < -a$.

$\therefore$ Abscissa of convergence, $\sigma_c = -a$.

When $\sigma > -a$, the $X(s)$ is given by,

Fig 2.28: ROC of $x(t) = e^{-at}\,u(t)$.

$$X(s) = -\frac{e^{-k\times\infty}e^{-j\Omega\times\infty}}{s+a} + \frac{1}{s+a} = -\frac{0\times e^{-j\Omega\times\infty}}{s+a} + \frac{1}{s+a} = \frac{1}{s+a} \qquad \boxed{e^{-k\times\infty} = e^{-\infty} = 0}$$

Therefore for a causal signal the ROC includes all points on the s-plane to the right of abscissa of convergence, $\sigma_c = -a$, as shown in Fig 2.28.

Case ii: *Left sided (anticausal) signal*

Let, $x(t) = e^{-bt}u(-t)$, where $b > 0$

$\qquad\qquad = e^{-bt}$ for $t \le 0$

Now, the Laplace transform of $x(t)$ is given by,

$$X(s) = \mathcal{L}\{x(t)\} = \int_{-\infty}^{+\infty} x(t)\,e^{-st}\ dt = \int_{-\infty}^{0} e^{-bt}\,e^{-st}\ dt$$

$$= \int_{-\infty}^{0} e^{-(s+b)t}\ dt = \left[\frac{e^{-(s+b)t}}{-(s+b)}\right]_{-\infty}^{0} = \frac{e^{0}}{-(s+b)} - \frac{e^{(\sigma + j\Omega + b)\infty}}{-(s+b)} \qquad \boxed{s = \sigma + j\Omega}$$

$$= -\frac{1}{s+b} + \frac{e^{(\sigma + b)\times\infty}e^{j\Omega\times\infty}}{s+b} = -\frac{1}{s+b} + \frac{e^{k\times\infty}e^{j\Omega\times\infty}}{s+b}$$

where, $k = \sigma + b = \sigma -(-b)$

When $\sigma > -b$, $k = \sigma -(-b) = $ Positive, $\therefore e^{k\infty} = e^{\infty} = \infty$

When $\sigma < -b$, $k = \sigma -(-b) = $ Negative, $\therefore e^{k\infty} = e^{-\infty} = 0$

Hence we can say that, $X(s)$ converges, when $\sigma < -b$, and does not converge for $\sigma > -b$.

$\therefore$ Abscissa of convergence, $\sigma_c = -b$.

When $\sigma < -b$, the $X(s)$ is given by,

Fig 2.29: ROC of $x(t) = e^{-bt}\,u(-t)$.

$$X(s) = -\frac{1}{s+b} + \frac{e^{k\times\infty}e^{j\Omega\times\infty}}{s+b} = -\frac{1}{s+b} + \frac{0\times e^{j\Omega\times\infty}}{s+b} = -\frac{1}{s+b} \qquad \boxed{e^{k\times\infty} = e^{-\infty} = 0}$$

Therefore for an anticausal signal the ROC includes all points on the s-plane to the left of abscissa of convergence, $\sigma_c = -b$, as shown in Fig 2.29.

Case iii: Two sided signal

Let, $x(t) = e^{-at} u(t) + e^{-bt} u(-t)$, where $a > 0$, $b > 0$, and $a > b$ (i.e., $-a < -b$)

Now, the Laplace transform of $x(t)$ is given by,

$$X(s) = \mathcal{L}\{x(t)\} = \int_{-\infty}^{+\infty} x(t)\, e^{-st}\, dt = \int_{-\infty}^{+\infty} \left[e^{-at} u(t) + e^{-bt} u(-t) \right] e^{-st}\, dt$$

$$= \int_{0}^{+\infty} e^{-at}\, e^{-st}\, dt + \int_{-\infty}^{0} e^{-bt}\, e^{-st}\, dt = \int_{0}^{+\infty} e^{-(s+a)t}\, dt + \int_{-\infty}^{0} e^{-(s+b)t}\, dt$$

$$= \left[\frac{e^{-(s+a)t}}{-(s+a)} \right]_{0}^{\infty} + \left[\frac{e^{-(s+b)t}}{-(s+b)} \right]_{-\infty}^{0} = \left[\frac{e^{-(\sigma+j\Omega+a)t}}{-(s+a)} \right]_{0}^{\infty} + \left[\frac{e^{-(\sigma+j\Omega+b)t}}{-(s+b)} \right]_{-\infty}^{0}$$

$$= \frac{e^{-(\sigma+j\Omega+a)\infty}}{-(s+a)} - \frac{e^{0}}{-(s+a)} + \frac{e^{0}}{-(s+b)} - \frac{e^{(\sigma+j\Omega+b)\infty}}{-(s+b)} \qquad \boxed{s = \sigma + j\Omega}$$

$$= -\frac{e^{-p\times\infty}\, e^{-j\Omega\times\infty}}{s+a} + \frac{1}{s+a} - \frac{1}{s+b} + \frac{e^{q\times\infty}\, e^{j\Omega\times\infty}}{s+b}$$

where, $p = \sigma + a = \sigma - (-a)$ and $q = \sigma + b = \sigma - (-b)$

When $\sigma > -a$, $p = \sigma - (-a) =$ Positive, $\therefore\ e^{-p\infty} = e^{-\infty} = 0$

When $\sigma < -a$, $p = \sigma - (-a) =$ Negative, $\therefore\ e^{-p\infty} = e^{+\infty} = \infty$

When $\sigma > -b$, $q = \sigma - (-b) =$ Positive, $\therefore\ e^{q\infty} = e^{\infty} = \infty$

When $\sigma < -b$, $q = \sigma - (-b) =$ Negative, $\therefore\ e^{q\infty} = e^{-\infty} = 0$

Hence we can say that, $X(s)$ converges, when σ lies between $-a$ and $-b$ (i.e., $-a < \sigma < -b$), and does not converge for $\sigma < -a$ and $\sigma > -b$.

$\therefore$ Abscissa of convergences, $\sigma_{c1} = -a$ and $\sigma_{c2} = -b$.

When $-a < \sigma < -b$, the $X(s)$ is given by,

$$\boxed{e^{-p\times\infty} = e^{-\infty} = 0}$$
$$\boxed{e^{q\times\infty} = e^{-\infty} = 0}$$

$$X(s) = -\frac{e^{-p\times\infty}\, e^{-j\Omega\times\infty}}{s+a} + \frac{1}{s+a} - \frac{1}{s+b} + \frac{e^{q\times\infty}\, e^{j\Omega\times\infty}}{s+b}$$

$$= -\frac{0\times e^{-j\Omega\times\infty}}{s+a} + \frac{1}{s+a} - \frac{1}{s+b} + \frac{0\times e^{j\Omega\times\infty}}{s+b}$$

$$= \frac{1}{s+a} - \frac{1}{s+b}$$

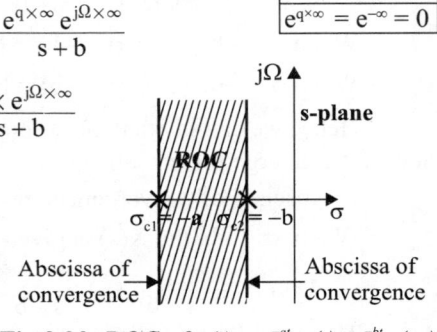

Therefore for a two sided signal the ROC includes all points on the s-plane in the region in between two abscissa of convergences, $\sigma_{c1} = -a$ and $\sigma_{c2} = -b$, as shown in Fig 2.30.

Fig 2.30: ROC of x(t) = e^{-at} u(t)+e^{-bt}u(−t).

2.8.2 ROC of Rational Function of s (ROC using Poles)

Case i: Right sided (causal) signal

Let x(t) be a right sided (causal) signal defined as,

$$x(t) = e^{-a_1 t} u(t) + e^{-a_2 t} u(t) + e^{-a_3 t} u(t) \quad ; \quad \text{where} - a_1 < -a_2 < -a_3$$

Now, the Laplace transform of x(t) is,

$$X(s) = \frac{1}{s+a_1} + \frac{1}{s+a_2} + \frac{1}{s+a_3} = \frac{N(s)}{(s+a_1)(s+a_2)(s+a_3)}$$

where, $N(s) = (s + a_2)(s + a_3) + (s + a_1)(s + a_3) + (s + a_1)(s + a_2)$

The poles of X(s) are,

$$p_1 = -a_1 \quad ; \quad p_2 = -a_2 \quad ; \quad p_3 = -a_3$$

Let, σ = Real part of s.

Now, the convergence criteria for X(s) are,

$$\sigma > -a_1 \quad ; \quad \sigma > -a_2 \quad ; \quad \sigma > -a_3$$

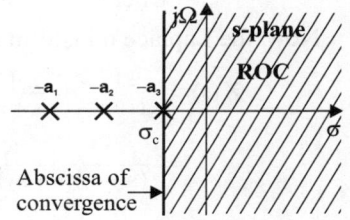

Since $-a_1 < -a_2 < -a_3$, the X(s) converges, when $\sigma > -a_3$, and does not converge for $\sigma < -a_3$.

$\therefore$ Abscissa of convergence, $\sigma_c = -a_3$.

Fig 2.31: ROC of a causal signal.

Therefore ROC of X(s) is all points to the right of abscissa of convergence (or right of the line passing through $-a_3$) in s-plane as shown in Fig 2.31. *In terms of poles of X(s) we can say that the ROC is right of right most pole of X(s), (i.e., right of the pole with largest real part).*

Case ii: Left sided (anticausal) signal

Let x(t) be a left sided (anticausal) signal defined as,

$$x(t) = e^{-a_1 t} u(-t) + e^{-a_2 t} u(-t) + e^{-a_3 t} u(-t) \quad ; \quad \text{where} - a_1 < -a_2 < -a_3$$

Now, the Laplace transform of x(t) is,

$$X(s) = -\frac{1}{s+a_1} - \frac{1}{s+a_2} - \frac{1}{s+a_3} = \frac{N(s)}{(s+a_1)(s+a_2)(s+a_3)}$$

where, $N(s) = -(s + a_2)(s + a_3) - (s + a_1)(s + a_3) - (s + a_1)(s + a_2)$

The poles of X(s) are,

$$p_1 = -a_1 \quad ; \quad p_2 = -a_2 \quad ; \quad p_3 = -a_3$$

Let, σ = Real part of s.

Now the convergence criteria for X(s) are,

$$\sigma < -a_1 \quad ; \quad \sigma < -a_2 \quad ; \quad \sigma < -a_3$$

Since $-a_1 < -a_2 < -a_3$, the X(s) converges, when $\sigma < -a_1$, and does not converge for $\sigma > -a_1$.

∴ Abscissa of convergence, $\sigma_c = -a_1$.

Therefore ROC of X(s) is all points to the left of abscissa of convergence (or left of the line passing through $-a_1$) in s-plane as shown in Fig 2.32. *In terms of poles of X(s) we can say that the ROC is left of left most pole of X(s), (i.e., left of the pole with smallest real part).*

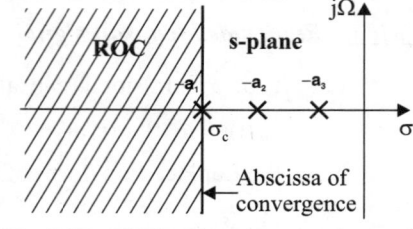

Fig 2.32: ROC of an anticausal signal.

Case iii: Two sided signal

Let x(t) be a two sided signal defined as,

$$x(t) = e^{-a_1 t} u(t) + e^{-a_2 t} u(t) + e^{-a_3 t} u(-t) + e^{-a_4 t} u(-t)$$

$$\text{where } -a_1 < -a_2 < -a_3 < -a_4$$

Now, the Laplace transform of x(t) is,

$$X(s) = \frac{1}{s+a_1} + \frac{1}{s+a_2} - \frac{1}{s+a_3} - \frac{1}{s+a_4}$$

$$= \frac{N(s)}{(s+a_1)(s+a_2)(s+a_3)(s+a_4)}$$

where, $N(s) = (s+a_2)(s+a_3)(s+a_4) + (s+a_1)(s+a_3)(s+a_4)$

$$- (s+a_1)(s+a_2)(s+a_4) - (s+a_1)(s+a_2)(s+a_3)$$

The poles of X(s) are,

$$p_1 = -a_1 \;\; ; \;\; p_2 = -a_2 \;\; ; \;\; p_3 = -a_3 \;\; ; \;\; p_4 = -a_4$$

Let, σ = Real part of s.

Now, the convergence criteria for X(s) are,

$$\sigma > -a_1 \;\; ; \;\; \sigma > -a_2 \;\; ; \;\; \sigma < -a_3 \;\; ; \;\; \sigma < -a_4$$

Since $-a_1 < -a_2 < -a_3 < -a_4$, the function X(s) converges, when σ lies between $-a_2$ and $-a_3$ (i.e., $-a_2 < \sigma < -a_3$), and does not converge for $\sigma < -a_2$ and $\sigma > -a_3$.

∴ Abscissa of convergences, $\sigma_{c1} = -a_2$ and $\sigma_{c2} = -a_3$.

Therefore ROC of X(s) is all points in the region in between two abscissa of convergences (or region inbetween the two lines passing through $-a_2$ and $-a_3$) in s-plane as shown in Fig 2.33.

Let X(s) be s-domain representation of a signal with causal and noncausal part. Let a_x be the magnitude of largest pole of causal part of the signal and let a_y be the magnitude of smallest pole of anticausal part of the signal and let $a_x < a_y$. *Now in term of poles of X(s) we can say that the ROC is the region in between two lines passing through a_x and a_y, where $a_x < a_y$.*

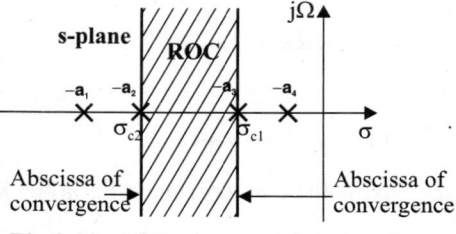

Fig 2.33: ROC of a two sided signal.

2.8.3 Properties of ROC   *(AU Jun'14, 2 Marks & Jun'13, 8 Marks)*

The various concepts of ROC that has been discussed in Section 2.8.1 and 2.8.2 are summarized as properties of ROC and given below.

Property-1 : The ROC of X(s) consists of strips parallel to the $j\Omega$ - axis in the s-plane.

Property-2 : If x(t) is of finite duration and is absolutely integrable, then the ROC is the entire s-plane.

Property-3 : If x(t) is right sided, and if the line passing through $Re(s) = \sigma_0$ is in ROC, then all values of s for which $Re(s) > \sigma_0$ will also be in ROC.

Property-4 : If x(t) is left sided, and if the line passing through $Re(s) = \sigma_0$ is in ROC, then all values of s for which $Re(s) < \sigma_0$ will also be in ROC.

Property-5 : If x(t) is two sided, and if the line passing through $Re(s) = \sigma_0$ is in ROC, then the ROC will consists of a strip in the s-plane that includes the line passing through $Re(s) = \sigma_0$.

Property-6 : If X(s) is rational, (where X(s) is Laplace transform of x(t)), then its ROC is bounded by poles or extends to infinity.

Property-7 : If X(s) is rational, (where X(s) is Laplace transform of x(t)), then ROC does not include any poles of X(s).

Property-8 : If X(s) is rational, (where X(s) is Laplace transform of x(t)), and if x(t) is right sided, then ROC is the region in s-plane to the right of the rightmost pole.

Property-9 : If X(s) is rational, (where X(s) is Laplace transform of x(t)), and if x(t) is left sided, then ROC is the region in s-plane to the left of the leftmost pole.

Example 2.25

Determine the Laplace transform of the following continuous time signals and their ROC.

a) $x(t) = A\ u(t)$ **b)** $x(t) = t\ u(t)$ **c)** $x(t) = e^{-3t}\ u(t)$

d) $x(t) = e^{-3t}\ u(-t)$ **e)** $x(t) = e^{-4|t|}$ **f)** $x(t) = t\ e^{-2|t|}$

Solution:

a) Given that, $x(t) = A\,u(t) = A$; $t \geq 0$ **(AU Dec'14, 2 Marks)**

By definition of Laplace transform,

$$X(s) = \mathcal{L}\{x(t)\} = \int_{-\infty}^{+\infty} x(t)\,e^{-st}\,dt = \int_{0}^{\infty} A\,e^{-st}\,dt = A\int_{0}^{\infty} e^{-st}\,dt \qquad \boxed{s = \sigma + j\Omega}$$

$$= A\left[\frac{e^{-st}}{-s}\right]_0^\infty = A\left[\frac{e^{-(\sigma+j\Omega)t}}{-s}\right]_0^\infty = A\left[\frac{e^{-(\sigma+j\Omega)\infty}}{-s} - \frac{e^0}{-s}\right] = A\left[\frac{e^{-\sigma\times\infty}\,e^{-j\Omega\times\infty}}{-s} + \frac{1}{s}\right]$$

When, $\sigma > 0$, (i.e., when σ is positive), $e^{-\sigma\infty} = e^{-\infty} = 0$

When, $\sigma < 0$, (i.e., when σ is negative), $e^{-\sigma\infty} = e^{\infty} = \infty$

Therefore we can say that, $X(s)$ converges when $\sigma > 0$.

When $\sigma > 0$, the $X(s)$ is given by,

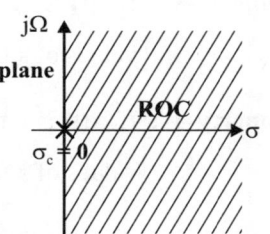

$$X(s) = A\left[\frac{e^{-\sigma\times\infty}\,e^{-j\Omega\times\infty}}{-s} + \frac{1}{s}\right] = A\left[\frac{0 \times e^{-j\Omega\times\infty}}{-s} + \frac{1}{s}\right] = \frac{A}{s}$$

$\therefore \mathcal{L}\{A\,u(t)\} = \dfrac{A}{s}$; with ROC as all points is s-plane to the right of line passing through $\sigma = 0$.(or ROC is right half of s-plane).

Fig 1: *ROC of x(t) = A u(t).*

---

b) Given that, $x(t) = t\,u(t) = t$; $t \geq 0$

By definition of Laplace transform, $\boxed{\int u\,dv = u\,v - \int v\,du}$

$$X(s) = \mathcal{L}\{x(t)\} = \int_{-\infty}^{+\infty} x(t)\,e^{-st}\,dt = \int_{0}^{\infty} t\,e^{-st}\,dt$$

$$\begin{aligned}u = t &\Rightarrow du = 1\\ dv = e^{-st} &\Rightarrow v = \frac{e^{-st}}{-s}\end{aligned}$$

$$= \left[t\,\frac{e^{-st}}{-s}\right]_0^\infty - \int_0^\infty \frac{e^{-st}}{-s}\times 1\,dt = \left[t\,\frac{e^{-st}}{-s}\right]_0^\infty - \left[\frac{e^{-st}}{s^2}\right]_0^\infty = \left[t\,\frac{e^{-(\sigma+j\Omega)t}}{-s}\right]_0^\infty - \left[\frac{e^{-(\sigma+j\Omega)t}}{s^2}\right]_0^\infty$$

$$= \left[\infty \times \frac{e^{-(\sigma+j\Omega)\infty}}{-s} - 0 \times \frac{e^0}{-s} - \frac{e^{-(\sigma+j\Omega)\infty}}{s^2} + \frac{e^0}{s^2}\right] \qquad \boxed{s = \sigma + j\Omega}$$

$$= \left[\infty \times \frac{e^{-\sigma\times\infty}\,e^{-j\Omega\times\infty}}{-s} - 0 - \frac{e^{-\sigma\times\infty}\,e^{-j\Omega\times\infty}}{s^2} + \frac{1}{s^2}\right]$$

When, $\sigma > 0$, (i.e., when σ is positive), $e^{-\sigma\infty} = e^{-\infty} = 0$

When, $\sigma < 0$, (i.e., when σ is negative), $e^{-\sigma\infty} = e^{\infty} = \infty$

Therefore we can say that, $X(s)$ converges when $\sigma > 0$.

When $\sigma > 0$, the $X(s)$ is given by,

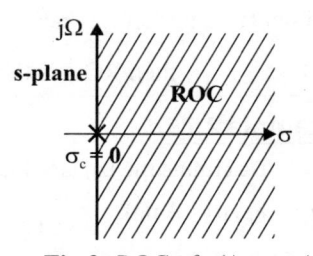

$$X(s) = \left[\infty \times \frac{e^{-\sigma\times\infty}\,e^{-j\Omega\times\infty}}{-s} - \frac{e^{-\sigma\times\infty}\,e^{-j\Omega\times\infty}}{s^2} + \frac{1}{s^2}\right]$$

$$= \left[\infty \times \frac{0 \times e^{-j\Omega\times\infty}}{-s} - \frac{0 \times e^{-j\Omega\times\infty}}{s^2} + \frac{1}{s^2}\right] = \frac{1}{s^2}$$

Fig 2: *ROC of x(t) = t u(t).*

$\therefore \mathcal{L}\{t\,u(t)\} = \dfrac{1}{s^2}$; with ROC as all points in s-plane to the right of line passing through $\sigma = 0$. (or ROC is right half of s-plane).

c) Given that, $x(t) = e^{-3t}\, u(t) = e^{-3t}$; $t \geq 0$

By definition of Laplace transform,

$$X(s) = \mathcal{L}\{x(t)\} = \int_{-\infty}^{+\infty} x(t)\, e^{-st}\, dt = \int_{0}^{\infty} e^{-3t} e^{-st}\, dt = \int_{0}^{\infty} e^{-(s+3)t}\, dt$$

$$= \left[\frac{e^{-(s+3)t}}{-(s+3)}\right]_{0}^{\infty} = \frac{e^{-(s+3)\infty}}{-(s+3)} - \frac{e^{0}}{-(s+3)} = -\frac{e^{-(\sigma+j\Omega+3)\infty}}{s+3} + \frac{1}{s+3} \qquad \boxed{s = \sigma + j\Omega}$$

$$= -\frac{e^{-(\sigma+3)\times\infty}\, e^{-j\Omega\times\infty}}{s+3} + \frac{1}{s+3} = -\frac{e^{-k\times\infty}\, e^{-j\Omega\times\infty}}{s+3} + \frac{1}{s+3}$$

where, $k = \sigma + 3 = \sigma - (-3)$

When, $\sigma > -3$, $k = \sigma - (-3) =$ Positive. $\quad \therefore e^{-k\infty} = e^{-\infty} = 0$

When, $\sigma < -3$, $k = \sigma - (-3) =$ Negative. $\therefore e^{-k\infty} = e^{\infty} = \infty$

Therefore we can say that, $X(s)$ converges when $\sigma > -3$.

When $\sigma > -3$, the $X(s)$ is given by,

$$X(s) = -\frac{e^{-k\times\infty}\, e^{-j\Omega\times\infty}}{s+3} + \frac{1}{s+3}$$

$$= -\frac{0 \times e^{-j\Omega\times\infty}}{s+3} + \frac{1}{s+3} = \frac{1}{s+3}$$

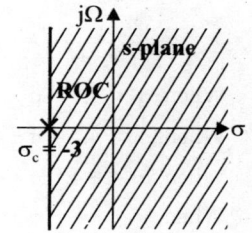

Fig 3: ROC of $x(t) = e^{-3t}\, u(t)$.

$\therefore \mathcal{L}\{e^{-3t}\, u(t)\} = \dfrac{1}{s+3}$; with ROC as all points in s-plane to the right of line passing through $\sigma = -3$.

---

d) Given that, $x(t) = e^{-3t}\, u(-t) = e^{-3t}$; $t \leq 0$

By definition of Laplace transform,

$$X(s) = \mathcal{L}\{x(t)\} = \int_{-\infty}^{+\infty} x(t)\, e^{-st}\, dt = \int_{-\infty}^{0} e^{-3t} e^{-st}\, dt = \int_{-\infty}^{0} e^{-(s+3)t}\, dt$$

$$= \left[\frac{e^{-(s+3)t}}{-(s+3)}\right]_{-\infty}^{0} = \frac{e^{0}}{-(s+3)} - \frac{e^{(s+3)\infty}}{-(s+3)} = -\frac{1}{s+3} + \frac{e^{(\sigma+j\Omega+3)\infty}}{s+3} \qquad \boxed{s = \sigma + j\Omega}$$

$$= -\frac{1}{s+3} + \frac{e^{(\sigma+3)\times\infty}\, e^{+j\Omega\times\infty}}{s+3} = -\frac{1}{s+3} + \frac{e^{k\times\infty}\, e^{j\Omega\times\infty}}{s+3}$$

where, $k = \sigma + 3 = \sigma - (-3)$

When, $\sigma > -3$, $k = \sigma - (-3) =$ Positive. $\quad \therefore e^{k\infty} = e^{\infty} = \infty$

When, $\sigma < -3$, $k = \sigma - (-3) =$ Negative. $\therefore e^{k\infty} = e^{-\infty} = 0$

Therefore we can say that, $X(s)$ converges when $\sigma < -3$.

When $\sigma < -3$, the $X(s)$ is given by,

$$X(s) = -\frac{1}{s+3} + \frac{e^{k\times\infty}\, e^{j\Omega\times\infty}}{s+3}$$

$$= -\frac{1}{s+3} + \frac{0 \times e^{j\Omega\times\infty}}{s+3} = -\frac{1}{s+3}$$

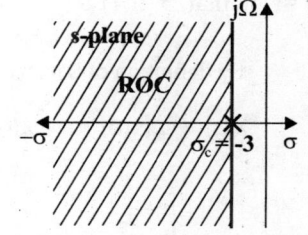

Fig 4: ROC of $x(t) = e^{-3t}\, u(-t)$.

$\therefore \mathcal{L}\{e^{-3t}\, u(-t)\} = -\dfrac{1}{s+3}$; with ROC as all points in s-plane to the left of line passing through $\sigma = -3$.

e) Given that, $x(t) = e^{-4|t|} = e^{4t}$; $t \leq 0$

$$= e^{-4t} \ ; \ t \geq 0$$

By definition of Laplace transform,

$$X(s) = \mathcal{L}\{x(t)\} = \int_{-\infty}^{+\infty} x(t) e^{-st} \, dt = \int_{-\infty}^{0} e^{4t} e^{-st} \, dt + \int_{0}^{\infty} e^{-4t} e^{-st} \, dt = \int_{-\infty}^{0} e^{-(s-4)t} \, dt + \int_{0}^{\infty} e^{-(s+4)t} \, dt$$

$$= \left[\frac{e^{-(s-4)t}}{-(s-4)}\right]_{-\infty}^{0} + \left[\frac{e^{-(s+4)t}}{-(s+4)}\right]_{0}^{\infty} = \left[\frac{e^{0}}{-(s-4)} - \frac{e^{-(s-4)\infty}}{-(s-4)}\right] + \left[\frac{e^{-(s+4)\infty}}{-(s+4)} - \frac{e^{0}}{-(s+4)}\right]$$

$$= -\frac{1}{s-4} + \frac{e^{(\sigma+j\Omega-4)\times\infty}}{s-4} - \frac{e^{-(\sigma+j\Omega+4)\times\infty}}{s+4} + \frac{1}{s+4} \qquad \boxed{s = \sigma + j\Omega}$$

$$= -\frac{1}{s-4} + \frac{e^{(\sigma-4)\times\infty} e^{j\Omega\times\infty}}{s-4} - \frac{e^{-(\sigma+4)\times\infty} e^{-j\Omega\times\infty}}{s+4} + \frac{1}{s+4}$$

When, $\sigma < 4$, $\sigma - 4 =$ Negative. $\therefore e^{(\sigma-4)\infty} = e^{-\infty} = 0$

When, $\sigma > 4$, $\sigma - 4 =$ Positive. $\therefore e^{(\sigma-4)\infty} = e^{\infty} = \infty$

When, $\sigma < -4$, $\sigma + 4 =$ Negative. $\therefore e^{-(\sigma+4)\infty} = e^{\infty} = \infty$

When, $\sigma > -4$, $\sigma + 4 =$ Positive. $\therefore e^{-(\sigma+4)\infty} = e^{-\infty} = 0$

Therefore we can say that, X(s) converges when σ lies between −4 and +4.

When σ lies between − 4 and + 4, the X(s) is given by,

$$X(s) = -\frac{1}{s-4} + \frac{e^{(\sigma-4)\times\infty} e^{j\Omega\times\infty}}{s-4} - \frac{e^{-(\sigma+4)\times\infty} e^{-j\Omega\times\infty}}{s+4} + \frac{1}{s+4}$$

$$= -\frac{1}{s-4} + \frac{e^{-\infty} e^{j\Omega\times\infty}}{s-4} - \frac{e^{-\infty} e^{-j\Omega\times\infty}}{s+4} + \frac{1}{s+4}$$

$$= -\frac{1}{s-4} + \frac{0 \times e^{j\Omega\times\infty}}{s-4} - \frac{0 \times e^{-j\Omega\times\infty}}{s+4} + \frac{1}{s+4}$$

$$= -\frac{1}{s-4} + 0 - 0 + \frac{1}{s+4}$$

Fig 5: *ROC of $x(t) = e^{-4|t|}$.*

$$= -\frac{1}{s+4} - \frac{1}{s-4} = \frac{s-4-(s+4)}{(s+4)(s-4)} = -\frac{8}{s^2-16}$$

$$\boxed{(a+b)(a-b) = a^2 - b^2}$$

$$\therefore \mathcal{L}\{e^{-4|t|}\} = -\frac{8}{s^2-16} \ ; \ \text{with ROC as all points in s-plane in between the lines passing through}$$
$$\sigma = -4 \text{ and } \sigma = 4 .$$

---

f) Given that, $x(t) = t\, e^{-2|t|} = t\, e^{2t}$; $-\infty \leq t \leq 0$ *(AU Dec'15, 16 Marks)*

$$= t\, e^{-2t} \ ; \quad 0 \leq t \leq +\infty$$

By definition of Laplace transform,

$$X(s) = \mathcal{L}\{x(t)\} = \int_{-\infty}^{+\infty} x(t) e^{-st} \, dt = \int_{-\infty}^{0} t\, e^{2t} e^{-st} \, dt + \int_{0}^{+\infty} t\, e^{-2t} e^{-st} \, dt = \int_{-\infty}^{0} t\, e^{-(s-2)t} \, dt + \int_{0}^{+\infty} t\, e^{-(s+2)t} \, dt$$

$$= \left[t \frac{e^{-(s-2)t}}{-(s-2)}\right]_{-\infty}^{0} - \int_{-\infty}^{0} \frac{e^{-(s-2)t}}{-(s-2)} \times 1 \, dt + \left[t \frac{e^{-(s+2)t}}{-(s+2)}\right]_{0}^{+\infty} - \int_{0}^{+\infty} \frac{e^{-(s+2)t}}{-(s+2)} \times 1 \, dt$$

$\int u \, dv = u\,v - \int v \, du$	
case i: $u = t$ $\Rightarrow du = 1$	**case ii:** $u = t$ $\Rightarrow du = 1$
$dv = e^{-(s-2)t}$ $\Rightarrow v = \dfrac{e^{-(s-2)t}}{-(s-2)}$	$dv = e^{-(s+2)t}$ $\Rightarrow v = \dfrac{e^{-(s+2)t}}{-(s+2)}$

$$\therefore \; X(s) = \left[\frac{-t\,e^{-(s-2)t}}{s-2}\right]_{-\infty}^{0} - \left[\frac{e^{-(s-2)t}}{(s-2)^2}\right]_{-\infty}^{0} + \left[\frac{-t\,e^{-(s+2)t}}{s+2}\right]_{0}^{\infty} - \left[\frac{e^{-(s+2)t}}{(s+2)^2}\right]_{0}^{\infty}$$

$$= 0 + \frac{\infty \times e^{(s-2)\infty}}{s-2} - \frac{e^0}{(s-2)^2} + \frac{e^{(s-2)\infty}}{(s-2)^2} - \frac{\infty \times e^{-(s+2)\infty}}{s+2} + 0 - \frac{e^{-(s+2)\infty}}{(s+2)^2} + \frac{e^0}{(s+2)^2}$$

$$= \frac{\infty \times e^{(\sigma+j\Omega-2)\infty}}{s-2} - \frac{1}{(s-2)^2} + \frac{e^{(\sigma+j\Omega-2)\infty}}{(s-2)^2} - \frac{\infty \times e^{-(\sigma+j\Omega+2)\infty}}{s+2} - \frac{e^{-(\sigma+j\Omega+2)\infty}}{(s+2)^2} + \frac{1}{(s+2)^2}$$

$$= \frac{\infty \times e^{(\sigma-2)\times\infty}\,e^{j\Omega\times\infty}}{s-2} - \frac{1}{(s-2)^2} + \frac{e^{(\sigma-2)\times\infty}\,e^{j\Omega\times\infty}}{(s-2)^2} - \frac{\infty \times e^{-(\sigma+2)\times\infty}\,e^{-j\Omega\times\infty}}{s+2} \qquad \boxed{s=\sigma+j\Omega}$$

$$- \frac{e^{-(\sigma+2)\times\infty}\,e^{-j\Omega\times\infty}}{(s+2)^2} + \frac{1}{(s+2)^2}$$

When, $\sigma < 2$, $\sigma - 2 =$ Negative, $\quad \therefore e^{(\sigma-2)\infty} = e^{-\infty} = 0$

When, $\sigma > 2$, $\sigma - 2 =$ Positive, $\quad \therefore e^{(\sigma-2)\infty} = e^{\infty} = \infty$

When, $\sigma < -2$, $\sigma + 2 =$ Negative, $\quad \therefore e^{-(\sigma+2)\infty} = e^{\infty} = \infty$

When, $\sigma > -2$, $\sigma + 2 =$ Positive, $\quad \therefore e^{-(\sigma+2)\infty} = e^{-\infty} = 0$

Therefore we can say that, X(s) converges when σ lies between − 2 and + 2. When σ lies between − 2 and + 2, the X(s) is given by,

$$\therefore \; X(s) = \frac{\infty\,e^{-\infty}\,e^{j\Omega\times\infty}}{s-2} - \frac{1}{(s-2)^2} + \frac{e^{-\infty}\,e^{j\Omega\times\infty}}{(s-2)^2} - \frac{\infty\,e^{-\infty}\,e^{j\Omega\times\infty}}{s+2} - \frac{e^{-\infty}\,e^{-j\Omega\times\infty}}{(s+2)^2} + \frac{1}{(s+2)^2}$$

$$= \frac{\infty \times 0 \times e^{j\Omega\times\infty}}{s-2} - \frac{1}{(s-2)^2} + \frac{0 \times e^{j\Omega\times\infty}}{(s-2)^2} - \frac{\infty \times 0 \times e^{-j\Omega\times\infty}}{s+2} - \frac{0 \times e^{-j\Omega\times\infty}}{(s+2)^2} + \frac{1}{(s+2)^2}$$

$$= -\frac{1}{(s-2)^2} + \frac{1}{(s+2)^2} = \frac{-(s+2)^2 + (s-2)^2}{(s-2)^2\,(s+2)^2}$$

$$= \frac{-(s^2+4s+4) - (s^2-4s+4)}{[(s-2)(s+2)]^2} = \frac{-8}{(s^2-4)^2}$$

Fig 6: ROC of x(t) = t e^{-2|t|}.

$$\therefore \mathcal{L}\{x(t)\} = -\frac{8}{(s^2-4)^2} \;; \text{with ROC as all points in s-plane in between the lines passing through } \sigma = -2 \text{ and } \sigma = 2.$$

Example 2.26

Determine the Laplace transform of the following signals.

a) $x(t) = \sin\Omega_0 t\, u(t)$ b) $x(t) = \cos\Omega_0 t\, u(t)$ c) $x(t) = \cosh\Omega_0 t\, u(t)$

d) $x(t) = e^{-at}\sin\Omega_0 t\, u(t)$ e) $x(t) = e^{-at}\cos\Omega_0 t\, u(t)$

Solution:

a) Given that, $x(t) = \sin\Omega_0 t\, u(t) = \sin\Omega_0 t\ ;\ t \geq 0$

$$\boxed{\sin\theta = \dfrac{e^{j\theta} - e^{-j\theta}}{2j}}$$

By definition of Laplace transform,

$$X(s) = \mathcal{L}\{x(t)\} = \int_{-\infty}^{\infty} x(t)\, e^{-st}\, dt = \int_{0}^{\infty} \sin\Omega_0 t\, e^{-st}\, dt = \int_{0}^{\infty} \frac{e^{j\Omega_0 t} - e^{-j\Omega_0 t}}{2j}\, e^{-st}\, dt$$

$$= \frac{1}{2j}\int_{0}^{\infty}\left(e^{-(s-j\Omega_0)t} - e^{-(s+j\Omega_0)t}\right) dt = \frac{1}{2j}\left[\frac{e^{-(s-j\Omega_0)t}}{-(s-j\Omega_0)} - \frac{e^{-(s+j\Omega_0)t}}{-(s+j\Omega_0)}\right]_{0}^{\infty}$$

$$= \frac{1}{2j}\left[\frac{e^{-\infty}}{-(s-j\Omega_0)} - \frac{e^{-\infty}}{-(s+j\Omega_0)} - \frac{e^{0}}{-(s-j\Omega_0)} + \frac{e^{0}}{-(s+j\Omega_0)}\right]$$

$$= \frac{1}{2j}\left[0 - 0 + \frac{1}{s-j\Omega_0} - \frac{1}{s+j\Omega_0}\right] = \frac{1}{2j}\left[\frac{(s+j\Omega_0) - (s-j\Omega_0)}{(s-j\Omega_0)(s+j\Omega_0)}\right]$$

$$= \frac{1}{2j}\left[\frac{s+j\Omega_0 - s + j\Omega_0}{s^2 - (j\Omega_0)^2}\right]$$

$$= \frac{1}{2j}\left[\frac{2j\Omega_0}{s^2 + \Omega_0^2}\right] = \frac{\Omega_0}{s^2 + \Omega_0^2}$$

$$\boxed{(a+b)(a-b) = a^2 - b^2}\quad \boxed{j^2 = -1}$$

$$\therefore\ \mathcal{L}\{\sin\Omega_0 t\, u(t)\} = \frac{\Omega_0}{s^2 + \Omega_0^2}$$

---

b) Given that, $x(t) = \cos\Omega_0 t\, u(t) = \cos\Omega_0 t\ ;\ t \geq 0$

By definition of Laplace transform,

$$\boxed{\cos\theta = \dfrac{e^{j\theta} + e^{-j\theta}}{2}}$$

$$X(s) = \mathcal{L}\{x(t)\} = \int_{-\infty}^{\infty} x(t)\, e^{-st}\, dt = \int_{0}^{\infty} \cos\Omega_0 t\, e^{-st}\, dt = \int_{0}^{\infty} \frac{e^{j\Omega_0 t} + e^{-j\Omega_0 t}}{2}\, e^{-st}\, dt$$

$$= \frac{1}{2}\int_{0}^{\infty}\left(e^{-(s-j\Omega_0)t} + e^{-(s+j\Omega_0)t}\right) dt = \frac{1}{2}\left[\frac{e^{-(s-j\Omega_0)t}}{-(s-j\Omega_0)} + \frac{e^{-(s+j\Omega_0)t}}{-(s+j\Omega_0)}\right]_{0}^{\infty}$$

$$= \frac{1}{2}\left[\frac{e^{-\infty}}{-(s-j\Omega_0)} + \frac{e^{-\infty}}{-(s+j\Omega_0)} - \frac{e^{0}}{-(s-j\Omega_0)} - \frac{e^{0}}{-(s+j\Omega_0)}\right]$$

$$= \frac{1}{2}\left[0 + 0 + \frac{1}{s-j\Omega_0} + \frac{1}{s+j\Omega_0}\right]$$

$$= \frac{1}{2}\left[\frac{s+j\Omega_0 + s - j\Omega_0}{(s-j\Omega_0)(s+j\Omega_0)}\right] = \frac{1}{2}\left[\frac{2s}{s^2 - (j\Omega_0)^2}\right] = \frac{s}{s^2 + \Omega_0^2}$$

$$\therefore\ \mathcal{L}\{\cos\Omega_0 t\, u(t)\} = \frac{s}{s^2 + \Omega_0^2}$$

$$\boxed{(a+b)(a-b) = a^2 - b^2}\quad \boxed{j^2 = -1}$$

---

c) Given that, $x(t) = \cosh\Omega_0 t\, u(t) = \cosh\Omega_0 t\ ;\ t \geq 0$

By definition of Laplace transform,

$$\boxed{\cosh\theta = \dfrac{e^{\theta} + e^{-\theta}}{2}}$$

$$X(s) = \mathcal{L}\{x(t)\} = \int_{-\infty}^{\infty} x(t)\, e^{-st}\, dt = \int_{0}^{\infty} \cosh\Omega_0 t\, e^{-st}\, dt = \int_{0}^{\infty} \frac{e^{\Omega_0 t} + e^{-\Omega_0 t}}{2}\, e^{-st}\, dt$$

$$= \frac{1}{2}\int_{0}^{\infty}\left(e^{-(s-\Omega_0)t} + e^{-(s+\Omega_0)t}\right) dt = \frac{1}{2}\left[\frac{e^{-(s-\Omega_0)t}}{-(s-\Omega_0)} + \frac{e^{-(s+\Omega_0)t}}{-(s+\Omega_0)}\right]_{0}^{\infty}$$

$$\therefore X(s) = \frac{1}{2}\left[\frac{e^{-\infty}}{-(s-\Omega_0)} + \frac{e^{-\infty}}{-(s+\Omega_0)} - \frac{e^0}{-(s-\Omega_0)} - \frac{e^0}{-(s+\Omega_0)}\right]$$

$$= \frac{1}{2}\left[0 + 0 + \frac{1}{s-\Omega_0} + \frac{1}{s+\Omega_0}\right]$$

$$= \frac{1}{2}\left[\frac{s+\Omega_0+s-\Omega_0}{(s-\Omega_0)(s+\Omega_0)}\right] = \frac{1}{2}\left[\frac{2s}{s^2-\Omega_0^2}\right] = \frac{s}{s^2-\Omega_0^2}$$

$$\boxed{(a+b)(a-b) = a^2 - b^2}$$

$$\therefore \mathcal{L}\{\cosh\Omega_0 t\, u(t)\} = \frac{s}{s^2-\Omega_0^2}$$

---

d) Given that, x(t) =$e^{-at} \sin\Omega_0 t\, u(t) = e^{-at} \sin\Omega_0 t$; $t \geq 0$ $\boxed{\sin\theta = \dfrac{e^{j\theta}-e^{-j\theta}}{2j}}$ **(AU Dec'13, 8 Marks)**

By definition of Laplace transform,

$$X(s) = \mathcal{L}\{x(t)\} = \int_{-\infty}^{\infty} x(t)\, e^{-st}\, dt = \int_{0}^{\infty} e^{-at}\sin\Omega_0 t\, e^{-st}\, dt = \int_{0}^{\infty} e^{-at}\frac{e^{j\Omega_0 t}-e^{-j\Omega_0 t}}{2j}\, e^{-st}\, dt$$

$$= \frac{1}{2j}\int_{0}^{\infty}(e^{-(s+a-j\Omega_0)t} - e^{-(s+a+j\Omega_0)t})\, dt = \frac{1}{2j}\left[\frac{e^{-(s+a-j\Omega_0)t}}{-(s+a-j\Omega_0)} - \frac{e^{-(s+a+j\Omega_0)t}}{-(s+a+j\Omega_0)}\right]_0^{\infty}$$

$$= \frac{1}{2j}\left[\frac{e^{-\infty}}{-(s+a-j\Omega_0)} - \frac{e^{-\infty}}{-(s+a+j\Omega_0)} - \frac{e^0}{-(s+a-j\Omega_0)} + \frac{e^0}{-(s+a+j\Omega_0)}\right]$$

$$= \frac{1}{2j}\left[0 - 0 + \frac{1}{s+a-j\Omega_0} - \frac{1}{s+a+j\Omega_0}\right] = \frac{1}{2j}\left[\frac{(s+a+j\Omega_0)-(s+a-j\Omega_0)}{(s+a-j\Omega_0)(s+a+j\Omega_0)}\right]$$

$$= \frac{1}{2j}\left[\frac{2j\Omega_0}{(s+a)^2-(j\Omega_0)^2}\right] = \frac{\Omega_0}{(s+a)^2+\Omega_0^2}$$

$$\therefore \mathcal{L}\{\sin\Omega_0 t\, u(t)\} = \frac{\Omega_0}{(s+a)^2+\Omega_0^2}$$
$$\boxed{(a+b)(a-b) = a^2 - b^2} \quad \boxed{j^2 = -1}$$

---

e) Given that, x(t) = $e^{-at}\cos\Omega_0 t\, u(t) = e^{-at}\cos\Omega_0 t$; $t \geq 0$ $\boxed{\cos\theta = \dfrac{e^{j\theta}+e^{-j\theta}}{2}}$ **(AU Jun'12, 8 Marks)**

By definition of Laplace transform,

$$X(s) = \mathcal{L}\{x(t)\} = \int_{-\infty}^{\infty} x(t)\, e^{-st}\, dt = \int_{0}^{\infty} e^{-at}\cos\Omega_0 t\, e^{-st}\, dt = \int_{0}^{\infty} e^{-at}\frac{e^{j\Omega_0 t}+e^{-j\Omega_0 t}}{2}\, e^{-st}\, dt$$

$$= \frac{1}{2}\int_{0}^{\infty}(e^{-(s+a-j\Omega_0)t} + e^{-(s+a+j\Omega_0)t})\, dt = \frac{1}{2}\left[\frac{e^{-(s+a-j\Omega_0)t}}{-(s+a-j\Omega_0)} + \frac{e^{-(s+a+j\Omega_0)t}}{-(s+a+j\Omega_0)}\right]_0^{\infty}$$

$$= \frac{1}{2}\left[\frac{e^{-\infty}}{-(s+a-j\Omega_0)} + \frac{e^{-\infty}}{-(s+a+j\Omega_0)} - \frac{e^0}{-(s+a-j\Omega_0)} - \frac{e^0}{-(s+a+j\Omega_0)}\right]$$

$$= \frac{1}{2}\left[0 - 0 + \frac{1}{s+a-j\Omega_0} + \frac{1}{s+a+j\Omega_0}\right] = \frac{1}{2}\left[\frac{(s+a+j\Omega_0)+(s+a-j\Omega_0)}{(s+a-j\Omega_0)(s+a+j\Omega_0)}\right]$$

$$= \frac{1}{2}\left[\frac{2(s+a)}{(s+a)^2-(j\Omega_0)^2}\right] = \frac{s+a}{(s+a)^2+\Omega_0^2}$$

$$\therefore \mathcal{L}\{e^{-at}\cos\Omega_0 t\, u(t)\} = \frac{s+a}{(s+a)^2+\Omega_0^2}$$
$$\boxed{(a+b)(a-b) = a^2 - b^2} \quad \boxed{j^2 = -1}$$

Example 2.27   **(AU Jun'11, 8 Marks)**

Determine the Laplace transform of the following signals.

a) $x(t) = u(t - 2)$ b) $x(t) = t^2 e^{-2t} u(t)$

Solution:

a) Given that, $x(t) = u(t - 2)$; $x(t) = 1$; $t \geq 2$ **(AU Jun'12, 2 Marks)**

By definition of Laplace transform,

$$X(s) = \mathcal{L}\{x(t)\} = \int_{-\infty}^{\infty} x(t)\, e^{-st}\, dt = \int_{2}^{\infty} e^{-st}\, dt$$

$$= \left[\frac{e^{-st}}{-s}\right]_{2}^{\infty} = \left[\frac{e^{-s\times\infty}}{-s} + \frac{e^{-s\times2}}{s}\right] = \frac{e^{-2s}}{s}$$

$$\therefore \mathcal{L}\{u(t-2)\} = \frac{e^{-2s}}{s}$$

b) Given that, $x(t) = t^2 e^{-2t} u(t) = t^2 e^{-2t}$; $t \geq 0$

By definition of Laplace transform,

$$X(s) = \mathcal{L}\{x(t)\} = \int_{-\infty}^{\infty} x(t)\, e^{-st}\, dt = \int_{0}^{\infty} t^2 e^{-2t} e^{-st}\, dt = \int_{0}^{\infty} t^2 e^{-(s+2)t}\, dt$$

$\int u\, dv = u\, v - \int v\, du$
$u = t^2$ $\Rightarrow$ $du = 2t$
$dv = e^{-(s+2)t} \Rightarrow$ $v = \dfrac{e^{-(s+2)t}}{-(s+2)}$

$$= \left[t^2 \frac{e^{-(s+2)t}}{-(s+2)}\right]_{0}^{\infty} - \int_{0}^{\infty} \frac{e^{-(s+2)t}}{-(s+2)} \times 2t\, dt$$

$$= \left[\frac{t^2 e^{-(s+2)t}}{-(s+2)}\right]_{0}^{\infty} + \int_{0}^{\infty} 2t \times \frac{e^{-(s+2)t}}{s+2}\, dt$$

$$= \left[\frac{t^2 e^{-(s+2)t}}{-(s+2)}\right]_{0}^{\infty} + \left[2t \times \frac{e^{-(s+2)t}}{-(s+2)^2}\right]_{0}^{\infty} - \int_{0}^{\infty} \frac{e^{-(s+2)t}}{-(s+2)^2} \times 2\, dt$$

$\int u\, dv = u\, v - \int v\, du$
$u = t$ $\Rightarrow$ $du = 1$
$dv = \dfrac{e^{-(s+2)t}}{s+2} \Rightarrow$ $v = \dfrac{e^{-(s+2)t}}{-(s+2)^2}$

$$= \left[\frac{t^2 e^{-(s+2)t}}{-(s+2)}\right]_{0}^{\infty} + \left[2t \times \frac{e^{-(s+2)t}}{-(s+2)^2}\right]_{0}^{\infty} - \left[2 \times \frac{e^{-(s+2)t}}{(s+2)^3}\right]_{0}^{\infty}$$

$$= \frac{\infty \times e^{-\infty}}{-(s+2)} - 0 + \frac{\infty \times e^{-\infty}}{-(s+2)^2} - 0 - \frac{e^{-\infty}}{(s+2)^3} + \frac{2e^0}{(s+2)^3} = \frac{2}{(s+2)^3}$$ $\boxed{e^{-\infty} = 0}$

$$\therefore \mathcal{L}\{t^2 e^{-2t} u(t)\} = \frac{2}{(s+2)^3}$$

Example 2.28   **(AU Jun'13, 10 Marks)**

Determine the Laplace transform of the following continous time signal $x(t) = e^{-b|t|}$ for

a) $b < 0$ b) $b > 0$

Solution:

Given that, $x(t) = e^{-b|t|}$ $= e^{bt}$; $t < 0$

$= e^{-bt}$; $t > 0$

By definition of Laplace transform,

$$X(s) = \mathcal{L}\{x(t)\} = \int_{-\infty}^{+\infty} x(t)\, e^{-st}\, dt = \int_{-\infty}^{0} e^{bt} e^{-st}\, dt + \int_{0}^{+\infty} e^{-bt} e^{-st}\, dt = \int_{-\infty}^{0} e^{-(s-b)t}\, dt + \int_{0}^{+\infty} e^{-(s+b)t}\, dt$$

$$\therefore X(s) = \left[\frac{e^{-(s-b)t}}{-(s-b)}\right]_{-\infty}^{0} + \left[\frac{e^{-(s+b)t}}{-(s+b)}\right]_{0}^{\infty} = \frac{e^0}{-(s-b)} - \frac{e^{(s-b)\infty}}{-(s-b)} + \frac{e^{-(s+b)\infty}}{-(s+b)} - \frac{e^0}{-(s+b)}$$

$$= -\frac{1}{s-b} + \frac{e^{(\sigma+j\Omega-b)\infty}}{s-b} - \frac{e^{-(\sigma+j\Omega+b)\infty}}{s+b} + \frac{1}{s+b} \qquad \boxed{s = \sigma + j\,\Omega}$$

$$= -\frac{1}{s-b} + \frac{e^{(\sigma-b)\infty}e^{j\Omega\infty}}{s-b} - \frac{e^{-(\sigma+b)\infty}e^{-j\Omega\infty}}{s+b} + \frac{1}{s+b}$$

a) b < 0

When $\sigma < -b$; $\sigma - b =$ Negative, $e^{(\sigma-b)\infty} = e^{-\infty} = 0$

When $\sigma > -b$; $\sigma - b =$ Positive, $e^{(\sigma-b)\infty} = e^{\infty} = \infty$

When $\sigma < -b$; $\sigma + b =$ Negative, $e^{-(\sigma+b)\infty} = e^{\infty} = \infty$

When $\sigma > -b$; $\sigma + b =$ Positive, $e^{-(\sigma+b)\infty} = e^{-\infty} = 0$

From the above analysis we can infer that there is no common area for ROC. Therefore, the Laplace transform does not exist for b < 0.

b) b > 0

When $q < b$; $\sigma - b =$ Negative, $e^{(\sigma-b)\infty} = e^{-\infty} = 0$

When $\sigma > b$; $\sigma - b =$ Positive, $e^{(\sigma-b)\infty} = e^{\infty} = \infty$

When $q < -b$; $\sigma + b =$ Negative, $e^{-(\sigma+b)\infty} = e^{\infty} = \infty$

When $\sigma > -b$; $\sigma + b =$ Positive, $e^{-(\sigma+b)\infty} = e^{-\infty} = 0$

Therefore we can say that, X(s) converges when σ lies between –b and +b.

When σ lies between -b and +b, the X(s) is given by,

$$X(s) = -\frac{1}{s-b} + \frac{e^{-\infty}e^{j\Omega\times\infty}}{s-b} - \frac{e^{-\infty}e^{j\Omega\times\infty}}{s+b} + \frac{1}{s+b} \qquad \boxed{e^{-\infty} = 0}$$

$$= -\frac{1}{s-b} + 0 + 0 + \frac{1}{s+b} = \frac{-(s+b)+(s-b)}{(s-b)(s+b)} = \frac{-2b}{s^2-b^2}$$

Example 2.29

Determine the Laplace transform of the signals shown below:

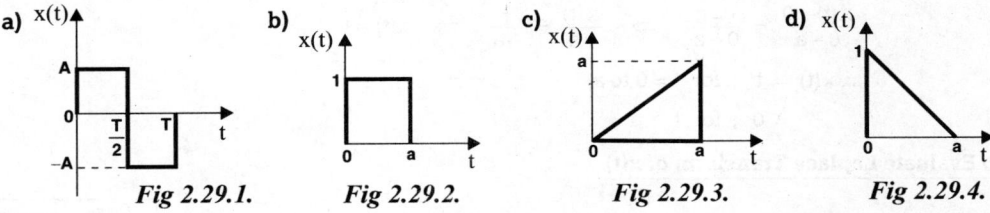

a) Fig 2.29.1.   b) Fig 2.29.2.   c) Fig 2.29.3.   d) Fig 2.29.4.

Solution:

a) The mathematical equation of the signal shown in Fig 2.29.1 is,

$$x(t) = A \quad ; \text{ for } 0 < t < T/2$$
$$= -A \quad ; \text{ for } T/2 < t < T$$

By definition of Laplace transform,

$$X(s) = \mathcal{L}\{x(t)\} = \int_{-\infty}^{\infty} x(t)\,e^{-st}\,dt = \int_{0}^{T} x(t)\,e^{-st}\,dt$$

$$\therefore \ X(s) = \int_{0}^{T/2} A\,e^{-st}\,dt + \int_{T/2}^{T}(-A)\,e^{-st}\,dt = \left[\dfrac{A\,e^{-st}}{-s}\right]_{0}^{T/2} + \left[\dfrac{-A\,e^{-st}}{-s}\right]_{T/2}^{T}$$

$$= \left[\dfrac{A\,e^{\frac{-sT}{2}}}{-s} - \dfrac{A\,e^{0}}{-s}\right] + \left[\dfrac{A\,e^{-sT}}{s} - \dfrac{A\,e^{\frac{-sT}{2}}}{s}\right]$$

$$= -\dfrac{A\,e^{\frac{-sT}{2}}}{s} + \dfrac{A}{s} + \dfrac{A\,e^{-sT}}{s} - \dfrac{A\,e^{\frac{-sT}{2}}}{s}$$

$$= \dfrac{A}{s}\left[1 + e^{-sT} - 2\,e^{\frac{-sT}{2}}\right] = \dfrac{A}{s}\left[1 - e^{\frac{-sT}{2}}\right]^{2} \qquad \boxed{(a-b)^2 = a^2 + b^2 - 2ab}$$

b)   The mathematical equation of the signal shown in Fig 2.29.2 is,

$$x(t) = 1 \ ; \ \text{for} \ 0 \le t \le a$$
$$= 0 \ ; \ \text{for} \ t > a$$

By definition of Laplace transform,

$$X(s) = \mathcal{L}\{x(t)\} = \int_{-\infty}^{\infty} x(t)\,e^{-st}\,dt = \int_{0}^{a} 1 \times e^{-st}\,dt = \int_{0}^{a} e^{-st}\,dt = \left[\dfrac{e^{-st}}{-s}\right]_{0}^{a}$$

$$= \dfrac{e^{-as}}{-s} - \dfrac{e^{0}}{-s} = -\dfrac{e^{-as}}{s} + \dfrac{1}{s} = \dfrac{1}{s}\left(1 - e^{-as}\right)$$

c)   **To Find Mathematical Equation for x(t)**

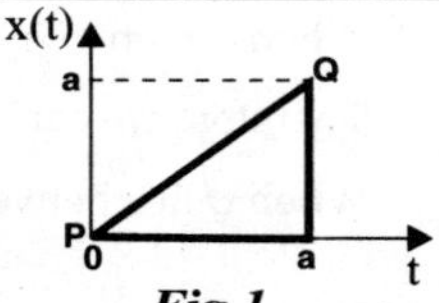

Consider the equation of straight line, $\dfrac{y - y_1}{y_1 - y_2} = \dfrac{x - x_1}{x_1 - x_2}$

Here, $y = x(t)$, $x = t$.

$\therefore$ The equation of straight line can be written as, $\dfrac{x(t) - x(t_1)}{x(t_1) - x(t_2)} = \dfrac{t - t_1}{t_1 - t_2}$ .....(1)

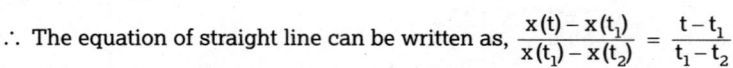

Fig 1.

Consider points P and Q as shown in Fig 1.

Coordinates of point - P = $[t_1, x(t_1)] = [0,0]$

Coordinates of point - Q = $[t_2, x(t_2)] = [a,a]$

On substituting the coordinates of points - P and Q in equation (1) we get,

$$\dfrac{x(t) - 0}{0 - a} = \dfrac{t - 0}{0 - a} \quad \Rightarrow \quad \dfrac{x(t)}{-a} = \dfrac{t}{-a} \quad \Rightarrow \quad x(t) = t$$

$$\therefore x(t) = t \ ; \ \text{for} \ t = 0 \ \text{to} \ a$$
$$= 0 \ ; \ \text{for} \ t > a$$

To Evaluate Laplace Transform of x(t)

$$X(s) = \mathcal{L}\{x(t)\} = \int_{-\infty}^{+\infty} x(t)\,e^{-st}\,dt = \int_{0}^{a} t\,e^{-st}\,dt$$

$$= \left[t \times \dfrac{e^{-st}}{-s} - \int \dfrac{e^{-st}}{-s} \times 1\,dt\right]_{0}^{a} = \left[-\dfrac{t\,e^{-st}}{s} - \dfrac{e^{-st}}{s^{2}}\right]_{0}^{a}$$

$$= \left[-\dfrac{a\,e^{-sa}}{s} - \dfrac{e^{-sa}}{s^{2}} + 0 + \dfrac{e^{0}}{s^{2}}\right] = \dfrac{1}{s^{2}} - \dfrac{e^{-sa}}{s^{2}} - \dfrac{a\,e^{-sa}}{s}$$

$$= \dfrac{1}{s^{2}}\left[1 - e^{-as} - as\,e^{-as}\right] = \dfrac{1}{s^{2}}\left[1 - e^{-as}(1 + as)\right]$$

$\int u\,dv = u\,v - \int v\,du$
$u = t \qquad \Rightarrow du = 1$
$dv = e^{-st} \Rightarrow v = \dfrac{e^{-st}}{-s}$

d) To Find Mathematical Equation for x(t)

Consider the equation of straight line, $\dfrac{y-y_1}{y_1-y_2} = \dfrac{x-x_1}{x_1-x_2}$

Here, $y = x(t)$, $x = t$.

∴ The equation of straight line can be written as, $\dfrac{x(t)-x(t_1)}{x(t_1)-x(t_2)} = \dfrac{t-t_1}{t_1-t_2}$ .....(1)

Consider points P and Q as shown in Fig 2.

Coordinates of point - P $= [t_1, x(t_1)] = [0,1]$

Coordinates of point - Q $= [t_2, x(t_2)] = [a,0]$

On substituting the coordinates of points-P and Q in equation (1) we get,

$$\frac{x(t)-1}{1-0} = \frac{t-0}{0-a} \;\Rightarrow\; x(t)-1 = -\frac{t}{a} \;\Rightarrow\; x(t) = 1-\frac{t}{a}$$

$$\therefore\; x(t) = 1-\frac{t}{a} \quad ; \text{ for } t = 0 \text{ to } a$$

$$= 0 \quad\quad ; \text{ for } t > a$$

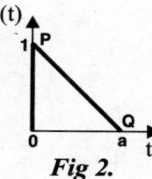

Fig 2.

To Evaluate Laplace transform of x(t)

$$X(s) = \mathcal{L}\{x(t)\} = \int\limits_{-\infty}^{+\infty} x(t)\, e^{-st}\, dt$$

$$= \int\limits_{0}^{a}\left(1-\frac{t}{a}\right)e^{-st}\, dt = \int\limits_{0}^{a} e^{-st}\, dt - \frac{1}{a}\int\limits_{0}^{a} t\, e^{-st}\, dt$$

$$= \left[\frac{e^{-st}}{-s}\right]_0^a - \frac{1}{a}\left[t\times\frac{e^{-st}}{-s} - \int\frac{e^{-st}}{-s}\times 1\, dt\right]_0^a$$

$$= \left[-\frac{e^{-st}}{s}\right]_0^a - \frac{1}{a}\left[-\frac{t\, e^{-st}}{s} - \frac{e^{-st}}{s^2}\right]_0^a$$

$$= \left[-\frac{e^{-as}}{s} + \frac{e^0}{s}\right] - \frac{1}{a}\left[-\frac{a\, e^{-as}}{s} - \frac{e^{-as}}{s^2} + 0 + \frac{e^0}{s^2}\right]$$

$$= -\frac{e^{-as}}{s} + \frac{1}{s} + \frac{e^{-as}}{s} + \frac{e^{-as}}{as^2} - \frac{1}{as^2}$$

$$= \frac{1}{s} + \frac{e^{-as}}{as^2} - \frac{1}{as^2} = \frac{1}{as^2}[e^{-as} + as - 1]$$

$\int u\, dv = u\, v - \int v\, du$
$u = t \qquad\qquad \Rightarrow du = 1$
$dv = e^{-st} \;\Rightarrow\; v = \dfrac{e^{-st}}{-s}$

Example 2.30

Determine the Laplace transform of the sine pulse shown in Fig 2.30.1.

Solution:

The mathematical equation of the sine pulse shown in Fig 2.30.1. is,

$$x(t) = A \sin t \;\; ; \text{ for } 0 < t < \pi$$

$$= 0 \quad\quad ; \text{ for } t > \pi$$

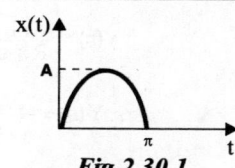

Fig 2.30.1.

$$X(s) = \mathcal{L}\{x(t)\} = \int_{-\infty}^{\infty} x(t) e^{-st} \, dt = \int_0^{\pi} A \sin t \, e^{-st} \, dt$$

$$= A \int_0^{\pi} \sin t \, e^{-st} \, dt = A \int_0^{\pi} \frac{e^{jt} - e^{-jt}}{2j} e^{-st} \, dt$$

$$\boxed{\sin\theta = \frac{e^{j\theta} - e^{-j\theta}}{2j}}$$

$$= \frac{A}{2j} \int_0^{\pi} \left[e^{-(s-j)t} - e^{-(s+j)t} \right] dt = \frac{A}{2j} \left[\frac{e^{-(s-j)t}}{-(s-j)} - \frac{e^{-(s+j)t}}{-(s+j)} \right]_0^{\pi}$$

$$= \frac{A}{2j} \left[-\frac{e^{-(s-j)\pi}}{(s-j)} + \frac{e^{-(s+j)\pi}}{(s+j)} + \frac{1}{s-j} - \frac{1}{s+j} \right]$$

$$= \frac{A}{2j} \left[-\frac{e^{-s\pi} e^{j\pi}}{s-j} + \frac{e^{-s\pi} e^{-j\pi}}{s+j} + \frac{1}{s-j} - \frac{1}{s+j} \right]$$

$$\boxed{\begin{aligned} e^{\pm j\pi} &= \cos\pi \pm j\sin\pi \\ &= -1 \pm j0 = -1 \end{aligned}}$$

$$= \frac{A}{2j} \left[\frac{e^{-s\pi}}{s-j} - \frac{e^{-s\pi}}{s+j} + \frac{1}{s-j} - \frac{1}{s+j} \right]$$

$$= \frac{A}{2j} \left[e^{-s\pi} \left(\frac{s+j-(s-j)}{(s-j)(s+j)} \right) + \frac{s+j-(s-j)}{(s-j)(s+j)} \right]$$

$$= \frac{A}{2j} \left[e^{-s\pi} \frac{2j}{s^2+1} + \frac{2j}{s^2+1} \right]$$

$$= \frac{A}{s^2+1} \left(e^{-s\pi} + 1 \right)$$

Example 2.31
(AU Jun'11, 8 Marks)

Determine the step response of the following circuit using Laplace transform.

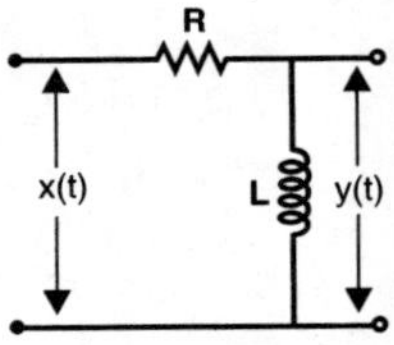

Solution:

Here the input $x(t)$ is the step input, $s(t) = A$.

On taking Laplace transform of the step input signal,

$$\mathcal{L}\{s(t)\} = \mathcal{L}\{A\} = \frac{A}{s}$$

The given circuit is represented in s-domain form as shown in Fig 1.

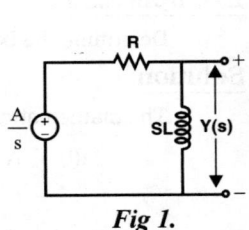

Fig 2.31.1.

By applying voltage division rule we get,

$$Y(s) = \frac{sL}{R+sL} \times \frac{A}{s} = \frac{AL}{L\left(\dfrac{R}{L}+s\right)}$$

$$\therefore Y(s) = \frac{A}{s+\dfrac{R}{L}}$$

On taking inverse Laplace transform of the above equation we get,

$$y(t) = A \, e^{-\frac{R}{L}t}$$

Fig 1.

Table 2.8: Laplace Transform of Some Standard Signals

Waveform	x(t)	$X(s) = \mathcal{L}\{x(t)\}$
	$\begin{aligned} x(t) &= A \quad; 0 < t < T \\ &= 0 \quad; \; t > T \end{aligned}$	$X(s) = \dfrac{A}{s}(1 - e^{-sT})$
	$\begin{aligned} x(t) &= \dfrac{At}{T} \quad; 0 < t < T \\ &= 0 \quad; \; t > T \end{aligned}$	$X(s) = \dfrac{A}{Ts^2}[1 - e^{-sT}(1 + sT)]$
	$\begin{aligned} x(t) &= \dfrac{2At}{T} \quad\quad\; ; 0 < t < \dfrac{T}{2} \\ &= 2A - \dfrac{2At}{T} \; ; \dfrac{T}{2} < 0 < T \end{aligned}$	$X(s) = \dfrac{2A}{Ts^2}\left(1 - e^{\frac{-sT}{2}}\right)^2$
	$\begin{aligned} x(t) &= A \quad; 0 < t < \dfrac{T}{2} \\ &= -A \; ; \dfrac{T}{2} < t < T \end{aligned}$	$X(s) = \dfrac{A}{s}\left(1 - e^{\frac{-sT}{2}}\right)^2$
	$\begin{aligned} x(t) &= A \sin t \; ; \; 0 < t < T \\ &= 0 \quad\quad\; ; \; t > T \end{aligned}$	$X(s) = \dfrac{A}{s^2 + 1}[1 - e^{-sT}(\cos T + s \sin T)]$
	$\begin{aligned} x(t) &= 1 \; ; \; 0 < t < 1 \\ &= -2 \; ; \; 1 < t < 2 \\ &= 2 \; ; \; 2 < t < 4 \\ &= -2 \; ; \; 4 < t < 5 \\ &= 0 \; ; \; t > 5 \end{aligned}$	$X(s) = \dfrac{1}{s}(1 - 3e^{-s} + 4e^{-2s} - 4e^{-4s} + 2e^{-5s})$

Table 2.8: Continued.....

Waveform	$x(t)$	$X(s)$
	$x(t) = A\,\sin t\,;\ 0 < t < T$ and $x(t+nT) = x(t)$	$X(s) = \dfrac{A}{s^2+1}\left(\dfrac{1-e^{-sT}(\cos T + s\sin T)}{1-e^{-sT}}\right)$
	$x(t) = A\,\sin t\,;\ 0 < t < \dfrac{T}{2}$ $= 0\qquad;\ \dfrac{T}{2} < t < T$ and $x(t+nT) = x(t)$	$X(s) = \dfrac{A\left[1 - e^{-\frac{sT}{2}}\left(\cos\dfrac{T}{2} + s\sin\dfrac{T}{2}\right)\right]}{(s^2+1)\left(1 - e^{-\frac{sT}{2}}\right)}$
	$x(t) = \dfrac{2At}{T}\qquad;\ 0 < t < \dfrac{T}{2}$ $= A - \dfrac{2At}{T}\ ;\ \dfrac{T}{2} < t < T$ and $x(t+nT) = x(t)$	$X(s) = \dfrac{2A\left[1 - \left(1 + \dfrac{Ts}{2}\right)e^{\frac{-sT}{2}}\right]}{Ts^2\left(1 - e^{\frac{-sT}{2}}\right)}$
	$x(t) = A - \dfrac{2At}{T}\,;\ 0 < t < T$ and $x(t+nT) = x(t)$	$X(s) = \dfrac{2A}{Ts}\left(\dfrac{T}{2}\,\dfrac{1+e^{-sT}}{1-e^{-sT}} - \dfrac{1}{s}\right)$
	$x(t) = A\ ;\ 0 < t < a$ $= 0\ ;\ a < t < T$ and $x(t+nT) = x(t)$	$X(s) = \dfrac{A}{s}\,\dfrac{1-e^{-as}}{1-e^{-sT}}$
	$x(t) = A\quad;\ 0 < t < \dfrac{T}{2}$ $= -A\ ;\ \dfrac{T}{2} < t < T$ and $x(t+nT) = x(t)$	$X(s) = \dfrac{A}{s}\left(\dfrac{1 - e^{\frac{-sT}{2}}}{1 + e^{\frac{-sT}{2}}}\right)$
	$x(t) = \dfrac{At}{T}\quad;\ 0 < t < T$ and $x(t+nT) = x(t)$	$X(s) = \dfrac{A}{Ts^2}\left[\dfrac{1 - e^{-sT}(1+sT)}{1 - e^{-sT}}\right]$

Table 2.9: Some Standard Laplace Transform Pairs

x(t)	X(s)	ROC
$\delta(t)$	1	Entire s-plane
$u(t)$	$\dfrac{1}{s}$	$\sigma > 0$
$t\,u(t)$	$\dfrac{1}{s^2}$	$\sigma > 0$
$\dfrac{t^{m-1}}{(m-1)!}u(t)$ where, $m = 1, 2, 3, \ldots$	$\dfrac{1}{s^m}$	$\sigma > 0$
$e^{-at}\,u(t)$	$\dfrac{1}{s+a}$	$\sigma > -a$
$-e^{-at}\,u(-t)$	$\dfrac{1}{s+a}$	$\sigma < -a$
$t^m\,u(t)$ where, $m = 1, 2, 3, \ldots$	$\dfrac{m!}{s^{m+1}}$	$\sigma > 0$
$t\,e^{-at}\,u(t)$	$\dfrac{1}{(s+a)^2}$	$\sigma > -a$
$\dfrac{1}{(m-1)!}\,t^{m-1}\,e^{-at}\,u(t)$ where, $m = 1, 2, 3, \ldots$	$\dfrac{1}{(s+a)^m}$	$\sigma > -a$
$t^m\,e^{-at}\,u(t)$ where, $m = 1, 2, 3, \ldots$	$\dfrac{m!}{(s+a)^{m+1}}$	$\sigma > -a$
$\sin\Omega_0 t\,u(t)$	$\dfrac{\Omega_0}{s^2 + \Omega_0^2}$	$\sigma > 0$
$\cos\Omega_0 t\,u(t)$	$\dfrac{s}{s^2 + \Omega_0^2}$	$\sigma > 0$
$\sinh\Omega_0 t\,u(t)$	$\dfrac{\Omega_0}{s^2 - \Omega_0^2}$	$\sigma > \Omega_0$
$\cosh\Omega_0 t\,u(t)$	$\dfrac{s}{s^2 - \Omega_0^2}$	$\sigma > \Omega_0$
$e^{-at}\sin\Omega_0 t\,u(t)$	$\dfrac{\Omega_0}{(s+a)^2 + \Omega_0^2}$	$\sigma > -a$
$e^{-at}\cos\Omega_0 t\,u(t)$	$\dfrac{s+a}{(s+a)^2 + \Omega_0^2}$	$\sigma > -a$

Example 2.32

Determine the poles and zeros of the rational function of s given below. Also sketch the pole-zero plot.

$$X(s) = \frac{(s+1)(s^2+10s+41)}{(s+4)(s^2+4s+13)}$$

Solution:

In the given function, both the numerator and denominator polynomials are third order polynomials. Hence the given function has three zeros and three ploes.

Let z_1, z_2, z_3, be zeros and p_1, p_2, p_3, be poles of X(s).

To Determine Zeros

Consider the numerator polynomial.

$(s + 1) (s^2 + 10s + 41) = 0$

The roots of quadratic $s^2 + 10s + 41 = 0$ are,

$$s = \frac{-10 \pm \sqrt{10^2 - 4 \times 41}}{2} = \frac{-10 \pm \sqrt{-64}}{2} = \frac{-10}{2} \pm \frac{j\sqrt{64}}{2} = -5+j4, \ -5-j4$$

$\therefore (s + 1) (s^2 + 10s + 41) = (s + 1) (s + 5 - j4) (s + 5 + j4) = 0$

$\therefore$ The roots of numerator polynomial are,

$s = -1 \ ; \ s = -5 + j4 \ ; \ s = -5 - j4$

$\therefore$ Zeros of X(s) are,

$z_1 = -1 \ ; \ z_2 = -5 + j4 \ ; \ z_3 = -5 - j4$

To Determine Poles

Consider the denominator polynomial.

$(s + 4) (s^2 + 4s + 13) = 0$

The roots of quadratic $s^2 + 4s + 13 = 0$ are,

$$s = \frac{-4 \pm \sqrt{4^2 - 4 \times 13}}{2} = \frac{-4 \pm \sqrt{-36}}{2}$$

$$= \frac{-4}{2} \pm \frac{j\sqrt{36}}{2} = -2 \pm j3 = -2+j3, \ -2-j3$$

$\therefore (s + 4) (s^2 + 4s + 13) = (s + 4) (s + 2 - j3) (s + 2 + j3) = 0$

$\therefore$ The roots of denominator polynomial are,

$s = -4 \ ; \ s = -2 + j3 \ ; \ s = -2 - j3$

$\therefore$ Poles of X(s) are,

$p_1 = -4 \ ; \ p_2 = -2 + j3 \ ; \ p_3 = -2 - j3$

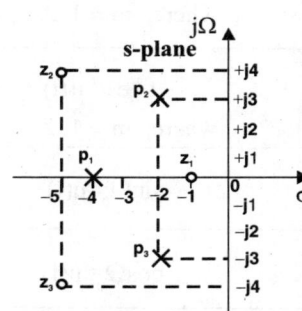

Fig 1: *Pole-zero plot.*

Pole-Zero Plot

The pole-zero plot of function X(s) is shown in Fig 1. The zeros are denoted by " **O** " and poles are denoted by "**x**".

2.9 Properties and Theorems of Laplace Transform

The properties and theorems of Laplace transform are listed in Table 2.7. The proof of properties and theorems are presented in this section.

1. Amplitude Scaling

In amplitude scaling, if the amplitude (or magnitude) of a time domain signal is multiplied by a constant A, then its Laplace transform is also multiplied by the same constant.

i.e. if $\mathcal{L}\{x(t)\} = X(s)$, then

$$\mathcal{L}\{A\,x(t)\} = A\,X(s)$$

Proof:

By definition of Laplace transform,

$$X(s) = \mathcal{L}\{x(t)\} = \int\limits_{-\infty}^{+\infty} x(t)\,e^{-st}\,dt \qquad\qquad .....(2.52)$$

$$\mathcal{L}\{A\,x(t)\} = \int\limits_{-\infty}^{+\infty} A\,x(t)\,e^{-st}\,dt$$

$$= A\int\limits_{-\infty}^{+\infty} x(t)\,e^{-st}\,dt$$

$$= A\,X(s) \qquad\qquad \boxed{\text{Using equation (2.52)}}$$

2. Linearity

The linearity property states that, Laplace transform of weighted sum of the two or more signals is equal to similar weighted sum of Laplace transform of the individual signals.

i.e., if $\mathcal{L}\{x_1(t)\} = X_1(s)$ and $\mathcal{L}\{x_2(t)\} = X_2(s)$, then

$$\mathcal{L}\{a_1\,x_1(t) + a_2\,x_2(t)\} = a_1\,X_1(s) + a_2\,X_2(s)$$

Proof:

By definition of Laplace transform,

$$X_1(s) = \mathcal{L}\{x_1(t)\} = \int\limits_{-\infty}^{+\infty} x_1(t)\,e^{-st}\,dt \qquad\qquad .....(2.53)$$

$$X_2(s) = \mathcal{L}\{x_2(t)\} = \int\limits_{-\infty}^{+\infty} x_2(t)\,e^{-st}\,dt \qquad\qquad .....(2.54)$$

$$\mathcal{L}\{a_1 x_1(t) + a_2 x_2(t)\} = \int\limits_{-\infty}^{+\infty} [a_1 x_1(t) + a_2 x_2(t)]\,e^{-st}\,dt$$

$$= a_1\int\limits_{-\infty}^{+\infty} x_1(t)\,e^{-st}\,dt + a_2\int\limits_{-\infty}^{+\infty} x_2(t)\,e^{-st}\,dt$$

$$= a_1 X_1(s) + a_2 X_2(s) \qquad \boxed{\begin{array}{l}\text{Using equations}\\ \text{(2.53) and (2.54)}\end{array}}$$

3. Time Differentiation

The time differentiation property states that if a causal signal x(t) is piecewise continuous, and Laplace transform of x(t) is X(s) then, Laplace transform of $\frac{d}{dt}$ x(t) is given by sX(s) – x(0).

i.e., If $\mathcal{L}\{x(t)\} = X(s)$, then

$$\mathcal{L}\left\{\frac{d}{dt} x(t)\right\} = sX(s) - x(0) \quad ; \quad \text{where, x(0) is value of x(t) at t = 0.}$$

Proof:

By definition of Laplace transform, the Laplace transform of a causal signal is given by,

$$X(s) = \mathcal{L}\{x(t)\} = \int_0^\infty x(t)\, e^{-st}\, dt \qquad .....(2.55)$$

$$\therefore \mathcal{L}\left\{\frac{d}{dt} x(t)\right\} = \int_0^\infty \frac{dx(t)}{dt}\, e^{-st}\, dt = \int_0^\infty e^{-st}\, \frac{dx(t)}{dt}\, dt$$

$$\boxed{\begin{array}{l} \int u\, dv = u\, v - \int v\, du \\ \hline u = e^{-st} \quad \Rightarrow\ du = -s\, e^{-st} \\ \hline dv = \frac{d}{dt}x(t) \Rightarrow\ v = x(t) \end{array}}$$

$$= [e^{-st} x(t)]_0^\infty - \int_0^\infty x(t)(-se^{-st})\, dt$$

$$\boxed{e^{-\infty} = 0 \text{ and } e^0 = 1}$$

$$= e^{-\infty} x(\infty) - e^0 x(0) + s \int_0^\infty x(t)\, e^{-st}\, dt$$

$$\boxed{\text{Using equation (2.55)}}$$

$$= s \int_0^\infty x(t)\, e^{-st}\, dt - x(0) = s\, X(s) - x(0)$$

4. Time Integration

The time integration property states that, if a causal signal x(t) is continuous and Laplace transform of x(t) is X(s), then the Laplace transform of $\int x(t)$ is given by, $\dfrac{X(s)}{s} + \dfrac{\left[\int x(t)\, dt\right]_{t=0}}{s}$

i.e.   If $\mathcal{L}\{x(t)\} = X(s)$, then

$$\mathcal{L}\left\{\int x(t)\right\} = \frac{X(s)}{s} + \frac{\left[\int x(t)\, dt\right]_{t=0}}{s}$$

Proof:

By definition of Laplace transform, the Laplace transform of a causal signal is given by,

$$X(s) = \mathcal{L}\{x(t)\} = \int_0^\infty x(t)\, e^{-st}\, dt \qquad .....(2.56)$$

$$\therefore \mathcal{L}\left\{\int x(t)\, dt\right\} = \int_0^\infty \left[\int x(t)\, dt\right] e^{-st}\, dt$$

$$\boxed{\begin{array}{l} \int u\, dv = u\, v - \int v\, du \\ \hline u = \int x(t)\, dt \ \Rightarrow\ du = x(t)\, dt \\ \hline dv = e^{-st} \qquad\ \Rightarrow\ v = \frac{e^{-st}}{-s} \end{array}}$$

$$= \left[\left[\int x(t)\, dt\right]\frac{e^{-st}}{-s}\right]_0^\infty - \int_0^\infty \frac{e^{-st}}{-s} \times x(t)\, dt$$

$$\therefore \mathcal{L}\left\{\int x(t)\, dt\right\} = \left[\int x(t)\ dt\right]\Bigg|_{t=\infty} \frac{e^{-\infty}}{-s} - \left[\int x(t)\ dt\right]\Bigg|_{t=0} \frac{e^0}{-s} + \frac{1}{s}\int_0^\infty x(t)\, e^{-st}\ dt$$

$$\boxed{e^{-\infty} = 0 \text{ and } e^0 = 1}$$

$$= \frac{1}{s}\left[\int x(t)\ dt\right]\Bigg|_{t=0} + \frac{1}{s}\int_0^\infty x(t)\, e^{-st}\ dt$$

$$= \frac{X(s)}{s} + \frac{\left[\int x(t)\ dt\right]\Big|_{t=0}}{s}$$

$$\boxed{\text{Using equation (2.56)}}$$

5. Frequency Shifting

The frequency shifting property of Laplace transform says that,

If, $\mathcal{L}\{x(t)\} = X(s)$, then

$$\mathcal{L}\{e^{\pm at}\, x(t)\} = X(s \mp a) \qquad [\text{i.e. } \mathcal{L}\{e^{at}\, x(t)\} = X(s-a) \text{ and } \mathcal{L}\{e^{-at}\, x(t)\} = X(s+a)]$$

Proof:

By definition of Laplace transform,

$$x(s) = \mathcal{L}\{x(t)\} = \int_{-\infty}^{+\infty} x(t)\, e^{-st}\ dt \qquad \qquad .....(2.57)$$

$$\therefore \mathcal{L}\{e^{\pm at}\, x(t)\} = \int_{-\infty}^{+\infty} e^{\pm at}\, x(t)\, e^{-st}\ dt$$

$$= \int_{-\infty}^{+\infty} x(t)\, e^{-(s \mp at)}\ dt$$

$$= X(s \mp a)$$

The term $\displaystyle\int_{-\infty}^{+\infty} x(t)\, e^{-(s \mp a)t}\ dt$ is similar to the form of definition of Laplace transform [equation(2.57)] except that s is replaced by $(s \mp a)$.

$$\therefore \int_{-\infty}^{+\infty} x(t)\, e^{-(s \mp a)t}\ dt = X(s \mp a)$$

6. Time Shifting

(AU Jun'12, 8 Marks)

The time shifting property of Laplace transform says that,

If, $\mathcal{L}\{x(t)\} = X(s)$, then

$$\mathcal{L}\{x(t \pm a)\} = e^{\pm as}\, X(s) \qquad [\text{i.e. } \mathcal{L}\{x(t+a)\} = e^{as}\, X(s) \text{ and } \mathcal{L}\{x(t-a)\} = e^{-as}\, X(s)]$$

Proof:

By definition of Laplace transform,

$$x(s) = \mathcal{L}\{x(t)\} = \int_{-\infty}^{+\infty} x(t)\, e^{-st}\, dt \qquad \qquad .....(2.58)$$

$$\therefore \mathcal{L}\{x(t \pm a)\} = \int_{-\infty}^{+\infty} x(t \pm a)\, e^{-st}\, dt = \int_{-\infty}^{+\infty} x(\tau)\, e^{-s(\tau \mp a)}\, d\tau$$

$$= \int_{-\infty}^{+\infty} x(\tau)\, e^{-s\tau} \times e^{\pm as}\ d\tau = e^{\pm as} \int_{-\infty}^{+\infty} x(\tau)\, e^{-s\tau}\ d\tau$$

$$= e^{\pm as} \int_{-\infty}^{+\infty} x(t)\, e^{-st}\ dt = e^{\pm as}\, X(s)$$

Let, $t \pm a = \tau$
$\therefore t = \tau \mp a$
On differentiating
$dt = d\tau$

Since τ is a dummy variable for integration we can change t to τ.

Using equation (2.58).

7. Frequency Differentiation

The frequency differentiation property of Laplace transform says that,

i.e. If $\mathcal{L}\{x(t)\} = X(s)$, then

$$\mathcal{L}\{t\,x(t)\} = -\frac{d}{ds}X(s)$$

Proof:

By definition of Laplace transform,

$$X(s) = \mathcal{L}\{x(t)\} = \int_{-\infty}^{+\infty} x(t)\,e^{-st}\,dt$$

On differentiating the above equation with respect to s we get,

$$\frac{d}{ds}X(s) = \frac{d}{ds}\left(\int_{-\infty}^{+\infty} x(t)\,e^{-st}\,dt\right)$$

$$= \int_{-\infty}^{+\infty} x(t)\left(\frac{d}{ds}e^{-st}\right)dt = \int_{-\infty}^{+\infty} x(t)(-te^{-st})\,dt \qquad \boxed{\text{Interchanging the order of integration and differentiation}}$$

$$= \int_{-\infty}^{+\infty} (-t\,x(t))e^{-st}\,dt = \mathcal{L}\{-t\,x(t)\} = -\mathcal{L}\{t\,x(t)\}$$

$$\therefore \mathcal{L}\{t\,x(t)\} = -\frac{d}{ds}X(s)$$

8. Frequency Integration

The frequency integration property of Laplace transform says that,

i.e., If $\mathcal{L}\{x(t)\} = X(s)$, then

$$\mathcal{L}\left\{\frac{1}{t}x(t)\right\} = \int_{s}^{\infty} X(s)\,ds$$

Proof:

By definition of Laplace transform,

$$X(s) = \mathcal{L}\{x(t)\} = \int_{-\infty}^{+\infty} x(t)\,e^{-st}\,dt$$

On integrating the above equation with respect to s between limits s to ∞ we get,

$$\int_{s}^{\infty} X(s)\,ds = \int_{s}^{\infty}\left[\int_{-\infty}^{+\infty} x(t)\,e^{-st}\,dt\right]ds \qquad \boxed{\text{Interchanging the order of integrations}}$$

$$= \int_{-\infty}^{\infty} x(t)\left[\int_{s}^{+\infty} e^{-st}\,ds\right]dt = \int_{-\infty}^{+\infty} x(t)\left[\frac{e^{-st}}{-t}\right]_{s}^{\infty}dt = \int_{-\infty}^{+\infty} x(t)\left[\frac{e^{-\infty}}{-t} - \frac{e^{-st}}{-t}\right]dt$$

$$= \int_{-\infty}^{+\infty} x(t)\left[0 + \frac{e^{-st}}{t}\right]dt = \int_{-\infty}^{+\infty}\left[\frac{1}{t}x(t)\right]e^{-st}\,dt = \mathcal{L}\left\{\frac{1}{t}x(t)\right\}$$

$$\therefore \mathcal{L}\left\{\frac{1}{t}x(t)\right\} = \int_{s}^{\infty} X(s)\,ds$$

9. Time Scaling   *(AU Jun'13, 2 Marks & Jun'12, 8 Marks)*

The time scaling property of Laplace transform says that,

If $\mathcal{L}\{x(t)\} = X(s)$, then

$$\mathcal{L}\{x(at)\} = \frac{1}{|a|} X\left(\frac{s}{a}\right)$$

Proof:

By definition of Laplace transform,

$$X(s) = \mathcal{L}\{x(t)\} = \int_{-\infty}^{+\infty} x(t)\, e^{-st}\, dt \qquad\qquad .....(2.59)$$

$$\therefore \mathcal{L}\{x(at)\} = \int_{-\infty}^{+\infty} x(at)\, e^{-st}\, dt = \int_{-\infty}^{+\infty} x(\tau)\, e^{-s\left(\frac{\tau}{a}\right)} \frac{d\tau}{a}$$

> Put, $at = \tau$
> $$\therefore t = \frac{\tau}{a}$$
> On differentiating
> $$dt = \frac{d\tau}{a}$$

$$= \frac{1}{a} \int_{-\infty}^{+\infty} x(\tau)\, e^{-\left(\frac{s}{a}\right)\tau}\, d\tau = \frac{1}{a} X\left(\frac{s}{a}\right)$$

The above transform is applicable for positive values of "a".

If "a" happens to be negative it can be proved that,

$$\mathcal{L}\{x(at)\} = -\frac{1}{a} X\left(\frac{s}{a}\right)$$

Hence in general,

$$\mathcal{L}\{x(at)\} = \frac{1}{|a|} X\left(\frac{s}{a}\right) \text{ for both positive and negative}$$

values of "a".

> The term $\int_{-\infty}^{+\infty} x(\tau)\, e^{-\left(\frac{s}{a}\right)\tau}\, d\tau$ is similar
> to the form of definition of Laplace transform (equation(2.59)) except that s is replaced by $\left(\frac{s}{a}\right)$.
> $$\therefore \int_{-\infty}^{+\infty} x(\tau)\, e^{-\left(\frac{s}{a}\right)\tau}\, d\tau = X\left(\frac{s}{a}\right)$$

---

10. Periodicity

The periodicity property of Laplace transform says that,

If $x(t) = x(t + nT)$, and $x_1(t)$ be one period of $x(t)$, and $\mathcal{L}\{x_1(t)\} = \int_{0}^{T} x_1(t)\, e^{-st}\, dt$, then

$$\mathcal{L}\{x(t+nT)\} = \frac{1}{1-e^{-sT}} \int_{0}^{T} x_1(t)\, e^{-st}\, dt$$

Proof:

By definition of Laplace transform,

$$\mathcal{L}\{x(t+nT)\} = \int_{0}^{\infty} x(t+nT)\, e^{-st}\, dt$$

$$= \int_{0}^{T} x_1(t)\, e^{-st}\, dt + \int_{T}^{2T} x_1(t-T)\, e^{-s(t+T)}\, dt + \int_{2T}^{3T} x_1(t-2T)\, e^{-s(t+2T)}\, dt + \cdots\cdots$$

$$\cdots\cdots + \int_{pT}^{(p+1)T} x_1(t-pT)\, e^{-s(t+pT)}\, dt + \cdots\cdots$$

$$\therefore\ \mathcal{L}\{x(t+nT)\} = \sum_{p=0}^{\infty} \int_{pT}^{(p+1)T} x_1(t-pT)\, e^{-s(t+pT)}\, dt$$

> The periodic signal will be identical in every period and so, $x_1(t + pT) = x_1(t)$.

$$= \sum_{p=0}^{\infty} \int_{0}^{T} x_1(t)\, e^{-st}\, e^{-psT}\, dt$$

> Interchanging the order of integration and summation

$$= \int_{0}^{T} x_1(t)\, e^{-st} \left(\sum_{p=0}^{\infty} e^{-psT} \right) dt$$

> Using infinite geometric series sum formula
> $$\sum_{n=0}^{\infty} C^n = \frac{1}{1-C}$$

$$= \int_{0}^{T} x_1(t)\, e^{-st} \left(\sum_{p=0}^{\infty} e^{-sT} \right)^{p} dt$$

$$= \int_{0}^{T} x_1(t)\, e^{-st} \left(\frac{1}{1-e^{-sT}} \right) dt$$

> The term $\dfrac{1}{1-e^{-sT}}$ is independent of t

$$= \left(\frac{1}{1-e^{-sT}} \right) \int_{0}^{T} x_1(t)\, e^{-st}\, dt$$

11. Initial Value Theorem   (AU Jun'11, 2 Marks)

The initial value theorem states that, if $x(t)$ and its derivative are Laplace transformable then, $\underset{t \to 0}{\text{Lt}}\ x(t) = \underset{s \to \infty}{\text{Lt}}\ s\, X(s)$

i.e. Initial value of signal, $x(0) = \underset{t \to 0}{\text{Lt}}\ x(t) = \underset{s \to \infty}{\text{Lt}}\ s\, X(s)$

Proof:

We know that, $\mathcal{L}\left\{ \dfrac{dx(t)}{dt} \right\} = s\, X(s) - x(0)$

On taking limit $s \to \infty$ on both sides of the above equation we get,

$$\underset{s \to \infty}{\text{Lt}}\ \mathcal{L}\left\{ \frac{dx(t)}{dt} \right\} = \underset{s \to \infty}{\text{Lt}}\ [s\, X(s) - x(0)]$$

> By definition of Laplace transform,
> $$\mathcal{L}\left\{ \frac{dx(t)}{dt} \right\} = \int_{0}^{\infty} \frac{dx(t)}{dt}\, e^{-st}\, dt$$

$$\underset{s \to \infty}{\text{Lt}} \int_{0}^{\infty} \frac{dx(t)}{dt}\, e^{-st}\, dt = \underset{s \to \infty}{\text{Lt}}\ [s\, X(s) - x(0)]$$

$$\int_{0}^{\infty} \frac{dx(t)}{dt} \left(\underset{s \to \alpha}{\text{Lt}}\ e^{-st} \right) dt = \left(\underset{s \to \alpha}{\text{Lt}}\ s\, X(s) \right) - x(0)$$

> Here $\dfrac{dx(t)}{dt}$ and $x(0)$ are not functions of s

$$0 = \underset{s \to \infty}{\text{Lt}}\ s\, X(s) - x(0)$$

> $\underset{s \to \infty}{\text{Lt}}\ e^{-st} = 0$

$$\therefore\ x(0) = \underset{s \to \infty}{\text{Lt}}\ s\, X(s)$$

> $x(0) = \underset{t \to 0}{\text{Lt}}\ x(t)$

$$\therefore\ \underset{t \to 0}{\text{Lt}}\ x(t) = \underset{s \to \infty}{\text{Lt}}\ s\, X(s)$$

12. Final Value Theorem   **(AU Jun'11, 2 Marks)**

The final value theorem states that if x(t) and its derivative are Laplace transformable then

$$\underset{t \to \infty}{Lt}\, x(t) = \underset{s \to 0}{Lt}\, s\, X(s)$$

i.e. Final value of signal, $x(\infty) = \underset{t \to \infty}{Lt}\, x(t) = \underset{s \to 0}{Lt}\, s\, X(s)$

Proof:

We know that, $\mathcal{L}\left\{\dfrac{dx(t)}{dt}\right\} = s\, X(s) - x(0)$

On taking limit s → 0 on both sides of the above equation we get,

$$\underset{s \to 0}{Lt}\, \mathcal{L}\left\{\dfrac{dx(t)}{dt}\right\} = \underset{s \to 0}{Lt}\, [s\, X(s) - x(0)]$$

$$\underset{s \to 0}{Lt}\, \int_0^\infty \dfrac{dx(t)}{dt}\, e^{-st}\, dt = \underset{s \to 0}{Lt}\, [s\, X(s) - x(0)]$$

$$\int_0^\infty \dfrac{dx(t)}{dt}\left(\underset{s \to 0}{Lt}\, e^{-st}\right) dt = \left(\underset{s \to 0}{Lt}\, s\, X(s)\right) - x(0)$$

$$\int_0^\infty \dfrac{dx(t)}{dt}\, dt = \underset{s \to 0}{Lt}\, s\, X(s) - x(0)$$

$$[x(t)]_0^\infty = \underset{s \to 0}{Lt}\, s\, X(s) - x(0)$$

$$x(\infty) - x(0) = \underset{s \to 0}{Lt}\, s\, X(s) - x(0)$$

$$\therefore\ x(\infty) = \underset{s \to 0}{Lt}\, s\, X(s)$$

$$\therefore\ \underset{t \to \infty}{Lt}\, x(t) = \underset{s \to 0}{Lt}\, s\, X(s)$$

> By definition of Laplace transform,
> $$\mathcal{L}\left\{\dfrac{dx(t)}{dt}\right\} = \int_0^\infty \dfrac{dx(t)}{dt}\, e^{-st}\, dt$$
>
> Here $\dfrac{dx(t)}{dt}$ and x(0) are not functions of s
>
> $$\underset{s \to 0}{Lt}\, e^{-st} = 1$$
>
> $$x(\infty) = \underset{t \to \infty}{Lt}\, x(t)$$

---

13. Convolution Theorem

The convolution theorem of Laplace transform says that, Laplace transform of convolution of two signals is given by the product of the Laplace transform of the individual signals.

i.e. if $\mathcal{L}\{x_1(t)\} = X_1(s)$ and $\mathcal{L}\{x_2(t)\} = X_2(s)$ then,

$$\mathcal{L}\{x_1(t) * x_2(t)\} = X_1(s)\, X_2(s) \qquad \qquad \dots\dots(2.60)$$

The equation (2.60) is also known as convolution property of Laplace transform.

According to the property of convolution,

$$x_1(t) * x_2(t) = \int_{-\infty}^{+\infty} x_1(\lambda)\, x_2(t - \lambda)\, d\lambda \qquad \qquad \dots\dots(2.61)$$

where, λ is a dummy variable used for integration.

Proof:

Let $x_1(t)$ and $x_2(t)$ be two time domain signals.

By definition of Laplace transform,

$$X_1(s) = \mathcal{L}\{x_1(t)\} = \int_{-\infty}^{+\infty} x_1(t)\, e^{-st}\, dt \qquad\qquad .....(2.62)$$

$$X_2(s) = \mathcal{L}\{x_2(t)\} = \int_{-\infty}^{+\infty} x_2(t)\, e^{-st}\, dt \qquad\qquad .....(2.63)$$

Now by definition of Laplace transform,

$$\mathcal{L}\{x_1(t) * x_2(t)\} = \int_{-\infty}^{+\infty} [x_1(t) * x_2(t)]\, e^{-st}\, dt \qquad\qquad .....(2.64)$$

$$= \int_{t=-\infty}^{t=+\infty} \left[\int_{\lambda=-\infty}^{\lambda=+\infty} [x_1(\lambda)\, x_2(t-\lambda)\, d\lambda] \right] e^{-st}\, dt \qquad \boxed{\text{Using equation (2.61)}}$$

$$= \int_{t=-\infty}^{t=+\infty} \int_{\lambda=-\infty}^{\lambda=+\infty} x_1(\lambda)\, x_2(t-\lambda)\, d\lambda\ e^{s\lambda}\, e^{-s\lambda} e^{-st}\, dt \qquad \boxed{e^{s\lambda}\, e^{-s\lambda} = e^0 = 1}$$

$$= \int_{t=-\infty}^{t=+\infty} \int_{\lambda=-\infty}^{\lambda=+\infty} x_1(\lambda)\, x_2(t-\lambda)\, e^{-s\lambda} e^{-s(t-\lambda)}\, d\lambda\, dt$$

$\boxed{\begin{array}{l} \text{Let,}\quad M = t-\lambda \\ \therefore\ dM = dt \\ \text{when } t=-\infty,\, M=-\infty-\lambda=-\infty \\ \text{when } t=+\infty,\, M=+\infty-\lambda=+\infty \end{array}}$

$$= \int_{M=-\infty}^{M=+\infty} \int_{\lambda=-\infty}^{\lambda=+\infty} x_1(\lambda)\, x_2(M)\, e^{-s\lambda}\, e^{-sM}\, d\lambda\, dM$$

$$= \int_{\lambda=-\infty}^{\lambda=+\infty} x_1(\lambda)\ e^{-s\lambda}\, d\lambda \times \int_{M=-\infty}^{M=+\infty} x_2(M)\, e^{-sM}\, dM$$

$\boxed{\begin{array}{l} \text{Here, } \lambda \text{ and } M \text{ are dummy} \\ \text{variables used for integration,} \\ \text{and so they can be changed to t.} \end{array}}$

$$= \int_{-\infty}^{+\infty} x_1(t)\, e^{-st}\, dt \times \int_{-\infty}^{+\infty} x_2(t)\, e^{-st}\, dt$$

$$= X_1(s)\, X_2(s) \qquad\qquad \boxed{\begin{array}{l} \text{Using equations} \\ \text{(2.62) and (2.63)} \end{array}}$$

$$\therefore\ \mathcal{L}\{x_1(t) * x_2(t)\} = X_1(s)\, X_2(s)$$

Table 2.10: Properties of Laplace Transform

Note: $\mathcal{L}\{x(t)\} = X(s); \quad \mathcal{L}\{x_1(t)\} = X_1(s); \quad \mathcal{L}\{x_2(t)\} = X_2(s)$

Property	Time Domain Signal	s-domain Signal
Amplitude scaling	$A\, x(t)$	$A\, X(s)$
Linearity	$a_1 x_1(t) \pm a_2 x_2(t)$	$a_1 X_1(s) \pm a_2 X_2(s)$
Time differentiation	$\dfrac{d}{dt} x(t)$	$s\, X(s) - x(0)$
	$\dfrac{d^m}{dt^m} x(t)$ where $m = 1, 2, 3 \dots$	$s^m X(s) - \displaystyle\sum_{K=1}^{m} s^{m-K}\, \dfrac{d^{(K-1)} x(t)}{dt^{K-1}}\Bigg|_{t=0}$
Time integration	$\displaystyle\int x(t)\, dt$	$\dfrac{X(s)}{s} + \dfrac{\left[\int x(t)\, dt\right]\big|_{t=0}}{s}$
	$\displaystyle\int \dots \int x(t)\, (dt)^m$ where $m = 1, 2, 3 \dots$	$\dfrac{X(s)}{s^m} + \displaystyle\sum_{K=1}^{m} \dfrac{1}{s^{m-K+1}} \left[\int \dots \int x(t)\, (dt)^k\right]\Bigg|_{t=0}$
Frequency shifting	$e^{\pm at} x(t)$	$X(s \mp a)$
Time shifting	$x(t \pm a)$	$e^{\pm as} X(s)$
Frequency differentiation	$t\, x(t)$	$-\dfrac{dX(s)}{ds}$
	$t^m x(t)$ where $m = 1, 2, 3 \dots$	$(-1)^m \dfrac{d^m}{ds^m} X(s)$
Frequency integration	$\dfrac{1}{t} x(t)$	$\displaystyle\int_s^{\infty} X(s)\, ds$
Time scaling	$x(at)$	$\dfrac{1}{|a|} X\!\left(\dfrac{s}{a}\right)$
Periodicity	$x(t + mT)$ where $m = 1, 2, 3 \dots$ $T = \text{Period}$	$\dfrac{1}{1 - e^{-sT}} \displaystyle\int_0^{T} x_1(t)\, e^{-st}\, dt$ where, $x_1(t)$ is one period of $x(t)$
Initial value theorem	$\underset{t \to 0}{\text{Lt}}\; x(t) = x(0)$	$\underset{s \to \infty}{\text{Lt}}\; s\, X(s)$
Final value theorem	$\underset{t \to \infty}{\text{Lt}}\; x(t) = x(\infty)$	$\underset{s \to 0}{\text{Lt}}\; s\, X(s)$
Convolution theorem	$x_1(t) * x_2(t)$ $= \displaystyle\int_{-\infty}^{+\infty} x_1(\lambda)\, x_2(t - \lambda)\, d\lambda$	$X_1(s)\, X_2(s)$

Example 2.33

Determine Laplace transform of periodic square wave shown in Fig 2.33.1.

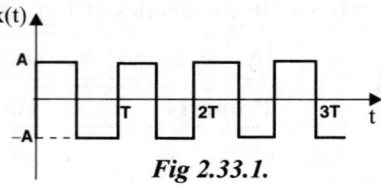

Fig 2.33.1.

Solution:

The given waveform satisfy the condition, $x(t + nT) = x(t)$, and so it is periodic.

Let $x_1(t)$ be one period of $x(t)$. The equation for one period of the periodic waveform of Fig 2.33.1 is,

$$x_1(t) = A \; ; \; \text{for } t = 0 \text{ to } \frac{T}{2}$$

$$= -A \; ; \; \text{for } t = \frac{T}{2} \text{ to } T$$

From periodicity property of Laplace transform,

If $X(s) = \mathcal{L}\{x(t)\}$, and if $x(t) = x(t + nT)$

Then, $X(s) = \dfrac{1}{1 - e^{-st}} \displaystyle\int_0^T x_1(t)\, e^{-st}\, dt$

where $x_1(t)$ is one period of $x(t)$.

$$\therefore \; X(s) = \mathcal{L}\{x(t)\} = \frac{1}{1 - e^{-st}} \int_0^T x_1(t)\, e^{-st}\, dt$$

Using the result of example 2.29(a), the above equation can be written as,

$$X(s) = \frac{1}{1 - e^{-sT}} \left[\frac{A}{s}\left(1 - e^{-\frac{sT}{2}}\right)^2 \right]$$

$$= \frac{1}{\left(1 + e^{-\frac{sT}{2}}\right)\left(1 + e^{-\frac{sT}{2}}\right)} \left[\frac{A}{s}\left(1 - e^{-\frac{sT}{2}}\right)^2 \right]$$

$$= \frac{A}{s} \left[\frac{1 - e^{-\frac{sT}{2}}}{1 + e^{-\frac{sT}{2}}} \right]$$

> From example 2.29 (a) we get,
>
> $$\int_0^T x_1(t)\, e^{-st}\, dt = \frac{A}{s}\left(1 - e^{-\frac{sT}{2}}\right)^2$$

> $a^2 - b^2 = (a + b)(a - b)$

Example 2.34

Determine the Laplace transform of the periodic signal whose waveform is shown in Fig 2.34.1.

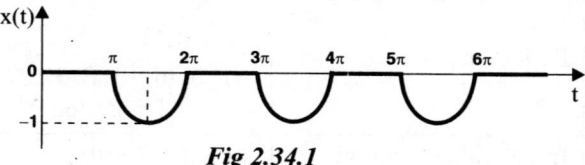

Fig 2.34.1

Solution:

The given waveform satisfies the condition, $x(t + nT) = x(t)$, and so it is periodic. Here, $T = 2\pi$.

Let $x_1(t)$ be one period of $x(t)$. The equation for one period of the periodic waveform of Fig 2.34.1 is,

$$x_1(t) = 0 \qquad ; \quad \text{for } t = 0 \text{ to } \pi$$

$$= \sin t \; ; \quad \text{for } t = \pi \text{ to } 2\pi$$

From periodicity property of Laplace transform,

If $X(s) = \mathcal{L}\{x(t)\}$, and if $x(t) = x(t + nT)$. Then, $X(s) = \dfrac{1}{1-e^{-st}} \displaystyle\int_0^T x_1(t)\,e^{-st}\,dt$, where $x_1(t)$ is one period of $x(t)$.

$$\therefore X(s) = \mathcal{L}\{x(t)\} = \frac{1}{1-e^{-sT}} \int_0^T x_1(t)\,e^{-st}\,dt = \frac{1}{1-e^{-2\pi s}} \int_0^{2\pi} x_1(t)\,e^{-st}\,dt \qquad \boxed{\text{Put, } T = 2\pi}$$

$$\boxed{\sin\theta = \frac{e^{j\theta} - e^{-j\theta}}{2j}}$$

$$= \frac{1}{1-e^{-2\pi s}}\left[\int_0^\pi 0\times e^{-st}\,dt + \int_\pi^{2\pi} \sin t\times e^{-st}\,dt\right] = \frac{1}{1-e^{-2\pi s}}\int_\pi^{2\pi}\sin t\times e^{-st}\,dt$$

$$= \frac{1}{1-e^{-2\pi s}}\int_\pi^{2\pi}\frac{e^{jt}-e^{-jt}}{2j}\,e^{-st}\,dt = \frac{1}{2j\left(1-e^{-2\pi s}\right)}\int_\pi^{2\pi}\left[e^{-(s-j)t}-e^{-(s+j)t}\right]dt$$

$$= \frac{1}{2j\left(1-e^{-2\pi s}\right)}\left[\frac{e^{-(s-j)t}}{-(s-j)} - \frac{e^{-(s+j)t}}{-(s+j)}\right]_\pi^{2\pi} = \frac{1}{2j\left(1-e^{-2\pi s}\right)}\left[-\frac{e^{jt}e^{-st}}{s-j} + \frac{e^{-jt}e^{-st}}{s+j}\right]_\pi^{2\pi}$$

$$= \frac{1}{2j\left(1-e^{-2\pi s}\right)}\left[-\frac{e^{j2\pi}e^{-2\pi s}}{s-j} + \frac{e^{-j2\pi}e^{-2\pi s}}{s+j} + \frac{e^{j\pi}e^{-\pi s}}{s-j} - \frac{e^{-j\pi}e^{-\pi s}}{s+j}\right]$$

$$= \frac{1}{2j\left(1-e^{-2\pi s}\right)}\left[-\frac{e^{-2\pi s}}{s-j} + \frac{e^{-2\pi s}}{s+j} - \frac{e^{-\pi s}}{s-j} + \frac{e^{-\pi s}}{s+j}\right]$$

$$\boxed{e^{\pm j\pi} = -1 \text{ and } e^{\pm j2\pi} = 1}$$

$$= \frac{1}{2j\left(1-e^{-2\pi s}\right)}\left[e^{-2\pi s}\frac{-(s+j)+(s-j)}{(s-j)(s+j)} + e^{-\pi s}\frac{-(s+j)+(s-j)}{(s-j)(s+j)}\right]$$

$$\boxed{a^2 - b^2 = (a+b)\,(a-b)}$$

$$\boxed{j^2 = -1}$$

$$= \frac{1}{2j\left(1-e^{-2\pi s}\right)}\left[e^{-2\pi s}\frac{-2j}{s^2+1} + e^{-\pi s}\frac{-2j}{s^2+1}\right]$$

$$= \frac{1}{2j(1+e^{-\pi s})(1-e^{-\pi s})}\left[\frac{-2je^{-\pi s}(1+e^{-\pi s})}{s^2+1}\right] = \frac{-e^{-\pi s}}{(1-e^{-\pi s})(s^2+1)}$$

Example 2.35

Determine the Laplace transform of the following signals using properties of Laplace transform.

a) $x(t) = (t^2 - 2t)\,u(t - 1)$　　**b)** Unit ramp signal starting at $t = a$　　**c)** $x(t)$ shown in Fig 2.35.1

Solution:

a) Given that, $x(t) = (t^2 - 2t)\,u(t-1) = t^2\,u(t-1) - 2t\,u(t-1)$

From table 2.7 we get,

$$\mathcal{L}\{t^2\,u(t)\} = \frac{2}{s^3}$$

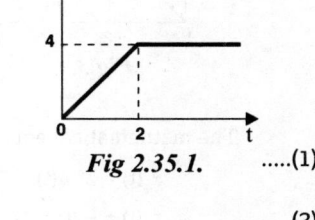

$x(t)$

4

0　　2　　　t

Fig 2.35.1.　　.....(1)

$$\mathcal{L}\{t\,u(t)\} = \frac{1}{s^2}\;;\;\therefore \mathcal{L}\{2t\,u(t)\} = \frac{2}{s^2} \qquad .....(2)$$

From time delay property of Laplace transform,

If $\mathcal{L}\{x(t)\,u(t)\} = X(s)$, then $\mathcal{L}\{x(t)\,u(t-a)\} = e^{-as}\,X(s)$

$$\therefore \; X(s) = \mathcal{L}\{x(t)\} = \mathcal{L}\{(t^2 - 2t)\, u(t-1)\} = \mathcal{L}\{t^2 u(t-1)\} - \mathcal{L}\{2t\, u(t-1)\}$$

$$= e^{-s}\, \mathcal{L}\{t^2 u(t)\} - e^{-s}\, \mathcal{L}\{2t\, u(t)\}$$

> Using time delay property

> Using equations (1) and (2)

$$= e^{-s}\frac{2}{s^3} - e^{-s}\frac{2}{s^2} = 2e^{-s}\left(\frac{1-s}{s^3}\right) = \frac{2e^{-s}(1-s)}{s^3}$$

---

b) Given that, Unit ramp signal starting at t = a.

The unit ramp starting at t = a, is unit ramp delayed by "a" units of time. The unit ramp waveform and the ramp waveform starting at t = a are shown in Fig 1 and Fig 2 respectively. The equation of unit ramp and delayed ramp are given below:

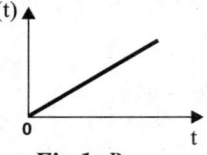

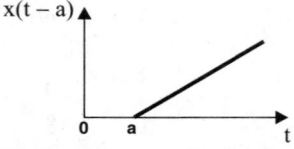

Fig 1: Ramp.   *Fig 2: Ramp starting at t = a.*

Unit ramp, x(t) = t u(t)

Delayed unit ramp, x(t − a) = (t − a) u(t − a)

From table 2.7 we get,

$$\mathcal{L}\{t\, u(t)\} = \frac{1}{s^2} \qquad\qquad .....(3)$$

From time delay property of Laplace transform,

If $\mathcal{L}\{x(t)\, u(t)\} = X(s)$, then $\mathcal{L}\{x(t)\, u(t-a)\} = e^{-as}\, X(s)$

Therefore Laplace transform of delayed ramp signal is,

$$\mathcal{L}\{x(t-a)\} = e^{-as}\, \mathcal{L}\{x(t)\}$$

$$= e^{-as}\, \mathcal{L}\{t\, u(t)\}$$

> Using time delay property

> Using equation (3)

$$= e^{-as}\frac{1}{s^2} = \frac{e^{-as}}{s^2}$$

---

c) The given signal can be decomposed into two signals as shown in Fig 4 and Fig 5.

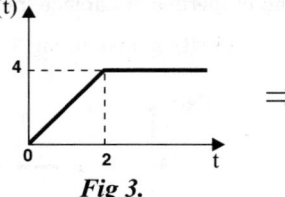

Fig 3.

$\Rightarrow$

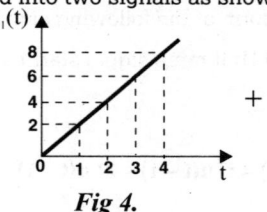

Fig 4.

$+$

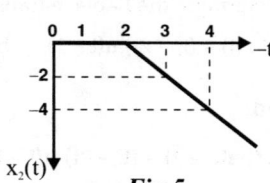

Fig 5.

The mathematical equation of signals, $x_1(t)$ and $x_2(t)$ are given below:

$$x_1(t) = 2t\, u(t)$$

$$x_2(t) = -2(t-2)\, u(t-2)$$

$$\therefore \; x(t) = x_1(t) + x_2(t) = 2t\, u(t) - 2(t-2)\, u(t-2)$$

From table 2.7 we get,

$$\mathcal{L}\{t\, u(t)\} = \frac{1}{s^2} \; ; \; \therefore \; \mathcal{L}\{2t\, u(t)\} = \frac{2}{s^2} \qquad\qquad .....(4)$$

$$\therefore \; X(s) = \mathcal{L}\{x(t)\} = \mathcal{L}\{x_1(t) + x_2(t)\} = \mathcal{L}\{2t\,u(t) - 2(t-2)u(t-2)\}$$

$$= \mathcal{L}\{2t\,u(t)\} - \mathcal{L}\{2(t-2)u(t-2)\}$$

$$= \mathcal{L}\{2t\,u(t)\} - e^{-2s}\,\mathcal{L}\{2t\,u(t)\}$$

Using time delay property

$$= \frac{2}{s^2} - e^{-2s}\frac{2}{s^2} = \frac{2(1 - e^{-2s})}{s^2}$$

Using equation (4)

Example 2.36

Let, $X(s) = \mathcal{L}\{x(t)\}$. Determine the initial value, $x(0)$ and the final value, $x(\infty)$, for the following signals using initial value and final value theorems.

a) $X(s) = \dfrac{1}{s(s-1)}$ **b)** $X(s) = \dfrac{s+1}{s^2 + 2s + 2}$ **c)** $X(s) = \dfrac{7s+6}{s(3s+5)}$

d) $X(s) = \dfrac{s^2 + 1}{s^2 + 6s + 5}$ **e)** $X(s) = \dfrac{s+5}{s^2(s+9)}$

Solution:

a) Given that, $X(s) = \dfrac{1}{s(s-1)}$

Initial value, $x(0) = \underset{s \to \infty}{\text{Lt}}\; s\,X(s) = \underset{s \to \infty}{\text{Lt}}\; s\dfrac{1}{s(s-1)} = \underset{s \to \infty}{\text{Lt}}\; s\dfrac{1}{s^2\left(1 - \dfrac{1}{s}\right)}$

$$= \underset{s \to \infty}{\text{Lt}}\; \frac{1}{s}\frac{1}{\left(1 - \dfrac{1}{s}\right)} = \frac{1}{\infty}\frac{1}{\left(1 - \dfrac{1}{\infty}\right)} = 0 \times \frac{1}{1-0} = 0$$

Final value, $x(\infty) = \underset{s \to 0}{\text{Lt}}\; s\,X(s) = \underset{s \to 0}{\text{Lt}}\; s\dfrac{1}{s(s-1)} = \underset{s \to 0}{\text{Lt}}\;\dfrac{1}{s-1} = \dfrac{1}{0-1} = -1$

b) Given that, $X(s) = \dfrac{s+1}{s^2 + 2s + 2}$

Initial value, $x(0) = \underset{s \to \infty}{\text{Lt}}\; s\,X(s) = \underset{s \to \infty}{\text{Lt}}\; s\dfrac{s+1}{s^2 + 2s + 2} = \underset{s \to \infty}{\text{Lt}}\; s\dfrac{s\left(1 + \dfrac{1}{s}\right)}{s^2\left(1 + \dfrac{2}{s} + \dfrac{2}{s^2}\right)}$

$$= \underset{s \to \infty}{\text{Lt}}\; \frac{1 + \dfrac{1}{s}}{1 + \dfrac{2}{s} + \dfrac{2}{s^2}} = \frac{1 + \dfrac{1}{\infty}}{1 + \dfrac{2}{\infty} + \dfrac{2}{\infty}} = \frac{1+0}{1+0+0} = 1$$

Final value, $x(\infty) = \underset{s \to 0}{\text{Lt}}\; s\,X(s) = \underset{s \to 0}{\text{Lt}}\; s\dfrac{s+1}{s^2 + 2s + 2}$

$$= 0 \times \frac{0+1}{0+0+2} = 0$$

c) Given that, $X(s) = \dfrac{7s+6}{s(3s+5)}$

Initial value, $x(0) = \underset{s \to \infty}{\text{Lt}}\; s\,X(s) = \underset{s \to \infty}{\text{Lt}}\; s\dfrac{7s+6}{s(3s+5)} = \underset{s \to \infty}{\text{Lt}}\; s\dfrac{s\left(7 + \dfrac{6}{s}\right)}{s^2\left(3 + \dfrac{5}{s}\right)}$

$$= \underset{s \to \infty}{\text{Lt}}\; \frac{7 + \dfrac{6}{s}}{3 + \dfrac{5}{s}} = \frac{7 + \dfrac{6}{\infty}}{3 + \dfrac{5}{\infty}} = \frac{7+0}{3+0} = \frac{7}{3}$$

Final value, $x(\infty) = \underset{s \to 0}{\text{Lt}}\ s\, X(s) = \underset{s \to 0}{\text{Lt}}\ s\,\dfrac{7s+6}{s(3s+5)}$

$$= \underset{s \to 0}{\text{Lt}}\ \dfrac{7s+6}{3s+5} = \dfrac{0+6}{0+5} = \dfrac{6}{5}$$

---

d) Given that, $X(s) = \dfrac{s^2+1}{s^2+6s+5}$

Initial value, $x(0) = \underset{s \to \infty}{\text{Lt}}\ s\, X(s) = \underset{s \to \infty}{\text{Lt}}\ s\,\dfrac{s^2+1}{s^2+6s+5)} = \underset{s \to \infty}{\text{Lt}}\ s\,\dfrac{s^2\left(1+\dfrac{1}{s^2}\right)}{s^2\left(1+\dfrac{6}{s}+\dfrac{5}{s^2}\right)}$

$$= \underset{s \to \infty}{\text{Lt}}\ s\ \dfrac{1+\dfrac{1}{s^2}}{1+\dfrac{6}{s}+\dfrac{1}{s^2}} = \infty \times \dfrac{1+\dfrac{1}{\infty}}{1+\dfrac{6}{\infty}+\dfrac{1}{\infty}} = \infty \times \dfrac{0+1}{0+0+5} = \infty$$

Final value, $x(\infty) = \underset{s \to 0}{\text{Lt}}\ s\, X(s) = \underset{s \to 0}{\text{Lt}}\ s\,\dfrac{s^2+1}{s^2+6s+5} = 0 \times \dfrac{0+1}{0+0+5} = 0$

---

e) Given that, $X(s) = \dfrac{s+5}{s^2(s+9)}$

Initial value, $x(0) = \underset{s \to \infty}{\text{Lt}}\ s\, X(s) = \underset{s \to \infty}{\text{Lt}}\ s\,\dfrac{s+5}{s^2(s+9)} = \underset{s \to \infty}{\text{Lt}}\ s\,\dfrac{s\left(1+\dfrac{5}{s}\right)}{s^3\left(1+\dfrac{9}{s}\right)}$

$$= \underset{s \to \infty}{\text{Lt}}\ \dfrac{1}{s}\ \dfrac{1+\dfrac{5}{s}}{1+\dfrac{9}{s}} = \dfrac{1}{\infty} \times \dfrac{1+\dfrac{1}{\infty}}{1+\dfrac{9}{\infty}} = 0 \times \dfrac{1+0}{1+0} = 0$$

Final value, $x(\infty) = \underset{s \to 0}{\text{Lt}}\ s\, X(s) = \underset{s \to 0}{\text{Lt}}\ s\,\dfrac{s+5}{s^2(s+9)} = \underset{s \to 0}{\text{Lt}}\ \dfrac{s+5}{s^2+9s} = \dfrac{0+5}{0+0} = \infty$

Example 2.37

Perform convolution of $x_1(t)$ and $x_2(t)$ using convolution theorem of Laplace transform.

a) $x_1(t) = u(t + 5)$, $x_2(t) = \delta(t - 7)$

b) $x_1(t) = u(t - 2)$, $x_2(t) = \delta(t + 6)$

c) $x_1(t) = u(t + 1)$, $x_2(t) = r(t - 2)$; where $r(t) = t\,u(t)$

Solution:

a) Given that, $x_1(t) = u(t + 5)$, $x_2(t) = \delta(t - 7)$

> $\mathcal{L}\{u(t)\} = \dfrac{1}{s}$ and $\mathcal{L}\{\delta(t)\} = 1$
>
> if $\mathcal{L}\{x(t)\} = X(s)$ then $\mathcal{L}\{x(t \pm a)\} = e^{\pm as}\,X(s)$

(AU Jun'12, 2 Marks)

$$\therefore\ X_1(s) = \mathcal{L}\{x_1(t)\} = \mathcal{L}\{u(t+5)\} = e^{5s}\,\mathcal{L}\{u(t)\} = e^{5s}\times\dfrac{1}{s} = \dfrac{e^{5s}}{s} \quad\quad .....(1)$$

$$\therefore\ X_2(s) = \mathcal{L}\{x_2(t)\} = \mathcal{L}\{\delta(t-7)\} = e^{-7s}\,\mathcal{L}\{\delta(t)\} = e^{-7s}\times 1 = e^{-7s} \quad\quad .....(2)$$

From convolution theorem of Laplace transform,

$$\mathcal{L}\{x_1(t) * x_2(t)\} = X_1(s)\,X_2(s)$$

$$= \dfrac{e^{5s}}{s} \times e^{-7s} = \dfrac{e^{5s-7s}}{s} = \dfrac{e^{-2s}}{s} \qquad \boxed{\text{Using equations (1) and (2)}}$$

$$\therefore\ x_1(t) * x_2(t) = \mathcal{L}^{-1}\left\{\dfrac{e^{-2s}}{s}\right\} = \mathcal{L}^{-1}\left\{\dfrac{1}{s}\right\}\Big|_{t=t-2} = u(t)|_{t=t-2} = u(t-2)$$

b) Given that, $x_1(t) = u(t-2)$, $x_2(t) = \delta(t+6)$

$$\boxed{\mathcal{L}\{u(t)\} = \frac{1}{s} \text{ and } \mathcal{L}\{\delta(t)\} = 1}$$
$$\boxed{\text{if } \mathcal{L}\{x(t)\} = X(s) \text{ then } \mathcal{L}\{x(t \pm a)\} = e^{\pm as} X(s)}$$

$$\therefore X_1(s) = \mathcal{L}\{x_1(t)\} = \mathcal{L}\{u(t-2)\}$$

$$= e^{-2s}\mathcal{L}\{u(t)\} = e^{-2s} \times \frac{1}{s} = \frac{e^{-2s}}{s} \qquad \qquad .....(1)$$

$$\therefore X_2(s) = \mathcal{L}\{x_2(t)\} = \mathcal{L}\{\delta(t+6)\} = e^{6s}\mathcal{L}\{\delta(t)\} = e^{6s} \times 1 = e^{6s} \qquad .....(2)$$

From convolution theorem of Laplace transform,

$$\mathcal{L}\{x_1(t) * x_2(t)\} = X_1(s) X_2(s)$$

$$= \frac{e^{-2s}}{s} \times e^{6s} = \frac{e^{-2s+6s}}{s} = \frac{e^{4s}}{s} \qquad \boxed{\text{Using equations (1) and (2)}}$$

$$\therefore x_1(t) * x_2(t) = \mathcal{L}^{-1}\left\{\frac{e^{4s}}{s}\right\} = \mathcal{L}^{-1}\left\{\frac{1}{s}\right\}\Big|_{t=t+4} = u(t)|_{t=t+4} = u(t+4)$$

c) Given that, $x_1(t) = u(t+1)$, $x_2(t) = r(t-2)$; where $r(t) = t\,u(t)$

$$\boxed{\mathcal{L}\{u(t)\} = \frac{1}{s} \text{ and } \mathcal{L}\{r(t)\} = \mathcal{L}\{t\,u(t)\} = \frac{1}{s^2}}$$
$$\boxed{\text{if } \mathcal{L}\{x(t)\} = X(s) \text{ then } \mathcal{L}\{x(t \pm a)\} = e^{\pm as} X(s)}$$

$$\therefore X_1(s) = \mathcal{L}\{x_1(t)\} = \mathcal{L}\{u(t+1)\}$$

$$= e^s \mathcal{L}\{u(t)\} = e^s \times \frac{1}{s} \qquad \qquad .....(1)$$

$$\therefore X_2(s) = \mathcal{L}\{x_2(t)\} = \mathcal{L}\{r(t-2)\} = e^{-2s}\mathcal{L}\{r(t)\} = e^{-2s}\mathcal{L}\{t\,u(t)\} = e^{-2s} \times \frac{1}{s^2} \qquad .....(2)$$

From convolution theorem of Laplace transform,

$$\mathcal{L}\{x_1(t) * x_2(t)\} = X_1(s) X_2(s)$$

$$= \frac{e^s}{s} \times \frac{e^{-2s}}{s^2} = \frac{e^{-s}}{s^3} \qquad \boxed{\text{Using equations (1) and (2)}}$$

$$\therefore x_1(t) * x_2(t) = \mathcal{L}^{-1}\left\{\frac{e^{-s}}{s^3}\right\} = \mathcal{L}^{-1}\left\{\frac{1}{s^3}\right\}\Big|_{t=t-1} = \left[\frac{t^2}{2} u(t)\right]\Big|_{t=t-1} \qquad \boxed{\mathcal{L}\left\{\frac{t^{m-1}}{(m-1)!} u(t)\right\} = \frac{1}{s^m}}$$

$$= \frac{(t-1)^2}{2} u(t-1)$$

Example 2.38

Perform convolution of $x_1(t)$ and $x_2(t)$ using convolution theorem of Laplace transform and sketch the resultant waveform, where, $x_1(t) = u(t) - u(t-1)$ and $x_2(t) = u(t) - u(t-2)$.

Solution:

$$x_1(t) = u(t) - u(t-1)$$

$$\therefore X_1(s) = \mathcal{L}\{x_1(t)\} = \mathcal{L}\{u(t) - u(t-1)\} = \mathcal{L}\{u(t)\} - \mathcal{L}\{u(t-1)\} = \mathcal{L}\{u(t)\} - e^{-s}\mathcal{L}\{u(t)\} = \frac{1}{s} - \frac{e^{-s}}{s} \qquad .....(1)$$

$$x_2(t) = u(t) - u(t-2)$$

$$\therefore X_2(s) = \mathcal{L}\{x_2(t)\} = \mathcal{L}\{u(t) - u(t-2)\} = \mathcal{L}\{u(t)\} - \mathcal{L}\{u(t-2)\} = \mathcal{L}\{u(t)\} - e^{-2s}\mathcal{L}\{u(t)\} = \frac{1}{s} - \frac{e^{-2s}}{s} \qquad .....(2)$$

From convolution theorem of Laplace transform,

$$\mathcal{L}\{x_1(t) * x_2(t)\} = X_1(s)\,X_2(s) = \left(\frac{1}{s} - \frac{e^{-s}}{s}\right)\left(\frac{1}{s} - \frac{e^{-2s}}{s}\right)$$

Using equations (1) and (2)

$$= \frac{1}{s^2} - \frac{e^{-2s}}{s^2} - \frac{e^{-s}}{s^2} + \frac{e^{-3s}}{s^2} = \frac{1}{s^2} - \frac{e^{-s}}{s^2} - \frac{e^{-2s}}{s^2} + \frac{e^{-3s}}{s^2}$$

$$\mathcal{L}\{t\,u(t)\} = \frac{1}{s^2}$$

$$\therefore \; x_1(t) * x_2(t) = \mathcal{L}^{-1}\left\{\frac{1}{s^2} - \frac{e^{-s}}{s^2} - \frac{e^{-2s}}{s^2} + \frac{e^{-3s}}{s^2}\right\}$$

$$\mathcal{L}\{(t-a)\,u(t-a)\} = \frac{e^{-as}}{s^2}$$

$$= \mathcal{L}^{-1}\left\{\frac{1}{s^2}\right\} - \mathcal{L}^{-1}\left\{\frac{e^{-s}}{s^2}\right\} - \mathcal{L}^{-1}\left\{\frac{e^{-2s}}{s^2}\right\} + \mathcal{L}^{-1}\left\{\frac{e^{-3s}}{s^2}\right\}$$

$$= t\,u(t) - (t-1)\,u(t-1) - (t-2)\,u(t-2) + (t-3)\,u(t-3)$$

To sketch the resultant waveform

For t = 0 to 1

When t = 0 to1, u(t) = 1, u(t – 1) = 0, u(t – 2) = 0, u(t – 3) = 0

$\therefore \; x_1(t) * x_2(t) = t \times 1 - (t-1) \times 0 - (t-2) \times 0 + (t-3) \times 0 = t$

Fig 1.

For t = 1 to 2

When t = 1 to 2, u(t) =1, u(t – 1) = 1, u(t – 2) = 0, u(t – 3) = 0

$\therefore \; x_1(t) * x_2(t) = t \times 1 - (t-1) \times 1 - (t-2) \times 0 + (t-3) \times 0 = 1$

For t = 2 to 3

When t = 2 to 3, u(t) =1, u(t – 1) = 1, u(t – 2) = 1, u(t – 3) = 0

$\therefore \; x_1(t) * x_2(t) = t \times 1 - (t-1) \times 1 - (t-2) \times 1 + (t-3) \times 0 = 3 - t$

For t > 3

When t > 3, u(t) = 1, u(t – 1) = 1, u(t – 2) = 1, u(t – 3) = 1

$\therefore \; x_1(t) * x_2(t) = t \times 1 - (t-1) \times 1 - (t-2) \times 1 + (t-3) \times 1 = 0$

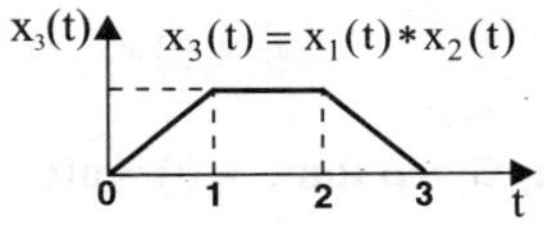

Using the calculated values, the resultant waveform is sketched as shown in Fig 1.

2.10 Inverse Laplace Transform

The *Inverse Laplace transform* of X(s) is defined as,

$$\mathcal{L}^{-1}\{X(s)\} = x(t) = \frac{1}{2\pi j} \int_{s=\sigma-j\Omega}^{s=\sigma+j\Omega} X(s)\,e^{st} \; ds$$

Performing inverse Laplace transform by using the above fundamental definition is tedious. But the inverse Laplace transform by partial fraction expansion method will be much easier. In this method the s-domain signal is expressed as a sum of first order and second order sections. Then the inverse Laplace transform is obtained by comparing each section with standard transform pair, (listed in table 2.8).

In the following section the inverse Laplace transform by partial fraction exapansion method is explained with example.

2.10.1 Inverse Laplace Transform by Partial Fraction Expansion Method

Let Laplace transform of x(t) be X(s). The s-domain signal X(s) will be a ratio of two polynomials in s (i.e., rational function of s). The roots of the denominator polynomial are called poles. The roots of numerator polynomials are called zeros. (For definition of poles and zeros please refer section 2.7.3). In signals and systems, three different types of s-domain signals are encountered. They are,

Case i : Signals with separate poles.

Case ii: Signals with multiple poles.

Case iii: Signals with complex conjugate poles.

The inverse Laplace transform by partial fraction expansion method of all the three cases are explained with an example.

Case i: **When s-Domain Signal X(s) has Distinct Poles**

Let, $X(s) = \dfrac{K}{s(s + p_1)(s + p_2)}$ .....(2.65)

By partial fraction expansion technique, the equation (2.65) can be expressed as,

$$X(s) = \frac{K}{s(s + p_1)(s + p_2)} = \frac{K_1}{s} + \frac{K_2}{s + p_1} + \frac{K_3}{s + p_2}$$ .....(2.66)

The residues K_1, K_2 and K_3 are given by,

$$K_1 = X(s) \times s\big|_{s = 0}$$

$$K_2 = X(s) \times (s + p_1)\big|_{s = -p_1}$$

$$K_3 = X(s) \times (s + p_2)\big|_{s = -p_2}$$

We know that, $\mathcal{L}\{u(t)\} = \dfrac{1}{s}$, $\mathcal{L}\{e^{-at}u(t)\} = \dfrac{1}{s + a}$

By using the above standard Laplace transform pairs the inverse Laplace transform of equation (2.66) can be obtained as shown below:

$$\therefore x(t) = \mathcal{L}^{-1}\{X(s)\} = \mathcal{L}^{-1}\left\{\frac{K_1}{s} + \frac{K_1}{s + p_1} + \frac{K_3}{s + p_2}\right\}$$

$$= K_1 \mathcal{L}^{-1}\left\{\frac{1}{s}\right\} + K_2 \mathcal{L}^{-1}\left\{\frac{1}{s + p_1}\right\} + K_3 \mathcal{L}^{-1}\left\{\frac{1}{s + p_2}\right\}$$

$$= K_1 u(t) + K_2 e^{-p_1 t} u(t) + K_3 e^{-p_2 t} u(t)$$

Example 2.39

Determine the inverse Laplace transform of $X(s) = \dfrac{2}{s(s+1)(s+2)}$.

Solution:

Given that, $X(s) = \dfrac{2}{s(s+1)(s+2)}$

By partial fraction expansion technique, X(s) can be expressed as,

$$X(s) = \frac{2}{s(s+1)(s+2)} = \frac{K_1}{s} + \frac{K_2}{s+1} + \frac{K_3}{s+2}$$

The residue K_1 is obtained by multiplying X(s) by s and letting s = 0.

$$K_1 = X(s) \times s\big|_{s=0} = \frac{2}{s(s+1)(s+2)} \times s\Big|_{s=0} = \frac{2}{(s+1)(s+2)}\Big|_{s=0} = \frac{2}{1 \times 2} = 1$$

The residue K_2 is obtained by multiplying X(s) by (s + 1) and letting s = –1.

$$K_2 = X(s) \times (s+1)\big|_{s=-1} = \frac{2}{s(s+1)(s+2)} \times (s+1)\Big|_{s=-1} = \frac{2}{s(s+2)}\Big|_{s=-1} = \frac{2}{-1(-1+2)} = -2$$

The residue K_3 is obtained by multiplying X(s) by (s + 2) and letting s = -2.

$$K_3 = X(s) \times (s+2)\big|_{s=-2} = \frac{2}{s(s+1)(s+2)} \times (s+2)\Big|_{s=-2} = \frac{2}{s(s+1)}\Big|_{s=-2} = \frac{2}{-2(-2+1)} = 1$$

$$\therefore X(s) = \frac{2}{s(s+1)(s+2)} = \frac{1}{s} - \frac{2}{s+1} + \frac{1}{s+2}$$

$$\therefore x(t) = \mathcal{L}^{-1}\{X(s)\} = \mathcal{L}^{-1}\Big\{\frac{1}{s} - \frac{2}{s+1} + \frac{1}{s+2}\Big\}$$

$$= u(t) - 2\,e^{-t}\,u(t) + e^{-2t}\,u(t)$$

$$= (1 - 2\,e^{-t} + e^{-2t})\,u(t) = (1 - e^{-t})^2\,u(t)$$

$$\boxed{\begin{aligned} \mathcal{L}\{u(t)\} &= \frac{1}{s} \\ \mathcal{L}\{e^{-at}\,u(t)\} &= \frac{1}{s+a} \end{aligned}}$$

$$\boxed{(x-y)^2 = x^2 - 2xy + y^2}$$

Case ii: *When s-Domain Signal X(s) has Multiple Poles*

Let, $X(s) = \dfrac{K}{s(s+p_1)(s+p_2)^2}$ .....(2.67)

By partial fraction expansion technique, the equation (2.67) can be expressed as,

$$X(s) = \frac{K}{s(s+p_1)(s+p_2)^2} = \frac{K_1}{s} + \frac{K_2}{s+p_1} + \frac{K_3}{(s+p_2)^2} + \frac{K_4}{(s+p_2)} \qquad \text{.....(2.68)}$$

The residues K_1, K_2, K_3, and K_4 are given by,

$$K_1 = X(s) \times s\big|_{s=0} \qquad ; \qquad K_2 = X(s) \times (s+p_1)\big|_{s=-p_1}$$

$$K_3 = X(s) \times (s+p_2)^2\big|_{s=-p_2} \quad ; \quad K_4 = \frac{d}{ds}[X(s) \times (s+p_2)^2]_{s=-p_2}$$

We know that, $\mathcal{L}\{u(t)\} = \dfrac{1}{s}$, $\mathcal{L}\{e^{-at}u(t)\} = \dfrac{1}{s+a}$, $\mathcal{L}\{t\,e^{-at}u(t)\} = \dfrac{1}{(s+a)^2}$

By using the above standard Laplace transform pairs the inverse Laplace transform of equation (2.68) can be obtained as shown below:

$$\therefore x(t) = \mathcal{L}^{-1}\{X(s)\} = \mathcal{L}^{-1}\Big\{\frac{K_1}{s} + \frac{K_2}{s+p_1} + \frac{K_3}{(s+p_2)^2} + \frac{K_4}{(s+p_2)}\Big\}$$

$$x(t) = K_1 \mathcal{L}^{-1}\left\{\frac{1}{s}\right\} + K_2 \mathcal{L}^{-1}\left\{\frac{1}{s+p_1}\right\} + K_3 \mathcal{L}^{-1}\left\{\frac{1}{(s+p_2)^2}\right\} + K_4 \mathcal{L}^{-1}\left\{\frac{1}{s+p_2}\right\}$$

$$= K_1 u(t) + K_2 e^{-p_1 t} u(t) + K_3 t e^{-p_2 t} u(t) + K_4 e^{-p_2 t} u(t)$$

In general if the pole p_2 in equation (2.67) has multiplicity of q, as shown below:

$$X(s) = \frac{K}{s(s+p_1)(s+p_2)^q}$$

Then the partial fraction of the above signal can be expressed as,

$$X(s) = \frac{K}{s(s+p_1)(s+p_2)^q} = \frac{K_1}{s} + \frac{K_2}{s+p_1} + \frac{K_3}{(s+p_2)^q} + \frac{K_{12}}{(s+p_2)^{q-1}} + \frac{K_{22}}{(s+p_2)^{q-2}} + \ldots + \frac{K_{(q-1)2}}{s+p_2}$$

The residue K_1, K_2, K_3 are evaluated as shown above. The $q-1$ residues K_{12}, K_{22}, ...$K_{(q-1)2}$ can be evaluated using the equation,

$$K_{r2} = \frac{1}{r!} \frac{d^r}{dq^r}[X(s) \times (s+p_2)^q] \; ; \; r = 1, 2, ...q-1$$

Example 2.40

Determine the inverse Laplace transform of $X(s) = \dfrac{2}{s(s+1)(s+2)^2}$.

Solution:

Given that, $X(s) = \dfrac{2}{s(s+1)(s+2)^2}$

By partial fraction expansion technique, X(s) can be expressed as,

$$X(s) = \frac{K}{s(s+1)(s+2)^2} = \frac{K_1}{s} + \frac{K_2}{(s+1)} + \frac{K_3}{(s+2)^2} + \frac{K_4}{(s+2)}$$

The residue K_1 is obtained by multiplying X(s) by s and letting s = 0.

$$K_1 = X(s) \times s \Big|_{s=0} = \frac{2}{s(s+1)(s+2)^2} \times s \Big|_{s=0} = \frac{2}{(s+1)(s+2)^2} \Big|_{s=0} = \frac{2}{1 \times 2^2} = 0.5$$

The residue K_2 is obtained by multiplying X(s) by (s + 1) and letting s = -1.

$$K_2 = X(s) \times (s+1) \Big|_{s=-1} = \frac{2}{s(s+1)(s+2)^2} \times (s+1) \Big|_{s=-1} = \frac{2}{-1(-1+2)^2} = -2$$

The residue K_3 is obtained by multiplying X(s) by $(s + 2)^2$ and letting s = -2.

$$K_3 = X(s) \times (s+2)^2 \Big|_{s=-2} = \frac{2}{s(s+1)(s+2)^2} \times (s+2)^2 \Big|_{s=-2} = \frac{2}{-2(-2+1)} = 1$$

The residue K_4 is obtained by differentiating the product X(s) × $(s + 2)^2$ with respect to s and then letting s = -2.

$$K_4 = \frac{d}{ds}[X(s) \times (s+2)^2]\Big|_{s=-2} = \frac{d}{ds}\left[\frac{2}{s(s+1)(s+2)^2} \times (s+2)^2\right]\Big|_{s=-2}$$

$$= \frac{d}{ds}\left[\frac{2}{s(s+1)}\right]\Big|_{s=-2} = \frac{-2(2s+1)}{s^2(s+1)^2}\Big|_{s=-2} = \frac{-2(2(-2)+1)}{(-2)^2(-2+1)^2} = 1.5$$

$\frac{d}{dx}\left(\frac{u}{v}\right) = \dfrac{\frac{du}{dx}v - u\frac{dv}{dx}}{v^2}$
$u = 2$ $v = s(s+1)$ $= s^2 + s$

$$\therefore X(s) = \frac{2}{s(s+1)(s+2)^2} = \frac{0.5}{s} - \frac{2}{s+1} + \frac{1}{(s+2)^2} + \frac{1.5}{s+2}$$

$$\therefore x(t) = \mathcal{L}^{-1}\{X(s)\} = \mathcal{L}^{-1}\left\{\frac{0.5}{s} - \frac{2}{s+1} + \frac{1}{(s+2)^2} + \frac{1.5}{s+2}\right\}$$

$\mathcal{L}\{u(t)\} = \frac{1}{s}$
$\mathcal{L}\{e^{-at}u(t)\} = \frac{1}{s+a}$
$\mathcal{L}\{te^{-at}u(t)\} = \frac{1}{(s+a)^2}$

$$= 0.5\,u(t) - 2e^{-t}u(t) + te^{-2t}u(t) + 1.5\,e^{-2t}u(t)$$

$$= (0.5 - 2e^{-t} + te^{-2t} + 1.5\,e^{-2t})\,u(t)$$

Case iii: *When s-Domain Signal X(s) has Complex Conjugate Poles*

Let, $X(s) = \dfrac{K}{(s+p_1)(s^2+bs+c)}$ .....(2.69)

By partial fraction expansion technique, the equation (2.69) can be expressed as,

$$X(s) = \frac{K}{(s+p_1)(s^2+bs+c)} = \frac{K}{s+p_1} + \frac{K_2s+K_3}{s^2+bs+c} \qquad .....(2.70)$$

The residue K_1 is given by, $K_1 = X(s) \times (s+p_1)\big|_{s=-p_1}$

The residues K_2 and K_3 are solved by cross multiplying the equation (2.70) and then equating the coefficients of like power of s.

Finally express $X(s)$ as shown below:

$$X(s) = \frac{K_1}{(s+p_1)} + \frac{K_2s+K_3}{s^2+bs+c}$$

$$= \frac{K_1}{(s+p_1)} + \frac{K_2s+K_3}{s^2+2\times\frac{b}{2}s+\left(\frac{b}{2}\right)^2+c-\left(\frac{b}{2}\right)^2}$$

Arranging, $s^2 + bs$, in the form of $(x+y)^2$
$(x+y)^2 = x^2 + 2xy + y^2$

$$= \frac{K_1}{(s+p_1)} + \frac{K_2s+K_3}{\left(s+\frac{b}{2}\right)^2+\left(c-\frac{b^2}{4}\right)}$$

$$= \frac{K_1}{s+p_1} + \frac{K_2s+K_3}{(s+a)^2+\Omega_0^2}$$

$$= \frac{K_1}{s+p_1} + K_2\frac{s+\frac{K_2}{K_3}}{(s+a)^2+\Omega_0^2} = \frac{K_1}{s+p_1} + K_2\frac{s+a+\frac{K_3}{K_2}-a}{(s+a)^2+\Omega_0^2}$$

Put, $\frac{b}{2} = a$ and $c - \frac{b^2}{4} = \Omega_0^2$
Put, $\frac{K_3}{K_2} - a = K_4$

$$\therefore X(s) = \frac{K_1}{s+p_1} + K_2 \frac{s+a+K_4}{(s+a)^2 + \Omega_0^2}$$

$$= \frac{K_1}{s+p_1} + K_2 \frac{s+a}{(s+a)^2 + \Omega_0^2} + \frac{K_2 K_4}{\Omega_0} \frac{\Omega_0}{(s+a)^2 + \Omega_0^2}$$ $\boxed{\text{Put, } \dfrac{K_2 K_4}{\Omega_0} = K_5}$

$$= \frac{K_1}{s+p_1} + K_2 \frac{s+a}{(s+a)^2 + \Omega_0^2} + K_5 \frac{\Omega_0}{(s+a)^2 + \Omega_0^2} \qquad \dots(2.71)$$

We know that, $\mathcal{L}\{e^{-at} u(t)\} = \dfrac{1}{s+a}$; $\mathcal{L}\{e^{-at} \cos\Omega_0 t \, u(t)\} = \dfrac{s+a}{(s+a)^2 + \Omega_0^2}$;

$$\mathcal{L}\{e^{-at} \sin\Omega_0 t \, u(t)\} = \frac{\Omega_0}{(s+a)^2 + \Omega_0^2}$$

By using the above standard Laplace transform pairs the inverse Laplace transform of equation (2.71) can be obtained as shown below:

$$\therefore x(t) = \mathcal{L}^{-1}\{X(s)\} = K_1 \mathcal{L}^{-1}\left\{ \frac{K_1}{s+p_1} + K_2 \frac{s+a}{(s+a)^2 + \Omega_0^2} + K_5 \frac{\Omega_0}{(s+a)^2 + \Omega_0^2} \right\}$$

$$= K_1 \mathcal{L}^{-1}\left\{ \frac{1}{s+p_1} \right\} + K_2 \mathcal{L}^{-1}\left\{ \frac{s+a}{(s+a)^2 + \Omega_0^2} \right\} + K_5 \mathcal{L}^{-1}\left\{ \frac{\Omega_0}{(s+a)^2 + \Omega_0^2} \right\}$$

$$= K_1 e^{-p_1 t} u(t) + K_2 e^{-at} \cos\Omega_0 t \, u(t) + K_5 e^{-at} \sin\Omega_0 t \, u(t)$$

Example 2.41

Determine the inverse Laplace transform of $X(s) = \dfrac{1}{(s+2)(s^2 + s + 1)}$.

Solution:

Given that, $X(s) = \dfrac{1}{(s+2)(s^2 + s + 1)}$

By partial fraction expansion technique, X(s) can be expressed as,

$$X(s) = \frac{1}{(s+2)(s^2 + s + 1)} = \frac{K_1}{s+2} + \frac{K_2 s + K_3}{s^2 + s + 1}$$

The residue K_1 is obtained by multiplying X(s) by (s + 2) and letting s = –2.

$$\therefore K_1 = X(s) \times (s+2) \Big|_{s=-2} = \frac{1}{(s+2)(s^2 + s + 1)} \times (s+2) \Big|_{s=-2} = \frac{1}{(-2)^2 - 2 + 1} = \frac{1}{3}$$

To solve K_2 and K_3, cross multiply the following equation and substitute the value of K_1. Then equate the coefficients of like power of s.

$$\frac{1}{(s+2)(s^2 + s + 1)} = \frac{K_1}{s+2} + \frac{K_2 s + K_3}{s^2 + s + 1}$$

$$1 = K_1(s^2 + s + 1) + (K_2 s + K_3)(s + 2)$$

$$1 = \frac{1}{3}(s^2 + s + 1) + K_2 s^2 + 2K_2 s + K_3 s + 2K_3 \qquad \boxed{K_1 = -\frac{1}{3}}$$

$$1 = \frac{s^2}{3} + \frac{s}{3} + \frac{1}{3} + K_2 s^2 + 2K_2 s + K_3 s + 2K_3$$

$$1 = \left(\frac{1}{3} + K_2\right)s^2 + \left(\frac{1}{3} + 2K_2 + K_3\right)s + \frac{1}{3} + 2K_3$$

On equating the coefficients of s^2 terms,

$$0 = \frac{1}{3} + K_2 \implies K_2 = -\frac{1}{3}$$

On equating the coefficients of s terms,

$$0 = \frac{1}{3} + 2K_2 + K_3 \implies K_3 = -\frac{1}{3} - 2K_2 = -\frac{1}{3} - 2 \times \left(-\frac{1}{3}\right) = \frac{1}{3} \qquad \boxed{\begin{array}{l}\text{Arranging, } s^2 + s, \text{ in}\\ \text{the form of } (x + y)^2\end{array}}$$

$$\therefore X(s) = \frac{1}{(s+2)(s^2+s+1)} = \frac{\frac{1}{3}}{s+2} + \frac{-\frac{1}{3}s + \frac{1}{3}}{s^2+s+1} = \frac{\frac{1}{3}}{s+2} + \frac{-\frac{1}{3}(s-1)}{s^2+s+1}$$

$$\boxed{(x + y)^2 = x^2 + 2xy + y^2}$$

$$= \frac{1}{3}\frac{1}{s+2} - \frac{1}{3}\frac{s-1}{(s^2+(2\times 0.5s)+0.5^2)+(1-0.5^2)} = \frac{1}{3}\frac{1}{s+2} - \frac{1}{3}\frac{s-1}{(s+0.5)^2+0.75}$$

$$= \frac{1}{3}\frac{1}{s+2} - \frac{1}{3}\frac{s+0.5-1-0.5}{(s+0.5)^2+(\sqrt{0.75})^2} = \frac{1}{3}\frac{1}{s+2} - \frac{1}{3}\frac{s+0.5-1.5}{(s+0.5)^2+0.886^2}$$

$$= \frac{1}{3}\frac{1}{s+2} - \frac{1}{3}\frac{s+0.5}{(s+0.5)^2+0.866^2} + \frac{1}{3}\times\frac{1.5}{0.866} - \frac{0.866}{(s+0.5)^2+0.866^2}$$

$$\therefore x(t) = \mathcal{L}^{-1}\{X(s)\} = \mathcal{L}^{-1}\left\{\frac{1}{3}\frac{1}{s+2} - \frac{1}{3}\frac{s+0.5}{(s+0.5)^2+0.866^2} + 0.557\frac{0.866}{(s+0.5)^2+0.866^2}\right\}$$

$$= \frac{1}{3}\mathcal{L}^{-1}\left\{\frac{1}{s+2}\right\} - \frac{1}{3}\mathcal{L}^{-1}\left\{\frac{s+0.5}{(s+0.5)^2+0.866^2}\right\} + 0.557\,\mathcal{L}^{-1}\left\{\frac{0.866}{(s+0.5)^2+0.866^2}\right\}$$

$$= \frac{1}{3}e^{-2t}u(t) - \frac{1}{3}e^{-0.5t}\cos 0.866t\, u(t) + 0.577\,e^{-0.5t}\sin 0.866t\, u(t)$$

$\mathcal{L}\{e^{-at}u(t)\} = \frac{1}{s+a}$	$\mathcal{L}\{e^{-at}\cos\Omega_0 t\, u(t)\} = \frac{s+a}{(s+a)^2+\Omega_0^2}$	$\mathcal{L}\{e^{-at}\sin\Omega_0\, u(t)\} = \frac{\Omega_0}{(s+a)^2+\Omega_0^2}$

2.10.2 Inverse Laplace Transform Using Convolution Theorem

The convolution theorem of equation (2.64) is useful to evaluate the inverse Laplace transform of complicated s-domain signals.

Let, $X(s) = X_1(s)\, X_2(s)$

Let, $x(t) = \mathcal{L}^{-1}\{X(s)\}$; $x_1(t) = \mathcal{L}^{-1}\{X_1(s)\}$; $x_2(t) = \mathcal{L}^{-1}\{X_2(s)\}$

Now, the inverse Laplace transform of $X_1(s)$ and $X_2(s)$ are computed separately to get $x_1(t)$ and $x_2(t)$, and then the inverse Laplace transform of $X(s)$ is obtained by convolution of $x_1(t)$ and $x_2(t)$, as shown below:

Here, $X(s) = X_1(s)\, X_2(s)$.

On taking inverse Laplace transform of the above equation we get,

$$\mathcal{L}^{-1}\{X(s)\} = \mathcal{L}^{-1}\{X_1(s)\, X_2(s)\}$$

$$\therefore \ x(t) = \mathcal{L}^{-1}\{X_1(s)\, X_2(s)\}$$

$$= x_1(t) * x_2(t)$$

> Using convolution property of Laplace transform.

$$= \int_{-\infty}^{+\infty} x_1(\lambda)\, x_2(t-\lambda)\, d\lambda$$

> Using equation (2.61).

Example 2.42

Find the inverse Laplace transform of the s-domain signal, $X(s) = \dfrac{4}{s^2(s^2+16)}$ using convolution theorem.

Solution:

Let, $X(s) = \dfrac{4}{s^2(s^2+16)} = \dfrac{4}{s^2(s^2+4^2)} = \dfrac{4}{s^2+4^2} \times \dfrac{1}{s^2} = X_1(s)\, X_2(s)$

where, $X_1(s) = \dfrac{4}{s^2+4^2}$ and $X_2(s) = \dfrac{1}{s^2}$

$$x_1(t) = \mathcal{L}^{-1}\{X_1(s)\} = \mathcal{L}^{-1}\left\{\dfrac{4}{s^2+4^2}\right\} = \sin 4t\, u(t) = \sin t \ ; \ t \geq 0$$

$$x_2(t) = \mathcal{L}^{-1}\{X_2(s)\} = \mathcal{L}^{-1}\left\{\dfrac{1}{s^2}\right\} = t\, u(t) = t \ ; \ t \geq 0$$

$$\therefore \ x(t) = \mathcal{L}^{-1}\{X(s)\} = \mathcal{L}^{-1}\{X_1(s)\, X_2(s)\}$$

> By convolution theorem
> $\mathcal{L}\{x_1(t) * x_2(t)\} = X_1(s)\, X_2(s)$

$$= x_1(t) * x_2(t)$$

> Since $x_1(t)$ and $x_2(t)$ are causal, limits of integration are changed to 0 to t.

$$= \int_{\lambda=-\infty}^{\lambda=+\infty} x_1(\lambda) x_1(t-\lambda)\, d\lambda = \int_{\lambda=0}^{\lambda=t} \sin 4\lambda\,(t-\lambda)\, d\lambda = \int_{\lambda=0}^{\lambda=t} (t-\lambda)\sin 4\lambda\, d\lambda$$

$$= \left[(t-\lambda)\left(\dfrac{-\cos 4\lambda}{4}\right) - \int \left(\dfrac{-\cos 4\lambda}{4}\right)(-1)\, d\lambda\right]_{\lambda=0}^{\lambda=t}$$

> $\int u\, dv = u\, v - \int v\, du$
>
> $u = t-\lambda \ \Rightarrow \ \dfrac{du}{d\lambda} = -1$
>
> $dv = \sin 4\lambda \ \Rightarrow \ v = \dfrac{-\cos 4\lambda}{4}$

$$\therefore \; x(t) = \left[-(t-\lambda)\left(\frac{\cos 4\lambda}{4}\right) - \frac{\sin 4\lambda}{16} \right]_{\lambda=0}^{\lambda=t}$$

$$= -(t-t)\left(\frac{\cos 4t}{4}\right) - \frac{\sin 4t}{16} + (t-0)\left(\frac{\cos 0}{4}\right) + \frac{\sin 0}{16} \qquad \boxed{\cos 0 = 1, \sin 0 = 0}$$

$$= 0 - \frac{\sin 4t}{16} + \frac{t}{4} + 0 = \frac{1}{4}\left(t - \frac{\sin 4t}{4}\right) \; ; \; t \geq 0 = \frac{1}{4}\left(t - \frac{\sin 4t}{4}\right)u(t)$$

Example 2.43

Find the inverse Laplace transform of the following s-domain signals.

a) $X(s) = \dfrac{8s+10}{(s+1)(s+2)}$ **b)** $X(s) = \dfrac{3s^2+8s+23}{(s+3)(s^2+2s+10)}$ **c)** $X(s) = \dfrac{8s^2+11s}{(s+2)(s+1)^2}$

Solution:

a) Given that, $X(s) = \dfrac{8s+10}{(s+1)(s+2)}$ ***(AU May'15, 10 Marks)***

By partial fraction expansion technique, X(s) can be expressed as

$$X(s) = \frac{8s+10}{(s+1)(s+2)} = \frac{K_1}{(s+1)} + \frac{K_2}{(s+2)}$$

The residual K_1 is obtained by multiplying X(s) by (s + 1) and letting s = –1.

$$\therefore \; K_1 = X(s) \times (s+1)\big|_{s=-1} = \frac{8s+10}{(s+1)(s+2)} \times (s+1)\big|_{s=-1} = \frac{8(-1)+10}{-1+2} = \frac{2}{1} = 2$$

The residual K_2 is obtained by multiplying X(s) by (s + 2) and letting s = –2.

$$\therefore \; K_2 = X(s) \times (s+2)\big|_{s=-2} = \frac{8s+10}{(s+1)(s+2)} \times (s+2)\big|_{s=-2} = \frac{8(-2)+10}{-2+1} = \frac{-6}{-1} = 6$$

$$\therefore \; X(s) = \frac{8s+10}{(s+1)(s+2)} = \frac{2}{(s+1)} + \frac{6}{(s+2)}$$

$$\therefore \; x(t) = \mathcal{L}^{-1}\{X(s)\} = \mathcal{L}^{-1}\left\{\frac{2}{s+1} + \frac{6}{s+2}\right\} \qquad\qquad \boxed{\mathcal{L}\{e^{-at}u(t)\} = \frac{1}{s+a}}$$

$$= 2\,\mathcal{L}^{-1}\left\{\frac{1}{s+1}\right\} + 6\,\mathcal{L}^{-1}\left\{\frac{1}{s+2}\right\} = 2e^{-t}u(t) + 6e^{-2t}u(t)$$

$$= [2e^{-t} + 6e^{-2t}]u(t)$$

b) Given that, $X(s) = \dfrac{3s^2+8s+23}{(s+3)(s^2+2s+10)}$

By partial fraction expansion technique, X(s) can be expressed as,

$$X(s) = \frac{3s^2+8s+23}{(s+3)(s^2+2s+10)} = \frac{K_1}{s+3} + \frac{K_2 s + K_3}{s^2+2s+10}$$

The residue K_1 is obtained by multiplying X(s) by (s + 3) and letting s = –3.

$$\therefore \; K_1 = X(s) \times (s+3)\big|_{s=-3} = \frac{3s^2+8s+23}{(s+3)(s^2+2s+10)} \times (s+3)\bigg|_{s=-3}$$

$$= \frac{3(-3)^2+8(-3)+23}{(-3)^2+2(-3)+10} = \frac{27-24+23}{9-6+10} = \frac{26}{13} = 2$$

To solve K_2 and K_3 cross multiply the following equation and substitute the value of K_1. Then equate the coefficients of like power of s.

$$\frac{3s^2+8s+23}{(s+3)(s^2+2s+10)} = \frac{K_1}{s+3} + \frac{K_2 s + K_3}{s^2+2s+10}$$

$$3s^2 + 8s + 23 = K_1(s^2 + 2s + 10) + (K_2 s + K_3)(s + 3)$$

$$3s^2 + 8s + 23 = K_1 s^2 + 2K_1 s + 10K_1 + K_2 s^2 + 3K_2 s + K_3 s + 3K_3$$

$$3s^2 + 8s + 23 = (K_1 + K_2)s^2 + (2K_1 + 3K_2 + K_3)s + 10K_1 + 3K_3$$

On equating the coefficients of s^2 terms, we get,

$$3 = K_1 + K_2 \implies K_2 = 3 - K_1 = 3 - 2 = 1$$

On equating the coefficients of s terms, we get,

$$8 = 2K_1 + 3K_2 + K_3 \implies K_3 = 8 - 2K_1 - 3K_2 = 8 - 2 \times 2 - 3 \times 1 = 1$$

$$\therefore X(s) = \frac{3s^2 + 8s + 23}{(s+3)(s^2 + 2s + 10)} = \frac{2}{s+3} + \frac{s+1}{s^2 + 2s + 10}$$

$$= \frac{2}{s+3} + \frac{s+1}{s^2 + 2s + 1 + 9} = \frac{2}{s+3} + \frac{s+1}{(s+1)^2 + 3^2}$$

$$\boxed{\begin{aligned} \mathcal{L}\{e^{-at} u(t)\} &= \frac{1}{s+a} \\ \mathcal{L}\{e^{-at} \cos\Omega_0 t\, u(t)\} &= \frac{s+a}{(s+a)^2 + \Omega_0^2} \end{aligned}}$$

$$\therefore x(t) = \mathcal{L}^{-1}\{X(s)\} = \mathcal{L}^{-1}\left\{ \frac{2}{s+3} + \frac{s+1}{(s+1)^2 + 3^2} \right\}$$

$$= 2\mathcal{L}^{-1}\left\{ \frac{1}{s+3} \right\} + \mathcal{L}^{-1}\left\{ \frac{s+1}{(s+1)^2 + 3^2} \right\} = 2e^{-3t} u(t) + e^{-t} \cos 3t\, u(t)$$

$$= (2e^{-3t} + e^{-t} \cos 3t)\, u(t)$$

c) Given that, $X(s) = \dfrac{8s^2 + 11s}{(s+2)(s+1)^3}$

By partial fraction expansion technique, X(s) can be expressed as,

$$X(s) = \frac{8s^2 + 11s}{(s+2)(s+1)^3} = \frac{K_1}{s+2} + \frac{K_2}{(s+1)^3} + \frac{K_3}{(s+1)^2} + \frac{K_4}{s+1}$$

The residue K_1 is obtained by multiplying X(s) by (s + 2) and letting s = –2.

$$\therefore K_1 = X(s) \times (s+2) \Big|_{s=-2} = \frac{8s^2 + 11s}{(s+2)(s+1)^3} \times (s+2) \Big|_{s=-2} = \frac{8 \times (-2)^2 + 11 \times (-2)}{(-2+1)^3} = -10$$

$$X(s) \times (s+1)^3 = \frac{8s^2 + 11s}{(s+2)(s+1)^3} \times (s+1)^3 = \frac{8s^2 + 11s}{s+2} \qquad \dots (1)$$

$$\frac{d}{ds}[X(s) \times (s+1)^3] = \frac{d}{ds}\left[\frac{8s^2 + 11s}{s+2} \right] = \frac{(16s+11)(s+2) - (8s^2+11s) \times 1}{(s+2)^2}$$

$$\boxed{\frac{d}{dx}\left(\frac{u}{v}\right) = \frac{\dfrac{du}{dx} v - u \dfrac{dv}{dx}}{v^2}}$$

$$= \frac{16s^2 + 32s + 11s + 22 - 8s^2 - 11s}{(s+2)^2} = \frac{8s^2 + 32s + 22}{(s+2)^2} \qquad \dots (2)$$

The residue K_2 is obtained by multiplying X(s) by $(s+1)^3$ and letting s = –1.

$$\therefore K_2 = X(s) \times (s+1)^3 \Big|_{s=-1} = \frac{8s^2 + 11s}{s+2} \Big|_{s=-1} = \frac{8 \times (-1)^2 + 11 \times (-1)}{-1+2} = -3 \quad \boxed{\text{Using equation (1)}}$$

The residue K_3 is obtained by differentiating the product $X(s) \times (s+1)^3$ with respect to s and then letting s = –1.

$$\therefore K_3 = \frac{d}{ds}[X(s) \times (s+1)^3] \Big|_{s=-1}$$

$$\therefore K_3 = \left.\frac{8s^2 + 32s + 22}{(s+2)^2}\right|_{s=-1} = \frac{8\times(-1)^2 + 32\times(-1) + 22}{(-1+2)^2} = -2 \qquad \boxed{\text{Using equation (2)}}$$

The residue K_4 is obtained by differentiating the product $X(s) \times (s+1)^3$ twice with respect to s and then dividing by 2! and letting $s = -1$.

$$\therefore K_4 = \left.\frac{1}{2!}\frac{d^2}{ds^2}[X(s)\times(s+1)^3]\right|_{s=-1} = \left.\frac{1}{2}\frac{d}{ds}\left[\frac{d}{ds}[X(s)\times(s+1)^3]\right]\right|_{s=-1} \qquad \boxed{\text{Using equation (2)}}$$

$$= \left.\frac{1}{2}\frac{d}{ds}\left[\frac{8s^2 + 32s + 22}{(s+2)^2}\right]\right|_{s=-1} = \left.\frac{1}{2}\frac{(16s+32)(s+2)^2 - (8s^2+32s+22)2(s+2)}{(s+2)^4}\right|_{s=-1}$$

$$= \frac{1}{2}\frac{[16\times(-1)+32](-1+2)^2 - [8\times(-1)^2 + 32\times(-1)+22]2(-1+2)}{(-1+2)^4} \qquad \boxed{\frac{d}{dx}\left(\frac{u}{v}\right) = \frac{\frac{du}{dx}v - u\frac{dv}{dx}}{v^2}}$$

$$= \frac{1}{2}\times\frac{-16+32-[8-32+22]\times2}{1} = 10$$

$$\therefore X(s) = \frac{8s^2+11s}{(s+2)(s+1)^3} = \frac{-10}{s+2} - \frac{3}{(s+1)^3} - \frac{2}{(s+1)^2} + \frac{10}{s+1} \qquad \boxed{\begin{array}{l}\mathcal{L}\{e^{-at}u(t)\} = \dfrac{1}{s+a}\\[2mm]\mathcal{L}\left\{\dfrac{t^{m-1}}{(m-1)!}e^{-at}u(t)\right\} = \dfrac{1}{(s+a)^m}\end{array}}$$

$$\therefore x(t) = \mathcal{L}^{-1}\{X(s)\} = \mathcal{L}^{-1}\left\{\frac{-10}{s+2} - \frac{3}{(s+1)^3} - \frac{2}{(s+1)^2} + \frac{10}{s+1}\right\}$$

$$= -10\,\mathcal{L}^{-1}\left\{\frac{1}{s+2}\right\} - 3\,\mathcal{L}^{-1}\left\{\frac{1}{(s+1)^3}\right\} - 2\,\mathcal{L}^{-1}\left\{\frac{1}{(s+1)^2}\right\} + 10\,\mathcal{L}^{-1}\left\{\frac{1}{s+1}\right\}$$

$$= -10\,e^{-2t}u(t) - 3\times\frac{1}{2!}t^2e^{-t}u(t) - 2t\,e^{-t}u(t) + 10\,e^{-t}u(t)$$

$$= (-10e^{-2t} - 1.5t^2e^{-t} - 2te^{-t} + 10e^{-t})\,u(t) = [(10 - 2t - 1.5t^2)\,e^{-t} - 10e^{-2t}]\,u(t)$$

Example 2.44   *(AU Dec'12, 8 Marks)*

Determine the inverse Laplace transform of $X(s) = \dfrac{1}{s^2 + 3s + 2}$; if ROC : $-2 < \text{Re}\{s\} < -1$

Solution:

Given that, $X(s) = \dfrac{1}{s^2 + 3s + 2}$

By partial fraction expansion technique, $X(s)$ can be expressed as,

$$X(s) = \frac{1}{(s+1)(s+2)} = \frac{K_1}{s+1} + \frac{K_2}{s+2}$$

The residual K_1 is obtained by multiplying $X(s)$ by $(s+1)$ and letting $s = -1$.

$$K_1 = \left.X(s)\times(s+1)\right|_{s=-1} = \left.\frac{1}{(s+1)(s+2)}(s+1)\right|_{s=-1} = \left.\frac{1}{s+2}\right|_{s=-1} = \frac{1}{-1+2} = 1$$

The residual K_2 is obtained by multiplying $X(s)$ by $(s+2)$ and letting $s = -2$.

$$K_2 = \left.X(s)\times(s+2)\right|_{s=-2} = \left.\frac{1}{(s+1)(s+2)}(s+2)\right|_{s=-2} = \left.\frac{1}{s+1}\right|_{s=-2} = \frac{1}{-2+1} = -1$$

$$\therefore x(t) = \frac{1}{(s+1)(s+2)} = \frac{1}{s+1} - \frac{1}{s+2} \qquad \boxed{\begin{array}{ll}\mathcal{L}\{e^{-at}u(t)\} = \dfrac{1}{s+a} & \text{; causal}\\[2mm]\mathcal{L}\{-e^{-at}u(-t)\} = \dfrac{1}{s+a} & \text{; noncausal}\end{array}}$$

ROC: $-2 < \text{Re}\{s\} < -1$:

Given that, ROC lies between lines passing through $s = -2$ and $s = -1$. Hence $x(t)$ will be two sided signal. The term corresponding to the pole, $p = -2$ will be causal signal and $p = -1$ will be anticausal sigal

$$\therefore x(t) = -e^{-t}u(-t) - e^{-2t}u(t)$$

Example 2.45

(AU Jun'14, 16 Marks)

Find the inverse Laplace transform of $X(s) = \dfrac{1}{(s+5)(s-3)}$ if the ROC is,

i) $-5 < \text{Re}\{s\} < 3$ ii) $\text{Re}\{s\} > 3$

Solution:

Given that, $X(s) = \dfrac{1}{(s+5)(s-3)}$

By partial fraction expansion technique, X(s) can be expressed as,

$$X(s) = \frac{1}{(s+5)(s-3)} = \frac{K_1}{s+5} + \frac{K_2}{s-3}$$

The residue K_1 is obtained by multiplying X(s) by (s + 5) and letting s = –5

$$\therefore K_1 = X(s) \times (s+5)\big|_{s=-5} = \frac{1}{(s+5)(s-3)} \times (s+5)\big|_{s=-5} = \frac{1}{-5-3} = -\frac{1}{8}$$

The residue K_2 is obtained by multiplying X(s) by (s – 3) and letting s = 3

$$\therefore K_2 = X(s) \times (s-3)\big|_{s=3} = \frac{1}{(s+5)(s-3)} \times (s-3)\big|_{s=3} = \frac{1}{3+5} = \frac{1}{8}$$

$$\therefore x(t) = \frac{1}{(s+5)(s-3)} = -\frac{1}{8}\left(\frac{1}{s+5}\right) + \frac{1}{8}\left(\frac{1}{s-3}\right)$$

Case i:

Given that ROC lies between lines passing through s = – 5 to s = 3. Hence x(t) will be two sided signal.

The term corresponding to the pole, p = – 5 will be causal signal and the term corresponding to the pole, p = 3 will be anticausal signal.

$$\therefore x(t) = -\frac{1}{8}\, e^{-5t}\, u(t) - \frac{1}{8}\, e^{3t}\, u(-t).$$

$$\boxed{\begin{aligned} \mathcal{L}\{e^{-at}\, u(t)\} &= \frac{1}{s+a} \quad ; \text{ causal} \\ \mathcal{L}\{-e^{-at}\, u(-t)\} &= \frac{1}{s+a} \quad ; \text{ anticausal} \end{aligned}}$$

Case ii:

Given that ROC is right of the line passing through s = 3, hence x(t) will be causal signal.

$$\therefore x(t) = -\frac{1}{8}\, e^{-5t}\, u(t) + \frac{1}{8}\, e^{3t}\, u(t).$$

Example 2.46

Find the inverse Laplace transform of $X(s) = \dfrac{4}{(s+2)(s+4)}$ if the ROC is,

i) $-2 > \text{Re}\{s\} > -4$ ii) $\text{Re}\{s\} < -4$ iii) $\text{Re}\{s\} > -2$

Solution:

Given that, $X(s) = \dfrac{4}{(s+2)(s+4)}$

By partial fraction expansion technique, X(s) can be expressed as,

$$X(s) = \frac{4}{(s+2)(s+4)} = \frac{K_1}{s+2} + \frac{K_2}{s+4}$$

The residue K_1 is obtained by multiplying X(s) by (s + 2) and letting s = –2.

$$\therefore K_1 = X(s) \times (s+2)\big|_{s=-2} = \frac{4}{(s+2)(s+4)} \times (s+2)\big|_{s=-2} = \frac{4}{-2+4} = 2$$

The residue K_2 is obtained by multiplying $X(s)$ by $(s + 4)$ and letting $s = -4$.

$$\therefore K_2 = X(s) \times (s+4)\Big|_{s=-4} = \frac{4}{(s+2)(s+4)} \times (s+4)\Big|_{s=-4} = \frac{4}{-4+2} = -2$$

$$\therefore x(t) = \frac{4}{(s+2)(s+4)} = \frac{2}{s+2} - \frac{2}{s+4}$$

Case i:

Given that ROC lies between lines passing through $s = -2$ to $s = -4$. Hence $x(t)$ will be two sided signal. The term corresponding to the pole, $p = -2$ will be anticausal signal and the term corresponding to the pole, $p = -4$ will be causal signal.

$$\therefore x(t) = -2e^{-2t}u(-t) -2e^{-4t}u(t)$$

$$\boxed{\begin{aligned} \mathcal{L}\{e^{-at}u(t)\} &= \frac{1}{s+a} \quad ; \text{ causal} \\ \mathcal{L}\{-e^{-at}u(-t)\} &= \frac{1}{s+a} \quad ; \text{ anticausal} \end{aligned}}$$

Case ii:

Given that ROC is left of the line passing through $s = -4$. Hence $x(t)$ will be anticausal signal.

$$\therefore x(t) = -2e^{-2t}u(-t) +2e^{-4t}u(-t) = 2[e^{-4t} - e^{-2t}]u(-t)$$

Case iii:

Given that ROC is right of the line passing through $s = -2$. Hence $x(t)$ will be causal signal.

$$\therefore x(t) = 2e^{-t}u(t) -2e^{-4t}u(t) = 2[e^{-t} - e^{-4t}]u(t)$$

2.11 Relation Between Fourier and Laplace Transform

(AU Dec'15, 2 Marks)

Let $x(t)$ be a continuous time signal, defined for all t.

The definition of Laplace tranform of $x(t)$ is,

$$\mathcal{L}\{x(t)\} = X(s) = \int_{-\infty}^{+\infty} x(t)\, e^{-st}\ dt$$

On substituting $s = \sigma + j\Omega$ in the above definition of Laplace tranform we get,

$$\mathcal{L}\{x(t)\} = X(s) = \int_{-\infty}^{+\infty} x(t)\, e^{-(\sigma + j\Omega)t}\ dt \qquad\qquad .....(2.72)$$

The definition of Fourier transform of $x(t)$ is,

$$\mathcal{F}\{x(t)\} = X(j\Omega) = \int_{-\infty}^{+\infty} x(t)\, e^{-j\Omega t}\ dt \qquad\qquad .....(2.73)$$

On comparing equations (2.72) and (2.73) we can say that, the Fourier transform of a continuous time signal, is obtained by letting $\sigma = 0$ (i.e., $s = j\Omega$) in the Laplace transform. .

$$\boxed{\therefore X(j\Omega) = X(s)\Big|_{s = j\Omega}}$$

Since $s = \sigma + j\Omega$, we can say that, the Laplace transform is a generalized transform and Fourier transform is a particular transform when $s = j\Omega$. Since $s = j\Omega$ represents the points on the imaginary axis in the s-plane, we can say that, the Fourier transform is an evaluation of the Laplace transform along the imaginary axis in the s-plane.

Since Fourier transform is evaluation of Laplace transform along imaginary axis, the ROC of $X(s)$ should include the imaginary axis. For all causal signals, the imaginary axis is included in ROC. Therefore for all causal signals the Fourier transform exist.

2.12 Summary of Important Concepts

1. The Fourier series is frequency domain representation of periodic signals.

2. The Fourier series exists only if Dirichlet's conditions are satisfied.

3. The signals with negative frequency are required for mathematical representation of real signals in terms of complex exponential signals.

4. In exponential form of Fourier series, $|c_n|$ represents the magnitude of nth harmonic component.

5. In exponential form of Fourier series, $\angle c_n$ represents the phase of the nth harmonic component.

8. The plot of magnitude versus n (or $n\Omega_0$) is called magnitude (line) spectrum.

9. The plot of phase versus n (or $n\Omega_0$) is called phase (line) spectrum.

10. For signals with even symmetry, the Fourier coefficients b_n are zero.

11. For signals with odd symmetry, the Fourier coefficients a_0 and a_n are zero.

12. For signals with half wave symmetry, the Fourier series will consist of odd harmonic terms alone.

13. A signal with half wave symmetry, if in addition has even/odd symmetry then it is said to have quarter wave symmetry.

14. For signals with quarter wave symmetry, the Fourier series will consist of either odd harmonics of sine terms or odd harmonics of cosine terms.

15. The Fourier transform has been developed from Fourier series by considering the fundamental period T as infinity.

16. The Fourier transform of a signal is also called analysis of the signal.

17. The inverse Fourier transform of a signal is also called synthesis of the signal.

18. The frequency spectrum of nonperiodic signals will be continuous, whereas frequency spectrum of periodic signals will be discrete.

19. The magnitude spectrum will have even symmetry and phase spectrum will have odd symmetry.

20. The Fourier transform of a periodic continuous time signal will have impulses located at the harmonic frequencies of the signal.

21. The ratio of Fourier transform of output and input signal of a system is called transfer function in frequency domain.

22. The Fourier transform of impulse response gives the frequency domain transfer function.

23. The Fourier transform is evaluation of Laplace transform along imaginary axis in s-plane.

24. The Laplace transform is used to transform a time domain signal to complex frequency domain.

25. The signal Ke^{st} can be thought of as an universal signal which represents all types of signals.

25. Complex frequency, s is defined as, $s = \sigma + j\Omega$.

26. The complex frequency plane or s-plane is a two dimensional complex plane with values of σ on horizontal axis and values of Ω on vertical axis.

27. In the definition of Laplace transform if the limits of integral is, 0 to $+\infty$, then the Laplace transform is called one sided Laplace transform or unilateral Laplace transform.

28. In the definition of Laplace transform if the limits of integral is, $-\infty$ to $+\infty$, then the Laplace transform is called two sided Laplace transform or bilateral Laplace transform.

29. The values of s for which the integral $\int\limits_{-\infty}^{+\infty} x(t)\, e^{-st}\, dt$ converges is called Region Of Convergence (ROC).

30. For a causal signal, the ROC includes all points on the s-plane to the right of abscissa of convergence.

31. For an anticausal signal, the ROC includes all points on the s-plane to the left of abscissa of convergence.

32. For a two sided signal the ROC includes all points on the s-plane in the region in between two abscissa of convergences.

33. The convolution theorem of Laplace transform says that, Laplace transform of convolution of two time domain signals is given by the product of the Laplace transform of the individual signals.

34. When X(s) is expressed as a ratio of two polynomials in s, then the s-domain signal X(s) is called a rational function of s.

35. The zeros and poles are two critical complex frequencies at which a rational function of s takes two extreme values zero and infinity respectively.

36. Since the signal X(s) attains infinte value at poles, the ROC of X(s) does not include poles.

37. In a realizable system, the number of zeros will be less than or equal to number of poles.

38. For a causal signal x(t), in terms of poles of X(s), the ROC is the region to the right of right most pole of X(s), (i.e., right of the pole with largest real part).

39. For an anticausal signal x(t), in terms of poles of X(s), the ROC is the region to the left of left most pole of X(s), (i.e., left of the pole with smallest real part).

2.13 Short-answer Questions

Q2.1 *Find the Fourier series coeffecients of continous time signal* cos πt   **(AU Jun'12, 2 Marks)**

Solution:

Given that, $x(t) = \cos \pi t$

Let, $x(t) = \cos \pi t = \dfrac{e^{j\pi t} + e^{-j\pi t}}{2} = \dfrac{1}{2}e^{j\pi t} + \dfrac{1}{2}e^{-j\pi t} = \dfrac{1}{2}e^{-j\pi t} + \dfrac{1}{2}e^{j\pi t}$ .....(1)

The exponential form of Fourier series is,

$$x(t) = \sum_{n=-\infty}^{+\infty} c_n\, e^{jn\Omega_0 t}$$

$$= \ldots + c_{-1}\, e^{-j\Omega_0 t} + c_0 + c_1\, e^{j\Omega_0 t} + \ldots \qquad \boxed{\Omega_0 = \pi} \qquad .....(2)$$

On comparing equations (1) and (2) we get,

$$c_0 = 0, \qquad c_1 = c_{-1} = \dfrac{1}{2}$$

Q2.2 *Determine the Fourier coefficients of the signal* $x(t) = 1 + \sin 2\Omega t + 2\cos 2\Omega t + \cos\left(3\Omega t + \dfrac{\pi}{3}\right)$

Solution:
                                                                    (AU May'15, 2 Marks)

Given that, $x(t) = 1 + \sin 2\Omega t + 2\cos 2\Omega t + \cos\left(3\Omega t + \dfrac{\pi}{3}\right)$

Let, $x(t) = 1 + \dfrac{e^{j2\Omega t} - e^{-j2\Omega t}}{2j} + 2\left(\dfrac{e^{j2\Omega t} + e^{-j2\Omega t}}{2}\right) + \dfrac{e^{j(3\Omega t + \frac{\pi}{3})} + e^{-j(3\Omega t + \frac{\pi}{3})}}{2}$

$= 1 + \dfrac{1}{2j}e^{j2\Omega t} - \dfrac{1}{2j}e^{-j2\Omega t} + e^{j2\Omega t} + e^{-j2\Omega t} + \dfrac{1}{2}\left(e^{j3\Omega t}\ e^{j\frac{\pi}{3}}\right) + \dfrac{1}{2}\left(e^{-j3\Omega t}\ e^{-j\frac{\pi}{3}}\right)$

$= 1 + \dfrac{1}{2j}e^{j2\Omega t} - \dfrac{1}{2j}e^{-j2\Omega t} + e^{j2\Omega t} + e^{-j2\Omega t} + \dfrac{1}{2}\left(e^{j3\Omega t}\left(\cos\dfrac{\pi}{3} + j\sin\dfrac{\pi}{3}\right)\right)$

$\qquad\qquad + \dfrac{1}{2}\left(e^{-j3\Omega t}\left(\cos\dfrac{\pi}{3} - j\sin\dfrac{\pi}{3}\right)\right)$

$= 1 + e^{j2\Omega t}\left(1 + \dfrac{1}{2j}\right) + e^{-j2\Omega t}\left(1 - \dfrac{1}{2j}\right) + \dfrac{1}{2}\left(e^{j3\Omega t}(0.5 + j0.866)\right) + \dfrac{1}{2}\left(e^{-j3\Omega t}(0.5 - j0.866)\right)$

$$\therefore \ x(t) = 1 + (1 - 0.5j)\,e^{j2\Omega t} + (1 + 0.5j)\,e^{-j2\Omega t} + (0.25 + j0.433)\,e^{j3\Omega t} + (0.25 - j0.433)\,e^{-j3\Omega t}$$

$$= (0.25 - j0.433)\,e^{-j3\Omega t} + (1 + j0.5)\,e^{-j2\Omega t} + 1 + (1 - j0.5)\,e^{j2\Omega t} + (0.25 + j0.433)\,e^{j3\Omega t} \qquad .....(1)$$

The exponential form of Fourier series is

$$x(t) = \sum_{n=-\infty}^{+\infty} c_n\,e^{-jn\Omega_0 t}$$

$$= ....+ c_{-3}\,e^{-3j\Omega_0 t} + c_{-2}\,e^{-2j\Omega_0 t} + c_{-1}\,e^{-j\Omega_0 t} + c_0 + c_1\,e^{j\Omega_0 t} + c_2\,e^{2j\Omega_0 t} + c_3\,e^{3j\Omega_0 t}....... \qquad ....(2)$$

On comparing equations (1) and (2) we get,

$c_{-3} = 0.25 - j\,0.433$	$c_0 = 1$
$c_{-2} = 1 + j\,05$	$c_1 = 0$
$c_{-1} = 0$	$c_2 = 1 - j\,0.5$
	$c_3 = 0.25 + j\,0.433$

Q2.3 *What is the value of x(t) at t = t₀ in the waveform shown in Fig Q2.3.*

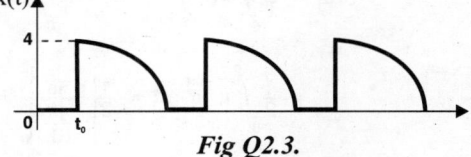

Fig Q2.3.

Solution:

In the waveform shown in Fig Q2.3, $t = t_0$ is a point of discontinuity. If $t = t_0$ is point of discontinuity, then the value of x(t) at $t = t_0$ is given by,

$$x(t_0) = \frac{x(t_0^+) + x(t_0^-)}{2}$$

Here, $x(t_0^+) = 4$, $\quad x(t_0^-) = 0$; $\quad \therefore x(t_0) = \frac{x(t_0^+) + x(t_0^-)}{2} = \frac{4 + 0}{2} = 2$

Q2.4 *Find the constant component of the periodic pulse signal shown in Fig Q2.4.*

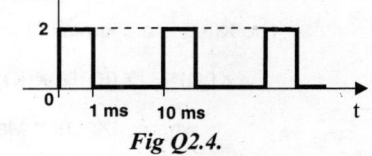

Fig Q2.4.

Solution:

The constant component of any periodic time domain signal is $a_0/2$, where,

$$a_0 = \frac{2}{T} \int_0^T x(t)\,dt$$

Here, $x(t) = 2$; for t = 0 to 1 ms

$\qquad = 0$; for t = 1 to 10 ms

$\qquad$ T = 10 ms

$$\therefore a_0 = \frac{2}{T} \int_0^T x(t)\,dt = \frac{2}{T} \int_0^1 x(t)\,dt = \frac{2}{10} \int_0^1 2\,dt = \frac{2}{10}[2t]_0^1 = \frac{2}{10}[2 - 0] = \frac{4}{10} = 0.4$$

$\therefore$ Constant component $= \dfrac{a_0}{2} = \dfrac{0.4}{2} = 0.2$

Q2.5 *Determine the magnitude of the fundamental frequency component of the periodic pulse signal shown in Fig Q2.5.*

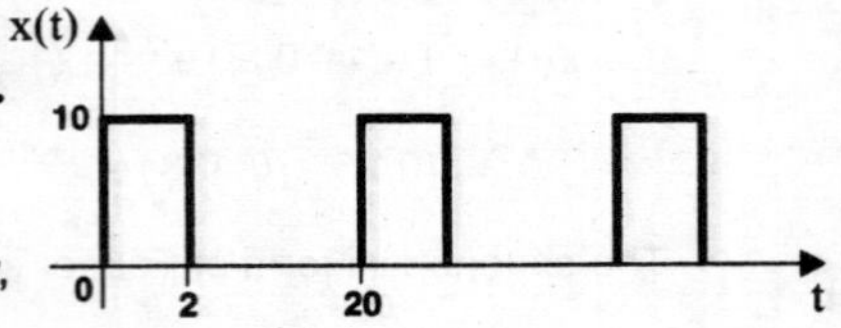

Fig Q2.5.

Solution:

The Fourier coefficient of nth harmonic component is given by,

$$c_n = \frac{1}{T}\int_0^T x(t)\,e^{-jn\Omega_0 t}\,dt \;;\; \text{where } \Omega_0 = \frac{2\pi}{T}$$

The Fourier coefficient of fundamental component is obtained when n = 1.

$$\therefore c_1 = \frac{1}{T}\int_0^T x(t)\,e^{-\frac{j2\pi t}{T}}\,dt$$

Here, x(t) = 10 ; for t = 0 to 2,

       = 0 ; for t = 2 to 10

T = 20

$$\therefore c_1 = \frac{1}{T}\int_0^T x(t)\,e^{-\frac{2\pi t}{T}}\,dt = \frac{1}{T}\int_0^2 x(t)\,e^{-\frac{2\pi t}{T}}\,dt = \frac{1}{20}\int_0^2 10\,e^{-\frac{j2\pi}{20}t}\,dt$$

$$= \frac{1}{2}\int_0^2 e^{-\frac{j\pi t}{10}}\,dt = \frac{1}{2}\left[\frac{e^{-\frac{j\pi t}{10}}}{\frac{-j\pi}{10}}\right]_0^2 = \frac{1}{2}\left[\frac{e^{-\frac{j\pi 2}{10}}}{\frac{-j\pi}{10}} - \frac{e^0}{\frac{-j\pi}{10}}\right]$$

$$= \frac{5}{-j\pi}\left(e^{-j\frac{\pi}{5}} - 1\right) = \frac{5}{-j\pi}\left(\cos\frac{\pi}{5} - j\sin\frac{\pi}{5} - 1\right)$$

$$= j1.5915\,(0.8090 - j0.5878 - 1) = 0.9355 - j0.304 = 0.9836\,\angle -0.314\text{ rad}$$

Magnitude of fundamental component, $|c_1| = 0.9836$

Q2.6 *What is magnitude and phase spectrum?*

Let x(t) be a time domain signal and $X(j\Omega)$ be the Fourier transform of x(t).

The $X(j\Omega)$ is a complex function of Ω and so it can be expressed as,

$$X(j\Omega) = |X(j\Omega)|\angle X(j\Omega)$$

where, $|X(j\Omega)|$ = Magnitude function or Magnitude spectrum.

          $\angle X(j\Omega)$ = Phase function or Phase spectrum.

Q2.7 *Determine the one sided spectrum of the signal* $x(t) = 7 + 10\,\cos(40\pi t + \frac{\pi}{2})$

Solution: **(AU Dec'11, 2 Marks)**

Given that, $x(t) = 7 + 10\cos(40\pi t + \frac{\pi}{2}) = 7 + 10\dfrac{e^{j\left(40\pi t + \frac{\pi}{2}\right)} + e^{-j\left(40\pi t + \frac{\pi}{2}\right)}}{2}$

$$\boxed{e^{\pm j\frac{\pi}{2}} = \cos\frac{\pi}{2} \pm j\sin\frac{\pi}{2} = \pm j}$$

$$= 7 + \frac{10}{2}e^{j40\pi t}e^{j\frac{\pi}{2}} + \frac{10}{2}e^{-j40\pi t}e^{-j\frac{\pi}{2}}$$

$$= 7 + j\frac{10}{2}e^{j40\pi t} - j\frac{10}{2}e^{-j40\pi t}$$

$$= -j5\,e^{-j40\pi t} + 7 + j5\,e^{j40\pi t} \qquad\qquad\qquad .....(1)$$

The Exponential Fourier series of x(t) is

$$x(t) = \sum_{n=-\infty}^{+\infty} c_n \, e^{-jn\Omega_0 t} = \dots + c_{-1} \, e^{-j\Omega_0 t} + c_0 + c_1 \, e^{j\Omega_0 t} + \dots \qquad \dots(2)$$

On comparing equations (1) and (2),

$$\boxed{\Omega_0 = 40\,\pi}$$

$$c_{-1} = -j5, \qquad c_0 = 7, \qquad c_1 = j5$$

Q2.8 *The Fourier transform of the signal shown in Fig Q2.8.1 is,* $X(j\Omega) = \dfrac{1}{\Omega^2}\left(e^{j\Omega} - j\Omega e^{j\Omega} - 1\right)$

Using the properties of Fourier transform find the Fourier transform of the signal shown in Fig Q2.8.2 and Q2.8.3.

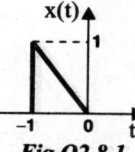

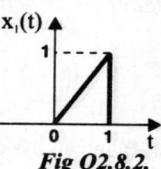

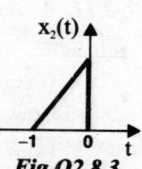

Fig Q2.8.1.   *Fig Q2.8.2.*   *Fig Q2.8.3.*

Solution:

The signal shown in fig Q2.8.2 is the folded version of the signal shown in fig Q2.8.1.

i.e., $x_1(t) = x(-t)$

Given that, $\mathcal{F}\{x(t)\} = X(j\Omega) = \dfrac{1}{\Omega^2}\left(e^{j\Omega} - j\Omega e^{j\Omega} - 1\right)$

Using time reversal property of Fourier transform we can write,

> By time reversal property,
> If $\mathcal{F}\{x(t)\} = X(j\Omega)$
> then $\mathcal{F}\{x(-t)\} = X(-j\Omega)$

$$\mathcal{F}\{x_1(t)\} = X_1(j\Omega) = X(-j\Omega) = \dfrac{1}{\Omega^2}\left(e^{-j\Omega} + j\Omega e^{-j\Omega} - 1\right)$$

The signal shown in fig Q2.8.3 is the shifted version of the signal shown in fig Q2.8.2.

i.e., $x_2(t) = x_1(t - t_0)$; and $t_0 = -1$

Using time shifting property of Fourier transform we can write,

> By time shifting property,
> If $\mathcal{L}\{x(t)\}$ $= X(s)$
> then $\mathcal{L}\{x(t - t_0)\} = e^{j\Omega t_0} X(j\Omega)$

$$\mathcal{F}\{x_2(t)\} = X_2(j\Omega) = e^{j\Omega t_0} X_1(j\Omega) = e^{-j\Omega} \dfrac{1}{\Omega^2}\left(e^{-j\Omega} + j\Omega e^{-j\Omega} - 1\right) = \dfrac{e^{-j\Omega}}{\Omega^2}\left(e^{-j\Omega} + j\Omega e^{-j\Omega} - 1\right)$$

Q2.9 *If Fourier transform of* $e^{-t} u(t)$ *is* $\dfrac{1}{1+j\Omega}$ *then find the Fourier transform of* $\dfrac{1}{1+t}$ *using duality property.*

Solution:

Given that, $x_1(t) = e^{-t} u(t)$, $X_1(j\Omega) = \dfrac{1}{1+j\Omega}$ and $x_2(t) = \dfrac{1}{1+t}$

Here, $x_2(t)$ and $X_1(j\Omega)$ are similar functions.

$\therefore$ By duality property,

> By duality property,
> If $\mathcal{F}\{x_1(t)\} = X_1(j\Omega)$
> and $x_2(t) \equiv X_1(j\Omega)$
> then $X_2(j\Omega) = 2\pi x_1(-j\Omega)$
> $= 2\pi\, x_1(t)\big|_{t=-j\Omega}$

$$X_2(j\Omega) = 2\pi\left[x_1(t)\big|_{t=-j\Omega}\right] = 2\pi\left[e^{-t} u(t)\big|_{t=-j\Omega}\right] = 2\pi e^{j\Omega} u(-j\Omega)$$

Q2.10 *Find the Fourier constant* a_0 *for the continuous time signal defined as,*

$$x(t) = kt \qquad ; \quad 0 \le t \le \dfrac{T}{2}$$

$$\qquad\quad = k(T-t) \; ; \quad \dfrac{T}{2} \le t \le T$$

Solution:

The Fourier constant a_0 is given by,

$$a_0 = \dfrac{2}{T}\int_0^T x(t)\; dt = \dfrac{2}{T}\int_0^{T/2} Kt\; dt + \dfrac{2}{T}\int_{T/2}^T K(T-t)\; dt$$

$$\therefore a_0 = \frac{2}{T}\left[\frac{Kt^2}{2}\right]_0^{T/2} + \frac{2}{T}\left[KTt - \frac{Kt^2}{2}\right]_{T/2}^{T} = \frac{2}{T}\left[\frac{KT^2}{8} - 0\right] + \frac{2}{T}\left[KT^2 - \frac{KT^2}{2} - \frac{KT^2}{2} + \frac{KT^2}{8}\right]$$

$$= \frac{2}{T}\left[\frac{KT^2}{8}\right] + \frac{2}{T}\left[\frac{KT^2}{8}\right] = \frac{4}{T}\frac{KT^2}{8} = \frac{KT}{2}$$

$$\therefore \frac{a_0}{2} = \frac{kT}{4}$$

Q2.11 A periodic signal x(t) is defined as x(t) = (1 – t)² ; 0 ≤ t ≤ T. Find the Fourier coefficient b_n.

Solution:

Given that, $x(t) = (1 - t)^2$; $0 \le t \le$ T.

Now, $x(-t) = (1-(-t)^2) = 1 - t^2$.

Here x(t) = x(–t) and so the given signal is even signal.

For even signals, b_n = 0.

Q2.12 A continuous time signal varies exponentially in the interval 0 to T. Find the Fourier constant $\frac{a_0}{2}$ of the signal.

Solution:

Given that, $x(t) = e^t$; $0 \le t \le$ T.

Now, $a_0 = \frac{2}{T}\int_0^T x(t)\ dt = \frac{2}{T}\int_0^T e^t\ dt = \frac{2}{T}[e^t]_0^T = \frac{2}{T}[e^T - e^0] = \frac{2}{T}(e^T - 1)$

$$\therefore \frac{a_0}{2} = \frac{e^T - 1}{T}$$

Q2.13 Find the Fourier transform of the signal $e^{-3|t|}$ u(t).

Solution:

$x(t) = e^{-3|t|} = e^{-3t}$ for t > 0
$\qquad\qquad = e^{3t}$ for t < 0

The fourier transform of x(t) is,

$$X(j\Omega) = \int_{-\infty}^{\infty} x(t)e^{-j\Omega t}\ dt = \int_{-\infty}^{0} e^{3t}e^{-j\Omega t}\ dt + \int_0^{\infty} e^{-3t}e^{-j\Omega t}\ dt = \int_{-\infty}^{0} e^{(3-j\Omega)t} + \int_0^{\infty} e^{-(3+j\Omega)t}\ dt$$

$$= \left[\frac{e^{(3-j\Omega)t}}{3-j\Omega}\right]_{-\infty}^{0} + \left[\frac{e^{-(3+j\Omega)t}}{-(3+j\Omega)}\right]_0^{\infty} = \frac{e^0}{3-j\Omega} - \frac{e^{-\infty}}{3-j\Omega} - \frac{e^{-\infty}}{3+j\Omega} + \frac{e^0}{3+j\Omega} \qquad \boxed{e^{-\infty}=0\ ;\ e^0=1}$$

$$= \frac{1}{3-j\Omega} + \frac{1}{3+j\Omega} = \frac{3+j\Omega+3-j\Omega}{3^2-(j\Omega)^2} = \frac{6}{3^2+\Omega^2} \qquad \boxed{(a+b)(a-b)=a^2-b^2\ \ |\ j^2=-1}$$

Q2.14 Determine the Fourier transform of x(t) using time shifting property, $x(t) = e^{-3|t-t_0|} + e^{-3|t+t_0|}$.

Solution:

$X(j\Omega) = \mathcal{F}\{x(t)\} = \mathcal{F}\left\{e^{-3|t-t_0|} + e^{3|t+t_0|}\right\}$

$\qquad\quad = \mathcal{F}\left\{e^{-3|t-t_0|}\right\} + \mathcal{F}\left\{e^{-3|t+t_0|}\right\}$

$\qquad\quad = \mathcal{F}\{e^{-3|t|}\}e^{-j\Omega t_0} + \mathcal{F}\{e^{-3|t|}\}e^{+j\Omega t_0}$

$\qquad\quad = \frac{3\times 2}{3^2+\Omega^2}\times e^{-j\Omega t_0} + \frac{3\times 2}{3^2+\Omega^2}\times e^{+j\Omega t_0} = \frac{6}{3^2+\Omega^2}(e^{-j\Omega t_0} + e^{j\Omega t_0})$

$\qquad\quad = \frac{12\cos\Omega t_0}{3^2+\Omega^2}$

By time shifting property,

$\mathcal{F}\{e^{-a|t-t_0|}\} = \frac{2}{a^2+\Omega^2}e^{-j\Omega t_0}$

$\mathcal{F}\{e^{-a|t|}\} = \frac{2a}{a^2+\Omega^2}$

$\cos\theta = \frac{e^{j\theta} + e^{-j\theta}}{2}$

Q2.15 For the signal shown in Fig Q2.15. Find:

   a) X(j0)   *b)* $\int\limits_{-\infty}^{+\infty} X(j\Omega)\, d\Omega$.

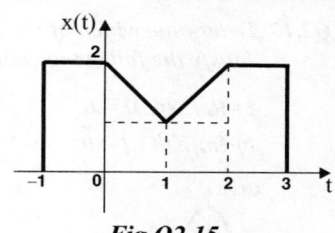

Fig Q2.15.

<u>Solution:</u>

a) By definition of Fourier transform,

$$X(j\Omega) = \int\limits_{-\infty}^{+\infty} x(t)\, e^{-j\Omega t}\, dt$$

By letting $\Omega = 0$ in the above equation we get,

$$X(j0) = \int\limits_{-\infty}^{+\infty} x(t)\, dt = \text{Area of the signal}$$

$\boxed{e^0 = 1}$

= Area of rectangle − Area of triangle

$$= 4 \times 2 - \frac{1}{2} \times 2 \times 1 = 8 - 1 = 7$$

b) By definition of inverse Fourier transform,

$$x(t) = \frac{1}{2\pi} \int\limits_{-\infty}^{+\infty} X(j\Omega)\, e^{j\Omega t}\, d\Omega$$

On letting $t = 0$ in the above equation we get,

$\boxed{e^0 = 1}$

$$x(0) = \frac{1}{2\pi} \int\limits_{-\infty}^{+\infty} X(j\Omega)\, e^0\, d\Omega$$

$\boxed{\text{From Fig Q2.15, } x(0) = 2}$

$$\therefore \int\limits_{-\infty}^{+\infty} X(j\Omega)\, d\Omega = 2\pi \times x(0) = 2\pi \times 2 = 4\pi$$

Q2.16 Find energy in frequency domain for the signal shown in FIg Q2.16.1.

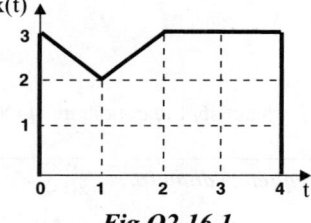

Fig Q2.16.1.

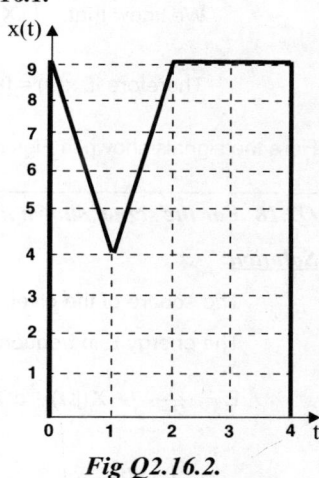

Fig Q2.16.2.

<u>Solution:</u>

    The square of the given signal is shown in Fig Q2.16.2.

    The energy E in frequency domain is given by,

$$E = \frac{1}{2\pi} \int\limits_{-\infty}^{+\infty} |X(j\Omega)|^2\, d\Omega$$

$$= \int\limits_{-\infty}^{+\infty} |x(t)|^2\, dt$$

$\boxed{\text{Using parseval's relation}}$

= Area of $x^2(t)$ = Area of rectangle - Area of triangle

$$= 9 \times 4 - \frac{1}{2} \times 2 \times 5 = 36 - 5 = 31\,\text{joules}$$

Q2.17 *Determine which of the following real signals shown in Fig Q2.17. have Fourier transforms that satisfy the following conditions.*

a) $Re[X(j\Omega)] = 0$

b) $Im[X(j\Omega)] = 0$

c) $\int_{-\infty}^{+\infty} X(j\Omega)\ d\Omega = 0$

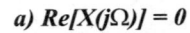 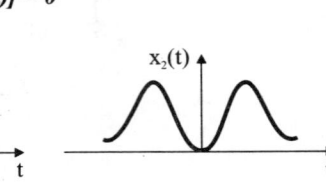 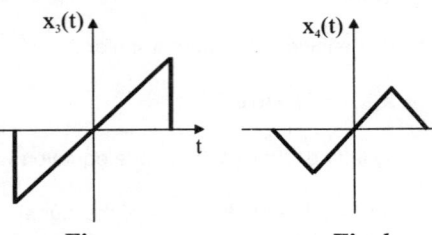

Fig a.   Fig b.   Fig c.   Fig d.

Fig Q2.17.

Solution:

a) Given that, $Re\{X(j\Omega)\} = 0$

In order to satisfy the given condition, the time domain signal should be real and odd.

∴ The signals shown in Figs c and d are real and odd, and so will satisfy the condition, $Re\{X(j\Omega)\} = 0$.

b) Given that, $Im\{j\Omega\} = 0$

In order to satisfy the given condition, the time domain signal should be real and even.

∴ The signals shown in Figs a and b are real and even, and so will satisfy the condition, $Im\{X(j\Omega)\} = 0$.

c) Given that $\int_{-\infty}^{+\infty} X(j\Omega)\,d\Omega = 0$

We know that, $\int_{-\infty}^{+\infty} X(j\Omega)\,d\Omega = 2\pi\,x(0)$

Therefore if, $x(0) = 0$, then $\int_{-\infty}^{+\infty} X(j\Omega)\,d\Omega = 0$

Here the signals shown in Figs b,c and d has $x(0) = 0$, and so they will satisfy the condition $\int_{-\infty}^{+\infty} X(j\Omega)\,d\Omega = 0$

Q2.18 *For the signal shown in Fig Q2.18.1., find energy in frequency domain.*

Solution:

The square of the given signal is shown in Fig Q2.18.2.

The energy E in frequency domain is given by,

$$E = \frac{1}{2\pi} \int_{-\infty}^{+\infty} |X(j\Omega)|^2\,d\Omega$$

$$= \int_{-\infty}^{+\infty} |x(t)|^2\,dt \qquad \boxed{\text{Using Parseval's relation}}$$

$$= \text{Area of } x2(t) = \text{Area of triangle in fig Q2.18.2.}$$

$$= \frac{1}{2} \times 2 \times 4 = 4 \text{ joules}$$

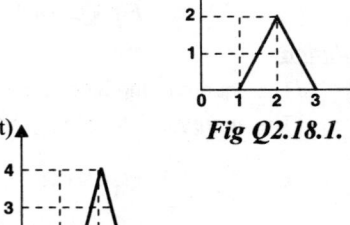

Fig Q2.18.1.

Fig Q2.18.2.

Q2.19 **Find the inverse Fourier transform of** $X(j\Omega) = \dfrac{1}{(4+j\Omega)^2}$**, using convolution property.**

Solution:

$$\text{Let,} \quad X(j\Omega) = \frac{1}{(4+j\Omega)^2} = \frac{1}{4+j\Omega} \times \frac{1}{4+j\Omega} = X_1(j\Omega) \times X_2(j\Omega)$$

$$\text{where,} \quad X_1(j\Omega) = \frac{1}{4+j\Omega} \quad \text{and} \quad X_2(j\Omega) = \frac{1}{4+j\Omega}$$

$$\therefore \; x_1(t) = \mathcal{F}^{-1}\{X_1(j\Omega)\} = \mathcal{F}^{-1}\left\{\frac{1}{4+j\Omega}\right\} = e^{-4t}u(t) = e^{-4t} \;;\; t \geq 0$$

Similarly, $x_2(t) = e^{-4t}u(t) = e^{-4t} \;;\; t \geq 0$

Here, $X(j\Omega) = X_1(j\Omega)\, X_2(j\Omega)$

On taking inverse Fourier transform of the above equation we get,

$$\mathcal{F}^{-1}\{X(j\Omega)\} = \mathcal{F}^{-1}\{X_1(j\Omega)X_2(j\Omega)\}$$

By convolution property of Fourier transform,
$\mathcal{F}\{x_1(t) * x_2(t)\} = X_1(j\Omega)\,X_2(j\Omega)$

$$\therefore \; x(t) = x_1(t) * x_2(t) = \int\limits_{-\infty}^{+\infty} x_1(\lambda)\, x_1(t-\lambda)\, d\lambda$$

$$= \int\limits_0^t e^{-4\lambda}\, e^{-4(t-\lambda)}\, d\lambda = \int\limits_0^t e^{-4t}\, e^{-4\lambda}\, e^{4\lambda}\, d\lambda = e^{-4t}\int\limits_0^t e^{-4\lambda+4\lambda}\, d\lambda$$

$$= e^{-4t}\int\limits_0^t d\lambda = e^{-4t}[\lambda]_0^t = e^{-4t}[t-0] = te^{-4t} \;;\; t \geq 0 \;\; = te^{-4t}u(t)$$

Q2.20 **If $x(t)$ and $X(j\Omega)$ are Fourier transform pair. Determine the Fourier transform of, $x(t)\sin\Omega_0 t$.**

Solution:

$$\mathcal{F}\{x(t)\sin\Omega_0 t\} = \mathcal{F}\left\{x(t)\frac{e^{j\Omega_0 t} - e^{-j\Omega_0 t}}{2j}\right\} = \frac{1}{2j}\mathcal{F}\{x(t)\, e^{j\Omega_0 t}\} - \frac{1}{2j}\mathcal{F}\{x(t)\, e^{-j\Omega_0 t}\}$$

$$= \frac{1}{2j}X(\Omega - \Omega_0) - \frac{1}{2j}X(\Omega + \Omega_0)$$

$$= \frac{1}{2j}[X(\Omega - \Omega_0) - X(\Omega + \Omega_0)]$$

By frequency shifting property,
If $\qquad \mathcal{F}\{x(t)\} = X(j\Omega)$
then $\mathcal{F}\{e^{j\Omega_0 t}\, x(t)\} = X(j(\Omega - \Omega_0))$

Q2.21 **Determine the exponential Fourier series representation of the following signals.**

     a) $x(t) = \cos 4t + \sin 6t$ *b) $x(t) = \sin^2 t$*

Solution:

  a) $x(t) = \cos 4t + \sin 6t$

$$= \frac{e^{j4t} + e^{-j4t}}{2} + \frac{e^{j6t} - e^{-j6t}}{2j} = -\frac{1}{2j}e^{-j6t} + \frac{1}{2}e^{-j4t} + \frac{1}{2}e^{j4t} + \frac{1}{2j}e^{j6t}$$

  b) $x(t) = \sin^2 t$

$$= \frac{1 - \cos 2t}{2} = \frac{1}{2} - \frac{1}{2}\cos 2t = \frac{1}{2} - \frac{1}{2}\left(\frac{e^{j2t} + e^{-j2t}}{2}\right)$$

$$= \frac{1}{2} - \frac{1}{4}e^{j2t} - \frac{1}{4}e^{-j2t} = \frac{1}{4}e^{-j2t} + \frac{1}{2} - \frac{1}{4}e^{j2t}$$

$\sin\theta = \dfrac{e^{j\theta} - e^{-j\theta}}{2j}$
$\cos\theta = \dfrac{e^{j\theta} + e^{-j\theta}}{2}$

Q2.22 **Find the Fourier series coefficients of complex exponential signal $x(t) = \sin 3\pi t + 2\cos 4\pi t$.**

Solution: **(AU Jun'12, 8 Marks)**

$$x(t) = \sin 3\pi t + 2\cos 4\pi t$$

$$= \frac{e^{j3\pi t} - e^{-j3\pi t}}{2j} + 2\frac{e^{j4\pi t} + e^{-j4\pi t}}{2}$$

$$\therefore\ x(t) = \frac{1}{2j}e^{j3\pi t} - \frac{1}{2j}e^{-j3\pi t} + e^{j4\pi t} + e^{-j4\pi t} = e^{-j4\pi t} - \frac{1}{2j}e^{-j3\pi t} + \frac{1}{2j}e^{j3\pi t} + e^{j4\pi t} \qquad \dots\dots(1)$$

The exponential form of Fourier series is,

$$x(t) = \sum_{n=-\infty}^{n=+\infty} c_n\, e^{jn\Omega_0 t} = \dots + c_{-4}\, e^{-j4\Omega_0 t} + c_{-3}\, e^{-j3\Omega_0 t} + c_{-2}\, e^{-j2\Omega_0 t} + c_{-1}\, e^{-j\Omega_0 t} + c_0$$

$$+ c_1\, e^{j\Omega_0 t} + c_2\, e^{j2\Omega_0 t} + c_3\, e^{j3\Omega_0 t} + c_4\, e^{j4\Omega_0 t} + \dots \qquad \dots\dots(2)$$

On comparing equations (1) and (2),

$$c_0 = 0,\ \ c_1 = c_{-1} = 0,\ \ c_2 = c_{-2} = 0,\ \ c_3 = \frac{1}{2j},\ \ c_{-3} = -\frac{1}{2j},\ \ c_4 = c_{-4} = 1 \qquad \boxed{\Omega_0 = \pi}$$

Q.2.23 Find the Laplace transform of $\sin^2 t$, $\delta(t)$ and $u(t)$.   **(AU Dec'11, 2 Marks)**

Solution:

$$\mathcal{L}\{\sin^2 t\} = \mathcal{L}\left\{\frac{1 - \cos 2t}{2}\right\} = \frac{1}{2}[\mathcal{L}\{1\} - \mathcal{L}\{\cos 2t\}] = \frac{1}{2}\left[\frac{1}{s} - \frac{s}{s^2 + 4}\right]$$

$$= \frac{1}{2}\left[\frac{s^2 + 4 - s^2}{s(s^2 + 4)}\right] = \frac{2}{s(s^2 + 4)}$$

$$u(t) = 1\ \ ;\ \ t \geq 0$$
$$\quad\ \ = 0\ \ ;\ \ t < 0$$

$$\mathcal{L}\{u(t)\} = \int_{-\infty}^{+\infty} u(t)\, e^{-st}\, dt = \int_0^\infty e^{-st}\, dt = \left[\frac{e^{-st}}{-s}\right]_0^\infty$$

$$= -\frac{1}{s}[e^{-\infty} - e^0] = \frac{1}{s}$$

$$\delta(t) = \infty\ ;\ t = 0$$
$$\quad\ \ = 0\ ;\ t \neq 0 \qquad \text{and} \qquad \int_{-\infty}^\infty \delta(t)\, dt = 1 \qquad\qquad\qquad \textbf{(AU May'15, 2 Marks)}$$

$$\mathcal{L}\{\delta(t)\} = \int_{-\infty}^\infty \delta(t)\, e^{-st}\, dt = e^{-st}\big|_{t=0} = e^0 = 1$$

Q2.24 Let $x(t)$ and $X(s)$ be Laplace transform pair. Given that, $X(s) = \dfrac{s}{s^2 + 1}$. Find Laplace transform of $y(t) = x\left(\dfrac{t}{5}\right) - x(2t)$.

Solution:

$$\mathcal{L}\left\{x\left(\frac{t}{5}\right)\right\} = 5\, X(5s) = 5\,\frac{5s}{(5s)^2 + 1} = \frac{25s}{25s^2 + 1} = \frac{s}{s^2 + \frac{1}{25}} = \frac{s}{s^2 + 0.04}$$

> By time scaling property,
> If $\quad \mathcal{L}\{x(t)\} = X(s)$
> then $\mathcal{L}\{x(at)\} = \dfrac{1}{|a|}X\left(\dfrac{s}{a}\right)$

$$\mathcal{L}\{x(2t)\} = \frac{1}{2}\, X\left(\frac{s}{2}\right) = \frac{1}{2}\,\frac{\frac{s}{2}}{\left(\frac{s}{2}\right)^2 + 1} = \frac{1}{4}\,\frac{s}{\frac{s^2}{4} + 1} = \frac{s}{s^2 + 4}$$

$$\text{Now,}\ \ \mathcal{L}\{y(t)\} = \mathcal{L}\left\{x\left(\frac{t}{5}\right) - x(2t)\right\} = \mathcal{L}\left\{x\left(\frac{t}{5}\right)\right\} - \mathcal{L}\{x(2t)\}$$

$$= \frac{s}{s^2 + 0.04} - \frac{s}{s^2 + 4} = \frac{s(s^2 + 4) - s(s^2 + 0.04)}{(s^2 + 0.04)(s^2 + 4)} = \frac{3.96s}{(s^2 + 0.04)(s^2 + 4)}$$

Q2.25 Find the inverse Laplace transform of $X(s) = \dfrac{1}{2}\left[\dfrac{1}{s} + \dfrac{s}{s^2+4}\right]$.

Solution:

$$x(t) = \mathcal{L}^{-1}\{X(s)\} = \mathcal{L}^{-1}\left\{\frac{1}{2}\left[\frac{1}{s} + \frac{s}{s^2+4}\right]\right\} = \frac{1}{2}\left[\mathcal{L}^{-1}\left\{\frac{1}{s}\right\} + \mathcal{L}^{-1}\left\{\frac{s}{s^2+4}\right\}\right]$$

$$= \frac{1}{2}[u(t) + \cos 2t\, u(t)] = \left[\frac{1+\cos 2t}{2}\right]u(t) = \cos^2 t\, u(t)$$

$$\boxed{\mathcal{L}\{u(t)\} = \frac{1}{s}}$$
$$\boxed{\mathcal{L}\{\cos\Omega t\, u(t)\} = \frac{s}{s^2+\Omega^2}}$$

$$\boxed{\cos^2\theta = \frac{1+\cos 2\theta}{2}}$$

Q2.26 Let $x(t)$ **and** $X(s)$ **be Laplace transform pair. Given that,** $x(t) = e^{-t}u(t)$. **Find inverse Laplace transform of** $e^{-3s}X(2s)$.

Solution:

$$X(s) = \mathcal{L}\{x(t)\} = \mathcal{L}\{e^{-t}u(t)\} = \frac{1}{s+1} \quad \Rightarrow \quad X(2s) = \frac{1}{2s+1} = \frac{1}{2}\,\frac{1}{s+\frac{1}{2}}$$

By time scaling property,
If $\mathcal{L}\{x(t)\} = X(s)$
then $\mathcal{L}\{x(at)\} = \dfrac{1}{|a|}X\left(\dfrac{s}{a}\right)$

$$\text{Now,} \quad \mathcal{L}^{-1}\{X(2s)\} = \mathcal{L}^{-1}\left\{\frac{1}{2}\,\frac{1}{s+\frac{1}{2}}\right\} = \frac{1}{2}\mathcal{L}^{-1}\left\{\frac{1}{s+\frac{1}{2}}\right\} = \frac{1}{2}e^{-\frac{1}{2}t}u(t)$$

$$\therefore \mathcal{L}^{-1}\{e^{-3s}X(2s)\} = \mathcal{L}^{-1}\{X(2s)\}\Big|_{t=t-3} = \frac{1}{2}e^{-\frac{1}{2}t}u(t)\Big|_{t=t-3}$$

$$= \frac{1}{2}e^{-\left(\frac{t-3}{2}\right)}u(t-3)$$

By time shifting property,
If $\mathcal{L}\{x(t)\} = X(s)$
then $\mathcal{L}\{x(t \pm a)\} = e^{\pm as}X(s)$

Q2.27 Find inverse Laplace transform of,

a) $X(s) = \dfrac{s}{(s^2+9)} \quad Re\{s\} > 0$ b) $X(s) = \dfrac{s}{(s^2+9)} \quad Re\{s\} < 0$

Solution:

a) When Re{s} > 0, the x(t) will be causal or right sided signal.

$$\therefore x(t) = \mathcal{L}^{-1}\{X(s)\} = \mathcal{L}^{-1}\left\{\frac{s}{s^2+9}\right\} = \mathcal{L}^{-1}\left\{\frac{s}{s^2+3^2}\right\} = \cos 3t\, u(t)$$

b) When Re{s} < 0, the x(t) will be anticausal and left sided signal.

$$\therefore x(t) = \mathcal{L}^{-1}\{X(s)\} = \mathcal{L}^{-1}\left\{\frac{s}{s^2+9}\right\} = \mathcal{L}^{-1}\left\{\frac{s}{s^2+3^2}\right\} = -\cos 3t\, u(-t)$$

Q2.28 Given that, $\mathcal{L}\{x(t)\} = X(s) = \dfrac{4s+1}{s^2+6s+3}$. **Determine the initial value,** $x(0)$.

Solution:

By initial value theorem,

Initial value, $x(0) = \underset{t \to 0}{\mathrm{Lt}}\, x(t) = \underset{s \to \infty}{\mathrm{Lt}}\, sX(s) = \underset{s \to \infty}{\mathrm{Lt}}\, s \times \dfrac{4s+1}{s^2+6s+3}$

$$= \underset{s \to \infty}{\mathrm{Lt}}\, s \times \frac{s\left(4+\frac{1}{s}\right)}{s^2\left(1+\frac{6}{s}+\frac{3}{s^2}\right)} = \frac{4+\frac{1}{\infty}}{1+\frac{6}{\infty}+\frac{3}{\infty}} = \frac{4+0}{1+0+0} = 4$$

Q2.29 *Given that,* $\mathcal{L}\{x(t)\} = X(s) = \dfrac{s+4}{s(s^2+7s+10)}$. *Determine the final value,* $x(\infty)$.

Solution:

By final value theorem,

Final value, $x(\infty) = \underset{t \to \infty}{Lt}\, x(\infty) = \underset{s \to 0}{Lt}\, sX(s)$

$= \underset{s \to 0}{Lt}\, s \times \dfrac{s+4}{s(s^2+7s+10)} = \underset{s \to 0}{Lt}\, \dfrac{s+4}{s^2+7s+10} = \dfrac{0+4}{0+0+10} = \dfrac{4}{10} = 0.4$

Q2.30 *Given that,* $X(s) = \dfrac{b_m s^m + b_{m-1}s^{m-1} + ... + b_1 s + b_0}{s(s^n + a_{n-1}s^{n-1} + ... + a_1 s + a_0)}$. *Find* $\underset{t \to \infty}{Lt}\, x(t)$.

Solution:

Using final value theorem,

$\underset{t \to \infty}{Lt}\, x(t) = \underset{s \to 0}{Lt}\, sX(s) = \underset{s \to 0}{Lt}\, s\, \dfrac{b_m s^m + b_{m-1}s^{m-1} + ... + b_1 s + b_0}{s(s^n + a_{n-1}s^{n-1} + ... + a_1 s + a_0)} = \dfrac{0+0+ ..... +0+b_0}{0+0+ ..... +0+a_0} = \dfrac{b_0}{a_0}$

Q2.31 *If* $X(s) = \dfrac{2}{s+3}$. *Find Laplace transform of* $\dfrac{d}{dt}x(t)$.

Solution:

Using initial value theorem,

$x(0) = \underset{s \to \infty}{Lt}\, s\,X(s) = \underset{s \to \infty}{Lt}\, s \times \dfrac{2}{s+3} = \underset{s \to \infty}{Lt}\, s \times \dfrac{2}{s\left(1+\dfrac{3}{s}\right)} = \dfrac{2}{1+\dfrac{3}{\infty}} = 2$

Using time differentiation property,

$\mathcal{L}\left\{\dfrac{d}{dt}x(t)\right\} = s\,X(s) - x(0) = s \times \dfrac{2}{s+3} - 2 = \dfrac{2s-2(s+3)}{s+3} = -\dfrac{6}{s+3}$

Q2.32 *If* $X(s) = \dfrac{0.4}{s+0.2}$. *Find Laplace transform of* $t\, x(t)$.

Solution:

Using frequency differentiation property,

> By frequency differentiation property,
> If $\mathcal{L}\{x(t)\} = X(s)$
> then $\mathcal{L}\{t\, x(t)\} = -\dfrac{dX(s)}{ds}$

$\mathcal{L}\{tx(t)\} = -\dfrac{d}{ds}X(s) = -\dfrac{d}{ds}\left[\dfrac{0.4}{s+0.2}\right] = -\dfrac{d}{ds}\left[0.4(s+0.2)^{-1}\right] = -\left[0.4 \times (-1) \times (s+0.2)^{-2}\right] = \dfrac{0.4}{(s+0.2)^2}$

2.14 MATLAB PROGRAMS

Program 2.1

Write a MATLAB program to find Fourier transform of the following signals.

a) A   b) u(t)   c) $Ae^{-t}u(t)$ d) $At\, e^{-bt}u(t)$ e) $A \cos\Omega_0 t$

```
% Program to find fourier transform of given time domain signals

    %Let t, A, b, o be any real variables
    syms t real; syms A real;  syms b real;  syms o real;

    %(a)
    x = A;
    disp('(a) Fourier transform of "A" is');
    X=fourier(x)
```

```
%(b)
x = heaviside(t);    %heaviside(t) is unit step signal
disp('(b) Fourier transform of  "u(t)" is');
X=fourier(x)


%(c)
x = A*exp(-t)*heaviside(t);
disp('(c) Fourier transform of  "A exp(-t) u(t)" is');
X=fourier(x)


%(d)
x=A*t*exp(-b*t)*heaviside(t);
disp('(d) Fourier transform of  "At exp(-b*t) u(t)" is');
X=fourier(x)


%(e)
x=A*cos(o*t);
disp('(e) Fourier transform of "A cos(o*t)" is');
X=fourier(x)
```

OUTPUT

    (a) Fourier transform of "A" is
```
        X =
            2*i*pi*dirac(1,w)
```
    (b) Fourier transform of "u(t)" is
```
        X =
            pi*dirac(w)-i/w
```
    (c) Fourier transform of "A exp(-t) u(t)" is
```
        X =
            A/(1+i*w)
```
    (d) Fourier transform of "At exp(-b*t) u(t)" is
```
        X =
            A/(b+i*w)^2
```
    (e) Fourier transform of "A cos(o*t)" is
```
        X =
            A*pi*(dirac(w-o)+dirac(w+o))
```

Program 2.2

    Write a MATLAB program to find inverse Fourier transform of the following frequency domain signals.

a) $X_1(j\Omega) = A\pi[\delta(\Omega - \Omega_0)]$ b) $X_2(j\Omega) = \dfrac{A\pi}{j}[\delta(\Omega - \Omega_0) - \delta(\Omega + \Omega_0)]$

c) $X_3(j\Omega) = \dfrac{A}{1 + j\Omega}$ d) $X_4(j\Omega) = \dfrac{3(j\Omega) + 14}{(j\Omega)^2 + 7(j\Omega) + 12}$

```
% Program to find inverse Fourier transform of frequency domain signals

%Let t, A, b, o, w be any real variables
syms t real; syms A real; syms o real; syms w real;

%(a)
X1=A*pi*(dirac(w-o)+dirac(w+o));
disp('(a) Inverse Fourier transform of X1 is');
x1=ifourier(X1,t)
```

```
%(b)
X2=A*pi*(dirac(w-o)-dirac(w+o))/i;
disp('(b) Inverse Fourier transform of X2 is');
x2=ifourier(X2,t)

%(c)
X3=A/(1+i*w);
disp('(c) Inverse Fourier transform of X3 is');
x3=ifourier(X3,t)

%(d)
X4=(3*w+14)/(w^2+7*w+12);
disp('(d) Inverse Fourier transform of X4 is');
x4=ifourier(X4,t)
```

OUTPUT

```
    (a) Inverse Fourier transform of X1 is
            x1 =
                A*cos(o*t)
    (b) Inverse Fourier transform of X2 is
            x2 =
                A*sin(o*t)
    (c) Inverse Fourier transform of X3 is
            x3 =
                A*exp(-t)*heaviside(t)
    (d) Inverse Fourier transform of X4 is
            x4 =
                1/2*i*(2*heaviside(t)-1)*(-2*exp(-4*i*t)+5*exp(-3*i*t))
```

Program 2.3

Write a MATLAB program to perform convolution of signals, $x_1(t) = e^{-2t}u(t)$ and $x_2(t) = e^{-6t}u(t)$, using Fourier transform.

```
% Program to perform convolution using Fourier transform

syms t real;
x1=exp(-2*t).*heaviside(t); %heaviside(t) is unit step signal
x2=exp(-6*t).*heaviside(t);

disp('Fourier transform of x1(t) is');
X1=fourier(x1)
disp('Fourier transform of x2(t) is');
X2=fourier(x2)
Y=X1*X2;

disp('Let x3 be convolution of x1(t) and x2(t).');
x3=ifourier(Y,t)
```

OUTPUT

```
Fourier transform of x1(t) is
    X1 =
        1/(2+i*w)
Fourier transform of x2(t) is
    X2 =
        1/(6+i*w)
Let x3 be convolution of x1(t) and x2(t).
    x3 =
        1/4*heaviside(t)*(exp(-2*t)-exp(-6*t))
```

Program 2.4

Write a MATLAB program to reconstruct the following periodic signal represented by its Fourier series, by considering only 3,5 and 59 terms.

$$x(t) = \frac{1}{2} + \sum_{n=1}^{\infty} b_n \sin n\Omega_0 t \; ; \; b_n = \frac{2}{\pi n} \; ; \; \Omega_0 = 2\pi F \; ; \; F = 1.$$

```
% Program to reconstruct periodic square pulse signal using
% partial sum of fourier series

syms t real;

N=input('Enter number of signals to reconstruct');
n_har=input('Enter no. of harmonics in each signal as array');

t=-1:.002:1;
omega_o=2*pi;
for  k=1:N
    n=[];
    n=[1:2:n_har(k)];
    b_n=2./(pi*n);
    L_n=length(n);
    x=0.5+b_n*sin(omega_o*n'*t);
    subplot(N,1,k),plot(t,x),xlabel('t'),ylabel('recons signal');
    axis( [-1 1 -0.5 1.5]);
    text(.55, 1.0,['no. of har. =',num2str(n_har(k))]);
    end
```

OUTPUT

```
Enter number of signals to reconstruct 3
Enter no. of harmonics in each signal as array [3 5 59]
```

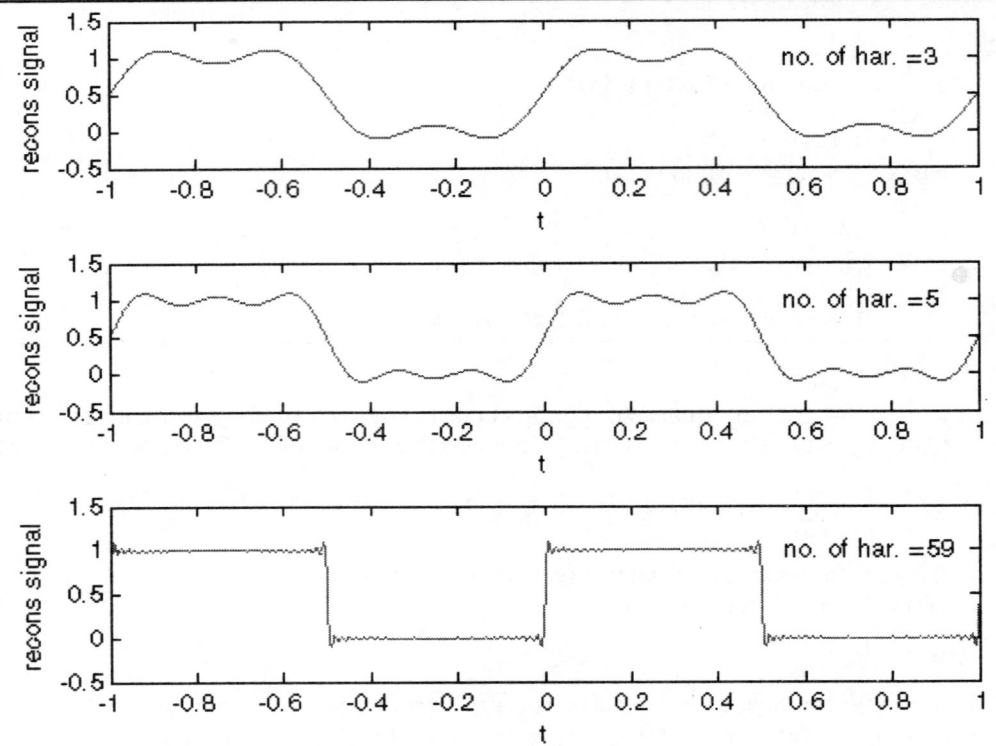

Fig P2.4: Reconstructed signals using only 3, 5 and 59 terms of Fourier series.

Program 2.5

Write a MATLAB program to find Laplace transform of the following standard causal signals.

a) t   b) e^{-at} c) te^{-at} d) $\cos\Omega t$ e) $\sinh\Omega t$ f) $e^{-at}\sin\Omega t$

```
%****Program to find Laplace transform of some standard signals

clear all
syms t real;      %Let a, t be any real variable
syms a real;
syms s complex;   %Let s be complex variable

%(a)
x=t;
disp('(a) Laplace transform of "t" is');
laplace(x)

%(b)
x=exp(-a*t);
disp('(b) Laplace transform of  "exp(-a*t)" is');
laplace(x)

%(c)
x=t*exp(-a*t);
disp('(c) Laplace transform of  "t*exp(-a*t)" is');
laplace(x)
```

```
%(d)
 sg=real(s);      %s - complex frequency ; sg - sigma ; o-omega
 o=imag(s);
 s=sg+(i*o);
 x=cos(o*t);
 disp('(d) Laplace transform of "cos(o*t)" is');
 laplace(x)

%(e)
 x=sinh(o*t);
 disp('(e) Laplace transform of " sinh(o*t)" is');
 laplace(x)

%(f)
 x=exp(-a*t)*sin(o*t);
 disp('(f) Laplace transform of  "exp(-a*t)*sin(o*t)" is');
 laplace(x)
```

OUTPUT

(a) Laplace transform of "t" is
     ans =
          $1/s^2$
(b) Laplace transform of "exp(-a*t)" is
     ans =
          $1/(s+a)$
(c) Laplace transform of "t*exp(-a*t)" is
     ans =
          $1/(s+a)^2$
(d) Laplace transform of "cos(o*t)" is
     ans =
          $s/(s^2-(1/2*s-1/2*conj(s))^2)$
(e) Laplace transform of " sinh(o*t)" is
     ans =
          $-i*(1/2*s-1/2*conj(s))/(s^2+(1/2*s-1/2*conj(s))^2)$
(f) Laplace transform of "exp(-a*t)*sin(o*t)" is
     ans =
          $-i*(1/2*s-1/2*conj(s))/((s+a)^2-(1/2*s-1/2*conj(s))^2)$

Program 2.6

Write a MATLAB program to find Laplace transform of the following standard causal signals.

a) t^n b) $t^{n-1}/(n-1)!$ c) $t^n e^{-at}$ d) $t^{n-1} e^{-at}/(n-1)!$

```
%****Program to find Laplace transform of some standard signals
 clear all
 syms t real;                      %Let t,a,n be any real variable
 syms a real;
 syms n real;

 n=input('Enter the value of n');  %get value of n from keyboard
```

```
%(a)
  x=t^n;
  disp('(a) Laplace transform of "t^n" is');
  laplace(x)

%(b)
  x=t^(n-1)/factorial(n-1);
  disp('(b) Laplace transform of "t^(n-1)/(n-1)!" is');
  laplace(x)

%(c)
  x=(t^n)*exp(-a*t);
  disp('(c) Laplace transform of "(t^n)*exp(-a*t)" is');
  laplace(x)

%(d)
  x=(t^(n-1)*exp(-a*t))/factorial(n-1);
  disp('(d) Laplace transform of "(t^(n-1)*exp(-a*t))/(n-1)!" is');
  laplace(x)
```

OUTPUT

```
    Enter the value of n4
      (a) Laplace transform of "t^n" is
          ans =
              24/s^5
      (b) Laplace transform of "t^(n-1)/(n-1)!" is
          ans =
              1/s^4
      (c) Laplace transform of "(t^n)*exp(-a*t)" is
          ans =
              24/(s+a)^5
      (d) Laplace transform of "(t^(n-1)*exp(-a*t))/(n-1)!" is
          ans =
              1/(s+a)^4
```

Program 2.7

Write a MATLAB program to find Laplace transform of the following causal signals.

a) $t^2 - 3t$ b) $1 + 0.4e^{-2t}\sin 3t$ c) $3\sin 2t + 3\cos 2t$

```
%****Program to find Laplace transform of given signals
  clear all
  syms t real; %Let t be real variable

%(a)
  x=t^2-(3*t);
  disp('(a) Laplace transform of "t^2-(3*t)" is');
  X=laplace(x);
  simplify(X)

%(b)
  x=1+0.4*exp(-(2*t))*sin(3*t);
  disp('(b) Laplace transform of "1+(0.4*exp(-(2*t)))*sin(3*t)" is');
  X=laplace(x);
  simplify(X)
```

```
%(c)
x=3*sin(2*t)+3*cos(2*t);
disp('(c) Laplace transform of "3*sin(2*t)+3*cos(2*t)" is');
X=laplace(x);
simplify(X)
```

OUTPUT

(a) Laplace transform of "t^2-(3*t)" is
      ans =
        -(-2+3*s)/s^3

(b) Laplace transform of "1+(0.4*exp(-(2*t)))*sin(3*t)" is
      ans =
        1/5*(5*s^2+26*s+65)/s/(s^2+4*s+13)

(c) Laplace transform of "3*sin(2*t)+3*cos(2*t)" is
      ans =
        3*(s+2)/(s^2+4)

Program 2.8

Write a MATLAB program to find inverse Laplace transform of the following s-domain signals.

a) $2/s(s+1)(s+2)$ b) $2/s(s+1)(s+2)^2$ c) $1/(s^2+s+1)(s+2)$

```
%****Program to determine inverse Laplace transform
clear all;
syms s complex;

X=2/(s*(s+1)*(s+2));
disp('Inverse Laplace transform of 2/(s(s+1)(s+2)) is');
x=ilaplace(X);
simplify(x)

X=2/(s*(s+1)*(s+2)^2);
disp('Inverse Laplace transform of 2/(s(s+1)(s+2)^2) is');
x=ilaplace(X);
simplify(x)

X=1/((s^2+s+1)*(s+2));
disp('Inverse Laplace transform of 1/((s^2+s+1)*(s+2)) is');
x=ilaplace(X);
simplify(x)
```

OUTPUT

Inverse Laplace transform of 2/(s(s+1)(s+2)) is
      ans =
        1+exp(-2*t)-2*exp(-t)

Inverse Laplace transform of 2/(s(s+1)(s+2)^2) is
      ans =
        t*exp(-2*t)-2*exp(-t)+3/2*exp(-2*t)+1/2

Inverse Laplace transform of 1/((s^2+s+1)*(s+2)) is
      ans =
        -1/3*exp(-1/2*t)*cos(1/2*3^(1/2)*t)
        +1/3*3^(1/2)*exp(-1/2*t)*sin(1/2*3^(1/2)*t)+1/3*exp(-2*t)

2.15 Exercises

I. Fill in the blanks with appropriate words

1. In Fourier series representation of a signal, if Ω_0 is the fundamental frequency then $n\Omega_0$ are called _____ frequencies.

2. Fourier series is useful for frequency domain analysis of _____ signals.

3. The plot of magnitude of Fourier coefficient c_n with respect to "n" is called_____.

4. For a signal x(t), the condition to be satisfied for half wave symmetry is_____.

5. For a signal x(t), the condition to be satisfied for quarter wave symmetry is_____ and _____.

6. If x(t) and X(jΩ) are Fourier transform pair then magnitude of X(jΩ) is called_____.

7. If x(t) and X(jΩ) are Fourier transform pair then phase of X(jΩ) is called_____.

8. The ROC of X(s) consists of stripes parallel to the ————— in s-plane.

9. The Fourier transform of impulse response will be equal to_____.

10. If x(t) is real and odd, then X(jΩ) will be_____.

11. Fourier transform of periodic continuous time signal consists of _____ located at _____ frequencies of the signal.

12. Fourier transform is evalution of Laplace transform along the _____ axis in s-plane.

13. The Laplace transform will transform a time domain signal to ————— signal.

14. The imaginary part of complex frequency is called ————— frequency.

15. The real part of complex frequency is called ————— frequency.

16. The ROC of Laplace transform of a causal signal includes all points on the s-plane to the ————— of abscissa of convergence.

17. A signal is said to be ————— if $e^{-\sigma t}|x(t)|$ approaches zero as t approaches infinity.

18. If Laplace transform of x(t) is X(s) then Laplace transform of Kx(t) is ————— .

19. If Laplace transform of x(t) is X(s) then Laplace transform of x(t – a) is ————— .

20. When X(s) is a ratio of two polynomials in s, then X(s) is called ————— of s.

21. The ————— are critical frequencies at which the signal X(s) become infinite.

22. The ————— are critical frequencies at which the signal X(s) become zero.

23. The ratio of Fourier transform of output and input of a system is called_____.

Answers

1. harmonic	7. phase spectrum	13. s-domain	19. $e^{-as}X(s)$
2. periodic	8. jΩ – axis	14. radian / real	20. rational function
3. magnitude line spectrum	9. transfer function	15. neper	21. poles
4. x(t ± T/2) = –x(t),	10. imaginary and odd	16. right	22. zeros
5. x(t ± T/2) = –x(t); x(–t) = ± x(t)	11. impulses, harmonic	17. exponential order	23. transfer function
6. magnitude spectrum	12. imaginary	18. K X(s)	

II. State whether the following statements are True/False.

1. The magnitude line spectrum is symmetric and phase line spectrum is antisymmetric.
2. For waveforms with even symmetry, the Fourier coefficient a_0 is always zero.
3. For waveforms with odd symmetry, the Fourier coefficients a_0 and a_n are always zero.
4. An alternating waveform will always have even harmonics only.
5. An even signal with half wave symmetry will always have even harmonics of cosine terms.
6. An odd signal with half wave symmetry will always have odd harmonics of sine terms.
7. The shifting of vertical axis of a waveform, will not affect the even/odd symmetry.
8. In frequency spectrum using Fourier transform, the magnitude spectrum will have even symmetry and phase spectrum will have odd symmetry.
9. The Fourier transform of a periodic signal will be summation of impulses.
10. The total average power in a periodic signal is equal to the sum of power in all of its harmonics.
11. Fourier transform is useful for frequency domain analysis of both periodic and nonperiodic signals.
12. Laplace transform is a generalized transform and Fourier transform is a particular transform.
13. The signal Ke^{st} is an universal signal which represents all types of signals.
14. The Laplace transform is used to transform a frequency domain signal to complex frequency domain.
15. The Laplace transform exists only if the signal is of exponential order.
16. The Laplace transform of the sum of two or more signals is equal to sum of transforms of individual signals.
17. Laplace transform of convolution of two signals is given by the product of the Laplace transform of the individual signals.
18. For a rational signal X(s), the ROC includes poles of X(s).
19. For a realizable system the number of finite poles and zeros should be equal.
20. For a causal rational signal X(s) the ROC is the region to the right of rightmost pole.
21. For an anticausal rational signal X(s) the ROC is the region to the left of leftmost pole.
22. If x(t) is finite and integrable then the ROC is entire s-plane.
23. Zeros and poles of transfer function are complex frequencies.

Answers

1. True	5. False	9. True	13. True	17. True	21. True
2. False	6. True	10. True	14. False	18. False	22. True
3. True	7. False	11. True	15. True	19. False	23. True
4. False	8. True	12. True	16. True	20. True	

III. Choose the right answer for the following questions.

1. *The Fourier coefficient 'a_n' can be evaluated as,*

a) $a_n = \dfrac{2}{T} \displaystyle\int_{-\infty}^{+\infty} x(t)\cos n\Omega_0 t \ dt$

b) $a_n = \dfrac{2}{T} \displaystyle\int_{0}^{T} x(t)\cos n\Omega_0 t \ dt$

c) $a_n = \dfrac{2}{T} \displaystyle\int_{-T/2}^{T/2} x(t)\sin n\Omega_0 t \ dt$

d) $a_n = \dfrac{2}{T} \displaystyle\int_{0}^{\infty} x(t)\cos n\Omega_0 t \ dt$

2. The exponential Fourier coefficient of $x(t) = \sin^2 t$ is,

   a) $C_1 = \dfrac{1}{\pi}$, $C_0 = 0$, $C_{-1} = -\dfrac{1}{\pi}$

   b) $C_{-1} = -\dfrac{1}{4}$, $C_0 = \dfrac{1}{2}$, $C_1 = -\dfrac{1}{4}$

   c) $C_{-1} = -\dfrac{1}{2}$, $C_0 = 1$, $C_1 = \dfrac{1}{2}$

   d) $C_{-1} = -2$, $C_0 = 1$, $C_1 = -2$

3. Fourier transform of the signum function, $x(t) = sgn(t)$ is,

   a) $2/\Omega$ **b)** $2/j\Omega$ **c)** $2/(j\Omega)^2$ **d)** $-2/j\Omega$

4. The Unit step signal $u(t)$ can also be expressed as,

   a) $1 + sgn(t)$ **b)** $1 + \delta(t)$ **c)** $\dfrac{1}{2} + \dfrac{1}{2} sgn(t)$ **d)** $2sgn(t)$

5. The necessary and sufficient condition for $x(t)$ to be real is,

   a) $X^*(j\Omega) = X(j\Omega)$ **b)** $X^*(j\Omega) = X(-j\Omega)$ **c)** $X(j\Omega) = X^*(-j\Omega)$ **d)** none of the above

6. Fourier series of any periodic signal $x(t)$ can only be obtained if,

   a) finite number of discontinuities within finite time interval, T.
   b) finite number of positive and negative maxima in the period, T.
   c) well defined at infinite number of points.
   d) both (a) and (b).

7. The inverse Fourier Transform of $X(j\Omega)$ is defined as,

   a) $\dfrac{1}{2\pi} \displaystyle\int_0^{\infty} X(j\Omega)\, e^{j\Omega t}\, d\Omega$

   b) $\dfrac{1}{2\pi} \displaystyle\int_{+\infty}^{-\infty} X(j\Omega)\, e^{j\Omega t}\, dt$

   c) $\dfrac{1}{2\pi} \displaystyle\int_{-\infty}^{\infty} X(j\Omega)\, e^{j\Omega t}\, d\Omega$

   d) $\dfrac{1}{2\pi} \displaystyle\int_0^{\infty} X(j\Omega)\, e^{-j\Omega t}\, d\Omega$

8. The frequency convolution property of Fourier transform says that, $\mathcal{F}\{x_1(t)\, x_2(t)\}$ is given by,

   a) $X_1(j\Omega)\, X_2(j\Omega)$

   b) $\dfrac{1}{2\pi} \displaystyle\int_{\lambda=-\infty}^{\lambda=+\infty} X_1(j\lambda)\, X_2(j(\Omega-\lambda))\, d\tau$

   c) $\dfrac{1}{2\pi} \displaystyle\int_{-\infty}^{+\infty} X_1(j\Omega)\, X_2(j\Omega)\, d\Omega$

   d) $\dfrac{1}{2\pi} \displaystyle\int_{-\infty}^{+\infty} |X_1(j\Omega)|^2\, |X_2(j\Omega)|^2\, d\Omega$

9. Fourier transform of a sinusoidal signal $2\sin\Omega_0 t$ is,

   a) $\dfrac{2\pi}{j}[\delta(\Omega-\Omega_0) - \delta(\Omega+\Omega_0)]$

   b) $\dfrac{2\pi}{j}[\delta(\Omega-\Omega_0) - \delta(\Omega+\Omega_0)]$

   c) $\dfrac{2\pi}{j}[\delta(\Omega+\Omega_0) - \delta(\Omega-\Omega_0)]$

   d) $\dfrac{2\pi}{j}[\delta(\Omega-\Omega_0) - \delta(\Omega+\Omega_0)]$

10. In the Fourier transform of a periodic signal, the magnitude of nth impulse is,

   a) $\dfrac{|C_n|}{2\pi}$ **b)** $2\pi|C_n|$ **c)** $|C_n|$ **d)** $\dfrac{|C_n|}{2}$

11. *Fourier transform of the causal signal, $x(t) = cos\Omega_0 t\ u(t)$ is,*

   a) $\dfrac{\Omega_0}{\Omega_0^2 - \Omega^2}$ b) $\dfrac{j\Omega}{\Omega_0^2 - \Omega^2}$ c) $\dfrac{-\Omega_0}{\Omega^2 + \Omega_0^2}$ d) $\dfrac{j\Omega}{\Omega_0^2 + \Omega^2}$

12. *If the signal x(t) has odd and half wave symmetry, then, the Fourier series will have only,*

   a) odd harmonics of sine terms.
   b) constant term and odd harmonics of cosine terms.
   c) even harmonics of sine terms.
   d) odd harmonics of cosine terms.

13. *Differentiation of signum function is, $\left(i.e., \dfrac{d}{dt} sgn(t)\right)$*

   a) $2\delta(t)$ b) $\delta(t)$ c) $\dfrac{1}{2}\delta(t)$ d) $2u(t)$

14. *Fourier transform of $cos\Omega_0 t$ is,*

   a) $X(\Omega - \Omega_0) + X(\Omega + \Omega_0)$ b) $\pi[\delta(\Omega - \Omega_0) + \delta(\Omega + \Omega_0)]$

   c) $\dfrac{1}{2}X(\Omega - \Omega_0) + \dfrac{1}{2}X(\Omega + \Omega_0)$ d) $\dfrac{\pi}{2}[X(\Omega - \Omega_0) + X(\Omega + \Omega_0)]$

15. *The Fourier transform of x(t) exists only if,*

   a) $\displaystyle\int_0^\infty x(t)\ dt < \infty$ b) $\displaystyle\int_{-\infty}^{+\infty} x(t)\ dt > 0$ c) $\displaystyle\int_{-\infty}^{+\infty} x(t)\ e^{j\Omega t}\ dt < \infty$ d) $\displaystyle\int_{-\infty}^{+\infty} x(t)\ dt < \infty$

16. *By convolution property of Fourier transform, $F\{x(t)* h(t)\}$ is,*

   a) $\dfrac{1}{2\pi}\displaystyle\int_{-\infty}^{+\infty} X(j\Omega)\ H(j\Omega)\ d\Omega$ b) $X(j\Omega)\ H(j\Omega)$

   c) $X(j\Omega) * H(j\Omega)$ d) None of the above

17. *A periodic signal x(t) of period T_0 is given by, $x(t) = 1\ ;\ |t| < T_1$.*

$$= 0\ ;\ T_1 < |t| < T_0/2$$

The constant component of x(t) is

   a) $\dfrac{T_0}{T_1}$ b) $\dfrac{T_1}{T_0}$ c) $\dfrac{2T_1}{T_0}$ d) $\dfrac{T_1}{2T_0}$

18. *For two periodic waveforms, square and triangular, the magnitude of the nth Fourier series coefficients, for n > 0, are respectively proportional to,*

   a) $|n^{-3}|$ and $|n^{-2}|$ b) $|n^{-2}|$ and $|n^{-3}|$ c) $|n^{-1}|$ and $|n^{-2}|$ d) $|n^{-4}|$ and $|n^{-2}|$

19. *When the waveform has parabolic structure/wiggles, the magnitude of higher harmonics,*

   a) increases rapidly
   b) decreases more rapidly
   c) remain same
   d) become zero

20. *The initial value of a continuous time signal in frequency domain is,*

   a) $X(0) = \displaystyle\int_0^\infty x(t)\,dt$ b) $X(0) = \dfrac{1}{2\pi}\displaystyle\int_{-\infty}^\infty x(j\Omega)\,dt$ c) $X(0) = \dfrac{1}{2\pi}\displaystyle\int_{-\infty}^\infty x(t)\,d\Omega$ d) $X(0) = \displaystyle\int_{-\infty}^\infty x(t)\,dt$

21. Which of the following cannot be the Fourier series expansion of a periodic signal?

a) $x(t) = 2 \cos t + 3 \cos 3t$ **b)** $x(t) = 2 \cos\pi t + 7 \cos t$

c) $x(t) = \cos t + 0.5$ **d)** $x(t) = 2 \cos 1.5\pi t + \sin 3.5\pi t$

22. Let C_n be a Fourier coefficient in exponential form. It is given that, $C_n = 2 + j7$ then C_{-n} is,

a) $-2 - j7$ **b)** $2 - j7$ **c)** $-2 + j7$ **d)** $2 + j7$

23. The Fourier transform of a function $x(t)$ is $X(j\Omega)$. Then, the Fourier transform of $\frac{d}{dt}x(t)$ will be,

a) $\dfrac{dX(\Omega)}{dF}$ **b)** $j2\pi\Omega\, X(\Omega)$ **c)** $j\Omega\, X(j\Omega)$ **d)** $\dfrac{X(\Omega)}{j\Omega}$

24. If Fourier transform of the signal $e^{-|t|}$ is $\dfrac{2}{1+\Omega^2}$ then, the Fourier transform of the signal $\dfrac{2}{1+t^2}$, using duality property is,

a) $2\pi\, e^{|j\Omega|}$ **b)** $2\pi\, e^{-|j\Omega|}$ **c)** $2\pi\,(-j\Omega)$ **d)** $2\pi\,(j\Omega)$

25. Fourier transform of gaussian pulse will be,

a) another gaussian pulse   **b)** squared sinc pulse

c) sinc pulse   **d)** impulse train

26. The signal $x(t) = u(6t)$ has a Fourier transform of,

a) $\dfrac{1}{j\Omega} + 6\pi\,\delta(\Omega)$ **b)** $\dfrac{6}{j\Omega} + 6\pi\,\delta(\Omega)$ **c)** $\dfrac{1}{j\Omega} + \pi\,\delta(\Omega)$ **d)** $\dfrac{1}{j6\Omega} + \pi\,\delta(6\Omega)$

27. Consider the signal, $x(t) = t(T-t); \ 0 \le t \le T$. The Fourier coefficient a_0 for this signal is,

a) $\dfrac{T^3}{4}$ **b)** $\dfrac{T^2}{3}$ **c)** $\dfrac{3}{T^2}$ **d)** $\dfrac{T^3}{12}$

28. The Fourier transform of the signal $x(t) = e^{7t}\, u(-t)$ is,

a) $\dfrac{1}{7+j\Omega}$ **b)** $\dfrac{7}{1+j\Omega}$ **c)** $\dfrac{7}{1-j\Omega}$ **d)** $\dfrac{1}{7-j\Omega}$

29. For a signal $x(t) = e^{-|t|}$, the Fourier transform is,

a) $\dfrac{2}{1+\Omega^2}$ **b)** $\dfrac{2}{1-\Omega^2}$ **c)** $\dfrac{1}{2-\Omega^2}$ **d)** $\dfrac{1}{2+\Omega^2}$

30. If $\mathcal{F}\{x_1(t)\} = \dfrac{a}{\Omega - a}$ and $\mathcal{F}\{x_2(t)\} = \dfrac{a}{\Omega + a}$, then $\mathcal{F}\{x_1(t) * x_2(t)\}$ is,

a) $\dfrac{\Omega - a}{\Omega + a}$ **b)** $\dfrac{\Omega + a}{\Omega - a}$ **c)** $\dfrac{a^2}{\Omega^2 - a^2}$ **d)** $\dfrac{a^2}{\Omega^2 + a^2}$

31. If Fourier transform of $x(t)$ is $X(\Omega)$ then Fourier transform of $e^{-j\Omega_0 t}\, x(t)$ is,

a) $\dfrac{X(\Omega)}{\Omega_0}$ **b)** $X(\Omega - \Omega_0)$ **c)** $X(\Omega + \Omega_0)$ **d)** $\Omega_0\, X(\Omega)$

32. If a signal $x(t)$ is differentiated 'm' times to produce an impulse then its Fourier coefficients will be proportional to,

a) n^m **b)** $\dfrac{1}{n^{m-1}}$ **c)** n^{m-1} **d)** $\dfrac{1}{n^m}$

33. *If Fourier transform of x(t) is* $\dfrac{2a}{a^2 + \Omega^2}$, *then Fourier transform of* $\dfrac{d^2}{dt^2} x(t)$ *is,*

a) $\dfrac{-2a\Omega^2}{a^2 + \Omega^2}$ b) $\dfrac{4a^2}{(a^2 + \Omega^2)^2}$ c) $\dfrac{2a\Omega}{(a^2 + \Omega^2)^2}$ d) $\dfrac{a^2 + \Omega^2}{2a}$

34. *The Fourier transform of the signal* $x(t) = e^{-4|t-t_0|}$ *is,*

a) $\left(\dfrac{4}{4^2 + \Omega^2}\right) e^{-j\Omega t_0}$ b) $\left(\dfrac{-4}{4^2 + \Omega^2}\right) e^{-j\Omega t_0}$ c) $\left(\dfrac{8}{4^2 + \Omega^2}\right) e^{-j\Omega t_0}$ d) $\left(\dfrac{-8}{4^2 + \Omega^2}\right) e^{-j\Omega t_0}$

35. *A signal x(t) has a magnitude of 5 at t =0. Then value of* $\displaystyle\int_{-\infty}^{\infty} X(j\Omega)\, d\Omega$ *will be,*

a) 10   b) 10π c) 5π d) $\dfrac{10}{\pi}$

36. *When* $\Omega = 0$ *in the signal x(t) = Ae^{st}, where s = $\sigma + j\Omega$, the signal x(t) represents,*

a) exponential signal   b) step signal   c) ramp signal   d) sinusoidal signal

37. *The mathematical operationi* $\displaystyle\int_{-\infty}^{+\infty} x(t)\, e^{-st}\, dt$ *is called,*

a) Laplace transform   b) two sided Laplace transform
c) one sided Laplace transform   d) generalized frequency domain transform

38. *The term* Ω *in s = $\sigma + j\Omega$ is called,*

a) frequency   b) real frequency   c) complex frequency   d) critical frequency

39. *A causal signal x(t) is said to be exponential order for any positive value of σ if,*

a) $\underset{t \to \infty}{\mathrm{Lt}}\, e^{+\sigma t} |x(t)| = \infty$ b) $\underset{t \to \infty}{\mathrm{Lt}}\, e^{-\sigma t} |x(t)| = \infty$ c) $\underset{t \to \infty}{\mathrm{Lt}}\, e^{+\sigma t} |x(t)| = 0$ d) $\underset{t \to \infty}{\mathrm{Lt}}\, e^{-\sigma t} |x(t)| = 0$

40. *The ROC of a causal signal x(t) is,*

a) entire s-plane   b) region in between two abscissa of convergence
c) right of abscissa of convergence   d) left of abscissa of convergence

41. *The Laplace transform of the causal signal $t^n u(t)$ is,*

a) $\dfrac{n!}{s^{n+1}}$ b) $\dfrac{n!}{s^n}$ c) $\dfrac{n}{s^{n+1}}$ d) $\dfrac{n}{s^n}$

42. *If x(t) and X(s) are Laplace transform pairs, then Laplace transform of $e^{-at} x(t)$ is,*

a) $e^{as} X(s)$ b) $e^{-as} X(s)$ c) $X(s - a)$ d) $X(s + a)$

43. *If x(t) and X(s) are Laplace transform pairs, then Laplace transform of* $\dfrac{x(t)}{t}$ *is,*

a) $\displaystyle\int_{0}^{\infty} X(s)\, ds$ b) $\displaystyle\int_{s}^{\infty} X(s)\, ds$ c) $\dfrac{1}{s} \displaystyle\int_{0}^{\infty} X(s)\, ds$ d) $\dfrac{1}{s} \displaystyle\int_{s}^{\infty} X(s)\, ds$

44. *If x(t) is periodic with period T, then Laplace transform of x(t) is defined as,*

a) $\dfrac{1}{1 - e^{-sT}} \displaystyle\int_{0}^{T} x(t)\, e^{-st}\, dt$ b) $\dfrac{1}{1 + e^{-sT}} \displaystyle\int_{0}^{T} x(t)\, e^{-st}\, dt$ c) $\dfrac{1}{1 - e^{sT}} \displaystyle\int_{0}^{T} x(t)\, e^{-st}\, dt$ d) $\dfrac{1}{1 + e^{sT}} \displaystyle\int_{0}^{T} x(t)\, e^{-st}\, dt$

45. **If x(t) and X(s) are Laplace transform pair, and X(s) is rational, then which of the following statements are true.**

   i) poles and zeros are critical frequencies.

   ii) the ROC of X(s) does not include poles.

   iii) the number of finite zeros will be less than or equal to number of poles.

   a) i) and iii) only   **b)** i) only   **c)** ii) only   **d)** all of the above

46. **If x(t) and X(s) are Laplace transform pair, and X(s) is rational, then which of the following statements are true regarding the ROC of X(s).**

   i) The ROC is bounded by poles or extends to infinity.

   ii) If x(t) is right sided, then ROC is right of right most pole.

   iii) If x(t) is left sided, then ROC is left of left most pole.

   a) i) only   **b)** ii) only   **c)** ii) and iii) only   **d)** all of the above

47. **The inverse Laplace transform of $X(s) = \dfrac{2}{s^2 + 2s + 5}$ is,**

   a) $x(t) = e^{-t} \cos 2t$ **b)** $x(t) = e^{-t} \sin 2t$ **c)** $x(t) = e^{-2t} \cos 5t$ **d)** $x(t) = e^{-2t} \sin 5t$

48. **The inverse Laplace transform of $X(s) = \dfrac{4}{s + 5}$ for ROC Re{s} > –5 and Re{s} < –5 are respectively.**

   a) $4\,e^{-5t}\,u(t)$ and $4\,e^{-5t}\,u(-t)$ **b)** $4\,e^{5t}\,u(t)$ and $4\,e^{5t}\,u(-t)$

   c) $4\,e^{-5t}\,u(t)$ and $-4\,e^{-5t}\,u(-t)$ **d)** $-4\,e^{-5t}\,u(t)$ and $-4\,e^{-5t}\,u(t)$

Answers								
1. b	7. c	13. a	19. b	25. a	31. b	37. b	43. b
2. b	8. b	14. b	20. d	26. c	32. d	38. b	44. a
3. b	9. a	15. d	21. b	27. d	33. a	39. d	45. d
4. c	10. b	16. b	22. b	28. d	34. c	40. c	46. d
5. a	11. b	17. c	23. c	29. a	35. b	41. a	47. b
6. d	12. a	18. c	24. b	30. c	36. a	42. d	48. c

IV. Answer the following questions.

1. Write the conditions for existence of Fourier series.
2. Write the trigonometric form of Fourier series representation of a periodic signal and explain.
3. Write the exponential form of Fourier series representation of a periodic signal and explain.
4. What is the relation between Fourier coefficients of trigonometric and exponential form?
5. Write a short note on negative frequency.
6. Define frequency spectrum or line spectrum.
7. Write short note on Fourier coefficients of signals with even symmetry and odd symmetry.
8. What is the effect of half wave symmetry on Fourier coefficients of a signal?
9. What is the effect of quarter wave symmetry on Fourier coefficients of a signal?
10. Write any two properties of Fourier series.

11. Write the Parseval's relation for continuous time periodic signal.
12. Explain how Fourier transform is obtained from Fourier series.
13. Define Fourier transform and inverse Fourier transform of a signal.
14. Write any two properties of Fourier transform.
15. State and prove time differentiation property of Fourier transform.
16. State and prove frequency differentiation property of Fourier transform.
17. Write the convolution theorem of Fourier transform.
18. What is the relation between Fourier transform and Laplace transform?
19. Show that Fourier transform of a periodic signal will be impulses.
20. Define complex frequency.
21. Define Laplace transform of a signal.
22. What is the condition to be satisfied for the existence of Laplace transform?
23. What is abscissa of convergence?
24. What is region of convergence(ROC)?
25. Define inverse Laplace transform.
26. Write any two properties of Laplace transform.
27. State and prove initial value theorem.
28. State and prove final value theorem.
29. Define the convolution theorem of Laplace transform.
30. Write any two properties of ROC of Laplace transform.
31. Write a procedure to determine inverse Laplace transform using convolution theorem.
32. Define transfer function of a continuous time system.
33. Define poles and zeros.
34. Write the procedure to perform convolution using Laplace transform.
35. Write the procedure to perform deconvolution using Laplace transform.

V. Solve the following problems

E2.1 Determine the trigonometric Fourier series representation of the signal shown in Fig E2.1.

E2.2 Determine the trigonometric Fourier series representation of the signal shown in Fig E2.2.

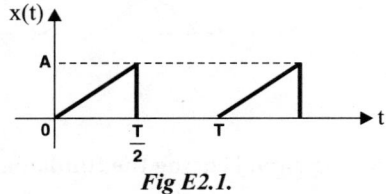

Fig E2.1.

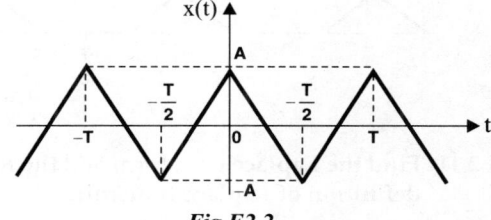

Fig E2.2.

E2.3 Determine the exponential Fourier series representation of the signal shown in Fig E2.3 and hence obtain the trigonometric Fourier series.

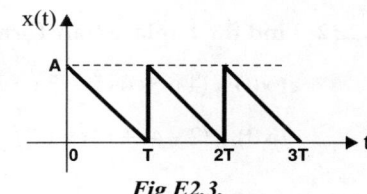

Fig E2.3.

E2.4   Determine the exponential Fourier series representation of the signal shown in Fig E2.4 and hence obtain the trigonometric Fourier series.

E2.5   Determine the exponential Fourier series representation of the signal shown in Fig E2.5.

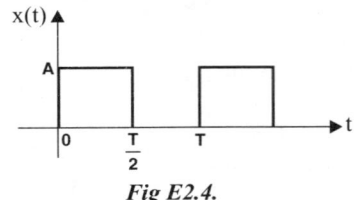

Fig E2.4.

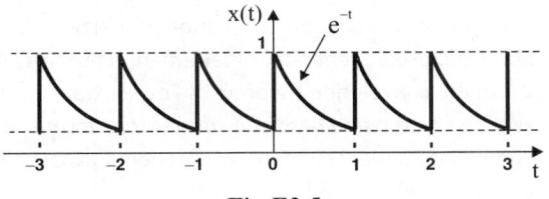

Fig E2.5.

E2.6   Find the Fourier transform of the time domain signals given below:

   a) $x(t) = t \sin\Omega_0 t$; $t = 0$ to $+\infty$

   b) $x(t) = t \cos \Omega_0 t$; $t = 0$ to $+\infty$

E2.7   Determine the Fourier transform of the rectangular pulse shown in Fig E2.7.

E2.8   Determine the Fourier transform of the triangular pulse shown in Fig E2.8.

E2.9   Determine the Fourier transform of the signal shown in Fig E2.9.

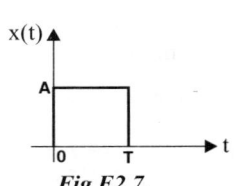

Fig E2.7.

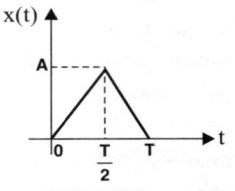

Fig E2.8.

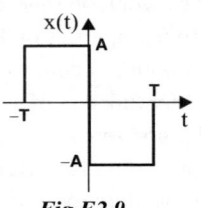

Fig E2.9.

E2.10 Determine Fourier transform of the periodic signal shown in Fig E2.10.

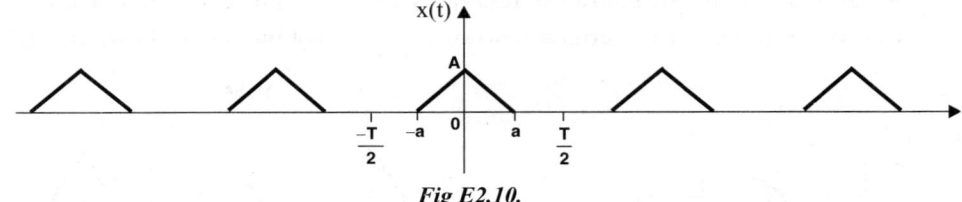

Fig E2.10.

E2.11 Find the Laplace transform and the ROC of the following signals using the fundamental definition of Laplace transform.

   a) $x(t) = (t^2 + 2t + 4)\, u(t)$ **b)** $x(t) = \cos^2 6t\, u(t)$ **c)** $x(t) = e^{-5t}\sin 7t\, u(t)$ **d)** $x(t) = e^{at+b}\, u(t)$

E2.12 Find the Laplace transform of the following signals.

   a) $x(t) = (4\, e^{-2t}\cos 5t - 3\, e^{-2t}\sin 5t)\, u(t)$ **b)** $x(t) = (t-2)^3\, u(t-2)$

   c) $x(t) = (2 - 4t + t^2)\, e^{-t}\, u(t)$ **d)** $x(t) = \delta(t) + (t\, e^{-3t} + 2 \cos 5t)\, u(t)$

E2.13 Find the Laplace transform of the following signals.

a)

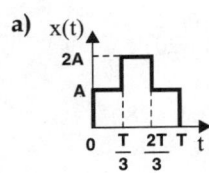

Fig E2.13.1.

b)

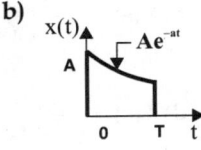

Fig E2.13.2.

c)

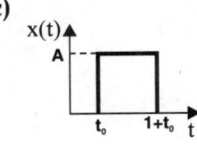

Fig E2.13.3.

d)

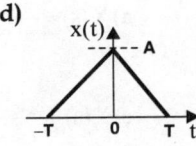

Fig E2.13.4.

E2.14 Find the Laplace transform of the signal shown in fig E2.14.

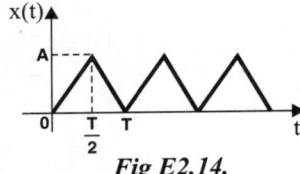

Fig E2.14.

E2.15 Find the initial value, $x(0)$ in time domain for the following s-domain signals.

a) $X(s) = \dfrac{10(s+1)}{(s+2)(s+6)}$

b) $X(s) = \dfrac{2s^2 + 2}{8s^3 + 4s^2 + 3s + 5}$

c) $X(s) = \dfrac{4}{s^3 + 2s^2 + 7s + 1}$

d) $X(s) = \dfrac{1}{s^3} + \dfrac{7}{s^2} + \dfrac{4}{s}$

E2.16 Find the final value, $x(\infty)$ in time domain for the following s-domain signals.

a) $X(s) = \dfrac{10(s+1)}{s(s+2)(s+4)}$ b) $X(s) = \dfrac{s}{s^2 + 4}$ c) $X(s) = \dfrac{(s+2)(s+3)}{s^2(s+5)}$ d) $X(s) = \dfrac{s(s+1)}{s^2 + 2s + 1}$

E2.17 Perform the convolution of $x_1(t)$ and $x_2(t)$ using Laplace transform.

a) $x_1(t) = t\,e^{-2t}u(t)$; $x_2(t) = 3\sin 4t\,u(t)$

b) $x_1(t) = (t-2)\,u(t-2)$; $x_2(t) = e^{-3t}u(t)$

c) $x_1(t) = 4t\,e^{-5t}u(t)$; $x_2(t) = u(t)$

d) $x_1(t) = t^2\,u(t)$; $x_2(t) = \delta(t-7)$

e) $x_1(t) = \sin t\,u(t)$; $x_2(t) = t\,e^{-t}u(t)$

E2.18 Determine the poles and zeros of the following s-domain signal and sketch the pole-zero plot.

$$X(s) = \dfrac{(s+1)(s^2 + 8s + 20)}{(s+3)(s^2 + 2s + 2)}$$

E2.19 Find the inverse Laplace transform of the following signals.

a) $X(s) = \dfrac{s}{s^2 + 4s + 3}$

b) $X(s) = \dfrac{2s+1}{s^3 + 7s^2 + 10s}$

c) $X(s) = \dfrac{1}{s^3 + 2s^2 + 2s + 1}$

d) $X(s) = \dfrac{s+1}{s^2 + 4s + 5}$

Signals and Systems-Simplified

E2.20 Find the inverse Laplace transform of the following signals.

a) $X(s) = \dfrac{s^2 + 2s - 2}{s(s+2)(s-3)}$

b) $X(s) = \dfrac{4s^2 + 15s + 62}{(s+1)(s^2 + 4s + 20)}$

c) $X(s) = \dfrac{2s^2 + 4s + 34}{(s+2)(s^2 + 6s + 25)}$

d) $X(s) = \dfrac{s^3 + 8s^2 + 23s + 28}{(s^2 + 4s + 13)(s^2 + 2s + 5)}$

E2.21 Find the inverse Laplace transform of $X(s) = \dfrac{1}{s^2(s^2 - a^2)}$ using convolution theorem.

Answers

E2.1 $x(t) = \dfrac{A}{4} - \dfrac{2A}{\pi^2}\left(\dfrac{\cos\Omega_0 t}{1^2} + \dfrac{\cos 3\Omega_0 t}{3^2} + \dfrac{\cos 5\Omega_0 t}{5^2} + \ldots\ldots\right) + \dfrac{A}{\pi}\left(\dfrac{\sin\Omega_0 t}{1} - \dfrac{\sin 2\Omega_0 t}{2} + \dfrac{\sin 3\Omega_0 t}{3} - \ldots\ldots\right)$

E2.2 $x(t) = \dfrac{8A}{\pi^2}\left(\dfrac{\cos\Omega_0 t}{1^2} + \dfrac{\cos 3\Omega_0 t}{3^2} + \dfrac{\cos 5\Omega_0 t}{5^2} + \ldots\ldots\right)$

E2.3 $x(t) = \dfrac{A}{j2\pi}\left(\ldots\ldots - \dfrac{e^{-j3\Omega_0 t}}{3} - \dfrac{e^{-j2\Omega_0 t}}{3} - \dfrac{e^{-j\Omega_0 t}}{3} + j\pi + \dfrac{e^{j\Omega_0 t}}{1} + \dfrac{e^{j2\Omega_0 t}}{2} + \dfrac{e^{j3\Omega_0 t}}{3} + \ldots\ldots\right)$

$x(t) = \dfrac{A}{2} + \dfrac{A}{\pi}\left(\dfrac{\sin\Omega_0 t}{1} + \dfrac{\sin 2\Omega_0 t}{2} + \dfrac{\sin 3\Omega_0 t}{3} + \ldots\ldots\right)$

E2.4 $x(t) = \dfrac{A}{j\pi}\left(\ldots\ldots - \dfrac{e^{-j5\Omega_0 t}}{5} - \dfrac{e^{-j3\Omega_0 t}}{3} - \dfrac{e^{-j\Omega_0 t}}{1} + \dfrac{j\pi}{2} + \dfrac{e^{j\Omega_0 t}}{1} + \dfrac{e^{j3\Omega_0 t}}{3} + \dfrac{e^{j5\Omega_0 t}}{5} + \ldots\ldots\right)$

$x(t) = \dfrac{A}{2} + \dfrac{2A}{\pi}\left(\dfrac{\sin\Omega_0 t}{1} + \dfrac{\sin 3\Omega_0 t}{3} + \dfrac{\sin 5\Omega_0 t}{5} + \ldots\ldots\right)$

E2.5 $x(t) = (1 - e^{-1})\left[\ldots\ldots + \dfrac{e^{-j3\Omega_0 t}}{1 - j6\pi} + \dfrac{e^{-j2\Omega_0 t}}{1 - j4\pi} + \dfrac{e^{-j\Omega_0 t}}{1 - j2\pi} + 1 + \dfrac{e^{j\Omega_0 t}}{1 + j2\pi} + \dfrac{e^{j2\Omega_0 t}}{1 + j4\pi} + \dfrac{e^{j3\Omega_0 t}}{1 + j6\pi} + \ldots\ldots\right]$

E2.6 i) $X(j\Omega) = \dfrac{-2\Omega\Omega_0}{(\Omega^2 + \Omega_0^2)^2}$ ii) $X(j\Omega) = \dfrac{\Omega_0^2 - \Omega^2}{(\Omega_0^2 + \Omega^2)^2}$

E2.7 $X(j\Omega) = \dfrac{A}{j\Omega}(1 - e^{-j\Omega T})$

E2.8 $X(j\Omega) = \dfrac{2A}{T\Omega^2}(2e^{-j\Omega T/2} - e^{-j\Omega T} - 1)$

E2.9 $X(j\Omega) = j\Omega A T^2 \sin c\left(\dfrac{\Omega T}{2\pi}\right)$

E2.10 $X(j\Omega) = \displaystyle\sum_{n=-\infty}^{+\infty} \dfrac{A}{2\pi^2 n^2}[1 - \cos 2n\pi]$

E2.11 a) $X(s) = \dfrac{4}{s^3}(s^2 + 0.5s + 0.5)$, ROC : $\sigma > 0$ **b)** $X(s) = \dfrac{s^2 + 72}{s(s^2 + 144)}(s^2 + 0.5s + 0.5)$, ROC : $\sigma > 0$

c) $X(s) = \dfrac{7}{(s+5)^2 + 49}$,   ROC : $\sigma > -5$ **d)** $X(s) = \dfrac{e^b}{s - a}$,   ROC : $\sigma > +a$

E2.12 **a)** $X(s) = \dfrac{4s - 7}{s^2 + 2s + 29}$

b) $X(s) = \dfrac{6\,e^{-2s}}{s^4}$

c) $X(s) = \dfrac{2s^2}{(s+1)^3}$

d) $X(s) = \dfrac{s^4 + 8s^3 + 47s^2 + 168s + 250}{(s+3)^2\,(s^2 + 25)}$

E2.13 **a)** $X(s) = \dfrac{A}{s}\big[1 + e^{-\frac{sT}{3}} - e^{-\frac{2sT}{3}} - e^{-sT}\big]$

b) $X(s) = \dfrac{A}{(s+a)}\big[1 - e^{-(s+a)T}\big]$

c) $X(s) = \dfrac{A}{s}\big[e^{-st_0}(1 - e^{-s})\big]$

d) $X(s) = \dfrac{2A}{Ts^2}(\cosh sT - 1)$

E2.14 $X(s) = \dfrac{2A}{Ts^2}\left(\dfrac{1 - e^{-\frac{sT}{2}}}{1 + e^{-\frac{sT}{2}}}\right)$

E2.15 **a)** $x(0) = 10$ **b)** $x(0) = 0.25$ **c)** $x(0) = 0$ **d)** $x(0) = 4$

E2.16 **a)** $x(\infty) = 1.25$ **b)** $x(\infty) = 0$ **c)** $x(\infty) = \infty$ **d)** $x(\infty) = 0$

E2.17 **a)** $x_1(t) * x_2(t) = (0.6\,t\,e^{-2t} + 0.12\,e^{-2t} - 0.12\cos 4t - 0.09\sin 4t)\,u(t)$

b) $x_1(t) * x_2(t) = \dfrac{1}{3}\big[(t-2) - \dfrac{1}{3} + \dfrac{1}{3}e^{-3(t-2)}\big]u(t-2)$

c) $x_1(t) * x_2(t) = 0.16\,[1 - 5t\,e^{-5t} - e^{-5t}]\,u(t)$

d) $x_1(t) * x_2(t) = (t - 7)^2\,u(t - 7)$

e) $x_1(t) * x_2(t) = 0.5\,[(t + 1)\,e^{-t} - \cos t]u(t)$

E2.18 $p_1 = -3,\quad p_2 = -1 + j1,\quad p_3 = -1 - j$

$z_1 = -2,\quad z_2 = -4 + j2,\quad z_3 = -4 - j2$

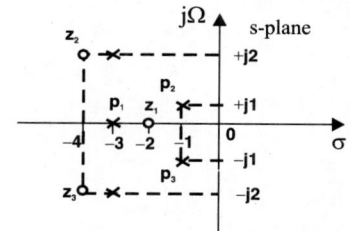

Fig E2.18: *Pole-zero plot.*

E2.19 **a)** $x(t) = [1.5\,e^{-3t} - 0.5t\,e^{-t}]\,u(t)$ **b)** $x(t) = [0.1 - 0.5\,e^{-2t} - 0.6\,e^{-5t}]\,u(t)$

c) $x(t) = [e^{-t} + e^{-0.5t}\,(\,0.577\sin 0.866t - \cos 0.866t\,)]\,u(t)$ **d)** $x(t) = e^{-2t}\cos t\,u(t)$

(or) $x(t) = [e^{-t} + 1.155\,e^{-0.5t}\sin 0.866t - 60^0\,)]\,u(t)$

E2.20 **a)** $x(t) = \left(\dfrac{1}{3} - \dfrac{1}{5}e^{-2t} + \dfrac{13}{15}e^{3t}\right)u(t)$ **b)** $x(t) = [\,3e^{-t} + e^{-2t}\cos 4t\,]\,u(t)$

c) $x(t) = 2[\,e^{-2t} - e^{-3t}\sin 4t\,]\,u(t)$ **d)** $x(t) = [\,e^{-2t}\sin 3t + e^{-t}\cos 2t\,]\,u(t)$

E2.21 $x(t) = \dfrac{1}{a^2}\left(\dfrac{\sinh at}{a} - t\right)u(t)$

CHAPTER 3

Linear Time Invariant Continuous Time System

3.1 Continuous Time System

A *continuous time system* (or analog system) is a physical device that operates on a continuous time signal (or an analog signal) called input or excitation, to produce another continuous time signal (or an analog signal) called output or response.

Linear Time Invariant (LTI) System

A continuous time system is linear if it obeys the principle of superposition and it is time invariant if its input-output relationship does not change with time. When a continuous time system satisfies the properties of linearity and time invariance then it is called an *LTI system* (Linear Time Invariant system).

Response of a Continuous Time System

In a continuous time system, the input signal x(t) is transformed by the system into a signal y(t), and the transformation can be expressed mathematically as shown in equation (3.1).

Response, $y(t) = \mathcal{H}\{x(t)\}$ .....(3.1)

where, $\mathcal{H}$ denotes the transformation (also called an operator).

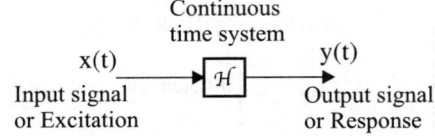

Fig 3.1: *Representation of continuous time system.*

3.1.1 Impulse Response   (AU Dec'12, 6 Marks & Jun'11, 2 Marks)

When the input to a continuous time system is a unit impulse signal $\delta(t)$ then the output is called an *impulse response* of the system and it is denoted by h(t).

∴ Impulse Response, $h(t) = \mathcal{H}\{\delta(t)\}$

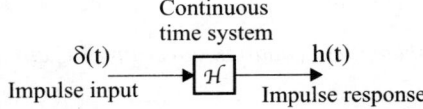

Fig 3.2: *Continuous time system with impulse input.*

3.1.2 Differential Equation Governing LTI Continuous Time System

The electric heaters, motors, generators, etc., are examples of electrical continuous time systems. The continuous time systems that operate on electrical energy can be modelled by three basic elements Resistor (R), Inductor (L) and Capacitor (C). The models constructed using these fundamental elements are called *electric circuits.*

In electric circuits the inputs and outputs are either voltage signals or current signals. The continuous time voltage signal is denoted by $v(t)$ and current signal by $i(t)$.

The basic RL, RC, and RLC circuits and their time domain KVL (Kirchoff's Voltage Law) equations are shown in Figs 3.2, 3.3 and 3.4 respectively. From these circuits it can be observed that the equations governing the continuous time systems are differential equations.

Also, it can be shown that all continuous time systems like Mechanical systems, Thermal systems, Hydraulic systems, etc., are all governed by differential equations.

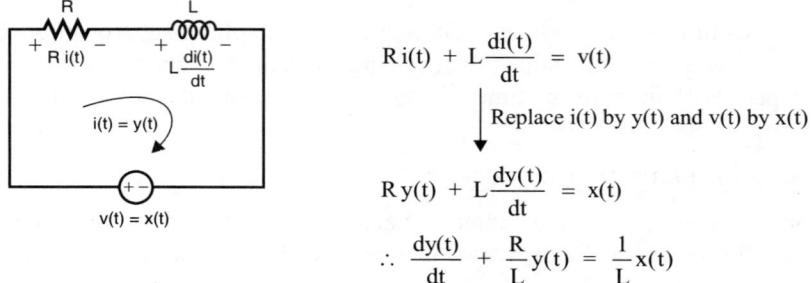

$$R\,i(t) + L\frac{di(t)}{dt} = v(t)$$

Replace $i(t)$ by $y(t)$ and $v(t)$ by $x(t)$

$$R\,y(t) + L\frac{dy(t)}{dt} = x(t)$$

$$\therefore \frac{dy(t)}{dt} + \frac{R}{L}y(t) = \frac{1}{L}x(t)$$

Fig 3.3: *RL circuit and the mathematical equation governing RL circuit.*

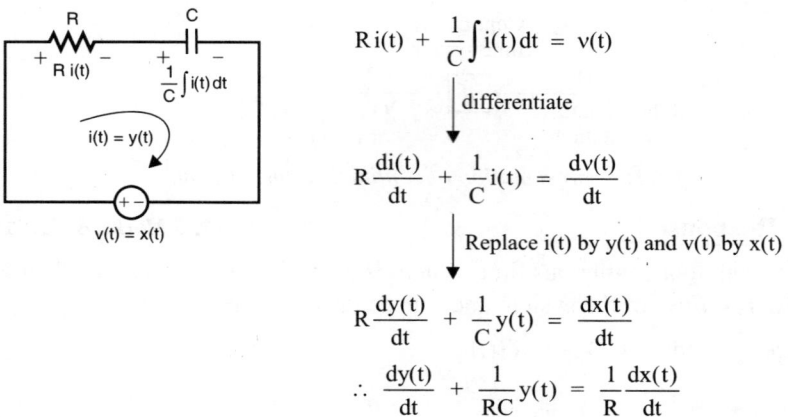

$$R\,i(t) + \frac{1}{C}\int i(t)\,dt = v(t)$$

differentiate

$$R\frac{di(t)}{dt} + \frac{1}{C}i(t) = \frac{dv(t)}{dt}$$

Replace $i(t)$ by $y(t)$ and $v(t)$ by $x(t)$

$$R\frac{dy(t)}{dt} + \frac{1}{C}y(t) = \frac{dx(t)}{dt}$$

$$\therefore \frac{dy(t)}{dt} + \frac{1}{RC}y(t) = \frac{1}{R}\frac{dx(t)}{dt}$$

Fig 3.4: *RC circuit and the mathematical equation governing RC circuit.*

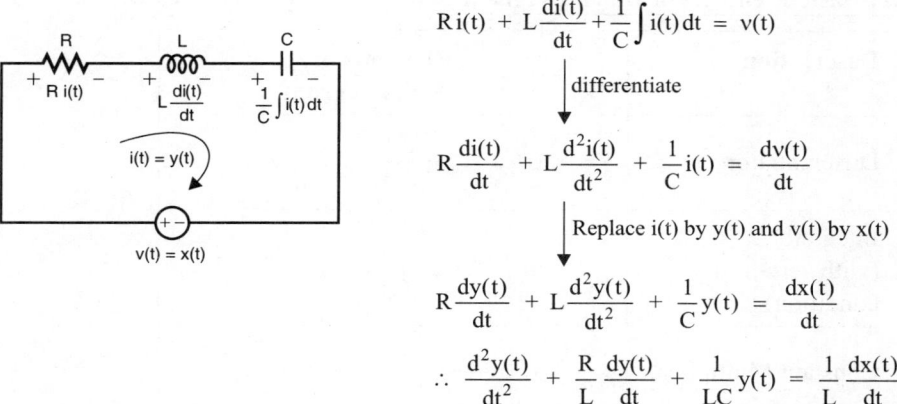

$$R\,i(t) + L\frac{di(t)}{dt} + \frac{1}{C}\int i(t)\,dt = v(t)$$

differentiate

$$R\frac{di(t)}{dt} + L\frac{d^2i(t)}{dt^2} + \frac{1}{C}i(t) = \frac{dv(t)}{dt}$$

Replace $i(t)$ by $y(t)$ and $v(t)$ by $x(t)$

$$R\frac{dy(t)}{dt} + L\frac{d^2y(t)}{dt^2} + \frac{1}{C}y(t) = \frac{dx(t)}{dt}$$

$$\therefore \frac{d^2y(t)}{dt^2} + \frac{R}{L}\frac{dy(t)}{dt} + \frac{1}{LC}y(t) = \frac{1}{L}\frac{dx(t)}{dt}$$

Fig 3.5: *RLC circuit and the mathematical equation governing RLC circuit.*

In general, the input-output relation of an LTI (Linear Time Invariant) continuous time system is represented by a constant coefficient ***differential equation*** shown below [equation (3.2)].

$$a_0\frac{d^N}{dt^N}y(t) + a_1\frac{d^{N-1}}{dt^{N-1}}y(t) + a_2\frac{d^{N-2}}{dt^{N-2}}y(t) + \ldots\ldots + a_{N-1}\frac{d}{dt}y(t) + a_N\,y(t) = b_0\frac{d^M}{dt^M}x(t)$$

$$+ b_1\frac{d^{M-1}}{dt^{M-1}}x(t) + b_2\frac{d^{M-2}}{dt^{M-2}}x(t) + \ldots\ldots + b_{M-1}\frac{d}{dt}x(t) + b_M\,x(t) \qquad\ldots\ldots(3.2)$$

where, N = Order of the system, $M \le N$, and $a_0 = 1$.

The solution of the above ***differential equation*** is the response $y(t)$ of the system, for the input $x(t)$.

3.2 Block Diagram Representation of Continuous Time System in Time Domain

Block Diagram

A ***block diagram*** of a system is a pictorial representation of the functions performed by the system. The block diagram of a system is constructed using the mathematical equation governing the system.

The basic elements of a block diagram are Differentiator, Integrator, Constant Multiplier and Signal Adder. The symbols used for the basic elements and their input-ouput relation are listed in Table 3.1.

Table 3.1: Basic Elements of Block Diagram   (AU Dec'13 & Dec'12, 2 Marks)

Description	Elements of Block Diagram
Differentiator	$x(t) \longrightarrow \boxed{\dfrac{d}{dt}} \longrightarrow \dfrac{d}{dt}x(t)$
Integrator (with zero initial condition)	$x(t) \longrightarrow \boxed{\int} \longrightarrow \int x(t)\,dt$
Constant Multiplier	$x(t) \longrightarrow \triangleright\ a \longrightarrow a\,x(t)$
Signal Adder	$x_1(t) \searrow \bigoplus \rightarrow x_1(t)+x_2(t)$ $\quad x_2(t) \nearrow$

Example 3.1

Construct the block diagram of the system described by the equation,

$$\frac{d^2y(t)}{dt^2} + 2\frac{dy(t)}{dt} + 3\,y(t) = 4\frac{dx(t)}{dt} + 5\,x(t)$$

Solution:

Case i: **Block diagram using differentiators**

Given that, $\dfrac{d^2y(t)}{dt^2} + 2\dfrac{dy(t)}{dt} + 3\,y(t) = 4\dfrac{dx(t)}{dt} + 5\,x(t)$

$$\therefore\ y(t) = -\frac{1}{3}\frac{d^2y(t)}{dt^2} - \frac{2}{3}\frac{dy(t)}{dt} + \frac{4}{3}\frac{dx(t)}{dt} + \frac{5}{3}x(t) \qquad\qquad .....(1)$$

The equation (1) is used to construct the block diagram using differentiators as shown in Fig 1.

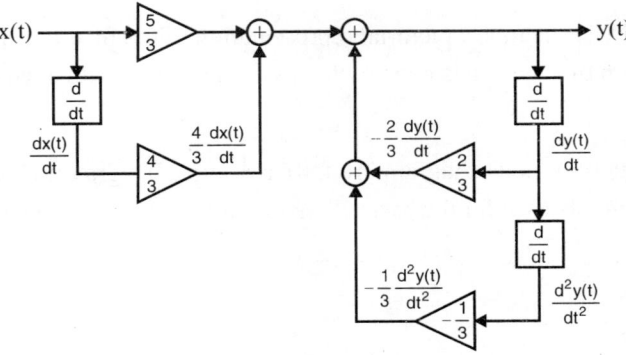

**Fig 1:** Block diagram using differentiators.

Case ii: **Block diagram using integrators**

Given that, $\dfrac{d^2 y(t)}{dt^2} + 2\,\dfrac{dy(t)}{dt} + 3y(t) = 4\,\dfrac{dx(t)}{dt} + 5x(t)$

$\downarrow$ Integrate with zero
initial conditions

$\dfrac{dy(t)}{dt} + 2\,y(t) + 3\int y(t)\,dt = 4\,x(t) + 5\int x(t)\,dt$

$\downarrow$ Integrate with zero
initial conditions

$y(t) + 2\int y(t)\,dt + 3\iint y(t)\,dt\,dt = 4\int x(t)\,dt + 5\iint x(t)\,dt\,dt$

$\therefore\ y(t) = -2\int y(t)\,dt - 3\iint y(t)\,dt\,dt + 4\int x(t)\,dt + 5\iint x(t)\,dt\,dt$ .....(2)

The equation (2) is used to construct the block diagram using integrators as shown in Fig 2.

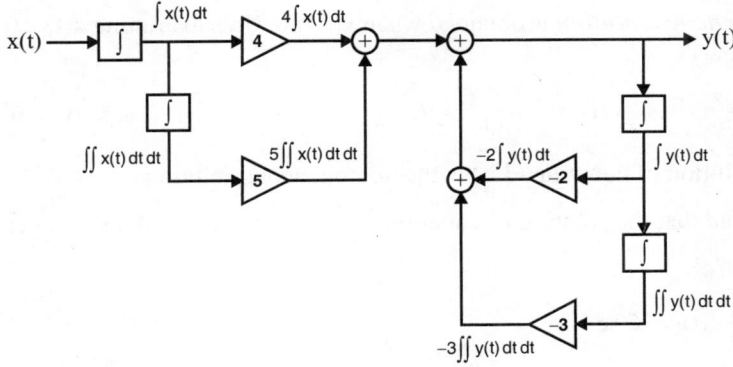

Fig 2: *Block diagram using integrators.*

3.3 Response of LTI Continuous Time System in Time Domain

The general equation governing an LTI continuous time system is,

$$a_0 \dfrac{d^N}{dt^N} y(t) + a_1 \dfrac{d^{N-1}}{dt^{N-1}} y(t) + a_2 \dfrac{d^{N-2}}{dt^{N-2}} y(t) + \ldots\ldots + a_{N-1} \dfrac{d}{dt} y(t) + a_N y(t) = b_0 \dfrac{d^M}{dt^M} x(t)$$

$$+ b_1 \dfrac{d^{M-1}}{dt^{N-1}} x(t) + b_2 \dfrac{d^{M-2}}{dt^{N-2}} x(t) + \ldots\ldots + b_{M-1} \dfrac{d}{dt} x(t) + b_M x(t) \qquad .....(3.3)$$

where, N = Order of the system, $M \le N$, and $a_0 = 1$.

The solution of the differential equation (3.3) is the **response** $y(t)$ of the LTI system, which consists of two parts. In mathematics, the two parts of the solution $y(t)$ are the homogeneous solution $y_h(t)$ and the particular solution $y_p(t)$.

$\boxed{\therefore\ \text{Response, } y(t) = y_h(t) + y_p(t)}$ .....(3.4)

The ***homogeneous solution*** is the response of the system when there is no input. The ***particular solution*** $y_p(t)$ is the solution of differential equation for specific input signal x(t) for $t \geq 0$.

In signals and systems the two parts of the solution y(t) are called zero-input response $y_{zi}(t)$ and zero-state response $y_{zs}(t)$.

$$\therefore \text{ Response, } y(t) = y_{zi}(t) + y_{zs}(t)$$ .....(3.5)

The ***zero input response*** is given by the homogeneous solution with constants evaluated using initial values of output (or initial conditions). The zero-input response is mainly due to initial output conditions (or initial stored energy) in the system. Hence, the zero-input response is also called *free response or natural response*.

The ***zero-state response*** is given by the sum of homogeneous solution and particular solution with zero initial conditions. The zero-state response is the response of the system due to input signal and with zero initial output condition. Hence, the zero-state response is also called *forced response*.

3.3.1 Homogeneous Solution

The ***homogeneous solution*** is obtained when x(t) = 0. On substituting x(t) = 0 in the system equation (3.3) we get,

$$\frac{d^N}{dt^N}y(t) + a_1\frac{d^{N-1}}{dt^{N-1}}y(t) + a_2\frac{d^{N-2}}{dt^{N-2}}y(t) + ... + a_{N-1}\frac{d}{dt}y(t) + a_N y(t) = 0$$ .....(3.6)

Now, the solution of equation (3.6) is the homogeneous solution.

Let us assume that the solution of equation (3.6) is in the form of an exponential.

i.e., $y(t) = Ce^{\lambda t}$

$$\therefore \frac{d}{dt}y(t) = C\lambda e^{\lambda t}$$

$$\frac{d^2}{dt^2}y(t) = C\lambda^2 e^{\lambda t}$$

$$\vdots$$

$$\frac{d^N}{dt^N}y(t) = C\lambda^N e^{\lambda t}$$

On substituting the above assumed solution in equation (3.6) we get,

$$C\lambda^N e^{\lambda t} + a_1 C\lambda^{N-1}e^{\lambda t} + a_2 C\lambda^{N-2}e^{\lambda t} + ... + a_{N-1}C\lambda e^{\lambda t} + a_N C\lambda e^{\lambda t} = 0$$

$$\therefore (\lambda^N + a_1 \lambda^{N-1} + a_2 \lambda^{N-2} + ... + a_{N-1}\lambda + a_N) Ce^{\lambda t} = 0$$

Now the ***characteristic polynomial*** of the system is given by,

$$\lambda^N + a_1 \lambda^{N-1} + a_2 \lambda^{N-2} + ... + a_{N-1}\lambda + a_N = 0$$

The characteristic polynomial has N roots, which are denoted as $\lambda_1, \lambda_2,...\lambda_N$.

The roots of the characteristic polynomial may be distinct real roots, repeated real roots or complex roots. The assumed solutions for various types of roots are given below:

Distinct Real Roots

Let the roots $\lambda_1, \lambda_2, \lambda_3, \dots \lambda_N$ be distinct real roots. Now the homogeneous solution will be in the form,

$$y_h(t) = C_1 e^{\lambda_1 t} + C_2 e^{\lambda_2 t} + \dots + C_{N-1} e^{\lambda_{N-1} t} + C_N e^{\lambda_N t}$$

where, $C_1, C_2, \dots C_N$ are constants that can be evaluated using initial conditions.

Repeated Real Roots

Let one of the real roots λ_1 repeats p times and the remaining $(N - p)$ roots are distinct real roots.

Now the homogeneous solution is in the form,

$$y_h(t) = (C_1 + C_2 t + C_3 t^2 + \dots + C_p t^{p-1}) e^{\lambda_1 t} + C_{p+1} e^{\lambda_{p+1} t} + \dots + C_N e^{\lambda_N t}$$

where, $C_1, C_2, C_3, \dots C_N$ are constants that can be evaluated using initial conditions.

Complex Roots

Let the characteristic polynomial have a pair of complex roots λ and λ^* and the remaining $(N - 2)$ roots be distinct real roots. Let, $\lambda = \alpha + j\beta$ and $\lambda^* = \alpha - j\beta$. Now, the homogeneous solution will be in the form,

$$y_h(t) = e^{\alpha t}(C_1 \cos \beta t + j C_2 \sin \beta t) + C_3 e^{\lambda_3 t} + \dots + C_{N-1} e^{\lambda_{N-1} t} + C_N e^{\lambda_N t}$$

where, $C_1, C_2, C_3 \dots C_N$ are constants that can be evaluated using initial conditions.

3.3.2 Particular Solution

The *particular solution*, $y_p(t)$ is the solution of Nth order differential equation governing the system for specific input signal $x(t)$ for $t \geq 0$. Since the input signal may have a different form, the particular solution depends on the form or type of the input signal $x(t)$.

If $x(t)$ is constant, then $y_p(t)$ is also a constant.

> **Example:**
> Let $x(t) = u(t)$, now $y_p(t) = K\, u(t)$

If $x(t)$ is exponential, then $y_p(t)$ is also an exponential.

> **Example:**
> Let $x(t) = e^{at}\, u(t)$, now $y_p(t) = K\, e^{at}\, u(t)$

If $x(t)$ is sinusoid, then $y_p(t)$ is also a sinusoid.

> **Example:**
> Let $x(t) = A \cos \Omega_0 t$, now $y_p(t) = K_1 \cos \Omega_0 t + K_2 \sin \Omega_0 t$

The general form of a particular solution for various types of inputs are listed in Table 3.2.

Table 3.2: Particular Solution

Input signal, x(t)	Particular solution, $y_p(t)$
$x(t) = A$	$y_p(t) = K$
$x(t) = Au(t)$	$y_p(t) = K\,u(t)$
$x(t) = A\,e^{\alpha t}$ (if $\alpha \neq \lambda_i$)	$y_p(t) = K\,e^{\alpha t}$
$x(t) = A\,e^{\alpha t}$ (if $\alpha = \lambda_i$)	$y_p(t) = K\,t\,e^{\alpha t}$
$x(t) = A\cos\Omega_0 t$ $\left.\rule{0pt}{12pt}\right\}$ $x(t) = A\sin\Omega_0 t$	$y_p(t) = K_1\cos\Omega_0 t + K_2\sin\Omega_0 t$

Note: λ_i *is one of the root of characteristic polynomial.*

3.3.3 Zero-Input and Zero-State Response

The ***zero-input response*** $y_{zi}(t)$ (or ***free response*** or ***natural response***) is obtained from the homogeneous solution $y_h(t)$ with constants evaluated using initial output (or initial conditions).

$\therefore$ **Zero – input response, $y_{zi}(t)$** $= \mathbf{y_h(t)}\big|_{\text{with constants evaluated using initial output condition}}$

The zero-state response $y_{zs}(t)$ (or forced response) is obtained from the sum of homogeneous solution and particular solution and evaluating the constants with zero initial conditions.

$\therefore$ **Zero – state response, $y_{zs}(t) = \left[\mathbf{y_h(t)} + \mathbf{y_p(t)}\right]\big|_{\text{with constants } C_1, C_2, ... C_N \text{ evaluated with zero initial output conditions}}}$

The sum of zero-input response and zero-state response will be the total response or complete response of a system.

3.3.4 Total Response

The total response y(t) of a continuous time system can be obtained by the following two methods.

Method -1

The ***total response*** y(t) is given by the sum of homogeneous solution and particular solution.

$\therefore$ Total response, $y(t) = y_h(t) + y_p(t)$

Procedure to Determine Total Response

1. Determine homogeneous solution $y_h(t)$ with constants $C_1, C_2, ...C_N$.

2. Determine particular solution $y_p(t)$ and estimate the constant K by evaluating the given system equation using $y_p(t)$ for any value of $t \geq 1$ so that no term of the system equation vanishes.

3. Now the total response y(t) is given by, the sum of $y_h(t)$ and $y_p(t)$

$\therefore$ Total response, $y(t) = y_h(t) + y_p(t)$

4. The total response y(t) will have N number of constants $C_1, C_2, ...C_N$. Evaluate the constants $C_1, C_2, ...C_N$ using initial outputs (or initial conditions).

Method -2

The *total response* y(t) is given by the sum of zero-input response and zero-state response.

$$\therefore \text{Total response, } y(t) = y_{zi}(t) + y_{zs}(t)$$

Procedure to Determine Total Response

1. Determine the homogeneous solution $y_h(t)$ with constants C_1, C_2, ...C_N.

2. Determine the zero-input response $y_{zi}(t)$ which is obtained from the homogeneous solution $y_h(t)$ by evaluating the constants C_1, C_2, ...C_N using initial outputs (or initial conditions).

3. Determine particular solution $y_p(t)$ and estimate the constant K by evaluating the given system equation using $y_p(t)$ for any value of $t \geq 1$ so that no term of the system equation vanishes.

4. Determine the zero-state response, $y_{zs}(t)$ which is given by sum of homogeneous solution and particular solution and evaluating the constants C_1, C_2, ...C_N with zero initial conditions.

5. Now the total response y(t) is given by the sum of zero-input response and zero-state response.

$$\therefore \text{Total response, } y(t) = y_{zi}(t) + y_{zs}(t)$$

Example 3.2

Determine the natural response of the system described by the equation,

$$\frac{d^2 y(t)}{dt^2} + 6\frac{dy(t)}{dt} + 5\,y(t) = \frac{dx(t)}{dt} + 4\,x(t); \quad y(0) = 1; \ \frac{dy(t)}{dt}\bigg|_{t=0} = -2$$

Solution:

The natural response is response of the system due to initial conditions and so it is given by zero-input response.

Zero-input response, $y_{zi}(t) = y_h(t)\big|_{\text{with constants evaluated using initial conditions}}$

where, $y_h(t)$ = Homogeneous solution

The given system equation is,

$$\frac{d^2 y(t)}{dt^2} + 6\frac{dy(t)}{dt} + 5\,y(t) = \frac{dx(t)}{dt} + 4\,x(t) \qquad \qquad .....(1)$$

Homogeneous Solution

The homogeneous solution is the solution of the system equation when x(t) = 0.

On substituting x(t) = 0 in system equation (equation (1)) we get,

$$\frac{d^2 y(t)}{dt^2} + 6\frac{dy(t)}{dt} + 5y(t) = 0 \qquad \qquad .....(2)$$

Let, $y(t) = C\,e^{\lambda t}$; $\quad \therefore \frac{d}{dt}y(t) = C\lambda e^{\lambda t}$; $\frac{d^2}{dt^2}y(t) = C\lambda^2 e^{\lambda t}$

On substituting the above terms in equation (2) we get,

$$C\lambda^2 e^{\lambda t} + 6C\lambda e^{\lambda t} + 5Ce^{\lambda t} = 0$$

$$\therefore (\lambda^2 + 6\lambda + 5)\,Ce^{\lambda t} = 0$$

The characteristic polynomial of the above equation is,

$$\lambda^2 + 6\lambda + 5 = 0 \quad \Rightarrow \quad (\lambda + 1)(\lambda + 5) = 0 \quad \Rightarrow \quad \lambda = -1, -5$$

Now the homogeneous solution is given by,

Homogeneous solution, $y_h(t) = C_1 e^{\lambda_1 t} + C_2 e^{\lambda_2 t} = C_1 e^{-t} + C_2 e^{-5t}$

Natural Response (or Zero-input Response)

Zero-input response, $y_{zi}(t) = y_h(t)\big|_{\text{with constants evaluated using initial conditions}}$

$$= C_1 e^{-t} + C_2 e^{-5t}\big|_{\text{with } C_1 \text{ and } C_2 \text{ evaluated using initial conditions}}$$

$$\therefore \frac{dy_{zi}(t)}{dt} = -C_1 e^{-t} - 5C_2 e^{-5t}$$

At $t = 0$, $y_{zi}(0) = C_1 e^0 + C_2 e^0 = C_1 + C_2$

Given that, $y_{zi}(0) = 1$, $\therefore C_1 + C_2 = 1$ .....(3)

At $t = 0$, $\frac{dy_{zi}(t)}{dt} = -C_1 e^0 - 5C_2 e^0 = -C_1 - 5C_2$

Given that, $\frac{dy_{zi}(t)}{dt}\big|_{t=0} = -2$, $\therefore -C_1 - 5C_2 = -2$ .....(4)

On adding equations (3) and (4) we get,

$$-4C_2 = -1 \quad \Rightarrow \quad C_2 = \frac{1}{4}$$

From equation (3), $C_1 = 1 - C_2 = 1 - \frac{1}{4} = \frac{3}{4}$

$\therefore$ Natural response, $y_{zi}(t) = \frac{3}{4} e^{-t} + \frac{1}{4} e^{-5t}$; $t \geq 0 = \frac{1}{4}(3e^{-t} + e^{-5t}) u(t)$

Example 3.3

Determine the forced response of the system described by the equation,

$$5 \frac{dy(t)}{dt} + 10 y(t) = 2 x(t), \quad \text{for the input, } x(t) = 2 u(t).$$

Solution:

The forced response is the response of the system due to input signal with zero initial conditions and so it is given by zero-state response.

Zero state response, $y_{zs}(t) = y_h(t) + y_p(t)\big|_{\text{with constants evaluated with zero initial conditions}}$

where, $y_h(t)$ = Homogeneous solution and $y_p(t)$ = Particular solution

The given system equation is,

$$5 \frac{dy(t)}{dt} + 10 y(t) = 2 x(t) \quad .....(1)$$

Homogeneous Solution

The homogeneous solution is the solution of the system equation when $x(t) = 0$.

On substituting $x(t) = 0$ in system equation [equation (1)] we get,

$$5 \frac{dy(t)}{dt} + 10 y(t) = 0 \quad .....(2)$$

Let, $y(t) = C\,e^{\lambda t}$; $\therefore \dfrac{d}{dt}\,y(t) = C\lambda e^{\lambda t}$

On substituting the above terms in equation (2) we get,

$$5\,C\lambda e^{\lambda t} + 10\,Ce^{\lambda t} = 0$$

$$\therefore (5\,\lambda + 10)\,Ce^{\lambda t} = 0$$

The characteristic polynomial of the above equation is,

$$5\lambda + 10 = 0 \qquad \Rightarrow \lambda + 2 = 0 \qquad \Rightarrow \qquad \lambda = -2$$

Now the homogeneous solution is given by,

Homogeneous solution, $y_h(t) = Ce^{\lambda t} = Ce^{-2t}$

Particular Solution

The particular solution is the solution of the system equation [equation (1)] for specific input.

Here input, $x(t) = 2\,u(t)$

Let the particular solution, $y_p(t)$ is of the form,

$$y_p(t) = K\ x(t)$$

$\therefore y_p(t) = 2K\,u(t)$; $\dfrac{dy_p(t)}{dt} = 2K\,\delta(t)$ $\boxed{\dfrac{d}{dt}\,u(t) = \delta(t)}$

On substituting the above terms and the input in system equation [equation (1)] we get,

$$5\,\frac{dy(t)}{dt} + 10\,y(t) = 2x(t)$$

$$\Downarrow$$

$$10\,K\,\delta(t) + 20\,K\,u(t) = 4\,u(t)$$

At $t = 1$, $10\,K\,\delta(1) + 20\,K\,u(1) = 4\,u(1)$ $\Rightarrow$ $20\,K = 4$ $\Rightarrow$ $K = 1/5$ $\boxed{\delta(1) = 0,\ u(1) = 1}$

$\therefore$ Particular solution, $y_p(t) = \dfrac{2}{5}\,u(t)$

Forced Response (or Zero-State Response)

Zero state response, $y_{zs}(t) = y_h(t) + y_p(t)\big|_{\text{with constants evaluated with zero initial conditions}}$

$$= Ce^{-2t} + \frac{2}{5}\,u(t)$$

At $t = 0$, $y_{zs}(t) = Ce^{0} + \dfrac{2}{5}\,u(0) = C + \dfrac{2}{5}$

Since, $y_{zs}(0) = 0$, $C + \dfrac{2}{5} = 0$ $\Rightarrow C = -\dfrac{2}{5}$

$\therefore$ Forced Response, $y_{zs}(t) = -\dfrac{2}{5}\,e^{-2t} + \dfrac{2}{5}\,u(t)$; for $t \ge 0 = \dfrac{2}{5}(1 - e^{-2t})\,u(t)$

Example 3.4

Determine the complete response of the system described by the equation,

$$\frac{d^2y(t)}{dt^2} + 5\frac{dy(t)}{dt} + 4\,y(t) = \frac{dx(t)}{dt}\ ;\ y(0) = 0\ ;\ \frac{dy(t)}{dt}\bigg|_{t=0} = 1,\ \ \text{for the input, } x(t) = e^{-2t}\,u(t)$$

Solution:

The given system equation is,

$$\frac{d^2y(t)}{dt^2} + 5\frac{dy(t)}{dt} + 4y(t) = \frac{dx(t)}{dt} \qquad \qquad ......(1)$$

Homogeneous Solution

The homogeneous solution is the solution of the system equation when x(t) = 0.

On substituting x(t) = 0 in system equation [equation (1)] we get,

$$\frac{d^2y(t)}{dt^2} + 5\frac{dy(t)}{dt} + 4y(t) = 0 \qquad \qquad .....(2)$$

Let, $y(t) = C\,e^{\lambda t}$; $\therefore \frac{d}{dt}y(t) = C\lambda e^{\lambda t}$ and $\frac{d^2}{dt^2}y(t) = C\lambda^2 e^{\lambda t}$

On substituting the above terms in equation (2) we get,

$$C\lambda^2 e^{\lambda t} + 5C\lambda e^{\lambda t} + 4Ce^{\lambda t} = 0$$

$$\therefore (\lambda^2 + 5\lambda + 4)Ce^{\lambda t} = 0$$

The characteristic polynomial of the above equation is,

$$\lambda^2 + 5\lambda + 4 = 0 \qquad \Rightarrow \qquad (\lambda + 4)(\lambda + 1) = 0 \qquad \Rightarrow \qquad \lambda = -4, -1$$

Now the homogeneous solution is given by,

Homogeneous solution, $y_h(t) = C_1 e^{\lambda_1 t} + C_2 e^{\lambda_2 t} = C_1 e^{-4t} + C_2 e^{-t}$

Particular Solution

The particular solution is the solution of the system equation [equation (1)] for specific input.

Here input, $x(t) = e^{-2t}u(t)$

$$\therefore x(t) = e^{-2t} \; ; \; \text{for} \; t \geq 0$$

$$\therefore \frac{dx(t)}{dt} = -2e^{-2t}$$

Let the particular solution, $y_p(t)$ is of the form,

$$y_p(t) = K\,x(t)$$

$$\therefore y_p(t) = K\,e^{-2t} ; \qquad \frac{dy_p(t)}{dt} = -2K\,e^{-2t} ; \qquad \frac{d^2y(t)}{dt^2} = 4K\,e^{-2t}$$

On substituting the above terms and the input in system equation [equation (1)] we get,

$$\frac{d^2y(t)}{dt^2} + 5\frac{dy(t)}{dt} + 4y(t) = \frac{dx(t)}{dt}$$

$$\Downarrow$$

$$4K\,e^{-2t} - 10K\,e^{-2t} + 4K\,e^{-2t} = -2\,e^{-2t}$$

On dividing throughout by e^{-2t} we get,

$$4K - 10K + 4K = -2 \qquad \Rightarrow \qquad -2K = -2 \qquad \Rightarrow \qquad K = 1$$

$\therefore$ Particular solution, $y_p(t) = e^{-2t}$

Total (or Complete) Response

Method - 1

Total response, $y(t) = y_h(t) + y_p(t)$

$\therefore y(t) = C_1 e^{-4t} + C_2 e^{-t} + e^{-2t}$

When $t = 0$, $y(t) = y(0) = C_1 e^0 + C_2 e^0 + e^0 = C_1 + C_2 + 1$

Given that $y(0) = 0$, $\therefore C_1 + C_2 + 1 = 0$ .....(3)

Here, $\dfrac{dy(t)}{dt} = -4C_1 e^{-4t} - C_2 e^{-t} - 2e^{-2t}$

Now, $\dfrac{dy(t)}{dt}\Big|_{t=0} = -4C_1 e^0 - C_2 e^0 - 2e^0 = -4C_1 - C_2 - 2$

Given that, $\dfrac{dy(t)}{dt}\Big|_{t=0} = 1$; $\therefore -4C_1 - C_2 - 2 = 1$ .....(4)

On adding equations (3) and (4) we get,

$-3C_1 - 1 = 1 \quad \Rightarrow \quad -3C_1 = 2 \quad \Rightarrow \quad C_1 = -\dfrac{2}{3}$

From equation (3), $C_2 = -1 - C_1 = -1 + \dfrac{2}{3} = -\dfrac{1}{3}$

$\therefore$ Total Response, $y(t) = -\dfrac{2}{3} e^{-4t} - \dfrac{1}{3} e^{-t} + e^{-2t}$; $t \geq 0$

$\qquad$ (or) $\qquad y(t) = \left(e^{-2t} - \dfrac{2}{3} e^{-4t} - \dfrac{1}{3} e^{-t}\right) u(t)$

Method - 2

Total response, $\qquad y(t) = y_{zi}(t) + y_{zs}(t)$

Zero-input response, $y_{zi}(t) = y_h(t)\big|_{\text{with constants evaluated using initial conditions}}$

$y_h(t) = C_1 e^{-4t} + C_2 e^{-t}$

$\therefore \dfrac{dy_h(t)}{dt} = -4C_1 e^{-4t} - C_2 e^{-t}$

At $t = 0$, $y_h(0) = C_1 e^0 + C_2 e^0 = C_1 + C_2$

Given that, $y_h(0) = 0$, $\therefore C_1 + C_2 = 0$ .....(5)

At $t = 0$, $\dfrac{dy_h(t)}{dt} = -4C_1 e^0 - C_2 e^0 = -4C_1 - C_2$

Given that, $\dfrac{dy_h(t)}{dt}\Big|_{t=0} = 1$, $\therefore -4C_1 - C_2 = 1$ .....(6)

On adding equations (5) and (6) we get,

$-3C_1 = 1 \quad \Rightarrow \quad C_1 = -\dfrac{1}{3}$

From equation (5), $C_2 = -C_1 = \dfrac{1}{3}$

$\therefore y_{zi}(t) = -\dfrac{1}{3} e^{-4t} + \dfrac{1}{3} e^{-t}$

Zero-state response, $y_{zs}(t) = y_h(t) + y_p(t)|_{\text{with constants evaluaed with zero initial conditions}}$

$\therefore y_{zs}(t) = C_1 e^{-4t} + C_2 e^{-t} + e^{-2t}$

$\therefore \dfrac{dy_{zs}(t)}{dt} = -4C_1 e^{-4t} - C_2 e^{-t} - 2e^{-2t}$

At $t = 0$, $y_{zs}(0) = C_1 e^0 + C_2 e^0 + e^0 = C_1 + C_2 + 1$

At $t = 0$, $\dfrac{dy_{zs}(t)}{dt} = -4C_1 e^0 - C_2 e^0 - 2e^0 = -4C_1 - C_2 - 2$

Since initial conditions are zero,

$C_1 + C_2 + 1 = 0$ .....(7)

$-4C_1 - C_2 - 2 = 0$ .....(8)

On adding equations (7) and (8) we get,

$-3C_1 - 1 = 0 \quad \Rightarrow \quad C_1 = -\dfrac{1}{3}$

From the equation (7), $C_2 = -1 - C_1 = -1 + \dfrac{1}{3} = -\dfrac{2}{3}$

$\therefore y_{zs}(t) = -\dfrac{1}{3} e^{-4t} - \dfrac{2}{3} e^{-t} + e^{-2t}$

$\therefore$ Total Response, $y(t) = y_{zi}(t) + y_{zs}(t)$

$= -\dfrac{1}{3} e^{-4t} + \dfrac{1}{3} e^{-t} - \dfrac{1}{3} e^{-4t} - \dfrac{2}{3} e^{-t} + e^{-2t}$

$= -\dfrac{2}{3} e^{-4t} - \dfrac{1}{3} e^{-t} + e^{-2t} \; ; \; t \geq 0 = \left(e^{-2t} - \dfrac{2}{3} e^{-4t} - \dfrac{1}{3} e^{-t}\right) u(t)$

3.4 Convolution Integral *(AU Dec'13, 8 Marks)* *(AU May'15, Jun'13, & Jun'11, 2 Marks)*

The *convolution* of two continuous time signals $x_1(t)$ and $x_2(t)$ is defined as,

$$\boxed{x_3(t) = \int_{-\infty}^{+\infty} x_1(\lambda) x_2(t-\lambda) \, d\lambda}$$ .....(3.7)

where, $x_3(t)$ is the signal obtained by convolving $x_1(t)$ and $x_2(t)$,

and λ is a dummy variable used for integration.

The convolution relation of equation (3.7) can be symbolically expressed as,

$x_3(t) = x_1(t) * x_2(t)$ .....(3.8)

where the symbol $*$ indicates convolution operation.

3.4.1 Procedure to Perform Convolution

The *convolution* of two continuous time signals $x_1(t)$ and $x_2(t)$ is defined as,

$$x_3(t) = x_1(t) * x_2(t) = \int_{-\infty}^{+\infty} x_1(\lambda) x_2(t-\lambda) \, d\lambda$$

where, $x_3(t)$ is the signal obtained by convolving $x_1(t)$ and $x_2(t)$,

λ is a dummy variable used for integration,

$*$ indicates convolution operation.

The computation of $x_3(t)$ using the above convolution equation for any value of t involves the following ,

1. Change of time index : The time index t in signals $x_1(t)$ and $x_2(t)$ is changed to λ to get $x_1(\lambda)$ and $x_2(\lambda)$.

3. Folding : The signal $x_2(\lambda)$ is folded to get $x_2(-\lambda)$.

3. Shifting : The signal $x_2(-\lambda)$ is shifted by t units of time to get $x_2(t - \lambda)$.

4. Multiplication : The signals $x_1(\lambda)$ and $x_2(t - \lambda)$ are multiplied to get a product signal.

5. Integration : The product signal is integrated to get $x_3(t)$. Let the product signal is nonzero in the interval $\lambda = \lambda_1$ to $\lambda = \lambda_2$, Now the signal $x_3(t)$ is given by,

$$x_3(t) = \int_{\lambda=\lambda_1}^{\lambda=\lambda_2} x_1(\lambda)\, x_2(t-\lambda)\ d\lambda$$

If both $x_1(t)$ and $x_2(t)$ are defined for $t > 0$, (i.e., both $x_1(t)$ and $x_2(t)$ are causal) then the product signal is nonzero in the interval $\lambda = 0$ to $\lambda = t$,

Now the signal $x_3(t)$ is given by,

$$x_3(t) = \int_{\lambda=0}^{\lambda=t} x_1(\lambda)\, x_2(t-\lambda)\ d\lambda$$

In order to determine the range of values of λ for product signal, graphical representation of signals will be very useful. The operations like folding, shifting and multiplication can be performed graphically to ascertain the range of values of λ over which the product signal is nonzero.

In the above convolution, if the signals $x_1(t)$ and $x_2(t)$ are defined by a single mathematical equation for $t = -\infty$ to $+\infty$, then time shift is valid for any value of t in the range $t = -\infty$ to $+\infty$. Therefore, the time shift, multiplication and integration are performed only once by taking general time shift t.

In the above convolution, if the signals $x_1(t)$ and $x_2(t)$ are defined by different mathematical equations in various intervals of time, then the time shift, multiplication and integration are performed in each interval of time by considering a time shift t in each interval.

3.4.2 Convolution of Impulse Signal

Case i: Convolution of impulse with x(t)

$$x(t) * \delta(t) = x(t)$$

Proof:

By definition of convolution,

$$x(t) * \delta(t) = \int_{\lambda=-\infty}^{\lambda=+\infty} x(\lambda)\, \delta(t-\lambda)\ d\lambda = x(\lambda)\big|_{\lambda=t}$$

$$\boxed{\delta(t-\lambda) \neq 0 \text{ only at } \lambda = t}$$

$$= x(t)$$

Case ii: Convolution of shifted impulse with x(t).

$$x(t) * \delta(t-t_0) = x(t-t_0)$$

Proof:

By definition of convolution,

$$x(t) * \delta(t-t_0) = \int_{\lambda=-\infty}^{\lambda=+\infty} x(\lambda)\,\delta(t-t_0-\lambda)\,d\lambda = x(\lambda)\big|_{\lambda=t-t_0}$$

$$\boxed{\delta(t-t_0-\lambda) \neq 0 \text{ only at } \lambda = t - t_0}$$

$$= x(t-t_0)$$

3.4.3 Response of LTI Continuous Time System Using Convolution

(AU Dec'13, 8 Marks, Dec'12, 6 Marks)

In an LTI continuous time system, the *response* y(t) of the system for an arbitrary input x(t) is given by convolution of input x(t) with impulse response h(t) of the system. It is expressed as,

$$y(t) = x(t) * h(t) = \int_{-\infty}^{+\infty} x(\lambda)\,h(t-\lambda)\,d\lambda \qquad\qquad .....(3.9)$$

where the symbol * represents convolution operation.

Proof:

Let y(t) be the response of system H for an input x(t).

$$\therefore y(t) = H\{x(t)\}$$

$$= H\left\{ \int_{-\infty}^{+\infty} x(\lambda)\,\delta(t-\lambda)\,d\lambda \right\}$$

$$\boxed{\text{From equation (1.4)} \quad x(t) = \int_{-\infty}^{+\infty} x(\lambda)\,\delta(t-\lambda)\,d\lambda}$$

$$= \int_{-\infty}^{+\infty} H\{x(\lambda)\,\delta(t-\lambda)\}\,d\lambda$$

$$\boxed{\text{In linear system, integration and system operation H can be interchanged.}}$$

$$\boxed{\text{The system H is a function of t and not a function of } \lambda.}$$

$$= \int_{-\infty}^{+\infty} x(\lambda)\,H\{\delta(t-\lambda)\}\,d\lambda$$

$$\boxed{\begin{array}{l}\text{By time invariant property}\\ \text{if } H\{\delta(t)\} = h(t)\\ \text{then, } H\{\delta(t-\lambda)\} = h(t-\lambda)\end{array}}$$

$$= \int_{-\infty}^{+\infty} x(\lambda)\,h(t-\lambda)\,d\lambda \qquad\qquad .....(3.10)$$

The equation (3.10) represents the convolution of input x(t) with the impulse response h(t) to yield the output y(t). Hence it is proved that the response y(t) of LTI continuous time system for an arbitrary input x(t) is given by convolution of input x(t) with impulse response h(t) of the system.

3.4.4 Unit Step Response Using Convolution

In an LTI system, if the input x(t) is a unit step signal, then the response is called a *unit step response*. In general the response y(t) of a system is given by convolution of input x(t) and impulse response h(t) of the system.

$$y(t) = x(t) * h(t) = \int_{\lambda=-\infty}^{\lambda=+\infty} x(\lambda)\,h(t-\lambda)\,d\lambda$$

Let the input x(t) be unit step input u(t), and the corresponding response be s(t). Now the unit step response s(t) is given by,

Unit Step Response, $s(t) = u(t) * h(t)$

$$= h(t) * u(t)$$

$$= \int_{\lambda = -\infty}^{\lambda = +\infty} h(\lambda)\, u(t - \lambda)\, d\lambda \qquad \boxed{\begin{aligned} u(t - \lambda) &= 1 \ ; \ -\infty < \lambda \le t \\ &= 0 \ ; \ \lambda > t \end{aligned}}$$

$$= \int_{\lambda = -\infty}^{\lambda = t} h(\lambda)\, d\lambda \qquad . \qquad\qquad\qquad ....(3.11)$$

If h(t) is defined only for $t \ge 0$ then the unit step response is given by,

$$s(t) = \int_{\lambda = 0}^{\lambda = t} h(\lambda)\, d\lambda$$

3.4.5 Properties of Convolution

The convolution of continuous time signals will satisfy the following properties.

Commutative property : $x_1(t) * x_2(t) = x_2(t) * x_1(t)$ .....(3.12)

Associative property : $[x_1(t) * x_2(t)] * x_3(t) = x_1(t) * [x_2(t) * x_3(t)]$ .....(3.13)

Distributive property : $x_1(t) * [x_2(t) + x_3(t)] = [x_1(t) * x_2(t)] + [x_1(t) * x_3(t)]$ .....(3.14)

Proof of Commutative Property:

Consider two continuous time signals, $x_1(t)$ and $x_2(t)$.

By Commutative property we can write,

$\quad x_1(t) * x_2(t) = x_2(t) * x_1(t)$

$\quad$ (LHS) $\qquad$ (RHS)

LHS $= x_1(t) * x_2(t)$

$$= \int_{m = -\infty}^{+\infty} x_1(m)\, x_2(t - m)\, dm \qquad\qquad .....(3.15)$$

$\qquad$ where m is a dummy variable used for convolution operation.

Let, $t - m = p$ $\qquad\qquad$ when $m = -\infty, \ p = t - m = t + \infty = +\infty$

$\quad \therefore m = t - p$ $\qquad\qquad$ when $m = +\infty, \ p = t - m = t - \infty = -\infty$

$\quad dm = -dp$

On replacing m by (t − p) and (t − m) by p in equation (3.15) we get,

$$\text{LHS} = -\int_{p = +\infty}^{-\infty} x_1(t - p)\, x_2(p)\, dp = \int_{p = -\infty}^{+\infty} x_2(p)\, x_1(t - p)\, dp \qquad \boxed{\begin{aligned}&\text{Here p is a dummy variable used} \\ &\text{for convolution operation.}\end{aligned}}$$

$$= x_2(t) * x_1(t)$$

$$= \text{RHS}$$

Proof of Associate Property:

Consider three continuous time signals $x_1(t)$, $x_2(t)$ and $x_3(t)$. By Associative property we can write,

$$[x_1(t) * x_2(t)] * x_3(t) = x_1(t) * [x_2(t) * x_3(t)]$$
$$\quad\quad\quad \text{LHS} \quad\quad\quad\quad\quad\quad \text{RHS}$$

Let, $y_1(t) = x_1(t) * x_2(t)$ .....(3.16)

Let us replace t by p.

$$\therefore y_1(p) = x_1(p) * x_2(p)$$

$$= \int_{m=-\infty}^{+\infty} x_1(m)\, x_2(p-m)\, dm \quad\quad\quad .....(3.17)$$

Let, $y_2(t) = x_2(t) * x_3(t)$ .....(3.18)

$$\therefore y_2(t) = \int_{q=-\infty}^{+\infty} x_2(q)\, x_3(t-q)\, dq$$

$$\therefore y_2(t-m) = \int_{q=-\infty}^{+\infty} x_2(q)\, x_3(t-q-m)\, dq \quad\quad .....(3.19)$$

where p, m and q are dummy variables used for convolution operation.

$$\text{LHS} = [x_1(t) * x_2(t)] * x_3(t)$$

$$= y_1(t) * x_3(t) \quad\quad\quad\quad\quad\quad\quad\quad \boxed{\text{Using equation (3.16).}}$$

$$= \int_{p=-\infty}^{+\infty} y_1(p)\, x_3(t-p)\, dp$$

$$= \int_{p=-\infty}^{+\infty} \int_{m=-\infty}^{+\infty} x_1(m)\, x_2(p-m)\, x_3(t-p)\, dm\, dp \quad \boxed{\text{Using equation (3.17).}}$$

$$= \int_{m=-\infty}^{+\infty} x_1(m)\, dm \int_{p=-\infty}^{+\infty} x_2(p-m)\, x_3(t-p)\, dp \quad\quad .....(3.20)$$

Let, $p - m = q$ $\quad\Big|\quad$ when $p = -\infty$, $q = p - m = -\infty - m = -\infty$

$\therefore p = q + m$ $\quad\Big|\quad$ when $p = +\infty$, $q = p - m = +\infty - m = +\infty$

$dp = dq$

On replacing $(p - m)$ by q and p by $(q + m)$ in the equation (3.20) we get,

$$\text{LHS} = \int_{m=-\infty}^{+\infty} x_1(m)\, dm \int_{q=-\infty}^{+\infty} x_2(q)\, x_2(t-q-m)\, dq$$

$$= \int_{m=-\infty}^{+\infty} x_1(m)\, y_2(t-m)\, dm \quad\quad\quad\quad \boxed{\text{Using equation (3.19).}}$$

$$= x_1(t) * y_2(t)$$

$$= x_1(t) * [x_2(t) * x_3(t)] \quad\quad\quad\quad\quad \boxed{\text{Using equation (3.18).}}$$

$$= \text{RHS}$$

Proof of Distributive Property:

Consider three continuous time signals $x_1(t)$, $x_2(t)$ and $x_3(t)$. By distributive property we can write,

$$x_1(t) * [x_2(t) + x_3(t)] = [x_1(t) * x_2(t)] + [x_1(t) * x_3(t)]$$

$$\underbrace{\qquad\qquad}_{\text{LHS}} \qquad \underbrace{\qquad\qquad\qquad}_{\text{RHS}}$$

$$\text{LHS} = x_1(t) * [x_2(t) + x_3(t)]$$

$$= x_1(t) * x_4(t)$$

$\boxed{x_4(t) = x_2(t) + x_3(t)}$

$$= \int_{m=-\infty}^{+\infty} x_1(m)\, x_4(t-m)\, dm$$

$\boxed{\begin{array}{l} m \text{ is a dummy variable} \\ \text{used for Integration.} \end{array}}$

$$= \int_{m=-\infty}^{+\infty} x_1(m)\, [x_2(t-m) + x_3(t-m)]\, dm$$

$\boxed{\begin{array}{l} \text{If, } x_4(t) = x_2(t) + x_3(t), \text{ then} \\ \quad x_4(t-m) = x_2(t-m) + x_3(t-m). \end{array}}$

$$= \int_{m=-\infty}^{+\infty} x_1(m)\, x_2(t-m)\, dm + \int_{m=-\infty}^{+\infty} x_1(m)\, x_3(t-m)\, dm$$

$$= [x_1(t) * x_2(t)] + [x_1(t) * x_3(t)]$$

$$= \text{RHS}$$

3.5 Interconnection of Continuous Time Systems

Smaller continuous time systems may be interconnected to form larger systems. Two possible basic ways of interconnection are ***cascade connection*** and ***parallel connection***. The cascade and parallel connections of two continuous time systems with impulse responses $h_1(t)$ and $h_2(t)$ are shown in Fig 3.6.

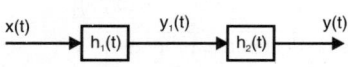

Fig 3.6a: Cascade connection.

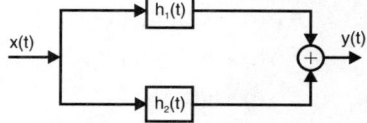

Fig 3.6b: Parallel connection.

Fig 3.6: Interconnection of continuous time systems.

Cascade Connected Continuous Time Systems

Two cascade connected continuous time systems with impulse response $h_1(t)$ and $h_2(t)$ can be replaced by a single equivalent continuous time system whose impulse response is given by convolution of individual impulse responses.

Fig 3.7: Cascade connected continuous time systems and their equivalent.

Proof:

With reference to Fig 3.7 we can write,

$$y_1(t) = x(t) * h_1(t) \qquad\qquad .....(3.21)$$

$$y(t) = y_1(t) * h_2(t) \qquad\qquad .....(3.22)$$

$$\therefore\ y(t) = [x(t) * h_1(t)] * h_2(t)$$ Using equation (3.21)

$$= x(t) * [h_1(t) * h_2(t)]$$ Using associative property.

$$= x(t) * h(t)$$ .....(3.23)

where, $h(t) = h_1(t) * h_2(t)$

From equation (3.23) we can say that the overall impulse response of two cascaded continuous time systems is given by convolution of individual impulse responses.

Parallel Connected Continuous Time Systems

Two parallel connected continuous time systems with impulse responses $h_1(t)$ and $h_2(t)$ can be replaced by a single equivalent continuous time system whose impulse response is given by the sum of individual impulse responses.

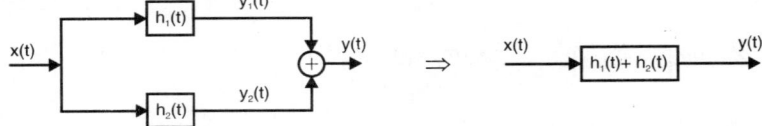

Fig 3.8: *Parallel connected continuous time systems and their equivalent.*

Proof:

With reference to Fig 3.8 we can write,

$$y_1(t) = x(t) * h_1(t)$$ .....(3.24)

$$y_2(t) = x(t) * h_2(t)$$ .....(3.25)

$$y(t) = y_1(t) + y_2(t)$$ .....(3.26)

$$= [x(t) * h_1(t)] + [x(t) * h_2(t)]$$ Using equation (3.24) and (3.25) .....(3.27)

$$= x(t) * [h_1(t) + h_2(t)]$$ Using distributive property.

$$= x(t) * h(t)$$ .....(3.28)

where, $h(t) = h_1(t) + h_2(t)$

From equation (3.28) we can say that the overall impulse response of two parallel connected continuous time systems is given by the sum of individual impulse responses.

Example 3.5
(AU Dec'14, 16 Marks)

Find the overall impulse response of the interconnected sytem. Given that $h_1(t) = e^{-2t}u(t)$, $h_2(t) = \delta(t) - \delta(t-1)$, $h_3(t) = \delta(t)$. Also find the output of the system for the input $x(t) = u(t)$ using convolution integral.

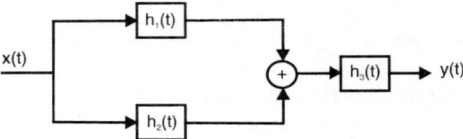

Solution:

Let, $h(t)$ = Overall impulse response.

Here, $h(t) = [h_1(t) + h_2(t)] * h_3(t)$

$$= [e^{-2t}u(t) + \delta(t) - \delta(t-1)] * \delta(t)$$

$$= e^{-2t}u(t) + \delta(t) - \delta(t-1)$$ $x(t) * \delta(t) = x(t)$

For LTI system the response is given by,

$$y(t) = x(t) * h(t)$$

Given that, $x(t) = u(t)$

$$\therefore \ y(t) = u(t) * [e^{-2t} u(t) + \delta(t) - \delta(t-1)]$$

$$= [u(t) * e^{-2t} u(t)] + [u(t) * \delta(t)] - [u(t) * \delta(t-1)]$$

$$= u(t) * e^{-2t}\big|_{t=0 \text{ to } \infty} + u(t) - u(t-1)$$

$$= \int_{\lambda=0}^{t} u(\lambda)\, e^{-2(t-\lambda)}\, d\lambda \,\bigg|_{t\geq0} + u(t) - u(t-1)$$

$$= \int_{\lambda=0}^{t} e^{-2t}\, e^{2\lambda}\, d\lambda \,\bigg|_{t\geq0} + u(t) + u(t-1)$$

$$= e^{-2t}\left[\frac{e^{2\lambda}}{2}\right]_0^t \bigg|_{t\geq0} + u(t) + u(t-1)$$

$$= e^{-2t}\left[\frac{e^{2t}}{2} - \frac{e^0}{2}\right]\bigg|_{t\geq0} + u(t) + u(t-1) = \left(\frac{1}{2} - \frac{e^{-2t}}{2}\right) u(t) + u(t) + u(t-1)$$

Using distributive property.

$$x(t) * \delta(t) = x(t)$$

$$x(t) * \delta(t-t_0) = x(t-t_0)$$

Here λ is dummy variable.

$$u(\lambda) = 1 \ ; \lambda \geq 0$$
$$\qquad\quad\ = 0 \ ; \lambda < 0$$

Example 3.6

Perform convolution of the following causal signals.

a) $x_1(t) = 2\, u(t), \ \ x_2(t) = u(t)$

b) $x_1(t) = e^{-2t}\, u(t), \ \ \ x_2(t) = e^{-5t}\, u(t)$

c) $x_1(t) = t\, u(t), \ \ x_2(t) = e^{-5t}\, u(t)$

d) $x_1(t) = \cos t\, u(t), \ x_2(t) = t\, u(t)$

e) $x_1(t) = e^{-2t}\, u(t-2), \ x_2(t) = e^{-2t}\, u(t)$

Solution:

a) Given that, $x_1(t) = 2\, u(t) = 2 \ ; \ t \geq 0$

$$x_2(t) = u(t) \ \ = 1 \ ; \ t \geq 0$$

Let, $x_3(t) = x_1(t) * x_2(t)$

By definition of convolution,

Since $x_1(t)$ and $x_2(t)$ are causal, the limits of integration is 0 to t.

$$x_3(t) = x_1(t) * x_2(t) = \int_{\lambda=0}^{\lambda=t} x_1(\lambda)\, x_2(t-\lambda)\, d\lambda = \int_{\lambda=0}^{\lambda=t} 2 \times 1\, d\lambda = 2 \int_{\lambda=0}^{\lambda=t} d\lambda$$

$$= 2[\lambda]_0^t = 2[t-0] = 2t \ ; \ \text{for } t \geq 0 = 2t\, u(t)$$

b) Given that, $x_1(t) = e^{-2t}\, u(t) \ = e^{-2t} \ ; \ t \geq 0$

$$x_2(t) = e^{-5t}\, u(t) \ = e^{-5t} \ ; \ t \geq 0$$

Let, $x_3(t) = x_1(t) * x_2(t)$

By definition of convolution,

> Since $x_1(t)$ and $x_2(t)$ are causal, the limits of integration is 0 to t.

$$x_3(t) = x_1(t) * x_2(t) = \int_{\lambda=0}^{\lambda=t} x_1(\lambda)\, x_2(t-\lambda)\, d\lambda = \int_{\lambda=0}^{\lambda=t} e^{-2\lambda}\, e^{-5(t-\lambda)}\, d\lambda = \int_{\lambda=0}^{\lambda=t} e^{-2\lambda}\, e^{-5t}\, e^{5\lambda}\, d\lambda$$

$$= e^{-5t} \int_{\lambda=0}^{\lambda=t} e^{-2\lambda+5\lambda}\, d\lambda = e^{-5t} \int_{\lambda=0}^{\lambda=t} e^{3\lambda}\, d\lambda = e^{-5t}\left[\frac{e^{3\lambda}}{3}\right]_0^t = e^{-5t}\left[\frac{e^{3t}}{3}-\frac{e^0}{3}\right]$$

$$= \frac{e^{-5t}}{3}(e^{3t}-1) = \frac{1}{3}(e^{-2t}-e^{-5t}) \; ; \text{ for } t \ge 0 = \frac{1}{3}(e^{-2t}-e^{-5t})\, u(t)$$

c) Given that, $x_1(t) = t\, u(t) = t \; ; \; t \ge 0$

$$x_2(t) = e^{-5t}\, u(t) = e^{-5t} \; ; \; t \ge 0$$

Let, $x_3(t) = x_1(t) * x_2(t)$

By definition of convolution,

> Since $x_1(t)$ and $x_2(t)$ are causal, the limits of integration is 0 to t.

$$x_3(t) = x_1(t) * x_2(t) = \int_{\lambda=0}^{\lambda=t} x_1(\lambda)\, x_2(t-\lambda)\, d\lambda$$

$$= \int_{\lambda=0}^{\lambda=t} \lambda\, e^{-5(t-\lambda)}\, d\lambda = \int_{\lambda=0}^{\lambda=t} \lambda\, e^{-5t}\, e^{5\lambda}\, d\lambda$$

$$= e^{-5t} \int_{\lambda=0}^{\lambda=t} \lambda\, e^{5\lambda}\, d\lambda \int = e^{-5t}\left[\lambda\,\frac{e^{5\lambda}}{5} - \int \frac{e^{5\lambda}}{5}\times 1\, d\lambda\right]_0^t$$

> $\int u\, dv = uv - \int v\, du$
>
> $u = \lambda \quad \Rightarrow du = 1$
>
> $dv = e^{5\lambda} \Rightarrow v = \dfrac{e^{5\lambda}}{5}$

$$= e^{-5t}\left[\lambda\,\frac{e^{5\lambda}}{5} - \frac{e^{5\lambda}}{25}\right]_0^t = e^{-5t}\left[t\,\frac{e^{5t}}{5} - \frac{e^{5t}}{25} - 0\times\frac{e^0}{5} + \frac{e^0}{25}\right]$$

$$= \frac{e^{-5t}}{25}(5te^{5t} - e^{5t} + 1) \; ; \text{ for } t \ge 0 = \frac{1}{25}(5t-1+e^{-5t})\, u(t)$$

$$= \frac{1}{25}(e^{-5t} + 5t - 1)\, u(t)$$

d) Given that, $x_1(t) = \cos t\, u(t) = \cos t \; ; \; t \ge 0$

$$x_2(t) = t\, u(t) \quad = t \quad ; \; t \ge 0$$

Let, $x_3(t) = x_1(t) * x_2(t)$

By definition of convolution,

> Since $x_1(t)$ and $x_2(t)$ are causal, the limits of integration is 0 to t.

$$x_3(t) = x_1(t) * x_2(t) = \int_{\lambda=0}^{\lambda=t} x_1(\lambda)\, x_2(t-\lambda)\, d\lambda$$

$$= \int_{\lambda=0}^{\lambda=t} \cos\lambda\, (t-\lambda)\, d\lambda = \int_{\lambda=0}^{\lambda=t} (t-\lambda)\cos\lambda\, d\lambda$$

> $\int u\, dv = uv - \int v\, du$
>
> $u = t-\lambda \quad \Rightarrow \dfrac{du}{d\lambda} = -1$
>
> $dv = \cos\lambda \Rightarrow \quad v = \sin\lambda$

$$= \left[(t-\lambda)\sin\lambda - \int \sin\lambda \times (-1)\, d\lambda\right]_{\lambda=0}^{\lambda=t}$$

$$\therefore \ x_3(t) = [(t-\lambda)\sin\lambda - \cos\lambda]_{\lambda=0}^{\lambda=t}$$

$$= [(t-t)\sin t - \cos t - (t-0)\sin0 + \cos0]$$

$$= 0 - \cos t - 0 + 1 = 1 - \cos t \ ; t \geq 0$$

$$= (1-\cos t)\,u(t)$$

---

e) Given that, $x_1(t) = e^{-2t}u(t-2)\ = e^{-2t}\ ; t \geq 2$

           $x_2(t) = e^{-2t}u(t)\ \ \ \ = e^{-2t}\ ; t \geq 0$ *(AU Dec'15, 16 Marks)*

Let, $x_3(t) = x_1(t) * x_2(t)$

By definition of convolution,

$$x_3(t) = x_1(t) * x_2(t) = \int_{\lambda=2}^{\lambda=t} x_1(\lambda)\,x_2(t-\lambda)\,d\lambda$$

> Since $x_1(t)$ and $x_2(t)$ are causal, the limits of integration is 2 to t.

$$= \int_{\lambda=2}^{\lambda=t} e^{-2\lambda}\,e^{-2(t-\lambda)}\,d\lambda = \int_{\lambda=2}^{\lambda=t} e^{-2\lambda}\,e^{-2t}e^{2\lambda}\,d\lambda$$

$$= e^{-2t}\int_{\lambda=2}^{\lambda=t} e^{-2\lambda}\,e^{2\lambda}\,d\lambda$$

$$= e^{-2t}\int_{\lambda=2}^{\lambda=t} e^{-2\lambda+2\lambda}\,d\lambda \qquad \boxed{e^{-2\lambda+2\lambda} = e^0 = 1}$$

$$= e^{-2t}\int_{\lambda=2}^{\lambda=t} d\lambda = e^{-2t}[\lambda]_2^t$$

$$= e^{-2t}(t-2) \ ; \ t \geq 2$$

$$= e^{-2t}(t-2)\,u(t-2)$$

Example 3.7

Determine the unit step response of the following systems whose impulse response are given below:

a) $h(t) = 3t\,u(t)$ **b)** $h(t) = e^{-5t}\,u(t)$ **c)** $h(t) = u(t+2)$

d) $h(t) = u(t-2)$ **e)** $h(t) = u(t+2) + u(t-2)$

Solution:

a) Given that, $h(t) = 3t\,u(t) = 3t\ ; \ t \geq 0$

$$\text{Unit Step Response, } s(t) = \int_{\lambda=-\infty}^{\lambda=t} h(\lambda)\,d\lambda = \int_{\lambda=0}^{\lambda=t} 3\lambda\,d\lambda = 3\int_{\lambda=0}^{\lambda=t} \lambda\,d\lambda$$

$$= 3\left[\frac{\lambda^2}{2}\right]_0^t = 3\left[\frac{t^2}{2} - \frac{0}{2}\right] = \frac{3}{2}t^2 \ ; \text{ for } t \geq 0 = \frac{3}{2}t^2\,u(t)$$

b) Given that, $h(t) = e^{-5t}\, u(t) = e^{-5t}$; $t \geq 0$

Unit Step Response, $s(t) = \int\limits_{\lambda=-\infty}^{\lambda=t} h(\lambda)\, d\lambda = \int\limits_{\lambda=0}^{\lambda=t} e^{-5\lambda}\, d\lambda = \left[\dfrac{e^{-5\lambda}}{-5}\right]_0^t$

$$= \left[\dfrac{e^{-5t}}{-5} - \dfrac{e^0}{-5}\right] = \dfrac{1}{5}\left(1 - e^{-5t}\right) \; ; \text{ for } t \geq 0 = \dfrac{1}{5}\left(1 - e^{-5t}\right) u(t)$$

c) Given that, $h(t) = u(t+2) = 1$; $t \geq -2$

Unit Step Response, $s(t) = \int\limits_{\lambda=-\infty}^{\lambda=t} h(\lambda)\, d\lambda = \int\limits_{\lambda=-2}^{\lambda=t} d\lambda = [\lambda]_{-2}^t = [t+2]$:

$$= t+2 \; ; \text{ for } t \geq -2 = (t+2)\, u(t+2)$$

d) Given that, $h(t) = u(t-2) = 1$; $t \geq 2$

Unit Step Response, $s(t) = \int\limits_{\lambda=-\infty}^{\lambda=t} h(\lambda)\, d\lambda = \int\limits_{\lambda=2}^{\lambda=t} d\lambda = [\lambda]_2^t = [t-2]$

$$= t-2 \; ; \text{ for } t \geq 2 = (t-2)\, u(t-2)$$

e) Given that, $h(t) = u(t+2) + u(t-2)$

Let, $h(t) = h_1(t) + h_2(t)$

where, $h_1(t) = u(t+2) = 1$; $t \geq -2$

$h_2(t) = u(t-2) = 1$; $t \geq 2$

Unit step response, $s(t) = h(t) * u(t) = [h_1(t) + h_2(t)] * u(t)$

$$= [h_1(t) * u(t)] + [h_2(t) * u(t)] = s_1(t) + s_2(t)$$

where, $s_1(t) = h_1(t) * u(t)$

$s_2(t) = h_2(t) * u(t)$

$s_1(t) = \int\limits_{\lambda=-\infty}^{\lambda=t} h_1(\lambda)\, d\lambda = \int\limits_{\lambda=-2}^{\lambda=t} d\lambda = [\lambda]_{-2}^t\, [t+2] = t+2 \; ; \text{ for } t \geq -2 = (t+2)\, u(t+2)$

$s_2(t) = \int\limits_{\lambda=-\infty}^{\lambda=t} h_2(\lambda)\, d\lambda = \int\limits_{\lambda=2}^{\lambda=t} d\lambda = [\lambda]_2^t = [t-2] = t-2 \; ; \text{ for } t \geq 2 = (t-2)\, u(t-2)$

Now, Unit step response, $s(t) = s_1(t) + s_2(t)$

$$= (t+2)\, u(t+2) + (t-2)\, u(t-2)$$

Example 3.8

Perform Convolution of the following signals, $x(t) = e^{-\alpha t}u(t)$, $h(t) = e^{-\beta t}u(t)$ if $|\alpha| < 1$, $|\beta| < 1$.

Solution:

Given that, $\mathbf{x(t)} = \mathbf{e^{-\alpha t}u(t)} = \mathbf{e^{-\alpha t}}$; $\mathbf{t \geq 0}$

$\qquad\mathbf{h(t)} = \mathbf{e^{-\beta t}u(t)} = \mathbf{e^{-\beta t}}$; $\mathbf{t \geq 0}$

Let, $y(t) = x(t) * h(t)$

By definition of convolution,

$$y(t) = x(t) * h(t) = \int_{\lambda=0}^{\lambda=t} e^{-\alpha\lambda} e^{-\beta(t-\lambda)} \, d\lambda = \int_{\lambda=0}^{\lambda=t} e^{-\alpha\lambda} e^{-\beta t} e^{\beta\lambda} \, d\lambda \qquad \boxed{\begin{array}{l}\text{Since } x_1(t) \text{ and } x_2(t) \text{ are causal,}\\ \text{the limits of integration is 0 to t.}\end{array}}$$

$$= e^{-\beta t} \int_{\lambda=0}^{\lambda=t} e^{(\beta-\alpha)\lambda} \, d\lambda$$

$$= e^{-\beta t}\left[\frac{e^{(\beta-\alpha)\lambda}}{\beta-\alpha}\right]_0^t = e^{-\beta t}\left[\frac{e^{(\beta-\alpha)t}}{\beta-\alpha} - \frac{e^0}{\beta-\alpha}\right]$$

$$= \frac{e^{-\beta t}}{\beta-\alpha}[e^{(\beta-\alpha)t} - 1] = \frac{e^{-\beta t}}{\beta-\alpha}[e^{\beta t} e^{-\alpha t} - 1]$$

$$= \frac{1}{\beta-\alpha}[e^{-\beta t} e^{\beta t} e^{-\alpha t} - e^{-\beta t}]$$

$$= \frac{1}{\beta-\alpha}(e^{-\alpha t} - e^{-\beta t}) \; ; \; t \geq 0 = \frac{1}{\beta-\alpha}(e^{-\alpha t} - e^{-\beta t})u(t)$$

Example 3.9

Perform convolution of the following signals, by graphical method.

a) $x_1(t) = e^{-3t}u(t)$, $x_2(t) = t\,u(t)$ **b)** $x_1(t) = e^{-at}$; $0 \leq t \leq T$, $x_2(t) = 1$; $0 \leq t \leq 2T$

Solution:

a) Given that, $x_1(t) = e^{-3t}u(t) = e^{-3t}$; $t \geq 0$

$\qquad\qquad x_2(t) = t\,u(t) = t$; $t \geq 0$

Let, $x_3(t) = x_1(t) * x_2(t)$

By definition of convolution,

$$x_3(t) = \int_{\lambda=-\infty}^{\lambda=+\infty} x_1(\lambda)\,x_2(t-\lambda)\,d\lambda$$

Let us change the time index t, in $x_1(t)$ and $x_2(t)$ to λ, to get $x_1(\lambda)$ and $x_2(\lambda)$, and then fold $x_2(\lambda)$ to get $x_2(-\lambda)$ graphically as shown below:

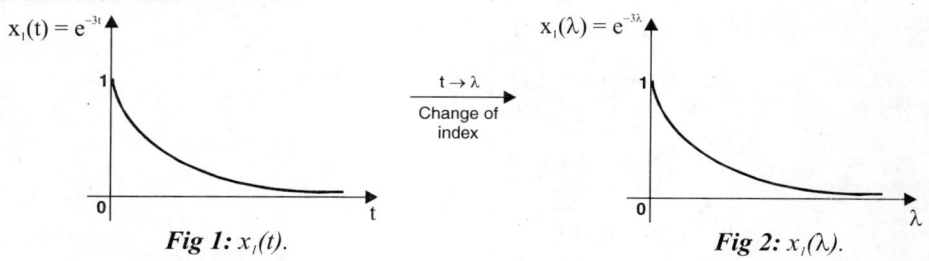

Fig 1: $x_1(t)$.   Fig 2: $x_1(\lambda)$.

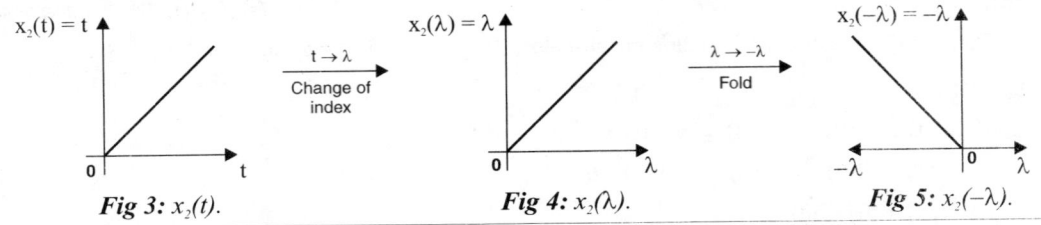

Fig 3: $x_2(t)$.   **Fig 4:** $x_2(\lambda)$.   **Fig 5:** $x_2(-\lambda)$.

Let us shift $x_2(-\lambda)$ by t units of time to get $x_2(t-\lambda)$ and then multiply with $x_1(\lambda)$ as shown below:

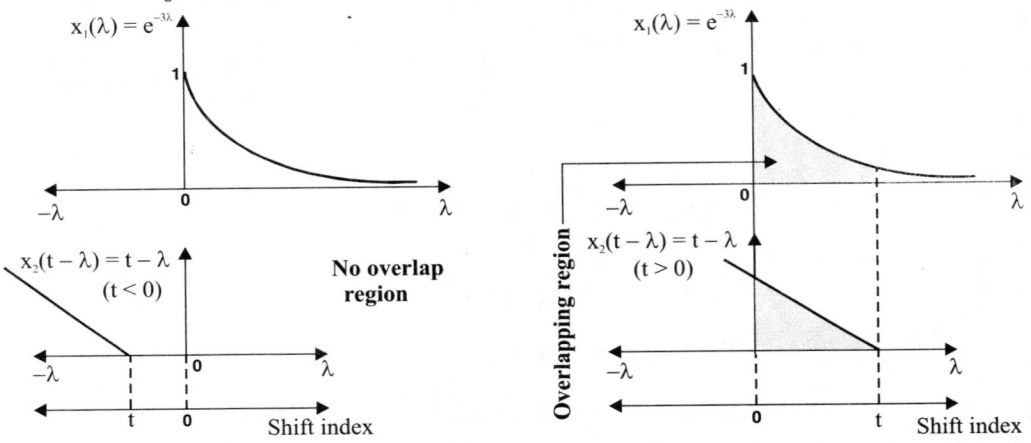

Fig 6: $x_1(\lambda)$ and $x_2(t-\lambda)$ when time shift, $t < 0$.   **Fig 7:** $x_1(\lambda)$ and $x_2(t-\lambda)$ when time shift, $t > 0$.

From Fig 6 and Fig 7, it is observed that the product of $x_1(\lambda)$ and $x_2(t-\lambda)$ is non-zero only for time shift $t > 0$.

For any time shift $t > 0$, the non-zero product exists in the overlapping region shown in Fig 7. Here the overlapping region is $\lambda = 0$ to $\lambda = t$. Hence integration of $x_1(\lambda)$ and $x_2(t-\lambda)$ is performed from $\lambda = 0$ to $\lambda = t$.

$$\therefore \ x_3(\lambda) = \int_{\lambda=0}^{\lambda=t} x_1(\lambda)\, x_2(t-\lambda)\, d\lambda \ ; \ t \geq 0$$

$$= \int_0^t e^{-3\lambda}(t-\lambda)\, d\lambda = \int_0^t (t-\lambda)\, e^{-3\lambda}\, d\lambda$$

$$= \left[(t-\lambda)\frac{e^{-3\lambda}}{-3} - \int \frac{e^{-3\lambda}}{-3} \times (-1)\, d\lambda \right]_0^t$$

$\int u\, dv = uv - \int v\, du$
$u = t - \lambda \ \Rightarrow \ \dfrac{du}{d\lambda} = -1$
$dv = e^{-3\lambda} \ \Rightarrow \ v = \dfrac{e^{-3\lambda}}{-3}$

$$= \left[(t-T)\frac{e^{-3\lambda}}{-3} + \frac{e^{-3\lambda}}{(-3)(-3)} \right]_0^t$$

$$= 0 + \frac{e^{-3t}}{9} - \frac{(t-0)\, e^0}{-3} - \frac{e^0}{9}$$

$$= \frac{e^{-3t}}{9} + \frac{t}{3} - \frac{1}{9} \ ; \ t \geq 0$$

$$= \frac{1}{9}[e^{-3t} + 3t - 1]\, u(t)$$

b) Given that, $x_1(t) = e^{-at}$; $0 \le t \le T$

$\qquad x_2(t) = 1$; $0 \le t \le 2T$

Let, $x_3(t) = x_1(t) * x_2(t)$

By definition of convolution,

$$x_3(t) = \int_{\lambda=-\infty}^{\lambda=+\infty} x_1(\lambda)\, x_2(t-\lambda)\, d\lambda$$

Let us change the time index t in $x_1(t)$ and $x_2(t)$ to λ, to get $x_1(\lambda)$ and $x_2(\lambda)$, and then fold $x_2(\lambda)$ to get $x_2(-\lambda)$ graphically as shown below:

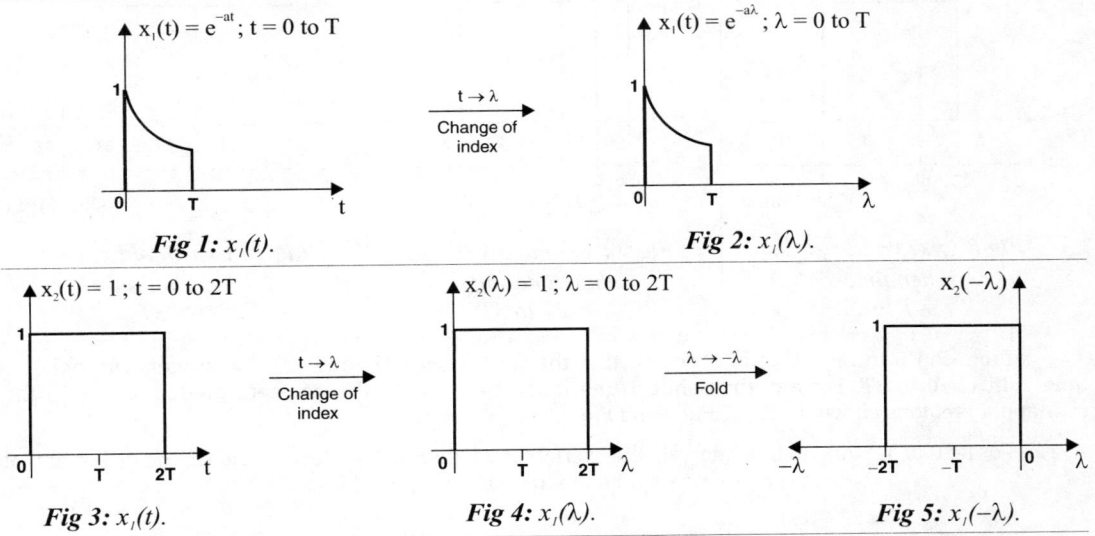

Fig 1: $x_1(t)$.   Fig 2: $x_1(\lambda)$.

Fig 3: $x_1(t)$.   Fig 4: $x_1(\lambda)$.   Fig 5: $x_1(-\lambda)$.

Let us shift $x_2(-\lambda)$ by t units of time to get $x_2(t-\lambda)$ and then multiply with $x_1(\lambda)$ as shown below:

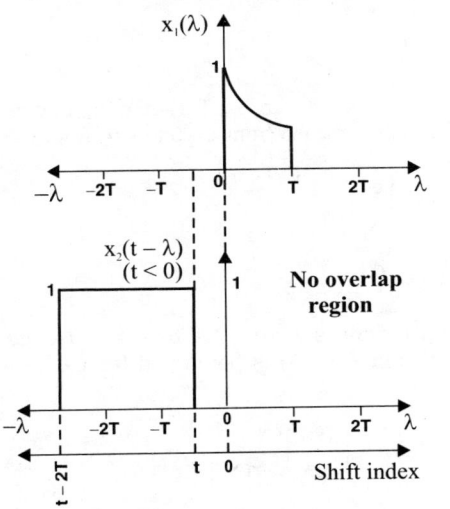

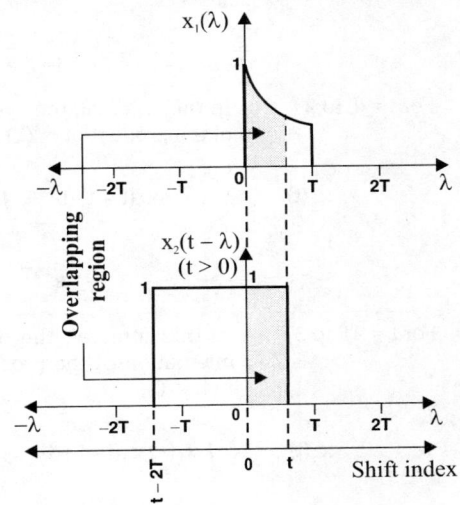

Fig 6: $x_1(\lambda)$ and $x_2(t-\lambda)$ when time shift, t < 0. | *Fig 7: $x_1(\lambda)$ and $x_2(t-\lambda)$ when time shift, t = 0 to T.*

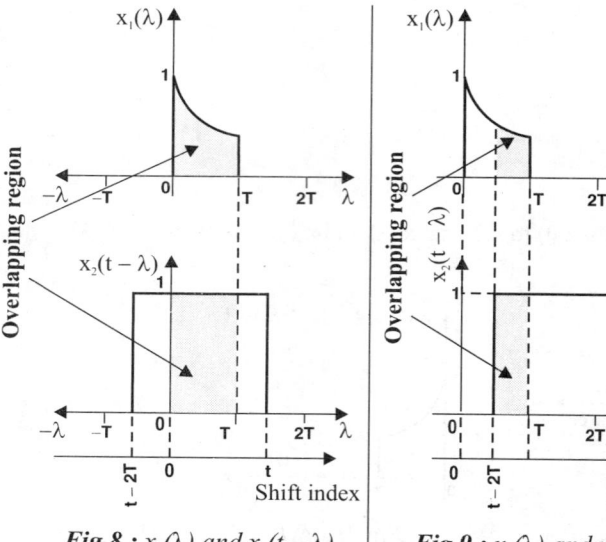

Fig 8 : $x_1(\lambda)$ *and* $x_2(t-\lambda)$
when time shift,
$t = T$ *to* $2T$.

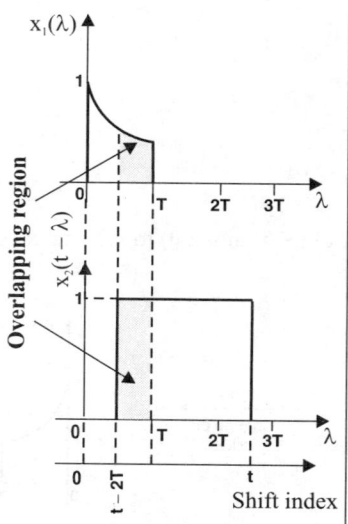

Fig 9 : $x_1(\lambda)$ *and* $x_2(t-\lambda)$
when time shift,
$t = 2T$ *to* $3T$.

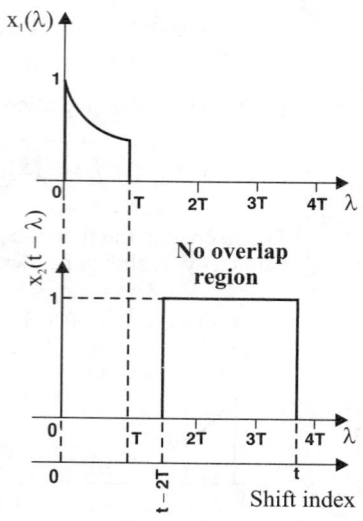

Fig 10 : $x_1(\lambda)$ *and* $x_2(t-\lambda)$
when time shift,
$t > 3T$.

From Fig 6 to Fig 10, it is observed that the product of $x_1(\lambda)$ and $x_2(t-\lambda)$ is non-zero only for time shift $t = 0$ to $3T$. For any time shift in the range $t = 0$ to $3T$, the non-zero product exists in the overlapping regions shown in Fig 7, Fig 8 and Fig 9.

For $t = 0$ to T : In this interval, the overlapping region is $\lambda = 0$ to $\lambda = t$. Hence the integration of the product of $x_1(\lambda)$ and $x_2(t-\lambda)$ is performed from $\lambda = 0$ to $\lambda = t$.

$$\therefore \; x_3(t) = \int_{\lambda=0}^{\lambda=t} x_1(\lambda)\,x_2(t-\lambda)\,d\lambda = \int_0^t e^{-a\lambda} \times 1\,d\lambda = \left[\frac{e^{-a\lambda}}{-a}\right]_0^t$$

$$= \frac{e^{-at}}{-a} - \frac{e^0}{-a} = -\frac{1}{a}e^{-at} + \frac{1}{a} = \frac{1}{a}\left(1 - e^{-at}\right)$$

For $t = T$ to $2T$: In this interval, the overlapping region is $\lambda = 0$ to $\lambda = T$. Hence the integration of the product of $x_1(\lambda)$ and $x_2(t-\lambda)$ is performed from $\lambda = 0$ to $\lambda = T$.

$$\therefore \; x_3(t) = \int_{\lambda=0}^{\lambda=T} x_1(\lambda)\,x_2(t-\lambda)\,d\lambda = \int_0^T e^{-a\lambda} \times 1\,d\lambda = \left[\frac{e^{-a\lambda}}{-a}\right]_0^T$$

$$= \frac{e^{-aT}}{-a} - \frac{e^0}{-a} = -\frac{1}{a}e^{-aT} + \frac{1}{a} = \frac{1}{a}\left(1 - e^{-aT}\right)$$

For $t = 2T$ to $3T$: In this interval, the overlapping region is $\lambda = t - 2T$ to $\lambda = T$. Hence the integration of the product of $x_1(\lambda)$ and $x_2(t-\lambda)$ is performed from $\lambda = t - 2T$ to $\lambda = T$.

$$\therefore \; x_3(t) = \int_{\lambda=t-2T}^{\lambda=T} x_1(\lambda)\,x_2(t-\lambda)\,d\lambda = \int_{t-2T}^T e^{-a\lambda} \times 1\,d\lambda = \left[\frac{e^{-a\lambda}}{-a}\right]_{t-2T}^T$$

$$= \frac{e^{-aT}}{-a} - \frac{e^{-a(t-2T)}}{-a} = -\frac{1}{a}e^{-aT} + \frac{1}{a}e^{-a(t-2T)} = \frac{1}{a}\left(e^{-a(t-2T)} - e^{-aT}\right)$$

$$\therefore \quad x_3(t) = 0 \qquad\qquad\qquad\qquad ; \quad t < 0$$

$$= \frac{1}{a}(1 - e^{-at}) \qquad\qquad ; \quad 0 < t < T$$

$$= \frac{1}{a}(1 - e^{-aT}) \qquad\qquad ; \quad T < t < 2T$$

$$= \frac{1}{a}(e^{-a(t-2T)} - e^{-aT}) \quad ; \quad 2T < t < 3T$$

$$= 0 \qquad\qquad\qquad\qquad\quad ; \quad t > 3T$$

Example 3.10

Perform convolution of the following signals, by graphical method and sketch the resultant signal.

a)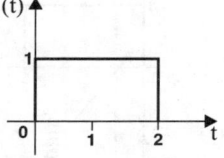

b)

Solution:

a) Let, $x_3(t) = x_1(t) * x_2(t)$

By definition of convolution,

$$x_3(t) = \int_{\lambda=-\infty}^{\lambda=+\infty} x_1(\lambda)\, x_2(t - \lambda)\, d\lambda$$

Let us change the time index t in $x_1(t)$ and $x_2(t)$ to λ, to get $x_1(\lambda)$ and $x_2(\lambda)$, and then fold $x_2(\lambda)$ to get $x_2(-\lambda)$ graphically as shown below:

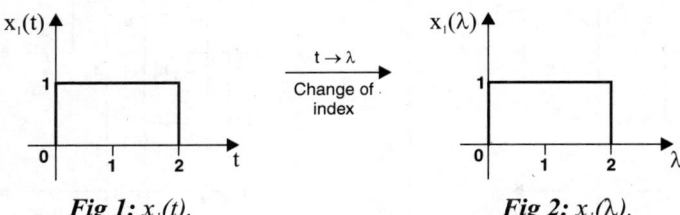

Fig 1: $x_1(t)$.   *Fig 2: $x_1(\lambda)$.*

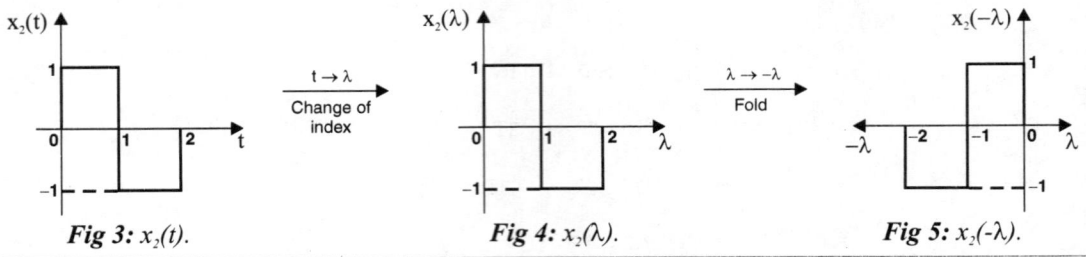

Fig 3: x₂(t).   *Fig 4: x₂(λ).*   *Fig 5: x₂(-λ).*

Let us shift $x_2(-\lambda)$ by t units of time to get $x_2(t - \lambda)$ and then multiply with $x_1(\lambda)$ as shown below:

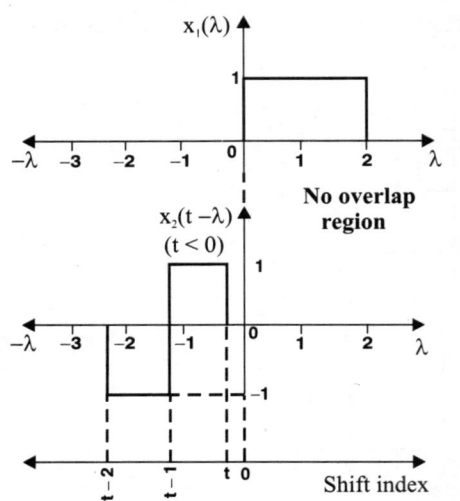

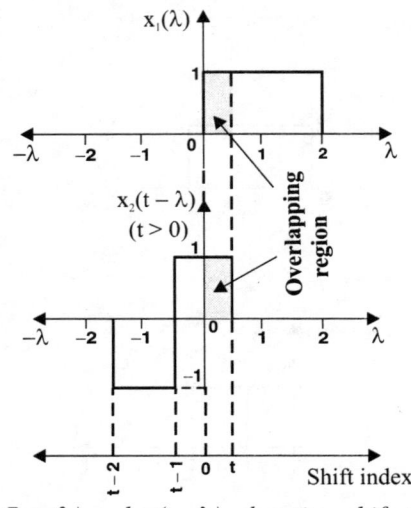

Fig 6: $x_1(\lambda)$ and $x_2(t - \lambda)$ when time shift, t < 0. | *Fig 7: $x_1(\lambda)$ and $x_2(t - \lambda)$ when time shift, t = 0 to 1.*

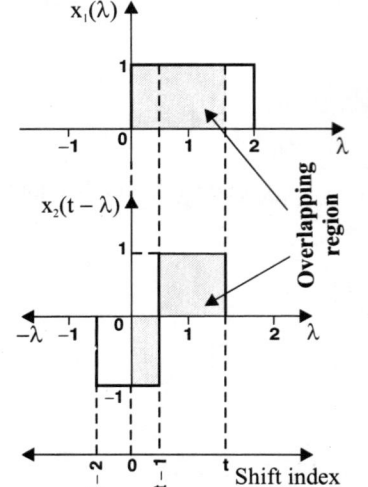

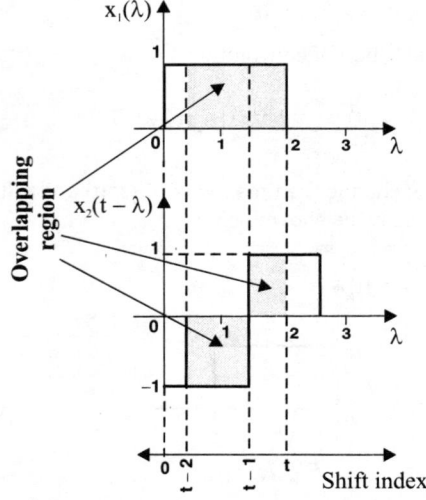

Fig 8: $x_1(\lambda)$ and $x_2(t - \lambda)$ when time shift, t = 1 to 2.

Fig 9: $x_1(\lambda)$ and $x_2(t - \lambda)$ when time shift, t = 2 to 3.

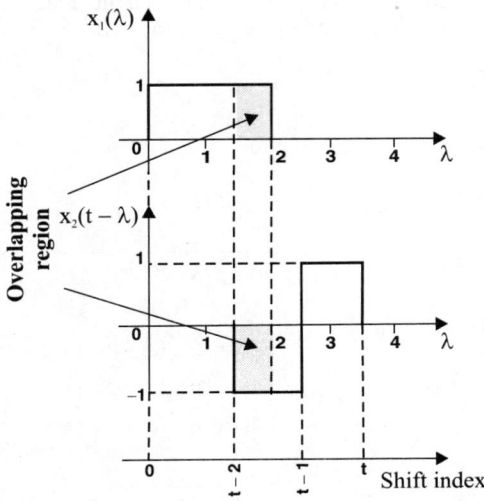

Fig 10: $x_1(\lambda)$ *and* $x_2(t-\lambda)$ *when time shift, t = 3 to 4.*

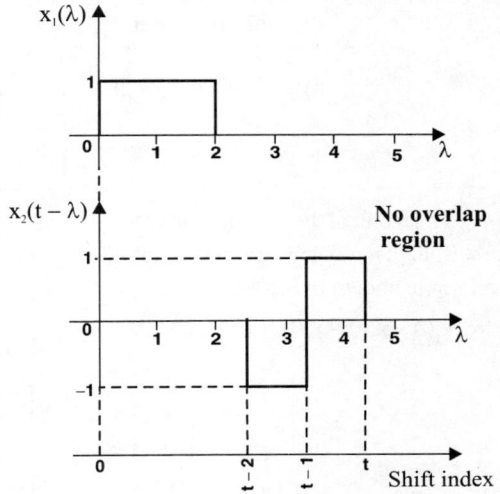

Fig 11: $x_1(\lambda)$ *and* $x_2(t-\lambda)$ *when time shift, t > 4.*

From Fig 6 to Fig 11, it is observed that the product of $x_1(\lambda)$ and $x_2(t-\lambda)$ is non-zero only for time shift t = 0 to 4. For any time shift in the range t = 0 to 4, the non-zero product exists in the overlapping regions shown in Fig 7 to Fig 10.

For t = 0 to 1 : In this interval, the overlapping region is $\lambda = 0$ to $\lambda = t$. Hence the integration of the product of $x_1(\lambda)$ and $x_2(t-\lambda)$ is performed from $\lambda = 0$ to $\lambda = t$.

$$\therefore x_3(t) = \int_{\lambda=0}^{\lambda=t} x_1(\lambda)\,x_2(t-\lambda)\,d\lambda = \int_0^t 1\times 1\,d\lambda = [\lambda]_0^t = t$$

For t = 1 to 2 : In this interval, the overlapping region is $\lambda = 0$ to $\lambda = t$. Hence the integration of the product of $x_1(\lambda)$ and $x_2(t-\lambda)$ is performed from $\lambda = 0$ to $\lambda = t$.

$$\therefore x_3(t) = \int_{\lambda=0}^{\lambda=t} x_1(\lambda)\,x_2(t-\lambda)\,d\lambda = \int_0^{t-1} 1\times(-1)\,d\lambda + \int_{t-1}^t 1\times 1\,d\lambda = [-\lambda]_0^{t-1} + [\lambda]_{t-1}^t$$

$$= -(t-1) + 0 + t - (t-1) = -t + 1 + t - t + 1 = 2 - t$$

For t = 2 to 3 : In this interval, the overlapping region is $\lambda = t-2$ to $\lambda = 2$. Hence the integration of the product of $x_1(\lambda)$ and $x_2(t-\lambda)$ is performed from $\lambda = t-2$ to $\lambda = 2$.

$$\therefore x_3(t) = \int_{\lambda=t-2}^{\lambda=2} x_1(\lambda)\,x_2(t-\lambda)\,d\lambda = \int_{t-2}^{t-1} 1\times(-1)\,d\lambda + \int_{t-1}^2 1\times 1\ d\lambda = [-\lambda]_{t-2}^{t-1} + [\lambda]_{t-1}^2$$

$$= -(t-1) + (t-2) + 2 - (t-1) = -t + 1 + t - 2 + 2 - t + 1 = 2 - t$$

For t = 3 to 4   :   In this interval, the overlapping region is $\lambda = t - 2$ to $\lambda = 2$. Hence the integration of the product of $x_1(\lambda)$ and $x_2(t - \lambda)$ is performed from $\lambda = t - 2$ to $\lambda = 2$.

$$\therefore \; x_3(t) = \int_{\lambda = t-2}^{\lambda = 2} x_1(\lambda) \, x_2(t - \lambda) \, d\lambda = \int_{t-2}^{2} 1 \times (-1) \, d\lambda = [-\lambda]_{t-2}^{2}$$

$$= -2 + (t - 2) = -2 + t - 2 = t - 4$$

The result of the convolution of $x_1(t)$ with $x_2(t)$ is given below and the resultant $x_3(t)$ waveform is shown in Fig 12.

$x_1(t) * x_2(t) = x_3(t) = 0$; $t < 0$

$= t$; $0 < t < 1$

$= 2 - t$; $1 < t < 3$

$= t - 4$; $3 < t < 4$

$= 0$; $t > 4$

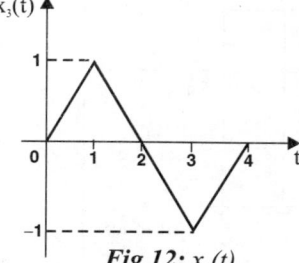

Fig 12: $x_3(t)$.

At t = 0, $x_3(0) = 0$
At t = 1, $x_3(1) = t = 1$
At t = 2, $x_3(2) = 2 - t = 2 - 2 = 0$
At t = 3, $x_3(3) = 2 - t = 2 - 3 = -1$
At t = 4, $x_3(4) = t - 4 = 4 - 4 = 0$

b)   Let, $x_3(t) = x_1(t) * x_2(t)$

By definition of convolution,

$$x_3(t) = \int_{\lambda = -\infty}^{\lambda = +\infty} x_1(\lambda) \, x_2(t - \lambda) \, d\lambda$$

Let us change the time index t in $x_1(t)$ and $x_2(t)$ to λ, to get $x_1(\lambda)$ and $x_2(\lambda)$, and then fold $x_2(\lambda)$ to get $x_2(-\lambda)$ graphically as shown below:

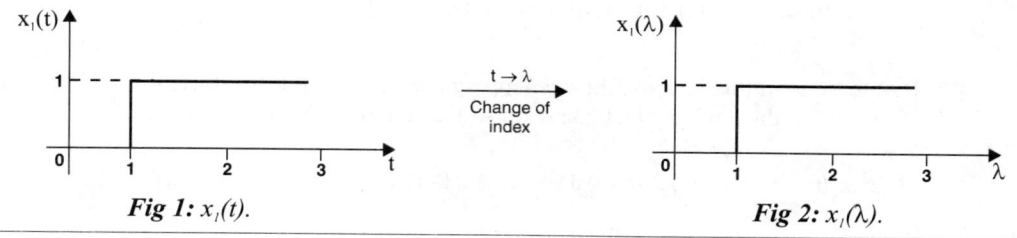

Fig 1: $x_1(t)$.   *Fig 2: $x_1(\lambda)$.*

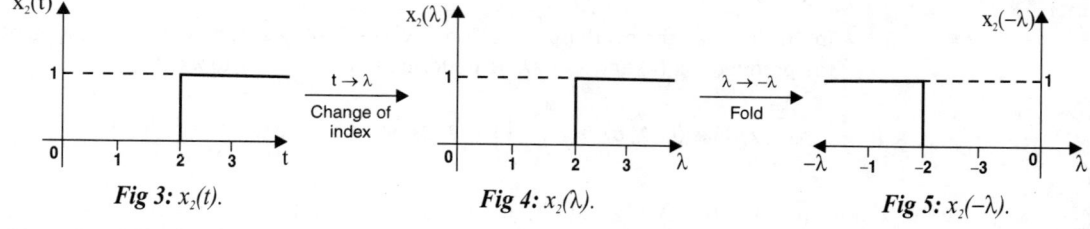

Fig 3: $x_2(t)$.   *Fig 4: $x_2(\lambda)$.*   *Fig 5: $x_2(-\lambda)$.*

Let us shift $x_2(-\lambda)$ by t units of time to get $x_2(t-\lambda)$ and then multiply with $x_1(\lambda)$ as shown below:

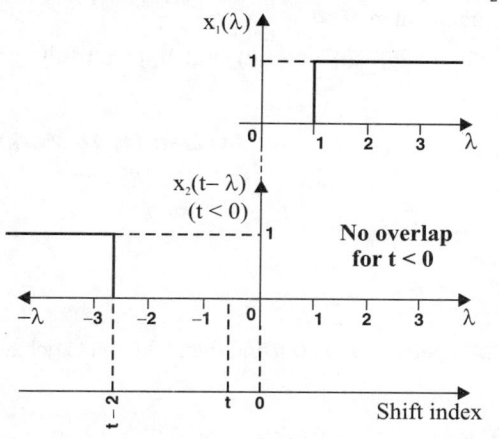

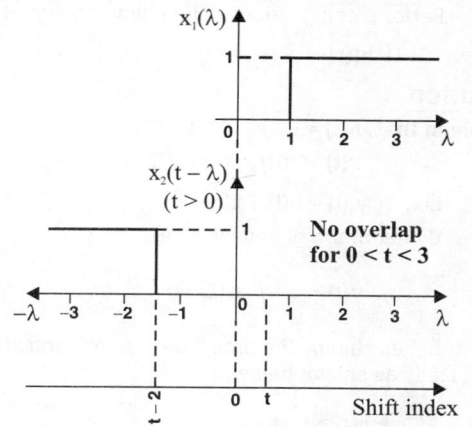

Fig 6: *$x_1(\lambda)$ and $x_2(t-\lambda)$ when time shift, $t < 0$.* | **Fig 7:** *$x_1(\lambda)$ and $x_2(t-\lambda)$ when time shift, $t > 0$.*

From Fig 6, Fig 7 and Fig 8, it is observed that the product of $x_1(\lambda)$ and $x_2(t-\lambda)$ is non-zero only for time shift $t \geq 3$. For any time shift $t > 3$, the non-zero product exists in the overlapping region shown in Fig 8.

For t > 3 :

$$x_3(t) = \int_{\lambda=1}^{\lambda=t-2} x_1(\lambda)\, x_2(t-\lambda)\, d\lambda$$

$$= \int_{1}^{t-2} 1 \times 1\, d\lambda = [\lambda]_1^{t-2}$$

$$= t - 2 - 1$$

$$= t - 3 \; ; \; t \geq 3$$

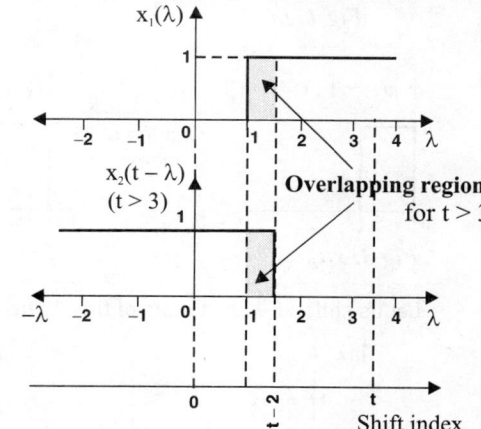

Fig 8: *$x_1(\lambda)$ and $x_2(t-\lambda)$ when time shift, $t > 3$.*

The result of the convolution of $x_1(t)$ with $x_2(t)$ is given below and the resultant waveform is shown in Fig 9.

$$x_1(t) * x_2(t) = x_3(t) = t - 3 \; ; \; t \geq 3$$

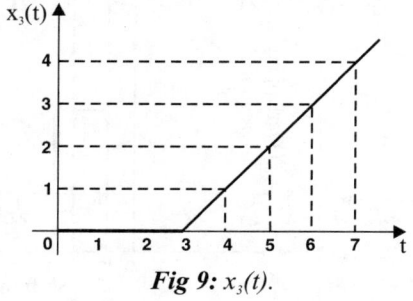

Fig 9: *$x_3(t)$.*

At t = 3, $x_3(t) = t - 3 = 3 - 3 = 0$	
At t = 4, $x_3(t) = t - 3 = 4 - 3 = 1$	
At t = 5, $x_3(t) = t - 3 = 5 - 3 = 2$	
At t = 6, $x_3(t) = t - 3 = 6 - 3 = 3$	
At t = 7, $x_3(t) = t - 3 = 7 - 3 = 4$	

Example 3.11

Perform convolution of the following signals, by graphical method.

    a) $h(t) = t$; $0 \leq t \leq T$ and $x(t) = u(t)$; $0 \leq t \leq T$ **b)** $x(t) = e^{-2t} u(t)$ and $h(t) = u(t + 2)$

Solution

a) Given that, $h(t) = t$, $0 \leq t \leq T$ **(AU Dec'11, 16 Marks)**

                  $x(t) = u(t)$, $0 \leq t \leq T$

Let, $y_3(t) = h(t) * x(t)$

By definition of convolution,

$$y(t) = \int_{\lambda = -\infty}^{\lambda = +\infty} h(\lambda)\, x(t - \lambda)\ d\lambda$$

Let us change the time index t in h(t) and x(t) to λ, to get $h(\lambda)$ and $x(\lambda)$, and then fold $x(\lambda)$ to get $x(-\lambda)$ graphically as shown below:

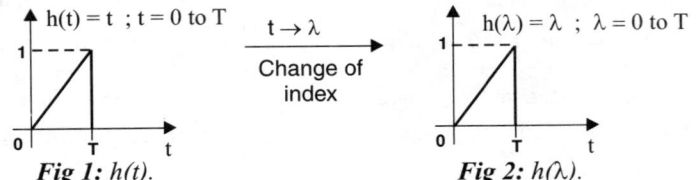

        Fig 1: *h(t).*   **Fig 2:** *h(λ).*

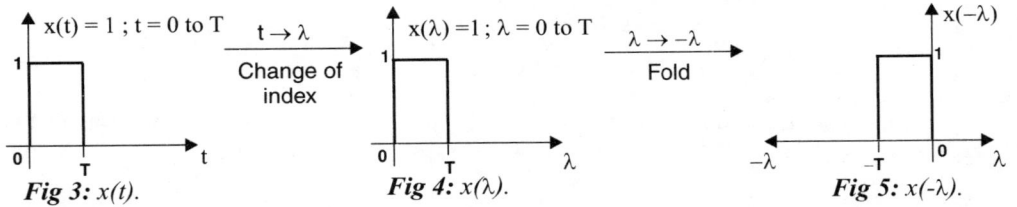

      Fig 3: *x(t).*   **Fig 4:** *x(λ).*   **Fig 5:** *x(-λ).*

Let us shift $x(-\lambda)$ by t units of time to get $x(t - \lambda)$ and then multiply with $h(\lambda)$ as shown below:

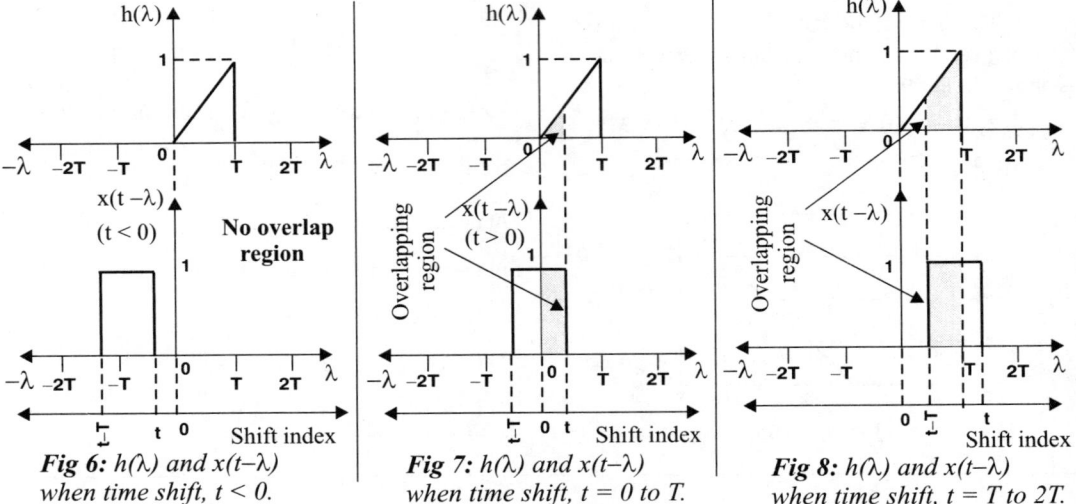

   Fig 6: *h(λ) and x(t-λ)*   **Fig 7:** *h(λ) and x(t-λ)*   **Fig 8:** *h(λ) and x(t-λ)*
   when time shift, t < 0.   *when time shift, t = 0 to T.*   *when time shift, t = T to 2T.*

From Fig 7 and Fig 8, it is observed that the product of $h(\lambda)$ and $x(t-\lambda)$ is non-zero only for time shift $t = 0$ to $2T$. For any time shift in the range $t = 0$ to $2T$, the non-zero product exists in the overlapping regions shown in Figs 7 and 8.

For $t = 0$ to T :   In this interval, the overlapping region is $\lambda = 0$ to $\lambda = t$. Hence the integration of the product of $h(\lambda)$ and $x(t-\lambda)$ is performed from $\lambda = 0$ to $\lambda = t$.

$$\therefore y(t) = \int\limits_{\lambda=0}^{\lambda=t} h(\lambda)\ x(t-\lambda)\ d\lambda = \int\limits_{0}^{t} 1\times 1\ d\lambda = [\lambda]_0^t = t$$

For $t = T$ to $2T$:   In this interval, the overlapping region is $\lambda = t-T$ to $\lambda = T$. Hence the integration of the product of $x_1(\lambda)$ and $x_2(t-\lambda)$ is performed from $\lambda = t-T$ to $\lambda = T$.

$$\therefore y(t) = \int\limits_{\lambda=t-T}^{\lambda=T} h(\lambda)\ x(t-\lambda)\ d\lambda = \int\limits_{t-T}^{T} 1\times 1\ d\lambda = [\lambda]_{t-T}^T = T-(t-T) = t$$

The result of the convolution of $h(t)$ with $x(t)$ is given below and the resultant waveform is shown in Fig 9.

$$h(t) * x(t) = y(t) = t\ ;\ 0 < t < 2T$$

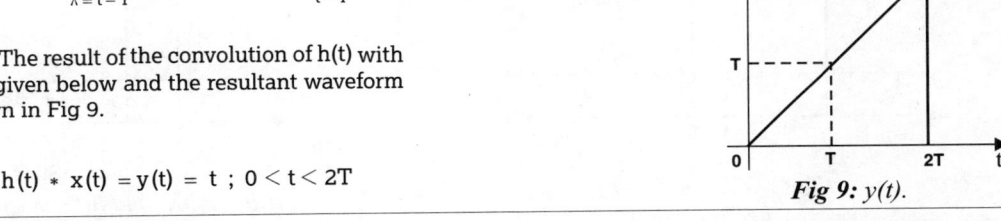

Fig 9: $y(t)$.

b) Given that, $x(t) = e^{-2t}\ u(t) = e^{-2t}\ ;\ t \geq 0$

         $h(t) = u(t+2) = 1\ \ ;\ t \geq -2$

                                                            (AU May'15, 6 Marks)

Let, $y(t) = x(t) * h(t)$

By definition of convolution,

$$y(t) = \int\limits_{\lambda=-\infty}^{\lambda=+\infty} x(\lambda)\ h(t-\lambda)\ d\lambda$$

Let us change the time index t, in $x(t)$ and $h(t)$ to λ, to get $x(\lambda)$ and $h(\lambda)$, and then fold $h(\lambda)$ to get $h(-\lambda)$ graphically as shown below:

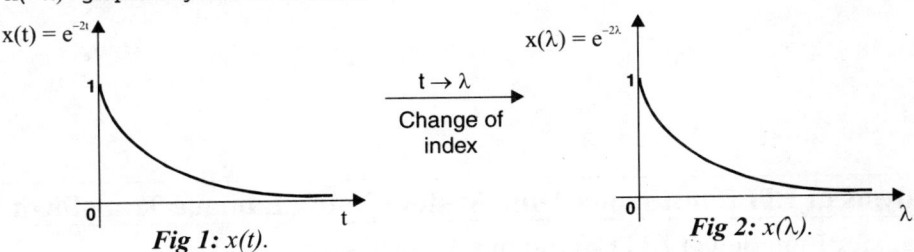

Fig 1: $x(t)$.   **Fig 2:** $x(\lambda)$.

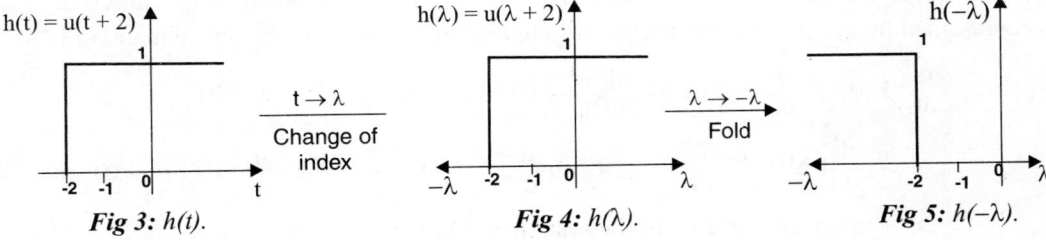

Fig 3: $h(t)$.   **Fig 4:** $h(\lambda)$.   **Fig 5:** $h(-\lambda)$.

Let us shift h($-\lambda$) by t units of time to get h(t $- \lambda$) and then multiply with x(λ) as shown below:

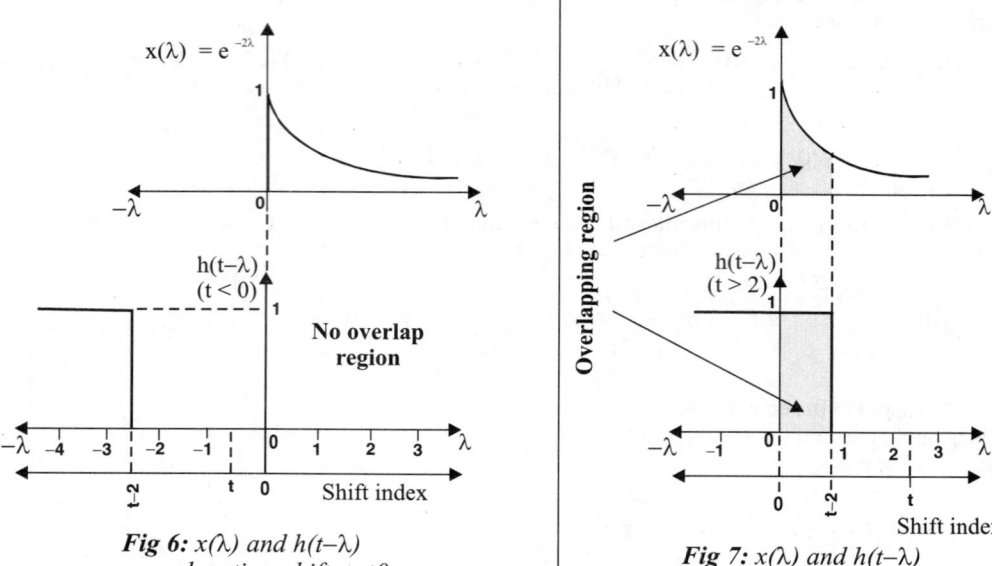

Fig 6: *x(λ) and h(t$-\lambda$)*
when time shift, t <0

Fig 7: *x(λ) and h(t$-\lambda$)*
when time shift, t > 2

From Figs 6 and 7, it is observed that the product of x(λ) and h(t $- \lambda$) is non - zero only for time shift t > 2.

For any time shift t > 2, the non - zero product exits in the overlapping region shown in Fig 7. Here the overlapping region is $\lambda = 0$ to $\lambda = t-2$. Hence integration of x(λ) and h(t $- \lambda$) is performed from $\lambda = 0$ to $\lambda = t-2$.

$$\therefore \ y(t) = \int_{\lambda=0}^{\lambda=t-2} x(\lambda)\, h(t-\lambda)\, d\lambda \ ; \ t \geq 0$$

$$= \int_{0}^{t-2} e^{-2\lambda} \times 1 \ d\lambda \ = \ \left[\frac{e^{-2\lambda}}{-2}\right]_{0}^{t-2} = -\frac{1}{2}[e^{-2(t-2)} - e^{0}] \ ; \ t \geq 2$$

$$= \frac{1}{2}[1 - e^{-2(t-2)}]\, u(t-2)$$

3.6 Analysis of LTI Continuous Time System Using Laplace Transform

3.6.1 Transfer Function of LTI Continuous Time System

In general, the input-output relation of a LTI (Linear Time Invariant) continuous time system is represented by the constant coefficient differential equation shown below, (equation (3.29)).

$$\frac{d^N}{dt^N}\, y(t) + a_1 \frac{d^{N-1}}{dt^{N-1}}\, y(t) + a_2 \frac{d^{N-2}}{dt^{N-2}}\, y(t) + ... + a_{N-1} \frac{d}{dt}\, y(t) + a_N\, y(t)$$

$$= \ b_0 \frac{d^M}{dt^M}\, x(t) + b_1 \frac{d^{M-1}}{dt^{M-1}}\, x(t) + b_2 \frac{d^{M-2}}{dt^{M-2}}\, x(t) + ... + b_{M-1} \frac{d}{dt}\, x(t) + b_M\, x(t) \quad .....(3.29)$$

where, N = Order of the system and M $\leq$ N.

On taking Laplace transform of equation (3.29) with zero initial conditions we get,

$$s^N Y(s) + a_1 s^{N-1} Y(s) + a_2 s^{N-2} Y(s) + \ldots + a_{N-1} s Y(s) + a_N Y(s)$$

$$= b_0 s^M X(s) + b_1 s^{M-1} X(s) + b_2 s^{M-2} X(s) + \ldots + b_{M-1} s X(s) + b_M X(s)$$

$$Y(s) (s^N + a_1 s^{N-1} + a_2 s^{N-2} + \ldots + a_{N-1} s + a_N)$$

$$= X(s) (b_0 s^M + b_1 s^{M-1} + b_2 s^{M-2} + \ldots + b_{M-1} s + b_M)$$

$$\therefore \ \frac{Y(s)}{X(s)} = \frac{b_0 s^M + b_1 s^{M-1} + b_2 s^{M-2} + \ldots + b_{M-1} s + b_M}{s^N + a_1 s^{N-1} + a_2 s^{N-2} + \ldots + a_{N-1} s + a_N} \qquad \ldots.(3.30)$$

The *transfer function* of a continuous time system is defined as the ratio of Laplace transform of output and Laplace transform of input. Hence the equation (3.30) is the transfer function of an LTI continuous time system.

The equation (3.30) is a rational function of s (i.e., ratio of two polynomials in s). The numerator and denominator polynomials of equation (3.29) can be expressed in the factorized form as shown in equation (3.31).

$$\frac{Y(s)}{X(s)} = G \frac{(s - z_1)(s - z_2)(s - z_3) \ldots (s - z_M)}{(s - p_1)(s - p_2)(s - p_3) \ldots (s - p_N)} \qquad \ldots.(3.31)$$

where, $z_1, z_2, z_3, \ldots z_M$ are roots of numerator polynomial

(or zeros of continuous time system)

$p_1, p_2, p_3, \ldots p_N$ are roots of denominator polynomial

(or poles of continuous time system)

3.6.2 Impulse Response and Transfer Function

Let, $x(t)$ = Input of a LTI continuous time system

$y(t)$ = Output / Response of the LTI continuous time system for the input $x(t)$

$h(t)$ = Impulse response (i.e., response for impulse input)

Now, the response $y(t)$ of the continuous time system is given by convolution of input and impulse response [Refer equation (3.9)].

$$\therefore \ y(t) = x(t) * h(t) = \int_{-\infty}^{+\infty} x(\lambda) h(t - \lambda) \, d\lambda \qquad \ldots (3.32)$$

On taking Laplace transform of equation (3.32) we get,

$$\mathcal{L}\{y(t)\} = \mathcal{L}\{x(t) * h(t)\}$$

Using convolution property of Laplace transform the above equation can be written as,

> If $\mathcal{L}\{x(t)\} = X(s)$
> and $\mathcal{L}\{h(t)\} = H(s)$
> then by convolution property
> $\mathcal{L}\{x(t) * h(t)\} = X(s) H(s)$.

$$Y(s) = X(s) H(s)$$

$$\therefore \ H(s) = \frac{Y(s)}{X(s)} \qquad \ldots.(3.33)$$

$$\therefore H(s) = \frac{Y(s)}{X(s)} = G \frac{(s-z_1)(s-z_2)(s-z_3)\,.....\,(s-z_M)}{(s-p_1)(s-p_2)(s-p_3)\,.....\,(s-p_N)}$$

Using equation (3.31).

From equation (3.33) we can conclude that *the transfer function of LTI continuous time system is also given by Laplace transform of the impulse response.*

Alternatively we can say that *the inverse Laplace transform of transfer function is the impulse response of the system.*

$$\therefore \text{Impulse response, } h(t) = \mathcal{L}^{-1}\{H(s)\} = \mathcal{L}^{-1}\left\{\frac{Y(s)}{X(s)}\right\}$$

Using equation (3.33).

3.6.3 Response of LTI Continuous Time System Using Laplace Transform

In general, the input-output relation of an LTI (Linear Time Invariant) continuous time system is represented by the constant coefficient differential equation shown below, (equation (3.34)).

$$\frac{d^N}{dt^N} y(t) + a_1 \frac{d^{N-1}}{dt^{N-1}} y(t) + a_2 \frac{d^{N-2}}{dt^{N-2}} y(t) + ... + a_{N-1} \frac{d}{dt} y(t) + a_N y(t)$$

$$= b_0 \frac{d^M}{dt^M} x(t) + b_1 \frac{d^{M-1}}{dt^{M-1}} x(t) + b_2 \frac{d^{M-2}}{dt^{M-2}} x(t) + ... + b_{M-1} \frac{d}{dt} x(t) + b_M x(t) \qquad .....(3.34)$$

The solution of the above differential equation (equation (3.34)) is the *(total) response* $y(t)$ of LTI system, which consists of two parts.

In signals and systems the two parts of the solution $y(t)$ are called zero-input response $y_{zi}(t)$ and zero-state response $y_{zs}(t)$.

$$\therefore \text{Response, } y(t) = y_{zi}(t) + y_{zs}(t)$$

Zero-Input Response (or Free Response or Natural Response) Using Laplace Transform

The *zero-input response* $y_{zi}(t)$ is mainly due to initial output (or initial stored energy) in the system. The zero-input response is obtained from system equation (equation (3.34)) when input $x(t) = 0$.

On substituting $x(t) = 0$ and $y(t) = y_{zi}(t)$ in equation (3.34) we get,

$$\frac{d^N}{dt^N} y_{zi}(t) + a_1 \frac{d^{N-1}}{dt^{N-1}} y_{zi}(t) + a_2 \frac{d^{N-2}}{dt^{N-2}} y_{zi}(t) + ... + a_{N-1} \frac{d}{dt} y_{zi}(t) + a_N y_{zi}(t) = 0$$

On taking Laplace transform of the above equation with non-zero initial conditions for output we can form an equation for $Y_{zi}(s)$.

The zero-input response $y_{zi}(t)$ is given by inverse Laplace transform of $Y_{zi}(s)$.

Zero-State Response (or Forced Response) Using Laplace Transform

The *zero-state response* $y_{zs}(t)$ is the response of the system due to input signal and with zero initial output. The zero-state response is obtained from the differential equation governing the system (equation (3.34)) for specific input signal $x(t)$ for $t \geq 0$ and with zero initial output.

On substituting $y(t) = y_{zs}(t)$ in equation (3.34) we get,

$$\frac{d^N}{dt^N} y_{zs}(t) + a_1 \frac{d^{N-1}}{dt^{N-1}} y_{zs}(t) + a_2 \frac{d^{N-2}}{dt^{N-2}} y_{zs}(t) + ... + a_{N-1} \frac{d}{dt} y_{zs}(t) + a_N y_{zs}(t)$$

$$= b_0 \frac{d^M}{dt^M} x(t) + b_1 \frac{d^{M-1}}{dt^{M-1}} x(t) + b_2 \frac{d^{M-2}}{dt^{M-2}} x(t) + ... + b_{M-1} \frac{d}{dt} x(t) + b_M x(t)$$

On taking Laplace transform of the above equation with zero initial conditions for output (i.e., $y_{zs}(t)$) and non-zero initial values for input (i.e., $x(t)$) we can form an equation for $Y_{zs}(s)$.

The zero-state response $y_{zs}(t)$ is given by inverse Laplace transform of $Y_{zs}(s)$.

Total Response

The *total response* $y(t)$ is the response of the system due to input signal and initial output (or intial stored energy). The total response is obtained from the differential equation governing the system (equation (3.34)) for specific input signal $x(t)$ for $t \geq 0$ and with non-zero initial conditions.

On taking Laplace transform of equation (3.34) with non-zero initial conditions for both input and output, and then substituting for $X(s)$ we can form an equation for $Y(s)$.

The total response $y(t)$ is given by inverse Laplace transform of $Y(s)$.

Alternatively the total response $y(t)$ is given by sum of zero-input response $y_{zi}(t)$ and zero-state response $y_{zs}(t)$.

$$\therefore \text{ Total response, } y(t) = y_{zi}(t) + y_{zs}(t)$$

3.6.4 Convolution and Deconvolution Using Laplace Transform

Convolution

The *convolution* operation is performed to find the response $y(t)$ of LTI continuous time system from the input $x(t)$ and impulse response $h(t)$.

$\therefore$ Response, $y(t) = x(t) * h(t)$

On taking Laplace transform of the above equation we get,

$$\mathcal{L}\{y(t)\} = \mathcal{L}\{x(t) * h(t)\}$$

$\therefore Y(s) = X(s) H(s)$ .....(3.35)

$\therefore$ Response, $y(t) = \mathcal{L}^{-1}\{Y(s)\}$

$$= \mathcal{L}^{-1}\{X(s) H(s)\}$$

> If $\mathcal{L}\{x(t)\} = X(s)$
> and $\mathcal{L}\{h(t)\} = H(s)$
> then by convolution property
> $\mathcal{L}\{x(t) * h(t)\} = X(s) H(s)$.

Procedure: 1. Take Laplace transform of $x(t)$ to get $X(s)$.

2. Take Laplace transform of $h(t)$ to get $H(s)$.

3. Compute the product of $X(s)H(s)$. Let, $X(s)H(s) = Y(s)$.

4. Take inverse Laplace transform of $Y(s)$ to get $y(t)$.

Deconvolution

The *deconvolution* operation is performed to extract the input $x(t)$ of an LTI continuous time system from the response $y(t)$ of the system.

From equation (3.35) get,

$$X(s) = \frac{Y(s)}{H(s)}$$

On taking inverse Laplace transform of the above equation we get,

$$\text{Input, } x(t) = \mathcal{L}^{-1}\{X(s)\} = \mathcal{L}^{-1}\left\{\frac{Y(s)}{H(s)}\right\}$$

Procedure: 1. Take Laplace transform of y(t) to get Y(s).

2. Take Laplace transform of h(t) to get H(s).

3. Divide Y(s) by H(s) to get X(s) (i.e., X(s) = Y(s) / H(s)).

4. Take inverse Laplace transform of X(s) to get x(t).

3.6.5 Stability in s-Domain

ROC of a Stable LTI System

Let H(s) be Laplace transform of h(t). Now by definition of Laplace transform we get,

$$H(s) \ = \ \int_{-\infty}^{+\infty} h(t)\, e^{-st}\, dt \ = \ \int_{-\infty}^{+\infty} h(t)\, e^{-(\sigma + j\Omega)t}\, dt$$

$$\boxed{s = \sigma + j\Omega}$$

$$\boxed{\int u\, dv = uv - \int v\, du}$$

$$\boxed{u = e^{-j\Omega t} \ \Rightarrow \ du = -j\Omega e^{-j\Omega t}}$$

$$\boxed{dv = h(t) \ \Rightarrow \ v = \int h(t)\, dt}$$

Let us evaluate H(s) for $\sigma = 0$.

$$\therefore H(s) = \int_{-\infty}^{+\infty} h(t)\, e^{-j\Omega t} dt = \int_{-\infty}^{+\infty} e^{-j\Omega t}\, h(t)\, dt = \left[e^{-j\Omega t}\left(\int h(t)\, dt \right) - \int h(t)\, dt \times (-j\Omega e^{-j\Omega t}) \right]_{-\infty}^{+\infty}$$

$$.....(3.36)$$

For a stable LTI system, $\int h(t)\, dt$ is constant.

Therefore, in equation (3.36), put, $\int h(t)\, dt = A$ where A constant.

$$\therefore \ H(s) = \left[e^{-j\Omega} A + j\Omega \int A\, e^{-j\Omega t} dt \right]_{-\infty}^{+\infty} = A\left[e^{-j\Omega} + j\Omega \int e^{-j\Omega t}\, dt \right]_{-\infty}^{+\infty} \qquad .....(3.37)$$

The evaluation of equation (3.37), is evaluation of Laplace transform for $s = j\Omega$, (i.e., evaluation of H(s) along imaginary axis) and so we can say that H(s) exists if the ROC includes the imaginary axis. Hence *for a stable LTI continuous time system the ROC should include the imaginary axis of s-plane.*

Location of Poles for Stability of Causal Systems

Let h(t) be impulse response of an LTI causal system. Now if h(t) satisfies the condition,

$$\boxed{\int_{0}^{\infty} h(t)\, dt < \infty}$$

$$.....(3.38)$$

then the LTI continuous time causal system is stable.

Let, $h(t) = e^{at}\, u(t)$

$$\text{Now,} \int_{0}^{\infty} h(t)\, dt = \int_{0}^{\infty} e^{at}\, u(t)\, dt = \int_{0}^{\infty} e^{at}\, dt = \left[\frac{e^{at}}{a} \right]_{0}^{\infty} = \frac{e^{a \times \infty}}{a} - \frac{e^{a \times 0}}{a} = \frac{e^{a \times \infty}}{a} - \frac{1}{a}$$

Let a be negative, and let k = − a, so that k is always positive.

$$\text{Now,} \int_{0}^{\infty} h(t)\, dt \ = \ \frac{e^{a \times \infty}}{a} - \frac{1}{a} = \frac{e^{-k\infty}}{-k} + \frac{1}{k} = \frac{e^{-\infty}}{-k} + \frac{1}{k}$$

$$\boxed{e^{-\infty} = 0}$$

$$= \ \frac{0}{-k} + \frac{1}{k} = \frac{1}{k} = \text{Constant, and so system is stable.} \qquad .....(3.39)$$

Let a be positive.

Now, $\displaystyle\int_{0}^{\infty} h(t) \, dt = \frac{e^{a\times\infty}}{a} - \frac{1}{a}$ $\boxed{e^{\infty} = \infty}$

$$= \frac{e^{a\infty}}{a} - \frac{1}{a} = \frac{\infty}{a} - \frac{1}{a} = \infty \text{, and so system is unstable. } \quad .....(3.40)$$

From the above discussion, the stability condition of equation (3.38) can be transformed as a condition on location of poles of transfer function of the LTI continuous time causal system in s-plane.

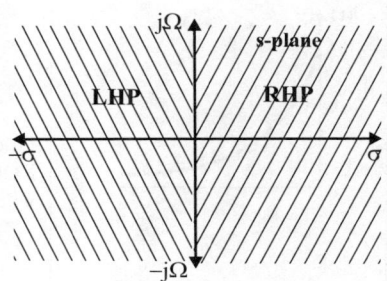

Consider the s-plane shown in Fig 3.9. The area to the right of vertical axis is called right half plane (RHP) and the area to the left of vertical axis is called left half plane (LHP).

The transfer function of a continuous time system is given by Laplace transform of its impulse response.

Fig 3.9 : s-plane.

Let, $h(t) = e^{at} u(t)$

$\therefore$ Transfer function, $H(s) = \mathcal{L}\{h(t)\} = \mathcal{L}\{e^{at} u(t)\} = \dfrac{1}{s-a}$

Here, the transfer function H(s) has pole at s = a.

> *If, a < 0, (i.e., if a is negative), then the pole will lie on the left half of s-plane, and from equation (3.39) we can say that the causal system is stable.*

> *If, a > 0, (i.e., if a is positive), then the pole will lie on the right half of s-plane. and from equation (3.39) we can say that the causal system is unstable.*

Therefore we can say that, *for a stable LTI continuous time causal system the poles should lie on the left half of s-plane.*

General Condition for Stability in s-Plane

On combining the condition for location of poles and the ROC we can say that,

1. *For a stable LTI continuous time causal system, the poles should lie on the left half of s-plane and the imaginary axis should be included in the ROC.*

2. *For a stable LTI continuous time noncausal system, the imaginary axis should be included in the ROC.*

The various types of impulse response of LTI continuous time system and their transfer functions and the locations of poles are summarized in Table 3.3.

Table 3.3: Impulse Response and Location of Poles of Transfer Function in s-Plane

Impulse Response h(t)	Transfer Function $H(s) = \mathcal{L}\{h(t)\}$	Location of Poles in s-plane and ROC
$h(t) = A\,e^{-at}\,u(t)\,;\,a > 0$ (Causal system)	$H(s) = \dfrac{A}{s+a}$ Pole at s = –a. ROC is σ > –a, where σ is real part of s.	The pole at s = –a, lies on left half of s-plane. ROC includes imaginary axis. Since pole lies on LHP and the imaginary axis is included in ROC, the system is stable.
$h(t) = A\,e^{-at}\,u(-t)\,;\,a > 0$ (Noncausal system)	$H(s) = -\dfrac{A}{s+a}$ Pole at s = –a. ROC is σ < –a, where σ is real part of s.	The pole at s = –a, lies on left half of s-plane. The ROC does not include imaginary axis. Since imaginary axis is not included in ROC, the system is unstable.
$h(t) = A\,e^{at}\,u(t)\,;\,a > 0$ (Causal system)	$H(s) = \dfrac{A}{s-a}$ Pole at s = +a. ROC is σ > +a, where σ is real part of s.	The pole at s = +a, lies on right half of s-plane. ROC does not include imaginary axis. Since pole lies on RHP and imaginary axis is not included in ROC, the system is unstable.

Table 3.3: Continued...

Impulse Response h(t)	Transfer Function $H(s) = \mathcal{L}\{h(t)\}$	Location of Poles in s-plane and ROC
$h(t) = A\,e^{at}\,u(-t)\,; a > 0$ (Noncausal system) 	$H(s) = -\dfrac{A}{s-a}$ Pole at $s = +a$. ROC is $\sigma < +a$, where σ is real part of s.	 The pole at $s = +a$, lies on right half of s-plane. The ROC includes imaginary axis. Since imaginary axis is included in ROC, the system is stable.
$h(t) = A\,e^{-a|t|}\,; a > 0$ (Noncausal system) 	$H(s) = \dfrac{A}{s+a} - \dfrac{A}{s-a}$ Poles at $s = -a, +a$. ROC is $-a < s < +a$, where s is real part of s.	 The pole at $s = +a$, lies on RHP and pole at $s = -a$, lies on LHP. ROC includes imaginary axis. Since the imaginary axis is included in ROC, the system is stable.
$h(t) = A\,e^{-at}\,u(t) + A\,e^{-bt}\,u(-t)$ where $a > 0, b > 0, a > b$ (Noncausal system) 	$H(s) = \dfrac{A}{s+a} - \dfrac{A}{s+b}$ Poles at $s = -a, -b$. ROC is $-a < s < -b$, where s is real part of s.	 The poles at $s = -a, -b$, lie on LHP. The ROC does not include imaginary axis. Since the imaginary axis is not included in ROC, the system is unstable.

Table 3.3: Continued...

Impulse Response h(t)	Transfer Function $H(s) = \mathcal{L}\{h(t)\}$	Location of Poles in s-plane and ROC
$h(t) = A\,u(t)$ (Causal system) 	$H(s) = \dfrac{A}{s}$ Pole at $s = 0$. ROC is $s > 0$, where s is real part of s.	 The pole at $s = 0$ lies on imaginary axis. The ROC does not include the imaginary axis. Since the imaginary axis is not included in ROC, the system is unstable.
$h(t) = A\,t\,u(t)$ (Causal system) 	$H(s) = \dfrac{A}{s^2}$ Double pole at $s = 0$. ROC is $s > 0$, where s is real part of s.	 The poles at $s = 0$ lie on imaginary axis. The ROC does not include the imaginary axis. Since the imaginary axis is not included in ROC, the system is unstable.
$h(t) = 2A \cos bt\, u(t)$ (Causal system) 	$H(s) = \dfrac{A}{s+jb} + \dfrac{A}{s-jb}$ Poles at $s = -jb, +jb$. ROC is $s > 0$, where s is real part of s.	 The poles at $s = -jb, +jb$, lie on imaginary axis. The ROC does not include the imaginary axis. Since the imaginary axis is not included in ROC, the system is unstable.

Table 3.3: Continued...

Impulse Response h(t)	Transfer Function $H(s) = \mathcal{L}\{h(t)\}$	Location of Poles in s-plane and ROC
$h(t) = 2A\,e^{-at}\cos bt\,u(t)$, where $a > 0$ (Causal system)	$H(s) = \dfrac{A}{s+a+jb} + \dfrac{A}{s+a-jb}$ Poles at $s = -a - jb$, $-a + jb$. ROC is $s > -a$, where s is real part of s.	The poles at $s = -a - jb$, $-a+jb$, lie on left half of s-plane. The ROC includes the imaginary axis. Since poles lie on LHP and the imaginary axis is included in ROC, the system is stable.
$h(t) = 2A\,e^{at}\cos bt\,u(t)$, where $a > 0$ (Causal system)	$H(s) = \dfrac{A}{s-a+jb} + \dfrac{A}{s-a-jb}$ Poles at $s = a - jb$, $a + jb$. ROC is $s > a$, where s is real part of s.	The poles at $s = a - jb$, $a + jb$, lie on right half of s-plane. The ROC does not include imaginary axis. Since poles lie on RHP and the imaginary axis is not included in ROC, the system is unstable.
$h(t) = 2A\,t\cos bt\,u(t)$ (Causal system)	$H(s) = \dfrac{A}{(s+jb)^2} + \dfrac{A}{(s-jb)^2}$ Double poles at $s = -jb$, $+jb$. ROC is $s > 0$, where s is real part of s.	The poles at $s = -jb$, $+jb$, lie on imaginary axis. The ROC does not include the imaginary axis. Since the imaginary axis is not included in ROC, the system is unstable.

Example 3.12

Using Laplace transform, determine the natural response of the system described by the equation,

$$\frac{d^2y(t)}{dt^2} + 6\frac{dy(t)}{dt^2} + 5\,y(t) = \frac{dx(t)}{dt} + 4\,x(t)\ ;\ y(0) = 1\ ;\ \frac{dy(t)}{dt}\bigg|_{t=0} = -2$$

Solution:

The natural response is the response of the system due to initial values of output alone. Hence for natural response the input x(t) is considered as zero. Therefore the natural response is also called zero-input response, $y_{zi}(t)$.

On substituting input x(t) = 0 and y(t) = $y_{zi}(t)$ in the given system equation we get,

$$\frac{d^2y_{zi}(t)}{dt^2} + 6\frac{dy_{zi}(t)}{dt^2} + 5\,y_{zi}(t) = 0$$

On taking Laplace transform of the above equation we get,

$$s^2\,Y_{zi}(s) - s\,y(0) - y'(0) + 6[s\,y_{zi}(s) - y(0)] + 5\,Y_{zi}(s) = 0$$

On substituting the given initial conditions of output in the above equation we get,

$$s^2\,Y_{zi}(s) - s \times 1 - (-2) + 6\,[s\,Y_{zi}(s) - 1] + 5\,Y_{zi}(s) = 0$$

$$s^2\,Y_{zi}(s) - s + 2 + 6s\,Y_{zi}(s) - 6 + 5\,Y_{zi}(s) = 0$$

$$(s^2 + 6s + 5)\,Y_{zi}(s) - s - 4 = 0$$

$$(s^2 + 6s + 5)\,Y_{zi}(s) = s + 4$$

$$\therefore Y_{zi}(s) = \frac{s+4}{s^2+6s+5} = \frac{s+4}{(s+1)(s+5)}$$

The roots of quadratic $s^2 + 6s + 5 = 0$ are,

$$s = \frac{-6 \pm \sqrt{6^2 - 4\times5}}{2} = \frac{-6\pm4}{2} = -1, -5$$

By partial fraction expansion technique, $Y_{zi}(s)$ can be expressed as shown below:

$$Y_{zi}(s) = \frac{s+4}{(s+1)(s+5)} = \frac{A}{s+1} + \frac{B}{s+5}$$

$$A = \frac{s+4}{(s+1)(s+5)} \times (s+1)\bigg|_{s=-1} = \frac{-1+4}{-1+5} = \frac{3}{4}$$

$$B = \frac{s+4}{(s+1)(s+5)} \times (s+5)\bigg|_{s=-5} = \frac{-5+4}{-5+1} = \frac{1}{4}$$

$$\therefore Y_{zi}(s) = \frac{3}{4}\frac{1}{s+1} + \frac{1}{4}\frac{1}{s+5}$$

On taking inverse Laplace transform of the above equation we get natural response.

$$\therefore \text{Natural response (or zero-input response)}\Big\}\ y_{zi}(t) = \mathcal{L}^{-1}\{Y_{zi}(s)\}$$

$$= \mathcal{L}^{-1}\Big\{\frac{3}{4}\frac{1}{s+1} + \frac{1}{4}\frac{1}{s+5}\Big\}$$

$$= \frac{3}{4}\mathcal{L}^{-1}\Big\{\frac{1}{s+1}\Big\} + \frac{1}{4}\mathcal{L}^{-1}\Big\{\frac{1}{s+5}\Big\}$$

$$\mathcal{L}\{e^{-at}u(t)\} = \frac{1}{s+a}$$

$$= \frac{3}{4}e^{-t}u(t) + \frac{1}{4}e^{-5t}u(t)$$

$$= \frac{1}{4}(3e^{-t} + e^{-5t})u(t)$$

Note: *Compare the above result with example 3.2.*

Example 3.13

Using Laplace transform determine the forced response of the system described by the equation,

$$5\frac{dy(t)}{dt} + 10\,y(t) = 2\,x(t) \text{ ; for the input, } x(t) = 2\,u(t)$$

Solution:

The forced response is the response of the system due to input alone. For forced response, the system equation is solved for the given input with zero initial output. (But initial values of input should be considered).

Input, $x(t) = 2\,u(t)$

$$\therefore X(s) = \mathcal{L}\{x(t)\} = \mathcal{L}\{2\,u(t)\} = \frac{2}{s} \qquad\qquad .....(1)$$

On substituting $y(t) = y_{zs}(t)$ in the given system equation we get,

$$5\frac{dy_{zs}(t)}{dt} + 10\,y_{zs}(t) = 2\,x(t)$$

On taking Laplace transform of the above equation with zero initial output conditions we get,

$$5s\,Y_{zs}(s) + 10\,Y_{zs}(s) = 2\,X(s)$$

$$5(s+2)\,Y_{zs}(s) = 2\,X(s)$$

$$\therefore Y_{zs}(s) = \frac{2}{5(s+2)}\,X(s)$$

$$= \frac{2}{5(s+2)} \times \frac{2}{s} = \frac{4}{5}\,\frac{1}{s(s+2)} \qquad \boxed{\text{Using equation (1).}}$$

By partial fraction expansion technique, $Y_{zs}(s)$ can be expressed as shown below:

$$Y_{zs}(s) = \frac{4}{5}\,\frac{1}{s(s+2)} = \frac{4}{5}\left[\frac{A}{s} + \frac{B}{s+2}\right]$$

$$A = \frac{1}{s(s+2)} \times s\Big|_{s=0} = \frac{1}{0+2} = \frac{1}{2}$$

$$B = \frac{1}{s(s+2)} \times (s+2)\Big|_{s=-2} = \frac{1}{-2} = -\frac{1}{2}$$

$$\therefore Y_{zs}(s) = \frac{4}{5}\left[\frac{1}{2}\frac{1}{s} - \frac{1}{2}\frac{1}{s+2}\right] = \frac{2}{5}\left[\frac{1}{s} - \frac{1}{s+2}\right]$$

On taking inverse Laplace transform of the above equation we get forced response.

$$\begin{array}{l}\therefore \text{ Forced response} \\ \text{(or zero - input response)}\end{array}\Big\} y_{zs}(t) = \mathcal{L}^{-1}\{Y_{zs}(s)\} = \mathcal{L}^{-1}\left\{\frac{2}{5}\left[\frac{1}{s} - \frac{1}{s+2}\right]\right\}$$

$$= \frac{2}{5}\left[\mathcal{L}^{-1}\left\{\frac{1}{s}\right\} - \mathcal{L}^{-1}\left\{\frac{1}{s+2}\right\}\right]$$

$$= \frac{2}{5}[u(t) - e^{-2t}\,u(t)]$$

$$= \frac{2}{5}(1 - e^{-2t})\,u(t)$$

$$\boxed{\begin{array}{l}\mathcal{L}\{u(t)\} = \dfrac{1}{s} \\[4pt] \mathcal{L}\{e^{-at}u(t)\} = \dfrac{1}{s+a}\end{array}}$$

> **Note:** *Compare the above result with example 3.3.*

Example 3.14   *(AU May'15, 10 Marks)*   *(AU Dec'11 & Dec'14, 16 Marks)*

Using Laplace transform determine the complete response of the system described by the equation,

$$\frac{d^2y(t)}{dt^2} + 5\,\frac{dy(t)}{dt} + 4\,y(t) = \frac{dx(t)}{dt} \; ; \; y(0) = 0 \; ; \; \frac{dy(t)}{dt}\Big|_{t=0} = 1, \text{ for the input, } x(t) = e^{-2t}\,u(t)$$

Solution:

Method - I

Input, $x(t) = e^{-2t}\,u(t)$

$$\therefore X(s) = \mathcal{L}\{x(t)\} = \mathcal{L}\{e^{-2t}\,u(t)\} = \frac{1}{s+2} \qquad\qquad .....(1)$$

Initial value of input, $x(0) = x(t)|_{t=0} = e^0\,u(0) = 1$

The given system equation is,

$$\frac{d^2y(t)}{dt^2} + 5\,\frac{dy(t)}{dt} + 4\,y(t) = \frac{dx(t)}{dt}$$

On taking Laplace transform of the above equation we get,

$$s^2\,Y(s) - s\,y(0) - y'(0) + 5[s\,Y(s) - y(0)] + 4\,Y(s) = s\,X(s) - x(0)$$

On substituting the initial values of output and input in the above equation we get,

$$s^2\,Y(s) - s \times 0 - 1 + 5\,[s\,Y(s) - 0] + 4\,Y(s) = s\,X(s) - 1$$

$$s^2\,Y(s) + 5s\,Y(s) + 4\,Y(s) = s\,X(s)$$

$$(s^2 + 5s + 4)\,Y(s) = s \times \frac{1}{s+2}$$

> Using equation (1).

$$\therefore Y(s) = \frac{s}{(s+2)(s^2 + 5s + 4)}$$

> The roots of quadratic $s^2 + 5s + 4 = 0$ are,
> $$s = \frac{-5 \pm \sqrt{5^2 - 4 \times 4}}{2}$$
> $$= \frac{-5 \pm 3}{2} = -1, -4$$

$$= \frac{s}{(s+2)(s+1)(s+4)}$$

$$= \frac{s}{(s+1)(s+2)(s+4)}$$

By partial fraction expansion technique, Y(s) can be expressed as shown below:

$$Y(s) = \frac{s}{(s+1)(s+2)(s+4)} = \frac{A}{(s+1)} + \frac{B}{(s+2)} + \frac{C}{(s+4)}$$

$$A = \frac{s}{\cancel{(s+1)}(s+2)(s+4)} \times \cancel{(s+1)}\Big|_{s=-1} = \frac{-1}{(-1+2)(-1+4)} = \frac{-1}{1 \times 3} = -\frac{1}{3}$$

$$B = \frac{s}{(s+1)\cancel{(s+2)}(s+4)} \times \cancel{(s+2)}\Big|_{s=-2} = \frac{-2}{(-2+1)(-2+4)} = \frac{-2}{-1 \times 2} = 1$$

$$C = \frac{s}{(s+1)(s+2)\cancel{(s+4)}} \times \cancel{(s+4)}\Big|_{s=-4} = \frac{-4}{(-4+1)(-4+2)} = \frac{-4}{-3 \times (-2)} = -\frac{2}{3}$$

$$\therefore Y(s) = -\frac{1}{3}\,\frac{1}{s+1} + \frac{1}{s+2} - \frac{2}{3}\,\frac{1}{s+4}$$

On taking inverse Laplace transform of the above equation we get complete / total response of the system.

$$\therefore \text{ Total response}, y(t) = \mathcal{L}^{-1}\{Y(s)\}$$

$$= \mathcal{L}^{-1}\left\{-\frac{1}{3}\frac{1}{s+1} + \frac{1}{s+2} - \frac{2}{3}\frac{1}{s+4}\right\}$$

$$= -\frac{1}{3}\mathcal{L}^{-1}\left\{\frac{1}{s+1}\right\} + \mathcal{L}^{-1}\left\{\frac{1}{s+2}\right\} - \frac{2}{3}\mathcal{L}^{-1}\left\{\frac{1}{s+4}\right\}$$

$$= -\frac{1}{3}e^{-t}u(t) + e^{-2t}u(t) - \frac{2}{3}e^{-4t}u(t) \qquad \boxed{\mathcal{L}\{e^{-at}u(t)\} = \frac{1}{s+a}}$$

$$= \left(-\frac{1}{3}e^{-t} + e^{-2t} - \frac{2}{3}e^{-4t}\right)u(t)$$

Method - II

Total response, $y(t) = y_{zi}(t) + y_{zs}(t)$

where, $y_{zi}(t)$ = Zero-input response

           $y_{zs}(t)$ = Zero-state response

Zero-input Response

On substituting $x(t) = 0$ and $y(t) = y_{zi}(t)$ in the system equation we get,

$$\frac{d^2y_{zi}(t)}{dt^2} + 5\frac{dy_{zi}(t)}{dt} + 4\,y_{zi}(t) = 0$$

On taking Laplace transform of the above equation we get,

$$s^2\,Y_{zi}(s) - s\,y(0) - y'(0) + 5\left[s\,Y_{zi}(s) - y(0)\right] + 4\,Y_{zi}(s) = 0$$

On substituting the initial values of output in the above equation we get,

$$s^2\,Y_{zi}(s) - s \times 0 - 1 + 5\,[s\,Y_{zi}(s) - 0] + 4\,Y_{zi}(s) = 0$$

$$\therefore \ (s^2 + 5s + 4)\ Y_{zi}(s) = 1$$

$$Y_{zi}(s) = \frac{1}{s^2 + 5s + 4} = \frac{1}{(s+1)(s+4)}$$

> The roots of quadratic $s^2 + 5s + 4 = 0$ are,
> $$s = \frac{-5 \pm \sqrt{5^2 - 4 \times 4}}{2}$$
> $$= \frac{-5 \pm 3}{2} = -1, -4$$

By partial fraction expansion technique, $Y_{zi}(s)$ can be expressed as shown below:

$$Y_{zi}(s) = \frac{1}{(s+1)(s+4)} = \frac{D}{s+1} + \frac{E}{s+4}$$

$$D = \frac{1}{(s+1)(s+4)} \times (s+1)\Big|_{s=-1} = \frac{1}{-1+4} = \frac{1}{3}$$

$$E = \frac{1}{(s+1)(s+4)} \times (s+4)\Big|_{s=-4} = \frac{1}{-4+1} = -\frac{1}{3}$$

$$\therefore \ Y_{zi}(s) = \frac{1}{3}\frac{1}{s+1} - \frac{1}{3}\frac{1}{s+4}$$

On taking inverse Laplace transform of $Y_{zi}(s)$ we get zero-input response, $y_{zi}(t)$

$$\therefore \text{ Zero - input response}, y_{zi}(t) = \mathcal{L}^{-1}\{Y_{zi}(s)\} = \mathcal{L}^{-1}\left\{\frac{1}{3}\frac{1}{s+1} - \frac{1}{3}\frac{1}{s+4}\right\}$$

$$= \frac{1}{3}\mathcal{L}^{-1}\left\{\frac{1}{s+1}\right\} - \frac{1}{3}\mathcal{L}^{-1}\left\{\frac{1}{s+4}\right\} \qquad \boxed{\mathcal{L}\{e^{-at}u(t)\} = \frac{1}{s+a}}$$

$$= \frac{1}{3}e^{-t}u(t) - \frac{1}{3}e^{-4t}u(t)$$

Zero-state Response

Given that, $x(t) = e^{-2t} u(t)$

$$\therefore \ x(0) = x(t)|_{t=0} = e^0 u(0) = 1 \qquad \qquad \qquad \text{.....(2)}$$

$$X(s) = \mathcal{L}\{x(t)\} = \mathcal{L}\{e^{-2t} u(t)\} = \frac{1}{s+2} \qquad \qquad \text{.....(3)}$$

On substituting $y(t) = y_{zs}(t)$ in the system equation we get,

$$\frac{d^2 y_{zs}(t)}{dt^2} + 5 \frac{dy_{zs}(t)}{dt} + 4 y_{zs}(t) = \frac{dx(t)}{dt}$$

On taking Laplace transform of the above equation with zero initial output we get,

$$s^2 \, Y_{zs}(s) + 5s \, Y_{zs}(s) + 4 \, Y_{zs}(s) = s \, X(s) - x(0)$$

$$\therefore \ (s^2 + 5s + 4) \, Y_{zs}(s) = s \times \frac{1}{s+2} - 1 \qquad \boxed{\text{Using equations (2) and (3).}}$$

$$(s+1)(s+4) \, Y_{zs}(s) = \frac{s-s-2}{s+2}$$

$$\therefore \ Y_{zs}(s) = \frac{-2}{(s+1)(s+2)(s+4)}$$

By partial fraction expansion technique, $Y_{zs}(s)$ can be expressed as shown below:

$$Y_{zs}(s) = \frac{-2}{(s+1)(s+2)(s+4)} = \frac{F}{s+1} + \frac{G}{s+2} + \frac{H}{s+4}$$

$$F = \frac{-2}{(s+1)(s+2)(s+4)} \times (s+1) \Big|_{s=-1} = \frac{-2}{(-1+2)(-1+4)} = \frac{-2}{1 \times 3} = -\frac{2}{3}$$

$$G = \frac{-2}{(s+1)(s+2)(s+4)} \times (s+2) \Big|_{s=-2} = \frac{-2}{(-2+1)(-2+4)} = \frac{-2}{-1 \times 2} = 1$$

$$H = \frac{-2}{(s+1)(s+2)(s+4)} \times (s+4) \Big|_{s=-4} = \frac{-2}{(-4+1)(-4+2)} = \frac{-2}{-3 \times (-2)} = -\frac{1}{3}$$

$$\therefore \ Y_{zs}(s) = -\frac{2}{3} \frac{1}{s+1} + \frac{1}{s+2} - \frac{1}{3} \frac{1}{s+4}$$

On taking inverse Laplace transform of $Y_{zs}(s)$ we get zero-state response, $y_{zs}(t)$

$$\therefore \ \text{Zero - state response}, y_{zs}(t) = \mathcal{L}^{-1}\{Y_{zs}(s)\} = \mathcal{L}^{-1}\left\{-\frac{2}{3} \frac{1}{s+1} + \frac{1}{s+2} - \frac{1}{3} \frac{1}{s+4}\right\}$$

$$= -\frac{2}{3} \mathcal{L}^{-1}\left\{\frac{1}{s+1}\right\} + \mathcal{L}^{-1}\left\{\frac{1}{s+2}\right\} - \frac{1}{3} \mathcal{L}^{-1}\left\{\frac{1}{s+4}\right\}$$

$$= -\frac{2}{3} e^{-t} u(t) + e^{-2t} u(t) - \frac{1}{3} e^{-4t} u(t) \qquad \boxed{\mathcal{L}\{e^{-at} u(t)\} = \frac{1}{s+a}}$$

Total / Complete Response

Total response, $y(t) = y_{zi}(t) + y_{zs}(t)$

$$= \left[\frac{1}{3} e^{-t} u(t) - \frac{1}{3} e^{-4t} u(t)\right] + \left[-\frac{2}{3} e^{-t} u(t) + e^{-2t} u(t) - \frac{1}{3} e^{-4t} u(t)\right]$$

$$= -\frac{1}{3} e^{-t} u(t) = e^{-2t} u(t) - \frac{2}{3} e^{-4t} u(t) = \left(-\frac{1}{3} e^{-t} + e^{-2t} - \frac{2}{3} e^{-4t}\right) u(t)$$

Note: *Compare the above result with example 3.4.*

Example 3.15

Determine the impulse response h(t) of the following system. Assume zero initial conditions.

a) $y(t) = x(t - t_0)$

b) $T_0 \dfrac{d^2 y(t)}{dt^2} + y(t) = x(t)$

c) $\dfrac{d^2 y(t)}{dt^2} + 4 \dfrac{dy(t)}{dt} + 3 y(t) = \dfrac{dx(t)}{dt} + 2 x(t)$

d) $\dfrac{d^2 y(t)}{dt^2} + 3 \dfrac{dy(t)}{dt} + 2 y(t) = x(t)$

Solution:

a) Given that, y(t) = x(t – t₀)

On taking Laplace transform with zero initial condition we get,

$$Y(s) = e^{-s t_0} X(s)$$

Input, $x(t) = \delta(t)$; $\therefore X(s) = \mathcal{L}\{x(t)\} = \mathcal{L}\{\delta(t)\} = 1$

When the input is impulse, the output is denoted by h(t). Let $\mathcal{L}\{h(t)\} = H(s)$.

$$\therefore Y(s) = e^{-s t_0} X(s) \implies H(s) = e^{-s t_0}$$

$\boxed{Y(s) = H(s) \; ; \; X(s) = 1.}$

Impulse response, $h(t) = \mathcal{L}^{-1}\{H(s)\}$

$$= \mathcal{L}^{-1}\{e^{-s t_0}\}$$

$$= \delta(t - t_0)$$

$\boxed{\begin{aligned} \mathcal{L}\{\delta(t)\} &= 1 \\ \mathcal{L}\{\delta(t - t_0)\} &= e^{-s t_0} \end{aligned}}$

---

b) Given that, $T_0 \dfrac{d^2 y(t)}{dt} + y(t) = x(t)$

On taking Laplace transform with zero initial condition we get,

$$T_0 \, s^2 \, Y(s) + Y(s) = X(s)$$

$$T_0 \left(s^2 + \dfrac{1}{T_0} \right) Y(s) = X(s)$$

Input, $x(t) = \delta(t)$; $\therefore X(s) = \mathcal{L}\{x(t)\} = \mathcal{L}\{\delta(t)\} = 1$

When the input is impulse, the output is denoted by h(t). Let $\mathcal{L}\{h(t)\} = H(s)$.

$$\therefore T_0 \left(s^2 + \dfrac{1}{T_0} \right) Y(s) = X(s) \implies T_0 \left(s^2 + \dfrac{1}{T_0} \right) H(s) = 1$$

$\boxed{Y(s) = H(s) \; ; \; X(s) = 1.}$

$$\therefore H(s) = \dfrac{1}{T_0 \left(s^2 + \dfrac{1}{T_0} \right)}$$

Impulse response, $h(t) = \mathcal{L}^{-1}\{H(s)\} = \mathcal{L}^{-1} \left\{ \dfrac{1}{T_0 \left(s^2 + \dfrac{1}{T_0} \right)} \right\}$

$\boxed{\mathcal{L}\{\sin \Omega_0 t \, u(t)\} = \dfrac{\Omega_0}{s^2 + \Omega_0^2}}$

$$= \mathcal{L}^{-1} \left\{ \dfrac{1}{\sqrt{T_0}} \dfrac{\frac{1}{\sqrt{T_0}}}{s^2 + \left(\frac{1}{\sqrt{T_0}} \right)^2} \right\} = \dfrac{1}{\sqrt{T_0}} \sin\left(\dfrac{1}{\sqrt{T_0}} t \right) u(t)$$

c) Given that, $\dfrac{d^2 y(t)}{dt^2} + 4 \dfrac{dy(t)}{dt} + 3\,y(t) = \dfrac{dx(t)}{dt} + 2\,x(t)$

On taking Laplace transform with zero initial condition we get,

$s^2\,Y(s) + 4s\,Y(s) + 3\,Y(s) = s\,X(s) + 2\,X(s)$

$(s^2 + 4s + 3)\,Y(s) = [s + 2]\,X(s)$

Input, $x(t) = \delta(t)$

$\therefore\ X(s) = \mathcal{L}\{x(t)\} = \mathcal{L}\{\delta(t)\} = 1$

When the input is impulse, the output is denoted by h(t). Let $\mathcal{L}\{h(t)\} = H(s)$.

$\therefore\ (s^2 + 4s + 3)\,Y(s) = (s + 2)\,X(s) \quad \Rightarrow \quad (s^2 + 4s + 3)\,H(s) = s + 2$

$\boxed{Y(s) = H(s)\ ;\ X(s) = 1.}$

$\therefore\ H(s) = \dfrac{s+2}{s^2 + 4s + 3} = \dfrac{s+2}{(s+1)(s+3)}$

By partial fraction expansion technique, H(s) can be expressed as,

$H(s) = \dfrac{s+2}{s^2 + 4s + 3} = \dfrac{A}{(s+1)} + \dfrac{B}{(s+3)}$

$A = \dfrac{s+2}{\cancel{(s+1)}(s+3)} \times \cancel{(s+1)}\Big|_{s=-1} = \dfrac{-1+2}{-1+3} = \dfrac{1}{2} = 0.5$

$B = \dfrac{s+2}{(s+1)\cancel{(s+3)}} \times \cancel{(s+3)}\Big|_{s=-3} = \dfrac{-3+2}{-3+1} = \dfrac{1}{2} = 0.5$

$\therefore\ H(s) = 0.5\,\dfrac{1}{s+1} + 0.5\,\dfrac{1}{s+3} = 0.5\left(\dfrac{1}{s+1} + \dfrac{1}{s+3}\right)$

$\therefore$ Impulse response, $h(t) = \mathcal{L}^{-1}\{H(s)\} = \mathcal{L}^{-1}\left\{0.5\left(\dfrac{1}{s+1} + \dfrac{1}{s+3}\right)\right\}$

$= 0.5\left[\mathcal{L}^{-1}\left\{\dfrac{1}{s+1}\right\} + \mathcal{L}^{-1}\left\{\dfrac{1}{s+3}\right\}\right]$

$\boxed{\mathcal{L}\{e^{-at}\,u(t)\} = \dfrac{1}{s+a}}$

$= 0.5\left[e^{-t}\,u(t) + e^{-3t}\,u(t)\right] = 0.5\left(e^{-t} + e^{-3t}\right)u(t)$

The roots of quadratic $s^2 + 4s + 3 = 0$ are,

$s = \dfrac{-4 \pm \sqrt{4^2 - 4 \times 3}}{2}$

$= \dfrac{-4 \pm 2}{2} = -1, -3$

---

d) Given that, $\dfrac{d^2 y(t)}{dt^2} + 3 \dfrac{dy(t)}{dt} + 2\,y(t) = x(t)$ *(AU Jun'11, Dec'12, & Jun'13, 8 Marks)*

On taking Laplace transform with zero initial conditions we get,

$s^2\,Y(s) + 3s\,Y(s) + 2\,Y(s) = X(s)$

$(s^2 + 3s + 2)\,Y(s) = X(s)$

Input, $x(t) = \delta(t)$; $\therefore\ X(s) = \mathcal{L}\{x(t)\} = \mathcal{L}\{\delta(t)\} = 1$

When the input is impulse, the output is denoted by h(t). Let $\mathcal{L}\{h(t)\} = H(s)$.

$\therefore\ (s^2 + 3s + 2)\,Y(s) = X(s) \quad \Rightarrow \quad (s^2 + 3s + 2)\,H(s) = 1$

$\boxed{Y(s) = H(s)\ ;\ X(s) = 1.}$

$\therefore\ H(s) = \dfrac{1}{s^2 + 3s + 2} = \dfrac{1}{(s+1)(s+2)}$

By partial fraction expansion technique, H(s) can be expressed as,

$H(s) = \dfrac{1}{(s+1)(s+2)} = \dfrac{A}{s+1} + \dfrac{B}{s+2}$

The roots of quadratic $s^2 + 3s + 2 = 0$ are,

$s = \dfrac{-3 \pm \sqrt{3^2 - 4 \times 2}}{2}$

$= \dfrac{-3 \pm 1}{2} = -1, -2$

$$A = \frac{1}{(s+1)(s+2)} \times (s+1)\Big|_{s=-1} = \frac{1}{-1+2} = 1$$

$$B = \frac{1}{(s+1)(s+2)} \times (s+2)\Big|_{s=-2} = \frac{1}{-2+1} = -1$$

$$\therefore H(s) = \frac{1}{s+1} - \frac{1}{s+2}$$

$$\boxed{\mathcal{L}\{e^{-at}u(t)\} = \frac{1}{s+a}}$$

$$\therefore \text{Impulse response, } h(t) = \mathcal{L}^{-1}\{H(s)\} = \mathcal{L}^{-1}\left\{\frac{1}{s+1} - \frac{1}{s+2}\right\} = \mathcal{L}^{-1}\left\{\frac{1}{s+1}\right\} - \mathcal{L}^{-1}\left\{\frac{1}{s+2}\right\}$$

$$= e^{-t}u(t) - e^{-2t}u(t) = (e^{-t} - e^{-2t})\,u(t)$$

Example 3.16

Perform convolution of $x_1(t) = e^{-2t}\cos 3t\, u(t)$ and $x_2(t) = 4\sin 3t\, u(t)$ using Laplace transform.

Solution:

Given that, $x_1(t) = e^{-2t}\cos 3t\, u(t)$ and $x_2(t) = 4\sin 3t\, u(t)$

$$\therefore X_1(s) = \mathcal{L}\{x_1(t)\} = \mathcal{L}\{e^{-2t}\cos 3t\, u(t)\} = \frac{s+2}{(s+2)^2 + 3^2} = \frac{s+2}{s^2+4s+13}$$

$$X_2(s) = \mathcal{L}\{x_2(t)\} = \mathcal{L}\{4\sin 3t\, u(t)\} = 4 \times \frac{3}{s^2+3^2} = \frac{12}{s^2+9}$$

Now by convolution theorem, $x_1(t) * x_2(t) = \mathcal{L}^{-1}\{X_1(s)\, X_2(s)\}$

Let, $X(s) = X_1(s)\, X_2(s)$

$$\therefore X(s) = \frac{s+2}{s^2+4s+13} \times \frac{12}{s^2+9} = \frac{12(s+2)}{(s^2+4s+13)(s^2+9)}$$

By partial fraction expansion, $X(s)$ can be expressed as,

$$X(s) = \frac{12(s+2)}{(s^2+4s+13)(s^2+9)} = \frac{As+B}{s^2+4s+13} + \frac{Cs+D}{s^2+9} \qquad \dots(1)$$

On cross-multiplying the equation (1) we get,

$$12(s + 2) = (A s + B)(s^2 + 9) + (C s + D)(s^2 + 4s + 13)$$

$$12s + 24 = A s^3 + 9A s + B s^2 + 9B + C s^3 + 4C s^2 + 13C s + D s^2 + 4D s + 13D$$

$$12s + 24 = (A + C)s^3 + (B + 4C + D)s^2 + (9A + 13C + 4D)s + (9B + 13D) \qquad \dots(2)$$

On equating the coefficients of s^3 terms of equation (2) we get,

$$A + C = 0 \quad \Rightarrow \quad A = -C \qquad \dots(3)$$

On equating the coefficients of s^2 terms of equation (2) we get,

$$B + 4C + D = 0 \qquad \dots(4)$$

On equating the coefficients of s terms of equation (2) we get,

$$9A + 13C + 4D = 12 \qquad \dots(5)$$

On equating constants of equation (2) we get,

$$9B + 13D = 24 \qquad \dots(6)$$

On substituting $A = -C$ in equation (5) we get,

$$9(-C) + 13C + 4D = 12$$

$$\therefore 4C + 4D = 12 \quad \Rightarrow \quad C + D = 3 \quad \Rightarrow \quad C = 3 - D \qquad \dots(7)$$

On substituting $C = 3 - D$ in equation (4) we get

$\qquad B + 4(3 - D) + D = 0$

$\qquad\qquad \therefore B - 3D = -12$ .....(8)

Equation (8) × –9 $\quad\Rightarrow\quad -9B + 27D = 108$

Equation (6) × 1 $\quad\Rightarrow\quad 9B + 13D = 24$

$$\overline{\qquad\qquad 40D = 132\qquad} \qquad \therefore D = \frac{132}{40} = 3.3$$

From equation (7), $\quad C = 3 - D = 3 - 3.3 = -0.3$

From equation (3), $\quad A = -C = 0.3$

From equation (6), $\quad B = \dfrac{24 - 13D}{9} = \dfrac{24 - 13 \times 3.3}{9} = -2.1$

$\therefore X(s) = \dfrac{As + B}{s^2 + 4s + 13} + \dfrac{Cs + D}{s^2 + 9} = \dfrac{0.3s - 2.1}{s^2 + 4s + 13} + \dfrac{-0.3s + 3.3}{s^2 + 9}$

> Arranging, $s^2 + 4s$, in the form of $(x + y)^2$.
>
> $(x + y)^2 = x^2 + 2xy + x^2.$

$\qquad = \dfrac{0.3\left(s - \dfrac{2.1}{0.3}\right)}{\left(s^2 + (2 \times 2s) + 2^2\right) + 3^2} + \dfrac{-0.3s + 3.3}{s^2 + 9}$

$\qquad = \dfrac{0.3(s - 7)}{(s + 2)^2 + 3^2} + \dfrac{-0.3s + 3.3}{s^2 + 9} = \dfrac{0.3(s + 2 - 9)}{(s + 2)^2 + 3^2.} + \dfrac{-0.3s + 3.3}{s^2 + 3^2}$

$\qquad = \dfrac{0.3(s + 2) - 0.3 \times 9}{(s + 2)^2 + 3^2} + \dfrac{-0.3s + 3.3}{s^2 + 3^2}$

$\qquad = 0.3\,\dfrac{s + 2}{(s + 2)^2 + 3^2} - 0.9\,\dfrac{3}{(s + 2)^2 + 3^2}$

$\qquad\qquad\qquad - 0.3\,\dfrac{s}{s^2 + 3^2} + 1.1\,\dfrac{3}{s^2 + 3^2}$

> $\mathcal{L}\{e^{-at}\cos\Omega_0 t\, u(t)\} = \dfrac{s + a}{(s + a)^2 + \Omega_0^2}$
>
> $\mathcal{L}\{e^{-at}\sin\Omega_0 t\, u(t)\} = \dfrac{\Omega_0}{(s + a)^2 + \Omega_0^2}$
>
> $\mathcal{L}\{\cos\Omega_0 t\, u(t)\} = \dfrac{s}{s^2 + \Omega_0^2}$
>
> $\mathcal{L}\{\sin\Omega_0 t\, u(t)\} = \dfrac{\Omega_0}{s^2 + \Omega_0^2}$

On taking inverse Laplace transform of the above equation we get,

$\qquad x(t) = 0.3\, e^{-2t} \cos 3t - 0.9\, e^{-2t} \sin 3t - 0.3 \cos 3t + 1.1 \sin 3t \ ; \ t \geq 0$

$\qquad\quad = e^{-2t}(0.3 \cos 3t - 0.9 \sin 3t) + (1.1 \sin 3t - 0.3 \cos 3t)\, u(t)$

Note: *The result of Example 3.16 can be further simplified as shown below:*

$\qquad x(t) = e^{-2t}(\cos 3t \times 0.3 - \sin 3t \times 0.9) + (\sin 3t \times 1.1 - \cos 3t \times 0.3) \ ; \ t \geq 0$

Let us construct two right angled triangles as shown below:

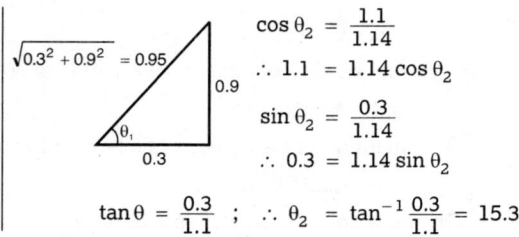

$\cos\theta_1 = \dfrac{0.3}{0.95}$

$\sqrt{0.3^2 + 0.9^2} = 0.95$

$\therefore 0.3 = 0.95 \cos\theta_1$

$\sin\theta_1 = \dfrac{0.9}{0.95}$

$\therefore 0.9 = 0.95 \sin\theta_1$

$\cos\theta_2 = \dfrac{1.1}{1.14}$

$\sqrt{0.3^2 + 0.9^2} = 0.95$

$\therefore 1.1 = 1.14 \cos\theta_2$

$\sin\theta_2 = \dfrac{0.3}{1.14}$

$\therefore 0.3 = 1.14 \sin\theta_2$

$\tan\theta = \dfrac{0.9}{0.3} \ ; \ \therefore \theta_1 = \tan^{-1}\dfrac{0.9}{0.3} = 71.6°$

$\tan\theta = \dfrac{0.3}{1.1} \ ; \ \therefore \theta_2 = \tan^{-1}\dfrac{0.3}{1.1} = 15.3°$

Using the relations obtained from right angled triangle, the $x(t)$ can be written as,

$\qquad x(t) = e^{-2t}(\cos 3t \times 0.95 \cos\theta_1 - \sin 3t \times 0.95 \sin\theta_1) + (\sin 3t \times 1.14 \cos\theta_2 - \cos 3t \times 1.14 \sin\theta_2)$

$\qquad\quad = 0.95\, e^{-2t}(\cos 3t \cos 71.6° - \sin 3t \sin 71.6°) + 1.14 (\sin 3t \cos 15.3° - \cos 3t \sin 15.3°)$

$\qquad\quad = 0.95\, e^{-2t} \cos(3t + 71.6°) + 1.14 \sin(3t - 15.3°) \ ; \ t \geq 0$

Example 3.17

Find the transfer function of the systems governed by the following differential equations.

a) $\dfrac{d^3y(t)}{dt^3} + 4\dfrac{d^2y(t)}{dt^2} + 3\dfrac{dy(t)}{dt} + 5\,y(t) = 2\dfrac{dx(t)}{dt} + x(t)$

b) $\dfrac{d^3y(t)}{dt^3} + 8\dfrac{d^2y(t)}{dt^2} + 6\dfrac{dy(t)}{dt} + 11\,y(t) = x(t-2)$

Solution:

a) Given that, $\dfrac{d^3y(t)}{dt^3} + 4\dfrac{d^2y(t)}{dt^2} + 3\dfrac{dy(t)}{dt} + 5\,y(t) = 2\dfrac{dx(t)}{dt} + x(t)$

On taking Laplace transform of the above equation with zero initial conditions we get,

$$s^3Y(s) + 4s^2Y(s) + 3sY(s) + 5Y(s) = 2sX(s) + X(s)$$

$$(s^3 + 8s^2 + 3s + 5)\,Y(s) = (2s + 1)\,X(s)$$

$\therefore$ Transfer Function, $\dfrac{Y(s)}{X(s)} = \dfrac{2s+1}{s^3 + 4s^2 + 3s + 5}$

if, $\mathcal{L}\{x(t)\} = X(s)$
then, $\mathcal{L}\left\{\dfrac{d^nx(t)}{dt^n}\right\} = s^nX(s)$
with zero initial conditions

b) Given that, $\dfrac{d^3y(t)}{dt^3} + 8\dfrac{d^2y(t)}{dt^2} + 6\dfrac{dy(t)}{dt} + 11\,y(t) = x(t-2)$

On taking Laplace transform of the above equation with zero initial conditions we get,

$$s^3Y(s) + 8s^2Y(s) + 6sY(s) + 11Y(s) = e^{-2s}X(s)$$

$$(s^3 + 8s^2 + 6s + 11)\,Y(s) = e^{-2s}X(s)$$

$\therefore$ Transfer Function, $\dfrac{Y(s)}{X(s)} = \dfrac{2s+1}{s^3 + 4s^2 + 6s + 11}$

if, $\mathcal{L}\{x(t)\} = X(s)$,
then, $\mathcal{L}\{x(t-a)\} = e^{-as}X(s)$

Example 3.18

Find the transfer function of the systems governed by the following impulse responses.

a) $h(t) = (2+t)\,e^{-3t}u(t)$ **b)** $h(t) = t^2\,u(t) - e^{-4t}\,u(t) + e^{-7t}\,u(t)$ **c)** $h(t) = u(t) + 0.5\,e^{-6t}\,u(t) + 0.2\,e^{-3t}\cos t\,u(t)$

Solution:

a) Impulse response, $h(t) = (2+t)\,e^{-3t}\,u(t)$

The transfer function is given by Laplace transform of Impulse response.

$\therefore$ Transfer function, $H(s) = \mathcal{L}\{h(t)\}$

$\mathcal{L}\{e^{-at}u(t)\} = \dfrac{1}{s+a}$; $\mathcal{L}\{t\,u(t)\} = \dfrac{1}{s^2}$
if, $\mathcal{L}\{x(t)\} = X(s)$,
then, $\mathcal{L}\{e^{-at}x(t)\} = X(s+a)$

$$= \mathcal{L}\{(2+t)\,e^{-3t}\,u(t)\} = \mathcal{L}\{2e^{-3t}\,u(t)\} + \mathcal{L}\{t\,e^{-3t}\,u(t)\}$$

$$= 2\mathcal{L}\{e^{-3t}\,u(t)\} + \mathcal{L}\{t\,u(t)\}|_{s=s+3} = 2\times\dfrac{1}{s+3} + \dfrac{1}{s^2}\Big|_{s=s+3}$$

$$= \dfrac{2}{s+3} + \dfrac{1}{(s+3)^2} = \dfrac{2(s+3)+1}{(s+3)^2} = \dfrac{2s+7}{s^2+6s+9}$$

b) Impulse response, $h(t) = t^2\, u(t) - e^{-4t}\, u(t) + e^{-7t}\, u(t)$

   The transfer function is given by Laplace transform of Impulse response.

$$\therefore \text{ Transfer function, } H(s) = \mathcal{L}\{h(t)\} = \mathcal{L}\{t^2 u(t) - e^{-4t} u(t) + e^{-7t} u(t)\}$$

$$= \mathcal{L}\{t^2 u(t)\} - \mathcal{L}\{e^{-4t} u(t)\} + \mathcal{L}\{e^{-7t} u(t)\}$$

$$= \frac{2}{s^3} - \frac{1}{s+4} + \frac{1}{s+7}$$

$$= \frac{2(s+4)(s+7) - s^3(s+7) + s^3(s+4)}{s^3(s+4)(s+7)}$$

$$= \frac{2(s^2+7s+4s+28) - s^4 - 7s^3 + s^4 + 4s^3}{s^3(s^2+11s+28)}$$

$$= \frac{-3s^2 + 2s^2 + 22s + 56}{s^5 + 11s^4 + 28s^3}$$

$$\boxed{\begin{aligned} \mathcal{L}\{t^m u(t)\} &= \frac{m!}{s^{m+1}} \\ \mathcal{L}\{e^{-at} u(t)\} &= \frac{1}{s+a} \end{aligned}}$$

c) Impulse response, $h(t) = u(t) + 0.5\, e^{-6t}\, u(t) + 0.2\, e^{-3t}\cos t\, u(t)$

   The transfer function is given by Laplace transform of Impulse response.

$$\therefore \text{ Transfer function, } H(s) = \mathcal{L}\{h(t)\}$$

$$= \mathcal{L}\{u(t) + 0.5\, e^{-6t} u(t) + 0.2\, e^{-3t}\cos t\, u(t)\}$$

$$= \mathcal{L}\{u(t)\} + 0.5 \times \mathcal{L}\{e^{-6t} u(t)\} + 0.2 \times \mathcal{L}\{\cos t\, u(t)\}\big|_{s=s+3}$$

$$= \frac{1}{s} + 0.5 \times \frac{1}{s+6} + 0.2 \times \frac{s}{s^2+1}\Big|_{s=s+3}$$

$$= \frac{1}{s} + \frac{0.5}{s+6} + \frac{0.2(s+3)}{(s+3)^2+1}$$

$$= \frac{1}{s} + \frac{0.5}{s+6} + \frac{0.2(s+3)}{s^2+6s+9+1}$$

$$= \frac{1}{s} + \frac{0.5}{s+6} + \frac{0.2(s+3)}{s^2+6s+10}$$

$$\boxed{\begin{aligned} \mathcal{L}\{u(t)\} &= \frac{1}{s} \\ \mathcal{L}\{e^{-at} u(t)\} &= \frac{1}{s+1} \\ \mathcal{L}\{\cos t\, u(t)\} &= \frac{s}{s^2+1} \end{aligned}}$$

$$\boxed{\begin{aligned} &\text{if, } \mathcal{L}\{x(t)\} = X(s), \\ &\text{then, } \mathcal{L}\{e^{-at} x(t)\} = X(s+a) \end{aligned}}$$

$$= \frac{(s+6)(s^2+6s+10) + 0.5s(s^2+6s+10) + 0.2s(s+3)(s+6)}{s(s+6)(s^2+6s+10)}$$

$$= \frac{s^3+6s^2+10s+6s^2+36s+60+0.5s^3+3s^2+5s+0.2s(s^2+9s+18)}{s(s^3+6s^2+10s+6s^2+36s+60)}$$

$$= \frac{s^3+6s^2+10s+6s^2+36s+60+0.5s^3+3s^2+5s+0.2s^3+1.8s^2+3.6s}{s(s^3+6s^2+10s+6s^2+36s+60)}$$

$$= \frac{1.7s^3+16.8s^2+54.6s+60}{s^4+12s^3+46s^2+60s}$$

Example 3.19

The unit step response of continuous time systems are given below. Determine the transfer function of the systems.

   a) $s(t) = u(t) + e^{-2t}\, u(t)$ **b)** $s(t) = t^2\, u(t) + t\, e^{-4t}\, u(t)$ **c)** $s(t) = t\, u(t) + \sin t\, u(t)$

Solution:

a) Unit step response, $s(t) = u(t) + e^{-2t}\, u(t)$

   On taking Laplace transform of unit step response we get,

$$S(s) = \mathcal{L}\{s(t)\} = \mathcal{L}\{u(t) + e^{-2t} u(t)\}$$

$$\therefore\ S(s) = \mathcal{L}\{u(t)+e^{-2t}u(t)\} = \mathcal{L}\{u(t)\} + \mathcal{L}\{e^{-2t}u(t)\}$$

$$\boxed{\mathcal{L}\{u(t)\}=\frac{1}{s}\ ;\ \mathcal{L}\{e^{-at}u(t)\}=\frac{1}{s+a}}$$

$$= \frac{1}{s} + \frac{1}{s+2} = \frac{s+2+s}{s(s+2)} = \frac{2s+2}{s(s+2)} \qquad \dots(1)$$

Let, $U(s) = \mathcal{L}\{u(t)\} = \frac{1}{s}$; where $u(t)$ is unit step signal. $\qquad \dots(2)$

Transfer function, $H(s) = \dfrac{\text{Laplace transform of output}}{\text{Laplace transform of input}} = \dfrac{S(s)}{U(s)}$

$$= S(s) \times \frac{1}{U(s)} = \frac{2s+2}{s(s+2)} \times s = \frac{2s+2}{s+2} \qquad \boxed{\text{Using equations (1) and (2).}}$$

b) Unit step response, $s(t) = t^2\,u(t) + t\,e^{-4t}\,u(t)$

On taking Laplace transform of unit step response we get,

$$\boxed{\mathcal{L}\{t^n\,u(t)\}=\frac{n!}{s^{n+1}}}$$

$$S(s) = \mathcal{L}\{s(t)\} = \mathcal{L}\{t^2\,u(t) = t\,e^{-4t}\,u(t)\}$$

$$\boxed{\begin{array}{l}\text{if, } \mathcal{L}\{x(t)\}=X(s),\\ \text{then, } \mathcal{L}\{e^{-at}x(t)\}=X(s+a)\end{array}}$$

$$= \mathcal{L}\{t^2\,u(t) + t\,e^{-4t}\,u(t)\} = \mathcal{L}\{t^2\,u(t)\} + \mathcal{L}\{t\,e^{-4t}\,u(t)\}$$

$$= \mathcal{L}\{t^2\,u(t)\} + \mathcal{L}\{t\,u(t)\}\big|_{s=s+4} = \frac{2}{s^3} + \frac{1}{s^2}\Big|_{s=s+4} = \frac{2}{s^3} + \frac{1}{(s+4)^2}$$

$$= \frac{2(s+4)^2+s^3}{s^3(s+4)^2} = \frac{2(s^2+8s+16)+s^3}{s^2(s^2+8s+16)} = \frac{s^2+2s^2+16s+32}{s^3(s^2+8s+16)} \qquad \dots(1)$$

Let, $U(s) = \mathcal{L}\{u(t)\} = \frac{1}{s}$; where $u(t)$ is step signal $\qquad \dots(2)$

Transfer function, $H(s) = \dfrac{\text{Laplace transform of output}}{\text{Laplace transform of input}} = \dfrac{S(s)}{U(s)}$ $\qquad \boxed{\text{Using equations (1) and (2).}}$

$$= S(s) \times \frac{1}{U(s)} = \frac{s^2+2s^2+16s+32}{s^3(s^2+8s+16)} \times s = \frac{s^2+2s^2+16s+32}{s^4+8s^3+16s^2}$$

c) Unit step response, $s(t) = t\,u(t) + \sin t\,u(t)$

On taking Laplace transform of unit step response we get,

$$\boxed{\begin{array}{l}\mathcal{L}\{t\,u(t)\}=\dfrac{1}{s^2}\\[2mm] \mathcal{L}\{\sin t\,u(t)\}=\dfrac{1}{s^2+1}\end{array}}$$

$$S(s) = \mathcal{L}\{s(t)\} = \mathcal{L}\{t\,u(t) + \sin t\,u(t)\}$$

$$= \mathcal{L}\{t\,u(t) + \sin t\,u(t)\} = \mathcal{L}\{t\,u(t)\} + \mathcal{L}\{\sin t\,u(t)\}$$

$$= \frac{1}{s^2} + \frac{1}{s^2+1} = \frac{s^2+1+s^2}{s^2(s^2+1)} = \frac{2s^2+1}{s^2(s^2+1)} \qquad \dots(1)$$

Let, $U(s) = \mathcal{L}\{u(t)\} = \frac{1}{s}$; where $u(t)$ is step signal $\qquad \dots(2)$

Transfer function, $H(s) = \dfrac{\text{Laplace transform of output}}{\text{Laplace transform of input}} = \dfrac{S(s)}{U(s)}$

$$= S(s) \times \frac{1}{U(s)} = \frac{2s^2+1}{s^2(s^2+1)} \times s = \frac{2s^2+1}{s^2+s} \qquad \boxed{\text{Using equations (1) and (2).}}$$

Example 3.20

Find the impulse response of continuous time systems governed by the following transfer functions.

a) $H(s) = \dfrac{1}{s^2(s-2)}$ **b)** $H(s) = \dfrac{1}{s(s+1)(s-2)}$ **c)** $H(s) = \dfrac{1}{s^2+s+1}$

Solution:

a) Transfer function, $H(s) = \dfrac{1}{s^2(s-2)}$

The impulse response is obtained by taking inverse Laplace transform of the transfer function.

$\therefore$ Impulse response, $h(t) = \mathcal{L}^{-1}\left\{\dfrac{1}{s^2(s-2)}\right\} = \mathcal{L}^{-1}\left\{\dfrac{A}{s^2} + \dfrac{B}{s} + \dfrac{C}{s-2}\right\}$ | Using partial fraction expansion technique. |

$A = \dfrac{1}{\cancel{s^2}(s-2)} \times \cancel{s^2}\Big|_{s=0} = \dfrac{1}{0-2} = -\dfrac{1}{2}$

$B = \dfrac{d}{ds}\left[\dfrac{1}{\cancel{s^2}(s-2)} \times \cancel{s^2}\right]\Big|_{s=0} = \dfrac{d}{ds}\left[\dfrac{1}{s-2}\right]\Big|_{s=0} = \dfrac{-1}{(s-2)^2}\Big|_{s=0} = \dfrac{1}{(0-2)^2} = -\dfrac{1}{4}$

$C = \dfrac{1}{s^2\cancel{(s-2)}} \times \cancel{(s-2)}\Big|_{s=2} = \dfrac{1}{2^2} = \dfrac{1}{4}$

$\therefore$ Impulse response, $h(t) = \mathcal{L}^{-1}\left\{-\dfrac{1}{2}\dfrac{1}{s^2} - \dfrac{1}{4}\dfrac{1}{s} + \dfrac{1}{4}\dfrac{1}{s-2}\right\}$

| $\mathcal{L}\{u(t)\} = \dfrac{1}{s}$ |
| $\mathcal{L}\{t\,u(t)\} = \dfrac{1}{s^2}$ |
| $\mathcal{L}\{e^{at}\,u(t)\} = \dfrac{1}{s-a}$ |

$\qquad = -\dfrac{1}{2}t\,u(t) - \dfrac{1}{4}u(t) + \dfrac{1}{4}e^{2t}u(t) = \dfrac{1}{4}\left(e^{2t} - 2t - 1\right)u(t)$

b) Transfer Function, $H(s) = \dfrac{1}{s(s+1)(s-2)}$

The impulse response is obtained by taking inverse Laplace transform of the transfer function.

$\therefore$ Impulse response, $h(t) = \mathcal{L}^{-1}\{H(s)\} = \mathcal{L}^{-1}\left\{\dfrac{1}{s(s+1)(s-2)}\right\} = \mathcal{L}^{-1}\left\{\dfrac{A}{s} + \dfrac{B}{s+1} + \dfrac{C}{s-2}\right\}$

$A = \dfrac{1}{\cancel{s}(s+1)(s-2)} \times \cancel{s}\Big|_{s=0} = \dfrac{1}{(0+1)(0-2)} = -\dfrac{1}{2}$ | Using partial fraction expansion technique. |

$B = \dfrac{1}{s\cancel{(s+1)}(s-2)} \times \cancel{(s+1)}\Big|_{s=-1} = \dfrac{1}{-1\times(-1-2)} = \dfrac{1}{3}$

$C = \dfrac{1}{s(s+1)\cancel{(s-2)}} \times \cancel{(s-2)}\Big|_{s=2} = \dfrac{1}{2\times(2+1)} = \dfrac{1}{6}$

$\therefore$ Impulse response, $h(t) = \mathcal{L}^{-1}\left\{-\dfrac{1}{2}\dfrac{1}{s} + \dfrac{1}{3}\dfrac{1}{s+1} + \dfrac{1}{6}\dfrac{1}{s-2}\right\}$

| $\mathcal{L}\{u(t)\} = \dfrac{1}{s}$ |
| $\mathcal{L}\{e^{\pm at}u(t)\} = \dfrac{1}{s \mp a}$ |

$\qquad = -\dfrac{1}{2}u(t) + \dfrac{1}{3}e^{-t}u(t) + \dfrac{1}{6}e^{2t}u(t) = \dfrac{1}{6}\left(e^{2t} + 2e^{-t} - 3\right)u(t)$

c) Transfer Function, $H(s) = \dfrac{1}{s^2+s+1}$

The impulse response is obtained by taking inverse Laplace transform of the transfer function.

$\therefore$ Impulse response, $h(t) = \mathcal{L}^{-1}\{H(s)\} = \mathcal{L}^{-1}\left\{\dfrac{1}{s^2+s+1}\right\}$

$\qquad = \mathcal{L}^{-1}\left\{\dfrac{1}{(s^2+2\times0.5\times s+0.5^2)+(1-0.5^2)}\right\}$

| Arranging, $s^2 + s$, in the form of $(x+y)^2$. |
| $(x+y)^2 = x^2 + 2xy + y^2$ |

$\therefore$ Impulse response, $h(t) = \mathcal{L}^{-1}\left\{\dfrac{1}{(s+0.5)^2+0.75}\right\} = \mathcal{L}^{-1}\left\{\dfrac{1}{(s+0.5)^2+\left(\sqrt{0.75}\,\right)^2}\right\}$

$\qquad = \mathcal{L}^{-1}\left\{\dfrac{1}{(s+0.5)^2+0.866^2}\right\} = \mathcal{L}^{-1}\left\{\dfrac{1}{0.866}\times\dfrac{0.866}{(s+0.5)^2+0.866^2}\right\}$

$\qquad = \dfrac{1}{0.866}\times\mathcal{L}^{-1}\left\{\dfrac{0.866}{(s+0.5)^2+0.866^2}\right\}$ $\boxed{\mathcal{L}\{e^{-at}\sin\Omega_0 t\, u(t)\} = \dfrac{\Omega_0}{(s+a)^2+\Omega_0^2}}$

$\qquad = \dfrac{1}{0.866}\, e^{-0.5t}\sin(0.866t)\, u(t)$

Example 3.21

The input and impulse response of continuous time systems are given below. Find the output of the continuous time systems.

a) $x(t) = \delta(t),\ h(t) = e^{-at}\, u(t)$ **b)** $x(t) = e^{-2t}\, u(t),\ h(t) = u(t)$ **c)** $x(t) = e^{-3t}\, u(t),\ h(t) = u(t-1)$

d) $x(t) = \cos 4t\, u(t) + \cos 7t\, u(t),\ h(t) = \delta(t-3)$ **e)** $x(t) = u(t-3) - u(t-5),\ h(t) = e^{-2t}\, u(t)$

Solution:

a) Given that, $x(t) = \delta(t)$ and $h(t) = e^{-at}\, u(t)$

$\boxed{\mathcal{L}\{\delta(t)\} = 1\,;\ \mathcal{L}\{e^{-at}\, u(t)\} = \dfrac{1}{s+a}}$

Let, $X(s) = \mathcal{L}\{x(t)\} = \mathcal{L}\{\delta(t)\} = 1$ .....(1)

$\qquad H(s) = \mathcal{L}\{h(t)\} = \mathcal{L}\{e^{-at}u(t)\} = \dfrac{1}{s+a}$ .....(2)

Response / Output, $y(t) = x(t) * h(t)$

On taking Laplace transform of the above equation we get,

$\qquad \mathcal{L}\{y(t)\} = \mathcal{L}\{x(t)*h(t)\}$

$\qquad\qquad = X(s)\, H(s)$ $\boxed{\begin{array}{l}\text{Using convolution theorem}\\ \text{of Laplace transform.}\end{array}}$

$\qquad\qquad = 1\times\dfrac{1}{s+a} = \dfrac{1}{s+a}$ $\boxed{\text{Using equations (1) and (2).}}$

Response / Output is given by inverse Laplace transform of the above equation.

$\qquad \therefore$ Response, $y(t) = \mathcal{L}^{-1}\left\{\dfrac{1}{s+a}\right\} = e^{-at}\, u(t)$

b) Given that, $x(t) = e^{-2t}\, u(t)$ and $h(t) = u(t)$ $\boxed{\mathcal{L}\{u(t)\} = \dfrac{1}{s}\,;\ \mathcal{L}\{e^{-at}\, u(t)\} = \dfrac{1}{s+a}}$ *(AU Jun'12, 8 Marks)*

Let, $X(s) = \mathcal{L}\{x(t)\} = \mathcal{L}\{e^{-at}\, u(t)\} = \dfrac{1}{s+2}$ .....(1)

$\qquad H(s) = \mathcal{L}\{h(t)\} = \mathcal{L}\{u(t)\} = \dfrac{1}{s}$ .....(2)

Response / Output, $y(t) = x(t) * h(t)$

On taking Laplace transform of the above equation we get,

$\qquad \mathcal{L}\{y(t)\} = \mathcal{L}\{x(t)*h(t)\}$

$\qquad\qquad = X(s)\, H(s)$ $\boxed{\begin{array}{l}\text{Using convolution theorem}\\ \text{of Laplace transform.}\end{array}}$

$\qquad\qquad = \dfrac{1}{s+2}\times\dfrac{1}{s} = \dfrac{1}{s(s+2)}$ $\boxed{\text{Using equations (1) and (2).}}$

Response / Output in time domain is given by inverse Laplace transform of the above equation.

$$\therefore \text{ Response, } y(t) = \mathcal{L}^{-1}\left\{\frac{1}{s(s+2)}\right\} = \mathcal{L}^{-1}\left\{\frac{A}{s} + \frac{B}{s+2}\right\}$$

Using partial fraction expansion technique.

$$A = \frac{1}{s(s+2)} \times s\bigg|_{s=0} = \frac{1}{0+2} = \frac{1}{2}$$

$$B = \frac{1}{s(s+2)} \times (s+2)\bigg|_{s=-2} = \frac{1}{-2} = -\frac{1}{2}$$

$$\therefore \text{ Response, } y(t) = \mathcal{L}^{-1}\left\{\frac{1}{2}\frac{1}{s} - \frac{1}{2}\frac{1}{s+2}\right\}$$

$$= \frac{1}{2}\mathcal{L}^{-1}\left\{\frac{1}{s}\right\} - \frac{1}{2}\mathcal{L}^{-1}\left\{\frac{1}{s+2}\right\} = \frac{1}{2}u(t) - \frac{1}{2}e^{-2t}u(t) = \frac{1}{2}(1-e^{-2t})\,u(t)$$

c) Given that, $x(t) = e^{-3t}u(t)$ and $h(t) = u(t-1)$

$$\text{Let, } X(s) = \mathcal{L}\{x(t)\} = \mathcal{L}\{e^{-3t}u(t)\} = \frac{1}{s+3} \quad .....(1)$$

$$H(s) = \mathcal{L}\{h(t)\} = \mathcal{L}\{u(t-1)\} = e^{-s} \times \frac{1}{s} = \frac{e^{-s}}{s} \quad .....(2)$$

Response / Output, $y(t) = x(t) * h(t)$

if $\mathcal{L}\{u(t)\} = \frac{1}{s}$; then $\mathcal{L}\{u(t-a)\} = \frac{e^{-as}}{s}$

On taking Laplace transform of the above equation we get,

$$\mathcal{L}\{y(t)\} = \mathcal{L}\{x(t) * h(t)\}$$

Using convolution theorem of Laplace transform.

$$= X(s)\,H(s)$$

$$= \frac{1}{s+3} \times \frac{e^{-s}}{s} = \frac{e^{-s}}{s(s+3)}$$

Using equations (1) and (2).

Response / Output in time domain is given by inverse Laplace transform of the above equation.

Using partial fraction expansion technique.

$$\therefore \text{ Response, } y(t) = \mathcal{L}\left\{\frac{e^{-s}}{s(s+3)}\right\} = \mathcal{L}^{-1}\left\{\frac{1}{s(s+3)}\right\}\bigg|_{t=t-1} = \mathcal{L}^{-1}\left\{\frac{A}{s} + \frac{B}{s+3}\right\}\bigg|_{t=t-1}$$

$$A = \frac{1}{s(s+3)} \times s\bigg|_{s=0} = \frac{1}{0+3} = \frac{1}{3}$$

if, $\mathcal{L}\{x(t)\} = X(s)$, then, $\mathcal{L}\{x(t-a)\} = e^{-as}X(s)$

$$B = \frac{1}{s(s+3)} \times (s+3)\bigg|_{s=-3} = \frac{1}{-3} = -\frac{1}{3}$$

$$\therefore \text{ Response, } y(t) = \mathcal{L}^{-1}\left\{\frac{1}{3}\frac{1}{s} - \frac{1}{3}\frac{1}{s+3}\right\}\bigg|_{t=t-1} = \left[\frac{1}{3}\right]\mathcal{L}^{-1}\left\{\frac{1}{s}\right\} - \frac{1}{3}\mathcal{L}^{-1}\left\{\frac{1}{s+3}\right\}\bigg|_{t=t-1}$$

$$= \left[\frac{1}{3}u(t) - \frac{1}{3}e^{-3t}u(t)\right]\bigg|_{t=t-1} = \left[\frac{1}{3}u(t-1) - \frac{1}{3}e^{-3(t-1)}u(t-1)\right]$$

$$= \frac{1}{3}[1-e^{-3(t-1)}]\,u(t-1)$$

d) Given that, $x(t) = \cos 4t\,u(t) + \cos 7t\,u(t)$ and $h(t) = \delta(t-3)$

$\mathcal{L}\{\cos\Omega_0 t\,u(t)\} = \frac{s}{s^2+\Omega_0^2}$

$$\text{Let, } X(s) = \mathcal{L}\{x(t)\} = \mathcal{L}\{\cos 4t\,u(t) + \cos 7t\,u(t)\} = \frac{s}{s^2+4^2} + \frac{s}{s^2+7^2} \quad .....(1)$$

$$H(s) = \mathcal{L}^{-1}\{h(t)\} = \mathcal{L}\{\delta(t-3)\} = e^{-3s} \times \mathcal{L}\{\delta(t)\} = e^{-3s} \times 1 = e^{-3s} \quad .....(2)$$

Response / Output, $y(t) = x(t) * h(t)$

if $\mathcal{L}\{\delta(t)\} = 1$; then $\mathcal{L}\{\delta(t-a)\} = e^{-as}$

On taking Laplace transform of the above equation we get,

$$\mathcal{L}\{y(t)\} = \mathcal{L}\{x(t) * h(t)\}$$

> Using convolution theorem of Laplace transform.

$$= X(s)\, H(s)$$

> Using equations (1) and (2).

$$= \left[\frac{s}{s^2+4^2} + \frac{s}{s^2+7^2}\right] \times e^{-3s} = e^{-3s} \times \left[\frac{s}{s^2+4^2} + \frac{s}{s^2+7^2}\right]$$

> if, $\mathcal{L}\{x(t)\} = X(s)$,
> then, $\mathcal{L}\{x(t-a)\} = e^{-as}\, X(s)$

Response / Output is given by inverse Laplace transform of the above equation.

$$\therefore \text{Response, } y(t) = \mathcal{L}^{-1}\left\{e^{-3s} \times \left[\frac{s}{s^2+4^2} + \frac{s}{s^2+7^2}\right]\right\} = \mathcal{L}^{-1}\left\{\frac{s}{s^2+4^2} + \frac{s}{s^2+7^2}\right\}\bigg|_{t=t-3}$$

$$= \left[\mathcal{L}^{-1}\left\{\frac{s}{s^2+4^2}\right\} + \mathcal{L}^{-1}\left\{\frac{s}{s^2+7^2}\right\}\right]\bigg|_{t=t-3} = \left[\cos 4t\, u(t) + \cos 7t\, u(t)\right]\big|_{t=t-3}$$

$$= \cos 4(t-3)\, u(t-3) + \cos 7(t-3)\, u(t-3) = \left[\cos 4(t-3) + \cos 7(t-3)\right]u(t-3)$$

e) Given that, $x(t) = u(t-3) - u(t-5)$ and $h(t) = e^{-3t}\, u(t)$ **(AU Jun'12, 8 Marks)**

Let, $X(s) = \mathcal{L}\{x(t)\} = \mathcal{L}\{u(t-3) - u(t-5)\} = \mathcal{L}\{u(t-3)\} - \mathcal{L}\{u(t-5)\}$

> $\mathcal{L}\{u(t-a)\} = \dfrac{e^{-as}}{s}$

$$= \frac{e^{-3s}}{3} - \frac{e^{-5s}}{5}$$ .....(1)

$$H(s) = \mathcal{L}\{h(t)\} = \mathcal{L}\{e^{-3t}\, u(t)\} = \frac{1}{s+3}$$ .....(2)

> $\mathcal{L}\{e^{-at}\, u(t)\} = \dfrac{1}{s+a}$

Response / Output, $y(t) = x(t) * h(t)$

On taking Laplace transform of the above equation we get,

$$\mathcal{L}\{y(t)\} = \mathcal{L}\{x(t) * h(t)\}$$

> Using convolution theorem of Laplace transform.

$$= X(s)\, H(s)$$

> Using equations (1) and (2).

$$Y(s) = \left(\frac{e^{-3s}}{s} - \frac{e^{-5s}}{s}\right) \times \frac{1}{s+3} = \frac{e^{-3s}}{s(s+3)} - \frac{e^{-5s}}{s(s+3)}$$

$$= e^{-3s}\left(\frac{A}{s} + \frac{B}{s+3}\right) - e^{-5s}\left(\frac{A}{s} + \frac{B}{s+3}\right)$$

Here, $\dfrac{1}{s(s+3)} = \dfrac{A}{s} + \dfrac{B}{s+3}$

$$A = \frac{1}{s(s+3)} \times s\bigg|_{s=0} = \frac{1}{0+3} = \frac{1}{3}$$

$$B = \frac{1}{s(s+3)} \times (s+3)\bigg|_{s=-3} = \frac{1}{-3} = -\frac{1}{3}$$

$$Y(s) = e^{-3s}\left(\frac{1}{3}\frac{1}{s} - \frac{1}{3}\frac{1}{(s+3)}\right) - e^{-5s}\left(\frac{1}{3}\frac{1}{s} - \frac{1}{3}\frac{1}{(s+3)}\right)$$

Response / Output in time domain is given by inverse Laplace transform of the above equation.

$$\therefore \text{Response, } y(t) = \mathcal{L}^{-1}\left\{e^{-3s} \times \left[\frac{1}{3}\frac{1}{s} - \frac{1}{3}\frac{1}{s+3}\right]\right\} - \mathcal{L}^{-1}\left\{e^{-5s} \times \left[\frac{1}{3}\frac{1}{s} - \frac{1}{3}\frac{1}{s+3}\right]\right\}$$

$$= \left[\frac{1}{3}\mathcal{L}^{-1}\left\{\frac{1}{s} - \frac{1}{s+3}\right\}\right]\bigg|_{t=t-3} - \left[\frac{1}{3}\mathcal{L}^{-1}\left\{\frac{1}{s} - \frac{1}{s+3}\right\}\right]\bigg|_{t=t-5}$$

$$= \left[\frac{1}{3}u(t) - \frac{1}{3}e^{-3t}\, u(t)\right]\bigg|_{t=t-3} - \left[\frac{1}{3}u(t) - \frac{1}{3}e^{-3t}\, u(t)\right]\bigg|_{t=t-5}$$

$$= \left[\frac{1}{3}u(t-3) - \frac{1}{3}e^{-3(t-3)}\, u(t-3)\right] - \left[\frac{1}{3}u(t-5) - \frac{1}{3}e^{-3(t-5)}\, u(t-5)\right]$$

$$= \frac{1}{3}\left[1 - e^{-3(t-3)}\right]u(t-3) - \frac{1}{3}\left[1 + e^{-3(t-5)}\right]u(t-5)$$

Example 3.22   *(AU Jun'12, 16 Marks)*

Using Laplace transform, find the differential equation and response of continuous time system governed by the following function.

$$H(s) = \frac{1}{s+10} \quad Re > -10; \text{ when the system input } x(t) = -2e^{-2t}\,u(t-1) - 3e^{-3t}\,u(t).$$

Solution:

Differential Equation of the System

Given that, $H(s) = \dfrac{1}{s+10}$

We know that, $H(s) = \dfrac{Y(s)}{X(s)}$

where, $Y(s) = $ Laplace transform of output

$X(s) = $ Laplace transform of input

$\therefore \dfrac{Y(s)}{X(s)} = \dfrac{1}{s+10}$

$\therefore (s+10)\,Y(s) = X(s)$ .....(1)

$s\,Y(s) + 10\,Y(s) = X(s)$

On taking inverse Laplace transform of above equation we get,

$\mathcal{L}^{-1}\{s\,Y(s)\} + \mathcal{L}^{-1}\{10\,Y(s)\} = \mathcal{L}^{-1}\{X(s)\}$

$$\dfrac{dy(t)}{dt} + 10\,y(t) = x(t) \quad\quad ....(2)$$

> if, $\mathcal{L}\{y(t)\} = Y(s)$,
>
> then, $\mathcal{L}\left\{\dfrac{d}{dt}\,y(t)\right\} = s\,Y(s)$

The equation (2) is the differential equation governing the continuous time system.

Response of the System

Input, $x(t) = -2e^{-2t}\,u(-t) - 3e^{-3t}\,u(t)$

$\therefore X(s) = \mathcal{L}\{x(t)\} = \mathcal{L}\{-2e^{-2t}\,u(-t) - 3e^{-3t}\,u(t)\}$

$$= \frac{2}{s+2} - \frac{3}{s+3} = \frac{2(s+3) - 3(s+2)}{(s+2)(s+3)} = \frac{-s}{s^2+5s+6} \quad .....(3)$$

> $\mathcal{L}\{e^{-at}u(t)\} = \dfrac{1}{s+a}$; causal
>
> $\mathcal{L}\{-e^{-at}u(-t)\} = \dfrac{1}{s+a}$; anticausal

From equation (1) we get,

$Y(s) = X(s) \times \dfrac{1}{s+10}$

$$= \frac{-s}{s^2+5s+6} \times \frac{1}{(s+10)} = \frac{-s}{(s+10)(s+2)(s+3)}$$

$\boxed{\text{Using equation (3)}}$

By partial expansion technique $Y(s)$ can be expressed as,

$$Y(s) = \frac{A}{(s+10)} + \frac{B}{(s+2)} + \frac{C}{(s+3)}$$

$$A = \left.\frac{-s}{(s+10)(s+2)(s+3)} \times (s+10)\right|_{s=-10} = \frac{+10}{(-10+2)(-10+3)} = \frac{10}{-8\times-7} = \frac{5}{28}$$

$$B = \left.\frac{-s}{(s+10)(s+2)(s+3)} \times (s+2)\right|_{s=-2} = \frac{+2}{(-2+10)(-2+3)} = \frac{2}{8\times1} = \frac{1}{4}$$

$$C = \left.\frac{-s}{(s+10)(s+2)(s+3)} \times (s+3)\right|_{s=-3} = \frac{+3}{(-3+10)(-3+2)} = \frac{3}{7\times-1} = -\frac{3}{7}$$

$$\therefore Y(s) = \frac{5}{28} \frac{1}{s+10} + \frac{1}{4} \frac{1}{s+2} - \frac{3}{7} \frac{1}{s+3}$$

Response / Output in time domain is given by inverse Laplace transform of the above equation.

$$\therefore \text{Response, } y(t) = \mathcal{L}^{-1}\left\{\frac{5}{28} \frac{1}{(s+10)} + \frac{1}{4} \frac{1}{(s+2)} - \frac{3}{7} \frac{1}{(s+3)}\right\}$$

$$= \frac{5}{28} \mathcal{L}^{-1}\left\{\frac{1}{(s+10)}\right\} + \frac{1}{4} \mathcal{L}^{-1}\left\{\frac{1}{(s+2)}\right\} - \frac{3}{7} \mathcal{L}^{-1}\left\{\frac{1}{(s+3)}\right\}$$

$$\boxed{\mathcal{L}\{e^{-at}u(t)\} = \frac{1}{s+a}}$$

$$= \frac{5}{28} e^{-10t}u(t) + \frac{1}{4} e^{-2t}u(t) - \frac{3}{7} e^{-3t}u(t)$$

Example 3.23   *(AU May'15, 10 Marks)*

Find the differential equation and complete respose y(t) of the system represented by the transfer function

$$H(s) = \frac{s+5}{s^2+4s+3} \text{ for the input } x(t) = e^{-2t} u(t).$$

Solution:

Differential Equation of the System

Given that, $H(s) = \dfrac{s+5}{s^2+4s+3}$

We know that, $H(s) = \dfrac{Y(s)}{X(s)}$

$$\therefore \frac{Y(s)}{X(s)} = \frac{s+5}{s^2+4s+3} \qquad\qquad\qquad .....(1)$$

On cross multiplying equation (1) we get,

$$Y(s)(s^2+4s+3) = X(s)(s+5)$$

$$s^2 Y(s) + 4 sY(s) + 3 Y(s) = s X(s) + 5 X(s)$$

$$\boxed{\begin{array}{l} \text{if, } \mathcal{L}\{x(t)\} = X(s), \\ \text{then, } \mathcal{L}\left\{\dfrac{d^m}{dt^m} x(t)\right\} = s^m X(s) \end{array}}$$

On taking inverse Laplace transform of the above equation we get,

$$\frac{d^2 y(t)}{dt^2} + 4 \frac{dy(t)}{dt} + 3y(t) = \frac{dx(t)}{dt} + 5x(t) \qquad\qquad .....(2)$$

The equation (2) is the differential equation governing the continuous time system.

Response of the System

Input, $x(t) = e^{-2t} u(t)$

$$\therefore X(s) = \mathcal{L}\{x(t)\} = \mathcal{L}\{ e^{-2t} u(t)\} = \frac{1}{s+2} \qquad\qquad .....(3)$$

From equation (1) we get,

$$Y(s) = \frac{s+5}{s^2+4s+3} \times X(s)$$

$$Y(s) = \frac{s+5}{s^2+4s+3} \times \frac{1}{s+2} = \frac{s+5}{(s^2+4s+3)(s+2)} \quad \boxed{\text{Using equation (3)}}$$

$$\boxed{\begin{array}{l} \text{The roots of quadratic} \\ s^2+4s+3 = 0 \\ s = \dfrac{-4 \pm \sqrt{4^2 - 4\times 3}}{2} \\ = \dfrac{-4 \pm 2}{2} = -1, -3 \end{array}}$$

$$Y(s) = \frac{s+5}{(s+1)(s+3)(s+2)} = \frac{s+5}{(s+1)(s+2)(s+3)}$$

By partial fraction expansion technique, Y(s) can be expressed as shown below:

$$Y(s) = \frac{s+5}{(s+1)(s+2)(s+3)} = \frac{A}{s+1} + \frac{B}{s+2} + \frac{C}{s+3}$$

$$A = \frac{s+5}{\cancel{(s+1)}(s+2)(s+3)} \times (s+1)\Big|_{s=-1} = \frac{(-1+5)}{(-1+2)(-1+3)} = \frac{4}{1 \times 2} = 2$$

$$B = \frac{s+5}{(s+1)\cancel{(s+2)}(s+3)} \times (s+2)\Big|_{s=-2} = \frac{(-2+5)}{(-2+1)(-2+3)} = \frac{3}{-1 \times 1} = -3$$

$$C = \frac{s+5}{(s+1)(s+2)\cancel{(s+3)}} \times (s+3)\Big|_{s=-3} = \frac{(-3+5)}{(-3+1)(-3+2)} = \frac{2}{-2 \times -1} = 1$$

$$\therefore Y(s) = \frac{2}{s+1} - \frac{3}{s+2} + \frac{1}{s+3}$$

On taking inverse Laplace transform of the above equation we get complete/total response of the system.

$$\therefore \text{Total response}, y(t) = \mathcal{L}^{-1}\{Y(s)\}$$

$$= \mathcal{L}^{-1}\left\{\frac{2}{s+1} - \frac{3}{s+2} + \frac{1}{s+3}\right\} = 2\mathcal{L}^{-1}\left\{\frac{1}{s+1}\right\} - 3\mathcal{L}^{-1}\left\{\frac{1}{s+2}\right\} + \mathcal{L}^{-1}\left\{\frac{1}{s+3}\right\}$$

$$= 2e^{-t}u(t) - 3e^{-2t}u(t) + e^{-3t}u(t) \qquad \boxed{\mathcal{L}\{e^{-at}u(t)\} = \frac{1}{s+a}}$$

$$= (2e^{-t} - 3e^{-2t} + e^{-3t})u(t)$$

Example 3.24

Perform convolution of the following causal signals, using Laplace transform.

a) $x_1(t) = 2\,u(t)$, $x_2(t) = u(t)$

b) $x_1(t) = e^{-2t}\,u(t)$, $x_2(t) = e^{-5t}\,u(t)$

c) $x_1(t) = t\,u(t)$, $x_2(t) = e^{-5t}\,u(t)$

d) $x_1(t) = \cos t\,u(t)$, $x_2(t) = t\,u(t)$

Solution:

a) Given that, $x_1(t) = 2\,u(t)$

$$x_2(t) = u(t)$$

Let, $X_1(s) = \mathcal{L}\{x_1(t)\} = \mathcal{L}\{2\,u(t)\} = 2\mathcal{L}\{u(t)\} = 2 \times \frac{1}{s} = \frac{2}{s}$

$$X_2(s) = \mathcal{L}\{x_2(t)\} = \mathcal{L}\{u(t)\} = \frac{1}{s}$$

$$\boxed{\begin{array}{l} \mathcal{L}\{u(t)\} = \frac{1}{s} \\[2mm] \mathcal{L}\{t\,u(t)\} = \frac{1}{s^2} \end{array}}$$

.....(1)

.....(2)

From convolution theorem of Laplace transform,

$$\mathcal{L}\{x_1(t) * x_2(t)\} = X_1(s)\,X_2(s)$$

$$= \frac{2}{s} \times \frac{1}{s} = \frac{2}{s^2}$$

$$\boxed{\text{Using equations (1) and (2).}}$$

$$\therefore x_1(t) * x_2(t) = \mathcal{L}^{-1}\left\{\frac{2}{s^2}\right\} = 2\mathcal{L}^{-1}\left\{\frac{1}{s^2}\right\}$$

$$= 2 \times t\,u(t) = 2t\,u(t)$$

b) Given that, $x_1(t) = e^{-2t}\,u(t)$

$$x_2(t) = e^{-5t}\,u(t)$$

Let, $X_1(s) = \mathcal{L}\{x_1(t)\} = \mathcal{L}\{e^{-2t}\,u(t)\} = \frac{1}{s+2}$

$$X_2(s) = \mathcal{L}\{x_2(t)\} = \mathcal{L}\{e^{-5t}\,u(t)\} = \frac{1}{s+5}$$

$$\boxed{\mathcal{L}\{e^{-at}u(t)\} = \frac{1}{s+a}}$$

.....(1)

.....(2)

From convolution theorem of Laplace transform,

$$\mathcal{L}\{x_1(t) * x_2(t)\} = X_1(s)\,X_2(s)$$

$$= \frac{1}{s+2} \times \frac{1}{s+5} = \frac{1}{(s+2)(s+5)}$$

$$\boxed{\text{Using equations (1) and (2).}}$$

$$\therefore x_1(t) * x_2(t) = \mathcal{L}^{-1}\left\{\frac{1}{(s+2)(s+5)}\right\} = \mathcal{L}^{-1}\left\{\frac{A}{s+2} + \frac{B}{s+5}\right\}$$

> Using partial fraction expansion technique.

$$A = \frac{1}{(s+2)(s+5)} \times (s+2)\Big|_{s=-2} = \frac{1}{-2+5} = \frac{1}{3}$$

$$B = \frac{1}{(s+2)(s+5)} \times (s+5)\Big|_{s=-5} = \frac{1}{-5+2} = -\frac{1}{3}$$

$$\therefore x_1(t) * x_2(t) = \mathcal{L}^{-1}\left\{\frac{1}{3}\frac{1}{s+2} - \frac{1}{3}\frac{1}{s+5}\right\} = \frac{1}{3}\mathcal{L}^{-1}\left\{\frac{1}{s+2}\right\} - \frac{1}{3}\mathcal{L}^{-1}\left\{\frac{1}{s+5}\right\}$$

$$= \frac{1}{3}e^{-2t}u(t) - \frac{1}{3}e^{-5t}u(t) = \frac{1}{3}(e^{-2t} - e^{-5t})u(t)$$

---

c) Given that, $x_1(t) = t\,u(t)$

   $x_2(t) = e^{-5t}\,u(t)$

> $\mathcal{L}\{u(t)\} = \frac{1}{s}$; $\mathcal{L}\{t\,u(t)\} = \frac{1}{s^2}$; $\mathcal{L}\{e^{-at}u(t)\} = \frac{1}{s+a}$

Let, $X_1(s) = \mathcal{L}\{x_1(t)\} = \mathcal{L}\{t\,u(t)\} = \dfrac{1}{s^2}$ .....(1)

$X_2(s) = \mathcal{L}\{x_2(t)\} = \mathcal{L}\{e^{-5t}u(t)\} = \dfrac{1}{s+5}$ .....(2)

From convolution theorem of Laplace transform,

$$\mathcal{L}\{x_1(t) * x_2(t)\} = X_1(s)\,X_2(s)$$

$$= \frac{1}{s^2} \times \frac{1}{s+5} = \frac{1}{s^2(s+5)}$$

> Using equations (1) and (2).

$$\therefore x_1(t) * x_2(t) = \mathcal{L}^{-1}\left\{\frac{1}{s^2(s+5)}\right\}$$

$$= \mathcal{L}^{-1}\left\{\frac{A}{s^2} + \frac{B}{s} + \frac{C}{s+5}\right\}$$

> Using partial fraction expansion technique.

$$A = \frac{1}{s^2(s+5)} \times s^2\Big|_{s=0} = \frac{1}{0+5} = \frac{1}{5}$$

$$B = \frac{d}{ds}\left[\frac{1}{s^2(s+5)} \times s^2\right]\Big|_{s=0} = \frac{d}{ds}\left[\frac{1}{s+5}\right]\Big|_{s=0} = \frac{-1}{(s+5)^2}\Big|_{s=0} = \frac{-1}{(0+5)^2} = -\frac{1}{25}$$

$$C = \frac{1}{s^2(s+5)} \times (s+5)\Big|_{s=-5} = \frac{1}{(-5)^2} = \frac{1}{25}$$

$$\therefore x_1(t) * x_2(t) = \mathcal{L}^{-1}\left\{\frac{1}{5}\frac{1}{s^2} - \frac{1}{25}\frac{1}{s} + \frac{1}{25}\frac{1}{s+5}\right\}$$

$$= \frac{1}{5}\mathcal{L}^{-1}\left\{\frac{1}{s^2}\right\} - \frac{1}{25}\mathcal{L}^{-1}\left\{\frac{1}{s}\right\} + \frac{1}{25}\mathcal{L}^{-1}\left\{\frac{1}{s+5}\right\}$$

$$= \frac{1}{5}t\,u(t) - \frac{1}{25}u(t) + \frac{1}{25}e^{-5t}u(t)$$

$$= \frac{1}{25}(e^{-5t} + 5t - 1)\,u(t)$$

---

d) Given that, $x_1(t) = \cos t\,u(t)$

   $x_2(t) = t\,u(t)$

> $\mathcal{L}\{t\,u(t)\} = \frac{1}{s^2}$; $\mathcal{L}\{\cos \Omega_0 t\,u(t)\} = \frac{s}{s^2 + \Omega_0^2}$

Let, $X_1(s) = \mathcal{L}\{x_1(t)\} = \mathcal{L}\{\cos t\,u(t)\} = \dfrac{s}{s^2+1}$ .....(1)

$X_2(s) = \mathcal{L}\{x_2(t)\} = \mathcal{L}\{t\,u(t)\} = \dfrac{1}{s^2}$ .....(2)

From convolution theorem of Laplace transform,

$$\mathcal{L}\{x_1(t) * x_2(t)\} = X_1(s)\, X_2(s) = \frac{1}{s^2} \times \frac{s}{s^2+1}$$ Using equations (1) and (2).

$$= \frac{1}{s(s^2+1)} = \frac{A}{s} + \frac{Cs+D}{s^2+1} \qquad .....(3)$$ Using partial fraction expansion technique.

On cross multiplying equation (3) we get,

$$1 = A\,(s^2+1) + (Cs+D)s$$

$$\therefore\ 1 = As^2 + A + Cs^2 + Ds \qquad\qquad\qquad .....(4)$$

On equating constants of equation (4) we get, $A = 1$

On equating coefficients of s of equation (4) we get, $D = 0$

On equating coefficients of s^2 of equation (4) we get, $A + C = 0 \ \Rightarrow\ C = -A = -1$

$$\therefore\ \mathcal{L}\{x_1(t) * x_2(t)\} = \frac{A}{s} + \frac{Cs+D}{s^2+1}$$

$$= \frac{1}{s} - \frac{s}{s^2+1}$$

$$\boxed{\begin{aligned}\mathcal{L}\{t\,u(t)\} &= \frac{1}{s} \\ \mathcal{L}\{\cos\Omega_0\, t\, u(t)\} &= \frac{s}{s^2+\Omega_0^2}\end{aligned}}$$

$$\therefore\ x_1(t) * x_2(t) = \mathcal{L}^{-1}\left\{\frac{1}{s} - \frac{1}{s^2+1}\right\} = \mathcal{L}^{-1}\left\{\frac{1}{s}\right\} - \mathcal{L}^{-1}\left\{\frac{s}{s^2+1}\right\}$$

$$= u(t) - \cos t\, u(t) = (1 - \cos t)\, u(t)$$

Note: *Compare the above result with example 3.6.*

Example 3.25
(AU Dec'12, 10 Marks)

Perform the convolution of the following signal.

$x(t) = e^{2t}\, u(-t),\ h(t) = u(t- 3)$

Solution:

Given that, $x(t) = e^{2t}\, u(-t)$ and $h(t) = u(t- 3)$

$$\boxed{\mathcal{L}\{-e^{-at}\, u(-t)\} = \frac{1}{s+a} \ ;\ \text{anticausal}}$$

Let, $X(s) = \mathcal{L}\{x(t)\} = \mathcal{L}\{e^{2t}\, u(-t)\} = \dfrac{-1}{s-2} \qquad .....(1)$

$$H(s) = \mathcal{L}\{h(t)\} = \mathcal{L}\{u(t-3)\} = e^{-3s} \times \frac{1}{s} = \frac{e^{-3s}}{s} \qquad .....(2)$$

$$\boxed{\begin{aligned}&\text{if } \mathcal{L}\{x(t)\} = X(s), \text{ then} \\ &\mathcal{L}\{x(t-a)\} = e^{-as}\, X(s)\end{aligned}}$$

Let, $y(t) = x(t) * h(t)$

On taking Laplace transform of the above equation we get,

$$\mathcal{L}\{y(t)\} = \mathcal{L}\{x(t) * h(t)\}$$ Using convolution property of Laplace transform.

$$= X(s)\, H(s)$$ Using equations (1) and (2)

$$= \frac{-1}{s-2} \times \frac{e^{-3s}}{s} = \frac{-e^{-3s}}{s(s-2)}$$

$$\therefore\ y(t) = \mathcal{L}^{-1}\left\{\frac{-e^{-3s}}{s(s-2)}\right\} = \mathcal{L}^{-1}\left\{\frac{-1}{s(s-2)}\right\}\Big|_{t=t-3} = \mathcal{L}^{-1}\left\{\frac{A}{s} + \frac{B}{(s-2)}\right\}\Big|_{t=t-3}$$ Using partial fraction expansion technique.

$$A = \frac{-1}{\cancel{s}(s-2)} \times \cancel{s}\Big|_{s=0} = \frac{-1}{0-2} = \frac{1}{2}$$

$$B = \frac{-1}{s(\cancel{s-2})} \times (\cancel{s-2})\Big|_{s=2} = \frac{-1}{2} = -\frac{1}{2}$$

$$\therefore \ y(t) = \mathcal{L}^{-1}\left\{\frac{1}{2}\frac{1}{s} - \frac{1}{2}\frac{1}{(s-2)}\right\}\Big|_{t=t-3} = \left\{\frac{1}{2}\mathcal{L}^{-1}\left\{\frac{1}{s}\right\} - \frac{1}{2}\mathcal{L}^{-1}\left\{\frac{1}{(s-2)}\right\}\right\}\Big|_{t=t-3}$$

$$= \left[\frac{1}{2}u(t) - \frac{1}{2}e^{-2t}u(t)\right]\Big|_{t=t-3} = \left[\frac{1}{2}u(t-3) - \frac{1}{2}e^{-2(t-3)}u(t-3)\right] = \frac{1}{2}(1 - e^{-2(t-3)})u(t-3)$$

Example 3.26

Perform deconvolution operation to extract the signal $x_1(t)$.

a) $x_1(t) * x_2(t) = 2t\,u(t)$; $x_2(t) = u(t)$

b) $x_1(t) * x_2(t) = \frac{1}{3}(e^{-2t} - e^{-5t})u(t)$; $x_2(t) = e^{-5t}u(t)$

c) $x_1(t) * x_2(t) = \frac{1}{25}(e^{-5t} + 5t - 1)u(t)$; $x_2(t) = e^{-5t}u(t)$

d) $x_1(t) * x_2(t) = u(t) - \cos t\,u(t)$; $x_2(t) = t\,u(t)$

Solution:

a) Given that, $x_1(t) * x_2(t) = 2t\,u(t)$; $x_2(t) = u(t)$

Let, $x_1(t) * x_2(t) = x_3(t)$ .....(1)

$\therefore x_3(t) = 2t\,u(t)$

Let, $X_3(s) = \mathcal{L}\{x_3(t)\} = \mathcal{L}\{2t\,u(t)\} = \dfrac{2}{s^2}$ .....(2)

$X_2(s) = \mathcal{L}\{x_2(t)\} = \mathcal{L}\{u(t)\} = \dfrac{1}{s}$ .....(3)

On taking Laplace transform of equation (1) we get,

$$\mathcal{L}\{x_1(t) * x_2(t)\} = \mathcal{L}\{x_3(t)\}$$

$$X_1(s)\,X_2(s) = X_3(s)$$

| Using convolution property of Laplace transform. |

$$\therefore X_1(s) = \frac{X_3(s)}{X_2(s)} = X_3(s) \times \frac{1}{X_2(s)} = \frac{2}{s^2} \times s = \frac{2}{s}$$

| Using equations (2) and (3). |

$$\therefore x_1(t) = \mathcal{L}^{-1}\{X_1(s)\} = \mathcal{L}^{-1}\left\{\frac{2}{s}\right\} = 2\,u(t)$$

b) Given that, $x_1(t) * x_2(t) = \frac{1}{3}(e^{-2t} - e^{-5t})u(t)$; $x_2(t) = e^{-5t}u(t)$

Let, $x_1(t) * x_2(t) = x_3(t)$ .....(1)

$$\therefore x_3(t) = \frac{1}{3}(e^{-2t} - e^{-5t})u(t) = \frac{e^{-2t}}{3}u(t) - \frac{e^{-5t}}{3}u(t)$$

Let, $X_3(s) = \mathcal{L}\{x_3(t)\} = \mathcal{L}\left\{\dfrac{e^{-2t}}{3}u(t) - \dfrac{e^{-5t}}{3}u(t)\right\} = \dfrac{1}{3(s+2)} - \dfrac{1}{3(s+5)}$ .....(2)

$$X_2(s) = \mathcal{L}\{x_2(t)\} = \mathcal{L}\{e^{-5t}u(t)\} = \frac{1}{s+5} \qquad \dots(3)$$

On taking Laplace transform of equation (1) we get,

$$\mathcal{L}\{x_1(t) * x_2(t)\} = \mathcal{L}\{x_3(t)\}$$

> Using convolution property of Laplace transform.

$$X_1(s)\,X_2(s) = X_3(s)$$

$$\therefore X_1(s) = \frac{X_3(s)}{X_2(s)} = X_3(s) \times \frac{1}{X_2(s)}$$

> Using equations (2) and (3).

$$= \left[\frac{1}{3(s+2)} - \frac{1}{3(s+5)}\right] \times (s+5)$$

$$= \frac{s+5}{3(s+2)} - \frac{1}{3} = \frac{s+5-(s+2)}{3(s+2)}$$

$$= \frac{3}{3(s+2)} = \frac{1}{s+2}$$

$$\therefore x_1(t) = \mathcal{L}^{-1}\{X_1(s)\} = \mathcal{L}^{-1}\left\{\frac{1}{s+2}\right\} = e^{-2t}u(t)$$

---

c) Given that, $x_1(t) * x_2(t) = \dfrac{1}{25}(e^{-5t} + 5t - 1)\,u(t)\;;\; x_2(t) = e^{-5t}\,u(t)$

Let, $x_1(t) * x_2(t) = x_3(t)$ $\qquad \dots(1)$

$$\therefore x_3(t) = \frac{1}{25}(e^{-5t} + 5t - 1)\,u(t) = \frac{e^{-5t}}{25}u(t) + \frac{t}{5}u(t) - \frac{1}{25}u(t)$$

$$\text{Let, } X_3(s) = \mathcal{L}\{x_3(t)\} = \mathcal{L}^{-1}\left\{\frac{e^{-5t}}{25}u(t) + \frac{t}{5}u(t) - \frac{1}{25}u(t)\right\} = \frac{1}{25(s+5)} + \frac{1}{5s^2} - \frac{1}{25s} \qquad \dots(2)$$

$$X_2(t) = \mathcal{L}\{x_2(t)\} = \mathcal{L}\{e^{-5t}u(t)\} = \frac{1}{s+5} \qquad \dots(3)$$

On taking Laplace transform of equation (1) we get,

$$\mathcal{L}\{x_1(t) * x_2(t)\} = \mathcal{L}\{x_3(t)\}$$

$$X_1(s)\,X_2(s) = X_3(s)$$

> Using convolution property of Laplace transform.

$$\therefore X_1(s) = \frac{X_3(s)}{X_2(s)} = X_3(s) \times \frac{1}{X_2(s)}$$

$$= \left[\frac{1}{25(s+5)} + \frac{1}{5s^2} - \frac{1}{25s}\right] \times (s+5) = \frac{1}{25} + \frac{s+5}{5s^2} - \frac{s+5}{25s}$$

> Using equations (2) and (3).

$$= \frac{s^2 + 5(s+5) - s(s+5)}{25s^2} = \frac{s^2 + 5s + 25 - s^2 - 5s}{25s^2} = \frac{25}{25s^2} = \frac{1}{s^2}$$

$$\therefore x_1(t) = \mathcal{L}^{-1}\{X_1(s)\} = \mathcal{L}^{-1}\left\{\frac{1}{s^2}\right\} = t\,u(t)$$

d) Given that, $x_1(t) * x_2(t) = u(t) - \cos t\, u(t)$; $x_2(t) = t\, u(t)$

Let, $x_1(t) * x_2(t) = x_3(t)$ .....(1)

$\therefore x_3(t) = u(t) - \cos t\, u(t)$

Let, $X_3(s) = \mathcal{L}\{x_3(t)\} = \mathcal{L}\{u(t) - \cos t\, u(t)\} = \dfrac{1}{s} - \dfrac{s}{s^2+1}$ .....(2)

$X_2(s) = \mathcal{L}\{x_2(t)\} = \mathcal{L}\{t\, u(t)\} = \dfrac{1}{s^2}$ .....(3)

On taking Laplace transform of equation (1) we get,

$\mathcal{L}\{x_1(t) * x_2(t)\} = \mathcal{L}\{x_3(t)\}$

$X_1(s)\, X_2(s) = X_3(s)$

> Using convolution property of Laplace transform.

$\therefore X_1(s) = \dfrac{X_3(s)}{X_2(s)} = X_3(s) \times \dfrac{1}{X_2(s)} = \left[\dfrac{1}{s} - \dfrac{s}{s^2+1}\right] \times s^2$

> Using equations (2) and (3).

$\qquad = s - \dfrac{s^3}{s^2+1} = \dfrac{s(s^2+1) - s^3}{s^2+1} = \dfrac{s^3 + s - s^3}{s^2+1} = \dfrac{s}{s^2+1}$

$\therefore x(t) = \mathcal{L}^{-1}\{X_1(s)\} = \mathcal{L}^{-1}\left\{\dfrac{s}{s^2+1}\right\} = \cos t\, u(t)$

> *Note: Compare the results of Example 3.26 with Example 3.24.*

Example 3.27

The impulse response of continuous time systems are given below. Determine the unit step response of the systems using convolution theorem of Laplace transform, and verify the result with Example 3.7.

a) $h(t) = 3t\, u(t)$ **b)** $h(t) = e^{-5t} u(t)$ **c)** $h(t) = u(t + 2)$

d) $h(t) = u(t - 2)$ **e)** $h(t) = u(t + 2) + u(t - 2)$

Solution:

a) Given that, $h(t) = 3t\, u(t)$

Let, $H(s) = \mathcal{L}\{h(t)\} = \mathcal{L}\{3t\, u(t)\} = 3\,\mathcal{L}\{t\, u(t)\} = 3 \times \dfrac{1}{s^2} = \dfrac{3}{s^2}$ .....(1)

$U(s) = \mathcal{L}\{u(t)\} = \dfrac{1}{s}$; where $u(t)$ is unit step signal .....(2)

Unit step response, $s(t) = h(t) * u(t)$

On taking Laplace transform of the above equation we get,

$\mathcal{L}\{s(t)\} = \mathcal{L}\{h(t) * u(t)\}$

> Using convolution theorem of Laplace transform.

$\qquad = H(s)\, U(s)$

$\qquad = \dfrac{3}{s^2} \times \dfrac{1}{s} = \dfrac{3}{s^3}$

> Using equations (2) and (3).

Unit step response is given by inverse Laplace transform of the above equation.

$\therefore$ Unit step response, $s(t) = \mathcal{L}^{-1}\left\{\dfrac{3}{s^3}\right\} = \mathcal{L}^{-1}\left\{\dfrac{3}{2} \times \dfrac{2}{s^{2+1}}\right\}$

> $\mathcal{L}\{t^m u(t)\} = \dfrac{m!}{s^{m+1}}$

$\qquad = \dfrac{3}{s^2}\,\mathcal{L}^{-1}\left\{\dfrac{2}{s^{2+1}}\right\} = \dfrac{3}{2}\, t^2 u(t)$

b) Given that, $h(t) = e^{-5t} u(t)$

Let, $H(s) = \mathcal{L}\{h(t)\} = \mathcal{L}\{e^{-5t} u(t)\} = \dfrac{1}{s+5}$ .....(1)

$U(s) = \mathcal{L}\{u(t)\} = \dfrac{1}{s}$; where $u(t)$ is unit step signal   .....(2)

Unit step response, $s(t) = h(t) * u(t)$

On taking Laplace transform of the above equation we get,

$$\mathcal{L}\{s(t)\} = \mathcal{L}\{h(t) * u(t)\} = H(s)\,U(s)$$

> Using convolution theorem of Laplace transform.

$$= \dfrac{1}{s+5} \times \dfrac{1}{s} = \dfrac{1}{s(s+5)}$$

> Using equations (1) and (2).

Unit step response is given by inverse Laplace transform of the above equation.

$\therefore$ Unit step response, $s(t) = \mathcal{L}^{-1}\left\{\dfrac{1}{s(s+5)}\right\} = \mathcal{L}^{-1}\left\{\dfrac{A}{s} + \dfrac{B}{s+5}\right\}$

> Using partial fraction expansion technique.

$$A = \dfrac{1}{\cancel{s}(s+5)} \times \cancel{s}\Big|_{s=0} = \dfrac{1}{0+5} = \dfrac{1}{5}$$

$$B = \dfrac{1}{s\cancel{(s+5)}} \times \cancel{(s+5)}\Big|_{s=-5} = \dfrac{1}{-5} = -\dfrac{1}{5}$$

> $\mathcal{L}\{u(t)\} = \dfrac{1}{s}$; $\mathcal{L}\{e^{-at} u(t)\} = \dfrac{1}{s+a}$

$\therefore$ Unit step response, $s(t) = \mathcal{L}^{-1}\left\{\dfrac{1}{5}\dfrac{1}{s} - \dfrac{1}{5}\dfrac{1}{s+5}\right\} = \dfrac{1}{5}\mathcal{L}^{-1}\left\{\dfrac{1}{s}\right\} - \dfrac{1}{5}\mathcal{L}^{-1}\left\{\dfrac{1}{s+5}\right\}$

$$= \dfrac{1}{5} u(t) - \dfrac{1}{5} e^{-5t} u(t) = \dfrac{1}{5}(1 - e^{-5t}) u(t)$$

---

c) Given that, $h(t) = u(t + 2)$

Let, $H(s) = \mathcal{L}\{h(t)\} = \mathcal{L}\{u(t+2)\} = e^{2s}\mathcal{L}\{u(t)\} = e^{2s} \times \dfrac{1}{s} = \dfrac{e^{2s}}{s}$ .....(1)

$U(s) = \mathcal{L}\{u(t)\} = \dfrac{1}{s}$; where $u(t)$ is unit step signal   .....(2)

Unit step response, $s(t) = h(t) * u(t)$

On taking Laplace transform of the above equation we get,

$$\mathcal{L}\{s(t)\} = \mathcal{L}\{h(t) * u(t)\}$$

> Using convolution theorem of Laplace transform.

$$= H(s)\,U(s)$$

$$= \dfrac{e^{2s}}{s} \times \dfrac{1}{s} = \dfrac{e^{2s}}{s^2}$$

> Using equations (1) and (2).

Unit step response is given by inverse Laplace transform of the above equation.

$\therefore$ Unit step response, $s(t) = \mathcal{L}^{-1}\left\{\dfrac{e^{2s}}{s^2}\right\} = \mathcal{L}^{-1}\left\{\dfrac{1}{s^2}\right\}\Big|_{t=t-2}$

> if $\mathcal{L}\{t\,u(t)\} = \dfrac{1}{s^2}$, then
>
> $\mathcal{L}\{(t+a)\,u(t+a)\} = \dfrac{e^{as}}{s^2}$

$$= t\,u(t)\Big|_{t=t+2} = (t+2)\,u(t+2)$$

d) Given that, h(t) = u(t − 2)

Let, $H(s) = \mathcal{L}\{h(t)\} = \mathcal{L}\{u(t-2)\} = e^{-2s}\mathcal{L}\{u(t)\} = e^{-2s} \times \dfrac{1}{s} = \dfrac{e^{-2s}}{s}$ .....(1)

$U(s) = \mathcal{L}\{u(t)\} = \dfrac{1}{s}$; where $u(t)$ is unit step signal .....(2)

Unit step response, $s(t) = h(t) * u(t)$

On taking Laplace transform of the above equation we get,

$\mathcal{L}\{s(t)\} = \mathcal{L}\{h(t) * u(t)\}$

$\qquad = H(s)\,U(s)$ | Using convolution theorem of Laplace transform. |

$\qquad = \dfrac{e^{-2s}}{s} \times \dfrac{1}{s} = \dfrac{e^{-2s}}{s^2}$ | Using equations (1) and (2). |

Unit step response is given by inverse Laplace transform of the above equation.

$\therefore$ Unit step response, $s(t) = \mathcal{L}^{-1}\left\{\dfrac{e^{-2s}}{s^2}\right\} = \mathcal{L}^{-1}\left\{\dfrac{1}{s^2}\right\}\Big|_{t=t-2}$

$\qquad\qquad = t\,u(t)\big|_{t=t-2} = (t-2)\,u(t-2)$

| if $\mathcal{L}\{t\,u(t)\} = \dfrac{1}{s^2}$, then $\mathcal{L}\{(t-a)\,u(t-a)\} = \dfrac{e^{-as}}{s^2}$ |

e) Given that, h(t) = u(t + 2) + u(t − 2)

Unit step response, $s(t) = h(t) * u(t)$

$\qquad\qquad = [u(t+2) + u(t-2)] * u(t)$

$\qquad\qquad = [u(t+2) * u(t)] + [u(t-2) * u(t)]$

$\qquad\qquad = (t+2)\,u(t+2) + (t-2)\,u(t-2)$ | Using the results of (c) and (d). |

Example 3.28

Perform convolution of $x_1(t)$ and $x_2(t)$ using convolution theorem of Laplace transform and verify the result with Example 3.10.

a)

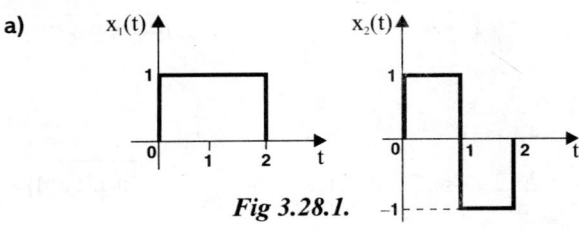

Fig 3.28.1.

b)

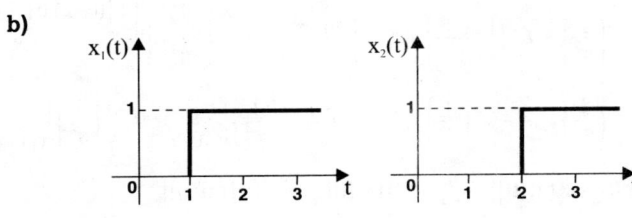

Fig 3.28.2.

Solution:

a)

The mathematical equation of the signal shown in Fig 1 is,

$$x_1(t) = u(t) - u(t-2)$$

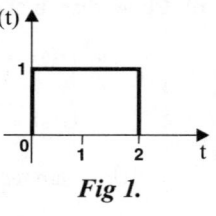

Fig 1.

On taking Laplace transform of above equation we get,

if $\mathcal{L}\{u(t)\} = \dfrac{1}{s}$, then

$\mathcal{L}\{u(t-a)\} = \dfrac{e^{-as}}{s}$

$$X_1(s) = \mathcal{L}\{x_1(t)\} = \mathcal{L}\{u(t) - u(t-2)\}$$

$$= \mathcal{L}\{u(t)\} - e^{-2s}\mathcal{L}\{u(t)\} = \frac{1}{s} - \frac{e^{-2s}}{s} \qquad \text{.....(1)}$$

The mathematical equation of the signal shown in Fig 2 is,

$$x_2(t) = 1 \quad ; \quad 0 < t < 1$$
$$= -1 \quad ; \quad 1 < t < 2$$

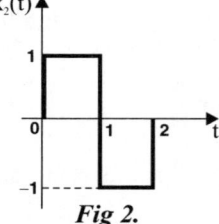

Fig 2.

Let, $X_2(s) = \mathcal{L}\{x_2(t)\}$

By the definition of Laplace transform,

$$X_2(s) = \int_{-\infty}^{+\infty} x_2(t)\, e^{-st}\, dt = \int_0^1 1 \times e^{-st}\, dt + \int_1^2 -1 \times e^{-st}\, dt = \int_0^1 e^{-st}\, dt - \int_1^2 e^{-st}\, dt$$

$$= \left[\frac{e^{-st}}{-s}\right]_0^1 - \left[\frac{e^{-st}}{-s}\right]_1^2 = \left[\frac{e^{-s}}{-s} - \frac{e^0}{-s}\right] - \left[\frac{e^{-2s}}{-s} - \frac{e^{-s}}{-s}\right]$$

$$= -\frac{e^{-s}}{s} + \frac{1}{s} + \frac{e^{-2s}}{s} - \frac{e^{-s}}{s}$$

$$= \frac{1}{s} - \frac{2e^{-s}}{s} + \frac{e^{-2s}}{s} \qquad \text{.....(2)}$$

By using convolution property of Laplace transform,

$$\mathcal{L}\{x_1(t) * x_2(t)\} = X_1(s)\, X_2(s)$$

$$\therefore\ x_1(t) * x_2(t) = \mathcal{L}^{-1}\{X_1(s)\, X_2(s)\}$$

$$= \mathcal{L}^{-1}\left\{\left(\frac{1}{s} - \frac{e^{-2s}}{s}\right)\left(\frac{1}{s} - \frac{2e^{-s}}{s} + \frac{e^{-2s}}{s}\right)\right\}$$

Using equations (1) and (2).

$$= \mathcal{L}^{-1}\left\{\frac{1}{s^2} - \frac{2e^{-s}}{s^2} + \frac{e^{-2s}}{s^2} - \frac{e^{-2s}}{s^2} + \frac{2e^{-3s}}{s^2} - \frac{e^{-4s}}{s^2}\right\}$$

$$\therefore\ x_1(t) * x_2(t) = \mathcal{L}^{-1}\left\{\frac{1}{s^2} - \frac{2e^{-s}}{s^2} + \frac{2e^{-3s}}{s^2} - \frac{e^{-4s}}{s^2}\right\}$$

if $\mathcal{L}\{t\, u(t)\} = \dfrac{1}{s^2}$, then

$\mathcal{L}\{(t-a)\, u(t-a)\} = \dfrac{e^{-as}}{s^2}$

$$= \mathcal{L}^{-1}\left\{\frac{1}{s^2}\right\} - \mathcal{L}^{-1}\left\{\frac{2e^{-s}}{s^2}\right\} + \mathcal{L}^{-1}\left\{\frac{2e^{-3s}}{s^2}\right\} - \mathcal{L}^{-1}\left\{\frac{e^{-4s}}{s^2}\right\}$$

$$= \mathcal{L}^{-1}\left\{\frac{1}{s^2}\right\} - 2\,\mathcal{L}^{-1}\left\{\frac{1}{s^2}\right\}\bigg|_{t=t-1} + 2\,\mathcal{L}^{-1}\left\{\frac{1}{s^2}\right\}\bigg|_{t=t-3} - \mathcal{L}^{-1}\left\{\frac{1}{s^2}\right\}\bigg|_{t=t-4}$$

$$= t\, u(t) - 2[t\, u(t)]\big|_{t=t-1} + 2[t\, u(t)]\big|_{t=t-3} - [t\, u(t)]\big|_{t=t-4}$$

$$= t\, u(t) - 2(t-1)\, u(t-1) + 2(t-3)\, u(t-3) - (t-4)\, u(t-4)$$

To verify the above result with Example 3.10(a)

<u>**For t = 0 to 1**</u>

When t = 0 to1, $u(t) = 1$, $u(t - 1) = 0$, $u(t - 3) = 0$, $u(t - 4) = 0$

     $\therefore$ $x_1(t) * x_2(t) = t \times 1 - 2(t - 1) \times 0 + 2(t - 3) \times 0 - (t - 4) \times 0 = t$

<u>**For t = 1 to 3**</u>

When t = 1 to 2, $u(t) = 1$, $u(t - 1) = 1$, $u(t - 3) = 0$, $u(t - 4) = 0$

     $\therefore$ $x_1(t) * x_2(t) = t \times 1 - 2(t - 1) \times 1 + 2(t - 3) \times 0 - (t - 4) \times 0 = t - 2t + 2 = 2 - t$

<u>**For t = 3 to 4**</u>

When t = 3 to 4, $u(t) = 1$, $u(t - 1) = 1$, $u(t - 3) = 1$, $u(t - 4) = 0$

     $\therefore$ $x_1(t) * x_2(t) = t \times 1 - 2(t - 1) \times 1 + 2(t - 3) \times 1 - (t - 4) \times 0 = t - 2t + 2 + 2t - 6 = t - 4$

<u>**For t > 4**</u>

When t > 4, $u(t) = 1$, $u(t - 1) = 1$, $u(t - 3) = 1$, $u(t - 4) = 1$

     $\therefore$ $x_1(t) * x_2(t) = t \times 1 - 2(t - 1) \times 1 + 2(t - 3) \times 1 - (t - 4) \times 1 = t - 2t + 2 + 2t - 6 - t + 4 = 0$

---

b)

The mathematical equation of the signal shown in Fig 1 is,

     $x_1(t) = u(t - 1)$

On taking Laplace transform of above equation we get,

     $X_1(s) = \mathcal{L}\{u(t - 1)\} = e^{-s}\mathcal{L}\{u(t)\} = \dfrac{e^{-s}}{s}$ .....(1)

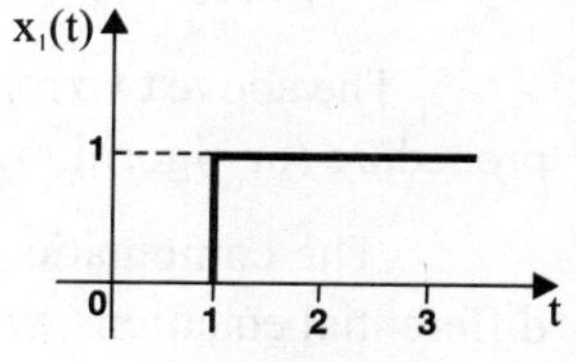

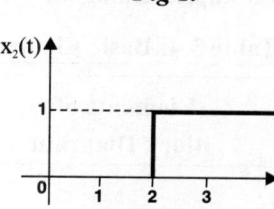

Fig 1.

The mathematical equation of the signal shown in Fig 2 is,

     $x_2(t) = u(t - 2)$

On taking Laplace transform of above equation we get,

     $X_1(s) = \mathcal{L}\{u(t - 2)\} = e^{-2s}\mathcal{L}\{u(t)\} = \dfrac{e^{-2s}}{s}$ .....(2)

By using convolution property of Laplace transform,

     $\mathcal{L}\{x_1(t) * x_2(t)\} = X_1(s)\, X_2(s)$

Fig 2.

     $\therefore$ $x_1(t) * x_2(t) = \mathcal{L}^{-1}\{X_1(s) X_2(s)\} = \mathcal{L}^{-1}\left\{\dfrac{e^{-s}}{s} \times \dfrac{e^{-2s}}{s}\right\} = \mathcal{L}^{-1}\left\{\dfrac{e^{-3s}}{s^2}\right\}$

> Using equations (1) and (2).

     $= \mathcal{L}^{-1}\left\{\dfrac{1}{s^2}\right\}\Big|_{t=t-3} = t\,u(t)\big|_{t=t-3} = (t - 3)\,u(t - 3)$

> if, $\mathcal{L}\{x(t)\} = X(s)$,
> then, $\mathcal{L}\{x(t - a)\} = e^{-as}X(s)$

To verify the above result with Example 3.10(b)

<u>**For t < 3**</u>

     When t < 3, u(t - 3) = 0, $\therefore$ $x_1(t) * x_2(t) = (t - 3) \times 0 = 0$

<u>**For t ≥ 3**</u>

     When t > 3, u(t - 3) = 1 $\therefore$ $x_1(t) * x_2(t) = (t - 3) \times 1 = t - 3$

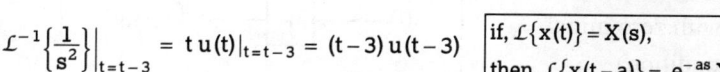

3.7 Block Diagram Representation of LTI Continuous Time Systems in s-Domain

In time domain, the input-output relation of a LTI (Linear Time Invariant) continuous time system is represented by constant coefficient differential equation shown in equation (3.41).

$$a_0 \frac{d^N}{dt^N} y(t) + a_1 \frac{d^{N-1}}{dt^{N-1}} y(t) + a_2 \frac{d^{N-2}}{dt^{N-2}} y(t) + \ldots + a_{N-1} \frac{d}{dt} y(t) + a_N y(t) = b_0 \frac{d^M}{dt^M} x(t)$$

$$+ b_1 \frac{d^{M-1}}{dt^{M-1}} x(t) + b_2 \frac{d^{M-2}}{dt^{M-2}} x(t) + \ldots b_{M-1} \frac{d}{dt} x(t) + b_M x(t) \qquad \ldots(3.41)$$

In s-domain, the input-output relation of a LTI (Linear Time Invariant) continuous time system is represented by the transfer function H(s), which is a rational function of s, as shown in equation (3.42).

$$H(s) = \frac{Y(s)}{X(s)} = \frac{b_0 s^M + b_1 s^{M-1} + b_2 s^{M-2} + \ldots + b_{M-1} s + b_M}{a_0 s^N + a_1 s^{N-1} + a_2 s^{N-2} + \ldots + a_{N-1} s + a_N} \qquad \ldots(3.42)$$

where, N = Order of the system, M ≤ N and $a_0 = 1$

The above two representations of continuous time system can be viewed as a computational procedure (or algorithm) to determine the output signal y(t) from the input signal x(t).

The computations in the above equation can be arranged into various equivalent sets of differential equations, with each set of equations defining a computational procedure or algorithm for implementing the system.

Table 3.4: Basic Elements of Block Diagram in Time Domain and s-Domain

Elements of Block Diagram	Time Domain Representation	s-domain Representation
Differentiator	$x(t) \longrightarrow \boxed{\frac{d}{dt}} \longrightarrow \frac{d}{dt} x(t)$	$X(s) \longrightarrow \boxed{s} \longrightarrow s\,X(s)$
Integrator (with zero initial condition)	$x(t) \longrightarrow \boxed{\int} \longrightarrow \int x(t)\,dt$	$X(s) \longrightarrow \boxed{1/s} \longrightarrow \frac{X(s)}{s}$
Constant Multiplier	$x(t) \longrightarrow \triangleright a \longrightarrow a\,x(t)$	$X(s) \longrightarrow \triangleright a \longrightarrow a\,X(s)$
Signal Adder	$x_1(t)$, $x_2(t) \longrightarrow \oplus \longrightarrow x_1(t) + x_2(t)$	$X_1(s)$, $X_2(s) \longrightarrow \oplus \longrightarrow X_1(s) + X_2(s)$

For each set of equations, we can construct a block diagram consisting of integrators, adders and multipliers. Such block diagrams are referred as realization of system or equivalently as structure for realizing system. The basic elements used to construct block diagrams are listed in Table 3.4.

(For block diagram representation of the continuous time system in time domain refer Chapter - 3, Section 3.2).

Some of the block diagram representations of the system gives a direct relation between time domain equation and s-domain equation.

The main advantage of rearranging the sets of differential equations is to reduce the computational complexity and memory requirements.

The different types of structures for realizing continuous time systems are,

1. Direct form-I structure
2. Direct form-II structure
3. Cascade structure
4. Parallel structure

3.7.1 Direct Form-I Structure

Consider the differential equation governing the continuous time system.

$$\frac{d^N}{dt^N}y(t) + a_1 \frac{d^{N-1}}{dt^{N-1}}y(t) + a_2 \frac{d^{N-2}}{dt^{N-2}}y(t) + \dots + a_{N-1}\frac{d}{dt}y(t) + a_N y(t) = b_0 \frac{d^M}{dt^M}x(t)$$

$$+ b_1 \frac{d^{M-1}}{dt^{M-1}}x(t) + b_2 \frac{d^{M-2}}{dt^{M-2}}x(t) + \dots + b_{M-1}\frac{d}{dt}x(t) + b_M x(t)$$

$$\therefore \frac{d^N}{dt^N}y(t) = -a_1 \frac{d^{N-1}}{dt^{N-1}}y(t) - a_2 \frac{d^{N-2}}{dt^{N-2}}y(t) - \dots - a_{N-1}\frac{d}{dt}y(t) - a_N y(t)$$

$$+ b_0 \frac{d^M}{dt^M}x(t) + b_1 \frac{d^{M-1}}{dt^{M-1}}x(t) + b_2 \frac{d^{M-2}}{dt^{M-2}}x(t) + \dots + b_{M-1}\frac{d}{dt}x(t) + b_M x(t)$$

On taking Laplace transform of the above equation with zero initial conditions we get,

$$s^N Y(s) = -a_1 s^{N-1}Y(s) - a_2 s^{N-2}Y(s) - \dots - a_{N-1}sY(s) - a_N Y(s)$$

$$+ b_0 s^M X(s) + b_1 s^{M-1}X(s) + b_2 s^{M-2}X(s) + \dots + b_{M-1}sX(s) + b_M X(s)$$

On dividing throughout by s^N and letting $M = N$ we get,

$$Y(s) = -a_1 \frac{Y(s)}{s} - a_2 \frac{Y(s)}{s^2} - \dots - a_{N-1}\frac{Y(s)}{s^{N-1}} - a_N \frac{Y(s)}{s^N}$$

$$+ b_0 X(s) + b_1 \frac{X(s)}{s} + b_2 \frac{X(s)}{s^2} + \dots + b_{N-1}\frac{X(s)}{s^{N-1}} + b_N \frac{X(s)}{s^N}$$

The above equation of Y(s) can be directly represented by a block diagram as shown in Fig 3.10 and this structure is called direct form-I structure. This structure uses separate integrators for input and output signals. Hence for realizing this structure more memory is required. The direct form structure provides a direct relation between time domain and s-domain equations.

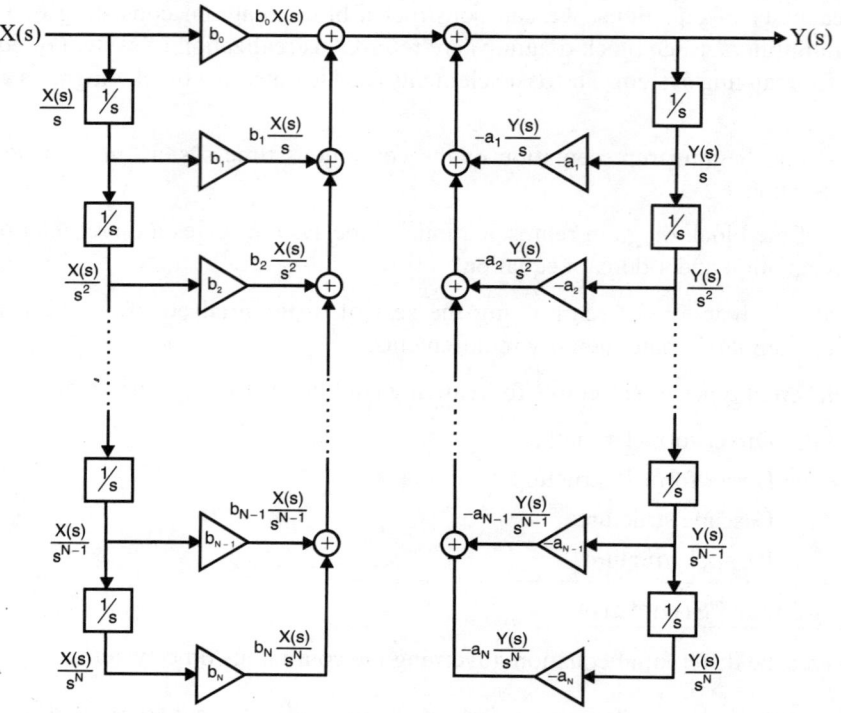

Fig 3.10: *Direct form-I structure of continuous time system.*

3.7.2 Direct Form-II Structure

An alternative structure called direct form-II structure can be realized which uses less number of integrators than the direct form-I structure.

Consider the differential equation governing the continuous time system.

$$\frac{d^N}{dt^N}y(t) + a_1\frac{d^{N-1}}{dt^{N-1}}y(t) + a_2\frac{d^{N-2}}{dt^{N-2}}y(t) + \ldots + a_{N-1}\frac{d}{dt}y(t) + a_N y(t) = b_0\frac{d^M}{dt^M}x(t)$$

$$+ b_1\frac{d^{M-1}}{dt^{M-1}}x(t) + b_2\frac{d^{M-2}}{dt^{M-2}}x(t) + \ldots + b_{M-1}\frac{d}{dt}x(t) + b_M x(t)$$

On taking Laplace transform of the above equation with zero initial conditions we get,

$$s^N Y(s) + a_1 s^{N-1}Y(s) + a_2 s^{N-2}Y(s) + \ldots + a_{N-1}sY(s) + a_N Y(s) = b_0 s^M X(s)$$

$$+ b_1 s^{M-1}X(s) + b_2 s^{M-2}X(s) + \ldots + b_{M-1}sX(s) + b_M X(s)$$

On dividing throughout by s^N and letting $M = N$ we get,

$$Y(s) + a_1\frac{Y(s)}{s} + a_2\frac{Y(s)}{s^2} + \ldots + a_{N-1}\frac{Y(s)}{s^{N-1}} + a_N\frac{Y(s)}{s^N} = b_0 X(s)$$

$$+ b_1\frac{X(s)}{s} + b_2\frac{X(s)}{s^2} + \ldots + b_{N-1}\frac{X(s)}{s^{N-1}} + b_N\frac{X(s)}{s^N}$$

$$Y(s)\left[1 + a_1 \frac{1}{s} + a_2 \frac{1}{s^2} + \dots + a_{N-1} \frac{1}{s^{N-1}} + a_N \frac{1}{s^N}\right]$$

$$= X(s)\left[b_0 + b_1 \frac{1}{s} + b_2 \frac{1}{s^2} + \dots + b_{N-1} \frac{1}{s^{N-1}} + b_N \frac{1}{s^N}\right]$$

$$\therefore \frac{Y(s)}{X(s)} = \frac{b_0 + b_1 \frac{1}{s} + b_2 \frac{1}{s^2} + \dots + b_{N-1} \frac{1}{s^{N-1}} + b_N \frac{1}{s^N}}{1 + a_1 \frac{1}{s} + a_2 \frac{1}{s^2} + \dots + a_{N-1} \frac{1}{s^{N-1}} + a_N \frac{1}{s^N}}$$

Let, $\dfrac{Y(s)}{X(s)} = \dfrac{W(s)}{X(s)} \times \dfrac{Y(s)}{W(s)}$

where, $\dfrac{W(s)}{X(s)} = \dfrac{1}{1 + a_1 \frac{1}{s} + a_2 \frac{1}{s^2} + \dots + a_{N-1} \frac{1}{s^{N-1}} + a_N \frac{1}{s^N}}$ .....(3.43)

and $\dfrac{Y(s)}{W(s)} = b_0 + b_1 \frac{1}{s} + b_2 \frac{1}{s^2} + \dots + b_{N-1} \frac{1}{s^{N-1}} + b_N \frac{1}{s^N}$ .....(3.44)

On cross multiplying equation (3.43) we get,

$$W(s) + a_1 \frac{W(s)}{s} + a_2 \frac{W(s)}{s^2} + \dots + a_{N-1} \frac{W(s)}{s^{N-1}} + a_N \frac{W(s)}{s^N} = X(s)$$

$$\therefore W(s) = -a_1 \frac{W(s)}{s} - a_2 \frac{W(s)}{s^2} - \dots - a_{N-1} \frac{W(s)}{s^{N-1}} - a_N \frac{W(s)}{s^N} + X(s) \qquad \text{.....(3.45)}$$

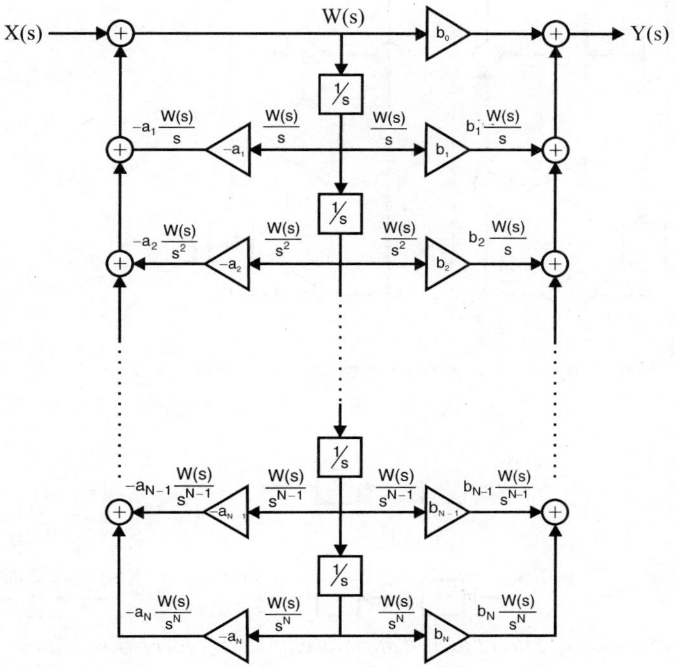

Fig 3.11: *Direct form-II structure of continuous time system for N = M.*

On cross multiplying equation (3.44) we get,

$$Y(s) = b_0 W(s) + b_1 \frac{W(s)}{s} + b_2 \frac{W(s)}{s^2} + \ldots\ldots + b_{N-1} \frac{W(s)}{s^{N-1}} + b_N \frac{W(s)}{s^N} \qquad \ldots\ldots(3.46)$$

The equations (3.45) and (3.46) represent the continuous time system in s-domain and can be realized by a direct structure called direct form-II structure as shown in Fig 3.11.

Conversion of Direct Form-I Structure to Direct Form-II Structure

The direct form-I structure can be converted to direct form-II structure by considering the direct form-I structure as cascade of two systems $\mathcal{H}_1$ and $\mathcal{H}_2$ as shown in Fig 3.12. By linearity property the order of cascading can be interchanged as shown in Fig 3.13 and Fig 3.14.

In Fig 3.14 we can observe that the input to the integrators in $\mathcal{H}_1$ and $\mathcal{H}_2$ are same and so the output of integrators in $\mathcal{H}_1$ and $\mathcal{H}_2$ are same. Therefore, instead of having separate integrators for $\mathcal{H}_1$ and $\mathcal{H}_2$, a single set of integrators can be used. Hence the integrators can be merged to combine the cascaded systems to a single system and the resultant structure will be direct form-II structure as that of Fig 3.11.

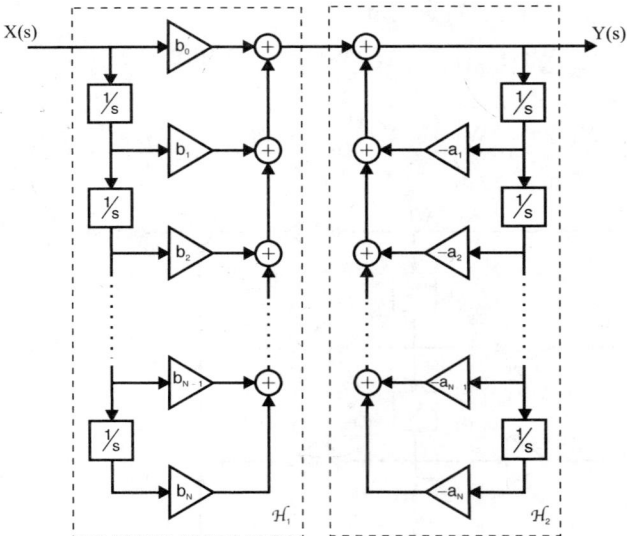

Fig 3.12: *Direct form-I structure as cascade of two systems.*

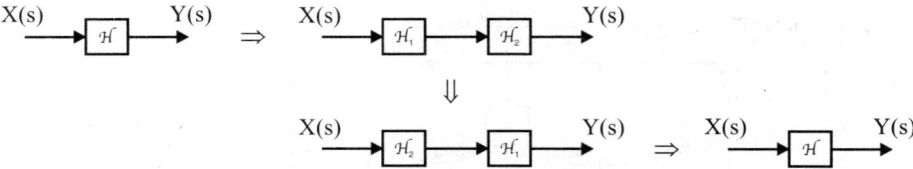

Fig 3.13: *Conversion of Direct form-I structure to Direct form-II structure.*

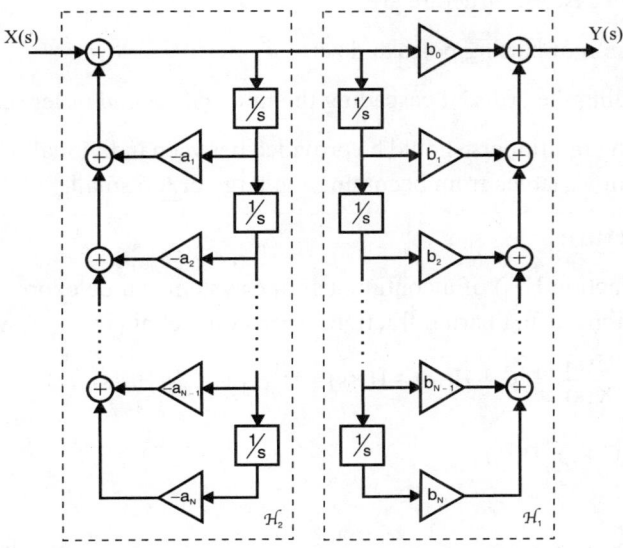

Fig 3.14: *Direct form-I structure after interchanging the order of cascade.*

3.7.3 Cascade Structure

The transfer function H(s) of a continuous time system can be expressed as a product of a number of second order or first order sections, as shown in equation (3.47).

$$H(s) = \frac{Y(s)}{X(s)} = H_1(s) \times H_2(s) \times H_3(s) \dots H_m(s)$$

$$= \prod_{i=1}^{m} H_i(s)$$

.....(3.47)

where, $H_i(s) = \dfrac{c_{0i} + c_{1i}\dfrac{1}{s} + c_{2i}\dfrac{1}{s^2}}{d_{0i} + d_{1i}\dfrac{1}{s} + d_{2i}\dfrac{1}{s^2}}$ $\boxed{\text{Second order section.}}$

or $H_i(s) = \dfrac{c_{0i} + c_{1i}\dfrac{1}{s}}{d_{0i} + d_{1i}\dfrac{1}{s}}$ $\boxed{\text{First order section.}}$

The individual second order or first order sections can be realized either in direct form-I or direct form-II structure. The overall system is obtained by cascading the individual sections as shown in Fig 3.15.

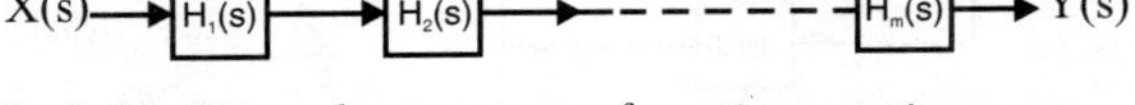

$$X(s) \rightarrow \boxed{H_1(s)} \rightarrow \boxed{H_2(s)} \rightarrow - - - - - \rightarrow \boxed{H_m(s)} \rightarrow Y(s)$$

Fig 3.15: *Cascade structure of continuous time system.*

The difficulty in cascade structure are,

1. Decision of pairing poles and zeros.

2. Deciding the order of cascading the first and second order sections.

3. Scaling multipliers should be provided between individual sections to prevent the system variables from becoming too large or too small.

3.7.4 Parallel Structure

The transfer function H(s) of a continuous time system can be expressed as a sum of first and second order sections, using partial fraction expansion technique as shown below:

$$H(s) = \frac{Y(s)}{X(s)} = C + H_1(s) + H_2(s) + H_3(s) + \ldots + H_m(s)$$

$$= C + \sum_{i=1}^{m} H_i(s)$$

$$\text{where, } H_i(s) = \frac{c_{0i} + c_{1i}\dfrac{1}{s}}{d_{0i} + d_{1i}\dfrac{1}{s} + d_{2i}\dfrac{1}{s^2}} \qquad \boxed{\text{Second order section.}}$$

$$\text{or} \qquad H_i(s) = \frac{c_{0i}}{d_{0i} + d_{1i}\dfrac{1}{s}} \qquad \boxed{\text{First order section.}}$$

The individual first and second order sections can be realized either in direct form-I or direct form-II structure. The overall system is obtained by connecting individual sections in parallel as shown in Fig 3.16.

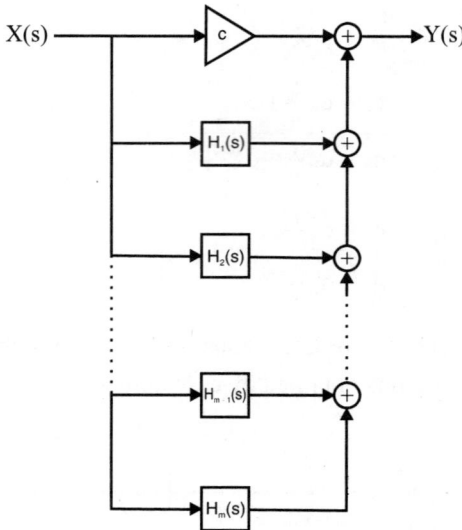

***Fig 3.16:** Parallel structure of continuous time system.*

Example 3.29
(AU Jun'13, 8 Marks)

Find the direct form-I structure of the continuous time system represented by the equation,

$$\frac{d^2y(t)}{dt^2} + 5\frac{dy(t)}{dt} + 4y(t) = \frac{dx(t)}{dt}$$

Solution:

Given that, $\dfrac{d^2y(t)}{dt^2} + 5\dfrac{dy(t)}{dt} + 4y(t) = \dfrac{dx(t)}{dt}$

On taking Laplace transform of the above equation we get,

$$s^2 Y(s) + 5s\, Y(s) + 4\, Y(s) = s\, X(s)$$

On dividing throughout by s^2 we get,

$$Y(s) + 5\,\frac{Y(s)}{s} + 4\,\frac{Y(s)}{s^2} = \frac{X(s)}{s}$$

$$\therefore\ Y(s) = -5\,\frac{Y(s)}{s} - 4\,\frac{Y(s)}{s^2} + \frac{X(s)}{s} \qquad \text{.....(1)}$$

The direct form-I structure of the given continuous time system is realized using equation (1) as shown in Fig 1.

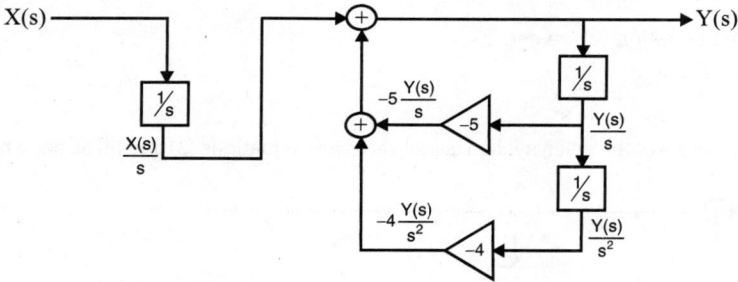

Fig 1: Direct form-I structure.

Example 3.30
(AU Jun'11, 8 Marks)

Find the direct form-II structure of the continuous time system represented by the equation,

$$\frac{d^3y(t)}{dt^3} + 4\frac{d^2y(t)}{dt^2} + 7\frac{dy(t)}{dt} + 8\,y(t) = 5\frac{d^2x(t)}{dt^2} + 4\frac{dx(t)}{dt} + 7\,x(t)$$

Solution:

Given that, $\dfrac{d^3y(t)}{dt^3} + 4\dfrac{d^2y(t)}{dt^2} + 7\dfrac{dy(t)}{dt} + 8\,y(t) = 5\dfrac{d^2x(t)}{dt^2} + 4\dfrac{dx(t)}{dt} + 7\,x(t)$

On taking Laplace transform of the above equation we get,

$$s^3\, Y(s) + 4s^2\, Y(s) + 7s\, Y(s) + 8\, Y(s) = 5s^2\, X(s) + 4s\, X(s) + 7\, X(s)$$

On dividing throughout by s^3 we get,

$$Y(s) + 4\,\frac{Y(s)}{s} + 7\,\frac{Y(s)}{s^2} + 8\,\frac{Y(s)}{s^3} = 5\,\frac{X(s)}{s} + 4\,\frac{X(s)}{s^2} + 7\,\frac{X(s)}{s^3}$$

$$\therefore\ Y(s)\left(1+\frac{4}{s}+\frac{7}{s^2}+\frac{8}{s^3}\right)\ =\ X(s)\left(\frac{5}{s}+\frac{4}{s^2}+\frac{7}{s^3}\right)$$

$$\therefore\ \frac{Y(s)}{X(s)}\ =\ \frac{\left(\dfrac{5}{s}+\dfrac{4}{s^2}+\dfrac{7}{s^3}\right)}{\left(1+\dfrac{4}{s}+\dfrac{7}{s^2}+\dfrac{8}{s^3}\right)}$$

Let, $\dfrac{Y(s)}{X(s)}\ =\ \dfrac{W(s)}{X(s)}\dfrac{Y(s)}{W(s)}$

where, $\dfrac{W(s)}{X(s)}\ =\ \dfrac{1}{\left(1+\dfrac{4}{s}+\dfrac{7}{s^2}+\dfrac{8}{s^3}\right)}$ .....(1)

$$\frac{Y(s)}{W(s)}\ =\ \left(\frac{5}{s}+\frac{4}{s^2}+\frac{7}{s^3}\right) \qquad\qquad .....(2)$$

On cross multiplying equation (1) we get,

$$W(s)+4\frac{W(s)}{s}+7\frac{W(s)}{s^2}+8\frac{W(s)}{s^3}\ =\ X(s)$$

$$\therefore\ W(s)\ =-4\frac{W(s)}{s}-7\frac{W(s)}{s^2}W(s)-8\frac{W(s)}{s^3}+X(s) \qquad .....(3)$$

On cross multiplying equation (2) we get,

$$Y(s)\ =\ 5\frac{W(s)}{s}+4\frac{W(s)}{s^2}+7\frac{W(s)}{s^3} \qquad\qquad .....(4)$$

The direct form-II structure of the given system is realized using equations (3) and (4) as shown in Fig 1.

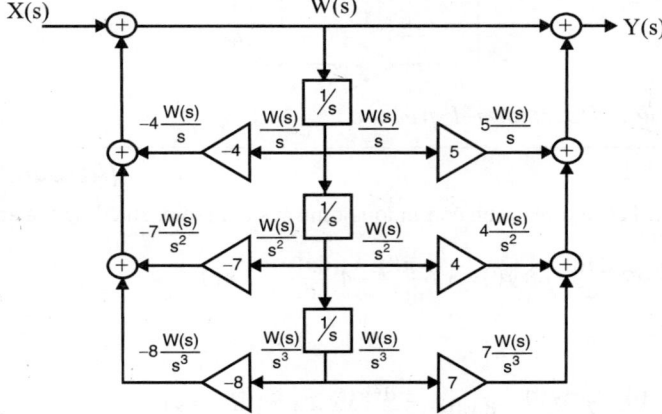

Fig 1: *Direct form–II structure.*

Example 3.31

Find the direct form-I and direct form-II structures of the continuous time system represented by the equation,

$$\frac{d^2y(t)}{dt^2}+0.6\frac{dy(t)}{dt}+0.7\,y(t)\ =\ \frac{d^2x(t)}{dt^2}+0.5\frac{dx(t)}{dt}+0.4\,x(t)$$

Solution:

Direct Form-I Structure

Given that, $\dfrac{d^2y(t)}{dt^2} + 0.6\dfrac{dy(t)}{dt} + 0.7\,y(t) = \dfrac{d^2x(t)}{dt^2} + 0.5\dfrac{dx(t)}{dt} + 0.4\,x(t)$

On taking Laplace transform of the above equation we get,

$$s^2\,Y(s) + 0.6s\,Y(s) + 0.7\,Y(s) = s^2\,X(s) + 0.5s\,X(s) + 0.4X(s)$$

On dividiing throughout by s^2 we get,

$$Y(s) + 0.6\dfrac{Y(s)}{s} + 0.7\dfrac{Y(s)}{s^2} = X(s) + 0.5\dfrac{X(s)}{s} + 0.4\dfrac{X(s)}{s^2} \qquad \dots(1)$$

$$\therefore\ Y(s) = -0.6\dfrac{Y(s)}{s} - 0.7\dfrac{Y(s)}{s^2} + X(s) + 0.5\dfrac{X(s)}{s} + 0.4\dfrac{X(s)}{s^2} \qquad \dots(2)$$

The direct form-I structure of the given continuous time system is realized using equation (2) as shown in Fig 1.

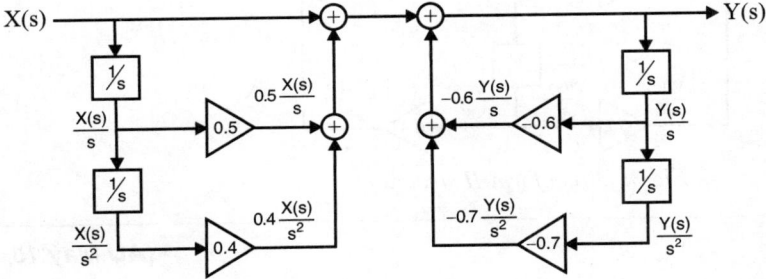

Fig 1: Direct form-I structure.

Direct Form-II Structure

From equation (1) we get,

$$Y(s)\left(1 + 0.6\,\dfrac{1}{s} + 0.7\,\dfrac{1}{s^2}\right) = X(s)\left(1 + 0.5\,\dfrac{1}{s} + 0.4\,\dfrac{1}{s^2}\right)$$

$$\therefore\ \dfrac{Y(s)}{X(s)} = \dfrac{1 + 0.5\,\dfrac{1}{s} + 0.4\,\dfrac{1}{s^2}}{1 + 0.6\,\dfrac{1}{s} + 0.7\,\dfrac{1}{s^2}}$$

Let, $\dfrac{Y(s)}{X(s)} = \dfrac{W(s)}{X(s)}\dfrac{Y(s)}{W(s)}$

where, $\dfrac{W(s)}{X(s)} = \dfrac{1}{1 + 0.6\,\dfrac{1}{s} + 0.7\,\dfrac{1}{s^2}}$ $\qquad \dots(3)$

$$\dfrac{Y(s)}{W(s)} = 1 + 0.5\,\dfrac{1}{s} + 0.4\,\dfrac{1}{s^2} \qquad \dots(4)$$

On cross multiplying equation (3) we get,

$$W(s) + 0.6\frac{W(s)}{s} + 0.7\frac{W(s)}{s^2} = X(s)$$

$$\therefore\ W(s) = -0.6\frac{W(s)}{s} - 0.7\frac{W(s)}{s^2} + X(s) \qquad\qquad .....(5)$$

On cross multiplying equation (4) we get,

$$Y(s) = W(s) + 0.5\frac{W(s)}{s} + 0.4\frac{W(s)}{s^2} \qquad\qquad .....(6)$$

The direct form-II structure of the given system is realized using equations (5) and (6) as shown in Fig 2.

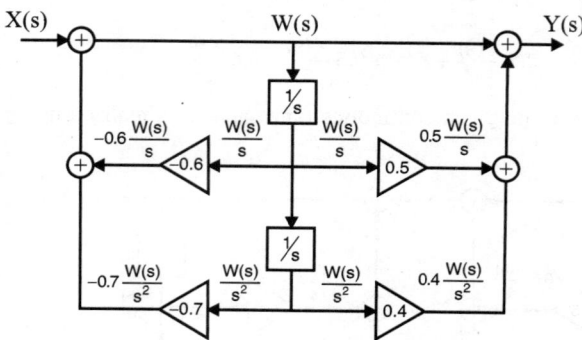

Fig 2: *Direct form-II structure.*

Example 3.32

(AU May'15, 6 Marks)

Find the direct form-I, and direct form-II structures of the continuous time system represented by the equation,

$$H(s) = \frac{4s + 28}{s^2 + 6s + 5}$$

Solution:

Direct Form-I Structure

Given that, $H(s) = \dfrac{4s + 28}{s^2 + 6s + 5}$

Let, $H(s) = \dfrac{Y(s)}{X(s)}$; where Y(s) = Output in s-domain and X(s) = Input in s-domain.

$$\therefore\ \frac{Y(s)}{X(s)} = \frac{4s + 28}{s^2 + 6s + 5} = \frac{s^2\left(\dfrac{4}{s} + \dfrac{28}{s}\right)}{s^2\left(1 + \dfrac{6}{s} + \dfrac{5}{s^2}\right)}$$

$$\therefore\ \frac{Y(s)}{X(s)} = \frac{\dfrac{4}{s} + \dfrac{28}{s^2}}{1 + \dfrac{6}{s}\ \dfrac{5}{s^2}} \qquad\qquad .....(1)$$

On cross multiplying equation (1) we get,

$$Y(s)\left(1+\frac{6}{s}+\frac{5}{s^2}\right) = X(s)\left(\frac{4}{s}+\frac{28}{s^2}\right)$$

$$Y(s) + 6\frac{Y(s)}{s} + 5\frac{Y(s)}{s^2} = 4\frac{X(s)}{s} + 28\frac{X(s)}{s^2}$$

$$\therefore Y(s) = 4\frac{X(s)}{s} + 28\frac{X(s)}{s^2} - 6\frac{Y(s)}{s} - 5\frac{Y(s)}{s^2} \qquad \dots\dots(2)$$

The direct form-I structure can be obtained from equation (2) as shown in Fig 1.

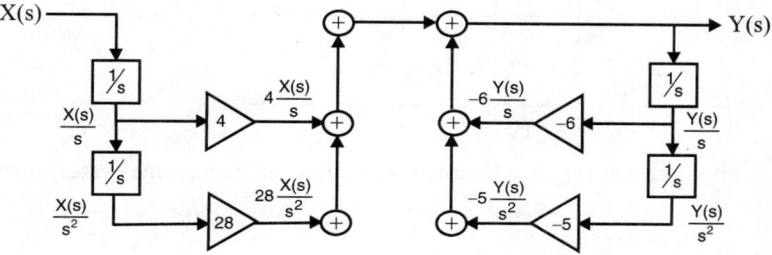

Fig 1: *Direct form–I structure.*

Direct Form-II Structure

Form equation (1) we get,

$$\frac{Y(s)}{X(s)} = \frac{\dfrac{4}{s}+\dfrac{28}{s^2}}{1+\dfrac{6}{s}+\dfrac{5}{s^2}}$$

Let, $\quad \dfrac{Y(s)}{X(s)} = \dfrac{W(s)}{X(s)}\dfrac{Y(s)}{W(s)}$

$$\text{where,} \quad \frac{W(s)}{X(s)} = \frac{1}{1+\dfrac{6}{s}+\dfrac{5}{s^2}} \qquad \dots\dots(3)$$

$$\frac{Y(s)}{W(s)} = \frac{4}{s} + \frac{28}{s^2} \qquad \dots\dots(4)$$

On cross multiplying equation (3) we get,

$$W(s) + 6\frac{W(s)}{s} + 5\frac{W(s)}{s^2} = X(s)$$

$$\therefore W(s) = X(s) - 6\frac{W(s)}{s} - 5\frac{W(s)}{s^2} \qquad \dots\dots(5)$$

On cross multiplying equation (4) we get,

$$Y(s) = 4\frac{W(s)}{s} + 28\frac{W(s)}{s^2} \qquad \dots\dots(6)$$

The equation (5) and (6) can be realized by a direct form-II structure as shown in Fig 2.

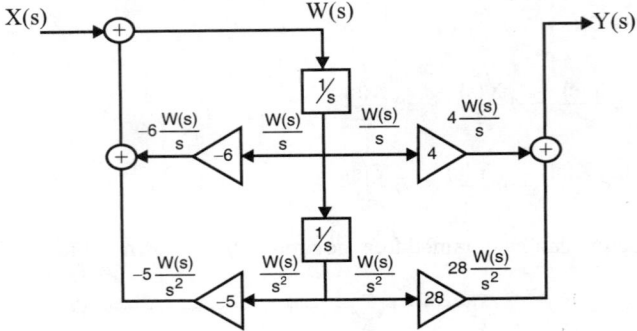

Fig 2: *Direct form–II structure.*

Example 3.33

Find the direct form-I and direct form-II structures of the continuous time system represented by transfer function,

$$H(s) = \frac{8s^3 - 4s^2 + 11s - 2}{\left(s - \frac{1}{4}\right)\left(s^2 - s + \frac{1}{2}\right)}$$

Solution:

Direct Form-I Structure

Given that, $H(s) = \dfrac{8s^3 - 4s^2 + 11s - 2}{\left(s - \frac{1}{4}\right)\left(s^2 - s + \frac{1}{2}\right)}$

Let, $H(s) = \dfrac{Y(s)}{X(s)}$; where Y(s) = Output in s-domain and X(s) = Input in s-domain.

$$\therefore \frac{Y(s)}{X(s)} = \frac{8s^3 - 4s^2 + 11s - 2}{\left(s - \frac{1}{4}\right)\left(s^2 - s + \frac{1}{2}\right)} = \frac{8s^3 - 4s^2 + 11s - 2}{s^3 - s^2 + \frac{1}{2}s - \frac{1}{4}s^2 + \frac{1}{4}s - \frac{1}{8}}$$

$$= \frac{8s^3 - 4s^2 + 11s - 2}{s^3 - \frac{5}{4}s^2 + \frac{3}{4}s - \frac{1}{8}} = \frac{s^3\left(8 - 4\frac{1}{s} + 11\frac{1}{s^2} - 2\frac{1}{s^3}\right)}{s^3\left(1 - \frac{5}{4}\frac{1}{s} + \frac{3}{4}\frac{1}{s^2} - \frac{1}{8}\frac{1}{s^3}\right)}$$

$$\therefore \frac{Y(s)}{X(s)} = \frac{8 - 4\frac{1}{s} + 11\frac{1}{s^2} - 2\frac{1}{s^3}}{1 - \frac{5}{4}\frac{1}{s} + \frac{3}{4}\frac{1}{s^2} - \frac{1}{8}\frac{1}{s^3}} \qquad \qquad .....(1)$$

On cross multiplying equation (1) we get,

$$Y(s) - \frac{5}{4}\frac{Y(s)}{s} + \frac{3}{4}\frac{Y(s)}{s^2} - \frac{1}{8}\frac{Y(s)}{s^3} = 8X(s) - 4\frac{X(s)}{s} + 11\frac{X(s)}{s^2} - 2\frac{X(s)}{s^3}$$

$$\therefore Y(s) = 8X(s) - 4\frac{X(s)}{s} + 11\frac{X(s)}{s^2} - 2\frac{X(s)}{s^3} + \frac{5}{4}\frac{Y(s)}{s} - \frac{3}{4}\frac{Y(s)}{s^2} + \frac{1}{8}\frac{Y(s)}{s^3} \qquad .....(2)$$

The direct form-I structure can be obtained from equation (2) as shown in Fig 1.

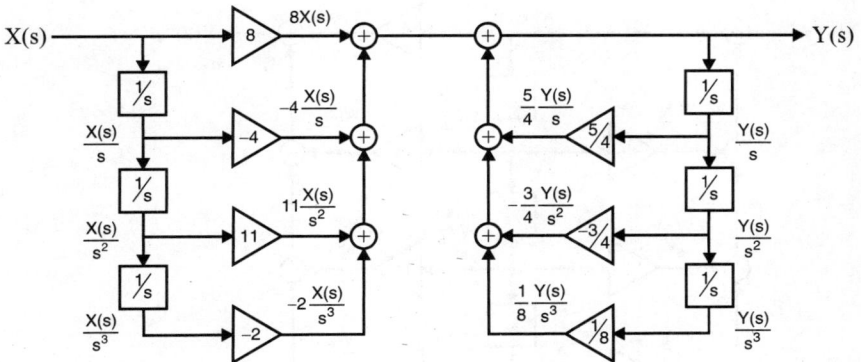

Fig 1: *Direct form-I structure.*

Direct Form-II Structure

From equation (1) we get,

$$\frac{Y(s)}{X(s)} = \frac{8 - 4\dfrac{1}{s} + 11\dfrac{1}{s^2} - 2\dfrac{1}{s^3}}{1 - \dfrac{5}{4}\dfrac{1}{s} + \dfrac{3}{4}\dfrac{1}{s^2} - \dfrac{1}{8}\dfrac{1}{s^3}}$$

Let, $\dfrac{Y(s)}{X(s)} = \dfrac{W(s)}{X(s)} \dfrac{Y(s)}{W(s)}$

where, $\dfrac{W(s)}{X(s)} = \dfrac{1}{1 - \dfrac{5}{4}\dfrac{1}{s} + \dfrac{3}{4}\dfrac{1}{s^2} - \dfrac{1}{8}\dfrac{1}{s^3}}$ .....(3)

$$\frac{Y(s)}{W(s)} = 8 - 4\frac{1}{s} + 11\frac{1}{s^2} - 2\frac{1}{s^3}$$ .....(4)

On cross multiplying equation (3) we get,

$$W(s) - \frac{5}{4}\frac{W(s)}{s} + \frac{3}{4}\frac{W(s)}{s^2} - \frac{1}{8}\frac{W(s)}{s^3} = X(s)$$

$$\therefore W(s) = X(s) + \frac{5}{4}\frac{W(s)}{s} - \frac{3}{4}\frac{W(s)}{s^2} + \frac{1}{8}\frac{W(s)}{s^3}$$ .....(5)

On cross multiplying equation (4) we get,

$$Y(s) = 8W(s) - 4\frac{W(s)}{s} + 11\frac{W(s)}{s^2} - 2\frac{W(s)}{s^3}$$ .....(6)

The equations (5) and (6) can be realized by a direct form-II structure as shown in Fig 2.

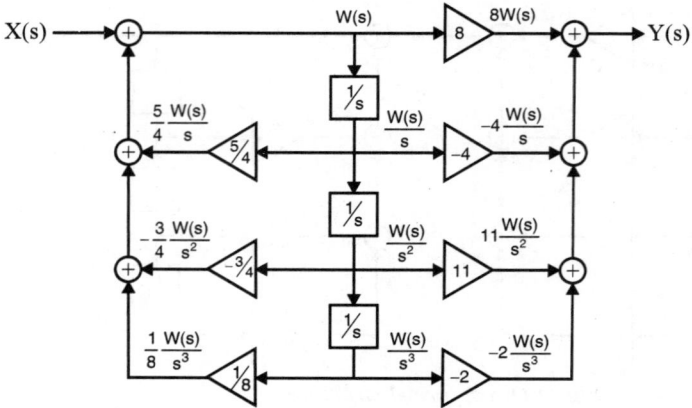

Fig 2: *Direct form-II structure.*

Example 3.34

Obtain the direct form-I, direct form-II, cascade and parallel structure of the continuous time LTI system governed by the equation,

$$\frac{d^3y(t)}{dt^3} + \frac{3}{8}\frac{d^2y(t)}{dt^2} - \frac{3}{32}\frac{dy(t)}{dt} - \frac{1}{64}y(t) = 3\frac{d^3x(t)}{dt^3} + 2\frac{d^2x(t)}{dt^2}$$

Solution:

Direct Form-I Structure

Given that, $\frac{d^3y(t)}{dt^3} + \frac{3}{8}\frac{d^2y(t)}{dt^2} - \frac{3}{32}\frac{dy(t)}{dt} - \frac{1}{64}y(t) = 3\frac{d^3x(t)}{dt^3} + 2\frac{d^2x(t)}{dt^2}$

On taking Laplace transform of the above equation we get,

$$s^3Y(s) + \frac{3}{8}s^2Y(s) - \frac{3}{32}sY(s) - \frac{1}{64}Y(s) = 3s^3X(s) + 2s^2X(s) \qquad .....(1)$$

On dividing equation (1) throughout by s^3 we get,

$$Y(s) + \frac{3}{8}\frac{Y(s)}{s} - \frac{3}{32}\frac{Y(s)}{s^2} - \frac{1}{64}\frac{Y(s)}{s^3} = 3X(s) + 2\frac{X(s)}{s} \qquad .....(2)$$

$$\therefore \ Y(s) = -\frac{3}{8}\frac{Y(s)}{s} + \frac{3}{32}\frac{Y(s)}{s^2} + \frac{1}{64}\frac{Y(s)}{s^3} + 3X(s) + 2\frac{X(s)}{s} \qquad .....(3)$$

The direct form-I structure of the given continuous time system is realized using equation (3) as sshown in Fig 1.

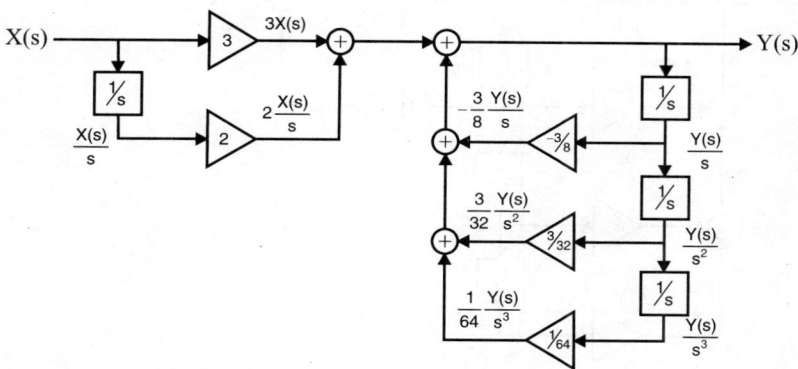

Fig 1: *Direct form-I structure.*

Direct Form-II Structure

From equation (2) we get,

$$Y(s)\left(1+\frac{3}{8}\frac{1}{s}-\frac{3}{32}\frac{1}{s^2}-\frac{1}{64}\frac{1}{s^3}\right) = X(s)\left(3+2\frac{1}{s}\right)$$

$$\therefore \frac{Y(s)}{X(s)} = \frac{3+2\frac{1}{s}}{1+\frac{3}{8}\frac{1}{s}-\frac{3}{32}\frac{1}{s^2}-\frac{1}{64}\frac{1}{s^3}}$$

Let, $\dfrac{Y(s)}{X(s)} = \dfrac{W(s)}{X(s)}\dfrac{Y(s)}{W(s)}$

where, $\dfrac{W(s)}{X(s)} = \dfrac{1}{1+\dfrac{3}{8}\dfrac{1}{s}-\dfrac{3}{32}\dfrac{1}{s^2}-\dfrac{1}{64}\dfrac{1}{s^3}}$ .....(4)

$$\frac{Y(s)}{W(s)} = 3+2\frac{1}{s}$$ .....(5)

On cross multiplying equation (4) we get,

$$W(s)+\frac{3}{8}\frac{W(s)}{s}-\frac{3}{32}\frac{W(s)}{s^2}-\frac{1}{64}\frac{W(s)}{s^3} = X(s)$$

$$\therefore W(s) = X(s)-\frac{3}{8}\frac{W(s)}{s}+\frac{3}{32}\frac{W(s)}{s^2}+\frac{1}{64}\frac{W(s)}{s^3}$$ .....(6)

On cross multiplying equation (5) we get,

$$Y(s) = 3W(s)+2\frac{W(s)}{s}$$ .....(7)

The equations (6) and (7) can be realized by a direct form-II structure as shown in Fig 2.

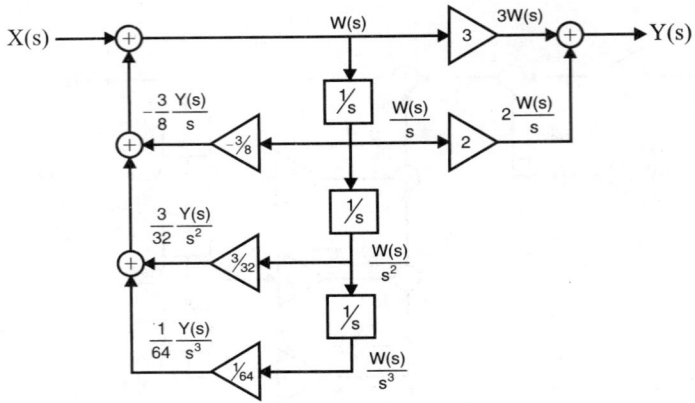

Fig 2: *Direct form-II structure.*

Cascade Structure

From equation (1) we get,

$$Y(s)\left(s^3 + \frac{3}{8}s^2 - \frac{3}{32}s - \frac{1}{64}\right) = X(s)\left(3s^3 + 2s^2\right)$$

$$\therefore \frac{Y(s)}{X(s)} = \frac{3s^3 + 2s^2}{s^3 + \frac{3}{8}s^2 - \frac{3}{32}s - \frac{1}{64}} \qquad \qquad .....(8)$$

For cascade realization, the numerator and denominator polynomials of equation (8) should be expressed in the factorized form, as shown below.

$$\therefore \frac{Y(s)}{X(s)} = \frac{3s^2\left(s + \frac{2}{3}\right)}{\left(s + \frac{1}{8}\right)\left(s^2 + \frac{2}{8}s - \frac{8}{64}\right)}$$

$$= \frac{3s^2\left(s + \frac{2}{3}\right)}{\left(s + \frac{1}{8}\right)\left(s^2 + \frac{1}{4}s - \frac{1}{8}\right)}$$

$$= \frac{3s^2\left(s + \frac{2}{3}\right)}{\left(s + \frac{1}{8}\right)\left(s + \frac{1}{2}\right)\left(s - \frac{1}{4}\right)} \qquad \qquad .....(9)$$

The roots of quadratic

$$s^2 + \frac{1}{4}s - \frac{1}{8} = 0$$

$$s = \frac{-\frac{1}{4} \pm \sqrt{\left(\frac{1}{4}\right)^2 + \frac{4}{8}}}{2}$$

$$= \frac{-\frac{1}{4} \pm \frac{3}{4}}{2} = \frac{1}{4}, -\frac{1}{2}$$

s = –1/8 is one of the root of denominator polynomial of equation (8).

–1/8	1	3/8	–3/32	–1/64
	↓	–1/8	–2/64	+1/64
	1	2/8	– 8/64	0

Since there are three first order factors in the denominator of equation (9), H(s) can be expressed as a product of three sections as shown in equation (10).

Let, $\dfrac{Y(s)}{X(s)} = H(s) = \dfrac{s}{s + \frac{1}{8}} \times \dfrac{s}{s + \frac{1}{2}} \times \dfrac{3s + 2}{s - \frac{1}{4}} = H_1(s) \times H_2(s) \times H_3(s)$

$$.....(10)$$

where, $H_1(s) = \dfrac{s}{s + \frac{1}{8}}$; $H_2(s) = \dfrac{s}{s + \frac{1}{2}}$ and $H_3(s) = \dfrac{3s + 2}{s - \frac{1}{4}}$

The transfer function $H_1(s)$ can be realized in direct form-II structure as shown below:

Let, $H_1(s) = \dfrac{Y_1(s)}{X(s)} = \dfrac{s}{s+\dfrac{1}{8}} = \dfrac{1}{1+\dfrac{1}{8}\dfrac{1}{s}}$

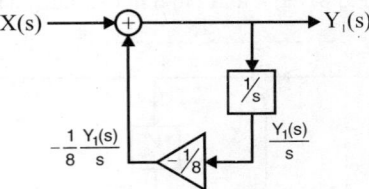

On cross multiplying the above function we get,

$$Y_1(s) + \dfrac{1}{8}\dfrac{Y_1(s)}{s} = X(s)$$

$$\therefore \; Y_1(s) = X(s) - \dfrac{1}{8}\dfrac{Y_1(s)}{s}$$

Fig 3: *Direct form-II structure of $H_1(s)$.*

The direct form-II structure of $H_1(s)$ is obtained using the above equation as shown in Fig 3.

The transfer function $H_2(s)$ can be realized in direct form-II structure as shown below:

Let, $H_2(s) = \dfrac{Y_2(s)}{Y_1(s)} = \dfrac{s}{s+\dfrac{1}{2}} = \dfrac{1}{1+\dfrac{1}{2}\dfrac{1}{s}}$

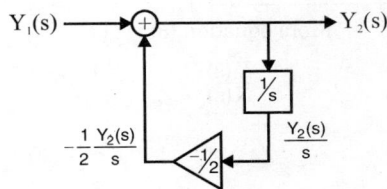

On cross multiplying the above function we get,

$$Y_2(s) + \dfrac{1}{2}\dfrac{Y_2(s)}{s} = Y_1(s)$$

$$\therefore \; Y_2(s) = Y_1(s) - \dfrac{1}{2}\dfrac{Y_2(s)}{s}$$

Fig 4: *Direct form-II structure of $H_2(s)$.*

The direct form-II structure of $H_2(s)$ is obtained using the above equation as shown in Fig 4.

The transfer function $H_3(s)$ can be realized in direct form-II structure as shown below:

$$H_3(s) = \dfrac{3s+2}{s-\dfrac{1}{4}} = \dfrac{3+2\dfrac{1}{s}}{1-\dfrac{1}{4}\dfrac{1}{s}}$$

Let, $H_3(s) = \dfrac{Y(s)}{Y_2(s)} = \dfrac{W(s)}{Y_2(s)}\dfrac{Y(s)}{W(s)} = \dfrac{3+2\dfrac{1}{s}}{1-\dfrac{1}{4}\dfrac{1}{s}}$

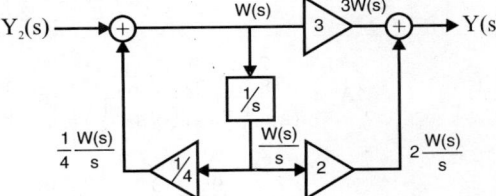

where, $\dfrac{W(s)}{Y_2(s)} = \dfrac{1}{1-\dfrac{1}{4}\dfrac{1}{s}}$

$$\dfrac{Y(s)}{W(s)} = 3+2\dfrac{1}{s}$$

Fig 5: *Direct form-II structure of $H_3(s)$.*

On cross multiplying the above functions we get,

$$W(s) = Y_2(s) + \dfrac{1}{4}\dfrac{W(s)}{s} \quad \text{and} \quad Y(s) = 3\,W(s) + 2\dfrac{W(s)}{s}$$

The direct form-II structure of $H_3(s)$ is obtained using the above equations as shown in Fig 5.

The cascade structure of the given system is obtained by connecting the individual sections shown in Fig 3, Fig 4 and Fig 5 in cascade as shown in Fig 6.

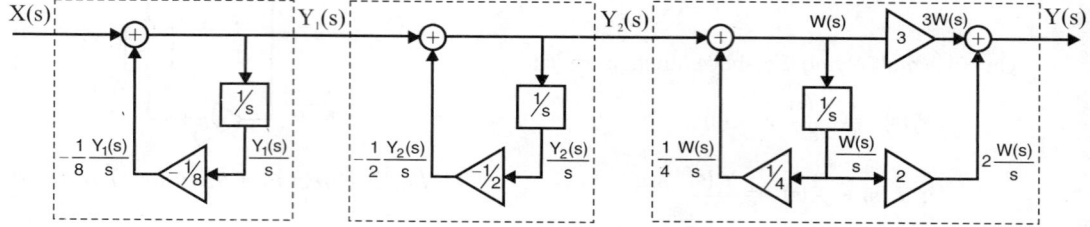

Fig 6: *Cascade structure of the system.*

Parallel Structure

From equation (8) we get,

$$\frac{Y(s)}{X(s)} = \frac{3s^3 + 2s^2}{s^3 + \frac{3}{8}s^2 - \frac{3}{32}s - \frac{1}{64}}$$

$$= 3 + \frac{\frac{7}{8}s^2 + \frac{9}{32}s + \frac{3}{64}}{s^3 + \frac{3}{8}s^2 - \frac{3}{32}s - \frac{1}{64}}$$

$$= 3 + \frac{\frac{7}{8}s^2 + \frac{9}{32}s + \frac{3}{64}}{\left(s + \frac{1}{8}\right)\left(s + \frac{1}{2}\right)\left(s - \frac{1}{4}\right)}$$

> Dividing numerator by denominator
>
> $$s^3 + \frac{3}{8}s^2 - \frac{3}{32}s - \frac{1}{64} \overline{\left.\begin{array}{r} 3 \\ \hline 3s^3 + 2s^2 \\ 3s^3 + \frac{9}{8}s^2 - \frac{9}{32}s - \frac{3}{64} \\ (-)\ (-)\ (+)\ (+) \\ \hline \frac{7}{8}s^2 + \frac{9}{32}s + \frac{3}{64} \end{array}\right.}$$

Using equation (9).

By partial fraction expansion technique the above function can be expressed as shown below:

$$\frac{Y(s)}{X(s)} = H(s) = 3 + \frac{\frac{7}{8}s^2 + \frac{9}{32}s + \frac{3}{64}}{\left(s + \frac{1}{8}\right)\left(s + \frac{1}{2}\right)\left(s - \frac{1}{4}\right)} = 3 + \frac{A}{s + \frac{1}{8}} + \frac{B}{s + \frac{1}{2}} + \frac{C}{s - \frac{1}{4}}$$

$$A = \frac{\frac{7}{8}s^2 + \frac{9}{32}s + \frac{3}{64}}{\left(s + \frac{1}{8}\right)\left(s + \frac{1}{2}\right)\left(s - \frac{1}{4}\right)} \times \left(s + \frac{1}{8}\right)\Bigg|_{s = -\frac{1}{8}} = \frac{\frac{7}{8}\left(-\frac{1}{8}\right)^2 + \frac{9}{32}\left(-\frac{1}{8}\right) + \frac{3}{64}}{\left(-\frac{1}{8} + \frac{1}{2}\right)\left(-\frac{1}{8} - \frac{1}{4}\right)}$$

$$= \frac{\frac{7}{512} - \frac{9}{256} + \frac{3}{64}}{\left(\frac{3}{8}\right)\left(-\frac{3}{8}\right)} = \frac{\frac{7 - 18 + 24}{512}}{-\frac{9}{64}} = \frac{\frac{13}{512}}{\frac{9}{64}} = \frac{13}{512} \times \left(-\frac{64}{9}\right) = -\frac{13}{72}$$

$$B = \frac{\frac{7}{8}s^2 + \frac{9}{32}s + \frac{3}{64}}{\left(s + \frac{1}{8}\right)\left(s + \frac{1}{2}\right)\left(s - \frac{1}{4}\right)} \times \left(s + \frac{1}{2}\right)\Bigg|_{s = -\frac{1}{2}} = \frac{\frac{7}{8}\left(-\frac{1}{2}\right)^2 + \frac{9}{32}\left(-\frac{1}{2}\right) + \frac{3}{64}}{\left(-\frac{1}{2} + \frac{1}{8}\right)\left(-\frac{1}{2} - \frac{1}{4}\right)}$$

$$= \frac{\frac{7}{32} - \frac{9}{64} + \frac{3}{64}}{\left(-\frac{3}{8}\right)\left(-\frac{3}{4}\right)} = \frac{\frac{14 - 9 + 3}{64}}{\frac{9}{32}} = \frac{\frac{8}{64}}{\frac{9}{32}} = \frac{8}{64} \times \frac{32}{9} = \frac{4}{9}$$

$$C = \frac{\frac{7}{8}s^2 + \frac{9}{32}s + \frac{3}{64}}{\left(s + \frac{1}{8}\right)\left(s + \frac{1}{2}\right)\left(s - \frac{1}{4}\right)} \times \left(s - \frac{1}{4}\right)\Bigg|_{s=\frac{1}{4}} = \frac{\frac{7}{8}\left(\frac{1}{4}\right)^2 + \frac{9}{32}\left(\frac{1}{4}\right) + \frac{3}{64}}{\left(\frac{1}{4} + \frac{1}{8}\right)\left(\frac{1}{4} + \frac{1}{2}\right)}$$

$$= \frac{\frac{7}{128} + \frac{9}{128} + \frac{3}{64}}{\left(\frac{3}{8}\right)\left(\frac{3}{4}\right)} = \frac{\frac{7+9+6}{128}}{\frac{9}{32}} = \frac{\frac{22}{128}}{\frac{9}{32}} = \frac{22}{128} \times \frac{32}{9} = \frac{11}{18}$$

$$\therefore \ H(s) = 3 + \frac{A}{s+\frac{1}{8}} + \frac{B}{s+\frac{1}{2}} + \frac{C}{s-\frac{1}{4}} = 3 + \frac{-\frac{13}{72}}{s+\frac{1}{8}} + \frac{\frac{4}{9}}{s+\frac{1}{2}} + \frac{\frac{11}{18}}{s-\frac{1}{4}}$$

Let, $H(s) = 3 + H_1(s) + H_2(s) + H_3(s)$

where, $H(s) = \dfrac{Y(s)}{X(s)}$; $H_1(s) = \dfrac{-\frac{13}{72}}{s+\frac{1}{8}}$; $H_2(s) = \dfrac{\frac{4}{9}}{s+\frac{1}{2}}$; $H_3(s) = \dfrac{\frac{11}{18}}{s-\frac{1}{4}}$

Now, $H(s) = \dfrac{Y(s)}{X(s)} = 3 + H_1(s) + H_2(s) + H_3(s)$

$$\therefore \ Y(s) = 3X(s) + H_1(s)X(s) + H_2(s)X(s) + H_3(s)X(s)$$

Let, $Y(s) = 3X(s) + Y_1(s) + Y_2(s) + Y_3(s)$ .....(11)

where, $Y_1(s) = H_1(s)X(s) = \dfrac{-\frac{13}{72}}{s+\frac{1}{8}}X(s) \Rightarrow \dfrac{Y_1(s)}{X(s)} = \dfrac{-\frac{13}{72}\frac{1}{s}}{1+\frac{1}{8}\frac{1}{s}}$

$$Y_2(s) = H_2(s)X(s) = \dfrac{\frac{4}{9}}{s+\frac{1}{2}}X(s) \Rightarrow \dfrac{Y_2(s)}{X(s)} = \dfrac{\frac{4}{9}\frac{1}{s}}{1+\frac{1}{2}\frac{1}{s}}$$

$$Y_3(s) = H_3(s)X(s) = \dfrac{\frac{11}{18}}{s-\frac{1}{4}}X(s) \Rightarrow \dfrac{Y_3(s)}{X(s)} = \dfrac{\frac{11}{18}\frac{1}{s}}{1-\frac{1}{4}\frac{1}{s}}$$

In equation (11), the output Y(s) is expressed as sum of four components. The first component is simply the input multiplied by a constant, and so it is realised as a constant multiplier in the parallel structure shown in Fig 10. The other three components involve first order section of transfer function, and so they are realized by direct form-II structures as shown below:

The transfer function $H_1(s)$ can be realized in direct form-II structure as shown below:

Let, $\dfrac{Y_1(s)}{X(s)} = \dfrac{W_1(s)}{X(s)}\dfrac{Y_1(s)}{W_1(s)} = \dfrac{-\frac{13}{72}\frac{1}{s}}{1+\frac{1}{8}\frac{1}{s}}$

where, $\dfrac{W_1(s)}{X(s)} = \dfrac{1}{1+\frac{1}{8}\frac{1}{s}}$ and $\dfrac{Y_1(s)}{W_1(s)} = -\frac{13}{72}\frac{1}{s}$ .....(12)

On cross multiplying and rearranging the transfer functions of equation (12) we get,

$$W_1(s) = X(s) - \frac{1}{8} \frac{W_1(s)}{s} \text{ and } Y_1(s) = -\frac{13}{72} \frac{W_1(s)}{s}$$

The direct form-II structure of $H_1(s)$ is obtained using the above equations as shown in Fig 7.

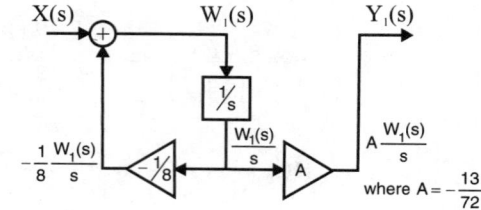

The transfer function $H_2(s)$ can be realized in direct form-II structure as shown below:

Let, $\dfrac{Y_2(s)}{X(s)} = \dfrac{W_2(s)}{X(s)} \dfrac{Y_2(s)}{W_2(s)} = \dfrac{\frac{4}{9}\frac{1}{s}}{1 + \frac{1}{2}\frac{1}{s}}$

Fig 7: *Direct form-II structure of $H_1(s)$.*

where, $\dfrac{W_2(s)}{X(s)} = \dfrac{1}{1 + \frac{1}{2}\frac{1}{s}}$ and $\dfrac{Y_2(s)}{W_2(s)} = \dfrac{4}{9}\dfrac{1}{s}$

On cross multiplying and rearranging the above functions we get,

$$W_2(s) = X(s) - \frac{1}{2} \frac{W_2(s)}{s} \text{ and } Y_2(s) = \frac{4}{9} \frac{W_2(s)}{s}$$

The direct form-II structure of $H_2(s)$ is obtained using the above equations as shown in Fig 8.

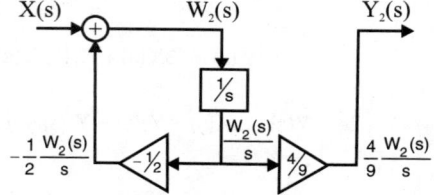

The transfer function $H_3(s)$ can be realized in direct form-II structure as shown below:

Let, $\dfrac{Y_3(s)}{X(s)} = \dfrac{W_3(s)}{X(s)} \dfrac{Y_3(s)}{W_3(s)} = \dfrac{\frac{11}{18}\frac{1}{s}}{1 - \frac{1}{4}\frac{1}{s}}$

Fig 8: *Direct form-II structure of $H_2(s)$.*

where, $\dfrac{W_3(s)}{X(s)} = \dfrac{1}{1 - \frac{1}{4}\frac{1}{s}}$ and $\dfrac{Y_3(s)}{W_3(s)} = \dfrac{11}{18}\dfrac{1}{s}$

On cross multiplying and rearranging the above functions we get,

$$W_3(s) = X(s) + \frac{1}{4} \frac{W_3(s)}{s} \text{ and } Y_3(s) = \frac{11}{18} \frac{W_3(s)}{s} \qquad \qquad .....(13)$$

The direct form-II structure of $H_3(s)$ is obtained using equation (13) as shown in Fig 9.

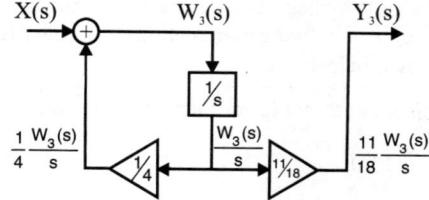

Fig 9: *Direct form-II structure of $H_3(s)$.*

The parallel structure of the given system is obtained by connecting the individual sections shown in Fig 7, Fig 8 and Fig 9 in parallel as shown in Fig 10.

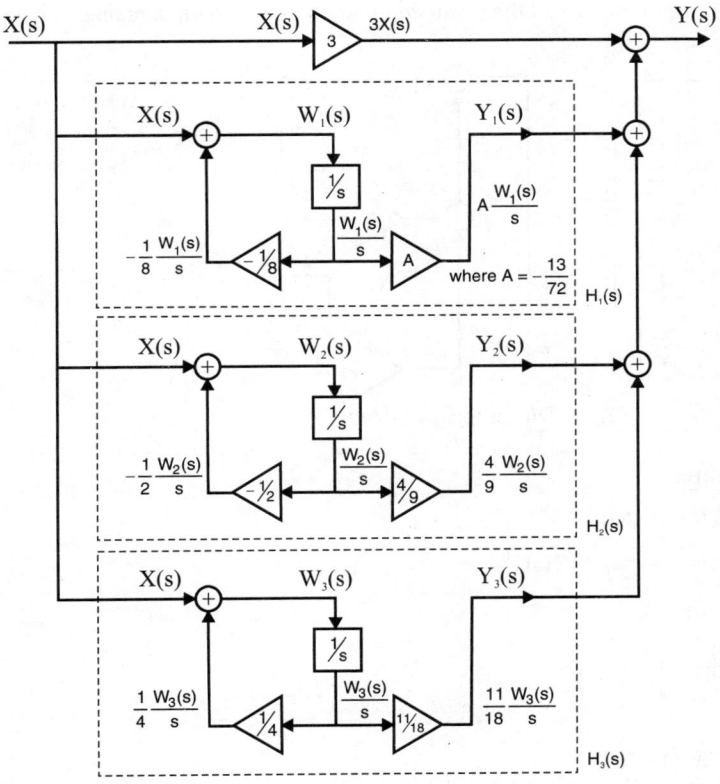

Fig 10: *Parallel structure of the system.*

Example 3.35   (AU Dec'13 & Dec'12, 8 Marks)

Obtain the direct form-I, direct form-II, cascade and parallel structure of the continuous time LTI system governed by the equation

$$\frac{d^2 y(t)}{dt} + 3\frac{dy(t)}{dt} + 2y(t) = x(t)$$

Solution:

Direct Form-I Structure

Given that, $\dfrac{d^2 y(t)}{dt} + 3\dfrac{dy(t)}{dt} + 2y(t) = x(t)$

On taking Laplace transform of the above equation we get,

$$s^2 Y(s) + 3\,sY(s) + 2Y(s) = X(s) \qquad\qquad .....(1)$$

On dividing equation (1) throughout by s^2 we get,

$$Y(s) + 3\frac{Y(s)}{s} + 2\frac{Y(s)}{s^2} = \frac{X(s)}{s^2} \qquad\qquad .....(2)$$

$$\therefore\ Y(s) = -3\frac{Y(s)}{s} - 2\frac{Y(s)}{s^2} + \frac{X(s)}{s^2} \qquad\qquad .....(3)$$

The direct form-I structure of the given continuous time system is realized using equation(3) as shown in Fig 1.

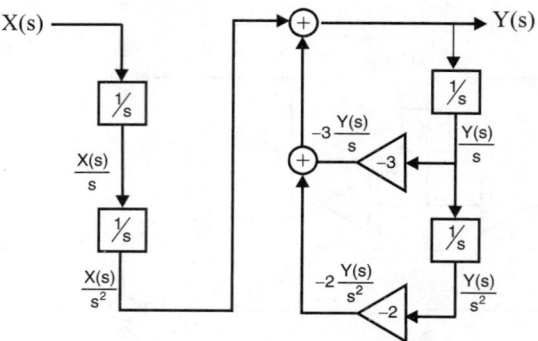

Fig 1: *Direct form-I structure.*

Direct Form-II Structure

From equation (2) we get,

$$Y(s)\left(1+\frac{3}{s}+\frac{2}{s^2}\right) = X(s)\frac{1}{s^2}$$

$$\therefore\ \frac{Y(s)}{X(s)} = \frac{\dfrac{1}{s^2}}{1+\dfrac{3}{s}+\dfrac{2}{s^2}}$$

Let, $\dfrac{Y(s)}{X(s)} = \dfrac{W(s)}{X(s)}\dfrac{Y(s)}{W(s)}$

where, $\dfrac{W(s)}{X(s)} = \dfrac{1}{1+\dfrac{3}{s}+\dfrac{2}{s^2}}$ .....(4)

$$\frac{Y(s)}{W(s)} = \frac{1}{s^2} \qquad\qquad .....(5)$$

On cross multiplying equation (4) we get,

$$W(s) + 3\frac{W(s)}{s} + 2\frac{W(s)}{s^2} = X(s)$$

$$\therefore W(s) = X(s) - 3\frac{W(s)}{s} - 2\frac{W(s)}{s^2} \qquad\qquad .....(6)$$

On cross multiplying equation (5) we get,

$$Y(s) = \frac{W(s)}{s^2} \qquad\qquad .....(7)$$

The equations (6) and (7) can be realized by a direct form-II structure as shown in Fig 2.

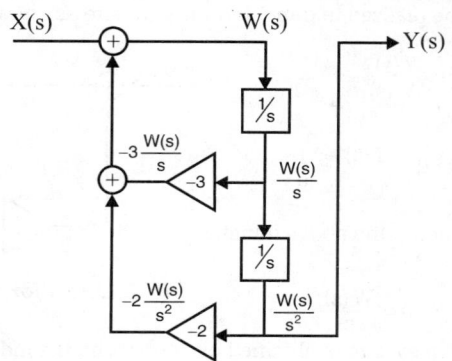

Fig 2: *Direct form-II structure.*

Cascade Structure

On rearranging equation (1) we get,

$$Y(s)(s^2 + 3s + 2) = X(s)$$

$$\therefore \frac{Y(s)}{X(s)} = \frac{1}{s^2 + 3s + 2} \qquad \qquad .....(8)$$

For cascade realization, the denominator polynomials of equation(8) should be expressed in the factorized form, as shown below:

$$\therefore \frac{Y(s)}{X(s)} = \frac{1}{(s^2 + 3s + 2)}$$

$$= \frac{1}{(s+2)(s+1)} \qquad \qquad .....(9)$$

The roots of quadratic
$s^2 + 3s + 3 = 0$
$s = \dfrac{-3 \pm \sqrt{3^2 - 4 \times 2}}{2}$
$= \dfrac{-3 \pm 1}{2} = -1, -2$

Since there are two first order factors in the dinominator of equation (9), H(s) can be expressed as a product of two sections as shown in equation (10).

$$\text{Let,} \quad \frac{Y(s)}{X(s)} = H(s) = \frac{1}{s+2} \times \frac{1}{s+1} = H_1(s) \times H_2(s) \qquad \qquad .....(10)$$

$$\text{where,} \quad H_1(s) = \frac{1}{s+2} \quad \text{and} \quad H_2(s) = \frac{1}{s+1}$$

The transfer function $H_1(s)$ can be realized in direct form-II structure as shown below:

$$\text{Let,} \quad H_1(s) = \frac{Y_1(s)}{X(s)} = \frac{W_1(s)}{X(s)} \times \frac{Y_1(s)}{W_1(s)} = = \frac{\frac{1}{s}}{1 + \frac{2}{s}}$$

$$\text{where,} \quad \frac{W_1(s)}{X_1(s)} = \frac{1}{1 + \frac{2}{s}} \quad \text{and} \quad \frac{Y_1(s)}{W_1(s)} = \frac{1}{s}$$

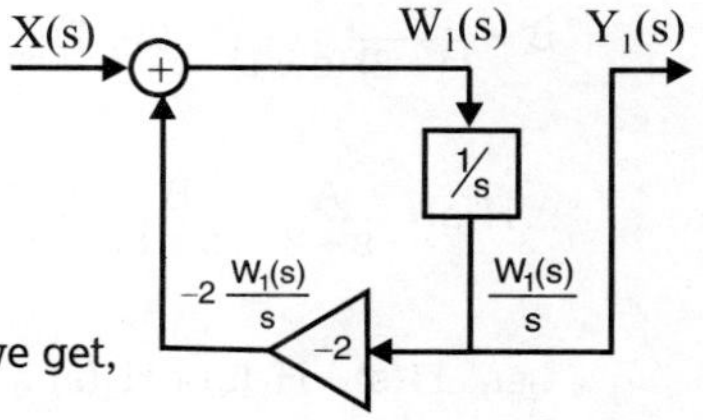

On cross multiplying and rearranging the above equations we get,

$$W_1(s) = X_1(s) - 2\frac{W_1(s)}{s} \quad \text{and} \quad Y_1(s) = \frac{W_1(s)}{s}$$

Fig 3: *Direct form -II structure of $H_1(s)$.*

The direct form-II structure of $H_1(s)$ is obtained using above equation as shown in Fig 3.

The transfer function $H_2(s)$ can be realized in direct form-II structure as shown below:

Let, $H_2(s) = \dfrac{Y(s)}{Y_1(s)} = \dfrac{W_2(s)}{Y_1(s)} \times \dfrac{Y(s)}{W_2(s)} = \dfrac{\frac{1}{s}}{1+\frac{1}{s}}$

where, $\dfrac{W_2(s)}{Y_1(s)} = \dfrac{1}{1+\frac{1}{s}}$ and $\dfrac{Y(s)}{W_1(s)} = \dfrac{1}{s}$

On cross multiplying and rearranging the above transfer functions we get,

$W_2(s) = Y_1(s) - \dfrac{W_2(s)}{s}$ and $Y(s) = \dfrac{W(s)}{s}$

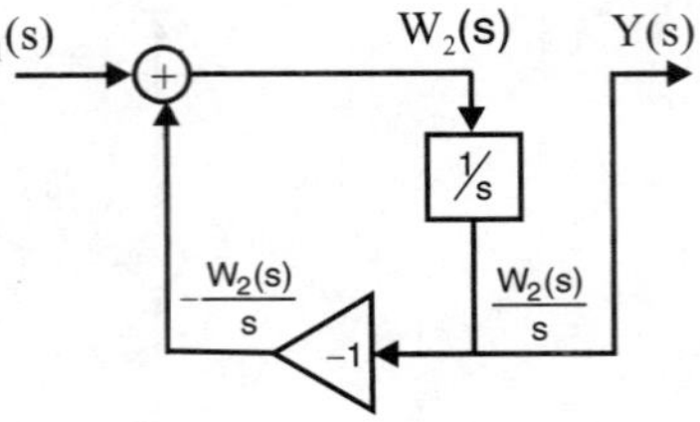

The cascade structure of the given system is obtained by connecting the individual sections shown in Fig 3 and Fig 4.

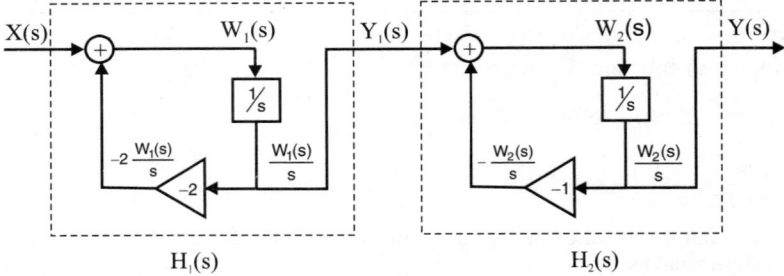

Fig 4: Direct form-II structure of $H_2(s)$.

Fig 5: Cascade structure of the system.

Parallel structure

From equation (9) we get,

$$\dfrac{Y(s)}{X(s)} = \dfrac{1}{s^2+3s+2} = \dfrac{1}{(s+2)(s+1)}$$

By partial fraction technique the above equation can be expressed as shown below:

$$\dfrac{Y(s)}{X(s)} = H(s) = \dfrac{1}{s^2+3s+2} = \dfrac{A}{s+2} + \dfrac{B}{s+1}$$

$A = \dfrac{1}{(s+2)(s+1)} \times (s+2)\Big|_{s=-2} = \dfrac{1}{-2+1} = -1$

$B = \dfrac{1}{(s+2)(s+1)} \times (s+1)\Big|_{s=-1} = \dfrac{1}{-1+2} = 1$

$\therefore H(s) = \dfrac{A}{s+2} + \dfrac{B}{s+1} = -\dfrac{1}{s+2} + \dfrac{1}{s+1}$

Let, $H(s) = H_1(s) + H_2(s)$

where, $H(s) = \dfrac{Y(s)}{X(s)}$; $H_1(s) = \dfrac{-1}{s+2}$; $H_2(s) = \dfrac{1}{s+1}$

Now, $H(s) = \dfrac{Y(s)}{X(s)} = H_1(s) + H_2(s)$

$\therefore Y(s) = H_1(s)\ X(s) + H_2(s)\ X(s)$

Let, $Y(s) = Y_1(s) + Y_2(s)$ .....(11)

where, $Y_1(s) = H_1(s)\ X(s) = -\dfrac{1}{s+2}\ X(s) \;\Rightarrow\; \dfrac{Y_1(s)}{X(s)} = \dfrac{-\dfrac{1}{s}}{1+\dfrac{2}{s}}$

$Y_2(s) = H_2(s)\ X(s) = \dfrac{1}{s+1}\ X(s) \Rightarrow \dfrac{Y_2(s)}{X(s)} = \dfrac{\dfrac{1}{s}}{1+\dfrac{1}{s}}$

In equation (11), the output $Y(s)$ is expressed as sum of two componenets. The two componenets are first order section of transfer function and they are realised by direct form-II structure as shown below:

The transfer function $H_1(s)$ can be realized in direct form-II structure as shown below:

Let, $\dfrac{Y_1(s)}{X(s)} = \dfrac{W_1(s)}{X(s)} \times \dfrac{Y_1(s)}{W_1(s)} = \dfrac{-\dfrac{1}{s}}{1+\dfrac{2}{s}}$

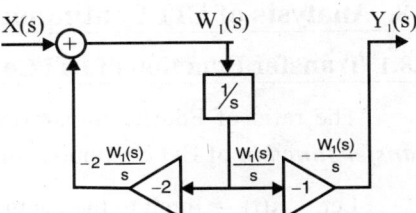

where, $\dfrac{W_1(s)}{X(s)} = \dfrac{1}{1+\dfrac{2}{s}}$ and $\dfrac{Y_1(s)}{W_1(s)} = -\dfrac{1}{s}$

On cross multiplying and rearranging the above equations we get,

$W_1(s) = X(s) - 2\,\dfrac{W_1(s)}{s}$ and $Y_1(s) = -\dfrac{W_1(s)}{s}$

Fig 6: *Direct form -II structure of $H_1(s)$.*

The direct form-II structure of $H_1(s)$ is obtained using the above equation as shown in Fig 6.

The transfer function $H_2(s)$ can be realized in direct form-II structure as shown below:

Let, $\dfrac{Y_2(s)}{X(s)} = \dfrac{W_2(s)}{X(s)} \times \dfrac{Y_2(s)}{W_2(s)} = \dfrac{\dfrac{1}{s}}{1+\dfrac{1}{s}}$

where, $\dfrac{W_2(s)}{X(s)} = \dfrac{1}{1+\dfrac{1}{s}}$ and $\dfrac{Y_2(s)}{W_2(s)} = \dfrac{1}{s}$

On cross multiplying and rearranging the above equations we get,

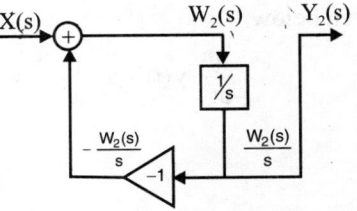

$W_2(s) = X(s) - \dfrac{W_2(s)}{s}$ and $Y_2(s) = \dfrac{W_2(s)}{s}$

Fig 7: *Direct form-II structure of $H_2(s)$.*

The direct form-II structure of $H_2(s)$ is obtained using the above equation as shown in Fig 7.

The parallel structure of the given system is obtained by connecting the individual sections shown in Fig 6 and Fig 7 in parallel as shown in Fig 8.

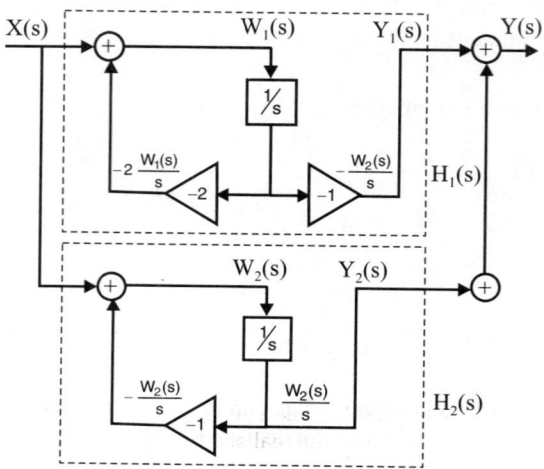

Fig 8: *Parallel structure of the system.*

3.8. Analysis of LTI Continuous Time System Using Fourier Transform

3.8.1 Transfer Function of LTI Continuous Time System in Frequency Domain

The ratio of Fourier transform of output and the Fourier transform of input is called *transfer function* of LTI continuous time system in frequency domain.

Let, $x(t)$ = Input to the continuous time system

   $y(t)$ = Output of the continuous time system

$\therefore$ $X(j\Omega)$ = Fourier transform of $x(t)$

   $Y(j\Omega)$ = Fourier transform of $y(t)$

Now, Transfer function $= \dfrac{Y(j\Omega)}{X(j\Omega)}$ .....(3.48)

The transfer function of LTI continuous time system in frequency domain can be obtained from the differential equation governing the input-output relation of an LTI continuous time system, given below:

$$\frac{d^N}{dt^N} y(t) + a_1 \frac{d^{N-1}}{dt^{N-1}} y(t) + a_2 \frac{d^{N-2}}{dt^{N-2}} y(t) + ... + a_{N-1} \frac{d}{dt} y(t) + a_N y(t) = b_0 \frac{d^M}{dt^M} x(t)$$

$$+ b_1 \frac{d^{M-1}}{dt^{M-1}} x(t) + b_2 \frac{d^{M-2}}{dt^{M-2}} x(t) + ... + b_{M-1} \frac{d}{dt} x(t) + b_M x(t)$$

On taking Fourier transform of the above equation and rearranging the resultant equation as a ratio of $Y(j\Omega)$ and $X(j\Omega)$, the transfer function of LTI continuous time system in frequency domain is obtained.

3.8.2 Impulse Response and Transfer Function

Consider an LTI continuous time system $\mathcal{H}$, shown in Fig 3.17. Let x(t) and y(t) be the input and output of the system respectively.

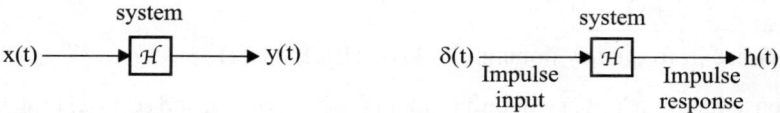

Fig 3.17: *Response of the system.*   **Fig 3.18:** *Impulse response of the system.*

For a continuous time system $\mathcal{H}$, if the input is impulse signal δ(t) as shown in Fig 3.18, then the output is called *impulse response*, which is denoted by h(t).

The importance of impulse response is that the response for any input to LTI system is given by convolution of input and impulse response.

Symbolically, the convolution operation is denoted as,

$$y(t) = x(t) * h(t) \qquad\qquad .....(3.49)$$

where, "∗" is the symbol for convolution.

Mathematically, the convolution operation is defined as,

$$y(t) \;=\; x(t) * h(t) \;=\; \int_{-\infty}^{+\infty} x(\tau)\, h(t - \tau)\, d\tau$$

where, τ is the dummy variable for integration.

Let,   H(jΩ) = Fourier transform of h(t)

       X(jΩ) = Fourier transform of x(t)

       Y(jΩ) = Fourier transform of y(t)

Now, by convolution property of Fourier transform we get,

$$\mathcal{F}\{x(t) * h(t)\} \;=\; X(j\Omega)\, H(j\Omega) \qquad\qquad .....(3.50)$$

Using equation (3.49), the above equation can be written as,

$$\mathcal{F}\{y(t)\} \;=\; X(j\Omega)\, H(j\Omega)$$

$$\therefore \; Y(j\Omega) \;=\; X(j\Omega)\, H(j\Omega)$$

$$\therefore \; H(j\Omega) \;=\; \frac{Y(j\Omega)}{X(j\Omega)} \qquad\qquad .....(3.51)$$

From equations (3.48) and (3.51) we can say that the *transfer function in frequency domain* is given by Fourier transform of impulse response.

$$\mathcal{F}\{h(t)\} \;=\; H(j\Omega) \;=\; \frac{Y(j\Omega)}{X(j\Omega)} \qquad\qquad .....(3.52)$$

3.8.3 Response of LTI Continuous Time System Using Fourier Transform

Consider the transfer function of LTI continuous time system, $H(j\Omega)$.

$$H(j\Omega) = \frac{Y(j\Omega)}{X(j\Omega)}$$

Now, response in frequency domain, $Y(j\Omega) = H(j\Omega)\, X(j\Omega)$

The response function $Y(j\Omega)$ will be a rational function of $j\Omega$, and so $Y(j\Omega)$ can be expressed as a ratio of two factorized polynomial in $j\Omega$ as shown below:

$$Y(j\Omega) = \frac{(j\Omega + z_1)(j\Omega + z_2)(j\Omega + z_3)\; .....}{(j\Omega + p_1)(j\Omega + p_2)(j\Omega + p_3)\; .....} \qquad .....(3.53)$$

By partial fraction expansion technique the equation (3.53) can be expressed as shown below:

$$Y(j\Omega) = \frac{k_1}{j\Omega + p_1} + \frac{k_2}{j\Omega + p_2} + \frac{k_3}{j\Omega + p_3} + ... \qquad .....(3.54)$$

where, $k_1, k_2, k_3, ...$ are residues.

Now the time domain response $y(t)$ can be obtained by taking inverse Fourier transform of equation (3.54). The inverse Fourier transform of each term in equation (3.54) can be obtained by comparing the terms with standard Fourier transform pair.

$$\mathcal{F}\{e^{-at}\, u(t)\} = \frac{1}{a + j\Omega} \qquad .....(3.55)$$

Using equation (3.55), the inverse Fourier transform of equation (3.54) can be obtained as shown below:

$$y(t) = k_1 e^{-p_1 t}\, u(t) + k_2 e^{-p_2 t}\, u(t) + k_3 e^{-p_3 t}\, u(t) + ... \qquad .....(3.56)$$

Since the transfer function is defined with zero initial conditions, the response obtained by using equation (3.56) is the time domain steady state (or forced) response of the LTI continuous time system.

> **Note:** *Only steady state or forced response can be computed via frequency domain.*

3.8.4 Frequency Response of LTI Continuous Time System

The output $y(t)$ of an LTI continuous time system is given by convolution of $h(t)$ and $x(t)$.

$$\therefore\; y(t) = x(t) * h(t) = h(t) * x(t) = \int_{-\infty}^{+\infty} h(\tau)\, x(t-\tau)\, d\tau \qquad .....(3.57)$$

Consider a special class of input (complex sinusoidal input), $\boxed{Ae^{j\Omega t} = A(\cos \Omega t + j\sin \Omega t)}$

$$x(t) = A\, e^{j\Omega t} \qquad .....(3.58)$$

where, A = Amplitude ; Ω = Angular frequency in rad/sec

$$\therefore\; x(t-\tau) = Ae^{j\Omega(t-\tau)} \qquad .....(3.59)$$

On substituting for x(t − τ) from equation (3.59) in equation (3.57) we get,

$$y(t) = \int_{-\infty}^{+\infty} h(\tau)\, Ae^{j\Omega(t-\tau)}\, d\tau = \int_{-\infty}^{+\infty} h(\tau)\, Ae^{j\Omega t}\, e^{-j\Omega\tau}\, d\tau$$

$$= Ae^{j\Omega t} \int_{-\infty}^{+\infty} h(\tau)\, e^{-j\Omega\tau}\, d\tau$$

$$= Ae^{j\Omega t} \int_{-\infty}^{+\infty} h(t)\, e^{-j\Omega t}\, dt$$

By the definition of Fourier transform,

$$H(j\Omega) = \int_{-\infty}^{+\infty} h(t)\, e^{-j\Omega t}\, dt$$

Replace τ by t

$$= x(t)\, H(j\Omega) \qquad\qquad .....(3.60)$$

From equation (3.60) we can say that if a complex sinusoidal signal is given as input signal to an LTI continuous time system, then the output is also a complex sinusoidal signal of the same frequency modified by $H(j\Omega)$. Hence, $H(j\Omega)$ is called the ***frequency response*** of the continuous time system.

Since the $H(j\Omega)$ is a complex function of Ω, the multiplication of $H(j\Omega)$ with input produces a change in the amplitude and phase of the input signal, and the modified input signal is the output signal. Therefore, an LTI system is characterized in the frequency domain by its frequency response.

The function $H(j\Omega)$ is a complex quantity and so it can be expressed as magnitude function and phase function.

$$\therefore H(j\Omega) = |H(j\Omega)|\, \angle H(j\Omega)$$

$$\text{where,}\quad |H(j\Omega)| = \text{Magnitude function}$$

$$\angle H(j\Omega) = \text{Phase function}$$

The sketch of magnitude function and phase function with respect to Ω will give the frequency response graphically.

$$\text{Let, } H(j\Omega) = H_r(j\Omega) + jH_i(j\Omega)$$

$$\text{where, } H_r(j\Omega) = \text{Real part of } H(j\Omega)$$

$$H_i(j\Omega) = \text{Imaginary part of } H(j\Omega)$$

The ***magnitude function*** is defined as,

$$|H(j\Omega)|^2 = H(j\Omega)\, H^*(j\Omega) = [H_r(j\Omega) + jH_i(j\Omega)]\,[H_r(j\Omega) - jH_i(j\Omega)]$$

$$\text{where, } H^*(j\Omega) \text{ is complex conjugate of } H(j\Omega)$$

$$\therefore |H(j\Omega)|^2 = H_r^2(j\Omega) + H_i^2(j\Omega)$$

$$\therefore |H(j\Omega)| = \sqrt{H(j\Omega)\, H^*(j\Omega)} \text{ or } |H(j\Omega)| = \sqrt{H_r^2(j\Omega) + H_i^2(j\Omega)}$$

The *phase function* is defined as,

$$\angle H(j\Omega) = Arg[H(j\Omega)] = \tan^{-1}\left[\frac{H_i(j\Omega)}{H_r(j\Omega)}\right]$$

From equation (3.52) we can say that the frequency response $H(j\Omega)$ of an LTI continuous time system is same as transfer function in frequency domain and so, the frequency response is also given by the ratio of Fourier transform of output to Fourier transform of input.

$$\therefore \text{ Frequency response, } H(j\Omega) = \frac{Y(j\Omega)}{X(j\Omega)}$$

.....(3.61)

Advantages of Frequency Response Analysis

1. The practical testing of systems can be easily carried with available sinusoidal signal generators and precise measurement equipments.

2. The transfer function of complicated systems can be determined experimentally by frequency response tests.

3. The design and parameter adjustment is carried out more easily in frequency domain.

4. In frequency domain, the effects of noise disturbance and parameter variations are relatively easy to visualize and incorporate corrective measures.

5. The frequency response analysis and designs can be extended to certain nonlinear systems.

Example 3.36   *(AU Dec'13, 8 Marks)*

Using Fourier transform find the impulse response and frequency response of the system described by the equation,

$$\frac{d^2y(t)}{dt^2} + 4\frac{dy(t)}{dt} + 3y(t) = \frac{dx(t)}{dt} + 2x(t)$$

Solution:

Given that, $\dfrac{d^2y(t)}{dt^2} + 4\dfrac{dy(t)}{dt} + 3y(t) = \dfrac{dx(t)}{dt} + 2x(t)$

On taking Fourier transform of above equation we get,

$$(j\Omega)^2 Y(j\Omega) + 4\,(j\Omega)\,Y(j\Omega) + 3\,Y(j\Omega) = j\Omega\,X(j\Omega) + 2\,X(j\Omega)$$

$$Y(j\Omega)[(j\Omega)^2 + 4\,j\Omega + 3] = X(j\Omega)\,(j\Omega + 2)$$

$$\therefore \frac{Y(j\Omega)}{X(j\Omega)} = \frac{j\Omega + 2}{(j\Omega)^2 + 4j\Omega + 3} = \frac{j\Omega + 2}{(j\Omega + 1)(j\Omega + 3)}$$

The frequency response, $H(j\Omega)$ is given by,

$$H(j\Omega) = \frac{Y(j\Omega)}{X(j\Omega)} = \frac{j\Omega + 2}{(j\Omega + 1)(j\Omega + 3)}$$

By partial fraction expansion technique the above equation can be written as,

$$H(j\Omega) = \frac{j\Omega + 2}{(j\Omega + 1)(j\Omega + 3)} = \frac{k_1}{j\Omega + 1} + \frac{k_2}{j\Omega + 3}$$

---

if, $\mathcal{F}\{x(t)\} = X(j\Omega)$,

then, $\dfrac{d^m}{dt^m}x(t) = (j\Omega)^m\,X(j\Omega)$

The roots of quadratic

$(j\Omega)^2 + 4j\Omega + 3 = 0$

$j\Omega = \dfrac{-4 \pm \sqrt{4^2 - 4 \times 3}}{2}$

$\quad = \dfrac{-4 \pm 2}{2} = -1, -3$

$$k_1 = \frac{j\Omega + 2}{(j\Omega + 1)(j\Omega + 3)} \times (j\Omega + 1) \Big|_{j\Omega = -1} = \frac{-1+2}{-1+3} = 0.5$$

$$k_2 = \frac{j\Omega + 2}{(j\Omega + 1)(j\Omega + 3)} \times (j\Omega + 3) \Big|_{j\Omega = -3} = \frac{-3+2}{-3+1} = 0.5$$

$$\therefore \ H(j\Omega) = \frac{0.5}{j\Omega + 1} + \frac{0.5}{j\Omega + 3}$$

The impulse response h(t), is given by inverse Fourier transform of transfer function, H(jΩ).

$$\therefore \text{ Impulse response, } h(t) = \mathcal{F}^{-1}\{H(j\Omega)\} = \mathcal{F}^{-1}\left\{\frac{0.5}{j\Omega + 1} + \frac{0.5}{j\Omega + 3}\right\} \qquad \boxed{\mathcal{F}\{e^{-at}\,u(t)\} = \frac{1}{j\Omega + a}}$$

$$= 0.5e^{-t}\,u(t) + 0.5e^{-3t}\,u(t) = 0.5\,(e^{-t} + e^{-3t})\,u(t)$$

Example 3.37

(AU Dec'15, 16 Marks)

Using Fourier transform find the impulse response and response of the system described by the equation,

$$\frac{d^2 y(t)}{dt^2} + 6\frac{dy(t)}{dt} + 8y(t) = 2x(t) \quad ; \text{ if } x(t) = u(t)$$

Solution:

Impulse Response of the System

Given that, $\dfrac{d^2 y(t)}{dt^2} + 6\dfrac{dy(t)}{dt} + 8y(t) = 2x(t)$

On taking Fourier transform of above equation we get,

$$(j\Omega)^2 Y(j\Omega) + 6\,(j\Omega)\,Y(j\Omega) + 8\,Y(j\Omega) = 2\,X(j\Omega)$$

$$Y(j\Omega)[(j\Omega)^2 + 6\,j\Omega + 8] = 2\,X(j\Omega)$$

$$\therefore \ \frac{Y(j\Omega)}{X(j\Omega)} = \frac{2}{(j\Omega)^2 + 6j\Omega + 8} = \frac{2}{(j\Omega + 2)(j\Omega + 4)}$$

The frequency response, H(jΩ) is given by

$$H(j\Omega) = \frac{Y(j\Omega)}{X(j\Omega)} = \frac{2}{(j\Omega + 2)(j\Omega + 4)} \qquad \qquad \dots\dots(1)$$

> if, $\mathcal{F}\{x(t)\} = X(j\Omega)$, then
>
> $\mathcal{F}\left\{\dfrac{d^m}{dt^m}x(t)\right\} = (j\Omega)^m\, X(j\Omega)$

> The roots of quadratic
> $(j\Omega)^2 + 6j\Omega + 8 = 0$
> $j\Omega = \dfrac{-6 \pm \sqrt{6^2 - 4\times 8}}{2}$
> $= \dfrac{-6 \pm 2}{2} = -2, -4$

By partial fraction expansion technique we can be write,

$$H(j\Omega) = \frac{2}{(j\Omega + 2)(j\Omega + 4)} = \frac{k_1}{j\Omega + 2} + \frac{k_2}{j\Omega + 4}$$

$$k_1 = \frac{2}{(j\Omega + 2)(j\Omega + 4)} \times (j\Omega + 2) \Big|_{j\Omega = -2} = \frac{2}{-2+4} = \frac{2}{2} = 1$$

$$k_2 = \frac{2}{(j\Omega + 2)(j\Omega + 4)} \times (j\Omega + 4) \Big|_{j\Omega = -4} = \frac{2}{-4+2} = \frac{2}{-2} = -1$$

$$\therefore \ H(j\Omega) = \frac{1}{j\Omega + 2} - \frac{1}{j\Omega + 4} \qquad \qquad \dots\dots(2)$$

On taking inverse Fourier transform of transfer function, H(jΩ) we get impulse response, h(t)

$$\therefore \ h(t) = \mathcal{F}^{-1}\{H(j\Omega)\} = \mathcal{F}^{-1}\left\{\frac{1}{j\Omega + 2} - \frac{1}{j\Omega + 4}\right\} \qquad \boxed{\mathcal{F}\{e^{-at}\,u(t)\} = \frac{1}{j\Omega + a}}$$

$$= e^{-2t}\,u(t) - e^{-4t}\,u(t) = [e^{-2t} - e^{-4t}]\,u(t)$$

Response of the System

Given that, $x(t) = u(t)$

$$\therefore \ X(j\Omega) = \mathcal{F}\{u(t)\} = \frac{1}{j\Omega} \qquad \qquad \qquad .....(2)$$

From equation (1), the frequency domain response $Y(j\Omega)$ is,

$$Y(j\Omega) = X(j\Omega) \times \frac{2}{(j\Omega + 2)(j\Omega + 4)}$$

$$= \frac{1}{j\Omega} \times \frac{2}{(j\Omega + 2)(j\Omega + 4)} = \frac{2}{j\Omega(j\Omega + 2)(j\Omega + 4)} \qquad \boxed{\text{Using equation (2).}}$$

By partial fraction expansion technique $Y(j\Omega)$ can be written as,

$$Y(j\Omega) = \frac{k_1}{j\Omega} + \frac{k_2}{j\Omega + 2} + \frac{k_3}{j\Omega + 4}$$

$$k_1 = \frac{2}{j\Omega(j\Omega + 2)(j\Omega + 4)} \times j\Omega \ \bigg|_{j\Omega = 0} = \frac{2}{(0+2)(0+4)} = \frac{1}{4} = 0.25$$

$$k_2 = \frac{2}{j\Omega(j\Omega + 2)(j\Omega + 4)} \times (j\Omega + 2) \ \bigg|_{j\Omega = -2} = \frac{2}{(-2)(-2+4)} = -\frac{1}{2} = -0.5$$

$$k_3 = \frac{2}{j\Omega(j\Omega + 2)(j\Omega + 4)} \times (j\Omega + 4) \ \bigg|_{j\Omega = -4} = \frac{2}{(-4)(-4+2)} = \frac{1}{4} = 0.25$$

$$\therefore \ Y(j\Omega) = \frac{0.25}{j\Omega} - \frac{0.5}{j\Omega + 2} + \frac{0.25}{j\Omega + 4}$$

On taking inverse Fourier transform of frequency domain response, $Y(j\Omega)$ we get time domain response, $y(t)$.

$$\therefore \ y(t) = \mathcal{F}^{-1}\{Y(j\Omega)\} = \mathcal{F}^{-1}\left\{ \frac{0.25}{j\Omega} - \frac{0.5}{j\Omega + 2} + \frac{0.25}{j\Omega + 4} \right\}$$

$$\boxed{\begin{array}{l} \mathcal{F}\{u(t)\} = \dfrac{1}{j\Omega} \\[2mm] \mathcal{F}\{e^{-at}u(t)\} = \dfrac{1}{j\Omega + a} \end{array}}$$

$$= 0.25 \ u(t) - 0.5e^{-2t}u(t) + 0.25 \ e^{-4t}u(t) = 0.25 \ (1 - 2e^{-2t} + e^{-4t}) \ u(t)$$

Example 3.38

The impulse response of an LTI system is $h(t) = 2 e^{-3t}u(t)$.

Find the response of the system for the input $x(t) = 2e^{-5t}u(t)$, using Fourier transform.

Solution:

Given that, $x(t) = 2 e^{-5t}u(t)$.

$$\therefore \ X(j\Omega) = \mathcal{F}\{x(t)\} = \mathcal{F}\{2 e^{-5t}u(t)\} = \frac{2}{j\Omega + 5} \qquad \qquad .....(1)$$

Given that, $h(t) = 2 e^{-3t}u(t)$.

$$\boxed{\mathcal{F}\{e^{-at}u(t)\} = \frac{1}{j\Omega + a}}$$

$$\therefore \ H(j\Omega) = \mathcal{F}\{h(t)\} = \mathcal{F}\{2 e^{-3t}u(t)\} = \frac{2}{j\Omega + 3} \qquad \qquad .....(2)$$

For LTI system, the response, $y(t) = x(t) * h(t)$ .....(3)

On taking Fourier transform of equation (3) we get,

$$\mathcal{F}\{y(t)\} = \mathcal{F}\{x(t) * h(t)\}$$

Let, $\mathcal{F}\{y(t)\} = Y(j\Omega)$

$$\therefore Y(j\Omega) = \mathcal{F}\{x(t) * h(t)\}$$

$$= X(j\Omega)\, H(j\Omega)$$

> Using convolution property of Fourier transform.

$$= \frac{2}{j\Omega + 5} \times \frac{2}{j\Omega + 3} = \frac{4}{(j\Omega + 5)(j\Omega + 3)}$$

> Using equations (1) and (2).

By partial fraction expansion technique, the above equation can be written as,

$$Y(j\Omega) = \frac{4}{(j\Omega + 5)(j\Omega + 3)} = \frac{k_1}{j\Omega + 5} + \frac{k_2}{j\Omega + 3}$$

$$k_1 = \left.\frac{4}{(j\Omega + 5)(j\Omega + 3)} \times (j\Omega + 5)\right|_{j\Omega = -5} = \frac{4}{-5 + 3} = -2$$

$$k_2 = \left.\frac{4}{(j\Omega + 5)(j\Omega + 3)} \times (j\Omega + 3)\right|_{j\Omega = -3} = \frac{4}{-3 + 5} = 2$$

$$\therefore Y(j\Omega) = -\frac{2}{j\Omega + 5} + \frac{2}{j\Omega + 3}$$

On taking inverse Fourier transform of frequency domain response, $Y(j\Omega)$ we get time domain response, $y(t)$.

$$\therefore y(t) = \mathcal{F}^{-1}\{Y(j\Omega)\} = \mathcal{F}^{-1}\left\{-\frac{2}{j\Omega + 5} + \frac{2}{j\Omega + 3}\right\}$$

> $\mathcal{F}\{e^{-at} u(t)\} = \dfrac{1}{j\Omega + a}$

$$= -2\,e^{-5t}\,u(t) + 2\,e^{-3t}\,u(t) = 2(e^{-3t} - e^{-5t})\,u(t)$$

Example 3.39   *(AU Jun'14, 16 Marks)*

Find the frequency response and impulse response of the system shown in Fig 3.39.1 using Fourier transform.

Solution:

With reference to Fig 1, by Kirchoff's voltage law we can write,

$$x(t) = R\, i(t) + L\frac{di(t)}{dt}$$

Here, $i(t) = y(t)$

$$\therefore x(t) = R\, y(t) + L\frac{dy(t)}{dt}$$

On taking Fourier transform of the above equation we get,

$$X(j\Omega) = R\, Y(j\Omega) + L\, j\Omega\, Y(j\Omega)$$

$$= Y(j\Omega)\,(R + L\, j\Omega)$$

$$\therefore \frac{Y(j\Omega)}{X(j\Omega)} = \frac{1}{R + L\, j\Omega}$$

The frequency response $H(j\Omega)$ is given by,

$$H(j\Omega) = \frac{Y(j\Omega)}{X(j\Omega)} = \frac{1}{R + j\Omega L} = \frac{1}{L\left(j\Omega + \dfrac{R}{L}\right)} = \frac{1}{L} \times \frac{1}{\left(j\Omega + \dfrac{R}{L}\right)}$$

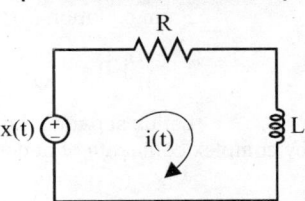

Fig 3.39.1.

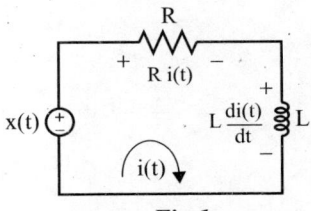

Fig 1.

On taking inverse Fourier transform of frequency response, $H(j\Omega)$ we get impulse response, $h(t)$.

$$\therefore h(t) = \mathcal{F}^{-1}\{H(j\Omega)\} = \mathcal{F}^{-1}\left\{\frac{1}{L}\,\frac{1}{j\Omega + \frac{R}{L}}\right\} = \frac{1}{L}e^{-\left(\frac{R}{L}\right)t}u(t)$$

Example 3.40

The output of the continuous time system is $y(t) = e^{-t}u(t)$ for an input $x(t) = e^{-2t}u(t)$. Find

a) Frequency response   **b)** Magnitude and phase function   **c)** Impulse response

Solution:

Given that, $y(t) = e^{-t}u(t)$ .....(1)

$$\therefore Y(j\Omega) = \mathcal{F}\{y(t)\} = \mathcal{F}\{e^{-t}u(t)\} = \frac{1}{j\Omega + 1}$$

Given that, $x(t) = e^{-2t}u(t)$

$$\therefore X(j\Omega) = \mathcal{F}\{x(t)\} = \mathcal{F}\{e^{-2t}u(t)\} = \frac{1}{j\Omega + 2}$$ .....(2)

a) Frequency Response

For LTI system, the frequency response $H(j\Omega)$ is given by,

$$H(j\Omega) = \frac{Y(j\Omega)}{X(j\Omega)}$$

$$\therefore H(j\Omega) = \frac{j\Omega + 2}{j\Omega + 1} \qquad \boxed{\text{Using equations (1) and (2).}}$$ .....(3)

b) Magnitude and Phase Function

From equation (3) the frequency response $H(j\Omega)$ is,

$$H(j\Omega) = \frac{j\Omega + 2}{j\Omega + 1}$$

Let us separate the real part and imaginary part by multiplying the numerator and denominator by complex conjugate of denominator.

$$\therefore H(j\Omega) = \frac{j\Omega + 2}{j\Omega + 1} \times \frac{-j\Omega + 1}{-j\Omega + 1} = \frac{\Omega^2 + j\Omega - j2\Omega + 2}{(1 + j\Omega)(1 - j\Omega)}$$

$$= \frac{\Omega^2 + 2 - j\Omega}{1 - (j\Omega)^2} = \frac{\Omega^2 + 2 - j\Omega}{1 + \Omega^2} = \frac{2 + \Omega^2}{1 + \Omega^2} - \frac{j\Omega}{1 + \Omega^2}$$

Let, $H(j\Omega) = H_r(j\Omega) + j\,H_i(j\Omega)$

$$\therefore H_i(j\Omega) = \frac{-\Omega}{1 + \Omega^2}$$ .....(4)

$$H_r(j\Omega) = \frac{2 + \Omega^2}{1 + \Omega^2}$$ .....(5)

The magnitude function $|H(j\Omega)|$ is defined as,

$$|H(j\Omega)| = \sqrt{H_r^2(j\Omega) + H_i^2(j\Omega)}$$

$$\therefore |H(j\Omega)| = \sqrt{\left(\frac{2+\Omega^2}{1+\Omega^2}\right)^2 + \left(\frac{-\Omega}{1+\Omega^2}\right)^2}$$

$$= \sqrt{\frac{(2+\Omega^2)^2}{(1+\Omega^2)^2} + \frac{\Omega^2}{(1+\Omega^2)^2}} = \sqrt{\frac{(2+\Omega^2)^2+\Omega^2}{(1+\Omega^2)^2}}$$

$$= \sqrt{\frac{4+\Omega^4+4\Omega^2+\Omega^2}{(1+\Omega^2)}} = \sqrt{\frac{\Omega^4+5\Omega^2+4}{(1+\Omega^2)^2}}$$

$$= \frac{\sqrt{(\Omega^2+1)(\Omega^2+4)}}{(\Omega^2+1)^2} = \sqrt{\frac{\Omega^2+4}{\Omega^2+1}}$$

The phase function $\angle H(j\Omega)$ is given as,

$$\angle H(j\Omega) = \tan^{-1}\left[\frac{H_i(j\Omega)}{H_r(j\Omega)}\right]$$

$$= \tan^{-1}\left(\frac{\dfrac{-\Omega}{1+\Omega^2}}{\dfrac{2+\Omega^2}{1+\Omega^2}}\right)$$

| Using equations (4) and (5). |

$$= \tan^{-1}\left(\frac{-\Omega}{2+\Omega^2}\right)$$

> Using equations (4) and (5).
>
> Let, $\Omega^2 = U$
> $\therefore \Omega^2 + 5\Omega^2 + 4 = U^2 + 5U + 4$
> The roots of quadratic
> $U^2 + 5U + 4 = 0$
> $$U = \frac{-5 \pm \sqrt{5^2 - 4 \times 4}}{2}$$
> $$= \frac{-5 \pm 3}{2} = -1, -4$$
> $\therefore U^2 + 5U + 4 = (U+1)(U+4)$
> $\qquad\qquad = (\Omega^2+1)(\Omega^2+4)$

c) Impulse response

From equation (3) the frequency response $H(j\Omega)$ is given by,

$$H(j\Omega) = \frac{j\Omega+2}{j\Omega+1} = \frac{j\Omega+1+1}{j\Omega+1}$$

$$= \frac{j\Omega+1}{j\Omega+1} + \frac{1}{j\Omega+1}$$

| $\mathcal{F}\{\delta(t)\} = 1 \; ; \; \mathcal{F}\{e^{-at}u(t)\} = \dfrac{1}{j\Omega+a}$ |

$$= 1 + \frac{1}{j\Omega+1}$$

On taking inverse Fourier transform of $H(j\Omega)$ we get impulse response, $h(t)$.

$$\therefore h(t) = \mathcal{F}^{-1}\{H(j\Omega)\} = \mathcal{F}^{-1}\left\{1 + \frac{1}{j\Omega+1}\right\} = \delta(t) + e^{-t}u(t)$$

3.9 Summary of Important Concepts

1. The homogenous solution is the response of a system when there is no input, whereas the particular solution is the response for specific input.

2. The free or natural response is the response of the system due to initial stored energy, whereas the forced response is the response due to a particular input when there is no initial energy.

3. The response (or total response) of an LTI system is the sum of natural and forced response.

4. The transfer function of a continuous time system is defined as the ratio of Laplace transform of output and Laplace transform of input.

5. The transfer function of an LTI continuous time system is also given by Laplace transform of the impulse response.(Alternatively, the inverse Laplace transform of transfer function is the impulse response of the system).

6. For a stable LTI continuous time system the ROC should include the imaginary axis of s-plane.

7. For a stable LTI continuous time causal system, the poles should lie on the left half of s-plane and the imaginary axis should be included in the ROC.

8.   For a stable LTI continuous time noncausal system, the imaginary axis should be included in the ROC.

9.   The response of an LTI system is given by convolution of input and impulse response.

10. The unit step response of an LTI system is given by the integral of its impulse response.

11. The convolution operation satisfies the commutative, associative and distributive properties.

12. When LTI systems are connected in cascade, the overall impulse response is given by the convolution of individual impulse responses.

13. When LTI systems are connected in parallel, the overall impulse response is given by the sum of its individual impulse responses.

14. The inverse system is used to recover the input from the response of a system.

15. The cascade of a system and its inverse is the identity system.

16. The deconvolution is the process of recovering input from the response of a system.

17. The ratio of Fourier transform of output and input signal of a system is called transfer function in frequency domain.

18. The Fourier transform of impulse response gives the frequency domain transfer function.

19. The Fourier transform is evaluation of Laplace transform along imaginary axis in s-plane.

3.10 Short-answer Questions

Q3.1   Determine natural response of the first order system governed by the equation,

$$\frac{dy(t)}{dt} + 3\,y(t) = x(t) \; ; \; y(0) = 2$$

Solution:

Homogeneous Solution, $y_h(t)$

Put, $x(t) = 0$ in the given equation.

$$\therefore \; \frac{dy(t)}{dt} + 3\,y(t) = 0 \qquad\qquad .....(1)$$

Let, $y_h(t) = Ce^{\lambda t}$; $\therefore \dfrac{dy_h(t)}{dt} = C\lambda e^{\lambda t}$

On substituting the above terms in equation (1) we get,

$$C\lambda e^{\lambda t} + 3Ce^{\lambda t} = 0 \quad \Rightarrow \quad C(\lambda + 3)e^{\lambda t} = 0$$

The characteristic equation is, $\lambda + 3 = 0 \quad \Rightarrow \quad \lambda = -3$

$$\therefore \; y_h(t) = Ce^{\lambda t} = Ce^{-3t}$$

Natural Response

Natutal response (or zero input response), $y_{zi}(t) = y_h(t)\big|_{\text{with constants evaluated using initial conditions}} = Ce^{-3t}$

At $t = 0$, $\; y_{zi}(0) = Ce^0 = C$

Given that, $\; y(0) = 2$, $\; \therefore C = 2$

$$\therefore \; \text{Natural response}, \; y_{zi}(t) = 2e^{-3t} \; ; \; t \geq 0 \; = 2e^{-3t}\,u(t)$$

Q3.2   Determine the unit step response of the first order system governed by the equation,

$$\frac{dy(t)}{dt} + 0.5y(t) = x(t) \text{ with zero intial conditions.}$$

Solution:

Homogeneous Solution, $y_h(t)$

Put, $x(t) = 0$ in the given equation.

$$\therefore \frac{dy(t)}{dt} + 0.5\,y(t) = 0 \qquad\qquad .....(1)$$

Let, $y_h(t) = Ce^{\lambda t}$; $\quad \therefore \frac{dy_h(t)}{dt} = C\lambda e^{\lambda t}$

On substituting the above terms in the given equation (1) we get,

$C\lambda e^{\lambda t} + 0.5Ce^{\lambda t} = 0 \quad \Rightarrow \quad C(\lambda + 0.5)e^{\lambda t} = 0$

The characteristic polynomial is, $\lambda + 0.5 = 0$ $\therefore \lambda = -0.5$

$\qquad \therefore y_h(t) = Ce^{-0.5t}$; $t \geq 0 = Ce^{-0.5t}\,u(t)$

Particular Solution, $y_p(t)$

Here, $x(t) = u(t)$

Let, $y_p(t) = K\,u(t)$; $\quad \therefore \frac{dy_p(t)}{dt} = K\,\delta(t)$

On substituting the above terms in the given equation we get,

$K\delta(t) + 0.5Ku(t) = u(t)$

At $t = 1$, $K\delta(t) + 0.5Ku(t) = u(1) \quad \Rightarrow \quad K \times 0 + 0.5K\,1 = 1 \quad \Rightarrow \quad K = \frac{1}{0.5} = 2$

$\qquad \therefore y_p(t) = 2\,u(t)$

Unit Step Response (or Total Response)

Unit step response (or Total Response), $y(t) = y_h(t) + y_p(t) = Ce^{-0.5t}\,u(t) + 2u(t)$

At $t = 0$, $y(0) = Ce^0 + 2 = C + 2$

Here, $y(0) = 0$, $\therefore C + 2 = 0 \quad \Rightarrow \quad C = -2$

$\therefore$ Unit step response, $y(t) = Ce^{-0.5t}\,u(t) + 2u(t) = -2e^{-0.5t}\,u(t) + 2u(t) = 2(1 - e^{-0.5t})\,u(t)$

Q3.3 *If two LTI systems with impulse responses, $h_1(t) = e^{-at}\,u(t)$ and $h_2(t) = e^{-bt}\,u(t)$ are connected in cascade, what will be the overall impulse response of cascaded system?*

Solution:

When LTI systems are in cascade, the overall impulse response is given by convolution of individual impulse responses.

Let, $h_o(t)$ = Overall impulse response of cascaded system.

Now, $h_o(t) = h_1(t) * h_2(t) = \displaystyle\int_{\lambda=0}^{t} h_1(\lambda)h_2(t-\lambda)\,d\lambda = \displaystyle\int_{\lambda=0}^{t} e^{-a\lambda}\,e^{-b(t-\lambda)}\,d\lambda$

$$= \int_{\lambda=0}^{t} e^{-a\lambda}\,e^{-bt}\,e^{b\lambda}\,d\lambda = e^{-bt}\int_{\lambda=0}^{t} e^{-(a-b)\lambda}\,d\lambda$$

$$= e^{-bt}\left[\frac{e^{-(a-b)\lambda}}{-(a-b)}\right]_0^t = e^{-bt}\left[\frac{e^{-(a-b)t}}{-(a-b)} - \frac{e^0}{-(a-b)}\right]$$

$$= \frac{1}{a-b}\left[-e^{-at} + e^{-bt}\right] \ ; \ t \geq 0 = \frac{1}{a-b}\left[-e^{-at} + e^{-bt}\right]u(t)$$

Q3.4 **Find the overall impulse response of the system shown in Fig Q3.4.**

Take, $h_1(t) = t\,u(t)$; $h_2(t) = 3\,u(t)$; $h_3(t) = 2\,u(t)$.

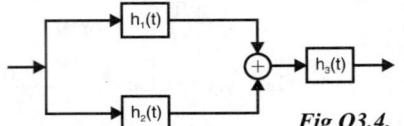

Fig Q3.4.

Solution:

Overall impulse response, $h_0(t) = [h_1(t) + h_2(t)] * h_3(t) = [h_1(t) * h_3(t)] + [h_2(t) * h_3(t)]$

$$= \int_{\lambda=0}^{t} h_1(\lambda)\,h_3(t-\lambda)\,d\lambda + \int_{\lambda=0}^{t} h_2(\lambda)\,h_3(t-\lambda)\,d\lambda$$

$\therefore$ Overall impulse response, $h_0(t) = \int_{\lambda=0}^{t} \lambda \times 2\,d\lambda + \int_{\lambda=0}^{t} 3 \times 2\,d\lambda = 2\int_{\lambda=0}^{t}\lambda\,d\lambda + 6\int_{\lambda=0}^{t} d\lambda = 2\left[\frac{\lambda^2}{2}\right]_0^t + 6[\lambda]_0^t$

$$= 2\left[\frac{t^2}{2} - 0\right] + 6[t - 0] = t^2 + 6t \ \ ; \ t \geq 0 = [t^2 + 6t]\,u(t)$$

Q3.5 **What is the importance of convolution?**

The convolution operation can be used to determine the response of an LTI system for any input from the knowledge of its impulse response. [The response of an LTI system is given by convolution of input and impulse response of the LTI system].

Q3.6 **Perform convolution of two sequences $x_1(t) = u(t-2)$ and $x_2(t) = \delta(t-1)$.**

(AU Dec'15, 2 Marks)

Solution:

Given that, $x_1(t) = u(t - 2)$; $x_2(t) = \delta(t - 1)$

By convolution, $x_1(t) * x_2(t) = u(t - 2) * \delta(t - 1)$

$= u(t - 2 - 1) = u(t - 3)$

$\boxed{\delta(t - t_0) * x(t) = x(t - t_0)}$

Q3.7 **The impulse response of a system is $e^{-4t}\,u(t)$, and for the input $x(t)$, the response is $(1 - e^{-4t})\,u(t)$. Find the input $x(t)$.**

Solution:

Given that, $h(t) = e^{-4t}\,u(t)$

$$\therefore H(s) = \mathcal{L}\{h(t)\} = \mathcal{L}\{e^{-4t}\,u(t)\} = \frac{1}{s+4}$$

Given that, $y(t) = (1 - e^{-4t})\,u(t) = u(t) - e^{-4t}\,u(t)$

$$\therefore Y(s) = \mathcal{L}\{y(t)\} = \mathcal{L}\{u(t) - e^{-4t}u(t)\} = \frac{1}{s} - \frac{1}{s+4}$$

We know that, $y(t) = x(t) * h(t)$

On taking Laplace transform of above equation we get,

$$\mathcal{L}\{y(t)\} = \mathcal{L}\{x(t) * h(t)\} \ \Rightarrow \ Y(s) = X(s)\,H(s)$$

Using convolution theorem of Laplace transform

$$\therefore X(s) = \frac{Y(s)}{H(s)} = Y(s) \times \frac{1}{H(s)} = \left(\frac{1}{s} - \frac{1}{s+4}\right) \times (s+4) = \frac{s+4}{s} - 1 = \frac{s+4-s}{s} = \frac{4}{s}$$

Now input, $x(t) = \mathcal{L}^{-1}\{X(s)\} = \mathcal{L}^{-1}\left\{\frac{4}{s}\right\} = 4\,u(t)$

Q3.8   *The impulse response of a system is $(4e^{-t} - 2e^{-3t})\, u(t)$. Determine the stability of the system.*

Solution:

Given that, $h(t) = (4e^{-t} - 2e^{-3t})\, u(t) = 4e^{-t}\, u(t) - 2e^{-3t}\, u(t)$

$\therefore\ H(s) = \mathcal{L}\{h(t)\} = \mathcal{L}\{4e^{-t}\, u(t) - 2e^{-3t}\, u(t)\}$

$$= \frac{4}{s+1} - \frac{2}{s+3} = \frac{4(s+3) - 2(s+1)}{(s+1)(s+3)} = \frac{2s+10}{(s+1)(s+3)} = \frac{2(s+5)}{(s+1)(s+3)}$$

The poles of $H(s)$ are lying at $s = -1$ and $s = -3$, and so they are lying on left half s-plane.

The ROC is the region to the right of right most pole and so imaginary axis is included in ROC.

Since the poles are lying on left half s-plane and the imaginary axis is included in ROC, the given causal system is stable.

Q3.9   *The impulse response of a system is $2e^{-5|t|}$. Determine the stability of the system.*

Solution:

Given that, $h(t) = 2e^{-5|t|} = 2e^{5t}$; $t \le 0$

$\hspace{6.5cm} = 2e^{-5t}$; $t \ge 0$

$\therefore\ h(t) = 2e^{5t}\, u(-t) + 2e^{-5t}\, u(t)$

$H(s) = \mathcal{L}\{h(t)\} = \mathcal{L}\{2e^{5t}\, u(-t) + 2e^{-5t}\, u(t)\}$

$$= -\frac{2}{s-5} + \frac{2}{s+5} = \frac{-2(s+5) + 2(s-5)}{(s-5)(s+5)} = \frac{-20}{(s-5)(s+5)}$$

The poles are lying at $s = 5$ and $s = -5$, and so one pole is lying on right half s-plane and another pole lying on left half s-plane.

The ROC is the region in between the lines passing through $s = 5$ and $s = -5$ and so imaginary axis is included in ROC.

Since imaginary axis is included in ROC, the given noncausal system is stable.

Q3.10   *Determine whether the given causal system with transfer function $H(s) = \dfrac{1}{s-2}$ is stable.*

Solution: ***(AU Dec'13, 2 Marks)***

Given that, $H(s) = \dfrac{1}{s-2}$

The impulse response is obtained by taking inverse Laplace transform of the transfer function.

$\therefore$ Impulse response, $h(t) = \mathcal{L}^{-1}\{H(s)\} = \mathcal{L}^{-1}\left\{\dfrac{1}{s-2}\right\} = e^{2t}\, u(t)$

The condition for stability is given by $\displaystyle\int_{-\infty}^{\infty} h(t)\, dt < \infty$.

$$\therefore\ \int_{-\infty}^{+\infty} |h(t)|\, dt = \int_{-\infty}^{+\infty} e^{2t}\, u(t)\, dt = \int_{0}^{+\infty} e^{2t}\, dt = \left[\frac{e^{2t}}{2}\right]_{0}^{\infty} = \frac{e^{\infty}}{2} - \frac{e^{0}}{2} = \infty - \frac{1}{2} = \infty$$

Here, $\displaystyle\int_{-\infty}^{\infty} h(t)\, dt = \infty$. Hence the system is unstable.

Q3.11 Determine the causality of the system described by impulse response $h(t) = e^{-t}u(t)$.

Solution: **(AU Dec'12, 2 Marks)**

Given that, $h(t) = e^{-t}u(t)$.

The condition for casuality is given by $h(t) = 0$ for $t < 0$

$\therefore\ h(t) = e^{-t}$; $t \geq 0$

$\quad\quad = 0$; $t < 0$

Therefore the condition is satisfied. Hence the system is casual.

Q3.12 Draw the direct form-I structure for the system represented by the transfer function

$$H(s) = \frac{2s^2 + 3s}{s^2 + 6s + 5}.$$

Solution:

Let, $H(s) = \dfrac{Y(s)}{X(s)} = \dfrac{2s^2 + 3s}{s^2 + 6s + 5}$

On cross multiplying the above equation we get,

$s^2Y(s) + 6sY(s) + 5Y(s) = 2s^2X(s) + 3sX(s)$

On dividing the above equation by s^2 we get,

$Y(s) + 6\dfrac{Y(s)}{s} + 5\dfrac{Y(s)}{s^2} = 2X(s) + 3\dfrac{X(s)}{s}$

$\therefore\ Y(s) = 2X(s) + 3\dfrac{X(s)}{s} - 6\dfrac{Y(s)}{s} - 5\dfrac{Y(s)}{s^2}$

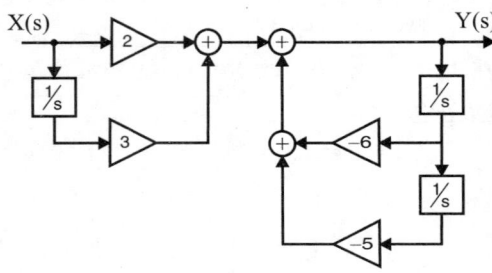

Fig Q3.12: Direct form-I structure.

The above equation can be used to construct the direct form-I structure as shown in Fig Q3.12.

Q3.13 Draw the direct form I-structure for the system described by $\dfrac{dy(t)}{dt} + y(t) = 0.1x(t)$.

Solution: **(AU Dec'14, 2 Marks)**

Given that, $\dfrac{dy(t)}{dt} + y(t) = 0.1x(t)$

On taking Lapalace transform of the above equation we get,

$s\,Y(s) + Y(s) = 0.1\,X(s)$

On dividing the above equation by s we get,

$Y(s) + \dfrac{Y(s)}{s} = 0.1\dfrac{X(s)}{s}$

$\therefore\ Y(s) = -\dfrac{Y(s)}{s} + 0.1\dfrac{X(s)}{s}$

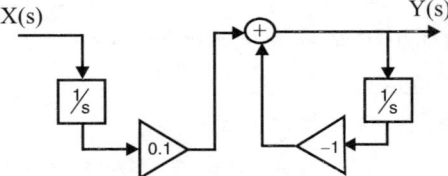

Fig Q3.13: Direct form-I structure.

The above equation can be used to construct the direct form-I structure as shown in Fig Q3.13.

Q3.14 Draw the direct form-II structure for the system represented by the transfer function

$$H(s) = \frac{0.4s^2 + 0.7s}{s^2 + 0.5s + 0.9}.$$

Solution:

Let, $H(s) = \dfrac{Y(s)}{X(s)} = \dfrac{0.4s^2 + 0.7s}{s^2 + 0.5s + 0.9} = \dfrac{s^2\left(0.4 + \dfrac{0.7}{s}\right)}{s^2\left(1 + \dfrac{0.5}{s} + \dfrac{0.9}{s^2}\right)} = \dfrac{0.4 + \dfrac{0.7}{s}}{1 + \dfrac{0.5}{s} + \dfrac{0.9}{s^2}}$

Let, $\dfrac{Y(s)}{X(s)} = \dfrac{W(s)}{X(s)} \times \dfrac{Y(s)}{W(s)}$

where, $\dfrac{W(s)}{X(s)} = \dfrac{1}{1 + \dfrac{0.5}{s} + \dfrac{0.9}{s^2}}$ .....(1)

$\dfrac{Y(s)}{W(s)} = 0.4 + \dfrac{0.7}{s}$ .....(2)

On cross multiplying and rearranging equation (1) we get,

$W(s) = X(s) - 0.5\,\dfrac{W(s)}{s} - 0.9\,\dfrac{W(s)}{s^2}$ .....(3)

On cross multiplying equation (2) we get,

$Y(s) = 0.4W(s) + 0.7\,\dfrac{W(s)}{s}$ .....(4)

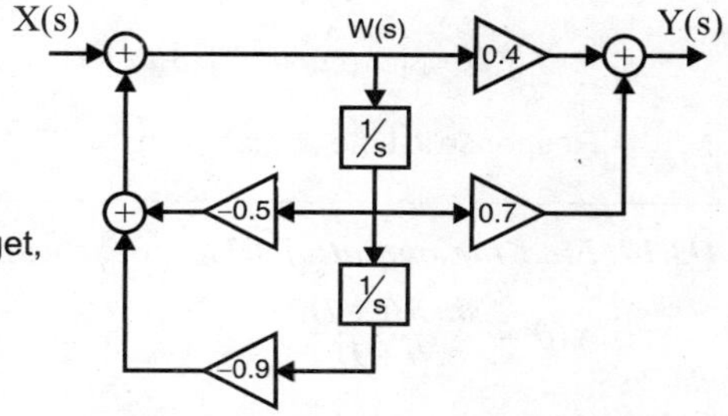

The direct form-II structure is constructed using equations (3) and (4) as shown in Fig Q3.14.

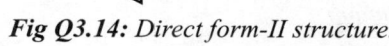

Fig Q3.14: *Direct form-II structure.*

Q3.15 *Given that, differential equation representation of a system* $\dfrac{d^2y(t)}{dt^2} + 2\dfrac{dy(t)}{dt} - 3y(t) = 2x(t)$. *Find the frequency response H(jΩ).*

(AU Dec'15, 2 Marks)

Solution:

Given that, $\dfrac{d^2y(t)}{dt^2} + 2\dfrac{dy(t)}{dt} - 3y(t) = 2x(t)$

On taking Fourier transform of the above equation we get,

$(j\Omega)^2\,Y(j\Omega) + 2\,j\Omega\,Y(j\Omega) - 3\,Y(j\Omega) = 2\,X(j\Omega)$

$Y(j\Omega)[(j\Omega)^2 + 2\,j\Omega - 3] = 2\,X(j\Omega)$

$\therefore\ \dfrac{Y(j\Omega)}{X(j\Omega)} = \dfrac{2}{(j\Omega)^2 + 2j\Omega - 3} = \dfrac{2}{(j\Omega + 3)(j\Omega - 1)}$

$\therefore$ Frequency response, $H(j\Omega) = \dfrac{Y(j\Omega)}{X(j\Omega)} = \dfrac{2}{(j3 + 3)(j\Omega - 1)}$

> The roots of quadratic $(j\Omega)^2 + 2j\Omega - 3 = 0$ are,
> $j\Omega = \dfrac{-2 \pm \sqrt{2^2 + 4 \times 3}}{2}$
> $= \dfrac{-2 \pm 4}{2} = 1, -3$

Q3.16 *The impulse response of a system in frequency domain is,*

$H(j\Omega) = +j\ ;\ \Omega < 0$

$\qquad\qquad = -j\ ;\ \Omega > 0$

Find the response for the input, x(t) = cos t.

Solution:

Given that, x(t) = cos t

$\therefore\ X(j\Omega) = \mathcal{F}\{x(t)\} = \mathcal{F}\{\cos t\} = \pi[\delta(\Omega + 1) + \delta(\Omega - 1)]$

$\therefore\ X(j\Omega) = \pi\delta(\Omega + 1)\ ;\ \Omega < 0$

$\qquad\qquad = \pi\delta(\Omega - 1)\ ;\ \Omega > 0$

Let y(t) be response and h(t) be impulse response in time domain.

We know that, $y(t) = x(t) * h(t)$

On taking Fourier transform of above equation we get,

$Y(j\Omega) = \mathcal{F}\{x(t) * h(t)\}$

$\qquad\quad = X(j\Omega)\,H(j\Omega)$

$\therefore$ $Y(j\Omega) = j \times \pi\delta(\Omega + 1)$; $\Omega < 0$

$\qquad$ $= -j \times \pi\delta(\Omega - 1)$; $\Omega > 0$

> Using convolution property of Fourier transform.

$\therefore$ $Y(j\Omega) = j\pi\delta(\Omega+1) - j\pi\delta(\Omega-1) = \dfrac{-\pi\,\delta(\Omega+1)}{j} + \dfrac{\pi\,\delta(\Omega-1)}{j} = \dfrac{\pi}{j}[\delta(\Omega-1) - \delta(\Omega+1)]$

Response in time domain, $y(t) = \mathcal{F}^{-1}\{Y(j\Omega)\} = \mathcal{F}^{-1}\left\{\dfrac{\pi}{j}[\delta(\Omega-1) - \delta(\Omega+1)]\right\} = \sin t$

Q3.17 *Find the output of LTI system with impulse response and input* $h(t) = \dfrac{\sin 5(t-1)}{\pi(t-1)}$ *and input*

$\qquad$ $x(t) = \dfrac{\sin 5(t+1)}{\pi(t+1)}$

Solution:

$\qquad$ $H(j\Omega) = \mathcal{F}\left\{\dfrac{\sin 5(t-1)}{\pi(t-1)}\right\} = \mathcal{F}\left\{\dfrac{\sin 5t}{\pi t}\right\} e^{-j\Omega}$

> Using time shifting property

$\qquad\qquad\quad = 1 \times e^{-j\Omega}$; $|\Omega| < 5 = e^{-j\Omega}$; $|\Omega| < 5$

> $\mathcal{F}\left\{\dfrac{\sin\Omega_0 t}{\pi t}\right\} = \mathcal{F}\left\{\dfrac{\Omega_0}{\pi}\dfrac{\sin\Omega_0 t}{\Omega_0 t}\right\}$
> $\qquad\qquad = \mathcal{F}\left\{\dfrac{\Omega_0}{\pi}\,\text{sinc}\,\Omega_0 t\right\}$
> $\qquad\qquad = 1$; $|\Omega| < \Omega_0$

Similarly, $X(j\Omega) = e^{-j\Omega}$; $|\Omega| < 5$

We know that,

Response, $y(t) = x(t) * h(t)$ $\qquad\qquad\qquad$ ......(1)

On taking Fourier transform of equation (1) we get,

$Y(j\Omega) = \mathcal{F}\{x(t) * h(t)\} = X(j\Omega)\,H(j\Omega)$

$\qquad\qquad = e^{j\Omega} \times e^{-j\Omega}$; $|\Omega| < 5 = 1$; $|\Omega| < 5$

> Using convolution property

$\therefore$ $y(t) = \mathcal{F}^{-1}\{Y(j\Omega)\} = \dfrac{5}{\pi}\,\text{sinc}\,5t = \dfrac{5}{\pi}\dfrac{\sin 5t}{5t} = \dfrac{\sin 5t}{\pi t}$

Q3.18 *Find the differential equation relating the input and output a continuous system represented by*

$\qquad$ $H(j\Omega) = \dfrac{4}{(j\Omega)^2 + 8j\Omega + 4}$ $\qquad\qquad$ **(AU Jun'14, 2 Marks)**

Solution:

Given that, $H(j\Omega) = \dfrac{4}{(j\Omega)^2 + 8j\Omega + 4}$

We know that, $H(j\Omega) = \dfrac{Y(j\Omega)}{X(j\Omega)} = \dfrac{4}{(j\Omega)^2 + 8j\Omega + 4}$

> if, $\mathcal{F}\{x(t)\} = X(j\Omega)$, then
> $\mathcal{F}\left\{\dfrac{d^m}{dt^m}x(t)\right\} = (j\Omega)^m\,X(j\Omega)$

On cross multiplying the above equation we get,

$(j\Omega)^2\,Y(j\Omega) + 8(j\Omega)\,Y(j\Omega) + 4\,Y(j\Omega) = 4\,X(j\Omega)$

On taking Inverse Fourier transform of the above equation we get,

$\dfrac{d^2 y(t)}{dt^2} + 8\dfrac{dy(t)}{dt} + 4y(t) = 4x(t)$

3.11 MATLAB Programs

Program 3.1

$\qquad$ Write a MATLAB program to perform convolution of the following two signals.

$\qquad$ x1(t)=1; 1<t<10 $\qquad\qquad$ x2(t)=1; 2<t<10

```
%*****************Program to perform convolution of two signals
%*****************x1(t)=1; t= 1 to 10 and x2(t)=1; t= 2 to 10

tmin=0; tmax=10; dt=0.01;
t=tmin:dt:tmax;                    %set time vector for given signal

x1=1.*(t>=1 & t<=10);              %generate signal x1(t)
x2=1.*(t>=2 & t<=10);              %generate signal x2(t)
x3=conv(x1,x2);                    %perform convolution of signals x1(t) and x2(t)

n3=length(x3);
t1=0:1:n3-1;                       %set time vector for x3(t) signal

subplot(3,1,1);plot(t,x1);
xlabel('t');ylabel('x1(t)');
title('signal x1(t)');

subplot(3,1,2);plot(t,x2);
xlabel('t');ylabel('x2(t)');
title('signal x2(t)');

subplot(3,1,3);plot(t1,x3); xlim ([0 600]);
xlabel('t / dt');ylabel('x3(t) / dt');
title('signal, x3(t) = x1(t)* x2(t)');
```

OUTPUT

The input and output waveforms of program 3.1 are shown in Fig P3.1.

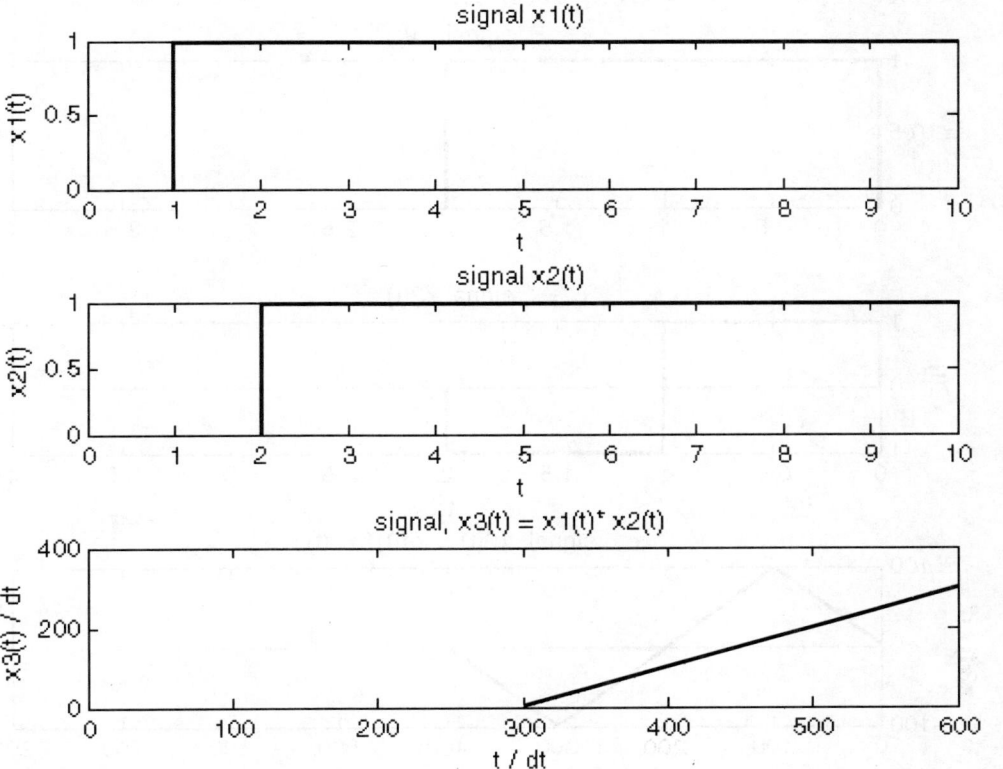

Fig P3.1: Input and Output waveforms of program 3.1.

Program 3.2

Write a MATLAB program to perform convolution of the following two signals.

x1(t)=1; 0<t<2   x2(t)= 1;   0<t<1
                              =-1;   1<t<2

```
%*****************Program to perform convolution of two signals
%*****************x1(t)=1; t= 0 to 2 and x2(t) = 1 ; for t= 0 to 1
%                                  = -1 ; for t= 1 to 2
tmin=0; tmax=4; dt=0.01;
t=tmin:dt:tmax;              %set time vector for given signal
x1=1.*(t>=0 & t<=2);        %generate signal x1(t)
xa=1;
xb=-1;
x2=xa.*(t>=0 & t<=1)+ xb.*(t>=1 & t<=2);    % generate signal x2(t)
x3=conv(x1,x2);                              % perform convolution of
                                             % x1(t) & x2(t)

n3=length(x3);
t1=0:1:n3-1;                %set time vector for signal x3(t)

subplot(3,1,1);plot(t,x1);
xlabel('t');ylabel('x1(t)');title('signal x1(t)');

subplot(3,1,2);plot(t,x2);
xlabel('t');ylabel('x2(t)');title('signal x2(t)');

subplot(3,1,3);plot(t1,x3);
xlabel('t / dt');ylabel('x3(t) / dt');
title('signal, x3(t) = x1(t)* x2(t)');
```

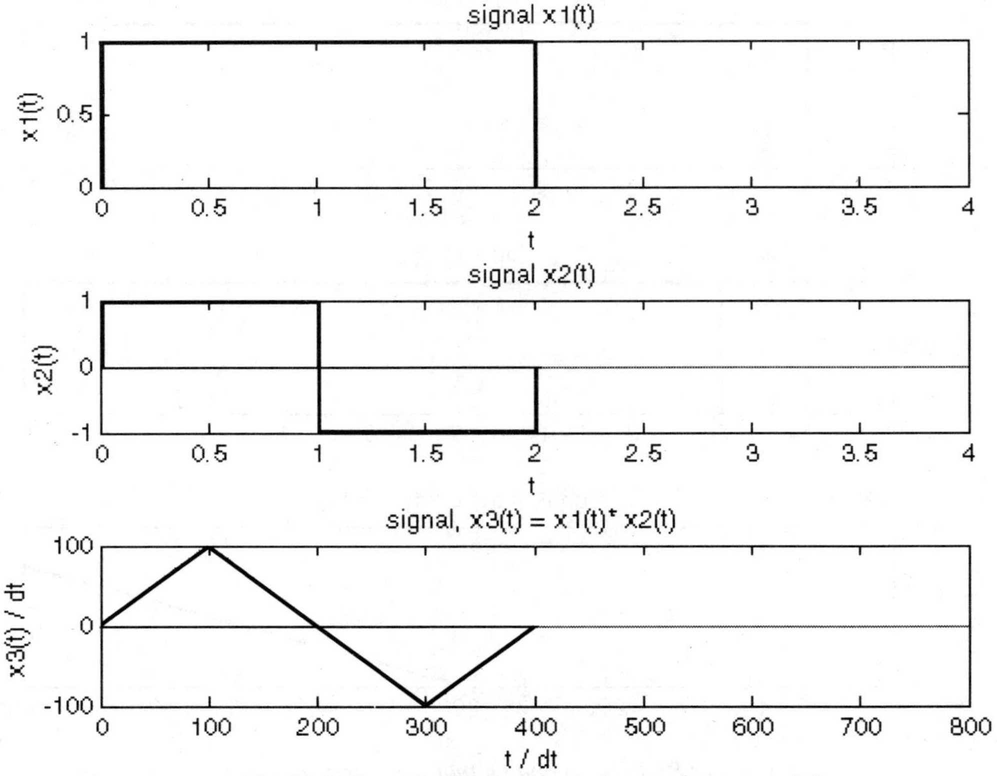

Fig P3.2: *Input and Output waveforms of program 3.2.*

OUTPUT

The input and output waveforms of program 3.2 are shown in Fig P3.2.

Program 3.3

Given that, x1(t) = e$^{-0.7t}$; 0<t<1   and   x2(t)=1;   0<t<2
Write a MATLAB program to perform following operations.
1. Convolution of x1(t) and x2(t).
2. Deconvolution of output of convolution with x2(t) to extract x1(t).
3. Deconvolution of output of convolution with x1(t) to extract x2(t).

```
%****************** Program to perform convolution and deconvolution
%****************** x1(t)=exp(-0.7t);0<=t<=T and x2(t)=1;0<=t<=T
T=2;
tmin = 0; tmax=2*T; dt=0.01;
t=tmin:dt:tmax;                        %set time vector for given signals

x1=exp(-0.7*t).*(t>=0 & t<T);          %generate signal x1(t)
x2=1.*(t>=0 & t<T);                    %generate signal x2(t)
x3=conv(x1,x2);                        %perform convolution of x1(t) and x2(t)

n3=length(x3);
```

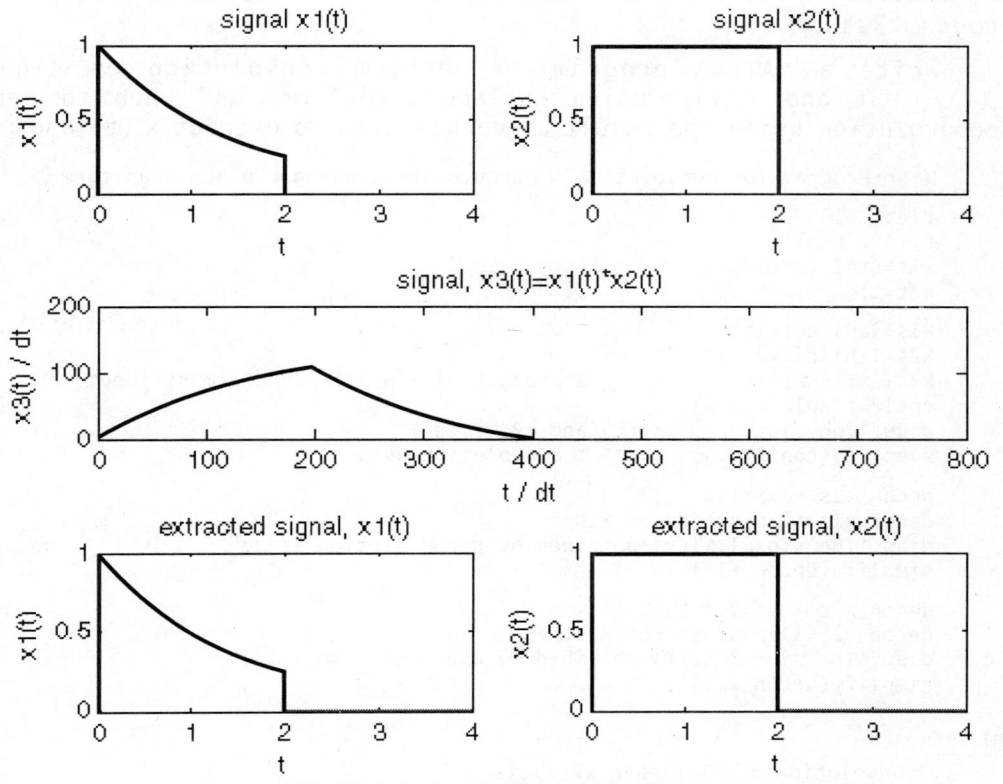

Fig P3.3: Input and Output waveforms of program 3.3.

```
t1=0:1:n3-1;          %set time vector for x3(t) signal
%plot the given signals and the signal obtained by convolution
subplot(3,2,1);plot(t,x1);title('signal x1(t)');
xlabel('t');ylabel('x1(t)');
subplot(3,2,2);plot(t,x2);title('signal x2(t)');
xlabel('t');ylabel('x2(t)');
subplot(3,2,3:4);plot(t1,x3);title('signal, x3(t)=x1(t)*x2(t)');
xlabel('t / dt');ylabel('x3(t) / dt');

x1d=deconv(x3,x2); %perform deconvolution to extract x1(t)
x2d=deconv(x3,x1); %perform deconvolution to extract x2(t)

ymin=min([min(x1d), min(x2d)]);
ymax=max([max(x1d), max(x2d)]);

%plot the signals obtained by deconvolution
subplot(3,2,5);plot(t,x1d);axis([tmin tmax ymin ymax]);
xlabel('t');ylabel('x1(t)');title('extracted signal, x1(t)');
subplot(3,2,6);plot(t,x2d);axis([tmin tmax ymin ymax]);
xlabel('t');ylabel('x2(t)');title('extracted signal, x2(t)');
```

OUTPUT

The input and output waveforms of program 3.3 are shown in Fig P3.3.

Program 3.4

Write a MATLAB program to perform convolution of signals, $x_1(t)=t^2-3t$ and $x_2(t)=t$, using Laplace transform, and then to perform deconvolution using the result of convolution to extract $x_1(t)$ and $x_2(t)$.

```
%****Program for convolution & deconvolution using Laplace transform

clear all;
syms t real;
x1t=(t^2-(2*t));
x2t=t;

X1s=laplace(x1t);
X2s=laplace(x2t);
X3s = X1s*X2s;              % product of laplace transform of inputs
con12=ilaplace(X3s);
disp('Convolution of x1(t) and x2(t) is');
simplify(con12)            % convolution output

decon_X1s=X3s/X1s;
decon_x1t=ilaplace(decon_X1s);
disp('The signal x1(t) obtained by deconvolution is');
simplify(decon_x1t)

decon_X2s=X3s/X2s;
decon_x2t=ilaplace(decon_X2s);
disp('The signal x2(t) obtained by deconvolution is');
simplify(decon_x2t)
```

OUTPUT

```
Convolution of x1(t) and x2(t) is
     ans =
         1/12*t^4-1/3*t^3
```

```
The signal x1(t) obtained by deconvolution is
      ans =
           t
The signal x2(t) obtained by deconvolution is
      ans =
         t^2-2*t
```

Program 3.5

Write a MATLAB program to find residues and poles of s-domain signal, $(1.5s^3+4.45s^2+4.25s+0.2)/(s^3+3.5s^2+3.5s+1)$

```
%*****Program to find residues and poles of s-domain signal

clear all
s=tf('s');
b=[1.5 4.45 4.25 0.2];      % Numerator coefficients
a=[1 3.5 3.5 1];            % Denominator coefficients
disp('The given transfer function is,');
f=tf([b], [a])

disp('The residues, poles and direct terms of given transfer function are,');
disp('(r - residue ; p - poles ; k - direct terms)');
[r,p,k]=residue(b,a)

disp('The numerator and denominator coefficients extracted from r,p,k are,');
[b,a]=residue(r,p,k)
```

OUTPUT

```
      The given transfer function is,
            Transfer function:
                1.5 s^3 + 4.45 s^2 + 4.25 s + 0.2
                ------------------
                s^3 + 3.5 s^2 + 3.5 s + 1
      The residues, poles and direct terms of given transfer function are,
      (r - residue ; p - poles ; k - direct terms)
          r =
             -1.6667
              2.2000
             -1.3333
          p =
             -2.0000
             -1.0000
             -0.5000
          k =
              1.5000
      The numerator and denominator coefficients extracted from r,p,k are,
          b =
             1.5000    4.4500    4.2500    0.2000
          a =
             1.0000    3.5000    3.5000    1.0000
```

Program 3.6

Write a MATLAB program to find the impulse response of the LTI system governed by the transfer function, $H(s)=1/(s^2+s+1)$.

```
%****Program to find impulse response of continuous time LTI system

syms s complex;
H=1/(s^2+s+1);
```

```
disp('Impulse response h(t) is');
h=ilaplace(H);
simplify(h)

s=tf('s');
H=1/(s^2+s+1);
t=0:0.01:20;              % set a time vector

h=impulse(H,t);           % impulse response of system H
plot(t,h);
xlabel('time in seconds');
ylabel('h(t)');
```

OUTPUT

```
Impulse response h(t) is
     ans =
          2/3*3^(1/2)*exp(-1/2*t)*sin(1/2*3^(1/2)*t)
```

The sketch of impulse response is shown in Fig P3.6.

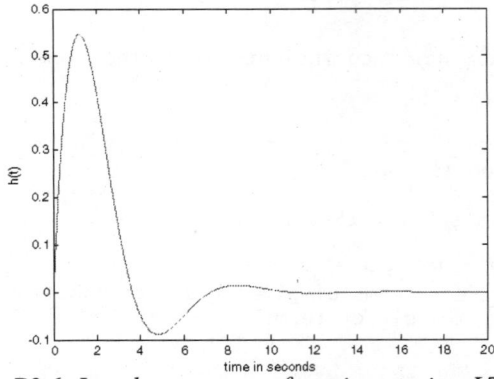

Fig P3.6: Impulse response of continuous time LTI system.

Program 3.7

Write a MATLAB program to find the step response of the first and second order LTI systems governed by the transfer functions,
$H(s)=1/(s+2)$ and $H(s)=1/(s^2+2.5s+25)$.

```
%********Program to find the step response of I and II order systems
syms s complex;
H1=1/(s+2);
disp('Step response of first order system is');
h1=ilaplace(H1);
simplify(h1)

H2=1/(s^2+2.5*s+25);
disp('Step response of second order system is');
h2=ilaplace(H2);
simplify(h2)

s=tf('s');
H1=1/(s+2);
```

```
H2=1/(s^2+2.5*s+25);
t1=0:0.0005:5;                    % set a time vector
s1=step(H1,t1);                   % step response of first order system
s2=step(H2,t1);                   % step response of second order system
subplot(2,1,1);plot(t1,s1);
xlabel('Time in seconds'); ylabel('s1(t)');
subplot(2,1,2);plot(t1,s2);
xlabel('Time in seconds');
ylabel('s2(t)');
```

OUTPUT

```
      Step response of first order system is
              ans =
                    exp(-2*t)
      Step response of second order system is
              ans =
                    4/75*15^(1/2)*exp(-5/4*t)*sin(5/4*15^(1/2)*t)
      The sketch of step response of first and second order system are shown in Fig P3.7.
```

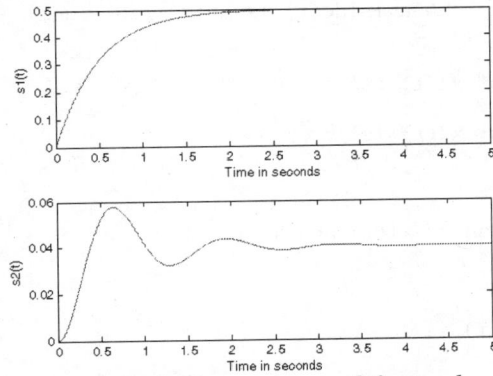

Fig P3.7: *Step response of first and*
second order system.

Program 3.8

Write a MATLAB program to determine response of LTI system whose impulse response is h(t)= 2e^{-3t}u(t) for the input is x(t)=2e^{-5t}u(t),using Fourier transform.

```
% Program to find the  response of LTI system using Fourier transform

syms t real;
h=2*exp(-3*t).*heaviside(t);
x=2*exp(-5*t).*heaviside(t);
disp('Fourier transform of the impulse response is');
H=fourier(h)

disp('Fourier transform of the input is');
X=fourier(x)

Y=H*X;
disp('The response of the given system is');
y_res=ifourier(Y,t)
```

OUTPUT

```
Fourier transform of the impulse response is
    H =
        2/(3+i*w)
Fourier transform of the input is
    X =
        2/(5+i*w)
The response of the given system is
    y_res =
        2*heaviside(t)*(exp(-3*t)-exp(-5*t))
```

Program 3.9

Write a MATLAB program to perform convolution of signals, $x_1(t)= e^{-2t}u(t)$ and $x_2(t)=e^{-6t}u(t)$, using Fourier transform.

```
% Program to perform convolution using Fourier transform

syms t real;
x1=exp(-2*t).*heaviside(t); %heaviside(t) is unit step signal
x2=exp(-6*t).*heaviside(t);

disp('Fourier transform of x1(t) is');
X1=fourier(x1)
disp('Fourier transform of x2(t) is');
X2=fourier(x2)
Y=X1*X2;

disp('Let x3 be convolution of x1(t) and x2(t).');
x3=ifourier(Y,t)
```

OUTPUT

```
Fourier transform of x1(t) is
    X1 =
        1/(2+i*w)
Fourier transform of x2(t) is
    X2 =
        1/(6+i*w)
Let x3 be convolution of x1(t) and x2(t).
    x3 =
        1/4*heaviside(t)*(exp(-2*t)-exp(-6*t))
```

3.12 Exercises

I. Fill in the blanks with appropriate words.

1. For———— of LTI system the integral of impulse response should be finite.

2. In ————— system the present output depends on past outputs.

3. The response of an LTI system is given by ————— of input and impulse response.

4. In ————— connection the output of a system becomes input for another system.

5. The ROC of X(s) consists of stripes parallel to the ————— in s-plane.

6. The ————— is the ratio of Laplace transform of output and input.

7. The Laplace transform of ————— gives the transfer function of the system.

8. For a stable LTI system the ROC should include ————— of s-plane.

9. For a stable causal system ——————— should lie on left half of s-plane.

10. Fourier transform of periodic continuous time signal consists of _____ located at _____ frequencies of the signal.

11. Fourier transform is evalution of Laplace transform along the _____ axis in s-plane.

Answers		
1. stability	5. $j\Omega$ – axis	9. poles
2. feedback	6. transfer function	10. impulses, harmonic
3. convolution	7. impulse response	11. imaginary
4. cascade	8. imaginary axis	

II. State whether the following statements are True/False.

1. All LTI systems are BIBO stable.

2. The stability of non-LTI systems can be tested by using its impulse response.

3. The nonfeedback systems depend on past inputs.

4. Convolution is a linear operation.

5. The convolution can be used to determine the response of non-LTI systems.

6. The convolution operation satisfies commutative property.

7. The inverse Laplace transform of the transfer function gives the impulse response.

8. The direct form-I structure uses less number of integrators than direct form-II structure.

9. Fourier transform is useful for frequency domain analysis of both periodic and non-periodic signals.

10. Laplace transform is a generalized transform and Fourier transform is a particular transform.

Answers				
1. False	3. True	5. False	7. True	9. True
2. False	4. True	6. True	8. False	10. True

III. Choose the right answer for the following questions.

1. *Which of the following responses of an LTI system does not depend on initial conditions?*

    a) natural response   **b)** free response   **c)** forced response   **d)** total response

2. *Which of the following statements are true ?*

    i)   an LTI system is always stable.

    ii) an LTI system is stable only if the integral of its impulse response is finite.

    iii) in a system, if the input is bounded then the output is always bounded.

    iv) in a system, even if the input is unbounded the output can be bounded

    a) ii only   **b)** ii and iv only   **c)** iii only   **d)** i and iv only

3. *Which of the following impulse responses of LTI systems represents an unstable system?*

    a) $h(t) = e^{at} u(t)$ **b)** $h(t) = e^{-at} u(t)$ **c)** $h(t) = t\, e^{-t} u(t)$ **d)** $h(t) = e^{-t} \sin t\, u(t)$

4. Which of the following impulse responses of LTI systems represents a stable system?

    a) h(t) = $e^t \cos t \, u(t)$ **b)** h(t) = $e^t \sin t \, u(t)$ **c)** h(t) = $e^{-t}\cos t \, u(t)$ **d)** h(t) = $t \sin t \, u(t)$

5. Which of the following represents the convolution of two causal signals $x_1(t)$ and $x_2(t)$?

    i) $\int_{\lambda=0}^{t} x_1(\lambda)x_2(t-\lambda)\, d\lambda$ ii) $\int_{\lambda=0}^{t} x_2(\lambda)x_1(t-\lambda)\, d\lambda$ iii) $\int_{t=0}^{\lambda} x_1(t)x_2(\lambda-t)\, dt$ iv) $\int_{t=0}^{\lambda} x_2(t)x_1(\lambda-t)\, dt$

    a) i only   **b)** ii only   **c)** i and ii only   **d)** all of the above

6. If $h_1(t)$ and $h_2(t)$ are impulse responses of two stable LTI systems in cascade, then overall impulse response of the cascaded system is,

    a) $h_1(t) * h_2(t)$ **b)** $h_1(t) \, h_2(t)$ **c)** $h_1(t) + h_2(t)$ **d)** $h_1(t) - h_2(t)$

7. If the impulse response of an LTI causal system is in the form e^{-at} then its response for step input of value A will be in the form,

    a) $A(1 - e^{-at})\, u(t)$ **b)** $Ae^{-at}\, u(t)$ **c)** $\dfrac{A}{a}(1-e^{-at})\, u(t)$ **d)** $(A - e^{-at})\, u(t)$

8. If x(t), y(t) and h(t) are input, output and impulse response of LTI continuous time system respectively then,

    a) $h(t) = x(t) * y(t)$ **b)** $x(t) = y(t) * h(t)$ **c)** $y(t) = x(t) * h(t)$ **d)** $y(t) = h(t) * h(t)$

9. If X(s), Y(s) and H(s) are Laplace transform of input, output and impulse response of LTI continuous time system respectively then,

    a) $x(t) = \mathcal{L}^{-1}\left\{\dfrac{H(s)}{Y(s)}\right\}$ **b)** $x(t) = \mathcal{L}^{-1}\left\{\dfrac{Y(s)}{H(s)}\right\}$

    c) $x(t) = \mathcal{L}^{-1}\left\{\dfrac{1}{Y(s)H(s)}\right\}$ **d)** $x(t) = \mathcal{L}^{-1}\{Y(s)H(s)\}$

10. The convolution of u(t) with u(t) will be equal to,

    a) $\delta(t)$ **b)** $u(t)$ **c)** $t\, u(t)$ **d)** $t^2\, u(t)$

11. The convolution of $e^{-at}\, u(t)$ with $e^{-at}\, u(t)$ will be equal to,

    a) $t\, u(t)$ **b)** $e^{-at}\, u(t)$ **c)** $e^{-2at}\, u(t)$ **d)** $te^{-at}\, u(t)$

12. Given that $H(s) = e^{-4s}$. What is the impulse response of the system?

    a) $\delta(t-4)$ **b)** $u(t-4)$ **c)** $e^{-4t}\, u(t)$ **d)** $e^{4t}\, u(t)$

13. Which of the following statements are true regarding the stability of continuous time system?

    i)   For stability of LTI system, the ROC should include imaginary axis.

    ii) For stability of causal system the poles should lie on left half s-plane.

    iii) For stability of noncausal system there is no restriction on location of poles in s-plane.

    a) i and ii only   **b)** i and iii only   **c)** ii and iii only   **d)** all of the above

Answers

1. c	4. c	7. c	10. c	13. d
2. b	5. d	8. c	11. d	
3. a	6. a	9. b	12. a	

IV. Answer the following questions.

1. What is the condition for stability of an LTI system?
2. Define feedback and nonfeedback systems.
3. Define convolution of two continuous time signals.
4. Show that the response of an LTI system can be obtained by convolution of input and impluse response?
5. List the properties of convolution.
6. State and prove the associative and commutative properties of convolution.
7. State and prove the distributive property of convolution.
8. What are the two ways of interconnection of LTI systems?
9. What is the relation between impulse response and unit step response of an LTI system?
10. Write the procedure to perform convolution using Laplace transform.
11. Write the procedure to perform deconvolution using Laplace transform.
12. What is the relation between impulse response and transfer function of the system in s-domain?
13. What is the condition for stability of a continuous time LTI system?
14. What is the condition for stability of LTI causal system?
15. What are the basic elements of block diagram?
16. Define frequency domain transfer function.
17. What is the relation between impulse response and transfer function in frequency domain.
18. How the response of a system can be determined via Fourier transform?

V. Solve the following problems.

E3.1 **Determine the natural response of the following systems.**

a) $\dfrac{d^3y(t)}{dt^3} + 8\dfrac{d^2y(t)}{dt^2} + 17\dfrac{dy(t)}{dt} + 10\,y(t) = \dfrac{dx(t)}{dt} + 0.5\,x(t)$

$y(0) = 2\,;\,\dfrac{dy(t)}{dt}\bigg|_{t=0} = -4\,;\,\dfrac{d^2y(t)}{dt^2}\bigg|_{t=0} = 10$

b) $\dfrac{d^2y(t)}{dt^2} + 1.6\dfrac{dy(t)}{dt} + 0.63y(t) = 0\,;\,\dfrac{dy(t)}{dt}\bigg|_{t=0} = -0.05\,;\,y(0) = 0.02$

c) $\dfrac{d^2y(t)}{dt^2} + 2\dfrac{dy(t)}{dt} + 0.75y(t) = 0\,;\,y(0) = 0.25\,;\,\dfrac{dy(t)}{dt}\bigg|_{t=0} = -1$

E3.2 **Determine the forced response of the following systems.**

a) $\dfrac{d^3y(t)}{dt^3} + 7\dfrac{d^2y(t)}{dt^2} + 14\dfrac{dy(t)}{dt} + 8\,y(t) = \dfrac{dx(t)}{dt} + 0.5\,x(t)\,;\,x(t) = e^{-3t}\,u(t)$

b) $\dfrac{d^2y(t)}{dt^2} + 1.8\dfrac{dy(t)}{dt} + 0.45\,y(t) = 0.5\dfrac{dx(t)}{dt} + 0.18\,x(t)\ ; x(t) = u(t)$

c) $\dfrac{d^2y(t)}{dt^2} + 0.7\dfrac{dy(t)}{dt} + 0.1\,y(t) = 0.4\,x(t)\ ; x(t) = 0.2\,e^{-0.3t}\,u(t)$

E3.3 Determine the total response of the following systems.

a) $\dfrac{d^2y(t)}{dt^2} + \dfrac{dy(t)}{dt} + 0.21y(t) = \dfrac{dx(t)}{dt} + x(t)\ ; x(t) = 0.2e^{-0.5t}u(t)\ ;$

$\left.\dfrac{dy(t)}{dt}\right|_{t=0} = -1\,; y(0) = 0.3$

b) $\dfrac{d^2y(t)}{dt^2} + 9\dfrac{dy(t)}{dt} + 14\,y(t) = 2x(t)\ ; x(t) = 0.7\,u(t)\ ; \left.\dfrac{dy(t)}{dt}\right|_{t=0} = -0.5\,; y(0) = 0.2$

c) $\dfrac{d^2y(t)}{dt^2} + 3.1\dfrac{dy(t)}{dt} + 2.2\,y(t) = 0.1x(t)\ ; x(t) = 0.7\,e^{-2.5t}\,u(t)\ ; \left.\dfrac{dy(t)}{dt}\right|_{t=0} = -0.5\,; y(0) = 0$

E3.4 Verify the stability of LTI systems whose impulse responses are given below:

a) $h(t) = t^2\,u(t)$ **b)** $h(t) = e^{-7t}\,u(t)$ **c)** $h(t) = e^{5t}\,u(t)$

d) $h(t) = t\sin t\,u(t)$ **e)** $h(t) = (A + Be^{-ct})\,u(t)$ **f)** $h(t) = 2e^{-3t}\cos t\,u(t)$

E3.5 Determine the overall impulse response of the following system.

a)   **b)**

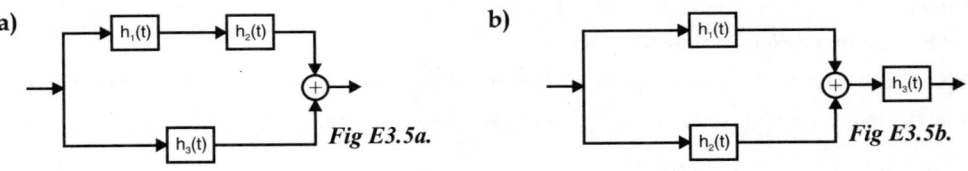

Fig E3.5a.   Fig E3.5b.

$h_1(t) = e^{-6t}\,u(t);\ h_2(t) = e^{2t}\,u(t);\ h_3(t) = t\,u(t)$ $h_1(t) = \cos t\,u(t);\ h_2(t) = \sin t\,u(t);\ h_3(t) = u(t)$

E3.6 Perform the convolution of the following signals.

a) $x_1(t) = t^2u(t)\ ;\ x_2(t) = u(t-1)$

b) $x_1(t) = t\,e^{-4t}u(t)\ ;\ x_2(t) = u(t)$

c) $x_1(t) = t\,u(t)\ ;\ x_2(t) = \sin t\,u(t)$

E3.7 Determine the unit step response of the following systems whose impulse responses are given below:

a) $h(t) = te^{-9t}\,u(t)$ **b)** $h(t) = e^{-t}\cos t\,u(t)$

c) $h(t) = e^{-5t}\,u(t-2)$ **d)** $h(t) = u(t-3) + (t+2)$

E3.8 Perform convolution of the following signals by graphical method and sketch the resultant waveform.

a)

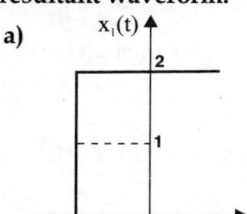 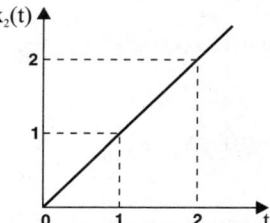

b)

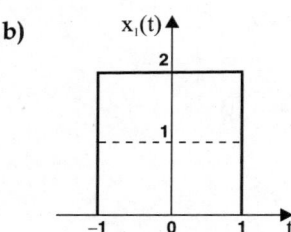 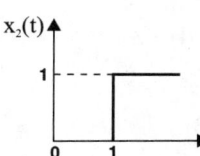

E3.9 Find the impulse response of the systems represented by following differential equations.

a) $\dfrac{dy(t)}{dt} = x(t - t_0)$

b) $3\dfrac{d^2y(t)}{dt^2} + y(t) = 0.5\,x(t)$

c) $\dfrac{d^2y(t)}{dt^2} + 0.8\dfrac{dy(t)}{dt} + 0.15y(t) = 0.2\dfrac{dx(t)}{dt} + x(t)$

E3.10 Using Laplace transform, determine the natural response of the system represented by following equations.

a) $\dfrac{d^2y(t)}{dt^2} + 0.15\dfrac{dy(t)}{dt} + 0.36y(t) = 0.1\dfrac{dx(t)}{dt} + 0.7x(t)$; $y(0) = 0.3$; $\dfrac{dy(t)}{dt}\bigg|_{t=0} = -0.2$

b) $\dfrac{d^2y(t)}{dt^2} + 10\dfrac{dy(t)}{dt} + 21y(t) = 8\,x(t)$; $y(0) = 2$; $\dfrac{dy(t)}{dt}\bigg|_{t=0} = -3$

E3.11 Using Laplace transform, determine the forced response of the system represented by following equations.

a) $\dfrac{d^2y(t)}{dt^2} + 9\dfrac{dy(t)}{dt} + 20\,y(t) = 0.2\dfrac{dx(t)}{dt} + 2\,x(t)$; Input $x(t) = 6\,u(t)$

b) $\dfrac{d^3y(t)}{dt^3} + 10\dfrac{d^2y(t)}{dt^2} + 27\dfrac{dy(t)}{dt} + 18\,y(t) = 12\,x(t)$; Input $x(t) = e^{-5t}\,u(t)$

E3.12 Using Laplace transform, determine the complete response of the system represented by following equations.

a) $\dfrac{d^2y(t)}{dt^2} + 1.4\dfrac{dy(t)}{dt} + 0.4\,y(t) = 0.2\dfrac{dx(t)}{dt} + 1.25\,x(t)\,;\,y(0) = 1\,;\,\dfrac{dy(t)}{dt}\bigg|_{t=0} = -2\,;$ for input $x(t) = 2e^{-3t}u(t)$

b) $\dfrac{d^2y(t)}{dt^2} + 11\dfrac{dy(t)}{dt} + 24\,y(t) = 3\,x(t)\;;\;y(0) = 2\quad;\dfrac{dy(t)}{dt}\bigg|_{t=0} = -0.5\,;$ for input $x(t) = 4\,u(t)$

E3.13 Find the transfer function of the system governed by the following differential equations.

a) $\dfrac{d^3y(t)}{dt^3} + 0.4\dfrac{d^2y(t)}{dt^2} + 0.3\dfrac{dy(t)}{dt} + 0.1y(t) = 0.3\dfrac{dx(t)}{dt} + x(t-1)$

b) $\dfrac{d^3y(t)}{dt^3} + 7\dfrac{d^2y(t)}{dt^2} + 5\dfrac{dy(t)}{dt} + 14y(t) = 3\dfrac{d^2x(t)}{dt^2} + 2\dfrac{dx(t)}{dt} + 4x(t)$

c) $4\dfrac{d^2y(t)}{dt^2} + 2\dfrac{dy(t)}{dt} + 8y(t) = x(t-3)$

E3.14 Find the transfer function of the system governed by the following impulse responses.

a) $h(t) = \delta(t-2) + e^{-4t}\,u(t)$ **b)** $h(t) = \delta(t) + t\,e^{-3t}\,u(t) + e^{-9t}\,u(t)$

c) $h(t) = 2\,u(t) + 3\,e^{-6t}\,u(t) + 4\,e^{-7t}\sin 2t\,u(t)$

E3.15 Determine the transfer function of the system, whose unit step responses are given below:

a) $s(t) = [t + \cosh 2t]\,u(t)$ **b)** $h(t) = [t^2 + 2\,e^{5t}\sin 3t]\,u(t)$

c) $s(t) = [3t^2 - 4e^{-7t} + 4e^{-t}]\,u(t)$

E3.16 Find the impulse response of the continuous time systems governed by the following transfer functions.

a) $H(s) = \dfrac{1}{s^2(s+\sqrt{2})}$ **b)** $H(s) = \dfrac{4}{s^2(s-16)}$ **c)** $H(s) = \dfrac{3}{s^2+18s+90}$

E3.17 Find the response of LTI systems whose input and impulse responses are given below:

a) $x(t) = 0.5e^{-2t}\,u(t)$; $h(t) = u(t)$

b) $x(t) = te^{-2t}\,u(t)$; $h(t) = \delta(t-2)$

c) $x(t) = 3e^{-2t}\cos4t\,u(t)$; $h(t) = e^{-6t}\,u(t)$

E3.18 Perform deconvolution operation to extract the signal $x_1(t)$.

a) $x_1(t) * x_2(t) = (1+t)\,u(t)$; $x_2(t) = \delta(t)$

b) $x_1(t) * x_2(t) = (te^{-t} + 3t^2)\,u(t)$; $x_2(t) = 2e^{-3t}u(t)$

c) $x_1(t) * x_2(t) = t^2\,u(t)$; $x_2(t) = 2\cosh t\,u(t)$

E3.19 Find the unit step response of the systems whose impulse responses are given below:

a) $h(t) = (t + \sin t)\, u(t)$ **b)** $h(t) = \delta(t - 3) + u(t - 3)$ **c)** $h(t) = (te^{-2t} + e^{-4t})\, u(t)$

E3.20 Draw the direct form-I and direct form-II structure of the continuous time systems respresented by the following equations.

a) $\dfrac{d^2 y(t)}{dt^2} + 5\dfrac{dy(t)}{dt} + 2\,y(t) = 3\dfrac{d^2 x(t)}{dt^2} + 4\dfrac{dx(t)}{dt} + 6x(t)$

b) $\dfrac{d^3 y(t)}{dt^3} + 0.9\dfrac{d^2 y(t)}{dt^2} + 1.2\dfrac{dy(t)}{dt} + 2.1\,y(t) = 3.1\dfrac{d^2 x(t)}{dt^2} + 0.6\dfrac{dx(t)}{dt} + 0.9x(t)$

E3.21 Draw the direct form-I and direct form-II structure of the continuous time systems governed by the following transfer functions.

a) $H(s) = \dfrac{4s^2 + 2s - 0.9}{s^3 + 2s^2 - 13s + 0.2}$ **b)** $H(s) = \dfrac{2s^3 + 3s^2 + 7s + 10}{s^3 - 9s^2 + 12s + 13}$

E3.22 Draw the cascade and parallel structure for the LTI system governed by the following transfer functions.

a) $H(s) = \dfrac{(s+2)(s-5)}{(s+4)(s^2+s+3)}$ **b)** $H(s) = \dfrac{s^2 + 2s}{(s+6)(s^2 - 7s + 10)}$

Answers

E3.1 **a)** $y_{zi}(t) = \left(\dfrac{1}{2}e^{-t} + \dfrac{4}{3}e^{-2t} + \dfrac{1}{6}e^{-5t}\right) u(t)$ **b)** $y_{zi}(t) = (-0.16e^{-0.7t} + 0.18e^{-0.9t})\, u(t)$

c) $y_{zi}(t) = (-0.625e^{-0.5t} + 0.875e^{-1.5t})\, u(t)$

E3.2 **a)** $y_{zs}(t) = \left(\dfrac{-5}{12}e^{-t} + \dfrac{5}{4}e^{-2t} + \dfrac{5}{12}e^{-4t} - \dfrac{5}{4}e^{-3t}\right) u(t)$

b) $y_{zs}(t) = (-0.5e^{-0.3t} + 0.1e^{-1.5t} + 0.4)\, u(t)$

c) $y_{zs}(t) = \left(\dfrac{8}{3}e^{-0.2t} + \dfrac{4}{3}e^{-0.5t} - 4e^{-0.3t}\right) u(t)$

E3.3 **a)** $y(t) = (4.275e^{-0.3t} + 8.525e^{-0.7t} - 12.5e^{-0.5t})\, u(t)$

b) $y(t) = (0.04e^{-2t} + 0.08e^{-7t} + 0.1)\, u(t)$

c) $y(t) = (-0.5e^{-1.1t} + 0.4e^{-2t} + 0.1e^{-2.5t})\, u(t)$

E3.4 **b)** **f)** – stable

a) **c)** **d)** **e)** – Unstable

E3.5 **a)** $h_o(t) = \left[h_1(t) * h_2(t)\right] + h_3(t) = \left(\dfrac{1}{8}e^{2t} - \dfrac{1}{8}e^{-6t} + t\right) u(t)$

b) $h_0(t) = \left[h_1(t) + h_2(t)\right] * h_3(t) = \left(\sin t - \cos t + 1\right) u(t)$

E3.6 **a)** $\left(\dfrac{t^3-1}{3}\right)u(t-1)$ **b)** $\dfrac{1}{16}\left(1-e^{-4t}-4te^{-4t}\right)u(t)$ c) $(t+\sin t)u(t)$

E3.7 **a)** $s(t)=\dfrac{1}{81}(1-e^{-9t}-9te^{-9t})\,u(t)$ **b)** $s(t)=\dfrac{1}{2}(1+e^{-t}\sin t-e^{-t}\cos t)\,u(t)$

   c) $s(t)=\dfrac{e^{-10}}{5}(1-e^{-5(t-2)})\,u(t-2)$ **d)** $s(t)=(t-3)\,u(t-3)+(t+2)\,u(t+2)$

E3.8 **a)** $x_1(t)*x_2(t)=(t+1)^2\,u(t+1)$ **b)** $x_1(t)*x_2(t)=2t-2(t-2)\,u(t-2)$
   (or) $x_1(t)*x_2(t)=(t+1)^2\;;\;t\ge-1$ (or) $x_1(t)*x_2(t)=2t\;;\;t=0\;\text{to}\;2$
                                                       $=4\;\;;\;t\ge2$

E3.9 **a)** $h(t)=u(t-t_0)$ **b)** $h(t)=\dfrac{0.5}{\sqrt{3}}\sin\left(\dfrac{1}{\sqrt{3}}t\right)u(t)$

   c) $h(t)=[\,4.7\,e^{-0.3t}-4.5\,e^{-0.5t}]\,u(t)$

E3.10 **a)** $y_{zi}(t)=[\,0.1778\,e^{-0.3t}+0.1222\,e^{-1.2t}]\,u(t)$ **b)** $y_{zi}(t)=\left(\dfrac{-3}{4}e^{-7t}+\dfrac{11}{4}e^{-3t}\right)u(t)$

E3.11 **a)** $y_{zi}(t)=(0.6-1.8e^{-4t}+1.2e^{-5t})\,u(t)$
   b) $y_{zi}(t)=(0.3e^{-t}-e^{-3t}+1.5e^{-5t}-0.8e^{-6t})\,u(t)$

E3.12 **a)** $y(t)=(0.25e^{-3t}+1.5833e^{-t}-0.8333e^{-0.4t})\,u(t)$
   b) $y(t)=(0.5-0.8e^{-8t}+2.3e^{-3t})\,u(t)$

E3.13 **a)** $\dfrac{Y(s)}{X(s)}=\dfrac{0.3s+e^{-s}}{s^3+0.4s^2+0.3s+0.1}$ **b)** $\dfrac{Y(s)}{X(s)}=\dfrac{3s^2+2s+4}{s^3+7s^2+5s+14}$ **c)** $\dfrac{Y(s)}{X(s)}=\dfrac{e^{-3s}}{4s^2+2s+8}$

E3.14 **a)** $H(s)=\dfrac{e^{-2s}(s+4)+1}{(s+4)}$

   b) $H(s)=\dfrac{s^3+16s^2+70s+99}{s^3+15s^2+63s+81}$

   c) $H(s)=\dfrac{5s^3+90s^2+481s+636}{s(s^3+20s^2+137s+318)}$

E3.15 **a)** $H(s)=\dfrac{2s^2-4}{s^2-4}$

   b) $H(s)=\dfrac{7s^2+10s+34}{s(s^2+10s+34)}$

   c) $H(s)=\dfrac{24s^3+6s^2+48s+42}{s^4+8s^3+7s^2}$

E3.16 a) $h(t) = \frac{1}{2} [\sqrt{2}\, t - 1 + e^{-\sqrt{2}\, t}]\, u(t)$ **b)** $h(t) = \frac{1}{4} [\cosh 4t - 1]\, u(t)$

     c) $h(t) = e^{-9t} \sin 3t\, u(t)$

E3.17 a) $y(t) = 0.25 [1 - e^{-2t}]\, u(t)$ **b)** $y(t) = (t - 2)\, e^{-(t-2)}\, u(t-2)$

     c) $y(t) = 0.375 [e^{-2t}(\sin 4t + \cos 4t) - e^{-6t}]\, u(t)$

E3.18 a) $x_1(t) = (t + 1)\, u(t)$ **b)** $x_1(t) = [(t + 0.5)\, e^{-t} + 3t + 4.5\, t^2]\, u(t)$

     c) $x_1(t) = \left(t - \dfrac{t^3}{6}\right) u(t)$

E3.19 a) $s(t) = [1 + t - \cos t]\, u(t)$ **b)** $s(t) = u(t-3) + (t-3)\, u(t-3)$

     c) $s(t) = 0.25 [2 - 2\, te^{-2t} - e^{-2t} - e^{-4t}]\, u(t)$

E3.20 a)

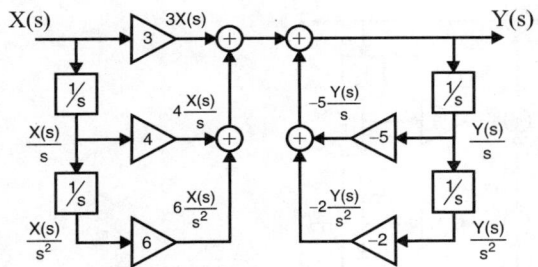 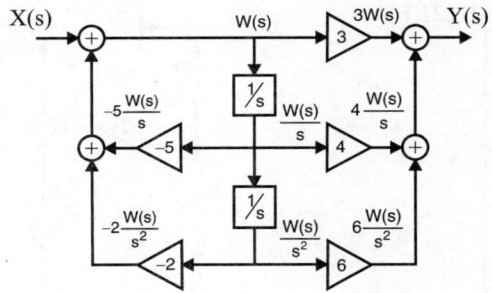

Fig E3.20a.1: Direct form-I structure.   *Fig E3.20a.2: Direct form-II structure.*

E3.20 b)

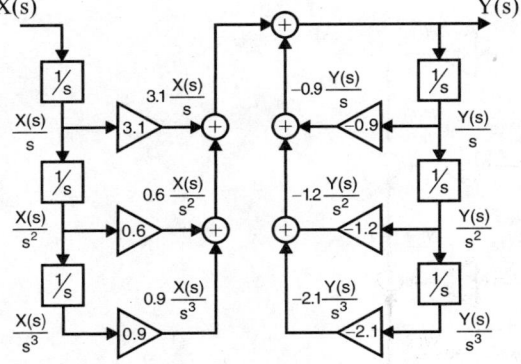 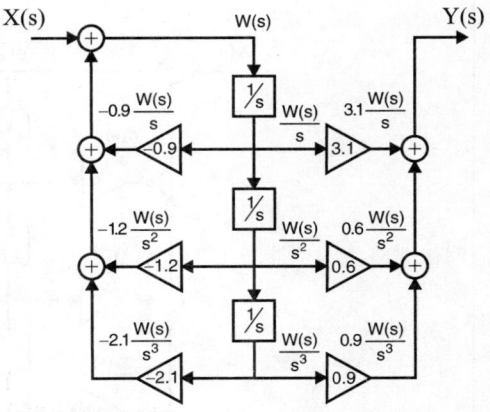

Fig E3.20b.1: Direct form-I structure.

Fig E3.20b.2: Direct form-II structure.

E3.21 a)

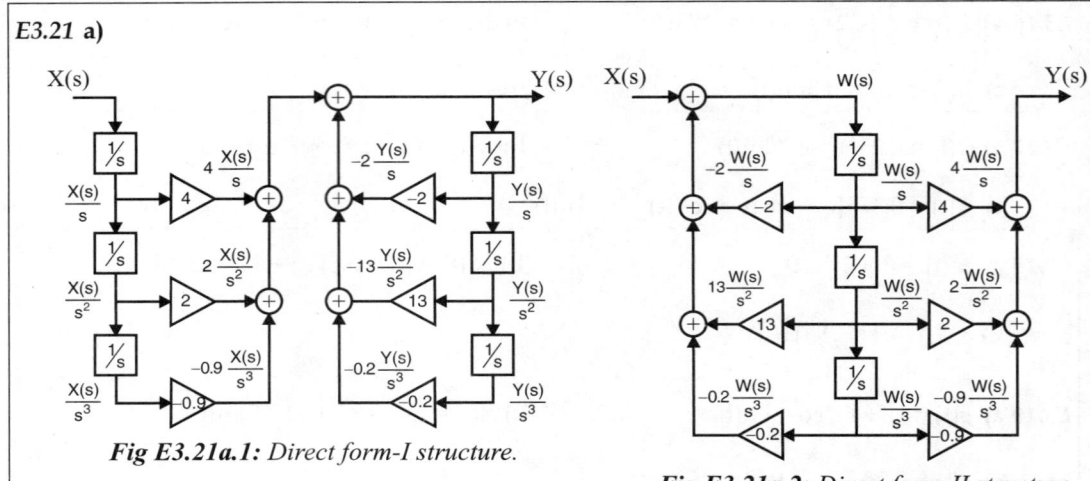

Fig E3.21a.1: *Direct form-I structure.*

Fig E3.21a.2: *Direct form-II structure.*

E3.21 b)

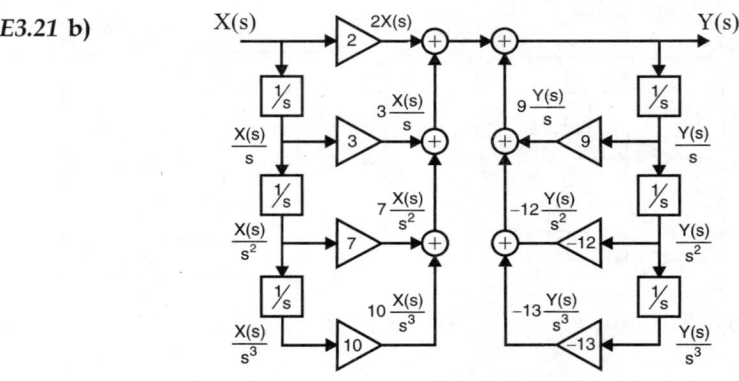

Fig E3.21b.1: *Direct form-I structure.*

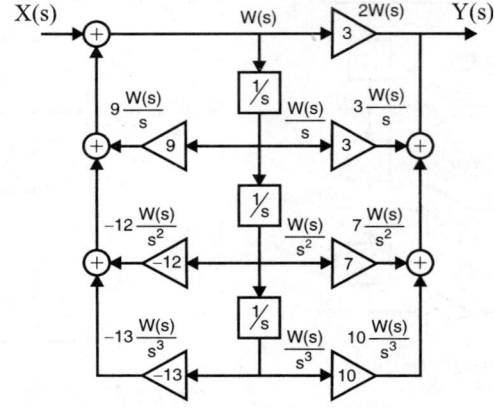

Fig E3.21b.2: *Direct form-II structure.*

E3.22 a)

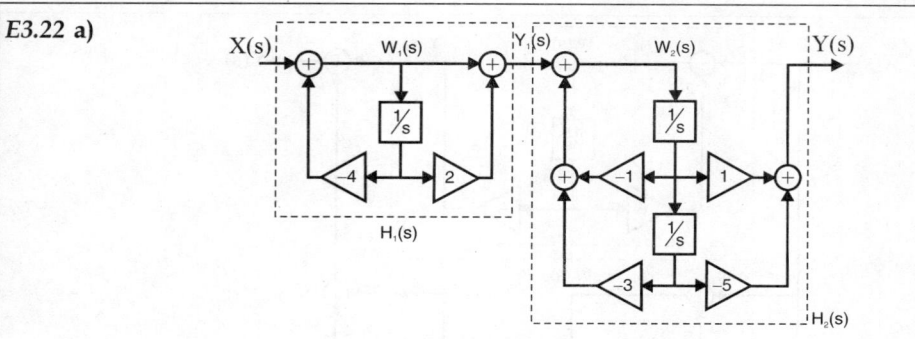

Fig E3.22a.1: *Cascade structure.*

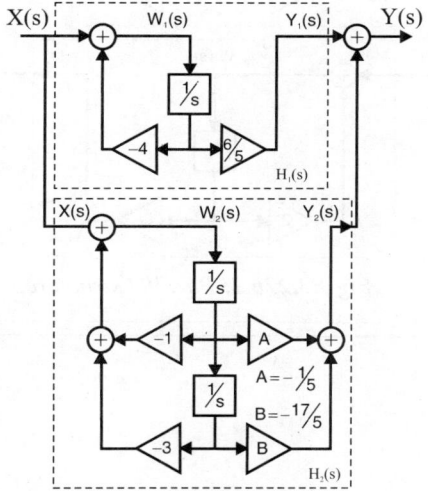

Fig E3.22a.2: *Parallel structure.*

E3.22 b)

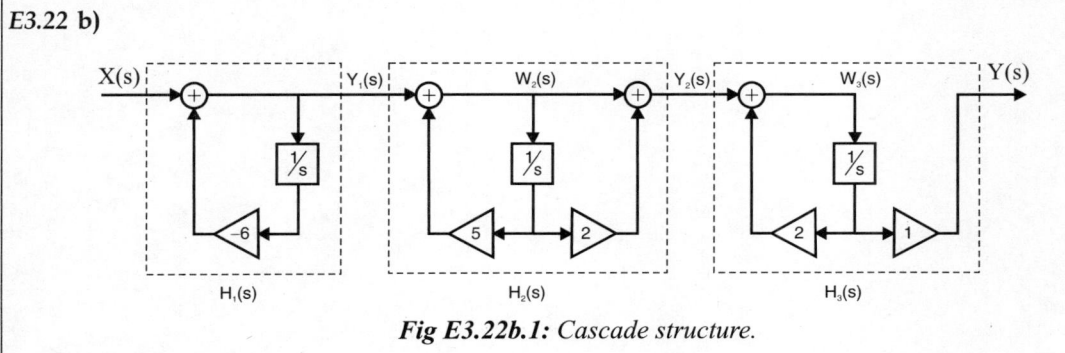

Fig E3.22b.1: *Cascade structure.*

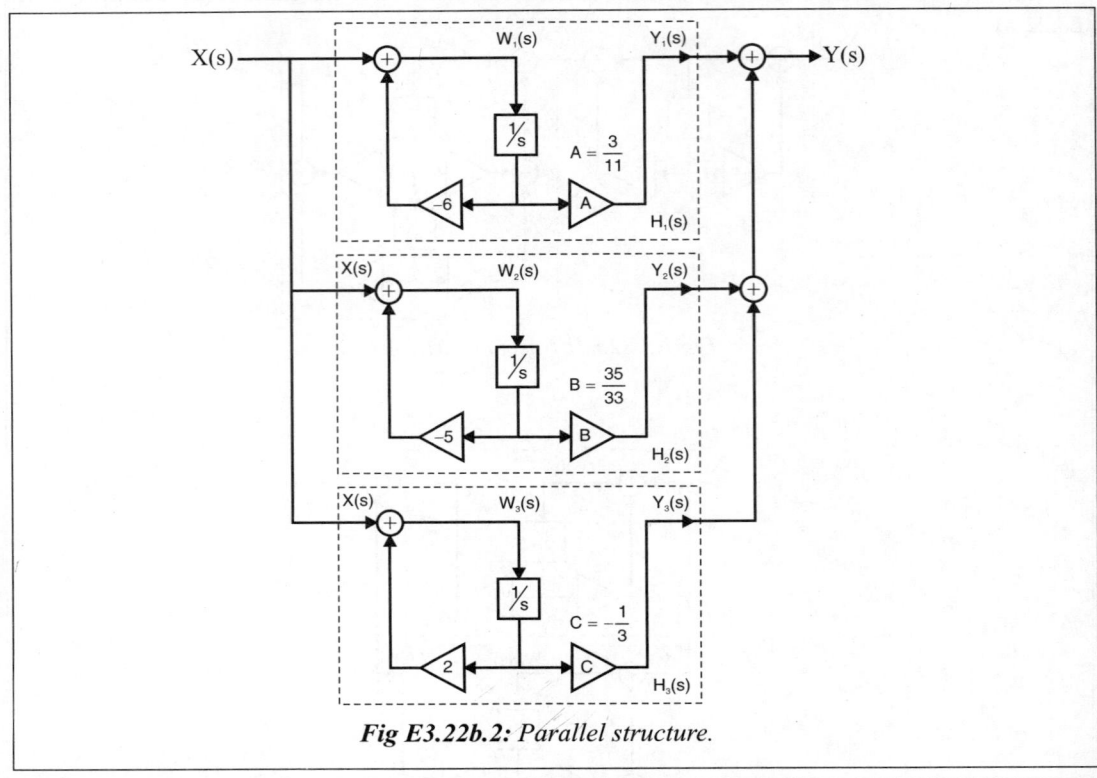

Fig E3.22b.2: Parallel structure.

CHAPTER 4

Analysis of Discrete Time Signals

4.1 Baseband Sampling   *(AU Dec'15 & 14, 16 Marks)*   *(AU Dec'12 & 11, 8 Marks)*

In communication engineering, the term baseband signal is used to indicate the unmodulated signal which has the original frequency components. Therefore a signal in its origina lform is called *baseband signal.* The sampling of unmouduled or original signal is called *baseband sampling.*

The *sampling* is the process of conversion of a continuous time signal into a discrete time signal. The sampling is performed by taking samples of continuous time signal at definite intervals of time. Usually, the time interval between two successive samples will be same and such type of sampling is called *periodic or uniform sampling*.

The time interval between successive samples is called *sampling time* (or sampling period or sampling interval), and it is denoted by "T". The unit of sampling period is second (s). The lower units are millisecond (ms) and microsecond (μs).

The inverse of sampling period is called *sampling frequency* (or sampling rate), and it is denoted by F_s. The unit of sampling frequency is hertz (Hz). The higher units are kHz and MHz.

Let, $x_a(t)$ = Analog / Continuous time signal.

$x(n)$ = Discrete time signal obtained by sampling $x_a(t)$.

Mathematically, the relation between $x(n)$ and $x_a(t)$ can be expressed as,

$$x(n) = x_a(t)\Big|_{t=nT} = x_a(nT) = x_a\left(\frac{n}{F_s}\right) ; \text{ for n in the range - } \infty < n < \infty$$

where, T = Sampling period or interval in seconds

$$F_s = \frac{1}{T} = \text{Sampling rate or sampling frequency in hertz}$$

Example: Let, $x_a(t) = A\cos(\Omega_0 t + \theta) = A\cos(2\pi F_0 t + \theta)$

where, Ω_0 = Frequency of analog signal in rad/sec

$F_0 = \dfrac{\Omega_0}{2\pi}$ = Frequency of analog signal in Hz

Let $x_a(t)$ be sampled at intervals of T seconds to get $x(n)$, where $T = \dfrac{1}{F_s}$

$$\therefore x(n) = x_a(t)\Big|_{t=nT} = A\cos(\Omega_0 t + \theta)\Big|_{t=nT}$$

$$= A\cos(\Omega_0 nT + \theta) = A\cos\left(\frac{2\pi F_0}{F_s}n + \theta\right) = A\cos(2\pi f_0 n + \theta) = A\cos(\omega_0 n + \theta)$$

where, $f_0 = \dfrac{F_0}{F_s}$ = Frequency of discrete sinusoid in cycles/sample

$\omega_0 = 2\pi f_0$ = Frequency of discrete sinusoid in radians/sample

4.1.1 Sampling and Aliasing *(AU Dec'11,'13 & June '11,'13 2 Marks) (AU Jun'11, 6 Marks)*

In Chapter - 1 Section 1.3.4, it is observed that any two sinusoid signals with frequencies in the range $-1/2 \leq f \leq +1/2$ are distinct and a discrete sinusoid with frequency, $f > |\pm 1/2|$ will be identical to another discrete sinusoid with frequency, $f < |\pm 1/2|$. Therefore, we can conclude that range of frequency of discrete time signal is $-1/2$ to $+1/2$. But the range of frequency of analog signal is $-\infty$ to $+\infty$. While sampling analog signals, the infinite frequency range continuous time signals are mapped (or converted) to finite frequency range discrete time signals.

The relation between frequency of analog and discrete time signal is,

$$f = \frac{F}{F_s} \qquad \qquad .....(4.1)$$

The range of frequency of discrete time signal is,

$$-\frac{1}{2} \leq f \leq \frac{1}{2} \qquad \qquad .....(4.2)$$

On substituting for f from equation (4.1) in equation (4.2) we get,

$$-\frac{1}{2} \leq \frac{F}{F_s} \leq \frac{1}{2} \qquad \qquad .....(4.3)$$

On multiplying equation (4.3) by F_s we get,

$$-\frac{F_s}{2} \leq F \leq \frac{F_s}{2} \qquad \qquad .....(4.4)$$

From equation (4.4) we can say that when an analog signal is sampled at a frequency F_s, the highest analog frequency that can be uniquely represented by a discrete time signal will be $F_s/2$. The continuous time signal with frequency above $F_s/2$ will be represented as a signal within the range $+ F_s/2$ to $- F_s/2$. Hence the signal with frequency above $F_s/2$ will have an identical signal with frequency below $F_s/2$ in the discrete form.

Hence infinite number of high frequency continuous time signals will be represented by a single discrete time signal. Such signals are called *alias.*

The phenomenon of high-frequency component getting the identity of low-frequency component during sampling is called *aliasing.*

Sampling an analog signal with frequency F by choosing a sampling frequency F_s such that $F_s/2 > F$ will not result in alias. But sampling frequency is selected such that $F_s/2 < F$ that the frequency above $F_s/2$ will have alias with frequency below $F_s/2$. Hence the point of reflection is $F_s/2$, and the frequency $F_s/2$ is called *folding frequency*.

The discrete time sinusoids, A sin $[(2\pi f_0 + 2\pi k)n]$, will be alias for integer values of k. It is also observed that, a sinusoidal signal with frequency F_1 will be an alias of sinusoidal signal with frequency F_2 if it is sampled at a frequency $F_s = F_1 - F_2$. In general, if the sampling frequency is any multiple of $F_1 - F_2$, [i.e., $F_s = k(F_1 - F_2)$ where $k = 1, 2, 3, ...$] the signal with frequency F_2 will be an alias of the signal with frequency F_1.

Let, F_{max} be maximum frequency of analog signal that can be uniquely represented as discrete time signal when sampled at a frequency F_s.

$$\text{Now, } F_{max} = \frac{F_s}{2} \qquad \qquad .....(4.5)$$

$$\therefore F_s = 2 F_{max} \qquad \qquad .....(4.6)$$

The equation (4.6) gives a choice for selecting sampling frequency. From equation (4.6) we can say that for unique representation of analog signal with maximum frequency F_{max}, the sampling frequency should be greater than $2F_{max}$.   **(AU Jun'12, 2 Marks)**

i.e., to avoid aliasing $F_s \geq 2F_{max}$ ..... (4.7)

When sampling frequency F_s is equal to $2F_{max}$, the sampling rate is called *Nyquist rate*.

It is observed that a nonshifted sinusoidal signal when sampled at Nyquist rate, will produce zero sample sequence (i.e., discrete sequence with all zeros), (because the sinusoidal signal is sampled at its zero crossings, Refer example 4.3). Hence to avoid zero sampling of sinewave, the sampling frequency F_s should be greater than $2F_{max}$, where F_{max} is the maximum frequency in the analog signal.

A discrete signal obtained by sampling can be reconstructed to an analog signal, only when it is sampled without aliasing. .The concepts discussed above are summarized as sampling theorem given below.   **(AU Jun'11, 2 Marks)**

Sampling Theorem : *A bandlimited continuous time signal with maximum frequency F_m hertz can be fully recovered from its samples provided that the sampling frequency F_s is greater than or equal to two times the maximum frequency F_m, (i.e., $F_s \geq 2F_m$).*

Note: The effects of aliasing in frequency spectrum are discussed in section 4.3.1.

Example 4.1

Consider the analog signals, $x_1(t) = 3 \cos 2\pi(20t)$ and $x_2(t) = 3 \cos 2\pi(70t)$.

Find a sampling frequency so that 70 Hz signal is an alias of the 20 Hz signal?

Solution:

Let, the sampling frequency, $F_s = 70 - 20 = 50\,\text{Hz}$.

$$\therefore x_1(n) = x_1(t)\Big|_{t=nT=\frac{n}{F_s}} = 3\cos 2\pi(20t)\Big|_{t=\frac{n}{F_s}} = 3\cos 2\pi\left(\frac{20 \times n}{50}\right) = 3\cos\frac{4\pi}{5}n$$

$$x_2(n) = x_2(t)\Big|_{t=nT=\frac{n}{F_s}} = 3\cos 2\pi(70t)\Big|_{t=\frac{n}{F_s}} = 3\cos 2\pi\left(\frac{70 \times n}{50}\right)$$

> For integer values of n
> $\cos (2\pi n + \theta) = \cos \theta$

$$= 3\cos\frac{14\pi}{5}n = 3\cos\left(2\pi n + \frac{4\pi}{5}n\right) = 3\cos\frac{4\pi}{5}n$$

From the above analysis, we observe that $x_1(n)$ and $x_2(n)$ are identical , and so $x_2(t)$ is an alias of $x_1(t)$ when sampled at a frequency of 50 Hz.

Example 4.2

Let an analog signal, $x_a(t) = 10 \cos 200\,\pi t$. If the sampling frequency is 150 Hz, find the discrete time signal $x(n)$. Also find an alias frequency corresponding to $F_s = 150\,\text{Hz}$.

Solution:

$$x(n) = x_a(t)\Big|_{t=nT=\frac{n}{F_s}} = 10\cos 200\pi t\Big|_{t=\frac{n}{F_s}} = 10\cos 200\pi \times \frac{n}{F_s}$$

$$= 10\cos\frac{200\pi \times n}{150} = 10\cos\frac{4\pi}{3}n = 10\cos\left(2\pi - \frac{2\pi}{3}\right)n = 10\cos\frac{2\pi}{3}n = 10\cos 2\pi\frac{1}{3}n$$

We know that the discrete time sinusoids whose frequencies are separated by integer multiples of 2π are identical.

$$\therefore 10\cos\frac{2\pi}{3}n = 10\cos\left(\frac{2\pi}{3}+2\pi\right)n = 10\cos\frac{8\pi}{3}n = 10\cos 2\pi\frac{4}{3}n$$

Now, $10\cos 2\pi\frac{4}{3}n$ is an alias of $10\cos 2\pi\frac{1}{3}n$

Here the frequency of the signal, $10\cos 2\pi\frac{4}{3}n$ is,

$$f = \frac{4}{3}\text{ cycles/sample}$$

We know that, $f = \dfrac{F}{F_s} \quad\Rightarrow\quad F = fF_s = \dfrac{4}{3}\times 150 = 200\text{Hz}$

$\therefore$ when, $F_s = 150$ Hz, $F = 200$ Hz is an alias frequency.

Example 4.3

Consider the analog signal, $x_a(t) = 6\cos 50\pi t + 3\sin 200\pi t - 3\cos 100\pi t$.

Determine the minimum sampling frequency and the sampled version of analog signal at this frequency. Sketch the waveform and show the sampling points. Comment on the result.

Solution:

The given analog signal can be written as shown below:

$$x_a(t) = 6\cos 50\pi t + 3\sin 200\pi t - 3\cos 100\pi t = 6\cos 2\pi F_1 t + 3\sin 2\pi F_2 t - 3\cos 2\pi F_3 t$$

$$\text{where, } 2\pi F_1 = 50\pi \quad;\quad F_1 = 25\text{Hz}$$

$$2\pi F_2 = 200\pi \quad;\quad F_2 = 100\text{Hz}$$

$$2\pi F_3 = 100\pi \quad;\quad F_3 = 50\text{Hz}$$

The maximum analog frequency in the signal is 100Hz. The sampling frequency should be twice that of this maximum analog frequency.

i.e., $F_s \geq 2F_{max} \quad\Rightarrow\quad F_s \geq 2\times 100$

Let, sampling frequency, $F_s = 200\text{Hz}$

Let, $x(n) =$ Sampled version of analog signal.

$$\therefore x(n) = x_a(t)\Big|_{t=nT} = x_a(t)\Big|_{t=\frac{n}{F_s}}$$

$$= 6\cos\frac{50\pi n}{200} + 3\sin\frac{200\pi n}{200} - 3\cos\frac{100\pi n}{200} = 6\cos\frac{\pi n}{4} + 3\sin\pi n - 3\cos\frac{\pi n}{2}$$

For integer values of n, $\sin\pi n = 0$.

$$\therefore x(n) = 6\cos\frac{\pi n}{4} - 3\cos\frac{\pi n}{2}$$

The components of analog waveform and the sampling points are shown in Fig1.

<u>Comment</u>: *In the sampled version of analog signal x(n), the component 3 sin 200πt will give always zero samples when sampled at 200Hz for any value of n. This is the drawback in sampling at Nyquist rate (i.e., sampling at $F_s = 2F_{max}$).*

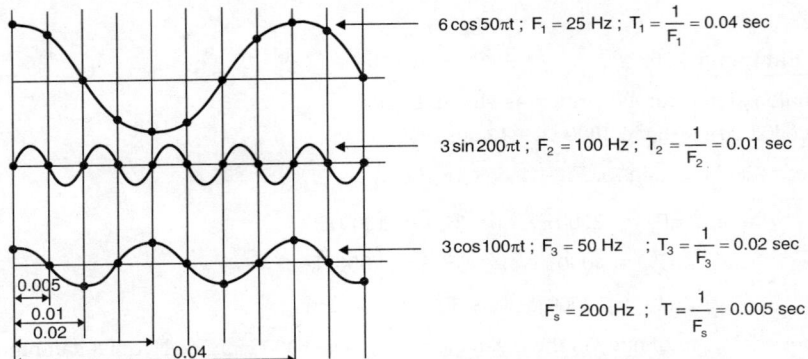

$6 \cos 50\pi t$; $F_1 = 25$ Hz ; $T_1 = \dfrac{1}{F_1} = 0.04$ sec

$3 \sin 200\pi t$; $F_2 = 100$ Hz ; $T_2 = \dfrac{1}{F_2} = 0.01$ sec

$3 \cos 100\pi t$; $F_3 = 50$ Hz   ; $T_3 = \dfrac{1}{F_3} = 0.02$ sec

$F_s = 200$ Hz ; $T = \dfrac{1}{F_s} = 0.005$ sec

Fig 1: Sampling points of the components of the signal $x_a(t)$.

(AU May'15, 8 Marks)

Example 4.4

A continuous time sinusoidal signal cos $(2\pi Ft + \theta)$ is sampled at a rate $F_s = 1000$ Hz. Determine the resulting signal samples, if the input signal frequency F is 400 Hz, 600 Hz and 1000 Hz respectively.

Solution:

Given that, $x_a(t) = \cos (2\pi Ft + \theta)$ and $F_s = 1000$ Hz

Let, $x(n)$ = Sampled version of continuous time signal.

$$\therefore x(n) = x_a(t)\Big|_{t=nT=\frac{n}{F_s}} = \cos(2\pi Ft + \theta)\Big|_{t=\frac{n}{F_s}}$$

$$= \cos\left(2\pi F \frac{n}{F_s} + \theta\right)$$

> For integer values of n
> $\cos (2\pi n + \theta) = \cos \theta$

$$= \cos\left(\frac{2\pi Fn}{1000} + \theta\right)$$

When F = 400Hz

$$x_1(n) = x(n)\big|_{F=400} = \cos\left(\frac{2\pi 400n}{1000} + \theta\right) = \cos\left(\frac{4\pi}{5}n + \theta\right)$$

When F = 600Hz

$$x_2(n) = x(n)\big|_{F=600} = \cos\left(\frac{2\pi 600n}{1000} + \theta\right) = \cos\left(\frac{6\pi}{5}n + \theta\right)$$

When F = 1000Hz

$$x_3(n) = x(n)\big|_{F=1000} = \cos\left(\frac{2\pi 1000n}{1000} + \theta\right) = \cos(2\pi n + \theta) = \cos\theta = \text{Constant independent of frequency.}$$

Example 4.5

Consider the analog signal, $x_a(t) = 2 \cos 2000\pi t + 5 \sin 4000\pi t + 12 \cos 12000\pi t$.

a) Determine the Nyquist sampling rate.

b) If the analog signal is sampled at $F_s = 5000$ Hz, determine the discrete time signal obtained by sampling.

Solution:

a) To Find Nyquist Sampling Rate

The given analog signal can be written as shown below:

$$x_a(t) = 2 \cos 2000 \pi t + 5 \sin 4000 \pi t - 12 \cos 12000 \pi t$$

$$= 2 \cos 2\pi \, F_1 t + 5 \sin 2\pi \, F_2 t - 12 \cos 2\pi \, F_3 t$$

where, $2\pi F_1 = 2000\pi \quad \Rightarrow \quad F_1 = 1000 \, Hz$

$\qquad\qquad 2\pi F_2 = 4000\pi \quad \Rightarrow \quad F_2 = 2000 \, Hz$

$\qquad\qquad 2\pi F_3 = 12000\pi \quad \Rightarrow \quad F_3 = 6000 Hz$

The maximum analog frequency in the given signal, F_{max} is 6000 Hz. The Nyquist sampling rate is twice that of this maximum analog frequency.

$\therefore$ Nyquist sampling rate, $F_s = 2\, F_{max} = 2 \times 6000 = 12000 \, Hz$

In order to avoid aliasing the sampling frequency, F_s should be greater than or equal to Nyquist rate.

b) To Determine the Discrete Time Signal Sampled at 5000 Hz

Let $x(n)$ be the discrete time signal obtained by sampling the given analog signal.

$$\therefore x(n) = x_a(t)\Big|_{t=nT} = x_a(t)\Big|_{t=\frac{n}{F_s}} = 2 \cos \frac{2000\pi n}{F_s} + 5 \sin \frac{4000\pi n}{F_s} + 12 \cos \frac{12000\pi n}{F_s}$$

$$= 2 \cos \frac{2000\pi n}{5000} + 5 \sin \frac{4000\pi n}{5000} + 12 \cos \frac{12000\pi n}{5000} \qquad \boxed{\begin{array}{l}\text{For integer values of n}\\ \cos (2\pi n + \theta) = \cos \theta \end{array}}$$

$$= 2 \cos \frac{2\pi n}{5} + 5 \sin \frac{4\pi n}{5} + 12 \cos \frac{12\pi n}{5} = 2 \cos \frac{2\pi n}{5} + 5 \sin \frac{4\pi n}{5} + 12 \cos\left(\frac{2\pi n}{5} + \frac{10\pi n}{5}\right)$$

$$= 2 \cos \frac{2\pi n}{5} + 5 \sin \frac{4\pi n}{5} + 12 \cos\left(\frac{2\pi n}{5} + 2\pi n\right) = 2 \cos \frac{2\pi n}{5} + 5 \sin \frac{4\pi n}{5} + 12 \cos \frac{2\pi n}{5}$$

$$= 14 \cos \frac{2\pi n}{5} + 5 \cos \frac{4\pi n}{5}$$

Comment: When sampled at 5000 Hz, the component $12 \cos 12000\pi t$ is an alias of the component $2 \cos 2000\pi \, t$.

4.2 Fourier Transform of Discrete Time Signals (Discrete Time Fourier Transform)

4.2.1 Definition of Discrete Time Fourier Transform   *(AU Dec'12, 2 Marks)*

The Fourier transform (FT) of discrete-time signals is called ***Discrete Time Fourier Transform*** (i.e., DTFT). But for convenience the DTFT is also referred as FT in this book.

Let, $x(n)$ = Discrete time signal

$\qquad X(e^{j\omega})$ = Fourier transform of $x(n)$

The Fourier transform of a finite energy discrete time signal, $x(n)$ is defined as,

$$X(e^{j\omega}) = \sum_{n=-\infty}^{+\infty} x(n) \, e^{-j\omega n}$$

Symbolically the Fourier transform of x(n) is denoted as,

$$\mathcal{F}\{x(n)\}$$

where, $\mathcal{F}$ is the operator that represents Fourier transform.

$$\therefore X(e^{j\omega}) = \mathcal{F}\{x(n)\} = \sum_{n=-\infty}^{+\infty} x(n)\, e^{-j\omega n}$$

The Fourier transform of a signal is said to exist if it can be expressed in a valid functional form. Since the computation of Fourier transform involves summing infinite number of terms, the Fourier transform exists only for the signals that are absolutely summable, i.e., given a signal x(n), the $X(e^{j\omega})$ exists only when,

$$\sum_{n=-\infty}^{+\infty} |x(n)| < \infty$$

Geometric Series

The Fourier transform and $\mathcal{Z}$-transform of a discrete time signal involves convergence of geometric series. Hence the following two geometric series sum formula will be useful in evaluating Fourier transform and $\mathcal{Z}$-transform.

1. Infinite geometric series sum formula.

   If C is a complex constant and $0 < |C| < 1$, then,

   $$\sum_{n=0}^{\infty} C^n = \frac{1}{1-C} \quad .$$

2. Finite geometric series sum formula.

   If C is a complex constant and,

   $$\text{When } C \neq 1, \quad \sum_{n=0}^{N-1} C^n = \frac{1-C^N}{1-C} = \frac{C^N-1}{C-1} \quad \text{or} \quad \sum_{n=0}^{N} C^n = \frac{C^{N+1}-1}{C-1}$$

   $$\text{When } C = 1, \quad \sum_{n=0}^{N-1} C^n = N \quad \text{or} \quad \sum_{n=0}^{N} C^n = N+1$$

Note: The infinite geometric series sum formula requires that the magnitude of C be strictly less than unity, but the finite geometric series sum formula is valid for any value of C.

4.2.2 Inverse Discrete Time Fourier Transform   *(AU Dec'12, 2 Marks)*

Let,   x(n)   =   Discrete time signal

        $X(e^{j\omega})$ =   Fourier transform of x(n)

The *inverse discrete time Fourier transform* of $X(e^{j\omega})$ is defined as,

$$x(n) = \frac{1}{2\pi} \int_{-\pi}^{\pi} X(e^{j\omega})\, e^{j\omega n}\, d\omega \quad ; \text{ for } n = -\infty \text{ to } +\infty$$

.....(4.8)

Symbolically the inverse Fourier transform can be expressed as,

$$\mathcal{F}^{-1}\{X(e^{j\omega})\},$$

where, $\mathcal{F}^{-1}$ is the operator that represents the inverse Fourier transform.

$$x(n) = \mathcal{F}^{-1}\{X(e^{j\omega})\} = \frac{1}{2\pi} \int_{-\pi}^{\pi} X(e^{j\omega})\, e^{j\omega n}\, d\omega \quad ; \text{ for } n = -\infty \text{ to } +\infty$$

Since $X(e^{j\omega})$ is periodic with period 2π, the limits of integral in the above definition of inverse Fourier transform can be either "$-\pi$ to $+\pi$", or "0 to 2π", or "any interval of 2π".

We also refer to $x(n)$ and $X(e^{j\omega})$ as a Fourier transform pair and this relation is expressed as,

$$x(n) \underset{\mathcal{F}^{-1}}{\overset{\mathcal{F}}{\rightleftharpoons}} X(e^{j\omega})$$

Alternate Method for Inverse Fourier Transform

The integral solution of equation (4.8) for the inverse Fourier transform is useful for analytic purpose, but sometimes it will be difficult to evaluate for typical functional forms of $X(e^{j\omega})$. An alternate and more useful method of determining the values of $x(n)$ follows directly from the definition of the Fourier transform.

Consider the definition of Fourier transform of $x(n)$.

$$X(e^{j\omega}) = \sum_{n=-\infty}^{+\infty} x(n)\, e^{-j\omega n}$$

Let us expand the above equation of $X(e^{j\omega})$ as shown below:

$$X(e^{j\omega}) = \sum_{n=-\infty}^{+\infty} x(n)\, e^{-j\omega n}$$

$$= \ldots + x(-2)e^{j2\omega} + x(-1)\, e^{j\omega} + x(0)\, e^{0} + x(1)\, e^{-j\omega} + x(2)\, e^{-j2\omega} + \ldots \qquad \ldots(4.9)$$

Let us express the given function of $X(e^{j\omega})$ as a power series of $e^{-j\omega}$ by long division as shown below:

$$X(e^{j\omega}) = \ldots + b_2\, e^{j2\omega} + b_1\, e^{j\omega} + a_0\, e^{0} + a_1\, e^{-j\omega} + a_2\, e^{-j2\omega} + \ldots \qquad \ldots(4.10)$$

On comparing the equations (4.9) and (4.10) we can say that the samples of signal $x(n)$ are simply the coefficients of $e^{-j\omega n}$.

$$\therefore \quad x(n) = \{ \ldots b_2, b_1, a_0, a_1, a_2 \ldots \}$$

$$\uparrow$$

4.2.3 Properties of Discrete Time Fourier Transform   *(AU Dec'12 & 11,8 Marks)*

1. Linearity Property

(AU Dec'15, 8 Marks)

The linearity property of Fourier transform states that the Fourier transform of a linear weighted combination of two or more signals is equal to the similar linear weighted combination of the Fourier transform of the individual signals.

Let, $\mathcal{F}\{x_1(n)\} = X_1(e^{j\omega})$ and $\mathcal{F}\{x_2(n)\} = X_2(e^{j\omega})$ then by linearity property,

$\mathcal{F}\{a_1 x_1(n) + a_2 x_2(n)\} = a_1 X_1(e^{j\omega}) + a_2 X_2(e^{j\omega})$; where a_1 and a_2 are constants.

Proof:

By the definition of Fourier transform,

$$X_1(e^{j\omega}) = \mathcal{F}\{x_1(n)\} = \sum_{n=-\infty}^{+\infty} x_1(n) e^{-j\omega n} \qquad \qquad .....(4.11)$$

$$X_2(e^{j\omega}) = \mathcal{F}\{x_2(n)\} = \sum_{n=-\infty}^{+\infty} x_2(n) e^{-j\omega n} \qquad \qquad .....(4.12)$$

$$\mathcal{F}\{a_1 x_1(n) + a_2 x_2(n)\} = \sum_{n=-\infty}^{+\infty} [a_1 x_1(n) + a_2 x_2(n)] e^{-j\omega n} = \sum_{n=-\infty}^{+\infty} [a_1 x_1(n) e^{-j\omega n} + a_2 x_2(n) e^{-j\omega n}]$$

$$= \sum_{n=-\infty}^{+\infty} a_1 x_1(n) e^{-j\omega n} + \sum_{n=-\infty}^{+\infty} a_2 x_2(n) e^{-j\omega n}$$

$$= a_1 \sum_{n=-\infty}^{+\infty} x_1(n) e^{-j\omega n} + a_2 \sum_{n=-\infty}^{+\infty} x_2(n) e^{-j\omega n}$$

$$= a_1 X_1(e^{j\omega}) + a_2 X_2(e^{j\omega}) \qquad \boxed{\text{Using equations (4.11) and (4.12)}}$$

2. Periodicity

Let, $\mathcal{F}\{x(n)\} = X(e^{j\omega})$, then $X(e^{j\omega})$ is periodic with period 2π.

$\therefore X(e^{j(\omega + 2\pi m)}) = X(e^{j\omega})$; where m is an integer

Proof:

$$X(e^{j(\omega + 2\pi m)}) = \sum_{n=-\infty}^{+\infty} x(n) e^{-j(\omega + 2\pi m)n}$$

$$= \sum_{n=-\infty}^{+\infty} x(n) e^{-j\omega n} e^{-j2\pi mn}$$

$$= \sum_{n=-\infty}^{+\infty} x(n) e^{-j\omega n} = X(e^{j\omega}) \qquad \boxed{\begin{array}{l}\text{Since m and n are} \\ \text{integers, } e^{-j2\pi mn} = 1\end{array}}$$

3. Time Shifting or Fourier Transform of Delayed Signal *(AU Jun'12,'11 & '13, 2 & 8 Marks)*

Let, $\mathcal{F}\{x(n)\} = X(e^{j\omega})$, then $\mathcal{F}\{x(n-m)\} = e^{-j\omega m} X(e^{j\omega})$

Also, $\mathcal{F}\{x(n + m)\} = e^{j\omega m} X(e^{j\omega})$

This relation means that if a signal is shifted in time domain by m samples, its magnitude spectrum remains unchanged. However, the phase spectrum is changed by an amount $\pm\omega m$. This result can be explained if we recall that the frequency content of a signal depends only on its shape. Mathematically, we can say that delaying by m units in time domain is equivalent to multiplying the spectrum by $e^{-j\omega m}$ in the frequency domain.

Proof:

By the definition of Fourier transform,

$$X(e^{j\omega}) = \mathcal{F}\{x(n)\} = \sum_{n=-\infty}^{+\infty} x(n)\, e^{-j\omega n} \qquad .....(4.13)$$

$$\therefore \mathcal{F}\{x(n-m)\} = \sum_{n=-\infty}^{+\infty} x(n-m)e^{-j\omega n}$$

> Let, $n - m = p$, $\therefore n = p + m$
> when $n \to -\infty$, $p \to -\infty$
> when $n \to +\infty$, $p \to +\infty$

$$= \sum_{p=-\infty}^{+\infty} x(p)e^{-j\omega(m+p)}$$

$$= \sum_{p=-\infty}^{+\infty} x(p)\, e^{-j\omega m}\, e^{-j\omega p}$$

$$= e^{-j\omega m}\sum_{p=-\infty}^{+\infty} x(p)e^{-j\omega p} = e^{-j\omega m}\sum_{n=-\infty}^{+\infty} x(n)e^{-j\omega n}$$

> Let, $p \to n$

> Using equation (4.13)

$$= e^{-j\omega m}\, X(e^{j\omega})$$

4. Time Reversal   *(AU Jun'11, 16 Marks)*

Let, $\mathcal{F}\{x(n)\} = X(e^{j\omega})$, then $\mathcal{F}\{x(-n)\} = X(e^{-j\omega})$

This means that if a signal is folded about the origin in time, its magnitude spectrum remains unchanged and the phase spectrum undergoes a change in sign (phase reversal).

Proof:

By the definition of Fourier transform

$$\mathcal{F}\{x(n)\} = \sum_{n=-\infty}^{+\infty} x(n)\, e^{-j\omega n} \qquad ....(4.14)$$

$$\mathcal{F}\{x(-n)\} = \sum_{n=-\infty}^{+\infty} x(-n)\, e^{-j\omega n} = \sum_{p=-\infty}^{+\infty} x(p)\, e^{j\omega p}$$

> Let, $p = -n$
> when $n \to -\infty$, $p \to +\infty$
> when $n \to +\infty$, $p \to -\infty$

$$= \sum_{p=-\infty}^{+\infty} x(p)(e^{-j\omega})^{-p} \qquad .....(4.15)$$

> The equation(4.15) is similar
> to the form of equation (4.14).

$$= X(e^{-j\omega})$$

5. Conjugation

If, $\mathcal{F}\{x(n)\} = X(e^{j\omega})$

then, $\mathcal{F}\{x^*(n)\} = X^*(e^{-j\omega})$

Proof:

By the definition of Fourier transform

$$X(e^{j\omega}) = \mathcal{F}\{x(n)\} = \sum_{n=-\infty}^{+\infty} x(n)\, e^{-j\omega n}$$

$$\mathcal{F}\{x^*(n)\} = \sum_{n=-\infty}^{+\infty} x^*(n)\, e^{-j\omega n}$$

$$= \left[\sum_{n=-\infty}^{+\infty} x(n)(e^{-j\omega})^{-n}\right]^* = \left[X(e^{-j\omega})\right]^*$$

$$= X^*(e^{-j\omega})$$

6. Frequency Shifting   (AU May'15, 4 Marks & June '13 '11,8 & 16 Marks)

Let, $\mathcal{F}\{x(n)\} = X(e^{j\omega})$, then $\mathcal{F}\{e^{j\omega_0 n} x(n)\} = X(e^{j(\omega - \omega_0)})$

According to this property, multiplication of a sequence $x(n)$ by $e^{j\omega_0 n}$ is equivalent to a frequency translation of the spectrum $X(e^{j\omega})$ by ω_0

Proof:

By the definition of Fourier transform,

$$X(e^{j\omega}) = \mathcal{F}\{x(n)\} = \sum_{n=-\infty}^{+\infty} x(n) e^{-j\omega n} \qquad .....(4.16)$$

$$\therefore \mathcal{F}\{e^{j\omega_0 n} x(n)\} = \sum_{n=-\infty}^{+\infty} e^{j\omega_0 n} x(n) e^{-j\omega n}$$

$$= \sum_{n=-\infty}^{+\infty} x(n) e^{-j(\omega - \omega_0)n} \qquad .....(4.17)$$

$$= X(e^{j(\omega - \omega_0)})$$

> The equation(4.17) is similar to the form of equation (4.16).

7. Fourier Transform of the Product of Two Signals   (AU Jun'14, 2 Marks)

Let, $\mathcal{F}\{x_1(n)\} = X_1(e^{j\omega})$

$\mathcal{F}\{x_2(n)\} = X_2(e^{j\omega})$

Now, $\mathcal{F}\{x_1(n) x_2(n)\} = \dfrac{1}{2\pi} \displaystyle\int_{-\pi}^{+\pi} X_1(e^{j\lambda}) X_2(e^{j(\omega - \lambda)}) \, d\lambda \qquad .....(4.18)$

The equation (4.18) is convolution of $X_1(e^{j\omega})$ and $X_2(e^{j\omega})$

This relation is the dual of time domain convolution. In other words, the Fourier transform of the product of two discrete time signals is equivalent to the convolution of their Fourier transform. On the other hand, the Fourier transform of the convolution of two discrete time signals is equivalent to the product of their Fourier transform.

Proof:

Let, $x_2(n) x_1(n) = x_3(n)$

Now, $\mathcal{F}\{x_2(n) x_1(n)\} = \mathcal{F}\{x_3(n)\} = \displaystyle\sum_{n=-\infty}^{+\infty} x_3(n) e^{-j\omega n}$

$$= \sum_{n=-\infty}^{+\infty} x_2(n) x_1(n) e^{-j\omega n} \qquad .....(4.19)$$

By the definition of inverse Fourier transform we get,

$$x_1(n) = \frac{1}{2\pi} \int_{-\pi}^{+\pi} X_1(e^{j\omega}) e^{j\omega n} \, d\omega \qquad \boxed{\text{Let, } \omega = \lambda}$$

$$= \frac{1}{2\pi} \int_{-\pi}^{+\pi} X_1(e^{j\lambda}) e^{j\lambda n} \, d\lambda \qquad .....(4.20)$$

On substituting for $x_1(n)$ from equation (4.20) in equation (4.19) we get,

$$\mathcal{F}\{x_1(n) x_2(n)\} = \sum_{n=-\infty}^{+\infty} x_2(n) \left[\frac{1}{2\pi} \int_{-\pi}^{+\pi} X_1(e^{j\lambda}) e^{j\lambda n} \, d\lambda \right] e^{-j\omega n}$$

On interchanging the order of summation and integration in the above equation we get,

$$\mathcal{F}\{x_1(n)\,x_2(n)\} = \frac{1}{2\pi} \int\limits_{-\pi}^{+\pi} \left[\sum_{n=-\infty}^{+\infty} x_2(n)\,e^{-j(\omega-\lambda)n} \right] X_1(e^{j\lambda})\,d\lambda$$

The term in the paranthesis in the above equation is similar to the definition of fourier transform of $x_2(n)$ but at a frequency argument of $(\omega-\lambda)$.

$$\therefore \mathcal{F}\{x_1(n)\,x_2(n)\} = \frac{1}{2\pi} \int\limits_{-\pi}^{+\pi} \left[X_2\left(e^{j(\omega-\lambda)n}\right) X_1(e^{j\lambda})\,d\lambda \right]$$

$$= \frac{1}{2\pi} \int\limits_{-\pi}^{+\pi} X_1(e^{j\lambda})\, X_2\left(e^{j(\omega-\lambda)n}\right) d\lambda$$

8. Differentiation in Frequency Domain

(AU May'15, 4 Marks)

If, $\mathcal{F}\{x(n)\} = X(e^{j\omega})$

then, $\mathcal{F}\{nx(n)\} = j\dfrac{d}{d\omega} X\left(e^{j\omega}\right)$

Proof:

By the definition of Fourier transform,

$$X(e^{j\omega}) = \mathcal{F}\{x(n)\} = \sum_{n=-\infty}^{+\infty} x(n)\,e^{-j\omega n} \qquad .....(4.21)$$

$$\mathcal{F}\{nx(n)\} = \sum_{n=-\infty}^{+\infty} nx(n)\,e^{-j\omega n} \qquad \boxed{\begin{array}{l}\text{Multiply by } j \text{ and } -j \\ j\times(-j) = 1\end{array}}$$

$$= \sum_{n=-\infty}^{+\infty} nx(n)\,j\times(-j)\,e^{-j\omega n}$$

$$= j \sum_{n=-\infty}^{+\infty} x(n)\left[-jne^{-j\omega n}\right]$$

$$= j \sum_{n=-\infty}^{+\infty} x(n)\left[\frac{d}{d\omega}e^{-j\omega n}\right] \qquad \boxed{\dfrac{d}{d\omega}\,e^{-j\omega n} = -jn\,e^{-j\omega n}}$$

$$= j\frac{d}{d\omega}\left[\sum_{n=-\infty}^{+\infty}x(n)\,e^{-j\omega n}\right] \qquad \boxed{\begin{array}{l}\text{Interchanging summation} \\ \text{and differentiation}\end{array}}$$

$$= j\frac{d}{d\omega}X(e^{j\omega}) \qquad \boxed{\text{Using equation (4.21)}}$$

9. Convolution Theorem

(AU Jun'11, 16 Marks)

If, $\mathcal{F}\{x_1(n)\} = X_1(e^{j\omega})$

and, $\mathcal{F}\{x_2(n)\} = X_2(e^{j\omega})$

then, $\mathcal{F}\{x_1(n) * x_2(n)\} = X_1(e^{j\omega})X_2(e^{j\omega})$

$$\text{where, } x_1(n) * x_2(n) = \sum_{m=-\infty}^{+\infty} x_1(m)\,x_2(n-m) \qquad\qquad .....(4.22)$$

The Fourier transform of the convolution of $x_1(n)$ and $x_2(n)$ is equal to the product of $X_1(e^{j\omega})$ and $X_2(e^{j\omega})$. It means that if we convolve two signals in time domain, it is equivalent to multiplying their spectra in frequency domain.

Proof:

By the definition of Fourier transform,

$$X_1(e^{j\omega}) = \mathcal{F}\{x_1(n)\} = \sum_{n=-\infty}^{+\infty} x_1(n)\,e^{-j\omega n} \qquad \text{.....(4.23)}$$

$$X_2(e^{j\omega}) = \mathcal{F}\{x_2(n)\} = \sum_{n=-\infty}^{+\infty} x_2(n)\,e^{-j\omega n} \qquad \text{.....(4.24)}$$

$$\mathcal{F}\{x_1(n)*x_2(n)\} = \sum_{n=-\infty}^{+\infty} \big[x_1(n)*x_2(n)\big]\,e^{-j\omega n}$$

$$= \sum_{n=-\infty}^{+\infty} \left[\sum_{m=-\infty}^{+\infty} x_1(m)\,x_2(n-m)\right]e^{-j\omega n} \qquad \boxed{\text{Using equation (4.22)}}$$

$$= \sum_{n=-\infty}^{+\infty}\sum_{m=-\infty}^{+\infty} x_1(m)\,x_2(n-m)e^{-j\omega n}e^{-j\omega m}e^{j\omega m} \qquad \boxed{\begin{array}{l}\text{Multiply by } e^{-j\omega m} \text{ and } e^{j\omega m}\\ e^{-j\omega m}\times e^{j\omega m}=1\end{array}}$$

$$= \sum_{m=-\infty}^{+\infty} x_1(m)\,e^{-j\omega m}\sum_{n=-\infty}^{+\infty} x_2(n-m)\,e^{-j\omega(n-m)} \qquad \boxed{\begin{array}{l}\text{Let, } n-m=p\\ \text{when } n\to-\infty,\ p\to-\infty\\ \text{when } n\to+\infty,\ p\to+\infty\end{array}}$$

$$= \sum_{m=-\infty}^{+\infty} x_1(m)\,e^{-j\omega m}\sum_{p=-\infty}^{+\infty} x_2(p)e^{-j\omega p}$$

$$= \left[\sum_{n=-\infty}^{+\infty} x_1(n)\,e^{-j\omega n}\right]\left[\sum_{n=-\infty}^{+\infty} x_2(n)\,e^{-j\omega n}\right] \qquad \boxed{\begin{array}{l}\text{Let } m=n, \text{ in first summation}\\ \text{Let } p=n, \text{ in second summation}\end{array}}$$

$$= X_1(e^{j\omega})\,X_2(e^{j\omega}) \qquad \boxed{\text{Using equations (4.23) and (4.24)}}$$

10. *Correlation*

If, $\mathcal{F}\{x(n)\} = X(e^{j\omega})$ and $\mathcal{F}\{y(n)\} = Y(e^{j\omega})$

then, $\mathcal{F}\{r_{xy}(m)\} = X(e^{j\omega})\,Y(e^{-j\omega})$

$$\text{where, } r_{xy}(m) = \sum_{n=-\infty}^{+\infty} x(n)y(n-m) \qquad \text{.....(4.25)}$$

Proof:

By the definition of Fourier transform,

$$X(e^{j\omega}) = \mathcal{F}\{x(n)\} = \sum_{n=-\infty}^{+\infty} x(n)\,e^{-j\omega n} \qquad \text{.....(4.26)}$$

$$Y(e^{j\omega}) = \mathcal{F}\{y(n)\} = \sum_{n=-\infty}^{+\infty} y(n)\,e^{-j\omega n} \qquad \text{.....(4.27)}$$

$$\mathcal{F}\{r_{xy}(m)\} = \sum_{m=-\infty}^{+\infty} r_{xy}(m)\,e^{-j\omega m} = \sum_{m=-\infty}^{+\infty}\left[\sum_{n=-\infty}^{+\infty} x(n)y(n-m)\right]e^{-j\omega m} \qquad \boxed{\text{Using equation (4.25)}}$$

$$= \sum_{m=-\infty}^{+\infty}\sum_{n=-\infty}^{+\infty} x(n)y(n-m)\,e^{-j\omega m}e^{-j\omega n}e^{j\omega n}$$

$$= \sum_{n=-\infty}^{+\infty} x(n)e^{-j\omega n}\sum_{m=-\infty}^{+\infty} y(n-m)\,e^{j\omega(n-m)} \qquad \boxed{\begin{array}{l}\text{Multiply by } e^{-j\omega m} \text{ and } e^{j\omega m}\\ e^{-j\omega m}\times e^{j\omega m}=1\end{array}}$$

$$\therefore \mathcal{F}\{r_{xy}(m)\} = \sum_{n=-\infty}^{+\infty} x(n)\, e^{-j\omega n} \sum_{p=-\infty}^{+\infty} y(p)\, e^{j\omega p}$$

> Let, $n - m = p$, $\therefore m = n - p$
> when $m \to -\infty$, $p \to +\infty$
> when $m \to +\infty$, $p \to -\infty$

$$= \left[\sum_{n=-\infty}^{+\infty} x(n)\, e^{-j\omega n} \right] \left[\sum_{p=-\infty}^{+\infty} y(p)\, (e^{-j\omega})^{-p} \right]$$

$$= X(e^{j\omega})\, Y(e^{-j\omega})$$

> Using equations (4.26) and (4.27)

11. Parseval's Relation

If, $\mathcal{F}\{x_1(n)\} = X_1(e^{j\omega})$ and, $\mathcal{F}\{x_2(n)\} = X_2(e^{j\omega})$,

then the Parseval's relation states that,

$$\sum_{n=-\infty}^{+\infty} x_1(n) x_2^*(n) = \frac{1}{2\pi} \int_{-\pi}^{+\pi} X_1(e^{j\omega})\, X_2^*(e^{j\omega}) \qquad \dots(4.28)$$

When $x_1(n) = x_2(n) = x(n)$, then Parseval's relation can be written as,

$$\sum_{n=-\infty}^{+\infty} |x(n)|^2 = \frac{1}{2\pi} \int_{-\pi}^{+\pi} |X(e^{j\omega})|^2\, d\omega$$

> $x(n)\, x^*(n) = |x(n)|^2$
> $X(e^{j\omega})\, X^*(e^{j\omega}) = |X(e^{j\omega})|^2$

The above equation is also called ***energy density spectrum*** of the signal $x(n)$.

Proof:

Let, $\mathcal{F}\{x_1(n)\} = X_1(e^{j\omega})$ and $\mathcal{F}\{x_2(n)\} = X_2(e^{j\omega})$

Now, by definition of Fourier transform,

$$\mathcal{F}\{x_1(n)\} = X_1(e^{j\omega}) = \sum_{n=-\infty}^{+\infty} x_1(n)\, e^{-j\omega n} \qquad \dots(4.29)$$

Now, by definition of inverse Fourier transform,

$$x_2(n) = \frac{1}{2\pi} \int_{-\pi}^{+\pi} X_2(e^{j\omega})\, e^{j\omega n}\, d\omega \qquad \dots(4.30)$$

Consider left-hand side of Parseval's relation [equation (4.28)],

$$\frac{1}{2\pi} \int_{-\pi}^{+\pi} X_1(e^{j\omega})\, X_2^*(e^{j\omega})\, d\omega$$

In the above expression, Let us substitute for $X_1(e^{j\omega})$ from equation (4.29),

$$\therefore \frac{1}{2\pi} \int_{-\pi}^{+\pi} X_1(e^{j\omega})\, X_2^*(e^{j\omega})\, d\omega = \frac{1}{2\pi} \int_{-\pi}^{+\pi} \left[\sum_{n=-\infty}^{+\infty} x_1(n) e^{-j\omega n} \right] X_2^*(e^{j\omega})\, d\omega$$

$$= \sum_{n=-\infty}^{+\infty} x_1(n) \left[\frac{1}{2\pi} \int_{-\pi}^{+\pi} X_2^*(e^{j\omega})\, e^{-j\omega n} d\omega \right]$$

> Interchanging summation and integration.

$$= \sum_{n=-\infty}^{+\infty} x_1(n) \left[\frac{1}{2\pi} \int_{-\pi}^{+\pi} X_2(e^{j\omega}) e^{j\omega n} d\omega \right]^*$$

$$= \sum_{n=-\infty}^{+\infty} x_1(n) x_2^*(n)$$

> Using equation (4.30).

Table 4.1: Properties of Discrete Time Fourier Transform

Note : $X(e^{jw}) = \mathcal{F}\{x(n)\}$; $X_1(e^{jw}) = \mathcal{F}\{x_1(n)\}$; $X_2(e^{jw}) = \mathcal{F}\{x_2(n)\}$; $Y(e^{jw}) = F\{y(n)\}$

Property	Discrete time signal	Fourier transform
Linearity	$a_1\, x_1(n) + a_2\, x_2(n)$	$a_1\, X_1(e^{j\omega}) + a_2\, X_2(e^{j\omega})$
Periodicity	$x(n)$	$X(e^{j\omega + 2\pi m}) = X(e^{j\omega})$
Time shifting	$x(n - m)$	$e^{-j\omega m}\, X(e^{j\omega})$
Time reversal	$x(-n)$	$X(e^{-j\omega})$
Conjugation	$x^*(n)$	$X^*(e^{-j\omega})$
Frequency shifting	$e^{j\omega_0 n}\, x(n)$	$X(e^{j(\omega - \omega_0)})$
Multiplication	$x_1(n)\, x_2(n)$	$\dfrac{1}{2\pi} \displaystyle\int_{-\pi}^{+\pi} X_1(e^{j\lambda})\, X_2(e^{j(\omega - \lambda)})\, d\lambda$
Differentiation in frequency domain	$n\, x(n)$	$j\dfrac{d}{d\omega}X(e^{j\omega})$
Convolution	$x_1(n) * x_2(n) = \displaystyle\sum_{n=-\infty}^{+\infty} x_1(m)\, x_2(n-m)$	$X_1(e^{j\omega})\, X_2(e^{j\omega})$
Correlation	$r_{xy}(m) = \displaystyle\sum_{m=-\infty}^{+\infty} x(n)y(n-m)$	$X(e^{j\omega})\, Y(e^{-j\omega})$
Symmetry of real signals	$x(n)$ is real	$X(e^{j\omega}) = X^*(e^{-j\omega})$ $\mathrm{Re}\{X(e^{j\omega})\} = \mathrm{Re}\{X(e^{-j\omega})\}$ $\mathrm{Im}\{X(e^{j\omega})\} = \mathrm{Im}\{X(e^{-j\omega})\}$ $\lvert X(e^{j\omega})\rvert = \lvert X(e^{-j\omega})\rvert,\ \angle X(e^{-j\omega}) = -\angle X(e^{-j\omega})$
Symmetry of real and even signal	$x(n)$ is real and even	$X(e^{j\omega})$ is real and even
Symmetry of real and odd signal	$x(n)$ is real and odd	$X(e^{j\omega})$ is imaginary and odd
Parseval's relation	$\displaystyle\sum_{n=-\infty}^{+\infty} x_1(n)\, x_2^*(n)$	$\dfrac{1}{2\pi}\displaystyle\int_{-\pi}^{+\pi} X_1(e^{j\omega})\, X_2^*(e^{j\omega})\, d\omega$
	Energy in time domain, $$E = \sum_{n=-\infty}^{+} \lvert x(n)\rvert^2$$	Energy in frequency domain, $$E = \frac{1}{2\pi}\int_{-\pi}^{+\pi} \lvert X(e^{j\omega})\rvert^2\, d\omega$$
	$$\sum_{n=-\infty}^{+\infty} \lvert x(n)\rvert^2 = \frac{1}{2\pi}\int_{-\pi}^{+\pi} \lvert X(e^{j\omega})\rvert^2\, d\omega$$	

Note: The term $\lvert X(e^{j\omega})\rvert^2$ represents the distribution of energy as a function of frequency and so it is called **energy density spectrum** or **energy spectral density**.

Table 4.2: Some Common Discrete Time Fourier Transform Pairs

$x(t)$	$x(n)$	$X(e^{j\omega})$	
		With Positive Power of $e^{j\omega}$	**With Negative Power of $e^{j\omega}$**
	$\delta(n)$	1	1
	$\delta(n-n_0)$	$\dfrac{1}{e^{j\omega n_0}}$	$e^{-j\omega n_0}$
	$u(n)$	$\dfrac{e^{j\omega}}{e^{j\omega}-1} + \displaystyle\sum_{m=-\infty}^{+\infty} \pi\delta(\omega-2\pi m)$	$\dfrac{1}{1-e^{-j\omega}} + \displaystyle\sum_{m=-\infty}^{+\infty} \pi\delta(\omega-2\pi m)$
	$a^n\,u(n)$	$\dfrac{e^{j\omega}}{e^{j\omega}-a}$	$\dfrac{1}{1-ae^{-j\omega}}$
	$n\,a^n\,u(n)$	$\dfrac{ae^{j\omega}}{\left(e^{j\omega}-a\right)^2}$	$\dfrac{ae^{-j\omega}}{\left(1-ae^{-j\omega}\right)^2}$
	$n^2\,a^n\,u(n)$	$\dfrac{ae^{j\omega}\left(e^{j\omega}+a\right)}{\left(e^{j\omega}-a\right)^3}$	$\dfrac{ae^{-j\omega}\left(1+ae^{-j\omega}\right)}{\left(1-ae^{-j\omega}\right)^3}$
$e^{-at}u(t)$	$e^{-anT}u(nT)$	$\dfrac{e^{j\omega}}{e^{j\omega}-e^{-aT}}$	$\dfrac{1}{1-e^{-j\omega}e^{-aT}}$
	1	$2\pi\displaystyle\sum_{m=-\infty}^{+\infty}\delta(\omega-2\pi m)$	
	$a^{|n|}$	$\dfrac{1-a^2}{1-2a\cos\omega+a^2}$	
	$\displaystyle\sum_{m=-\infty}^{+\infty}\delta(n-mN)$	$\dfrac{2\pi}{N}\displaystyle\sum_{m=-\infty}^{+\infty}\delta\left(\omega-\dfrac{2\pi m}{N}\right)$	
$e^{j\Omega_0 t}$	$e^{j\Omega_0 nt}=e^{j\omega_0 n}$ where, $\omega_0=\Omega_0 T$	$2\pi\displaystyle\sum_{m=-\infty}^{+\infty}\delta(\omega-\omega_0-2\pi m)$	
$\sin\Omega_0 t$	$\sin\Omega_0 nT$ $=\sin\omega_0 n$ where, $\omega_0=\Omega_0 T$	$\dfrac{\pi}{j}\displaystyle\sum_{m=-\infty}^{+\infty}\left[\delta(\omega-\omega_0-2\pi m)-\delta(\omega+\omega_0-2\pi m)\right]$	
$\cos\Omega_0 t$	$\cos\Omega_0 nT$ $=\cos\omega_0 n$ where, $\omega_0=\Omega_0 T$	$\pi\displaystyle\sum_{m=-\infty}^{+\infty}\left[\delta(\omega-\omega_0-2\pi m)+\delta(\omega+\omega_0-2\pi m)\right]$	

4.3 Frequency Spectrum of Discrete Time Signal

The Fourier transform $X(e^{j\omega})$ of a signal $x(n)$ represents the frequency content of $x(n)$. We can say that, by taking Fourier transform, the signal $x(n)$ is decomposed into its frequency components. Hence $X(e^{j\omega})$ is also called *frequency spectrum* of discrete time signal or *signal spectrum*.

Since $X(e^{j\omega})$ is periodic function of ω with period 2π, the frequency spectrum of discrete time signal will be periodic with period 2π. For analysis purpose one period of ω is considered from 0 to 2π or $-\pi$ to π.

Since $X(e^{j\omega})$ is a complex function of ω the frequency spectrum can be divided into two components: Magnitude spectrum and Phase spectrum.

The $X(e^{j\omega})$ is a complex valued function of ω, and so it can be expressed in polar form as,

$$X(e^{j\omega}) = |X(e^{j\omega})| \angle X(e^{j\omega})$$

$$\text{where, } |X(e^{j\omega})| \ = \text{Magnitude} \qquad\qquad .....(4.31)$$

$$\angle X(e^{j\omega}) \ = \text{Phase} \qquad\qquad\qquad .....(4.32)$$

The Magnitude of $X(e^{j\omega})$ is called magnitude spectrum and phase of $X(e^{j\omega})$ is called phase spectrum.

$$\therefore \ \text{Magnitude spectrum} \ = \ |\,X(e^{j\omega})\,| \qquad\qquad .....(4.33)$$

$$\text{Phase spectrum} \ = \ \angle X(e^{j\omega}) \qquad\qquad .....(4.34)$$

Magnitude spectrum

Let, $x(n) \ = \ $ Discrete time signal

$X(e^{j\omega}) = \mathcal{F}\{x(n)\} = $ Frequency spectrum of $x(n)$

$X^*(e^{j\omega}) = $ Conjugate of $X(e^{j\omega})$

We know that, $\quad X(e^{j\omega}) \, X^*(e^{j\omega}) \ = \ |X\,(e^{j\omega})|^2$

$\therefore$ Magnitude spectrum, $|X(e^{j\omega})| = \sqrt{X(e^{j\omega})\, X^*(e^{j\omega})}$ .....(4.35)

Alternatively, $X(e^{j\omega})$ can be expressed in the rectangular form as shown below:

$$X(e^{j\omega}) \ = \ X_r(e^{j\omega}) + j\, X_i(e^{j\omega})$$

$$\text{where, } X_r(e^{j\omega}) \ = \ \text{Real part of } X(e^{j\omega})$$

$$X_i(e^{j\omega}) = \text{Imaginary part of } X(e^{j\omega})$$

Now, the magnitude spectrum can be defined as,

$$\text{Magnitude spectrum, } |X(e^{j\omega})| \ = \ \sqrt{X_r^2(e^{j\omega}) + X_i^2(e^{j\omega})} \qquad .....(4.36)$$

Phase spectrum

From the rectangular form of $X(e^{j\omega})$, the phase of $X(e^{j\omega})$ can be expressed as,

$$\text{Phase of } X(e^{j\omega}) = \angle X(e^{j\omega}) \ = \ \tan^{-1} \frac{X_i(e^{j\omega})}{X_r(e^{j\omega})}$$

$$\therefore \ \text{Phase spectrum, } \angle X(e^{j\omega}) \ = \ \tan^{-1} \frac{X_i(e^{j\omega})}{X_r(e^{j\omega})} \qquad\qquad .....(4.37)$$

4.3.1 Aliasing in Frequency Spectrum Due to Sampling

Let x(t) be an analog signal and $X(j\Omega)$ be Fourier transform of x(t).

Now by definition of continuous time inverse Fourier transform,

$$x(t) = \frac{1}{2\pi} \int_{-\infty}^{+\infty} X(j\Omega) \, e^{j\Omega t} \, d\Omega \qquad \qquad \qquad ....(4.38)$$

Let x(nT) be a discrete time signal obtained by sampling x(t) with sampling period, T.

$$\therefore x(nT) = x(t)\big|_{t=nT}$$

$$= \frac{1}{2\pi} \int_{-\infty}^{\infty} X(j\Omega) \, e^{j\Omega t} \, d\Omega \big|_{t=nT} \qquad \boxed{\text{Using equation (4.38).}}$$

$$= \frac{1}{2\pi} \int_{-\infty}^{+\infty} X(j\Omega) \, e^{j\Omega nT} \, d\Omega \qquad \boxed{\begin{array}{l}\text{Expressing the integration} \\ \text{as summation of infinite} \\ \text{number of integrals.}\end{array}}$$

$$= \frac{1}{2\pi} \sum_{m=-\infty}^{+\infty} \int_{\frac{(2m-1)\pi}{T}}^{\frac{(2m+1)\pi}{T}} X\Big(j\Big(\Omega + \frac{2\pi m}{T}\Big)\Big) \, e^{j\big(\Omega + \frac{2\pi m}{T}\big)nT} \, d\Omega \qquad \boxed{\begin{array}{l}X(j\Omega) \text{ in the interval} \\ \dfrac{(2m-1)\pi}{T} \text{ to } \dfrac{(2m+1)\pi}{T} \\ \text{is identical with } X(j\Omega) \\ \text{in the interval } -\dfrac{\pi}{T} \text{ to } +\dfrac{\pi}{T}\end{array}}$$

$$= \frac{1}{2\pi} \sum_{m=-\infty}^{+\infty} \int_{-\pi/T}^{+\pi/T} X\Big(j\Big(\Omega + \frac{2\pi m}{T}\Big)\Big) \, e^{j\Omega nT} \, e^{j2\pi mn} \, d\Omega \qquad \boxed{\begin{array}{l}\text{Since m and n are} \\ \text{integers } e^{j2\pi mn} = 1\end{array}}$$

$$= \frac{1}{2\pi} \sum_{m=-\infty}^{+\infty} \int_{-\pi/T}^{+\pi/T} X\Big(j\Big(\frac{\omega}{T} + \frac{2\pi m}{T}\Big)\Big) \, e^{j\omega n} \, d\omega \qquad \boxed{\begin{array}{l}\text{The relation between analog and} \\ \text{digital frequency is } \Omega = \dfrac{\omega}{T}\end{array}}$$

$$= \frac{1}{2\pi} \sum_{m=-\infty}^{+\infty} \frac{1}{T} \int_{-\pi}^{+\pi} X\Big(j\Big(\frac{\omega}{T} + \frac{2\pi m}{T}\Big)\Big) \, e^{j\omega n} \, d\omega$$

$$= \frac{1}{2\pi} \int_{-\pi}^{+\pi} \frac{1}{T} \sum_{m=-\infty}^{+\infty} X\Big(j\Big(\frac{\omega}{T} + \frac{2\pi m}{T}\Big)\Big) \, e^{j\omega n} \, d\omega \qquad \qquad .....(4.39)$$

By the definition of inverse Fourier transform of a discrete time signal, the x(nT) can be written as,

$$x(nT) = \frac{1}{2\pi} \int_{-\pi}^{+\pi} X(e^{j\omega}) \, e^{j\omega n} \, d\omega \qquad \qquad .....(4.40)$$

On comparing equations (4.39) and (4.40) we can write,

$$X(e^{j\omega}) = \frac{1}{T} \sum_{m=-\infty}^{+\infty} X\Big(j\Big(\frac{\omega}{T} + \frac{2\pi m}{T}\Big)\Big) \qquad \qquad .....(4.41)$$

$$= \frac{1}{T} \sum_{m=-\infty}^{+\infty} X\Big(j\Big(\Omega + \frac{2\pi m}{T}\Big)\Big) \qquad \qquad ....(4.42)$$

In equation (4.42) if $X(j\Omega)$ is the original spectrum of analog signal, then $X\left(j\left(\Omega + \frac{2\pi m}{T}\right)\right)$ is the frequency shifted version of $X(j\Omega)$, shifted by $\frac{2\pi m}{T}$. In equation (4.42) the term $\frac{1}{T}$ will scale the amplitude of the spectrum $X\left(j\left(\Omega + \frac{2\pi m}{T}\right)\right)$ by a factor $\frac{1}{T}$.

Therefore from equation (4.42) we can say that $X(e^{j\omega})$ is sum of frequency shifted and amplitude scaled version of $X(j\Omega)$. In general we can say that the *frequency spectrum of a discrete time signal obtained by sampling continuous time signal will be sum of frequency shifted and amplitude scaled spectrum of continuous time signal*. This concept is illustrated in Fig 4.1.

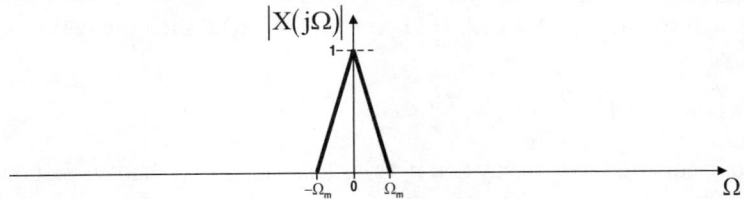

Fig 4.1a: Spectrum of a continuous time signal x(t), with maximum frequency Ω_m.

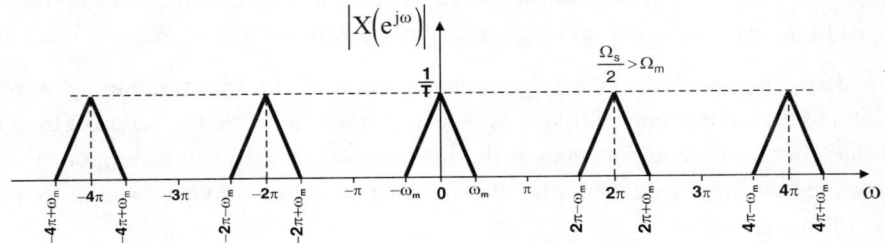

Fig 4.1b: Spectrum of sampled version of x(t), with $\Omega_s/2 > \Omega_m$

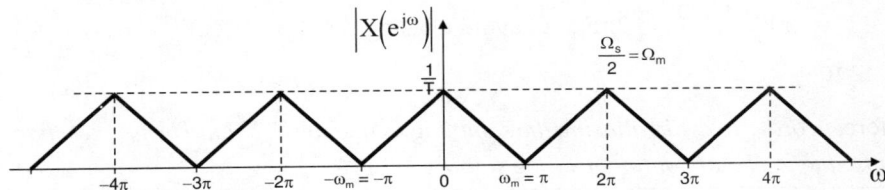

Fig 4.1c: Spectrum of sampled version of x(t), with $\Omega_s/2 = \Omega_m$.

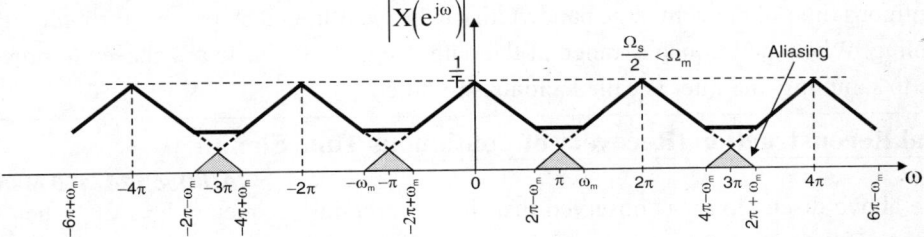

Fig 4.1d: Spectrum of sampled version of x(t), with $\Omega_s/2 < \Omega_m$.

Fig 4.1: Spectrum of a continuous time signal and its sampled version, sampled at various sampling rates.

The frequency Ω of a continuous time signal can be converted to frequency ω of a discrete time signal by choosing the transformation, $\omega = \Omega T$, where T is the sampling time, $1/T = F_s$ is the sampling cyclic frequency, and $2\pi F_s = \Omega_s$ is the radian sampling frequency.

In this transformation, the radian frequency ω of sampled version of discrete time signal is unique in the interval $-\pi$ to $+\pi$, and the cyclic frequency f of sampled version of discrete time signal is unique in the interval $-1/2$ to $+1/2$.

The maximum frequency in the spectrum shown in Fig 4.1a is Ω_m. Let ω_m be the corresponding maximum frequency of the sampled version of the discrete time signal when the spectrum of Fig 4.1a is sampled at a frequency of $\Omega_s/2$. If Ω_m is equal to $\Omega_s/2$, then the corresponding value of ω_m is given by,

$$\omega_m = \Omega_m T = \frac{\Omega_s}{2} T = \frac{2\pi F_s}{2} T = \frac{\pi}{T} T = \pi$$

From the above equation we can say that if Ω_m is less than $\Omega_s/2$, then corresponding ω_m will be less than π and if Ω_m is greater than $\Omega_s/2$ then corresponding ω_m will be greater than π. From Fig 4.1b and Fig 4.1c it is observed that, as long as Ω_m is less than $\Omega_s/2$, then corresponding ω_m is less than or equal to π, and so there is no overlapping of the components of frequency spectrum.

From Fig 4.1d it is observed, when Ω_m is greater than $\Omega_s/2$, then corresponding ω_m will be greater than π, and so the components of frequency spectrum overlaps. Due to overlap of frequency spectrum, the high frequency components get the identity of low frequency components. This phenomenon is called *aliasing*. Due to aliasing the information shifts from one band of frequency to another band of frequency.

Therefore in order to avoid aliasing, $\Omega_s/2$, should be greater than or equal to Ω_m.

Since, $\Omega_m = 2\pi F_m$ and $\Omega_s = 2\pi F_s$, to avoid aliasing, $2\pi F_s/2 > 2\pi F_m$

$\therefore$ $F_s > 2F_m$

Therefore, *in order to avoid aliasing the sampling frequency F_s should be greater than twice the maximum frequency F_m of continuous time signal* .

4.3.2 Antialaising filter   *(AU Jun'14, 2 Marks)*

A continous time signal with large bandwidth can be bandlimited by passing through a filter before sampling. When the frequency range of the output signal of the filter is chosen to prevent alaising due to sampling, the filter is called antialiasing filter.

4.3.3 Signal Reconstruction (Recovery of Continuous Time Signal)

(AU Dec'12&11, 8 Marks)

In the above discussion it is observed that, if the sampling frequency $F_s > 2F_m$, then the spectrum $X(e^{j\omega})$ of the sampled continuous time signal will have aliased components of the spectrum $X(j\Omega)$ of original continuous time signal. The aliasing of spectral components prevents the recovery of original signal x(t) from the sampled signal x(n).

When the spectrum of sampled signal has no aliasing then it is possible to recover the original signal from the sampled signal. When there is no aliasing, the spectrum $X(e^{j\omega})$ can be passed through a low pass filter with cut-off frequency, ω_s/π. Now the equation of spectrum $X(e^{j\omega})$ [equation(4.42)]can be written as shown below:

$$X(e^{j\omega}) = \frac{1}{T}X(j\Omega) \quad \Rightarrow \quad X(j\Omega) = TX(e^{j\omega}) \qquad \qquad .....(4.43)$$

On taking inverse Fourier transform of $X(j\Omega)$ we get x(t). Hence by definition of inverse Fourier transform of continuous time signal we get,

$$x(t) = \frac{1}{2\pi}\int_{-\infty}^{+\infty} X(j\Omega)e^{j\Omega t}d\Omega = \frac{1}{2\pi}\int_{-\pi/T}^{+\pi/T} X(j\Omega)e^{j\Omega t}d\Omega$$

$\quad$ Because $X(j\Omega)$ is zero outside the interval $-\pi/T$ to π/T

$$= \frac{1}{2\pi}\int_{-\pi/T}^{+\pi/T} TX(e^{j\omega})e^{j\Omega t}d\Omega$$

$\quad$ Substituting for $X(j\Omega)$ from equation (4.43).

$$= \frac{1}{2\pi}\int_{-\pi/T}^{+\pi/T} T\sum_{n=-\infty}^{+\infty}x(nT)e^{-j\omega n}e^{j\Omega t}d\Omega$$

$\quad$ Using the definition of Fourier transform of discrete time signal.

$$= \frac{1}{2\pi}\int_{-\pi/T}^{+\pi/T} T\sum_{n=-\infty}^{+\infty}x(nT)e^{-j\Omega Tn}e^{j\Omega t}d\Omega = \frac{T}{2\pi}\sum_{n=-\infty}^{+\infty}x(nT)\int_{-\pi/T}^{+\pi/T}e^{j\Omega(t-nT)}d\Omega$$

$$= \frac{T}{2\pi}\sum_{n=-\infty}^{+\infty}x(nT)\left[\frac{e^{j\Omega(t-nT)}}{j(t-nT)}\right]_{-\pi/T}^{+\pi/T} = \frac{T}{2\pi}\sum_{n=-\infty}^{+\infty}x(nT)\left[\frac{e^{j(\pi/T)(t-nT)}}{j(t-nT)} - \frac{e^{j(-\pi/T)(t-nT)}}{j(t-nT)}\right]$$

$$= \sum_{n=-\infty}^{+\infty}x(nT)\frac{1}{(\pi/T)(t-nT)}\left[\frac{e^{j(\pi/T)(t-nT)} - e^{-j(\pi/T)(t-nT)}}{2j}\right]$$

$$= \sum_{n=-\infty}^{+\infty}x(nT)\frac{\sin[(\pi/T)(t-nT)]}{(\pi/T)(t-nT)} \qquad \qquad .....(4.44)$$

The equation (4.44) can be used to reconstruct the original continuous time signal x(t) from its samples and the equation (4.44) is also called *ideal interpolation formula*.

4.3.4 Sampling of Bandpass Signal

A continuous time signal is called *bandpass signal* if its frequency spectrum lies in a narrow band of frequencies. Let the lower and upper value of this narrow band of frequency be F_1 and F_2 respectively. Now the *bandwidth*, "$B = F_1 - F_2$". Let F_c be a frequency corresponding to centre of bandwidth. The frequency spectrum of some of the bandpass signals are shown in Fig 4.2.

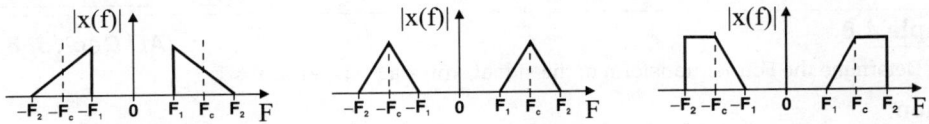

Fig 4.2: Sample frequency spectrum of continuous time bandpass signals.

The maximum frequency in the bandpass signal is F_2. According to sampling theorem, to avoid aliasing the bandpass signal has to be sampled at a sampling frequency greater than $2F_2$. When F_2 happens to be a very high frequency, then sampling rate will be very high. In order to avoid high sampling rates the bandpass signals can be shifted in frequency to an equivalent lowpass signal and the equivalent lowpass signal can be sampled at a lower rate.

A bandpass signal can be shifted in frequency by an amount F_c to convert the signal to an equivalent lowpass signal, and when the upper cutoff frequency F_2 is an integer multiple of bandwidth B, then the equivalent lowpass signal can be sampled at a rate of 2B samples per second. When the upper cutoff frequency F_2 is not an integer multiple of bandwidth B, then the sampling rate has to be slightly increased and go upto 4B.

In general, the bandpass signals with a bandwidth of B Hz can be sampled at a rate of 2B to 4B Hz.

Example 4.6

Find the Fourier transform of $x(n)$, where $x(n) = 1$; $0 \leq n \leq 5$

$= 0$; otherwise

Solution:

By the definition of Fourier transform,

$$X(e^{j\omega}) = \mathcal{F}\{x(n)\} = \sum_{n=-\infty}^{+\infty} x(n) e^{-j\omega n} = \sum_{n=0}^{5} x(n) e^{-j\omega n} = \frac{1 - e^{-j6\omega}}{1 - e^{-j\omega}}$$

> Using finite geometric series sum formula,
> $$\sum_{n=0}^{N} C^n = \frac{1 - C^{N+1}}{1 - C}$$

$$= \frac{1 - e^{\frac{-j6\omega}{2}} e^{\frac{-j6\omega}{2}}}{1 - e^{\frac{-j\omega}{2}} e^{\frac{-j\omega}{2}}} = \frac{\left(e^{\frac{j6\omega}{2}} - e^{\frac{-j6\omega}{2}}\right) e^{\frac{-j6\omega}{2}}}{\left(e^{\frac{j\omega}{2}} - e^{\frac{-j\omega}{2}}\right) e^{\frac{-j\omega}{2}}}$$

> $$\boxed{e^{j\theta} \, e^{-j\theta} = 1}$$

$$= \left(\frac{2j \sin \frac{6\omega}{2}}{2j \sin \frac{\omega}{2}}\right) e^{\frac{-j6\omega}{2} + \frac{j\omega}{2}} = \frac{\sin \frac{6\omega}{2}}{\sin \frac{\omega}{2}} e^{\frac{-j5\omega}{2}} = \frac{\sin 3\omega}{\sin \frac{\omega}{2}} e^{\frac{-j5\omega}{2}}$$

> $$\boxed{\sin \theta = \frac{e^{j\theta} - e^{-j\theta}}{2j}}$$

Example 4.7

(AU Dec'11, 8 Marks)

Find the Fourier transform of the discrete time signal, $x(n) = u(n - 2)$.

Solution:

Given that, $x(n) = u(n - 2) = 1$; $n \geq 2$

By the definition of Fourier transform,

$$X(e^{j\omega}) = \mathcal{F}\{x(n)\} = \sum_{n=-\infty}^{\infty} x(n) e^{-j\omega n} = \sum_{n=2}^{\infty} e^{-j\omega n} = \sum_{n=0}^{\infty} e^{-j\omega n} - e^{-j\omega n}\Big|_{n=0} - e^{-j\omega n}\Big|_{n=1}$$

$$= \frac{1}{1 - e^{-j\omega}} - 1 - e^{-j\omega} = \frac{1 - (1 - e^{-j\omega}) - e^{-j\omega}(1 - e^{-j\omega})}{1 - e^{-j\omega}}$$

> Using infinite geometric series sum formula
> $$\sum_{n=0}^{\infty} C^n = \frac{1}{1 - C} \; ; \text{ if, } 0 < |C| < 1$$

$$= \frac{1 - 1 + e^{-j\omega} - e^{-j\omega} + e^{-j2\omega}}{1 - e^{-j\omega}} = \frac{e^{-j2\omega}}{1 - e^{-j\omega}}$$

Example 4.8

(AU Dec'13, 8 Marks)

Determine the Fourier transform of the signal, $x(n) = a^{|n|}$; $-1 < a < 1$.

Solution:

Given that, $x(n) = a^{|n|} = a^{-n}$; $n < 0$

$= a^n$; $n \geq 0$

By definition of Fourier transform,

$$X(e^{j\omega}) = \mathcal{F}\{x(n)\} = \sum_{n=-\infty}^{+\infty} x(n)e^{-j\omega n}$$

$$= \sum_{n=-\infty}^{-1} a^{-n} e^{-j\omega n} + \sum_{n=0}^{\infty} a^{n} e^{-j\omega n}$$

$$= \sum_{n=1}^{\infty} a^{n} e^{j\omega n} + \sum_{n=0}^{\infty} a^{n} e^{-j\omega n}$$

$$= -(ae^{j\omega})^{0} + (ae^{j\omega})^{0} + \sum_{n=1}^{\infty} a^{n} e^{j\omega n} + \sum_{n=0}^{\infty} a^{n} e^{-j\omega n}$$

$$= -1 + \sum_{n=0}^{\infty} a^{n} e^{j\omega n} + \sum_{n=0}^{\infty} a^{n} e^{-j\omega n} \qquad \boxed{(ae^{j\omega})^{0} = 1}$$

$$= -1 + \sum_{n=0}^{\infty} (ae^{j\omega})^{n} + \sum_{n=0}^{\infty} (ae^{-j\omega})^{n} \quad = -1 + \frac{1}{1 - ae^{j\omega}} + \frac{1}{1 - ae^{-j\omega}}$$

$$= \frac{-(1 - ae^{j\omega})(1 - ae^{-j\omega}) + 1 - ae^{-j\omega} + 1 - ae^{j\omega}}{(1 - ae^{j\omega})(1 - ae^{-j\omega})}$$

$$= \frac{\cancel{-1} + \cancel{ae^{-j\omega}} + \cancel{ae^{j\omega}} - a^{2} + \cancel{1} - \cancel{ae^{-j\omega}} + 1 - \cancel{ae^{j\omega}}}{1 - ae^{-j\omega} - ae^{+j\omega} + a^{2}} = \frac{1 - a^{2}}{1 - a(e^{j\omega} + e^{-j\omega}) + a^{2}}$$

$$= \frac{1 - a^{2}}{1 - 2a\cos\omega + a^{2}}$$

> Using infinite geometric series sum formula
> $$\sum_{n=0}^{\infty} C^{n} = \frac{1}{1 - C}$$
> if, $0 < |C| < 1$

$$\boxed{\cos\theta = \frac{e^{j\theta} + e^{-j\theta}}{2}}$$

Example 4.9

Find $X(e^{j\omega})$, if $x(n) = \frac{1}{2}\left[\left(\frac{1}{3}\right)^{n} + \left(\frac{1}{5}\right)^{n}\right]u(n)$.

Solution:

Given that, $x(n) = \frac{1}{2}\left[\left(\frac{1}{3}\right)^{n} + \left(\frac{1}{5}\right)^{n}\right]u(n)$; for all $n = \frac{1}{2}\left(\frac{1}{3}\right)^{n} + \frac{1}{2}\left(\frac{1}{5}\right)^{n}$; for $n \geq 0$

By definition of Fourier transform,

$$X(e^{j\omega}) = \mathcal{F}\{x(n)\} = \sum_{n=-\infty}^{+\infty} x(n)e^{-j\omega n} = \sum_{n=0}^{+\infty}\left[\frac{1}{2}\left(\frac{1}{3}\right)^{n} + \frac{1}{2}\left(\frac{1}{5}\right)^{n}\right]e^{-j\omega n}$$

$$= \frac{1}{2}\sum_{n=0}^{\infty}\left(\frac{1}{3}\right)^{n}e^{-j\omega n} + \frac{1}{2}\sum_{n=0}^{\infty}\left(\frac{1}{5}\right)^{n}e^{-j\omega n} = \frac{1}{2}\sum_{n=0}^{\infty}\left(\frac{1}{3}e^{-j\omega}\right)^{n} + \frac{1}{2}\sum_{n=0}^{\infty}\left(\frac{1}{5}e^{-j\omega}\right)^{n}$$

$$= \frac{1}{2}\frac{1}{1 - \frac{1}{3}e^{-j\omega}} + \frac{1}{2}\frac{1}{1 - \frac{1}{5}e^{-j\omega}}$$

> Using infinite geometric series sum formula
> $$\sum_{n=0}^{\infty} C^{n} = \frac{1}{1 - C} \; ; \text{ if, } 0 < |C| < 1$$

$$= \frac{1}{2}\left[\frac{1 - \frac{1}{5}e^{-j\omega} + 1 - \frac{1}{3}e^{-j\omega}}{\left(1 - \frac{1}{3}e^{-j\omega}\right)\left(1 - \frac{1}{5}e^{-j\omega}\right)}\right]$$

$$= \frac{1}{2}\frac{\left[2 - \left(\frac{1}{5} + \frac{1}{3}\right)e^{-j\omega}\right]}{\left[1 - \left(\frac{1}{5} + \frac{1}{3}\right)e^{-j\omega} + \frac{1}{15}e^{-j2\omega}\right]}$$

$$= \frac{1}{2}\frac{2 - \frac{3+5}{5\times 3}e^{-j\omega}}{1 - \frac{3+5}{5\times 3}e^{-j\omega} + \frac{1}{15}e^{-j2\omega}} = \frac{1 - \frac{4}{15}e^{j\omega}}{1 - \frac{8}{15}e^{-j\omega} + \frac{1}{15}e^{-j2\omega}}$$

$$= \frac{15 - 4e^{-j\omega}}{15 - 8e^{-j\omega} + e^{-j2\omega}}$$

Example 4.10   *(AU May'15, 5 Marks)*

Find the DTFT of $x(n) = \left(\frac{1}{2}\right)^{n-1} u(n-1)$.

Solution:

Given that, $x(n) = \left(\frac{1}{2}\right)^{n-1} u(n-1) = \left(\frac{1}{2}\right)^{n}$; $n \geq 1$

By definition of Fourier transform

$$X(e^{j\omega}) = \mathcal{F}\{x(n)\} = \sum_{n=-\infty}^{+\infty} x(n)\, e^{-j\omega n}$$

$$= \sum_{n=1}^{\infty} \left(\frac{1}{2}\right)^{n} e^{-j\omega n} = -\left(\frac{1}{2}\right)^{0} + \left(\frac{1}{2}\right)^{0} + \sum_{n=1}^{\infty} \left(\frac{1}{2}\right)^{n} e^{-j\omega n}$$

$$= -1 + \sum_{n=0}^{\infty} \left(\frac{1}{2}\right)^{n} e^{-j\omega n} = -1 + \sum_{n=0}^{\infty} \left(\frac{1}{2} e^{-j\omega}\right)^{n} = -1 + \frac{1}{1 - \frac{1}{2} e^{-j\omega}}$$

$$= \frac{-\left(1 - \frac{1}{2} e^{j\omega}\right) + 1}{1 - \frac{1}{2} e^{-j\omega}} = \frac{\frac{1}{2} e^{-j\omega}}{1 - \frac{1}{2} e^{-j\omega}}$$

$$= \frac{e^{-j\omega}}{2 - e^{-j\omega}}$$

> Using infinite geometric series sum formula
> $$\sum_{n=0}^{\infty} C^{n} = \frac{1}{1-C}$$
> if $0 < |C| < 1$

Example 4.11

Determine the Fourier transform of the following discrete time signals

a) $x_1(n) = x(1-n)$ **b)** $x_2(n) = (n-1)^2 x(n)$

Solution: *(AU Jun'12, 8 Marks)*

a) Given that, $x_1(n) = x(1-n)$

Let, $\mathcal{F}\{x(n)\} = X(e^{j\omega})$

$\therefore \mathcal{F}\{x(-n)\} = X(e^{j\omega})\big|_{\omega = -\omega} = X(e^{-j\omega})$

$\therefore \mathcal{F}\{x(-n+1)\} = e^{j\omega}\, \mathcal{F}\{x(-n)\} = e^{j\omega} X(e^{-j\omega})$

$\therefore \mathcal{F}\{x_1(n)\} = e^{j\omega} X(e^{-j\omega})$

b) Given that, $x_2(n) = (n-1)^2 x(n)$ *(AU Jun'12, 8 Marks)*

$= (n^2 - 2n + 1)\, x(n) = n^2 x(n) - 2n\, x(n) + x(n)$

Let, $\mathcal{F}\{x(n)\} = X(e^{j\omega})$

$\therefore \mathcal{F}\{n\, x(n)\} = j\dfrac{d}{d\omega} X(e^{j\omega})$

$\therefore \mathcal{F}\{n^2 x(n)\} = j\dfrac{d}{d\omega} \mathcal{F}\{n\, x(n)\} = j\dfrac{d}{d\omega} j\dfrac{d}{d\omega} X(e^{j\omega}) = -\dfrac{d^2}{d\omega^2} X(e^{j\omega})$

$\therefore \mathcal{F}\{x_2(n)\} = \mathcal{F}\{(n-1)^2 x(n)\} = \mathcal{F}\{n^2 x(n) - 2n\, x(n) + x(n)\}$

$= \mathcal{F}\{n^2 x(n)\} - 2\, \mathcal{F}\{n\, x(n)\} + \mathcal{F}\{x(n)\} = -\dfrac{d^2}{d\omega^2} X(e^{j\omega}) - 2j\dfrac{d}{d\omega} X(e^{j\omega}) + X(e^{j\omega}).$

Example 4.12

Compute the Fourier transform and sketch the magnitude and phase spectrum of causal three sample sequence given by,

$$x(n) = \frac{1}{3} \; ; 0 \le n \le 2$$

$$= 0 \; ; \text{else}$$

Solution:

Let, $X(e^{j\omega})$ be Fourier transform of x(n).

Now by definition of Fourier transform,

$$X(e^{j\omega}) = \sum_{n=-\infty}^{+\infty} x(n)e^{-j\omega n} = \sum_{n=0}^{2} x(n)e^{-j\omega n}$$

$$= x(0)e^{0} + x(1)e^{-j\omega} + x(2)e^{-j2\omega} = \frac{1}{3} + \frac{1}{3}e^{-j\omega} + \frac{1}{3}e^{-j2\omega}$$

$$= \frac{1}{3} + \frac{1}{3}(\cos\omega - j\sin\omega) + \frac{1}{3}(\cos 2\omega - j\sin 2\omega)$$

$\boxed{e^{\pm j\theta} = \cos\theta \pm \sin\theta}$

$$= \frac{1}{3}(1 + \cos\omega + \cos 2\omega) - j\frac{1}{3}(\sin\omega + \sin 2\omega)$$

The $X(e^{j\omega})$ is evaluated for various values of ω and tabulated in table 1. The magnitude and phase of $X(e^{j\omega})$ for various values of ω are also listed in table 1. Using the values listed in table 1, the magnitude and phase spectrum are sketched as shown in Fig 1 and Fig 2 respectively.

Table 1: Magnitude and Phase of X(e^{jw})

ω	$X(e^{j\omega})$	$\lvert X(e^{j\omega}) \rvert$	$\angle X(e^{j\omega})$ in rad
0	$1 + j0 \quad = 1\angle 0$	1	0
$\frac{\pi}{8}$	$0.877 - j0.363 = 0.949\angle -0.392 = 0.949\angle -0.125\pi$	0.949	-0.125π
$\frac{2\pi}{8}$	$0.569 - j0.569 = 0.805\angle -0.785 = 0.805\angle -0.25\pi$	0.805	-0.25π
$\frac{3\pi}{8}$	$0.225 - j0.544 = 0.587\angle -1.179 = 0.587\angle -0.375\pi$	0.587	-0.375π
$\frac{4\pi}{8} = \frac{\pi}{2}$	$0 - j0.333 \quad = 0.333\angle -1.571 = 0.333\angle -0.5\pi$	0.333	-0.5π
$\frac{5\pi}{8}$	$-0.03 - j0.072 = 0.078\angle -1.966 = 0.078\angle -0.625\pi$	0.078	-0.625π
$\frac{6\pi}{8}$	$0.098 - j0.098 = 0.139\angle -0.785 = 0.139\angle -0.25\pi$	0.139	-0.25π
$\frac{7\pi}{8}$	$0.261 + j0.108 = 0.282\angle 0.392 = 0.282\angle 0.125\pi$	0.282	0.125π
$\frac{8\pi}{8} = \pi$	$0.333 + j0 \quad = 0.333\angle 0 \quad = 0.333\angle 0$	0.333	0

Table 1: Continued...

ω	$X(e^{j\omega})$	$\lvert X(e^{j\omega})\rvert$	$\angle X(e^{j\omega})$ in rad
$\dfrac{9\pi}{8}$	$0.261 - j0.108 = 0.282\angle 0.392 = 0.282\angle -0.125\pi$	0.282	-0.125π
$\dfrac{10\pi}{8}$	$0.098 + j0.098 = 0.139\angle 0.785 = 0.139\angle 0.25\pi$	0.139	0.25π
$\dfrac{11\pi}{8}$	$-0.03 + j0.072 = 0.078\angle 1.966 = 0.078\angle 0.625\pi$	0.078	0.625π
$\dfrac{12\pi}{8} = \dfrac{3\pi}{2}$	$0 + j0.333 \quad = 0.333\angle 1.571 = 0.333\angle 0.5\pi$	0.333	0.5π
$\dfrac{13\pi}{8}$	$0.225 + j0.544 = 0.587\angle 1.179 = 0.587\angle 0.375\pi$	0.589	0.375π
$\dfrac{14\pi}{8}$	$0.569 + j0.569 = 0.805\angle 0.785 = 0.805\angle 0.25\pi$	0.805	0.25π
$\dfrac{15\pi}{8}$	$0.877 + j0.363 = 0.949\angle 0.392 = 0.949\angle 0.125\pi$	0.949	0.125π
$\dfrac{16\pi}{8} = 2\pi$	$1 + j0 \qquad = 1\angle 0$	1	0

Note : The function X(e^{jw}) is calculated using complex mode of calculator. The magnitude and phase are calculated using rectangular to polar conversion technique.

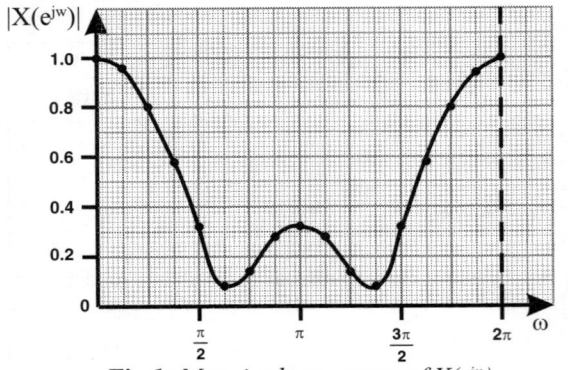

Fig 1: *Magnitude spectrum of X(e$^{j\omega}$).*   **Fig 2:** *Phase spectrum of X(e$^{j\omega}$).*

Example 4.13

Find the convolution of the sequences, $x_1(n) = x_2(n) = \{1, 3, 5\}$
$\uparrow$

Solution:

By definition of Fourier transform,

$$X_1(e^{j\omega}) = \mathcal{F}\{x_1(n)\} = \sum_{n=-\infty}^{+\infty} x_1(n)e^{-j\omega n} = \sum_{n=-1}^{+1} x_1(n)e^{-j\omega n} = x(-1)\,e^{j\omega} + x(0)\,e^{0} + x(1)e^{-j\omega} = e^{j\omega} + 3 + 5e^{-j\omega}$$

Since, $x_1(n) = x_2(n)$, $X_2(e^{j\omega}) = X_1(e^{j\omega}) = e^{j\omega} + 3 + 5e^{-j\omega}$

Let, $x_3(n) = x_1(n) * x_2(n)$

$\therefore\ X_3(e^{j\omega}) = \mathcal{F}\{x_3(n)\} = \mathcal{F}\{x_1(n) * x_2(n)\}$

$$\therefore X_3(e^{j\omega}) = X_1(e^{j\omega})X_2(e^{j\omega}) = (e^{j\omega} + 3 + 5e^{-j\omega})(e^{j\omega} + 3 + 5e^{-j\omega})$$

Using convolution property of Fourier transform.

$$= e^{j2\omega} + 3e^{j\omega} + 5 + 3e^{j\omega} + 9 + 15e^{-j\omega} + 5 + 15e^{-j\omega} + 25e^{-j2\omega}$$

$$= e^{j2\omega} + 6e^{j\omega} + 19 + 30e^{-j\omega} + 25e^{-j2\omega} \qquad .....(1)$$

By definition of Fourier transform,

$$X_3(e^{j\omega}) = \sum_{n=-\infty}^{+\infty} x_3(n)e^{-j\omega n} = ...x_3(-2)e^{j2\omega} + x_3(-1)e^{j\omega} + x_3(0)$$

$$+ x_3(1)e^{-j\omega} + x_3(2)e^{-j2\omega} + ... \qquad .....(2)$$

On comparing the coefficient of $e^{j\omega n}$ in equations (1) and (2) we get,

$$x_3(n) = x_1(n) * x_2(n) = \{1, 6, 19, 30, 25\}$$
$$\uparrow$$

Example 4.14

If $X(e^{j\omega}) = \frac{1}{5}(1 + 3\cos\omega)$, find x(n).

Solution:

Given that, $X(e^{j\omega}) = \frac{1}{5}(1 + 3\cos\omega) = \frac{1}{5} + \frac{3}{5}\frac{e^{j\omega} + e^{-j\omega}}{2} = \frac{3}{10}e^{j\omega} + \frac{1}{5} + \frac{3}{10}e^{-j\omega}$

$$\boxed{\cos\theta = \frac{e^{j\theta} + e^{-j\theta}}{2}}$$

$$= 0.3e^{j\omega} + 0.2 + 0.3e^{-j\omega} \qquad .....(1)$$

Let, $x(n) = \mathcal{F}^{-1}\{X(e^{j\omega})\}$

By definition of Fourier transform we get,

$$X(e^{j\omega}) = \sum_{n=-\infty}^{+\infty} x(n)e^{-j\omega n} = ... + x(-2)e^{j2\omega} + x(-1)e^{j\omega} + x(0) + x(1)e^{-j\omega} + x(2)e^{-j2\omega} + ... \qquad .....(2)$$

On comparing the equations (1) and (2) we can say that the samples of x(n) are the coefficients of $e^{j\omega n}$.

$$\therefore x(-1) = 0.3 \; ; \; x(0) = 0.2 \; ; \; x(1) = 0.3 \; ; \; \text{and} \; x(n) = 0, \text{for } n < -1 \text{ and } n > 1$$

$$\therefore x(n) = \{0.3, \ 0.2, \ 0.3\}$$
$$\uparrow$$

Example 4.15

Find the inverse Fourier transform of, $X(e^{j\omega}) = (1 - ae^{-j\omega})^{-1}$.

Solution:

Given that, $X(e^{j\omega}) = (1 - ae^{-j\omega})^{-1} = \dfrac{1}{1 - ae^{-j\omega}}$

Using Taylor series expansion, the above equation of $X(e^{j\omega})$ can be expanded as shown below:

$$X(e^{j\omega}) = 1 + ae^{-j\omega} + a^2 e^{-j2\omega} + ... + a^k e^{-jk\omega} + ... \qquad .....(1)$$

Let, $x(n) = \mathcal{F}^{-1}\{X(e^{j\omega})\}$.

By definition of Fourier transform we get,

$$X(e^{j\omega}) = \sum_{n=-\infty}^{+\infty} x(n)e^{-j\omega n}$$

$$= ...... + x(-2)e^{j2\omega} + x(-1)e^{j\omega} + x(0) + x(1)e^{-j\omega} + x(2)e^{-j2\omega} + ... \qquad .....(2)$$

On comparing the equations (1) and (2) we can say that the samples of x(n) are the coefficients of $e^{-j\omega n}$.

$$\therefore x(n) = \{1, \ a, \ a^2, ..., \ a^k, ...\}$$

$$\therefore x(n) = a^n \ ; \ n \geq 0 \qquad \text{or } x(n) = a^n \ u(n)$$
$$\qquad = 0 \ \ ; \ n < 0$$

Example 4.16

Determine the discrete time sequence from the spectrum $X(e^{j\omega})$, where $X(e^{j\omega}) = \dfrac{1}{2}\dfrac{e^{j\omega}+1+e^{-j\omega}}{1-ae^{-j\omega}}$

Solution:

The discrete time sequence y(n) is obtained by taking inverse Fourier transform of $X(e^{j\omega})$.

$$X(e^{j\omega}) = \frac{1}{2}\frac{e^{j\omega}+1+e^{-j\omega}}{1-ae^{-j\omega}} = \frac{1}{2}\left[\frac{e^{j\omega}}{1-ae^{-j\omega}} + \frac{1}{1-ae^{-j\omega}} + \frac{e^{-j\omega}}{1-ae^{-j\omega}}\right]$$

$$X(e^{j\omega}) = \frac{1}{2}\left[X_1(e^{j\omega}) + X_2(e^{j\omega}) + X_3(e^{j\omega})\right]$$

where, $X_1(e^{j\omega}) = \dfrac{e^{j\omega}}{1-ae^{-j\omega}}$; $X_2(e^{j\omega}) = \dfrac{1}{1-ae^{-j\omega}}$ and $X_3(e^{j\omega}) = \dfrac{e^{-j\omega}}{1-ae^{-j\omega}}$

Let, $x_1(n) = \mathcal{F}^{-1}\{X_1(e^{j\omega})\}$; $x_2(n) = \mathcal{F}^{-1}\{X_2(e^{j\omega})\}$; $x_3(n) = \mathcal{F}^{-1}\{X_3(e^{j\omega})\}$

> $u(n) = 1$ for $n \geq 0$
> $\qquad = 0$ for $n < 0$

By Taylor's series expansion we get,

$$X_2(e^{j\omega}) = \frac{1}{1-ae^{-j\omega}} = 1 + ae^{-j\omega} + a^2 e^{-j2\omega} + a^3 e^{-j3\omega} + ...$$

$$= \sum_{n=0}^{+\infty} a^n e^{-j\omega n} = \sum_{n=-\infty}^{+\infty} a^n u(n) e^{-j\omega n} \qquad\qquad .....(1)$$

By definition of Fourier transform we can write,

$$X_2(e^{j\omega}) = \sum_{n=-\infty}^{+\infty} x_2(n) e^{-j\omega n} \qquad\qquad\qquad .....(2)$$

By comparing equations (1) and (2) we can write,

$$x_2(n) = a^n \ u(n)$$

> By time shifting property
> If $\mathcal{F}\{x(n)\} = X(e^{j\omega})$
> then, $\mathcal{F}\{x(n+m)\} = e^{\pm j\omega m} X(e^{j\omega})$

Here, $X_1(e^{j\omega}) = \dfrac{e^{j\omega}}{1-ae^{-j\omega}} = e^{j\omega} X_2(e^{j\omega})$

$$\therefore x_1(n) = a^{(n+1)} u(n+1)$$

> Using shifting property

Here, $X_3(e^{j\omega}) = \dfrac{e^{-j\omega}}{1-ae^{-j\omega}} = e^{-j\omega} X_2(e^{j\omega})$

$$\therefore x_3(n) = a^{(n-1)} u(n-1)$$

> Using shifting property

Let, x(n) = Inverse Fourier transform of $X(e^{j\omega})$.

$$\therefore x(n) = \mathcal{F}^{-1}\{X(e^{j\omega})\} = \mathcal{F}^{-1}\left\{\frac{1}{2}\left[X_1(e^{j\omega}) + X_2(e^{j\omega}) + X_3(e^{j\omega})\right]\right\}$$

$$= \frac{1}{2}\left[\mathcal{F}^{-1}\{X_1(\omega)\} + \mathcal{F}^{-1}\{X_2(\omega)\} + \mathcal{F}^{-1}\{X_3(\omega)\}\right] = = \frac{1}{2}\left[x_1(n) + x_2(n) + x_3(n)\right]$$

$$= \frac{1}{2}\left[a^{(n+1)} u(n+1) + a^n u(n) + a^{(n-1)} u(n-1)\right]$$

Example 4.17

If $X(e^{j\omega}) = e^{-j3}$; $|\omega| \le 1$

     $= 0$; $1 < |\omega| \le \pi$,   Find x(n) and plot.

Solution:

The x(n) is obtained by taking inverse Fourier transform of $X(e^{j\omega})$.

By definition of inverse Fourier transform,

$$x(n) = \mathcal{F}^{-1}\{X(e^{j\omega})\} = \frac{1}{2\pi}\int_{-\pi}^{+\pi} X(e^{j\omega})e^{j\omega n}d\omega = \frac{1}{2\pi}\int_{-1}^{+1} e^{-j3\omega}e^{j\omega n}d\omega$$

$$= \frac{1}{2\pi}\int_{-1}^{+1} e^{j\omega(n-3)}d\omega = \frac{1}{2\pi}\left[\frac{e^{j\omega(n-3)}}{j(n-3)}\right]_{-1}^{+1} = \frac{1}{j2\pi(n-3)}[e^{j(n-3)} - e^{-j(n-3)}]$$

$$= \frac{1}{\pi(n-3)}\left[\frac{e^{j(n-3)} - e^{-j(n-3)}}{2j}\right] = \frac{1}{\pi(n-3)}\sin(n-3)$$

$$\boxed{\sin\theta = \frac{e^{j\theta} - e^{-j\theta}}{2j}}$$

$$= \frac{\sin(n-3)}{\pi(n-3)} \quad ; \quad \text{for all n , except n = 3.}$$

When $n = 3$, $x(n) = \underset{(n-3)\to 0}{Lt}\frac{\sin(n-3)}{\pi(n-3)} = \frac{1}{\pi}\underset{(n-3)\to 0}{Lt}\frac{\sin(n-3)}{(n-3)} = \frac{1}{\pi}$

$$\boxed{\begin{array}{c}\text{L' Hospital rule}\\[4pt]\underset{\theta\to 0}{Lt}\dfrac{\sin\theta}{\theta} = 1\end{array}}$$

The signal x(n) is an infinite duration signal and can be evaluated for all integer values of n in the range $n = -\infty$ to $+\infty$. Here x(n) is evaluated for n = -2 to + 8 and plotted.

$$x(n) = \frac{\sin(n-3)}{\pi(n-3)}$$

$$\boxed{\begin{array}{l}\textbf{\textit{Note}} : \text{Evaluate sin(n–3) by keeping}\\ \text{calculator in radian mode.}\end{array}}$$

When $n = -2$; $x(-2) = \dfrac{\sin(-2-3)}{\pi(-2-3)} = -0.061$

When $n = -1$; $x(-1) = \dfrac{\sin(-1-3)}{\pi(-1-3)} = -0.06$

When $n = 0$; $x(0) = \dfrac{\sin(0-3)}{\pi(0-3)} = 0.015$

When $n = 1$; $x(1) = \dfrac{\sin(1-3)}{\pi(1-3)} = 0.145$

When $n = 2$; $x(2) = \dfrac{\sin(2-3)}{\pi(2-3)} = 0.268$

When $n = 3$; $x(3) = \dfrac{1}{\pi} = 0.318$

When $n = 4$; $x(4) = \dfrac{\sin(4-3)}{\pi(4-3)} = 0.268$

When $n = 5$; $x(5) = \dfrac{\sin(5-3)}{\pi(5-3)} = 0.145$

When $n = 6$; $x(6) = \dfrac{\sin(6-3)}{\pi(6-3)} = 0.015$

When $n = 7$; $x(7) = \dfrac{\sin(7-3)}{\pi(7-3)} = -0.06$

When $n = 8$; $x(8) = \dfrac{\sin(8-3)}{\pi(8-3)} = -0.061$

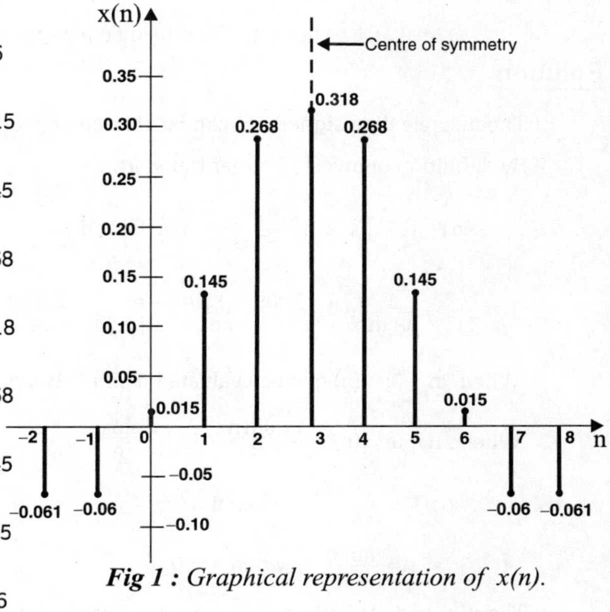

Fig 1 : *Graphical representation of x(n).*

Here x(n) is a symmetrical signal with centre of symmetry at n = 3.

Example 4.18

Find $x(n)$, if $X(e^{j\omega}) = \dfrac{1}{1 - \dfrac{1}{8}e^{-j\omega}}$

Solution

Given that, $X(e^{j\omega}) = \dfrac{1}{1 - \dfrac{1}{8}e^{-j\omega}}$

By Taylor's series expansion we can write,

$X(e^{j\omega}) = \dfrac{1}{1 - \dfrac{1}{8}e^{-j\omega}} = 1 + \dfrac{1}{8}e^{-j\omega} + \left(\dfrac{1}{8}e^{-j\omega}\right)^2 + \left(\dfrac{1}{8}e^{-j\omega}\right)^3 + \dots$

$$= \sum_{n=0}^{\infty}\left(\dfrac{1}{8}e^{-j\omega}\right)^n = \sum_{n=0}^{\infty}\left(\dfrac{1}{8}\right)^n e^{-j\omega n} \qquad \dots (1)$$

> Using infinite geometric series sum formula
> $$\sum_{n=0}^{\infty} C^n = \dfrac{1}{1-C}$$
> if $0 < |C| < 1$

By definition of Fourier transform we can write,

$$X(e^{j\omega}) = \sum_{n=0}^{\infty} x(n)e^{-j\omega n} \; ; \text{ for } n \geq 0 \qquad \dots (2)$$

On comparing equations (1) and (2) we get,

$$x(n) = \left(\dfrac{1}{8}\right)^n \; ; \text{ for } n \geq 0$$

$$\therefore x(n) = \left(\dfrac{1}{8}\right)^n u(n) \; ; \text{ for all } n$$

Example 4.19

If $X(e^{j\omega}) = 1 \; ; \quad |\omega| \leq 1$

$= 0 \; ; \quad 1 < |\omega| \leq \pi$, Find the discrete time signal $x(n)$, and plot.

Solution

The discrete time signal $x(n)$ can be obtained by taking inverse Fourier transform of $X(e^{j\omega})$.

By definition of inverse Fourier transform,

$$x(n) = F^{-1}\{X(e^{j\omega})\} = \dfrac{1}{2\pi}\int_{-\pi}^{+\pi} X(e^{j\omega})e^{j\omega n}d\omega = \dfrac{1}{2\pi}\int_{-1}^{+1} 1 \times e^{j\omega n}d\omega = \dfrac{1}{2\pi}\left[\dfrac{e^{j\omega n}}{jn}\right]_{-1}^{+1}$$

> $\sin\theta = \dfrac{e^{j\theta} - e^{-j\theta}}{2j}$

$$= \dfrac{1}{j2\pi n}[e^{jn} - e^{-jn}] = \dfrac{1}{\pi n}\left[\dfrac{e^{jn} - e^{-jn}}{2j}\right] = \dfrac{2\sin n}{\pi n} \; ; \text{ for all } n \text{ , except when } n = 0$$

When $n = 0$; $x(n)$ can be evaluated using L' Hospital rule.

When $n = 0$; $x(n) = \underset{n\to 0}{\text{Lt}} \dfrac{\sin n}{\pi n} = \dfrac{1}{\pi}\underset{n\to 0}{\text{Lt}}\dfrac{\sin n}{n} = \dfrac{1}{\pi}$

> L' Hospital rule
> $\underset{\theta\to 0}{\text{Lt}} \dfrac{\sin\theta}{\theta} = 1$

$$\therefore x(n) = \dfrac{1}{\pi} \qquad ; \text{ when } n = 0$$

$$= \dfrac{\sin n}{\pi n} \; ; \text{ when } n \neq 0$$

The $x(n)$ is an infinite duration signal and can be evaluated for all integer values of n in the range $n = -\infty$ to $+\infty$. Here $x(n)$ is evaluated for $n = -5$ to $+5$ and plotted.

When $n = -5 \; ; \; x(-5) = \dfrac{\sin(-5)}{\pi(-5)} \quad = -0.061$

When $n = -4$; $x(-4) = \dfrac{\sin(-4)}{\pi(-4)} = -0.06$

When $n = -3$; $x(-3) = \dfrac{\sin(-3)}{\pi(-3)} = 0.015$

When $n = -2$; $x(-2) = \dfrac{\sin(-2)}{\pi(-2)} = 0.145$

When $n = -1$; $x(-1) = \dfrac{\sin(-1)}{\pi(-1)} = 0.268$

When $n = 0$; $x(0) = \dfrac{1}{\pi} = 0.318$

When $n = 1$; $x(1) = \dfrac{\sin(1)}{\pi(1)} = 0.268$

When $n = 2$; $x(2) = \dfrac{\sin(2)}{\pi(2)} = 0.145$

When $n = 3$; $x(3) = \dfrac{\sin(3)}{\pi(3)} = 0.015$

When $n = 4$; $x(4) = \dfrac{\sin(4)}{\pi(4)} = -0.06$

When $n = 5$; $x(5) = \dfrac{\sin(5)}{\pi(5)} = -0.061$

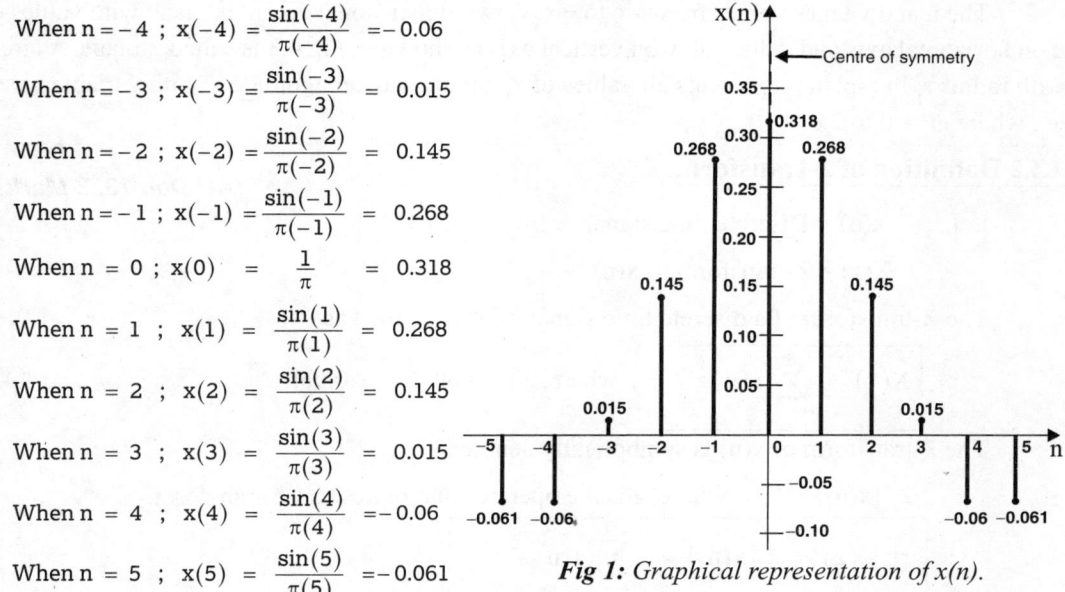

Fig 1: Graphical representation of x(n).

Here $x(n)$ is a symmetrical signal with centre of symmetry at $n = 0$.

4.4 Z-Transform

    Transform techniques are an important tool in the analysis of signals and systems. The Laplace transforms are popularly used for analysis of continuous time signals and systems. Similarly Z-transform plays an important role in analysis and representation of discrete time signals and systems. The Z-transform provides a method for the analysis of discrete time signals and systems in the frequency domain which is generally more efficient than its time domain analysis.

    A transform of a sampled signal or sequence was defined in 1947 by W. Hurewicz as,

$$z[f(kT)] = \sum_{k=0}^{\infty} f(kT) \, z^{-k}$$

which was later denoted in 1952 as Z-transform by a sampled-data control group at Columbia University led by professor John R. Raggazini and including L.A. Zadeh, E.I. Jury, R.E. Kalman, J.E. Bertram, B. Friedland and G.F. Franklin, (Source : www.ling.upenn.edu).

4.4.1 Z-Plane

    The Z-transform of $x(n)$ will convert the time domain signal $x(n)$ to z-domain signal $X(z)$, where the signal becomes a function of complex variable z.

    The complex variable z is defined as,

$$z = u + jv = r \, e^{j\omega}$$

where, u = Real part of z ; v = Imaginary part of z

$r = \sqrt{u^2 + v^2}$ = Magnitude of z

$\omega = \tan^{-1}\dfrac{v}{u}$ = Phase or Argument of z

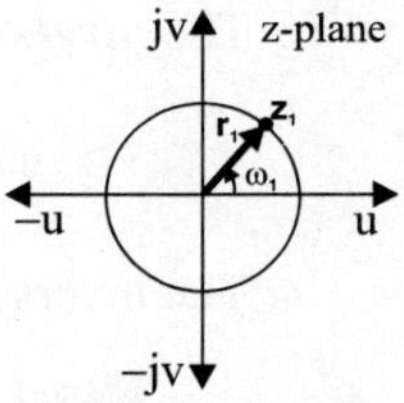

Fig 4.3: z-plane.

The u and v takes values from $-\infty$ to $+\infty$. A two dimensional complex plane with values of u on horizontal axis and values of v on vertical axis as shown in Fig 4.3 is called z-plane. A circle with radius r_1 in z-plane represents all values of z_1 having same magnitude r_1 with variable phase ω_1, where $\omega_1 = 0$ to 2π.

4.4.2 Definition of Z-Transform   *(AU Dec'13, 2 Marks)*

Let, $x(n)$ = Discrete time signal

$X(z)$ = Z-transform of $x(n)$

The Z-transform of a discrete time signal, $x(n)$ is defined as,

$$\boxed{X(z) = \sum_{n=-\infty}^{\infty} x(n) z^{-n}}$$; where, z is a complex variable   .....(4.45)

The Z-transform of $x(n)$ is symbolically denoted as,

$Z\{x(n)\}$;   where, z is the operator that represents Z-transform.

$$\boxed{\therefore\ X(z) = Z\{x(n)\} = \sum_{n=-\infty}^{+\infty} x(n) z^{-n}}$$

Since the time index n is defined for both positive and negative values, the discrete time signal $x(n)$ in equation (4.42) is considered to be two-sided and the transform is called *two-sided Z-transform*. If the signal $x(n)$ is one-sided signal, [i.e., $x(n)$ is defined only for positive value of n] then the Z-transform is called one-sided Z-transform.

The *one-sided Z-transform* of $x(n)$ is defined as,

$$\boxed{X(z) = Z\{x(n)\} = \sum_{n=0}^{+\infty} x(n) z^{-n}}$$ .....(4.46)

The computation of $X(z)$ involves summation of infinite terms which are functions of z. Hence it is possible that the infinite series may not converge to finite value for certain values of z. Therefore for every $X(z)$ there will be a set of values of z for which $X(z)$ can be computed. Such a set of values will lie in a particular region of z-plane and this region is called (*Region Of Convergence* (ROC)) of $X(z)$.

4.4.3 Inverse Z-Transform

Let, $X(z)$ be Z-transform of $x(n)$. Now the signal $x(n)$ can be uniquely determined from $X(z)$ and its region of convergence (ROC).

The *inverse Z-transform* of $X(z)$ is defined as,

$$x(n) = \frac{1}{2\pi j} \oint_c X(z) z^{n-1} dz$$ .....(4.47)

The inverse Z-transform of $X(z)$ is symbolically denoted as,

$Z^{-1}\{X(z)\}$;   where, Z^{-1} is the operator that represents the inverse Z-transform

$$\therefore\ x(n)\ =\ \mathcal{Z}^{-1}\{X(z)\}\ =\ \frac{1}{2\pi j}\oint_c X(z)\,z^{n-1}\,dz$$

We also refer x(n) and X(z) as a $\mathcal{Z}$-transform pair and this relation is expressed as,

$$x(n) \underset{\mathcal{Z}^{-1}}{\overset{\mathcal{Z}}{\rightleftarrows}} X(z)$$

Proof:

Consider the definition of $\mathcal{Z}$-transform of x(n),

$$X(z)\ =\ \sum_{n=-\infty}^{+\infty} x(n)\,z^{-n}\ =\ \sum_{k=-\infty}^{+\infty} x(k)\,z^{-k} \qquad \boxed{\text{Let } n \to k}$$

$$\therefore\ X(z)\,z^{n-1}\ =\ \sum_{k=-\infty}^{+\infty} x(k)\,z^{-k}\,z^{n-1} \qquad \boxed{\text{Multiply both sides by } z^{n-1}}$$

Let us integrate the above equation on both sides over a closed contour "C" within the ROC of X(z) which encloses the origin.

$$\therefore\ \oint_c X(z)\,z^{n-1}\,dz\ =\ \oint_c \sum_{k=-\infty}^{+\infty} x(k)\,z^{n-1-k}\,dz$$

$$=\ \sum_{k=-\infty}^{+\infty} x(k) \oint_c z^{n-1-k}\,dz \qquad \boxed{\begin{array}{l}\text{Interchanging the order of}\\ \text{summation and integration.}\end{array}}$$

$$=\ 2\pi j \sum_{k=-\infty}^{+\infty} x(k)\,\frac{1}{2\pi j}\oint_c z^{n-1-k}\,dz \qquad \boxed{\text{Multiply and divide by } 2\pi j.}$$
.....(4.48)

By Cauchy integral theorem,

$$\frac{1}{2\pi j}\oint_c z^{n-1-k}\,dz\ =\ 1 \quad ;\ k=n$$
$$=\ 0 \quad ;\ k \neq n$$

On applying Cauchy integral theorem the equation (4.48) reduces to,

$$\oint_c X(z)\,z^{n-1}\,dz\ =\ 2\pi j\,x(n)$$

$$\therefore\ x(n)\ =\ \frac{1}{2\pi j}\oint_c X(z)\,z^{n-1}\,dz \qquad \boxed{\left.\sum_{n=-\infty}^{+\infty} x(k)\right|_{n=k}\ =\ x(n)}$$

4.4.4 Poles and Zeros of Rational Function of z

Let, X(z) be $\mathcal{Z}$-transform of x(n). When X(z) is expressed as a ratio of two polynomials in z or z^{-1}, then X(z) is called a *rational function* of z.

Let X(z) be expressed as a ratio of two polynomials in z, as shown below:

$$X(z)\ =\ \frac{N(z)}{D(z)}\ =\ \frac{b_0 + b_1 z^{-1} + b_2 z^{-2} + b_3 z^{-3} + \ldots + b_M z^{-M}}{a_0 + a_1 z^{-1} + a_2 z^{-2} + a_3 z^{-3} + \ldots + a_N z^{-N}} \qquad \ldots(4.49)$$

where, N(z) = Numerator polynomial of X(z)

D(z) = Denominator polynomial of X(z)

In equation (4.49) let us scale the coefficients of numerator polynomial by b_0 and that of denominator polynomial by a_0, and then convert the polynomials to positive power of z as shown below:

$$X(z) = \frac{b_0\left(1 + \dfrac{b_1}{b_0}z^{-1} + \dfrac{b_2}{b_0}z^{-2} + \dfrac{b_3}{b_0}z^{-3} + ... + \dfrac{b_M}{b_0}z^{-M}\right)}{a_0\left(1 + \dfrac{a_1}{a_0}z^{-1} + \dfrac{a_2}{a_0}z^{-2} + \dfrac{a_3}{a_0}z^{-3} + ... + \dfrac{a_N}{a_0}z^{-N}\right)}$$

$$= G\,\frac{z^{-M}\left(z^M + \dfrac{b_1}{b_0}z^{M-1} + \dfrac{b_2}{b_0}z^{M-2} + \dfrac{b_3}{b_0}z^{M-3} + ... + \dfrac{b_M}{b_0}\right)}{z^{-N}\left(z^N + \dfrac{a_1}{a_0}z^{N-1} + \dfrac{a_2}{a_0}z^{N-2} + \dfrac{a_3}{a_0}z^{N-3} + ... + \dfrac{a_N}{a_0}\right)}$$

$\boxed{\text{Let, M = N}}$

$$= G\,\frac{(z - z_1)(z - z_2)(z - z_3)..... (z - z_N)}{(z - p_1)(z - p_2)(z - p_3)..... (z - p_N)} \qquad .....(4.50)$$

where, z_1, z_2, z_3, ...z_N are roots of numerator polynomial

p_1, p_2, p_3, ...p_N are roots of denominator polynomial

G is a scaling factor.

In equation (4.50) if the value of z is equal to one of the roots of the numerator polynomial, then the function X(z) will become zero.

Therefore, the roots of numerator polynomial z_1, z_2, z_3, ...z_N are called zeros of X(z). Hence the *zeros* are defined as values z at which the function X(z) become zero.

In equation (4.50) if the value of z is equal to one of the roots of the denominator polynomial then the funcion X(z) will become infinite. Therefore the roots of denominator polynomial p_1, p_2, p_3, ...p_N are called poles of X(z). Hence the *poles* are defined as values of z at which the function X(z) become infinite.

Since the function X(z) attains infinite values at poles, the ROC of X(z) does not include poles.

In a realizable system, the number of zeros will be less than or equal to number of poles. Also for every zero, we can associate one pole (the missing zeros are assumed to exist at infinity).

Let z_i be the zero associated with the pole p_i. If we evaluate |X(z)| for various values of z, then |X(z)| will be zero for z = z_i and infnite for z = p_i. Hence the plot of |X(z)| in a three-dimensional plane will look like a pole (or pillar-like structure) and so the point z = p_i is called a pole.

Representation of Poles and Zeros in z-Plane

The complex variable, z is defined as,

$z = u + jv$

where, u = Real part of z

v = Imaginary part of z

Hence the z-plane is a complex plane, with u on real axis and v on imaginary axis (Refer Fig 4.3 in section 4.4.1). In the z-plane, the zeros are marked by small circle "**o**" and the poles are marked by letter "**x**".

For example consider a rational function of z shown below:

$$X(z) = \frac{0.5 - 0.4\,z^{-1} + 0.06\,z^{-2}}{2 + 1.6\,z^{-1} + 0.64\,z^{-2}}$$

$$= \frac{0.5\left(1 - \dfrac{0.4}{0.5}\,z^{-1} + \dfrac{0.06}{0.5}\,z^{-2}\right)}{2\left(1 + \dfrac{1.6}{2}\,z^{-1} + \dfrac{0.64}{2}\,z^{-2}\right)} = \frac{0.25\,(1 - 0.8\,z^{-1} + 0.12\,z^{-2})}{1 + 0.8\,z^{-1} + 0.32\,z^{-2}}$$

$$= \frac{0.25\,z^{-2}\,(z^2 - 0.8\,z + 0.12)}{z^{-2}\,(z^2 + 0.8\,z + 0.32)} = \frac{0.25\,(z - 0.2)\,(z - 0.6)}{(z + 0.4 - j0.4)\,(z + 0.4 + j0.4)} \qquad .....(4.51)$$

The roots of quadratic, $z^2 - 0.8z + 0.12 = 0$ are,

$$z = \frac{0.8 \pm \sqrt{0.8^2 - 4 \times 0.12}}{2} = \frac{0.8 \pm 0.4}{2} = 0.6,\ 0.2$$

$$\therefore\ z^2 + 0.8z + 0.12 = (z - 0.6)\,(z - 0.2)$$

The roots of quadratic, $z^2 + 0.8z + 0.32 = 0$ are,

$$z = \frac{-0.8 \pm \sqrt{0.8^2 - 4 \times 0.32}}{2} = \frac{0.8 \pm \sqrt{-0.64}}{2} = \frac{0.8 \pm j0.8}{2} = -0.4 \pm j0.4$$

$$\therefore\ z^2 + 0.8z + 0.32 = (z + 0.4 - j0.4)\,(z + 0.4 + j0.4)$$

The zeros of X(z) are roots of numerator polynomial, which has two roots.

Therefore, the zeros of X(z) are,

$$z_1 = 0.6,\ \ z_2 = 0.2$$

The poles of X(z) are roots of denominator polynomial, which has two roots.

Therefore, the poles of X(z) are,

$$p_1 = -0.4 + j0.4,\ \ p_2 = -0.4 - j0.4$$

The pole-zero plot of X(z) is shown in Fig 4.4.

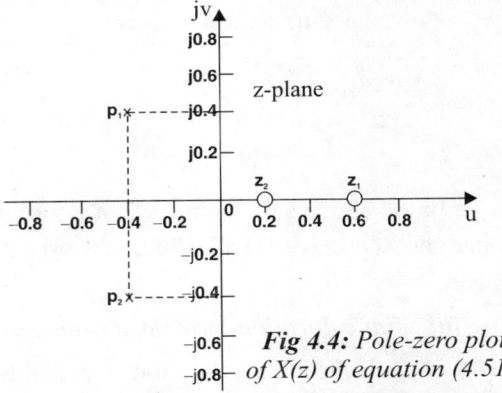

Fig 4.4: Pole-zero plot of X(z) of equation (4.51).

4.5 Region of Convergence of Ƶ-transform   *(AU Jun'11, 2 Marks)*

Since the Ƶ-transform is an infinite power series, it exists only for those values of z for which the series converges. The region of convergence, (ROC) of X(z) is the set of all values of z, for which X(z) attains a finite value. The ROC for the following six types of signals are discussed here.

4.5.1 ROC of Discrete Time Signal

Case i: Finite duration, right-sided (causal) signal

Case ii: Finite duration, left-sided (anticausal) signal

Case iii: Finite duration, two-sided (noncausal) signal

Case iv: Infinite duration, right-sided (causal) signal

Case v: Infinite duration, left-sided (anticausal) signal

Case vi: Infinite duration, two-sided (noncausal) signal

Case i: *Finite duration, right-sided (causal) signal*

Let, x(n) be a finite duration signal with N-samples, defined in the range $0 \le n \le (N-1)$.

$\therefore$ x(n) = {x(0), x(1), x(2),.....x(N–1)}

Now, the $\mathbb{Z}$-transform of x(n) is,

$$X(z) = \sum_{n=0}^{N-1} x(n) z^{-n}$$

$$= x(0) + x(1)z^{-1} + x(2)z^{-2} +........+ x(N-1)z^{-(N-1)}$$

$$= x(0) + \frac{x(1)}{z} + \frac{x(2)}{z^2} +........+ \frac{x(N-1)}{z^{(N-1)}}$$

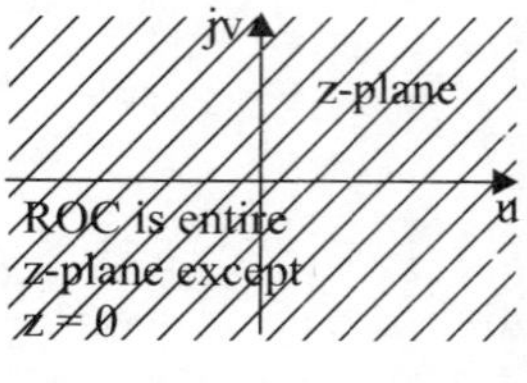

Fig 4.5: *ROC of finite duration causal signal.*

In the above summation, when z = 0, all the terms except the first term become infinite. Hence, the X(z) exists for all values of z, except z = 0. *Therefore, the ROC of finite duration right-sided (or causal signal) is entire z-plane except z = 0.*

Case ii: *Finite duration, left-sided (anticausal) signal*

Let, x(n) be a finite duration signal with N-samples, defined in the range $-(N-1) \le n \le 0$.

$\therefore$ x(n) = {x(–(N–1)),.....,x(–2), x(–1), x(0)}

Now, the $\mathbb{Z}$-transform of x(n) is,

$$X(z) = \sum_{n=-(N-1)}^{0} x(n) z^{-n}$$

$$= x(-(N-1)) z^{(N-1)} +........+x(-2)z^2 +x(-1)z + x(0)$$

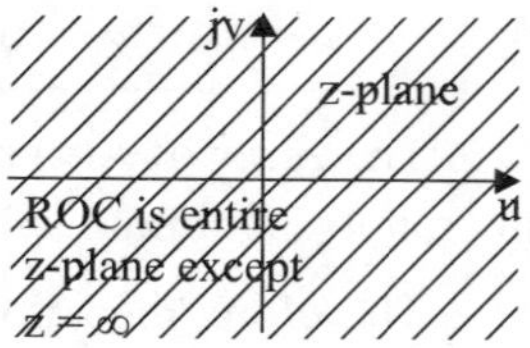

Fig 4.6: *ROC of finite duration anticausal signal.*

In the above summation, when z = ∞, all the terms except the last term become infinite. Hence the X(z) exists for all values of z, except, z = ∞. *Therefore, the ROC of X(z) is entire z-plane, except z = ∞.*

Case iii: *Finite duration, two-sided (noncausal) signal*

Let, x(n) be a finite duration signal with N-samples, defined in the range $-M \le n \le + M$,

where, M = $\dfrac{N-1}{2}$

$\therefore$ x(n) = {x(-M),........, x(-2), x(-1), x(0), x(1), x(2), ...x(M)}

Now, the $\mathbb{Z}$-transform of x(n) is,

$$X(z) = \sum_{n=-M}^{+M} x(n) z^{-n}$$

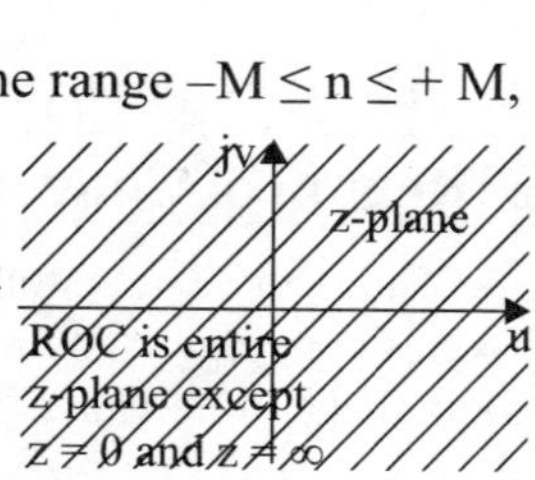

Fig 4.7: *ROC of finite duration two-sided signal.*

$$= x(-M)z^M +........+x(-2)z^2 +x(-1)z + x(0) + x(1)z^{-1} + x(2)z^{-2} +......+ x(M)z^{-M}$$

$$= x(-M)z^M +........+ x(-2)z^2 + x(-1)z + x(0) + \frac{x(1)}{z} + \frac{x(2)}{z^2} +......+ \frac{x(M)}{z^M}$$

In the above summation, when z = 0, the terms with negative power of z attain infinity and when z = ∞, the terms with positive power of z attain infinity. Hence X(z) converges for all values of z, except z = 0 and z = ∞. *Therefore, the ROC is entire z-plane, except z = 0 and z = ∞.*

Case iv: Infinite duration, right-sided (causal) signal

Let, $x(n) = r_1^n$; $n \ge 0$

Now, the $\mathcal{Z}$-transform of $x(n)$ is,

$$X(z) = \sum_{n=-\infty}^{+\infty} x(n) z^{-n} = \sum_{n=0}^{\infty} r_1^n z^{-n} = \sum_{n=0}^{\infty} (r_1 z^{-1})^n$$

If, $0 < |r_1 z^{-1}| < 1$, then $\displaystyle\sum_{n=0}^{\infty} (r_1 z^{-1})^n = \dfrac{1}{1 - r_1 z^{-1}}$

$$\therefore X(z) = \frac{1}{1 - r_1 z^{-1}}$$

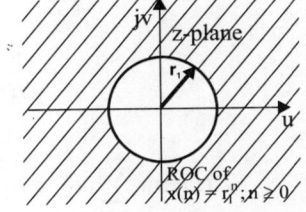

> Using infinite geometric series sum formula
> $$\sum_{n=0}^{\infty} C^n = \frac{1}{1-C}$$
> If, $0 < |C| < 1$

Here the condition to be satisfied for the convergence of X(z) is,

$$0 < |r_1 z^{-1}| < 1$$

$$\therefore |r_1 z^{-1}| < 1 \implies \left|\frac{r_1}{z}\right| < 1 \implies |z| > |r_1|$$

Fig 4.8: ROC of infinite duration right-sided signal.

The term $|r_1|$ represents a circle of radius r_1 in z-plane as shown in Fig 4.8. From the above analysis we can say that, X(z) converges for all points external to the circle of radius r_1 in z-plane. *Therefore, the ROC of X(z) is exterior of the circle of radius r_1 in z-plane as shown in Fig 4.8.*

Case v: Infinite duration, left-sided (anticausal) signal

Let, $x(n) = r_2^n$; $n \le 0$

Now, the $\mathcal{Z}$-transform of $x(n)$ is,

$$X(z) = \sum_{n=-\infty}^{+\infty} x(n) z^{-n} = \sum_{n=-\infty}^{0} r_2^n z^{-n} = \sum_{n=0}^{+\infty} r_2^{-n} z^n = \sum_{n=0}^{+\infty} (r_2^{-1} z)^n$$

If, $0 < |r_2^{-1} z| < 1$, then $\displaystyle\sum_{n=0}^{\infty} (r_2^{-1} z)^n = \dfrac{1}{1 - r_2^{-1} z}$

$$\therefore X(z) = \frac{1}{1 - r_2^{-1} z}$$

> Using infinite geometric series sum formula
> $$\sum_{n=0}^{\infty} C^n = \frac{1}{1-C}$$
> If, $0 < |C| < 1$

Here the condition to be satisfied for the convergence of X(z) is,

$$0 < |r_2^{-1} z| < 1$$

$$\therefore |r_2^{-1} z| < 1 \implies \left|\frac{z}{r_2}\right| < 1 \implies |z| < |r_2|$$

The term $|r_2|$ represents a circle of radius r_2 in z-plane as shown in Fig 4.9. From the above analysis we can say that X(z) converges for all points internal to the circle of radius r_2 in z-plane. *Therefore, the ROC of X(z) is interior of the circle of radius r_2 as shown in Fig 4.9.*

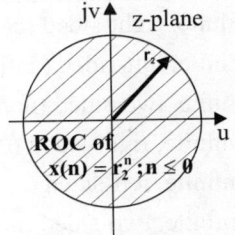

Fig 4.9: ROC of infinite duration left-sided signal.

Case vi: *Infinite duration, two-sided (noncausal) signal*

Let, $x(n) = r_1^n u(n) + r_2^n u(-n)$

Now, the $\mathcal{Z}$-transform of $x(n)$ is,

$$X(z) = \sum_{n=-\infty}^{\infty} x(n) z^{-n} = \sum_{n=-\infty}^{0} r_2^n z^{-n} + \sum_{n=0}^{+\infty} r_1^n z^{-n} = \sum_{n=0}^{+\infty} r_2^{-n} z^n + \sum_{n=0}^{+\infty} r_1^n z^{-n}$$

$$= \sum_{n=0}^{+\infty} (r_2^{-1} z)^n + \sum_{n=0}^{+\infty} (r_1 z^{-1})^n$$

> Infinite geometric series sum formula
> $$\sum_{n=0}^{\infty} C^n = \frac{1}{1-C}; \quad \text{If, } 0 < |C| < 1$$

$$= \frac{1}{1 - r_2^{-1} z} + \frac{1}{1 - r_1 z^{-1}}$$

> Using infinite geometric series sum formula
> if, $0 < |r_2^{-1} z| < 1$, and, $0 < |r_1 z^{-1}| < 1$

The term $\displaystyle\sum_{n=0}^{\infty} (r_2^{-1} z)^n$ converges if,

$$0 < |r_2^{-1} z| < 1$$

$$\therefore \quad |r_2^{-1} z| < 1 \Rightarrow \frac{|z|}{|r_2|} < 1 \Rightarrow |z| < |r_2|$$

The term $\displaystyle\sum_{n=0}^{\infty} (r_1 z^{-1})^n$ converges if,

$$0 < |r_1 z^{-1}| < 1$$

$$\therefore \quad |r_1 z^{-1}| < 1 \Rightarrow \frac{|r_1|}{|z|} < 1 \Rightarrow |z| > |r_1|$$

The term $|r_2|$ represents a circle of radius r_2 and $|r_1|$ represents a circle of radius r_1 in z-plane. If $|r_2| > |r_1|$ then there will be a region between two circles as shown in Fig 4.10. Now the $X(z)$ will converge for all points in the region between two circles (because the first term of $X(z)$ converges for $|z| < |r_2|$ and the second term of $X(z)$ converges for $|z| > |r_1|$). Hence the *ROC is the region between two circles of radius r_1 and r_2 as shown in Fig 4.10.*

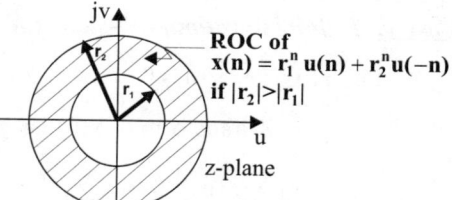

Fig 4.10: *ROC of infinite duration two-sided signal.*

Table 4.3: Summary of ROC of Discrete Time Signals

Sequence	ROC
Finite, right-sided (causal)	Entire z-plane except $z = 0$
Finite, left-sided (anticausal)	Entire z-plane except $z = \infty$
Finite, two-sided (noncausal)	Entire z-plane except $z = 0$ and $z = \infty$
Infinite, right-sided (causal)	Exterior of circle of radius r_1, where $|z| > r_1$
Infinite, left-sided (anticausal)	Interior of circle of radius r_2, where $|z| < r_2$
Infinite, two-sided (noncausal)	The area between two circles of radius r_2 and r_1 where, $r_2 > r_1$, and $r_1 < |z| < r_2$, (i.e., $|z| > r_1$, and, $|z| < r_2$)

Table 4.3: Characteristic Families of Signals and Corresponding ROC

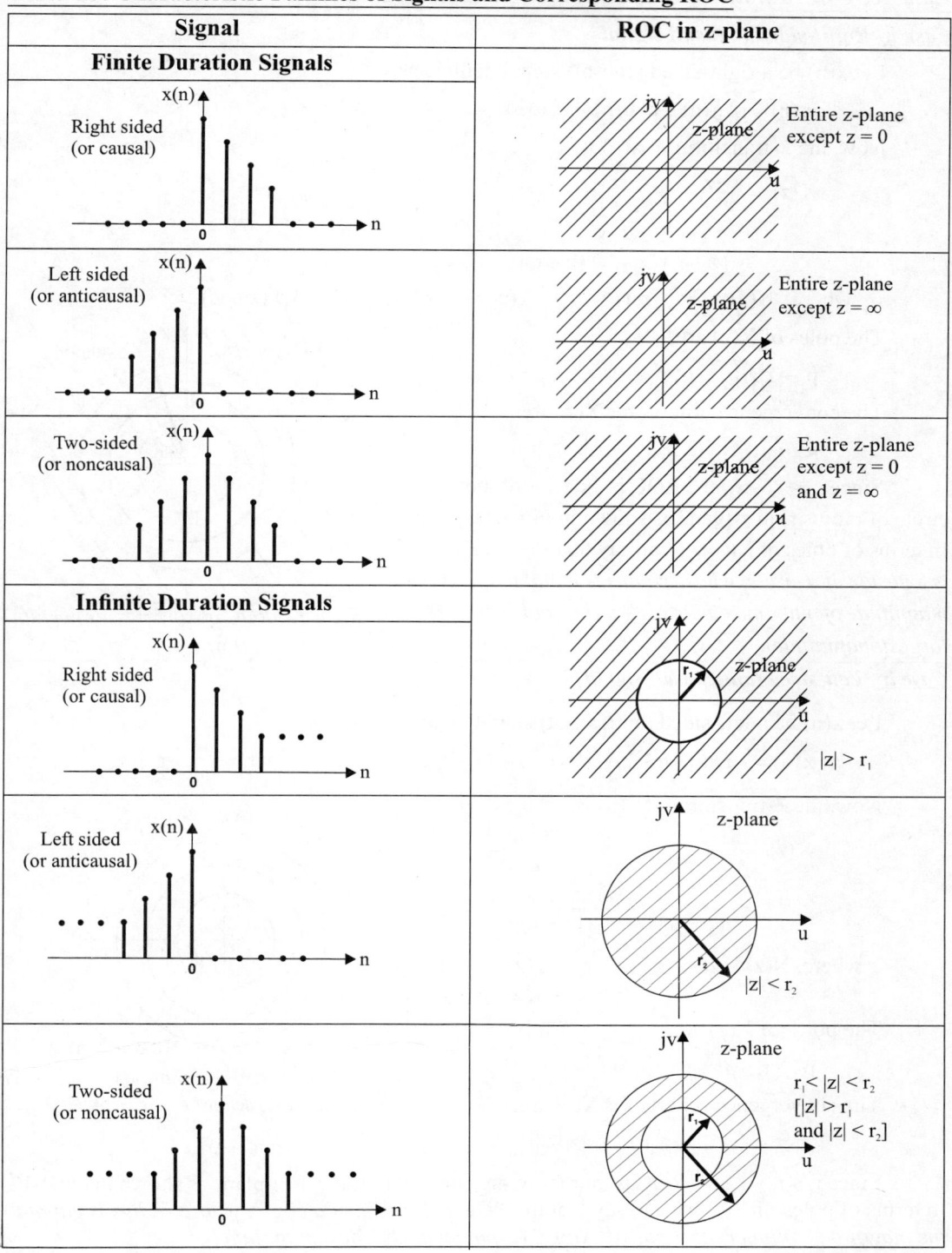

4.5.2 ROC of Rational Function of z (ROC using Poles)

Case i: *Right-sided (causal) signal*

Let x(n) be a right-sided (causal) signal defined as,

$$x(n) = r_1^n u(n) + r_2^n u(n) + r_3^n u(n) \quad ; \quad \text{where } r_1 < r_2 < r_3$$

Now, the $\mathcal{Z}$-transform of x(n) is,

$$X(z) = \frac{z}{z - r_1} + \frac{z}{z - r_2} + \frac{z}{z - r_3}$$

$$= \frac{N(z)}{(z - r_1)(z - r_2)(z - r_3)}$$

where, $N(z) = z(z - r_2)(z - r_3) + z(z - r_1)(z - r_3) + z(z - r_1)(z - r_2)$

The poles of X(z) are,

$$p_1 = r_1, \ p_2 = r_2, \ p_3 = r_3$$

The convergence criteria for X(z) are,

$$|z| > |r_1| \quad ; \quad |z| > |r_2| \quad ; \quad |z| > |r_3|$$

Since $r_1 < r_2 < r_3$, the ROC is exterior of the circle of radius r_3 in z-plane as shown in Fig.4.11. In terms of poles of X(z) we can say that the *ROC is exterior of a circle, whose radius is equal to the magnitude of outer most pole of X(z) (i.e., pole with largest magnitude).*

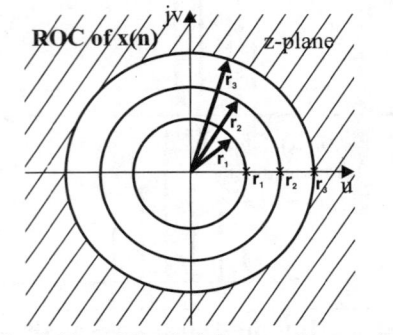

Fig 4.11: *ROC of x(n) = $r_1 u(n) + r_2 u(n)$ $+ r_3 u(n)$ where $r_1 < r_2 < r_3$.*

Case ii: *Left-sided (anticausal) signal*

Let x(n) be a left-sided (anticausal) signal defined as,

$$x(n) = -r_1^n u(-n-1) - r_2^n u(-n-1) - r_3^n u(-n-1) \ ; \quad \text{where } r_1 < r_2 < r_3$$

Now, the $\mathcal{Z}$-transform of x(n) is,

$$X(z) = \frac{z}{z - r_1} + \frac{z}{z - r_2} + \frac{z}{z - r_3}$$

$$= \frac{N(z)}{(z - r_1)(z - r_2)(z - r_3)}$$

where, $N(z) = z(z - r_2)(z - r_3) + z(z - r_1)(z - r_3)$
$$+ z(z - r_1)(z - r_2)$$

The poles of X(z) are,

$$p_1 = r_1, \ p_2 = r_2, \ p_3 = r_3$$

The convergence criteria for X(z) are,

$$|z| < |r_1| \quad ; \quad |z| < |r_2| \quad ; \quad |z| < |r_3|$$

Fig 4.12: *ROC of x(n) = $-r_1 u(-n-1) - r_2 u(-n-1)$ $-r_3 u(-n-1)$ where $r_1 < r_2 < r_3$.*

Since $r_1 < r_2 < r_3$, the ROC is interior of the circle of radius r_1 in z-plane as shown in Fig.4.12. In terms of poles of X(z) we can say that the *ROC is interior of a circle, whose radius is equal to the magnitude of inner most pole of X(z) (i.e., pole with smallest magnitude).*

Case iii: *Two-sided (noncausal) signal*

Let $x(n)$ be two-sided signal defined as,

$$x(n) = r_1^n \, u(n) + r_2^n \, u(n) - r_3^n \, u(-n-1) - r_4^n \, u(-n-1); \qquad \text{where } r_1 < r_2 < r_3 < r_4$$

Now, the $\mathcal{Z}$-transform of $x(n)$ is,

$$X(z) = \frac{z}{z - r_1} + \frac{z}{z - r_2} + \frac{z}{z - r_3} + \frac{z}{z - r_4} = \frac{N(z)}{(z - r_1)(z - r_2)(z - r_3)(z - r_4)}$$

$$\text{where, } N(z) = z(z - r_2)(z - r_3)(z - r_4) + z(z - r_1)(z - r_3)(z - r_4)$$
$$+ z(z - r_1)(z - r_2)(z - r_4) + z(z - r_1)(z - r_2)(z - r_3)$$

The poles of $X(z)$ are,

$$p_1 = r_1 \; ; \; p_2 = r_2 \; ; \; p_3 = r_3 \; ; \; p_4 = r_4$$

The convergence criteria for $X(z)$ are,

$$|z| > |r_1| \; ; \; |z| > |r_2| \; ; \; |z| < |r_3| \; ; \; |z| < |r_4|$$

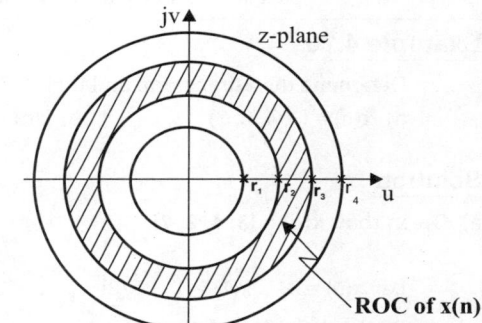

Since $r_1 < r_2 < r_3 < r_4$, the ROC is the region inbetween the circles of radius r_2 and r_3 as shown in Fig 4.13. Let r_x be the magnitude of largest pole of causal signal and let r_y be the magnitude of smallest pole of anticausal signal and let $r_x < r_y$. Now in terms of poles of $X(z)$ we can say that the *ROC is the region in between two circles of radius r_x and r_y, where $r_x < r_y$.*

Fig 4.13: ROC of $x(n) = r_1 u(n) + r_2 u(n)$
$-r_3 u(-n-1) - r_4 u(-n-1)$.

4.5.3 Properties of ROC

The various concepts of ROC that has been discussed above are summarized as properties of ROC and given below:

Property - 1 : The ROC of $X(z)$ is a ring or disk in z-plane, with centre at origin.

Property - 2 : If $x(n)$ is finite duration right-sided (causal) signal, then the ROC is entire z-plane except $z = 0$.

Property - 3 : If $x(n)$ is finite duration left-sided (anticausal) signal, then the ROC is entire z-plane except $z = \infty$.

Property - 4 : If $x(n)$ is finite duration two-sided (noncausal) signal, then the ROC is entire z-plane except $z = 0$ and $z = \infty$.

Property - 5 : If $x(n)$ is infinite duration right-sided (causal) signal, then the ROC is exterior of a circle of radius r_1.

Property - 6 : If $x(n)$ is infinite duration left-sided (anticausal) signal, then the ROC is interior of a circle of radius r_2.

Property - 7 : If $x(n)$ is infinite duration two-sided (noncausal) signal, then the ROC is the region in between two circles of radius r_1 and r_2.

Property - 8 : If $X(z)$ is rational, [where $X(z)$ is $\mathcal{Z}$-transform of $x(n)$], then the ROC does not include any poles of $X(z)$.

Property - 9 : If $X(z)$ is rational, [where $X(z)$ is $\mathcal{Z}$-transform of $x(n)$], and if $x(n)$ is right-sided then the ROC is exterior of a circle whose radius corresponds to the pole with largest magnitude.

Property - 10 : If $X(z)$ is rational, [where $X(z)$ is $\mathcal{Z}$-transform of $x(n)$], and if $x(n)$ is left-sided, then the ROC is interior of a circle whose radius corresponds to the pole with smallest magnitude.

Property-11 : If $X(z)$ is rational, [where $X(z)$ is $\mathcal{Z}$-transform of $x(n)$], and if $x(n)$ is two-sided then the ROC is region in between two circles whose radius corresponds to the pole of causal part with largest magnitude and the pole of anticausal part with smallest magnitude.

Example 4.20

Determine the $\mathcal{Z}$-transform and their ROC of the following discrete time signals.

a) $x(n) = \{3, 4, 2, 7\}$ **b)** $x(n) = \{6, 8, 9, 3\}$ **c)** $x(n) = \{2, 4, 6, 8, 10\}$
    ↑   ↑   ↑

Solution:

a) Given that, $x(n) = \{3, 4, 2, 7\}$
        ↑

i.e., $x(0) = 3$; $x(1) = 4$; $x(2) = 2$; $x(3) = 7$; and $x(n) = 0$ for $n < 0$ and for $n > 3$.

By the definition of $\mathcal{Z}$-transform,

$$X(z) = \mathcal{Z}\{x(n)\} = \sum_{n=-\infty}^{\infty} x(n)\, z^{-n}$$

> The $x(n)$ is defined in the range $n = 0$ to 3, hence the limits of summation is changed to $n = 0$ to $n = 3$.

$$= \sum_{n=0}^{3} x(n)\, z^{-n}$$

$$= x(0)\, z^{0} + x(1)\, z^{-1} + x(2)\, z^{-2} + x(3)\, z^{-3}$$

$$= 3 + 4z^{-1} + 2z^{-2} + 7z^{-3}$$

$$= 3 + \frac{4}{z} + \frac{2}{z^2} + \frac{7}{z^3}$$

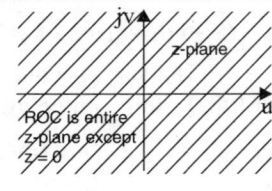

In $X(z)$, when $z = 0$, except the first term all other terms will become infinite. Hence $X(z)$ will be finite for all values of z, except $z = 0$. Therefore, the ROC is entire z-plane except $z = 0$.

b) Given that, $x(n) = \{6, 8, 9, 3\}$
              ↑

i.e, $x(-3) = 6$; $x(-2) = 8$; $x(-1) = 9$; $x(0) = 3$; and $x(n) = 0$ for $n < -3$ and for $n > 0$.

By the definition of $\mathcal{Z}$-transform,

$$X(z) = \mathcal{Z}\{x(n)\} = \sum_{n=-\infty}^{\infty} x(n)\, z^{-n}$$

> The $x(n)$ is defined in the range $n = -3$ to 0, hence the limits of summation is changed to $n = -3$ to $n = 0$.

$$= \sum_{n=-3}^{0} x(n)\, z^{-n}$$

$$= x(-3)\, z^{3} + x(-2)\, z^{2} + x(-1)\, z + x(0)$$

$$= 6z^{3} + 8z^{2} + 9z + 3$$

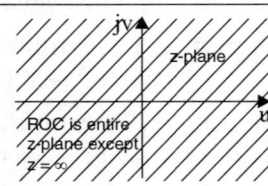

In $X(z)$, when $z = \infty$, except the last term all other terms become infinite. Hence $X(z)$ will be finite for all values of z, except $z = \infty$. Therefore, the ROC is entire z-plane except $z = \infty$.

c) Given that, $x(n) = \{2, 4, 6, 8, 10\}$

$$\uparrow$$

i.e, $x(-2) = 2$; $x(-1) = 4$; $x(0) = 6$; $x(1) = 8$; $x(2) = 10$ and $x(n) = 0$ for $n < -2$ and for $n > 2$.

By the definition of $\mathcal{Z}$-transform,

$$X(z) = \mathcal{Z}\{x(n)\} = \sum_{n=-\infty}^{\infty} x(n) z^{-n}$$

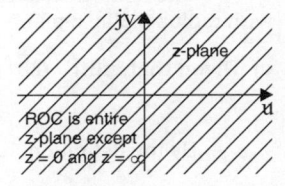

The $x(n)$ is defined in the range $n = -2$ to $+2$, hence the limits of summation is changed to $n = -2$ to $n = +2$.

$$= \sum_{n=-2}^{2} x(n) z^{-n}$$

$$= x(-2) z^2 + x(-1) z^1 + x(0) z^0 + x(1) z^{-1} + x(2) z^{-2}$$

$$= 2z^2 + 4z + 6 + 8z^{-1} + 10z^{-2}$$

$$= 2z^2 + 4z + 6 + \frac{8}{z} + \frac{10}{z^2}$$

ROC is entire z-plane except $z = 0$ and $z = \infty$

In $X(z)$, when $z = 0$, the terms with negative power of z will become infinite and when $z = \infty$, the terms with positive power of z will become infinite. Hence $X(z)$ will be finite for all values of z except when $z = 0$ and $z = \infty$. Therefore, the ROC is entire z-plane except $z = 0$ and $z = \infty$.

Example 4.21

Determine the $\mathcal{Z}$-transform and their ROC of the following discrete time signals.

a) $x(n) = u(n)$ **b)** $x(n) = 0.3^n u(n)$ **c)** $x(n) = 0.8^n u(-n-1)$

d) $x(n) = 0.3^n u(n) + 0.8^n u(-n-1)$ **(e)** $x(n) = a^{|n|}$; $0 < a < 1$

Solution:

a) Given that, $x(n) = u(n) = 1$; for $n \geq 0$

$$= 0 \text{ ; for } n < 0$$

By the definition of $\mathcal{Z}$-transform,

$$X(z) = \mathcal{Z}\{x(n)\} = \sum_{n=-\infty}^{\infty} x(n) z^{-n} = \sum_{n=0}^{\infty} u(n) z^{-n}$$

Using infinite geometric series sum formula

$$\sum_{n=0}^{\infty} C^n = \frac{1}{1-C}$$

If, $0 < |C| < 1$

$$= \sum_{n=0}^{\infty} z^{-n} = \sum_{n=0}^{\infty} (z^{-1})^n = \frac{1}{1-z^{-1}}$$

$$= \frac{1}{1-1/z} = \frac{z}{z-1}$$

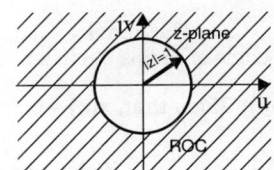

Here the condition for convergence is, $0 < |z^{-1}| < 1$.

$$\therefore |z^{-1}| < 1 \quad \Rightarrow \quad \frac{1}{|z|} < 1 \quad \Rightarrow \quad |z| > 1$$

The term $|z| = 1$ represents a circle of unit radius in z-plane. Therefore, the ROC is exterior of unit circle in z-plane.

b) Given that, $x(n) = 0.3^n u(n) = 0.3^n$; for $n \geq 0$

$$= 0 \text{ ; for } n < 0$$

By the definition of $\mathcal{Z}$-transform,

$$X(z) = \mathcal{Z}\{x(n)\} = \sum_{n=-\infty}^{\infty} x(n) z^{-n} = \sum_{n=0}^{\infty} 0.3^n z^{-n}$$

$$= \sum_{n=0}^{\infty} (0.3 z^{-1})^n = \frac{1}{1-0.3 z^{-1}}$$

Using infinite geometric series sum formula.

$$\therefore\ X(z) = \frac{1}{1 - 0.3\frac{1}{z}} = \frac{z}{z - 0.3}$$

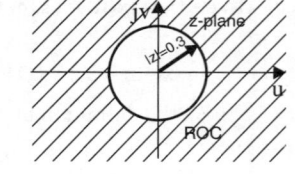

Here the condition for convergence is, $0 < |0.3\ z^{-1}| < 1.$

$$\therefore\ |0.3\ z^{-1}| < 1 \quad \Rightarrow \quad \frac{0.3}{|z|} < 1 \quad \Rightarrow \quad |z| > 0.3$$

The term $|z| = 0.3$ represents a circle of radius 0.3 in z-plane. Therefore, the ROC is exterior of circle with radius 0.3 in z-plane.

---

c) **Given that, $x(n) = 0.8^n\ u(-n-1) = 0 \quad$; for $n > -1$**

$$= 0.8^n\ ;\ \text{for}\ n \le -1$$

By the definition of $\mathbb{Z}$-transform,

$$X(z) = \mathbb{Z}\{x(n)\} = \sum_{n=-\infty}^{\infty} x(n)\ z^{-n} = \sum_{n=-\infty}^{-1} 0.8^n\ z^{-n}$$

$$= \sum_{n=1}^{\infty} 0.8^{-n}\ z^n = -(0.8^{-1}z)^0 + (0.8^{-1}z)^0 + \sum_{n=1}^{\infty} (0.8^{-1}z)^n$$

$$\boxed{(0.8^{-1}z)^0 = 1}$$

$$= \sum_{n=0}^{\infty} (0.8^{-1}z)^n - 1 = \frac{1}{1 - 0.8^{-1}z} - 1$$

$$\boxed{\text{Using infinite geometric series sum formula.}}$$

$$= \frac{1 - (1 - 0.8^{-1}z)}{1 - 0.8z^{-1}z} = \frac{0.8^{-1}z}{1 - 0.8^{-1}z} = \frac{1}{\frac{1}{0.8^{-1}z} - 1}$$

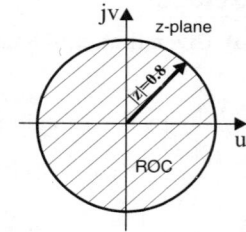

$$= -\frac{1}{1 - 0.8z^{-1}} = -\frac{1}{1 - \frac{0.8}{z}} = -\frac{z}{z - 0.8}$$

Here the condition for convergence is, $0 < |0.8^{-1}\ z| < 1.$

$$\therefore\ |0.8^{-1}z| < 1 \quad \Rightarrow \quad \frac{|z|}{0.8} < 1 \quad \Rightarrow \quad |z| < 0.8$$

The term $|z| = 0.8$, represents a circle of radius 0.8 in z-plane. Therefore, the ROC is interior of the circle of radius 0.8 in z-plane.

---

d) **Given that, $x(n) = 0.3^n\ u(n) + 0.8^n\ u(-n-1)$**

$$X(z) = \mathbb{Z}\{x(n)\} = \mathbb{Z}\{0.3^n\ u(n) + 0.8^n\ u(-n-1)\}$$

$$\boxed{\text{Using linearity property.}}$$

$$= \mathbb{Z}\{0.3^n\ u(n)\} + \mathbb{Z}\{0.8^n\ u(-n-1)\}$$

$$\boxed{\text{Using the results of (b) and (c).}}$$

$$= \frac{z}{z - 0.3} - \frac{z}{z - 0.8}$$

$$= \frac{z(z - 0.8) - z(z - 0.3)}{(z - 0.3)(z - 0.8)} = \frac{z^2 - 0.8z - z^2 + 0.3z}{z^2 - 0.8z - 0.3z + 0.24} = \frac{-0.5z}{z^2 - 1.1z + 0.24}$$

Here the condition for convergence of $0.3^n u(n)$ is,

$$0 < |0.3\, z^{-1}| < 1 \quad \Rightarrow \quad |z| > 0.3$$

and the condition for convergence of $0.8^n u(-n-1)$ is,

$$0 < |0.8^{-1}\, z| < 1 \quad \Rightarrow \quad |z| < 0.8$$

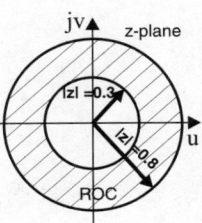

The term $|z| = 0.8$, represents a circle of radius 0.8 in z-plane and the term $|z| = 0.3$ represents a circle of radius 0.3 in z-plane. Hence the common region of convergence for both the terms of x(n) is the region in between the circles of radius $|z| = 0.8$ and $|z| = 0.3$ in the z-plane.

---

e) **Given that, $x(n) = a^{|n|}$; $0 < a < 1$**   *(AU May'15, 5 Marks)*

$$\therefore\ x(n) = a^{-n} ;\quad n < 0$$

$$= a^n \ \vdots\quad n \geq 0$$

By the definition of z - transform

$$X(z) = \mathcal{Z}\{x(n)\} = \sum_{n=-\infty}^{\infty} x(n)\, z^{-n} = \sum_{n=-\infty}^{-1} a^{-n} z^{-n} + \sum_{n=0}^{\infty} a^n z^{-n}$$

$$= \sum_{n=1}^{\infty} a^n z^n + \sum_{n=0}^{\infty} a^n z^{-n}$$

$$= \sum_{n=1}^{\infty} (az)^n + \sum_{n=0}^{\infty} (az^{-1})^n = -(az)^0 + (az)^0 + \sum_{n=1}^{\infty} (az)^n + \sum_{n=0}^{\infty} (az^{-1})^n$$

$$= -1 + \sum_{n=0}^{\infty}(az)^n + \sum_{n=0}^{\infty}(az^{-1})^n = -1 + \frac{1}{1-az} + \frac{1}{1-az^{-1}}$$

$$= \frac{-(1-az)(1-az^{-1}) + 1 - az^{-1} + 1 - az}{(1-az)(1-az^{-1})} \qquad \boxed{\text{Using infinite geometric series sum formula.}}$$

$$= \frac{-\cancel{1} + \cancel{az^{-1}} + \cancel{az} - a^2 + \cancel{1} - \cancel{az^{-1}} + 1 - \cancel{az}}{1 - az^{-1} - az + a^2}$$

$$= \frac{1 - a^2}{1 - a(z + z^{-1}) + a^2}$$

$$= \frac{1 - a^2}{1 - 2a\cosh z + a^2} \qquad \boxed{\cosh\theta = \dfrac{e^{\theta} + e^{-\theta}}{2}}$$

Here the term $\displaystyle\sum_{n=0}^{\infty} (az)^n$ converges if

$$0 < |az| < 1$$

$$\therefore\ |az| < 1$$

Since $0 < a < 1$, $|z| < 1$

Here the term $\displaystyle\sum_{n=0}^{\infty} (az^{-1})^n$ converges if

$$0 < |az^{-1}| < 1$$

$$\therefore\ |az^{-1}| < 1 \quad \Rightarrow \quad \frac{|a|}{|z|} < 1$$

$$\therefore\ |z| > |a|$$

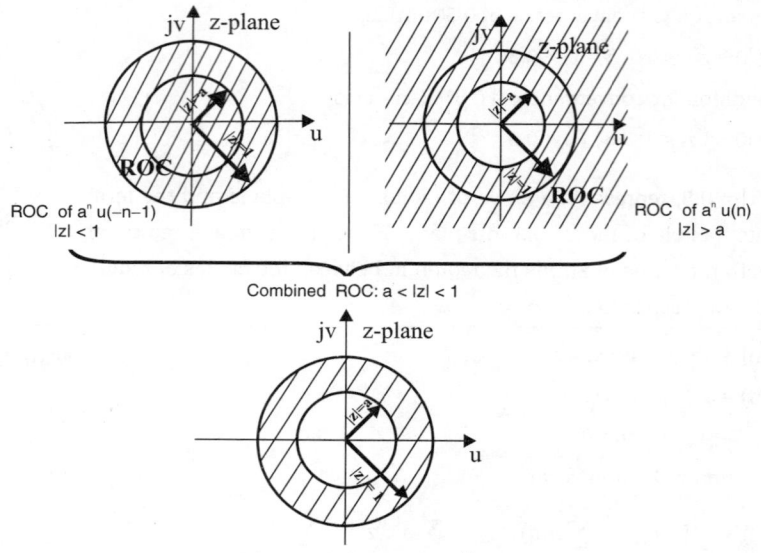

Fig 1: *ROC of $x(n) = a^{|n|}$.*

Example 4.22

Determine the z - transform and sketch the ROC along with location of poles for each of the following discrete time signals.

a) $x(n) = \left(\frac{1}{2}\right)^n u(n) - \left(\frac{1}{3}\right)^n u(n)$ **b)** $x(n) = \left(\frac{1}{2}\right)^n u(n) + \left(\frac{1}{3}\right)^n u(n-1)$

Solution:

a) Given that, $x(n) = \left(\frac{1}{2}\right)^n u(n) - \left(\frac{1}{3}\right)^n u(n) = \left(\frac{1}{2}\right)^n - \left(\frac{1}{3}\right)^n$; $n \geq 0$ *(AU Jun'12, 8 Marks)*

By definition of z- transform,

$u(n) = 1$;	for $n \geq 0$
$= 0$;	for $n < 0$

$$X(z) = Z\{x(n)\} = \sum_{n=-\infty}^{\infty} x(n) z^{-n} = \sum_{n=0}^{\infty} \left(\frac{1}{2}\right)^n z^{-n} - \sum_{n=0}^{\infty} \left(\frac{1}{3}\right)^n z^{-n}$$

$$= \sum_{n=0}^{\infty} \left(\frac{1}{2} z^{-1}\right)^n - \sum_{n=0}^{\infty} \left(\frac{1}{3} z^{-1}\right)^n$$

Using infinite geometric series sum formula
$$\sum_{n=0}^{\infty} C^n = \frac{1}{1-C} \ ;$$
$$\text{if, } 0 < |C| < 1$$

$$= \frac{1}{1 - \frac{1}{2} z^{-1}} - \frac{1}{1 - \frac{1}{3} z^{-1}} = \frac{z}{z - \frac{1}{2}} - \frac{z}{z - \frac{1}{3}}$$

$$= \frac{z\left(z - \frac{1}{3}\right) - z\left(z - \frac{1}{2}\right)}{\left(z - \frac{1}{2}\right)\left(z - \frac{1}{3}\right)} = \frac{z^2 - \frac{1}{3} z - z^2 + \frac{1}{2} z}{\left(z - \frac{1}{2}\right)\left(z - \frac{1}{3}\right)}$$

$$= \frac{\left(-\frac{1}{3} + \frac{1}{2}\right) z}{\left(z - \frac{1}{2}\right)\left(z - \frac{1}{3}\right)} = \frac{\left(\frac{-2+3}{6}\right) z}{\left(z - \frac{1}{2}\right)\left(z - \frac{1}{3}\right)} = \frac{\frac{1}{6} z}{\left(z - \frac{1}{2}\right)\left(z - \frac{1}{3}\right)}$$

Poles:

Poles are roots of denominator polynomial.

$\therefore$ Poles are at, $z = \dfrac{1}{2}$ and $z = \dfrac{1}{3}$

The location of poles in z-plane are denoted by the symbol "**x**" in the sketch of ROC.

ROC:

Here the term $\displaystyle\sum_{n=0}^{\infty} \left(\dfrac{1}{2}z^{-1}\right)^n$ converges if

$\left|\dfrac{1}{2}z^{-1}\right| < 1 \quad \Rightarrow \quad \dfrac{\frac{1}{2}}{|z|} < 1$

$\therefore |z| > \dfrac{1}{2}$

Here the term $\displaystyle\sum_{n=0}^{\infty} \left(\dfrac{1}{3}z^{-1}\right)^n$ converges if

$\left|\dfrac{1}{3}z^{-1}\right| < 1 \quad \Rightarrow \quad \dfrac{\frac{1}{3}}{|z|} < 1$

$\therefore |z| > \dfrac{1}{3}$

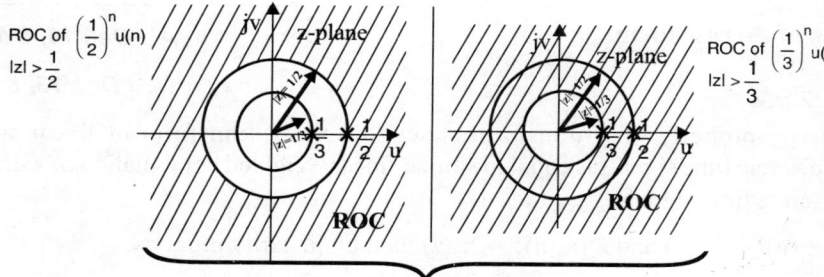

Combined ROC : $|z| > \dfrac{1}{2}$

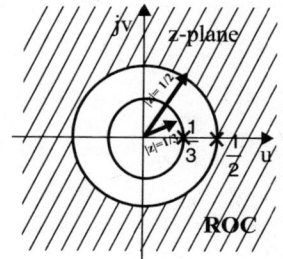

Fig 1: *ROC of* $\left(\dfrac{1}{2}\right)^n u(n) - \left(\dfrac{1}{3}\right)^n u(n).$

---

b) Given that, $x(n) = \left(\dfrac{1}{2}\right)^n u(n) + \left(\dfrac{1}{3}\right)^n u(n-1)$ **(AU Jun'12, 8 Marks)**

By definition of $\mathcal{Z}$- transform

$$X(z) = \mathcal{Z}\{x(n)\} = \sum_{n=-\infty}^{\infty} x(n)\, z^{-n} = \sum_{n=-\infty}^{\infty} \left[\left(\dfrac{1}{2}\right)^n u(n) + \left(\dfrac{1}{3}\right)^n u(n-1)\right] z^{-n}$$

$$= \sum_{n=-\infty}^{\infty} \left(\dfrac{1}{2}\right)^n u(n)\, z^{-n} + \sum_{n=-\infty}^{\infty} \left(\dfrac{1}{3}\right)^n u(n-1)\, z^{-n}$$

$$= \sum_{n=0}^{\infty} \left(\dfrac{1}{2}\right)^n z^{-n} + \sum_{n=1}^{\infty} \left(\dfrac{1}{3}\right)^n z^{-n} = \sum_{n=0}^{\infty} \left(\dfrac{1}{2}z^{-1}\right)^n + \sum_{n=1}^{\infty} \left(\dfrac{1}{3}z^{-1}\right)^n$$

$$= \sum_{n=0}^{\infty} \left(\dfrac{1}{2}z^{-1}\right)^n - \left(\dfrac{1}{3}z^{-1}\right)^0 + \left(\dfrac{1}{3}z^{-1}\right)^0 + \sum_{n=1}^{\infty} \left(\dfrac{1}{3}z^{-1}\right)^n = \sum_{n=0}^{\infty} \left(\dfrac{1}{2}z^{-1}\right)^{-n} + \sum_{n=0}^{\infty} \left(\dfrac{1}{3}z^{-1}\right)^{-n} - 1$$

$u(n) = 1$;	for $n \geq 0$
$= 0$;	for $n < 0$
$u(n-1) = 1$;	for $n \geq 1$
$= 0$;	for $n < 1$

$\boxed{\left(\dfrac{1}{3}z^{-1}\right)^0 = 1}$

$$\therefore \ X(z) = \cfrac{1}{1 - \frac{1}{2}z^{-1}} + \cfrac{1}{1 - \frac{1}{3}z^{-1}} - 1 = \cfrac{z}{z - \frac{1}{2}} + \cfrac{z}{z - \frac{1}{3}} - 1$$

Using infinite geometric series sum formula

$$\sum_{n=0}^{\infty} C^n = \frac{1}{1-C} \ ;$$

if, $0 < |C| < 1$

$$= \cfrac{z\left(z - \frac{1}{3}\right) + z\left(z - \frac{1}{2}\right) - \left(z - \frac{1}{2}\right)\left(z - \frac{1}{3}\right)}{\left(z - \frac{1}{2}\right)\left(z - \frac{1}{3}\right)}$$

$$= \cfrac{z^2 - \frac{1}{3}z + z^2 - \frac{1}{2}z - z^2 + \frac{1}{3}z + \frac{1}{2}z - \frac{1}{6}}{\left(z - \frac{1}{2}\right)\left(z - \frac{1}{3}\right)} = \cfrac{z^2 - \frac{1}{6}}{\left(z - \frac{1}{2}\right)\left(z - \frac{1}{3}\right)}$$

Poles and ROC:

The poles and ROC will be same as that of Example 4.23 (a).

4.6 Properties of $\mathcal{Z}$-Transform   *(AU Dec'15, 8 Marks)*

1. Linearity Property   *(AU Dec'13, 8 Marks)*

The linearity property of $\mathcal{Z}$-transform states that the $\mathcal{Z}$-transform of linear weighted combination of discrete time signals is equal to similar linear weighted combination of $\mathcal{Z}$-transform of individual discrete time signals.

Let, $\mathcal{Z}\{x_1(n)\} = X_1(z)$ and $\mathcal{Z}\{x_2(n)\} = X_2(z)$ then by linearity property,

$$\mathcal{Z}\{a_1x_1(n) + a_2x_2(n)\} = a_1X_1(z) + a_2X_2(z) \quad ; \quad \text{where, } a_1 \text{ and } a_2 \text{ are constants.}$$

Proof:

By definition of $\mathcal{Z}$-transform,

$$X_1(z) = \mathcal{Z}\{x_1(n)\} = \sum_{n=-\infty}^{+\infty} x_1(n)z^{-n} \qquad \qquad \dots\dots(4.52)$$

$$X_2(z) = \mathcal{Z}\{x_2(n)\} = \sum_{n=-\infty}^{+\infty} x_2(n)z^{-n} \qquad \qquad \dots\dots(4.53)$$

$$\therefore \ \mathcal{Z}\{a_1x_1(n) + a_2x_2(n)\} = \sum_{n=-\infty}^{+\infty} [a_1x_1(n) + a_2x_2(n)]z^{-n} = \sum_{n=-\infty}^{+\infty} [a_1x_1(n)z^{-n} + a_2x_2(n)z^{-n}]$$

$$= \sum_{n=-\infty}^{+\infty} a_1x_1(n)z^{-n} + \sum_{n=-\infty}^{+\infty} a_2x_2(n)z^{-n} = a_1 \sum_{n=-\infty}^{+\infty} x_1(n)z^{-n} + a_2 \sum_{n=-\infty}^{+\infty} x_2(n)z^{-n}$$

$$= a_1 X_1(z) + a_2 X_2(z) \qquad \boxed{\text{Using equations (4.52) and (4.53).}}$$

---

2. Shifting Property   *(AU Dec'13, 8 Marks)*

Case i: Two-sided $\mathcal{Z}$-transform

The shifting property of $\mathcal{Z}$-transform states that, $\mathcal{Z}$-transform of a shifted signal shifted by m-units of time is obtained by multiplying z^m to $\mathcal{Z}$-transform of unshifted signal.

Let, $\mathcal{Z}\{x(n)\} = X(z)$

Now, by shifting property,

$$\mathcal{Z}\{x(n-m)\} = z^{-m}\,X(z)$$

$$\mathcal{Z}\{x(n+m)\} = z^{m}\,X(z)$$

Proof:

By definition of $\mathcal{Z}$-transform,

$$X(z) = \mathcal{Z}\{x(n)\} = \sum_{n=-\infty}^{+\infty} x(n)\,z^{-n} \qquad \qquad \dots(4.54)$$

$$\therefore \ \mathcal{Z}\{x(n-m)\} = \sum_{n=-\infty}^{+\infty} x(n-m)\,z^{-n}$$

> Let, $n - m = p,\ \therefore\ n = p + m$
> when $n \to -\infty$, $p \to -\infty$
> when $n \to +\infty$, $p \to +\infty$

$$= \sum_{p=-\infty}^{+\infty} x(p)\,z^{-(m+p)}$$

$$= \sum_{p=-\infty}^{+\infty} x(p)\,z^{-m}\,z^{-p}$$

$$= z^{-m} \sum_{p=-\infty}^{+\infty} x(p)\,z^{-p} = z^{-m}\sum_{n=-\infty}^{+\infty} x(n)\,z^{-n}$$

> Let, $p \to n$

$$= z^{-m}\,X(z)$$

> Using equation (4.54).

By definition of $\mathcal{Z}$-transform,

$$\mathcal{Z}\{x(n+m)\} = \sum_{n=-\infty}^{+\infty} x(n+m)\,z^{-n}$$

> Let, $n + m = p,\ \therefore\ n = p - m$
> when $n \to -\infty$, $p \to -\infty$
> when $n \to +\infty$, $p \to +\infty$

$$= \sum_{p=-\infty}^{+\infty} x(p)\,z^{-(p-m)}$$

$$= \sum_{p=-\infty}^{+\infty} x(p)\,z^{-p}\,z^{m}$$

$$= z^{m} \sum_{p=-\infty}^{+\infty} x(p)\,z^{-p} = z^{m}\sum_{n=-\infty}^{+\infty} x(n)\,z^{-n}$$

> Let, $p \to n$
>
> Using equation (4.54).

$$= z^{m}\,X(z)$$

Case ii: One-sided $\mathcal{Z}$-transform

Let $x(n)$ be a discrete time signal defined in the range $0 < n < \infty$.

Let, $\mathcal{Z}\{x(n)\} = X(z)$

Now by shifting property,

$$\mathcal{Z}\{x(n-m)\} = z^{-m}\,X(z) + \sum_{i=1}^{m} x(-i)\,z^{-(m-i)}$$

$$\mathcal{Z}\{x(n+m)\} = z^{m}\,X(z) + \sum_{i=1}^{m} x(i)\,z^{(m-i)}$$

Proof:

By definition of one-sided z-transform,

$$X(z) = \mathcal{Z}\{x(n)\} = \sum_{n=0}^{+\infty} x(n) z^{-n} \qquad \ldots(4.55)$$

$$\therefore \ \mathcal{Z}\{x(n-m)\} = \sum_{n=0}^{+\infty} x(n-m) z^{-n} = \sum_{n=0}^{+\infty} x(n-m) z^{-n} z^{m} z^{-m} \qquad \boxed{\text{Multiply by } z^m \text{ and } z^{-m}}$$

$$= z^{-m} \sum_{n=0}^{+\infty} x(n-m) z^{-(n-m)} \qquad \boxed{\begin{array}{l}\text{Let, } n-m = p, \\ \text{when } n \to 0 \ , \ p \to -m \\ \text{when } n \to +\infty, \ p \to +\infty\end{array}}$$

$$= z^{-m} \sum_{p=-m}^{+\infty} x(p) z^{-p}$$

$$= z^{-m} \sum_{p=0}^{+\infty} x(p) z^{-p} + z^{-m} \sum_{p=-m}^{-1} x(p) z^{-p}$$

$$= z^{-m} \sum_{p=0}^{+\infty} x(p) z^{-p} + z^{-m} \sum_{p=1}^{m} x(-p) z^{p} \qquad \boxed{\begin{array}{l}\text{Let } p = n, \text{ in first summation.} \\ \text{Let } p = i, \text{ in second summation.}\end{array}}$$

$$= z^{-m} \sum_{n=0}^{+\infty} x(n) z^{-n} + z^{-m} \sum_{i=1}^{m} x(-i) z^{i} \qquad \boxed{\text{Using equation (4.55).}}$$

$$= z^{-m} X(z) + \sum_{i=1}^{m} x(-i) z^{-(m-i)} \qquad \ldots(4.56)$$

Note: *In equation (4.56) if x(–i) for i = 1 to m are zero then the shifting property of one-sided z-transform for delayed signal will be same as that for two-sided z-transform.*

By definition of one-sided z-transform,

$$\mathcal{Z}\{x(n+m)\} = \sum_{n=0}^{+\infty} x(n+m) z^{-n} = \sum_{n=0}^{+\infty} x(n+m) z^{-n} z^{m} z^{-m} \qquad \boxed{\text{Multiply by } z^m \text{ and } z^{-m}}$$

$$= z^{m} \sum_{n=0}^{+\infty} x(n+m) z^{-(n+m)} \qquad \boxed{\begin{array}{l}\text{Let, } n+m = p, \\ \text{when } n \to 0 \ , \ p \to m \\ \text{when } n \to +\infty, \ p \to +\infty\end{array}}$$

$$= z^{m} \sum_{p=m}^{+\infty} x(p) z^{-p}$$

$$= z^{m} \sum_{p=0}^{+\infty} x(p) z^{-p} - z^{m} \sum_{p=0}^{m-1} x(p) z^{-p} \qquad \boxed{\begin{array}{l}\text{Let } p = n, \text{ in first summation.} \\ \text{Let } p = i, \text{ in second summation.}\end{array}}$$

$$= z^{m} \sum_{n=0}^{+\infty} x(n) z^{-n} - z^{m} \sum_{i=0}^{m-1} x(i) z^{-i} \qquad \boxed{\text{Using equation (4.55).}}$$

$$= z^{m} X(z) - \sum_{i=0}^{m-1} x(i) z^{m-i} \qquad \ldots(4.57)$$

Note: *In equation (4.57) if x(i) for i = 0 to m–1 are zero then the shifting property of one-side z-transform for advanced signal will be same as that for two-sided z-transform.*

3. Multiplication by n (or Differentiation in z-domain)   (AU Dec'13, 8 Marks)

If $\mathcal{Z}\{x(n)\} = X(z)$

then $\mathcal{Z}\{x(n)\} = -z\dfrac{d}{dz}X(z)$

In general,

$$\mathcal{Z}\{n^m\,x(n)\} = \left(-z\dfrac{d}{dz}\right)^m X(z)$$

$$= \underbrace{-z\dfrac{d}{dz}\left(-z\dfrac{d}{dz}\left(.....\left(-z\dfrac{d}{dz}\left(-z\dfrac{d}{dz}\,X(z)\right)\right).....\right)\right)}_{m-\text{times}}$$

Proof:

By definition of $\mathcal{Z}$-transform,

$$X(z) = \mathcal{Z}\{x(n)\} = \sum_{n=-\infty}^{+\infty} x(n)\,z^{-n} \qquad \qquad .....(4.58)$$

$$\therefore \mathcal{Z}\{n\,x(n)\} = \sum_{n=-\infty}^{+\infty} n\,x(n)\,z^{-n}$$

$$= \sum_{n=-\infty}^{+\infty} n\,x(n)\,z^{-n}\,z\,z^{-1} \qquad \boxed{\text{Multiply by } z \text{ and } z^{-1}}$$

$$= -z\sum_{n=-\infty}^{+\infty} x(n)\left[-n\,z^{-n-1}\right]$$

$$= -z\sum_{n=-\infty}^{+\infty} x(n)\left[\dfrac{d}{dz}z^{-n}\right] \qquad \boxed{\dfrac{d}{dz}z^{-n} = -nz^{-n-1}}$$

$$= -z\dfrac{d}{dz}\sum_{n=-\infty}^{+\infty} x(n)\,z^{-n} \qquad \boxed{\begin{array}{l}\text{Interchanging summation}\\\text{and differentiation.}\end{array}}$$

$$= -z\dfrac{d}{dz}X(z) \qquad \boxed{\text{Using equation(4.58).}}$$

4. Multiplication by an Exponential Sequence, a^n (or Scaling in z-domain)

If $\mathcal{Z}\{x(n)\} = X(z)$

then $\mathcal{Z}\{a^n\,x(n)\} = X(a^{-1}z)$

Proof:

By definition of $\mathcal{Z}$-transform,

$$\mathcal{Z}\{x(n)\} = \sum_{n=-\infty}^{+\infty} x(n)\,z^{-n} \qquad \qquad ....(4.59)$$

$$\therefore \mathcal{Z}\{a^n\,x(n)\} = \sum_{n=-\infty}^{+\infty} a^n\,x(n)\,z^{-n}$$

$$= \sum_{n=-\infty}^{+\infty} x(n)\,(a^{-1}z)^{-n} \qquad \qquad .....(4.60)$$

$$= X(a^{-1}z) \qquad \boxed{\begin{array}{l}\text{The equation (4.60) is similar}\\\text{to the form of equation (4.49).}\end{array}}$$

5. Time Reversal

If $\ \mathcal{Z}\{x(n)\} = X(z)$

then $\mathcal{Z}\{x(-n)\} = X(z^{-1})$

Proof:

By definition of z-transform,

$$\mathcal{Z}\{x(n)\} = \sum_{n=-\infty}^{+\infty} x(n)\, z^{-n} \qquad\qquad .....(4.61)$$

$$\therefore\ \mathcal{Z}\{x(-n)\} = \sum_{n=-\infty}^{+\infty} x(-n)\, z^{-n} = \sum_{p=-\infty}^{+\infty} x(p)\, z^{p}$$

Let, p = −n
when n → −∞, p → +∞
when n → +∞, p → −∞

$$= \sum_{p=-\infty}^{+\infty} x(p)\, (z^{-1})^{-p} \qquad\qquad .....(4.62)$$

The equation (4.62) is similar to the form of equation (4.61).

$$= X(z^{-1})$$

6. Conjugation

If $\ \mathcal{Z}\{x(n)\} = X(z)$

then $\mathcal{Z}\{x^*(n)\} = X^*(z^*)$

Proof:

By definition of z-transform,

$$X(z) = \mathcal{Z}\{x(n)\} = \sum_{n=-\infty}^{+\infty} x(n)\, z^{-n} \qquad\qquad .....(4.63)$$

$$\therefore\ \mathcal{Z}\{x^*(n)\} = \sum_{n=-\infty}^{+\infty} x^*(n)\, z^{-n}$$

$$= \left[\sum_{n=-\infty}^{+\infty} x(n)\, (z^*)^{-n} \right]^* \qquad\qquad .....(4.64)$$

The equation (4.64) is similar to the form of equation (4.63).

$$= [X(z^*)]^*$$

$$= X^*(z^*)$$

7. Convolution Theorem

If $\ \mathcal{Z}\{x_1(n)\} = X_1(z)$

and $\ \mathcal{Z}\{x_2(n)\} = X_2(z)$

then $\mathcal{Z}\{x_1(n) * x_2(n)\} = X_1(z)\, X_2(z)$

where, $x_1(n) * x_2(n) = \sum_{m=-\infty}^{+\infty} x_1(m)\, x_2(n-m)$ $\qquad\qquad .....(4.65)$

Proof:

By definition of z-transform,

$$X_1(z) = \mathcal{Z}\{x_1(n)\} = \sum_{n=-\infty}^{+\infty} x_1(n)\, z^{-n} \qquad\qquad .....(4.66)$$

$$X_2(z) = \mathcal{Z}\{x_2(n)\} = \sum_{n=-\infty}^{+\infty} x_2(n)\, z^{-n} \qquad\qquad .....(4.67)$$

$$\therefore\ \mathcal{Z}\{x_1(n) * x_2(n)\} = \sum_{n=-\infty}^{+\infty} [x_1(n) * x_2(n)]\, z^{-n}$$

$$\therefore \; Z\{x_1(n) * x_2(n)\} = \sum_{n=-\infty}^{+\infty} \left[\sum_{m=-\infty}^{+\infty} x_1(m) \, x_2(n-m) \right] z^{-n}$$

Using equation (4.65).

$$= \sum_{n=-\infty}^{+\infty} \sum_{m=-\infty}^{+\infty} x_1(m) \, x_2(n-m) \, z^{-n} \, z^{-m} \, z^{m}$$

Multiply by z^{m} and z^{-m}

$$= \sum_{m=-\infty}^{+\infty} x_1(m) \, z^{-m} \sum_{n=-\infty}^{+\infty} x_2(n-m) \, z^{-(n-m)}$$

Let, $n + m = p$,
when $n \to -\infty$, $p \to -\infty$
when $n \to +\infty$, $p \to +\infty$

$$= \sum_{m=-\infty}^{+\infty} x_1(m) \, z^{-m} \sum_{p=-\infty}^{+\infty} x_2(p) \, z^{-p}$$

Let $m = n$, in first summation.
Let $p = n$, in second summation.

$$= \left[\sum_{n=-\infty}^{+\infty} x_1(n) \, z^{-n} \right]\left[\sum_{n=-\infty}^{+\infty} x_2(n) \, z^{-n} \right]$$

$$= X_1(z) \, X_2(z)$$

Using equations (4.66) and (4.67)

8. Correlation Property   *(AU Dec'13, 8 Marks)*

If $Z\{x(n)\} = X(z)$ and $Z\{y(n)\} = Y(z)$

then $Z\{r_{xy}(m)\} = X(z) \, Y(z^{-1})$

where, $r_{xy}(m) = \displaystyle\sum_{n=-\infty}^{+\infty} x(n) \, y(n-m)$ .....(4.68)

Proof:

By definition of Z-transform,

$$X(z) = Z\{x(n)\} = \sum_{n=-\infty}^{+\infty} x(n) \, z^{-n} \qquad .....(4.69)$$

$$Y(z) = Z\{y(n)\} = \sum_{n=-\infty}^{+\infty} y(n) \, z^{-n} \qquad .....(4.70)$$

$$\therefore \; Z\{r_{xy}(m)\} = \sum_{m=-\infty}^{+\infty} r_{xy}(m) \, z^{-m}$$

$$= \sum_{n=-\infty}^{+\infty} \left[\sum_{n=-\infty}^{+\infty} x(n) \, y(n-m) \right] z^{-m}$$

Using equation (4.68).

$$= \sum_{n=-\infty}^{+\infty} \sum_{n=-\infty}^{+\infty} x(n) \, y(n-m) \, z^{-m} \, z^{-n} \, z^{n}$$

Multiply by z^{m} and z^{-m}

$$= \sum_{n=-\infty}^{+\infty} x(n) \, z^{-n} \sum_{m=-\infty}^{+\infty} y(n-m) \, z^{(n-m)}$$

Let, $n - m = p$, $\therefore m = n - p$
when $m \to -\infty$, $p \to +\infty$
when $m \to +\infty$, $p \to -\infty$

$$= \sum_{n=-\infty}^{+\infty} x(n) \, z^{-n} \sum_{p=-\infty}^{+\infty} y(p) \, z^{p}$$

$$= \left[\sum_{n=-\infty}^{+\infty} x(n) \, z^{-n} \right]\left[\sum_{p=-\infty}^{+\infty} y(p) \, (z^{-1})^{-p} \right]$$

Using equations (4.69) and (4.70)

$$= X(z) \, Y(z^{-1})$$

9. Initial Value Theorem

Let $x(n)$ be an one-sided signal defined in the range $0 \leq n \leq \infty$.

Now, if $\mathcal{Z}\{x(n)\} = X(z)$,

then the initial value of $x(n)$ [i.e., $x(0)$] is given by,

$$x(0) = \underset{z \to \infty}{\text{Lt}}\ X(z)$$

Proof:

By definition of one-sided $\mathcal{Z}$ - transfrom,

$$X(z) = \sum_{n=0}^{+\infty} x(n)\, z^{-n}$$

On expanding the above summation we get,

$$X(z) = x(0) + x(1)\, z^{-1} + x(2)\, z^{-2} + x(3)\, z^{-3} + \ldots\ldots$$

$$\therefore\ X(z) = x(0) + \frac{x(1)}{z} + \frac{x(2)}{z^2} + \frac{x(3)}{z^3} + \ldots\ldots$$

On taking limit $z \to \infty$ in the above equation we get,

$$\underset{z \to \infty}{\text{Lt}}\ X(z) = \underset{z \to \infty}{\text{Lt}}\ \left[x(0) + \frac{x(1)}{z} + \frac{x(2)}{z^2} + \frac{x(3)}{z^3} + \ldots\ldots \right]$$

$$= x(0) + 0 + 0 + 0 + \ldots\ldots$$

$$\therefore\ x(0) = \underset{z \to \infty}{\text{Lt}}\ X(z)$$

10. Final Value Theorem

(AU Jun'12, 2 Marks)

Let $x(n)$ be a one-sided signal defined in the range $0 \leq n \leq \infty$.

Now, if $\mathcal{Z}\{x(n)\} = X(z)$,

then the final value of $x(n)$ [i.e., $x(\infty)$] is given by,

$$x(\infty) = \underset{z \to 1}{\text{Lt}}\ (1 - z^{-1})\, X(z) \quad \text{or} \quad x(\infty) = \underset{z \to 1}{\text{Lt}}\ \left[\frac{z-1}{z} \right] X(z)$$

Proof:

By definition of one-sided $\mathcal{Z}$-transfrom,

$$\mathcal{Z}\{x(n)\} = X(z) = \sum_{n=0}^{+\infty} x(n)\, z^{-n} \qquad \ldots\ldots(4.71)$$

$$\therefore\ \mathcal{Z}\{x(n-1) - x(n)\} = \sum_{n=0}^{+\infty} [x(n-1) - x(n)]\, z^{-n}$$

$$\text{(LHS)} \qquad\qquad \text{(RHS)}$$

$$\text{LHS} = \mathcal{Z}\{x(n-1) - x(n)\}$$

$$= \mathcal{Z}\{x(n-1)\} - \mathcal{Z}\{x(n)\} \qquad \boxed{\text{Using linearity property.}}$$

$$= z^{-1} X(z) + x(-1) - X(z) \qquad \boxed{\text{Using shifting property and equation (4.71).}}$$

$$= x(-1) - (1 - z^{-1})\, X(z)$$

$$= \underset{z \to 1}{\text{Lt}}\ [x(-1) - (1 - z^{-1})\, X(z)] \qquad \boxed{\text{Taking limit } z \to 1}$$

$$= x(-1) - \underset{z \to 1}{\text{Lt}}\ (1 - z^{-1})\, X(z) \qquad \ldots\ldots(4.72)$$

$$\text{RHS} = \sum_{n=0}^{+\infty} [x(n-1) - x(n)] z^{-n}$$

$$= \underset{z \to 1}{\text{Lt}} \sum_{n=0}^{+\infty} [x(n-1) - x(n)] z^{-n} \qquad \boxed{\text{Taking limit } z \to 1}$$

$$= \sum_{n=0}^{+\infty} [x(n-1) - x(n)] \qquad \boxed{\begin{array}{l}\text{On applying limit } z \to 1, \text{ the term } z^{-n} \\ \text{becomes unity.}\end{array}}$$

$$= \underset{p \to \infty}{\text{Lt}} \sum_{n=0}^{p} [x(n-1) - x(n)] \qquad \boxed{\begin{array}{l}\text{Changing the summation index from} \\ 0 \text{ to } p \text{ and then taking limit } p \to \infty.\end{array}}$$

$$= \underset{p \to \infty}{\text{Lt}} \left[\begin{array}{l} [x(-1) - x(0)] + [x(0) - x(1)] + [x(1) - x(2)] + \ldots \\ \ldots + [x(p-2) - x(p-1)] + [x(p-1) - x(p)] \end{array} \right] = \underset{p \to \infty}{\text{Lt}} [x(-1) - x(p)]$$

$$= x(-1) - x(\infty) \qquad \qquad \ldots(4.73)$$

On equating equation (4.72) with (4.73) we get,

$$x(-1) - x(\infty) = x(-1) - \underset{z \to 1}{\text{Lt}} (1 - z^{-1}) X(z)$$

$$\therefore x(\infty) = \underset{z \to 1}{\text{Lt}} (1 - z^{-1}) X(z)$$

11. Complex Convolution Theorem (or Multiplication in Time Domain)

Let, $\mathcal{Z}\{x_1(n)\} = X_1(z)$ and $\mathcal{Z}\{x_2(n)\} = X_2(z)$. **(AU Jun'14, 2 Marks)**

Now, the complex convolution theorem states that,

$$\mathcal{Z}\{x_1(n) x_2(n)\} = \frac{1}{2\pi j} \oint_C X_1(v) X_2\left(\frac{z}{v}\right) v^{-1} dv$$

where, v is a dummy variable used for contour integration

Proof:

Let, $\mathcal{Z}\{x_1(n)\} = X_1(z)$ and $\mathcal{Z}\{x_2(n)\} = X_2(z)$.

Now, by definition of inverse $\mathcal{Z}$-transform,

$$x_1(n) = \frac{1}{2\pi j} \oint_C X_1(z) z^{n-1} dz = \frac{1}{2\pi j} \oint_C X_1(v) v^{n-1} dv \qquad \boxed{\text{Let, } z = v} \qquad \ldots(4.74)$$

Now, by definition of $\mathcal{Z}$-transform,

$$X_2(z) = \sum_{n=-\infty}^{+\infty} x_2(n) z^{-n} \qquad \ldots(4.75)$$

Using the definition of $\mathcal{Z}$-transform, the $\mathcal{Z}\{x_1(n) x_2(n)\}$ can be written as,

$$\mathcal{Z}\{x_1(n) x_2(n)\} = \sum_{n=-\infty}^{+\infty} x_1(n) x_2(n) z^{-n}$$

$$= \sum_{n=-\infty}^{+\infty} \left[\frac{1}{2\pi j} \oint_C X_1(v) v^{n-1} dv \right] x_2(n) z^{-n} \qquad \boxed{\text{Using equation (4.74).}}$$

$$= \frac{1}{2\pi j} \oint_C X_1(v) \sum_{n=-\infty}^{+\infty} x_2(n) z^{-n} v^n v^{-1} dv$$

$$= \frac{1}{2\pi j} \oint_C X_1(v) \left[\sum_{n=-\infty}^{+\infty} x_2(n) \left(\frac{z}{v}\right)^{-n} \right] v^{-1} dv \qquad \boxed{\begin{array}{l}\text{Interchanging the order of} \\ \text{summation and integration.}\end{array}}$$

$$= \frac{1}{2\pi j} \oint_C X_1(v) X_2\left(\frac{z}{v}\right) v^{-1} dv \qquad \boxed{\text{Using equation (4.75).}}$$

12. Parseval's Relation

If $\mathcal{Z}\{x_1(n)\} = X_1(z)$ and $\mathcal{Z}\{x_2(n)\} = X_2(z)$.

Then the Parseval's relation states that,

$$\sum_{n=-\infty}^{+\infty} x_1(n)\, x_2^*(n) = \frac{1}{2\pi j} \oint_C X_1(z)\, X_2^*\left(\frac{1}{z^*}\right) z^{-1}\, dz$$

Proof:

Let, $\mathcal{Z}\{x_1(n)\} = X_1(z)$ and $\mathcal{Z}\{x_2(n)\} = X_2(z)$.

Now, by definition of inverse $\mathcal{Z}$-transform,

$$x_1(n) = \frac{1}{2\pi j} \oint_C X_1(z)\, z^{n-1}\, dz = \frac{1}{2\pi j} \oint_C X_1(v)\, v^{n-1}\, dv \qquad \boxed{\text{Let, } z = v} \quad .....(4.76)$$

Now, by definition of $\mathcal{Z}$-transform,

$$\mathcal{Z}\{x_2(n)\} = \sum_{n=-\infty}^{+\infty} x_2(n)\, z^{-n} \qquad\qquad .....(4.77)$$

Using the definition of $\mathcal{Z}$-transform, the $\mathcal{Z}\{x_1(n)\,x_2{}^*(n)\}$ can be written as,

$$\mathcal{Z}\{x_1(n)\, x_2^*(n)\} = \sum_{n=-\infty}^{+\infty} x_1(n)\, x_2^*(n)\, z^{-n} \qquad .....(4.78)$$

On substituting for $x_1(n)$ from equation (4.76) in equation (4.77) we can write,

$$\sum_{n=-\infty}^{+\infty} x_1(n)\, x_2^*(n)\, z^{-n} = \sum_{n=-\infty}^{+\infty} \left[\frac{1}{2\pi j} \oint_C X_1(v)\, v^{n-1}\, dv\right] x_2^*(n)\, z^{-n}$$

$$= \frac{1}{2\pi j} \oint_C X_1(v)\left[\sum_{n=-\infty}^{+\infty} x_2^*(n)\, z^{-n}\, v^n\right] v^{-1}\, dv \qquad \boxed{\begin{array}{l}\text{Interchanging the}\\ \text{order of summation}\\ \text{and integration.}\end{array}}$$

$$= \frac{1}{2\pi j} \oint_C X_1(v)\left[\sum_{n=-\infty}^{+\infty} x_2^*(n)\left(\frac{z}{v}\right)^{-n}\right] v^{-1}\, dv$$

$$= \frac{1}{2\pi j} \oint_C X_1(v)\left[\sum_{n=-\infty}^{+\infty} x_2(n)\left(\frac{z^*}{v^*}\right)^{-n}\right]^* v^{-1}\, dv$$

$$= \frac{1}{2\pi j} \oint_C X_1(v)\, X_2^*\left(\frac{z^*}{v^*}\right) v^{-1}\, dv \qquad \boxed{\text{Using equation (4.77).}}$$

Let us take limit $z \to 1$ in the above equation,

$$\therefore \lim_{z\to 1} \sum_{n=-\infty}^{+\infty} x_1(n)\, x_2^*(n)\, z^{-n} = \lim_{z\to 1} \frac{1}{2\pi j} \oint_C X_1(v)\, X_2^*\left(\frac{z^*}{v^*}\right) v^{-1}\, dv$$

$$\sum_{n=-\infty}^{+\infty} x_1(n)\, x_2^*(n) = \frac{1}{2\pi j} \oint_C X_1(v)\, X_2^*\left(\frac{1}{v^*}\right) v^{-1}\, dv$$

$$\therefore \sum_{n=-\infty}^{+\infty} x_1(n)\, x_2^*(n) = \frac{1}{2\pi j} \oint_C X_1(z)\, X_2^*\left(\frac{1}{z^*}\right) z^{-1}\, dz \qquad \boxed{\text{Let, } v = z}$$

Table 4.4: Summary of Properties of $\mathbb{Z}$-Transform

Note: $X(z) = \mathbb{z}\{x(n)\}$; $X_1(z) = \mathbb{z}\{x_1(n)\}$; $X_2(z) = \mathbb{z}\{x_2(n)\}$; $Y(z) = \mathbb{z}\{y(n)\}$

Property		Discrete Time Signal	$\mathbb{Z}$-transform
Linearity		$a_1 x_1(n) + a_2 x_2(n)$	$a_1 X_1(z) + a_2 X_2(z)$
Shifting $(m \geq 0)$	$x(n)$; for $n \geq 0$	$x(n-m)$	$z^{-m} X(z) + \sum\limits_{i=1}^{m} x(-i) z^{-(m-i)}$
		$x(n+m)$	$z^m X(z) - \sum\limits_{i=0}^{m-1} x(i) z^{m-i}$
	$x(n)$; for all n	$x(n-m)$	$z^{-m} X(z)$
		$x(n+m)$	$z^m X(z)$
Multiplication by n^m (or differentiation in z-domain)		$n^m x(n)$	$\left[-z \dfrac{d}{dz}\right]^m X(z)$
Scaling in z-domain (or multiplication by a^n)		$a^n x(n)$	$X(a^{-1} z)$
Time reversal		$x(-n)$	$X(z^{-1})$
Conjugation		$x^*(n)$	$X^*(z^*)$
Convolution		$x_1(n) * x_2(n) = \sum\limits_{m=-\infty}^{+\infty} x_1(m) x_2(n-m)$	$X_1(z) X_2(z)$
Corrrelation		$r_{xy}(m) = \sum\limits_{m=-\infty}^{+\infty} x(m) y(n-m)$	$X(z) Y(z^{-1})$
Initial value		$x(0) = \underset{z \to \infty}{Lt}\ X(z)$	
Final value		$x(\infty) = \underset{z \to 1}{Lt}\ (1 - z^{-1}) X(z)$ $= \underset{z \to 1}{Lt}\ \left(\dfrac{z-1}{z}\right) X(z)$ if $X(z)$ is analytic for $|z| > 1$	
Complex convolution theorem		$x_1(n) x_2(n)$	$\dfrac{1}{2\pi j} \oint\limits_{C} X_1(v) X_2\left[\dfrac{z}{v}\right] v^{-1} dv$
Parseval's relation		$\sum\limits_{n=-\infty}^{+\infty} x_1(n) x_2^*(n) = \dfrac{1}{2\pi j} \oint\limits_{C} X_1(z) X_2\left[\dfrac{1}{z^*}\right] z^{-1} dz$	

Table 4.5: Some Common 𝑍-transform Pairs

x(t)	x(n)	X(z) With Positive Power of z	X(z) With Negative Power of z	ROC
	$\delta(n)$	1	1	Entire z-plane
	u(n) or 1	$\dfrac{z}{z-1}$	$\dfrac{1}{1-z^{-1}}$	$|z|>1$
	$a^n\,u(n)$	$\dfrac{z}{z-a}$	$\dfrac{1}{1-az^{-1}}$	$|z|>|a|$
	$n\,a^n\,u(n)$	$\dfrac{az}{(z-a)^2}$	$\dfrac{az^{-1}}{(1-az^{-1})^2}$	$|z|>|a|$
	$n^2\,a^n\,u(n)$	$\dfrac{az\,(z+a)}{(z-a)^3}$	$\dfrac{az^{-1}\,(1+az^{-1})}{(1-az^{-1})^3}$	$|z|>|a|$
	$-\,a^n\,u(-n-1)$	$\dfrac{z}{z-a}$	$\dfrac{1}{1-az^{-1}}$	$|z|<|a|$
	$-\,na^n\,u(-n-1)$	$\dfrac{az}{(z-a)^2}$	$\dfrac{az^{-1}}{(1-az^{-1})^2}$	$|z|<|a|$
$t\,u(t)$	$nT\,u(nT)$	$\dfrac{Tz}{(z-1)^2}$	$\dfrac{Tz^{-1}}{(1-z^{-1})^2}$	$|z|>1$
$t^2\,u(t)$	$(nT)^2\,u(nT)$	$\dfrac{T^2z\,(z+1)}{(z-1)^3}$	$\dfrac{T^2z^{-1}\,(1+z^{-1})}{(1-z^{-1})^3}$	$|z|>1$
$e^{-at}\,u(t)$	$e^{-anT}\,u(nT)$	$\dfrac{z}{z-e^{-aT}}$	$\dfrac{1}{1-e^{-aT}\,z^{-1}}$	$|z|>|e^{-aT}|$
$t\,e^{-at}\,u(t)$	$nT\,e^{-anT}\,u(nT)$	$\dfrac{zT\,e^{-aT}}{(z-e^{-aT})^2}$	$\dfrac{z^{-1}T\,e^{-aT}}{(1-e^{-aT}\,z^{-1})^2}$	$|z|>|e^{-aT}|$
$\sin\Omega_0 t\,u(t)$	$\sin\Omega_0 nT\,u(nT)$ $=\sin\omega n\,u(nT)$ where, $\omega=\Omega_0 T$	$\dfrac{z\sin\omega}{z^2-2z\cos\omega+1}$	$\dfrac{z^{-1}\sin\omega}{1-2z^{-1}\cos\omega+z^{-2}}$	$|z|>1$
$\cos\Omega_0 t\,u(t)$	$\cos\Omega_0 nT\,u(nT)$ $=\cos\omega n\,u(nT)$ where, $\omega=\Omega_0 T$	$\dfrac{z\,(z-\cos\omega)}{z^2-2z\cos\omega+1}$	$\dfrac{1-z^{-1}\cos\omega}{1-2z^{-1}\cos\omega+z^{-2}}$	$|z|>1$

Note: 1. *The signals multiplied by u(n) are causal signals (defined for n ≥ 0).*
2. *The signals multiplied by u(-n-1) are anticausal signals (defined for n ≤ 0).*

Example 4.23

Find the one-sided z-transform of the following discrete time signals.

a) $x(n) = n\, a^{(n-1)}$ **b)** $x(n) = n^3$

Solution:

a) Given that, $x(n) = n\, a^{(n-1)}$; $n \geq 0$

Let, $x_1(n) = a^n$

By definition of one-sided z-transform,

$$X_1(z) = z\{x_1(n)\} \sum_{n=0}^{\infty} x_1(n)\, z^{-n} = \sum_{n=0}^{\infty} a^n z^{-n} = \sum_{n=0}^{\infty} (a\, z^{-1})^n = \frac{1}{1 - a z^{-1}} = \frac{z}{z - a}$$

> Using infinite geometric series sum formula.

Let, $x_1(n-1) = a^{n-1}$

By shifting property,

$$z\{x_1(n-1)\} = z^{-1} X_1(z) = z^{-1}\frac{z}{z-a} = \frac{1}{z-a}$$

> If $z\{x(n)\} = X(z)$
> then $z\{n\, x(n)\} = -z\frac{d}{dz}X(z)$

Given that, $x(n) = n\, a^{n-1}$

$$\therefore X(z) = z\{x(n)\} = z\{n\, a^{n-1}\} = z\{n\, x_1(n-1)\} = -z\frac{d}{dz}X_1(z)$$

$$= -z\frac{d}{dz}\frac{1}{z-a} = -z \times \frac{-1}{(z-a)^2} = \frac{z}{(z-a)^2}$$

---

b) Given that, $x(n) = n^3$; $n \geq 0$

Let us multiply the given discrete time signal by a discrete unit step signal,

$$\therefore \ x(n) = n^3\, u(n)$$

> *Note: Multiplying a one-sided sequence by u(n) will not alter its value.*

By the differentiation in z-domain property of z-transform, we get,

$$z\{n^m\, u(n)\} = \left(-z\frac{d}{dz}\right)^m U(z)$$

where, $U(z) = z\{u(n)\} = \dfrac{z}{z-1}$

$$\therefore -z\frac{d}{dz}U(z) = -z\left[\frac{d}{dz}\frac{z}{z-1}\right] = -z\left[\frac{z-1-z}{(z-1)^2}\right] = \frac{z}{(z-1)^2}$$

> $d\dfrac{u}{v} = \dfrac{v\, du - u\, dv}{v^2}$

$$\therefore \left(-z\frac{d}{dz}\right)^2 U(z) = -z\frac{d}{dz}\left[-z\frac{d}{dz}U(z)\right] = -z\frac{d}{dz}\left[\frac{z}{(z-1)^2}\right] = -z\left(\frac{(z-1)^2 - z \times 2(z-1)}{(z-1)^4}\right)$$

$$= -z\left(\frac{(z-1)(z-1-2z)}{(z-1)^4}\right) = -z\left(\frac{-(z+1)}{(z-1)^3}\right) = \frac{z(z+1)}{(z-1)^3}$$

$$\therefore \left(-z\frac{d}{dz}\right)^3 U(z) = -z\frac{d}{dz}\left[\left(-z\frac{d}{dz}\right)^2 U(z)\right] = -z\frac{d}{dz}\frac{z(z+1)}{(z-1)^3} = -z\frac{d}{dz}\frac{z^2+z}{(z-1)^3}$$

$$= -z\left[\frac{(z-1)^3(2z+1) - (z^2+z)3(z-1)^2}{(z-1)^6}\right] = -z\left[\frac{(z-1)^2[(z-1)(2z+1) - 3(z^2+z)]}{(z-1)^6}\right]$$

$$= -z\frac{(2z^2+z-2z-1-3z^2-3z)}{(z-1)^4} = -z\frac{(-z^2-4z-1)}{(z-1)^4} = \frac{z(z^2+4z+1)}{(z-1)^4}$$

$$\therefore X(z) = z\{n^3 u(n)\} = \left(-z\frac{d}{dz}\right)^3 U(z) = \frac{z(z^2+4z+1)}{(z-1)^4}$$

Example 4.24

Find the one-sided z-transform of the discrete time signals generated by mathematically sampling the following continuous time signals.

     a) t^2 **b)** $\sin \Omega_0 t$ **c)** $\cos \Omega_0 t$

Solution:

a) Given that, $x(t) = t^2$

The discrete time signal is generated by replacing t by nT, where T is the sampling time period.

$$\therefore \; x(n) = t^2\big|_{t=nT} = (nT)^2 = n^2\, T^2 = n^2\, g(n)$$

$$\text{where, } g(n) = T^2$$

By the definition of one-sided z-transform we get,

$$\boxed{\begin{array}{l} \text{If } z\{g(n)\} = G(z) \\[4pt] \text{then } z\{n^m\, g(n)\} = \left(-z\dfrac{d}{dz}\right)^m G(z) \end{array}}$$

$$G(z) = z\{g(n)\} = z\{T^2\} = \sum_{n=0}^{\infty} T^2\, z^{-n} = T^2 \sum_{n=0}^{\infty} (z^{-1})^n = T^2\left(\frac{1}{1-z^{-1}}\right) = \frac{T^2\, z}{z-1}$$

By the multiplication by n^m property of z-transform we get,

$$X(z) = z\{x(n)\} = z\{n^2\, g(n)\} = \left(-z\frac{d}{dz}\right)^2 G(z) = -z\frac{d}{dz}\left(-z\frac{d}{dz}\, G(z)\right) \quad \boxed{d\,\frac{u}{v} = \frac{v\,du - u\,dv}{v^2}}$$

$$= -z\frac{d}{dz}\left(-z\frac{d}{dz}\,\frac{T^2\, z}{z-1}\right) = -z\frac{d}{dz}\left(-z \times \frac{(z-1)\,T^2 - T^2 z}{(z-1)^2}\right)$$

$$= -z\frac{d}{dz}\left(\frac{z\,T^2}{(z-1)^2}\right) = -z \times \frac{(z-1)^2\, T^2 - zT^2 \times 2(z-1)}{(z-1)^4}$$

$$= -z \times \frac{(z-1)(zT^2 - T^2 - 2zT^2)}{(z-1)^4} = -z \times \frac{-zT^2 - T^2}{(z-1)^3} = \frac{z\,T^2(z+1)}{(z-1)^3}$$

---

b) Given that, $x(t) = \sin \Omega_0 t$

The discrete time signal is generated by replacing t by nT, where T is the sampling time period.

$$\therefore \; x(n) = \sin \Omega_0 t\big|_{t=nT} = \sin(\Omega_0 nT) = \sin \omega n \; ; \quad \text{where } \omega = \Omega_0 T$$

By the definition of one-sided z-transform,

$$X(z) = z\{x(n)\} = \sum_{n=0}^{\infty} x(n)\, z^{-n} = \sum_{n=0}^{\infty} \sin \omega n \times z^{-n}$$

$$\boxed{\sin \theta = \frac{e^{j0} - e^{-j0}}{2j}}$$

$$= \sum_{n=0}^{\infty} \frac{e^{j\omega n} - e^{-j\omega n}}{2j}\, z^{-n} = \frac{1}{2j} \sum_{n=0}^{\infty} e^{j\omega n}\, z^{-n} - \frac{1}{2j} \sum_{n=0}^{\infty} e^{-j\omega n}\, z^{-n}$$

$$= \frac{1}{2j} \sum_{n=0}^{\infty} (e^{j\omega}\, z^{-1})^n - \frac{1}{2j} \sum_{n=0}^{\infty} (e^{-j\omega}\, z^{-1})^n$$

$$= \frac{1}{2j}\,\frac{1}{1 - e^{j\omega}\, z^{-1}} - \frac{1}{2j}\,\frac{1}{1 - e^{-j\omega}\, z^{-1}}$$

$$\boxed{\begin{array}{l}\text{Using infinite geometric} \\ \text{series sum formula.}\end{array}}$$

$$= \frac{1}{2j}\,\frac{z}{z - e^{j\omega}} - \frac{1}{2j}\,\frac{z}{z - e^{-j\omega}}$$

$$= \frac{z(z - e^{-j\omega}) - z(z - e^{j\omega})}{2j\,(z - e^{j\omega})(z - e^{-j\omega})} = \frac{z^2 - z e^{-j\omega} - z^2 + z e^{j\omega}}{2j\,(z^2 - z e^{-j\omega} - z e^{j\omega} + e^{j\omega} e^{-j\omega})}$$

$$= \frac{z(e^{j\omega} - e^{-j\omega})/2j}{z^2 - z(e^{j\omega} + e^{-j\omega}) + 1}$$

$$\boxed{\sin \theta = \frac{e^{j0} - e^{-j0}}{2j}}$$

$$= \frac{z \sin \omega}{z^2 - 2z \cos \omega + 1} \quad ; \quad \text{where } \omega = \Omega_0 T$$

$$\boxed{\cos \theta = \frac{e^{j0} + e^{-j0}}{2}}$$

c) Given that, $x(t) = \cos\Omega_0 t$ *(AU Dec'12, 8 Marks)*

The discrete time signal is generated by replacing t by nT, where T is the sampling time period.

$$\therefore\ x(n) = \cos\Omega_0 t\big|_{t=nT} = \cos(\Omega_0 nT) = \cos\omega n\ ;\ \text{where } \omega = \Omega_0 T$$

By the definition of one-sided z-transform,

$$\mathcal{Z}\{x(n)\} = X(z) = \sum_{n=0}^{\infty} x(z)\, z^{-n} = \sum_{n=0}^{\infty} \cos\omega n \times z^{-n}$$

$$\boxed{\cos\theta = \frac{e^{j\theta} + e^{-j\theta}}{2}}$$

$$= \sum_{n=0}^{\infty} \frac{e^{j\omega n} + e^{-j\omega n}}{2}\, z^{-n} = \frac{1}{2}\sum_{n=0}^{\infty} e^{j\omega n}\, z^{-n} + \frac{1}{2}\sum_{n=0}^{\infty} e^{-j\omega n}\, z^{-n}$$

$$= \frac{1}{2}\sum_{n=0}^{\infty} (e^{j\omega} z^{-1})^n + \frac{1}{2}\sum_{n=0}^{\infty} (e^{-j\omega} z^{-1})^n$$

$$= \frac{1}{2}\frac{1}{1 - e^{j\omega} z^{-1}} + \frac{1}{2}\frac{1}{1 - e^{-j\omega} z^{-1}}$$

$$\boxed{\text{Using infinite geometric series sum formula.}}$$

$$= \frac{1}{2}\frac{z}{z - e^{j\omega}} + \frac{1}{2}\frac{z}{z - e^{-j\omega}}$$

$$= \frac{z(z - e^{-j\omega}) + z(z - e^{j\omega})}{2(z - e^{j\omega})(z - e^{-j\omega})} = \frac{z^2 - z e^{-j\omega} + z^2 - z e^{j\omega}}{2(z^2 - z e^{-j\omega} - z e^{j\omega} + e^{j\omega} e^{-j\omega})}$$

$$= \frac{2z^2 - z(e^{j\omega} + e^{-j\omega})}{2[z^2 - z(e^{j\omega} + e^{-j\omega}) + 1]} = \frac{z^2 - z(e^{j\omega} + e^{-j\omega})/2}{z^2 - z(e^{j\omega} + e^{-j\omega}) + 1}$$

$$= \frac{z(z - \cos\omega)}{z^2 - 2z\cos\omega + 1}\ ;\quad \text{where } \omega = \Omega_0 T$$

$$\boxed{\cos\theta = \frac{e^{j\theta} + e^{-j\theta}}{2}}$$

Example 4.25   *(AU Dec'13, June'13, 8 Marks)*

Determine the z- transform and their ROC of the discrete time signal, $x(n) = r^n \cos(\omega n)\, u(n)$.

Solution:

Given that, $x(n) = r^n \cos(\omega n)\, u(n) = r^n \cos(\omega n)\ ;\ n \geq 0$

$$\boxed{\begin{aligned} u(n) &= 1\ ;\quad \text{for } n \geq 0 \\ &= 0\ ;\quad \text{for } n < 0 \end{aligned}}$$

By definition of z-transform,

$$X(z) = \mathcal{Z}\{x(n)\} = \sum_{n=-\infty}^{\infty} x(n) z^{-n} = \sum_{n=0}^{\infty} r^n \cos\omega n\, z^{-n}$$

$$\boxed{\cos\theta = \frac{e^{j\theta} + e^{-j\theta}}{2}}$$

$$= \sum_{n=0}^{\infty} r^n \left(\frac{e^{j\omega n} + e^{-j\omega n}}{2}\right) z^{-n} = \frac{1}{2}\sum_{n=0}^{\infty} r^n e^{j\omega n}\, z^{-n} + \frac{1}{2}\sum_{n=0}^{\infty} r^n e^{-j\omega n}\, z^{-n}$$

$$= \frac{1}{2}\sum_{n=0}^{\infty} (r\, e^{j\omega}\, z^{-1})^n + \frac{1}{2}\sum_{n=0}^{\infty} (r\, e^{-j\omega}\, z^{-1})^n \qquad \qquad .....(1)$$

$$= \frac{1}{2}\frac{1}{1 - re^{j\omega} z^{-1}} + \frac{1}{2}\frac{1}{1 - re^{-j\omega} z^{-1}}$$

$$\boxed{\begin{aligned} &\text{Using infinite geometric series sum formula} \\ &\sum_{n=0}^{\infty} C^n = \frac{1}{1-C}\ ;\ \text{if, } 0 < |C| < 1 \end{aligned}}$$

$$= \frac{1}{2}\left[\frac{1 - re^{-j\omega} z^{-1} + 1 - re^{j\omega} z^{-1}}{(1 - re^{j\omega} z^{-1})(1 - re^{-j\omega} z^{-1})}\right]$$

$$= \frac{1}{2}\left[\frac{2 - r z^{-1}(e^{j\omega} + e^{-j\omega})}{1 - re^{-j\omega} z^{-1} - re^{j\omega} z^{-1} + r^2 z^{-2}}\right]$$

$$\therefore X(z) = \frac{1}{2}\left[\frac{2 - r\,z^{-1}\,2\cos\omega}{1 - r\,z^{-1}(e^{j\omega} + e^{-j\omega}) + r^2\,z^{-2}}\right]$$

$$= \frac{1 - r\cos\omega\,z^{-1}}{1 - r\,z^{-1}\,2\cos\omega + r^2\,z^{-2}}$$

$$= \frac{1 - r\cos\omega\,z^{-1}}{z^{-2}\,(z^2 - 2r\cos\omega\,z + r^2)}$$

$$= \frac{z^2 - r\cos\omega\,z}{z^2 - 2r\cos\omega\,z + r^2}$$

$$= \frac{z(z - r\cos\omega)}{z^2 - 2r\cos\omega\,z + r^2}$$

> Note: Here the term $e^{j\omega}$ and $e^{-j\omega}$ contribute only phase

Here the infinite series of equation (1) converges only if $|\,r e^{j\omega}\,z^{-1}| < 1$ and $|\,r e^{-j\omega}\,z^{-1}| < 1$

$$|\,r e^{j\omega}\,z^{-1}| < 1 \quad \Rightarrow \quad \left|\frac{r\,e^{j\omega}}{z}\right| < 1 \quad \Rightarrow \quad \frac{r}{|z|} < 1 \quad \Rightarrow \quad |z| > r$$

$$r e^{-j\omega}\,z^{-1}| < 1 \quad \Rightarrow \quad \left|\frac{r\,e^{-j\omega}}{z}\right| < 1 \quad \Rightarrow \quad \frac{r}{|z|} < 1 \quad \Rightarrow \quad |z| > r$$

$\therefore$ The ROC is $|z| > |r|$

Example 4.26
                                                                          (AU May'15, 6 Marks)

Using suitable z - transform properties, Find X(z) of $x(n) = (n-2)\left(\frac{1}{2}\right)^{(n-2)} u(n-2)$.

Solution:

Let, $x_1(n) = \left(\frac{1}{2}\right)^n u(n) = \left(\frac{1}{2}\right)^n \; ; \; n \ge 0$

By the definition of z - transform,

$$X_1(z) = z\{x_1(n)\} = \sum_{n=-\infty}^{+\infty} x_1(n)\,z^{-n}$$

$$= \sum_{n=0}^{\infty} \left(\frac{1}{2}\right)^n z^{-n} = \sum_{n=0}^{\infty} \left(\frac{1}{2}z^{-1}\right)^n$$

$$= \frac{1}{1 - \frac{1}{2}z^{-1}} = \frac{z}{z - \frac{1}{2}} \qquad\qquad\qquad .....(1)$$

By differntiation in z-domain property of z-transform,

$$z\left\{n\left(\frac{1}{2}\right)^n u(n)\right\} = z\{nx_1(n)\} = -z\frac{d}{dz}X_1(z)$$

$$= -z\frac{d}{dz}\left(\frac{z}{z - \frac{1}{2}}\right) = -z\left[\frac{\left(z - \frac{1}{2}\right)\times 1 - z\times 1}{\left(z - \frac{1}{2}\right)^2}\right]$$

> Using equation (1).

$$= -z\left[\frac{z - \frac{1}{2} - z}{\left(z - \frac{1}{2}\right)^2}\right] = \frac{\frac{1}{2}z}{\left(z - \frac{1}{2}\right)^2} \qquad\qquad .....(2)$$

By shifting property of z-transform,

$$z\left\{(n-2)\left(\frac{1}{2}\right)^{(n-2)}u(n-2)\right\} = z^{-2}z\left\{n\left(\frac{1}{2}\right)^{n}u(n)\right\}$$

$$= z^{-2}\frac{\frac{1}{2}z}{\left(z-\frac{1}{2}\right)^{2}} \qquad \boxed{\text{Using equation (2).}}$$

$$= \frac{\frac{1}{2}}{z\left(z-\frac{1}{2}\right)^{2}}$$

$$\therefore X(z) = z\{x(n)\} = z\left\{(n-2)\left(\frac{1}{2}\right)^{(n-2)}u(n-2)\right\} = \frac{\frac{1}{2}}{z\left(z-\frac{1}{2}\right)^{2}}$$

Example 4.27

Find the one-sided z-transform of the discrete time signals generated by mathematically sampling the following continuous time signals.

a) $e^{-at}\cos\Omega_0 t$ **b)** $e^{-at}\sin\Omega_0 t$

Solution:

a) Given that, $x(t) = e^{-at}\cos\Omega_0 t$

The discrete time signal $x(n)$ is generated by replacing t by nT, where T is the sampling time period.

$$\therefore x(n) = e^{-at}\cos\Omega_0 t\Big|_{t=nT} = e^{-anT}\cos\Omega_0 nT = e^{-anT}\cos\omega n \; ; \text{ where } \omega = \Omega_0 T$$

By the definition of one-sided z-transform we get,

$$X(z) = z\{x(n)\} = \sum_{n=0}^{\infty}e^{-anT}\cos\omega n\,z^{-n} = \sum_{n=0}^{\infty}e^{-anT}\left[\frac{e^{j\omega n}+e^{-j\omega n}}{2}\right]z^{-n} \qquad \boxed{\cos\theta = \frac{e^{j\theta}+e^{-j\theta}}{2}}$$

$$= \frac{1}{2}\sum_{n=0}^{\infty}(e^{-aT}e^{j\omega}z^{-1})^{n} + \frac{1}{2}\sum_{n=0}^{\infty}(e^{-aT}e^{-j\omega}z^{-1})^{n}$$

$$= \frac{1}{2}\frac{1}{1-e^{-aT}e^{j\omega}z^{-1}} + \frac{1}{2}\frac{1}{1-e^{-aT}e^{-j\omega}z^{-1}}$$

Using infinite geometric series sum formula

$$\sum_{n=0}^{\infty}C^{n} = \frac{1}{1-C} \;;$$

$$\text{if, } 0<|C|<1$$

$$= \frac{1}{2}\frac{1}{1-e^{j\omega}/z\,e^{aT}} + \frac{1}{2}\frac{1}{1-e^{-j\omega}/z\,e^{aT}}$$

$$= \frac{1}{2}\left[\frac{z\,e^{aT}}{z\,e^{aT}-e^{j\omega}} + \frac{z\,e^{aT}}{z\,e^{aT}-e^{-j\omega}}\right]$$

$$= \frac{1}{2}\left[\frac{z\,e^{aT}(z\,e^{aT}-e^{-j\omega}) + z\,e^{aT}(z\,e^{aT}-e^{j\omega})}{(z\,e^{aT}-e^{j\omega})(z\,e^{aT}-e^{-j\omega})}\right]$$

$$= \frac{z\,e^{aT}}{2}\left[\frac{z\,e^{aT}-e^{-j\omega}+z\,e^{aT}-e^{j\omega}}{(z\,e^{aT})^{2}-z\,e^{aT}e^{-j\omega}-z\,e^{aT}e^{j\omega}+e^{j\omega}e^{-j\omega}}\right]$$

$$\therefore \; X(z) = \frac{z\,e^{aT}}{2}\left[\frac{2z\,e^{aT} - (e^{j\omega} + e^{-j\omega})}{z^2\,e^{2aT} - z\,e^{aT}(e^{j\omega} + e^{-j\omega}) + 1}\right]$$

$$= \left[\frac{z\,e^{aT}(z\,e^{aT} - \cos\omega)}{z^2\,e^{2aT} - 2z\,e^{aT}\cos\omega + 1}\right] \quad ; \quad \text{where } \omega = \Omega_0 T \qquad \boxed{\cos\theta = \frac{e^{j\theta} + e^{-j\theta}}{2}}$$

b) Given that, $x(t) = e^{-at}\sin\Omega_0 t$

The discrete time signal $x(n)$ is generated by replacing t by nT, where T is the sampling time period.

$$\therefore \; x(n) = e^{-at}\sin\Omega_0 t\Big|_{t=nT} = e^{-anT}\sin\Omega_0 nT = e^{-anT}\sin\omega n \; ; \; \text{where } \omega = \Omega_0 T$$

By the definition of one-sided z-transform we get,

$$X(z) = Z\{x(n)\} = \sum_{n=0}^{\infty} e^{-anT}\sin\omega n\, z^{-n} = \sum_{n=0}^{\infty} e^{-anT}\left(\frac{e^{j\omega n} - e^{-j\omega n}}{2j}\right)z^{-n}$$

$$= \frac{1}{2j}\sum_{n=0}^{\infty}(e^{-aT}e^{j\omega}z^{-1})^n - \frac{1}{2j}\sum_{n=0}^{\infty}(e^{-aT}e^{-j\omega}z^{-1})^n \qquad \boxed{\sin\theta = \frac{e^{j\theta} - e^{-j\theta}}{2j}}$$

$$= \frac{1}{2j}\frac{1}{1 - e^{-aT}e^{j\omega}z^{-1}} - \frac{1}{2j}\frac{1}{1 - e^{-aT}e^{-j\omega}z^{-1}}$$

$$\boxed{\begin{array}{l}\text{Using infinite}\\\text{geometric series}\\\text{sum formula,}\\[4pt]\displaystyle\sum_{n=0}^{\infty} C^n = \frac{1}{1-C}\end{array}}$$

$$= \frac{1}{2j}\frac{1}{1 - e^{j\omega}/z\,e^{aT}} - \frac{1}{2j}\frac{1}{1 - e^{-j\omega}/z\,e^{aT}}$$

$$= \frac{1}{2j}\frac{z\,e^{aT}}{z\,e^{aT} - e^{j\omega}} - \frac{1}{2j}\frac{z\,e^{aT}}{z\,e^{aT} - e^{-j\omega}}$$

$$= \frac{1}{2j}\left[\frac{z\,e^{aT}(z\,e^{aT} - e^{-j\omega}) - z\,e^{aT}(z\,e^{aT} - e^{j\omega})}{(z\,e^{aT} - e^{j\omega})(z\,e^{aT} - e^{-j\omega})}\right]$$

$$= \frac{1}{2j}\left[\frac{z\,e^{aT}[z\,e^{aT} - e^{-j\omega} - z\,e^{aT} - e^{j\omega}]}{(z\,e^{aT})^2 - z\,e^{aT}e^{-j\omega} - z\,e^{aT}e^{j\omega} + e^{j\omega}e^{-j\omega}}\right]$$

$$= \left[\frac{z\,e^{aT}(e^{j\omega} - e^{-j\omega})/2j}{z^2\,e^{2aT} - z\,e^{aT}(e^{j\omega} + e^{-j\omega}) + 1}\right] \qquad \boxed{\sin\theta = \frac{e^{j\theta} - e^{-j\theta}}{2j}}$$

$$= \left[\frac{z\,e^{aT}\sin\omega}{z^2\,e^{2aT} - 2z\,e^{aT}\cos\omega + 1}\right] \quad ; \quad \text{where } \omega = \Omega_0 T \qquad \boxed{\cos\theta = \frac{e^{j\theta} + e^{-j\theta}}{2}}$$

Example 4.28

Find the initial value, $x(0)$ and final value, $x(\infty)$ of the following z-domain signals.

a) $X(z) = \dfrac{1}{1 - z^2}$ **b)** $\dfrac{2 - 4z^{-1}}{1 + 2z^{-1} - 3z^{-2}}$ **c)** $X(z) = \dfrac{1 - 3z^{-1}}{1 - 3.6z^{-1} + 1.8z^{-2}}$

Solution:

a) Given that, $X(z) = \dfrac{1}{1 - z^{-2}}$

By initial value theorem of z-transform we get,

$$x(0) = \underset{z\to\infty}{Lt}\; X(z) = \underset{z\to\infty}{Lt}\frac{1}{1 - z^{-2}} = \underset{z\to\infty}{Lt}\frac{1}{1 - \dfrac{1}{z^2}} = \frac{1}{1 - \dfrac{1}{\infty}} = \frac{1}{1 - 0} = 1$$

By final value theorem of z-transform we get,

$$x(\infty) = \underset{z \to 1}{\text{Lt}} \; (1 - z^{-1}) \, X(z) = \underset{z \to 1}{\text{Lt}} \; (1 - z^{-1}) \frac{1}{1 - z^{-2}}$$

$$\boxed{a^2 - b^2 = (a+b)(a-b)}$$

$$= \underset{z \to 1}{\text{Lt}} \; (1 - z^{-1}) \frac{1}{(1 - z^{-1})(1 + z^{-1})} = \underset{z \to 1}{\text{Lt}} \; \frac{1}{(1 + z^{-1})} = \frac{1}{1 + 1^{-1}} = \frac{1}{2}$$

b) Given that, $\quad X(z) = \dfrac{2 - 4z^{-1}}{1 + 2z^{-1} - 3z^{-2}}$

By initial value theorem of z-transform we get,

$$x(0) = \underset{z \to \infty}{\text{Lt}} \; X(z) = \underset{z \to \infty}{\text{Lt}} \; \frac{2 - 4z^{-1}}{1 + 2z^{-1} - 3z^{-2}} = \underset{z \to \infty}{\text{Lt}} \; \frac{2 - \dfrac{4}{z}}{1 + \dfrac{2}{z} - \dfrac{3}{z^2}}$$

$$= \frac{2 - \dfrac{4}{\infty}}{1 + \dfrac{2}{\infty} - \dfrac{3}{\infty}} = \frac{2 - 0}{1 + 0 - 0} = 2$$

> The root of quadratic
> $z^2 + 2z - 3 = 0$ are,
> $$z = \frac{-2 \pm \sqrt{2^2 - 4 \times (-3)}}{2} = \frac{-2 \pm 4}{2} = 1, -3$$

By final value theorem of z-transform we get,

$$x(\infty) = \underset{z \to 1}{\text{Lt}} \; (1 - z^{-1}) \, X(z) = \underset{z \to 1}{\text{Lt}} \; (1 - z^{-1}) \, \frac{2 - 4z^{-1}}{1 + 2z^{-1} - 3z^{-2}}$$

$$= \underset{z \to 1}{\text{Lt}} \; \frac{2z^{-2}(z-1)(z-2)}{z^{-2}(z^2 + 2z - 3)} = \underset{z \to 1}{\text{Lt}} \; \frac{2(z-1)(z-2)}{(z-1)(z+3)} = \underset{z \to 1}{\text{Lt}} \; \frac{2(z-2)}{(z+3)} = \frac{2(1-2)}{(1+3)} = \frac{-2}{4} = -0.5$$

c) Given that, $\quad X(z) = \dfrac{1 - 3z^{-1}}{1 - 3.6z^{-1} + 1.8z^{-2}}$

By initial value theorem of z-transform we get,

$$x(0) = \underset{z \to \infty}{\text{Lt}} \; X(z) = \underset{z \to \infty}{\text{Lt}} \; \frac{1 - 3z^{-1}}{1 - 3.6z^{-1} + 1.8z^{-2}} = \underset{z \to \infty}{\text{Lt}} \; \frac{1 - \dfrac{3}{z}}{1 - \dfrac{3.6}{z} + \dfrac{1.8}{z^2}}$$

$$= \frac{1 - \dfrac{3}{\infty}}{1 - \dfrac{3.6}{\infty} + \dfrac{1.8}{\infty}} = \frac{1 - 0}{1 - 0 + 0} = 1$$

By final value theorem of z-transform we get,

> The root of quadratic $z^2 - 3.6z + 1.8 = 0$ are,
> $$z = \frac{3.6 \pm \sqrt{3.6^2 - 4 \times 1.8}}{2} = \frac{3.6 \pm 2.4}{2} = 3, 0.6$$

$$x(\infty) = \underset{z \to 1}{\text{Lt}} \; (1 - z^{-1}) \, X(z)$$

$$= \underset{z \to 1}{\text{Lt}} \; (1 - z^{-1}) \frac{1 - 3z^{-1}}{1 - 3.6z^{-1} + 1.8z^{-2}} = \underset{z \to 1}{\text{Lt}} \; \frac{z^{-2}(z-1)(z-3)}{z^{-2}(z^2 - 3.6z + 1.8)}$$

$$= \underset{z \to 1}{\text{Lt}} \; \frac{(z-1)(z-3)}{(z-3)(z-0.6)} = \underset{z \to 1}{\text{Lt}} \; \frac{z-1}{z-0.6} = \frac{1-1}{1-0.6} = \frac{0}{0.4} = 0$$

4.7 Relation Between $\mathcal{Z}$-Transform and Discrete Time Fourier Transform

The $\mathcal{Z}$-transform of a discrete time signal x(n) is defined as,   **(AU Jun'11, 4 Marks)**

$$X(z) = \sum_{n=-\infty}^{+\infty} x(n)z^{-n}$$
..... (4.79)

where, z is a complex variable.

The Fourier transform of a discrete time signal x(n) is given by,

$$X(e^{j\omega}) = \sum_{n=-\infty}^{+\infty} x(n)e^{-j\omega n}$$
..... (4.80)

From equation (4.79) and (4.80) we can say that if we replace **z** by $e^{j\omega}$ in the $\mathcal{Z}$-transform of x(n) we get Fourier transform of x(n).

The X(z) can be viewed as a unique representation of the signal x(n) in the complex z-plane. In z-plane, the point $z = e^{j\omega}$, represents a point with unit magnitude and having a phase of ω. The range of frequency of discrete time signal ω is 0 to 2π. Hencewe can say that, the points on unit circle in z-plane are given by $z = e^{j\omega}$, when ω is varied from 0 to 2π. Therefore the Fourier transform of a discrete time signal x(n) can be obtained by evaluating the $\mathcal{Z}$- transform on a circle of unit radius as shown in the following equation.

$$X(e^{j\omega}) = X(z)\Big|_{z=e^{j\omega}} = \sum_{n=-\infty}^{+\infty} x(n)z^{-n}\Big|_{z=e^{j\omega}} = \sum_{n=-\infty}^{+\infty} x(n)e^{-j\omega n}$$

It is important to note that X(z) exists for $z = e^{j\omega}$ if unit circle is included in ROC of X(z). Therefore the Fourier transform can be obtained from $\mathcal{Z}$-tranform by evaluating X(z) at $z = e^{j\omega}$, if and only if ROC of X(z) includes the unit circle. Fourier transform of some of the common signals that can be obtained from $\mathcal{Z}$-transform are listed in table 4.8

Table 4.6: Some Common $\mathcal{Z}$-transform and Fourier Transform Pairs

x(t)	x(n)	X(z)	$X(e^{j\omega})$
	$\delta(n)$	1	1
	$a^n u(n)$; $|a| < 1$	$\dfrac{z}{z-a}$	$\dfrac{e^{j\omega}}{e^{j\omega}-a}$
	$n\, a^n u(n)$; $|a| < 1$	$\dfrac{az}{(z-a)^2}$	$\dfrac{ae^{j\omega}}{(e^{j\omega}-a)^2}$
	$n^2 a^n u(n)$; $|a| < 1$	$\dfrac{az(z+a)}{(z-a)^3}$	$\dfrac{ae^{j\omega}(e^{j\omega}+a)}{(e^{j\omega}-a)^3}$
$e^{-at}u(t)$	$e^{-anT}u(nT)$; $|e^{-aT}| < 1$	$\dfrac{z}{z-e^{-aT}}$	$\dfrac{e^{j\omega}}{e^{j\omega}-e^{-aT}}$
$te^{-at}u(t)$	$nTe^{-anT}u(nT)$; $|e^{-aT}| < 1$	$\dfrac{zTe^{-aT}}{(z-e^{-aT})^2}$	$\dfrac{e^{j\omega}Te^{-aT}}{(e^{j\omega}-e^{-aT})^2}$

4.8 Various Methods of Computing Inverse $\mathbb{Z}$-Transform   *(AU May'15, 2 Marks)*

Let $X(z)$ be $\mathbb{Z}$-transform of the discrete time signal $x(n)$. The inverse $\mathbb{Z}$-transform is the process of recovering the discrete time signal $x(n)$ from its $\mathbb{Z}$-transform $X(z)$. The signal $x(n)$ can be uniquely determined from $X(z)$ and its ROC.

The inverse $\mathbb{Z}$-transform can be determined by the following three methods.

      1. Direct evaluation by contour integration (or residue method).

      2. Partial fraction expansion method.

      3. Power series expansion method.

4.8.1 Inverse $\mathbb{Z}$-Transform by Contour Integration or Residue Method

Let, $X(z)$ be $\mathbb{Z}$-transform of $x(n)$.

Now by definition of inverse $\mathbb{Z}$-transform,

$$x(n) = \frac{1}{2\pi j} \oint_C X(z)\, z^{n-1}\, dz \qquad\qquad .....(4.81)$$

Using partial fraction expansion technique the function $X(z)\, z^{n-1}$ can be expressed as shown below:

$$X(z)\, z^{n-1} = \frac{A_1}{z - p_1} + \frac{A_2}{z - p_2} + \frac{A_3}{z - p_3} + ... + \frac{A_N}{z - p_N} \qquad .....(4.82)$$

where, $p_1, p_2, p_3, ... p_N$ are poles of $X(z)\, z^{n-1}$ and $A_1, A_2, A_3, ... A_N$ are residues.

The residue A_1 is obtained by multiplying the equation (4.83) by $(z - p_1)$ and letting $z = p_1$.

Similarly other residues are evaluated.

$$\therefore\ A_1 = (z - p_1)\, X(z)\, z^{n-1}\big|_{z = p_1} \qquad\qquad .....(4.83.1)$$

$$A_2 = (z - p_2)\, X(z)\, z^{n-1}\big|_{z = p_2} \qquad\qquad .....(4.83.2)$$

$$A_3 = (z - p_3)\, X(z)\, z^{n-1}\big|_{z = p_3} \qquad\qquad .....(4.83.3)$$

$$\vdots$$

$$A_N = (z - p_N)\, X(z)\, z^{n-1}\big|_{z = p_N} \qquad\qquad .....(4.83.N)$$

Using equation (4.82) the equation (4.81) can be written as,

$$x(n) = \frac{1}{2\pi j} \oint_C \left[\frac{A_1}{z - p_1} + \frac{A_2}{z - p_2} + \frac{A_3}{z - p_3} + ... + \frac{A_N}{z - p_N} \right] dz$$

$$= \frac{1}{2\pi j} \left[A_1 \oint_C \frac{dz}{z - p_1} + A_2 \oint_C \frac{dz}{z - p_2} + A_3 \oint_C \frac{dz}{z - p_3} + ... + A_N \oint_C \frac{dz}{z - p_N} \right]$$

$$\qquad\qquad\qquad\qquad\qquad\qquad\qquad\qquad\qquad\qquad\qquad\qquad\qquad\qquad .....(4.84)$$

If, $G(z) = \dfrac{1}{z - p_0}$, then by **Cauchy's integral theorem,**

$$\oint_C G(z)\, dz = \oint_C \frac{1}{z - p_0}\, dz = 2\pi j \quad ; \text{ if } p_0 \text{ is a point inside the counter C in } z - \text{plane.}$$

$$= 0 \quad ; \text{ if } p_0 \text{ is a point outside the counter C in } z - \text{plane.}$$

Using Cauchy's integral theorem, the equation (4.84) can be written as shown below.

$$x(n) = \frac{1}{2\pi j}[A_1\, 2\pi j + A_2\, 2\pi j + A_3\, 2\pi j + \dots + A_N\, 2\pi j] \qquad \dots (4.85)$$

$$= A_1 + A_2 + A_3 + \dots + A_N$$

$$= \text{Sum of residues of } X(z)\, z^{n-1}$$

On substituting for residues from equation (4.83.1) to (4.83.N) in equation (4.85), we get,

$$x(n) = (z - p_1)\, X(z)\, z^{n-1}\big|_{z = p_1} + (z - p_2)\, X(z)\, z^{n-1}\big|_{z = p_2}$$

$$+ (z - p_3)\, X(z)\, z^{n-1}\big|_{z = p_3} + \dots + (z - p_N)\, X(z)\, z^{n-1}\big|_{z = p_N}$$

$$\therefore \; x(n) = \sum_{i=1}^{N}\left[(z - p_i)\, X(z)\, z^{n-1}\big|_{z = p_i}\right]. \qquad \dots (4.86)$$

where, N = Number or poles of $X(z)\, z^{n-1}$ lying inside the contour C.

Using equation (4.86), by considering only the poles lying inside the contour C, the inverse $\mathcal{Z}$-transform can be evaluated. For a stable system the contour C is the unit circle in z-plane.

4.8.2 Inverse $\mathcal{Z}$-Transform by Partial Fraction Expansion Method

Let $X(z)$ be $\mathcal{Z}$-transform of $x(n)$, and $X(z)$ be a rational function of z. Now the function $X(z)$ can be expressed as a ratio of two polynomials in z as shown below: (Refer equation 4.49).

$$X(z) = \frac{N(z)}{D(z)} \qquad \dots (4.87)$$

where, N(z) = Numerator polynomial of X(z)

D(z) = Denominator polynomial of X(z)

Let us divide both sides of equation (4.87) by z and express equation (4.88) as shown below:

$$\frac{X(z)}{z} = \frac{N(z)}{z\, D(z)}$$

$$\therefore \; \frac{X(z)}{z} = \frac{Q(z)}{D(z)} \qquad \dots (4.88)$$

where, $Q(z) = \dfrac{N(z)}{z}$

Note: *It is convenient, if we consider $\dfrac{X(z)}{z}$ rather than X(z) for inverse $\mathcal{Z}$-transform by partial fraction expansion method.*

On factorizing the denominator polynomial of equation (4.88) we get,

$$\frac{X(z)}{z} = \frac{Q(z)}{D(z)} = \frac{Q(z)}{(z-p_1)(z-p_2)(z-p_3)\,.....\,(z-p_N)} \qquad .....(4.89)$$

where, $p_1, p_2, p_3, \,.....\,p_N$ are roots of denominator polynomial [as well as poles of X(z)].

The equation (4.89) can be expressed as a series of sum terms by partial fraction expansion technique as shown below:

$$\frac{X(z)}{z} = \frac{A_1}{z-p_1} + \frac{A_2}{z-p_2} + \frac{A_3}{z-p_3} +.....+ \frac{A_N}{z-p_N}$$

where, $A_1, A_2, A_3, \,.....\,A_N$ are residues.

$$\therefore X(z) = A_1 \frac{z}{z-p_1} + A_2 \frac{z}{z-p_2} + A_3 \frac{z}{z-p_3} +.....+ A_N \frac{z}{z-p_N} \qquad .....(4.90)$$

Now, the inverse $\mathbb{Z}$-transform of equation (4.90) is obtained by comparing each term with standard $\mathbb{Z}$-transform pair. The two popular $\mathbb{Z}$-transform pairs useful for inverse $\mathbb{Z}$-transform of equation (4.90) are given below:

If a^n is a causal (or right-sided) signal then,

$$\mathbb{Z}\{a^n\,u(n)\} = \frac{z}{z-a} \ ; \ \text{with ROC}\,|z|>|a|$$

If a^n is an anticausal (or left-sided) signal then,

$$\mathbb{Z}\{-a^n\,u(-n-1)\} = \frac{z}{z-a} \ ; \ \text{with ROC}\,|z|<|a|$$

Let r_1 be the magnitude of the largest pole and let the ROC be $|z|>r_1$ (where r_1 is radius of a circle in z-plane), then each term of equation (4.90) gives rise to a causal sequence, and so the inverse $\mathbb{Z}$-transform of equation (4.90) will be as shown in equation (4.91).

$$x(n) = A_1\,p_1^n\,u(n) + A_2\,p_2^n\,u(n) + A_3\,p_3^n\,u(n) + \,.....\, + A_N\,p_N^n\,u(n) \qquad .....(4.91)$$

Let r_2 be the magnitude of the smallest pole and let ROC be $|z| < r_2$ (where r_2 is radius of a circle in z-plane), then each term of equation (4.90) give rise to an anticausal sequence, and so the inverse $\mathbb{Z}$-transform of equation (4.90) will be as shown in equation (4.92).

$$x(n) = -A_1\,p_1^n\,u(-n-1) - A_2\,p_2^n\,u(-n-1) - A_3\,p_3^n\,u(-n-1) - \,.....\, - A_N\,p_N^n\,u(-n-1) \qquad .....(4.92)$$

Sometimes the specified ROC will be in between two circles of radius r_x and r_y, where $r_x < r_y$. [i.e., ROC is $r_x < |z| < r_y$]. Now in this case, the terms with magnitude of pole less than r_x will give rise to causal signal and the terms with magnitude of pole greater than r_y will give rise to anticausal signal so that the inverse $\mathbb{Z}$-transform of X(z) will give a two-sided signal. [Refer Section 4.5.2, case iii].

Evaluation of Residues

The coefficients of the denominator polynomial D(z) are assumed real and so the roots of the denominator polynomial are real and/or complex conjugate pairs (i.e., complex roots will occur only in conjugate pairs). Hence on factorizing the denominator polynomial we get the following cases. [The roots of the denominator polynomial are poles of X(z)].

Case i : *When roots (or poles) are real and distinct.*

Case ii: *When roots (or poles) have multiplicity.*

Case iii: *When roots (or poles) are complex conjugate.*

Case i: When roots (or poles) are real and distinct

In this case $\dfrac{X(z)}{z}$ can be expressed as,

$$\frac{X(z)}{z} = \frac{Q(z)}{D(z)} = \frac{Q(z)}{(z-p_1)(z-p_2)\dots(z-p_N)}$$

$$= \frac{A_1}{(z-p_1)} + \frac{A_2}{(z-p_2)} + \dots + \frac{A_N}{(z-p_N)}$$

where, $A_1, A_2 \dots A_N$ are residues and $p_1, p_2, \dots p_N$ are poles.

The residue A_1 is evaluated by multiplying both sides of $\dfrac{X(z)}{z}$ by $(z–p_1)$ and letting $z = p_1$. Similarly other residues are evaluated.

$$\therefore \; A_1 = (z-p_1)\frac{X(z)}{z}\bigg|_{z=p_1}$$

$$A_2 = (z-p_2)\frac{X(z)}{z}\bigg|_{z=p_2}$$

$$\vdots$$

$$A_N = (z-p_N)\frac{X(z)}{z}\bigg|_{z=p_N}$$

Case ii: When roots (or poles) have multiplicity

Let one pole have a multiplicity of q (i.e., repeats q times). In this case $\dfrac{X(z)}{z}$ is expressed as,

$$\frac{X(z)}{z} = \frac{Q(z)}{D(z)} = \frac{Q(z)}{(z-p_1)(z-p_2)\dots(z-p_x)^q\dots(z-p_N)}$$

$$= \frac{A_1}{(z-p_1)} + \frac{A_2}{(z-p_2)} + \dots$$

$$+ \frac{A_{x0}}{(z-p_x)^q} + \frac{A_{x1}}{(z-p_x)^{q-1}} + \dots + \frac{A_{x(q-1)}}{(z-p_x)} + \dots + \frac{A_N}{(z-p_N)}$$

where, $A_{x0}, A_{x1}, \dots A_{x(q-1)}$ are residues of repeated root (or pole), $z = p_x$

The residues of distinct real roots are evaluated as explained in case i.

The residue A_{xr} of repeated root is obtained as shown below:

$$A_{xr} = \frac{1}{r!}\frac{d^r}{dz^r}\left[(z-p_x)\frac{X(z)}{z}\right]\bigg|_{z=p_x} \; ; \; \text{where, } r = 0,1,2,\dots(q-1)$$

Case iii: When roots (or poles) are complex conjugate

Let $\dfrac{X(z)}{z}$ has one pair of complex conjugate pole. In this case $\dfrac{X(z)}{z}$ can be expressed as,

$$\frac{X(z)}{z} = \frac{Q(z)}{D(z)} = \frac{Q(z)}{(z-p_1)(z-p_2)\ldots(z^2+az+b)\ldots(z-p_N)}$$

$$= \frac{A_1}{z-p_1} + \frac{A_2}{z-p_2} + \ldots + \frac{A_x}{z-(x+jy)} + \frac{A_x^*}{z-(x-jy)} + \ldots + \frac{A_N}{z-p_N}$$

The residues of real and nonrepeated roots are evaluated as explained in case i.

The residue A_x is evaluated as that of case i and the residue A_x^* is the conjugate of A_x.

4.8.3 Inverse 𝒵-Transform by Power Series Expansion Method

Let $X(z)$ be 𝒵-transform of $x(n)$, and $X(z)$ be a rational function of z as shown below:

$$X(z) = \frac{N(z)}{D(z)} = \frac{b_0 + b_1 z^{-1} + b_2 z^{-2} + b_3 z^{-3} + \ldots + b_M z^{-M}}{a_0 + a_1 z^{-1} + a_2 z^{-2} + a_3 z^{-3} + \ldots + a_N z^{-N}} \qquad \ldots(4.93)$$

On dividing the numerator polynomial $N(z)$ by denominator polynomial $D(z)$ we can express $X(z)$ as a power series of z. It is possible to express $X(z)$ as positive power of z or as negative power of z or with both positive and negative power of z as shown below:

Case i: $X(z) = \dfrac{N(z)}{D(z)} = c_0 + c_1 z^{-1} + c_2 z^{-2} + c_3 z^{-3} + \ldots$ $\ldots(4.93.1)$

Case ii: $X(z) = \dfrac{N(z)}{D(z)} = d_0 + d_1 z^{1} + d_2 z^{2} + d_3 z^{3} + \ldots$ $\ldots(4.93.2)$

Case iii: $X(z) = \dfrac{N(z)}{D(z)} = \ldots + e_{-3} z^{3} + e_{-2} z^{2} + e_{-1} z + e_0$

$$+ \; e_1 z^{-1} + e_2 z^{-2} + e_3 z^{-3} + \ldots \qquad \ldots(4.93.3)$$

The case-i power series of z is obtained when the ROC is exterior of a circle of radius r in z-plane (i.e., ROC is $|z| > r$).

The case-ii power series of z is obtained when the ROC is interior of a circle of radius r in z-plane (i.e., ROC is $|z| < r$).

The case-iii power series of z is obtained when the ROC is in between two circles of radius r_1 and r_2 in z-plane (i.e., ROC is $r_1 < |z| < r_2$).

By the definition of 𝒵-transform, we get,

$$X(z) = \sum_{n=-\infty}^{\infty} x(n) z^{-n}$$

On expanding the summation we get,

$$X(z) = \ldots\ldots x(-3) z^{3} + x(-2) z^{2} + x(-1) z^{1} + x(0) z^{0}$$

$$+ \; x(1) z^{-1} + x(2) z^{-2} + x(3) z^{-3} + \ldots\ldots \qquad \ldots(4.94)$$

On comparing the coefficients of z of equations (4.93) and (4.94), the samples of x(n) are determined. [i.e., the coefficient of z^i is the ith sample, x(i) of the signal x(n)].

> **Note:** *The different methods of evaluation of inverse $\mathbb{Z}$-transform of a function X(z) will result in different type of mathematical expressions. But the inverse $\mathbb{Z}$-transform is unique for a specified ROC and so on evaluating the expressions for each value of n, we may get a same signal.*

Example 4.29

Determine the inverse $\mathbb{Z}$-transform of the function, $X(z) = \dfrac{3 + 2z^{-1} + z^{-2}}{1 - 3z^{-1} + 2z^{-2}}$ by the following three methods and prove that the inverse $\mathbb{Z}$-transform is unique.

1. Residue Method
2. Partial Fraction Expansion Method
3. Power Series Expansion Method

> The roots of quadratic
> $z^2 - 3z + 2 = 0$ are,
> $z = \dfrac{3 \pm \sqrt{3^2 - 4 \times 2}}{2} = \dfrac{3 \pm 1}{2} = 2, 1$

Solution:

Method-1 : Residue Method

Given that, $X(z) = \dfrac{3 + 2z^{-1} + z^{-2}}{1 - 3z^{-1} + 2z^{-2}} = \dfrac{z^{-2}(3z^2 + 2z + 1)}{z^{-2}(z^2 - 3z + 2)} = \dfrac{3z^2 + 2z + 1}{z^2 - 3z + 2}$

Let us divide the numerator polynomial by denominator polynomial and express X(z) as shown below:

$$X(z) = \frac{3z^2 + 2z + 1}{z^2 - 3z + 2} = 3 + \frac{11z - 5}{z^2 - 3z + 2}$$

$$= 3 + \frac{11z - 5}{(z - 1)(z - 2)}$$

$$
\begin{array}{r}
3 \\
z^2 - 3z + 2 \enclose{longdiv}{3z^2 + 2z + 1} \\
3z^2 - 9z + 6 \\
\hline
(-) \quad (+) \quad (-) \\
\hline
11z - 5
\end{array}
$$

Let, $X_1(z) = 3$ and $X_2(z) = \dfrac{11z - 5}{z^2 - 3z + 2}$; $\therefore X(z) = X_1(z) + X_2(z)$

$x(n) = \mathbb{Z}^{-1}\{X(z)\} = \mathbb{Z}^{-1}\{X_1(z)\} + \mathbb{Z}^{-1}\{X_2(z)\}$

$\qquad = \mathbb{Z}^{-1}\{3\} + \mathbb{Z}^{-1}\{X_2(z)\}$

$\qquad = 3\,\delta(n) + \displaystyle\sum_{i=1}^{N} \left[(z - p_i)\, X_2(z)\, z^{n-1}\big|_{z = p_i}\right]$ Using residue theorem.

$\qquad = 3\,\delta(n) + (z - 1)\dfrac{11z - 5}{(z - 1)(z - 2)}\, z^{n-1}\bigg|_{z=1} + (z - 2)\dfrac{11z - 5}{(z - 1)(z - 2)}\, z^{n-1}\bigg|_{z=2}$

$\qquad = 3\,\delta(n) + \dfrac{11 - 5}{1 - 2}(1)^{n-1} + \dfrac{11 \times 2 - 5}{2 - 1}\, 2^{n-1}$

$\qquad = 3\,\delta(n) - 6\,u(n-1) + 17\,(2)^{n-1}u(n-1) = 3\,\delta(n) + [-6 + 17\,(2)^{n-1}]\,u(n-1)$

When n = 0, x(0) = 3 - 0 + 0   = 3

When n = 1, x(1) = 0 - 6 + 17 × 2^0 = 11

When n = 2, x(2) = 0 - 6 + 17 × 2^1 = 28

When n = 3, x(3) = 0 - 6 + 17 × 2^2 = 62

When n = 4,   x(4) = 0 - 6 + 17 × 2^3 = 130

$\therefore$ x(n) = {3, 11, 28, 62, 130, ...}
$\qquad\quad \uparrow$

Method-2 : Partial Fraction Expansion Method

Given that, $X(z) = \dfrac{3 + 2z^{-1} + z^{-2}}{1 - 3z^{-1} + 2z^{-2}} = \dfrac{z^{-2}(3z^2 + 2z + 1)}{z^{-2}(z^2 - 3z + 2)} = \dfrac{3z^2 + 2z + 1}{(z - 1)(z - 2)}$

$$\frac{X(z)}{z} = \frac{3z^2 + 2z + 1}{z(z - 1)(z - 2)}$$

Let, $\dfrac{X(z)}{z} = \dfrac{3z^2 + 2z + 1}{z(z - 1)(z - 2)} = \dfrac{A_1}{z} + \dfrac{A_2}{z - 1} + \dfrac{A_3}{z - 2}$

Now, $A_1 = z \dfrac{X(z)}{z}\bigg|_{z=0} = (z) \dfrac{3z^2 + 2z + 1}{z(z - 1)(z - 2)}\bigg|_{z=0} = \dfrac{0 + 0 + 1}{(0 - 1)(0 - 2)} = 0.5$

$A_2 = (z - 1) \dfrac{X(z)}{z}\bigg|_{z=1} = (z-1) \dfrac{3z^2 + 2z + 1}{z(z-1)(z - 2)}\bigg|_{z=1} = \dfrac{3 + 2 + 1}{1 \times (1 - 2)} = -6$

$A_3 = (z - 2) \dfrac{X(z)}{z}\bigg|_{z=2} = (z-2) \dfrac{3z^2 + 2z + 1}{z(z - 1)(z-2)}\bigg|_{z=2} = \dfrac{3 \times 2^2 + 2 \times 2 + 1}{2 \times (2 - 1)} = 8.5$

$$\frac{X(z)}{z} = \frac{0.5}{z} - \frac{6}{z-1} + \frac{8.5}{z-2}$$

$\therefore X(z) = 0.5 - 6\dfrac{z}{z-1} + 8.5\dfrac{z}{z-2}$

$z\{\delta(n)\} = 1$
$z\{u(n)\} = \dfrac{z}{z-1}$
$z\{a^n u(n)\} = \dfrac{z}{z-a}$

On taking inverse z-transform of X(z) we get,

$$x(n) = 0.5\,\delta(n) - 6\,u(n) + 8.5\,(2)^n\,u(n) = 0.5\,\delta(n) + [-6 + 8.5(2)^n]\,u(n)$$

When n = 0, x(0) = 0.5 − 6 + 8.5 × 2^0 = 3 | When n = 3, x(3) = 0 − 6 + 8.5 × 2^3 = 62

When n = 1, x(1) = 0 − 6 + 8.5 × 2^1 = 11 | When n = 4, x(4) = 0 − 6 + 8.5 × 2^4 = 130

When n = 2, x(2) = 0 − 6 + 8.5 × 2^2 = 28

$\therefore$ x(n) = {3, 11, 28, 62, 130, ...}
$\uparrow$

Method-3 : Power Series Expansion Method

Given that, $X(z) = \dfrac{3 + 2z^{-1} + z^{-2}}{1 - 3z^{-1} + 2z^{-2}}$

Let us divide the numerator polynomial by denominator polynomial as shown below:

$$
\begin{array}{r}
3 + 11z^{-1} + 28z^{-2} + 62z^{-3} + 130z^{-4} + \cdots\cdots\cdots \\[4pt]
1 - 3z^{-1} + 2z^{-2}\,\overline{\big)\ 3 + 2z^{-1} + z^{-2}} \\
3 - 9z^{-1} + 6z^{-2} \\
\hline
(-)\ \ (+)\ \ \ \ (-) \\
11z^{-1} - 5z^{-2} \\
11z^{-1} - 33z^{-2} + 22z^{-3} \\
\hline
(-)\ \ \ \ (+)\ \ \ \ \ (-) \\
28z^{-2} - 22z^{-3} \\
28z^{-2} - 84z^{-3} + 56z^{-4} \\
\hline
(-)\ \ \ \ (+)\ \ \ \ \ (-) \\
62z^{-3} - 56z^{-4} \\
62z^{-3} - 186z^{-4} + 124z^{-5} \\
\hline
(-)\ \ \ \ (+)\ \ \ \ \ (-) \\
130z^{-4} - 124z^{-5}
\end{array}
$$

$$\therefore X(z) = \frac{3 + 2z^{-1} + z^{-2}}{1 - 3z^{-1} + 2z^{-2}} = 3 + 11z^{-1} + 28z^{-2} + 62z^{-3} + 130z^{-4} + ...$$ .....(1)

Let, x(n) be inverse z-transform of X(z).

Now, by definition of z-transform,

$$X(z) = \sum_{n=-\infty}^{+\infty} x(n)z^{-n}$$

$$= ..... + x(0) + x(1)z^{-1} + x(2)z^{-2} + x(3)z^{-3} + x(4)z^{-4} + ...$$ .....(2)

On comparing equations (1) and (2) we get,

x(0) = 3

x(1) = 11

x(2) = 28

x(3) = 62

x(4) = 130 and so on.

$\therefore$ x(n) = {3, 11, 28, 62, 130, ...}

Conclusion: It is observed that the results of all the three methods are same.

Example 4.30

Determine the inverse z-transform of the following z-domain functions.

a) $X(z) = \dfrac{3z^2 + 2z + 1}{z^2 + 4z + 3}$ b) $X(z) = \dfrac{z - 0.6}{z^2 + z + 2}$ c) $X(z) = \dfrac{2z - 4}{(z - 1)(z + 2)^2}$

Solution:

a) Given that, $X(z) = \dfrac{3z^2 + 2z + 1}{z^2 + 4z + 3}$

On dividing the numerator by denominator, the X(z) can be expressed as shown below:

$$X(z) = \frac{3z^2 + 2z + 1}{z^2 + 4z + 3} = 3 + \frac{-10z - 8}{z^2 + 4z + 3} = 3 + \frac{-10z - 8}{(z + 1)(z + 3)}$$

By partial fraction expansion we get, $X(z) = 3 + \dfrac{A_1}{z + 1} + \dfrac{A_2}{z + 3}$

$$A_1 = (z+1)\frac{-10z - 8}{(z+1)(z + 1)}\Big|_{z=-1} = \frac{-10z - 8}{z + 3}\Big|_{z=-1}$$

$$= \frac{10 - 8}{-1 + 3} = \frac{2}{2} = 1$$

$$A_2 = (z+1)\frac{-10z - 8}{(z + 1)(z+3)}\Big|_{z=-3} = \frac{-10z - 8}{z + 1}\Big|_{z=-3}$$

$$= \frac{-10 \times (-3) - 8}{-3 + 1} = -11$$

$$\therefore X(z) = 3 + \frac{1}{z + 1} - \frac{11}{z + 3} = 3 + \frac{1}{z}\frac{z}{z - (-1)} - 11\frac{1}{z}\frac{z}{z - (-3)}$$

$$= 3 + z^{-1}\frac{z}{z - (-1)} - 11z^{-1}\frac{z}{z - (-3)}$$

Side box (division):

$z^2 + 4z + 3 \overline{)\; 3z^2 + 2z + 1}$ quotient 3

$3z^2 + 12z + 9$

(−) (−) (−)

$-10z - 8$

The roots of quadratic

$z^2 + 4z + 3 = 0$

$= \dfrac{-4 \pm \sqrt{4^2 - 4 \times 3}}{2}$

$= \dfrac{-4 \pm 2}{2} = -1, -3$

Multiply and divide by z.

On taking inverse z-transform of $X(z)$ we get,

$$x(n) = 3\,\delta(n) + (-1)^{n-1} u(n-1) - 11(-3)^{n-1} u(n-1)$$

$$= 3\,\delta(n) + [(-1)^{n-1} - 11(-3)^{n-1}] u(n-1)$$

When $n = 0$,	$x(0)$	$= 3 + 0 + 0$	$=$	3
When $n = 1$,	$x(1)$	$= 0 + 1 - 11$	$= -10$	
When $n = 2$,	$x(2)$	$= 0 - 1 + 33$	$=$	32
When $n = 3$,	$x(3)$	$= 0 + 1 - 99$	$= -98$	
When $n = 4$,	$x(4)$	$= 0 - 1 + 297$	$=$	296

$\therefore x(n) = \{3, -10, 32, -98, 296, \ldots\}$
$\uparrow$

$z\{\delta(n)\} = 1$
$z\{a^n u(n)\} = \dfrac{z}{z - a}$
If $z\{a^n u(n)\} = \dfrac{z}{z - a}$
then by time shifting property,
$z\{a^{(n-1)} u(n-1)\} = z^{-1}\,\dfrac{z}{z - a}$

Alternate Method

$$X(z) = \frac{3z^2 + 2z + 1}{z^2 + 4z + 3}$$

$$\therefore \frac{X(z)}{z} = \frac{3z^2 + 2z + 1}{z(z^2 + 4z + 3)} = \frac{3z^2 + 2z + 1}{z(z + 1)(z + 3)}$$

By partial fraction expansion technique $\dfrac{X(z)}{z}$ can be expressed as,

$$\therefore \frac{X(z)}{z} = \frac{3z^2 + 2z + 1}{z(z + 1)(z + 3)} = \frac{A_1}{z} + \frac{A_2}{z + 1} + \frac{A_3}{z + 3}$$

$$A_1 = z\,\frac{X(z)}{z}\bigg|_{z=0} = \not{z}\,\frac{3z^2 + 2z + 1}{\not{z}(z + 1)(z + 3)}\bigg|_{z=0}$$

$$= \frac{0 + 0 + 1}{(0 + 1)(0 + 3)} = \frac{1}{3}$$

$$A_2 = (z + 1)\,\frac{X(z)}{z}\bigg|_{z=-1} = (\not{z + 1})\,\frac{3z^2 + 2z + 1}{z(\not{z + 1})(z + 3)}\bigg|_{z=-1}$$

$$= \frac{3(-1)^2 + 2(-1) + 1}{-1 \times (-1 + 3)} = -1$$

$$A_3 = (z + 3)\,\frac{X(z)}{z}\bigg|_{z=-3} = (\not{z + 3})\,\frac{3z^2 + 2z + 1}{z(z + 1)(\not{z + 3})}\bigg|_{z=-3}$$

$$= \frac{3(-3)^2 + 2(-3) + 1}{-3 \times (-3 + 1)} = \frac{22}{6} = \frac{11}{3}$$

$$\therefore \frac{X(z)}{z} = \frac{1}{3}\frac{1}{z} - \frac{1}{z + 1} + \frac{11}{3}\frac{1}{z + 3}$$

$$\therefore X(z) = \frac{1}{3} - \frac{z}{z + 1} + \frac{11}{3}\frac{z}{z + 3}$$

$$= \frac{1}{3} - \frac{z}{z - (-1)} + \frac{11}{3}\frac{z}{z - (-3)}$$

$z\{\delta(n)\} = 1$
$z\{a^n u(n)\} = \dfrac{z}{z - a}$

On taking inverse $\mathbb{Z}$-transform of X(z) we get,

$$x(n) = \mathbb{Z}^{-1}\{X(z)\} = \frac{1}{3}\delta(n) - (-1)^n u(n) + \frac{11}{3}(-3)^n u(n) = \frac{1}{3}\delta(n) + \left[-(-1)^n + \frac{11}{3}(-3)^n\right]u(n)$$

When n = 0, $x(0) = \frac{1}{3} - 1 + \frac{11}{3} = 3$

When n = 1, $x(1) = 0 + 1 + \frac{11}{3} \times -3 = -10$

When n = 2, $x(2) = 0 - 1 + \frac{11}{3} \times (-3)^2 = 32$

When n = 3, $x(3) = 0 + 1 + \frac{11}{3} \times (-3)^3 = -98$

When n = 4, $x(4) = 0 - 1 + \frac{11}{3} \times (-3)^4 = 296$

Note: The closed form expression of x(n) in the two methods look different, but on evaluating x(n) for various values of n we get same signal x(n).

$\therefore$ x(n) = {3, –10, 32, –98, 296, ...}

b) Given that, $X(z) = \dfrac{z - 0.6}{z^2 + z + 2}$

$$X(z) = \frac{z - 0.6}{z^2 + z + 2} = \frac{z - 0.6}{(z + 0.5 - j1.323)(z + 0.5 + j1.323)}$$

By partial fraction expansion we get,

$$X(z) = \frac{A}{(z + 0.5 - j1.323)} + \frac{A^*}{(z + 0.5 + j1.323)}$$

The roots of quadratic $z^2 + z + 2 = 0$ are,
$$z = \frac{-1 \pm \sqrt{1 - 4 \times 2}}{2} = \frac{-1 \pm j\sqrt{7}}{2} = -0.5 \pm j1.323$$

$$A = (z + 0.5 - j1.323)\frac{z - 0.6}{(z + 0.5 - j1.323)(z + 0.5 + j1.323)}\bigg|_{z=-0.5+j1.323}$$

$$= \frac{-0.5 + j1.323 - 0.6}{-0.5 + j1.323 + 0.5 + j1.323}$$

$$= \frac{-1.1 + j1.323}{j2.646} = \frac{-1.1}{j2.646} + \frac{j1.323}{j2.646}$$

$$= 0.5 + j0.416$$

$\therefore$ $A^* = (0.5 + j0.416)^* = 0.5 - j0.416$

$\therefore$ $X(z) = \dfrac{0.5 + j0.416}{z + 0.5 - j1.323} + \dfrac{0.5 - j0.416}{z + 0.5 + j1.323}$ $\boxed{z^{-1}z = 1.}$

$$= (0.5 + j0.416)z^{-1}\frac{z}{z - (-0.5 + j1.323)} + (0.5 - j0.416)z^{-1}\frac{z}{z - (-0.5 - j1.323)}$$

On taking inverse $\mathbb{Z}$-transform of X(z) we get,

$$x(n) = \mathbb{Z}^{-1}\{X(z)\} = (0.5 + j0.416)(-0.5 + j1.323)^{(n-1)}u(n-1)$$
$$+ (0.5 - j0.416)(-0.5 - j1.323)^{(n-1)}u(n-1)$$

If $\mathbb{Z}\{a^n u(n)\} = \dfrac{z}{z - a}$
then by time shifting property,
$\mathbb{Z}\{a^{(n-1)} u(n - 1)\} = z^{-1}\dfrac{z}{z - a}$

Alternatively the above result can be expressed as shown below:

$$0.5 + j0.416 = 0.5 + j0.416 = 0.65 \ \angle 39.7° = 0.65 \ \angle 0.22\pi$$

$$0.5 - j0.416 = 0.5 - j0.416 = 0.65 \ \angle -39.7° = 0.65 \ \angle -0.22\pi$$

$$-0.5 + j1.323 = 1.414 \ \angle 110.7° = 1.414 \ \angle 0.61\pi$$

$$-0.5 - j1.323 = 1.414 \ \angle -110.7° = 1.414 \ \angle -0.61\pi$$

$180° = \pi$ rad
$\therefore \ 1° = \dfrac{\pi}{180}$ rad
$\therefore \ 39.7° = \dfrac{39.7}{180} \pi = 0.22\pi$ rad
$110.7° = \dfrac{110.7}{180} \pi = 0.61\pi$ rad

Now, x(n) for n ≥ 1 can be written as shown below:

$$\therefore \ x(n) = [0.65 \ \angle 0.22\pi] [1.414 \ \angle 0.61\pi]^{(n-1)} + [0.65 \ \angle -0.22\pi] [1.414 \ \angle -0.61\pi]^{(n-1)}$$

$$= [0.65 \ \angle 0.22\pi] [1.414^{(n-1)} \angle 0.61\pi \ (n-1)]$$

$$\qquad + [0.65 \ \angle -0.22\pi] [1.414^{(n-1)} \ \angle -0.61\pi \ (n-1)]$$

$$= 0.65 \ (1.414)^{(n-1)} \ \angle (0.22\pi + 0.61\pi n - 0.61\pi)$$

$$\qquad + 0.65 \ (1.414)^{(n-1)} \ \angle (-0.22\pi - 0.61\pi n + 0.61\pi)$$

$$= 0.65 \ (1.414)^{(n-1)} \ [1 \ \angle (0.61n - 0.39)\pi + 1 \ \angle -(0.61n - 0.39)\pi]$$

$$= 0.65 \ (1.414)^{(n-1)} \ [\cos ((0.61n - 0.39)\pi) + j \ \sin((0.61n - 0.39)\pi) + \cos((0.61n - 0.39)\pi)$$

$$\qquad -j \ \sin((0.61n - 0.39)\pi) \ ]$$

$$= 0.65 \ (1.414)^{(n-1)} \ 2 \cos ((0.61n - 0.39)\pi) \ ; \text{ for } n \geq 1$$

$$= 1.3 \ (1.414)^{(n-1)} \cos ((0.61n - 0.39)\pi) \ u(n - 1)$$

c) Given that, $X(z) = \dfrac{2z - 4}{(z - 1)(z + 2)^2}$

By partial fraction expansion we get,

$$X(z) = \frac{2z - 4}{(z - 1)(z + 2)^2} = \frac{A_1}{z - 1} + \frac{A_2}{(z + 2)^2} + \frac{A_3}{(z + 2)}$$

$$A_1 = (z - 1) \frac{2z - 4}{(z - 1)(z + 2)^2} \bigg|_{z=1} = \frac{2z - 4}{(z + 2)^2} \bigg|_{z=1} = \frac{2 - 4}{(1 + 2)^2} = \frac{-2}{9} = -0.22$$

$$A_2 = (z + 2)^2 \frac{2z - 4}{(z - 1)(z + 2)^2} \bigg|_{z=-2} = \frac{2z - 4}{z - 1} \bigg|_{z=-2} = \frac{2 \times -2 - 4}{-2 - 1} = \frac{-8}{-3} = 2.67$$

$$A_3 = \frac{d}{dz} \left[(z + 2)^2 \frac{2z - 4}{(z - 1)(z + 2)^2} \right] \bigg|_{z=-2} = \frac{d}{dz} \left[\frac{2z - 4}{z - 1} \right] \bigg|_{z=-2}$$

$$= \frac{2(z - 1) - (2z - 4)}{(z - 1)^2} \bigg|_{z=-2} = \frac{2(-2 - 1) - (2 \times -2 - 4)}{(-2 - 1)^2} = \frac{2}{9} = 0.22$$

$d\dfrac{u}{v} = \dfrac{v \, du - u \, dv}{v^2}$
Multiply and divide by z.

$$\therefore \ X(z) = \frac{-0.22}{z - 1} + \frac{2.67}{(z + 2)^2} + \frac{0.22}{z + 2}$$

$$= -0.22 \, z^{-1} \frac{z}{z - 1} + \frac{2.67}{-2} z^{-1} \frac{-2z}{(z - (-2))^2} + 0.22 \, z^{-1} \frac{z}{z - (-2)}$$

$z^{-1}z = 1.$

$$\mathcal{Z}\{u(n)\} = \frac{z}{z - 1} \ ; \ \mathcal{Z}\{a^n u(n)\} = \frac{z}{z - a} \ ; \ \mathcal{Z}\{na^n u(n)\} = \frac{z}{(z - a)^2}$$

If $\mathcal{Z}\{x(n)\} = X(z)$ then by time shifting property $\mathcal{Z}\{x(n - 1)\} = z^{-1} X(z)$

$$\therefore \mathcal{Z}\{u(n - 1)\} = z^{-1} \frac{z}{z - 1} \ ; \ \mathcal{Z}\{a^{(n-1)} u(n - 1)\} = z^{-1} \frac{z}{z - a}$$

$$\text{and} \ \mathcal{Z}\{(n - 1)a^{(n-1)} u(n - 1)\} = z^{-1} \frac{az}{(z - a)^2}$$

On taking inverse z-transform of $X(z)$ using standard transform and shifting property we get,

$$x(n) = z^{-1}\{X(z)\} = -0.22\, u(n-1) - 1.335\,(n-1)\,(-2)^{n-1}\,u(n-1) + 0.22\,(-2)^{n-1}\,u(n-1)$$

$$= [\,-0.22 + [-0.335\,(n-1) + 0.22]\,(-2)^{n-1}\,]\,u(n-1)$$

Example 4.34

Determine the inverse z-transform of the following function.

a) $X(z) = \dfrac{1}{1 + 4.5\,z^{-1} + 3.5\,z^{-2}}$ b) $X(z) = \dfrac{z^2}{z^2 - z + 0.5}$

c) $X(z) = \dfrac{2}{(1 + z^{-1})(1 - z^{-1})^2}$ d) $X(z) = \dfrac{1}{z^2 - 1.2z + 0.2}$

Solution:

(AU Jun'13, 8 Marks)

a) Given that, $X(z) = \dfrac{1}{1 + 4.5\,z^{-1} + 3.5\,z^{-2}}$

> The roots of quadratic
> $z^2 - 4.5z + 3.5 = 0$ are,
> $$z = \frac{-4.5 \pm \sqrt{4.5^2 - 4 \times 3.5}}{2}$$
> $$= \frac{-4.5 \pm 2.5}{2} = -1, -3.5$$

$$X(z) = \frac{1}{1 + 4.5\,z^{-1} + 3.5\,z^{-2}} = \frac{1}{1 + \dfrac{4.5}{z} + \dfrac{3.5}{z^2}}$$

$$= \frac{z^2}{z^2 + 4.5z + 3.5} = \frac{z^2}{(z+1)(z+3.5)}$$

$$\therefore \frac{X(z)}{z} = \frac{z}{(z+1)(z+3.5)}$$

By partial fraction expansion, $X(z)/z$ can be expressed as,

$$\frac{X(z)}{z} = \frac{A_1}{z+1} + \frac{A_2}{z+3.5}$$

$$A_1 = (z+1)\frac{X(z)}{z}\Big|_{z=-1} = (z+1)\frac{z}{(z+1)(z+3.5)}\Big|_{z=-1} = \frac{-1}{-1+3.5} = -0.4$$

$$A_2 = (z+3.5)\frac{X(z)}{z}\Big|_{z=-3.5} = (z+3.5)\frac{z}{(z+1)(z+3.5)}\Big|_{z=-3.5} = \frac{-3.5}{-3.5+1} = 1.4$$

$$\therefore \frac{X(z)}{z} = \frac{-0.4}{z+1} + \frac{1.4}{z+3.5}$$

> $z\{a^n u(n)\} = \dfrac{z}{z-a}$; $\text{ROC}\,|z| > |a|$

$$\therefore X(z) = \frac{-0.4z}{z+1} + \frac{1.4z}{z+3.5} = \frac{-0.4z}{z-(-1)} + \frac{1.4z}{z-(-3.5)}$$

On taking inverse z-transform of $X(z)$, we get,

$$x(n) = z^{-1}\{X(z)\} = -0.4(-1)^n\,u(n) + 1.4(-3.5)^n\,u(n) = [-0.4(-1)^n + 1.4(-3.5)^n]\,u(n)$$

b) Given that, $X(z) = \dfrac{z^2}{z^2 - z + 0.5}$

$$X(z) = \frac{z^2}{z^2 - z + 0.5} = \frac{z^2}{(z - 0.5 - j0.5)(z - 0.5 + j0.5)}$$

$$\therefore \frac{X(z)}{z} = \frac{z}{(z - 0.5 - j0.5)(z - 0.5 + j0.5)}$$

> The roots of quadratic
> $z^2 - z + 0.5 = 0$ are,
> $$z = \frac{1 \pm \sqrt{1 - 4 \times 0.5}}{2}$$
> $$= 0.5 \pm j0.5$$

By partial fraction expansion, we can write,

$$\frac{X(z)}{z} = \frac{A}{z - 0.5 - j0.5} + \frac{A^*}{z - 0.5 + j0.5}$$

$$A = (z - 0.5 - j0.5) \frac{X(z)}{z}\bigg|_{z = 0.5 + j0.5}$$

$$= (\cancel{z - 0.5 - j0.5}) \frac{z}{(\cancel{z - 0.5 - j0.5})(z - 0.5 + j0.5)}\bigg|_{z = 0.5 + j0.5}$$

$$= \frac{0.5 + j0.5}{0.5 + j0.5 - 0.5 + j0.5} = \frac{0.5 + j0.5}{j1.0} = -j(j0.5 + 0.5) = 0.5 - j0.5$$

$$\therefore A^* = (0.5 - j0.5)^* = 0.5 + j0.5$$

$$\therefore \frac{X(z)}{z} = \frac{0.5 - j0.5}{z - 0.5 - j0.5} + \frac{0.5 + j0.5}{z - 0.5 + j0.5}$$

$$X(z) = \frac{(0.5 - j0.5)z}{z - (0.5 + j0.5)} + \frac{(0.5 + j0.5)z}{z - (0.5 - j0.5)}$$

On taking inverse z-transform of $X(z)$ we get,

$$\boxed{\begin{array}{l} Z\{a^n u(n)\} = \dfrac{z}{z-a} \; ; \\ ROC: |z| > |a| \end{array}}$$

$$x(n) = Z^{-1}\{X(z)\} = (0.5 - j0.5)(0.5 + j0.5)^n u(n) + (0.5 + j0.5)(0.5 - j0.5)^n u(n)$$

Alternatively the above result can be expressed as shown below:

Here, $0.5 + j0.5 = 0.707\angle 45° = 0.707\angle 0.25\pi$

$0.5 - j0.5 = 0.707\angle -45° = 0.707\angle -0.25\pi$

$$\boxed{\begin{array}{l} 180° = \pi \text{ rad} \; ; \; \therefore 1° = \dfrac{\pi}{180} \text{ rad} \\ \therefore 45° = \dfrac{45}{180}\pi = 0.25\pi \text{ rad} \end{array}}$$

$$\therefore x(n) = [0.707\angle -0.25\pi][0.707 \angle 0.25\pi]^n u(n) + [0.707\angle 0.25\pi][0.707\angle -0.25\pi]^n u(n)$$

$$= [0.707\angle -0.25\pi][0.707^n \angle 0.25\pi n] u(n) + [0.707\angle 0.25\pi][0.707^n \angle -0.25\pi n] u(n)$$

$$= 0.707^{(n+1)} \angle (0.25\pi (n-1)) u(n) + 0.707^{(n+1)} \angle (-0.25\pi (n-1)) u(n)$$

$$= 0.707^{(n+1)} [1\angle 0.25\pi(n-1) + 1\angle -0.25\pi(n-1)] u(n)$$

$$= 0.707^{(n+1)} [\cos(0.25\pi(n-1)) + j\,\cancel{\sin(0.25\pi(n-1))} + \cos(0.25\pi(n-1)) - j\,\cancel{\sin(0.25\pi(n-1))}] u(n)$$

$$= 0.707^{(n+1)} \, 2\cos(0.25\pi (n-1)) u(n)$$

---

c) Given that, $X(z) = \dfrac{2}{(1 + z^{-1})(1 - z^{-1})^2}$

$$X(z) = \frac{2}{(1 + z^{-1})(1 - z^{-1})^2} = \frac{2}{z^{-1}(z + 1)z^{-2}(z - 1)^2} = \frac{2z^3}{(z + 1)(z - 1)^2}$$

$$\therefore \frac{X(z)}{z} = \frac{2z^2}{(z + 1)(z - 1)^2}$$

By partial fraction expansion, we can write,

$$\frac{X(z)}{z} = \frac{2z^2}{(z + 1)(z - 1)^2} = \frac{A_1}{z + 1} + \frac{A_2}{(z - 1)^2} + \frac{A_3}{z - 1}$$

$$A_1 = (z+1) \left. \frac{X(z)}{z} \right|_{z=-1} = \cancel{(z+1)} \left. \frac{2z^2}{\cancel{(z+1)}(z-1)^2} \right|_{z=-1} = \left. \frac{2z^2}{(z-1)^2} \right|_{z=-1} = \frac{2(-1)^2}{(-1-1)^2} = \frac{2}{4} = 0.5$$

$$A_2 = (z-1)^2 \left. \frac{X(z)}{z} \right|_{z=1} = \cancel{(z-1)^2} \left. \frac{2z^2}{(z+1)\cancel{(z-1)^2}} \right|_{z=1} = \left. \frac{2z^2}{z+1} \right|_{z=1} = \frac{2}{1+1} = 1$$

$$A_3 = \frac{d}{dz} \left[(z-1)^2 \frac{X(z)}{z} \right]_{z=1} = \frac{d}{dz} \left[\cancel{(z-1)^2} \frac{2z^2}{(z+1)\cancel{(z-1)^2}} \right]_{z=1}$$

$$\boxed{d\frac{u}{v} = \frac{v\,du - u\,dv}{v^2}}$$

$$= \frac{d}{dz} \left[\frac{2z^2}{z+1} \right]_{z=1} = \left. \frac{(z+1)\,4z - 2z^2}{(z+1)^2} \right|_{z=1} = \frac{(1+1)\times 4 - 2}{(1+1)^2} = \frac{6}{4} = 1.5$$

$$\therefore \frac{X(z)}{z} = \frac{0.5}{z+1} + \frac{1}{(z-1)^2} + \frac{1.5}{z-1}$$

$$\therefore X(z) = 0.5 \frac{z}{z-(-1)} + \frac{z}{(z-1)^2} + 1.5 \frac{z}{z-1}$$

On taking inverse z-transform of X(z) we get,

$$\boxed{\begin{array}{l} z\{a^n u(n)\} = \dfrac{z}{z-a} \\[2mm] z\{n\,u(n)\} = \dfrac{z}{(z-1)^2} \\[2mm] z\{u(n)\} = \dfrac{z}{z-1} \end{array}}$$

$$x(n) = z^{-1}\{X(z)\} = 0.5\,(-1)^n\,u(n) + n\,u(n) + 1.5\,u(n)$$

$$= [0.5\,(-1)^n + n + 1.5]\,u(n)$$

---

d) Given that, $X(z) = \dfrac{1}{z^2 - 1.2z + 0.2}$

(AU Jun'12, 8 Marks)

$$\therefore X(z) = \frac{1}{(z-1)(z-0.2)}$$

By partial fraction expansion, X(z) can be expressed as,

$$X(z) = \frac{A_1}{z-1} + \frac{A_2}{z-0.2}$$

$$A_1 = (z-1)\,X(z)\big|_{z=1} = \cancel{(z-1)} \left. \frac{1}{\cancel{(z-1)}(z-0.2)} \right|_{z=1} = \frac{1}{1-0.2} = 1.25$$

$$A_2 = (z-0.2)\,X(z)\big|_{z=0.2} = \cancel{(z-0.2)} \left. \frac{1}{(z-1)\cancel{(z-0.2)}} \right|_{z=0.2} = \frac{1}{0.2-1} = -1.25$$

$$\therefore X(z) = \frac{1.25}{z-1} - \frac{1.25}{z-0.2}$$

$$= 1.25 \times z^{-1} \frac{z}{z-1} - 1.25 \times z^{-1} \frac{z}{z-0.2}$$

$$\boxed{z^{-1}z = 1.}\quad \boxed{z\{a^n u(n)\} = \frac{z}{z-a}}$$

On taking inverse z-transform we get,

$$x(n) = z^{-1}\{X(z)\} = \left[1.25u(n) - 1.25(0.2)^n u(n) \right]\Big|_{n=n-1}$$

$$\boxed{\begin{array}{l} \text{if} \quad z\{x(n)\} = X(z) \\ \text{then } z\{x(n-1)\} = z^{-1}X(z) \end{array}}$$

The roots of quadratic
$z^2 - 1.2z + 0.2 = 0$ are
$$z = \frac{1.2 \pm \sqrt{1.2^2 - 4 \times 0.2}}{2}$$
$$= \frac{1.2 \pm 0.8}{2} = 1,\,0.2$$

$$= 1.25\,u(n-1) - 1.25\,(0.2)^{n-1}\,u(n-1)$$

Example 4.32

Determine the inverse z-transform of $X(z) = \dfrac{1}{1 - 4.5z^{-1} + 3.5z^{-2}}$ by long division method.

(a) if ROC : $|z| > 3.5$ **(b)** if ROC : $|z| < 1.0$

Solution:

Given that, $X(z) = \dfrac{1}{1 - 4.5z^{-1} + 3.5z^{-2}}$

$$= \frac{1}{z^{-2}(z^{-2} - 4.5z + 3.5)} = \frac{z^2}{(z - 3.5)(z - 1)}$$

The poles of $X(z)$ are, $z = 3.5$ and $z = 1.0$.

The roots of quadratic
$z^2 - 4.5z + 3.5 = 0$ are,
$z = \dfrac{4.5 \pm \sqrt{4.5^2 - 4 \times 3.5}}{2}$
$= \dfrac{4.5 \pm 2.5}{2} = 3.5, 1$

a) When ROC is $|z| > 3.5$

In this case, the ROC is exterior of circle whose radius corresponds to largest pole. Hence x(n) will be a causal signal. (Refer section 4.5.2).

Let us express $X(z)$ as a power series expansion in negative powers of z, by dividing the numerator of $X(z)$ by its denominator as shown below:

$$
\begin{array}{r}
1 + 4.5\,z^{-1} + 16.75\,z^{-2} + 59.625\,z^{-3} + 209.6875\,z^{-4} + \ldots\ldots\ldots
\end{array}
$$

$1 - 4.5\,z^{-1} + 3.5\,z^{-2}$) 1

$\qquad\qquad\qquad\qquad 1 - 4.5\,z^{-1} + 3.5\,z^{-2}$

$\qquad\qquad\qquad (-)\ (+)\qquad\ (-)$

$\qquad\qquad\qquad\qquad 4.5\,z^{-1} - 3.5\,z^{-2}$

$\qquad\qquad\qquad\qquad 4.5\,z^{-1} - 20.25\,z^{-2} + 15.75\,z^{-3}$

$\qquad\qquad\qquad\quad (-)\qquad\quad (+)\qquad\qquad (-)$

$\qquad\qquad\qquad\qquad\qquad 16.75\,z^{-2} - 15.75\,z^{-3}$

$\qquad\qquad\qquad\qquad\qquad 16.75\,z^{-2} - 75.375\,z^{-3} + 58.625\,z^{-4}$

$\qquad\qquad\qquad\qquad (-)\qquad\qquad (+)\qquad\qquad\quad (-)$

$\qquad\qquad\qquad\qquad\qquad\qquad 59.625\,z^{-3} - 58.625\,z^{-4}$

$\qquad\qquad\qquad\qquad\qquad\qquad 59.625\,z^{-3} - 268.3125\,z^{-4} + 208.6875\,z^{-5}$

$\qquad\qquad\qquad\qquad\qquad (-)\qquad\qquad (+)\qquad\qquad\qquad (-)$

$\qquad\qquad\qquad\qquad\qquad\qquad\qquad 209.6875\,z^{-4} + 208.6875\,z^{-5}$

$\qquad\qquad\qquad\qquad\qquad\qquad\qquad\qquad\qquad \vdots$

$\therefore\ X(z) = \dfrac{1}{1 - 4.5z^{-1} + 3.5z^{-2}}$

$$= 1 + 4.5\,z^{-1} + 16.75\,z^{-2} + 59.625\,z^{-3} + 209.6875\,z^{-4} + \ldots \qquad\qquad \ldots\ldots(1)$$

If $X(z)$ is z-transform of $x(n)$ then, by the definition of z-transform we get,

$$X(z) = z\{x(n)\} = \sum_{n=-\infty}^{\infty} x(n)\,z^{-n}$$

For a causal signal,

$$X(z) = \sum_{n=0}^{\infty} x(n)\,z^{-n}$$

On expanding the summation we get,

$$X(z) = x(0)\,z^0 + x(1)\,z^{-1} + x(2)\,z^{-2} + x(3)\,z^{-3} + x(4)\,z^{-4} + \ldots\ldots \qquad\qquad \ldots\ldots(2)$$

On comparing the two power series of X(z) [equations (1) and (2)], we get,

$x(0) = 1$; $x(1) = 4.5$; $x(2) = 16.75$; $x(3) = 59.625$; $x(4) = 209.6875$; ....

$x(n) = \{1,\ 4.5,\ 16.75,\ 59.625,\ 209.6875,\ ....\}$

↑

b) When ROC is |z| < 1.0

In this case, the ROC is interior of circle whose radius corresponds to smallest pole. Hence x(n) will be an anticausal signal. (Refer section 4.5.2).

Let us express X(z) as a power series expansion in positive powers of z. Therefore, rewrite the denominator polynomial of X(z) in the reverse order and then the numerator, is divided by the denominator as shown below.

$$
\begin{array}{r}
0.286\,z^2 + 0.368\,z^3 + 0.391\,z^4 + 0.398\,z^5 + 1.4\,z^6 + \;.....
\end{array}
$$

$3.5\,z^{-2} - 4.5\,z^{-1} + 1$)

$\quad 1$

$\quad 1 - 1.287\,z + 0.286\,z^2$

$\quad (-)\ (+)\qquad (-)$

$\qquad\quad 1.287\,z - 0.286\,z^2$

$\qquad\quad 1.287\,z - 1.656\,z^2 + 0.368\,z^3$

$\qquad (-)\qquad (+)\qquad (-)$

$\qquad\qquad\quad 1.37\,z^2 - 0.368\,z^3$

$\qquad\qquad\quad 1.37\,z^2 - 1.76\,z^3 + 0.391\,z^4$

$\qquad\qquad (-)\qquad (+)\qquad (-)$

$\qquad\qquad\qquad\quad 1.392\,z^3 - 0.391\,z^4$

$\qquad\qquad\qquad\quad 1.392\,z^3 - 1.791\,z^4 + 0.398\,z^5$

$\qquad\qquad\qquad (-)\qquad (+)\qquad (-)$

$\qquad\qquad\qquad\qquad\quad 1.4\,z^4 - 0.398\,z^5$

$\qquad\qquad\qquad\qquad\qquad\qquad \vdots$

$$\therefore\ X(z) = \frac{1}{1 - 4.5\,z^{-1} + 3.5\,z^{-2}} = \frac{1}{3.5\,z^{-2} - 4.5\,z^{-1} + 1}$$

$$= 0.286\,z^2 + 0.368\,z^3 + 0.391\,z^4 + 0.398\,z^5 + 1.4\,z^6 + \;..... \qquad .....(3)$$

If X(z) is z-transform of x(n) then, by the definition of z-transform we get,

$$X(z) = z\{x(n)\} = \sum_{n=-\infty}^{\infty} x(n)\,z^{-n}$$

For an anticausal signal,

$$X(z) = \sum_{n=-\infty}^{0} x(n)\,z^{-n}$$

On expanding the summation we get,

$$X(z) = ... \; x(-6)\,z^6 + x(-5)\,z^5 + x(-4)\,z^4 + x(-3)\,z^3 + x(-2)\,z^2 + x(-1)\,z + x(0) \qquad .....(4)$$

On comparing the two power series of X(z) [equations (3) and (4)], we get,

$x(0) = 0$; $x(-1) = 0$; $x(-2) = 0.286$; $x(-3) = 0.368$; $x(-4) = 0.391$;

$x(-5) = 0.398$; $x(-6) = 1.4$; ....

$x(n) = \{........,\ 1.4,\ 0.398,\ 0.391,\ 0.368,\ 0.286,\ 0,\ 0\}$

↑

Example 4.33   *(AU Jun'14, 16 Marks)*

Find the inverse z-transform of, $X(z) = \dfrac{z^2}{(z-0.5)(z-1)^2}$ $|z| > 1$.

Solution:

Given that, $X(z) = \dfrac{z^2}{(z-0.5)(z-1)^2}$

$$\therefore \frac{X(z)}{z} = \frac{z}{(z-0.5)(z-1)^2}$$

By partial fraction expansion technique we get,

$$\frac{X(z)}{z} = \frac{K_1}{(z-0.5)} + \frac{K_2}{(z-1)^2} + \frac{K_3}{(z-1)}$$

$$K_1 = (z-0.5)\frac{X(z)}{z}\bigg|_{z=0.5} = (\cancel{z-0.5})\frac{z}{(\cancel{z-0.5})(z-1)^2}\bigg|_{z=0.5} = \frac{0.5}{(0.5-1)^2} = 2$$

$$K_2 = (z-1)\frac{X(z)}{z}\bigg|_{z=1} = (\cancel{z-1})^2\frac{z}{(z-0.5)\cancel{(z-1)^2}}\bigg|_{z=1} = \frac{1}{1-0.5} = 2$$

$$K_3 = \frac{d}{dz}(z-1)^2\frac{X(z)}{z}\bigg|_{z=1} = \frac{d}{dz}\left[\cancel{(z-1)^2}\frac{z}{(z-0.5)\cancel{(z-1)^2}}\right]\bigg|_{z=1} = \frac{(z-0.5)-z}{(z-0.5)^2}\bigg|_{z=1}$$

$$= \frac{z-0.5-z}{(z-0.5)^2}\bigg|_{z=1} = \frac{-0.5}{(1-0.5)^2} = -2$$

$$\therefore \frac{X(z)}{z} = \frac{2}{z-0.5} + \frac{2}{(z-1)^2} - \frac{2}{z-1}$$

$$X(z) = \frac{2z}{z-0.5} + \frac{2z}{(z-1)^2} - \frac{2z}{z-1}$$

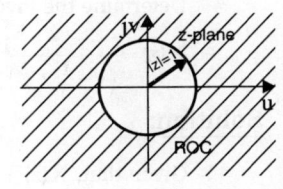

$$Z\{a^n u(n)\} = \frac{z}{z-a}$$

$$Z\{n\, u(n)\} = \frac{z}{(z-1)^2}$$

$$Z\{u(n)\} = \frac{z}{z-1}$$

Given that ROC is $|z| > 1$.

The poles of $X(z)$ are at, $z = 1$ amd $z = 0.5$

The specified ROC is exterior of the circle whose radius corresponds to the largest pole. Hence x(n) will be a causal signal.

Let us take inverse z-transform of $X(z)$ to get x(n) as causal signal.

$$\therefore x(n) = Z^{-1}\{X(z)\} = 2(0.5)^n\, u(n) + 2n\, u(n) - 2\, u(n)$$

$$= [\, 2(0.5)^n + 2n - 2\,]\, u(n)$$

Example 4.34   *(AU Dec'13, 8 Marks)*

Determine the inverse z - transform of, $X(z) = \dfrac{1+z^{-1}}{\left(1-\frac{2}{3}z^{-1}\right)^2}$ if ROC is $|z| > \frac{2}{3}$

Solution:

Given that, $X(z) = \dfrac{1+z^{-1}}{\left(1-\frac{2}{3}z^{-1}\right)^2} = \dfrac{1+z^{-1}}{z^{-2}\left(z-\frac{2}{3}\right)^2}$

$$= \frac{z^2\,(1+z^{-1})}{\left(z-\frac{2}{3}\right)^2} = \frac{z^2+z}{\left(z-\frac{2}{3}\right)^2} = \frac{z(z+1)}{\left(z-\frac{2}{3}\right)^2}$$

$$\therefore \frac{X(z)}{z} = \frac{z+1}{\left(z - \frac{2}{3}\right)^2}$$

By partial fraction expansion we can write,

$$\frac{X(z)}{z} = \frac{A_1}{\left(z - \frac{2}{3}\right)^2} + \frac{A_2}{\left(z - \frac{2}{3}\right)}$$

$$A_1 = \left(z - \frac{2}{3}\right)^2 \frac{X(z)}{z}\Bigg|_{z=\frac{2}{3}} = \left(z - \frac{2}{3}\right)^2 \frac{(z+1)}{\left(z - \frac{2}{3}\right)^2}\Bigg|_{z=\frac{2}{3}} = \frac{2}{3} + 1 = \frac{5}{3}$$

$$A_2 = \frac{d}{dz}\left[\left(z - \frac{2}{3}\right)^2 \frac{X(z)}{z}\right]\Bigg|_{z=\frac{2}{3}} = \frac{d}{dz}\left[\left(z - \frac{2}{3}\right)^2 \frac{z+1}{\left(z - \frac{2}{3}\right)^2}\right]\Bigg|_{z=\frac{2}{3}} = \frac{d}{dz}(z+1) = 1$$

$$\therefore \frac{X(z)}{z} = \frac{\frac{5}{3}}{\left(z - \frac{2}{3}\right)^2} + \frac{1}{\left(z - \frac{2}{3}\right)}$$

$$X(z) = \frac{5}{3}\frac{z}{\left(z - \frac{2}{3}\right)^2} + \frac{z}{\left(z - \frac{2}{3}\right)}$$

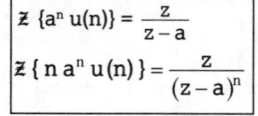

Given that, ROC is $|z| > \frac{2}{3}$. Therefore the x(n) will be causal.

Let us take inverse $\mathcal{Z}$-transform of X(z) to get x(n) as causal signal.

$$\therefore x(n) = \mathcal{Z}^{-1}\{X(z)\} = \frac{5}{3}n\left(\frac{2}{3}\right)^n u(n) + \left(\frac{2}{3}\right)^n u(n)$$

$$= \left[\frac{5}{3}n\left(\frac{2}{3}\right)^n + \left(\frac{2}{3}\right)^n\right]u(n)$$

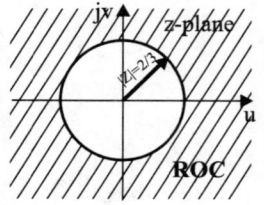

Example 4.35   *(AU Dec'12, 8 Marks)*

Determine the inverse $\mathcal{Z}$-transform of the following z-domain function.

$$X(z) = \frac{1}{\left(1 - \frac{1}{3}z^{-1}\right)\left(1 - \frac{1}{6}z^{-1}\right)}; \quad \text{when } ROC: |z| > \frac{1}{3}$$

Solution:

Given that, $X(z) = \dfrac{1}{\left(1 - \frac{1}{3}z^{-1}\right)\left(1 - \frac{1}{6}z^{-1}\right)} = \dfrac{1}{z^{-1}\left(z - \frac{1}{3}\right)z^{-1}\left(z - \frac{1}{6}\right)} = \dfrac{z^2}{\left(z - \frac{1}{3}\right)\left(z - \frac{1}{6}\right)}$

$$\therefore \frac{X(z)}{z} = \frac{z}{\left(z - \frac{1}{3}\right)\left(z - \frac{1}{6}\right)}$$

By partial fraction expansion, we can write,

$$\frac{X(z)}{z} = \frac{A}{\left(z - \frac{1}{3}\right)} + \frac{B}{\left(z - \frac{1}{6}\right)}$$

$$A = \left(z - \frac{1}{3}\right)\frac{X(z)}{z}\Bigg|_{z=\frac{1}{3}} = \left(z - \frac{1}{3}\right)\frac{z}{\left(z - \frac{1}{3}\right)\left(z - \frac{1}{6}\right)}\Bigg|_{z=\frac{1}{3}} = \frac{\frac{1}{3}}{\frac{1}{3} - \frac{1}{6}} = \frac{\frac{1}{3}}{\frac{1}{6}} = 2$$

$$B = \left(z - \frac{1}{6}\right)\frac{X(z)}{z}\bigg|_{z=\frac{1}{6}} = \left(z - \frac{1}{6}\right)\frac{z}{\left(z-\frac{1}{3}\right)\left(z-\frac{1}{6}\right)}\bigg|_{z=\frac{1}{6}} = \frac{\frac{1}{6}}{\frac{1}{6}-\frac{1}{3}} = \frac{\frac{1}{6}}{\frac{-1}{6}} = -1$$

$$\therefore \frac{X(z)}{z} = \frac{2}{z-\frac{1}{3}} + \frac{-1}{z-\frac{1}{6}}$$

$$\boxed{Z\{a^n u(n)\} = \frac{z}{z-a}}$$

$$X(z) = \frac{2z}{z-\frac{1}{3}} - \frac{z}{z-\frac{1}{6}}$$

Given that, ROC is $|z| > \frac{1}{3}$.

The poles of X(z) are at, $z = \frac{1}{3}$ and $z = \frac{1}{6}$.

The specified ROC is exterior of circle whose radius corresponds to the largest pole. Hence x(n) will be causal.

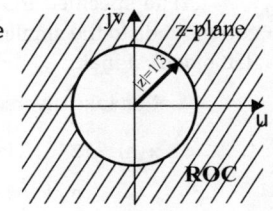

Let us take inverse Z-transform of X(z) to get x(n) as causal.

$$\therefore x(n) = Z^{-1}\{X(z)\} = 2\left(\frac{1}{3}\right)^n u(n) - \left(\frac{1}{6}\right)^n u(n)$$

$$= \left[2\left(\frac{1}{3}\right)^n - \left(\frac{1}{6}\right)^n\right]u(n)$$

Example 4.36   *(AU Dec'14, 16 Marks)*

Determine inverse Z- transform of, $X(z) = \dfrac{z^{-1}}{1-0.25z^{-1}-0.375z^{-2}}$

(a) if ROC is $|z| > 0.75$ **(b)** if ROC is $|z| < 0.5$

$$\boxed{\begin{array}{l}\text{The roots of quadratic}\\ z^2 - 0.25z - 0.375 = 0\\[4pt] z = \dfrac{0.25 \pm \sqrt{0.25^2 + 4 \times 0.375}}{2}\\[6pt] = \dfrac{0.25 \pm 1.25}{2} = 0.75, -0.5\end{array}}$$

Solution:

Given that, $X(z) = \dfrac{z^{-1}}{1-0.25z^{-1}-0.375z^{-2}}$

$$\therefore X(z) = \frac{z^{-1}}{z^{-2}(z^2 - 0.25z - 0.375)} = \frac{z}{z^2 - 0.25z - 0.375} = \frac{z}{(z-0.75)(z+0.5)}$$

$$\therefore \frac{X(z)}{z} = \frac{1}{(z-0.75)(z+0.5)}$$

By partial fraction expansion technique we get,

$$\frac{X(z)}{z} = \frac{1}{(z-0.75)(z+0.5)} = \frac{A_1}{z-0.75} + \frac{A_2}{z+0.5}$$

$$A_1 = (z-0.75)\frac{X(z)}{z}\bigg|_{z=0.75} = (z-0.75)\frac{1}{(z-0.75)(z+0.5)}\bigg|_{z=0.75} = \frac{1}{0.75+0.5} = 0.8$$

$$A_2 = (z+0.5)\frac{X(z)}{z}\bigg|_{z=-0.5} = (z+0.5)\frac{1}{(z-0.75)(z+0.5)}\bigg|_{z=-0.5} = \frac{1}{-0.5-0.75} = -0.8$$

$$\frac{X(z)}{z} = \frac{0.8}{z-0.75} - \frac{0.8}{z+0.5}$$

$$\therefore X(z) = \frac{0.8z}{z-0.75} - \frac{0.8z}{z+0.5}$$

Now, the poles of X(z) are at, z = 0.75 and z = -0.5

a) ROC is $|z| > 0.75$

$$\boxed{Z\{a^n u(n)\} = \frac{z}{z-a} \; ; ROC|z|>|a|}$$

The specified ROC is exterior of the circle whose radius corresponds to the largest pole, hence x(n) will be a causal (right sided) signal.

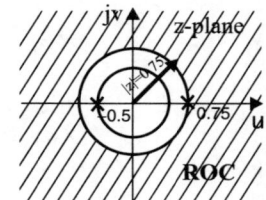

Let us take inverse Z-transform of X(z) to get x(n) as causal.

$$\therefore \; x(n) = Z^{-1}\{X(z)\} = 0.8\,(0.75)^n \, u(n) - 0.8\,(-0.5)^n \, u(n)$$

$$= 0.8\big[(0.75)^n - (0.5)^n\big]\, u(n)$$

b) ROC is $|z| < 0.5$

$$\boxed{Z\{-a^n u(-n-1)\} = \frac{z}{z-a} \; ; ROC:|z|<|a|}$$

The specified ROC is interior of the circle whose radius corresponds to the smallest pole, hence x(n) will be an anticausal (left sided) signal.

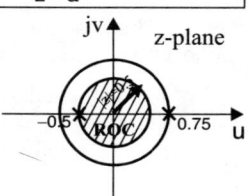

Let us take inverse Z-transform of X(z) to get x(n) as anticausal.

$$\therefore x(n) = Z^{-1}\{X(z)\} = -0.8(0.75)^n\, u(-n-1) + 0.8(-0.5)^n\, u(-n-1)$$

$$= 0.8\big[-(0.75)^n + (-0.5)^n\big]\, u(-n-1)$$

Example 4.37

(AU Jun'11 & Dec 11, 10,8 Marks)

Determine the inverse Z - transform of $X(z) = \dfrac{z^2}{z^2 - 1.5z + 0.5}$

a) if ROC $|z| > 1$ **b)** if ROC $|z| < 0.5$ **c)** if ROC $0.5 < |z| < 1$

Solution:

Given that, $X(z) = \dfrac{z^2}{z^2 - 1.5z + 0.5} = \dfrac{z^2}{(z-1)(z-0.5)}$

$$\therefore \; \frac{X(z)}{z} = \frac{z}{(z-1)(z-0.5)}$$

The roots of quadratic
$z^2-1.5\,z-0.5 = 0$ are
$z = \dfrac{1.5 \pm \sqrt{1.5^2 - 4\times 0.5}}{2}$
$= \dfrac{1.5 \pm 0.5}{2} = 1, 0.5$

By partial fraction expansion technique we get,

$$\frac{X(z)}{z} = \frac{z}{(z-1)(z-0.5)} = \frac{A_1}{z-1} + \frac{A_2}{z-0.5}$$

$$A_1 = (z-1) \frac{X(z)}{z}\Big|_{z=1} = (z-1)\frac{z}{(z-1)(z-0.5)}\Big|_{z=1} = \frac{1}{1-0.5} = 2$$

$$A_2 = (z-0.5) \frac{X(z)}{z}\Big|_{z=0.5} = (z-0.5)\frac{z}{(z-1)(z-0.5)}\Big|_{z=0.5} = \frac{0.5}{0.5-1} = -1$$

$$\therefore \; \frac{X(z)}{z} = \frac{2}{z-1} - \frac{1}{z-0.5}$$

$$\therefore \; X(z) = 2\frac{z}{z-1} - \frac{z}{z-0.5}$$

$\therefore$ The poles of X(z) are at, z = 1 and z = 0.5

a) ROC is $|z| > 1$

The specified ROC is exterior of the circle whose radius corresponds to the largest pole, hence x(n) will be causal signal.

Let us take inverse Z-transform of X(z) to get x(n) as causal.

$$\therefore \; x(n) = Z^{-1}\{X(z)\} = 2\,u(n) - (0.5)^n\, u(n)$$

$$= [2 - 0.5^n]\, u(n)$$

$$\boxed{Z\{a^n u(n)\} = \frac{z}{z-a} \; ; ROC:|z|>|a|}$$

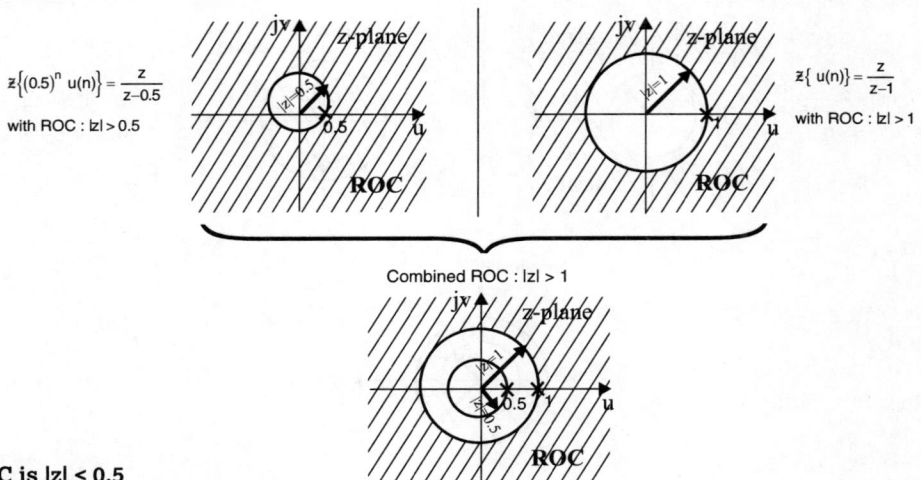

$$\mathcal{Z}\{(0.5)^n\,u(n)\} = \frac{z}{z-0.5}$$

with ROC : |z| > 0.5

$$\mathcal{Z}\{u(n)\} = \frac{z}{z-1}$$

with ROC : |z| > 1

Combined ROC : |z| > 1

b) ROC is |z| < 0.5

The specified ROC is interior of the circle whose radius corresponds to the smallest pole, hence x(n) will be anticausal signal.

Let us take inverse $\mathcal{Z}$-transform of X(z) to get x(n) as anticausal.

$$\therefore \; x(n) = \mathcal{Z}^{-1}\{X(z)\} = -2\,u(-n-1) - (0.5)^n\,u(-n-1)$$

$$\mathcal{Z}\{-a^n u(-n-1)\} = \frac{z}{z-a}\; ;\; \mathrm{ROC}:|z|<|a|$$

$$= -\,[2 + 0.5^n\,]\,u(-n-1)$$

$$\mathcal{Z}\{-0.5^n\,u(-n-1)\} = \frac{z}{z-0.5}$$

with ROC : |z| < 0.5

$$\mathcal{Z}\{-u(-n-1)\} = \frac{z}{z-1}$$

with ROC : |z| < 1

ROC

Combined ROC : |z| < 0.5

c) ROC is 0.5 < |z| < 1

The specified ROC is region between the two circles of radius 0.5 and 1. Hence the term correspond to the pole at z =1 will be anticausal signal (because |z| < 1) and the term corresponds to the pole at z = 0.5 will be causal signal (because |z| > 0.5).

Let us take inverse $\mathcal{Z}$-transform of $\frac{z}{z-1}$ to get anticausal signal and $\frac{z}{z-0.5}$ to get causal signal.

$$\therefore \; x(n) = \mathcal{Z}^{-1}\{X(z)\} = -2\,u(-n-1) - (0.5)^n\,u(n)$$

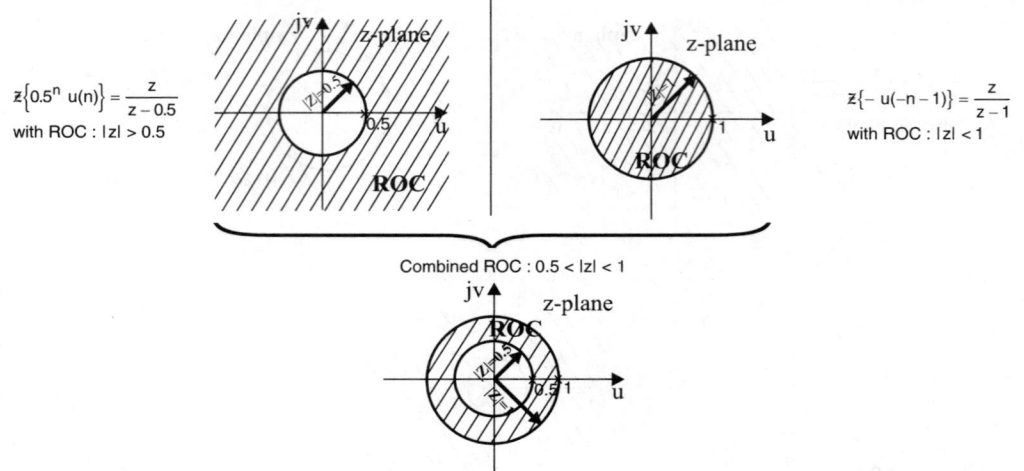

$$\bar{z}\{0.5^n\, u(n)\} = \frac{z}{z - 0.5}$$
with ROC : |z| > 0.5

$$\bar{z}\{-\, u(-n-1)\} = \frac{z}{z - 1}$$
with ROC : |z| < 1

Combined ROC : 0.5 < |z| < 1

Example 4.38

Determine the inverse $\bar{z}$-transform of $X(z) = \dfrac{1}{1 - 0.8\, z^{-1} + 0.12\, z^{-2}}$

a) if ROC is, |z| > 0.6   **b)** if ROC is, |z| < 0.2   **c)** if ROC is, 0.2 < |z| < 0.6

Solution:

Given that, $X(z) = \dfrac{1}{1 - 0.8\, z^{-1} + 0.12\, z^{-2}}$

$$= \frac{1}{z^{-2}(z^2 - 0.8\, z + 0.12)} = \frac{z^2}{(z - 0.6)(z - 0.2)}$$

$$\therefore \frac{X(z)}{z} = \frac{z}{(z - 0.6)(z - 0.2)}$$

> The roots of quadratic $z^2 - 0.8z + 0.12 = 0$ are,
> $$z = \frac{0.8 \pm \sqrt{0.8^2 - 4 \times 0.12}}{2} = \frac{0.8 \pm 0.4}{2} = 0.6, 0.2$$

By partial fraction expansion technique we get,

$$\frac{X(z)}{z} = \frac{z}{(z - 0.6)(z - 0.2)} = \frac{A_1}{z - 0.6} + \frac{A_2}{z - 0.2}$$

$$A_1 = (z - 0.6)\frac{X(z)}{z}\bigg|_{z=0.6} = (z - 0.6)\frac{z}{(z - 0.6)(z - 0.2)}\bigg|_{z=0.6} = \frac{0.6}{0.6 - 0.2} = 1.5$$

$$A_2 = (z - 0.2)\frac{X(z)}{z}\bigg|_{z=0.2} = (z - 0.2)\frac{z}{(z - 0.6)(z - 0.2)}\bigg|_{z=0.2} = \frac{0.2}{0.2 - 0.6} = -0.5$$

$$\therefore \frac{X(z)}{z} = \frac{1.5}{z - 0.6} - \frac{0.5}{z - 0.2}$$

$$\therefore X(z) = 1.5\frac{z}{z - 0.6} - 0.5\frac{z}{z - 0.2}$$

Now, the poles of $X(z)$ are at, $z = 0.6$ and $z = 0.2$

a) ROC is |z| > 0.6

The specified ROC is exterior of the circle whose radius corresponds to the largest pole, hence x(n) will be a causal (or right-sided) signal. (Refer section 4.5.2).

Let us take inverse $\bar{z}$-transform of $X(z)$ to get x(n) as causal.

$$\therefore x(n) = \bar{z}^{-1}\{X(z)\} = 1.5(0.6)^n\, u(n) - 0.5\,(0.2)^n\, u(n)$$

$$= [1.5\,(0.6)^n - (0.2)^n]\, u(n)$$

$$\boxed{\bar{z}\{a^n\, u(n)\} = \frac{z}{z - a}\; ; \text{ROC:}\, |z| > |a|}$$

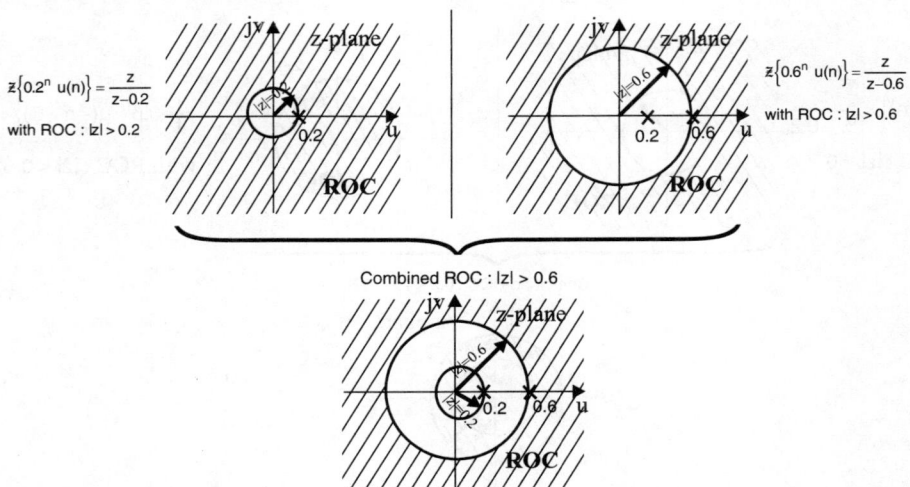

$$\mathcal{Z}\{0.2^n\, u(n)\} = \frac{z}{z-0.2}$$
with ROC : |z| > 0.2

$$\mathcal{Z}\{0.6^n\, u(n)\} = \frac{z}{z-0.6}$$
with ROC : |z| > 0.6

Combined ROC : |z| > 0.6

b) ROC is |z| < 0.2

The specified ROC is interior of the circle whose radius corresponds to the smallest pole, hence x(n) will be an anticausal (or left-sided) signal. (Refer section 4.5.2).

Let us take inverse $\mathcal{Z}$-transform of X(z) to get x(n) as anticausal.

$$\therefore\ x(n) = \mathcal{Z}^{-1}\{X(z)\} = 1.5(-(0.6)^n\, u(-n-1)) - 0.5\,[-(0.2)^n\, u(-n-1)]$$

$$= -[1.5\,(0.6)^n + 0.5\,(0.2)^n]\, u(-n-1)$$

$$\boxed{\begin{array}{l}\mathcal{Z}\{-a^n\, u(-n-1)\} = \dfrac{z}{z-a}\\[4pt] \text{ROC}: |z| < |a|\end{array}}$$

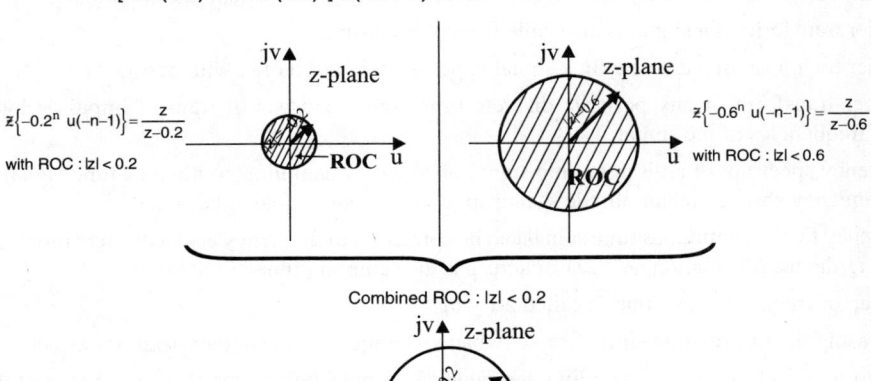

$$\mathcal{Z}\{-0.2^n\, u(-n-1)\} = \frac{z}{z-0.2}$$
with ROC : |z| < 0.2

$$\mathcal{Z}\{-0.6^n\, u(-n-1)\} = \frac{z}{z-0.6}$$
with ROC : |z| < 0.6

Combined ROC : |z| < 0.2

c) ROC is 0.2 < |z| < 0.6

The specified ROC is the region in between two circles of radius 0.2 and 0.6. Hence the term corresponds to the pole, at z = 0.6 will be anticausal signal (because |z| < 0.6) and the term corresponds to the pole, at z = 0.2, will be a causal signal (because |z| > 0.2). (Refer section 4.5.2).

Let us take inverse $\mathcal{Z}$-transform of $\dfrac{z}{z-0.6}$ to get anticausal signal and $\dfrac{z}{z-0.2}$ to get causal signal.

$$\therefore\ x(n) = \mathcal{Z}^{-1}\{X(z)\} = 1.5(-(0.6)^n\, u(-n-1)) - 0.5\,(0.2)^n\, u(n)$$

$$= -1.5(0.6)^n\, u(-n-1)) - 0.5\,(0.2)^n\, u(n)$$

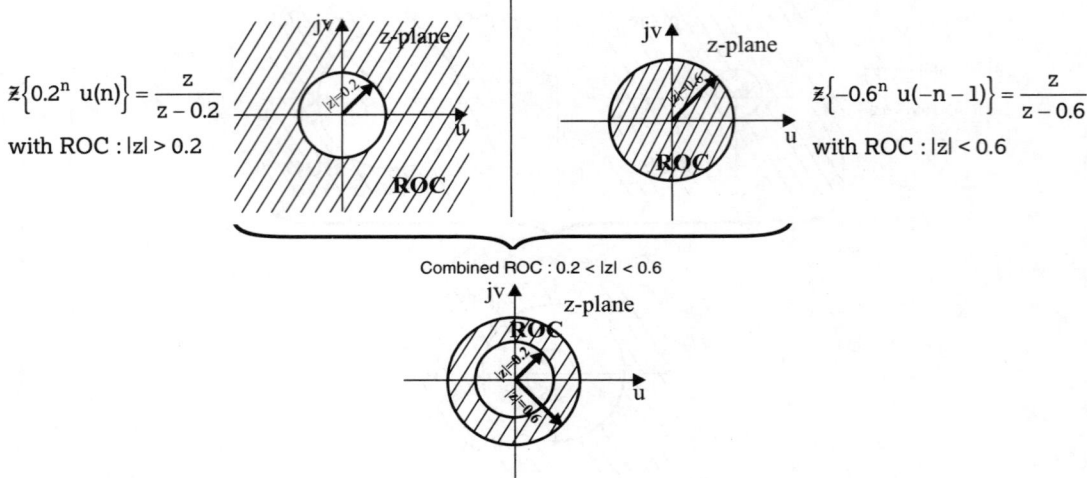

$$\mathbb{Z}\{0.2^n \ u(n)\} = \frac{z}{z - 0.2}$$

with ROC : $|z| > 0.2$

$$\mathbb{Z}\{-0.6^n \ u(-n - 1)\} = \frac{z}{z - 0.6}$$

with ROC : $|z| < 0.6$

Combined ROC : $0.2 < |z| < 0.6$

4.9 Summary of Important Concepts

1. The Fourier transform technique can be applied to both periodic and nonperiodic discrete time signals
2. The Fourier transform is also called analysis of discrete time signal $x(n)$.
3. The inverse Fourier transform is also called synthesis of discrete time signal $x(n)$.
4. The Fourier transform exists only for the discrete time signals that are absolutely summable.
5. The Fourier transform of a signal is also called signal spectrum.
6. The Fourier transform of a discrete time signal is periodic function of ω with period 2π.
7. The Fourier transform of any periodic discrete time signal consists of train of impulses located at harmonic frequencies of the signal.
8. The frequency spectrum of a discrete time signal obtained by sampling continuous time signal will be sum of frequency shifted and amplitude scaled spectrum of continuous time signal.
9. The frequency Ω of a continuous time signal can be converted to frequency ω of a discrete time signal by choosing the transformation, $\omega = \Omega T$, where T is the sampling time.
10. The overlap of frequency spectrum is called aliasing.
11. Due to aliasing the information shifts from one band of frequency to another band of frequency.
12. In order to avoid aliasing, the sampling frequency F_s should be greater than or equal to twice the maximum frequency F_m of continuous time signal.
13. When the spectrum of sampled signal has no aliasing then it is possible to recover the original signal from the sampled signal.
14. The bandpass signals with a bandwidth of B Hz can be sampled at a rate of 2B to 4B Hz.
15. The $\mathbb{Z}$-transform provides a method for analysis of discrete time signals and systems in frequency domain.
16. The ROC of $X(z)$ is a set of all values of z, for which $X(z)$ attains a finite value.
17. Since ROC is a set of values of z, it will be a ring or disk in z-plane, with centre at origin.
18. The zeros are defined as values of z at which the function $X(z)$ becomes zero.
19. The poles are defined as values of z at which the function $X(z)$ becomes infinite.
20. In a realizable system, the number of zeros will be less than or equal to number of poles.

21. If x(n) is finite duration right-sided (causal) signal, then the ROC is entire z-plane except z = 0.

22. If x(n) is finite duration left-sided (anticausal) signal, then the ROC is entire z-plane except z = ∞.

23. If x(n) is finite duration two-sided (noncausal) signal, then the ROC is entire z-plane except z = 0 and z = ∞.

24. If x(n) is infinite duration right-sided (causal) signal, then the ROC is exterior of a circle of radius r_1.

25. If x(n) is infinite duration left-sided (anticausal) signal, then the ROC is interior of a circle of radius r_2.

26. If x(n) is infinite duration two-sided (noncausal) signal, then the ROC is the region in between two circles of radius r_1 and r_2.

27. If X(z) is rational, [where X(z) is $\mathcal{Z}$-transform of x(n)], then the ROC does not include any poles of X(z).

28. If X(z) is rational, [where X(z) is $\mathcal{Z}$-transform of x(n)], and if x(n) is right-sided, then ROC is exterior of a circle whose radius corresponds to pole with largest magnitude.

29. If X(z) is rational, [where X(z) is $\mathcal{Z}$-transform of x(n)], and if x(n) is left-sided, then ROC is interior of a circle whose radius corresponds to pole with smallest magnitude.

30. If X(z) is rational, [where X(z) is $\mathcal{Z}$-transform of x(n)], and if x(n) is two-sided, then ROC is region in between two circles whose radii corresponds to pole of causal part with largest magnitude and pole of anticausal part with smallest magnitude.

31. The inverse $\mathcal{Z}$-transform is the process of recovering the discrete time signal x(n) from its $\mathcal{Z}$-transform X(z).

4.10 Short-answer Questions   *(AU Dec'15, 2 Marks)*

Q4.1   State the need for sampling.

Solution:

The sampling is needed for processing of continous time signal using its sampled version of signal in digital systems.

Q4.2   Determine the Nyquist samping rate for x(t) = sin (200 πt) + 3 sin² (120 πt).

Solution: *(AU May'15, 2 Marks)*

Given that, x(t) = sin (200 πt) + 3 sin² (120 πt)

$$= \sin(200\,\pi t) + 3\left(\frac{1 - \cos(2 \times 120\,\pi t)}{2}\right)$$

$$\boxed{\sin^2 = \frac{1 - \cos 2\theta}{2}}$$

$$= \sin(200\,\pi t) + \frac{3}{2} - \frac{3}{2}\cos(240\,\pi t)$$

The analog signal x(t) can be written as shown below:

$$x(t) = \sin 2\pi F_1 t + \frac{3}{2}\cos 2\pi F_2 t - \frac{3}{2}\cos 2\pi F_3 t$$

where, $2\pi F_1 = 200\,\pi \Rightarrow F_1 = 100$ Hz

$2\pi F_2 = 0 \qquad \Rightarrow F_2 = 0$ Hz

$2\pi F_3 = 240\,\pi \Rightarrow F_3 = 120$ Hz

The maximum analog signal frequency in the signal is 120 Hz. The Nyquist sampling rate is twice that of the maximum analog frequency.

∴ Nyquist sampling rate, $F_s = 2\,F_{max} = 2 \times 120$ Hz = 240 Hz.

Q4.3   *Find the discrete time Fourier transform of the signal,* $x(n) = (0.2)^n u(n) + (0.2)^{-n} u(-n-1)$.

Solution:

Given that, $x(n) = (0.2)^n u(n) + (0.2)^{-n} u(-n-1) = (0.2)^n$; $n \geq 0$

$$= (0.2)^{-n} ; \; n \leq -1$$

By definition of Fourier transform,

$$X(e^{j\omega}) = \mathcal{F}\{x(n)\} = \sum_{n=-\infty}^{\infty} x(n)e^{-j\omega n} = \sum_{n=0}^{\infty} (0.2)^n e^{-j\omega n} + \sum_{n=-\infty}^{-1} (0.2)^{-n} e^{-j\omega n} \qquad \boxed{(0.2e^{j\omega})^0 = 1}$$

$$= \sum_{n=0}^{\infty} (0.2e^{-j\omega})^n + \sum_{n=-\infty}^{-1} (0.2e^{j\omega})^{-n} = \sum_{n=0}^{\infty} (0.2e^{-j\omega})^n + \sum_{n=1}^{\infty} (0.2e^{j\omega})^n + (0.2e^{j\omega})^0 - (0.2e^{j\omega})^0$$

$$= \sum_{n=0}^{\infty} (0.2e^{-j\omega})^n + \sum_{n=0}^{\infty} (0.2e^{j\omega})^n - 1 = \frac{1}{1-0.2e^{-j\omega}} + \frac{1}{1-0.2e^{j\omega}} - 1$$

$$= \frac{1-0.2e^{j\omega} + 1 - 0.2e^{-j\omega} - (1-0.2e^{-j\omega})(1-0.2e^{j\omega})}{(1-0.2e^{-j\omega}) + (1-0.2e^{j\omega})}$$

$$= \frac{1-0.2e^{j\omega} + 1 - 0.2e^{-j\omega} - (1-0.2e^{j\omega} - 0.2e^{-j\omega} + 0.04)}{1-0.2e^{j\omega} - 0.2e^{-j\omega} + 0.04}$$

$$= \frac{1-0.04}{1-0.2(e^{j\omega} + e^{-j\omega}) + 0.04} = \frac{0.96}{1.04 - 0.4\cos\omega}$$

> Using infinite geometric series sum formula
> $$\sum_{n=0}^{\infty} C^n = \frac{1}{1-C}$$
> when $|C| < 1$

> $$\cos\theta = \frac{e^{j\theta} + e^{-j\theta}}{2}$$

Q4.4   *Find the Discrete time Fourier transform of the signal,*   **(AU Dec'14, 2 Marks)**

     $x(n) = \delta(n) + \delta(n-1)$

Solution:

Given that, $x(n) = \delta(n) + \delta(n-1)$

By definition of Fourier transform,

$$X(e^{j\omega}) = \mathcal{F}\{x(n)\} = \sum_{n=-\infty}^{+\infty} x(n)e^{-j\omega n}$$

$$= \sum_{n=-\infty}^{+\infty} [\delta(n) + \delta(n-1)]e^{-j\omega n}$$

$$= \sum_{n=-\infty}^{+\infty} \delta(n)e^{-j\omega n} + \sum_{n=-\infty}^{+\infty} \delta(n-1)e^{-j\omega n}$$

$$= 1 \times e^0 + 1 \times e^{-j\omega}$$

$$= 1 + e^{-j\omega} = 1 + \frac{1}{e^{j\omega}} = \frac{e^{j\omega}+1}{e^{j\omega}}$$

> $\delta(n) = 1$; $n = 0$
> $\quad\quad = 0$; $n \neq 0$
>
> $\delta(n-1) = 1$; $n = 1$
> $\quad\quad\quad\quad = 0$; $n \neq 1$

Q4.5   *Find the inverse Fourier transform of the rectangular pulse spectrum defined as,*

     $X(e^{j\omega}) = 1$; $|\omega| \leq W$

           $= 0$; $W \leq |\omega| \leq \pi$

Solution:

By definition inverse Fourier transform,

$$x(n) = \mathcal{F}^{-1}\{X(e^{j\omega})\} = \frac{1}{2\pi} \int_{-\pi}^{\pi} X(e^{j\omega})e^{j\omega n}\, d\omega = \frac{1}{2\pi} \int_{-W}^{W} e^{j\omega n}\, d\omega$$

$$\therefore \; x(n) = \frac{1}{2\pi}\left[\frac{e^{j\omega n}}{jn}\right]_{-W}^{W} = \frac{1}{2\pi}\left[\frac{e^{jWn}}{jn} - \frac{e^{-jWn}}{jn}\right] = \frac{1}{\pi n}\left[\frac{e^{jWn} - e^{-jWn}}{2j}\right]$$

$$= \frac{\sin Wn}{\pi n} = \frac{W}{\pi}\frac{\sin Wn}{Wn} = \frac{W}{\pi}\text{ sinc } Wn$$

$\sin\theta = \dfrac{e^{j\theta} - e^{-j\theta}}{2j}$
$\dfrac{\sin\theta}{\theta} = \sin c\theta$

Q4.6   *Determine the inverse Fourier transform of $X(e^{j\omega}) = 2\pi\delta(\omega - \omega_0)$, $|\omega_0| \leq \pi$.*

Solution:

The inverse Fourier transform of $X(e^{j\omega})$ is

$$x(n) = \frac{1}{2\pi}\int_{-\pi}^{\pi} X(e^{j\omega})\, e^{j\omega n}\, d\omega = \frac{1}{2\pi}\int_{-\pi}^{\pi} 2\pi\delta(\omega - \omega_0)e^{j\omega n}\, d\omega$$

$$= \int_{-\pi}^{\pi} \delta(\omega - \omega_0)e^{j\omega n}\, d\omega = e^{j\omega n}\Big|_{\omega = \omega_0} = e^{j\omega_0 n}$$

> **Note :** Here the integral limit is $-\pi$ to $+\pi$, and in this range there is only one impulse located at ω_0.

Q4.7   *Determine the sampling period for the signal $X(j\Omega) = U(j\Omega + j\Omega_0) - U(j\Omega - j\Omega_0)$, to sample without aliasing.*

Solution:

The frequency spectrum of the given signal can be plotted as shown in Fig Q4.7.

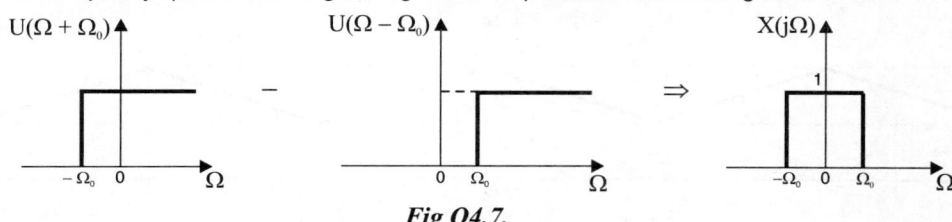

Fig Q4.7.

From the frequency spectrum of Fig Q4.7, it is observed that the maximum frequency, Ω_{max} is Ω_0

$$\therefore \; \Omega_{max} = \Omega_0 \; \Rightarrow \; 2\pi F_{max} = \Omega_0 \; \Rightarrow \; F_{max} = \frac{\Omega_0}{2\pi}$$

In order to avoid aliasing the sampling frequency, F_s should be greater than $2F_{max}$.

$$\therefore \; F_s \geq 2F_{max}$$

Since sampling period, $T = \dfrac{1}{F_s}$ in order to avoid aliasing,

$$T \leq \frac{1}{2F_{max}}$$

$$\therefore \; \text{Minimum sampling period, } T = \frac{1}{F_s} = \frac{1}{2F_{max}} = \frac{1}{\frac{2\Omega_0}{2\pi}} = \frac{\pi}{\Omega_0}$$

$\therefore$ In order to avoid aliasing the sampling period T should be less than $\dfrac{\pi}{\Omega_0}$ $\left(\text{i.e., } T < \dfrac{\pi}{\Omega_0}\right)$.

Q4.8   *A signal x(t) whose spectrum is shown in Fig Q4.8.1 is sampled at a rate of 300 samples / sec. Sketch the spectrum of the sampled discrete time signal*

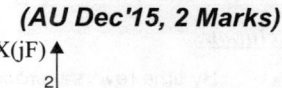

(AU Dec'15, 2 Marks)

Solution:

From the spectrum shown in Fig Q4.8.1 it is observed that the maximum frequency, F_m in the signal is 100 Hz. Given that, Sampling frequency, F_s is 300 Hz, which is greater than $2 F_m$, and so the signal is sampled without aliasing.

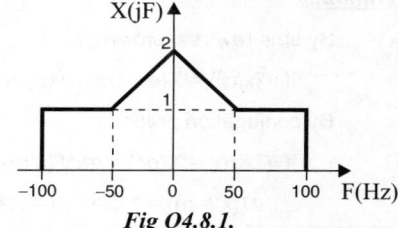

Fig Q4.8.1.

Frequency "f" of sampled discrete time signal corresponding to any frequency "F" of continuous time signal is given by, $f = F / F_s$.

The magnitude of the spectrum of discrete time signal will be scaled by 1/T, where $T = 1/F_s$.

The frequency spectrum of a discrete time signal will be periodic with periodicity of - 0.5 to + 0.5.

Therefore the frequency spectrum of sampled discrete time signal will be as shown in Fig Q4.8.2.

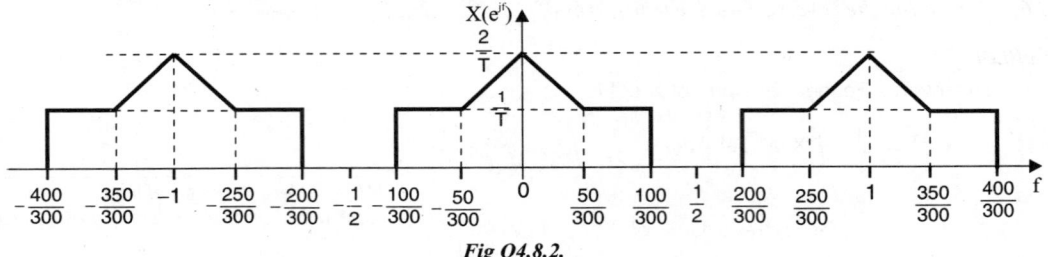

Fig Q4.8.2.

Q4.9 *If the spectrum shown in Fig Q4.8.1 is sampled at a rate of 100 samples / sec. Sketch the spectrum of the sampled discrete time signal.*

Since the sampling frequency is less than $2F_m$, the spectrum of the sampled signal will have aliasing as shown in Fig Q4.9.1.

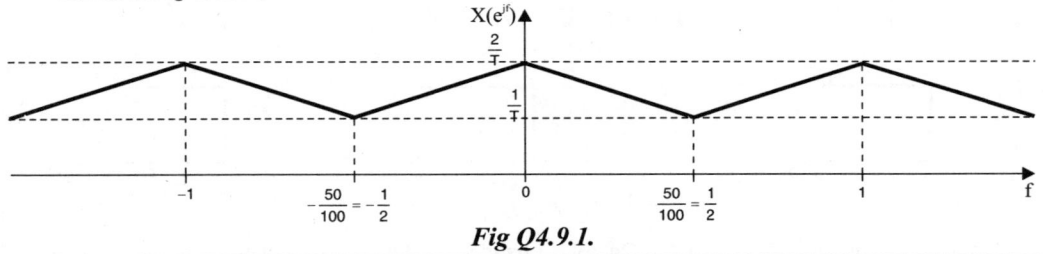

Fig Q4.9.1.

Q4.10 *Consider the sampling of the bandpass signal whose frequency spectrum is shown in Fig Q4.10. Determine the minimum sampling rate F_s to avoid aliasing.*

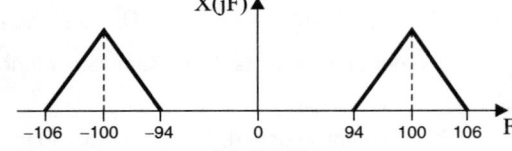

Fig Q4.10.

Solution:

The given signal is a bandpass signal. The bandwidth, B = 106 – 94 = 12 Hz.

Here the upper cutoff frequency (106 Hz) is not an integer multiple of bandwidth, B. Hence the minimum sampling rate should be 4B, in order to avoid aliasing.

∴ Minimum sampling rate = 4B = 4 × 12 = 48 Hz.

Q4.11 *If $X(e^{j\omega})$ is DTFT of x(n), what is the DTFT of x*(−n)?*   **(AU Jun'13, 2 Marks)**

Solution:

By time reversal property,

if $\mathcal{F}\{x(n)\} = X(e^{j\omega})$ then $\mathcal{F}\{x(-n)\} = X(e^{-j\omega})$

By conjugation property,

if $\mathcal{F}\{x(n)\} = X(e^{j\omega})$ then $\mathcal{F}\{x^*(n)\} = X^*(e^{-j\omega})$

∴ $\mathcal{F}\{\{x^*(-n)\} = X^*((e^{-j\omega})^*) = X^*(e^{j\omega})$

Q4.12 Find the Z-transform and their ROC of the discrete time signal. x(n) = {1, −1, 2, 3,4 }

$\qquad\qquad\qquad\qquad\qquad\qquad\qquad\qquad\qquad\qquad\qquad\qquad\qquad\uparrow$

Solution: **(AU Dec'15, 2 Marks)**

From the given sequence we can write,

$x(-3) = 1, x(-2) = -1, x(-1) = 2, x(0) = 3, x(1) = 4$

By definition of Z-transform,

> The given sequence is a finite duration sequence defined in the range n = − 3 to 1, hence the limits of summation is changed to n = −3 to 1.

$$X(z) = Z\{x(n)\} = \sum_{n=-\infty}^{\infty} x(n)\,z^{-n} = \sum_{n=-3}^{1} x(n)\,z^{-n}$$

$$\therefore \; X(z) = x(-3)\,z^3 + x(-2)\,z^2 + x(-1)\,z^1 + x(0)\,z^0 + x(z)\,z^{-1}$$

$$= 1 \times z^3 + (-1)\,z^2 + 2 \times z^1 + 3 \times z^0 + 4 \times z^{-1}$$

$$= z^3 - z^2 + 2z + 3 + 4z^{-1}$$

In X(z), when z = 0, the terms with negative power of z will become infinite and when z = ∞, the terms with positive power of z will become infinite. Hence X(z) will be finite for all values of z except when z = 0 and z = ∞. Therefore, the ROC is entire z- plane except z = 0 and z = ∞.

Q4.13 Determine the Z- transform of δ(n + k) **(AU Jun'13, 2 Marks)**

Solution:

Given that, x(n) = δ(n + k)

By definition of Z- transform,

$$X(z) = Z\{x(n)\} = \sum_{n=-\infty}^{+\infty} x(n)\,z^{-n} = \sum_{n=-\infty}^{+\infty} \delta(n+k)\,z^{-n} = 1 \times z^{-(-k)} = z^k$$

> $\delta(n+k) = 1 \;\; ; \quad n = -k$
> $\qquad\qquad = 0 \;\; ; \quad n \neq -k$

Q4.14 Find the Z-transform of aⁿ u(n) , |a| < 1 **(AU Dec'11, 2 Marks)**

Solution:

By the definition of Z-transform,

$$Z\{a^n\,u(n)\} = \sum_{n=0}^{\infty} a^n z^{-n} = \sum_{n=0}^{\infty} (a\,z^{-1})^n = \frac{1}{1 - a z^{-1}} = \frac{1}{1 - a/z} = \frac{z}{z-a}$$

> $u(n) = 1 \,; n \geq 0$
> $\qquad\quad = 0 \,; n < 0$

> Infinite geometric series sum formula,
> $$\sum_{n=0}^{\infty} C^n = \frac{1}{1-C} \; ;$$
> if, $0 < |C| < 1$

Q4.15 Find the Z-transform of e⁻ᵃⁿᵀ u(n).

Solution:

By the definition of Z-transform,

$$Z\{e^{-anT}\,u(n)\} = \sum_{n=0}^{\infty} e^{-anT} z^{-n} = \sum_{n=0}^{\infty} (e^{-aT}\,z^{-1})^n = \frac{1}{1 - e^{-aT}\,z^{-1}} = \frac{1}{1 - e^{-aT}/z} = \frac{z}{z - e^{-aT}}$$

Q4.16 Find the Z-transform of x(n) defined as,

$x(n) = b^n \;\; ; \;\; 0 \leq n \leq N-1$

$\qquad = 0 \;\;\; ; \;\;\; otherwise$

Solution:

By the definition of Z-transform,

$$Z\{x(n)\} = \sum_{n=-\infty}^{+\infty} x(n)z^{-n} = \sum_{n=0}^{N-1} b^n z^{-n}$$

> Finite geometric series sum formula,
> $$\sum_{n=0}^{N-1} C^n = \frac{1-C^N}{1-C}$$

$$\therefore \; \mathcal{Z}\{x(n)\} =_{=} \sum_{n=0}^{N-1} (bz^{-1})^n = \frac{1-(bz^{-1})^N}{1-bz^{-1}} = \frac{1-b^N z^{-N}}{1-bz^{-1}} = \frac{z^{-N}(z^N - b^N)}{z^{-1}(z-b)} = \frac{z^{-N+1}(z^N - b^N)}{z-b}$$

Q4.17 Find the Z-transform of $x(n) = a^{n+1} u(n+1)$.

Solution:

Given that, $x(n) = a^{n+1} u(n+1) = a^{n+1}$; $n \geq -1$

By the definition of Z-transform,

$$\mathcal{Z}\{x(n)\} = \sum_{n=-\infty}^{+\infty} x(n)z^{-n} = \sum_{n=-1}^{+\infty} a^{n+1} z^{-n} = a^{n+1} z^{-n}\big|_{n=-1} + \sum_{n=0}^{+\infty} a^{n+1} z^{-n} = a^0 z + \sum_{n=0}^{+\infty} a^n a z^{-n}$$

$$= z + a\sum_{n=0}^{+\infty} (az^{-1})^n = z + a\frac{1}{1-az^{-1}} = z + \frac{az}{z-a} = \frac{z(z-a)+az}{z-a} = \frac{z^2}{z-a}$$

Q4.18 Determine the inverse Z-transform of $X(z) = \log(1 + az^{-1})$; $|z| > |a|$

Solution:

Given that, $X(z) = \log(1 + az^{-1})$; $|z| > |a|$

Let, $x(n) = \mathcal{Z}^{-1}\{X(z)\}$

By differentiation property of Z-transform we get,

> Since ROC is exterior of a circle of radius "a", the $x(n)$ should be a causal signal.

$$\mathcal{Z}\{nx(n)\} = -z\frac{d}{dz}X(z)$$

$$= -z\frac{d}{dz}[\log(1+az^{-1})] = -z\frac{1}{1+az^{-1}}(-az^{-2}) = \frac{az^{-1}}{1+az^{-1}}$$

> $X(z) = \log(1 + az^{-1})$

$$= \frac{az^{-1}}{z^{-1}(z+a)} = az^{-1}\frac{z}{z-(-a)}$$

> If $\mathcal{Z}\{x(n)\} = X(z)$
> then by shifting property
> $\mathcal{Z}\{x(n-m)\} = z^{-m}X(z)$

$$\therefore \; nx(n) = \mathcal{Z}^{-1}\left\{az^{-1}\frac{z}{z-(-a)}\right\} = a\mathcal{Z}^{-1}\left\{\frac{z}{z-(-a)}\right\}\bigg|_{n=n-1} = a(-a)^n u(n)\big|_{n=n-1}$$

$$= a(-a)^{n-1} u(n-1)$$

$$\therefore \; x(n) = \frac{a}{n}(-a)^{n-1} u(n-1)$$

Q4.19 Determine $x(0)$ if the Z-transform of $x(n)$ is $X(z) = \dfrac{2z^2}{(z+3)(z-4)}$.

Solution:

By initial value theorem of Z-transform,

$$x(0) = \underset{z\to\infty}{Lt}\, X(z) = \underset{z\to\infty}{Lt}\, \frac{2z^2}{(z+3)(z-4)}$$

$$= \underset{z\to\infty}{Lt}\, \frac{2z^2}{z^2\left(1+\frac{3}{z}\right)\left(1-\frac{4}{z}\right)} = \frac{2}{\left(1+\frac{3}{\infty}\right)\left(1-\frac{4}{\infty}\right)} = \frac{2}{(1+0)(1-0)} = 2$$

Q4.20 Determine the Z-transform of $x(n) = (n-3)u(n)$.

Solution:

Let, $\mathcal{Z}\{u(n)\} = U(z) = \dfrac{z}{z-1}$

$$\therefore \; \mathcal{Z}\{x(n)\} = \mathcal{Z}\{(n-3)\, u(n)\} = \mathcal{Z}\{n\, u(n) - 3\, u(n)\}$$

$$= \mathcal{Z}\{n\, u(n)\} - 3\, \mathcal{Z}\{u(n)\} = -z\frac{d}{dz}U(z) - 3\,U(z)$$

$$= -z\frac{d}{dz}\left(\frac{z}{z-1}\right) - 3\frac{z}{z-1} = -z\frac{z-1-z}{(z-1)^2} - \frac{3z}{z-1}$$

$$= \frac{z}{(z-1)^2} - \frac{3z}{z-1} = \frac{z - 3z(z-1)}{(z-1)^2} = \frac{z - 3z^2 + 3z}{(z-1)^2} = \frac{-3z^2 + 4z}{(z-1)^2} = \frac{z(4-3z)}{(z-1)^2}$$

$$\boxed{\mathcal{Z}\{u(n)\} = \frac{z}{z-1}}$$

$$\boxed{\mathcal{Z}\{n\,x(n)\} = -z\frac{d}{dz}X(z)}$$

$$\boxed{d\frac{u}{v} = \frac{v\,du - u\,dv}{v^2}}$$

4.11 MATLAB Programs

Program 4.1

Write a MATLAB program to find one-sided $\mathcal{Z}$-transform of the following standard causal signals.

a) n   b) a^n c) na^n d) e^{-anT}

```
%Program to find the Z-transform of some standard signals

    clear all
    syms n T a real;      %Let n, T, a be real variable
    syms z complex;       %Let z be complex variable

%(a)
    x = n;
    disp('(a) z-transform of "n" is');
    ztrans(x)

%(b)
    x = a^n;
    disp('(b) z-transform of  "a^n" is');
    ztrans(x)

%(c)
    x=n*(a^n);
    disp('(c) z-transform of  "n(a^n)" is');
    ztrans(x)

%(d)
    x=exp(-a*n*T);
    disp('(d) z-transform of "exp(-a*n*T)" is');
    ztrans(x)
```

OUTPUT

```
    (a) z-transform of "n" is
          ans =
                z/(z-1)^2
    (b) z-transform of  "a^n" is
          ans =
                z/a/(z/a-1)
    (c) z-transform of  "n(a^n)" is
          ans =
                z*a/(z-a)^2
    (d) z-transform of "exp(-a*n*T)" is
          ans =
                z/exp(-a*T)/(z/exp(-a*T)-1)
```

Program 4.2

Write a MATLAB program to find ƶ-transform of the following causal signals.

 a) 0.5^n b) $1+n(0.4)^{(n-1)}$

```
%********** program to determine z-transform of given signals
clear all
syms n real; %Let n be real variable

%(a)
x1=0.5^n;
disp('(a) z-transform of "0.5^n" is');
X1=ztrans(x1)

%(b)
x2=1+n*(0.4^(n-1));
disp('(b) z-transform of "1+n*(0.4^(n-1))" is');
X2=ztrans(x2)
```

OUTPUT
```
    (a) z-transform of "0.5^n" is
        X1 =
            2*z/(2*z-1)
    (b) z-transform of "1+n*(0.4^(n-1))" is
        X2 =
            z/(z-1)+25*z/(5*z-2)^2
```

Program 4.3

Write a MATLAB program to find inverse ƶ-transform of the following z-domain signals.

 a) $1/(1-1.5z^{-1}+0.5z^{-2})$ b) $1/((1+z^{-1})(1-z^{-1})^2)$

```
%***********Program to determine the inverse z-transform
syms n z

X=1/(1-1.5*(z^(-1))+0.5*(z^(-2)));
disp('Inverse z-transform of 1/(1-1.5z^-1+0.5z^-2)is');
x=iztrans(X,z,n);
simplify(x)

X=1/((1+(z^(-1)))*((1-(z^(-1))^2)));
disp('Inverse z-transform of 1/((1+z^-1)*(1-z^-1)^2))is');
x=iztrans(X,z,n);
simplify(x)
```

OUTPUT
```
    Inverse z-transform of 1/(1-1.5z^-1+0.5z^-2) is
        ans =
            2-2^(-n)
    Inverse z-transform of 1/((1+z^-1)*(1-z^-1)^2)) is
        ans =
            3/4*(-1)^n+1/2*(-1)^n*n+1/4
```

4.12 Exercises

I. Fill in the blanks with appropriate words.

1. A bandlimited signal with maximum frequency F_m can be fully recovered from its samples if sampled at a frequency greater than or equal to _____.

2. The sampling rate for a bandpass signal with bandwidth "B" is _____.

3. The frequency spectrum of a discrete time signal is periodic with period _____.

4. The ————— of X(z) is the set of all values of z, for which X(z) attains a finite value.

5. The phenomena of high frequency components acquiring the identity of low frequency components is called _____.

6. For a causal signal the ROC should be _____ the circle of radius whose value corresponds to pole with _____ magnitude.

7. If X(z) is rational, then the ROC does not include _____ of X(z).

8. The sequences multiplied by u(−n) are _____ and defined for _____.

9. The Z-transform of a shifted signal, shifted by 'q' units of time is obtained by _____ to Z-transform of unshifted signal.

10. The Fourier transform of a discrete time signal can be obtained by evaluating the Z-transform along _____.

Answers

1. $2F_m$	5. aliasing	9. multiplying z^q
2. 2B to 4B	6. outside, largest	10. unit circle
3. 2π	7. poles	
4. region of convergence	8. anticausal sequences, $n \leq 0$	

II. State whether the following statements are True/False.

1. The Fourier transform of discrete signal is a discrete function of ω.

2. The frequency response is periodic with a periodicity of 2π.

3. Convolving two signals in time domain is equivalent to multiplying their spectra in frequency domain.

4. Multiplication of a sequence x(n) by $e^{j\omega_0 n}$ is same as frequency translation of the spectrum $X(e^{j\omega})$ by ω_0.

5. The spectrum of sampled version of a discrete time signal is sum of frequency shifted and amplitude scaled version of original spectrum of continuous time signal, $X(j\Omega)$.

6. If a discrete time signal is shifted in time by 'n_0' samples, then its magnitude spectrum shifts by ωn_0.

7. The Fourier transform can be obtained from Z-transform only if ROC of X(z) includes unit circle.

8. The Z-transform is unique only for a specified ROC.

9. The ROC of Z-transform of a causal signal always includes unit circle.

10. The ROC of Z-transform will be always a circular region with centre at origin of z-plane.

Answers

1. False	3. True	5. True	7. True	9. False
2. True	4. True	6. False	8. True	10. True

III. Choose the right answer for the following questions.

1. *The Fourier transform of x(n) = 1, for all 'n' is,*

   a) $2\pi \sum_{m=-\infty}^{+\infty} \delta(\omega - 2\pi m)$ **b)** $\pi \sum_{m=-\infty}^{+\infty} \delta(\omega - 2\pi m)$ **c)** $2\pi \sum_{m=-\infty}^{+\infty} \delta(\omega - m)$ **d)** $2\pi \sum_{m=-\infty}^{+\infty} \delta(\omega - \pi m)$

2. *If $\mathcal{F}\{x(n)\} = X(e^{j\omega})$, then $\mathcal{F}\{x(n-3)\}$ will be,*

   a) $e^{-j3\omega} X(e^{-j\omega})$ **b)** $e^{j3\omega} X(e^{-j\omega})$ **c)** $e^{-j3\omega} X(e^{j\omega})$ **d)** $e^{j3\omega} X(e^{j\omega})$

3. *If a signal is folded about the origin in time then its,*

   a) magnitude spectrum undergoes change in sign
   b) phase spectrum undergoes change in sign
   c) magnitude remains unchanged
   d) both c and b

4. *The Fourier transform of correlation sequence of two discrete time signals $x_1(n)$ and $x_2(n)$ is given by,*

   a) $X_1(e^{j\omega}) X_2(e^{j\omega})$ **b)** $X_1(e^{j\omega}) X_2(e^{-j\omega})$ **c)** $X_1(e^{-j\omega}) X_2(e^{-j\omega})$ **d)** none of the above

5. *A bandlimited continuous time signal with maximum frequency F_m, sampled at a frequency F_s, can be fully recovered from its samples, provided that,*

   a) $F_s \geq 2F_m$ **b)** $F_s = 2F_m$ **c)** $F_m \geq 2F_s$ **d)** $F_s = F_m$

6. *If the bandwidth of a bandpass signal x(t) is 2F, then the minimum sampling rate for bandpass signal must be,*

   a) 2F samples/sec   **b)** 4F samples/sec   **c)** $\frac{F}{2}$ samples/sec   **d)** $\frac{F}{4}$ samples/sec

7. *The discrete time Fourier transform of the signal, $x(n) = 0.5^{(n-1)} u(n-1)$ is,*

   a) $\frac{e^{-j\omega}}{1-0.5e^{-j\omega}}$ **b)** $e^{-j\omega}(1-0.5e^{-j\omega})$ **c)** $\frac{0.5e^{-j\omega}}{1-0.5e^{-j\omega}}$ **d)** $\frac{0.5e^{j\omega}}{1-0.5e^{-j\omega}}$

8. *The Z-transform of $a^{-n} u(-n-1)$ is,*

   a) $\frac{-z}{z-1/a}$ **b)** $\frac{z}{z-1/a}$ **c)** $\frac{z}{z-a}$ **d)** $\frac{-z}{z-a}$

9. *The inverse Z-transform of $\frac{3}{z-4}$, $|z| > 4$ is,*

   a) $3(4)^n u(n-1)$ **b)** $3(4)^{n-1} u(n)$ **c)** $3(4)^{n-1} u(n+1)$ **d)** $3(4)^{n-1} u(n-1)$

10. *ROC of x(n) contains,*

    a) poles   **b)** zeros   **c)** no poles   **d)** no zeros

11. *The inverse Z-transform of $X(z) = e^{a/z}$, $|z| > 0$ is,*

    a) $x(n) = \frac{-a^n}{n!} u(n)$ **b)** $x(n) = \frac{a^n}{n!} u(n)$ **c)** $x(n) = \frac{a^{n-1}}{n!} u(n)$ **d)** none of the above

12. *The Z-transform of x(n) = [u(n) – u(n – 3)], for ROC |z| > 1 is,*

   a) $X(z) = \dfrac{z - z^{-3}}{z - 1}$ b) $X(z) = \dfrac{z^{-2}}{(z-1)^2}$ c) $X(z) = \dfrac{z - 4z^{-2} + 3z^{-3}}{(z-1)^2}$ d) $\dfrac{z - z^{-2}}{z - 1}$

13. *If x(n) = {0.5, –0.25, 1}, then Z-transform of the signal is,*

   a) $\dfrac{z^2}{0.5z^2 - 0.25z + 1}$ b) $\dfrac{z^2}{z^2 - 0.5z + 0.25}$ c) $\dfrac{0.5z^2 - 0.25z + 1}{z^2}$ d) $\dfrac{2z^2 + 4z + 1}{z^2}$

14. *If Z-transform of x(n) is X(z) then Z-transform of x(–n) is,*

   a) $-X(z)$ b) $X(-z)$ c) $-X(z^{-1})$ d) $X(z^{-1})$

15. *The inverse Z-transform of X(z) can be defined as,*

   a) $x(n) = \dfrac{1}{2\pi} \oint_c X(z) z^{n-1} dz$ b) $x(n) = \dfrac{1}{2j} \oint_c X(z) z^{n-1} dz$

   c) $x(n) = \dfrac{1}{2\pi j} \oint_c X(z) z^{n-1} dz$ d) $x(n) = \dfrac{1}{2\pi j} \oint_c X(z) z^{-n} dz$

16. *The Z-transform is a,*
   a) finite series   b) infinite power series
   c) geometric series   d) both a and c

17. *If the Z-transform of x(n) is X(z), then Z-transform of (0.5)^n x(n) is,*

   a) $X(0.5\,z)$ b) $X(0.5^{-1}\,z)$ c) $X(2^{-1}\,z)$ d) $X(2z)$

18. *The Z-transform of correlation of the sequences x(n) and y(n) is,*

   a) $X^*(z)\,Y^*(z^{-1})$ b) $X(z)\,Y(z^{-1})$ c) $X(z) * Y(z)$ d) $X(z^{-1})\,Y(z^{-1})$

Answers				
1. a	5. a	9. d	13. c	17. b
2. c	6. b	10. c	14. d	18. b
3. d	7. a	11. b	15. c	
4. b	8. a	12. d	16. b	

IV. Answer the following questions.

1. Define Fourier transform of a discrete time signal.
2. State and prove any two properties of Fourier transform.
3. State and prove the time delay property of Fourier transform.
4. Give the significance of Parseval's relation.
5. Define inverse Fourier transform.
6. What is aliasing of frequency spectrum?
7. Explain how a bandlimited signal can be sampled without aliasing?
8. What is ideal interpolation formula? What is its significance?
9. Write a short note on sampling of bandpass signals.

10. Define one-sided and two-sided Z-transform.
11. What is region of convergence (ROC)?
12. State the final value theorem with regard to Z-transform.
13. State the initial value theorem with regard to Z-transform.
14. Define Z-transform of unit step signal.
15. What are the different methods available for inverse Z-transform?

V. Solve the following problems.

E4.1 Determine the Fourier transform of the following signals.

a) $x(n) = 3\cos\dfrac{2\pi}{5}n$

b) $x(n) = \{-3,\ 4,\ -1,\ 2\}$
$\qquad\qquad\quad\uparrow$

c) $x(n) = (-1)^n\ ;\ 0\le n\le 7$
$\qquad = 0\qquad ;\ \text{otherwise}$

d) $x(n) = 0.5\left[\left(\dfrac{1}{0.4}\right)^n - \left(\dfrac{1}{0.8}\right)^n\right]u(n)$

E4.2 Determine the convolution of the following sequences, using Fourier transform.

a) $x_1(n) = \{2,\ -2,\ 2\},\quad x_2(n) = \{-2,\ 2,\ -2\}$
$\qquad\qquad\ \uparrow\qquad\qquad\qquad\qquad\ \uparrow$

b) $x_1(n) = \{-2,\ -1,\ 0\},\quad x_2(n) = \{-3,\ 5,\ -7\}$
$\qquad\qquad\quad\uparrow\qquad\qquad\qquad\qquad\ \uparrow$

E4.3 Determine the inverse Fourier transform of the following functions of ω.

a) $X(e^{j\omega}) = 2j\omega$

b) $X(e^{j\omega}) = \dfrac{1}{(1-ae^{-j\omega})^2}\ ;\ |a|<1$

c) $Y(e^{j\omega}) = \dfrac{1+\frac{1}{7}e^{-j\omega}}{1-\frac{1}{7}e^{-j\omega}}$

d) $H(e^{j\omega}) = \dfrac{1}{\left(1-\frac{1}{6}e^{-j\omega}\right)\left(1-\frac{1}{5}e^{-j\omega}\right)}$

E4.4 Determine the Z-transform and their ROC of the following discrete time signals.

a) $x(n) = \{4,\ 2,\ 8,\ 5\}$
$\qquad\qquad\ \uparrow$

b) $x(n) = \{3,\ 0,\ 0,\ 4,\ 45,\ 1\}$
$\qquad\qquad\qquad\qquad\quad\uparrow$

c) $x(n) = \{2,\ 1,\ 1,\ 2,\ 5,\ 8,\ 2\}$
$\qquad\qquad\qquad\ \uparrow$

d) $x(n) = -0.2^n\,u(n-1)$

e) $x(n) = (0.6)^n\,u(n) + (0.7)^n\,u(-n-1)$

f) $x(n) = (0.9)^{|n|}$

E4.5 Find the one-sided Z-transform of the following discrete time signals.

a) $x(n) = n^2\,5^n\,u(n)$

b) $x(n) = n(0.5)^{n+4}$

c) $x(n) = (0.5)^{n-2}[u(n) - u(n-2)]$

E4.6 Find the one-sided Z-transform of the discrete signals generated by mathematically sampling the following continuous time signals.

a) $x(t) = 4t\,e^{-0.6t}\,u(t)$

b) $x(t) = 2t^3\,u(t)$

E4.7 Find the time domain initial value $x(0)$ and final value $x(\infty)$ of the following z-domain functions.

a) $X(z) = \dfrac{0.5}{(1-z^{-1})^2(1+z^{-1})}$

b) $X(z) = \dfrac{z^3}{(z-1)(z^2-0.2)}$

E4.8 Determine the inverse Z-transform of the following functions using contour integral method.

a) $X(z) = \dfrac{(2z-1)z}{(z-1)^2}$ b) $X(z) = \dfrac{z^2+z}{(z-2)^2}$ c) $X(z) = \dfrac{(1-e^{-a})z}{(z-1)(z-e^{-a})}$

E4.9 Determine the inverse Z-transform of the following functions by partial fraction method.

a) $X(z) = \dfrac{z^2}{(z+1)(z+2)^2}$ b) $X(z) = \dfrac{2z^2-z}{z^3-5z^2+8z-4}$ c) $X(z) = \dfrac{z(z^2+3)}{(z^2+1)^2}$

E4.10 Determine the inverse Z-transform of the function, $X(z) = \dfrac{2-z^{-1}}{\left[1-\frac{1}{4}z^{-1}\right]\left[1-\frac{1}{3}z^{-1}\right]}$

a) $\text{ROC}: |z| > \frac{1}{3}$, b) $\text{ROC}: |z| < \frac{1}{4}$, c) $\text{ROC}: \frac{1}{4} < |z| > \frac{1}{3}$

E4.11 Determine the inverse Z-transform of the following function using power series method

$X(z) = \dfrac{z}{2z^2-3z+1}$

a) $\text{ROC}: |z| < 0.5$, b) $\text{ROC}: |z| > 1$

E4.12 Determine the inverse Z-transform for the following functions using power series method.

a) $X(z) = \dfrac{z^2+z}{z^2-2z+1}$; $\text{ROC}:|z|>1$ b) $X(z) = \dfrac{1-\frac{1}{3}z^{-1}}{1+\frac{1}{3}z^{-1}}$; $\text{ROC}:|z|>\frac{1}{3}$

Answers

E4.1 a) $X(e^{j\omega}) = 3\pi \displaystyle\sum_{m=-\infty}^{+\infty} \left[\delta\left(\omega - \frac{2\pi}{5} - 2\pi m\right) + \delta\left(\omega + \frac{2\pi}{5} - 2\pi m\right)\right]$

b) $X(e^{j\omega}) = -3 + 4e^{-j\omega} - e^{-j2\omega} + 2e^{-j3\omega}$ c) $X(e^{j\omega}) = \dfrac{\sin 4\omega}{\cos(\omega/2)} e^{j\left(\frac{\pi - 7\omega}{2}\right)}$

d) $X(e^{j\omega}) = \dfrac{0.625e^{-j\omega}}{1 - 3.75e^{-j\omega} + 3.125e^{-j2\omega}}$

E4.2 a) $x(n) = \{-4,\ 8,\ -12,\ -8,\ -4\}$ b) $x(n) = \{6,\ -7,\ 9,\ 7\}$
          $\uparrow$ $\uparrow$

E4.3 a) $x(n) = \dfrac{2\cos \pi n}{n}$ b) $x(n) = (n+1)a^n u(n)$; $|a| < 1$

c) $y(n) = \left(\frac{1}{7}\right)^n [u(n) + u(n-1)]$ d) $h(n) = \left[6\left(\frac{1}{5}\right)^n - 5\left(\frac{1}{6}\right)^n\right]u(n)$

E4.4 a) $X(z) = 4 + \dfrac{2}{z} + \dfrac{8}{z^2} + \dfrac{5}{z^3}$ b) $X(z) = 3z^5 + 4z^2 + 45z + 1$

ROC is entire z-plane except at z = 0.   ROC is entire z-plane except at z = ∞.

c) $X(z) = 2z^3 + 1z^2 + z + 2 + 5z^{-1} + 8z^{-2} + 2z^{-3}$

ROC is entire z-plane except at z = 0 and z = ∞.

d) $X(z) = \dfrac{-0.2}{z - 0.2}$; ROC is exterior of the circle of radius 0.2 in z-plane.

e) $X(z) = \dfrac{-0.1z}{(z - 0.6)(z - 0.7)}$; ROC is $0.6 < |z| < 0.7$

f) $X(z) = \dfrac{-0.21z}{(z - 0.9)(z - 1.11)}$; ROC is $0.9 < |z| < 1.11$

E4.5   **a)** $X(z) = \dfrac{5z(z + 5)}{(z - 5)^3}$ **b)** $X(z) = \dfrac{0.5^5 z}{(z - 0.5)^2}$ **c)** $X(z) = \dfrac{4z^2 - 1}{z(z - 0.5)}$

E4.6   **a)** $X(z) = \dfrac{4zTe^{-0.6T}}{(z - e^{-0.6T})^2}$ **b)** $X(z) = \dfrac{2T^3 z(z^2 + 4z + 1)}{(z - 1)^4}$

E4.7   **a)** Initial value, $x(0) = 0.5$ **b)** Initial value, $x(0) = 1$

Final value, $x(\infty) = \infty$ Final value, $x(\infty) = 1.25$

E4.8   **a)** $x(n) = [n + 2]\, u(n)$ **b)** $x(n) = (n + 1)\, 2^n\, u(n) + n\, 2^{(n-1)}\, u(n - 1)$

c) $x(n) = (1 - e^{-an})\, u(n)$

E4.9   **a)** $x(n) = [(-2)^n - (-1)^n - n(-2)^n]\, u(n)$ **b)** $x(n) = [1 + (1.5n - 1)2^n]\, u(n)$

c) $x(n) = \left[[j(-j)^n - j^n] + \dfrac{n}{2j}[(-j)^n - j^n] \right] u(n)$

E4.10   **a)** $x(n) = \left[6\left(\dfrac{1}{4}\right)^n - 4\left(\dfrac{1}{3}\right)^n \right] u(n)$ **b)** $x(n) = \left[-6\left(\dfrac{1}{4}\right)^n + 4\left(\dfrac{1}{3}\right)^n \right] u(-n - 1)$

c) $x(n) = 6\left(\dfrac{1}{4}\right)^n u(n) + 4\left(\dfrac{1}{3}\right)^n u(-n - 1)$

E4.11   **a)** $x(n) = \{.....31,\ 15,\ 7,\ 3,\ 1\}$ **b)** $x(n) = \left\{ 0,\ \dfrac{1}{2},\ \dfrac{3}{4},\ \dfrac{7}{8},\ \dfrac{15}{16},\ \dfrac{31}{32},\ ..... \right\}$

                                      $\uparrow$ $\uparrow$

E4.12   **a)** $x(n) = \{1,\ 3,\ 5,\ 7,.....\}$ **b)** $x(n) = \left\{ 1,\ -\dfrac{2}{3},\ \dfrac{2}{9},\ -\dfrac{2}{27},\ \dfrac{2}{81},\ ..... \right\}$

                    $\uparrow$ $\uparrow$

CHAPTER 5

Linear Time Invariant Discrete Time System

5.1 Discrete Time System

A *discrete time system* is a device or algorithm that operates on a discrete time signal, called the input or excitation, to produce another discrete time signal called the output or the response of the system.

Linear Time Invariant (LTI) System

A discrete time system is linear if it obeys the principle of superposition and it is time invariant if its input-output relationship does not change with time. When a discrete time system satisfies the properties of linearity and time invariance then it is called an *LTI system* (Linear Time Invariant system).

Response of Discrete Time System

In a discrete time system the input signal x(n) is transformed by the system into a signal y(n), and the transformation can be expressed mathematically as shown in equation (5.1). The diagrammatic representation of discrete time system is shown in Fig 5.1.

Response, $y(n) = \mathcal{H}\{x(n)\}$ .....(5.1)

where, $\mathcal{H}$ denotes the transformation (also called an operator).

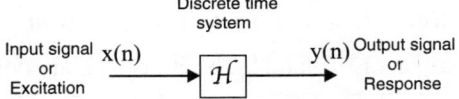

Fig 5.1: Representation of discrete time system.

5.1.1 Impulse Response

When the input to a discrete time system is a unit impulse $\delta(n)$ then the output is called an *impulse response* of the system and is denoted by h(n).

∴ Impulse Response, $h(n) = \mathcal{H}\{\delta(n)\}$ .....(5.2)

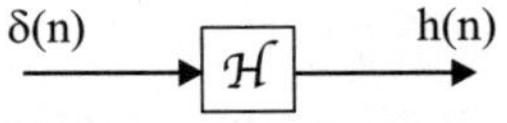

$$\delta(n) \xrightarrow{\quad} \boxed{\mathcal{H}} \xrightarrow{h(n)}$$

Fig 5.2: Discrete time system with impulse input.

5.1.2 Difference Equation Governing Discrete Time System

The mathematical equation governing the discrete time system can be developed as shown below:

The response of a discrete time system at any time instant depends on the present input, past inputs and past outputs.

Let us consider the response at n = 0. Let us assume a relaxed system and so at n = 0, there is no past input or output. Therefore the response at n = 0, is a function of present input alone.

$$\therefore \; y(0) = F[x(0)]$$

Let us consider the response at n =1. Now the present input is $x(1)$, the past input is $x(0)$ and past output is $y(0)$. Therefore the response at n = 1, is a function of $x(1)$, $x(0)$, $y(0)$.

$$\therefore \; y(1) = F[y(0), x(1), x(0)]$$

Let us consider the response at n = 2. Now the present input is $x(2)$, the past inputs are $x(1)$ and $x(0)$, and past outputs are $y(1)$ and $y(0)$. Therefore the response at n = 2, is a function of $x(2)$, $x(1)$, $x(0)$, $y(1)$, $y(0)$.

$$\therefore \; y(2) = F[y(1), y(0), x(2), x(1), x(0)]$$

Similarly, at n = 3, $y(3) = F[y(2), y(1), y(0), x(3), x(2), x(1), x(0)]$

at n = 4, $y(4) = F[y(3), y(2), y(1), y(0), x(4), x(3), x(2), x(1), x(0)]$ and so on.

In general, at any time instant n,

$$y(n) = F[y(n-1), y(n-2), y(n-3), \ldots y(1), y(0), x(n), x(n-1),$$
$$x(n-2), x(n-3) \ldots x(1), x(0)]$$

For an LTI system, the response $y(n)$ can be expressed as a weighted summation of dependent terms. Therefore, the above equation can be written as,

$$y(n) = -a_1 \, y(n-1) - a_2 \, y(n-2) - a_3 \, y(n-3) - \ldots\ldots$$
$$+ b_0 \, x(n) + b_1 \, x(n-1) + b_2 \, x(n-2) + b_3 \, x(n-3) + \ldots\ldots \quad \ldots\ldots (5.3)$$

where, $a_1, a_2, a_3, \ldots$ and $b_0, b_1, b_2, b_3, \ldots$ are constants.

> *Note:* *Negative constants are inserted for output signals, because output signals are fedback from output to input. Positive constants are inserted for input signals, because input signals are fed forward from input to output.*

Practically, the response $y(n)$ at any time instant n, may depend on N number of past outputs, present input and M number of past inputs where $M \leq N$. Hence, the equation (5.3) can be written as,

$$y(n) = -a_1 \, y(n-1) - a_2 \, y(n-2) - a_3 \, y(n-3) - \ldots\ldots - a_N \, y(n-N)$$
$$+ b_0 \, x(n) + b_1 \, x(n-1) + b_2 \, x(n-2) + b_3 \, x(n-3) + \ldots\ldots + b_M \, x(n-M)$$

$$\therefore \quad y(n) = -\sum_{m=1}^{N} a_m \, y(n-m) + \sum_{m=0}^{M} b_m \, x(n-m) \qquad \ldots\ldots(5.4)$$

The equation (5.4) is a constant coefficient ***difference equation***, governing the input-output relation of an LTI discrete time system.

In equation (5.4) the value of "N" gives the *order* of the system.

If $N = 1$, the discrete time system is called 1st order system

If $N = 2$, the discrete time system is called 2nd order system

If $N = 3$, the discrete time system is called 3rd order system , and so on.

The general difference equation governing 1st order discrete time LTI system is,

$$y(n) = -a_1 \, y(n-1) + b_0 \, x(n) + b_1 \, x(n-1)$$

The general difference equation governing 2nd order discrete time LTI system is,

$$y(n) = -a_2 \, y(n-2) - a_1 \, y(n-1) + b_0 \, x(n) + b_1 \, x(n-1) + b_2 \, x(n-2)$$

5.2 Block Diagram Representation of Discrete Time System

The discrete time system can be represented diagrammatically by *block diagram*. These diagrammatic representations are useful for physical implementation of discrete time system in hardware or software.

The basic elements employed in block diagram are adder, constant multiplier, unit delay element and unit advance element.

Adder : An adder is used to represent addition of two discrete time signals.

Constant Multiplier : A constant multiplier is used to represent multiplication of a scaling factor (constant) to a discrete time signal.

Unit Delay Element : A unit delay element is used to represent the delay of samples of a discrete time signal by one sampling time.

Unit Advance Element : A unit advance element is used to represent the advance of samples of a discrete time signal by one sampling time.

The symbolic representation of the basic elements of block diagram are listed in table 5.1.

Table 5.1: Basic Elements of Block Diagram   **(AU May'15, 2 Marks)**

Element	Block Diagram Representation
Adder	$x_1(n)$ $\quad$ $x_1(n) + x_2(n)$ $\oplus$ $\quad$ $x_2(n)$
Constant multiplier	$x(n)$ $\quad$ $\triangleright$ a $\quad$ $ax(n)$
Unit delay element	$x(n)$ $\quad$ z^{-1} $\quad$ $x(n-1)$
Unit advance element	$x(n)$ $\quad$ z $\quad$ $x(n+1)$

Example 5.1

Construct the block diagram of the discrete time systems whose input-output relations are described by the following difference equations.

a) $y(n) = 0.7 \, x(n) + 0.7 \, x(n-1)$ **b)** $y(n) = 0.4 \, y(n-1) + x(n) - 3 \, x(n-2)$

c) $y(n) = 0.2 \, y(n-1) + 0.7 \, x(n) + 0.9 \, x(n-1)$

Solution:

a) Given that, $y(n) = 0.7 \, x(n) + 0.7 \, x(n-1)$

The individual terms of the given equation are $0.7 \, x(n)$ and $0.7 \, x(n-1)$. They are represented by basic elements as shown below:

The input to the system is $x(n)$ and the output of the system is $y(n)$. The above elements are connected as shown below to get the output $y(n)$.

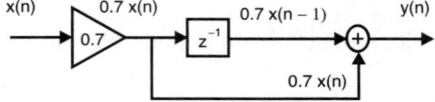

Fig 1: *Block diagram of the system described by the equation*
$$y(n) = 0.7 \, x(n) + 0.7 \, x(n-1).$$

b) Given that, $y(n) = 0.4 \, y(n-1) + x(n) - 3 \, x(n-2)$

The individual terms of the given equation are $0.4 \, y(n-1)$ and $-3 \, x(n-2)$. They are represented by basic elements as shown below:

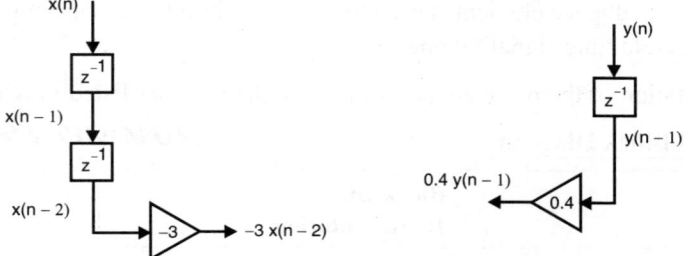

The input to the system is $x(n)$ and the output of the system is $y(n)$. The above elements are connected as shown below to get the output $y(n)$.

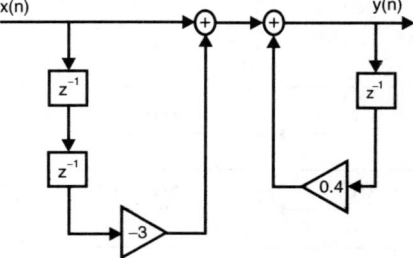

Fig 2: *Block diagram of the system described by the equation*
$$y(n) = 0.4 \, y(n-1) + x(n) - 3 \, x(n-2).$$

c) Given that, y(n) = 0.2 y(n – 1) + 0.7 x(n) + 0.9 x(n – 1)

The individual terms of the given equation are 0.2 y(n – 1), 0.7 x(n) and 0.9 x(n – 1). They are represented by basic elements as shown below:

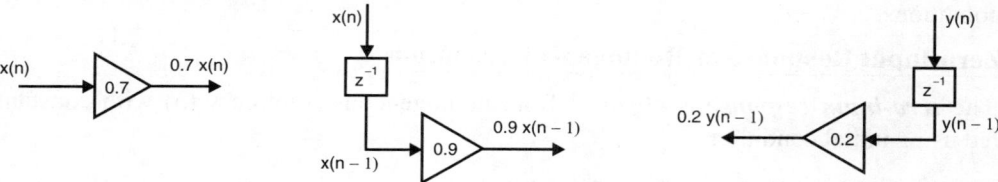

The input to the system is x(n) and the output of the system is y(n). The above elements are connected as shown below to get the output y(n).

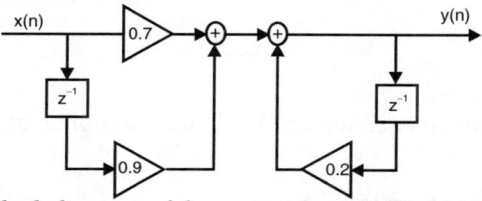

Fig 3: *Block diagram of the system described by the equation*
$$y(n) = 0.2\, y(n-1) + 0.7\, x(n) + 0.9\, x(n-1).$$

5.3 Response of LTI Discrete Time System in Time Domain

The general equation governing an LTI discrete time system is,

$$y(n) = -\sum_{m=1}^{N} a_m\, y(n-m) + \sum_{m=0}^{M} b_m\, x(n-m)$$

$$\therefore\ y(n) + \sum_{m=1}^{N} a_m\, y(n-m) = \sum_{m=0}^{M} b_m\, x(n-m)$$

$$(\text{or})\ \sum_{m=0}^{N} a_m\, y(n-m) = \sum_{m=0}^{M} b_m\, x(n-m) \text{ with } a_0 = 1 \qquad \text{.......(5.5)}$$

The solution of the difference equation (5.5) is the ***response*** y(n) of LTI system, which consists of two parts. In mathematics, the two parts of the solution y(n) are homogeneous solution $y_h(n)$ and particular solution $y_p(n)$.

$$\boxed{\therefore\ \text{Response, } y(n) = y_h(n) + y_p(n)} \qquad \text{.....(5.6)}$$

The ***homogeneous solution*** is the response of the system when there is no input. The ***particular solution*** $y_p(n)$ is the solution of difference equation for specific input signal x(n) for n $\geq$ 0.

In signals and systems, the two parts of the solution y(n) are called zero-input response $y_{zi}(n)$ and zero-state response $y_{zs}(n)$.

$$\boxed{\therefore\ \text{Response, } y(n) = y_{zi}(n) + y_{zs}(n)} \qquad \text{.....(5.7)}$$

The ***zero-input response*** is mainly due to initial conditions (or initial stored energy) in the system. Hence zero-input response is also called ***free response*** or ***natural response***. The ***zero-input response*** is given by homogeneous solution with constants evaluated using initial conditions.

The *zero-state response* is the response of the system due to input signal and with zero initial condition. Hence the zero-state response is called forced response. The *zero-state response* or *forced response* is given by the sum of homogeneous solution and particular solution with zero initial conditions.

5.3.1 Zero-Input Response or Homogeneous Solution

The *zero-input response* is obtained from homogeneous solution $y_h(n)$ with constants evaluated using initial condition.

$$\therefore \text{ Zero-input response, } y_{zi}(n) = y_h(n)\big|_{\text{with constants evaluated using initial conditions}}$$

The *homogeneous solution* is obtained when $x(n) = 0$. Therefore the homogeneous solution is the solution of the equation,

$$\sum_{m=0}^{N} a_m \, y(n-m) = 0 \qquad\qquad .....(5.8)$$

Let us assume that the solution of equation (5.8) is in the form of an exponential.

$$\therefore y(n) = \lambda^n$$

On substituting $y(n) = \lambda^n$ in equation (5.8) we get,

$$\sum_{m=0}^{N} a_m \, \lambda^{(n-m)} = 0$$

On expanding the above equation (by taking $a_0 = 1$), we get,

$$\lambda^n + a_1 \lambda^{n-1} + a_2 \lambda^{n-2} + ... + a_{N-1} \lambda^{n-(N-1)} + a_N \lambda^{n-N} = 0$$
$$\lambda^{n-N} (\lambda^N + a_1 \lambda^{N-1} + a_2 \lambda^{N-2} + ... + a_{N-1} \lambda + a_N) = 0$$

Now, the *characteristic polynomial* of the system is given by,

$$\lambda^N + a_1 \lambda^{N-1} + a_2 \lambda^{N-2} + ... + a_{N-1} \lambda + a_N = 0 \qquad\qquad .....(5.9)$$

The characteristic polynomial has N roots, which are denoted as $\lambda_1, \lambda_2, ... \lambda_N$.

The roots of the characteristic polynomial may be distinct real roots, repeated real roots or complex. The assumed solutions for various types of roots are given below:

Distinct Real Roots

Let the roots $\lambda_1, \lambda_2, \lambda_3, ... \lambda_N$ be distinct real roots. Now the homogeneous solution will be in the form,

$$y_h(n) = C_1\lambda_1^n + C_2\lambda_2^n + C_3\lambda_3^n + ... + C_N\lambda_N^n$$

where, $C_1, C_2, C_3, ...C_N$ are constants that can be evaluated using initial conditions.

Repeated Real Roots

Let one of the real roots λ_1 repeats p times and the remaining $(N-p)$ roots are distinct real roots. Now, the homogeneous solution is in the form,

$$y_h(n) = \left(C_1 + C_2 n + C_3 n^2 + ... + C_p n^{p-1}\right)\lambda_1^n + C_{p+1}\lambda_{p+1}^n + C_N\lambda_N^n$$

where, $C_1, C_2, C_3, ...C_N$ are constants that can be evaluated using initial conditions.

Complex Roots

Let the characteristic polynomial has a pair of complex roots λ and λ^* and the remaining $(N-2)$ roots be distinct real roots. Now, the homogeneous solution will be in the form,

$$y_h(n) = r^n [C_1 \cos n\theta + C_2 \sin n\theta] + C_3 \lambda_3{}^n + C_4 \lambda_4{}^n + ... + C_N \lambda_N{}^n$$

$$\text{where, } \lambda = a+jb, \ \lambda^* = a-jb, \ r = \sqrt{a^2+b^2}, \ \theta = \tan^{-1}\frac{b}{a}$$

$C_1, C_2, C_3 ... C_N$ are constants that can be evaluated using initial conditions.

5.3.2 Particular Solution

The *particular solution*, $y_p(n)$ is the solution of the difference equation for specific input signal $x(n)$ for $n \geq 0$. Since the input signal may have different form, the particular solution depends on the form or type of the input signal $x(n)$.

If $x(n)$ is constant, then $y_p(n)$ is also a constant.

> **Example:**
> Let, $x(n) = u(n)$; now, $y_p(n) = K \, u(n)$

If $x(n)$ is exponential, then $y_p(n)$ is also an exponential.

> **Example:**
> Let, $x(n) = a^n u(n)$; now, $y_p(n) = K \, a^n u(n)$

If $x(n)$ is sinusoid, then $y_p(n)$ is also a sinusoid.

> **Example:**
> Let, $x(n) = A \cos \omega_0 n$; now, $y_p(n) = K_1 \cos \omega_0 n + K_2 \sin \omega_0 n$

The general form of particular solution for various types of inputs are listed in Table 5.2.

Table 5.2: Particular Solution

Input Signal, $x(n)$	Particular Solution, $y_p(n)$
A	K
$A B^n$	$K B^n$
$A n^B$	$K_0 n^B + K_1 n^{(B-1)} + ... + K_B$
$A^n n^B$	$A^n (K_0 n^B + K_1 n^{(B-1)} + ...+ K_B)$
$A \cos \omega_0 n$	$K_1 \cos \omega_0 n + K_2 \sin \omega_0 n$
$A \sin \omega_0 n$	$K_1 \cos \omega_0 n + K_2 \sin \omega_0 n$

5.3.3 Zero-State Response

The *zero-state response* or *forced response* is obtained from the sum of homogeneous solution and particular solution and evaluating the constants with zero initial conditions.

$$\therefore \text{Zero} - \text{state response}, \ y_{zs}(n) = y_h(n) + y_p(n)\big|_{\text{with constants } C_1, C_2, ...C_N \text{ evaluated with zero initial conditions}}$$

5.3.4 Total Response

The total response of discrete time system can be obtained by the following two methods.

Method-1

The *total response* is given by sum of homogeneous solution and particular solution.

$\therefore$ Total response, $y(n) = y_h(n) + y_p(n)$

Procedure to Determine Total Response by Method-1

1. Determine the homogeneous solution $y_h(n)$ with constants $C_1, C_2, ...C_N$.

2. Determine the particular solution $y_p(n)$ and evaluate the constants K for any value of $n \geq 1$ so that no term of $y(n)$ vanishes.

3. Now the total response is given by the sum of $y_h(n)$ and $y_p(n)$.

$\therefore$ Total response, $y(n) = y_h(n) + y_p(n)$

4. The total response will have N number of constants $C_1, C_2, ...C_N$. Evaluate the given difference equation for $n = 0, 1, 2, ...N-1$ and form one set of N number of equations. Then evaluate the total response for $n = 0, 1, 2, ... N-1$ and form another set of N number of equations. Now solve the constants $C_1, C_2, ...C_N$ using the two sets of N number of equations.

Method-2

The *total response* is given by sum of zero-input response and zero-state response.

$\therefore$ Total response, $y(n) = y_{zi}(n) + y_{zs}(n)$

Procedure to Determine Total Response by Method-2

1. Determine the homogeneous solution $y_h(n)$ with constants $C_1, C_2, ...C_N$.

2. Determine the zero-input response, which is obtained from the homogeneous solution $y_h(n)$ and evaluating the constants $C_1, C_2, ...C_N$ using the initial conditions.

3. Determine the particular solution $y_p(n)$ and evaluate the constants K for any value of $n \geq 1$ so that no term of $y(n)$ vanishes.

4. Determine the zero-state response, $y_{zs}(n)$ which is given by sum of homogeneous solution and particular solution and evaluating the constants $C_1, C_2, ...C_N$ with zero initial conditions.

5. Now, the total response is given by sum of zero input response and zero state response.

$\therefore$ Total response, $y(n) = y_{zi}(n) + y_{zs}(n)$

Example 5.2

Determine the response of first order discrete time system governed by the difference equation,

$y(n) = -0.8\, y(n-1) + x(n)$

When the input is unit step, and with initial condition a) $y(-1) = 0$ b) $y(-1) = 2/9$.

Solution:

Given that, $y(n) = -0.8\, y(n-1) + x(n)$

$$\therefore \quad y(n) + 0.8\, y(n-1) = x(n) \qquad\qquad .....(1)$$

Homogeneous Solution

The homogeneous equation is the solution of equation (1) when $x(n) = 0$.

$$\therefore \quad y(n) + 0.8\, y(n-1) = 0 \qquad\qquad .....(2)$$

Put, $y(n) = \lambda^n$ in equation (2).

$$\therefore \quad \lambda^n + 0.8\, \lambda^{(n-1)} = 0$$

$$\lambda^{(n-1)}\,(\lambda + 0.8) = 0 \quad\Rightarrow\quad \lambda = -0.8$$

The homogeneous solution $y_h(n)$ is given by,

$$y_h(n) = C\,\lambda^n = C\,(-0.8)^n \ ; \quad \text{for } n \geq 0 \qquad\qquad .....(3)$$

Particular Solution

Given that the input is unit step and so the particular solution will be in the form,

$$y(n) = K\,u(n) \qquad\qquad .....(4)$$

On substituting for $y(n)$ from equation (4) in equation (1) we get,

$$y(n) + 0.8\, y(n-1) = x(n) \quad\Rightarrow\quad K\,u(n) + 0.8\,K\,u(n-1) = u(n) \qquad .....(5)$$

In order to determine the value of K, let us evaluate equation (5) for $n = 1$, ($\because$ we have to evaluate equation (5) for any $n \geq 1$, such that none of the term vanishes).

From equation (5) when $n = 1$, we get,

$$K + 0.8\,K = 1 \quad\Rightarrow\quad 1.8\,K = 1 \quad\Rightarrow\quad K = \frac{1}{1.8} = \frac{10}{18} = \frac{5}{9}$$

The particular solution $y_p(n)$ is given by,

$$y_p(n) = K\,u(n) = \frac{5}{9}\,u(n)\,; \quad \text{for all } n$$

$$= \frac{5}{9} \quad ; \ \text{for } \ n \geq 0$$

Total Response

The total response $y(n)$ of the system is given by sum of homogeneous and particular solution.

$$\therefore \ \text{Response, } y(n) = y_h(n) + y_p(n)$$

$$\therefore \quad y(n) = C(-0.8)^n + \frac{5}{9} \ ; \ \text{for } \ n \geq 0 \qquad\qquad .....(6)$$

When $n = 0$, from equation (1), we get, $y(0) + 0.8\, y(-1) = 1$

$$\therefore \quad y(0) = 1 - 0.8\, y(-1) \qquad\qquad .....(7)$$

When $n = 0$, from equation (6), we get, $y(0) = C + \dfrac{5}{9}$ $\qquad\qquad .....(8)$

On equating (7) and (8) we get, $C + \dfrac{5}{9} = 1 - 0.8\, y(-1)$

$$\therefore \quad C = 1 - 0.8\, y(-1) - \frac{5}{9}$$

$$= \frac{4}{9} - 0.8\, y(-1) \qquad\qquad .....(9)$$

On substituting for C from equation (9) in equation (6) we get,

$$y(n) = \left(\frac{4}{9} - 0.8\, y(-1)\right)(-0.8)^n + \frac{5}{9}$$

a) When $y(-1) = 0$

$$y(n) = \frac{4}{9}(-0.8)^n + \frac{5}{9} \; ; \; \text{for } n \geq 0$$

$$= \left[\frac{4}{9}(-0.8)^n + \frac{5}{9}\right] u(n)$$

b) When $y(-1) = 2/9$

$$y(n) = \left(\frac{4}{9} - 0.8 \times \frac{2}{9}\right)(-0.8)^n + \frac{5}{9} = \frac{2.4}{9}(-0.8)^n + \frac{5}{9} = \frac{24}{90}(-0.8)^n + \frac{5}{9}$$

$$\therefore \; y(n) = \frac{5}{9} + \frac{12}{45}(-0.8)^n \; ; \quad \text{for } n \geq 0$$

$$= \left[\frac{5}{9} + \frac{12}{45}(-0.8)^n\right] u(n)$$

Example 5.3

Determine the response $y(n)$, $n \geq 0$ of the system described by the second order difference equation,

$$y(n) - 0.2\, y(n-1) - 0.03\, y(n-2) = x(n) + 0.4\, x(n-1),$$

when the input signal is, $x(n) = 0.2^n\, u(n)$ and with initial conditions $y(-2) = 0$, $y(-1) = 0.5$.

Solution:

Given that, $y(n) - 0.2\, y(n-1) - 0.03\, y(n-2) = x(n) + 0.4\, x(n-1)$ .....(1)

Homogeneous Solution

The homogeneous solution is the solution of equation (1) when $x(n) = 0$.

$$\therefore \; y(n) - 0.2\, y(n-1) - 0.03\, y(n-2) = 0 \qquad\qquad\qquad\qquad .....(2)$$

Put $y(n) = \lambda^n$ in equation (2).

$$\therefore \; \lambda^n - 0.2\, \lambda^{n-1} - 0.03\, \lambda^{n-2} = 0$$

$$\lambda^{n-2}(\lambda^2 - 0.2\lambda - 0.03) = 0$$

The characteristic equation is,

$$\lambda^2 - 0.2\lambda - 0.03 = 0 \quad \Rightarrow \quad (\lambda - 0.3)(\lambda + 0.1) = 0$$

$$\therefore \text{ The roots are, } \lambda = 0.3, -0.1$$

> The roots of quadratic,
> $\lambda^2 - 0.2\lambda - 0.03 = 0$ are,
> $$\lambda = \frac{0.2 \pm \sqrt{0.2^2 + 4 \times 0.03}}{2}$$
> $$= \frac{0.2 \pm 0.4}{2} = 0.3, -0.1.$$

The homogeneous solution, $y_h(n)$ is given by,

$$y_h(n) = C_1\lambda_1^n + C_2\lambda_2^n$$

$$= C_1(0.3)^n + C_2(-0.1)^n \; ; \; \text{for } n \geq 0 \qquad\qquad\qquad .....(3)$$

Particular Solution

Given that the input is an exponential signal, $0.2^n\, u(n)$ and so the particular solution will be in the form,

$$y(n) = K\, 0.2^n\, u(n) \qquad\qquad\qquad\qquad\qquad\qquad\qquad\qquad .....(4)$$

On substituting for $y(n)$ from equation (4) in equation (1) we get,

$$K0.2^n\, u(n) - 0.2\, K\, 0.2^{(n-1)}\, u(n-1) - 0.03\, K\, 0.2^{(n-2)}\, u(n-2) = 0.2^n\, u(n) + 0.4 \times 0.2^{(n-1)}\, u(n-1) \quad .....(5)$$

In order to determine the value of K, let us evaluate equation (5) for $n = 2$, ($\because$ we have to evaluate equation (5) for any $n \geq 1$, such that none of the term vanishes).

From equation (5) when n = 2, we get,

$$K\ 0.2^2 - 0.2K \times 0.2^1 - 0.03K \times 0.2^0 = 0.2^2 + 0.4 \times 0.2^1$$

$$0.04K - 0.04K - 0.03K = 0.04 + 0.08$$

$$-0.03K = 0.12$$

$$\therefore\ K\ =\ -\frac{0.12}{0.03}\ =\ -4$$

The particular solution $y_p(n)$ is given by,

$$y_p\ (n) = K\ 0.2^n\ u(n) = (-4)\ 0.2^n\ u(n)$$

Total Response

The total response y(n) of the system is given by sum of homogeneous and particular solution.

$$\therefore\ \text{Response,}\ \ y(n) = y_h\ (n) + y_p\ (n)$$

$$= C_1\ 0.3^n + C_2\ (-0.1)^n + (-4)\ 0.2^n\ ;\ \ \text{for } n \geq 0 \qquad\qquad .....(6)$$

To find y(0) and y(1)

When $n = 0$,

From equation (1) we get,

$$y(0)\ -\ 0.2\ y(-1) - 0.03\ y(-2) = x(0) + 0.4\ x(-1) \qquad\qquad .....(7)$$

Given that, $\ y(-1) = 0.5,\ y(-2) = 0$

$$x(n) = 0.2^n\ u(n), \qquad\qquad \therefore\ \ x(0) = 0.2^0 = 1,\ \ \ x(-1) = 0$$

On substituting the above conditions in equation (7) we get,

$$y(0) - 0.2 \times 0.5 - 0.03 \times 0 = 1 + 0$$

$$\therefore\ \ y(0) = 1.1 \qquad\qquad .....(8)$$

When $\ n = 1$,

From equation (1) we get,

$$y(1) - 0.2\ y(0) - 0.03\ y(-1) = x(1) + 0.4\ x(0) \qquad\qquad .....(9)$$

We know that, $y(0) = 1.1,\ \ \ y(-1) = 0.5,\ \ \ y(-2) = 0$

Given that, $\quad x(n) = 0.2^n\ u(n),\quad\ \therefore\ \ x(0) = 0.2^0 = 1$

$$x(1) = 0.2^1 = 0.2$$

On substituting the above conditions in equation (9) we get,

$$y(1) - 0.2 \times 1.1 - 0.03 \times 0.5 = 0.2 + 0.4 \times 1$$

$$\therefore\ \ y(1) =\ 0.6\ +\ 0.235\ = 0.835 \qquad\qquad .....(10)$$

To solve constants C_1 and C_2

When n = 0,

From equation (6) we get,

$$y(0) = C_1\ 0.3^0 + C_2\ (-0.1)^0 + (-4)\ 0.2^0 = C_1 + C_2 - 4 \qquad\qquad .....(11)$$

From equations (8) and (11) we can write,

$$C_1 + C_2 - 4 = 1.1$$

$$\therefore\ \ C_1 + C_2 = 5.1 \qquad\qquad .....(12)$$

When n = 1,

From equation (6) we get,

$$y(1) = C_1 \times 0.3 + C_2 (-0.1) + (-4)\, 0.2 = 0.3\, C_1 - 0.1 C_2 - 0.8 \qquad .....(13)$$

From equations (10) and (13) we can write,

$$0.3\, C_1 - 0.1 C_2 - 0.8 = 0.835$$

$$\therefore \quad 0.3\, C_1 - 0.1 C_2 = 1.635 \qquad\qquad .....(14)$$

Equation (12) × 0.1 $\Rightarrow$ $0.1 C_1 + 0.1 C_2 = 0.51$

Equation (13) $\Rightarrow$ $0.3 C_1 - 0.1 C_2 = 1.635$

                Add $0.4 C_1 \qquad\quad = 2.145$

$$\therefore \; C_1 \; = \; \frac{2.145}{0.4} \; = \; 5.3625$$

From equation(12),

$$C_2 = 5.1 - C_1 = 5.1 - 5.3625$$
$$= -0.2625$$

Total Response

$$y(n) = [5.3625(0.3)^n - 0.2625(-0.1)^n + (-4)\, 0.2^n]\, u(n) \quad ; \quad \text{for all } n$$

5.4 Convolution Sum

                                                  (AU Dec'13, 2 Marks)

5.4.1 Discrete or Linear Convolution

The ***Discrete*** or ***Linear convolution*** of two discrete time sequences $x_1(n)$ and $x_2(n)$ is defined as,

$$\boxed{x_3(n) = \sum_{m=-\infty}^{\infty} x_1(m)\, x_2(n-m)} \quad \text{or} \quad \boxed{x_3(n) = \sum_{m=-\infty}^{+\infty} x_2(m)\, x_1(n-m)} \qquad .....(5.10)$$

        where, $x_3(n)$ is the sequence obtained by convolving $x_1(n)$ and $x_2(n)$

                      m is a dummy variable

The convolution relation of equation (5.10) can be symbolically expressed as,

$$x_3(n) = x_1(n) * x_2(n) = x_2(n) * x_1(n) \qquad ..... (5.11)$$

        where, the symbol * indicates convolution operation.

In linear convolution, the sequences $x_1(n)$ and $x_2(n)$ are nonperiodic sequences and the sequence $x_3(n)$ obtained by convolution is also nonperiodic. Hence, this convolution is also called *aperiodic convolution.*

Procedure for Evaluating Linear Convolution

Let $x_1(n)$ and $x_2(n)$ be two discrete time sequences.

Let $x_3(n)$ be the sequence obtained by convolution of $x_1(n)$ and $x_2(n)$.

$$\therefore \; x_3(n) = x_1(n) * x_2(n) = \sum_{m=-\infty}^{+\infty} x_1(m)\, x_2(n-m) \; ; \; -\infty < n < +\infty$$

Now, each sample of $x_3(n)$ can be computed using the above equation.

The value of $x_3(n)$ at n = q is obtained by replacing n by q, in the above equation.

$$\therefore \; x_3(q) = \sum_{m=-\infty}^{+\infty} x_1(m)\, x_2(q-m) \qquad .....(5.12)$$

The evaluation of equation (5.12) to determine the value of $x_3(n)$ at $n = q$, involves the following five steps.

1. Change of index : Change the index n in the sequences $x_1(n)$ and $x_2(n)$, to get the sequences $x_1(m)$ and $x_2(m)$.

2. Folding : Fold $x_2(m)$ about $m = 0$, to obtain $x_2(-m)$.

3. Shifting : Shift $x_2(-m)$ by q to the right if q is positive, shift $x_2(-m)$ by q to the left if q is negative to obtain $x_2(q-m)$.

4. Multiplication : Multiply $x_1(m)$ by $x_2(q-m)$ to get a product sequence. Let the product sequence be $v_q(m)$. Now, $v_q(m) = x_1(m) \times x_2(q-m)$.

5. Summation : Sum all the values of the product sequence $v_q(m)$ to obtain the value of $x_3(n)$ at $n = q$. [i.e., $x_3(q)$].

The above procedure will give the value of $x_3(n)$ at a single time instant say $n = q$. In general, we are interested in evaluating the values of the sequence $x_3(n)$ over all the time instants in the range $-\infty < n < \infty$. Hence the steps 3, 4 and 5 given above must be repeated, for all possible time shifts in the range $-\infty < n < \infty$.

Convolution of Finite Duration Sequences

In convolution of finite duration sequences it is possible to predict the length of resultant sequence.

If the sequence $x_1(n)$ has N_1 samples and sequence $x_2(n)$ has N_2 samples then the output sequence $x_3(n)$ will be a finite duration sequence consisting of "$N_1 + N_2 - 1$" samples.

i.e., if, Length of $x_1(n) = N_1$

Length of $x_2(n) = N_2$

then, Length of $x_3(n) = N_1 + N_2 - 1$

In the convolution of finite duration sequences it is possible to predict the start and end of the resultant sequence. If $x_1(n)$ starts at $n = n_1$ and $x_2(n)$ starts at $n = n_2$ then, the initial value of n for $x_3(n)$ is "$n = n_1 + n_2$". The value of $x_1(n)$ for $n < n_1$ and the value of $x_2(n)$ for $n < n_2$ are then assumed to be zero. The final value of n for $x_3(n)$ is "$n = (n_1 + n_2) + (N_1 + N_2 - 2)$".

i.e., if, $x_1(n)$ start at $n = n_1$

$x_2(n)$ start at $n = n_2$

then, $x_3(n)$ start at $n = n_1 + n_2$

and $x_3(n)$ end at $n = (n_1 + n_2) + (N_1 + N_2 - 1) - 1$

$= (n_1 + n_2) + (N_1 + N_2 - 2)$

5.4.2 Representation of Discrete Time Signal as Summation of Impulses

A discrete time signal can be expressed as summation of impulses and this concept will be useful to prove that the response of discrete time LTI system can be determined using discrete convolution.

Let, $x(n)$ = Discrete time signal

$\delta(n)$ = Unit impulse signal

$\delta(n - m)$ = Delayed impulse signal

We know that, $\delta(n) = 1$; at $n = 0$

$\qquad = 0$; when $n \neq 0$

and, $\delta(n - m) = 1$; at $n = m$

$\qquad = 0$; when $n \neq m$

If we multiply the signal $x(n)$ with the delayed impulse $\delta(n - m)$ then the product is nonzero only at $n = m$ and zero for all other values of n. Also at $n = m$, the value of product signal is m^{th} sample $x(m)$ of the signal $x(n)$.

$\therefore x(n) \, \delta(n - m) = x(m)$

Each multiplication of the signal $x(n)$ by an unit impulse at some delay m, in essence picks out the single value $x(m)$ of the signal $x(n)$ at $n = m$, where the unit impulse is nonzero. Consequently if we repeat this multiplication for all possible delays in the range $-\infty < m < \infty$ and add all the product sequences, the result will be a sequence that is equal to the sequence $x(n)$.

For example, $x(n) \, \delta(n - (-2)) \quad = \quad x(-2)$

$x(n) \, \delta(n - (-1)) \quad = \quad x(-1)$

$x(n) \, \delta(n) \qquad = \quad x(0)$

$x(n) \, \delta(n - 1) \quad = \quad x(1)$

$x(n) \, \delta(n - 2) \quad = \quad x(2)$

From the above products we can say that each sample of $x(n)$ can be expressed as a product of the sample and delayed impulse, as shown below.

$\therefore \quad x(-2) = x(-2) \, \delta(n - (-2))$

$x(-1) = x(-1) \, \delta(n - (-1))$

$x(0) = x(0) \, \delta(n)$

$x(1) = x(1) \, \delta(n - 1)$

$x(2) = x(2) \, \delta(n - 2)$

$\therefore x(n) = \quad \cdots + x(-2) + x(-1) + x(0) + x(1) + x(2) + \cdots$

$$\therefore \ x(n) \ = \ \cdots\cdots + x(-2)\,\delta(n-(-2)) + x(-1)\,\delta(n-(-1)) + x(0)\,\delta(n) + x\,(1)\,\delta(n-1)$$

$$+ x(2)\,\delta(n-2) + \cdots\cdots$$

$$\therefore \ x(n) \ = \ \sum_{m=-\infty}^{+\infty} x(m)\,\delta(n-m) \qquad\qquad .....(5.13)$$

In equation (5.13) each product $x(m)\,\delta(n-m)$ is an impulse and the summation of impulses gives the sequence $x(n)$.

5.4.3 Response of LTI Discrete Time System Using Discrete Convolution

In an LTI system, the response $y(n)$ of the system for an arbitrary input $x(n)$ is given by convolution of input $x(n)$ with impulse response $h(n)$ of the system. It is expressed as,

$$y(n) \ = \ x(n) * h(n) \ = \ \sum_{m=-\infty}^{+\infty} x(m)\,h(n-m) \qquad\qquad .....(5.14)$$

where, the symbol $*$ represents convolution operation.

Proof:

Let $y(n)$ be the response of system H for an input $x(n)$.

$$\therefore \ y(n) = \mathcal{H}\{x(n)\} \qquad\qquad .....(5.15)$$

$$= \mathcal{H}\left\{ \sum_{m=-\infty}^{+\infty} x(m)\,\delta(n-m) \right\}$$

From equation (5.13)

$$\therefore \ x(n) \ = \ \sum_{m=-\infty}^{+\infty} x(m)\,\delta(n-m)$$

In linear system, summation and system operation $\mathcal{H}$ can be interchanged.

$$= \sum_{m=-\infty}^{+\infty} \mathcal{H}\{x(m)\,\delta(n-m)\}$$

The system $\mathcal{H}$ is a function of n and not a function of m.

$$= \sum_{m=-\infty}^{+\infty} x(m)\,\mathcal{H}\{\delta(n-m)\}$$

By time invariant property
if $\mathcal{H}\{\delta(n)\} = h(n)$
then, $\mathcal{H}\{\delta(n-m)\} = h(n-m)$

$$= \sum_{m=-\infty}^{+\infty} x(m)\,h(n-m) \qquad\qquad .....(5.16)$$

The equation (5.16) represents the convolution of input $x(n)$ with the impulse response $h(n)$ to yield the output $y(n)$. Hence it is proved that the response $y(n)$ of LTI discrete time system for an arbitrary input $x(n)$ is given by convolution of input $x(n)$ with impulse response $h(n)$ of the system.

5.4.4 Properties of Linear Convolution

The Discrete or Linear convolution will satisfy the following properties.

Commutative property : $x_1(n) * x_2(n) \ = \ x_2(n) * x_1(n)$ $\qquad\qquad .....(5.17)$

Associative property : $[x_1(n) * x_2(n)] * x_3(n) = x_1(n) * [x_2(n) * x_3(n)]$ $\qquad .....(5.18)$

Distributive property : $x_1(n) * [x_2(n) + x_3(n)] = [x_1(n) * x_2(n)] + [x_1(n) * x_3(n)]$ $\quad .....(5.19)$

Proof of Commutative Property:

Consider convolution of $x_1(n)$ and $x_2(n)$.

By commutative property we can write,

$x_1(n) * x_2(n) = x_2(n) * x_1(n)$

$\quad$ (LHS) $\qquad$ (RHS)

LHS $= x_1(n) * x_2(n)$

$$= \sum_{m=-\infty}^{+\infty} x_1(m)\, x_2(n-m) \qquad \qquad .....(5.20)$$

where, m is a dummy variable used for convolution operation.

Let, $\quad n - m = p$ $\qquad$ when $m = -\infty$, $p = n - m = n + \infty = +\infty$

$\therefore m = n - p$ $\qquad$ when $m = +\infty$, $p = n - m = n - \infty = -\infty$

On replacing m by $(n - p)$ and $(n - m)$ by p in equation (5.20) we get,

$$\text{LHS} = \sum_{p=-\infty}^{+\infty} x_1(n-p)\, x_2(p) = \sum_{p=-\infty}^{+\infty} x_2(p)\, x_1(n-p)$$

$$= x_2(n) * x_1(n)$$

$$= \text{RHS}$$

$\boxed{\text{p is a dummy variable used for convolution operation.}}$

Proof of Associative Property:

Consider the discrete time signals $x_1(n)$, $x_2(n)$ and $x_3(n)$.

Let, $\quad y_1(n) = x_1(n) * x_2(n)$ $\qquad\qquad\qquad .....(5.21)$

Let us replace n by p

$$\therefore y_1(p) = x_1(p) * x_2(p) = \sum_{m=-\infty}^{+\infty} x_1(m)\, x_2(p-m) \qquad .....(5.22)$$

Let, $\quad y_2(n) = x_2(n) * x_3(n)$ $\qquad\qquad\qquad .....(5.23)$

$$\therefore y_2(n) = \sum_{q=-\infty}^{+\infty} x_1(q)\, x_2(n-q)$$

$$\therefore y_2(n-m) = \sum_{q=-\infty}^{+\infty} x_1(q)\, x_2(n-q-m) \qquad .....(5.24)$$

where p, m and q are dummy variables used for convolution operation.

By associative property we can write,

$[x_1(n) * x_2(n)] * x_3(n) = x_1(n) * [x_2(n) * x_3(n)]$

$\qquad\quad$ LHS $\qquad\qquad\qquad$ RHS

LHS $= [x_1(n) * x_2(n)] * x_3(n)$

$$= y_1(n) * x_3(n) = \sum_{p=-\infty}^{+\infty} y_1(p)\, x_3(n-p) \qquad \boxed{\text{Using equation (5.21).}}$$

$$= \sum_{p=-\infty}^{+\infty} \sum_{m=-\infty}^{+\infty} x_1(m)\, x_2(p-m)\, x_3(n-p) \qquad \boxed{\text{Using equation (5.22).}}$$

$$= \sum_{m=-\infty}^{+\infty} x_1(m) \sum_{p=-\infty}^{+\infty} x_2(p-m)\, x_3(n-p) \qquad .....(5.25)$$

Let, $p - m = q$ | when $p = -\infty$, $q = p - m = -\infty - m = -\infty$

$\therefore p = q + m$ | when $p = +\infty$, $q = p - m = +\infty - m = +\infty$

On replacing $(p - m)$ by q, and p by $(q + m)$ in the equation (5.25) we get,

$$\text{LHS} = \sum_{m=-\infty}^{+\infty} x_1(m) \sum_{q=-\infty}^{+\infty} x_2(q)\, x_3(n - q - m)$$

$$= \sum_{m=-\infty}^{+\infty} x_1(m)\, y_2(n - m)$$

Using equation (5.24).

$$= x_1(n) * y_2(n)$$

$$= x_1(n) * [x_2(n) * x_3(n)]$$

Using equation (5.23).

$$= \text{RHS}$$

Proof of Distributive Property:

Consider the discrete time signals $x_1(n)$, $x_2(n)$ and $x_3(n)$. By distributive property we can write,

$$x_1(n) * [x_2(n) + x_3(n)] = [x_1(n) * x_2(n)] + [x_1(n) * x_3(n)]$$

$\qquad\qquad$ LHS $\qquad\qquad\qquad\qquad$ RHS

$$\text{LHS} = x_1(n) * [x_2(n) + x_3(n)]$$

$$= x_1(n) * x_4(n)$$

$x_4(n) = x_2(n) + x_3(n)$

$$= \sum_{m=-\infty}^{+\infty} x_1(m)\, x_4(n - m)$$

m is a dummy variable used for convolution opearation.

$$= \sum_{m=-\infty}^{+\infty} x_1(m)\, [x_2(n - m) + x_3(n - m)]$$

$x_4(n - m) = x_2(n - m) + x_3(n - m)$

$$= \sum_{m=-\infty}^{+\infty} x_1(m)\, x_2(n - m) + \sum_{m=-\infty}^{+\infty} x_1(m)\, x_3(n - m)$$

$$= [x_1(n) * x_2(n)] + [x_1(n) * x_3(n)]$$

$$= \text{RHS}$$

5.4.5 Methods of Performing Linear Convolution

Method 1: Graphical Method

Let $x_1(n)$ and $x_2(n)$ be the input sequences and $x_3(n)$ be the output sequence.

1. Change the index "n" of input sequences to "m" to get $x_1(m)$ and $x_2(m)$.

2. Sketch the graphical representation of the input sequences $x_1(m)$ and $x_2(m)$.

3. Let us fold $x_2(m)$ to get $x_2(-m)$. Sketch the graphical representation of the folded sequence $x_2(-m)$.

4. Shift the folded sequence $x_2(-m)$ to the left graphically so that the product of $x_1(m)$ and shifted $x_2(-m)$ gives only one nonzero sample. Now multiply $x_1(m)$ and shifted $x_2(-m)$ to get a product sequence, and then sum up the samples of product sequence, which is the first sample of output sequence.

5. To get the next sample of output sequence, shift $x_2(-m)$ of previous step to one position right and multiply the shifted sequence with $x_1(m)$ to get a product sequence. Now the sum of the samples of product sequence gives the second sample of output sequence.

6. To get subsequent samples of output sequence, the step 5 is repeated until we get a non-zero product sequence.

Method 2: Tabular Method

The tabular method is same as that of graphical method, except that the tabular representation of the sequences are employed instead of graphical representation. In tabular method, every input sequence, folded and shifted sequence is represented by a row in a table.

Method 3: Matrix Method

Let $x_1(n)$ and $x_2(n)$ be the input sequences and $x_3(n)$ be the output sequence. In matrix method one of the sequences is represented as a row and the other as a column as shown below:

Multiply each column element with row elements and fill up the matrix array.

Now the sum of the diagonal elements gives the samples of output sequence $x_3(n)$. (The sum of the diagonal elements are shown below for reference).

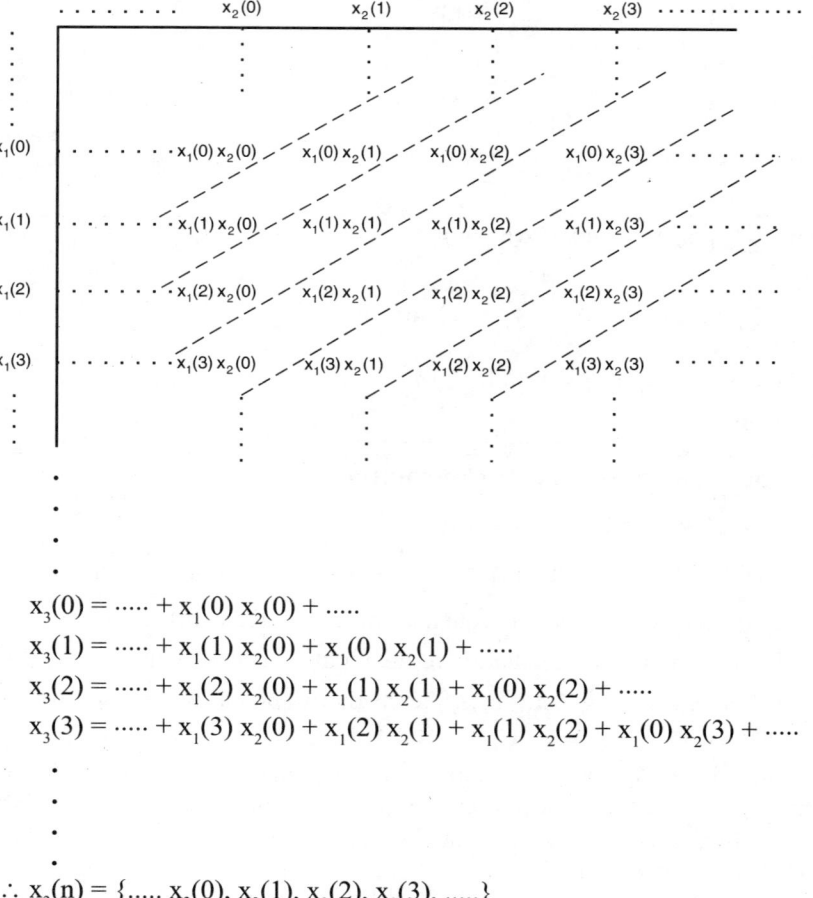

$$x_3(0) = \cdots + x_1(0)\, x_2(0) + \cdots$$
$$x_3(1) = \cdots + x_1(1)\, x_2(0) + x_1(0)\, x_2(1) + \cdots$$
$$x_3(2) = \cdots + x_1(2)\, x_2(0) + x_1(1)\, x_2(1) + x_1(0)\, x_2(2) + \cdots$$
$$x_3(3) = \cdots + x_1(3)\, x_2(0) + x_1(2)\, x_2(1) + x_1(1)\, x_2(2) + x_1(0)\, x_2(3) + \cdots$$

$$\therefore\ x_3(n) = \{\cdots\ x_3(0),\ x_3(1),\ x_3(2),\ x_3(3),\ \cdots\}$$

Example 5.4   *(AU May'15, 6 Marks)*

Determine the response of the LTI system whose input x(n) and impulse response h(n) are given by,

x(n) = {1, 2, 0.5, 1} and h(n) = {1, 2, 1, −1}
    ↑   ↑

Solution:

The response y(n) of the system is given by convolution of x(n) and h(n).

$$y(n) = x(n) * h(n) = \sum_{m=-\infty}^{\infty} x(m)\,h(n-m)$$

In this example the convolution operation is performed by three methods.

The Input sequence starts at n = 0 and the impulse response sequence starts at n = −1. Therefore the output sequence starts at n = 0 + (−1) = −1.

The input and impulse response consists of 4 samples, so the output consists of 4 + 4 − 1 = 7 samples.

Method 1 : Graphical Method

The graphical representation of x(n) and h(n) after replacing n by m are shown below. The sequence h(m) is folded with respect to m = 0 to obtain h(−m).

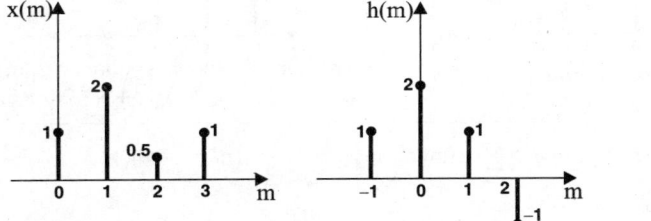

Fig 1: Input sequence.   *Fig 2: Impulse response.*   *Fig 3: Folded impulse response.*

The samples of y(n) are computed using the convolution formula,

$$y(n) = \sum_{m=-\infty}^{\infty} x(m)\,h(n-m) = \sum_{m=-\infty}^{\infty} x(m)\,h_n(m)\ ;\ \text{where } h_n(m) = h(n-m)$$

The computation of each sample using the above equation are graphically shown in Fig 4 to Fig 10. The graphical representation of output sequence is shown in Fig 11.

When n = − 1 ; $y(-1) = \sum_{m=-\infty}^{+\infty} x(m)\,h(-1-m) = \sum_{m=-\infty}^{+\infty} x(m)\,h_{-1}(m) = \sum_{m=-\infty}^{+\infty} v_{-1}(m)$

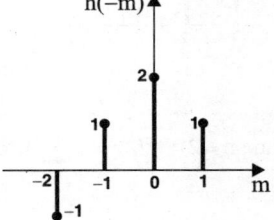

Fig 4: Computation of y(−1).   The sum of product sequence $v_{-1}(m)$ gives y(−1). ∴ y(−1) = 1

When n = 0 ; $y(0) = \sum\limits_{m=-\infty}^{+\infty} x(m) \, h(0-m) = \sum\limits_{m=-\infty}^{+\infty} x(m) \, h_0(m) = \sum\limits_{m=-\infty}^{+\infty} v_0(m)$

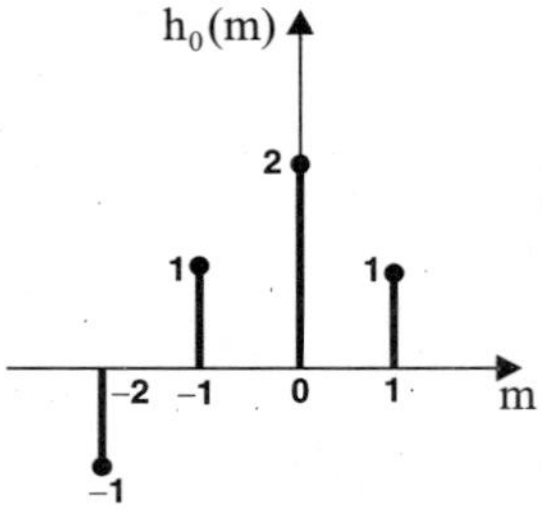

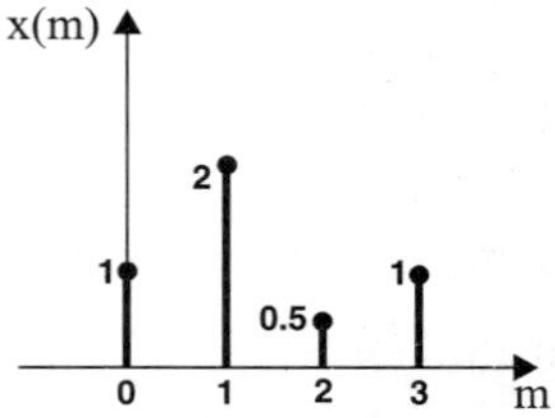

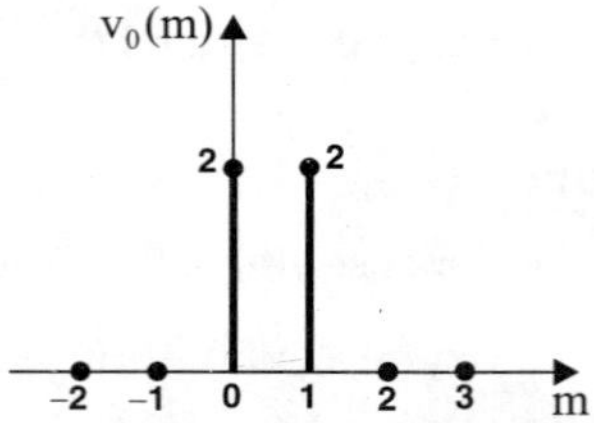

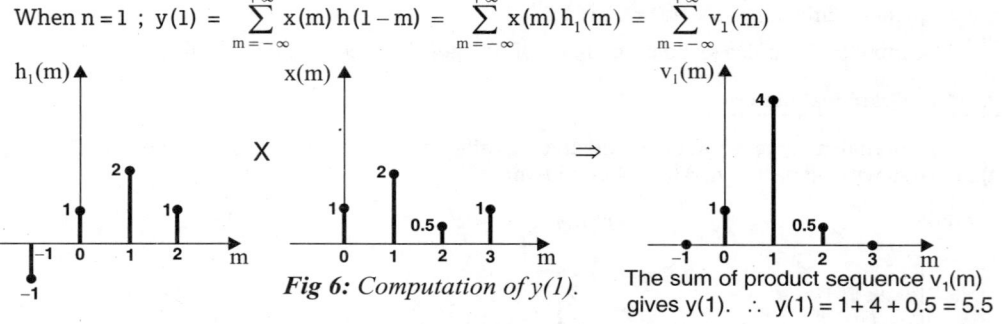

Fig 5: *Computation of y(0).*

The sum of product sequence $v_0(m)$ gives y(0). ∴ y(0) = 2 + 2 = 4

When n = 1 ; $y(1) = \sum\limits_{m=-\infty}^{+\infty} x(m) \, h(1-m) = \sum\limits_{m=-\infty}^{+\infty} x(m) \, h_1(m) = \sum\limits_{m=-\infty}^{+\infty} v_1(m)$

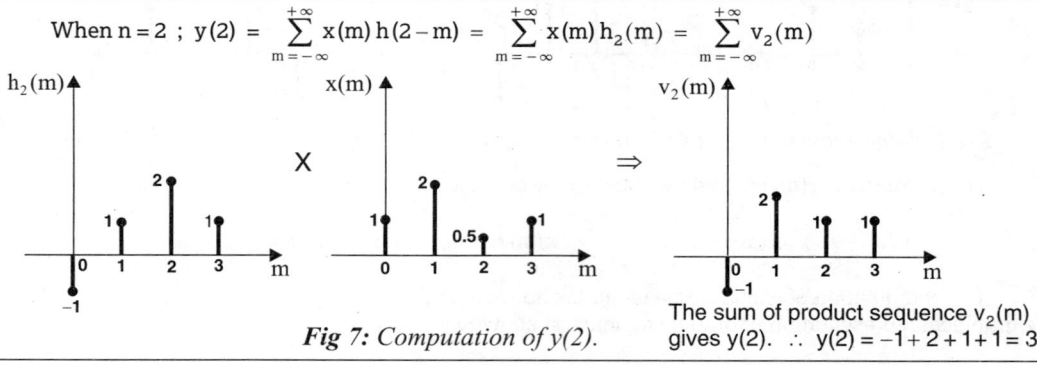

Fig 6: *Computation of y(1).*

The sum of product sequence $v_1(m)$ gives y(1). ∴ y(1) = 1 + 4 + 0.5 = 5.5

When n = 2 ; $y(2) = \sum\limits_{m=-\infty}^{+\infty} x(m) \, h(2-m) = \sum\limits_{m=-\infty}^{+\infty} x(m) \, h_2(m) = \sum\limits_{m=-\infty}^{+\infty} v_2(m)$

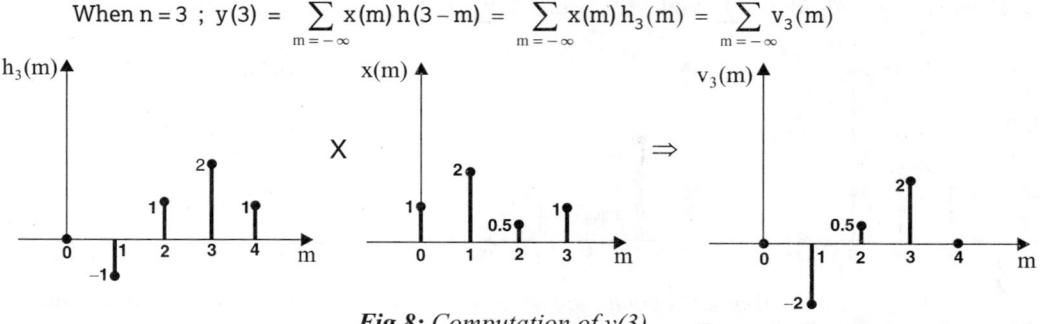

Fig 7: *Computation of y(2).*

The sum of product sequence $v_2(m)$ gives y(2). ∴ y(2) = −1 + 2 + 1 + 1 = 3

When n = 3 ; $y(3) = \sum\limits_{m=-\infty}^{+\infty} x(m) \, h(3-m) = \sum\limits_{m=-\infty}^{+\infty} x(m) \, h_3(m) = \sum\limits_{m=-\infty}^{+\infty} v_3(m)$

Fig 8: *Computation of y(3).*

The sum of product sequence $v_3(m)$ gives y(3). ∴ y(3) = −2 + 0.5 + 2 = 0.5

When $n = 4$; $y(4) = \sum_{m=-\infty}^{+\infty} x(m)\, h(4-m) = \sum_{m=-\infty}^{+\infty} x(m)\, h_4(m) = \sum_{m=-\infty}^{+\infty} v_4(m)$

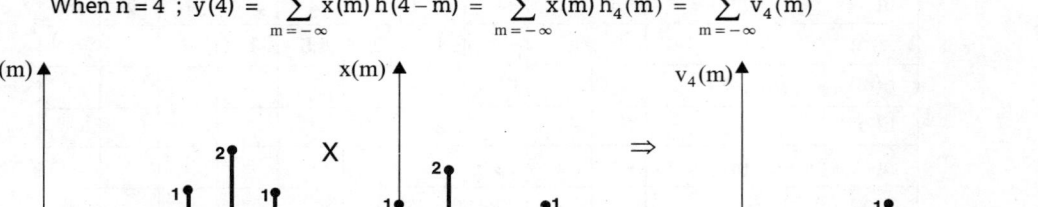

Fig 9: Computation of y(4).

The sum of product sequence $v_4(m)$ gives $y(4)$. $\therefore y(4) = -0.5 + 1 = 0.5$

When $n = 5$; $y(5) = \sum_{m=-\infty}^{+\infty} x(m)\, h(5-m) = \sum_{m=-\infty}^{+\infty} x(m)\, h_5(m) = \sum_{m=-\infty}^{+\infty} v_5(m)$

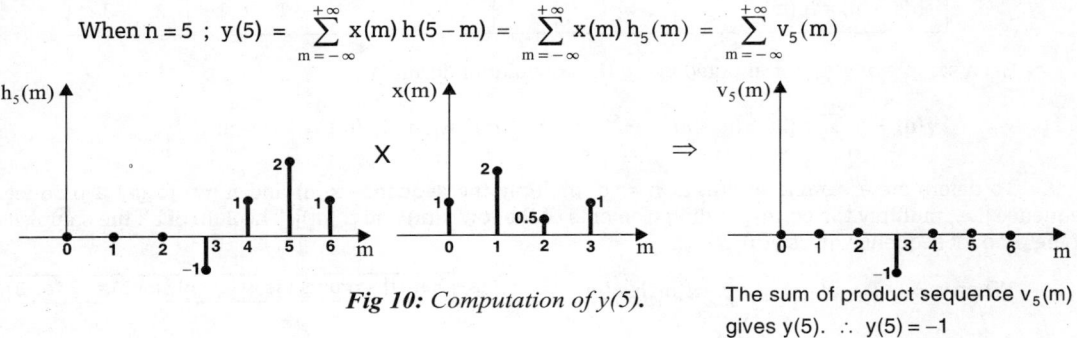

Fig 10: Computation of y(5).

The sum of product sequence $v_5(m)$ gives $y(5)$. $\therefore y(5) = -1$

The output sequence, $y(n) = \{1, 4, 5.5, 3, 0.5, 0.5, -1\}$

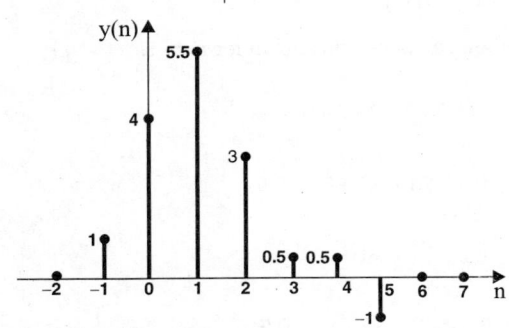

Fig 11: Graphical representation of y(n).

Method 2 : Tabular Method

The given sequences and the shifted sequences can be represented in the tabular array as shown below:

> **Note:** *The unfilled boxes in the table are considered as zeros.*

m	-3	-2	-1	0	1	2	3	4	5	6
x(m)				1	2	0.5	1			
h(m)			1	2	1	-1				
h(-m)		-1	1	2	1					
$h(-1-m) = h_{-1}(m)$	-1	1	2	1						
$h(0-m) = h_0(m)$		-1	1	2	1					
$h(1-m) = h_1(m)$			-1	1	2	1				
$h(2-m) = h_2(m)$				-1	1	2	1			
$h(3-m) = h_3(m)$					-1	1	2	1		
$h(4-m) = h_4(m)$						-1	1	2	1	
$h(5-m) = h_5(m)$							-1	1	2	1

Each sample of y(n) is computed using the convolution formula,

$$y(n) = \sum_{m=-\infty}^{+\infty} x(m)\, h(n-m) = \sum_{m=-\infty}^{+\infty} x(m)\, h_n(m), \text{ where } h_n(m) = h(n-m)$$

To determine a sample of y(n) at n = q, multiply the sequence x(m) and $h_q(m)$ to get a product sequence [i.e., multiply the corresponding elements of the row x(m) and $h_q(m)$]. The sum of all the samples of the product sequence gives y(q).

When n = -1 ; $y(-1) = \sum_{m=-3}^{3} x(m)\, h_{-1}(m)$ | Here the product is valid only for m = -3 to +3. |

$$= x(-3)\, h_{-1}(-3) + x(-2) h_{-1}(-2) + x(-1) h_{-1}(-1) + x(0)\, h_{-1}(0) + x(1)\, h_{-1}(1)$$
$$+ x(2)\, h_{-1}(2) + x(3)\, h_{-1}(3)$$

$$= 0 + 0 + 0 + 1 + 0 + 0 + 0 = 1$$

The samples of y(n) for other values of n are calculated as shown for n = -1.

When n = 0 ; $y(0) = \sum_{m=-2}^{3} x(m)\, h_0(m) = 0+0+2+2+0+0 = 4$

When n = 1 ; $y(1) = \sum_{m=-1}^{3} x(m)\, h_1(m) = 0+1+4+0.5+0 = 5.5$

When n = 2 ; $y(2) = \sum_{m=0}^{3} x(m)\, h_2(m) = -1+2+1+1 = 3$

When n = 3 ; $y(3) = \sum_{m=0}^{4} x(m)\, h_3(m) = 0-2+0.5+2+0 = 0.5$

When n = 4 ; $y(4) = \sum_{m=0}^{5} x(m)\, h_4(m) = 0+0-0.5+1+0+0 = 0.5$

When n = 5 ; $y(5) = \sum_{m=0}^{6} x(m)\, h_5(m) = 0+0+0-1+0+0+0 = -1$

The output sequence , y(n) = { 1, 4, 5.5, 3, 0.5, 0.5, -1}

↑

Method 3 : Matrix Method

(AU May'15, 6 Marks)

The input sequence x(n) is arranged as a column and the impulse response is arranged as a row as shown below. The elements of the two-dimensional array are obtained by multiplying the corresponding row element with the column element. The sum of the diagonal elements gives the samples of y(n).

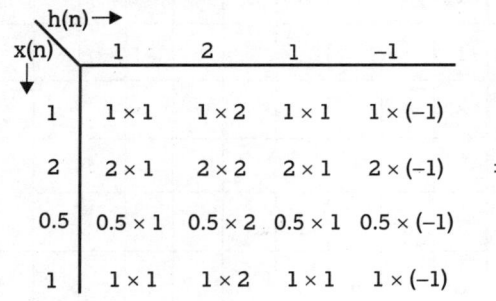 ⟹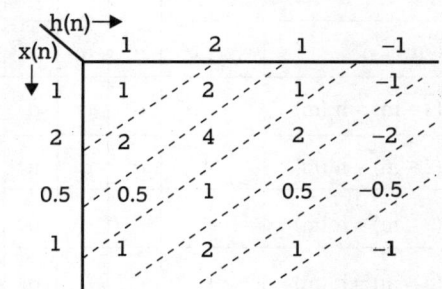

$$y(-1) = 1$$
$$y(0) = 2 + 2 = 4$$
$$y(1) = 0.5 + 4 + 1 = 5.5$$
$$y(2) = 1 + 1 + 2 + (-1) = 3$$

$$y(3) = 2 + 0.5 + (-2) = 0.5$$
$$y(4) = 1 + (-0.5) = 0.5$$
$$y(5) = -1$$

$$\therefore y(n) = \{1, 4, 5.5, 3, 0.5, 0.5, -1\}$$
$$\uparrow$$

Example 5.5

(AU Dec'13, 10 Marks)

Determine the response of the LTI system whose input x(n) and impulse response h(n) are given by,

$$x(n) = 1, \quad 0 \le n \le 4$$
$$= 0, \quad \text{otherwise}$$

and $h(n) = \alpha^n, 0 \le n \le 6$

$$= 0, \quad \text{otherwise}$$

Solution:

The input sequence starts at n = 0 and the impulse response sequence starts at n = 0. Therefore the output sequence starts at n = 0 + 0 = 0.

The input sequence consist of 5 samples and the impulse response sequence consists of 7 samples so the output consists of 5 + 7 −1 = 11 samples.

The given sequences and the shifted sequences can be represented in the tabular array as shown below:

m	−6	−5	−4	−3	−2	−1	0	1	2	3	4	5	6	7	8	9	10
x(m)							1	1	1	1	1						
h(m)							1	α	α^2	α^3	α^4	α^5	α^6				
h(−m)	α^6	α^5	α^4	α^3	α^2	α	1										
h(1 − m) = h_1(m)		α^6	α^5	α^4	α^3	α^2	α	1									
h(2 − m) = h_2(m)			α^6	α^5	α^4	α^3	α^2	α	1								
h(3 − m) = h_3(m)				α^6	α^5	α^4	α^3	α^2	α	1							
h(4 − m) = h_4(m)					α^6	α^5	α^4	α^3	α^2	α	1						
h(5 − m) = h_5(m)						α^6	α^5	α^4	α^3	α^2	α	1					
h(6 − m) = h_6(m)							α^6	α^5	α^4	α^3	α^2	α	1				
h(7 − m) = h_7(m)								α^6	α^5	α^4	α^3	α^2	α	1			
h(8 − m) = h_8(m)									α^6	α^5	α^4	α^3	α^2	α	1		
h(9 − m) = h_9(m)										α^6	α^5	α^4	α^3	α^2	α	1	
h(10 − m) = h_{10}(m)											α^6	α^5	α^4	α^3	α^2	α	1

Each sample of y(n) is computed using the convolution formula,

$$y(n) = \sum_{m=-\infty}^{+\infty} x(m)\, h(n-m) = \sum_{m=-\infty}^{+\infty} x(m)\, h_n(m), \text{ where } h_n(m) = h(n-m)$$

To determine a sample of y(n) at n = q, multiply the sequence x(m) and h_q(m) to get a product sequence [i.e., multiply the corresponding elements of the row x(m) and h_q(m)]. The sum of all the samples of the product sequence gives y(q).

When n = 0 ; $y(0) = \sum_{m=-6}^{4} x(m)\, h_0(m)$ | Here the product is valid only for m = −6 to +4.

$$= x(-6)\, h_0(-6) + x(-5)h_0(-5) + x(-4)\, h_0(-4) + x(-3)\, h_0(-3) + x(-2)h_0(-2)$$
$$+ x(-1)h_0(-1) + x(0)\, h_0(0) + x(1)\, h_0(1) + x(2)\, h_0(2) + x(3)\, h_0(3) + x(4)\, h_0(4)$$

$$= 0 \times \alpha^6 + 0 \times \alpha^5 + 0 \times \alpha^4 + 0 \times \alpha^3 + 0 \times \alpha^2 + 0 \times \alpha + 1 \times 1 + 1 \times 0 + 1 \times 0 + 1 \times 0 + 1 \times 0 = 1$$

The samples of y(n) for other values of n are calculated as shown for n = 0.

When n = 1 ; $y(1) = \sum_{m=-5}^{4} x(m)\, h_1(m) = 0 + 0 + 0 + 0 + 0 + \alpha + 1 + 0 + 0 + 0 = \alpha + 1$

When n = 2 ; $y(2) = \sum_{m=-4}^{4} x(m)\, h_2(m) = 0 + 0 + 0 + 0 + \alpha^2 + \alpha + 1 + 0 + 0 = \alpha^2 + \alpha + 1$

When $n = 3$; $y(3) = \sum\limits_{m=-3}^{4} x(m)\, h_3(m) = 0+0+0+\alpha^3+\alpha^2+\alpha+1+0 = \alpha^3+\alpha^2+\alpha+1$

When $n = 4$; $y(4) = \sum\limits_{m=-2}^{4} x(m)\, h_4(m) = 0+0+\alpha^4+\alpha^3+\alpha^2+\alpha+1 = \alpha^4+\alpha^3+\alpha^2+\alpha+1$

When $n = 5$; $y(5) = \sum\limits_{m=-1}^{5} x(m)\, h_5(m) = 0+\alpha^5+\alpha^4+\alpha^3+\alpha^2+\alpha+0 = \alpha^5+\alpha^4+\alpha^3+\alpha^2+\alpha$

When $n = 6$; $y(6) = \sum\limits_{m=0}^{6} x(m)\, h_6(m) = \alpha^6+\alpha^5+\alpha^4+\alpha^3+\alpha^2+0+0 = \alpha^6+\alpha^5+\alpha^4+\alpha^3+\alpha^2$

When $n = 7$; $y(7) = \sum\limits_{m=0}^{7} x(m)\, h_7(m) = 0+\alpha^6+\alpha^5+\alpha^4+\alpha^3+0+0+0 = \alpha^6+\alpha^5+\alpha^4+\alpha^3$

When $n = 8$; $y(8) = \sum\limits_{m=0}^{8} x(m)\, h_8(m) = 0+0+\alpha^6+\alpha^5+\alpha^4+0+0+0+0 = \alpha^6+\alpha^5+\alpha^4$

When $n = 9$; $y(9) = \sum\limits_{m=0}^{9} x(m)\, h_9(m) = 0+0+0+\alpha^6+\alpha^5+0+0+0+0+0 = \alpha^6+\alpha^5$

When $n = 10$; $y(10) = \sum\limits_{m=0}^{10} x(m)\, h_{10}(m) = 0+0+0+0+\alpha^6+0+0+0+0+0+0 = \alpha^6$

Example 5.6

Determine the output y(n) of a relaxed LTI system with impulse response,

$h(n) = a^n\, u(n)$; where $|a| < 1$ and

When input is a unit step sequence, i.e., $x(n) = u(n)$.

Solution:

The graphical representation of x(n) and h(n) after replacing n by m are shown below. Also the sequence x(m) is folded to get x(–m).

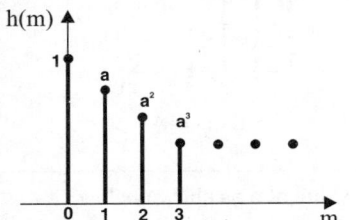

Fig 1: *Impulse response.*

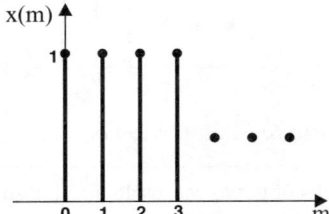

Fig 2: *Impulse sequence.*

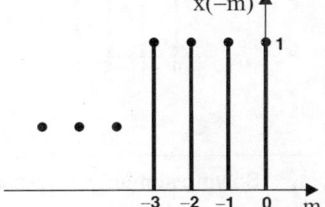

Fig 3: *Folded input sequence.*

Here both h(m) and x(m) are infinite duration sequences starting at n = 0. Hence the output sequence y(n) will also be an infinite duration sequence starting at n = 0.

By convolution formula,

$$y(n) = \sum_{m=-\infty}^{+\infty} h(m)\, x(n-m) = \sum_{m=0}^{\infty} h(m)\, x_n(m), \text{ where } x_n(m) = x(n-m)$$

The computation of some samples of y(n) using the above equation are graphically shown below:

When $n = 0$; $y(0) = \sum\limits_{m=0}^{\infty} h(m)\, x(0-m) = \sum\limits_{m=0}^{\infty} h(m)\, x_0(m) = \sum\limits_{m=0}^{\infty} v_0(m)$

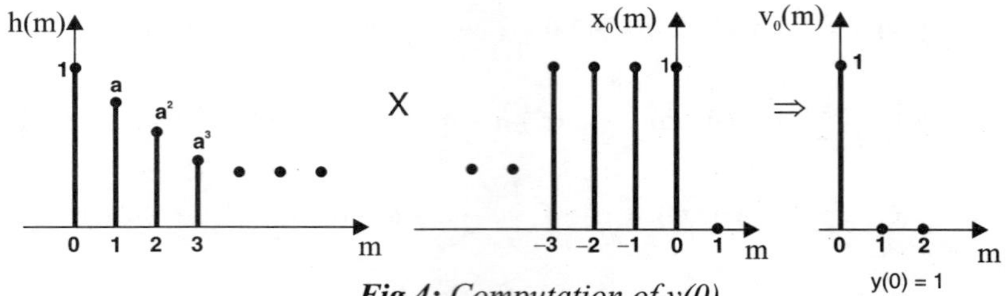

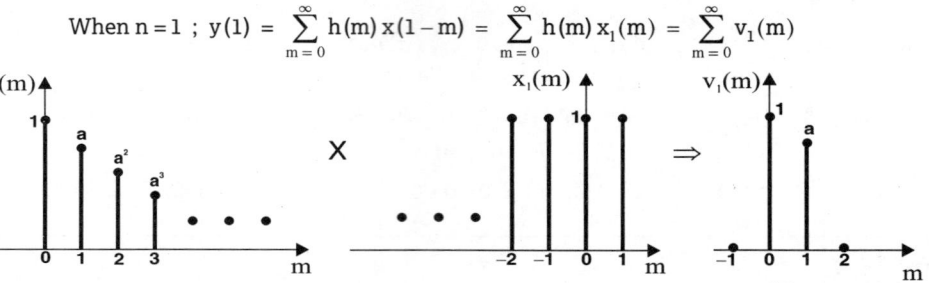

Fig 4: Computation of y(0).

When $n = 1$; $y(1) = \sum\limits_{m=0}^{\infty} h(m)\, x(1-m) = \sum\limits_{m=0}^{\infty} h(m)\, x_1(m) = \sum\limits_{m=0}^{\infty} v_1(m)$

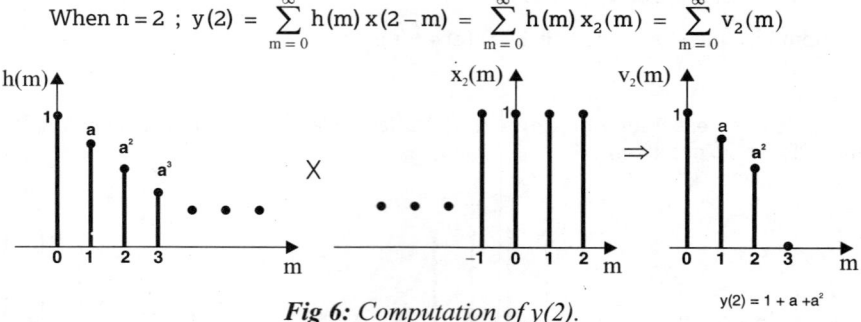

Fig 5: Computation of y(1).

When $n = 2$; $y(2) = \sum\limits_{m=0}^{\infty} h(m)\, x(2-m) = \sum\limits_{m=0}^{\infty} h(m)\, x_2(m) = \sum\limits_{m=0}^{\infty} v_2(m)$

Fig 6: Computation of y(2).

Solving similarly for other values of n, we can write y(n) for any value of n as shown below.

$$y(n) = 1 + a + a^2 + \ldots\ldots + a^n = \sum\limits_{p=0}^{n} a^p \ ; \ \text{for } n \geq 0$$

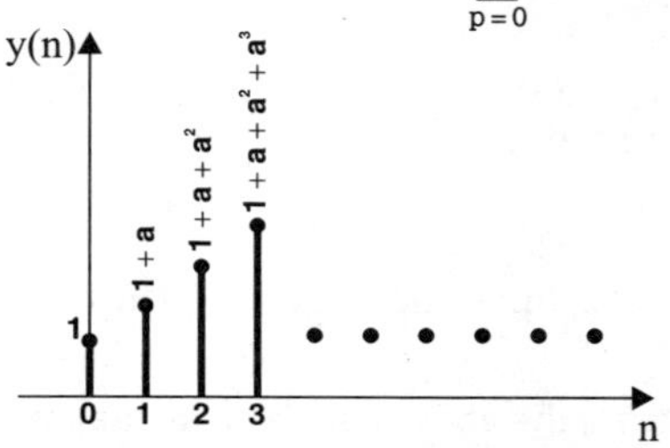

Fig 7: Graphical representation of y(n).

Example 5.7 (AU Jun'14, 16 Marks)

Find the inverse Fourier transform of $X(e^{j\omega}) = \dfrac{1}{(1-\alpha e^{-j\omega})^2}$, $|\alpha| < 1$ using convolution property.

Solution:

Given that, $X(e^{j\omega}) = \dfrac{1}{(1-\alpha e^{-j\omega})^2}$

Let, $X(e^{j\omega}) = X_1(e^{j\omega}) \, X_2(e^{j\omega})$

where, $X_1(e^{j\omega}) = X_2(e^{j\omega}) = \dfrac{1}{1-\alpha e^{-j\omega}}$

$\therefore x_1(n) = \mathcal{F}^{-1}\{X_1(e^{j\omega})\} = \mathcal{F}^{-1}\left\{\dfrac{1}{1-\alpha e^{-j\omega}}\right\} = \alpha^n \, u(n)$

$x_2(n) = \mathcal{F}^{-1}\{X_2(e^{j\omega})\} = \mathcal{F}^{-1}\left\{\dfrac{1}{1-\alpha e^{-j\omega}}\right\} = \alpha^n \, u(n)$

By convolution property of Fourier transform we can write,

$\mathcal{F}\{x_1(n) * x_2(n)\} = X_1(e^{j\omega}) \, X_2(e^{j\omega})$

$\therefore x_1(n) * x_2(n) = \mathcal{F}^{-1}\{X_1(e^{j\omega}) \, X_2(e^{j\omega})\}$

$\therefore x_1(n) * x_2(n) = \mathcal{F}^{-1}\{X(e^{j\omega})\}$

$\boxed{x(n) = \mathcal{F}^{-1}\{X(e^{j\omega})\}}$

$\therefore x(n) = x_1(n) * x_2(n)$

Here, $x_1(n) = \alpha^n \, u(n) = \alpha^n$; $n \geq 0$ $\Rightarrow$ $x_1(m) = \alpha^m$; $m \geq 0$

$x_2(n) = \alpha^n \, u(n) = \alpha^n$; $n \geq 0$ $\Rightarrow$ $x_2(m) = \alpha^m$; $m \geq 0$

$\therefore x_2(n-m) = \alpha^{(n-m)}$; $n > 0$ and $0 < m < n$

$\therefore x(n) = x_1(n) * x_2(n)$

$= \displaystyle\sum_{m=-\infty}^{+\infty} x_1(m) \, x_2(n-m)$; for $n \geq 0$

$\boxed{\begin{array}{l}\text{The product } x_1(m) \, x_2(n-m) \text{ is}\\ \text{nonzero in the range } 0 < m < n\end{array}}$

$= \displaystyle\sum_{m=0}^{n} \alpha^m \, \alpha^{(n-m)} = \sum_{m=0}^{n} \alpha^m \, \alpha^n \alpha^{-m}$

$= \alpha^n \displaystyle\sum_{m=0}^{n} \alpha^m \, \alpha^{-m}$

When $n = 0$; $x(0) = \alpha^0 \displaystyle\sum_{m=0}^{} \alpha^m \, \alpha^{-m} = \alpha^0 \, \alpha^0 \, \alpha^0 = 1$

When $n = 1$; $x(1) = \alpha^1 \displaystyle\sum_{m=0}^{1} \alpha^m \, \alpha^{-m} = \alpha^1[\alpha^0 \, \alpha^0 + \alpha^1 \, \alpha^{-1}] = (1+1)\alpha$

When $n = 2$; $x(2) = \alpha^2 \displaystyle\sum_{m=0}^{2} \alpha^m \, \alpha^{-m} = \alpha^2[\alpha^0 \, \alpha^0 + \alpha^1 \, \alpha^{-1} + \alpha^2 \, \alpha^{-2}] = \alpha^2[1+1+1] = (2+1)\alpha^2$

When $n = 3$; $x(3) = \alpha^3 \displaystyle\sum_{m=0}^{3} \alpha^m \, \alpha^{-m} = \alpha^3[\alpha^0 \, \alpha^0 + \alpha^1 \, \alpha^{-1} + \alpha^2 \, \alpha^{-2} + \alpha^3 \, \alpha^{-3}] = \alpha^3[1+1+1+1] = (3+1)\alpha^3$

When $n = 4$; $x(4) = \alpha^4 \displaystyle\sum_{m=0}^{4} \alpha^m \, \alpha^{-m} = \alpha^4[\alpha^0 \alpha^0 + \alpha^1 \alpha^{-1} + \alpha^2 \alpha^{-2} + \alpha^3 \alpha^{-3} + \alpha^4 \alpha^{-4}] = \alpha^4[1+1+1+1+1] = (4+1)\alpha^4$

From the above analysis the nth term of $x(n)$ can be written as,

$x(n) = (n+1)\,\alpha^n$; $n \geq 0$

$\therefore x(n) = \mathcal{F}^{-1}\{X(e^{j\omega})\} = (n+1)\,\alpha^n \, u(n)$

5.5 Interconnections of Discrete Time Systems

Smaller discrete time systems may be interconnected to form larger systems. Two possible basic ways of interconnection are ***cascade connection*** and ***parallel connection***. The cascade and parallel connections of two discrete time systems with impulse responses $h_1(n)$ and $h_2(n)$ are shown in Fig 5.3.

Fig 5.3 a: *Cascade connection.*   ***Fig 5.3 b:*** *Parallel connection.*

Fig 5.3: *Interconnection of discrete time systems.*

5.5.1 Cascade Connected Discrete Time Systems

Two cascade connected discrete time systems with impulse response $h_1(n)$ and $h_2(n)$ can be replaced by a single equivalent discrete time system whose impulse response is given by convolution of individual impulse responses.

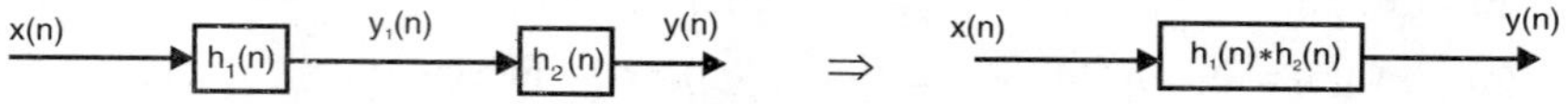

Fig 5.4: *Cascade connected discrete time system and its equivalent*

Proof:

With reference to Fig 5.4 we can write,

$$y_1(n) = x(n) * h_1(n) \qquad \qquad \qquad .....(5.26)$$

$$y(n) = y_1(n) * h_2(n) \qquad \qquad \qquad .....(5.27)$$

Using equation (5.26), the equation (5.27) can be written as,

$$y(n) = x(n) * h_1(n) * h_2(n)$$

$$= x(n) * [h_1(n) * h_2(n)]$$

$$= x(n) * h(n) \qquad \qquad \qquad .....(5.28)$$

where, $h(n) = h_1(n) * h_2(n)$

From equation (5.28) we can say that the overall impulse response of two cascaded discrete time systems is given by convolution of individual impulse responses.

5.5.2 Parallel Connected Discrete Time Systems

Two parallel connected discrete time systems with impulse responses $h_1(n)$ and $h_2(n)$ can be replaced by a single equivalent discrete time system whose impulse response is given by sum of individual impulse responses.

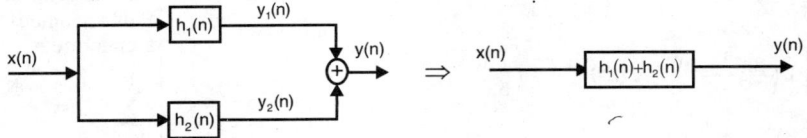

Fig 5.5: *Parallel connected discrete time system and its equivalent.*

Proof:

With reference to Fig 5.5 we can write,

$$y_1(n) = x(n) * h_1(n) \qquad\qquad .....(5.29)$$

$$y_2(n) = x(n) * h_2(n) \qquad\qquad .....(5.30)$$

$$y(n) = y_1(n) + y_2(n) \qquad\qquad .....(5.31)$$

On substituting for $y_1(n)$ and $y_2(n)$ from equations (5.29) and (5.30) in equation (5.31) we get,

$$y(n) = [x(n) * h_1(n)] + [x(n) * h_2(n)] \qquad\qquad .....(5.32)$$

By using distributive property of convolution, the equation (5.32) can be written as shown below:

$$y(n) = x(n) * [h_1(n) + h_2(n)]$$

$$= x(n) * h(n) \qquad\qquad .....(5.33)$$

where, $h(n) = h_1(n) + h_2(n)$

From equation (5.33) we can say that the overall impulse response of two parallel connected discrete time systems is given by sum of individual impulse responses.

Example 5.8

Determine the impulse response for the cascade of two LTI systems having impulse responses,

$$h_1(n) = \left(\tfrac{2}{5}\right)^n u(n) \text{ and } h_2(n) = \left(\tfrac{1}{5}\right)^n u(n).$$

Solution:

Given that, $h_1(n) = \left(\tfrac{2}{5}\right)^n u(n) = \left(\tfrac{2}{5}\right)^n \ ; \ n \geq 0 \quad \Rightarrow \quad h_1(m) = \left(\tfrac{2}{5}\right)^m \ ; \ m \geq 0$

$$h_2(n) = \left(\tfrac{1}{5}\right)^n u(n) = \left(\tfrac{1}{5}\right)^n \ ; \ n \geq 0 \quad \Rightarrow \quad h_2(m) = \left(\tfrac{1}{5}\right)^m \ ; \ m \geq 0$$

$$\therefore h_2(m-n) = \left(\tfrac{1}{5}\right)^{m-n} \ ; \ n > 0 \text{ and } 0 < m < n$$

Let $h(n)$ be the impulse response and given by convolution of $h_1(n)$ and $h_2(n)$.

$$\therefore h(n) = h_1(n) * h_2(n) = \sum_{m=-\infty}^{+\infty} h_1(m)\, h_2(n-m) \ ; \ n \geq 0$$

where, m is a dummy variable used for convolution operation

The product $h_1(m) h_2(n-m)$ will be nonzero in the range $0 \le m \le n$. Therefore the summation index in the above equation is changed to $m = 0$ to n.

$$\therefore h(n) = \sum_{m=0}^{n} h_1(m) h_2(n-m) = \sum_{m=0}^{n} \left(\frac{2}{5}\right)^m \left(\frac{1}{5}\right)^{n-m} = \sum_{m=0}^{n} \left(\frac{2}{5}\right)^m \left(\frac{1}{5}\right)^n \left(\frac{1}{5}\right)^{-m} = \left(\frac{1}{5}\right)^n \sum_{m=0}^{n} \left(\frac{2}{5}\right)^m 5^m$$

$$= \left(\frac{1}{5}\right)^n \sum_{m=0}^{n} \left(\frac{2\times 5}{5}\right)^m = \left(\frac{1}{5}\right)^n \sum_{m=0}^{n} 2^m$$

> Finite geometric series sum formula.
> $$\sum_{n=0}^{N} C^n = \frac{C^{N+1} - 1}{C - 1}$$

$$= \left(\frac{1}{5}\right)^n \left(\frac{2^{n+1}-1}{2-1}\right) = 2(2^n)\left(\frac{1}{5}\right)^n - \left(\frac{1}{5}\right)^n$$

$$= 2\left(\frac{2}{5}\right)^n - \left(\frac{1}{5}\right)^n \; ; \; n \ge 0$$

$$= \left[2\left(\frac{2}{5}\right)^n - \left(\frac{1}{5}\right)^n\right] u(n)$$

Example 5.9

Determine the overall impulse response of the interconnected discrete time systems shown below:

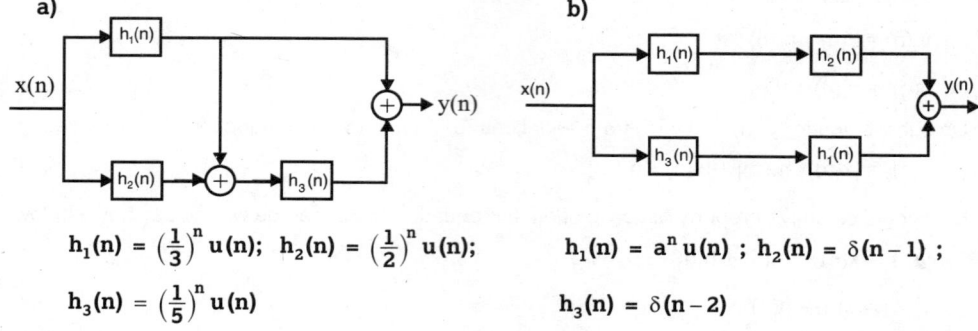

$$h_1(n) = \left(\frac{1}{3}\right)^n u(n); \quad h_2(n) = \left(\frac{1}{2}\right)^n u(n); \qquad h_1(n) = a^n u(n) \; ; \; h_2(n) = \delta(n-1) \; ;$$

$$h_3(n) = \left(\frac{1}{5}\right)^n u(n) \qquad\qquad\qquad\qquad h_3(n) = \delta(n-2)$$

Solution:

a)   The given system can be redrawn as shown below:

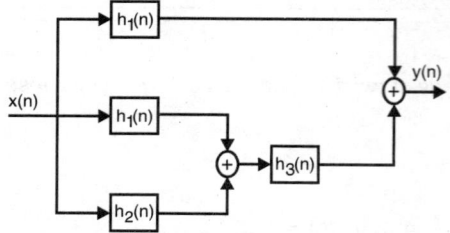

The above system can be reduced to single equivalent system as shown below:

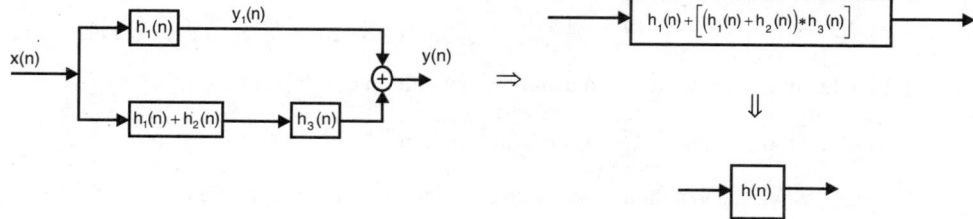

Here, $h(n) = h_1(n) + [(h_1(n) + h_2(n)) * h_3(n)]$

$\qquad = h_1(n) + [h_1(n) * h_3(n)] + [h_2(n) * h_3(n)]$

Given that, $h_1(n) = \left(\frac{1}{3}\right)^n u(n) = \left(\frac{1}{3}\right)^n$; $n \geq 0 \quad \Rightarrow \quad h_1(m) = \left(\frac{1}{3}\right)^m$; $m \geq 0$

$\qquad h_2(n) = \left(\frac{1}{2}\right)^n u(n) = \left(\frac{1}{2}\right)^n$; $n \geq 0 \quad \Rightarrow \quad h_2(m) = \left(\frac{1}{2}\right)^m$; $m \geq 0$

$\qquad h_3(n) = \left(\frac{1}{5}\right)^n u(n) = \left(\frac{1}{5}\right)^n$; $n \geq 0 \quad \Rightarrow \quad h_3(m) = \left(\frac{1}{5}\right)^m$; $m \geq 0$

$\qquad\qquad\qquad\qquad\qquad \therefore\ h_3(m-n) = \left(\frac{1}{5}\right)^{m-n}$; $n > 0$ and $0 < m < n$

Let us evaluate the convolution of $h_1(n)$ and $h_3(n)$.

$\therefore\ h_1(n) * h_3(n) = \displaystyle\sum_{m=-\infty}^{\infty} h_1(m)\, h_3(n-m)$; $n \geq 0$

> The product of $h_1(m)\, h_3(n-m)$ is nonzero in the range $0 \leq m \leq n$. Therefore the summation index is changed to $m = 0$ to n.

$\qquad = \displaystyle\sum_{m=0}^{n} h_1(m)\, h_3(n-m)$

$\qquad = \displaystyle\sum_{m=0}^{n} \left(\frac{1}{3}\right)^m \left(\frac{1}{5}\right)^{n-m} = \displaystyle\sum_{m=0}^{n} \left(\frac{1}{3}\right)^m \left(\frac{1}{5}\right)^n \left(\frac{1}{5}\right)^{-m}$

$\qquad = \left(\frac{1}{5}\right)^n \displaystyle\sum_{m=0}^{n} \left(\frac{1}{3}\right)^m 5^m = \left(\frac{1}{5}\right)^n \displaystyle\sum_{m=0}^{n} \left(\frac{5}{3}\right)^m$

$\qquad = \left(\frac{1}{5}\right)^n \dfrac{\left(\frac{5}{3}\right)^{n+1} - 1}{\frac{5}{3} - 1}$

> Finite geometric series sum formula.
> $$\sum_{n=0}^{N} C^n = \frac{C^{N+1} - 1}{C - 1}$$

$\qquad = \left(\frac{1}{5}\right)^n \dfrac{\left(\frac{5}{3}\right)^n \frac{5}{3} - 1}{\frac{5-3}{3}} = \left(\frac{1}{5}\right)^n \left[\frac{3}{2}\left(\frac{5}{3}\right)^n \frac{5}{3} - \frac{3}{2}\right]$

$\qquad = \frac{5}{2}\left(\frac{1}{5}\right)^n \left(\frac{5}{3}\right)^n - \frac{3}{2}\left(\frac{1}{5}\right)^n = \frac{5}{2}\left(\frac{1}{3}\right)^n - \frac{3}{2}\left(\frac{1}{5}\right)^n$; $n \geq 0$

$\qquad = \frac{5}{2}\left(\frac{1}{3}\right)^n u(n) - \frac{3}{2}\left(\frac{1}{5}\right)^n u(n)$

Let us evaluate the convolution of $h_2(n)$ and $h_3(n)$.

$\therefore\ h_2(n) * h_3(n) = \displaystyle\sum_{m=-\infty}^{+\infty} h_2(m)\, h_3(n-m)$

> The product of $h_2(m)$ and $h_3(n-m)$ is nonzero in the range $0 \leq m \leq n$. Therefore the summation index is changed to $m = 0$ to n.

$\qquad = \displaystyle\sum_{m=0}^{n} h_2(m)\, h_3(n-m)$

$\qquad = \displaystyle\sum_{m=0}^{n} \left(\frac{1}{2}\right)^m \left(\frac{1}{5}\right)^{n-m} = \displaystyle\sum_{m=0}^{n} \left(\frac{1}{2}\right)^m \left(\frac{1}{5}\right)^n \left(\frac{1}{5}\right)^{-m}$

> Finite geometric series sum formula.
> $$\sum_{n=0}^{N} C^n = \frac{C^{N+1} - 1}{C - 1}$$

$\qquad = \left(\frac{1}{5}\right)^n \displaystyle\sum_{m=0}^{n} \left(\frac{1}{2}\right)^m 5^m = \left(\frac{1}{5}\right)^n \displaystyle\sum_{m=0}^{n} \left(\frac{5}{2}\right)^m$

$\qquad = \left(\frac{1}{5}\right)^n \dfrac{\left(\frac{5}{2}\right)^{n+1} - 1}{\frac{5}{2} - 1} = \left(\frac{1}{5}\right)^n \dfrac{\left(\frac{5}{2}\right)^n \frac{5}{2} - 1}{\frac{5-2}{2}} = \left(\frac{1}{5}\right)^n \left[\frac{2}{3}\left(\frac{5}{2}\right)^n \frac{5}{2} - \frac{2}{3}\right]$

$\qquad = \frac{5}{3}\left(\frac{1}{5}\right)^n \left(\frac{5}{2}\right)^n - \frac{2}{3}\left(\frac{1}{5}\right)^n = \frac{5}{3}\left(\frac{1}{2}\right)^n - \frac{2}{3}\left(\frac{1}{5}\right)^n$; $n \geq 0$

$\qquad = \frac{5}{3}\left(\frac{1}{2}\right)^n u(n) - \frac{2}{3}\left(\frac{1}{5}\right)^n u(n)$

Now, the overall impulse response h(n) is given by,

$$h(n) = h_1(n) + [h_1(n) * h_3(n)] + [h_2(n) * h_3(n)]$$

$$= \left(\frac{1}{3}\right)^n u(n) + \frac{5}{2}\left(\frac{1}{3}\right)^n u(n) - \frac{3}{2}\left(\frac{1}{5}\right)^n u(n) + \frac{5}{3}\left(\frac{1}{2}\right)^n u(n) - \frac{2}{3}\left(\frac{1}{5}\right)^n u(n)$$

$$= \left(1+\frac{5}{2}\right)\left(\frac{1}{3}\right)^n u(n) - \left(\frac{3}{2}+\frac{2}{3}\right)\left(\frac{1}{5}\right)^n u(n) + \frac{5}{3}\left(\frac{1}{2}\right)^n u(n)$$

$$= \left[\frac{7}{2}\left(\frac{1}{3}\right)^n - \frac{13}{6}\left(\frac{1}{5}\right)^n + \frac{5}{3}\left(\frac{1}{2}\right)^n\right] u(n)$$

b)   The given system can be reduced to single equivalent system as shown below:

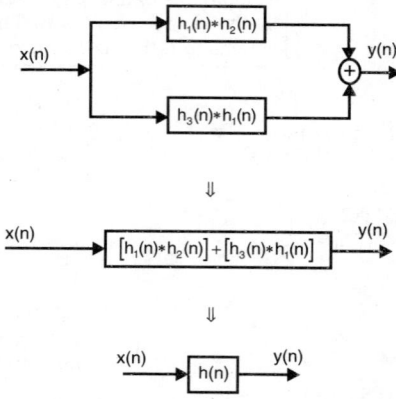

Here, $h(n) = [h_1(n) * h_2(n)] + [h_3(n) * h_1(n)]$

Given that, $h_1(n) = a^n u(n) = a^n$; $n \geq 0$ $\Rightarrow$ $h_1(m) = a^m$; $m \geq 0$

$$\therefore \; h_1(m-n) = a^{m-n} \; ; \; n>0 \text{ and } 0 \leq m \leq n$$

$$h_2(n) = \delta(n-1) \qquad\qquad \Rightarrow \quad \begin{aligned} h_2(m) &= \delta(m-1) = 1 \; ; \; m = 1 \\ &= 0 \; ; \; m \neq 1 \end{aligned}$$

$$h_3(n) = \delta(n-2) \qquad\qquad \Rightarrow \quad \begin{aligned} h_3(m) &= \delta(m-2) = 1 \; ; \; m = 2 \\ &= 0 \; ; \; m \neq 2 \end{aligned}$$

Let us evaluate the convolution of $h_1(n)$ and $h_2(n)$.

$$h_1(n) * h_2(n) = \sum_{m=-\infty}^{\infty} h_1(m) h_2(n-m) \; ; \; n \geq 1$$

Using commutative property.

$$= \sum_{m=-\infty}^{\infty} h_2(m) h_1(n-m)$$

$$= \sum_{m=-\infty}^{\infty} \delta(m-1) a^{(n-m)} = \sum_{m=-\infty}^{\infty} \delta(m-1) a^n a^{-m}$$

$$= a^n \sum_{m=-\infty}^{\infty} \delta(m-1) a^{-m}$$

$$= a^n a^{-1} = a^{n-1} \; ; \; \text{for } n \geq 1$$

The product $\delta(m-1)\alpha^{-m}$ will be nonzero only when m = 1.

$$= a^{n-1} u(n-1) \; ; \; \text{for all n.}$$

Let us evaluate the convolution of $h_3(n)$ and $h_1(n)$.

$$h_3(n) * h_1(n) = \sum_{m=-\infty}^{\infty} h_3(m) h_1(n-m) \; ; \; n \geq 2$$

$$= \sum_{m=-\infty}^{\infty} \delta(m-2) a^{(n-m)} = \sum_{m=-\infty}^{\infty} \delta(m-2) a^n a^{-m}$$

> The product $\delta(m-2)a^{-m}$ will be nonzero only when m = 2.

$$= a^n \sum_{m=-\infty}^{\infty} \delta(m-2) a^{-m}$$

$$= a^n a^{-2} = a^{n-2} \; ; \; \text{for } n \geq 2$$

$$= a^{n-2} u(n-2) \; ; \; \text{for all } n$$

Now, the overall impulse response h(n) is given by,

$$h(n) = [h_1(n) * h_2(n)] + [h_3(n) * h_1(n)]$$

$$= a^{(n-1)} u(n-1) + a^{(n-2)} u(n-2)$$

Example 5.10   *(AU Dec'15 & Dec'14, 16 Marks)*

Perform convolution of the following signal

$$x(n) = \left(\tfrac{1}{2}\right)^{n-2} u(n-2) \text{ and } h(n) = u(n+2)$$

Solution:

Given that, $x(n) = \left(\tfrac{1}{2}\right)^{n-2} u(n-2) = \left(\tfrac{1}{2}\right)^{n-2} \; ; \; n \geq 2 \Rightarrow x(m) = \left(\tfrac{1}{2}\right)^{m-2} \; ; \; m \geq 2$

$$\begin{aligned} h(n) = u(n+2) &= 1 \; ; n \geq -2 \\ &= 0 \; ; n < -2 \end{aligned} \qquad \Rightarrow \qquad \begin{aligned} h(m) = u(m+2) &= 1 \; ; m \geq -2 \\ &= 0 \; ; m < -2 \end{aligned}$$

$$\therefore \; h(m-n) = u(n-m+2) = 1 \; ; \; n > 2 \text{ and } 2 < m < n$$

$$x(n) * h(n) = \sum_{m=-\infty}^{+\infty} x(m) h(n-m) \; ; \; n \geq 2$$

> The product x(m) h(n – m) is nonzero in the range 2 ≤ m ≤ n. Therefore, summation index is changed to m = 2 to n

$$= \sum_{m=-\infty}^{+\infty} \left(\tfrac{1}{2}\right)^{m-2} u(n-m+2)$$

$$= \sum_{m=2}^{n} \left(\tfrac{1}{2}\right)^{m} \left(\tfrac{1}{2}\right)^{-2} = \left(\tfrac{1}{2}\right)^{-2} \sum_{m=2}^{n} \left(\tfrac{1}{2}\right)^{m}$$

$$= \frac{1}{2^{-2}} \left[\sum_{m=0}^{n} \left(\tfrac{1}{2}\right)^{m} - \left(\tfrac{1}{2}\right)^{0} - \left(\tfrac{1}{2}\right)^{1} \right]$$

$$= 2^2 \left[\sum_{m=0}^{n} \left(\tfrac{1}{2}\right)^{m} - 1 - \tfrac{1}{2} \right]$$

$$= 2^2 \left[\frac{1 - \left(\tfrac{1}{2}\right)^{n+1}}{1 - \tfrac{1}{2}} - \frac{3}{2} \right]$$

> Finite geometric series sum formula
> $$\sum_{n=0}^{N} C^n = \frac{1 - C^{N+1}}{1 - C}$$

$$\therefore\ x(n) * h(n) = 2^2 \left[2\left(1 - \left(\tfrac{1}{2}\right)^n \left(\tfrac{1}{2}\right)\right) - \tfrac{3}{2} \right]$$

$$= 2^2 \times 2 - 2^2 \left(\tfrac{1}{2}\right)^n - 2^3\, \tfrac{3}{2} = 8 - \frac{1}{2^{-2} 2^n} - 6 = 2 - \frac{1}{2^{n-2}}$$

$$= 2 - \left(\tfrac{1}{2}\right)^{n-2} \ ; \ n \geq 2 = \left[2 - \left(\tfrac{1}{2}\right)^{n-2} \right] u(n-2)$$

Example 5.11   (AU Jun'14, 16 Marks)

Perform the convolution of sum of the two sequences, $x(n) = r(n)$ and $h(n) = u(n)$.

Solution:

Given that, $x(n) = r(n) = n$ $\Rightarrow$ $x(m) = r(m) = m$

$h(n) = u(n) = 1 \ ; \ n \geq 0$ $\Rightarrow$ $h(m) = 1 \ ; \ m \geq 0$

$\qquad\qquad = 0 \ ; \ n < 0$ $= 0 \ ; \ m < 0$

$$\therefore\ h(n-m) = 1 \ ; \ n > 0 \text{ and } 0 < m < n$$

$\therefore\ x(n) * h(n)$

$$= \sum_{m = -\infty}^{+\infty} x(m)\, h(n-m)$$

$$= \sum_{m=0}^{n} r(m) \times 1$$

The product $x(m)\, h(n-m)$ is valid in the range $0 \leq m \leq n$. Therefore, the summation index is changed to $m = 0$ to n.

$$= \sum_{m=0}^{n} r(m)$$

$$= \sum_{m=0}^{n} m$$

5.6 Analysis of LTI Discrete Time System Using Discrete Time Fourier Transform

5.6.1 Transfer Function of LTI Discrete Time System in Frequency Domain

The ratio of Fourier transform of output and the Fourier transform of input is called *transfer function* of LTI discrete time system in frequency domain.

Let, $x(n)$ = Input to the discrete time system

$\qquad y(n)$ = Output of the discrete time system

$\therefore\ X(e^{j\omega})$ = Fourier transform of $x(n)$

$Y(e^{j\omega})$ = Fourier transform of $y(n)$

Now, Transfer function $= \dfrac{Y(e^{j\omega})}{X(e^{j\omega})}$ .....(5.34)

The transfer function of an LTI discrete time system in frequency domain can be obtained from the difference equation governing the input-output relation of the LTI discrete time system given below, [refer equation (5.4)].

$$y(n) = -\sum_{m=1}^{N} a_m y(n-m) + \sum_{m=0}^{M} b_m x(n-m)$$

On taking Fourier transform of above equation we get,

$$\mathcal{F}\{y(n)\} = \mathcal{F}\left\{-\sum_{m=1}^{N} a_m \, y(n-m) + \sum_{m=0}^{M} b_m \, x(n-m)\right\}$$

$$Y(e^{j\omega}) = -\sum_{m=1}^{N} a_m \, \mathcal{F}\{y(n-m)\} + \sum_{m=0}^{M} b_m \, \mathcal{F}\{x(n-m)\}$$

$$Y(e^{j\omega}) = -\sum_{m=1}^{N} a_m \, e^{-j\omega m} \, Y(e^{j\omega}) + \sum_{m=0}^{M} b_m \, e^{-j\omega m} \, X(e^{j\omega})$$

$$\therefore \; Y(e^{j\omega}) + \sum_{m=1}^{N} a_m \, e^{-j\omega m} \, Y(e^{j\omega}) = \sum_{m=0}^{M} b_m \, e^{-j\omega m} \, X(e^{j\omega})$$

$$Y(e^{j\omega})\left[1 + \sum_{m=1}^{N} a_m \, e^{-j\omega n}\right] = X(e^{j\omega}) \sum_{m=1}^{M} b_m \, e^{-j\omega n}$$

$$\therefore \; \frac{Y(e^{j\omega})}{X(e^{j\omega})} = \frac{\displaystyle\sum_{m=1}^{M} b_m e^{-j\omega m}}{1 + \displaystyle\sum_{n=1}^{N} a_m e^{-j\omega m}} = \frac{b_1 e^{-j\omega} + b_2 e^{-j2\omega} + \ldots\ldots + b_m e^{-j\omega m}}{1 + a_1 e^{-j\omega} + a_2 e^{-j2\omega} + \ldots\ldots + a_N e^{-j\omega N}} \qquad \ldots(5.35)$$

The above equation is frequency domain transfer function of Nth order LTI discrete time system.

5.6.2 Impulse Response and Transfer Function

Let, $h(n)$ = Impulse response (ie., response for impulse input)

$H(e^{j\omega})$ = Fourier transform of $h(n)$

Now, the response $y(n)$ of the discrete time system is given by convolution of input $x(n)$ and impulse response, $h(n)$.

$$\therefore y(n) = x(n) * h(n) \qquad\qquad \ldots(5.36)$$

On taking Fourier transform of above equation we get,

$$\mathcal{F}\{y(n)\} = \mathcal{F}\{x(n) * h(n)\}$$

$$\therefore \; \mathcal{F}\{y(n)\} = X(e^{j\omega}) \, H(e^{j\omega})$$

$$\therefore Y(e^{j\omega}) = X(e^{j\omega}) \, H(e^{j\omega})$$

> Convolution theorem of Fourier transform,
> $$\mathcal{F}\{x(n) * h(n)\} = X(e^{j\omega}) H(e^{j\omega})$$

$$\boxed{\therefore H(e^{j\omega}) = \frac{Y(e^{j\omega})}{X(e^{j\omega})}} \qquad\qquad \ldots(5.37)$$

From equations (5.34) and (5.37) we can say that the ***transfer function*** of a discrete time system in frequency domain is also given by discrete time Fourier transform of impulse response.

5.6.3 Response of LTI Discrete Time System using Discrete Time Fourier Transform

Let, $y(n)$ = Response of LTI discrete time system

$Y(e^{j\omega})$ = Fourier transform of $y(n)$

Consider the transfer function of LTI discrete time system.

$$H(e^{j\omega}) = \frac{Y(e^{j\omega})}{X(e^{j\omega})}$$

$\therefore\ Y(e^{j\omega}) = X(e^{j\omega})\, H(e^{j\omega})$

On taking inverse Fourier transform of above equation we get,

$$\mathcal{F}^{-1}\{Y(e^{j\omega})\} = \mathcal{F}^{-1}\{X(e^{j\omega})\, H(e^{j\omega})\}$$

$\therefore$ Response, $y(n) = \mathcal{F}^{-1}\{X(e^{j\omega})\, H(e^{j\omega})\}$ .....(5.38)

From the equation (5.39) we can say that the response $y(n)$ is given by the inverse Fourier transform of the product of $X(e^{j\omega})$ and $H(e^{j\omega})$.

Since the transfer function is defined with zero initial conditions, the response obtained by using equation (5.39) is the forced response or steady state response of discrete time system.

5.6.4 Frequency Response of LTI Discrete Time System

Consider a special class of input (complex sinusoidal input), $\boxed{Ae^{j\omega n} = A(\cos \omega n + j \sin \omega n)}$

$$x(n) = A\, e^{j\omega n}\ ;\ \ -\infty < n < \infty$$ .....(5.39)

where, A = Amplitude

ω = Arbitrary frequency in the interval $-\pi$ to $+\pi$.

$\therefore x(n-m) = Ae^{j\omega(n-m)}$ .....(5.40)

The output $y(n)$ of LTI system is given by convolution of $h(n)$ and $x(n)$.

$$y(n) = x(n)*h(n) = h(n)*x(n) = \sum_{m=-\infty}^{+\infty} h(m)x(n-m)$$

$$= \sum_{m=-\infty}^{+\infty} h(m)Ae^{j\omega(n-m)} = \sum_{m=-\infty}^{+\infty} h(m)Ae^{j\omega n}e^{-j\omega m}$$ $\boxed{\text{Using equation (5.40)}}$

$$= Ae^{j\omega n}\sum_{m=-\infty}^{+\infty} h(m)e^{-j\omega m}$$ $\boxed{\text{Replace } m \text{ by } n}$

$$= Ae^{j\omega n}\sum_{n=-\infty}^{+\infty} h(n)\, e^{-j\omega n}$$ $\boxed{\text{By definition of Fourier transform}}$
$$\boxed{H(e^{j\omega}) = \sum_{n=-\infty}^{+\infty} h(n)\, e^{-j\omega n}}$$

$$= Ae^{j\omega n}\, H(e^{j\omega})$$ $\boxed{\text{Using equation (5.39)}}$

$$= x(n)\, H(e^{j\omega})$$

$\therefore$ Output, $y(n) = x(n)\, H(e^{j\omega})$ .....(5.41)

From equation (5.41), we can say that if a complex sinusoidal signal is given as input signal to an LTI system, then the output is also a sinusoidal signal of the same frequency modified by $H(e^{j\omega})$. Hence $H(e^{j\omega})$ is called the *frequency response* of the system. An LTI system is characterized in the frequency domain by its frequency response.

The function $H(e^{j\omega})$ is a complex quantity. Therefore, $H(e^{j\omega})$ produces a change in the amplitude and phase of the input signal.

Let us express $H(e^{j\omega})$ as magnitude function and phase function.

$$\therefore H(e^{j\omega}) = |H(e^{j\omega})| \angle H(e^{j\omega})$$

$$\text{where,} \quad |H(e^{j\omega})| = \text{Magnitude function} \qquad \qquad .....(5.42)$$

$$\angle H(e^{j\omega}) = \text{Phase function} \qquad \qquad .....(5.43)$$

The sketch of magnitude function and phase function with respect to ω will give the frequency response graphically.

$$\text{Let, } H(e^{j\omega}) = H_r(e^{j\omega}) + jH_i(e^{j\omega})$$

$$\text{where, } H_r(e^{j\omega}) = \text{Real part of } H(e^{j\omega})$$

$$H_i(e^{j\omega}) = \text{Imaginary part of } H(e^{j\omega})$$

| $H^*(e^{j\omega})$ is complex conjugate of $H(e^{j\omega})$ |

$$\therefore |H(e^{j\omega})|^2 = H(e^{j\omega}) H^*(e^{j\omega})$$

$$= \left[H_r(e^{j\omega}) + jH_i(e^{j\omega})\right]\left[H_r(e^{j\omega}) - jH_i(e^{j\omega})\right]$$

$$= H_r^2(e^{j\omega}) + H_i^2(e^{j\omega})$$

The ***magnitude function*** is defined as,

$$\text{Magnitude function, } |H(e^{j\omega})| = \sqrt{H(e^{j\omega}) H^*(e^{j\omega})} \text{ or } \sqrt{H_r^2(e^{j\omega}) + H_i^2(e^{j\omega})} \qquad .....(5.44)$$

The ***phase function*** is defined as,

$$\text{Phase function, } \angle H(e^{j\omega}) = \tan^{-1}\left[\frac{H_i(e^{j\omega})}{H_r(e^{j\omega})}\right] \qquad .....(5.45)$$

The drawback in frequency response analysis using Fourier transform is that the frequency response is a continuous function of ω and so it cannot be processed by digital systems.

From equation (5.37) we can say that the frequency response $H(e^{j\omega})$ of an LTI system is same as transfer function in frequency domain and so, the frequency response is also given by the ratio of Fourier transform of output to Fourier transform of input.

$$\therefore \text{ Frequency response, } H(e^{j\omega}) = \frac{Y(e^{j\omega})}{X(e^{j\omega})} \qquad .....(5.46)$$

5.6.5 Properties of Frequency Response

1. The frequency response is periodic function of ω with a period of 2π.

2. If h(n) is real, then the magnitude of $H(e^{j\omega})$ is symmetric and phase of $H(e^{j\omega})$ is antisymmetric over the interval $0 \leq \omega \leq 2\pi$.

3. If h(n) is complex, then the real part of $H(e^{j\omega})$ is symmetric and the imaginary part of $H(e^{j\omega})$ is antisymmetric over the interval $0 \leq \omega \leq 2\pi$.

4. The impulse response h(n) is discrete, whereas the frequency response $H(e^{j\omega})$ is continuous function of ω.

5.6.6 Frequency Response of First-Order Discrete Time System

A first-order discrete time system is characterized by the difference equation,

$$y(n) = x(n) + a\,y(n-1) \qquad\qquad .....(5.47)$$

On taking Fourier transform of equation(5.47) we get,

$$Y(e^{j\omega}) = X(e^{j\omega}) + a\,e^{-j\omega}\,Y(e^{j\omega})$$

$$\therefore\;\; Y(e^{j\omega}) - a\,e^{-j\omega}\,Y(e^{j\omega}) = X(e^{j\omega})$$

$$\therefore\;\; Y(e^{j\omega})\,[1 - a\,e^{-j\omega}] = X(e^{j\omega})$$

$$\therefore\;\;\; H(e^{j\omega}) = \frac{Y(e^{j\omega})}{X(e^{j\omega})} = \frac{1}{1 - ae^{-j\omega}} \qquad\qquad .....(5.48)$$

$$\therefore\;\; |H(e^{j\omega})|^2 = H(e^{j\omega})H^*(e^{j\omega}) = \frac{1}{(1-ae^{-j\omega})}\frac{1}{(1-ae^{j\omega})} = \frac{1}{1 - ae^{j\omega} - ae^{-j\omega} + a^2 e^{-j\omega}e^{j\omega}}$$

$$= \frac{1}{1 - a(e^{j\omega}+e^{-j\omega})+a^2} = \frac{1}{1 - 2a\cos\omega + a^2} \qquad \boxed{\cos\theta = \frac{e^{j\theta}+e^{-j\theta}}{2}} \;\;.....(5.49)$$

The equation(5.48) is the frequency response of first-order system. The frequency response can be expressed graphically as two functions: Magnitude function and Phase function.

The magnitude function of $H(e^{j\omega})$ is defined as,

$$\text{Magnitude function, } |H(e^{j\omega})| = \sqrt{H(e^{j\omega})\,H^*(e^{j\omega})} = \frac{1}{\sqrt{1 - 2a\cos\omega + a^2}}$$

Let us separate the real part and imaginary part of $H(e^{j\omega})$, by multiplying the numerator and denominator of $H(e^{j\omega})$ [equation (5.48)], by the complex conjugate of the denominator as shown below:

$$\therefore\;\; H(e^{j\omega}) = \frac{1}{1 - ae^{-j\omega}} \times \frac{1 - ae^{+j\omega}}{1 - ae^{+j\omega}} = \frac{1 - ae^{j\omega}}{1 - 2a\cos\omega + a^2} = \frac{1 - a(\cos\omega + j\sin\omega)}{1 - 2a\cos\omega + a^2}$$

$$= \frac{1 - a\cos\omega}{1 - 2a\cos\omega + a^2} + j\frac{-a\sin\omega}{1 - 2a\cos\omega + a^2}$$

$$\therefore\; \text{Real part, } H_r(e^{j\omega}) = \frac{1 - a\cos\omega}{1 - 2a\cos\omega + a^2} \qquad \boxed{\begin{array}{l}\text{Using equation (5.49)}\\[4pt] e^{j\omega} = \cos\omega + j\sin\omega\end{array}}$$

$$\text{Imaginary part, } H_i(e^{j\omega}) = \frac{-a\sin\omega}{1 - 2a\cos\omega + a^2}$$

The phase function of $H(e^{j\omega})$ is defined as,

$$\text{Phase function, } \angle H(e^{j\omega}) = \tan^{-1}\!\left[\frac{H_i(e^{j\omega})}{H_r(e^{j\omega})}\right] = \tan^{-1}\!\left[\frac{-a\sin\omega}{1 - a\cos\omega}\right]$$

The Magnitude and Phase responses are calculated for $a = 0.5, 0.8, -0.5$ and -0.8 and tabulated in Table-5.3. Using the calculated values, the $|H(e^{j\omega})|$ and $\angle H(e^{j\omega})$ are sketched graphically for $a = 0.5, 0.8, -0.5$ and -0.8 in Fig 5.6, 5.7, 5.8 and 5.9 respectively. From the plots it is inferred that the first-order system behaves as a lowpass filter when "a" is in the range of "$0 < a < 1$" and behaves as a highpass filter when "a" is in the range of "$-1 < a < 0$".

Table 5.3: Frequency Response of First-Order Discrete Time System

$\left|H(e^{j\omega})\right| = \dfrac{1}{\sqrt{1 - 2a\cos\omega + a^2}}$	$\angle H(e^{j\omega}) = \tan^{-1}\left(\dfrac{-a\sin\omega}{1 - a\cos\omega}\right)$ $= \left[\dfrac{1}{\pi}\tan^{-1}\left(\dfrac{-a\sin\omega}{1 - a\cos\omega}\right)\right]\pi$

ω	a = 0.5		a = 0.8		a = –0.5		a = –0.8	
	$\left\|H(e^{j\omega})\right\|$	$\angle H(e^{j\omega})$	$\left\|H(e^{j\omega})\right\|$	$\angle H(e^{j\omega})$	$\left\|H(e^{j\omega})\right\|$	$\angle H(e^{j\omega})$	$H(e^{j\omega})\|$	$\angle H(e^{j\omega})$
$\dfrac{-8\pi}{8} = -\pi$	0.667	0	0.556	0	2	0	5	0
$\dfrac{-7\pi}{8}$	0.678	0.04π	0.566	0.06π	1.751	-0.11π	2.486	-0.28π
$\dfrac{-6\pi}{8}$	0.715	0.08π	0.601	0.11π	1.357	-0.16π	1.402	-0.29π
$\dfrac{-5\pi}{8}$	0.783	0.12π	0.666	0.16π	1.074	-0.17π	0.986	-0.26π
$\dfrac{-4\pi}{8} = \dfrac{-\pi}{2}$	0.894	0.15π	0.781	0.21π	0.894	-0.15π	0.781	-0.21π
$\dfrac{-3\pi}{8}$	1.074	0.17π	0.986	0.26π	0.783	-0.12π	0.666	-0.16π
$\dfrac{-2\pi}{8}$	1.357	0.16π	1.402	0.29π	0.715	-0.08π	0.601	-0.11π
$\dfrac{-\pi}{8}$	1.751	0.11π	2.486	0.28π	0.678	-0.04π	0.566	-0.06π
0	2	0	5	0	0.667	0	0.556	0
$\dfrac{\pi}{8}$	1.751	-0.11π	2.486	-0.28π	0.678	0.04π	0.566	0.06π
$\dfrac{2\pi}{8}$	1.357	-0.16π	1.402	-0.29π	0.715	0.08π	0.601	0.11π
$\dfrac{3\pi}{8}$	1.074	-0.17π	0.986	-0.26π	0.783	0.12π	0.666	0.16π
$\dfrac{4\pi}{8} = \dfrac{\pi}{2}$	0.894	-0.15π	0.781	-0.21π	0.894	0.15π	0.781	0.21π
$\dfrac{5\pi}{8}$	0.783	-0.12π	0.666	-0.16π	1.074	0.17π	0.986	0.26π
$\dfrac{6\pi}{8}$	0.715	-0.08π	0.601	-0.11π	1.357	0.16π	1.402	0.29π
$\dfrac{7\pi}{8}$	0.678	-0.04π	0.566	-0.06π	1.751	0.11π	2.486	0.28π
$\dfrac{8\pi}{8} = \pi$	0.667	0	0.556	0	2	0	5	0

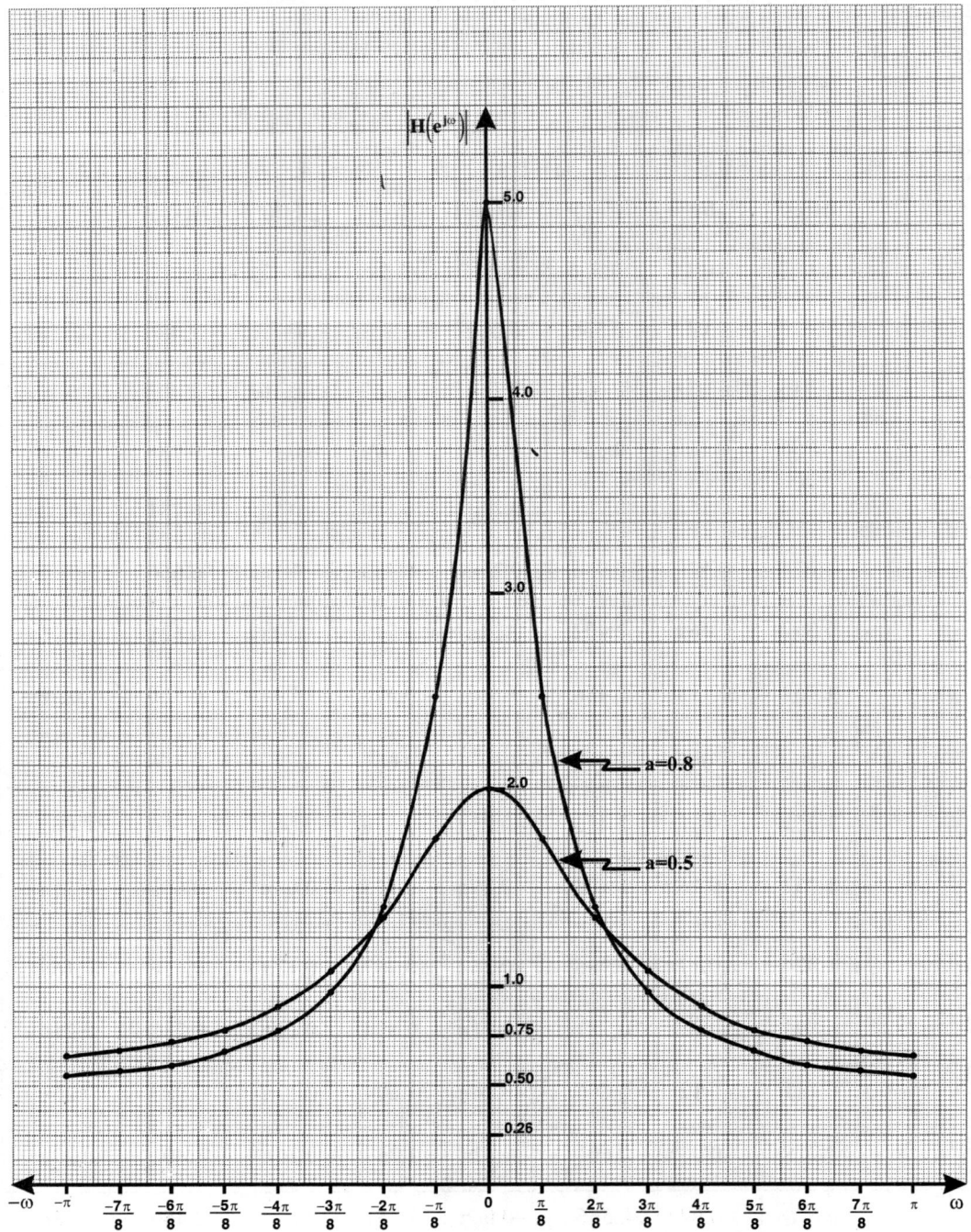

Fig 5.6: *Magnitude response of 1ˢᵗ order discrete time system when a = 0.5 and 0.8.*

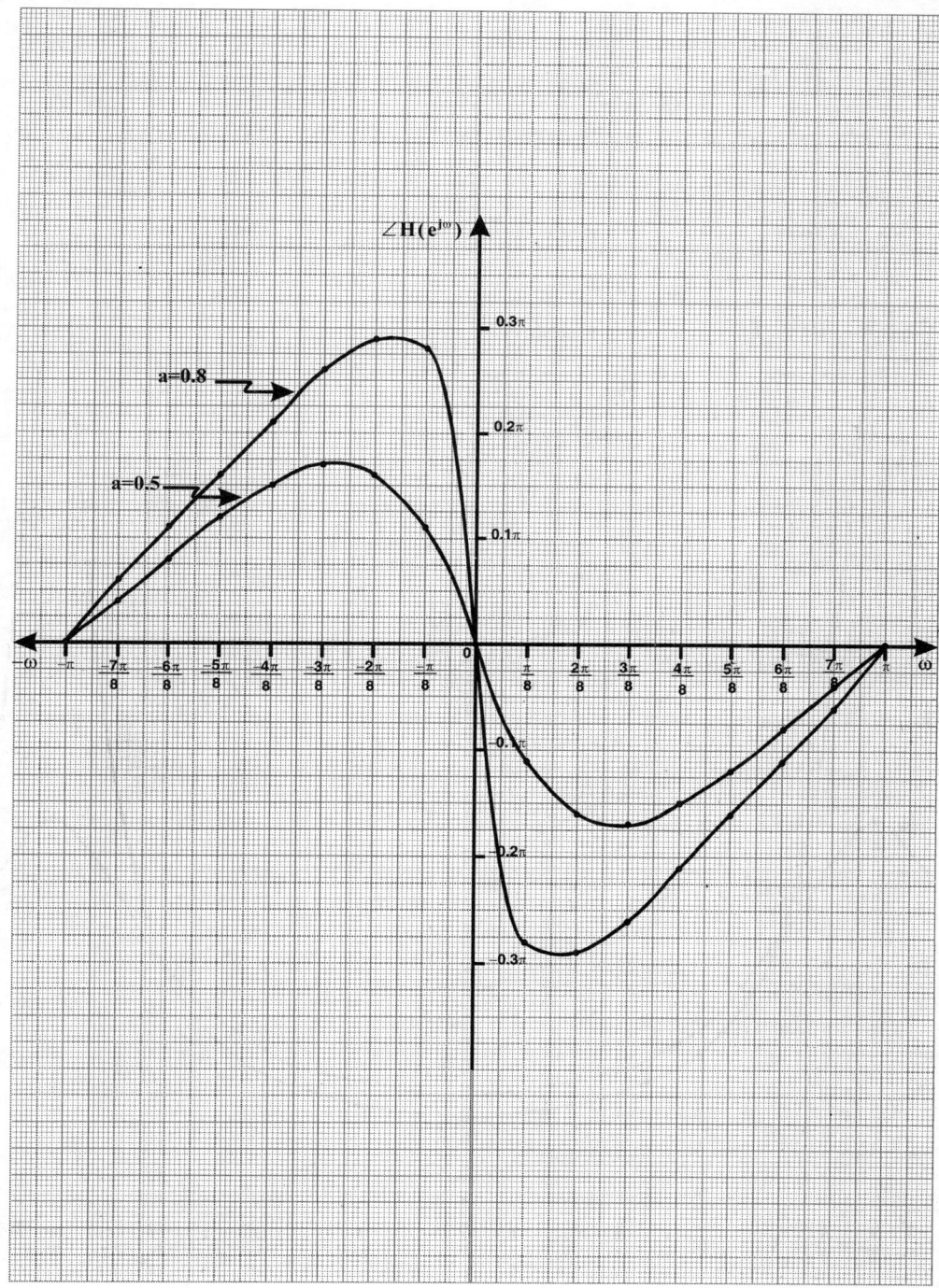

Fig 5.7: *Phase response of 1ˢᵗ order discrete time system when a = 0.5 and 0.8.*

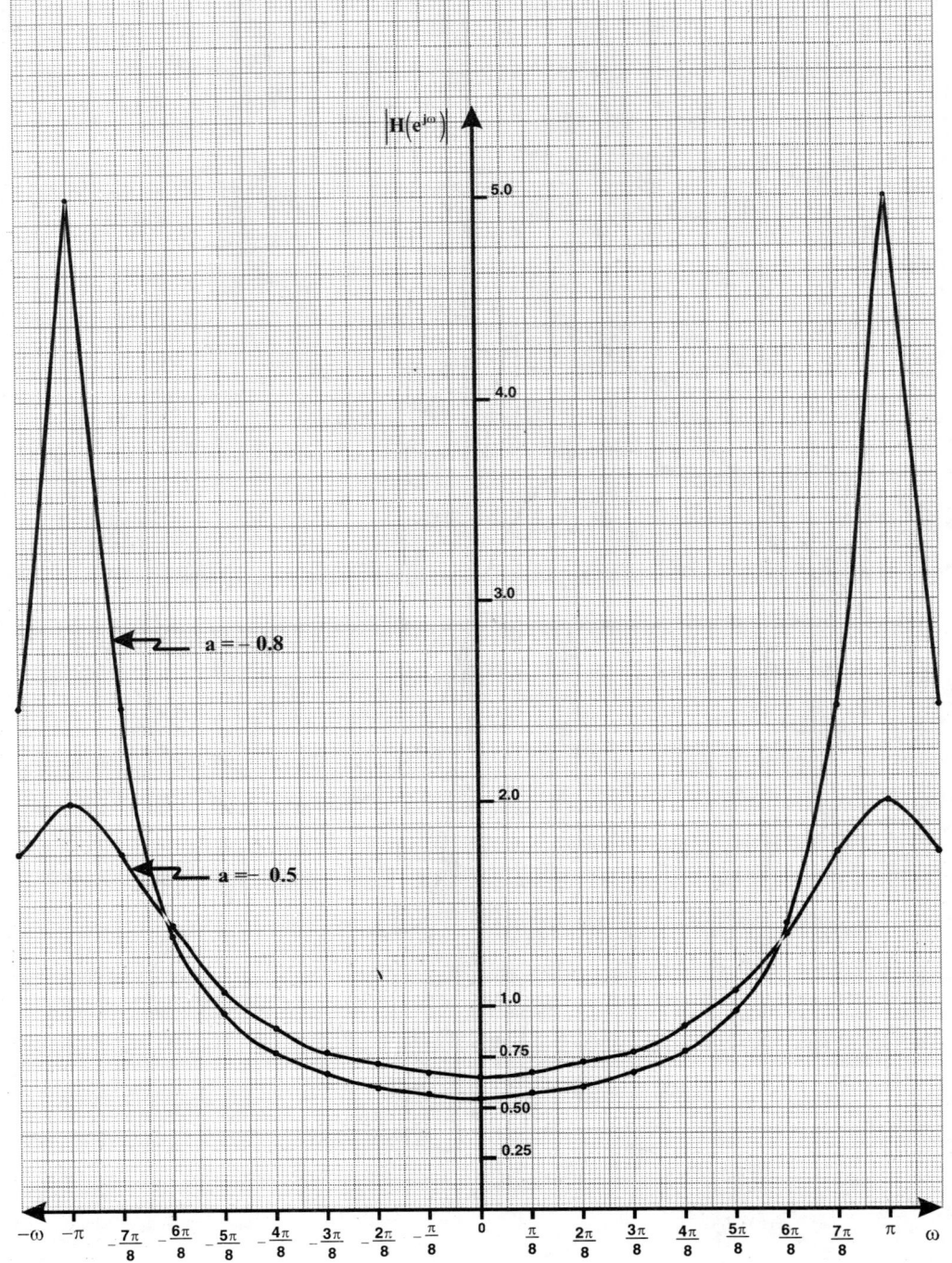

Fig 5.8: *Magnitude response of 1st order discrete time system when a = −0.5 and −0.8.*

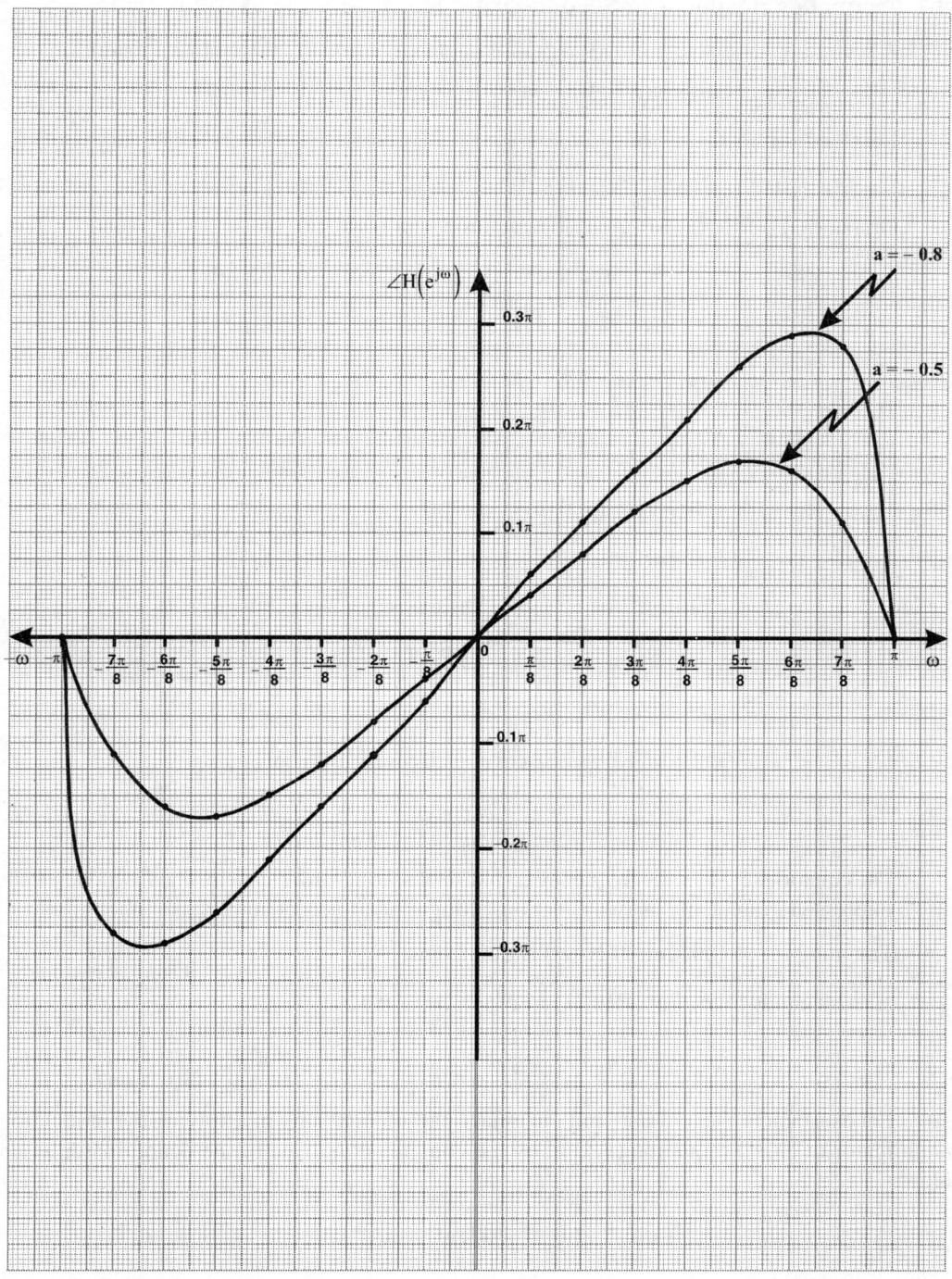

Fig 5.9: *Phase response of 1ˢᵗ order discrete time system when a = −0.5 and −0.8.*

5.6.7 Frequency Response of Second-Order Discrete Time System

A second order discrete time system is characterized by the difference equation,

$$y(n) = 2r \cos\omega_0 \, y(n-1) - r^2 \, y(n-2) + x(n) - r \cos\omega_0 \, x(n-1)$$

Let $a = -r \cos\omega_0$; $\alpha = -2r \cos\omega_0$; $\beta = r^2$

$$\therefore \; y(n) = -\alpha \, y(n-1) - \beta \, y(n-2) + x(n) + a \, x(n-1) \qquad\qquad .....(5.50)$$

On taking Fourier transform of the equation (5.50) we get,

$$Y(e^{j\omega}) = -\alpha e^{-j\omega} Y(e^{j\omega}) - \beta e^{-j2\omega} Y(e^{j\omega}) + X(e^{j\omega}) + a e^{-j\omega} X(e^{j\omega})$$

$$Y(e^{j\omega}) + \alpha e^{-j\omega} Y(e^{j\omega}) + \beta e^{-j2\omega} Y(e^{j\omega}) = X(e^{j\omega}) + a e^{-j\omega} X(e^{j\omega})$$

$$Y(e^{j\omega}) \left[1 + \alpha e^{-j\omega} + \beta e^{-j2\omega}\right] = X(e^{j\omega}) \left[1 + a e^{-j\omega}\right]$$

$$\therefore H(e^{j\omega}) = \frac{Y(e^{j\omega})}{X(e^{j\omega})} = \frac{1 + a e^{-j\omega}}{1 + \alpha e^{-j\omega} + \beta e^{-j2\omega}} \qquad\qquad .....(5.51)$$

$$\therefore \; |H(e^{j\omega})|^2 = H(e^{j\omega}) H^*(e^{j\omega}) = \frac{1 + a e^{-j\omega}}{1 + \alpha e^{-j\omega} + \beta e^{-j2\omega}} \, \frac{1 + a e^{+j\omega}}{1 + \alpha e^{+j\omega} + \beta e^{+j2\omega}}$$

$$= \frac{1 + a e^{j\omega} + a e^{-j\omega} + a^2}{1 + \alpha e^{j\omega} + \beta e^{j2\omega} + \alpha e^{-j\omega} + \alpha^2 + \alpha\beta e^{j\omega} + \beta e^{-j2\omega} + \alpha\beta e^{-j\omega} + \beta^2}$$

$$= \frac{1 + a(e^{j\omega} + e^{-j\omega}) + a^2}{1 + \alpha^2 + \beta^2 + \alpha\beta(e^{j\omega} + e^{-j\omega}) + \beta(e^{j2\omega} + e^{-j2\omega}) + \alpha(e^{j\omega} + e^{-j\omega})}$$

$$= \frac{1 + 2a \cos\omega + a^2}{1 + \alpha^2 + \beta^2 + 2\alpha\beta \cos\omega + 2\beta \cos 2\omega + 2\alpha \cos\omega}$$

$$= \frac{1 + a^2 + 2a \cos\omega}{1 + \alpha^2 + \beta^2 + 2\alpha(1 + \beta)\cos\omega + 2\beta \cos 2\omega} \qquad\qquad .....(5.52)$$

The equation (5.51) is the frequency response of second-order system. The frequency response can be expressed graphically as two functions: Magnitude function and Phase function.

The magnitude function of $H(e^{j\omega})$ is defined as,

Magnitude function, $|H(e^{j\omega})| = \sqrt{H(e^{j\omega}) H^*(e^{j\omega})}$

$\boxed{\text{Using equation (5.52).}}$

$$= \left[\frac{1 + a^2 + 2a \cos\omega}{1 + \alpha^2 + \beta^2 + 2\alpha(1 + \beta)\cos\omega + 2\beta \cos 2\omega} \right]^{\frac{1}{2}}$$

Let us separate the real part and imaginary part of $H(e^{j\omega})$, by multiplying the numerator and denominator of $H(e^{j\omega})$ [equation (5.51)], by the complex conjugate of the denominator as shown below:

$$\therefore H(e^{j\omega}) = \frac{1 + a e^{-j\omega}}{1 + \alpha e^{-j\omega} + \beta e^{-j2\omega}} \, \frac{1 + \alpha e^{j\omega} + \beta e^{j2\omega}}{1 + \alpha e^{j\omega} + \beta e^{j2\omega}}$$

$$= \frac{1 + \alpha e^{j\omega} + \beta e^{j2\omega} + a e^{-j\omega} + a\alpha + a\beta e^{j\omega}}{1 + \alpha^2 + \beta^2 + 2\alpha(1 + \beta)\cos\omega + 2\beta \cos 2\omega}$$

$\boxed{\text{Using equation (5.52).}}$

$$\therefore \; H(e^{j\omega}) = \frac{1 + a\alpha + ae^{-j\omega} + (a\beta + \alpha)e^{j\omega} + \beta e^{j2\omega}}{1 + \alpha^2 + \beta^2 + 2\alpha(1+\beta)\cos\omega + 2\beta\cos 2\omega} \qquad \boxed{e^{\pm j\theta} = \cos\theta \pm j\sin\theta}$$

$$= \frac{1 + a\alpha + a(\cos\omega - j\sin\omega) + (a\beta + \alpha)(\cos\omega + j\sin\omega) + \beta(\cos 2\omega + j\sin 2\omega)}{1 + \alpha^2 + \beta^2 + 2\alpha(1+\beta)\cos\omega + 2\beta\cos 2\omega}$$

Real part, $H_r(e^{j\omega}) = \dfrac{1 + a\alpha + (a + a\beta + \alpha)\cos\omega + \beta\cos 2\omega}{1 + \alpha^2 + \beta^2 + 2\alpha(1+\beta)\cos\omega + 2\beta\cos 2\omega}$

Imaginary part, $H_i(e^{j\omega}) = \dfrac{(a\beta + \alpha - a)\sin\omega + \beta\sin 2\omega}{1 + \alpha^2 + \beta^2 + 2\alpha(1+\beta)\cos\omega + 2\beta\cos 2\omega}$

The phase function of $H(e^{j\omega})$ is defined as,

Phase function, $\angle H(e^{j\omega}) = \tan^{-1}\dfrac{H_i(e^{j\omega})}{H_r(e^{j\omega})} = \tan^{-1}\dfrac{(a\beta + \alpha - a)\sin\omega + \beta\sin 2\omega}{1 + a\alpha + (a + a\beta + \alpha)\cos\omega + \beta\cos 2\omega}$

The magnitude and phase response are calculated for r=0.5 and 0.9 and $\omega_0 = \pi/4$, and tabulated in Table 5.4. Using the calculated values, the $|H(e^{j\omega})|$ and $\angle H(e^{j\omega})$ are sketched graphically for r = 0.5 and 0.8 and $\omega_0 = \pi/4$ as shown in Fig 5.10, 5.11. From the plots it can be inferred that the second-order system behaves as a resonant filter (or bandpass filter). The magnitude response shows a sharp peak close to the frequency $\omega = \omega_0 = \pi/4$, which is called resonant frequency.

Table 5.4: Frequency Response of Second-Order Discrete Time System

$$|H(e^{j\omega})| = \left(\frac{1 + a^2 + 2a\cos\omega}{1 + \alpha^2 + \beta^2 + 2\alpha(1+\beta)\cos\omega + 2\beta\cos 2\omega}\right)^{\frac{1}{2}}$$

$$\angle H(e^{j\omega}) = \tan^{-1}\left(\frac{(a\beta + \alpha - a)\sin\omega + \beta\sin 2\omega}{1 + a\alpha + (a + a\beta + \alpha)\cos\omega + \beta\cos 2\omega}\right)$$

$$= \left[\frac{1}{\pi}\tan^{-1}\left(\frac{(a\beta + \alpha - a)\sin\omega + \beta\sin 2\omega}{1 + a\alpha + (a + a\beta + \alpha)\cos\omega + \beta\cos 2\omega}\right)\right]\pi$$

Case - i

$r = 0.5, \; \omega_0 = \dfrac{\pi}{4}$

$\therefore a = -r\cos\omega_0 = -0.5\cos\dfrac{\pi}{4} = -0.3536$

$\alpha = -2r\cos\omega_0 = -2 \times 0.5\cos\dfrac{\pi}{4} = -0.7071$

$\beta = r^2 = 0.5^2 = 0.25$

$|H(e^{j\omega})| = \left(\dfrac{1.125 - 0.7072\cos\omega}{1.5625 - 1.7678\cos\omega + 0.5\cos 2\omega}\right)^{1/2}$

$\angle H(e^{j\omega}) = \left[\dfrac{1}{\pi}\tan^{-1}\left(\dfrac{-0.4419\sin\omega + 0.25\sin 2\omega}{1.25 - 1.1491\cos\omega + 0.25\cos 2\omega}\right)\right]\pi$

Case - ii

$r = 0.9, \; \omega_0 = \dfrac{\pi}{4}$

$\therefore a = -r\cos\omega_0 = -0.9\cos\dfrac{\pi}{4} = -0.6364$

$\alpha = -2r\cos\omega_0 = -2 \times 0.9\cos\dfrac{\pi}{4} = -1.2728$

$\beta = r^2 = 0.9^2 = 0.81$

$|H(e^{j\omega})| = \left(\dfrac{1.405 - 1.2728\cos\omega}{3.2761 - 4.6075\cos\omega + 1.62\cos 2\omega}\right)^{1/2}$

$\angle H(e^{j\omega}) = \left[\dfrac{1}{\pi}\tan^{-1}\left(\dfrac{-1.1519\sin\omega + 0.81\sin 2\omega}{1.81 - 2.4247\cos\omega + 0.81\cos 2\omega}\right)\right]\pi$

Table 5.4: Continued...

ω	r = 0.5		r = 0.9	
	$\|H(e^{j\omega})\|$	$\angle H(e^{j\omega})$	$\|H(e^{j\omega})\|$	$\angle H(e^{j\omega})$
$\dfrac{-8\pi}{8} = -\pi$	0.69	0	0.53	0
$\dfrac{-7\pi}{8}$	0.71	0.04π	0.55	0.07π
$\dfrac{-6\pi}{8}$	0.76	0.08π	0.59	0.14π
$\dfrac{-5\pi}{8}$	0.86	0.12π	0.7	0.21π
$\dfrac{-4\pi}{8} = \dfrac{-\pi}{2}$	1.03	0.13π	0.92	0.27π
$\dfrac{-3\pi}{8}$	1.27	0.11π	1.58	0.32π
$\dfrac{-2\pi}{8}$	1.41	0.05π	5.28	0.02π
$\dfrac{-\pi}{8}$	1.29	-0.01π	1.18	-0.24π
0	1.19	0	0.68	0
$\dfrac{\pi}{8}$	1.29	0.01π	1.18	0.24π
$\dfrac{2\pi}{8}$	1.41	-0.05π	5.28	-0.02π
$\dfrac{3\pi}{8}$	1.27	-0.11π	1.58	-0.32π
$\dfrac{4\pi}{8} = \dfrac{\pi}{2}$	1.03	-0.13π	0.92	-0.27π
$\dfrac{5\pi}{8}$	0.86	-0.12π	0.7	-0.21π
$\dfrac{6\pi}{8}$	0.76	-0.08π	0.59	-0.14π
$\dfrac{7\pi}{8}$	0.71	-0.04π	0.55	-0.07π
$\dfrac{8\pi}{8} = \pi$	0.69	0	0.53	0

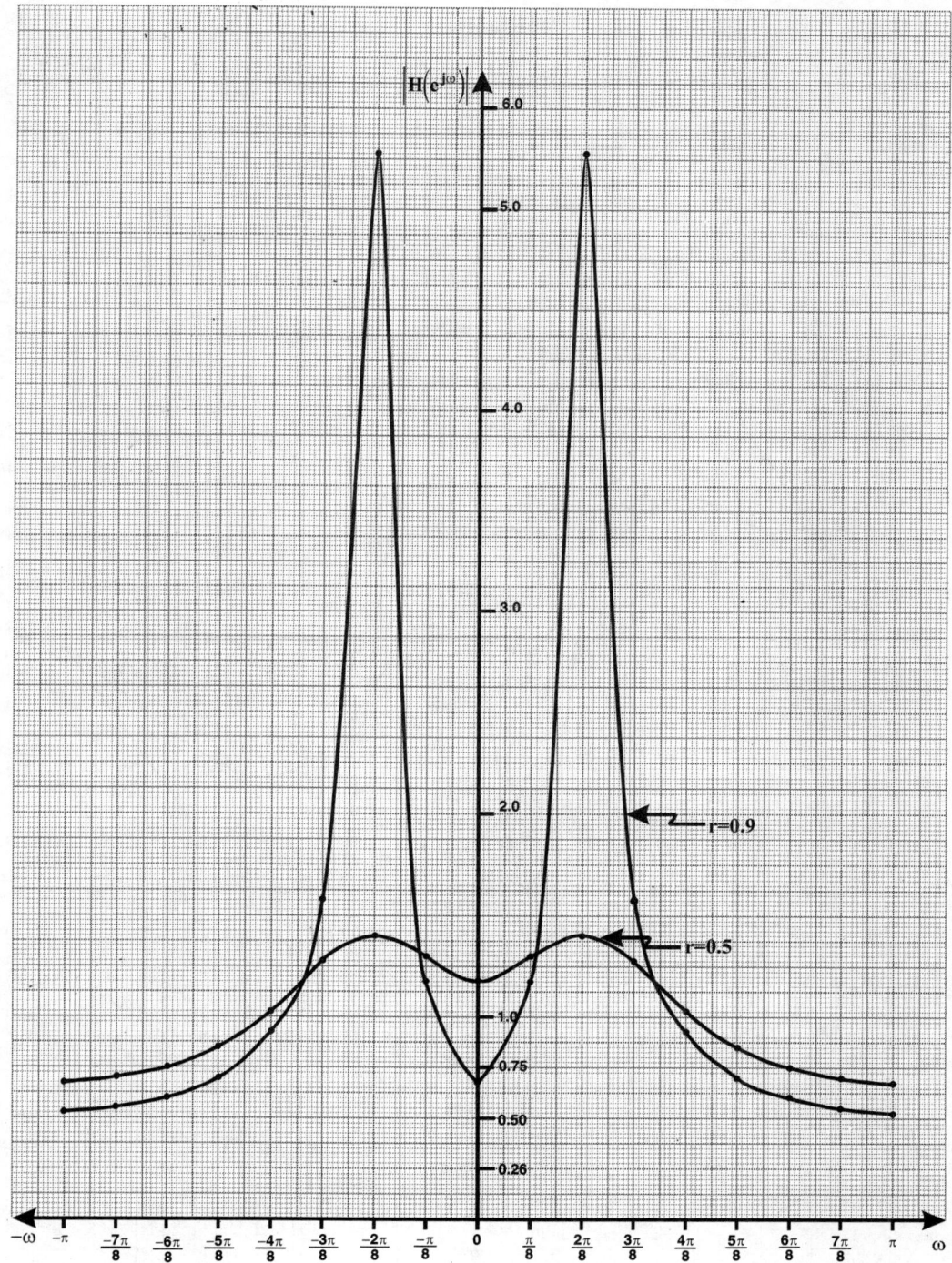

Fig 5.10: *Magnitude response of 2ⁿᵈ order discrete time system.*

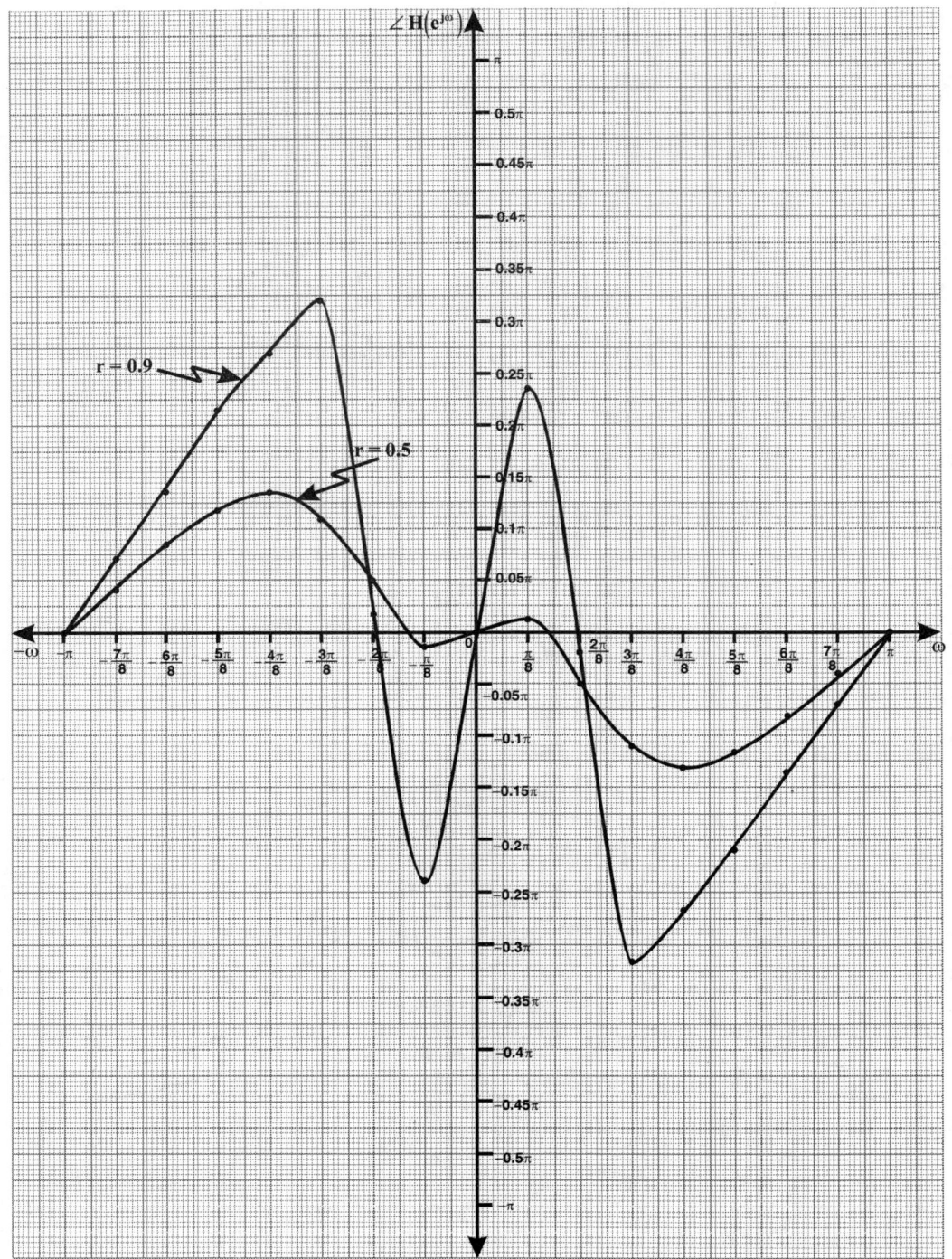

Fig 5.11: *Phase response of 2nd order discrete time system.*

Example 5.12

The impulse response of an LTI system is $h(n) = \{1, 2, 2, 1\}$. Find the response of the system for the input $x(n) = \{1, 2, 3, 4\}$.

Solution:

Let, $y(n)$ = Response of the system

$Y(e^{j\omega})$ = Fourier transform of $y(n)$

The response $y(n)$ of the system is given by convolution of $x(n)$ and $h(n)$.

$\therefore \ y(n) = x(n) * h(n)$

On taking Fourier transform of above equation we get,

$\mathcal{F}\{y(n)\} = \mathcal{F}\{x(n) * h(n)\}$

$\therefore \ Y(e^{j\omega}) = X(e^{j\omega})H(e^{j\omega})$

> Convolution theorem of Fourier transform,
>
> $\mathcal{F}\{x(n) * h(n)\} = X(e^{j\omega})\,H(e^{j\omega})$

By definition of Fourier transform, we can write,

$$X(e^{j\omega}) = \sum_{n=-\infty}^{+\infty}x(n)e^{-j\omega n} = \sum_{n=0}^{3}x(n)e^{-j\omega n}$$

$$= x(0)e^{0}+x(1)e^{-j\omega}+x(2)e^{-j2\omega}+x(3)e^{-j3\omega}$$

$$= 1+2e^{-j\omega}+3e^{-j2\omega}+4e^{-j3\omega}$$

By definition of Fourier transform we can write,

$$H(e^{j\omega}) = \sum_{n=-\infty}^{+\infty}h(n)e^{-j\omega n} = \sum_{n=0}^{3}h(n)e^{-j\omega n}$$

$$= h(0)e^{0}+h(1)e^{-j\omega}+h(2)e^{-j2\omega}+h(3)e^{-j3\omega}$$

$$= 1+2e^{-j\omega}+2e^{-j2\omega}+e^{-j3\omega}$$

$$X(e^{j\omega})\,H(e^{j\omega}) = (1 + 2\,e^{-j\omega} + 3\,e^{-j2\omega}+ 4e^{-j3\omega})\,(1 + 2\,e^{-j\omega} + 2e^{-j2\omega} + e^{-j3\omega})$$

$$= 1 + 2\,e^{-j\omega} + 2e^{-j2\omega} + e^{-j3\omega}$$
$$+ 2\,e^{-j\omega} + 4\,e^{-j2\omega} + 4\,e^{-j3\omega} + 2\,e^{-j4\omega}$$
$$+ 3\,e^{-j2\omega} + 6\,e^{-j3\omega} + 6\,e^{-j4\omega} + 3\,e^{-j5\omega}$$
$$+ 4e^{-j3\omega} + 8\,e^{-j4\omega} + 8e^{-j5\omega} + 4e^{-j6\omega}$$

$$= 1 + 4\,e^{-j\omega}+ 9\,e^{-j2\omega} + 15\,e^{-j3\omega} + 16\,e^{-j4\omega} + 11\,e^{-j5\omega} + 4e^{-j6\omega} \qquad ..... \ (1)$$

By definition of Fourier transform we get,

$$Y(e^{j\omega}) = \sum_{n=-\infty}^{+\infty}y(n)e^{-j\omega n}$$

$$= ..... \ y(0)\ e^{0}+ y(1)\ e^{-j\omega}+ y(2)\ e^{-j2\omega}+ y(3)\ e^{-j3\omega}+ y(4)\ e^{-j4\omega}+ y(5)\ e^{-j5\omega}+ y(6)\ e^{-j6\omega}+ ..... \ ..... \ (2)$$

On comparing equations (1) and (2) we get,

$$y(n) = \{1, \ 4, \ 9, \ 15, \ 16, \ 11, \ 4\}$$
$$\uparrow$$

Example 5.13   *(AU Dec'15, 16 Marks)*

The impulse response of an LTI system is $h(n) = \alpha^n\, u(n)$. Find the response of the system for the input $x(n) = \beta^n\, u(n)$ with $|\alpha|$ and $|\beta| < 1$ using DTFT.

a) $\alpha = \beta$ b) $\alpha \neq \beta$

Solution:

Let, $y(n)$ = Response of LTI system

$Y(e^{j\omega})$ = Fourier transform of $y(n)$

The response y(n) of the system is given by the convolution of x(n) and h(n).

$$\therefore y(n) = x(n) * y(n)$$

On taking Fourier transform of above equation we get,

$$\mathcal{F}\{y(n)\} = \mathcal{F}\{x(n) * y(n)\}$$

By definition of Fourier transform, we can write

> Convolution theorem of Fourier transform,
> $$\mathcal{F}\{x(n) * h(n)\} = X(e^{j\omega})\, H(e^{j\omega})$$

$$X(e^{j\omega}) = \sum_{n=-\infty}^{+\infty} x(n)\, e^{-j\omega n} = \sum_{n=0}^{\infty} \beta^n e^{-j\omega n} = \sum_{n=0}^{\infty} (\beta e^{-j\omega})^n = \frac{1}{1 - \beta e^{-j\omega}}$$

$$H(e^{j\omega}) = \sum_{n=-\infty}^{+\infty} h(n)\, e^{-j\omega n} = \sum_{n=0}^{\infty} \alpha^n e^{-j\omega n} = \sum_{n=0}^{\infty} (\alpha e^{-j\omega})^n = \frac{1}{1 - \alpha e^{-j\omega}}$$

$$\therefore Y(e^{j\omega}) = X(e^{j\omega})\, H(e^{j\omega}) = \frac{1}{1 - \beta e^{-j\omega}} \times \frac{1}{1 - \alpha e^{-j\omega}} = \frac{1}{e^{-j\omega}(e^{j\omega} - \beta)(e^{j\omega} - \alpha)}$$

$$\therefore Y(e^{j\omega}) = \frac{e^{j\omega} e^{j\omega}}{(e^{j\omega} - \beta)(e^{j\omega} - \alpha)} \qquad \qquad .....(1)$$

a) $\alpha = \beta$

From equation (1) when $\alpha = \beta$ we get,

$$Y(e^{j\omega}) = \frac{e^{j\omega} e^{j\omega}}{(e^{j\omega} - \beta)^2}$$

By partial fraction expansion technique we can write,

$$\frac{Y(e^{j\omega})}{e^{j\omega}} = \frac{e^{j\omega}}{(e^{j\omega} - \beta)^2} = \frac{A}{(e^{j\omega} - \beta)^2} + \frac{B}{e^{j\omega} - \beta}$$

$$A = (e^{j\omega} - \beta)^2 \frac{Y(e^{j\omega})}{e^{j\omega}} \bigg|_{e^{j\omega} = \beta} = (e^{j\omega} - \beta)^2 \frac{e^{j\omega}}{(e^{j\omega} - \beta)^2} \bigg|_{e^{j\omega} = \beta} = \beta$$

$$B = \frac{d}{d(e^{j\omega})}\left((e^{j\omega} - \beta)^2 \frac{Y(e^{j\omega})}{e^{j\omega}}\right)\bigg|_{e^{j\omega} = \beta} = \frac{d}{d(e^{j\omega})} e^{j\omega} \bigg|_{e^{j\omega} = \beta} = 1$$

$$\therefore \frac{Y(e^{j\omega})}{e^{j\omega}} = \frac{A}{(e^{j\omega} - \beta)^2} + \frac{B}{e^{j\omega} - \beta} = \frac{\beta}{(e^{j\omega} - \beta)^2} + \frac{1}{e^{j\omega} - \beta}$$

> $$\mathcal{F}\{a^n u(n)\} = \frac{1}{1 - a e^{-j\omega}}$$
> $$\mathcal{F}\{n a^n u(n)\} = \frac{a e^{-j\omega}}{(1 - a e^{-j\omega})^2}$$

$$\therefore Y(e^{j\omega}) = \frac{\beta}{(e^{j\omega})^2 (1 - \beta e^{-j\omega})^2} + \frac{e^{j\omega}}{e^{j\omega}(1 - \beta e^{-j\omega})}$$

$$= \frac{\beta e^{-j\omega}}{(1 - \beta e^{-j\omega})^2} + \frac{1}{1 - \beta e^{-j\omega}}$$

The response y(n) is obtained by taking inverse Fourier transform of $Y(e^{j\omega})$

$$\therefore y(n) = \mathcal{F}\{Y(e^{j\omega})\} = n\beta^n u(n) + \beta^n u(n) = (n+1)\beta^n u(n)$$

b) $\alpha \neq \beta$

From equation (1) we get,

$$Y(e^{j\omega}) = \frac{e^{j\omega} e^{j\omega}}{(e^{j\omega} - \beta)(e^{j\omega} - \alpha)}$$

By partial fraction expansion technique we can write,

$$\frac{Y(e^{j\omega})}{e^{j\omega}} = \frac{e^{j\omega}}{(e^{j\omega} - \beta)(e^{j\omega} - \alpha)} = \frac{A}{e^{j\omega} - \beta} + \frac{B}{e^{j\omega} - \alpha}$$

$$A = (e^{j\omega} - \beta) \frac{Y(e^{j\omega})}{e^{j\omega}} \bigg|_{e^{j\omega} = \beta} = (\cancel{e^{j\omega}} - \beta) \frac{e^{j\omega}}{(\cancel{e^{j\omega}} - \beta)(e^{j\omega} - \alpha)} \bigg|_{e^{j\omega} = \beta} = \frac{\beta}{\beta - \alpha}$$

$$B = (e^{j\omega} - \alpha) \frac{Y(e^{j\omega})}{e^{j\omega}} \bigg|_{e^{j\omega} = \alpha} = (\cancel{e^{j\omega}} - \alpha) \frac{e^{j\omega}}{(e^{j\omega} - \beta)(\cancel{e^{j\omega}} - \alpha)} \bigg|_{e^{j\omega} = \alpha} = \frac{\alpha}{\alpha - \beta}$$

$$\therefore \frac{Y(e^{j\omega})}{e^{j\omega}} = \frac{A}{e^{j\omega} - \beta} + \frac{B}{e^{j\omega} - \alpha}$$

$$\therefore Y(e^{j\omega}) = \frac{\beta}{\beta - \alpha} \frac{e^{j\omega}}{(e^{j\omega} - \beta)} + \frac{\alpha}{\alpha - \beta} \frac{e^{j\omega}}{(e^{j\omega} - \alpha)}$$

$$= \frac{\beta}{\beta - \alpha} \frac{1}{1 - \beta e^{-j\omega}} + \frac{\alpha}{\alpha - \beta} \frac{1}{1 - \alpha e^{-j\omega}}$$

The response y(n) is obtained by taking inverse Fourier transform of $Y(e^{j\omega})$.

$$\therefore y(n) = \mathcal{F}^{-1}\{Y(e^{j\omega})\} = \frac{\beta}{\beta - \alpha} \beta^n u(n) + \frac{\alpha}{\alpha - \beta} \alpha^n u(n) = -\frac{\beta}{\alpha - \beta} \beta^n u(n) + \frac{\alpha}{\alpha - \beta} \alpha^n u(n)$$

$$= \frac{1}{\alpha - \beta} [\alpha^{n+1} - \beta^{n+1}] u(n)$$

Example 5.14

The impulse response of an LTI system is given by, h(n) = 0.8^n u(n). Find the frequency response.

Solution:

The frequency response $H(e^{j\omega})$ is obtained by taking Fourier transform of the impulse response h(n).

Given that, impulse response, h(n) = 0.8^n u(n) for all

On taking Fourier transform we get,

$$H(e^{j\omega}) = \mathcal{F}\{h(n)\} = \sum_{n=-\infty}^{+\infty} h(n) e^{-j\omega n}$$

$$\boxed{\begin{aligned} u(n) &= 1 \text{ for } n \geq 0 \\ &= 0 \text{ for } n < 0 \end{aligned}}$$

$$= \sum_{n=-\infty}^{+\infty} 0.8^n u(n) e^{-j\omega n} = \sum_{n=0}^{\infty} 0.8^n e^{-j\omega n} = \sum_{n=0}^{\infty} (0.8 \, e^{-j\omega})^n$$

$$= \frac{1}{1 - 0.8 e^{-j\omega}}$$

$$\boxed{\begin{aligned} &\text{Using infinite geometric series sum formula,} \\ &\sum_{n=0}^{\infty} C^n = \frac{1}{1-C} \quad \text{where} \, |C| < 1 \end{aligned}}$$

The frequency response has two functions: Magnitude function and phase function,

The magnitude function is defined as,

$$\text{Magnitude function, } |H(e^{j\omega})| = \sqrt{H(e^{j\omega}) H^*(e^{j\omega})} = \sqrt{\frac{1}{1 - 0.8 e^{-j\omega}} \times \frac{1}{1 - 0.8 e^{j\omega}}}$$

$$\boxed{\cos\theta = \frac{e^{j\theta} + e^{-j\theta}}{2}}$$

$$= \frac{1}{\sqrt{1 - 0.8 e^{j\omega} - 0.8 e^{-j\omega} + 0.8^2}} = \frac{1}{\sqrt{1 - 0.8(e^{j\omega} + e^{-j\omega}) + 0.64}}$$

$$= \frac{1}{\sqrt{1.64 - 0.8(2\cos\omega)}} = \frac{1}{\sqrt{1.64 - 1.6\cos\omega}} \qquad \dots\dots(1)$$

Let us separate the real and imaginary parts of $H(e^{j\omega})$, by multiplying the numerator and denominator by complex conjugate of the denominator.

$$\therefore H(e^{j\omega}) = \frac{1}{1 - 0.8 e^{-j\omega}} \times \frac{1 - 0.8 e^{j\omega}}{1 - 0.8 e^{j\omega}} = \frac{1 - 0.8 e^{j\omega}}{1.64 - 1.6\cos\omega}$$

$$\boxed{\text{Using equation (1).}}$$

$$\therefore\ H_i(e^{j\omega}) = \frac{1 - 0.8(\cos\omega + j\sin\omega)}{1.64 - 1.6\cos\omega} = \frac{1 - 0.8\cos\omega}{1.64 - 1.6\cos\omega} - \frac{j0.8\sin\omega}{1.64 - 1.6\cos\omega}$$

$$\therefore\ H_i(e^{j\omega}) = \frac{-0.8\sin\omega}{1.64 - 1.6\cos\omega}$$

$$H_r(e^{j\omega}) = \frac{1 - 0.8\cos\omega}{1.64 - 1.6\cos\omega}$$

$$\boxed{e^{j\theta} = \cos\theta + j\sin\theta}$$

$$\therefore\ \frac{H_i(e^{j\omega})}{H_r(e^{j\omega})} = \frac{\dfrac{-0.8\sin\omega}{1.64 - 1.6\cos\omega}}{\dfrac{1 - 0.8\cos\omega}{1.64 - 1.6\cos\omega}} = \frac{-0.8\sin\omega}{1 - 0.8\cos\omega} \qquad\qquad .....(2)$$

The phase function is defined as,

$$\text{Phase function,}\ \angle H(e^{j\omega}) = \tan^{-1}\left[\frac{H_i(e^{j\omega})}{H_r(e^{j\omega})}\right] = \tan^{-1}\left[\frac{-0.8\sin\omega}{1 - 0.8\cos\omega}\right] \qquad \boxed{\text{Using equation (2).}}$$

Example 5.15

A system has impulse response h(n) given by, $h(n) = -0.25\,\delta(n+1) + 0.5\,\delta(n) - 0.75\,\delta(n-1)$.

a) Is the system BIBO stable? **b)** Is the system causal? Justify your answer. **c)** Find the frequency response.

Solution:

We know that, $\delta(n-k) = 1$; when $n = k$
$\qquad\qquad\qquad\quad = 0$; when $n \ne k$

Let us evaluate h(n) for different values of n.

When $n = -2$; $h(n) = h(-2) = -0.25\delta(-1) + 0.5\delta(-2) - 0.75\delta(-3) = 0\quad +0\quad +0\quad = 0$

When $n = -1$; $h(n) = h(-1) = -0.25\delta(0)\ +0.5\delta(-1) - 0.75\delta(-2) = -0.25 + 0\quad +0\quad = -0.25$

When $n = \ 0$; $h(n) = h(0)\ = -0.25\delta(1)\ +0.5\delta(0)\ - 0.75\delta(-1) = 0\quad +0.5 + 0\quad = 0.5$

When $n = \ 1$; $h(n) = h(1)\ = -0.25\delta(2)\ +0.5\delta(1)\ - 0.75\delta(0)\ = 0\quad +0\quad -0.75 = -0.25$

When $n = \ 2$; $h(n) = h(2)\ = -0.25\delta(3)\ +0.5\delta(2)\ - 0.75\delta(1)\ = 0\quad +0\quad +0\quad = 0$

From the above analysis, we can infer that $h(n) = 0$ for $n < -1$ and $n > 1$, and $h(n) \ne 0$ only for $n = -1, 0, 1$.
Here, $h(-1) = -0.25,\ h(0) = 0.5,\ h(1) = -0.75$

$$\therefore\ \text{Impulse response, } h(n) = \{-0.25,\ \underset{\uparrow}{0.5},\ -0.75\}$$

a) Check for Stability

For stability of a system, $\displaystyle\sum_{n=-\infty}^{+\infty}|h(n)| < \infty$

$$\therefore\ \sum_{n=-\infty}^{+\infty}|h(n)| = |h(-1)| + |h(0)| + |h(1)| = 0.25 + 0.5 + 0.75 = 1.5$$

Since $\displaystyle\sum_{n=-\infty}^{+\infty}|h(n)| < \infty$, the system is BIBO stable.

b) Check for Causality

In a causal system the present output should depend only on present and past inputs or outputs, and should not depend on future inputs or outputs. In the given system, the response h(n) depends on the future input $\delta(n+1)$. Hence the system is non causal.

c) Frequency Response

The frequency response, $H(e^{j\omega})$ is given by the Fourier transform of h(n).

By definition of Fourier transform,

Frequency response, $H(e^{j\omega}) = \mathcal{F}\{h(n)\} = \displaystyle\sum_{n=-\infty}^{+\infty} h(n)e^{-j\omega n} = h(-1)e^{j\omega} + h(0) + h(1)e^{-j\omega}$

$$= -0.25e^{j\omega} + 0.5 - 0.75e^{-j\omega}$$

$$= -0.25(\cos\omega + j\sin\omega) + 0.5 - 0.75(\cos\omega - j\sin\omega)$$

$$= -0.25\cos\omega - j0.25\sin\omega + 0.5 - 0.75\cos\omega + j0.75\sin\omega$$

$$= 0.5 - \cos\omega + j0.5\sin\omega$$

$\therefore$ Magnitude function, $|H(e^{j\omega})| = \sqrt{H_i^2(e^{j\omega}) + H_r^2(e^{j\omega})} = \sqrt{(0.5-\cos\omega)^2 + (0.5\sin\omega)^2}$

Phase function, $\angle H(e^{j\omega}) = \tan^{-1}\left[\dfrac{H_i(e^{j\omega})}{H_r(e^{j\omega})}\right] = \tan^{-1}\left[\dfrac{0.5\sin\omega}{0.5-\cos\omega}\right]$

Example 5.16

(AU Jun'13, 8 Marks)

Find the Fourier transform of $x(n) = \left(\frac{1}{2}\right)^n u(n)$ and sketch the magnitude and phase function.

Solution:

Given that, $x(n) = \left(\frac{1}{2}\right)^n u(n) = \left(\frac{1}{2}\right)^n$; $n \geq 0$

By definition of Fourier transform,

$$X(e^{j\omega}) = \mathcal{F}\{x(n)\} = \sum_{n=-\infty}^{\infty} x(n)e^{-j\omega n} = \sum_{n=0}^{\infty} \left(\frac{1}{2}\right)^n e^{-j\omega n} = \sum_{n=0}^{\infty} \left(\frac{1}{2}e^{-j\omega}\right)^n = \frac{1}{1 - \frac{1}{2}e^{-j\omega}}$$

The magnitude function is defined as,

Magnitude function, $|X(e^{j\omega})| = \sqrt{X(e^{j\omega})X^*(e^{j\omega})}$

Using infinite geometric series sum formula,
$\displaystyle\sum_{n=0}^{\infty} C^n = \dfrac{1}{1-C}$
where $0 < |C| < 1$

$$= \sqrt{\frac{1}{(1 - 0.5\,e^{-j\omega}) \times (1 - 0.5\,e^{j\omega})}}$$

$$= \sqrt{\frac{1}{1 - 0.5\,e^{j\omega} - 0.5\,e^{-j\omega} + 0.5^2}}$$

$$= \frac{1}{\sqrt{1 - 0.5(e^{j\omega} + e^{-j\omega}) + 0.5^2}}$$

$$= \frac{1}{\sqrt{1 - 0.5 \times 2\cos\omega + 0.5^2}}$$

$\cos\theta = \dfrac{e^{j\theta} + e^{-j\theta}}{2}$

$$= \frac{1}{\sqrt{1.25 - \cos\omega}}$$

.....(1)

Let us separate the real part and imaginary parts of $X(e^{j\omega})$, by multiplying the numerator and denominator of $X(e^{j\omega})$ by the conjugate of denominator $X(e^{j\omega})$.

$$\therefore X(e^{j\omega}) = \frac{1}{1-\dfrac{1}{2}e^{-j\omega}} \times \frac{1-\dfrac{1}{2}e^{j\omega}}{1-\dfrac{1}{2}e^{j\omega}} = \frac{1-\dfrac{1}{2}(\cos\omega + j\sin\omega)}{1.25 - \cos\omega}$$

$$\boxed{e^{\pm j\theta} = \cos\theta \pm j\sin\theta}$$

$$\boxed{\text{Using equation (1)}}$$

$$= \frac{1 - 0.5\cos\omega - j0.5\sin\omega}{1.25 - \cos\omega}$$

$$= \frac{1 - 0.5\cos\omega}{1.25 - \cos\omega} - j\frac{0.5\sin\omega}{1.25 - \cos\omega}$$

$$\therefore H_r(e^{j\omega}) = \frac{1 - 0.5\cos\omega}{1.25 - \cos\omega} \text{ and } H_i(e^{j\omega}) = \frac{0.5\sin\omega}{1.25 - \cos\omega}$$

Phase function, $\angle X(e^{j\omega}) = \tan^{-1}\left[\dfrac{X_i(e^{j\omega})}{X_r(e^{j\omega})}\right] = \tan^{-1}\left[\dfrac{\dfrac{0.5\sin\omega}{1.25-\cos\omega}}{\dfrac{1-0.5\cos\omega}{1.25-\cos\omega}}\right] = \tan^{-1}\left[\dfrac{-0.5\sin\omega}{1-0.5\cos\omega}\right]$

Refer Table 5.3 and Fig 5.7 for plot of magnitude and phase function (Substitute a = 0.5)

Example 5.17

Find the frequency response of the LTI system, governed by the difference equation,

$$y(n) + a_1 y(n-1) + a_2 y(n-2) = x(n)$$

Solution:

Let, $\mathcal{F}\{x(n)\} = X(e^{j\omega})$, $\mathcal{F}\{y(n)\} = Y(e^{j\omega})$, $\mathcal{F}\{y(n-k)\} = e^{-j\omega k}Y(e^{j\omega})$

Given that, $y(n) + a_1 y(n-1) + a_2 y(n-2) = x(n)$

On taking Fourier transform we get,

$$Y(e^{j\omega}) + a_1 e^{-j\omega}Y(e^{j\omega}) + a_2 e^{-j2\omega}Y(e^{j\omega}) = X(e^{j\omega})$$

$$\therefore (1 + a_1 e^{-j\omega} + a_2 e^{-j2\omega})Y(e^{j\omega}) = X(e^{j\omega})$$

$$\therefore \text{ Frequency response, } H(e^{j\omega}) = \frac{Y(e^{j\omega})}{X(e^{j\omega})} = \frac{1}{1 + a_1 e^{-j\omega} + a_2 e^{-j2\omega}}$$

The magnitude function of $H(e^{j\omega})$ is defined as,

Magnitude function, $|H(e^{j\omega})| = [H(e^{j\omega})\,H^*(e^{j\omega})]^{1/2}$; where $H^*(e^{j\omega})$ is conjugate of $H(e^{j\omega})$

$$\therefore |H(e^{j\omega})| = \left[\frac{1}{1+a_1 e^{-j\omega}+a_2 e^{-j2\omega}} \times \frac{1}{1+a_1 e^{j\omega}+a_2 e^{j2\omega}}\right]^{\frac{1}{2}}$$

$$= \left[\frac{1}{1+a_1 e^{j\omega}+a_2 e^{j2\omega}+a_1 e^{-j\omega}+a_1^2+a_1 a_2 e^{j\omega}+a_2 e^{-j2\omega}+a_1 a_2 e^{-j\omega}+a_2^2}\right]^{\frac{1}{2}}$$

$$\therefore |H(e^{j\omega})| = \left[\frac{1}{1+a_1^2+a_2^2+a_1(e^{j\omega}+e^{-j\omega})+a_2(e^{j2\omega}+e^{-j2\omega})+a_1a_2(e^{j\omega}+e^{-j\omega})}\right]^{\frac{1}{2}}$$

$$= \left[\frac{1}{1+a_1^2+a_2^2+2a_1\cos\omega+2a_2\cos 2\omega+2a_1a_2\cos\omega}\right]^{\frac{1}{2}} \qquad \boxed{\cos\theta = \frac{e^{j\theta}+e^{-j\theta}}{2}}$$

$$= \left[\frac{1}{1+a_1^2+a_2^2+2a_1(a_2+1)\cos\omega+2a_2\cos 2\omega}\right]^{\frac{1}{2}} \qquad \text{.....(1)}$$

Let us separate the real and imaginary parts, by multiplying the numerator and denominator of $H(e^{j\omega})$ by the conjugate of the denominator of $H(e^{j\omega})$.

$$\therefore H(e^{j\omega}) = \frac{1}{1+a_1e^{-j\omega}+a_2e^{-j2\omega}} \times \frac{1+a_1e^{j\omega}+a_2e^{j2\omega}}{1+a_1e^{j\omega}+a_2e^{j2\omega}}$$

$$= \frac{1+a_1e^{j\omega}+a_2e^{j2\omega}}{1+a_1^2+a_1^2+2a_1(a_2+1)\cos\omega+2a_2\cos 2\omega} \qquad \boxed{\text{Using equation (1)}}$$
$$\boxed{e^{\pm j\theta} = \cos\theta \pm j\sin\theta}$$

$$= \frac{1+a_1(\cos\omega+j\sin\omega)+a_2(\cos 2\omega+j\sin 2\omega)}{1+a_1^2+a_2^2+2a_1(a_2+1)\cos\omega+2a_2\cos 2\omega}$$

$$= \frac{1+a_1\cos\omega+a_2\cos 2\omega}{1+a_1^2+a_2^2+2a_1(a_2+1)\cos\omega+2a_2\cos 2\omega}$$

$$+j\frac{a_1\sin\omega+a_2\sin 2\omega}{1+a_1^2+a_2^2+2a_1(a_2+1)\cos\omega+2a_2\cos 2\omega}$$

$$\therefore H_r(e^{j\omega}) = \frac{1+a_1\cos\omega+a_2\cos 2\omega}{1+a_1^2+a_2^2+2a_1(a_2+1)\cos\omega+2a_2\cos 2\omega}$$

$$H_i(e^{j\omega}) = \frac{a_1\sin\omega+a_2\sin 2\omega}{1+a_1^2+a_2^2+2a_1(a_2+1)\cos\omega+2a_2\cos 2\omega}$$

$$\therefore \text{Phase function}, \ \angle H(e^{j\omega}) = \tan^{-1}\frac{H_i(e^{j\omega})}{H_r(e^{j\omega})} = \tan^{-1}\left[\frac{a_1\sin\omega+a_2\sin 2\omega}{1+a_1\cos\omega+a_2\cos 2\omega}\right]$$

Example 5.18

The impulse response of an LTI system is given by $h(n) = r^n\cos(\omega_0 n)\,u(n)$. Find the frequency response of the system.

Solution:

The frequency response $H(e^{j\omega})$ is obtained by taking Fourier transform of $h(n)$.

By definition of Fourier transform,

$$\boxed{\begin{array}{l} u(n) = 1 \text{ for } n \geq 0 \\ \qquad\quad = 0 \text{ for } n < 0 \end{array}}$$

$$H(e^{j\omega}) = \sum_{n=-\infty}^{+\infty} h(n)\, e^{-j\omega n} = \sum_{n=0}^{+\infty} r^n \cos \omega_0 n\, e^{-j\omega n}$$

$$\therefore\ H(e^{j\omega}) = \sum_{n=0}^{+\infty} r^n \left[\frac{e^{j\omega_0 n} + e^{-j\omega_0 n}}{2} \right] e^{-j\omega n} = \frac{1}{2} \sum_{n=0}^{+\infty} \left[r^n e^{j\omega_0 n} e^{-j\omega n} + r^n e^{-j\omega_0 n} e^{-j\omega n} \right]$$

$$= \frac{1}{2} \sum_{n=0}^{+\infty} \left[r\, e^{j\omega_0} e^{-j\omega} \right]^n + \frac{1}{2} \sum_{n=0}^{\infty} \left[r\, e^{-j\omega_0} e^{-j\omega} \right]^n$$

$$= \frac{1}{2} \frac{1}{1 - r\, e^{j\omega_0} e^{-j\omega}} + \frac{1}{2} \frac{1}{1 - r\, e^{-j\omega_0} e^{-j\omega}} = \frac{1}{2} \left[\frac{1 - r\, e^{-j\omega_0} e^{-j\omega} + 1 - r\, e^{j\omega_0} e^{-j\omega}}{\left(1 - r\, e^{j\omega_0} e^{-j\omega}\right)\left(1 - r\, e^{-j\omega_0} e^{-j\omega}\right)} \right]$$

$$= \frac{1}{2} \frac{2 - r\, e^{-j\omega}\left(e^{-j\omega_0} + e^{j\omega_0}\right)}{1 - r\, e^{-j\omega_0} e^{-j\omega} - r\, e^{j\omega_0} e^{-j\omega} + r^2 e^{-j2\omega}} = \frac{1}{2} \frac{2 - r\, e^{-j\omega}\left(e^{j\omega_0} + e^{-j\omega_0}\right)}{1 - r\, e^{-j\omega}\left(e^{j\omega_0} + e^{-j\omega_0}\right) + r^2 e^{-j2\omega}}$$

$$= \frac{1}{2} \frac{2 - r\, e^{-j\omega} 2 \cos \omega_0}{1 - r\, e^{-j\omega} 2 \cos \omega_0 + r^2 e^{-j2\omega}} = \frac{1 - r \cos \omega_0\, e^{-j\omega}}{1 - 2r \cos \omega_0\, e^{-j\omega} + r^2 e^{-j2\omega}}$$

$$\therefore\ \text{Frequency responce, } H(e^{j\omega}) = \frac{1 - r \cos \omega_0\, e^{-j\omega}}{1 - 2r \cos \omega_0\, e^{-j\omega} + r^2 e^{-j2\omega}}$$

Let, $-r \cos \omega_0 = a$; $-2r \cos \omega_0 = \alpha$; and $r^2 = \beta$.

$$\therefore\ H(e^{j\omega}) = \frac{1 + a\, e^{-j\omega}}{1 + \alpha\, e^{-j\omega} + \beta\, e^{-j2\omega}}$$

The function $H(e^{j\omega})$ is same as frequency response [equation (5.51)] of standard second order system discussed in section 5.6.7. Hence, refer section 5.6.7 for derivation of magnitude and phase function.

Example 5.19

Determine the frequency response of an LTI system governed by the difference equation,

$$y(n) = x(n) + 0.81\, x(n - 1) + 0.81\, x(n - 2) - 0.45\, y(n - 2)$$

Solution:

Let, $\mathcal{F}\{y(n)\} = Y(e^{j\omega})$ $\therefore$ $\mathcal{F}\{y(n - k)\} = e^{-j\omega k} Y(e^{j\omega})$

Let, $\mathcal{F}\{x(n)\} = X(e^{j\omega})$ $\therefore$ $\mathcal{F}\{x(n - k)\} = e^{-j\omega k} X(e^{j\omega})$

Given that, $y(n) = x(n) + 0.81\, x(n - 1) + 0.81\, x(n - 2) - 0.45\, y(n - 2)$

On taking Fourier transform we get,

$$Y(e^{j\omega}) = X(e^{j\omega}) + 0.81\, e^{-j\omega} X(e^{j\omega}) + 0.81\, e^{-j2\omega} X(e^{j\omega}) - 0.45\, e^{-j2\omega} Y(e^{j\omega})$$

$$\therefore\ Y(e^{j\omega}) + 0.45 e^{-j2\omega} Y(e^{j\omega}) = X(e^{j\omega}) + 0.81\, e^{-j\omega} X(e^{j\omega}) + 0.81\, e^{-j2\omega} X(e^{j\omega})$$

$$\therefore\ (1 + 0.45\, e^{-j2\omega})\, Y(e^{j\omega}) = (1 + 0.81 e^{-j\omega} + 0.81 e^{-j2\omega})\, X(e^{j\omega})$$

$\therefore$ Frequency response, $H\left(e^{j\omega}\right) = \dfrac{Y\left(e^{j\omega}\right)}{X\left(e^{j\omega}\right)} = \dfrac{1 + 0.81\,e^{-j\omega} + 0.81\,e^{-j2\omega}}{1 + 0.45\,e^{-j2\omega}}$

Magnitude function, $\left|H\left(e^{j\omega}\right)\right| = \left[H\left(e^{j\omega}\right)H^{*}\left(e^{j\omega}\right)\right]^{\frac{1}{2}}$

$$= \left[\frac{1 + 0.81\,e^{-j\omega} + 0.81\,e^{-j2\omega}}{1 + 0.45\,e^{-j2\omega}} \times \frac{1 + 0.81\,e^{j\omega} + 0.81\,e^{j2\omega}}{1 + 0.45\,e^{j2\omega}}\right]^{\frac{1}{2}}$$

$$= \left[\frac{\begin{array}{c}1 + 0.81\,e^{j\omega} + 0.81\,e^{j2\omega} + 0.81\,e^{-j\omega} + 0.81^{2} + 0.81^{2}\,e^{j\omega} \\ + 0.81\,e^{-j2\omega} + 0.81^{2}\,e^{-j\omega} + 0.81^{2}\end{array}}{1 + 0.45\,e^{j2\omega} + 0.45\,e^{-j2\omega} + 0.45^{2}}\right]^{\frac{1}{2}}$$

$$= \left[\frac{2.3 + 0.81\left(e^{j\omega} + e^{-j\omega}\right) + 0.66\left(e^{j\omega} + e^{-j\omega}\right) + 0.81\left(e^{j2\omega} + e^{-j2\omega}\right)}{1.2 + 0.45\left(e^{j2\omega} + e^{-j2\omega}\right)}\right]^{\frac{1}{2}}$$

$$= \left[\frac{2.31 + 1.62\cos\omega + 1.32\cos\omega + 1.62\cos 2\omega}{1.2 + 0.9\cos 2\omega}\right]^{\frac{1}{2}} \qquad \dots\dots(1)$$

Let us separate real part and imaginary parts of $H(e^{j\omega})$, by multiplying the numerator and denominator of $H(e^{j\omega})$ by the complex conjugate of $H(e^{j\omega})$.

$$\therefore\ H\left(e^{j\omega}\right) = \frac{1 + 0.81\,e^{-j\omega} + 0.81\,e^{-j2\omega}}{1 + 0.45\,e^{-j2\omega}} \times \frac{1 + 0.45\,e^{j2\omega}}{1 + 0.45\,e^{j2\omega}}$$

$$= \frac{1 + 0.45\,e^{j2\omega} + 0.81\,e^{-j\omega} + 0.36\,e^{j\omega} + 0.81\,e^{-j2\omega} + 0.36}{1.2 + 0.9\cos 2\omega} \qquad \boxed{\text{Using equation (1)}}$$

$$= \frac{\begin{array}{c}1.36 + 0.45(\cos 2\omega + j\sin 2\omega) + 0.81(\cos\omega - j\sin\omega) \\ + 0.36(\cos\omega + j\sin\omega) + 0.81(\cos 2\omega - j\sin 2\omega)\end{array}}{1.2 + 0.9\cos 2\omega} \qquad \boxed{e^{\pm j\theta} = \cos\theta \pm j\sin\theta}$$

$$= \frac{1.36 + 1.17\cos\omega + 1.26\cos 2\omega}{1.2 + 0.9\cos 2\omega} + j\,\frac{-0.45\sin\omega - 0.36\sin 2\omega}{1.2 + 0.9\cos 2\omega} = H_{r}(e^{j\omega}) + jH_{i}(e^{j\omega})$$

Phase function, $\angle H\left(e^{j\omega}\right) = \tan^{-1}\left[\dfrac{H_{i}\left(e^{j\omega}\right)}{H_{r}\left(e^{j\omega}\right)}\right] = \tan^{-1}\left[\dfrac{-0.45\sin\omega - 0.36\sin 2\omega}{1.36 + 1.17\cos\omega + 1.26\cos 2\omega}\right]$

5.7 Analysis of LTI Discrete Time System Using z-transform

5.7.1 Transfer Function of LTI Discrete Time System

Let, $x(n)$ = Input of LTI discrete time system

$y(n)$ = Output of LTI discrete time system

$X(z)$ = z-transform of $x(n)$

$Y(z)$ = z-transform of $y(n)$

The *transfer function* of a n LTI discrete time system is defined as the ratio of z-transform of output and z-transform of input with zero initial conditions.

$$\therefore \text{ Transfer function} = \frac{Y(z)}{X(z)}\Bigg|_{\text{with zero initial conditions}} \qquad\qquad .....(5.53)$$

The mathematical equation governing the input-output relation of an LTI discrete time system is given by,

$$y(n) = -\sum_{m=1}^{N} a_m \, y(n-m) + \sum_{m=0}^{M} b_m \, x(n-m) \qquad\qquad .....(5.54)$$

The equation (5.54) is a constant coefficient difference equation and N is the order of the system.

Let us take z-transform of equation (5.54) with zero initial conditions.

$$\therefore z\{y(n)\} = z\left\{-\sum_{m=1}^{N} a_m \, y(n-m) + \sum_{m=0}^{M} b_m \, x(n-m)\right\}$$

$$\therefore Y(z) = z\left\{-\sum_{m=1}^{N} a_m \, y(n-m)\right\} + z\left\{\sum_{m=0}^{M} b_m \, x(n-m)\right\}$$

$$= -\sum_{m=1}^{N} a_m \, z\{y(n-m)\} + \sum_{m=0}^{M} b_m \, z\{x(n-m)\}$$

$$= -\sum_{m=1}^{N} a_m \, z^{-m} \, Y(z) + \sum_{m=0}^{M} b_m \, z^{-m} \, X(z)$$

$$\therefore Y(z) + \sum_{m=1}^{N} a_m \, z^{-m} \, Y(z) = \sum_{m=0}^{M} b_m \, z^{-m} \, X(z)$$

> By time shift property
> if $z\{x(n)\} = X(z)$
> then $z\{x(n-m)\} = z^{-m} \, X(z)$

$$Y(z)\left[1 + \sum_{m=1}^{N} a_m \, z^{-m}\right] = \sum_{m=0}^{M} b_m \, z^{-m} \, X(z)$$

$$\therefore \quad \frac{Y(z)}{X(z)} = \frac{\displaystyle\sum_{m=0}^{M} b_m \, z^{-m}}{1 + \displaystyle\sum_{m=1}^{N} a_m \, z^{-m}}$$

On expanding the summations in the above equation we get,

$$\frac{Y(z)}{X(z)} = \frac{b_0 + b_1 z^{-1} + b_2 z^{-2} + b_3 z^{-3} + ... + b_M z^{-M}}{1 + a_1 z^{-1} + a_2 z^{-2} + a_3 z^{-3} + ... + a_N z^{-N}}$$
    .....(5.55)

The equation (5.55) is the transfer function of an LTI discrete time system expressed as a rational function of z.

The numerator and denominator polynomials of equation (5.55) are converted to positive power of z and then expressed in the factorized form as shown in equation (5.56). $\boxed{\text{Let } M = N}$

$$\frac{Y(z)}{X(z)} = G \frac{(z - z_1)(z - z_2)(z - z_3) ..... (z - z_N)}{(z - p_1)(z - p_2)(z - p_3) ..... (z - p_N)}$$
    .....(5.56)

where, $z_1, z_2, z_3, ... z_N$ are roots of numerator polynomial (or zeros of discrete time system)

$p_1, p_2, p_3, ... p_N$ are roots of denominator polynomial (or poles of discrete time system).

5.7.2 Impulse Response and Transfer Function

Let, h(n) = Impulse response (i.e., response for impulse input)

H(z) = $\mathcal{Z}$-transform of h(n)

Now, the response y(n) of the discrete time system is given by convolution of input x(n) and impulse response, h(n).

$$\therefore \ y(n) = x(n) * h(n)$$

On taking $\mathcal{Z}$-transform of the above equation we get,

$$\mathcal{Z}\{y(n)\} = \mathcal{Z}\{x(n) * h(n)\}$$

$$\therefore \ \mathcal{Z}\{y(n)\} = X(z) H(z)$$

$$\therefore \ Y(z) = X(z) H(z)$$

$$\therefore \ H(z) = \frac{Y(z)}{X(z)}$$
    .....(5.57)

> If $\mathcal{Z}\{x(n)\} = X(z)$
> and $\mathcal{Z}\{h(n)\} = H(z)$
> then by convolution property,
> $\mathcal{Z}\{x(n) * h(n)\} = X(z) H(z)$

From equation (5.57) we can conclude that the transfer function of an LTI discrete time system is also given by $\mathcal{Z}$-transform of the impulse response.

Alternatively, we can say that the inverse $\mathcal{Z}$-transform of transfer function is the impulse response of the system.

$$\therefore \ \text{Impluse response, } h(n) = \mathcal{Z}^{-1}\{H(z)\} = \mathcal{Z}^{-1}\left\{\frac{Y(z)}{X(z)}\right\}$$

5.7.3 Response of LTI Discrete Time System Using $\mathcal{Z}$-Transform

In general, the input-output relation of an LTI (Linear Time Invariant) discrete time system is represented by the constant coefficient difference equation shown below, [equation (5.54)].

$$y(n) = - \sum_{m=1}^{N} a_m \, y(n-m) + \sum_{m=0}^{M} b_m \, x(n-m)$$

(or) $\qquad \displaystyle\sum_{m=0}^{N} a_m\, y(n-m) = \sum_{m=0}^{M} b_m\, x(n-m)$ with $a_0 = 1$ .....(5.58)

The solution of the above difference equation [equation (5.58)] is the (total) response $y(n)$ of LTI discrete time system, which consists of two parts. In signals and systems the two parts of the solution $y(n)$ are called zero-input response $y_{zi}(n)$ and zero-state response $y_{zs}(n)$.

$\boxed{\therefore \text{Response, } y(n) = y_{zi}(n) + y_{zs}(n)}$ .....(5.59)

Zero-input Response (or Free Response or Natural Response) Using Z-Transform

The *zero-input response* $y_{zi}(n)$ is mainly due to initial output (or initial stored energy) in the system. The zero-input response is obtained from system equation [equation (5.58)] when input $x(n) = 0$.

On substituting $x(n) = 0$ and $y(n) = y_{zi}(n)$ in equation (5.58) we get,

$$\sum_{m=0}^{M} a_m\, y_{zi}(n-m) = 0 \;;\; \text{with } a_0 = 1$$

On taking Z-transform of the above equation with non-zero initial conditions for output we can form an equation for $Y_{zi}(z)$. The zero-input response $y_{zi}(n)$ of a discrete time system is given by inverse Z-transform of $Y_{zi}(z)$.

Zero-State Response (or Forced Response) Using Z-Transform

The zero-state response $y_{zs}(n)$ is the response of the system due to input signal and with zero initial output. The zero-state response is obtained from the difference equation governing the system [equation(5.58)] for specific input signal $x(n)$ for $n \geq 0$ and with zero initial output.

On substituting $y(n) = y_{zs}(n)$ in equation (5.58) we get,

$$\sum_{m=0}^{N} a_m\, y_{zs}(n-m) = \sum_{m=0}^{M} b_m\, x(n-m) \;;\; \text{with } a_0 = 1$$

On taking Z-transform of the above equation with zero initial conditions for output [i.e., $y_{zs}(n)$] and nonzero initial values for input [i.e., $x(n)$] we can form an equation for $Y_{zs}(z)$. The zero-state response $y_{zs}(n)$ of a discrete time system is given by inverse Z-transform of $Y_{zs}(z)$.

Total Response

The total response $y(n)$ is the response of the system due to input signal and initial output (or intial stored energy). The total response is obtained from the difference equation governing the system [equation(5.58)] for specific input signal $x(n)$ for $n \geq 0$ and with nonzero initial conditions.

On taking Z-transform of equation (5.58) with nonzero initial conditions for both input and output, and then substituting for $X(z)$ we can form an equation for $Y(z)$. The total response $y(n)$ is given by inverse Z-transform of $Y(z)$. Alternatively, the total response $y(n)$ is given by sum of zero-input response $y_{zi}(n)$ and zero-state response $y_{zs}(n)$.

$\boxed{\therefore \text{Total response, } y(n) = y_{zi}(n) + y_{zs}(n)}$ .....(5.60)

5.7.4 Convolution and Deconvolution Using z-Transform

Convolution

The convolution operation is performed to find the response y(n) of an LTI discrete time system from the input x(n) and impulse response h(n).

∴ Response, $y(n) = x(n) * h(n)$

On taking z-transform of the above equation we get,

$$z\{y(n)\} = z\{x(n) * h(n)\}$$

$$\therefore Y(z) = X(z)\, H(z)$$

∴ Response, $y(n) = z^{-1}\{Y(z)\} = z^{-1}\{X(z)\, H(z)\}$

> If $z\{x(n)\} = X(z)$
> and $z\{h(n)\} = H(z)$
> then by convolution property,
> $z\{x(n)\} * h(n)\} = X(z)\, H(z)$

Procedure:
1. Take z-transform of x(n) to get X(z).
2. Take z-transform of h(n) to get H(z).
3. Get the product X(z) H(z).
4. Take inverse z-transform of the product X(z) H(z).

Deconvolution

The deconvolution operation is performed to extract the input x(n) of an LTI system from the response y(n) of the system.

From equation (5.57) get,

$$X(z) = \frac{Y(z)}{H(z)}$$

On taking inverse z-transform of the above equation we get,

$$\text{Input, } x(n) = z^{-1}\{X(z)\} = z^{-1}\left\{\frac{Y(z)}{H(z)}\right\}$$

Procedure:
1. Take z-transform of y(n) to get Y(z).
2. Take z-transform of h(n) to get H(z).
3. Divide Y(z) by H(z) to get X(z), [i.e., X(z) = Y(z) / H(z)].
4. Take inverse z-transform of X(z) to get x(n).

5.7.5 Stability in z-Domain

Location of Poles for Stability

Let, h(n) be the impulse response of an LTI discrete time system. Now, if h(n) satisfies the condition,

$$\sum_{n=-\infty}^{+\infty} |h(n)| < \infty \qquad\qquad .....(5.61)$$

then the LTI discrete time system is stable.

The stability condition of equation (5.61) can be transformed as a condition on location of poles of transfer function of the LTI discrete time system in z-plane.

Let, $h(n) = a^n u(n)$

Now, $\sum_{n=-\infty}^{+\infty} |h(n)| = \sum_{n=-\infty}^{+\infty} |a^n u(n)| = \sum_{n=0}^{\infty} a^n$

If $|a|$ is such that, $0 < |a| < 1$, then $\sum_{n=0}^{\infty} a^n = \dfrac{1}{1-a}$ = constant, and so the system is stable.

If $|a| > 1$, then $\sum_{n=0}^{\infty} a^n = \infty$ and so the system is unstable.

Now, $H(z) = \mathcal{Z}\{h(n)\} = \mathcal{Z}\{a^n u(n)\} = \dfrac{z}{z-a}$

Here $H(z)$ has pole at $z = a$.

If $|a| < 1$, then the pole will lie inside the unit circle and if $|a| > 1$, then the pole will lie outside the unit circle. Therefore we can say that, *for a stable discrete time system the poles should lie inside the unit circle.* The various types of impulse response of LTI discrete time system and their transfer functions and the locations of poles are summarized in table 5.5.

ROC of a Stable System   (AU Dec'14, 2 Marks)

Let, $H(z)$ be $\mathcal{Z}$-transform of $h(n)$. Now, by definition of $\mathcal{Z}$-transform we get,

$$H(z) = \sum_{n=-\infty}^{+\infty} h(n) z^{-n}$$

Let us evaluate $H(z)$ for $z = 1$.

$$\therefore H(z) = \sum_{n=-\infty}^{+\infty} h(n)$$

On taking absolute value on both sides we get,

$$|H(z)| = \left| \sum_{n=-\infty}^{+\infty} h(n) \right| \qquad \Rightarrow \qquad |H(z)| = \sum_{n=-\infty}^{+\infty} |h(n)|$$

For a stable LTI discrete time system,

$$\sum_{n=-\infty}^{+\infty} |h(n)| < \infty \qquad \Rightarrow \qquad |H(z)| < \infty$$

Therefore, we can conclude that $z = 1$ will be a point in the ROC of a stable system. Hence *for a stable discrete time system the ROC of impulse response should include the unit circle.*

General Condition for Stability in z-plane

On combining the condition for location of poles and the ROC we can say that *for a stable LTI discrete time system the poles should lie inside the unit circle and the unit circle should be included in ROC of impulse response of the system.*

Table 5.5: Impluse Response and Location of Poles

Impluse Response h(n)	Transfer Function H(z)	Location of Poles in z-plane and ROC
$h(n) = a^n u(n)$; $0 < a < 1$ h(n) graph $\sum_{n=0}^{+\infty} |h(n)| < \infty$; Stable system	$H(z) = \dfrac{z}{z-a}$ ROC is $|z| > a$ pole at $z = a$	z-plane plot Since $0 < a < 1$, the pole $z = a$, lies inside the unit circle. The ROC contains the unit circle.
$h(n) = (-a)^n u(n)$; $0 < |-a| < 1$ h(n) graph $\sum_{n=0}^{+\infty} |h(n)| < \infty$; Stable system	$H(z) = \dfrac{z}{z+a}$ ROC is $|z| > |-a|$ pole at $z = -a$	z-plane plot Since $0 < |-a| < 1$, the pole $z = -a$, lies inside the unit circle. The ROC contains the unit circle.
$h(n) = a^n u(n)$; $a > 1$ h(n) graph $\sum_{n=0}^{+\infty} |h(n)| < \infty$; Unstable system	$H(z) = \dfrac{z}{z-a}$ ROC is $|z| > a$ pole at $z = a$	z-plane plot Since $a > 1$, the pole at $z = a$, lies outside the unit circle. The ROC does not contain the unit circle.
$h(n) = (-a)^n u(n)$; $|-a| > 1$ h(n) graph $\sum_{n=0}^{+\infty} |h(n)| \infty$; Unstable system	$H(z) = \dfrac{z}{z+a}$ ROC is $|z| > |-a|$ pole at $z = -a$	z-plane plot Since $|-a| > 1$, the pole at $z = -a$, lies outside the unit circle. The ROC does not contain the unit circle.

Table 5.5: Continued....

Impulse Response h(n)	Transfer Function H(z)	Location of Poles in z-plane and ROC
$h(n) = a\,u(n)$; $a > 0$ (i.e., a is positive) [plot of h(n)] $\sum_{n=0}^{+\infty}|h(n)| = \infty$; Unstable system	$H(z) = \dfrac{az}{z-1}$ ROC is $|z| > 1$ pole at $z = 1$	[z-plane plot] The pole z = 1 lies on the unit circle. The ROC does not contain the unit circle.
$h(n) = a(-1)^n u(n)$; $a > 0$ (i.e., a is positive) [plot of h(n)] $\sum_{n=0}^{+\infty}|h(n)| = \infty$; Unstable system	$H(z) = \dfrac{az}{z+1}$ ROC is $|z| > 1$ pole at $z = -1$	[z-plane plot] The pole z = -1 lies on the unit circle. The ROC does not contain the unit circle.
$h(n) = n\,a^n u(n)$; $0 < a < 1$ [plot of h(n)] $\sum_{n=0}^{+\infty}|h(n)| < \infty$; Stable system	$H(z) = \dfrac{az}{(z-a)^2}$ ROC is $|z| > a$ Two poles at $z = a$	[z-plane plot] Since 0 < a < 1, the pole z = a, lie inside the unit circle. The ROC contains the unit circle.
$h(n) = n\,(-a)^n u(n)$; $0 < |-a| < 1$ [plot of h(n)] $\sum_{n=0}^{+\infty}|h(n)| < \infty$; Stable system	$H(z) = \dfrac{az}{(z+a)^2}$ ROC is $|z| > a$ Two poles at $z = -a$	[z-plane plot] Since 0 < |-a| < 1, the pole z = -a, lies inside the unit circle. The ROC contains the unit circle.

Table 5.5 : Continued....

Impulse Response h(n)	Transfer Function H(z)	Location of Poles in z-plane and ROC
$h(n) = a^n u(n) \; ; \; a > 1$ $\sum\limits_{n=0}^{+\infty} |h(n)| < \infty$; Unstable system	$H(z) = \dfrac{az}{(z - a)^2}$ ROC is $|z| > a$ Two poles at $z = a$	Since a> 1, the two poles at z = a lie outside the unit circle. The ROC does not contain the unit circle.
$h(n) = n\,(-a)^n u(n) \; ; \; |-a| > 1$ $\sum\limits_{n=0}^{+\infty} |h(n)| = \infty$; Unstable system	$H(z) = \dfrac{az}{(z + a)^2}$ ROC is $|z| > |{-a}|$ Two poles at $z = -a$	Since I-al >1, the two poles at z = -a lie outside the unit circle. The ROC does not contain the unit circle.
$h(n) = n\, u(n)$ $\sum\limits_{n=0}^{+\infty} |h(n)| = \infty$; Unstable system	$H(z) = \dfrac{az}{(z - 1)^2}$ ROC is $|z| > 1$ Two poles at $z = 1$	The two poles at z = 1, lie on the unit circle. The ROC does not contain the unit circle.

Table 5.5 : Continued....

Impulse Response h(n)	Transfer Function H(z)	Location of Poles in z-plane and ROC
$h(n) = n(-1)^n u(n)$; $|-a| > 1$ $\sum\limits_{n=0}^{\infty} |h(n)| = \infty$; Unstable system	$H(z) = \dfrac{az}{(z+1)^2}$ ROC is $|z| > 1$ Two poles at $z = -1$	The two poles at $z = -1$, lie on the unit circle. The ROC does not contain the unit circle.
$h(n) = r^n \cos \omega_0 n \, u(n)$; $0 < r < 1$ $\sum\limits_{n=0}^{\infty} |h(n)| < \infty$; Stable system	$H(z) = \dfrac{z(z - r\cos\omega_0)}{(z - r\cos\omega_0 - jr\sin\omega_0)(z - r\cos\omega_0 + jr\sin\omega_0)}$ ROC is $|z| > r$ A pair of conjugate poles at $z = p_1 = r\cos\omega_0 + jr\sin\omega_0$ $z = p_2 = r\cos\omega_0 - jr\sin\omega_0$	Since $0 < r < 1$, the conjugate pole pairs lie inside the unit circle. The ROC contains the unit circle.
$h(n) = r^n \cos \omega_0 n \, u(n)$; $r > 1$ $\sum\limits_{n=0}^{\infty} |h(n)| = \infty$; Unstable system	$H(z) = \dfrac{z(z - r\cos\omega_0)}{(z - r\cos\omega_0 - jr\sin\omega_0)(z - r\cos\omega_0 + jr\sin\omega_0)}$ ROC is $|z| > r$ A pair of conjugate poles at $z = p_1 = r\cos\omega_0 + jr\sin\omega_0$ $z = p_2 = r\cos\omega_0 - jr\sin\omega_0$	Since $r > 1$, the conjugate pole pairs lie outside the unit circle. The ROC does not contain the unit circle.
$h(n) = \cos \omega_0 n \, u(n)$ $\sum\limits_{n=0}^{\infty} |h(n)| = 0$; Stable system	$H(z) = \dfrac{z(z - \cos\omega_0)}{(z - \cos\omega_0 - j\sin\omega_0)(z - \cos\omega_0 + j\sin\omega_0)}$ ROC is $|z| > r$ A pair of conjugate poles at $z = p_1 = \cos\omega_0 + j\sin\omega_0$ $z = p_2 = \cos\omega_0 - j\sin\omega_0$	Since conjugate pole pairs lie on the circle. The ROC does not contain the unit circle.

Example 5.20

Determine the impulse response h(n) for the system described by the second-order difference equation,

$y(n) + 4y(n - 1) + 3y(n - 2) = x(n - 1)$.

Solution:

The difference equation governing the system is,

$$y(n) + 4y(n - 1) + 3y(n - 2) = x(n - 1)$$

Let us take $\mathcal{Z}$-transform of the difference equation governing the system with zero initial conditions.

$\therefore \mathcal{Z}\{y(n) + 4y(n - 1) + 3y(n - 2)\} = \mathcal{Z}\{x(n - 1)\}$

$\mathcal{Z}\{y(n)\} + 4\,\mathcal{Z}\{y(n - 1)\} + 3\,\mathcal{Z}\{y(n - 2)\} = \mathcal{Z}\{x(n - 1)\}$

$Y(z) + 4z^{-1}Y(z) + 3z^{-2}Y(z) = z^{-1}X(z)$

$(1 + 4z^{-1} + 3z^{-2})\,Y(z) = z^{-1}\,X(z)$

$\therefore \dfrac{Y(z)}{X(z)} = \dfrac{z^{-1}}{1 + 4z^{-1} + 3z^{-2}}$

> If $\mathcal{Z}\{x(n)\} = X(z)$
> then by shifting property
> $\mathcal{Z}\{x(n - m)\} = z^{-m}X(z)$
> If $\mathcal{Z}\{y(n)\} = Y(z)$
> then by shifting property
> $\mathcal{Z}\{y(n - m)\} = z^{-m}Y(z)$

> The roots of quadratic
> $z^2 + 4z + 3$ are,
> $z = \dfrac{-4 \pm \sqrt{4^2 - 4 \times 3}}{2}$
> $= \dfrac{-4 \pm 2}{2} = -1, -3$

We know that, $\dfrac{Y(z)}{X(z)} = H(z)$

$\therefore H(z) = \dfrac{z^{-1}}{1 + 4z^{-1} + 3z^{-2}} = \dfrac{z^{-1}}{z^{-2}(z^2 + 4z + 3)} = \dfrac{z}{(z + 1)(z + 3)}$

Using partial fraction expansion technique, we can write,

$\dfrac{H(z)}{z} = \dfrac{1}{(z + 1)(z + 3)} = \dfrac{A}{z + 1} + \dfrac{B}{z + 3}$

$A = (z + 1)\,\dfrac{1}{(z + 1)(z + 3)}\Big|_{z=-1} = \dfrac{1}{-1 + 3} = \dfrac{1}{2} = 0.5$

$B = (z + 3)\,\dfrac{1}{(z + 1)(z + 3)}\Big|_{z=-3} = \dfrac{1}{-3 + 1} = -\dfrac{1}{2} = -0.5$

$\therefore \dfrac{H(z)}{z} = \dfrac{0.5}{z + 1} - \dfrac{0.5}{z + 3} \quad \Rightarrow \quad H(z) = \dfrac{0.5z}{z + 1} - \dfrac{0.5z}{z + 3}$

> $\mathcal{Z}\{a^n\,u(n)\} = \dfrac{z}{z - a}$

The impulse response h(n) is given by inverse $\mathcal{Z}$-transform of H(z).

Impulse response, $h(n) = \mathcal{Z}^{-1}\{H(z)\} = \mathcal{Z}^{-1}\left\{\dfrac{0.5z}{z + 1} - \dfrac{0.5z}{z + 3}\right\} = 0.5\mathcal{Z}^{-1}\left\{\dfrac{z}{z - (-1)}\right\} - 0.5\mathcal{Z}^{-1}\left\{\dfrac{z}{z - (-3)}\right\}$

$= 0.5(-1)^n\,u(n) - 0.5\,(-3)^n\,u(n) = 0.5[(-1)^n - (-3)^n]\,u(n)$

Example 5.21 *(AU Jun'12, 8 Marks)*

Determine the impulse response h(n) for the system described by the second-order difference equation,

$y(n) + y(n - 1) - 2y(n - 2) + 3y(n - 3) = x(n) + 2x(n - 1)$.

Solution:

The difference equation governing the LTI system is,

$y(n) + y(n - 1) - 2y(n - 2) + 3y(n - 3) = x(n) + 2x(n - 1)$

Let us take z-transform of the difference equation governing the system with zero initial coditions.

$$\therefore z\{y(n) + y(n-1) - 2y(n-2) + 3y(n-3)\} = z\{x(n) + 2x(n-1)\}$$

$$z\{y(n)\} + z\{y(n-1)\} - 2\,z\{y(n-2)\}\} + 3\,z\{y(n-3)\} = z\{x(n)\} + 2\,z\{x(n-1)\}$$

$$Y(z) + z^{-1}Y(z) - 2z^{-2}Y(z) + 3z^{-3}Y(z) = X(z) + 2z^{-1}X(z)$$

$$(1 + z^{-1} - 2z^{-2} + 3z^{-3})\,Y(z) = (1 + 2z^{-1})\,X(z)$$

If $z\{x(n)\} = X(z)$
then $z\{x(n-m)\} = z^{-m}\,X(z)$

$$\therefore \frac{Y(z)}{X(z)} = \frac{1 + 2z^{-1}}{1 + z^{-1} - 2z^{-2} + 3z^{-3}} = \frac{z^{-1}(z+2)}{z^{-3}(z^3 + z^2 - 2z + +3)} = \frac{z^2(z+2)}{z^3 + z^2 - 2z + 3}$$

By trial and error, one of the root of $z^3 + z^2 - 2z + 3 = 0$ is found to be $z = -2.375$

$$\therefore z^3 + z^2 - 2z + 3 = (z + 2.375)(z^2 - 1.375z + 1.266)$$

$$\therefore \frac{Y(z)}{X(z)} = \frac{z^2(z+2)}{(z+2.375)(z^2 - 1.375z + 1.266)}$$

$$= \frac{z^2(z+2)}{(z+2.375)(z - 0.688 - j0.891)(z - 0.688 + j0.891)}$$

$$\begin{array}{r|l} & z^2 - 1.375z + 1.2666 \\ \hline z + 2.375 & z^3 + z^2 - 2z + 3 \\ & z^3 + 2.375\,z^2 \\ & \underline{(-)\ \ (-)} \\ & -1.375\,z^2 - 2z \\ & -1.375\,z^2 - 3.266z \\ & \underline{(+)\qquad (+)} \\ & 1.266z + 3 \\ & 1.266z + 3 \\ & \underline{(-)\qquad (-)} \\ & 0 \end{array}$$

We know that, $\dfrac{Y(z)}{X(z)} = H(z)$

$$\therefore H(z) = \frac{z^2(z+2)}{(z+2.375)(z - 0.688 - j0.891)(z - 0.688 + j0.891)}$$

Using partial fraction expansion technique, we can write,

$$\frac{H(z)}{z} = \frac{z^2 + 2z}{(z+2.375)(z - 0.688 - j0.891)(z - 0.688 + j0.891)}$$

The roots of quadratic equation $z^2 - 1.375z + 1.266 = 0$ are

$$z = \frac{1.375 \pm \sqrt{1.375^2 - 4 \times 1.266}}{2}$$

$$= 0.688 \pm j0.891$$

$$= \frac{A}{z+2.375} + \frac{B}{z - 0.688 - j0.891} + \frac{B^*}{z - 0.688 + j0.891}$$

$$A = (z+2.375)\frac{H(z)}{z}\Big|_{z=-2.375} = \cancel{(z+2.375)}\frac{z^2 + 2z}{\cancel{(z+2.375)}(z - 0.688 - j0.891)(z - 0.688 + j0.891)}\Big|_{z=-2.375}$$

$$= \frac{(-2.375)^2 + 2(-2.375)}{(-2.375 - 0.688 - j0.891)(-2.375 - 0.688 + j0.891)} = 0.088$$

$$B = (z - 0.688 - j0.891)\frac{H(z)}{z}\Big|_{z=0.688 + j0.891}$$

$$= \cancel{(z - 0.688 - j0.891)}\frac{z^2 + 2z}{(z+2.375)\cancel{(z - 0.688 - j0.891)}(z - 0.688 + j0.891)}\Big|_{z=0.688 + j0.891}$$

$$= \frac{(0.688 + j0.891)^2 + 2(0.688 + j0.891)}{(0.688 + j0.891 + 2.375)(0.688 + j0.891 - 0.688 + j0.891)} = 0.456 - j0.326$$

$$\therefore B^* = 0.456 + j0.326$$

$$\therefore \frac{H(z)}{z} = \frac{A}{z+2.375} + \frac{B}{z - 0.688 - j0.891} + \frac{B^*}{z - 0.688 + j0.891}$$

$$\therefore H(z) = 0.088\frac{z}{z - (-2.375)} + (0.456 - j0.326)\frac{z}{z - (0.688 + j0.891)}$$

$$+ (0.456 + j0.326)\frac{z}{z - (0.688 - j0.891)}$$

The impulse response $h(n)$ is given by inverse z-transform of $H(z)$.

Impulse response, $h(n) = z^{-1}\{H(z)\} = 0.088(-2.375)^n\,u(n) + (0.456 - j0.326)(0.688 + j0.891)^n\,u(n)$

$$+ (0.456 + j0.326)(0.688 - j0.891)^n\,u(n)$$

Example 5.22   *(AU Dec'12, 6 Marks)*

Determine the impulse response h(n) and transfer function of the system described by the difference equation,

$$y(n) = \frac{1}{3}y(n-1) + 3x(n).$$

Solution:

The difference equation governing the system is,

$$y(n) = \frac{1}{3}y(n-1) + 3x(n)$$

Let us take $\mathcal{Z}$-transform of the difference equation governing the system with zero initial coditions.

$$\therefore \mathcal{Z}\{y(n)\} = \mathcal{Z}\left\{\frac{1}{3}y(n-1) + 3x(n)\right\}$$

$$\mathcal{Z}\{y(n)\} = \mathcal{Z}\left\{\frac{1}{3}y(n-1)\right\} + 3\mathcal{Z}\{x(n)\}$$

$$Y(z) = \frac{1}{3}z^{-1}Y(z) + 3X(z)$$

$$Y(z) - \frac{1}{3}z^{-1}Y(z) = 3X(z)$$

$$Y(z)\left[1 - \frac{1}{3}z^{-1}\right] = 3X(z)$$

$\mathcal{Z}\{x(n) = X(z)$
$\mathcal{Z}\{y(n) = Y(z)$
$\mathcal{Z}\{y(n-1) = z^{-1} Y(z)$

$$\therefore \frac{Y(z)}{X(z)} = \frac{3}{1 - \frac{1}{3}z^{-1}}$$

We know that, $\dfrac{Y(z)}{X(z)} = H(z)$

$$\therefore H(z) = \frac{3}{1 - \frac{1}{3}z^{-1}}$$

The impulse response h(n) is given by inverse $\mathcal{Z}$-transform of H(z).

$$\therefore \text{Impulse response, } h(n) = \mathcal{Z}^{-1}\{H(z)\} = \mathcal{Z}^{-1}\left\{\frac{3}{1 - \frac{1}{3}z^{-1}}\right\}$$

$$= 3\left(\frac{1}{3}\right)^{n} u(n)$$

Example 5.23   *(AU Jun'14 & '13, 16 & 8 Marks)*

Determine the transfer function and impulse response h(n) for the system governed by the difference equation, $y(n) - \frac{5}{6}y(n-1) + \frac{1}{6}y(n-2) = x(n).$

Solution:

The difference equation governing the LTI system is,

$$y(n) - \frac{5}{6}y(n-1) + \frac{1}{6}y(n-2) = x(n)$$

On taking $\mathcal{Z}$-transform of the difference equation governing the system with zero initial conditions,

$$\mathcal{Z}\left\{y(n) - \frac{5}{6}y(n-1) + \frac{1}{6}y(n-2)\right\} = \mathcal{Z}\{x(n)\}$$

$$\mathcal{Z}\{y(n)\} - \frac{5}{6}\mathcal{Z}\{y(n-1)\} + \frac{1}{6}\mathcal{Z}\{y(n-2)\} = \mathcal{Z}\{x(n)\}$$

$$Y(z) - \frac{5}{6}z^{-1}Y(z) + \frac{1}{6}z^{-2}Y(z) = X(z)$$

$$Y(z)\left(1 - \frac{5}{6}z^{-1} + \frac{1}{6}z^{-2}\right) = X(z)$$

$\mathcal{Z}\{y(n)\} = Y(z),$	$\mathcal{Z}\{x(n)\} = X(z),$
$\mathcal{Z}\{y(n-1)\} = z^{-1} Y(z)$	$\mathcal{Z}\{y(n-2)\} = z^{-2} Y(z)$

∴ Transfer function, $H(z) = \dfrac{Y(z)}{X(z)} = \dfrac{1}{1 - \frac{5}{6}z^{-1} + \frac{1}{6}z^{-2}} = \dfrac{z^2}{z^2 - \frac{5}{6}z + \frac{1}{6}}$

We know that, $\dfrac{Y(z)}{X(z)} = H(z)$; ∴ $H(z) = \dfrac{z^2}{z^2 - \frac{5}{6}z + \frac{1}{6}}$

The roots of quadratic
$z^2 - \frac{5}{6}z + \frac{1}{6} = 0$ are,
$$z = \frac{\frac{5}{6} \pm \sqrt{\left(\frac{5}{6}\right)^2 - 4 \times \frac{1}{6}}}{2}$$
$$= \frac{1}{2}, \frac{1}{3}$$

Using partial fraction expansion technique, we can write,

$$\frac{H(z)}{z} = \frac{z}{z^2 - \frac{5}{6}z + \frac{1}{6}} = \frac{z}{\left(z - \frac{1}{2}\right)\left(z - \frac{1}{3}\right)} = \frac{A}{z - \frac{1}{2}} + \frac{B}{z - \frac{1}{3}}$$

$$A = \left(z - \frac{1}{2}\right)\frac{H(z)}{z}\bigg|_{z = \frac{1}{2}} = \left(z - \frac{1}{2}\right)\frac{z}{\left(z - \frac{1}{2}\right)\left(z - \frac{1}{3}\right)}\bigg|_{z = \frac{1}{2}} = \frac{\frac{1}{2}}{\frac{1}{2} - \frac{1}{3}} = \frac{\frac{1}{2}}{\frac{3-2}{2 \times 3}} = 3$$

$$B = \left(z - \frac{1}{3}\right)\frac{H(z)}{z}\bigg|_{z = \frac{1}{3}} = \left(z - \frac{1}{3}\right)\frac{z}{\left(z - \frac{1}{2}\right)\left(z - \frac{1}{3}\right)}\bigg|_{z = \frac{1}{3}} = \frac{\frac{1}{3}}{\frac{1}{3} - \frac{1}{2}} = \frac{\frac{1}{3}}{\frac{2-3}{3 \times 2}} = -2$$

$$\therefore \frac{H(z)}{z} = \frac{3}{z - \frac{1}{2}} - \frac{2}{z - \frac{1}{3}}$$

$$\therefore H(z) = 3\frac{z}{z - \frac{1}{2}} - 2\frac{z}{z - \frac{1}{3}}$$

∴ Impulse response, $h(n) = \mathcal{Z}^{-1}\{H(z)\} = \mathcal{Z}^{-1}\left\{3\dfrac{z}{z - \frac{1}{2}} - 2\dfrac{z}{z - \frac{1}{3}}\right\} = 3\mathcal{Z}^{-1}\left\{\dfrac{z}{z - \frac{1}{2}}\right\} - 2\mathcal{Z}^{-1}\left\{\dfrac{z}{z - \frac{1}{3}}\right\}$

$$= 3\left(\frac{1}{2}\right)^n u(n) - 2\left(\frac{1}{3}\right)^n u(n)$$

$$= \left[3\left(\frac{1}{2}\right)^n - 2\left(\frac{1}{3}\right)^n\right]u(n)$$

Example 5.24 *(AU Dec'13, 8 Marks)*

Determine the transfer function and impulse response for the system governed by the difference equation, $y(n) - \frac{1}{4}y(n-1) - \frac{3}{8}y(n-2) = -x(n) + 2x(n-1)$.

Solution:

The difference equation governing the LTI system is,

$$y(n) - \frac{1}{4}y(n-1) - \frac{3}{8}y(n-2) = -x(n) + 2x(n-1)$$

Let us take $\mathcal{Z}$-transform of the difference equation governing the system with zero initial conditions,

$$\therefore \mathcal{Z}\left\{y(n) - \frac{1}{4}y(n-1) - \frac{3}{8}y(n-2)\right\} = \mathcal{Z}\{-x(n) + 2x(n-1)\}$$

$$\mathcal{Z}\{y(n)\} - \frac{1}{4}\mathcal{Z}\{y(n-1)\} - \frac{3}{8}\mathcal{Z}\{y(n-2)\} = -\mathcal{Z}\{x(n)\} + 2\mathcal{Z}\{x(n-1)\}$$

$$Y(z) - \frac{1}{4}z^{-1}Y(z) - \frac{3}{8}z^{-2}Y(z) = -X(z) + 2z^{-1}X(z)$$

$$Y(z)\left(1 - \frac{1}{4}z^{-1} - \frac{3}{8}z^{-2}\right) = X(z)\left(-1 + 2z^{-1}\right)$$

$\therefore$ Transfer function, $H(z) = \dfrac{Y(z)}{X(z)} = \dfrac{2z^{-1}-1}{1-\dfrac{1}{4}z^{-1}-\dfrac{3}{8}z^{-2}} = \dfrac{z^{-1}(2-z)}{z^{-2}\left(z^2-\dfrac{1}{4}z-\dfrac{3}{8}\right)}$

$\qquad\qquad = \dfrac{z(2-z)}{z^2-0.25z-0.375} = \dfrac{z(2-z)}{(z-0.75)(z+0.5)}$

$\therefore \dfrac{H(z)}{z} = \dfrac{2-z}{(z-0.75)(z+0.5)}$

Using partial franction expansion technique, we can write,

$\dfrac{H(z)}{z} = \dfrac{2-z}{(z-0.75)(z+0.5)} = \dfrac{A}{z-0.75} + \dfrac{B}{z+0.5}$

The roots of quadratic equation
$z^2-0.25z-0.375 = 0$ are,
$z = \dfrac{0.25 \pm \sqrt{0.25^2-4\times(-0.375)}}{2}$
$= \dfrac{0.25 \pm 1.25}{2} = 0.75,\,-0.5$

$A = (z-0.75)\dfrac{H(z)}{z}\bigg|_{z=0.75} = (\cancel{z-0.75})\dfrac{2-z}{(\cancel{z-0.75})(z+0.5)}\bigg|_{z=0.75} = \dfrac{2-0.75}{0.75+0.5} = 1$

$B = (z+0.5)\dfrac{H(z)}{z}\bigg|_{z=-0.5} = (\cancel{z+0.5})\dfrac{2-z}{(z-0.75)(\cancel{z+0.5})}\bigg|_{z=-0.5} = \dfrac{2+0.5}{-0.5-0.75} = -2$

$\therefore \dfrac{H(z)}{z} = \dfrac{1}{z-0.75} - \dfrac{2}{z+0.5}$

$\therefore H(z) = \dfrac{z}{z-0.75} - \dfrac{2z}{z+0.5}$

The impulse response h(n) is given by inverse z-transform of H(z).

$\therefore$ Impulse response, $h(n) = z^{-1}\{H(z)\} = z^{-1}\left\{\dfrac{z}{z-0.75} - \dfrac{2z}{z+0.5}\right\} = z^{-1}\left\{\dfrac{z}{z-(0.75)}\right\} - 2z^{-1}\left\{\dfrac{z}{z-(-0.5)}\right\}$

$\qquad\qquad = (0.75)^n\,u(n) - 2\,(-0.5)^n\,u(n) = [0.75^n - 2(-0.5)^n]\,u(n)$

Example 5.25

Find the transfer function and unit sample response of the second-order difference equation with zero initial condition, $y(n) = x(n) - 0.25y(n-2)$.

Solution:

The difference equation governing the system is,

$\qquad y(n) = x(n) - 0.25\,y(n-2)$

Let us take z-transform of the difference equation governing the system with zero initial condition.

$\qquad z\{y(n)\} = z\{x(n) - 0.25\,y(n-2)\}$

$\qquad z\{y(n)\} = z\{x(n)\} - 0.25\,z\{y(n-2)\}$

$\boxed{z\{x(n)\} = X(z)}$

$\qquad Y(z) = X(z) - 0.25\,z^{-2}\,Y(z)$

$\qquad Y(z) + 0.25z^{-2}\,Y(z) = X(z)$

$\qquad [1+0.25z^{-2}]\,Y(z) = X(z)$

$\boxed{\begin{array}{l}z\{y(n)\} = Y(z)\\ z\{y(n-2)\} = z^{-2}\,Y(z)\\ \text{(Using shifting property)}\end{array}}$

$\therefore$ Trasfer function, $\dfrac{Y(z)}{X(z)} = \dfrac{1}{1+0.25z^{-2}}$

We know that, $\dfrac{Y(z)}{X(z)} = H(z)$

$\boxed{\begin{array}{l}z^2+0.25=0\\ \therefore z^2=-0.25\\ \therefore z=\pm\sqrt{-0.25}=\pm j0.5\end{array}}$

$\therefore H(z) = \dfrac{1}{1+0.25z^{-2}} = \dfrac{1}{z^{-2}(z^2+0.25)} = \dfrac{z^2}{(z+j0.5)(z-j0.5)}$

Using partial fraction expansion technique we can write,

$\dfrac{H(z)}{z} = \dfrac{z}{(z+j0.5)(z-j0.5)} = \dfrac{A}{z+j0.5} + \dfrac{A^*}{z-j0.5}$; where A^* is conjugate of A

$$A = (z+j0.5)\frac{H(z)}{z}\Big|_{z=-0.5} = \cancel{(z+j0.5)}\frac{z}{\cancel{(z+j0.5)}(z-j0.5)}\Big|_{z=-j0.5}$$

$$= \frac{-j0.5}{-j0.5-j0.5} = \frac{-j0.5}{2(-j0.5)} = \frac{1}{2} = 0.5$$

$$\therefore A^* = 0.5$$

$$\frac{H(z)}{z} = \frac{A}{z+j0.5} + \frac{A^*}{z-j0.5} = \frac{0.5}{z+j0.5} + \frac{0.5}{z-j0.5}$$

$$\therefore H(z) = \frac{0.5z}{z+j0.5} + \frac{0.5z}{z-j0.5} = \frac{0.5z}{z-(-j0.5)} + \frac{0.5z}{z-j0.5}$$

The impulse response is obtained by taking inverse z-transform of H(z).

$$\therefore \text{ Impulse response, } h(n) = z^{-1}\{H(z)\} = z\left\{\frac{0.5z}{z-(-j0.5)} + \frac{0.5z}{z-j0.5)}\right\}$$

$$= 0.5\left[z^{-1}\left\{\frac{z}{z-(-j0.5)}\right\} + z^{-1}\left\{\frac{z}{z-j0.5)}\right\}\right] \qquad \boxed{z\{a^n u(n)\} = \frac{z}{z-a}}$$

$$= 0.5[(-j0.5)^n u(n) + (j0.5)^n u(n)] = 0.5[(-j0.5)^n + (j0.5)^n] u(n)$$

Alternatively the above result can be expressed as shown below:

Here, $-j0.5 = 0.5\angle-90° = 0.5\angle-0.5\pi$

$+j0.5 = 0.5\angle 90° = 0.5\angle 0.5\pi$

$\boxed{180° = \pi \text{ rad} \quad ; \quad \therefore 1° = \frac{\pi}{180}\text{ rad}}$

$\therefore h(n) = 0.5[(0.5\angle-0.5\pi)^n + (0.5\angle 0.5\pi)^n] u(n)$

$\therefore 90^0 = 90\times\frac{\pi}{180} = 0.5\pi \text{ rad}$

$= 0.5[0.5^n\angle-0.5n\pi + 0.5^n\angle 0.5n\pi] u(n)$

$= 0.5(0.5)^n[1\angle-0.5n\pi + 1\angle 0.5n\pi] u(n)$

$= 0.5(0.5^n)[\cos 0.5n\pi - j\sin 0.5n\pi + \cos 0.5n\pi + j\sin 0.5n\pi] u(n)$

$= 0.5(0.5^n)[2\cos 0.5n\pi] u(n)$

$= 0.5^n \cos(0.5n\pi) u(n)$

Example 5.26

Determine the impulse response sequence of the discrete time LTI system defined by,

$y(n) - 4y(n-1) + 4y(n-2) = x(n) - 5x(n-3)$.

Solution:

The difference equation governing the LTI system is,

$$y(n) - 4y(n-1) + 4y(n-2) = x(n) - 5x(n-3)$$

$\boxed{z\{x(n)\} = X(z), \therefore z\{ax(n-m)\} = az^{-m}X(z)}$

On taking z-transform of the difference equation governing the system with zero initial conditions we get,

$\boxed{z\{y(n)\} = Y(z), \therefore z\{ay(n-m)\} = az^{-m}Y(z)}$

$$z\{y(n) - 4y(n-1) + 4y(n-2)\} = z\{x(n) - 5x(n-3)\}$$

$$z\{y(n)\} - 4z\{y(n-1)\} + 4z\{y(n-2)\} = z\{x(n)\} - 5z\{x(n-3)\}$$

$$Y(z) - 4z^{-1}Y(z) + 4z^{-2}Y(z) = X(z) - 5z^{-3}X(z)$$

$$[1 - 4z^{-1} + 4z^{-2}]Y(z) = [1 - 5z^{-3}]X(z)$$

$$\therefore \frac{Y(z)}{X(z)} = \frac{1 - 5z^{-3}}{1 - 4z^{-1} + 4z^{-2}}$$

We know that, $\dfrac{Y(z)}{X(z)} = H(z)$

$$\therefore H(z) = \frac{1 - 5z^{-3}}{1 - 4z^{-1} + 4z^{-2}} = \frac{1 - 5z^{-3}}{z^{-2}(z^2 - 4z + 4)} = \frac{z^2 - 5z^{-1}}{(z - 2)^2}$$

$$= \frac{z^2}{(z - 2)^2} - \frac{5z^{-1}}{(z - 2)^2} = \frac{1}{2}z\frac{2z}{(z - 2)^2} - \frac{5}{2}z^{-2}\frac{2z}{(z - 2)^2}$$

$z\{a^n u(n)\} = \dfrac{z}{z - a}$
$z\{na^n u(n)\} = \dfrac{az}{(z - a)^2}$
if $z\{x(n)\} = X(z)$ then by shifting property $z\{x(n \pm m)\} = Z^{\pm m}X(z)$

$$\boxed{(a - b)^2 = a^2 - 2ab + b^2}$$

The impulse response is obtained by taking inverse z-transform of $H(z)$.

$$\therefore \text{Impulse response, } h(n) = z^{-1}\{H(z)\} = z^{-1}\left\{\frac{1}{2}z\frac{2z}{(z - 2)^2} - \frac{5}{2}z^{-2}\frac{2z}{(z - 2)^2}\right\}$$

$$= \frac{1}{2}z^{-1}\left\{z\frac{2z}{(z - 2)^2}\right\} - \frac{5}{2}z^{-1}\left\{z^{-2}\frac{2z}{(z - 2)^2}\right\}$$

$$= \frac{1}{2}z^{-1}\left\{\frac{2z}{(z - 2)^2}\right\}\Big|_{n = n+1} - \frac{5}{2}z^{-1}\left\{\frac{2z}{(z - 2)^2}\right\}\Big|_{n = n-2}$$

$$= \frac{1}{2}[n\,2^n u(n)]\Big|_{n = n+1} - \frac{5}{2}[n\,2^n u(n)]\Big|_{n = n-2}$$

$$= \frac{1}{2}(n + 1)(2)^{n+1}u(n + 1) - \frac{5}{2}(n - 2)(2)^{n - 2}u(n - 2)$$

Example 5.27   *(AU Dec'14, 16 Marks)*

Determine the transfer function and impulse response of LTI discrete time system governed by the difference equation $y(n) = \frac{3}{2}y(n - 1) - \frac{1}{2}y(n - 2) + x(n) + x(n - 1)$ and the input $x(n) = u(n)$.

Solution:

Given that, $y(n) = \frac{3}{2}y(n - 1) - \frac{1}{2}y(n - 2) + x(n) + x(n - 1)$

On taking z- transform of the above equation we get,

$$Y(z) = \frac{3}{2}z^{-1}Y(z) - \frac{1}{2}z^{-2}Y(z) + X(z) + z^{-1}X(z)$$

$$Y(z)\left(1 - \frac{3}{2}z^{-1} + \frac{1}{2}z^{-2}\right) = X(z)(1 + z^{-1})$$

$z\{y(n)\} = Y(z),\ \therefore z\{y(n - m)\} = z^{-m}Y(z)$
$z\{x(n)\} = X(z),\ \therefore z\{x(n - m)\} = z^{-m}X(z)$

$$\therefore \frac{Y(z)}{X(z)} = \frac{1 + z^{-1}}{1 - \frac{3}{2}z^{-1} + \frac{1}{2}z^{-2}}$$

We know that $\dfrac{Y(z)}{X(z)} = H(z)$

The roots of the quadratic, $z^2 - 1.5z + 0.5 = 0$ are,
$z = \dfrac{1.5 \pm \sqrt{1.5^2 - 4 \times 0.5}}{2} = 1 \text{ or } 0.5$

$$\therefore H(z) = \frac{1 + z^{-1}}{1 - \frac{3}{2}z^{-1} + \frac{1}{2}z^{-2}} = \frac{z^{-1}(z + 1)}{z^{-2}(z^2 - 1.5z + 0.5)} = \frac{z(z + 1)}{(z - 1)(z - 0.5)}$$

By partial fraction expansion technique,

$$\frac{H(z)}{z} = \frac{z + 1}{(z - 1)(z - 0.5)} = \frac{A}{z - 1} + \frac{B}{z - 0.5}$$

$$A = (z - 1)\frac{H(z)}{z}\Big|_{z = 1} = (z - 1)\frac{(z + 1)}{(z - 1)(z - 0.5)}\Big|_{z = 1} = \frac{1 + 1}{1 - 0.5} = 4$$

$$B = (z-0.5)\frac{H(z)}{z}\Big|_{z=0.5} = (z-0.5)\frac{(z+1)}{(z-1)(z-0.5)}\Big|_{z=0.5} = \frac{0.5+1}{0.5-1} = -3$$

$$\therefore \frac{H(z)}{z} = \frac{A}{z-1} + \frac{B}{z-0.5} = \frac{4}{z-1} - \frac{3}{z-0.5}$$

$$\therefore H(z) = 4\frac{z}{z-1} - 3\frac{z}{z-0.5}$$

The impulse response is obtained by taking inverse z- transform of H(z).

$$h(n) = z^{-1}\{H(z)\} = \left\{\frac{4z}{z-1} - \frac{3z}{z-0.5}\right\}$$

$$= 4z^{-1}\left\{\frac{z}{z-1}\right\} - 3z^{-1}\left\{\frac{z}{z-0.5}\right\}$$

$$= 4u(n) - 3(0.5)^n u(n) = [4 - 3(0.5)^n] u(n)$$

$$\boxed{z\left\{\frac{z}{z-a}\right\} = a^n}$$

Example 5.28   *(AU Dec'11, 16 Marks)*

Determine the system function and impulse response for the system governed by the difference equation $y(n) - \frac{1}{2}y(n-1) + \frac{1}{4}y(n-2) = x(n)$ and the input $x(n) = u(n)$.

Solution:

Given that, $y(n) - \frac{1}{2}y(n-1) + \frac{1}{4}y(n-2) = x(n)$

On taking z- transform of the above equation we get,

$$Y(z) - \frac{1}{2}z^{-1}Y(z) + \frac{1}{4}z^{-2}Y(z) = X(z)$$

$$Y(z)\left(1 - \frac{1}{2}z^{-1} + \frac{1}{4}z^{-2}\right) = X(z)$$

$$\therefore \frac{Y(z)}{X(z)} = \frac{1}{1 - \frac{1}{2}z^{-1} + \frac{1}{4}z^{-2}}$$

Let, H(z) = System function or Transfer function

We know that $\frac{Y(z)}{X(z)} = H(z)$

$\therefore$ System function, $H(z) = \dfrac{1}{1 - \frac{1}{2}z^{-1} + \frac{1}{4}z^{-2}}$

The system function can also be expressed in positive power of z as shown below:

$$\therefore H(z) = \frac{1}{z^{-2}\left(z^2 - \frac{1}{2}z + \frac{1}{4}\right)} = \frac{z^2}{z^2 - \frac{1}{2}z + \frac{1}{4}}$$

By partial fraction expansion technique,

$$\frac{H(z)}{z} = \frac{z}{\left(z - \frac{1}{4} - j\frac{\sqrt{3}}{4}\right)\left(z - \frac{1}{4} + j\frac{\sqrt{3}}{4}\right)}$$

$$= \frac{A}{z - \frac{1}{4} - j\frac{\sqrt{3}}{4}} + \frac{A^*}{z - \frac{1}{4} + j\frac{\sqrt{3}}{4}}$$

$$A = \left(z - \frac{1}{4} - j\frac{\sqrt{3}}{4}\right)\frac{H(z)}{z}\Big|_{z=\frac{1}{4}+j\frac{\sqrt{3}}{4}} = \left(z - \frac{1}{4} - j\frac{\sqrt{3}}{4}\right) \times \frac{z}{\left(z - \frac{1}{4} - j\frac{\sqrt{3}}{4}\right)\left(z - \frac{1}{4} + j\frac{\sqrt{3}}{4}\right)}$$

> The roots of quadratic equation $z^2 - \frac{1}{2}z + \frac{1}{4} = 0$ are
> $$z = \frac{\frac{1}{2} \pm \sqrt{\frac{1}{4} - 4 \times 1 \times \frac{1}{4}}}{2}$$
> $$= \frac{1}{2}\left(\frac{1}{2} \pm j\frac{\sqrt{3}}{2}\right) = \frac{1}{4} \pm j\frac{\sqrt{3}}{4}$$

$$\therefore A = \frac{\frac{1}{4}+j\frac{\sqrt{3}}{4}}{\frac{1}{4}+j\frac{\sqrt{3}}{4}-\frac{1}{4}+j\frac{\sqrt{3}}{4}} = \frac{\frac{1}{4}+j\frac{\sqrt{3}}{4}}{j\frac{\sqrt{3}}{2}} = -j\frac{2}{\sqrt{3}}\left(\frac{1}{4}+j\frac{\sqrt{3}}{4}\right) = \frac{1}{2}-j\frac{1}{2\sqrt{3}}$$

$$\therefore A^* = \frac{1}{2}+j\frac{1}{2\sqrt{3}}$$

$$\therefore \frac{H(z)}{z} = \frac{\frac{1}{2}-j\frac{1}{2\sqrt{3}}}{z-\frac{1}{4}-j\frac{\sqrt{3}}{4}} + \frac{\frac{1}{2}+j\frac{1}{2\sqrt{3}}}{z-\frac{1}{4}+j\frac{\sqrt{3}}{4}}$$

$$\therefore H(z) = \left(\frac{1}{2}-j\frac{1}{2\sqrt{3}}\right)\frac{z}{z-\left(\frac{1}{4}+j\frac{\sqrt{3}}{4}\right)} + \left(\frac{1}{2}+j\frac{1}{2\sqrt{3}}\right)\frac{z}{z-\left(\frac{1}{4}-j\frac{\sqrt{3}}{4}\right)}$$

The impulse response is obtained by taking inverse z-transform of H(z).

$$\therefore \text{ Impulse response, } h(n) = z^{-1}\{H(z)\} = z^{-1}\left\{\left(\frac{1}{2}-j\frac{1}{2\sqrt{3}}\right)\frac{z}{z-\left(\frac{1}{4}+j\frac{\sqrt{3}}{4}\right)} + \left(\frac{1}{2}+j\frac{1}{2\sqrt{3}}\right)\frac{z}{z-\left(\frac{1}{4}-j\frac{\sqrt{3}}{4}\right)}\right\}$$

$$= \left(\frac{1}{2}-j\frac{1}{2\sqrt{3}}\right)\left(\frac{1}{4}+j\frac{\sqrt{3}}{4}\right)^n u(n) + \left(\frac{1}{2}+j\frac{1}{2\sqrt{3}}\right)\left(\frac{1}{4}-j\frac{\sqrt{3}}{4}\right)^n u(n)$$

Example 5.29   *(AU Jun'13,16 Marks)*

Determine the response of the system governed by the difference equation and also determine its stability.

$$y(n) = 0.7y(n-1) - 0.12y(n-2) + x(n-1) + x(n-2) \text{ for input } x(n) = n\,u(n)$$

Solution:

Given that, $x(n) = n\,u(n)$

$$\therefore X(z) = z\{x(n)\} = z\{n\,u(n)\} = \frac{z}{(z-1)^2}$$

The difference eqution governing the LTI system is,

$$y(n) = 0.7y(n-1) - 0.12y(n-2) + x(n-1) + x(n-2)$$

On taking z- transform we get,

$$\therefore Y(z) = 0.7z^{-1}Y(z) - 0.12z^{-2}Y(z) + z^{-1}X(z) + z^{-2}X(z)$$

$$Y(z) - 0.7z^{-1}Y(z) + 0.12z^{-2}Y(z) = z^{-1}X(z) + z^{-2}X(z)$$

$$Y(z)(1 - 0.7z^{-1} + 0.12z^{-2}) = X(z)(z^{-1} + z^{-2})$$

> The roots of quadratic
> $$z^2 - 0.7z + 0.12$$
> $$= \frac{0.7 \pm \sqrt{(0.7)^2 - 4\times 0.12}}{2}$$
> $$= \frac{0.7 \pm 0.1}{2} = 0.4, 0.3$$

$$Y(z) = X(z)\,\frac{z^{-1}+z^{-2}}{1-0.7z^{-1}+0.12z^{-2}} \qquad\qquad .....(1)$$

$$Y(z) = \frac{z}{(z-1)^2} \times \frac{z^{-2}(z+1)}{z^{-2}(z^2-0.7z+0.12)} = \frac{z(z+1)}{(z-1)^2(z-0.4)(z-0.3)}$$

By partial fraction expansion technique,

$$\frac{Y(z)}{z} = \frac{z+1}{(z-1)^2(z-0.4)(z-0.3)} = \frac{A}{(z-1)^2} + \frac{B}{z-1} + \frac{C}{z-0.4} + \frac{D}{z-0.3}$$

$$A = (z-1)^2\left.\frac{Y(z)}{z}\right|_{z=1} = (z-1)^2\left.\frac{(z+1)}{(z-1)^2(z-0.4)(z-0.3)}\right|_{z=1} = \frac{1+1}{(1-0.4)(1-0.3)} = 4.762$$

$$B = \frac{d}{dz}(z-1)^2 \frac{Y(z)}{z}\bigg|_{z=1} = \frac{d}{dz}\left(\frac{z+1}{z^2-0.7z+0.12}\right)\bigg|_{z=1}$$

$$= \frac{(z^2-0.7z+0.12)-(z+1)(2z-0.7)}{(z^2-0.7z+0.12)^2}\bigg|_{z=1} = \frac{(1-0.7+0.12)-(1+1)(2-0.7)}{(1-0.7+0.12)^2} = -12.358$$

$$C = (z-0.4)\frac{Y(z)}{z}\bigg|_{z=0.4} = (z-0.4)\frac{(z+1)}{(z-1)^2\,(z-0.4)(z-0.3)}\bigg|_{z=0.4} = \frac{0.4+1}{(0.4-1)^2(0.4-0.3)} = 38.889$$

$$D = (z-0.3)\frac{Y(z)}{z}\bigg|_{z=0.3} = (z-0.3)\frac{(z+1)}{(z-1)^2\,(z-0.4)(z-0.3)}\bigg|_{z=0.3} = \frac{0.3+1}{(0.3-1)^2(0.3-0.4)} = -26.531$$

$$\frac{Y(z)}{z} = \frac{4.762}{(z-1)^2} - \frac{12.358}{z-1} + \frac{38.889}{z-0.4} - \frac{26.531}{z-0.3}$$

$$\therefore\; Y(z) = 4.762\frac{z}{(z-1)^2} - 12.358\frac{z}{z-1} + 38.889\frac{z}{z-0.4} - 26.531\frac{z}{z-0.3}$$

The steady state response is obtained by taking inverse z-transform of $Y(z)$.

$$\therefore\; y(n) = z^{-1}\{Y(z)\} = z^{-1}\left\{4.762\frac{z}{(z-1)^2} - 12.358\frac{z}{z-1} + 38.889\frac{z}{z-0.4} - 26.531\frac{z}{z-0.3}\right\}$$

$$= 4.762\,nu(n) - 12.358\,u(n) + 38.889(0.4)^n\,u(n) - 26.531(0.3)^n\,u(n)$$

Stability

From equation (1) we get,

$$\frac{Y(z)}{X(z)} = \frac{z^{-1}+z^{-2}}{1-0.7z^{-1}+0.12z^{-2}} = \frac{z^{-2}(z+1)}{z^{-2}(z^2-0.7z+0.12)} = \frac{z+1}{(z-0.4)(z-0.3)}$$

We know that the transfer function of the system is,

$$H(z) = \frac{Y(z)}{X(z)} = \frac{z+1}{(z-0.4)(z-0.3)}$$

$\therefore$ The poles of $H(z)$ are at $z = 0.4$ and $z = 0.3$

For a causal system the ROC is exterior of circle corresponds to largest pole. Therefore, the ROC is $|z| > 0.4$. Here the system is causal system and the ROC includes unit circle and so the system is stable.

Example 5.30

Obtain and sketch the impulse response of shift invariant system described by,
$y(n) = 0.4\,x(n) + x(n-1) + 0.2\,x(n-2) + x(n-3) + 0.6\,x(n-4)$.

Solution:

The difference equation governing the system is,

$$y(n) = 0.4\,x(n) + x(n-1) + 0.2\,x(n-2) + x(n-3) + 0.6\,x(n-4)$$

On taking z-transform we get,

$$Y(z) = 0.4X(z) + z^{-1}X(z) + 0.2z^{-2}X(z) + z^{-3}X(z) + 0.6z^{-4}X(z)$$

$$Y(z) = [0.4 + z^{-1} + 0.2z^{-2} + z^{-3} + 0.6z^{-4}]\,X(z)$$

$$\therefore\; \frac{Y(z)}{X(z)} = [0.4 + z^{-1} + 0.2z^{-2} + z^{-3} + 0.6z^{-4}]$$

If $z\{x(n)\} = X(z)$ then by shifting property
$z\{x(n-k)\} = z^{-k}X(n)$

We know that, $\dfrac{Y(z)}{X(z)} = H(z)$

$\therefore H(z) = 0.4 + z^{-1} + 0.2z^{-2} + z^{-3} + 0.6z^{-4}$ .....(1)

By the definition of one sided z-transform we get,

$H(z) = \displaystyle\sum_{n=0}^{+\infty} h(n) z^{-n}$

$= h(0) z^0 + h(1) z^{-1} + h(2) z^{-2} + h(3) z^{-3} + h(4) z^{-4} + ...$ ....(2)

On comparing equations (1) and (2) we get,

$h(0) = 0.4$ $h(3) = 1$

$h(1) = 1$ $h(4) = 0.6$

$h(2) = 0.2$ $h(n) = 0$; for $n < 0$ and $n > 4$

$\therefore$ Impulse response, $h(n) = \{0.4, \quad 1.0, \quad 0.2, \quad 1.0, \quad 0.6\}$
$\qquad\qquad\qquad\qquad\qquad\quad \uparrow$

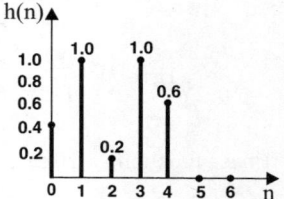

Fig 1: Graphical representation of impulse response h(n).

Example 5.31

An LTI system is described by the difference equation, $y(n) = ay(n-1) + bx(n)$. Find the impulse response, magnitude function and phase function. Solve b, if $|H(e^{j\omega})| = 1$. Sketch the magnitude and phase response for $a = 0.7$.

Solution:

a) To Find Impulse Response

Given that, $y(n) = ay(n-1) + bx(n)$.

On taking z-transform we get,

$Y(z) = az^{-1}Y(z) + b\,X(z)$

$\boxed{\begin{aligned} &z\{x(n)\} = X(z), \\ &z\{y(n)\} = Y(z), \\ \therefore\;&z\{y(n-1)\} = z^{-1}Y(z). \end{aligned}}$

$\therefore Y(z) - az^{-1}Y(z) = b\,X(z) \quad\Rightarrow\quad (1 - az^{-1})\,Y(z) = b\,X(z)$

$\therefore H(z) = \dfrac{Y(z)}{X(z)} = \dfrac{b}{1 - az^{-1}}$

The impulse response is obtained by taking inverse z-transform of $H(z)$.

$\therefore$ Impulse response, $h(n) = z^{-1}\{H(z)\} = z^{-1}\left\{\dfrac{b}{1-az^{-1}}\right\}$

$= b\,z^{-1}\left\{\dfrac{1}{1-az^{-1}}\right\} = b\,a^n\,u(n)$

$\boxed{z\{a^n u(n)\} = \dfrac{1}{1-az^{-1}}}$

b) To Find Frequency Response

The frequency response $H(e^{j\omega})$ is obtained by evaluating $H(z)$ at, $z = e^{j\omega}$.

$\therefore$ Frequency response, $H\left(e^{j\omega}\right) = H(z)\big|_{z = e^{j\omega}} = \dfrac{b}{1-az^{-1}}\bigg|_{z=e^{j\omega}} = \dfrac{b}{1-a\,e^{-j\omega}}$

Magnitude function, $\left|H\left(e^{j\omega}\right)\right| = \left[H\left(e^{j\omega}\right) \times H^*\left(e^{j\omega}\right)\right]^{\frac{1}{2}} = \left[\dfrac{b}{1-a\,e^{-j\omega}} \times \dfrac{b}{1-a\,e^{j\omega}}\right]^{\frac{1}{2}} = \left[\dfrac{b^2}{1-a\,e^{j\omega}-a\,e^{-j\omega}+a^2}\right]^{\frac{1}{2}}$

$= \left[\dfrac{b^2}{1+a^2-a\left(e^{j\omega}+e^{-j\omega}\right)}\right]^{\frac{1}{2}} = \left[\dfrac{b^2}{1+a^2-2a\cos\omega}\right]^{\frac{1}{2}} = \dfrac{b}{\sqrt{1+a^2-2a\cos\omega}}$

Let us seprate real and imaginary parts of H(e^jω), by multiplying the numerator and denominator of H(e^jω) by the complex conjugate of the denominator.

$$\therefore\ H\left(e^{j\omega}\right)=\frac{b}{1-a\,e^{-j\omega}}\times\frac{1-a\,e^{j\omega}}{1-a\,e^{j\omega}}=\frac{b-ab\,e^{j\omega}}{1-a\,e^{j\omega}-a\,e^{-j\omega}+a^{2}}=\frac{b-ab\left(\cos\omega+j\sin\omega\right)}{1+a^{2}-a\left(e^{j\omega}+e^{-j\omega}\right)}$$

$$=\frac{b-ab\cos\omega-jab\sin\omega}{1+a^{2}-2a\cos\omega}=\frac{b(1-a\cos\omega)}{1+a^{2}-2a\cos\omega}+j\,\frac{-ab\sin\omega}{1+a^{2}-2a\cos\omega}$$

$$\therefore\ H_{r}\left(e^{j\omega}\right)=\frac{b(1-a\cos\omega)}{1+a^{2}-2a\cos\omega}\ \text{and}\ H_{i}\left(e^{j\omega}\right)=\frac{-ab\sin\omega}{1+a^{2}-2a\cos\omega}$$

Phase function, $\angle H\left(e^{j\omega}\right)=\tan^{-1}\left[\dfrac{H_{i}\left(e^{j\omega}\right)}{H_{r}\left(e^{j\omega}\right)}\right]=\tan^{-1}\left[\dfrac{-ab\sin\omega}{b(1-a\cos\omega)}\right]=\tan^{-1}\left[\dfrac{-a\sin\omega}{1-a\cos\omega}\right]$

c) To Evaluate b and Sketch Frequency Response

Given that, $|H(e^{j\omega})| = 1$

$$\therefore\ \frac{b}{\sqrt{1+a^{2}-2a\cos\omega}}=1\quad\text{or}\quad b=\sqrt{1+a^{2}-2a\cos\omega}$$

When $a = 0.7$, $\angle H\left(e^{j\omega}\right)=\tan^{-1}\left[\dfrac{-a\sin\omega}{1-a\cos\omega}\right]=\tan^{-1}\left[\dfrac{-0.7\sin\omega}{1-0.7\cos\omega}\right]$

The phase function is periodic in the range $-\pi$ to $+\pi$. Hence the phase function is evaluated for various values of ω in the range $-\pi$ to $+\pi$.

When $\omega=\dfrac{-4\pi}{4}$, $\angle H(e^{j\omega})=\tan^{-1}\dfrac{-0.7\sin(-\pi)}{1-0.7\cos(-\pi)}\ =0$

When $\omega=\dfrac{-3\pi}{4}$, $\angle H(e^{j\omega})=\tan^{-1}\dfrac{-0.7\sin\left(\frac{-3\pi}{4}\right)}{1-0.7\cos\left(\frac{-3\pi}{4}\right)}\ =0.32\ =\dfrac{0.32}{\pi}\times\pi\ =0.1\pi\ \text{rad}$

When $\omega=\dfrac{-2\pi}{4}$, $\angle H(e^{j\omega})=\tan^{-1}\dfrac{-0.7\sin\left(\frac{-\pi}{2}\right)}{1-0.7\cos\left(\frac{-\pi}{2}\right)}\ =0.61\ =\dfrac{0.61}{\pi}\times\pi\ =0.19\pi\ \text{rad}$

When $\omega=\dfrac{-\pi}{4}$, $\angle H(e^{j\omega})=\tan^{-1}\dfrac{-0.7\sin\left(\frac{-\pi}{4}\right)}{1-0.7\cos\left(\frac{-\pi}{4}\right)}\ =0.775=\dfrac{0.775}{\pi}\times\pi\ =0.25\pi\ \text{rad}$

When $\omega=0$, $\angle H(e^{j\omega})=\tan^{-1}\dfrac{-0.7\sin(0)}{1-0.7\cos(0)}\ =0$

When $\omega=\dfrac{\pi}{4}$, $\angle H(e^{j\omega})=\tan^{-1}\dfrac{-0.7\sin\left(\frac{\pi}{4}\right)}{1-0.7\cos\left(\frac{\pi}{4}\right)}\ =-0.775\ =\dfrac{-0.775}{\pi}\times\pi=-0.25\pi\ \text{rad}$

When $\omega=\dfrac{2\pi}{4}$, $\angle H(e^{j\omega})=\tan^{-1}\dfrac{-0.7\sin\left(\frac{\pi}{2}\right)}{1-0.7\cos\left(\frac{\pi}{2}\right)}\ =-0.61\ =\dfrac{-0.61}{\pi}\times\pi\ =-0.19\pi\ \text{rad}$

When $\omega=\dfrac{3\pi}{4}$, $\angle H(e^{j\omega})=\tan^{-1}\dfrac{-0.7\sin\left(\frac{3\pi}{4}\right)}{1-0.7\cos\left(\frac{3\pi}{4}\right)}\ =-0.32\ =\dfrac{-0.32}{\pi}\times\pi\ =-0.1\pi\ \text{rad}$

When $\omega=\dfrac{4\pi}{4}$, $\angle H(e^{j\omega})=\tan^{-1}\dfrac{-0.7\sin(\pi)}{1-0.7\cos(\pi)}\ =0$

The phase function of Fig 2 is sketched using the above calculated values.

The magnitude function is a straight line, passing through "1" as shown in Fig 1.

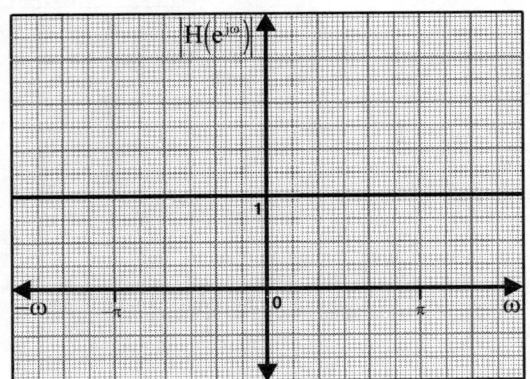

Fig 1: *Magnitude function.*

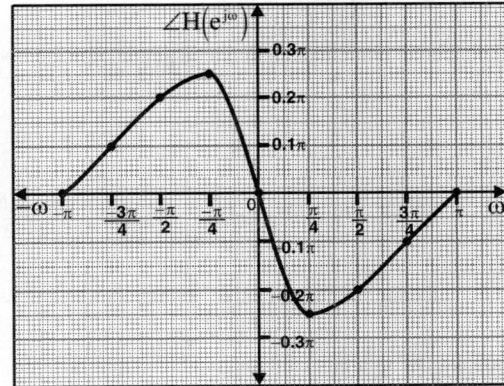

Fig 2: *Phase function.*

Example 5.32

The impulse response of system is $h(n) = 1 \quad ; \quad 0 \le n \le (N-1)$

$\qquad\qquad\qquad\qquad\qquad = 0 \quad ; \quad$ otherwise

Find the transfer function and frequency response.

Solution:

The transfer function H(z) is obtained by taking z-transform of the impulse response,

$\therefore$ Transfer function, $H(z) = z\{h(n)\} = \displaystyle\sum_{n=0}^{\infty} h(n)\, z^{-n} = \sum_{n=0}^{N-1} z^{-n}$

> Using finite geometric series sum formula,
>
> $\displaystyle\sum_{n=0}^{N-1} \dfrac{1-C^N}{1-C}$

$$= \frac{1-\left(z^{-1}\right)^N}{1-z^{-1}} = \frac{1-z^{-N}}{1-z^{-1}}$$

The frequency response $H(e^{j\omega})$ is obtained by evaluating H(z) at $z = e^{j\omega}$.

$\therefore$ Frequency function, $H\left(e^{j\omega}\right) = H(z)\big|_{z=e^{j\omega}} = \dfrac{1-z^{-N}}{1-z^{-1}}\bigg|_{z=e^{j\omega}} = \dfrac{1-e^{-j\omega N}}{1-e^{-j\omega}}$

$\therefore$ Magnitude function, $\left|H\left(e^{j\omega}\right)\right| = \left[H\left(e^{j\omega}\right) H^*\left(e^{j\omega}\right)\right]^{\frac{1}{2}}$

> $\cos\theta = \dfrac{e^{j\theta}+e^{-j\theta}}{2}$

$$= \left[\frac{1-e^{-j\omega N}}{1-e^{-j\omega}} \times \frac{1-e^{j\omega N}}{1-e^{j\omega}}\right]^{\frac{1}{2}} = \left[\frac{1-e^{j\omega N}-e^{-j\omega N}+1}{1-e^{j\omega}-e^{-j\omega}+1}\right]^{\frac{1}{2}}$$

$$= \left[\frac{2-\left(e^{j\omega N}+e^{-j\omega N}\right)}{2-\left(e^{j\omega}+e^{-j\omega}\right)}\right]^{\frac{1}{2}} = \left[\frac{2-2\cos\omega N}{2-2\cos\omega}\right]^{\frac{1}{2}} = \left[\frac{1-\cos\omega N}{1-\cos\omega}\right]^{\frac{1}{2}}$$

Let us seprate real and imaginary parts of $H(e^{j\omega})$, by multiplying the numerator and denominator of $H(e^{j\omega})$ by the complex conjugate of the denominator.

$\therefore H(e^{j\omega}) = \dfrac{1-e^{-j\omega N}}{1-e^{-j\omega}} \times \dfrac{1-e^{j\omega}}{1-e^{j\omega}}$

$$\therefore \; H(e^{j\omega}) = \frac{1 - e^{j\omega} - e^{-j\omega N} + e^{-j\omega N}e^{j\omega}}{1 - e^{j\omega} - e^{-j\omega} + 1} = \frac{1 - e^{j\omega} - e^{-j\omega N} + e^{-j\omega(N-1)}}{2 - 2\cos\omega} \qquad \boxed{e^{\pm j\theta} = \cos\theta \pm j\sin\theta}$$

$$= \frac{1 - (\cos\omega + j\sin\omega) - (\cos\omega N - j\sin\omega N) + (\cos\omega(N-1) - j\sin\omega(N-1))}{2 - 2\cos\omega}$$

Re al part, $\qquad H_r(e^{j\omega}) = \dfrac{1 - \cos\omega - \cos\omega N + \cos\omega(N-1)}{2 - 2\cos\omega}$

Im aginary part, $\quad H_i(e^{j\omega}) = \dfrac{-\sin\omega + \sin\omega N - \sin\omega(N-1)}{2 - 2\cos\omega}$

Phase function, $\angle H(e^{j\omega}) = \tan^{-1}\left[\dfrac{H_i(e^{j\omega})}{H_r(e^{j\omega})}\right] = \tan^{-1}\left[\dfrac{-\sin\omega + \sin\omega N - \sin\omega(N-1)}{1 - \cos\omega - \cos\omega N + \cos\omega(N-1)}\right]$

Example 5.33

A causal system is represented by the following difference equation.

$$y(n) + \frac{1}{4} y(n-1) = x(n) + \frac{1}{2} x(n-1)$$

Find the system transfer function H(z), the impulse response and frequency response of the system.

Solution:

a) System Transfer Function

The system transfer function, $H(z) = \dfrac{Y(z)}{X(z)}$ $\qquad\boxed{Z\{y(n)\} = Y(z)\;;Z\{y(n-1)\} = z^{-1}Y(z)}$

Given that, $y(n) + \dfrac{1}{4} y(n-1) = x(n) + \dfrac{1}{2} x(n-1)$ $\qquad\boxed{Z\{x(n)\} = X(z)\;;Z\{x(n-1)\} = z^{-1}X(z)}$

On taking Z-transform of the difference equation governing the system we get,

$$Y(z) + \frac{1}{4}z^{-1}Y(z) = X(z) + \frac{1}{2}z^{-1}X(z)$$

$$Y(z)\left(1 + \frac{1}{4}z^{-1}\right) = X(z)\left(1 + \frac{1}{2}z^{-1}\right)$$

$$\therefore \text{System transfer function, } H(z) = \frac{Y(z)}{X(z)} = \frac{1 + \dfrac{1}{2}z^{-1}}{1 + \dfrac{1}{4}z^{-1}}$$

b) Impulse Response

The impulse response h(n) is given by inverse Z-transform of H(z).

$$H(z) = \frac{1 + \dfrac{1}{2}z^{-1}}{1 + \dfrac{1}{4}z^{-1}}$$
$\qquad\qquad\boxed{Z\{a^n u(n)\} = \dfrac{1}{1 - az^{-1}}}$

$$= \frac{1}{1 - \left(-\dfrac{1}{4}\right)z^{-1}} + \frac{1}{2}\frac{z^{-1}}{1 - \left(-\dfrac{1}{4}\right)z^{-1}}$$
$\qquad\boxed{\begin{array}{l}\text{If } Z\{x(n)\} = X(z) \text{ then by time shifting}\\ \text{property } Z\{x(n-1)\} = z^{-1}X(z).\end{array}}$

The impulse response is obtained by taking inverse Z-transform of H(z).

$$\therefore \; \text{Impulse response, } h(n) = Z^{-1}\{H(z)\} = \left(-\frac{1}{4}\right)^n u(n) + \frac{1}{2}\left(-\frac{1}{4}\right)^{(n-1)} u(n-1)$$

c) Frequency Response

The frequency response $H(e^{j\omega})$ is the Fourier transform of h(n), or $H(e^{j\omega})$ is obtained by evaluating H(z) at $z = e^{j\omega}$, or $H(e^{j\omega})$ is given by $\dfrac{Y(e^{j\omega})}{X(e^{j\omega})}$.

Method 1

By definition of Fourier transform,

$$H(e^{j\omega}) = \mathcal{F}\{h(n)\} = \sum_{n=-\infty}^{+\infty} h(n)e^{-j\omega n}$$

$$= \sum_{n=-\infty}^{+\infty} \left[\left(-\frac{1}{4}\right)^n u(n) + \frac{1}{2}\left(-\frac{1}{4}\right)^{n-1} u(n-1)\right] e^{-j\omega n}$$

$$= \sum_{n=-\infty}^{+\infty} \left(-\frac{1}{4}\right)^n u(n)e^{-j\omega n} + \frac{1}{2}\sum_{n=-\infty}^{+\infty} \left(-\frac{1}{4}\right)^{n-1} u(n-1)e^{-j\omega n}$$

$$= \sum_{n=-\infty}^{+\infty} \left(-\frac{1}{4}\right)^n u(n)e^{-j\omega n} + \frac{1}{2}\sum_{m=-\infty}^{+\infty} \left(-\frac{1}{4}\right)^{m} u(m)e^{-j\omega(m+1)}$$

$$= \sum_{n=0}^{+\infty} \left(-\frac{1}{4}\right)^n e^{-j\omega n} + \frac{1}{2}\sum_{m=0}^{+\infty} \left(-\frac{1}{4}\right)^{m} e^{-j\omega m}e^{-j\omega}$$

$$= \sum_{n=0}^{+\infty} \left(-\frac{1}{4}e^{-j\omega}\right)^n + \frac{e^{-j\omega}}{2}\sum_{m=0}^{+\infty} \left(-\frac{1}{4}e^{-j\omega}\right)^m$$

$$= \frac{1}{1-\left(-\frac{1}{4}e^{-j\omega}\right)} + \frac{e^{-j\omega}}{2}\frac{1}{1-\left(-\frac{1}{4}e^{-j\omega}\right)}$$

$$= \frac{1}{1+\frac{1}{4}e^{-j\omega}} + \frac{\frac{1}{2}e^{-j\omega}}{1+\frac{1}{4}e^{-j\omega}} = \frac{1+\frac{1}{2}e^{-j\omega}}{1+\frac{1}{4}e^{-j\omega}}$$

> Let, $n - 1 = m$
> $\therefore n = m + 1$
> When $n = -\infty, m = -\infty$
> When $n = +\infty, m = +\infty$

> Using infinite geometric series sum formula,
> $$\sum_{n=0}^{\infty} C^n = \frac{1}{1-C}$$

Method 2

The frequency response, $H(e^{j\omega}) = H(z)\Big|_{z=e^{j\omega}} = \dfrac{1+\frac{1}{2}z^{-1}}{1+\frac{1}{4}z^{-1}}\Bigg|_{z=e^{j\omega}} = \dfrac{1+\frac{1}{2}e^{-j\omega}}{1+\frac{1}{4}e^{-j\omega}}$

Method 3

Given that, $y(n)+\frac{1}{4}y(n-1) = x(n)+\frac{1}{2}x(n-1)$

On taking Fourier transform,

$$Y(e^{j\omega})+\frac{1}{4}e^{-j\omega}Y(e^{j\omega}) = X(e^{j\omega})+\frac{1}{2}e^{-j\omega}X(e^{j\omega})$$

$$Y(e^{j\omega})\left[1+\frac{1}{4}e^{-j\omega}\right] = X(e^{j\omega})\left[1+\frac{1}{2}e^{-j\omega}\right]$$

$\therefore$ Frequency response, $H(e^{j\omega}) = \dfrac{Y(e^{j\omega})}{X(e^{j\omega})} = \dfrac{1+\frac{1}{2}e^{-j\omega}}{1+\frac{1}{4}e^{-j\omega}}$

Magnitude and Phase Function

Magnitude function, $|H(e^{j\omega})| = [H(e^{j\omega})H^*(e^{j\omega})]^{\frac{1}{2}}$;   where $H^*(e^{j\omega})$ is conjugate of $H(e^{j\omega})$

$$= \left[\frac{1+\frac{1}{2}e^{-j\omega}}{1+\frac{1}{4}e^{-j\omega}} \times \frac{1+\frac{1}{2}e^{j\omega}}{1+\frac{1}{4}e^{j\omega}}\right]^{\frac{1}{2}} = \left[\frac{1+\frac{1}{2}e^{j\omega}+\frac{1}{2}e^{-j\omega}+\frac{1}{4}}{1+\frac{1}{4}e^{j\omega}+\frac{1}{4}e^{-j\omega}+\frac{1}{16}}\right]^{\frac{1}{2}}$$

$$= \left[\frac{1+\frac{1}{2}(e^{j\omega}+e^{-j\omega})+\frac{1}{4}}{1+\frac{1}{4}(e^{j\omega}+e^{-j\omega})+\frac{1}{16}}\right]^{\frac{1}{2}} = \left[\frac{\frac{5}{4}+\cos\omega}{\frac{17}{16}+\frac{1}{2}\cos\omega}\right]^{\frac{1}{2}} \qquad \boxed{\cos\theta = \frac{e^{j\theta}+e^{-j\theta}}{2}}$$

Let us separate the real part and imaginary parts of $H(e^{j\omega})$, by multiplying the numerator and denominator of $H(e^{j\omega})$ by the conjugate of denominator of $H(e^{j\omega})$.

$$\therefore H(e^{j\omega}) = \frac{1+\frac{1}{2}e^{-j\omega}}{1+\frac{1}{4}e^{-j\omega}} \times \frac{1+\frac{1}{4}e^{j\omega}}{1+\frac{1}{4}e^{j\omega}} = \frac{1+\frac{1}{4}e^{j\omega}+\frac{1}{2}e^{-j\omega}+\frac{1}{8}}{1+\frac{1}{4}e^{j\omega}+\frac{1}{4}e^{-j\omega}+\frac{1}{16}}$$

$$= \frac{\frac{9}{8}+\frac{1}{4}(\cos\omega+j\sin\omega)+\frac{1}{2}(\cos\omega-j\sin\omega)}{\frac{17}{16}+\frac{1}{4}(e^{j\omega}+e^{-j\omega})} \qquad \boxed{e^{\pm j\theta} = \cos\theta \pm j\sin\theta}$$

$$= \frac{\frac{9}{8}+\frac{1}{4}\cos\omega+\frac{1}{2}\cos\omega+j\frac{1}{4}\sin\omega-j\frac{1}{2}\sin\omega}{\frac{17}{16}+\frac{1}{2}\cos\omega}$$

$$= \frac{\frac{9}{8}+\frac{3}{4}\cos\omega}{\frac{17}{16}+\frac{1}{2}\cos\omega} + j\frac{\left(-\frac{1}{4}\sin\omega\right)}{\frac{17}{16}+\frac{1}{2}\cos\omega}$$

$$\therefore H_r(e^{j\omega}) = \frac{\frac{9}{8}+\frac{3}{4}\cos\omega}{\frac{17}{16}+\frac{1}{2}\cos\omega} \quad \text{and} \quad H_i(e^{j\omega}) = \frac{-\frac{1}{4}\sin\omega}{\frac{17}{16}+\frac{1}{2}\cos\omega}$$

Phase function, $\angle H(e^{j\omega}) = \tan^{-1}\dfrac{H_i(e^{j\omega})}{H_r(e^{j\omega})} = \tan^{-1}\left[\dfrac{-\frac{1}{4}\sin\omega}{\frac{9}{8}+\frac{3}{4}\cos\omega}\right] = \tan^{-1}\left[\dfrac{-2\sin\omega}{9+6\cos\omega}\right]$

Example 5.34

Determine the response of discrete time LTI system governed by the difference equation, $y(n) = -0.8\,y(n-1) + x(n)$, when the input is unit step and initial condition, **a)** $y(-1) = 0$ and **b)** $y(-1) = 2/9$.

Solution:

Given that, $x(n) = u(n)$; $\therefore X(z) = \mathcal{Z}\{x(n)\} = \mathcal{Z}\{u(n)\} = \dfrac{z}{z-1}$ .....(1)

Given that, $y(n) = -0.8\,y(n-1) + x(n)$

$\quad\quad \therefore y(n) + 0.8\,y(n-1) = x(n)$

$\boxed{\begin{array}{l}\text{If } \mathcal{Z}\{y(n)\} = Y(z) \text{ then}\\ \mathcal{Z}\{y(n-1)\} = z^{-1}Y(z) - y(-1)\end{array}}$

On taking $\mathcal{Z}$-transform of above equation we get,

$\quad Y(z) + 0.8[z^{-1}Y(z) + y(-1)] = X(z)$

$\boxed{\text{Using equation (1).}}$

$$Y(z)[1 + 0.8z^{-1}] + 0.8y(-1) = \frac{z}{z-1}$$

$$Y(z)\left(1 + \frac{0.8}{z}\right) = \frac{z}{z-1} - 0.8y(-1)$$

$$Y(z)\left(\frac{z + 0.8}{z}\right) = \frac{z}{z-1} - 0.8y(-1)$$

$$\therefore Y(z) = \frac{z^2}{(z-1)(z+0.8)} - 0.8\frac{z\,y(-1)}{z+0.8}$$

Let, $P(z) = \dfrac{z^2}{(z-1)(z+0.8)}$ $\Rightarrow$ $\dfrac{P(z)}{z} = \dfrac{z}{(z-1)(z+0.8)}$

Let, $\dfrac{z}{(z-1)(z+0.8)} = \dfrac{A}{z-1} + \dfrac{B}{z+0.8}$

$$A = \frac{z}{(z-1)(z+0.8)} \times (z-1)\Big|_{z=1} = \frac{1}{1+0.8} = \frac{1}{1.8} = \frac{10}{18} = \frac{5}{9}$$

$$B = \frac{z}{(z-1)(z+0.8)} \times (z+0.8)\Big|_{z=-0.8} = \frac{-0.8}{-0.8-1} = \frac{-0.8}{-1.8} = \frac{8}{18} = \frac{4}{9}$$

$$\therefore \frac{P(z)}{z} = \frac{5}{9}\frac{1}{z-1} + \frac{4}{9}\frac{1}{z+0.8} \quad \Rightarrow \quad P(z) = \frac{5}{9}\frac{z}{z-1} + \frac{4}{9}\frac{z}{z+0.8}$$

$$\therefore Y(z) = \frac{5}{9}\frac{z}{z-1} + \frac{4}{9}\frac{z}{z+0.8} - 0.8\frac{z\,y(-1)}{z+0.8} \qquad \dots\dots(2)$$

a) When y(−1) = 0

From equation (2), when y(−1) = 0, we get,

$$\therefore Y(z) = \frac{5}{9}\frac{z}{z-1} + \frac{4}{9}\frac{z}{z+0.8}$$

$$\therefore \text{Response, } y(n) = z^{-1}\{Y(z)\} = z^{-1}\left\{\frac{5}{9}\frac{z}{z-1} + \frac{4}{9}\frac{z}{z+0.8}\right\} = \frac{5}{9}z^{-1}\left\{\frac{z}{z-1}\right\} + \frac{4}{9}z^{-1}\left\{\frac{z}{z-(-0.8)}\right\}$$

$$= \frac{5}{9}u(n) + \frac{4}{9}(-0.8)^n u(n) = \frac{1}{9}[5 + 4(-0.8)^n]\,u(n)$$

b) When y(−1) = 2/9

From equation (2), when y(−1) = 2/9, we get,

$$\therefore Y(z) = \frac{5}{9}\frac{z}{z-1} + \frac{4}{9}\frac{z}{z+0.8} - 0.8 \times \frac{2}{9}\frac{z}{z+0.8} = \frac{5}{9}\frac{z}{z-1} + \frac{2.4}{9}\frac{z}{z+0.8}$$

$$= \frac{5}{9}\frac{z}{z-1} + \frac{24}{90}\frac{z}{z+0.8} = \frac{5}{9}\frac{z}{z-1} + \frac{12}{45}\frac{z}{z+0.8}$$

$$\therefore \text{Response, } y(n) = z^{-1}\{Y(z)\} = z^{-1}\left\{\frac{5}{9}\frac{z}{z-1} + \frac{12}{45}\frac{z}{z+0.8}\right\} = \frac{5}{9}z^{-1}\left\{\frac{z}{z-1}\right\} + \frac{12}{45}z^{-1}\left\{\frac{z}{z-(-0.8)}\right\}$$

$$= \frac{5}{9}u(n) + \frac{12}{45}(-0.8)^n\,u(n) = \left[\frac{5}{9} + \frac{12}{45}(-0.8)^n\right]u(n)$$

Note: *Compare the result with example 5.2.*

Example 5.35

Determine the response of LTI discrete time system governed by the difference equation, $y(n) - 0.2\,y(n - 1) - 0.03\,y(n - 2) = x(n) + 0.4\,x(n - 1)$ for the input, $x(n) = 0.2^n\,u(n)$ and with initial condition, $y(-2) = 0,\ y(-1) = 0.5$.

Solution:

Given that, $x(n) = 0.2^n\,u(n)$; $\therefore X(z) = Z\{x(n)\} = Z\{0.2^n u(n)\} = \dfrac{z}{z - 0.2}$ .....(1)

Given that, $y(n) - 0.2\,y(n - 1) - 0.03\,y(n - 2) = x(n) + 0.4\,x(n - 1)$

On taking Z-transform of above equation we get,

$$Y(z) - 0.2\,[z^{-1}Y(z) + y(-1)] - 0.03\,[z^{-2}Y(z) + z^{-1}y(-1) + y(-2)\,] = X(z) + 0.4\,[\,z^{-1}X(z) + x(-1)] \quad .....(2)$$

If $Z\{y(n)\} = Y(z)$, then $Z\{y(n-1)\} = z^{-1}Y(z) + y(-1)$

and $Z\{y(n-2)\} = z^{-2}Y(z) + z^{-1}y(-1) + y(-2)$

Given that, $y(-2) = 0,\quad y(-1) = 0.5$

$$x(n) = 0.2^n\,u(n) = 0.2^n\,;\ \text{ for } n \ge 0 \qquad \Rightarrow \qquad x(-1) = 0$$

$$= 0\ ;\ \text{ for } n < 0$$

On substituting the above initial conditions in equation (2) we get,

$$Y(z) - 0.2\,z^{-1}Y(z) - 0.2 \times 0.5 - 0.03\,z^{-2}Y(z) - 0.03\,z^{-1} \times 0.5 + 0 = X(z) + 0.4\ z^{-1}X(z) + 0$$

$$Y(z) - \frac{0.2}{z}Y(z) - 0.1 - \frac{0.03}{z^2}Y(z) - \frac{0.015}{z} = X(z) + \frac{0.4}{z}X(z)$$

$$\therefore Y(z)\left(1 - \frac{0.2}{z} - \frac{0.03}{z^2}\right) - \left(\frac{0.015}{z} + 0.1\right) = X(z)\left(1 + \frac{0.4}{z}\right)$$

$$Y(z)\left(\frac{z^2 - 0.2z - 0.03}{z^2}\right) - \left(\frac{0.015 + 0.1z}{z}\right) = \left(\frac{z}{z - 0.2}\right)\left(\frac{z + 0.4}{z}\right) \qquad \boxed{\text{Using equation (1).}}$$

$$Y(z)\,\frac{(z - 0.3)(z + 0.1)}{z^2} = \frac{z + 0.4}{z - 0.2} + \frac{0.015 + 0.1z}{z}$$

$$Y(z)\,\frac{(z - 0.3)(z + 0.1)}{z^2} = \frac{z(z + 0.4) + (0.015 + 0.1z)(z - 0.2)}{z(z - 0.2)}$$

$$Y(z)\,\frac{(z - 0.3)(z + 0.1)}{z^2} = \frac{z^2 + 0.4z + 0.015z - 0.003 + 0.1z^2 - 0.02z}{z(z - 0.2)}$$

$$Y(z)\,\frac{(z - 0.3)(z + 0.1)}{z^2} = \frac{1.1z^2 + 0.395z - 0.003}{z(z - 0.2)}$$

$$Y(z) = \frac{1.1z^2 + 0.395z - 0.003}{z(z - 0.2)} \times \frac{z^2}{(z - 0.3)(z + 0.1)}$$

The roots of the quadratic, $z^2 - 0.2z - 0.03 = 0$ are,

$$z = \frac{0.2 \pm \sqrt{0.2^2 - 4 \times (-0.03)}}{2}$$

$$= \frac{0.2 \pm 0.4}{2} = 0.3,\, -0.1$$

$$= \frac{z(1.1z^2 + 0.395z - 0.003)}{(z - 0.2)(z - 0.3)(z + 0.1)}$$

Let, $\dfrac{Y(z)}{z} = \dfrac{1.1z^2 + 0.395z - 0.003}{(z - 0.2)(z - 0.3)(z + 0.1)} = \dfrac{A}{z - 0.2} + \dfrac{B}{z - 0.3} + \dfrac{C}{z + 0.1}$

$$A = \frac{1.1z^2 + 0.395z - 0.003}{(z-0.2)(z-0.3)(z+0.1)} \times (z-0.2)\Big|_{z=0.2} = \frac{1.1\times 0.2^2 + 0.395\times 0.2 - 0.003}{(0.2-0.3)(0.2+0.1)} = -4$$

$$B = \frac{1.1z^2 + 0.395z - 0.003}{(z-0.2)(z-0.3)(z+0.1)} \times (z-0.3)\Big|_{z=0.3} = \frac{1.1\times 0.3^2 + 0.395\times 0.3 - 0.003}{(0.3-0.3)(0.3+0.1)} = 5.3625$$

$$C = \frac{1.1z^2 + 0.395z - 0.003}{(z-0.2)(z-0.3)(z+0.1)} \times (z+0.1)\Big|_{z=-0.1} = \frac{1.1\times (-0.1)^2 + 0.395\times (-0.1) - 0.003}{(-0.1-0.2)(-0.1-0.3)} = -0.2625$$

$$\therefore \frac{Y(z)}{z} = \frac{-4}{z-0.2} + \frac{5.3625}{z-0.3} - \frac{0.2625}{z+0.1}$$

$$\therefore Y(z) = -4\frac{z}{z-0.2} + 5.3625\frac{z}{z-0.3} - 0.2625\frac{z}{z-(-0.1)}$$

$$\therefore \text{Response, } y(n) = \mathcal{Z}^{-1}\{Y(z)\} = \mathcal{Z}^{-1}\left\{-4\frac{z}{z-0.2} + 5.3625\frac{z}{z-0.3} - 0.2625\frac{z}{z-(-0.1)}\right\}$$

$$= -4(0.2)^n u(n) + 5.3625(0.3)^n u(n) - 0.2625(-0.1)^n u(n)$$

$$= [-4(0.2)^n + 5.3625(0.3)^n - 0.2625(-0.1)^n]u(n)$$

Note: Compare the result with example 5.3.

Example 5.36

Find the response of the time invariant system with impulse response, h(n) = {1, 2, –1, –2} to an input signal, x(n) = {1, 2, 3, 4}.

Solution:

Let, y(n) = Response or Output of an LTI system.

The response of an LTI system is given by the convolution of input signal and impulse response.

$$\therefore y(n) = x(n) * h(n)$$

By convolution property
$$\mathcal{Z}\{x(n) * h(n)\} = X(z)H(z)$$

On taking $\mathcal{Z}$-transform we get,

$$\mathcal{Z}\{y(n)\} = \mathcal{Z}\{x(n) * h(n)\}$$

$$\therefore Y(z) = X(z)\,H(z)$$

Given that, x(n) = {1, 2, 3, 4}

By definition of $\mathcal{Z}$-transform,

$$X(z) = \sum_{n=-\infty}^{\infty} x(n)z^{-n} = \sum_{n=0}^{3} x(n)z^{-n} = x(0)z^0 + x(1)z^{-1} + x(2)z^{-2} + x(3)z^{-3}$$

$$= 1 + 2z^{-1} + 3z^{-2} + 4z^{-3}$$

Given that, h(n) = { 1, 2, –1, –2}

By definition of $\mathcal{Z}$-transform,

$$H(z) = \sum_{n=-\infty}^{\infty} h(n)z^{-n} = \sum_{n=0}^{3} h(n)z^{-n} = h(0)z^0 + h(1)z^{-1} + h(2)z^{-2} + h(3)z^{-3}$$

$$= 1 + 2z^{-1} - z^{-2} - 2z^{-3}$$

$$\therefore Y(z) = X(z)\, H(z)$$

$$= [\,1 + 2z^{-1} + 3z^{-2} + 4z^{-3}\,]\,[\,1 + 2z^{-1} - z^{-2} - 2z^{-3}\,]$$

$$= 1 + 2z^{-1} - z^{-2} - 2z^{-3}$$

$$+ 2z^{-1} + 4z^{-2} - 2z^{-3} - 4z^{-4}$$

$$+ 3z^{-2} + 6z^{-3} - 3z^{-4} - 6z^{-5}$$

$$+ 4z^{-3} + 8z^{-4} - 4z^{-5} - 8z^{-6}$$

$$= 1 + 4z^{-1} + 6z^{-2} + 6z^{-3} + z^{-4} - 10z^{-5} - 8z^{-6} \qquad \dots(1)$$

By definition of z-transform we get,

$$Y(z) = \sum_{n=-\infty}^{\infty} y(n)z^{-n}$$

$$= y(0)\,z^{0} + y(1)\,z^{-1} + y(2)\,z^{-2} + y(3)\,z^{-3} + y(4)\,z^{-4} + y(5)\,z^{-5} + y(6)\,z^{-6} + \dots \qquad \dots(2)$$

On comparing equations (1) and (2) we get,

$y(0) = 1$	$y(2) = 6$	$y(4) = 1$	$y(6) = -8$
$y(1) = 4$	$y(3) = 6$	$y(5) = -10$	

$$\therefore \text{ Response, } y(n) = \{1, \quad 4, \quad 6, \quad 6, \quad 1, \quad -10, \quad -8\}$$
$$\uparrow$$

Example 5.37

(AU June'12, 8 Marks)

Find the input of the time invariant system with impulse response, $h(n) = \{1, 2, 3\}$ to an output signal, $y(n) = \{3, 8, 14, 8, 3\}$.

Solution:

Given that, $y(n) = \{3, 8, 14, 8, 3\}$

$$\therefore Y(z) = \mathcal{Z}\{y(n)\} = \sum_{n=-\infty}^{\infty} y(n)z^{-n} = \sum_{n=0}^{4} y(n)z^{-n}$$

$$= y(0) + y(1)\,z^{-1} + y(2)\,z^{-2} + y(3)\,z^{-3} + y(4)\,z^{-4}$$

$$= 3 + 8z^{-1} + 14z^{-2} + 8z^{-3} + 3z^{-4}$$

Given that, $h(n) = \{1, 2, 3\}$

$$\therefore H(z) = \mathcal{Z}\{h(n)\} = \sum_{n=-\infty}^{\infty} h(n)z^{-n} = \sum_{n=0}^{2} h(n)z^{-n} = h(0) + h(1)z^{-1} + h(2)z^{-2}$$

$$= 1 + 2z^{-1} + 3z^{-2}$$

We know that, $H(z) = \dfrac{Y(z)}{X(z)}$

$$\therefore X(z) = \frac{Y(z)}{H(z)} = \frac{3 + 8z^{-1} + 14z^{-2} + 8z^{-3} + 3z^{-4}}{1 + 2z^{-1} + 3z^{-2}}$$

$$
\begin{array}{r}
3 + 2z^{-1} + z^{-2} \\
\hline
\end{array}
$$

$$
1 + 2z^{-1} + 3z^{-2} \,) \, \begin{array}{l}
3 + 8z^{-1} + 14z^{-2} + 8z^{-3} + 3z^{-4} \\
3 + 6z^{-1} + 9z^{-2} \\
\underline{(-)\ (-)\quad\ (-)} \\
\quad\ 2z^{-1} + 5z^{-2} + 8z^{-3} \\
\quad\ 2z^{-1} + 4z^{-2} + 6z^{-3} \\
\underline{\quad (-)\quad\ (-)\quad\ (-)} \\
\qquad\qquad z^{-2} + 2z^{-3} + 3z^{-4} \\
\qquad\qquad z^{-2} + 2z^{-3} + 3z^{-4} \\
\underline{\qquad\qquad (-)\quad (-)\quad\ (-)} \\
\qquad\qquad\qquad\qquad 0
\end{array}
$$

$\therefore \ X(z) \ = \ 3 + 2z^{-1} + z^{-2}$ .....(1)

By the definition of z-transform,

$$X(z) \ = \ z\{x(n)\} \ = \ \sum_{n=-\infty}^{\infty} x(n)z^{-n} \ = x(0)z^{0} + x(1)z^{-1} + x(2)z^{-2} + x(3)z^{-3} \qquad \text{.....(2)}$$

On comparing equations (1) and (2) we get,

$x(0) = 1 \quad ; \quad x(1) = 2 \quad ; \quad x(2) = 3$

$\therefore$ Input, $x(n) = \{ \ 3, \ 2, \ 1 \ \}$

                   ↑

Example 5.38

Using z-transform, perform deconvolution of the response, $y(n) = \{ 1, 4, 6, 6, 1, -10, -8 \}$ and impulse response $h(n) = \{ 1, 2, -1, -2 \}$ to extract the input $x(n)$.

Solution:

Given that, $y(n) = \{ 1, 4, 6, 6, 1, -10, -8 \}$

$$\therefore Y(z) \ = \ z\{y(n)\} \ = \ \sum_{n=-\infty}^{\infty} y(n)z^{-n} \ = \ \sum_{n=0}^{6} y(n)z^{-n}$$

$$= y(0) + y(1) \, z^{-1} + y(2) \, z^{-2} + y(3) \, z^{-3} + y(4) \, z^{-4} + y(5) \, z^{-5} + y(6) \, z^{-6}$$

$$= 1 + 4z^{-1} + 6z^{-2} + 6z^{-3} + z^{-4} - 10z^{-5} - 8z^{-6}$$

Given that, $h(n) = \{ 1, 2, -1, -2 \}$

$$\therefore H(z) \ = \ z\{h(n)\} \ = \ \sum_{n=-\infty}^{\infty} h(n)z^{-n} \ = \ \sum_{n=0}^{3} h(n)z^{-n} \ = \ h(0) + h(1)z^{-1} + h(2)z^{-2} + h(3)z^{-3}$$

$$= 1 + 2z^{-1} - z^{-2} - 2z^{-3}$$

We know that, $H(z) \ = \ \dfrac{Y(z)}{X(z)}$

$$\therefore X(z) \ = \ \frac{Y(z)}{H(z)} \ = \ \frac{1 + 4z^{-1} + 6z^{-2} + 6z^{-3} + z^{-4} - 10z^{-5} - 8z^{-6}}{1 + 2z^{-1} - z^{-2} - 2z^{-3}}$$

$$
\begin{array}{r}
1 + 2z^{-1} + 3z^{-2} + 4z^{-3} \\
\hline
1 + 2z^{-1} - z^{-2} - 2z^{-3} \,\big)\; 1 + 4z^{-1} + 6z^{-2} + 6z^{-3} + z^{-4} - 10z^{-5} - 8z^{-6} \\
\end{array}
$$

$$1 + 2z^{-1} - z^{-2} - 2z^{-3}\quad \underline{1 + 2z^{-1} - z^{-2} - 2z^{-3}}$$

$$2z^{-1} + 7z^{-2} + 8z^{-3} + z^{-4}$$

$$\underline{2z^{-1} + 4z^{-2} - 2z^{-3} - 4z^{-4}}$$

$$3z^{-2} + 10z^{-3} + 5z^{-4} - 10z^{-5}$$

$$\underline{3z^{-2} + 6z^{-3} - 3z^{-4} - 6z^{-5}}$$

$$4z^{-3} + 8z^{-4} - 4z^{-5} - 8z^{-6}$$

$$\underline{4z^{-3} + 8z^{-4} - 4z^{-5} - 8z^{-6}}$$

$$0$$

$\therefore X(z) = 1 + 2z^{-1} + 3z^{-2} + 4z^{-3}$ .....(1)

By the definition of z-transform,

$$X(z) = \mathcal{Z}\{x(n)\} = \sum_{n=-\infty}^{\infty} x(n)z^{-n}$$

On expanding the above summation we get,

$$X(z) = \ldots\ldots + x(0) + x(1)\,z^{-1} + x(2)\,z^{-2} + x(3)\,z^{-3} + \ldots\ldots$$ .....(2)

On comparing equations (1) and (2) we get,

$$x(0) = 1 \quad ; \quad x(1) = 2 \quad ; \quad x(2) = 3 \quad ; \quad x(3) = 4$$

$\therefore$ Input, $x(n) = \{\, 1, 2, 3, 4 \,\}$
$\uparrow$

Example 5.39

An LTI system is described by the equation, $y(n) = x(n) + 0.8\,x(n-1) + 0.8\,x(n-2) - 0.49\,y(n-2)$. Determine the transfer function of the system. Sketch the poles and zeros on the z-plane.

Solution:

Given that, $y(n) = x(n) + 0.8\,x(n-1) + 0.8\,x(n-2) - 0.49\,y(n-2)$

On taking z-transform we get,

$$Y(z) = X(z) + 0.8z^{-1}X(z) + 0.8z^{-2}X(z) - 0.49z^{-2}Y(z)$$

$$Y(z) + 0.49z^{-2}Y(z) = X(z) + 0.8z^{-1}X(z) + 0.8z^{-2}X(z)$$

$$[1 + 0.49z^{-2}]\,Y(z) = [1 + 0.8z^{-1} + 0.8z^{-2}]\,X(z)$$

$$\boxed{\begin{aligned} \mathcal{Z}\{y(n)\} &= Y(z); \; \therefore \; \mathcal{Z}\{y(n-m)\} = z^{-m}Y(z) \\ \mathcal{Z}\{x(n)\} &= X(z); \; \therefore \; \mathcal{Z}\{x(n-m)\} = z^{-m}X(z) \end{aligned}}$$

$$\therefore \frac{Y(z)}{X(z)} = \frac{1 + 0.8z^{-1} + 0.8z^{-2}}{1 + 0.49z^{-2}}$$ .....(1)

The equation(1) is the transfer function of the LTI system.

$$\therefore \text{ Transfer function, } H(z) = \frac{Y(z)}{X(z)} = \frac{1 + 0.8z^{-1} + 0.8z^{-2}}{1 + 0.49z^{-2}}$$

The transfer function can also be expressed as a rational function in positive power of z as shown below:

$$\therefore H(z) = \frac{z^{-2}(z^2 + 0.8z + 0.8)}{z^{-2}(z^2 + 0.49)} = \frac{z^2 + 0.8z + 0.8}{z^2 + 0.49}$$

The poles are the roots of the denominator polynomial,

$z^2 + 0.49 = 0$

$\therefore z^2 = -0.49$

$z = \pm\sqrt{-0.49} = \pm j0.7$

$\therefore$ The poles are, $p_1 = j0.7$, $p_2 = -j0.7$

The zeros are the roots of the numerator polynomial,

$z^2 + 0.8z + 0.8 = 0$

$z = \dfrac{-0.8 \pm \sqrt{0.8^2 - 4 \times 0.8}}{2} = \dfrac{-0.8 \pm \sqrt{-2.56}}{2}$

$= \dfrac{-0.8 \pm j1.6}{2} = -0.4 \pm j0.8$

$\therefore$ The zeros are, $z_1 = -0.4 + j0.8$ and $z_2 = -0.4 - j0.8$

$\therefore H(z) = \dfrac{z^2 + 0.8z + 0.8}{z^2 + 0.49} = \dfrac{(z + 0.4 - j0.8)(z + 0.4 + j0.8)}{(z - j0.7)(z + j0.7)}$

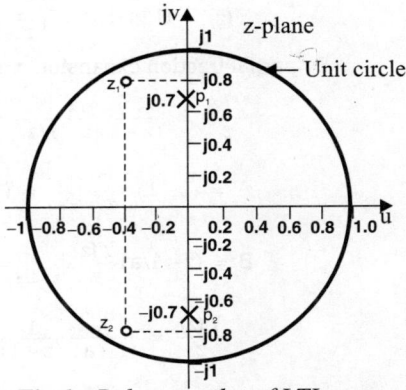

Fig 1 : Pole-zero plot of LTI system.

The Fig1 shows the location of poles and zeros on the z-plane. The poles are marked as "X" and zeros as " 0 ".

Example 5.40

Determine the step response of an LTI system whose impulse response h(n) is given by,

$h(n) = a^{-n} u(-n); \quad 0 < a < 1.$

Solution:

On taking $\mathcal{Z}$-transform of impulse response h(n) we get,

$$H(z) = \mathcal{Z}\{h(n)\} = \sum_{n=-\infty}^{\infty} h(n)z^{-n} = \sum_{n=-\infty}^{0} a^{-n}z^{-n} = \sum_{n=0}^{\infty} a^n z^n = \sum_{n=0}^{\infty}(az)^n$$

$$\boxed{\begin{array}{l} \because u(-n) = 1 \ ; \ n \le 0 \\ = 0 \ ; \ n > 0 \end{array}}$$

If $|az| < 1$, then using infinite geometric series sum formula,

$$H(z) = \sum_{n=0}^{\infty}(az)^n = \frac{1}{1-az} = \frac{-\dfrac{1}{a}}{z - \dfrac{1}{a}} \ ; \quad ROC: |z| < \left|\dfrac{1}{a}\right| \qquad \boxed{\begin{array}{l} |az| < 1 \\ \therefore |z| < \dfrac{1}{|a|} \end{array}} \quad ....(1)$$

The step input, $u(n) = 1 \ ; \ n \ge 0$

$= 0 \ ; \ n < 0$

On taking $\mathcal{Z}$-transform of unit step signal we get,

$$U(z) = \mathcal{Z}\{u(n)\} = \frac{z}{z-1} \ ; \quad ROC \ |z| > 1 \qquad \qquad .....(2)$$

Let y(n) be step response. Now the step response is given by convolution of step input, u(n) and impulse response, h(n).

$\therefore y(n) = u(n) * h(n)$

On taking $\mathcal{Z}$-transform we get,

$\mathcal{Z}\{y(n)\} = \mathcal{Z}\{u(n) * h(n)\}$

$\boxed{\begin{array}{l} \text{By convolution property,} \\ \mathcal{Z}\{u(n) * h(n)\} = U(z)H(z) \end{array}}$

$\therefore Y(z) = U(z) H(z)$

On substituting for U(z) and H(z) from equations (1) and (2) respectively we get,

$$Y(z) = U(z)H(z) = \left(\frac{z}{z-1}\right)\left(\frac{-1/a}{z-1/a}\right)$$

By partial fraction expansion we can write,

$$\frac{Y(z)}{z} = \frac{-1/a}{(z-1)(z-1/a)} = \frac{A}{z-1} + \frac{B}{z-1/a}$$

$$A = (z-1)\frac{Y(z)}{z}\Big|_{z=1} = (z-1)\frac{-1/a}{(z-1)(z-1/a)}\Big|_{z=1} = \frac{-1/a}{1-1/a} = \frac{-1}{a-1} = \frac{1}{1-a}$$

$$B = (z-1/a)\frac{Y(z)}{z}\Big|_{z=1/a} = (z-1/a)\frac{-1/a}{(z-1)(z-1/a)}\Big|_{z=1/a} = \frac{-1/a}{1/a-1} = \frac{-1}{1-a}$$

$$\therefore \frac{Y(z)}{z} = \frac{1}{(1-a)}\frac{1}{(z-1)} - \frac{1}{(1-a)}\frac{1}{(z-1/a)}$$

$$\therefore Y(z) = \frac{1}{(1-a)}\frac{z}{(z-1)} - \frac{1}{(1-a)}\frac{z}{(z-1/a)}$$

$\mathcal{Z}\{-u(-n-1)\} = \dfrac{z}{z-1}$; anticausal
$\mathcal{Z}\{-b^n u(-n-1)\} = \dfrac{z}{z-b}$; anticausal

Note: *Since impulse response is anticausal, the step response is also anticausal.*

On taking inverse $\mathcal{Z}$-transform of Y(z) we get step response.

$$\therefore \text{Step response, } y(n) = \mathcal{Z}^{-1}\{Y(z)\} = -\frac{1}{1-a}u(-n-1) + \frac{1}{1-a}\left(\frac{1}{a}\right)^n u(-n-1)$$

$$= \left[\left(\frac{1}{a}\right)^n - 1\right]\left(\frac{1}{1-a}\right)u(-n-1)$$

Example 5.41   *(AU May'15, 10 Marks)*

Determine the impulse response and step response for the system described by the second order difference equation,

$$y(n) + y(n-1) - 2y(n-2) = x(n-1) + 2x(n-2)$$

Solution:

a) Impulse response

The difference equation governing the system is,

$$y(n) + y(n-1) - 2y(n-2) = x(n-1) + 2x(n-2)$$

Let us take the $\mathcal{Z}$ - transform of the difference equation governing the system with zero initial conditions

$$\therefore \mathcal{Z}\{y(n) + y(n-1) - 2y(n-2)\} = \mathcal{Z}\{x(n-1) + 2x(n-2)\}$$

$$\mathcal{Z}\{y(n)\} + \mathcal{Z}\{y(n-1)\} - 2\mathcal{Z}\{y(n-2)\} = \mathcal{Z}\{x(n-1)\} + 2\mathcal{Z}\{x(n-2)\}$$

$$Y(z) + z^{-1}Y(z) - 2z^{-2}Y(z) = z^{-1}X(z) + 2z^{-2}X(z)$$

$$(1 + z^{-1} - 2z^{-2})Y(z) = (z^{-1} + 2z^{-2})X(z)$$

The roots of the quadratic,
$z^2 + z - 2 = 0$ are,
$z = \dfrac{-1 \pm \sqrt{1 + 4 \times 2}}{2}$
$= \dfrac{-1 \pm 3}{2} = 1, -2$

$$\therefore \frac{Y(z)}{X(z)} = \frac{z^{-1} + 2z^{-2}}{1 + z^{-1} - 2z^{-2}} = \frac{z^{-2}(z+2)}{z^{-2}(z^2 + z - 2)} = \frac{z+2}{(z-1)(z+2)} = \frac{1}{z-1} \qquad \text{.....(1)}$$

We know that the transfer function, $H(z) = \dfrac{Y(z)}{X(z)}$

$$\therefore \text{Transfer function, } H(z) = \frac{1}{z-1}$$

The impulse response h(n) is given by inverse z-transform of H(z).

∴ Impulse response, $h(n) = z^{-1}\{H(z)\} = z^{-1}\left\{\dfrac{1}{z-1}\right\}$

$\boxed{z\{u(n)\} = \dfrac{z}{z-1}}$

$\qquad\qquad\qquad = z^{-1}\left\{z^{-1}\dfrac{z}{z-1}\right\}$

$\boxed{\begin{array}{l}\text{Using shifting property}\\ z\{x(n-m)\} = z^{-m}\,X(z)\end{array}}$

$\qquad\qquad\qquad = u(n)\big|_{n=n-1} = u(n-1)$

$\boxed{z^{-1}z = 1}$

b) Step response

Now, the input is unit step input u(n)

∴ Input, $x(n) = u(n)$

∴ $X(z) = z\{u(n)\} = \dfrac{z}{z-1}$ .....(2)

From equation (1) we get,

∴ $Y(z) = \dfrac{1}{z-1}\,X(z)$

$\boxed{\text{Using equation (2).}}$

$\qquad = \dfrac{z}{z-1} \times \dfrac{1}{z-1} = \dfrac{z}{(z-1)^2}$

The step response y(n) is obtained by taking inverse z-transform.

∴ Step respone, $y(n) = z^{-1}\{Y(z)\} = z^{-1}\left\{\dfrac{z}{(z-1)^2}\right\} = n\,u(n)$

$\boxed{z\{n\,a^n u(n)\} = \dfrac{az}{(z-a)^2}}$

Example 5.42

(AU May'15, 12 Marks)

A causal system has $x(n) = \delta(n) + \dfrac{1}{4}\delta(n-1) - \dfrac{1}{8}\delta(n-2)$ and $y(n) = \delta(n) - \dfrac{3}{4}\delta(n-1)$. Find the impulse response and output if $x(n) = \left(\dfrac{1}{2}\right)^n u(n)$

Solution:

Impulse response

Given that, $x(n) = \delta(n) + \dfrac{1}{4}\delta(n-1) - \dfrac{1}{8}\delta(n-2)$

Let us take z - transform of the above equation with zero initial conditions.

$z\{x(n)\} = z\left\{\delta(n) + \dfrac{1}{4}\delta(n-1) - \dfrac{1}{8}\delta(n-2)\right\}$

$\boxed{\begin{array}{l}z\{\delta(n)\} = 1\\ z\{x(n-m)\} = z^{-m}\,X(z)\end{array}}$

$z\{x(n)\} = z\{\delta(n)\} + \dfrac{1}{4}\,z\{\delta(n-1)\} - \dfrac{1}{8}\,z\{\delta(n-2)\}$

∴ $X(z) = 1 + \dfrac{1}{4}z^{-1} - \dfrac{1}{8}z^{-2}$ .....(1)

Given that, $y(n) = \delta(n) - \dfrac{3}{4}\delta(n-1)$

Let us take z-transform of the above equation with zero initial conditions.

$z\{y(n)\} = z\left\{\delta(n) - \dfrac{3}{4}\delta(n-1)\right\}$

$z\{y(n)\} = z\{\delta(n)\} - \dfrac{3}{4}\,z\{\delta(n-1)\}$

∴ $Y(z) = 1 - \dfrac{3}{4}z^{-1}$ .....(2)

We know that the transfer function, $H(z) = \dfrac{Y(z)}{X(z)}$

∴ $H(z) = \dfrac{1 - \dfrac{3}{4}z^{-1}}{1 + \dfrac{1}{4}z^{-1} - \dfrac{1}{8}z^{-2}} = \dfrac{z^2\left(1 - \dfrac{3}{4}z^{-1}\right)}{z^2\left(1 + \dfrac{1}{4}z^{-1} - \dfrac{1}{8}z^{-2}\right)}$

$\boxed{\text{Using equations (1) and (2).}}$

$$\therefore \ H(z) = \frac{z\left(z - \frac{3}{4}\right)}{z^2 + \frac{1}{4}z - \frac{1}{8}} \qquad \qquad \qquad .....(3)$$

The roots of the quadratic,
$$z^2 + \frac{1}{4}z - \frac{1}{8} = 0 \text{ are,}$$
$$z = \frac{-\frac{1}{4} \pm \sqrt{\left(\frac{1}{4}\right)^2 + 4 \times \frac{1}{8}}}{2}$$
$$= \frac{1}{2}\left(-\frac{1}{4} \pm \frac{3}{4}\right) = \frac{1}{4}, -\frac{1}{2}$$

$$\therefore \ \frac{H(z)}{z} = \frac{z - \frac{3}{4}}{z^2 + \frac{1}{4}z - \frac{1}{8}} = \frac{\left(z - \frac{3}{4}\right)}{\left(z - \frac{1}{4}\right)\left(z + \frac{1}{2}\right)}$$

By using partial fraction technique

$$\frac{H(z)}{z} = \frac{A}{z - \frac{1}{4}} + \frac{B}{z + \frac{1}{2}}$$

$$A = \frac{H(z)}{z}\left(z - \frac{1}{4}\right)\bigg|_{z = \frac{1}{4}} = \frac{\left(z - \frac{3}{4}\right)}{\left(z - \frac{1}{4}\right)\left(z + \frac{1}{2}\right)}\left(z - \frac{1}{4}\right)\bigg|_{z = \frac{1}{4}} = \frac{\frac{1}{4} - \frac{3}{4}}{\frac{1}{4} + \frac{1}{2}} = \frac{-\frac{2}{4}}{\frac{3}{4}} = -\frac{2}{3}.$$

$$B = \frac{H(z)}{z}\left(z + \frac{1}{2}\right)\bigg|_{z = -\frac{1}{2}} = \frac{\left(z - \frac{3}{4}\right)}{\left(z - \frac{1}{4}\right)\left(z + \frac{1}{2}\right)}\left(z + \frac{1}{2}\right)\bigg|_{z = -\frac{1}{2}} = \frac{-\frac{1}{2} - \frac{3}{4}}{-\frac{1}{2} - \frac{1}{4}} = \frac{-\frac{5}{4}}{-\frac{3}{4}} = \frac{5}{3}.$$

$$\frac{H(z)}{z} = \frac{-\frac{2}{3}}{z - \frac{1}{4}} + \frac{\frac{5}{3}}{z + \frac{1}{2}}$$

$$\therefore \ H(z) = -\frac{2}{3}\frac{z}{z - \frac{1}{4}} + \frac{5}{3}\frac{z}{z + \frac{1}{2}}$$

The impulse response h(n) is obtained by taking inverse z-transform of H(z).

$$h(n) = z^{-1}\{H(z)\} = z^{-1}\left[\frac{-2}{3}\frac{z}{z - \frac{1}{4}} + \frac{5}{3}\frac{z}{z + \frac{1}{2}}\right]$$

$$= -\frac{2}{3}z^{-1}\left\{\frac{z}{z - \frac{1}{4}}\right\} + \frac{5}{3}z^{-1}\left\{\frac{z}{z - \left(-\frac{1}{2}\right)}\right\}$$

$$= -\frac{2}{3}\left(\frac{1}{4}\right)^n u(n) + \frac{5}{3}\left(-\frac{1}{2}\right)^n u(n) = \left[-\frac{2}{3}\left(\frac{1}{4}\right)^n + \frac{5}{3}\left(-\frac{1}{2}\right)^n\right]u(n)$$

Output response

Given that, $x(n) = \left(\frac{1}{2}\right)^n u(n)$

Let us take z-transform of the above equation we get,

$$z\{x(n)\} = z\left\{\left(\frac{1}{2}\right)^n u(n)\right\}$$

$$X(z) = \frac{z}{z - \frac{1}{2}} \qquad \qquad .....(4)$$

We know that, $Y(z) = X(z)H(z)$

$$\therefore \ Y(z) = \frac{z}{z - \frac{1}{2}} \frac{z\left(z - \frac{3}{4}\right)}{\left(z - \frac{1}{4}\right)\left(z + \frac{1}{2}\right)} \quad \Rightarrow \quad \frac{Y(z)}{z} = \frac{z\left(z - \frac{3}{4}\right)}{\left(z - \frac{1}{4}\right)\left(z - \frac{1}{2}\right)\left(z + \frac{1}{2}\right)}$$

By using partial fraction technique,

$$\frac{Y(z)}{z} = \frac{A}{z - \frac{1}{4}} + \frac{B}{z - \frac{1}{2}} + \frac{C}{z + \frac{1}{2}}$$

Using equations (3) and (4).

$$A = \frac{Y(z)}{z}\left(z - \frac{1}{4}\right)\bigg|_{z=\frac{1}{4}} = \frac{z\left(z - \frac{3}{4}\right)}{\left(z - \frac{1}{4}\right)\left(z - \frac{1}{2}\right)\left(z + \frac{1}{2}\right)}\left(z - \frac{1}{4}\right)\bigg|_{z=\frac{1}{4}} = \frac{\frac{1}{4}\left(\frac{1}{4} - \frac{3}{4}\right)}{\left(\frac{1}{4} - \frac{1}{2}\right)\left(\frac{1}{4} + \frac{1}{4}\right)} = \frac{\frac{1}{4}\left(-\frac{1}{2}\right)}{\left(-\frac{1}{4}\right)\frac{1}{2}} = 1$$

$$B = \frac{Y(z)}{z}\left(z - \frac{1}{2}\right)\bigg|_{z=\frac{1}{2}} = \frac{z\left(z - \frac{3}{4}\right)}{\left(z - \frac{1}{4}\right)\left(z - \frac{1}{2}\right)\left(z + \frac{1}{2}\right)}\left(z - \frac{1}{2}\right)\bigg|_{z=\frac{1}{2}} = \frac{\frac{1}{2}\left(\frac{1}{2} - \frac{3}{4}\right)}{\left(\frac{1}{2} - \frac{1}{4}\right)\left(\frac{1}{2} + \frac{1}{2}\right)} = \frac{\frac{1}{2}\left(-\frac{1}{4}\right)}{\frac{1}{4} \times 1} = -\frac{1}{2}$$

$$C = \frac{Y(z)}{z}\left(z + \frac{1}{2}\right)\bigg|_{z=-\frac{1}{2}} = \frac{z\left(z - \frac{3}{4}\right)}{\left(z - \frac{1}{4}\right)\left(z - \frac{1}{2}\right)\left(z + \frac{1}{2}\right)}\left(z + \frac{1}{2}\right)\bigg|_{z=-\frac{1}{2}} = \frac{-\frac{1}{2}\left(-\frac{1}{2} - \frac{3}{4}\right)}{\left(-\frac{1}{2} - \frac{1}{4}\right)\left(-\frac{1}{2} - \frac{1}{2}\right)}$$

$$= \frac{-\frac{1}{2}\left(-\frac{5}{4}\right)}{-\frac{3}{4}\left(-\frac{1}{4}\right)} = \frac{5}{8} \times \frac{16}{3} = \frac{10}{3}$$

$$\therefore \frac{Y(z)}{z} = \frac{1}{z - \frac{1}{4}} - \frac{1}{2}\frac{1}{z - \frac{1}{2}} + \frac{10}{3}\frac{1}{z + \frac{1}{2}}$$

$$Y(z) = \frac{z}{z - \frac{1}{4}} - \frac{1}{2}\frac{z}{z - \frac{1}{2}} + \frac{10}{3}\frac{z}{z + \frac{1}{2}}$$

The output y(n) is obtained by using inverse z-transform of $Y(z)$.

$$y(n) = z^{-1}\{Y(z)\} = z^{-1}\left\{\frac{z}{z - \frac{1}{4}} - \frac{1}{2}\frac{z}{z - \frac{1}{2}} + \frac{10}{3}\frac{z}{z + \frac{1}{2}}\right\}$$

$$= z^{-1}\left\{\frac{z}{z - \frac{1}{4}}\right\} - \frac{1}{2}z^{-1}\left\{\frac{z}{z - \frac{1}{2}}\right\} + \frac{10}{3}z^{-1}\left\{\frac{z}{z - \left(-\frac{1}{2}\right)}\right\}$$

$$= \left(\frac{1}{4}\right)^n u(n) - \frac{1}{2}\left(\frac{1}{2}\right)^n u(n) + \frac{10}{3}\left(-\frac{1}{2}\right)^n u(n) = \left[\left(\frac{1}{4}\right)^n - \left(\frac{1}{2}\right)^{n+1} + \frac{10}{3}\left(-\frac{1}{2}\right)^n\right]u(n)$$

Example 5.43

Test the stability of the first-order system governed by the equation, $y(n) = x(n) + b\,y(n-1)$, where $|b| < 1$.

Solution:

Given that, $y(n) = x(n) + b\,y(n-1)$ $\boxed{z\{y(n)\} = Y(z); \; z\{y(n-1)\} = z^{-1}Y(z); \; z\{x(n)\} = X(z)}$

On taking z-transform we get,

$$Y(z) = X(z) + b\,z^{-1}\,Y(z) \qquad \Rightarrow \qquad Y(z) - b\,z^{-1}\,Y(z) = X(z) \qquad \Rightarrow \qquad (1 - b\,z^{-1})\,Y(z) = X(z)$$

$$\therefore \frac{Y(z)}{X(z)} = \frac{1}{1 - bz^{-1}}$$

We know that, $Y(z)/X(z)$ is equal to $H(z)$.

$$\therefore H(z) = \frac{1}{1 - bz^{-1}} = \frac{1}{z^{-1}(z - b)} = \frac{z}{z - b}$$

On taking inverse z-transform of $H(z)$ we get the impulse response h(n). $\boxed{z\{a^n u(n)\} = \frac{z}{z - a}}$

$$\therefore \text{Impulse response, } h(n) = b^n\,u(n)$$

The condition to be satisfied for the stability of the system is, $\sum\limits_{n=-\infty}^{\infty} |h(n)| < \infty$

$$\sum_{n=-\infty}^{\infty} |h(n)| = \sum_{n=0}^{\infty} |b^n| = \sum_{n=0}^{\infty} |b|^n$$

Since |b| < 1, using the infinite geometric series sum formula we can write,

$$\sum_{n=0}^{\infty} |b|^n = \frac{1}{1-|b|}$$

$$\therefore \sum_{n=-\infty}^{\infty} |h(n)| = \frac{1}{1-|b|}$$

> **Infinite geometric series sum formula**
> $$\sum_{n=0}^{\infty} c^n = \frac{1}{1-C} \; ; \text{if}, \, 0 < |c| < 1$$

The term 1/(1–|b|) is less than infinity and so the system is stable.

Example 5.44   *(AU Jun'13, 8 Marks)*

Determine the range of values of "a" for the stability of LTI system with impulse response, h(n) = a^n u(n).

Solution:

The condition to be satisfied for the stability of the system is, $\sum\limits_{n=-\infty}^{\infty} |h(n)| < \infty$

Given that, h(n) = a^n u(n)

$$\therefore \sum_{n=-\infty}^{\infty} |h(n)| = \sum_{n=0}^{\infty} |a^n|$$

The summation of infinite terms in the above equation converges if, 0 < |a| < 1. Hence by using infinite geometric series formula,

$$\sum_{n=-\infty}^{\infty} |h(n)| = \frac{1}{1-|a|}$$
$$= \text{Constant}$$

> **Infinite geometric series sum formula**
> $$\sum_{n=0}^{\infty} c^n = \frac{1}{1-C} \; ; \text{if}, \, 0 < |c| < 1$$

Therefore, the system is stable if |a| < 1.

Example 5.45   *(AU Dec'12, 10 Marks)*

Determine the impulse response of the LTI system defined by,

$$H(z) = \frac{3 - 4z^{-1}}{1 - 3.5z^{-1} + 1.5z^{-2}}.$$

Specify the ROC of H(z) and determine h(n) for the following conditions

   a) The system is stable
   b) The system is causal
   c) The system is anticausal

Solution

Given that, $H(z) = \dfrac{3 - 4z^{-1}}{1 - 3.5z^{-1} + 1.5z^{-2}}$

$$= \frac{z^{-1}(3z - 4)}{z^{-2}(z^2 - 3.5z + 1.5)} = \frac{z(3z - 4)}{(z - 3)(z - 0.5)}$$

> The roots of the quadratic,
> $z^2 - 3.5z + 1.5 = 0$ are,
> $$z = \frac{3.5 \pm \sqrt{3.5^2 - 4 \times 1.5}}{2}$$
> $$= \frac{3.5 \pm 2.5}{2} = 3, 0.5$$

$$\therefore \frac{H(z)}{z} = \frac{3z-4}{(z-3)(z-0.5)}$$

By partial fraction expansion, we can write,

$$\frac{H(z)}{z} = \frac{A}{z-3} + \frac{B}{z-0.5}$$

$$A = (z-3)\frac{H(z)}{z}\bigg|_{z=3} = (z-3)\times\frac{3z-4}{(z-3)(z-0.5)}\bigg|_{z=3} = \frac{3\times3-4}{3-0.5} = 2$$

$$B = (z-0.5)\frac{H(z)}{z}\bigg|_{z=0.5} = (z-0.5)\times\frac{3z-4}{(z-3)(z-0.5)}\bigg|_{z=0.5} = \frac{3\times0.5-4}{0.5-3} = 1$$

$$\therefore H(z) = \frac{2z}{z-3} + \frac{z}{z-0.5}$$

The poles of H(z) are at z = 3, z = 0.5.

a) The system is stable

The ROC os stable system should include the unit circle. Therefore, ROC is 0.5 < |z| < 3.

Let us take inverse z-transform of $\frac{z}{z-3}$ to get anticausal signal and $\frac{z}{z-0.5}$ to get causal signal.

$$\therefore \text{Impulse response, } h(n) = z\{H(z)\} = z^{-1}\left\{\frac{2z}{z-3} + \frac{z}{z-0.5}\right\}$$

$$= -2(3)^n u(-n-1) + 0.5^n u(n)$$

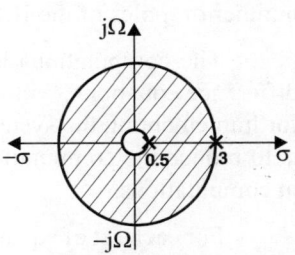

Fig 1: ROC: 0.5<|z|<3.

b) The system is causal

For a causal system the ROC should be exterior of a circle, whose radius is equal to the magnitude of outermost pole (i.e., pole with largest magnitude of H(z)).

Here the outermost pole(largest magnitude) pole is p = 3. Therefore ROC is |z| > 3.

Let us take inverse z-transform of H(z) to get h(n) as causal.

$$\therefore \text{Impulse response, } h(n) = z\{H(z)\} = z^{-1}\left\{\frac{2z}{z-3} + \frac{z}{z-0.5}\right\}$$

$$= 2(3)^n u(n) + 0.5^n u(n)$$

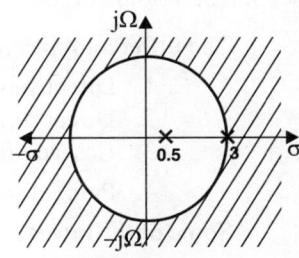

Fig 2: ROC: |z|>3.

c) The system is anticausal

For an anticausal system, the ROC should be interior of a circle, whose radius is equal to the magnitude of innermost pole (i.e., pole with smallest magnitude of H(z)).

Here the innermost pole (smallest magnitude pole) is p = 0.5. Therefore ROC is |z| < 0.5.

Let us take inverse z-transform of H(z) to get h(n) as anticausal.

$$\therefore \text{Impulse response, } h(n) = z\{H(z)\} = z^{-1}\left\{\frac{2z}{z-3} + \frac{z}{z-0.5}\right\}$$

$$= -2(3)^n u(-n-1) - 0.5^n u(-n-1)$$

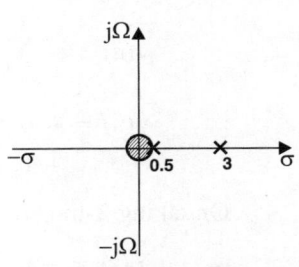

Fig 3: ROC: |z|<0.5.

5.8 Structures for Realization of IIR Systems　　　*(AU May'15, 2 Marks)*

In general, the time domain representation of an Nth order IIR system is,

$$y(n) = -\sum_{m=1}^{N} a_m\, y(n-m) + \sum_{m=1}^{N} b_m\, x(n-m)$$

and the z-domain representation of an Nth order IIR system is,

$$H(z) = \frac{Y(z)}{X(z)} = \frac{b_0 + b_1 z^{-1} + b_2 z^{-2} + ... + b_M z^{-M}}{1 + a_1 z^{-1} + a_2 z^{-2} + a_N z^{-N}}$$

The above two representations of IIR system can be viewed as a computational procedure (or algorithm) to determine the output sequence $y(n)$ from the input sequence $x(n)$. Also, in the above representations the value of M gives the number of zeros and the value of N gives the number of poles of the IIR system.

The computations in the above equation can be arranged into various equivalent sets of difference equations, with each set of equations defining a computational procedure or algorithm for implementing the system. The main advantage of rearranging the sets of difference equations is to reduce the computational complexity, memory requirements and finite-word-length effects in computations.

For each set of equations, we can construct a block diagram consisting of delays, adders and multipliers. Such block diagrams are referred as realization of system or equivalently as a structure for realizing system. The basic block diagram representation of discrete system is discussed in Section 5.2. Some of the standard block diagram structures are discussed in this section.

The different types of structures for realizing the IIR systems are,

　　　1. Direct form-I structure
　　　2. Direct form-II structure
　　　3. Cascade form structure
　　　4. Parallel form structure

Some of the block diagram representation of the system gives a direct relation between the time domain equation and the z-domain equation.

5.8.1 Direct Form-I Structure of IIR System

Consider the difference equation governing an IIR system.

$$y(n) = -\sum_{m=1}^{N} a_m\, y(n-m) + \sum_{m=0}^{M} b_m\, x(n-m)$$

$$y(n) = -a_1\, y(n-1) - a_2\, y(n-2) - ..... - a_N\, y(n-N)$$
$$+ b_0\, x(n) + b_1\, x(n-1) + b_2\, x(n-2) + ... + b_M\, x(n-M)$$

On taking $\mathbb{Z}$-transform of the above equation we get,

$$Y(z) = -a_1 z^{-1} Y(z) - a_2 z^{-2} Y(z) - ..... - a_N z^{-N} Y(z)$$
$$+ b_0\, X(z) + b_1 z^{-1} X(z) + b_2 z^{-2} X(z) + ... + b_M z^{-M} X(z) \quad .....(5.65)$$

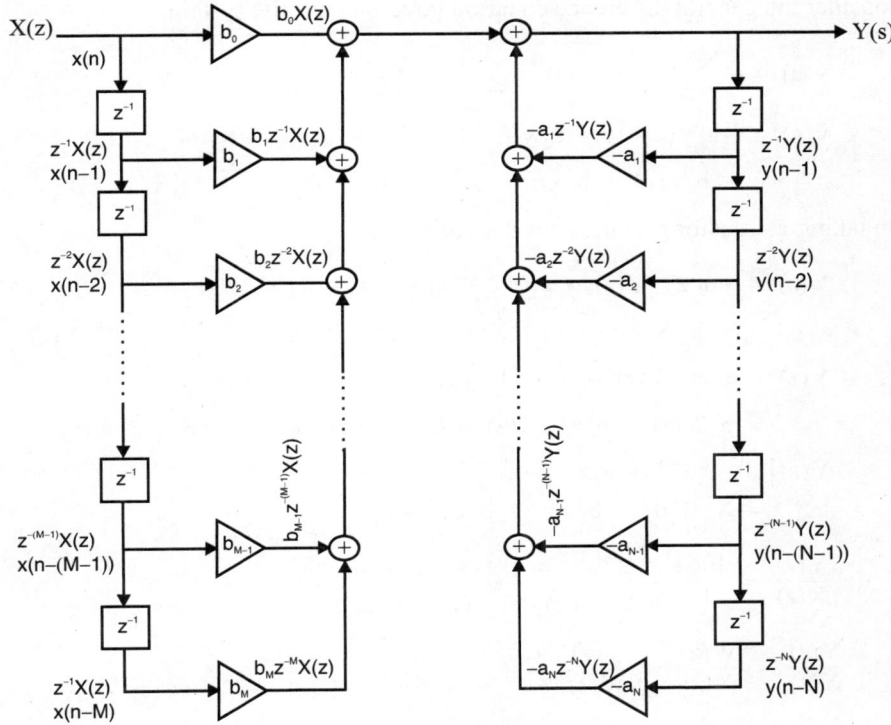

Fig 5.12: *Direct form-I structure of IIR system.*

The equation of Y(z) [equation (5.65)] can be directly represented by a block diagram as shown in Fig 5.12 and this structure is called direct form-I structure. The direct form-I structure provides a direct relation between time domain and z-domain equations. The direct form-I structure uses separate delays (z^{-1}) for input and output samples. Hence for realizing direct form-I structure more memory is required.

From the direct form-I structure it is observed that the realization of an Nth order discrete time system with M number of zeros and N number of poles, involves M+N+1 number of multiplications and M+N number of additions. Also this structure involves M+N delays and so M+N memory locations are required to store the delayed signals.

When the number of delays in a structure is equal to the order of the system, the structure is called *canonic structure*. In direct form-I structure the number of delays is not equal to order of the system and so direct form-I structure is noncanonic structure.

5.8.2 Direct Form-II Structure of IIR System

An alternative structure called direct form-II structure can be realized which uses less number of delay elements than the direct form-I structure.

Consider the general difference equation governing an IIR system.

$$y(n) = -\sum_{m=1}^{N} a_m\, y(n-m) + \sum_{m=0}^{M} b_m\, x(n-m)$$

$$y(n) = -a_1\, y(n-1) - a_2\, y(n-2) - \,..... \, - a_N\, y(n-N)$$

$$+\, b_0\, x(n) + b_1\, x(n-1) + b_2\, x(n-2) + \,..... \, + b_M\, x(n-M)$$

On taking $\mathcal{Z}$-transform of the above equation we get,

$$Y(z) = -a_1\, z^{-1}\, Y(z) - a_2\, z^{-2}\, Y(z) - \,..... \, - a_N\, z^{-N}\, Y(z)$$

$$+\, b_0\, X(z) + b_1\, z^{-1}\, X(z) + b_2\, z^{-2}\, X(z) + \,...... \, + b_M\, z^{-M}\, X(z)$$

$$Y(z) + a_1\, z^{-1}\, Y(z) + a_2\, z^{-2}\, Y(z) + \,..... \, + a_N\, z^{-N}\, Y(z)$$

$$= b_0\, X(z) + b_1\, z^{-1}\, X(z) + b_2\, z^{-2}\, X(z) + \,...... \, + b_M\, z^{-M}\, X(z)$$

$$Y(z)\left[1 + a_1 z^{-1} + a_2 z^{-2} + \,..... \, + a_N z^{-N}\right]$$

$$= X(z)\left[b_0 + b_1\, z^{-1} + b_2\, z^{-2} + \,..... \, + b_M\, z^{-M}\right]$$

$$\frac{Y(z)}{X(z)} = \frac{b_0 + b_1\, z^{-1} + b_2\, z^{-2} + \,..... \, + b_M\, z^{-M}}{1 + a_1\, z^{-1} + a_2\, z^{-2} + \,..... \, + a_N\, z^{-N}}$$

Let, $\dfrac{Y(z)}{X(z)} = \dfrac{W(z)}{X(z)} \times \dfrac{Y(z)}{W(z)}$

where, $\dfrac{W(z)}{X(z)} = \dfrac{1}{1 + a_1\, z^{-1} + a_2\, z^{-2} + \,..... \, + a_N\, z^{-N}}$ .....(5.66)

$\dfrac{Y(z)}{W(z)} = b_0 + b_1\, z^{-1} + b_2\, z^{-2} + \,..... \, + b_M\, z^{-M}$ .....(5.67)

On cross multiplying equation (5.66) we get,

$$W(z) + a_1\, z^{-1}\, W(z) + a_2\, z^{-2}\, W(z) + \,..... \, + a_N\, z^{-N}\, W(z) = X(z)$$

$$\therefore\ W(z) = X(z) - a_1\, z^{-1}\, W(z) - a_2\, z^{-2}\, W(z) - \,..... \, - a_N\, z^{-N}\, W(z) \qquad ..... \,(5.68)$$

On cross multiplying equation (5.67) we get,

$$Y(z) = b_0\, W(z) + b_1\, z^{-1}\, W(z) + b_2\, z^{-2}\, W(z) + \,..... \, + b_M\, z^{-M}\, W(z) \qquad ..... \,(5.69)$$

The equations (5.68) and (5.69) represent the IIR system in z-domain and can be realized by a direct structure called direct form-II structure as shown in Fig 5.13. In direct form-II structure the number of delays is equal to order of the system and so the direct form-II structure is *canonic structure.*

From the direct form-II structure it is observed that the realization of an Nth order discrete time system with M number of zeros and N number of poles, involves M+N+1 number of multiplications and M+N number of additions. In a realizable system, $N \geq M$, and so the number of delays in direct form-II structure will be equal to N. Hence, when a system is realized using direct form-II structure, N memory locations are required to store the delayed signals.

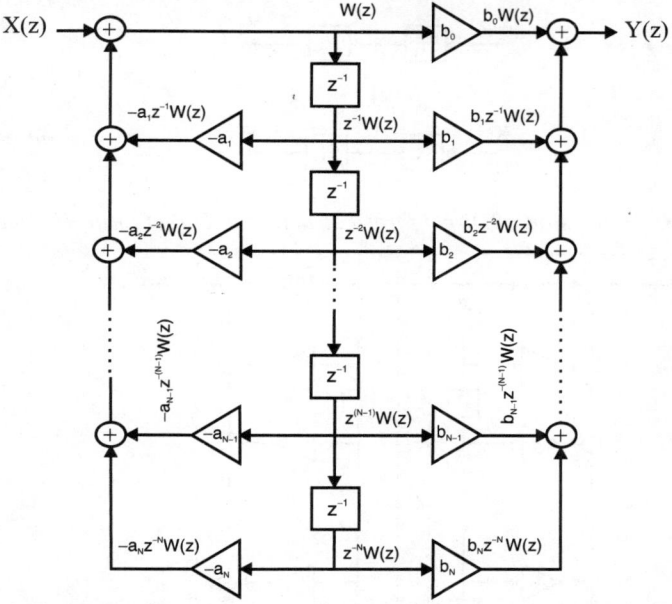

Fig 5.13: *Direct form-II structure of IIR system for N=M.*

Conversion of Direct Form-I Structure to Direct Form-II Structure

The direct form-I structure can be converted to direct form-II structure by considering the direct form-I structure as cascade of two systems $\mathcal{H}_1$ and $\mathcal{H}_2$ as shown in Fig 5.14

By linearity property the order of cascading can be interchanged as shown in Fig 5.15 and Fig 5.16.

In Fig 5.16 we can observe that the input to the delay elements in $\mathcal{H}_1$ and $\mathcal{H}_2$ are same and so the output of delay elements in $\mathcal{H}_1$ and $\mathcal{H}_2$ are same. Therefore instead of having separate delays for $\mathcal{H}_1$ and $\mathcal{H}_2$, a single set of delays can be used. Hence the delays can be merged to combine the cascaded systems to a single system and the resultant structure will be direct form-II structure as that of Fig 5.13

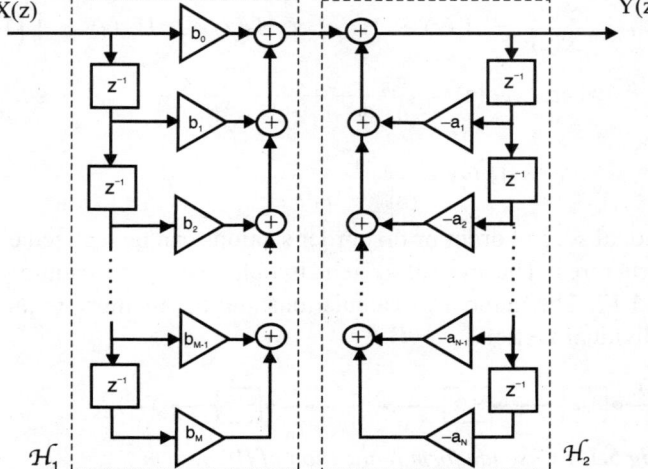

Fig 5.14: *Direct form-I structure as cascade of two systems.*

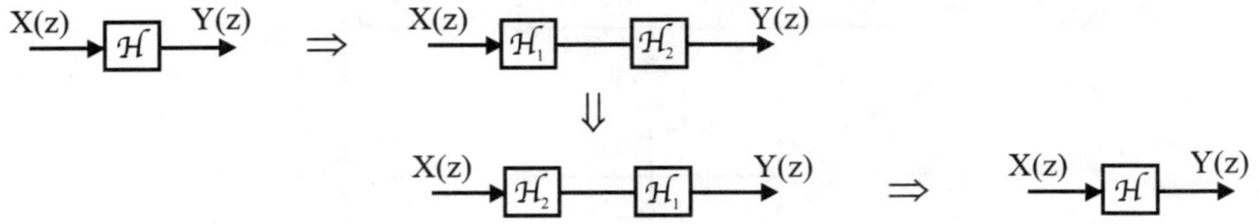

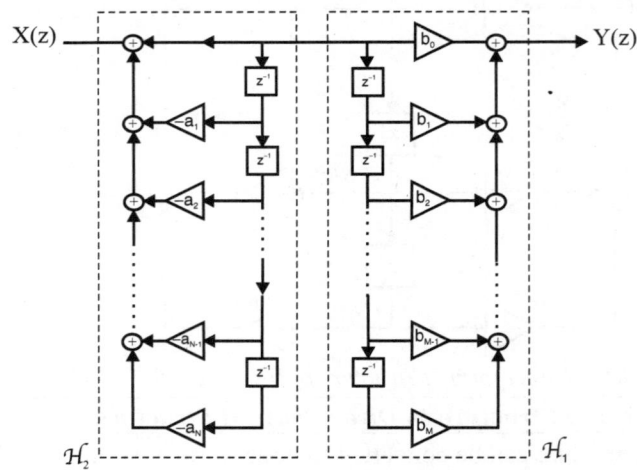

Fig 5.15: *Conversion of Direct form-I structure to Direct form -II structure.*

Fig 5.16: *Direct form-I structure after*
interchanging the order of cascading.

5.8.3 Cascade Form Realization of IIR System

The transfer function $H(z)$ can be expressed as a product of a number of second order or first order sections, as shown in equation (5.70).

$$H(z) = \frac{Y(z)}{X(z)} = H_1(z) \times H_2(z) \times H_3(z) \; ..... \; H_m(z) = \prod_{i=1}^{m} H_i(z) \qquad .....(5.70)$$

$$\text{where,} \quad H_i(z) = \frac{c_{0i} + c_{1i}\, z^{-1} + c_{2i}\, z^{-2}}{d_{0i} + d_{1i}\, z^{-1} + d_{2i}\, z^{-2}} \qquad \boxed{\text{Second order section}}$$

$$\text{or,} \quad H_i(z) = \frac{c_{0i} + c_{1i}\, z^{-1}}{d_{0i} + d_{1i}\, z^{-1}} \qquad \boxed{\text{First order section}}$$

The individual second order or first order sections can be realized either in direct form-I or direct form-II structures. The overall system is obtained by cascading the individual sections as shown in Fig 5.17. The number of calculations and the memory requirement depends on the realization of individual sections.

Fig 5.17: *Cascade form realization of IIR system.*

The difficulty in cascade structure are,

1.   Decision of pairing poles and zeros.

2.   Deciding the order of cascading the first and second order sections.

3.   Scaling multipliers should be provided between individual sections to prevent the system variables from becoming too large or too small.

5.8.4 Parallel Form Realization of IIR System

The transfer function H(z) of a discrete time system can be expressed as a sum of first and second order sections, using partial fraction expansion technique as shown in equation (5.71).

$$H(z) = \frac{Y(z)}{X(z)} = C + H_1(z) + H_2(z) + \dots + H_m(z) \qquad \dots(5.71)$$

$$= C + \sum_{i=1}^{m} H_i(z)$$

where, $\quad H_i(z) = \dfrac{c_{0i} + c_{1i}\, z^{-1}}{d_{0i} + d_{1i}\, z^{-1} + d_{2i}\, z^{-2}}$

Second order section

or, $\quad H_i(z) = \dfrac{c_{0i}}{d_{0i} + d_{1i}\, z^{-1}}$

First order section

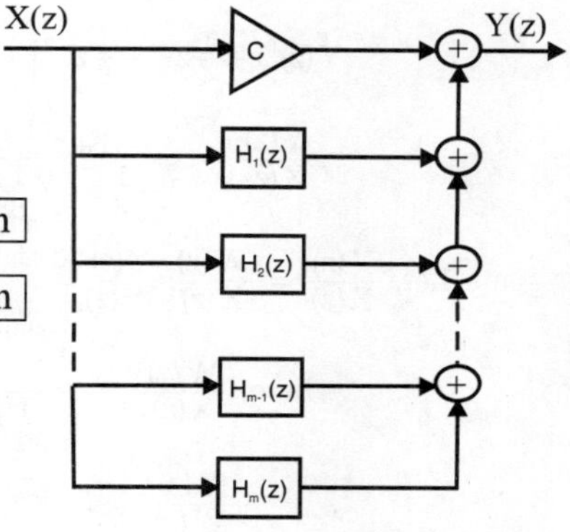

The individual first and second order sections can be realized either in direct form-I or direct form-II structures. The overall system is obtained by connecting the individual sections in parallel as shown in Fig 5.18. The number of calculations and the memory requirement depends on the realization of individual sections.

Fig 5.18: Parallel form realization of IIR system.

Example 5.46   *(AU Dec'13, 6 Marks)*

Find the direct form-I and direct form-II realizations of a discrete time system governed by the equation,

$$y(n) + \tfrac{1}{4}y(n-1) + \tfrac{1}{8}y(n-2) = x(n) + x(n-1).$$

Solution:

Direct Form-I

Given that, $y(n) + \tfrac{1}{4}y(n-1) + \tfrac{1}{8}y(n-2) = x(n) + x(n-1)$ .....(1)

On taking z-transform of equation(1) we get,

$$Y(z) + \tfrac{1}{4}z^{-1}Y(z) + \tfrac{1}{8}z^{-2}Y(z) = X(z) + z^{-1}X(z) \qquad \dots(2)$$

$$\therefore\ Y(z) = -\tfrac{1}{4}z^{-1}Y(z) - \tfrac{1}{8}z^{-2}Y(z) + X(z) + z^{-1}X(z) \qquad \dots(3)$$

The direct form-I structure can be obtained from equation (3), as shown in Fig 1.

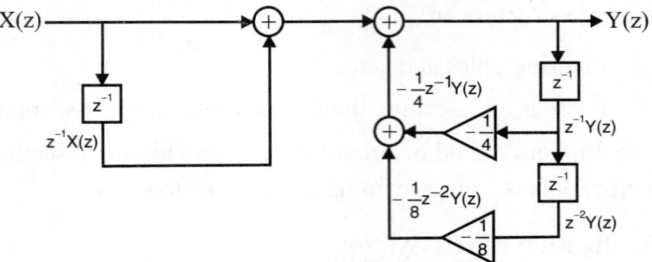

Fig 1: *Direct form-I realization structure.*

Direct Form-II

Consider equation (2).

$$Y(z) + \frac{1}{4}z^{-1}Y(z) + \frac{1}{8}z^{-2}Y(z) = X(z) + z^{-1}X(z)$$

$$Y(z)\left[1 + \frac{1}{4}z^{-1} + \frac{1}{8}z^{-2}\right] = X(z)\left[1 + z^{-1}\right]$$

$$\therefore \frac{Y(z)}{X(z)} = \frac{1 + z^{-1}}{1 + \frac{1}{4}z^{-1} + \frac{1}{8}z^{-2}} \qquad\qquad .....(4)$$

Let, $\dfrac{Y(z)}{X(z)} = \dfrac{W(z)}{X(z)} \dfrac{Y(z)}{W(z)}$

where, $\dfrac{W(z)}{X(z)} = \dfrac{1}{1 + \frac{1}{4}z^{-1} + \frac{1}{8}z^{-2}}$ $\qquad\qquad .....(5)$

$$\frac{Y(z)}{W(z)} = 1 + z^{-1} \qquad\qquad .....(6)$$

On cross multiplying equation (5) we get,

$$W(z) + \frac{1}{4}z^{-1}W(z) + \frac{1}{8}z^{-2}W(z) = X(z)$$

or $W(z) = X(z) - \frac{1}{4}z^{-1}W(z) - \frac{1}{8}z^{-2}W(z)$ $\qquad\qquad .....(7)$

On cross multiplying equation (6) we get,

$$Y(z) = W(z) + z^{-1}W(z) \qquad\qquad .....(8)$$

The equations (7) and (8) can be realized by a direct form-II structure as shown in Fig 2.

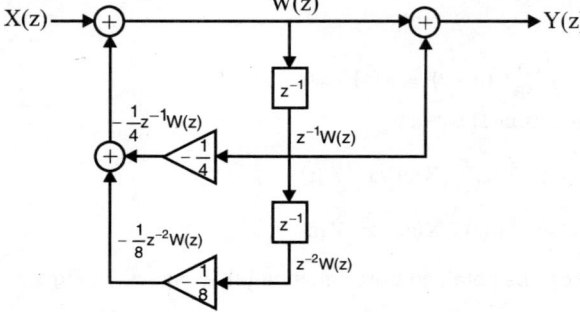

Fig 2: *Direct form-II realization structure.*

Example 5.47

Find the direct form-I and direct form-II realizations of a discrete time system represented by transfer

function, $H(z) = \dfrac{8z^3 - 4z^2 + 11z - 2}{\left(z - \frac{1}{4}\right)\left(z^2 - z + \frac{1}{2}\right)}$

Solution:

Direct Form-I

Let, $H(z) = \dfrac{Y(z)}{X(z)}$;   where, $Y(z) = $ Output   and $X(z) = $ Input

$$\therefore \frac{Y(z)}{X(z)} = \frac{8z^3 - 4z^2 + 11z - 2}{\left(z - \frac{1}{4}\right)\left(z^2 - z + \frac{1}{2}\right)} = \frac{8z^3 - 4z^2 + 11z - 2}{z^3 - z^2 + \frac{1}{2}z - \frac{1}{4}z^2 + \frac{1}{4}z - \frac{1}{8}}$$

$$= \frac{8z^3 - 4z^2 + 11z - 2}{z^3 - \frac{5}{4}z^2 + \frac{3}{4}z - \frac{1}{8}} = \frac{z^3\left(8 - 4z^{-1} + 11z^{-2} - 2z^{-3}\right)}{z^3\left(1 - \frac{5}{4}z^{-1} + \frac{3}{4}z^{-2} - \frac{1}{8}z^{-3}\right)}$$

$$= \frac{8 - 4z^{-1} + 11z^{-2} - 2z^{-3}}{1 - \frac{5}{4}z^{-1} + \frac{3}{4}z^{-2} - \frac{1}{8}z^{-3}} \qquad\qquad .....(1)$$

On cross multiplying equation (1) we get,

$$Y(z) - \frac{5}{4}z^{-1}Y(z) + \frac{3}{4}z^{-2}Y(z) - \frac{1}{8}z^{-3}Y(z) = 8X(z) - 4z^{-1}X(z) + 11z^{-2}X(z) - 2z^{-3}X(z)$$

$$\therefore Y(z) = 8X(z) - 4z^{-1}X(z) + 11z^{-2}X(z) - 2z^{-3}X(z)$$

$$+ \frac{5}{4}z^{-1}Y(z) - \frac{3}{4}z^{-2}Y(z) + \frac{1}{8}z^{-3}Y(z) \qquad\qquad .....(2)$$

The direct form-I structure can be obtained from equation (2) as shown in Fig 1.

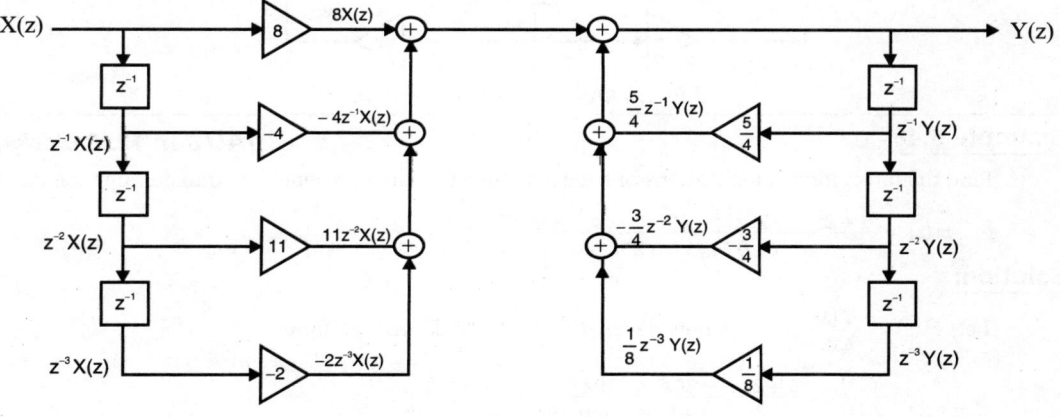

Fig 1: *Direct form-I realization.*

Direct Form-II

From equation (1) we get,

$$\frac{Y(z)}{X(z)} = \frac{8 - 4z^{-1} + 11z^{-2} - 2z^{-3}}{1 - \frac{5}{4}z^{-1} + \frac{3}{4}z^{-2} - \frac{1}{8}z^{-3}}$$

Let, $\dfrac{Y(z)}{X(z)} = \dfrac{W(z)}{X(z)} \dfrac{Y(z)}{W(z)}$

Where, $\dfrac{W(z)}{X(z)} = \dfrac{1}{1 - \dfrac{5}{4}z^{-1} + \dfrac{3}{4}z^{-2} - \dfrac{1}{8}z^{-3}}$ .....(3)

$\dfrac{Y(z)}{W(z)} = 8 - 4z^{-1} + 11z^{-2} - 2z^{-3}$ .....(4)

On cross multiplying equation (3) we get,

$$W(z) - \frac{5}{4}z^{-1}W(z) + \frac{3}{4}z^{-2}W(z) - \frac{1}{8}z^{-3}W(z) = X(z)$$

$$\therefore W(z) = X(z) + \frac{5}{4}z^{-1}W(z) - \frac{3}{4}z^{-2}W(z) + \frac{1}{8}z^{-3}W(z) \quad .....(5)$$

On cross multiplying equation (4) we get,

$$Y(z) = 8W(z) - 4z^{-1}W(z) + 11z^{-2}W(z) - 2z^{-3}W(z) \quad .....(6)$$

The equations (5) and (6) can be realized by a direct form-II structure as shown in Fig 2.

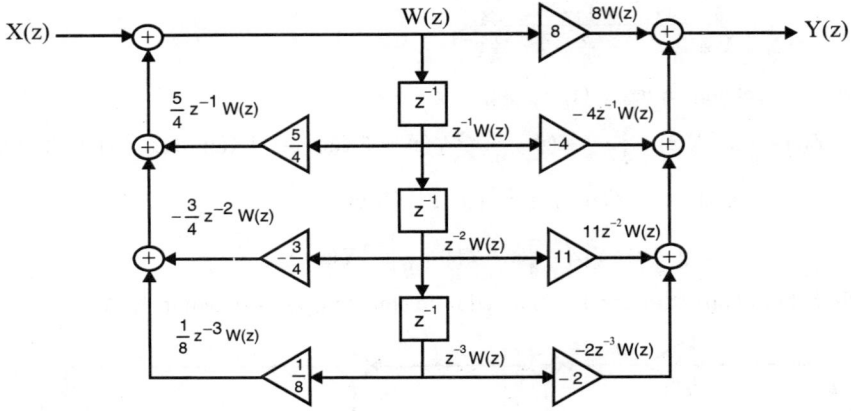

Fig 2: Direct form-II realization.

Example 5.48

(AU Jun'12, 8 Marks)

Find the direct form-II realizations of a discrete time system represented by transfer function,

$$H(z) = \frac{1 + 2z^{-1} - 20z^{-2} - 20z^{-3} - 5z^{-4} + 6z^{-6}}{1 + 0.5z^{-1} - 0.25z^{-2}}$$

Solution:

Let, $H(z) = \dfrac{Y(z)}{X(z)}$;   where, $Y(z)$ = Output   and $X(z)$ = Input

$$\therefore \frac{Y(z)}{X(z)} = \frac{1 + 2z^{-1} - 20z^{-2} - 20z^{-3} - 5z^{-4} + 6z^{-6}}{1 + 0.5z^{-1} - 0.25z^{-2}}$$

Let, $\dfrac{Y(z)}{X(z)} = \dfrac{W(z)}{X(z)} \dfrac{Y(z)}{W(z)}$

$$\frac{W(z)}{X(z)} = \frac{1}{1 + 0.5z^{-1} - 0.25z^{-2}} \quad .....(1)$$

$$\frac{Y(z)}{W(z)} = 1 + 2z^{-1} - 20z^{-2} - 20z^{-3} - 5z^{-4} + 6z^{-6} \quad .....(2)$$

On cross multiplying equation (1) we get,

$$W(z) + 0.5z^{-1}W(z) - 0.25z^{-2}W(z) = X(z)$$

$$\therefore W(z) = X(z) - 0.5z^{-1}W(z) + 0.25z^{-2}W(z) \qquad .....(3)$$

On cross multiplying equation (2) we get,

$$Y(z) = W(z) + 2z^{-1}W(z) - 20z^{-2}W(z) - 20z^{-3}W(z) - 5z^{-4}W(z) + 6z^{-6}W(z) \qquad .....(4)$$

The equations (3) and (4) can be realized by a direct form-II structure as shown in Fig 1.

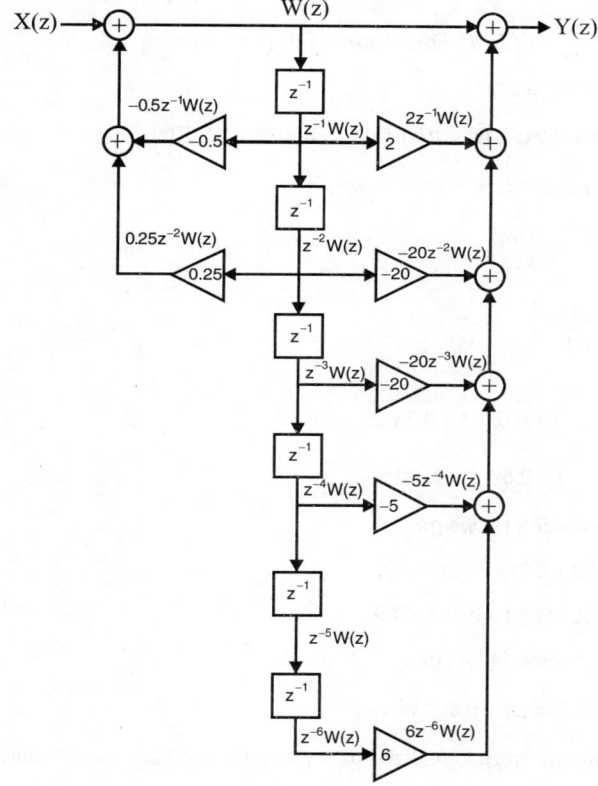

Fig 1: *Direct form-II realization.*

Example 5.49

Find the digital network in direct form-I and II for the system described by the difference equation,

$$y(n) = x(n) + 0.5\,x(n-1) + 0.4\,x(n-2) - 0.6\,y(n-1) - 0.7\,y(n-2).$$

Solution:

Given that, $y(n) = x(n) + 0.5\,x(n-1) + 0.4\,x(n-2) - 0.6\,y(n-1) - 0.7\,y(n-2)$

On taking z-transform we get,

$$Y(z) = X(z) + 0.5z^{-1}X(z) + 0.4z^{-2}X(z) - 0.6z^{-1}Y(z) - 0.7z^{-2}\,Y(z) \qquad .....(1)$$

The direct form-I digital network can be realized using equation (1) as shown in Fig 1.

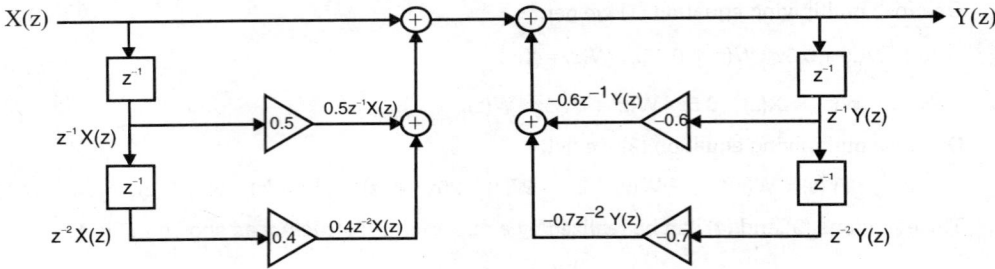

Fig 1: *Direct form-I digital network.*

On rearranging equation (1) we get,

$$Y(z) + 0.6z^{-1}Y(z) + 0.7z^{-2}Y(z) = X(z) + 0.5z^{-1}X(z) + 0.4z^{-2}X(z)$$

$$(1+0.6z^{-1} + 0.7z^{-2})\,Y(z) = (1 + 0.5z^{-1} + 0.4z^{-2})\,X(z)$$

$$\frac{Y(z)}{X(z)} = \frac{1 + 0.5z^{-1} + 0.4z^{-2}}{1 + 0.6z^{-1} + 0.7z^{-2}} \qquad\qquad .....(2)$$

Let, $\dfrac{Y(z)}{X(z)} = \dfrac{W(z)}{X(z)}\,\dfrac{Y(z)}{W(z)}$

Where, $\dfrac{W(z)}{X(z)} = \dfrac{1}{1 + 0.6\,z^{-1} + 0.7\,z^{-2}}$ $\qquad\qquad .....(3)$

$$\frac{Y(z)}{W(z)} = 1 + 0.5z^{-1} + 0.4z^{-2} \qquad\qquad .....(4)$$

On cross multiplying equation (3) we get,

$$W(z) + 0.6z^{-1}W(z) + 0.7z^{-2}W(z) = X(z)$$

$$\therefore\ W(z) = X(z) - 0.6z^{-1}W(z) - 0.7z^{-2}W(z) \qquad\qquad .....(5)$$

On cross multiplying equation (4) we get,

$$Y(z) = W(z) + 0.5z^{-1}W(z) + 0.4z^{-2}W(z) \qquad\qquad .....(6)$$

The direct form-II digital network is realized using equations (5) and (6) as shown in Fig 2.

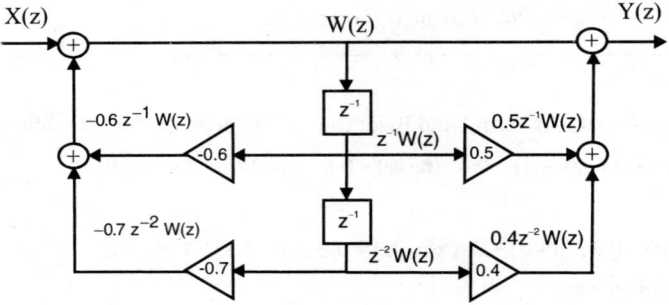

Fig 2: *Direct form-II digital network.*

Example 5.50

Realize the digital network described by H(z) in two ways. $H(z) = \dfrac{1 - r\cos\omega_0\, z^{-1}}{1 - 2r\cos\omega_0\, z^{-1} + r^2 z^{-2}}$

Solution:

Let, $H(z) = \dfrac{Y(z)}{X(z)} = \dfrac{1 - r\cos\omega_0\, z^{-1}}{1 - 2r\cos\omega_0\, z^{-1} + r^2 z^{-2}}$

On cross multiplying we get,

$$Y(z) - 2r\cos\omega_0 z^{-1} Y(z) + r^2 z^{-2} Y(z) = X(z) - r\cos\omega_0 z^{-1} X(z)$$

$$\therefore\ Y(z) = X(z) - r\cos\omega_0 z^{-1} X(z) + 2r\cos\omega_0 z^{-1} Y(z) - r^2 z^{-2} Y(z)$$

Let, $r\cos\omega_0 = a$. $\therefore\ Y(z) = X(z) - az^{-1} X(z) + 2az^{-1} Y(z) - r^2 z^{-2} Y(z)$ ..... (1)

The equation (1) can be used to construct direct form-I structure of H(z) as shown in Fig 1.

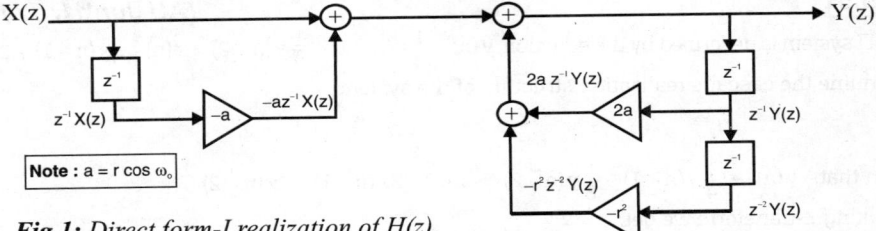

Fig 1: *Direct form-I realization of H(z).*

Consider the direct form-I structure as cascade of two systems $H_1(z)$ and $H_2(z)$ as shown in Fig 2.

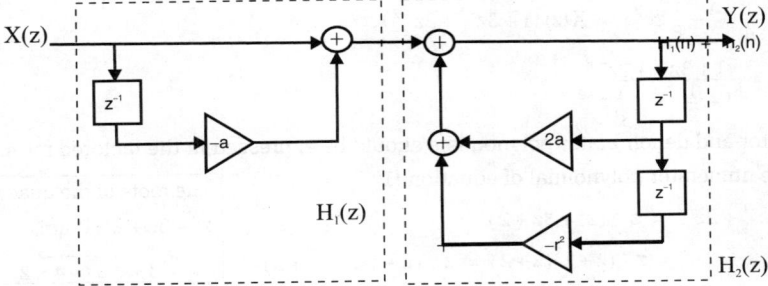

Fig 2: *Direct form-I structure as cascade of two systems.*

In an LT1 system, by linearity property, the order of cascading can be changed. Hence the systems $H_1(z)$ and $H_2(z)$ are interchanged and the Fig 2 is redrawn as shown in Fig 3.

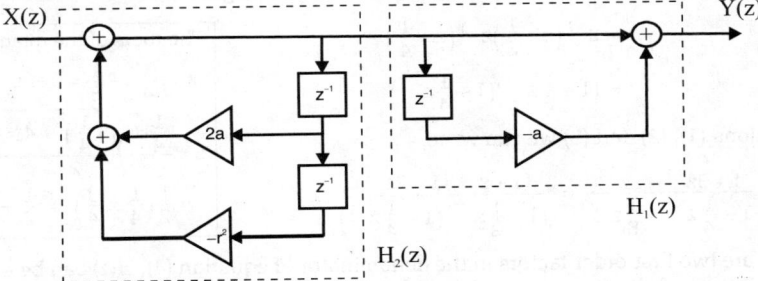

Fig 3: *Direct form-I structure with $H_1(z)$ and $H_2(z)$ interchanged.*

Since the input to delay elements in both the systems $H_1(z)$ and $H_2(z)$ are same, the outputs will also be same. Hence the delays can be combined and the resultant structure is direct form-II structure, which is shown in Fig 4.

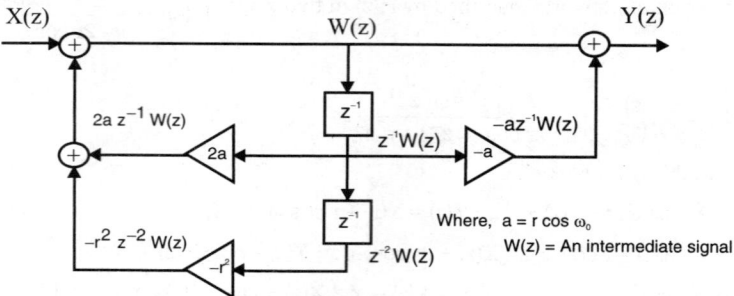

Fig 4: *Direct form-II structure of H(z).*

Example 5.51

(AU Jun'11, 12 Marks)

An LTI system is described by the equation, $y(n) - \frac{1}{4}y(n-1) - \frac{1}{8}y(n-2) = x(n) + 3x(n-1) + 2x(n-2)$.

Determine the cascade realization structure of the system.

Solution:

Given that, $y(n) - \frac{1}{4}y(n-1) - \frac{1}{8}y(n-2) = x(n) + 3x(n-1) + 2x(n-2)$

On taking z-transform we get,

$$Y(z) - \frac{1}{4}z^{-1}Y(z) - \frac{1}{8}z^{-2}Y(z) = X(z) + 3z^{-1}X(z) + 2z^{-2}X(z)$$

$$Y(z)\left(1 - \frac{1}{4}z^{-1} - \frac{1}{8}z^{-2}\right) = X(z)(1 + 3z^{-1} + 2z^{-2})$$

$$\therefore \quad \frac{Y(z)}{X(z)} = \frac{1 + 3z^{-1} + 2z^{-2}}{1 - \frac{1}{4}z^{-1} - \frac{1}{8}z^{-2}} \qquad \dots\dots(1)$$

The numerator and denominator polynomials should be expressed in the factored form.

Consider the numerator polynomial of equation (1).

$$1 + 3z^{-1} + 2z^{-2} = z^{-2}(z^2 + 3z + 2)$$

$$= z^{-2}(z+1)(z+2) = z^{-1}(z+1)z^{-1}(z+2)$$

$$= (1 + z^{-1})(1 + 2z^{-1})$$

Consider the denominator polynomial of equation (1)

$$1 - \frac{1}{4}z^{-1} - \frac{1}{8}z^{-2} = z^{-2}\left(z^2 - \frac{1}{4}z - \frac{1}{8}\right) = z^{-2}\left(z + \frac{1}{4}\right)\left(z - \frac{1}{2}\right) \qquad \dots\dots(3)$$

$$= z^{-1}\left(z - \frac{1}{2}\right)z^{-1}\left(z + \frac{1}{4}\right)$$

$$= \left(1 - \frac{1}{2}z^{-1}\right)\left(1 + \frac{1}{4}z^{-1}\right)$$

From equations (1), (2) and (3) we can write,

$$H(z) = \frac{1 + 3z^{-1} + 2z^{-2}}{1 - \frac{1}{4}z^{-1} + \frac{1}{8}z^{-2}} = \frac{(1 + z^{-1})(1 + 2z^{-1})}{\left(1 - \frac{1}{2}z^{-1}\right)\left(1 + \frac{1}{4}z^{-1}\right)}$$

The roots of the quadratic,
$z^2 + 3z + 2 = 0$ are,
$z = \dfrac{-3 \pm \sqrt{3^2 - 4 \times 2}}{2}$
$= \dfrac{-3 \pm 1}{2} = -1, -2$ $\quad\dots\dots(2)$

The roots of the quadratic,
$z^2 - \dfrac{1}{4}z - \dfrac{1}{8} = 0$ are,
$z = \dfrac{\dfrac{1}{4} \pm \sqrt{\left(\dfrac{1}{4}\right)^2 + 4 \times \dfrac{1}{8}}}{2}$
$= \dfrac{1}{2}\left(\dfrac{1}{4} \pm \dfrac{3}{4}\right) = \dfrac{1}{2}, -\dfrac{1}{4}$ $\quad\dots\dots(4)$

Since there are two first-order factors in the denominator of equation (4), H(z) can be expressed as a product of two sections as shown in equation (5).

Let, $H(z) = \dfrac{1+z^{-1}}{1-\frac{1}{2}z^{-1}} \times \dfrac{1+2z^{-1}}{1+\frac{1}{4}z^{-1}} = H_1(z) \times H_2(z)$ ..... (5)

where, $H_1(z) = \dfrac{1+z^{-1}}{1-\frac{1}{2}z^{-1}}$; $H_2(z) = \dfrac{1+2z^{-1}}{1+\frac{1}{4}z^{-1}}$

Let, $H_1(z) = \dfrac{Y_1(z)}{X(z)} = \dfrac{W_1(z)}{X(z)}\dfrac{Y_1(z)}{W_1(z)} = \dfrac{1+z^{-1}}{1-\frac{1}{2}z^{-1}}$

where, $\dfrac{W_1(z)}{X(z)} = \dfrac{1}{1-\frac{1}{2}z^{-1}}$ and $\dfrac{Y_1(z)}{W_1(z)} = 1+z^{-1}$

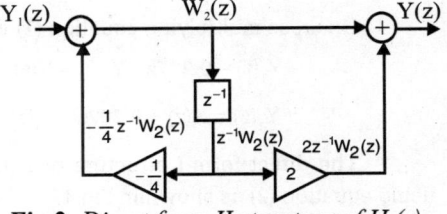

$\therefore W_1(z) = X(z) + \frac{1}{2}z^{-1}W_1(z)$ ..... (6)

$Y_1(z) = W_1(z) + z^{-1}W_1(z)$ ..... (7)

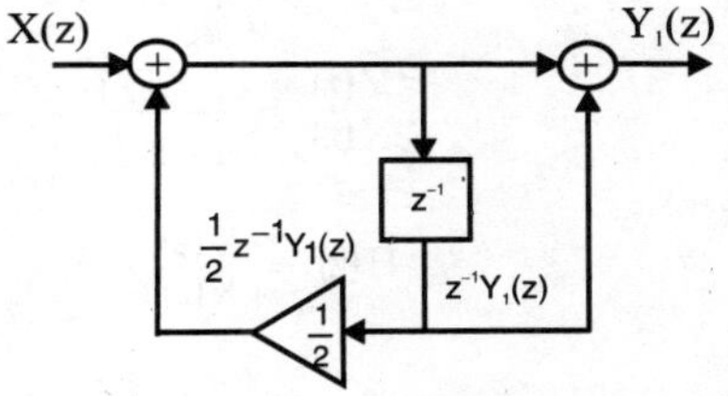

Fig 1: *Direct form-II structure of $H_1(z)$.*

Using equations (6) and (7) the transfer function $H_1(z)$ can be realized in direct form-II structure as shown in Fig 1.

Let, $H_2(z) = \dfrac{Y(z)}{Y_1(z)} = \dfrac{W_2(z)}{Y_1(z)}\dfrac{Y(z)}{W_2(z)} = \dfrac{1+2z^{-1}}{1+\frac{1}{4}z^{-1}}$

where, $\dfrac{W_2(z)}{Y_1(z)} = \dfrac{1}{1+\frac{1}{4}z^{-1}}$ and $\dfrac{Y(z)}{W_2(z)} = 1+2z^{-1}$

$\therefore W_2(z) = Y_1(z) - \frac{1}{4}z^{-1}W_2(z)$ ..... (8)

$Y(z) = W_2(z) + 2z^{-1}W_2(z)$ ..... (9)

Fig 2: *Direct form-II structure of $H_2(z)$.*

Using equations (8) and (9) the transfer function $H_2(z)$ can be realized in direct form-II structure as shown in Fig 2.

The cascade structure of the given LTI system is obtained by connecting the individual sections shown in Fig 1, Fig 2 in cascade as shown in Fig 3.

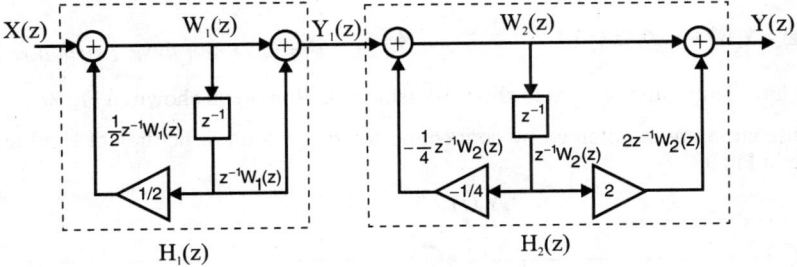

Fig 3: *Cascade realization of the system.*

Example 5.52

An LTI System is described by the equation, $y(n) + y(n-1) - \frac{1}{4}y(n-2) = x(n)$.
Determine the cascade realization structure of the system.

Solution:

Given that, $y(n) + y(n-1) - \frac{1}{4}y(n-2) = x(n)$

On taking z-transform we get,

$$Y(z) + z^{-1}Y(z) - \frac{1}{4}z^{-2}Y(z) = X(z)$$

$$\left(1 + z^{-1} - \frac{1}{4}z^{-2}\right)Y(z) = X(z)$$

$$\therefore \frac{Y(z)}{X(z)} = \frac{1}{1 + z^{-1} - \frac{1}{4}z^{-2}}$$

> The roots of the quadratic,
> $$z^2 + z - \frac{1}{4} = 0 \text{ are,}$$
> $$z = \frac{-1 \pm \sqrt{1^2 + 4 \times \frac{1}{4}}}{2}$$
> $$= \frac{-1 \pm \sqrt{2}}{2} = 0.207, -1.207$$

$$\therefore H(z) = \frac{Y(z)}{X(z)} = \frac{1}{1 + z^{-1} - \frac{1}{4}z^{-2}} = \frac{1}{z^{-2}\left(z^2 + z - \frac{1}{4}\right)}$$

$$= \frac{1}{z^{-2}(z - 0.207)(z + 1.207)} = \frac{1}{(1 - 0.207z^{-1})(1 + 1.207z^{-1})}$$

Let, $H(z) = H_1(z)\,H_2(z)$

where, $H_1(z) = \dfrac{1}{1 - 0.207z^{-1}}$; $H_2(z) = \dfrac{1}{1 + 1.207z^{-1}}$

Let, $H_1(z) = \dfrac{Y_1(z)}{X(z)} = \dfrac{1}{1 - 0.207z^{-1}}$ .....(1)

On cross multiplying equation (1) we get,

$$Y_1(z) - 0.207z^{-1}\,Y_1(z) = X(z)$$

$$\therefore Y_1(z) = X(z) + 0.207z^{-1}\,Y_1(z) \qquad ..... (2)$$

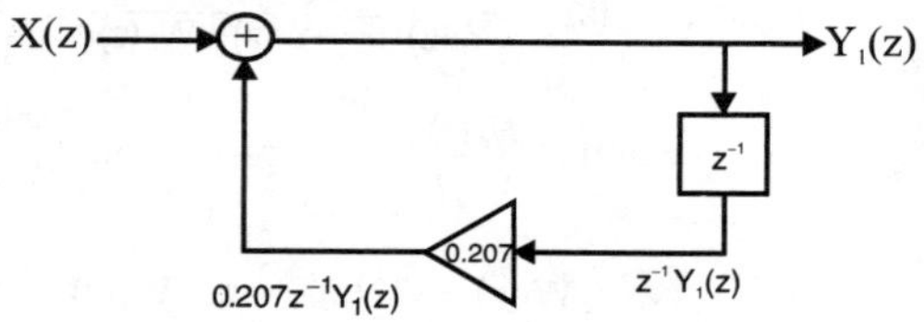

The direct form-I structure of $H_1(z)$ is obtained using equation (2) as shown in Fig 1.

Let, $H_2(z) = \dfrac{Y(z)}{Y_1(z)} = \dfrac{1}{1 + 1.207z^{-1}}$ ..... (3)

On cross multiplying equation (3) we get,

$$Y(z) + 1.207\,z^{-1}\,Y(z) = Y_1(z)$$

$$Y(z) = Y_1(z) - 1.207\,z^{-1}\,Y(z) \qquad ..... (4)$$

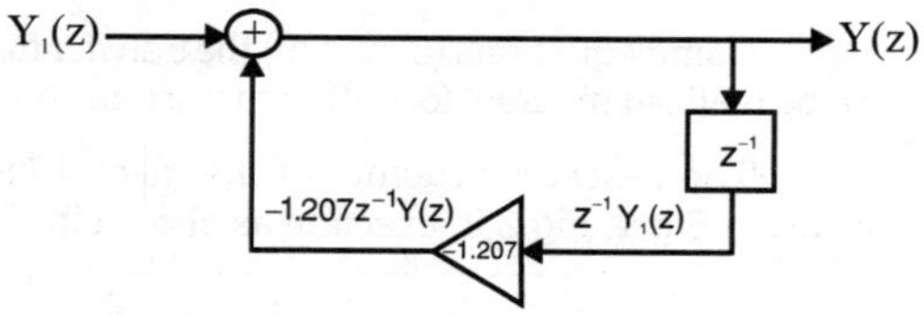

The direct form-I structure of $H_2(z)$ is obtained using equation (4) as shown in Fig 2.

Fig 1: Direct form-I structure of $H_1(z)$.

Fig 2: Direct form-I structure of $H_2(z)$.

The cascade structure is obtained by connecting the direct form structures of $H_1(z)$ and $H_2(z)$ in cascade as shown in Fig 3.

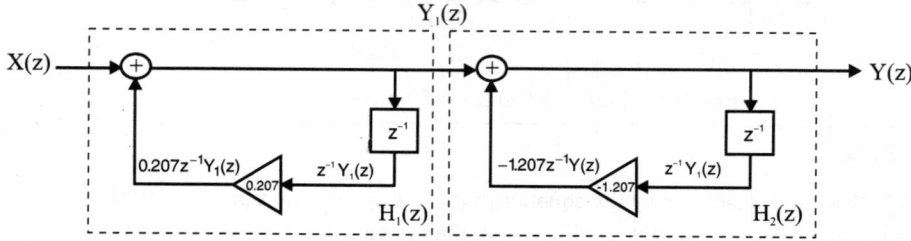

Fig 3: Cascade structure.

Example 5.53

Obtain the cascade realization of the system, $H(z) = \dfrac{2 + z^{-1} + z^{-2}}{\left(1 + \frac{1}{2}z^{-1}\right)\left(1 - \frac{1}{2}z^{-1}\right)\left(1 + \frac{1}{2}z^{-1}\right)}$.

Solution:

Given that, $H(z) = \dfrac{2 + z^{-1} + z^{-2}}{\left(1 + \frac{1}{2}z^{-1}\right)\left(1 - \frac{1}{2}z^{-1}\right)\left(1 + \frac{1}{2}z^{-1}\right)}$

On examining the roots of numerator polynomial it is found that the roots are complex conjugate. Hence H(z) can be realized as cascade of one first order and one second order system.

$$\therefore H(z) = \frac{1}{1 - \frac{1}{2}z^{-1}} \times \frac{2 + z^{-1} + z^{-2}}{\left(1 + \frac{1}{2}z^{-1}\right)\left(1 + \frac{1}{2}z^{-1}\right)} = \frac{1}{1 - \frac{1}{2}z^{-1}} \times \frac{2 + z^{-1} + z^{-2}}{1 + z^{-1} + \frac{1}{4}z^{-2}}$$

Let, $H(z) = H_1(z) \times H_2(z)$

where, $H_1(z) = \dfrac{1}{1 - \frac{1}{2}z^{-1}}$ and $H_2(z) = \dfrac{2 + z^{-1} + z^{-2}}{1 + z^{-1} + \frac{1}{4}z^{-2}}$

Let, $H_1(z) = \dfrac{Y_1(z)}{X(z)} = \dfrac{1}{1 - \frac{1}{2}z^{-1}}$ .....(1)

On cross multiplying equation (1) we get,

$$Y_1(z) - \frac{1}{2}z^{-1}Y_1(z) = X(z) \; ; \quad \therefore Y_1(z) = X(z) + \frac{1}{2}z^{-1}Y_1(z) \qquad \text{.....(2)}$$

The direct form-II structure of $H_1(z)$ can be obtained from equation (2) as shown in Fig 1.

Let, $H_2(z) = \dfrac{Y(z)}{Y_1(z)} = \dfrac{2 + z^{-1} + z^{-2}}{1 + z^{-1} + \frac{1}{4}z^{-2}}$

Let, $\dfrac{Y(z)}{Y_1(z)} = \dfrac{W_2(z)}{Y_1(z)} \dfrac{Y(z)}{W_2(z)}$

where, $\dfrac{W_2(z)}{Y_1(z)} = \dfrac{1}{1 + z^{-1} + \frac{1}{4}z^{-2}}$ .....(3)

$\dfrac{Y(z)}{W_2(z)} = 2 + z^{-1} + z^{-2}$ .....(4)

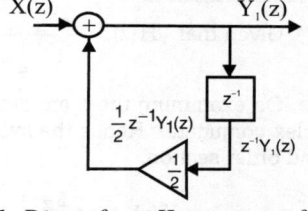

Fig 1: *Direct form-II structure of $H_1(z)$.*

On cross multiplying equation (3) we get,

$$W_2(z) + z^{-1}W_2(z) + \frac{1}{4}z^{-2}W_2(z) = Y_1(z)$$

$$\therefore W_2(z) = Y_1(z) - z^{-1}W_2(z) - \frac{1}{4}z^{-2}W_2(z) \qquad \text{.....(5)}$$

On cross multiplying equation (4) we get,

$$Y(z) = 2W_2(z) + z^{-1}W_2(z) + z^{-2}W_2(z) \qquad \text{..... (6)}$$

The direct form-II structure of $H_2(z)$ can be obtained using equations (5) and (6) as shown in Fig 2.

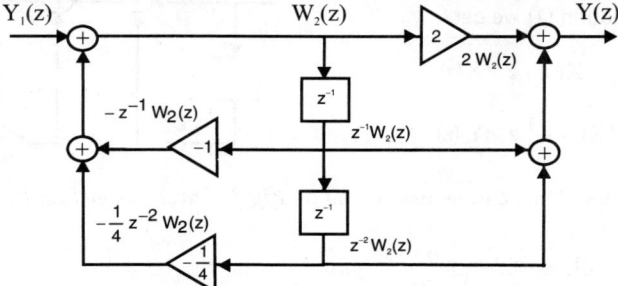

Fig 2: *Direct form-II structure of $H_2(z)$.*

The cascade realization of H(z) is obtained by connecting the direct form-II structures of $H_1(z)$ and $H_2(z)$ in cascade as shown in Fig 3.

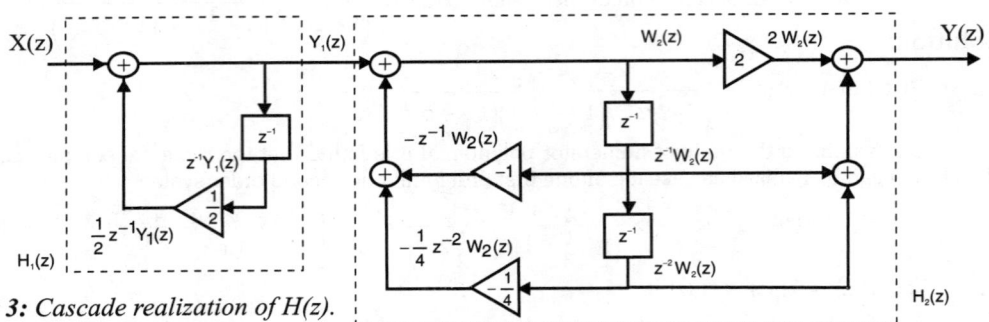

Fig 3: *Cascade realization of H(z).*

Example 5.54

The transfer function of a system is given by, $H(z) = \dfrac{\left(1+z^{-1}\right)^3}{\left(1-\dfrac{1}{4}z^{-1}\right)\left(1-z^{-1}+\dfrac{1}{2}z^{-2}\right)}$

Realize the system in cascade and parallel structures.

Solution:

Cascade Realization

Given that $H(z) = \dfrac{\left(1+z^{-1}\right)^3}{\left(1-\dfrac{1}{4}z^{-1}\right)\left(1-z^{-1}+\dfrac{1}{2}z^{-2}\right)}$

On examining the roots of the quadratic factor in the denominator it is observed that the roots are complex conjugate. Hence the system has to be realized as cascading of one first order section and one second order section.

$$\therefore H(z) = \frac{1+z^{-1}}{1-\frac{1}{4}z^{-1}} \times \frac{\left(1+z^{-1}\right)^2}{1-z^{-1}+\frac{1}{2}z^{-2}} = \frac{1+z^{-1}}{1-\frac{1}{4}z^{-1}} \times \frac{1+2z^{-1}+z^{-2}}{1-z^{-1}+\frac{1}{2}z^{-2}}$$

Let, $H(z) = H_1(z) \times H_2(z)$

Where, $H_1(z) = \dfrac{1+z^{-1}}{1-\dfrac{1}{4}z^{-1}}$ and $H_2(z) = \dfrac{1+2z^{-1}+z^{-2}}{1-z^{-1}+\dfrac{1}{2}z^{-2}}$

Let, $H_1(z) = \dfrac{Y_1(z)}{X(z)} = \dfrac{1+z^{-1}}{1-\dfrac{1}{4}z^{-1}}$ ..... (1)

On cross multiplying equation (1) we get,

$$Y_1(z) - \frac{1}{4}z^{-1}Y_1(z) = X(z) + z^{-1}X(z)$$

$$\therefore Y_1(z) = X(z) + z^{-1}X(z) + \frac{1}{4}z^{-1}Y_1(z) \qquad ..... (2)$$

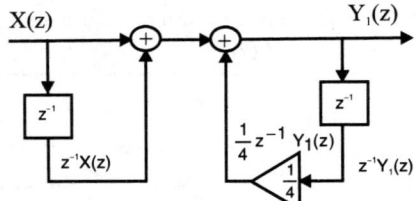

The direct form-I structure of $H_1(z)$ can be drawn using   **Fig 1:** *Direct form-I realization of $H_1(z)$.*
equation (2) as shown in Fig 1.

Let, $H_2(z) = \dfrac{Y(z)}{Y_1(z)} = \dfrac{1 + 2z^{-1} + z^{-2}}{1 - z^{-1} + \dfrac{1}{2}z^{-2}}$ .....(3)

On cross multiplying equation (3) we get

$$Y(z) - z^{-1}Y(z) + \frac{1}{2}z^{-2}Y(z) = Y_1(z) + 2z^{-1}Y_1(z) + z^{-2}Y_1(z)$$

$$\therefore Y(z) = Y_1(z) + 2z^{-1}Y_1(z) + z^{-2}Y_1(z) + z^{-1}Y(z) - \frac{1}{2}z^{-2}Y(z) \qquad .....(4)$$

The direct form-I structure of $H_2(z)$ can be drawn using equation (4) as shown in Fig 2.

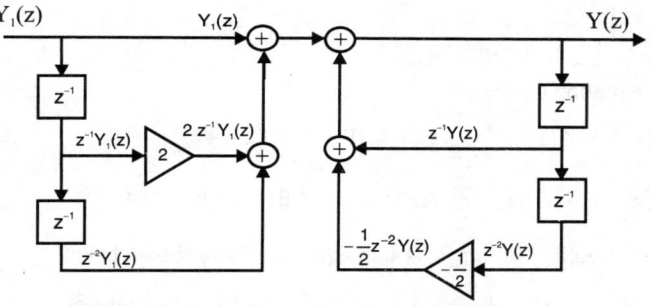

Fig 2: *The direct form-I structure of* $H_2(z)$.

The cascade realization of H(z) is obtained by connecting the direct form - I structures of $H_1(z)$ and $H_2(z)$ in cascade as shown in Fig 3.

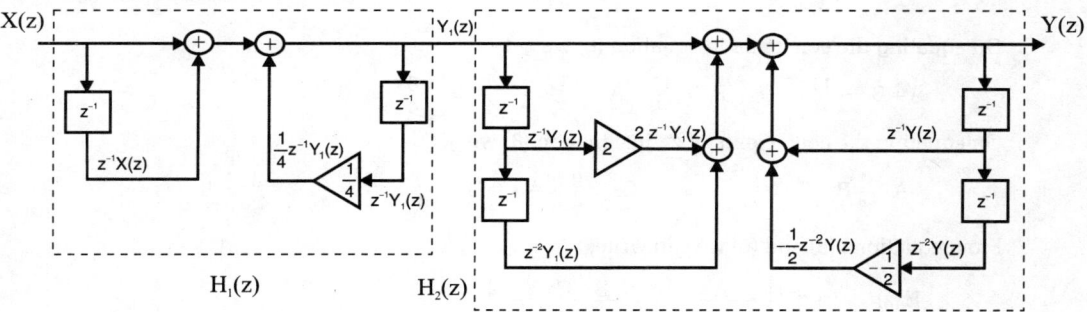

Fig 3: *Cascade realization of H(z).*

Parallel Realization

Given that, $H(z) = \dfrac{(1+z^{-1})^3}{\left(1-\frac{1}{4}z^{-1}\right)\left(1-z^{-1}+\frac{1}{2}z^{-2}\right)} = \dfrac{(1+z^{-1})(1+z^{-1})^2}{\left(1-\frac{1}{4}z^{-1}\right)\left(1-z^{-1}+\frac{1}{2}z^{-2}\right)}$

$= \dfrac{(1+z^{-1})(1+2z^{-1}+z^{-2})}{\left(1-\frac{1}{4}z^{-1}\right)\left(1-z^{-1}+\frac{1}{2}z^{-2}\right)} = \dfrac{1+2z^{-1}+z^{-2}+z^{-1}+2z^{-2}+z^{-3}}{1-z^{-1}+\frac{1}{2}z^{-2}-\frac{1}{4}z^{-1}+\frac{1}{4}z^{-2}-\frac{1}{8}z^{-3}}$

$= \dfrac{1+3z^{-1}+3z^{-2}+z^{-3}}{1-\frac{5}{4}z^{-1}+\frac{3}{4}z^{-2}-\frac{1}{8}z^{-3}}$

$= 1 + \dfrac{\frac{17}{4}z^{-1}+\frac{9}{4}z^{-2}+\frac{9}{8}z^{-3}}{1-\frac{5}{4}z^{-1}+\frac{3}{4}z^{-2}-\frac{1}{8}z^{-3}}$

$$
\begin{array}{r}
1 \\
1-\frac{5}{4}z^{-1}+\frac{3}{4}z^{-2}-\frac{1}{8}z^{-3} \overline{\Big)\ 1+3z^{-1}\ \ +3z^{-2}\ \ +z^{-3}} \\
\underline{1-\frac{5}{4}z^{-1}+\frac{3}{4}z^{-2}-\frac{1}{8}z^{-3}} \\
{\scriptstyle (-)\,(+)\qquad (-)\qquad (+)} \\
\frac{17}{4}z^{-1}+\frac{9}{4}z^{-2}+\frac{9}{8}z^{-3}
\end{array}
$$

$$\therefore H(z) = 1 + \frac{\frac{17}{4}z^{-1} + \frac{9}{4}z^{-2} + \frac{9}{8}z^{-3}}{\left(1 - \frac{1}{4}z^{-1}\right)\left(1 - z^{-1} + \frac{1}{2}z^{-2}\right)} = 1 + z^{-1}\left[\frac{\frac{17}{4} + \frac{9}{4}z^{-1} + \frac{9}{8}z^{-2}}{\left(1 - \frac{1}{4}z^{-1}\right)\left(1 - z^{-1} + \frac{1}{2}z^{-2}\right)}\right] \qquad .....(5)$$

By partial fraction expansion we can write,

$$\frac{\frac{17}{4} + \frac{9}{4}z^{-1} + \frac{9}{8}z^{-2}}{\left(1 - \frac{1}{4}z^{-1}\right)\left(1 - z^{-1} + \frac{1}{2}z^{-2}\right)} = \frac{A}{\left(1 - \frac{1}{4}z^{-1}\right)} + \frac{B + Cz^{-1}}{\left(1 - z^{-1} + \frac{1}{2}z^{-2}\right)} \qquad .....(6)$$

On cross multiplying equation (6) we get,

$$\frac{17}{4} + \frac{9}{4}z^{-1} + \frac{9}{8}z^{-2} = A\left(1 - z^{-1} + \frac{1}{2}z^{-2}\right) + (B + Cz^{-1})\left(1 - \frac{1}{4}z^{-1}\right) \qquad .....(7)$$

$$\frac{17}{4} + \frac{9}{4}z^{-1} + \frac{9}{8}z^{-2} = A - Az^{-1} + \frac{1}{2}Az^{-2} + B - \frac{1}{4}Bz^{-1} + Cz^{-1} - \frac{1}{4}Cz^{-2} \qquad .....(8)$$

The residue A can be solved by putting, $z^{-1} = 4$, in equation (7) as shown below.

$$\frac{17}{4} + \frac{9}{4} \times 4 + \frac{9}{8} \times 4^2 = A\left(1 - 4 + \frac{1}{2} \times 4^2\right) \quad \Rightarrow \quad \frac{17}{4} + 9 + 18 = A(1 - 4 + 8)$$

$$\therefore \frac{17 + 36 + 72}{4} = 5A \quad \Rightarrow \quad \frac{125}{4} = 5A$$

$$\therefore A = \frac{125}{4} \times \frac{1}{5} = \frac{25}{4}$$

On equating the constants in equation (8) we get,

$$A + B = \frac{17}{4} \quad \Rightarrow \quad B = \frac{17}{4} - A = \frac{17}{4} - \frac{25}{4} = -\frac{8}{4} = -2$$

On equating the coefficients of z^{-1} in equation (8) we get,

$$-A - \frac{1}{4}B + C = \frac{9}{4} \quad \Rightarrow \quad C = \frac{9}{4} + A + \frac{1}{4}B = \frac{9}{4} + \frac{25}{4} - \frac{2}{4} = \frac{32}{4} = 8$$

From equations (5) and (6) we can write,

$$H(z) = 1 + z^{-1}\left[\frac{A}{1 - \frac{1}{4}z^{-1}} + \frac{B + Cz^{-1}}{1 - z^{-1} + \frac{1}{2}z^{-2}}\right]$$

$$\therefore H(z) = 1 + \frac{\frac{25}{4}z^{-1}}{1 - \frac{1}{4}z^{-1}} + \frac{-2z^{-1} + 8z^{-2}}{1 - z^{-1} + \frac{1}{2}z^{-2}}$$

$$\text{Let,} \quad H(z) = 1 + \frac{\frac{25}{4}z^{-1}}{1 - \frac{1}{4}z^{-1}} + \frac{-2z^{-1} + 8z^{-2}}{1 - z^{-1} + \frac{1}{2}z^{-2}} = 1 + H_1(z) + H_2(z)$$

$$\text{where,} \quad H_1(z) = \frac{\frac{25}{4}z^{-1}}{1 - \frac{1}{4}z^{-1}} \; ; \qquad H_2(z) = \frac{-2z^{-1} + 8z^{-2}}{1 - z^{-1} + \frac{1}{2}z^{-2}}$$

$$\text{Let,} \quad H(z) = \frac{Y(z)}{X(z)} \; ; \quad H_1(z) = \frac{Y_1(z)}{X(z)} \; ; \quad H_2(z) = \frac{Y_2(z)}{X(z)}$$

$$\therefore H(z) = 1 + H_1(z) + H_2(z) \qquad \Rightarrow \qquad \frac{Y(z)}{X(z)} = 1 + \frac{Y_1(z)}{X(z)} + \frac{Y_2(z)}{X(z)}$$

$$\therefore Y(z) = X(z) + Y_1(z) + Y_2(z)$$

Realization of $H_1(z)$

$$H_1(z) = \frac{Y_1(z)}{X(z)} = \frac{\frac{25}{4}z^{-1}}{1 - \frac{1}{4}z^{-1}}$$

On cross multiplying the above equation we get,

$$Y_1(z) - \frac{1}{4}z^{-1}Y_1(z) = \frac{25}{4}z^{-1}X(z)$$

$$Y_1(z) = \frac{25}{4}z^{-1}X(z) + \frac{1}{4}z^{-1}Y_1(z) \qquad \qquad \dots\dots(9)$$

Using equation (9) the direct form-I structure of $H_1(z)$ is drawn as shown in Fig 4.

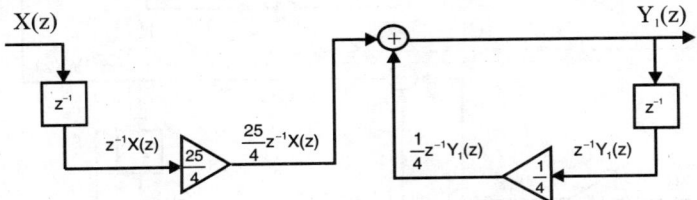

Fig 4: *Direct form-I structure of $H_1(z)$.*

Realization of $H_2(z)$

$$H_2(z) = \frac{Y_2(z)}{X(z)} = \frac{-2z^{-1} + 8z^{-2}}{1 - z^{-1} + \frac{1}{2}z^{-2}}$$

On cross multiplying the above equation we get,

$$Y_2(z) - z^{-1}Y_2(z) + \frac{1}{2}z^{-2}Y_2(z) = -2z^{-1}X(z) + 8z^{-2}X(z)$$

$$\therefore \ Y_2(z) = -2z^{-1}X(z) + 8z^{-2}X(z) + z^{-1}Y_2(z) - \frac{1}{2}z^{-2}Y_2(z) \qquad \qquad \dots\dots(10)$$

Using equation (10) the direct form-I structure of $H_2(z)$ is drawn as shown in Fig 5.

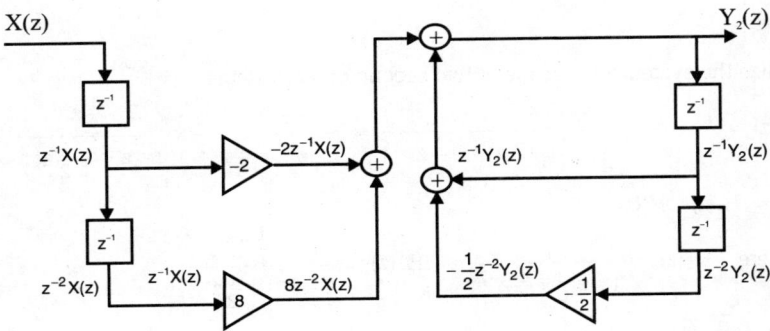

Fig 5: *Direct form-I structure of $H_2(z)$.*

Parallel Structure

The parallel structure of H(z) is obtained by connecting the direct form-I structure of $H_1(z)$ and $H_2(z)$ as shown in Fig 6.

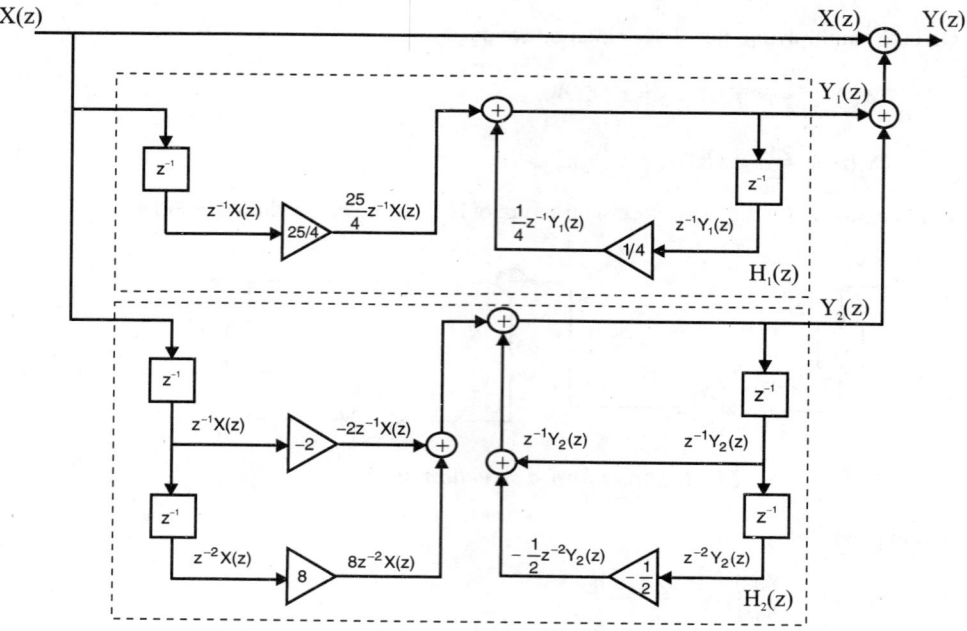

Fig 6: *Parallel structure of H(z).*

Example 5.55

Realize the given system in cascade and parallel forms.

$$H(z) = \frac{1 + \frac{1}{2}z^{-1}}{\left(1 - z^{-1} + \frac{1}{4}z^{-2}\right)\left(1 - z^{-1} + \frac{1}{2}z^{-2}\right)}$$

Solution:

Cascade Form

Let us realize the system as cascade of two second order systems.

$$H(z) = \frac{1 + \frac{1}{2}z^{-1}}{\left(1 - z^{-1} + \frac{1}{4}z^{-2}\right)\left(1 - z^{-1} + \frac{1}{2}z^{-2}\right)} = \frac{1}{1 - z^{-1} + \frac{1}{4}z^{-2}} \times \frac{1 + \frac{1}{2}z^{-1}}{1 - z^{-1} + \frac{1}{2}z^{-2}}$$

Let, $H(z) = H_1(z) \times H_2(z)$

where, $H_1(z) = \dfrac{1}{1 - z^{-1} + \frac{1}{4}z^{-2}}$; $H_2(z) = \dfrac{1 + \frac{1}{2}z^{-1}}{1 - z^{-1} + \frac{1}{2}z^{-2}}$

Let, $H_1(z) = \dfrac{Y_1(z)}{X(z)} = \dfrac{1}{1 - z^{-1} + \frac{1}{4}z^{-2}}$ .....(1)

On cross multiplying equation (1) we get,

$$Y_1(z) - z^{-1}Y_1(z) + \frac{1}{4}z^{-2}Y_1(z) = X(z)$$

$$\therefore Y_1(z) = X(z) + z^{-1}Y_1(z) - \frac{1}{4}z^{-2}Y_1(z) \qquad .....(2)$$

The equation (2) can be realized in direct form-II structure as shown in Fig 1.

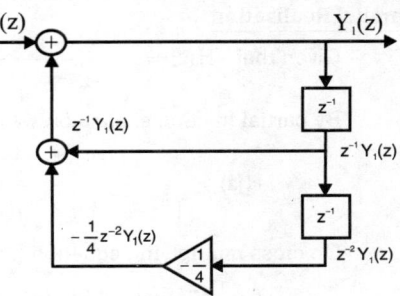

Let, $H_2(z) = \dfrac{Y(z)}{Y_1(z)} = \dfrac{1+\frac{1}{2}z^{-1}}{1-z^{-1}+\frac{1}{2}z^{-2}}$

Let, $\dfrac{Y(z)}{Y_1(z)} = \dfrac{W_2(z)}{Y_1(z)} \dfrac{Y(z)}{W_2(z)}$

where, $\dfrac{W_2(z)}{Y_1(z)} = \dfrac{1}{1-z^{-1}+\frac{1}{2}z^{-2}}$ .....(3)   **Fig 1:** *Direct form-II structure of $H_1(z)$.*

$$\dfrac{Y(z)}{W_2(z)} = 1+\frac{1}{2}z^{-1} \qquad\qquad .....(4)$$

On cross multiplying equation (3) we get

$$W_2(z) - z^{-1}W_2(z) + \frac{1}{2}z^{-2}W_2(z) = Y_1(z)$$

$$\therefore W_2(z) = Y_1(z) + z^{-1}W_2(z) - \frac{1}{2}z^{-2}W_2(z) \qquad\qquad .....(5)$$

On cross multiplying equation (4) we get,

$$Y(z) = W_2(z) + \frac{1}{2}z^{-1}W_2(z) \qquad\qquad .....(6)$$

Using equations (5) and (6) the system $H_2(z)$ can be realized in direct form-II structure as shown in Fig 2

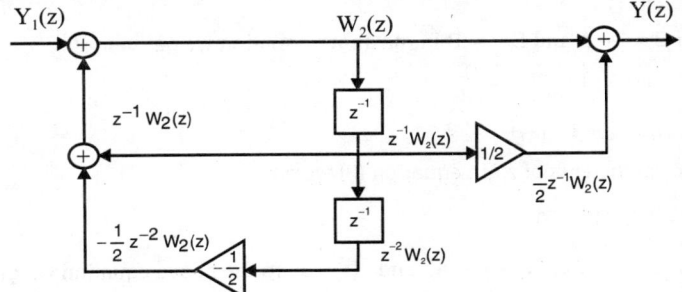

Fig 2: *Direct form-II structure of system $H_2(z)$.*

Cascade structure of H(z) is obtained by connecting structures of $H_1(z)$ and $H_2(z)$ in cascade as shown in Fig 3.

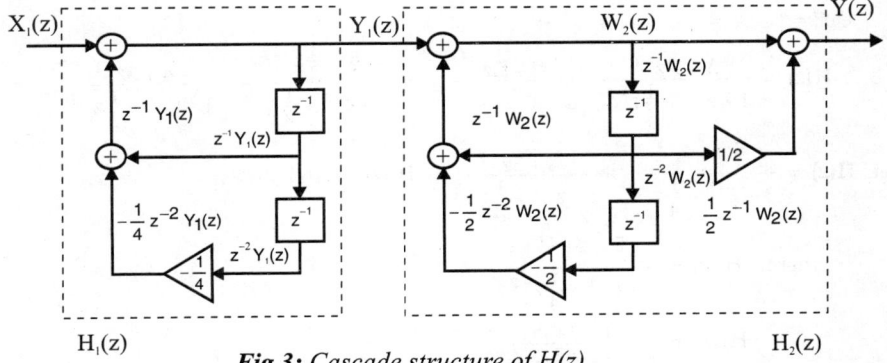

Fig 3: *Cascade structure of H(z).*

Parallel Realization

Given that, $H(z) = \dfrac{1 + \frac{1}{2}z^{-1}}{\left(1 - z^{-1} + \frac{1}{4}z^{-2}\right)\left(1 - z^{-1} + \frac{1}{2}z^{-2}\right)}$

By partial fraction expansion we can write,

$$H(z) = \frac{1 + \frac{1}{2}z^{-1}}{\left(1 - z^{-1} + \frac{1}{4}z^{-2}\right)\left(1 - z^{-1} + \frac{1}{2}z^{-2}\right)} = \frac{A + Bz^{-1}}{1 - z^{-1} + \frac{1}{4}z^{-2}} + \frac{C + Dz^{-1}}{1 - z^{-1} + \frac{1}{2}z^{-2}} \qquad .....(7)$$

On cross multiplying equation (7) we get,

$$1 + \frac{1}{2}z^{-1} = (A + Bz^{-1})\left(1 - z^{-1} + \frac{1}{2}z^{-2}\right) + (C + Dz^{-1})\left(1 - z^{-1} + \frac{1}{4}z^{-2}\right)$$

$$1 + \frac{1}{2}z^{-1} = A - Az^{-1} + \frac{1}{2}Az^{-2} + Bz^{-1} - Bz^{-2} + \frac{1}{2}Bz^{-3}$$

$$+ C - Cz^{-1} + \frac{1}{4}Cz^{-2} + Dz^{-1} - Dz^{-2} + \frac{1}{4}Dz^{-3}$$

$$1 + \frac{1}{2}z^{-1} = (A + C) + (-A + B - C + D)z^{-1} + \left(\frac{1}{2}A - B + \frac{1}{4}C - D\right)z^{-2} + \left(\frac{1}{2}B + \frac{1}{4}D\right)z^{-3} \qquad .....(8)$$

On equating the constants in equation (8) we get,

$$A + C = 1 \quad \Rightarrow \quad C = 1 - A$$

On equating the coefficients of z^{-3} in equation (8) we get,

$$\frac{1}{2}B + \frac{1}{4}D = 0 \quad \therefore \frac{1}{4}D = -\frac{1}{2}B \quad \Rightarrow \quad D = -2B$$

On equating the coefficients of z^{-1} in equation (8) we get,

$$-A + B - C + D = \frac{1}{2}$$

On substituting $C = 1 - A$ and $D = -2B$ in the above equation we get,

$$-A + B - (1 - A) + (-2B) = \frac{1}{2} \quad \Rightarrow \quad -B = \frac{1}{2} + 1 \quad \Rightarrow \quad B = -\frac{3}{2}$$

$$\therefore D = -2B = -2 \times \left(\frac{-3}{2}\right) = 3$$

On equating the coefficients of z^{-2} in equation (8) we get,

$$\frac{1}{2}A - B + \frac{1}{4}C - D = 0$$

On substituting $B = -3/2$, $C = 1 - A$, and $D = 3$ in the above equation we get,

$$\frac{1}{2}A - \left(-\frac{3}{2}\right) + \frac{1}{4}(1 - A) - 3 = 0 \quad \Rightarrow \quad \frac{1}{2}A - \frac{1}{4}A = -\frac{3}{2} - \frac{1}{4} + 3$$

$$\therefore \frac{2A - A}{4} = \frac{-6 - 1 + 12}{4} \quad \Rightarrow \quad \frac{A}{4} = \frac{5}{4} \quad \Rightarrow \quad A = 5$$

$$\therefore C = 1 - A = 1 - 5 = -4$$

$$\therefore H(z) = \frac{A + Bz^{-1}}{1 - z^{-1} + \frac{1}{4}z^{-2}} + \frac{C + Dz^{-1}}{1 - z^{-1} + \frac{1}{2}z^{-2}} = \frac{5 - \frac{3}{2}z^{-1}}{1 - z^{-1} + \frac{1}{4}z^{-2}} + \frac{-4 + 3z^{-1}}{1 - z^{-1} + \frac{1}{2}z^{-2}}$$

Let, $H(z) = \dfrac{5 - \frac{3}{2}z^{-1}}{1 - z^{-1} + \frac{1}{4}z^{-2}} + \dfrac{-4 + 3z^{-1}}{1 - z^{-1} + \frac{1}{2}z^{-2}} = H_1(z) + H_2(z)$

where, $H_1(z) = \dfrac{5 - \frac{3}{2}z^{-1}}{1 - z^{-1} + \frac{1}{4}z^{-2}}$

$H_2(z) = \dfrac{-4 + 3z^{-1}}{1 - z^{-1} + \frac{1}{2}z^{-2}}$

Let, $H(z) = \dfrac{Y(z)}{X(z)}$; $H_1(z) = \dfrac{Y_1(z)}{X(z)}$; $H_2(z) = \dfrac{Y_2(z)}{X(z)}$

$\therefore H(z) = H_1(z) + H_2(z)$

$\therefore \dfrac{Y(z)}{X(z)} = \dfrac{Y_1(z)}{X(z)} + \dfrac{Y_2(z)}{X(z)} \quad \Rightarrow \quad Y(z) = Y_1(z) + Y_2(z)$

Realization of $H_1(z)$
_____

$$H_1(z) = \dfrac{Y_1(z)}{X(z)} = \dfrac{5 - \dfrac{3}{2}z^{-1}}{1 - z^{-1} + \dfrac{1}{4}z^{-2}}$$

Let, $\dfrac{Y_1(z)}{X(z)} = \dfrac{W_1(z)}{X(z)} \dfrac{Y_1(z)}{W_1(z)}$

where, $\dfrac{W_1(z)}{X(z)} = \dfrac{1}{1 - z^{-1} + \dfrac{1}{4}z^{-2}}$ .....(9)

$\dfrac{Y_1(z)}{W_1(z)} = 5 - \dfrac{3}{2}z^{-1}$ .....(10)

On cross multiplying equation (9) we get,

$$W_1(z) - z^{-1}W_1(z) + \dfrac{1}{4}z^{-2}W_1(z) = X(z)$$

$\therefore W_1(z) = X(z) + z^{-1}W_1(z) - \dfrac{1}{4}z^{-2}W_1(z)$ .....(11)

On cross multiplying equation (10) we get,

$$Y_1(z) = 5W_1(z) - \dfrac{3}{2}z^{-1}W_1(z)$$ .....(12)

    The direct form-II structure of system $H_1(z)$ can be realized using equations (11) and (12) as shown in Fig 4.

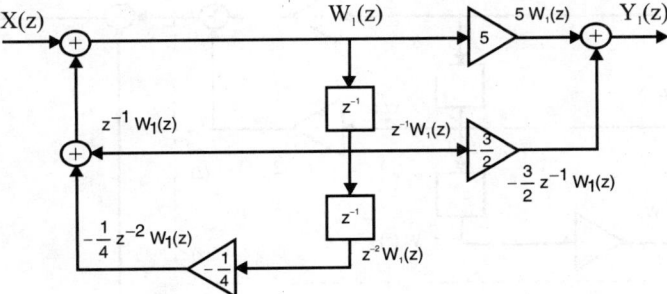

Fig 4: *Direct form-II structure of system $H_1(z)$.*

Realization of $H_2(z)$
_____

$$H_2(z) = \dfrac{Y_2(z)}{X(z)} = \dfrac{-4 + 3z^{-1}}{1 - z^{-1} + \dfrac{1}{2}z^{-2}}$$

Let, $\dfrac{Y_2(z)}{X(z)} = \dfrac{W_2(z)}{X(z)} \dfrac{Y_2(z)}{W_2(z)}$

where, $\dfrac{W_2(z)}{X(z)} = \dfrac{1}{1 - z^{-1} + \dfrac{1}{2}z^{-2}}$ .....(13)

$\dfrac{Y_2(z)}{W_2(z)} = -4 + 3z^{-1}$ .....(14)

On cross multiplying the equation (13) we get,

$$W_2(z) - z^{-1}W_2(z) + \frac{1}{2}z^{-2}W_2(z) = X(z)$$

$$\therefore W_1(z) = X(z) + z^{-1}W_1(z) - \frac{1}{4}z^{-2}W_1(z) \qquad \qquad .....(15)$$

On cross multiplying equation (14) we get,

$$Y_2(z) = -4\,W_2(z) + 3z^{-1}\,W_2(z) \qquad \qquad ..... (16)$$

The direct form-II structure of system $H_2(z)$ can be realized using equations (15) and (16) as shown in Fig 5.

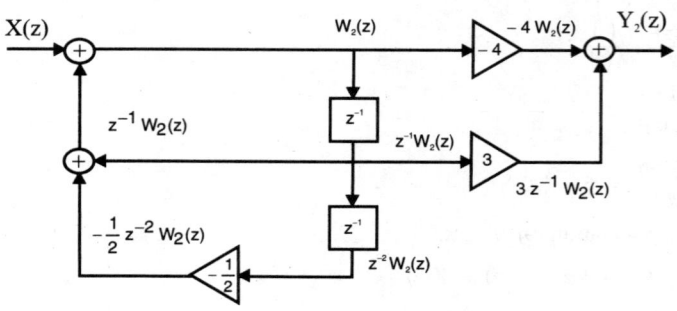

Fig 5: *Direct form-II structure of* $H_2(z)$.

The parallel form structure of H(z) is obtained by connecting the direct form-II structure of $H_1(z)$ and $H_2(z)$ in parallel as shown in Fig 6.

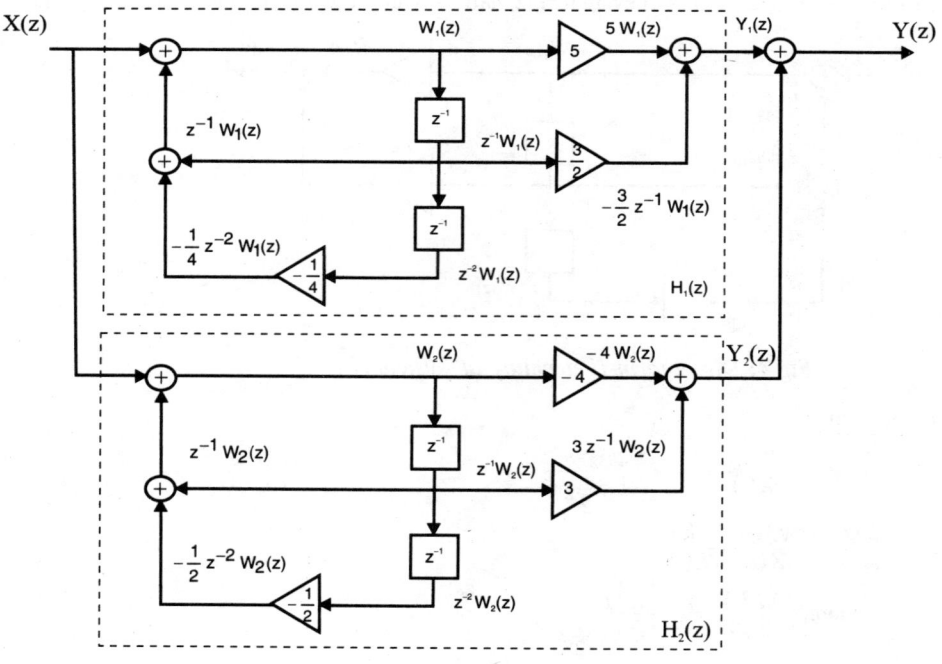

Fig 6: *Parallel form realization of system H(z).*

Example 5.56

Obtain the direct form-I, direct form-II, cascade and parallel form realizations of the LTI system governed by the equation,

$$y(n) = -\frac{3}{8} y(n-1) + \frac{3}{32} y(n-2) + \frac{1}{64} y(n-3) + x(n) + 3x(n-1) + 2x(n-2).$$

Solution:

Direct Form-I

Given that,

$$y(n) = -\frac{3}{8} y(n-1) + \frac{3}{32} y(n-2) + \frac{1}{64} y(n-3) + x(n) + 3x(n-1) + 2x(n-2) \quad \text{.....(1)}$$

On taking z-transform of equation(1) we get,

$$Y(z) = -\frac{3}{8} z^{-1} Y(z) + \frac{3}{32} z^{-2} Y(z) + \frac{1}{64} z^{-3} Y(z) + X(z) + 3 z^{-1} X(z) + 2 z^{-2} X(z) \quad \text{.....(2)}$$

The direct form-I structure can be obtained from equation (2), as shown in Fig 1.

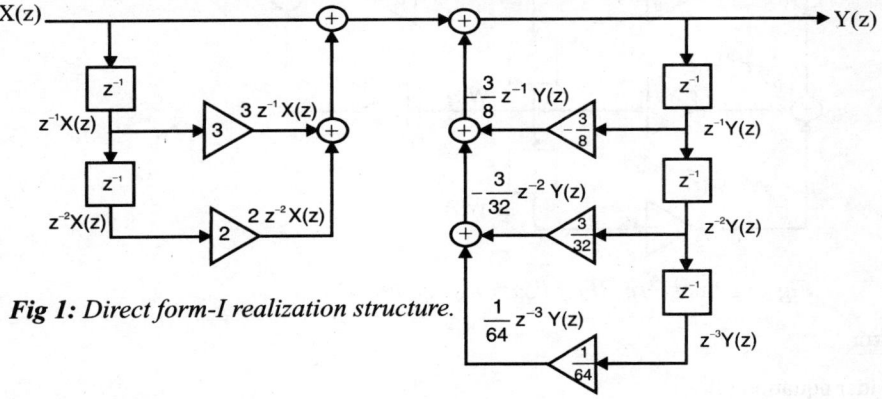

Fig 1: *Direct form-I realization structure.*

Direct Form-II

Consider equation (2).

$$Y(z) = -\frac{3}{8} z^{-1} Y(z) + \frac{3}{32} z^{-2} Y(z) + \frac{1}{64} z^{-3} Y(z) + X(z) + 3z^{-1} X(z) + 2z^{-2} X(z)$$

$$Y(z) + \frac{3}{8} z^{-1} Y(z) - \frac{3}{32} z^{-2} Y(z) - \frac{1}{64} z^{-3} Y(z) = X(z) + 3z^{-1} X(z) + 2z^{-2} X(z)$$

$$Y(z)\left[1 + \frac{3}{8} z^{-1} - \frac{3}{32} z^{-2} - \frac{1}{64} z^{-3}\right] = X(z)\left[1 + 3z^{-1} + 2z^{-2}\right]$$

$$\therefore \frac{Y(z)}{X(z)} = \frac{1 + 3z^{-1} + 2z^{-2}}{1 + \frac{3}{8} z^{-1} - \frac{3}{32} z^{-2} - \frac{1}{64} z^{-3}} \quad \text{.....(3)}$$

Let, $\dfrac{Y(z)}{X(z)} = \dfrac{W(z)}{X(z)} \dfrac{Y(z)}{W(z)}$

where, $\dfrac{W(z)}{X(z)} = \dfrac{1}{1 + \frac{3}{8} z^{-1} - \frac{3}{32} z^{-2} - \frac{1}{64} z^{-3}} \quad \text{.....(4)}$

$$\frac{Y(z)}{W(z)} = 1 + 3z^{-1} + 2z^{-2} \quad \text{.....(5)}$$

On cross multiplying equation (4) we get,

$$W(z)\left(1 + \frac{3}{8}z^{-1} - \frac{3}{32}z^{-2} - \frac{1}{64}z^{-3}\right) = X(z) \qquad .....(6)$$

or $W(z) = X(z) - \frac{3}{8}z^{-1}W(z) + \frac{3}{32}z^{-2}W(z) + \frac{1}{64}z^{-3}W(z)$

On cross multiplying equation (5) we get,

$$Y(z) = W(z) + 3z^{-1}W(z) + 2z^{-2}W(z) \qquad .....(7)$$

The equations (6) and (7) can be realized by a direct form-II structure as shown in Fig 2.

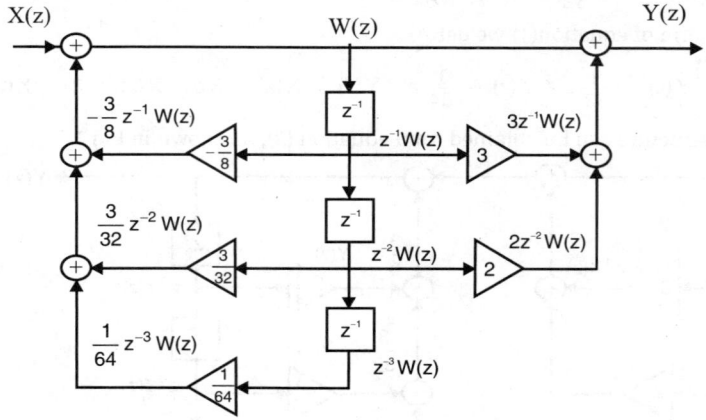

Fig 2: *Direct form-II realization structure.*

Cascade Form

Consider equation (3).

$$\frac{Y(z)}{X(z)} = H(z) = \frac{1 + 3z^{-1} + 2z^{-2}}{1 + \frac{3}{8}z^{-1} - \frac{3}{32}z^{-2} - \frac{1}{64}z^{-3}} \qquad .....(8)$$

The numerator and denominator polynomials should be expressed in the factored form.

Consider the numerator polynomial of equation (8).

$$1 + 3z^{-1} + 2z^{-2} = z^{-2}(z^2 + 3z + 2) = z^{-1}(z+1)z^{-1}(z+2)$$

$$= (1 + z^{-1})(1 + 2z^{-1}) \qquad .....(9)$$

Consider the denominator polynomial of equation (8)

$$1 + \frac{3}{8}z^{-1} - \frac{3}{32}z^{-2} - \frac{1}{64}z^{-3} = z^{-3}\left(z^3 + \frac{3}{8}z^2 - \frac{3}{32}z - \frac{1}{64}\right) \qquad .....(10)$$

$$= z^{-3}\left(z + \frac{1}{8}\right)\left(z^2 + \frac{2}{8}z - \frac{8}{64}\right) \qquad \boxed{\begin{array}{l} z = -1/8 \text{ is one of the root} \\ \text{of the equation (10).} \end{array}}$$

$$\therefore 1 + \frac{3}{8}z^{-1} - \frac{3}{32}z^{-2} - \frac{1}{64}z^{-3} = z^{-3}\left(z+\frac{1}{8}\right)\left(z+\frac{1}{2}\right)\left(z-\frac{1}{4}\right)$$

$-1/8$	1	3/8	$-3/32$	$-1/64$
	$\downarrow$	$-1/8$	$-2/64$	$+1/64$
	1	2/8	$-8/64$	0

$$= z^{-1}\left(z+\frac{1}{8}\right)z^{-1}\left(z+\frac{1}{2}\right)z^{-1}\left(z-\frac{1}{4}\right)$$

$$= \left(1+\frac{1}{8}z^{-1}\right)\left(1+\frac{1}{2}z^{-1}\right)\left(1-\frac{1}{4}z^{-1}\right) \qquad .....(11)$$

From equations (8), (9) and (11) we can write,

$$H(z) = \frac{1 + 3z^{-1} + 2z^{-2}}{1 + \frac{3}{8}z^{-1} - \frac{3}{32}z^{-2} - \frac{1}{64}z^{-3}} = \frac{(1+z^{-1})(1+2z^{-1})}{\left(1+\frac{1}{8}z^{-1}\right)\left(1+\frac{1}{2}z^{-1}\right)\left(1-\frac{1}{4}z^{-1}\right)} \qquad .....(12)$$

Since there are three first order factors in the denominator of equation (12), H(z) can be expressed as a product of three sections as shown in equation (13).

$$\text{Let, } H(z) = \frac{1 + z^{-1}}{1 + \frac{1}{8}z^{-1}} \times \frac{1 + 2z^{-1}}{1 + \frac{1}{2}z^{-1}} \times \frac{1}{1 - \frac{1}{4}z^{-1}} = H_1(z) \times H_2(z) \times H_3(z) \qquad .....(13)$$

$$\text{where, } H_1(z) = \frac{1+z^{-1}}{1+\frac{1}{8}z^{-1}} \;\; ; H_2(z) = \frac{1+2z^{-1}}{1+\frac{1}{2}z^{-1}} \text{ and } H_3(z) = \frac{1}{1-\frac{1}{4}z^{-1}}$$

The transfer function $H_1(z)$ can be realized in direct form-II structure as shown in Fig 3.

$$\text{Let, } H_1(z) = \frac{Y_1(z)}{X(z)} = \frac{W_1(z)}{X(z)} \frac{Y_1(z)}{W_1(z)} = \frac{1+z^{-1}}{1+\frac{1}{8}z^{-1}}$$

$$\text{Where, } \frac{W_1(z)}{X(z)} = \frac{1}{1+\frac{1}{8}z^{-1}} \text{ and } \frac{Y_1(z)}{W_1(z)} = 1 + z^{-1}$$

$$\therefore W_1(z) = X(z) - \frac{1}{8}z^{-1}W_1(z)$$

$$Y_1(z) = W_1(z) + z^{-1}W_1(z)$$

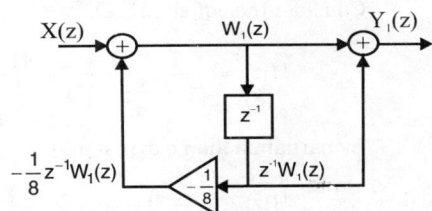

Fig 3: Direct form-II structure of $H_1(z)$.

The transfer function $H_2(z)$ can be realized in direct form-II structure as shown in Fig 4.

$$\text{Let, } H_2(z) = \frac{Y_2(z)}{Y_1(z)} = \frac{W_2(z)}{Y_1(z)} \frac{Y_2(z)}{W_2(z)} = \frac{1+2z^{-1}}{1+\frac{1}{2}z^{-1}}$$

$$\text{where, } \frac{W_2(z)}{Y_1(z)} = \frac{1}{1+\frac{1}{2}z^{-1}} \text{ and } \frac{Y_2(z)}{W_2(z)} = 1 + 2z^{-1}$$

$$\therefore W_2(z) = Y_1(z) - \frac{1}{2}z^{-1}W_2(z)$$

$$Y_2(z) = W_2(z) + 2z^{-1}W_2(z)$$

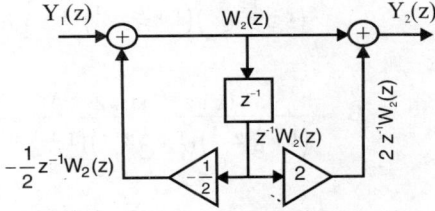

Fig 4: Direct form-II structure of $H_2(z)$.

The transfer function $H_3(z)$ can be realized in direct form-II structure as shown in Fig 5.

$$\text{Let, } H_3(z) = \frac{Y(z)}{Y_2(z)} = \frac{1}{1-\frac{1}{4}z^{-1}}$$

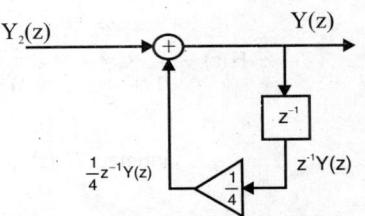

Fig 5: Direct form-II structure of $H_3(z)$.

$$\therefore\ Y(z) - \frac{1}{4} z^{-1} Y(z) = Y_2(z)$$

$$Y(z) = Y_2(z) + \frac{1}{4} z^{-1} Y(z)$$

The cascade structure of the given system is obtained by connecting the individual sections shown in Fig 3, Fig 4 and Fig 5 in cascade as shown in Fig 6.

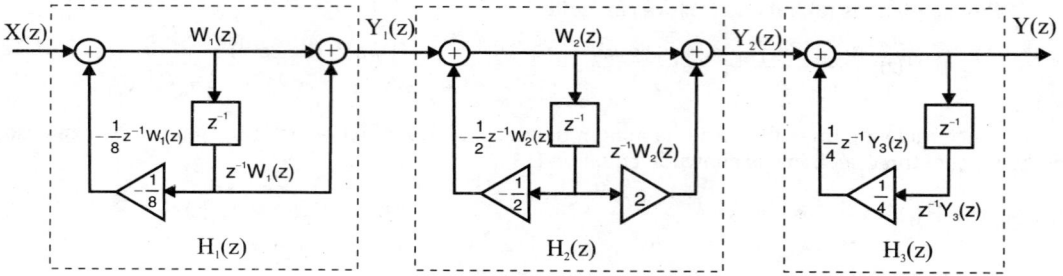

Fig 6: *Cascade realization of the system.*

Parallel Form

Consider the equation (12).

$$H(z) = \frac{(1 + z^{-1})(1 + 2z^{-1})}{\left(1 + \frac{1}{8}z^{-1}\right)\left(1 + \frac{1}{2}z^{-1}\right)\left(1 - \frac{1}{4}z^{-1}\right)}$$

By partial fraction expansion,

$$H(z) = \frac{A}{1 + \frac{1}{8}z^{-1}} + \frac{B}{1 + \frac{1}{2}z^{-1}} = \frac{C}{1 - \frac{1}{4}z^{-1}}$$

$$A = \frac{(1+z^{-1})(1+2z^{-1})}{\left(1+\frac{1}{8}z^{-1}\right)\left(1+\frac{1}{2}z^{-1}\right)\left(1-\frac{1}{4}z^{-1}\right)} \times \left(1+\frac{1}{8}z^{-1}\right)\Bigg|_{z^{-1}=-8} = \frac{(1-8)(1-16)}{(1-4)(1+2)} = -\frac{35}{3}$$

$$B = \frac{(1+z^{-1})(1+2z^{-1})}{\left(1+\frac{1}{8}z^{-1}\right)\left(1+\frac{1}{2}z^{-1}\right)\left(1-\frac{1}{4}z^{-1}\right)} \times \left(1+\frac{1}{2}z^{-1}\right)\Bigg|_{z^{-1}=-2} = \frac{(1-2)(1-4)}{\left(1-\frac{1}{4}\right)\left(1+\frac{1}{2}\right)} = \frac{(-1)\times(-3)}{\frac{3}{4}\times\frac{3}{2}} = \frac{8}{3}$$

$$C = \frac{(1+z^{-1})(1+2z^{-1})}{\left(1+\frac{1}{8}z^{-1}\right)\left(1+\frac{1}{2}z^{-1}\right)\left(1-\frac{1}{4}z^{-1}\right)} \times \left(1-\frac{1}{4}z^{-1}\right)\Bigg|_{z^{-1}=4} = \frac{(1+4)(1+8)}{\left(1+\frac{1}{2}\right)(1+2)} = \frac{5\times9}{\frac{3}{2}\times3} = 10$$

$$\therefore\ H(z) = \frac{-\frac{35}{3}}{1 + \frac{1}{8}z^{-1}} + \frac{\frac{8}{3}}{1 + \frac{1}{2}z^{-1}} + \frac{10}{1 - \frac{1}{4}z^{-1}} = H_1(z) + H_2(z) + H_3(z)$$

$$\text{where,}\ \ H_1(z) = \frac{-\frac{35}{3}}{1 + \frac{1}{8}z^{-1}}\ ; \qquad H_2(z) = \frac{\frac{8}{3}}{1 + \frac{1}{2}z^{-1}}\ ; \qquad H_3(z) = \frac{10}{1 - \frac{1}{4}z^{-1}}$$

$$\text{Let,}\ \ H(z) = \frac{Y(z)}{X(z)}\ ;\ \ H_1(z) = \frac{Y_1(z)}{X(z)}\ ;\ \ H_2(z) = \frac{Y_2(z)}{X(z)}\ ;\ \ H_3 = \frac{Y_3(z)}{X(z)}$$

$$\therefore \ H(z) = H_1(z) + H_2(z) + H_3(z) \qquad \Rightarrow \qquad \frac{Y(z)}{X(z)} = \frac{Y_1(z)}{X(z)} + \frac{Y_2(z)}{X(z)} + \frac{Y_3(z)}{X(z)}$$

$$\therefore \ Y(z) = Y_1(z) + Y_2(z) + Y_3(z)$$

The transfer function $H_1(z)$ can be realized in direct form-I structure as shown in Fig 7.

Let, $H_1(z) = \dfrac{Y_1(z)}{X(z)} = \dfrac{-\dfrac{35}{3}}{1+\dfrac{1}{8}z^{-1}}$

On cross multiplying and rearranging we get,

$$Y_1(z) = -\frac{1}{8}z^{-1}Y_1(z) - \frac{35}{3}X(z)$$

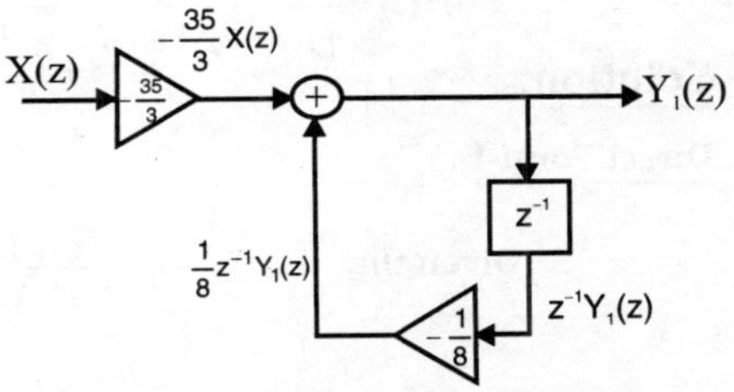

The transfer function $H_2(z)$ can be realized in direct form-I structure as shown in Fig 8.

Let, $H_2(z) = \dfrac{Y_2(z)}{X(z)} = \dfrac{\dfrac{8}{3}}{1+\dfrac{1}{2}z^{-1}}$

On cross multiplying and rearranging we get,

$$Y_2(z) = -\frac{1}{2}z^{-1}Y_2(z) + \frac{8}{3}X(z)$$

The transfer function $H_3(z)$ can be realized in direct form-I structure as shown in Fig 9.

Let, $H_3(z) = \dfrac{Y_3(z)}{X(z)} = \dfrac{10}{1-\dfrac{1}{4}z^{-1}}$

On cross multiplying and rearranging we get,

$$Y_3(z) = \frac{1}{4}z^{-1}Y_3(z) + 10\,X(z)$$

The overall structure is obtained by connecting the individual sections shown in Fig 7, Fig 8 and Fig 9 in parallel as shown in Fig 10.

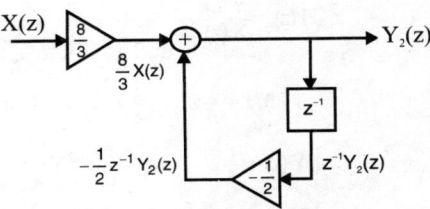

Fig 7: Direct form-I structure of $H_1(z)$.

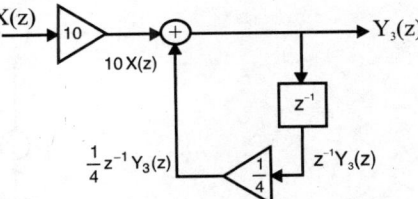

Fig 8: Direct form-I structure of $H_2(z)$.

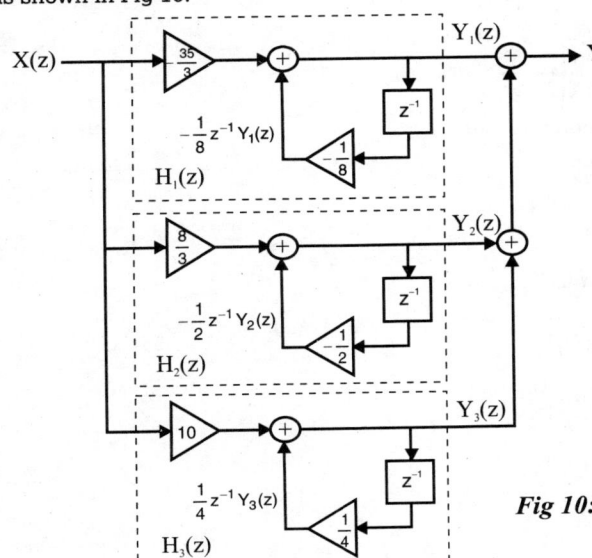

Fig 9: Direct form-I structure of $H_3(z)$.

Fig 10: Parallel form realization.

Example 5.57

Obtain the direct form-I, direct form-II, cascade and parallel form realizations of a discrete time system represented by transfer function,

$$H(z) = \frac{1}{\left(1 + \frac{1}{2}z^{-1}\right)\left(1 - \frac{1}{4}z^{-1}\right)}$$

Solution:

Direct Form-I

Given that, $H(z) = \dfrac{1}{\left(1 + \frac{1}{2}z^{-1}\right)\left(1 - \frac{1}{4}z^{-1}\right)}$

We know that, $H(z) = \dfrac{Y(z)}{X(z)}$

$$\therefore H(z) = \frac{Y(z)}{X(z)} = \frac{1}{\left(1 + \frac{1}{2}z^{-1}\right)\left(1 - \frac{1}{4}z^{-1}\right)} = \frac{1}{1 + \frac{1}{4}z^{-1} - \frac{1}{8}z^{-2}}$$

$$Y(z)\left(1 + \frac{1}{4}z^{-1} - \frac{1}{8}z^{-2}\right) = X(z)$$

$$Y(z) + \frac{1}{4}z^{-1}Y(z) - \frac{1}{8}z^{-2}Y(z) = X(z) \qquad\qquad .....(1)$$

$$Y(z) = X(z) - \frac{1}{4}z^{-1}Y(z) + \frac{1}{8}z^{-2}Y(z) \qquad\qquad .....(2)$$

The direct form-I structure can be obtained from equation (2), as shown in Fig 1.

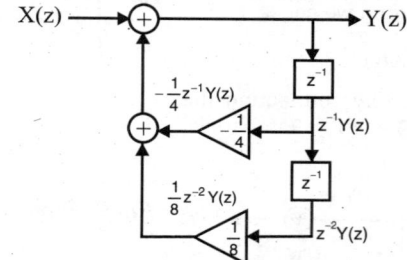

Fig 1: *Direct form-I realization structure.*

Direct Form-II

Consider equation (2).

$$Y(z) = X(z) - \frac{1}{4}z^{-1}Y(z) + \frac{1}{8}z^{-2}Y(z)$$

$$Y(z) + \frac{1}{4}z^{-1}Y(z) - \frac{1}{8}z^{-2}Y(z) = X(z)$$

$$Y(z)\left(1 + \frac{1}{4}z^{-1} - \frac{1}{8}z^{-2}\right) = X(z)$$

$$\therefore \frac{Y(z)}{X(z)} = \frac{1}{1 + \frac{1}{4}z^{-1} - \frac{1}{8}z^{-2}} \qquad\qquad .....(3)$$

Let, $\dfrac{Y(z)}{X(z)} = \dfrac{W(z)}{X(z)} \dfrac{Y(z)}{W(z)}$

where, $\dfrac{W(z)}{X(z)} = \dfrac{1}{1+\dfrac{1}{4}z^{-1}-\dfrac{1}{8}z^{-2}}$ .....(4)

$\dfrac{Y(z)}{W(z)} = 1$ .....(5)

On cross multiplying equation (4) we get,

$$W(z)+\dfrac{1}{4}z^{-1}W(z)-\dfrac{1}{8}z^{-2}W(z) = X(z)$$

$$\therefore W(z) = X(z)-\dfrac{1}{4}z^{-1}W(z)+\dfrac{1}{8}z^{-2}W(z) \qquad .....(6)$$

On cross multiplying equation (5) we get,

$$Y(z) = W(z) \qquad .....(7)$$

The equations (6) and (7) can be realized by a direct form-II structure as shown in Fig 2.

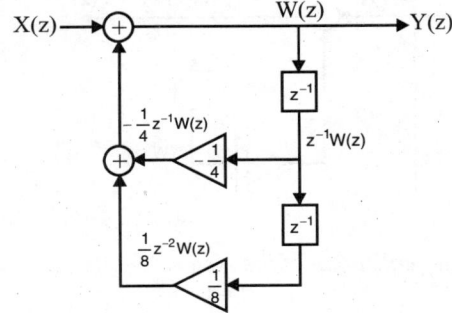

Fig 2: *Direct form-II realization structure.*

Cascade Form

Given that, $H(z) = \dfrac{1}{\left(1+\dfrac{1}{2}z^{-1}\right)\left(1-\dfrac{1}{4}z^{-1}\right)}$ .....(8)

Since there are two first-order factors in the denominator of equation (8), H(z) can be expressed as a product of two sections as shown in equation (9).

Let, $H(z) = \dfrac{1}{1+\dfrac{1}{2}z^{-1}} \times \dfrac{1}{1-\dfrac{1}{4}z^{-1}} = H_1(z) \times H_2(z)$ .....(9)

where, $H_1(z) = \dfrac{1}{1+\dfrac{1}{2}z^{-1}}$; $H_2(z) = \dfrac{1}{1-\dfrac{1}{4}z^{-1}}$

The transfer function $H_1(z)$ can be realized in direct form-II structure using equations (10) and (11), as shown in Fig 3.

Let, $H_1(z) = \dfrac{Y_1(z)}{X(z)} = \dfrac{W_1(z)}{X(z)}\dfrac{Y_1(z)}{W_1(z)} = \dfrac{1}{1+\dfrac{1}{2}z^{-1}}$

where, $\dfrac{W_1(z)}{X(z)} = \dfrac{1}{1+\dfrac{1}{2}z^{-1}}$ and $\dfrac{Y_1(z)}{W_1(z)} = 1$

$\therefore W_1(z) = X(z)-\dfrac{1}{2}z^{-1}W_1(z)$ .....(10)

$Y_1(z) = W_1(z)$ .....(11)

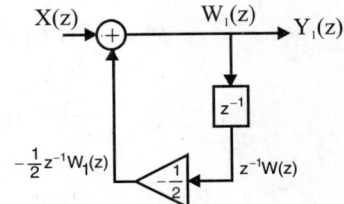

Fig 3: *Direct form-II structure of $H_1(z)$.*

The transfer function $H_2(z)$ can be realized in direct form-II structure using equations (12) and (13), as shown in Fig 4.

Let, $H_2(z) = \dfrac{Y(z)}{Y_1(z)} = \dfrac{W_2(z)}{Y_1(z)} \dfrac{Y(z)}{W_2(z)} = \dfrac{1}{1 - \frac{1}{4}z^{-1}}$

where, $\dfrac{W_2(z)}{Y_1(z)} = \dfrac{1}{1 - \frac{1}{4}z^{-1}}$ and $\dfrac{Y(z)}{W_2(z)} = 1$

$\therefore W_2(z) = Y_1(z) + \frac{1}{4}z^{-1}W_2(z)$ .....(12)

$Y_2(z) = W_2(z)$ .....(13)

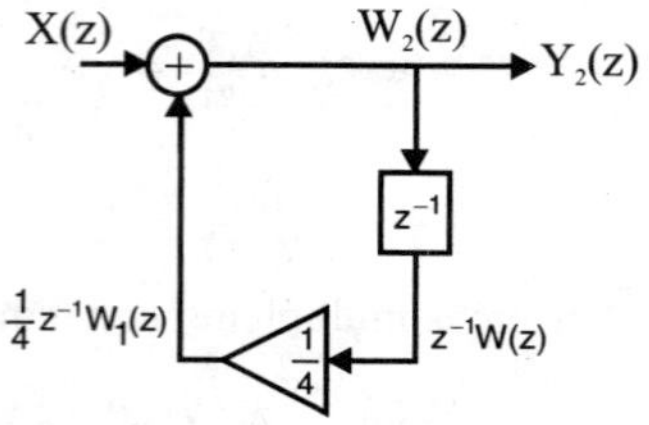

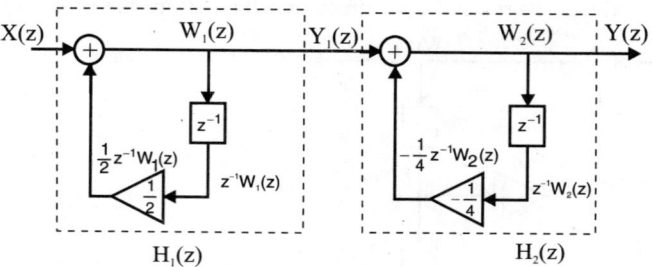

Fig 4: *Direct form-II structure of $H_2(z)$.*

The cascade structure of the given LTI system is obtained by connecting the individual sections shown in Fig 3, Fig 4 in cascade as shown in Fig 5.

Fig 5: *Cascade realization of the system.*

Parallel Form

Given that, $H(z) = \dfrac{1}{\left(1 + \frac{1}{2}z^{-1}\right)\left(1 - \frac{1}{4}z^{-1}\right)}$

By partial fraction expansion,

$H(z) = \dfrac{1}{\left(1 + \frac{1}{2}z^{-1}\right)\left(1 - \frac{1}{4}z^{-1}\right)} = \dfrac{A}{\left(1 + \frac{1}{2}z^{-1}\right)} + \dfrac{B}{\left(1 - \frac{1}{4}z^{-1}\right)}$

$A = \dfrac{1}{\left(1 + \frac{1}{2}z^{-1}\right)\left(1 - \frac{1}{4}z^{-1}\right)} \times \left(1 + \frac{1}{2}z^{-1}\right)\Big|_{z^{-1}=-2} = \dfrac{1}{1 + \frac{1}{2}} = \dfrac{1}{\frac{3}{2}} = \dfrac{2}{3}$

$B = \dfrac{1}{\left(1 + \frac{1}{2}z^{-1}\right)\left(1 - \frac{1}{4}z^{-1}\right)} \times \left(1 - \frac{1}{4}z^{-1}\right)\Big|_{z^{-1}=4} = \dfrac{1}{1 + \frac{4}{2}} = \dfrac{1}{3}$

$\therefore H(z) = \dfrac{\frac{2}{3}}{1 + \frac{1}{2}z^{-1}} + \dfrac{\frac{1}{3}}{1 - \frac{1}{4}z^{-1}} = H_1(z) + H_2(z)$

Where, $H_1(z) = \dfrac{\frac{2}{3}}{1 + \frac{1}{2}z^{-1}}$; $H_2(z) = \dfrac{\frac{1}{3}}{1 - \frac{1}{4}z^{-1}}$

Let, $H(z) = \dfrac{Y(z)}{X(z)}$; $H_1(z) = \dfrac{Y_1(z)}{X(z)}$; $H_2(z) = \dfrac{Y_2(z)}{X(z)}$

$\therefore H(z) = H_1(z) + H_2(z) \implies \dfrac{Y(z)}{X(z)} = \dfrac{Y_1(z)}{X(z)} + \dfrac{Y_2(z)}{X(z)}$

$\therefore Y(z) = Y_1(z) + Y_2(z)$

The transfer function $H_1(z)$ can be realized in direct form-I structure using equation (14) as shown in Fig 6.

Let, $H_1(z) = \dfrac{Y_1(z)}{X(z)} = \dfrac{\dfrac{2}{3}}{1 + \dfrac{1}{2}z^{-1}}$

On cross multiplying and rearranging we get,

$$Y_1(z) = -\frac{1}{2}z^{-1}Y_1(z) + \frac{2}{3}X(z) \qquad .....(14)$$

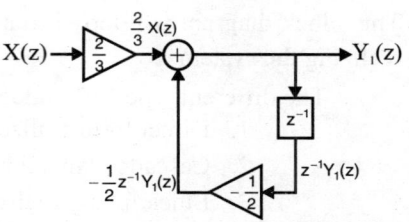

Fig 6: Direct form-I structure of $H_1(z)$.

The transfer function $H_2(z)$ can be realized in direct form-I structure using equation (15) as shown in Fig 7.

Let, $H_2(z) = \dfrac{Y_2(z)}{X(z)} = \dfrac{\dfrac{1}{3}}{1 - \dfrac{1}{4}z^{-1}}$

On cross multiplying and rearranging we get,

$$Y_2(z) = \frac{1}{4}z^{-1}Y_2(z) + \frac{1}{3}X(z) \qquad .....(15)$$

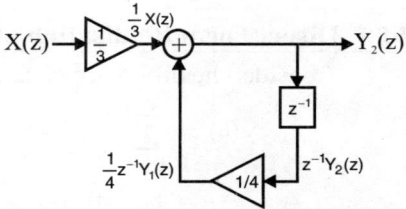

Fig 7: Direct form-I structure of $H_2(z)$.

The overall structure is obtained by connecting the individual sections shown in Fig 6, Fig 7 in parallel as shown in Fig 8.

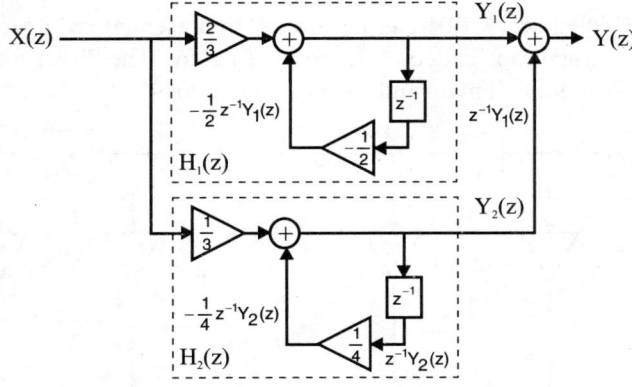

Fig 8: Parallel form realization.

5.9 Structures for Realization of FIR Systems

In general, the time domain representation of an Nth order FIR system is,

$$y(n) = \sum_{m=0}^{N-1} b_m x(n-m) = b_0 x(n) + b_1 x(n-1) + b_2 x(n-2) + ... + b_{N-1} x(n-(N-1))$$

and the z-domain representation of a FIR system is,

$$H(z) = \frac{Y(z)}{X(z)} = b_0 + b_1 z^{-1} + b_2 z^{-2} + ... + b_{N-1} z^{-(N-1)}$$

The above two representations of FIR system can be viewed as a computational procedure (or algorithm) to determine the output sequence $y(n)$ from the input sequence $x(n)$. These equations can be used to construct the block diagram of the FIR system using delays, adders and multipliers.

This block diagram is referred to as realization of the system or equivalently as a structure for realizing the system.

The different types of standard structures for realizing FIR systems are,

      1. Direct form realization
      2. Cascade realization
      3. Linear phase realization

Some of the block diagram representation of the system gives a direct relation between time domain equation and z-domain equation.

5.9.1 Direct Form Realization of FIR System

Consider the difference equation governing a FIR system,

$$y(n) = \sum_{m=0}^{N-1} b_m x(n-m)$$

> If $\mathcal{Z}\{x(n)\} = X(z)$ then,
> $\mathcal{Z}\{x(n-k)\} = z^{-k} X(z)$

$$= b_0 x(n) + b_1 x(n-1) + b_2 x(n-2) + ... + b_{N-1} x(n-(N-1))$$

On taking $\mathcal{Z}$-transform of the above equation we get,.

$$\therefore Y(z) = b_0 X(z) + b_1 z^{-1} X(z) + b_2 z^{-2} X(z) + b_3 z^{-3} X(z) + \qquad .....(5.72)$$
$$..... + b_{N-2} z^{-(N-2)} X(z) + b_{N-1} z^{-(N-1)} X(z)$$

The equation of $Y(z)$ [equation (5.72)] can be directly represented by a block diagram as shown in Fig 5.19 and this structure is called direct form structure. The direct form structure provides a direct relation between time domain and z-domain equations.

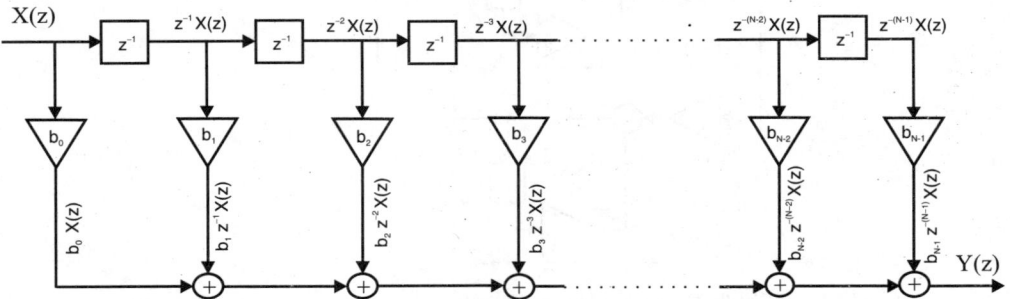

Fig 5.19: Direct form structure of FIR system.

From the direct form structure it is observed that the realization of an Nth order FIR discrete time system involves N number of multiplications and N−1 number of additions. Also the structure involves N−1 delays and so N−1 memory locations are required to store the delayed signals.

5.9.2 Cascade Form Realization of FIR System

Consider the transfer function of a FIR system,

$$H(z) = \frac{Y(z)}{X(z)} = b_0 + b_1 z^{-1} + b_2' z^{-2} + ... + b_{N-1} z^{-(N-1)}$$

The transfer function of FIR system is $(N-1)^{th}$ order polynomial in z. This polynomial can be factorized into first and second order factors and the transfer function can be expressed as a product of first and second order factors or sections as shown in equation (5.73).

$$H(z) = \frac{Y(z)}{X(z)} = H_1(z) \times H_2(z) \times H_3(z) \ldots H_m(z) = \prod_{i=1}^{m} H_i(z) \qquad \ldots(5.73)$$

where, $H_i(z) = c_{0i} + c_{1i} z^{-1} + c_{2i} z^{-2}$ | Second order section |

or, $H_i(z) = c_{0i} + c_{1i} z^{-1}$ | First order section |

The individual second order or first order sections can be realized either in direct form structure or linear phase structure. The overall system is obtained by cascading the individual sections as shown in Fig 5.20 The number of calculations and the memory requirement depends on the realization of individual sections.

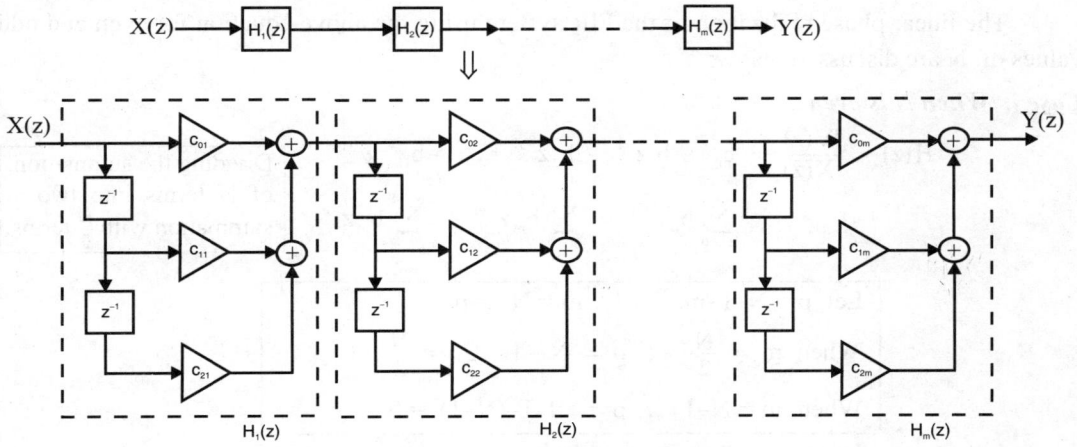

Fig 5.20: *Cascade structure of FIR system.*

5.9.3 Linear Phase Realization of FIR System

Consider the impulse response, h(n) of FIR system,

$$h(n) = \{b_0, \ b_1, \ b_2, \ \ldots\ldots\ldots\ldots b_{N-1}\}$$
     ↑

In FIR system, for linear phase response the impulse response should be symmetrical.

The condition for symmetry is,

     h(n) = h(N−1−n)

Proof :

Let, N =7, $\therefore$ h(n) = h(N−1−n) = h(6−n)	Let, N =8, $\therefore$ h(n)= h(N−1−n) = h(7−n)
n = 0, 1, 2, 3, 4, 5, 6	n = 0, 1, 2, 3, 4, 5, 6, 7
When n = 0; h(0) = h(6)	When n = 0; h(0) = h(7)
When n = 1; h(1) = h(5)	When n = 1; h(1) = h(6)
When n = 2; h(2) = h(4)	When n = 2; h(2) = h(5)
When n = 3; h(3) = h(3)	When n = 3; h(3) = h(4)

When the impulse response is symmetric, the samples of impulse response will satisfy the condition,

$$b_n = b_{N-1-n}$$

By using the above symmetry condition it is possible to reduce the number of multipliers required for the realization of FIR system. Hence, the linear phase realization is also called *realization with minimum number of multipliers.*

Consider the transfer function of a FIR system,

$$H(z) = \frac{Y(z)}{X(z)} = b_0 + b_1 z^{-1} + b_2 z^{-2} + ... + b_{N-1} z^{-(N-1)}$$

The linear phase realization of the FIR system using the above equation for even and odd values of N are discussed below:

Case i: **When N is even**

$$H(z) = \frac{Y(z)}{X(z)} = b_0 + b_1 z^{-1} + b_2 z^{-2} + ... + b_{N-1} z^{-(N-1)}$$

$$= \sum_{m=0}^{N-1} b_m z^{-m} = \sum_{m=0}^{\frac{N}{2}-1} b_m z^{-m} + \sum_{m=\frac{N}{2}}^{N-1} b_m z^{-m}$$

> Dividing the summation of N terms into two summation with $\frac{N}{2}$ terms.

> Let, $p = N-1-m$, $\quad \therefore m = N-1-p$
>
> When $m = \dfrac{N}{2}$; $p = N-1-\dfrac{N}{2} = \dfrac{N}{2}-1$
>
> When $m = N-1$; $p = N-1-(N-1) = 0$

$$\therefore \frac{Y(z)}{X(z)} = \sum_{m=0}^{\frac{N}{2}-1} b_m z^{-m} + \sum_{p=0}^{\frac{N}{2}-1} b_{N-1-p} z^{-(N-1-p)}$$

$$= \sum_{m=0}^{\frac{N}{2}-1} b_m z^{-m} + \sum_{m=0}^{\frac{N}{2}-1} b_{N-1-m} z^{-(N-1-m)}$$

> Let, $p = m$

$$= \sum_{m=0}^{\frac{N}{2}-1} b_m z^{-m} + \sum_{m=0}^{\frac{N}{2}-1} b_m z^{-(N-1-m)}$$

> $b_m = b_{N-1-m}$
> (Symmetry condition)

$$= \sum_{m=0}^{\frac{N}{2}-1} b_m [z^{-m} + z^{-(N-1-m)}]$$

$$\therefore Y(z) = b_0 [X(z) + z^{-(N-1)}X(z)] + b_1 [z^{-1}X(z) + z^{-(N-2)}X(z)] + ...$$

$$+ b_{\frac{N}{2}-2} \left[z^{-(\frac{N}{2}-2)}X(z) + z^{-(\frac{N}{2}+1)}X(z)\right] + b_{\frac{N}{2}-1} \left[z^{-(\frac{N}{2}-1)}X(z) + z^{-\frac{N}{2}}X(z)\right]$$

When N is even, the above equation can be used to construct the direct form structure of linear phase FIR system with minimum number of multipliers, as shown in Fig 5.21. From the direct form linear phase structure it is observed that the realization of an Nth order FIR discrete time system for even values of N involves N/2 number of multiplications and N−1 number of additions. Also the structure involves N−1 delays and so N−1 memory locations are required to store the delayed signals.

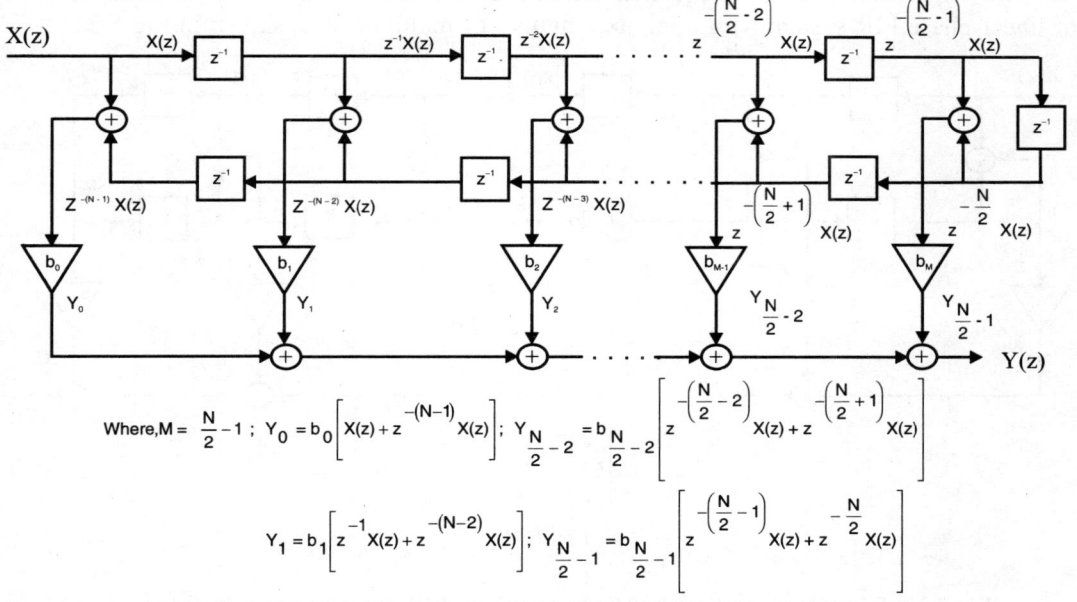

Where, $M = \dfrac{N}{2} - 1$; $Y_0 = b_0\left[X(z) + z^{-(N-1)} X(z)\right]$; $Y_{\frac{N}{2}-2} = b_{\frac{N}{2}-2}\left[z^{-\left(\frac{N}{2}-2\right)}X(z) + z^{-\left(\frac{N}{2}+1\right)}X(z)\right]$

$Y_1 = b_1\left[z^{-1}X(z) + z^{-(N-2)}X(z)\right]$; $Y_{\frac{N}{2}-1} = b_{\frac{N}{2}-1}\left[z^{-\left(\frac{N}{2}-1\right)}X(z) + z^{-\frac{N}{2}}X(z)\right]$

Fig 5.21: *Direct form realization of a linear phase FIR system when N is even.*

Case ii: **When N is odd**

$$H(z) = \frac{Y(z)}{X(z)} = b_0 + b_1 z^{-1} + b_2 z^{-2} + \dots + b_{N-1} z^{-(N-1)} = \sum_{m=0}^{N-1} b_m z^{-m}$$

$$= \sum_{m=0}^{\frac{N-3}{2}} b_m z^{-m} + b_{\frac{N-1}{2}} z^{-\left(\frac{N-1}{2}\right)} + \sum_{m=\frac{N+1}{2}}^{N-1} b_m z^{-m}$$

| Dividing the summation of N terms into two summation with $\frac{N-1}{2}$ terms. |

Let, $p = N-1-m$, $\quad \therefore m = N-1-p$

When, $m = \dfrac{N+1}{2}$; $p = N - 1 - \dfrac{N+1}{2} = \dfrac{N-3}{2}$

When, $m = N-1$; $p = N-1-(N-1) = 0$

$$\therefore \frac{Y(z)}{X(z)} = \sum_{m=0}^{\frac{N-3}{2}} b_m z^{-m} + b_{\frac{N-1}{2}} z^{-\left(\frac{N-1}{2}\right)} + \sum_{p=0}^{\frac{N-3}{2}} b_{N-1-p} z^{-(N-1-p)}$$

$$= \sum_{m=0}^{\frac{N-3}{2}} b_m z^{-m} + b_{\frac{N-1}{2}} z^{-\left(\frac{N-1}{2}\right)} + \sum_{m=0}^{\frac{N-3}{2}} b_{N-1-m} z^{-(N-1-m)} \qquad \boxed{\text{Let, } p = m}$$

$$= \sum_{m=0}^{\frac{N-3}{2}} b_m z^{-m} + b_{\frac{N-1}{2}} z^{-\left(\frac{N-1}{2}\right)} + \sum_{m=0}^{\frac{N-3}{2}} b_m z^{-(N-1-m)} \qquad \boxed{\begin{array}{l} b_m = b_{N-1-m} \\ \text{(Symmetry condition)} \end{array}}$$

$$= b_{\frac{N-1}{2}} z^{-\left(\frac{N-1}{2}\right)} + \sum_{m=0}^{\frac{N-3}{2}} b_m\left[z^{-m} + z^{-(N-1-m)}\right]$$

$$\therefore Y(z) = b_{\frac{N-1}{2}} z^{-\left(\frac{N-1}{2}\right)} X(z) + b_0\left[X(z) + z^{-(N-1)} X(z)\right] + b_1\left[z^{-1}X(z) + z^{-(N-2)} X(z)\right] +$$

$$\dots + b_{\frac{N-5}{2}}\left[z^{-\left(\frac{N-5}{2}\right)} X(z) + z^{-\left(\frac{N+3}{2}\right)} X(z)\right] + b_{\frac{N-3}{2}}\left[z^{-\left(\frac{N-3}{2}\right)} X(z) + z^{-\left(\frac{N+1}{2}\right)} X(z)\right]$$

When N is odd, the above equation can be used to construct the direct form structure of linear phase FIR system with minimum number of multipliers, as shown in Fig 5.22.

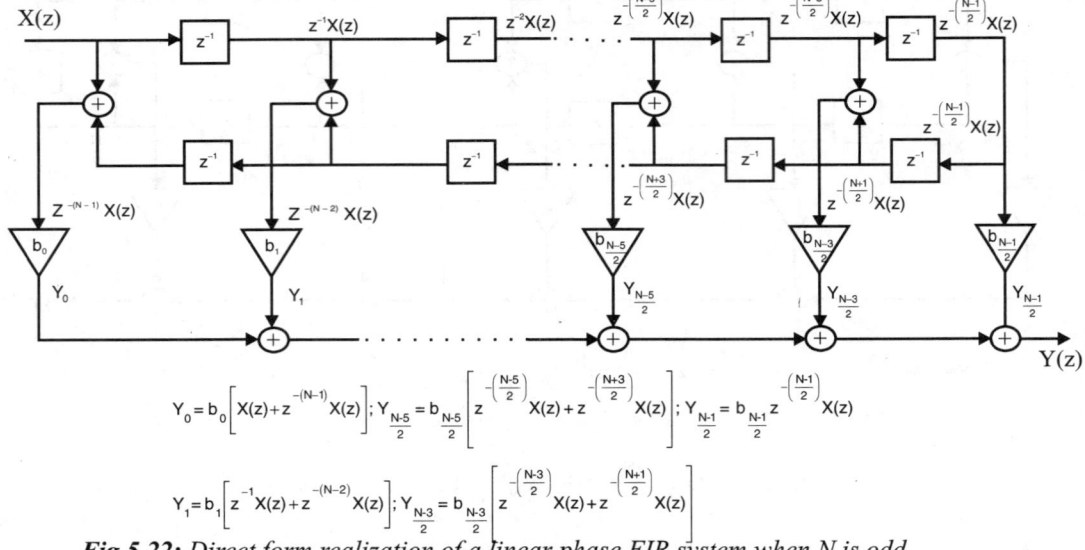

$$Y_0 = b_0 \left[X(z) + z^{-(N-1)} X(z) \right]; \; Y_{\frac{N-5}{2}} = b_{\frac{N-5}{2}} \left[z^{-\left(\frac{N-5}{2}\right)} X(z) + z^{-\left(\frac{N+3}{2}\right)} X(z) \right]; \; Y_{\frac{N-1}{2}} = b_{\frac{N-1}{2}} z^{-\left(\frac{N-1}{2}\right)} X(z)$$

$$Y_1 = b_1 \left[z^{-1} X(z) + z^{-(N-2)} X(z) \right]; \; Y_{\frac{N-3}{2}} = b_{\frac{N-3}{2}} \left[z^{-\left(\frac{N-3}{2}\right)} X(z) + z^{-\left(\frac{N+1}{2}\right)} X(z) \right]$$

Fig 5.22: Direct form realization of a linear phase FIR system when N is odd.

From the direct form linear phase structure it is observed that the realization of an Nth order FIR discrete time system for odd values of N involves (N+1)/2 number of multiplications and N−1 number of additions. Also the structure involves N−1 delays and so N−1 memory locations are required to store the delayed signals.

Example 5.58

Draw the direct form structure of the FIR system described by the transfer function

$$H(z) = 1 + \frac{1}{2} z^{-1} + \frac{3}{4} z^{-2} + \frac{1}{4} z^{-3} + \frac{1}{2} z^{-4} + \frac{1}{8} z^{-5}$$

Solution:

Let, $H(z) = \dfrac{Y(z)}{X(z)} = 1 + \dfrac{1}{2} z^{-1} + \dfrac{3}{4} z^{-2} + \dfrac{1}{4} z^{-3} + \dfrac{1}{2} z^{-4} + \dfrac{1}{8} z^{-5}$

$\therefore Y(z) = X(z) + \dfrac{1}{2} z^{-1} X(z) + \dfrac{3}{4} z^{-2} X(z) + \dfrac{1}{4} z^{-3} X(z) + \dfrac{1}{2} z^{-4} X(z) + \dfrac{1}{8} z^{-5} X(z)$ .....(1)

The direct form structure of FIR system can be obtained directly from equation (1).

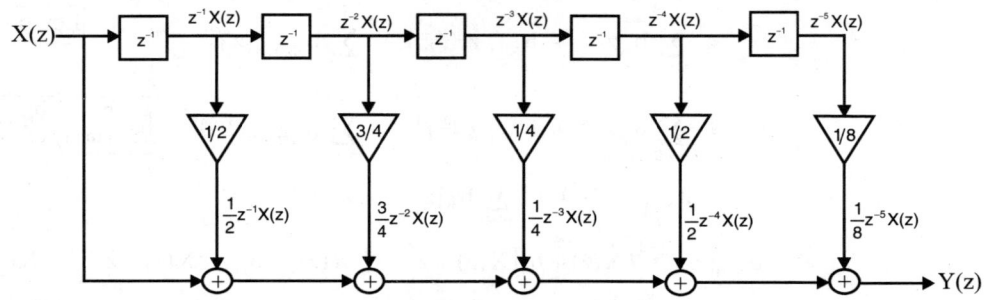

Fig 1: Direct form structure of H(z).

Example 5.59

Realize the following system with minimum number of multipliers.

a) $H(z) = \frac{1}{4} + \frac{1}{2}z^{-1} + \frac{3}{4}z^{-2} + \frac{1}{2}z^{-3} + \frac{1}{4}z^{-4}$

b) $H(z) = 1 + \frac{1}{2}z^{-1} + \frac{1}{2}z^{-2} + z^{-3}$

c) $H(z) = \left(1 + \frac{1}{2}z^{-1} + z^{-2}\right)\left(1 + \frac{1}{4}z^{-1} + z^{-2}\right)$

Solution:

a) Given that, $H(z) = \frac{1}{4} + \frac{1}{2}z^{-1} + \frac{3}{4}z^{-2} + \frac{1}{2}z^{-3} + \frac{1}{4}z^{-4}$ .....(1)

By the definition of z-transform we get,

$$H(z) = \sum_{n=0}^{\alpha} h(n)\,z^{-n} = h(0) + h(1)\,z^{-1} + h(2)\,z^{-2} + h(3)\,z^{-3} + ... \qquad .....(2)$$

On comparing equations (1) and (2) we get,

Impulse response, $h(n) = \left\{\frac{1}{4}, \frac{1}{2}, \frac{3}{4}, \frac{1}{2}, \frac{1}{4}\right\}$

Here h(n) satisfies the condition h(n) = h(N – 1 – n) and so impulse response is symmetrical. Hence the system has linear phase and can be realized with minimum number of multipliers.

Let, $H(z) = \frac{Y(z)}{X(z)} = \frac{1}{4} + \frac{1}{2}z^{-1} + \frac{3}{4}z^{-2} + \frac{1}{2}z^{-3} + \frac{1}{4}z^{-4}$

$\therefore Y(z) = \frac{1}{4}X(z) + \frac{1}{2}z^{-1}X(z) + \frac{3}{4}z^{-2}X(z) + \frac{1}{2}z^{-3}X(z) + \frac{1}{4}z^{-4}X(z)$

$\qquad = \frac{1}{4}[X(z)+z^{-4}X(z)] + \frac{1}{2}[z^{-1}X(z)+z^{-3}X(z)] + \frac{3}{4}z^{-2}X(z)$ .....(3)

The direct form structure of linear phase FIR system is constructed using equation (3) as shown in Fig 1.

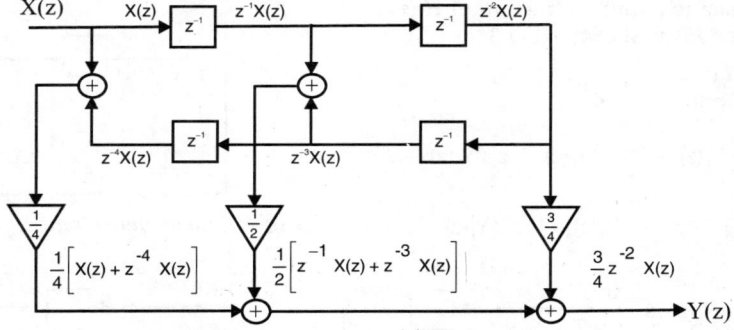

Fig 1: *Linear phase realization of H(z).*

b) Given that, $H(z) = 1 + \frac{1}{2}z^{-1} + \frac{1}{2}z^{-2} + z^{-3}$

Let, $H(z) = \frac{Y(z)}{X(z)} = 1 + \frac{1}{2}z^{-1} + \frac{1}{2}z^{-2} + z^{-3}$

$\therefore Y(z) = X(z) + \frac{1}{2}z^{-1}X(z) + \frac{1}{2}z^{-2}X(z) + z^{-3}X(z)$

$\qquad = [X(z) + z^{-3}X(z)] + \frac{1}{2}[z^{-1}X(z) + z^{-2}X(z)]$ .....(4)

The direct form realization of H(z) with minimum number of multipliers (i.e., linear phase realization) is obtained using equation (4) as shown in Fig 2.

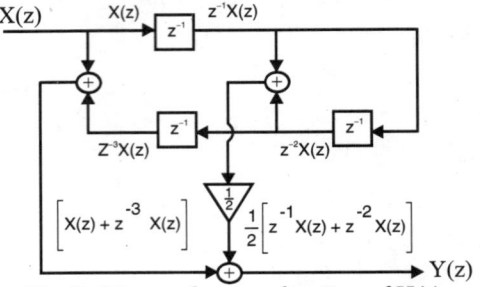

Fig 2: *Linear phase realization of H(z).*

c) Given that, $H(z) = \left(1 + \frac{1}{2}z^{-1} + z^{-2}\right)\left(1 + \frac{1}{4}z^{-1} + z^{-2}\right)$

The given system can be realized as cascade of two second order systems. Each system can be realized with minimum number of multipliers.

Let, $H(z) = H_1(z)\, H_2(z)$

where, $H_1(z) = 1 + \frac{1}{2}z^{-1} + z^{-2}$; $H_2(z) = 1 + \frac{1}{4}z^{-1} + z^{-2}$

Let, $H_1(z) = \dfrac{Y_1(z)}{X(z)} = 1 + \frac{1}{2}z^{-1} + z^{-2}$

$\therefore Y_1(z) = X(z) + \frac{1}{2}z^{-1}X(z) + z^{-2}X(z)$

$\qquad = [X(z) + z^{-2}X(z)] + \frac{1}{2}z^{-1}X(z)$ ......(5)

The linear phase realization structure of $H_1(z)$ is obtained using equation (5) as shown in Fig 3.

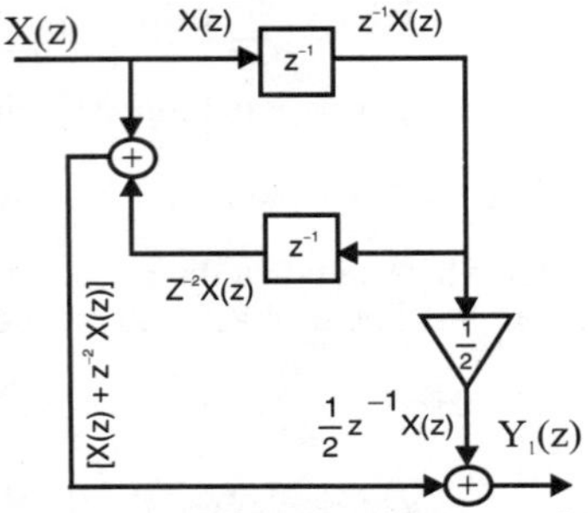

Fig 3: *Linear phase realization of $H_1(z)$.*

Let, $H_2(z) = \dfrac{Y(z)}{Y_1(z)} = 1 + \frac{1}{4}z^{-1} + z^{-2}$

$\therefore Y(z) = Y_1(z) + \frac{1}{4}z^{-1}Y_1(z) + z^{-2}Y_1(z)$

$\qquad = [Y_1(z) + z^{-2}Y_1(z)] + \frac{1}{4}z^{-1}Y_1(z)$ ......(6)

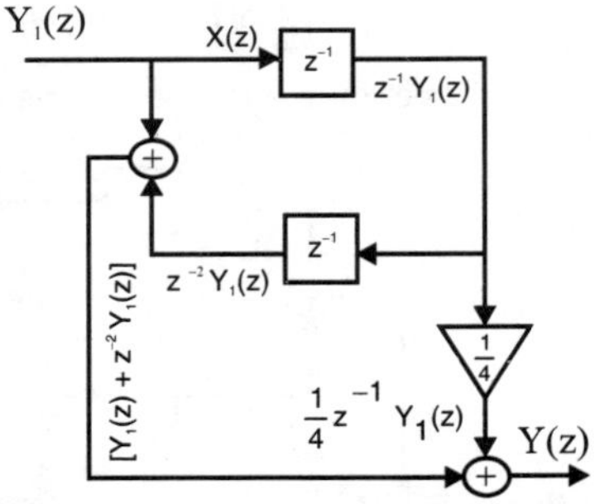

Fig 4: *Linear phase realization of $H_2(z)$.*

The linear phase realization structure of $H_2(z)$ is obtained using equation (6) as shown in Fig 4.

The linear phase structure of H(z) is obtained by connecting the linear phase realization structures of $H_1(z)$ and $H_2(z)$ in cascade as shown in Fig 5.

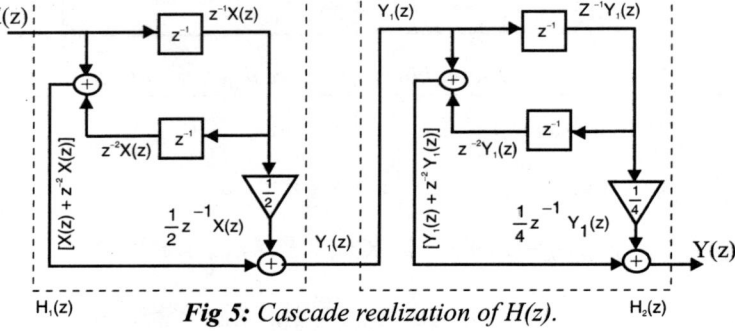

Fig 5: *Cascade realization of H(z).*

5.10 Summary of Important Concepts

1.   The convolution of N_1 and N_2 sample sequences produce a sequence consisting of N_1+N_2-1 samples.

2.   In an LTI system, response for an arbitrary input is given by convolution of input with impulse response.

3.   The ratio of Fourier transform of output and input of an LTI discrete time system is called transfer function of the LTI discrete time system in frequency domain.

4.   The frequency domain transfer function is also given by Fourier transform of impulse response.

5.   The Fourier transform of impulse response is called frequency response of the system.

6.   The frequency response of discrete time system is periodic continuous function of ω with period 2π.

7.   The first order discrete time system behaves as either lowpass filter or highpass filter.

8.   The second order discrete time system behaves as a resonant filter or bandpass filter.

9.   The transfer function of an LTI discrete time system is defined as the ratio of Z-transform of output and Z-transform of input.

10. The transfer function of an LTI discrete time system is also given by Z-transform of the impulse response.

11. The inverse Z-transform of transfer function is the impulse response of the system.

12. The zero-input response $y_{zi}(n)$ is mainly due to initial output (or initial stored energy) in the system.

13. The zero-state response $y_{zs}(n)$ is the response of the system due to input signal and with zero initial output.

14. The total response y(n) is the response of the system due to input signal and initial output (or initial stored energy).

15. The convolution operation is performed to find the response y(n) of an LTI discrete time system from the input x(n) and impulse response h(n).

16. The deconvolution operation is performed to extract the input x(n) of an LTI system from the response y(n) and impulse response h(n) of the system.

17. Mathematically, a discrete time system is represented by a difference equation.

18. Physically, a discrete time system is realized or implemented either as a digital hardware or as a software running on a digital hardware.

19. The processing of the discrete time signal by the digital hardware involves mathematical operations like addition, multiplication, and delay.

20. The time taken to process the discrete time signal and the computational complexity, depends on number of calculations involved and the type of arithmetic used for computation.

21. The various structures proposed for IIR and FIR systems, attempt to reduce the computational complexity, errors in computation and the memory requirement of the system.

22. When a discrete time system is designed by considering all the infinite samples of the impulse response, then the system is called IIR (Infinite Impulse Response) system.

23. When a discrete time system is designed by choosing only finite samples (usually N-samples) of the impulse response, then the system is called FIR (Finite Impulse Response) system.

24. The IIR systems are recursive systems, whereas the FIR systems are nonrecursive systems.

25. The direct form-I structure of IIR system offers a direct relation between time domain and z-domain equations.

26. Since separate delays are employed for input and output samples, realizing IIR system using direct form-I structure require more memory.

27. The direct form-I and II structure realization of an Nth order IIR discrete time system involves M+N+1 number of multiplications and M+N number of additions.

28. The direct form-I structure realization of an Nth order IIR discrete time system involves M+N delays and so M+N memory locations are required to store the delayed signals.

29. In a realizable N^{th} order IIR discrete time system, the direct form-II structure realization involves N delays and so N memory locations are required to store the delayed signals.

30. In canonic structure, the number of delays will be equal to the order of the system.

31. The direct form-II structure of IIR system is canonic whereas the direct form-I structure is noncanonic.

32. In cascade realization of IIR system, the Nth order transfer function is divided into first and second order sections and they are realized in direct form-I or II structure and then connected in cascade.

33. In parallel realization of IIR system, the Nth order transfer function is divided into first and second order sections and they are realized in direct form-I or II structure and then connected in parallel.

34. In cascade and parallel realization of IIR systems, the number of calculations and the memory requirement depends on the realization of individual sections.

35. Direct form structure of FIR system provides a direct relation between time domain and z-domain equations.

36. The realization of an Nth order FIR discrete time system using direct form structure involves N number of multiplications and N−1 number of additions.

37. The realization of an Nth order FIR discrete time system using direct form structure involves N−1 delays and so N−1 memory locations are required to store the delayed signals.

38. The condition for symmetry of impulse response of FIR system is, $h(n) = h(N-1-n)$.

39. The linear phase realization is also called realization with minimum number of multipliers.

40. In cascade realization of FIR system, the Nth order transfer function is divided into first and second order sections and they are realized in direct form or linear phase structure and then connected in cascade.

41. The direct form linear phase realization structure of an Nth order FIR discrete time system for even values of N involves N/2 number of multiplications, and N−1 number of additions.

42. The direct form linear phase realization structure of an Nth order FIR discrete time system for odd values of N involves (N+1)/2 number of multiplications, and N−1 number of additions.

5.11 Short-answer and Questions

Q5.1 *What are the basic elements used to construct the block diagram of discrete time system?*

The basic elements used to construct the block diagram of discrete time system are adder, constant multiplier and unit delay element.

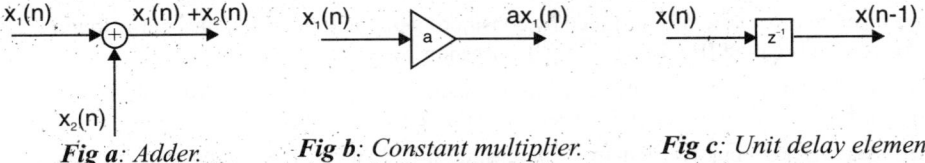

Fig a: Adder.   *Fig b: Constant multiplier.*   *Fig c: Unit delay element.*

Q5.2 *Why linear convolution is important in digital signal processing ?*

The response or output of an LTI discrete time system for any input x(n) is given by linear convolution of the input x(n) and the impulse response h(n) of the system. (This means that if the impulse response of a system is known, then the response of the system for any input can be determined by convolution operation.)

Q5.3 *In y(n) = x(n)∗ h(n), how will you determine the start and end point of y(n)? What will be the length of y(n)?*

Let, length of x(n) be N_1 and starts at n = n_x. Let, length of h(n) be N_2 and starts at n = n_h.

Now, y(n) will start at n = $n_x + n_h$

y(n) will end at n = $(n_x + n_h) + (N_1 + N_2 - 2)$

The length of y(n) is $N_1 + N_2 - 1$.

Q5.4 *Perform convolution of the two sequences, x(n) = {1,1,1,1} and h(n) = {2,2}*

Solution: **(AU Jun'12, 2 Marks) (AU Dec'12 &11, 2 Marks)**

The input sequences x(n) is arranged as a column and the impulse response is arranged as a row as shown below. The elements of the two-dimensional array are obtained by multiplying the corresponding row element with the column element. The sum of the diagonal elements gives the samples of y(n).

The input sequence starts at n = 0 and the impulse response sequences starts at n = 0. Therefore the output sequence starts at n = 0 + 0 = 0.

The input consits of 4 samples and impulse response consists of 2 samples, so the output consists of 4 + 2 − 1 = 5

∴ y(0) = 2   y(3) = 2 + 2 = 4

y(1) = 2 + 2 = 4   y(4) = 2

y(2) = 2 + 2 = 4

∴ y(n) = {2, 4, 4, 4, 2}
                 ↑

Q5.5 *Perform convolution of the two sequences, x(n) = {1,1,3} and h(n) = {1,4,−1}*

Solution: **(AU Dec'15, 2 Marks)**

The input sequences x(n) is arranged as a column and the impulse response is arranged as a row as shown below. The elements of the two-dimensional array are obtained by multiplying the corresponding row element with the column element. The sum of the diagonal elements gives the samples of y(n).

The input sequence starts at n = 0 and the impulse response sequences starts at n = 0. Therefore the output sequence starts at n = 0 + 0 = 0.

The input and impulse response consists of 3 samples, so the output consists of 3 + 3 − 1 = 5

∴ y(0) = 1   y(3) = 12 − 1 = 11

y(1) = 1 + 4 = 5   y(4) = −3

y(2) = 3 + 4 − 1 = 6

∴ y(n) = {1, 5, 6, 11, −3}
                    ↑

Q5.6 *Perform the following convolutions,*   **(AU Jun'11, 2 Marks)**

   *a) x(n) * δ(n) b) δ(n) * [h₁(n) + h₂(n)]*

<u>Solution:</u>

a) $x(n) * \delta(n) = \sum\limits_{m=-\infty}^{+\infty} x(n)\,\delta(n-m)$ $\boxed{\begin{array}{l}\delta(n-m) = 1\ ;\ m = n \\ \qquad\qquad = 0\ ;\ m \neq n\end{array}}$

   $= x(m)\,\delta(n-m)\big|_{m=n} = x(n)$

b) $\delta(n) * [h_1(n) + h_2(n)] = \delta(n) * h_1(n) + \delta(n) * h_2(n)$ $\boxed{\text{Using solution of (a)}}$

   $= h_1(n) + h_2(n)$

Q5.7 *A causal discrete time LTI system has a system function* $H(z) = \dfrac{1 - 2az^{-1}}{2b + z^{-1}}$. *Here 'a' is real and*

$|a| < 1$. Find the value of 'b' so that the frequency response $H(e^{j\omega})$ of the system satisfies the condition $|H(e^{j\omega})| = 1$ for all ω.

<u>Solution:</u>

The frequency response of the system can be obtained by putting, $z = e^{j\omega}$ in H(z).

   $\therefore H(e^{j\omega}) = H(z)\big|_{z=e^{j\omega}} = \dfrac{1 - 2ae^{-j\omega}}{2b + e^{-j\omega}}$

Here, $|H(e^{j\omega})| = 1$; $\therefore \left|\dfrac{1 - 2ae^{-j\omega}}{2b + e^{-j\omega}}\right| = 1 \Rightarrow |1 - 2ae^{-j\omega}| = |2b + e^{-j\omega}|$ $\boxed{e^{\pm j\theta} = \cos\theta \pm j\sin\theta}$

   $\therefore |1 - 2a\cos\omega + j2a\sin\omega| = |2b + \cos\omega - j\sin\omega|$

   $(1 - 2a\cos\omega)^2 + (2a\sin\omega)^2 = (2b + \cos\omega)^2 + (\sin\omega)^2$ $\boxed{\sin^2\theta + \cos^2\theta = 1}$

   $1 + 4a^2\cos^2\omega - 4a\cos\omega + 4a^2\sin^2\omega = 4b^2 + 4b\cos\omega + \cos^2\omega + \sin^2\omega$

   $1 + 4a^2 - 4a\cos\omega = 4b^2 + 4b\cos\omega + 1$

   The above equation is true, when b = – a .

   Hence to satisfy the condition $|H(e^{j\omega})| = 1$ for all ω, b = – a .

Q5.8 *Determine the transfer function of the LTI system defined by the equation,*

   y(n) – 0.5 y(n– 1) = x(n) + 0.4 x(n– 1)   **(AU Jun'12, 2 Marks)**

<u>Solution:</u>

Given that,   y(n) – 0.5 y(n – 1) = x(n) + 0.4 x(n – 1)

On taking z-transform we get,

   $Y(z) - 0.5\,z^{-1}Y(z) = X(z) + 0.4\,z^{-1}X(z)$ $\Rightarrow$ $Y(z)\,[1 - 0.5\,z^{-1}] = X(z)\,[1 + 0.4\,z^{-1}]$

$\therefore$ Transfer function, $\dfrac{Y(z)}{X(z)} = \dfrac{1 + 0.4\,z^{-1}}{1 - 0.5\,z^{-1}}$

Q5.9 *Find the value of y(n) = x(n–1) * δ(n–2)*

<u>Solution:</u>   **(AU Dec'14, 2 Marks)**

   $y(n) = x(n-1) * \delta(n-2) = \sum\limits_{m=-\infty}^{+\infty} x(m-1)\,\delta(n-m+2)$ $\boxed{\begin{array}{l}\delta(n-m+2) = 1\ ;\ m = n+2 \\ \qquad\qquad\qquad = 0\ ;\ m \neq n+2\end{array}}$

   $= x(m-1)\,\delta(n-m+2)\big|_{m=n+2} = x(n+2-1) = x(n+1)$

Q5.10 *An LTI system is governed by equation, y(n) = –2 y(n – 2) – 0.5 y(n – 1) + 3 x(n – 1) + 5 x(n). Determine the transfer function of the system.*

Solution:

Given that, $y(n) = -2\,y(n-2) - 0.5\,y(n-1) + 3\,x(n-1) + 5\,x(n)$

On taking z-transform of above equation we get,

$Y(z) + 2\,z^{-2}Y(z) + 0.5\,z^{-1}Y(z) = 3\,z^{-1}X(z) + 5\,X(z)$

$Y(z)\,[\,1 + 2\,z^{-2} + 0.5\,z^{-1}\,] = [\,3\,z^{-1} + 5\,]\,X(z)$

$\therefore$ Transfer function, $H(z) = \dfrac{Y(z)}{X(z)} = \dfrac{3z^{-1}+5}{1+2z^{-2}+0.5z^{-1}} = \dfrac{3z^{-1}+5}{z^{-2}(z^{2}+2+0.5z)} = \dfrac{5z^{2}+3z}{z^{2}+0.5z+2}$

Q5.11 *Determine the impulse response for the cascaded two LTI systems having impulse responses $h_1(n) = u(n)$ and $h_2(n) = \delta(n) + 2\delta(n-1)$.*   **(AU Jun'14, 2 Marks)**

Solution:

Let h(n) be the impulse response of the cascade system. Now h(n) is given by convolution of $h_1(n)$ and $h_2(n)$.

$\therefore$ $h(n) = h_1(n) * h_2(n)$

$\qquad = u(n) * \{\,\delta(n) + 2\,\delta(n-1)\,\}$

$\qquad = u(n) * \delta(n) + 2\,u(n) * \delta(n-1)$

$\qquad = u(n) + 2\,u(n-1)$

$x(n) * \delta(n)\quad = x(n)$
$x(n) * \delta(n-1) = x(n-1)$

Q5.12 *What will be the overall impulse response h(n) if the two system having their impulse responses as $h_1(n)$ and $h_2(n)$ are connected in series and parallel?*   **(AU Dec'11, 2 Marks)**

Solution:

Let, h(n) = Overall impulse response.

For parallel combination, $h(n) = h_1(n) + h_2(n)$

For series combination, $h(n) = h_1(n) * h_2(n)$.

Q5.13 *The transfer function of an LTI system is $H(z) = \dfrac{z-1}{(z-2)(z+3)}$. Determine the impulse response.*

Solution:

$H(z) = \dfrac{z-1}{(z-2)(z+3)} = \dfrac{A}{z-2} + \dfrac{B}{z+3}$

$A = \dfrac{z-1}{(z-2)\,(z+3)} \times (z-2)\Big|_{z=2} = \dfrac{z-1}{z-2}\Big|_{z=2} = \dfrac{2-1}{2+3} = \dfrac{1}{5}$

$B = \dfrac{z-1}{(z-2)\,(z+3)} \times (z+3)\Big|_{z=-3} = \dfrac{z-1}{z-2}\Big|_{z=-3} = \dfrac{-3-1}{-3-2} = \dfrac{-4}{-5} = \dfrac{4}{5}$

$z\{a^{n}u(n)\} = \dfrac{z}{z-a}$

$z\{a^{(n-1)}u(n-1)\} = z^{-1}\dfrac{z}{z-a}$

$\therefore H(z) = \dfrac{1}{5}\dfrac{1}{z-2} + \dfrac{4}{5}\dfrac{1}{z+3}$

Impulse response, $h(n) = z^{-1}\{H(z)\} = z^{-1}\left\{\dfrac{1}{5}\dfrac{1}{z-2} + \dfrac{4}{5}\dfrac{1}{z+3}\right\}$

$\qquad\qquad = z^{-1}\left\{\dfrac{1}{5}z^{-1}\dfrac{z}{z-2} + \dfrac{4}{5}z^{-1}\dfrac{z}{z-(-3)}\right\}$

$\qquad\qquad = \dfrac{1}{5}2^{(n-1)}u(n-1) + \dfrac{4}{5}(-3)^{(n-1)}u(n-1)$

$\qquad\qquad = \dfrac{1}{5}\big[2^{(n-1)} + 4(-3)^{(n-1)}\big]u(n-1)$

Q5.14 *Determine the response of LTI system governed by the equation, $y(n) - 0.5\ y(n - 1) = x(n)$, for input $x(n) = 5^n\ u(n)$, and initial condition $y(-1) = 2$.*

Solution:

Given that, $x(n) = 5^n\ u(n)$; $X(z) = Z\{u(n)\} = \dfrac{z}{z-5}$

Given that, $y(n) - 0.5\ y(n - 1) = x(n)$,

On taking Z-transform of above equation we get,

$$Y(z) - 0.5\,[\,z^{-1}Y(z) + y(-1)\,] = X(z)$$

$$Y(z) - 0.5[z^{-1}Y(z) + 2] = \dfrac{z}{z-5}$$

$$Y(z) - 0.5z^{-1}Y(z) - 1 = \dfrac{z}{z-5} \quad \Rightarrow \quad Y(z)\left[1 - \dfrac{0.5}{z}\right] = \dfrac{z}{z-5} + 1 \quad \Rightarrow \quad Y(z)\left[\dfrac{z - 0.5}{z}\right] = \dfrac{z + z - 5}{z-5}$$

$$\therefore Y(z) = \dfrac{z(2z - 5)}{(z - 0.5)(z - 5)} \quad \Rightarrow \quad \dfrac{Y(z)}{z} = \dfrac{2z - 5}{(z - 0.5)(z - 5)}$$

Let, $\dfrac{Y(z)}{z} = \dfrac{2z - 5}{(z - 0.5)(z - 5)} = \dfrac{A}{z - 0.5} + \dfrac{B}{z - 5}$

$A = \dfrac{2z - 5}{(z - 0.5)(z - 5)} \times (z - 0.5)\Big|_{z = 0.5} = \dfrac{2 \times 0.5 - 5}{0.5 - 5} = \dfrac{-4}{-4.5} = \dfrac{40}{45} = \dfrac{8}{9}$

$B = \dfrac{2z - 5}{(z - 0.5)(z - 5)} \times (z - 5)\Big|_{z = 5} = \dfrac{2 \times 5 - 5}{5 - 0.5} = \dfrac{5}{4.5} = \dfrac{50}{45} = \dfrac{10}{9}$

$\therefore \dfrac{Y(z)}{z} = \dfrac{8}{9}\dfrac{1}{z - 0.5} + \dfrac{10}{9}\dfrac{1}{z - 5} \quad \Rightarrow \quad Y(z) = \dfrac{8}{9}\dfrac{z}{z - 0.5} + \dfrac{10}{9}\dfrac{z}{z - 5}$ $\boxed{Z\{a^n u(n)\} = \dfrac{z}{z - a}}$

$\therefore$ Re sponse, $y(n) = Z^{-1}\{Y(z)\} = Z^{-1}\left\{\dfrac{8}{9}\dfrac{z}{z - 0.5} + \dfrac{10}{9}\dfrac{z}{z - 5}\right\}$

$$= \dfrac{8}{9}\,0.5^n\,u(n) + \dfrac{10}{9}\,5^n\,u(n) = \left[\dfrac{8}{9}\,0.5^n + \dfrac{10}{9}\,5^n\right]u(n)$$

Q5.15 *A signal $x(t) = a^t$ is sampled at a frequency of $1/T$ Hz in the range $-\infty < t < 0$. Find the Z-transform of the sampled version of the signal.*

Solution:

Given that, $x(t) = a^t$; $-\infty < t < 0$

The sampled version of the signal $x(nT)$ is given by, $x(nT) = a^{nT}$; $-\infty < nT < 0$

Now the Z - transform of $x(nT)$ is,

$$Z\{x(nT)\} = \sum_{n = -\infty}^{+\infty} x(nT)z^{-n} = \sum_{n = -\infty}^{0} a^{nT}z^{-n} = \sum_{n = 0}^{\infty} a^{-nT}z^{n}$$

$$= \sum_{n = 0}^{\infty} (a^{-T}z)^{n} = \dfrac{1}{1 - a^{-T}z} = \dfrac{1}{1 - z/a^{T}} = \dfrac{a^{T}}{a^{T} - z}$$

Q5.16 *The transfer function of a system is given by, $H(z) = \dfrac{1}{1 - 0.5z^{-1}} + \dfrac{1}{1 - 2z^{-1}}$. Determine the stability and causality of the system for a) ROC : $|z| > 2$;   b) ROC : $|z| < 0.5$.*

Solution: **(AU Dec'13, 2 Marks)**

a) ROC is $|z| > 2$

When ROC is $|z| > 2$, the impulse response $h(n)$ should be right-sided signal.

$\therefore$ Impulse response, $h(n) = Z^{-1}\{H(z)\} = Z^{-1}\left\{\dfrac{1}{1-0.5z^{-1}} + \dfrac{1}{1-2z^{-1}}\right\} = (0.5^n + 2^n)u(n)$

1. The ROC does not include unit circle. Hence the system is unstable.
2. The impulse response is right-sided signal. Hence the system is causal.

<u>b) ROC is |z| < 0.5</u>

When ROC is |z| < 0.5, the impulse response h(n) should be left-sided signal.

$\therefore$ Impulse response, $h(n) = Z^{-1}\{H(z)\} = Z^{-1}\left\{\dfrac{1}{1-0.5z^{-1}} + \dfrac{1}{1-2z^{-1}}\right\} = (-0.5^n - 2^n)u(n-1)$

1. The ROC does not include unit circle. Hence the system is unstable.
2. The impulse response is left-sided sequence. Hence the system is anticausal.

Q5.17 Determine the stability of the LTI system with impulse response h(n) = sin πn
Solution:

Given that, h(n) = sin πn. **(AU Dec'14, 2 Marks)**

The condition for the stability of a system is $\displaystyle\sum_{n=-\infty}^{+\infty} |h(n)| < \infty$

$\displaystyle\sum_{n=-\infty}^{+\infty} |h(n)| = \sum_{n=-\infty}^{+\infty} |\sin \pi n|$

Sin θ lies between −1 and +1 for all θ.The output is bounded for any value of input and therfore the given system is stable.

Q5.18 Determine the stability of the LTI system described by the transfer function,

$H(z) = \dfrac{z}{z-\frac{1}{2}} + \dfrac{2z}{z-3}$ *for ROC ;* $\dfrac{1}{2} < |z| < 3$ **(AU Jun'14, 2 Marks)**

Solution:

Given that, ROC is $\dfrac{1}{2} < |z| < 3$

When ROC is $\dfrac{1}{2} < |z| < 3$, the impulse response h(n) is two sided signal. Since |z| > 1/2, the term with pole z = 1/2 corresponds to right sided signal.Since |z| <3, the term with pole z = 3 corresopnds to left sided signal.

The impulse response, $h(n) = Z^{-1}\{H(z)\} = Z^{-1}\left\{\dfrac{z}{z-\frac{1}{2}} + \dfrac{2z}{z-3}\right\} = \left(\dfrac{1}{2}\right)^n u(n) - 2(3)^n u(-n-1)$

Here the ROC includes the unit circle but one of the pole lie outside the unit circle and hence the system is unstable.

Q5.19 Determine the stability and causality of the system described by the transfer function,

$H(z) = \dfrac{1+z^{-1}}{(1-0.5z^{-1})(1+0.25z^{-1})}$ **(AU Dec'12, 2 Marks)**

Solution:

Given that, $H(z) = \dfrac{1+z^{-1}}{(1-0.5z^{-1})(1+0.25z^{-1})} = \dfrac{1+z^{-1}}{z^{-1}(z-0.5)z^{-1}(z+0.25)} = \dfrac{z^2(1+z^{-1})}{(z-0.5)(z+0.25)}$

$= \dfrac{z^2+z}{(z-0.5)(z+0.25)}$

The poles of H(z) lie at z = 0.5 and z = −0.25

When ROC is |z| > 0.5, the impulse response h(n) is right-sided signal.

1. The ROC includesthe unit circles. Hence the system is stable.
2. The impulse response is right-sided signal. Hence the system is causal

Q5.20 *Determine the stability and causality of the system described by the transfer function,*

$$H(z) = \frac{1}{1 - 0.25z^{-1}} + \frac{1}{1 - 2z^{-1}} \text{ for ROC : } 0.25 < |z| < 2.$$

Solution:

Given that, ROC is $0.25 < |z| < 2$

When ROC is $0.25 < |z| < 2$, the impulse response h(n) is two-sided signal. Since $|z| > 0.25$,

the term with pole $z = 0.25$ corresponds to right-sided signal. Since $|z| < 2$, the term with pole $z = 2$ corresponds to left-sided signal.

$\therefore$ Impulse response, $h(n) = Z^{-1}\{H(z)\} = Z^{-1}\left\{\dfrac{1}{1 - 0.5z^{-1}} + \dfrac{1}{1 - 2z^{-1}}\right\} = 0.5^n u(n) - 2^n u(-n-1)$

1. The ROC includes the unit circle but one of the pole lie outside the unit circle. Hence the system is unstable.

2. The impulse response is two-sided noncausal signal. Hence the system is noncausal.

Q5.21 *Using Z-transform, determine the response of the LTI system with impulse response,*

$$h(n) = \{1, -1, 1\}, \text{ for an input } x(n) = \{-2, 3, 1\}.$$

Solution:

Given that, $x(n) = \{-2, 3, 1\}$

$\therefore X(z) = Z\{x(n)\} = \displaystyle\sum_{n=-\infty}^{+\infty} x(n)z^{-n} = \displaystyle\sum_{n=0}^{2} x(n)z^{-n} = x(0) + x(1)z^{-1} + x(2)z^{-2} = -2 + 3z^{-1} + z^{-2}$

Given that, $h(n) = \{1, -1, 1\}$

$\therefore H(z) = Z\{h(n)\} = \displaystyle\sum_{n=-\infty}^{+\infty} h(n)z^{-n} = \displaystyle\sum_{n=0}^{2} h(n)z^{-n} = h(0) + h(1)z^{-1} + h(2)z^{-2} = 1 - z^{-1} + z^{-2}$

We know that, $H(z) = \dfrac{Y(z)}{X(z)}$

$\therefore Y(z) = X(z)\, H(z) = (-2 + 3z^{-1} + z^{-2})(1 - z^{-1} + z^{-2})$

$= -2 + 2z^{-1} - 2z^{-2} + 3z^{-1} - 3z^{-2} + 3z^{-3} + z^{-2} - z^{-3} + z^{-4}$

$= -2 + 5z^{-1} - 4z^{-2} + 2z^{-3} + z^{-4}$ .....(1)

By the definition of Z-transform,

$\therefore Y(z) = Z\{y(n)\} = \displaystyle\sum_{n=-\infty}^{+\infty} y(n)z^{-n}$

On expanding the above summation we get,

$Y(z) = \,..... + y(0) + y(1)z^{-1} + y(2)z^{-2} + y(3)z^{-3} + y(4)z^{-4} + \,.....$ .....(2)

On comparing equations (1) and (2) we get,

$y(0) = -2$; $y(1) = 5$; $y(2) = -4$; $y(3) = 2$; $y(4) = 1$

$\therefore$ Response, $y(n) = \{-2, 5, -4, 2, 1\}$

Q5.22 *Using Z-transform, perform deconvolution of response y(n) = {−2, 5, −4, 2, 1} and impulse response h(n) = { 1, −1, 1 }, to extract the input x(n).*

Solution:

Given that, $y(n) = \{ -2, 5, -4, 2, 1 \}$

$$Y(z) = Z\{y(n)\} = \sum_{n=-\infty}^{+\infty} y(n)z^{-n} = \sum_{n=0}^{4} y(n)z^{-n}$$

$= y(0) + y(1) z^{-1} + y(2) z^{-2} + y(3) z^{-3} + y(4) z^{-4} = -2 + 5 z^{-1} - 4 z^{-2} + 2 z^{-3} + z^{-4}$

Given that, $h(n) = \{ 1, -1, 1 \}$

$$H(z) = Z\{h(n)\} = \sum_{n=-\infty}^{+\infty} h(n)z^{-n} = \sum_{n=0}^{2} h(n)z^{-n} = h(0) + h(1)z^{-1} + h(2)z^{-2} = 1 - z^{-1} + z^{-2}$$

We know that, $H(z) = \dfrac{Y(z)}{x(z)}$

$\therefore X(z) = \dfrac{Y(z)}{H(z)} = \dfrac{-2 + 5z^{-1} - 4z^{-2} + 2z^{-3} + z^{-4}}{1 - z^{-1} + z^{-2}}$

$= -2 + 3z^{-1} + z^{-2}$ .....(1)

By the definition of Z-transform,

$$X(z) = Z\{x(n)\} = \sum_{n=-\infty}^{+\infty} x(n)z^{-n}$$

On expanding the above summation we get,

$X(z) = ... + x(0) + x(1) z^{-1} + x(2) z^{-2} .....$ .....(2)

On comparing equation (1) and (2) we get,

$x(0) = -2$; $x(1) = 3$; $x(2) = 1$

$\therefore$ Input, $x(n) = \{ -2, 3, 1 \}$

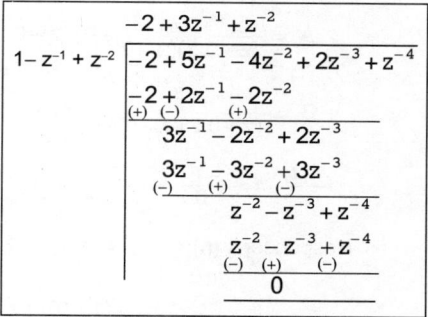

Q5.23 *In an LTI system the impulse response, h(n) = C ^{n} for n ≤ 0. Determine the range of values of C, for which the system is stable.*

Solution:

Given that, $h(n) = C^n$ for $n \le 0$.

$\therefore \displaystyle\sum_{n=-\infty}^{+\infty} h(n) = \sum_{n=-\infty}^{0} C^n = \sum_{n=0}^{+\infty} C^{-n} = \sum_{n=0}^{+\infty} (C^{-1})^n$

If, $0 < |C^{-1}| < 1$, then $\displaystyle\sum_{n=0}^{+\infty} (C^{-1})^n = \dfrac{1}{1 - C^{-1}}$

If, $|C^{-1}| > 1$, then $\displaystyle\sum_{n=0}^{+\infty} (C^{-1})^n = \infty$

$\therefore$ For stability, $|C^{-1}| < 1 \Rightarrow \dfrac{1}{|C|} < 1 \Rightarrow |C| > 1$

Q5.24 *Using Z-transform, determine the response of the LTI system with impulse response $h(n) = 0.4^n u(n)$ for an input $x(n) = 0.2^n u(n)$.*

Solution:

Given that, $x(n) = 0.2^n u(n)$.

$$\therefore X(z) = Z\{x(n)\} = Z\{0.2^n u(n)\} = \frac{z}{z-0.2}$$

Given that, $h(n) = 0.4^n u(n)$.

$$\therefore H(z) = Z\{h(n)\} = Z\{0.4^n u(n)\} = \frac{z}{z-0.4}$$

We know that, $H(z) = \dfrac{Y(z)}{x(z)}$

$$\therefore Y(z) = X(z)H(z) = \frac{z}{z-0.2} \times \frac{z}{z-0.4} = \frac{z^2}{(z-0.2)(z-0.4)}$$

Let, $\dfrac{Y(z)}{z} = \dfrac{z}{(z-0.2)(z-0.4)} = \dfrac{A}{z-0.2} + \dfrac{B}{z-0.4}$

$$A = \frac{z}{(z-0.2)(z-0.4)} \times (z-0.2)\big|_{z=0.2} = \frac{0.2}{0.2-0.4} = \frac{0.2}{-0.2} = -1$$

$$B = \frac{z}{(z-0.2)(z-0.4)} \times (z-0.4)\big|_{z=0.4} = \frac{0.4}{0.4-0.2} = \frac{0.4}{0.2} = 2$$

$$\therefore \frac{Y(z)}{z} = \frac{-1}{z-0.2} + \frac{2}{z-0.4} \quad \Rightarrow \quad Y(z) = -\frac{z}{z-0.2} + 2\frac{z}{z-0.4}$$

Response, $y(n) = Z^{-1}\{Y(z)\} = Z^{-1}\left\{-\dfrac{z}{z-0.2} + 2\dfrac{z}{z-0.4}\right\}$

$$= -(0.2)^n u(n) + 2(0.4)^n u(n) = [2(0.4)^n - (0.2)^n]u(n)$$

Q5.25 *Using Z-transform perform deconvolution of response, $y(n) = 2 (0.4)^n u(n) - (0.2)^n u(n)$ and impulse response, $h(n) = 0.4^n u(n)$, to extract the input $x(n)$.*

Solution:

Given that, $y(n) = 2 (0.4)^n u(n) - (0.2)^n u(n)$

$$\therefore Y(z) = Z\{y(n)\} = Z\{2(0.4)^n u(n) - (0.2)^n u(n)\}$$

$$= \frac{2z}{z-0.4} - \frac{z}{z-0.2} = \frac{2z(z-0.2)-z(z-0.4)}{(z-0.4)(z-0.2)} = \frac{2z^2 - 0.4z - z^2 + 0.4z}{(z-0.4)(z-0.2)} = \frac{z^2}{(z-0.4)(z-0.2)}$$

Given that, $h(n) = 0.4^n u(n)$

$$\therefore H(z) = Z\{h(n)\} = Z\{0.4^n u(n)\} = \frac{z}{z-0.4}$$

We know that, $H(z) = \dfrac{Y(z)}{x(z)}$

$$\therefore X(z) = \frac{Y(z)}{H(z)} = Y(z) \times \frac{1}{H(z)} = \frac{z^2}{(z-0.4)(z-0.2)} \times \frac{z-0.4}{z} = \frac{z}{z-0.2}$$

$$\therefore \text{Input, } x(n) = Z^{-1}\{X(z)\} = Z^{-1}\left\{\frac{z}{z-0.2}\right\} = 0.2^n u(n)$$

Q5.26 *Obtain the transfer function for the structure shown in Fig Q5.26.*

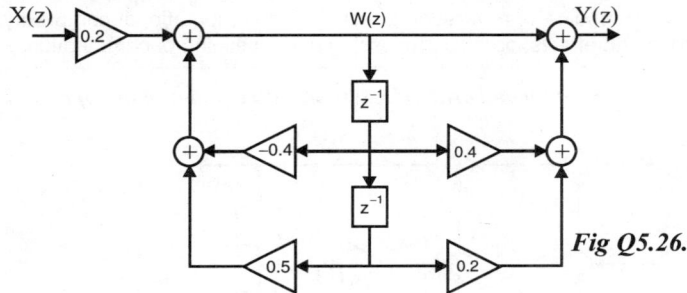

Fig Q5.26.

Solution:

The following z-domain equations can be obtained from the given direct form-II structure.

$$W(z) = -0.4 \, z^{-1} W(z) + 0.5 \, z^{-2} W(z) + 0.2 \, X(z)$$

$$\therefore W(z) + 0.4 \, z^{-1} W(z) - 0.5 \, z^{-2} W(z) = 0.2 \, X(z) \quad \Rightarrow \quad \frac{W(z)}{X(z)} = \frac{0.2}{1 + 0.4z^{-1} - 0.5z^{-2}}$$

$$Y(z) = W(z) + 0.4 \, z^{-1} W(z) + 0.2 \, z^{-2} W(z) \quad \Rightarrow \quad \frac{Y(z)}{W(z)} = 1 + 0.4z^{-1} + 0.2z^{-2}$$

The given direct form-II digital network can be realized by the transfer function,

$$\frac{Y(z)}{X(z)} = \frac{W(z)}{X(z)} \times \frac{Y(z)}{W(z)} = \frac{0.2(1 + 0.4z^{-1} + 0.2z^{-2})}{1 + 0.4z^{-1} - 0.5z^{-2}}$$

Q5.27 *Realize the following FIR system with minimum number of multipliers.*

$$h(n) = \{-0.5, \ 0.8, \ -0.5\}$$

Solution:

Given that, $h(n) = \{-0.5, \ 0.8, \ -0.5\}$

On taking z- transform,

$$H(z) = \sum_{n=0}^{\infty} h(n) z^{-n} = \sum_{n=0}^{2} h(n) z^{-n}$$

$$= h(0) + h(1) z^{-1} + h(2) z^{-2}$$

$$= -0.5 + 0.8 z^{-1} - 0.5 z^{-2}$$

Let, $H(z) = \dfrac{Y(z)}{X(z)} = -0.5 + 0.8 z^{-1} - 0.5 z^{-2}$

$$\therefore Y(z) = -0.5 \, X(z) + 0.8 \, z^{-1} X(z) - 0.5 \, z^{-2} X(z)$$

$$= -0.5 \left[X(z) + z^{-2} X(z) \right] + 0.8 \, z^{-1} X(z)$$

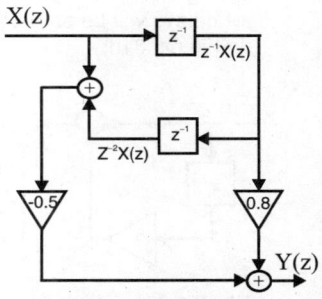

Fig Q5.27: Linear phase realization.

The linear phase structure which has minimum number of multipliers is drawn using the above equation as shown in Fig Q5.27.

Q5.28 *The transfer function of an IIR system has 'Z' number of zeros and 'P' number of poles. How many number of additions, multiplications and memory locations are required to realize the system in direct form-I and direct form-II.*

The realization of IIR system with Z zeros and P poles in direct form-I and II structure, involves $Z+P$ number of additions and $Z+P+1$ number of multiplications. The direct form-I structure requires $Z+P$ memory locations whereas the direct form-II structure requires only P number of memory locations.

Q5.29 *What are the factors that influence the choice of structure for realization of an LTI system?*

The factors that influence the choice of realization structure are computational complexity, memory requirements, finite word length effects, parallel processing and pipelining of computations.

Q5.30 *Draw the direct form-I structure of second order IIR system with equal number of poles and zeros.*

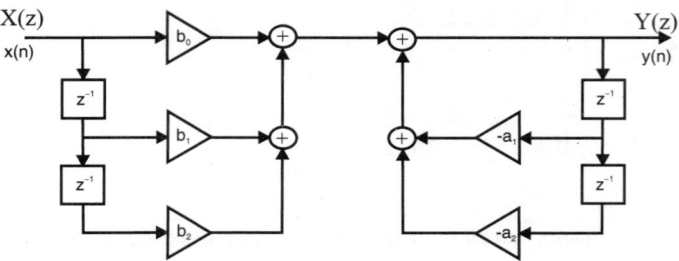

Fig Q5.30: *Direct form-I structure of second order IIR system.*

Q5.31 *An LTI system is described by the difference equation, $y(n) = a_1 y(n-1) + x(n) + b_1 x(n-1)$. Realize it in direct form-I structure and convert to direct form-II structure.*

Solution:

Given that, $y(n) = a_1 y(n-1) + x(n) + b_1 x(n-1)$. Using the given equation the direct form-I structure is drawn as shown in Fig Q5.31a.

Direct form-I structure can be considered as cascade of two systems H_1 and H_2 as shown in Fig Q5.31b. By linearity property, order of cascading can be changed as shown in Fig Q5.31c.

In Fig Q5.31c, we can observe that the input to the delay in H_1 and H_2 are same and so the output of delays will be same. Hence the delays can be combined to get direct form-II structure as shown in Fig Q5.31d.

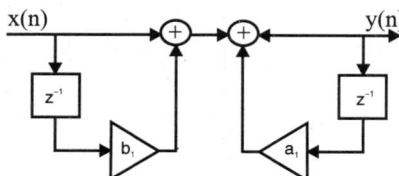

Fig Q5.31a: *Direct form-I structure.*

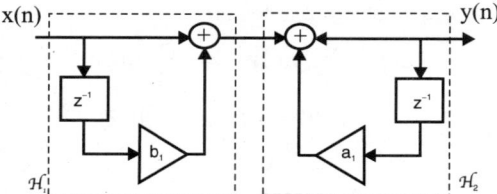

Fig Q5.31b: *Direct form-I structure as cascade of two systems.*

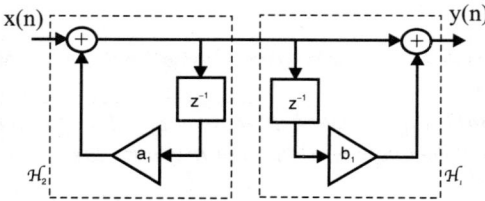

Fig Q5.31c: *Direct form-I structure after interchanging the order of cascading.*

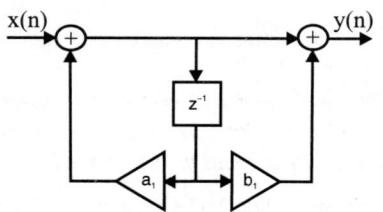

Fig Q5.31d: *Direct form-II structure.*

Q5.32 What is the advantage in cascade and parallel realization of IIR systems ?

In digital implementation of LTI system the coefficients of the difference equation governing the system are quantized. While quantizing the coefficients the value of poles may change. This will end up in a frequency response different to that of desired frequency response.

These effects can be avoided or minimized, if the LTI system is realized in cascade or parallel structure. [i.e, The sensitivity of frequency response characteristics to quantization of the coefficients is minimized].

Q5.33 Compare the direct form-I and II structures of an IIR systems, with M zeros and N poles.

Direct form-I	Direct form-II
1. Separate delay for input and output.	1. Same delay for input and output.
2. M + N + 1 multiplications are involved.	2. M + N + 1 multiplications are involved.
3. M + N additions are involved.	3. M + N additions are involved.
4. M + N delays are involved.	4. N delays are involved.
5. M + N memory locations are required.	5. N memory locations are required.
6. Noncanonical structure.	6. Canonical structure.

Q5.34 Compare the direct form and linear phase structures of an Nth order FIR system.

Direct form	Linear phase
1. Impulse response need not be symmetric.	1. Impulse response should be symmetric.
2. N multiplications are involved.	2. N/2 or (N+1)/2 multiplications are involved.
3. N−1 additions and delays are involved.	3. N−1 additions and delays are involved.
4. N−1 memory locations are required.	4. N−1 memory locations are required.

Q5.35 What is the advantage in linear phase realization of FIR systems ?

The advantage in the linear phase realization structure is that it involves minimum number of multiplications. In linear phase realization of Nth order FIR system, the number of multiplications for even values of N will be N/2 and for odd values of N will be (N+1)/2, whereas the direct form realization involves N multiplications.

5.12 MATLAB Programs

Program 5.1

Write a MATLAB program to perform convolution of the following two discrete time signals.

x1(n)=1; 1<n<10   x2(n)=1;   2<n<10

```
%*******************Program to perform convolution of two signals
%*******************x1(N)=1; n= 1 to 10 and x2(n)=1; n= 2 to 10

n = 0 : 1 : 15;              %specify range of n

x1=1.*(n>=1 & n<=10);       %generate signal x1(n)
x2=1.*(n>=2 & n<=10);       %generate signal x2(n)
N1=length(x1);
N2=length(x2);
```

```
x3=conv(x1,x2);                    %perform convolution of signals x1(n) and x2(n)
n1=0 : 1 : N1+N2-2;                %specify range of n for x3(n)
subplot(3,1,1);stem(n,x1);
xlabel('n');ylabel('x1(n)');
title('signal x1(n)');

subplot(3,1,2);stem(n,x2);
xlabel('n');ylabel('x2(n)');
title('signal x2(n)');

subplot(3,1,3);stem(n1,x3);
xlabel('n');ylabel('x3(n)');
title('signal, x3(n) = x1(n)*x2(n)');
```

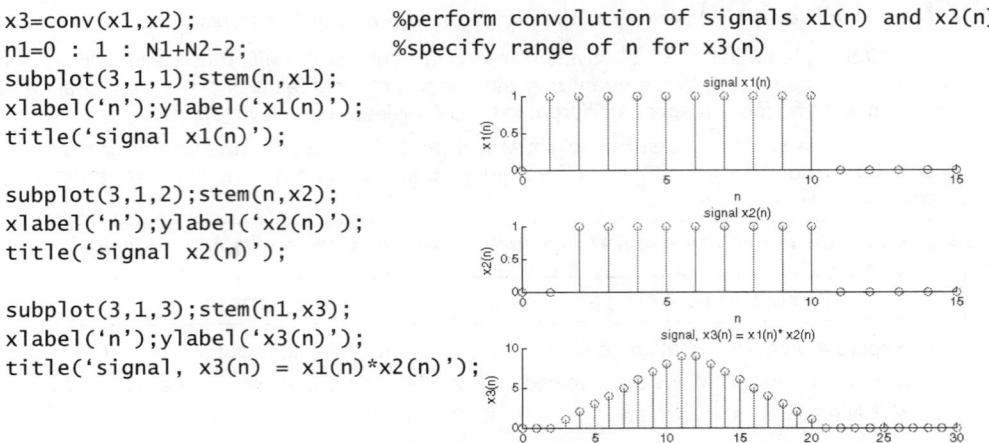

OUTPUT

Fig P5.1: Output waveforms of program 5.1.

The input and output waveforms of program 5.1 are shown in Fig P5.1.

Program 5.2

Write a MATLAB program to sketch the magnitude and phase spectrum of discrete time systems represented by the following transfer functions.

a) $H(e^{j\omega})=(1-e^{-j3\omega})/3(1-e^{-j\omega})$ b) $H(e^{j\omega})=2e^{-j\omega/2}\cos(\omega/2)$

c) $H(e^{j\omega})=2e^{-j\omega/2}\sin(\omega/2)$

```
% Program to sketch the magnitude and phase spectrum
% of the given discrete time systems

clear all

MagH1=[]; MagH2=[]; MagH3=[]; PhaH1=[]; PhaH2=[]; PhaH3=[]; w1=[];

for w=-2*pi:0.01:2*pi
H1=(1/3)*(1-exp(-3*i*w))/(1-exp(-i*w));
H2=2*(exp(-i*w/2))*(cos(w/2));
H3=2*(exp(-i*w/2))*(sin(w/2));

H1_M=abs(H1);       H2_M=abs(H2);   H3_M=abs(H3);
H1_P=angle(H1); H2_P=angle(H2); H3_P=angle(H3);

MagH1=[MagH1,H1_M];      %store the magnitude as an array
MagH2=[MagH2,H2_M];
MagH3=[MagH3,H3_M];

PhaH1=[PhaH1,H1_P];      %store the phase as an array
PhaH2=[PhaH2,H2_P];
PhaH3=[PhaH3,H3_P];

w1=[w1,w];               %store the frequency as an array
end

subplot(3,2,1),plot(w1,MagH1);
xlabel('w in rad.'),ylabel('Mag. of H1');
subplot(3,2,2),plot(w1,PhaH1);
xlabel('w in rad.'),ylabel('Pha. of H1');
```

```
subplot(3,2,3),plot(w1,MagH2);
xlabel('w in rad.'),ylabel('Mag. of H2');
subplot(3,2,4),plot(w1,PhaH2);
xlabel('w in rad.'),ylabel('Pha. of H2');

subplot(3,2,5),plot(w1,MagH3);
xlabel('w in rad.'),ylabel('Mag. of H3');
subplot(3,2,6),plot(w1,PhaH3);
xlabel('w in rad.'),ylabel('Pha. of H3');
```

OUTPUT

The magnitude and phase spectrum of program 5.2 are shown in Fig P5.2.

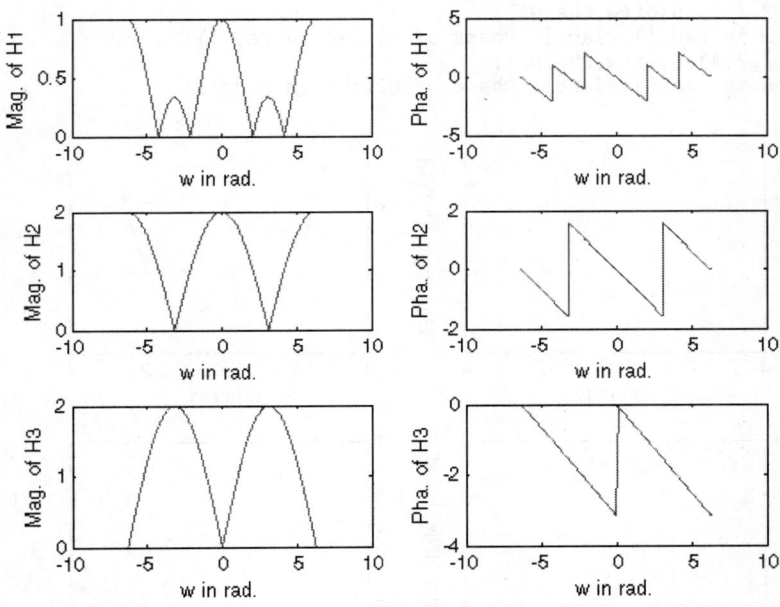

Fig P5.2*: Magnitude and phase spectrum of program 5.2.*

Program 5.3

Write a MATLAB program to sketch the frequency response of the first-order discrete time system governed by the transfer function,

$H(e^{j\omega})=1/(1-ae^{-j\omega})$ for a=0.5 and a=-0.5.

```
% Program to sketch frequency response of first-order discrete
  time system

clear all
j=sqrt(-1);w=[];Mag_H1=[];Pha_H1=[];Mag_H2=[];Pha_H2=[];

for w1=-pi:0.01:pi
    H1 = 1/(1-0.5*exp(-j*w1));
    H2 = 1/(1+0.5*exp(-j*w1));
    H1_M = abs(H1);
    H2_M = abs(H2);
```

```
     H1_P = angle(H1);
        H2_P = angle(H2);
        Mag_H1=[Mag_H1, H1_M];
        Mag_H2=[Mag_H2, H2_M];
        Pha_H1=[Pha_H1,H1_P];
        Pha_H2=[Pha_H2,H2_P];
        w=[w,w1];
     end

     subplot(2,2,1),plot(w,Mag_H1);
     xlabel('w in rad.'),ylabel('Magnitude of H1(jw)');
     subplot(2,2,2),plot(w,Mag_H2);
     xlabel('w in rad.'),ylabel('Magnitude of H2(jw)');
     subplot(2,2,3),plot(w,Pha_H1);
     xlabel('w in rad.'),ylabel('Phase of H1(jw) in rad.');
     subplot(2,2,4),plot(w,Pha_H2);
     xlabel('w in rad.'),ylabel('Phase of H2(jw) in rad.');
```

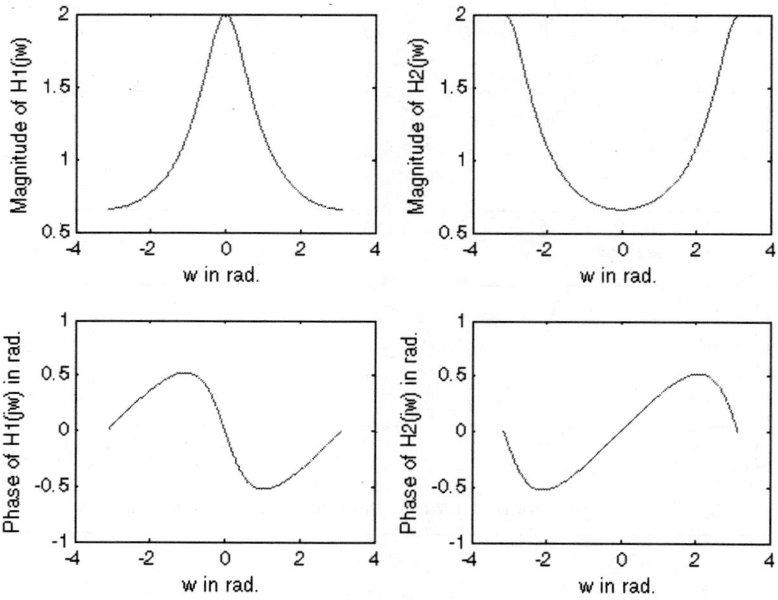

Fig P5.3: *Magnitude and phase spectrum of first-order discrete time system.*

OUTPUT

The frequency response consists of two parts : Magnitude spectrum and Phase spectrum. The magnitude and phase spectrum of first-order discrete time system for a=0.5 and for a=-0.5 are shown in Fig P5.3.

Program 5.4

Write a MATLAB program to sketch the frequency response of the second-order discrete time system governed by the transfer function,

$H(e^{j\omega})=(1+ae^{-j\omega})/(1+\alpha e^{-j\omega}+\beta e^{-j2\omega})$

where, $a=-r\cos\omega_0$; $\alpha=2a$; $\beta=r^2$; $r=0.9$; $\omega_0=\pi/2$.

```
% Program to sketch frequency response of second-order
% discrete time system
```

```
clear all

j=sqrt(-1);w=[];Mag_H=[];Pha_H=[];
r=0.9; wo=pi/2;
a=(-1*r*cos(wo));
alpha=2*a;
Beta=r^2;

for  w1=-pi:0.01:pi
     Num_of_H=(1+a*exp(-j*w1));
     Den_of_H=(1+((alpha)*exp(-j*w1))+((Beta)*exp(-j*2*w1)));
     H=Num_of_H / Den_of_H;
     H_M=abs(H);
     H_P=angle(H);
     Mag_H=[Mag_H,H_M];
     Pha_H=[Pha_H,H_P];
     w=[w,w1];
end
subplot(2,1,1),plot(w,Mag_H);
xlabel('w in radians'),ylabel('Magnitude of H(jw)');
subplot(2,1,2),plot(w,Pha_H);
xlabel('w in radians'),ylabel('Phase of H(jw)');
```

OUTPUT

The frequency response consists of two parts : Magnitude spectrum and Phase spectrum. The magnitude and phase spectrum of the given second-order discrete time system are shown in Fig P5.4.

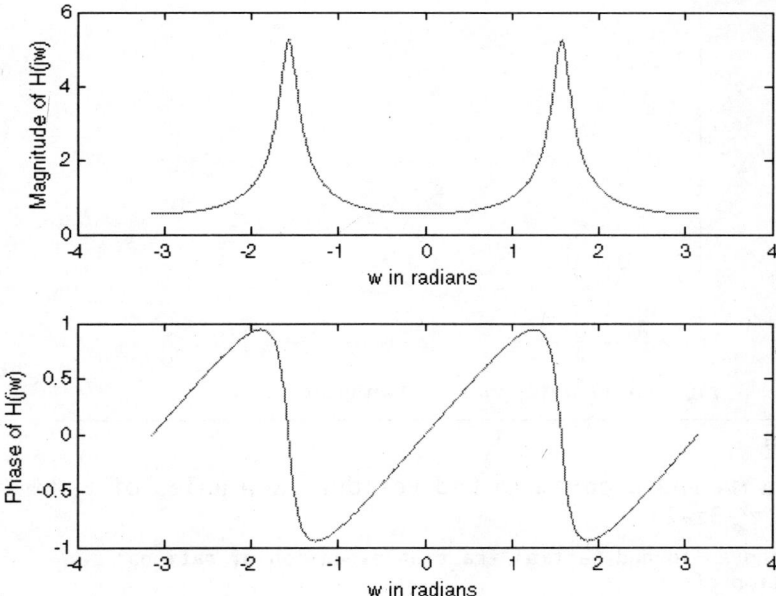

Fig P5.4*: Magnitude and phase spectrum of second-order discrete time system.*

Program 5.5

Write a MATLAB program to compute and sketch the impulse response of discrete time system governed by transfer function, $H(z)=1/(1-0.8z^{-1}+0.16z^2)$.

```
%******* Program to find impulse response of a discrete time system
clear all
syms z n

H=1/(1-0.8*(z^(-1))+0.16*(z^(-2)));
disp('Impulse response h(n) is');
h=iztrans(H);              %compute impulse response
simplify(h)

N=15;
b=[0 0 1];                 %numerator coefficients
a=[1 -0.8 0.16];           %denominator coefficients
[H,n]=impz(b,a,N);         %compute N samples of impulse response

stem(n,H);                 %sketch impulse response
xlabel('n');
ylabel('h(n)');
```

OUTPUT

```
Impulse response h(n) is
      ans =
            2^n*5^(-n)+2^n*5^(-n)*n
```

The sketch of impulse response is shown in Fig P5.5.

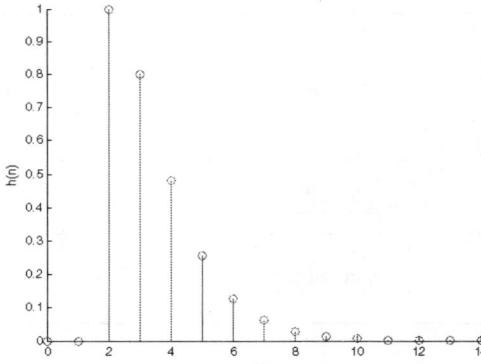

***Fig P5.5**: Impulse response of program 5.5.*

Program 5.6

Write a MATLAB program to find residues and poles of z-domain signal, $(3z^2+2z+1)/(z^2-3z+2)$

```
%*** Program to find partial fraction expansion of rational
% function of z

clear all
H=tf('z');
Ts=0.1;
```

```
b=[3 2 1 ];        %Numerator coefficients
a=[1 -3 2];        %Denominator coefficients

disp('The given transfer function is,');
H=tf([b], [a],Ts)

disp('The residues, poles and direct terms of given TF are,');
disp('(r - residue ; p - poles ; k - direct terms)');
[r,p,k]=residue(b,a)

disp('The num. and den. coefficients extracted from r,p,k,');
[b,a]=residue(r,p,k)
```

OUTPUT

```
The given transfer function is,
        Transfer function:
                        3 z^2 + 2 z + 1
                        ---------------
                         z^2 - 3 z + 2
Sampling time: 0.1
The residues, poles and direct terms of given TF are,
(r - residue ; p - poles ; k - direct terms)
        r =
                17
                -6
        p =
                2
                1
        k =
                3
The num. and den. coefficients extracted from r,p,k are,
        b =
                3       2       1
        a =
                1      -3       2
```

Program 5.7

Write a MATLAB program to perform convolution of signals, $x_1(n) = (0.4)^n u(n)$ and $x_2(n) = (0.5)^n u(n)$, using z-transform, and then to perform deconvolution using the result of convolution to extract $x_1(n)$ and $x_2(n)$.

```
%*** Program to perform convolution and deconvolution using z-transform

clear all;
syms n z
x1n=0.4^n;
x2n=0.5^n;

X1z=ztrans(x1n);
X2z=ztrans(x2n);
X3z=X1z*X2z;                        %product of z-transform of inputs
con12=iztrans(X3z);
disp('Convolution of x1(n) and x2(n) is');
simplify(con12)                    % convolution output

decon_X1z=X3z/X1z;
decon_x1n=iztrans(decon_X1z);
disp('The signal x1(n) obtained by deconvolution is');
simplify(decon_x1n)
```

```
decon_x2z=X3z/X2z;
decon_x2n=iztrans(decon_X2z);
disp('The signal x2(n) obtained by deconvolution is');
simplify(decon_x2n)
```

OUTPUT

```
    Convolution of x1(n) and x2(n) is
        ans =
            5*2^(-n)-4*2^n*5^(-n)
    The signal x1(n) obtained by deconvolution is
        ans =
            2^(-n)
    The signal x2(n) obtained by deconvolution is
        ans =
            2^n*5^(-n)
```

Program 5.8

Write a MATLAB program to find poles and zeros of z-domain signal, $(z^2+0.8z+0.8)/(z^2+0.49)$, and sketch the pole zero plot.

```
% Program to determine poles and zeros of rational function of Z and
% to plot the poles and zeros in z-plane

clear all
syms z

num_coeff=[1 0.8 0.8];                    %find the factors of z^2+0.8z+0.8
disp('Roots of numerator polynomial z^2+0.8z+0.8 are zeros.');
zeros=roots(num_coeff)

den_coeff=[1 0 0.49];                      %find the factors of z^2+0.49
disp('Roots of denominator polynomial z^2+0.49 are poles.');
poles=roots(den_coeff)

H=tf('z');
Ts=0.1;

H=tf([num_coeff],[den_coeff],Ts);
zgrid on;
pzmap(H);                                  %Pole-zero plot
```

OUTPUT

```
    Roots of numerator polynomial z^2+0.8z+0.8 are zeros.
        zeros =
                -0.4000 + 0.8000i
                -0.4000 - 0.8000i

    Roots of denominator polynomial z^2+0.49 are poles.
        poles =
                        0 + 0.7000i
                        0 - 0.7000i

    The pole-zero plot is shown in Fig P5.8.
```

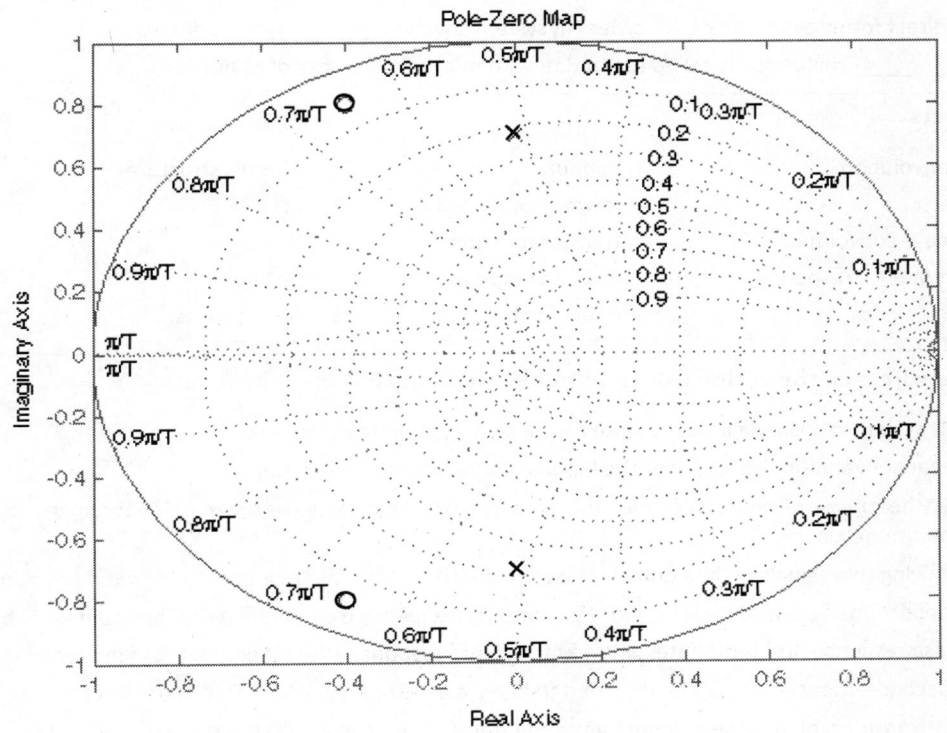

Fig P5.8 : *Pole-Zero plot of program 5.8.*

5.13 Exercises

I. Fill in the blanks with appropriate words.

1. The response of an LTI system is given by _____ of input and impulse response.

2. An LTI system is causal if and only if its impulse response is _____ for negative values of n.

3. The _____ is called aperiodic convolution.

4. The Fourier transform of the impulse response of an LTI system is called _____ .

5. The Fourier transform of the discrete signal can be obtained by evaluating the Z-transform along _____ .

6. A second-order LTI system will behave as a _____ filter.

7. A first-order LTI system will behave as a _____ filter.

8. The ratio of Z-transform of output to Z-transform of input is called _____ of the system.

9. The inverse Z-transform of transfer function is _____ of the system.

10. In IIR systems, the _____ structure will give direct relation between time domain and z-domain.

11. When number of delays is equal to order of the system, the structure is called _____ .

12. The direct form realization of IIR system with M zeros and N poles involves _____ multiplications.
13. The direct form-II realization of Nth order IIR system requires _____ delays and memory locations.
14. The direct form realization of Nth order FIR system involves _____ additions.
15. _____ realization is called realization with minimum number of multipliers.

Answers

1.	convolution	6.	bandpass	11.	canonic structure
2.	zero	7.	lowpass or highpass	12.	M+N+1
3.	linear convolution	8.	transfer function	13.	N
4.	frequency response	9.	impulse response	14.	N−1
5.	unit circle	10.	direct form-I	15.	Linear phase

II. State whether the following statements are True/False.

1. An LTI system is unstable if the impulse response is absolutely summable.
2. In linear convolution the length of the input sequences should be same.
3. When the impulse response is complex, the real part of frequency response is symmetric and imaginary part is antisymmetric.
4. Convolving two signals in time domain is equivalent to multiplying their spectra in frequency domain.
5. Multiplication of a sequence x(n) by $e^{j\omega_0 n}$ is same as frequency translation of the spectrum $X(e^{j\omega})$ by ω_0.
6. Impulse response h(n) is discrete, whereas frequency response $H(e^{j\omega})$ is continuous function of ω.
7. The second-order system can be designed to behave as either low pass or high pass filter.
8. The Z-transform of impulse response gives the transfer function of LTI system.
9. If X(z) and H(z) are Z-transform of input and impulse response respectively, then the response of LTI system is given by inverse Z-transform of the product X(z) H(z).
10. For a stable LTI discrete time system the poles should lie on the unit circle.
11. The direct form-I structure of IIR system employs same delay for input and output samples.
12. In direct form-II realization of IIR system, N memory locations are required to store delayed signals.
13. In parallel or cascade realization, the memory requirement depends on realization of individual sections.
14. Scaling multipliers has to be provided between individual sections of cascade structure.
15. The linear phase realization of Nth order FIR system for odd values of N involves N/2 multiplications.
16. For linear phase realization of FIR system, the impulse response should be symmetric.

Answers

1.	False	5.	True	9.	True	13.	True
2.	False	6.	True	10.	False	14.	True
3.	True	7.	False	11.	False	15.	False
4.	True	8.	True	12.	True	16.	True

III. Choose the right answer for the following questions.

1. *Two parallel connected discrete time systems with impulse responses $h_1(n)$ and $h_2(n)$ can be replaced by a single equivalent discrete time system with impulse response,*

   a) $h_1(n) * h_2(n)$ **b)** $h_1(n) + h_2(n)$ **c)** $h_1(n) - h_2(n)$ **d)** $h_1(n) * [h_1(n) + h_2(n)]$

2. *If $h(n)$ is real, then magnitude of $H(e^{j\omega})$ is _____ and phase of $H(e^{j\omega})$ is _____.*

   a) symmetric, antisymmetric   b) antisymmetric, symmetric

   c) symmetric, symmetric   d) antisymmetric, antisymmetric

3. *The second order LTI discrete time system behaves as,*

   a) low pass filter   b) high pass filter   c) resonant filter   d) all pass filter

4. *For a stable LTI discrete time system poles should lie —— and unit circle should be ——.*

   a) outside unit circle, included in ROC   b) inside unit circle, outside of ROC

   c) inside unit circle, included in ROC   d) outside unit circle, outside of ROC

5. *An LTI system with impulse response, $h(n) = (-a)^n u(n)$ and $-a < -1$ will be,*

   a) stable system   b) unstable system

   c) anticausal system   d) neither stable nor causal

6. *If $X(z)$ has a single pole on the unit circle, on negative real axis then, $x(n)$ is,*

   a) signed constant sequence   b) signed decaying sequence

   c) signed growing sequence   d) constant sequence

7. *The factor that influence the choice of realization of strurcture is,*

   a) memory requirements   b) computational complexity

   c) parallel processing and pipelining   d) all the above.

8. *The structure that uses separate delays for input and output samples is,*

   a) direct form-II   b) direct form-I   c) cascade form   d) parallel form

9. *The linear phase realization sturcture is used to represent,*

   a) FIR systems   b) IIR systems

   c) both FIR and IIR systems   d) all discrete time systems

10. *The effect of quantization of coefficients on the frequency response is minimized in,*

    a) cascade realization   b) parallel realization

    c) direct form structure   d) both a and b

11. *The direct form-I and II structures of IIR system will be identical in,*

    a) all pole system   b) all zero system

    c) both a and b   d) first order and second order systems.

12. *The condition for symmetry of impulse response of FIR system is,*

    a) $h(n) = h(N-1)$ b) $h(n) = h(N+1)$ c) $h(n) = h(N-n)$ d) $h(n) = h(N-1-n)$

13. *The linear phase realization is used in FIR systems, mainly to minimize,*

   a) multipliers   b) memory   c) delays   d) adders

14. *Which one of the following FIR system has linear phase response?*

   a) y(n) = 0.4 x(n) + 0.1 x(n–1) + 0.5 x(n–2)   b) y(n) = 0.3 x(n) + x(n–1) + 3.0 x(n–2)

   c) y(n) = 0.5 x(n) + 0.7 x(n–1)   d) y(n) = 0.6x(n) + 0.6 x(n–1)

15. *The quantization error increases, when the order of the system 'N' increases in case of,*

   a) direct form realization   b) cascade or parallel form realization

   c) all IIR systems   d) all FIR systems

16. *The number of memory locations required to realize the system,* $H(z) = \dfrac{1+z^{-2}+2z^{-3}}{1+z^{-2}+z^{-4}}$

   a) 8   b) 7   c) 2   d) 10

17. *Number of multipliers and adders required for direct form realization of Nth order FIR system are,*

   a) N, N+1   b) N, N–1   c) N+1, N   d) N–1, N+1

18. *The realization of linear phase FIR system for odd values of 'N' needs,*

   a) $\dfrac{N}{2}$ multipliers   b) $\dfrac{N+1}{2}$ multipliers   c) N–1 multipliers   d) N multipliers

Answers

1. b	6. a	11. c	16. b
2. a	7. d	12. d	17. b
3. c	8. b	13. a	18. b
4. c	9. a	14. d	
5. a	10. d	15. a	

IV. Answer the following questions.

1.   What is discrete time system?

2.   What is impulse response? Explain its significance.

3.   Write the difference equation governing the Nth order LTI system.

4.   Write the expression for discrete convolution.

5.   What are FIR and IIR systems?

6.   Write the convolution sum formula for FIR and IIR systems.

7.   Write the properties of linear convolution.

8.   Prove the distributive property of linear convolution.

9.   What are the two ways of interconnecting LTI systems?

10. What is the importance of linear convolution in signals and systems?

11. Define the frequency spectrum of a discrete time signal in terms of Fourier transform.

12. Write the properties of frequency response of an LTI system.
13. What is frequency response of an LTI system?
14. Define the transfer function of an LTI system.
15. Write the transfer function of Nth order LTI system.
16. Give the importance of convolution and deconvolution operations using Z-transform.
17. Give the conditions for stability of an LTI discrete time system in z-plane.
18. What are the various issues that are addressed by realization structures?
19. What are the basic elements used to construct the realization structures of discrete time system?
20. List the different types of structures for realization of IIR systems.
21. Draw the direct form-I structure of an Nth order IIR system with equal number of poles and zeros.
22. Draw the direct form-II structure of an Nth order IIR system with equal number of poles and zeros.
23. Explain the conversion of direct form-I structure to direct form-II structure with an example.
24. What are the difficulties in cascade realization?
25. Explain the realization of cascade structure of an IIR system.
26. Explain the realization of parallel structure of an IIR system.
27. What are the different types of structure for realization of FIR systems?
28. Draw the direct form structure of an Nth order FIR system.
29. What is the necessary condition for Linear phase realization of FIR system?
30. Draw the linear phase realization structure of an Nth order FIR system when 'N' is even.
31. Draw the linear phase realization structure of an Nth order FIR system when 'N' is odd.
32. Explain the realization of cascade structure of a FIR system.

V. Solve the following problems.

E5.1 **Construct the block diagram of the discrete time systems whose input-output relations are described by the following difference equations.**

a) $y(n) = 2y(n-1) + 2.1\,x(n-1) + 0.5\,x(n-2)$

b) $y(n) = 1.6\,x(n-2) + 0.7\,x(n) + 3y(n-1) + 0.3y(n-2)$

E5.2 **Determine the response of the discrete time systems governed by the following difference equations.**

a) $y(n) = 0.1y(n-1) + x(n-1) + 0.7x(n)$; $x(n) = 2^{-n}\,u(n)$; $y(-1) = -1$

b) $y(n) + 2.1y(n-1) + 0.2y(n-2) = x(n) + 0.56x(n-1)$; $x(n) = u(n)$; $y(-2) = 1$; $y(-1) = -3$

E5.3 **a) Determine the impulse response for the cascade of two LTI systems having impulse responses,**

$$h_1(n) = \left(\frac{1}{7}\right)^n u(n) \ \text{ and } \ h_2(n) = \delta(n-3)$$

b) Determine the overall impulse response of the interconnected discrete time system shown in Fig E5.3.

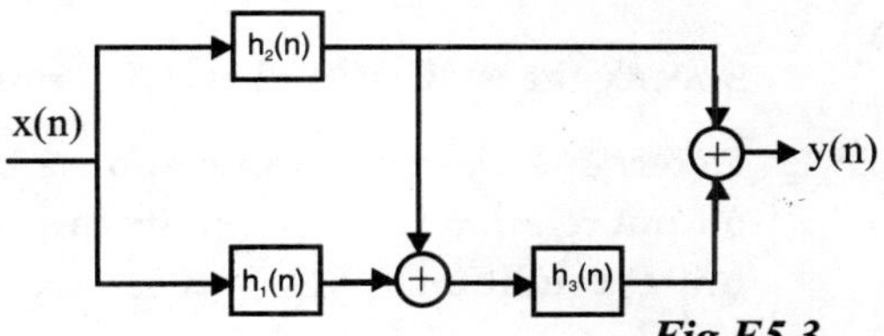

Fig E5.3.

Take, $h_1(n) = \left(\frac{1}{3}\right)^n u(n)$; $h_2(n) = \left(\frac{1}{6}\right)^n u(n)$; $h_3(n) = \left(\frac{1}{9}\right)^n u(n)$

E5.4 Determine the response of an LTI system whose impulse response h(n) and input x(n) are given by,

a) $h(n) = \{1, \ 4, \ 1, \ -2, \ 1\}$, $x(n) = \{1, \ 3, \ 5, \ -1, \ -2\}$
        $\uparrow$ $\uparrow$

b) $h(n) = \begin{cases} 1 & ; \ 0 \leq n \leq 2 \\ 0 & ; \quad\ n \geq 3 \end{cases}$, $x(n) = a^n u(n); \ |a| < 1$

E5.5 The input x(n) and impulse response h(n) of an LTI system are given by,

$x(n) = \{-1, \ 1, \ -1, \ 1, \ -1, \ 1\}$; $h(n) = \{-0.5, \ 0.5, \ -1, \ 0.5, \ -1, \ -2\}$
        $\uparrow$ $\uparrow$

Find the response of the system using linear convolution.

E5.6 A discrete LTI system is described by a difference equation, $y(n) = x(n) - x(n-1)$. Determine the frequency response $H(e^{j\omega})$, impulse response h(n). Sketch the magnitude function and phase function.

E5.7 Sketch the magnitude and phase function of the discrete time LTI system described by the equation $y(n) = x(n) + x(n-1)$.

E5.8 The impulse response of an LTI system is $h(n) = \{-2, \ -1, \ 1, \ -2\}$. Find the response of the system $x(n) = \{2, \ 2, \ 4, \ 1\}$ for the input, using convolution property of Fourier transform.

E5.9 A causal system is represented by the following difference equation,
$y(n) - 0.2y(n-1) = x(n) - 0.6 \times (n-1)$

Find the system transfer function H(z), impulse response and frequency response of the system. Also determine the magnitude and phase function.

E5.10 Determine the transfer function and impulse response for the systems described by the following equations.

a) $y(n) + 2y(n-1) - 3y(n-2) = x(n-1)$ b) $y(n) - \frac{7}{4}y(n-1) + \frac{5}{8}y(n-2) = 2x(n)$

c) $y(n) = 0.2x(n) - 5x(n-1) + 0.6y(n-1) - 0.08y(n-2)$ d) $y(n) - \frac{3}{2}y(n-1) = x(n) + \frac{2}{3}x(n-1)$

E5.11 A discrete time LTI system is characterized by the transfer function, $H(z) = \dfrac{z(6z-8)}{\left(z - \frac{1}{2}\right)(z-3)}$.

Specify the ROC of H(z) and determine h(n) for the system to be, (i) stable, (ii) causal.

E5.12 Determine the unit step response of the discrete time LTI system, whose input and output relation is described by the difference equation, $y(n) + 7\,y(n-1) = x(n)$, where the initial condition is, $y(-1) = 1$.

E5.13 Determine the response of discrete time LTI system governed by the following difference equation, $4\,y(n) + 5\,y(n-1) + y(n-2) = x(n)$; with initial conditions, $y(-2) = -2$ and $y(-1) = 1$, for the input $x(n) = (0.5)^n u(n)$.

E5.14 An LTI system has the impulse response h(n) defined by $h(n) = x_1(n-1) * x_2(n)$. The Z-transform of the two signals $x_1(n)$ and $x_2(n)$ are $X_1(z) = 2 - 4z^{-1}$ and $X_2(z) = 1 + 5z^{-2}$ respectively. Determine the output of the system for the input $\delta(n-1)$.

E5.15 Obtain the direct form-I, direct form-II, cascade and parallel form realizations of the LTI system governed by the equation,

$$y(n) = -\frac{5}{4}y(n-1) + \frac{1}{8}y(n-2) + \frac{1}{16}y(n-3) + x(n) + 5x(n-1) + 6x(n-2)$$

E5.16 Realize the direct form-I and direct form-II structures of the IIR system represented by the transfer function,

$$H(z) = \frac{2(z+2)}{(z-0.1)(z+0.5)(z+0.4)}$$

E5.17 Determine the direct form-I, II, cascade and parallel realization of the following LTI system.

$$H(z) = \frac{z^3 - 2z^2 + 2z - 1}{(z-0.5)(z^2+z-0.5)}$$

E5.18 Realize the cascade and parallel structures of the system governed by the difference equation,

$$y(n) - \frac{3}{4}y(n-1) + \frac{1}{8}y(n-2) = x(n) + \frac{1}{2}x(n-1)$$

E5.19 Draw the direct form structure of the FIR systems described by the following equations,

a) $y(n) = x(n) + \frac{1}{3}x(n-1) + \frac{1}{4}x(n-2) + \frac{1}{5}x(n-3) + \frac{2}{7}x(n-4)$

b) $y(n) = 0.35x(n) + 0.3x(n-1) + 0.125x(n-2) - 0.25x(n-3)$
$- 0.35x(n-4) - 0.3x(n-5) - 0.125x(n-6)$

E5.20 Realize the following FIR systems with minimum number of multipliers.

a) $H(z) = 0.2 + 0.5z^{-1} + 0.3z^{-2} + 0.5z^{-3} + 0.2z^{-4}$

b) $H(z) = \left(0.1 + \frac{1}{8}z^{-1} + z^{-2}\right)\left(2 - \frac{1}{9}z^{-1} + 2z^{-2}\right)$

c) $y(n) = -\frac{1}{2}x(n) + \frac{3}{5}x(n-1) + \frac{3}{8}x(n-2) + \frac{3}{5}x(n-3) - \frac{1}{2}x(n-4)$

Answers

E5.1 a)

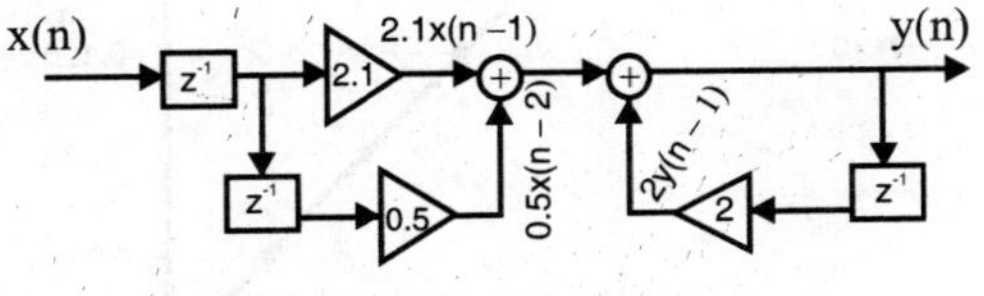

Fig E5.1a.1: Block diagram.

E5.1 b)

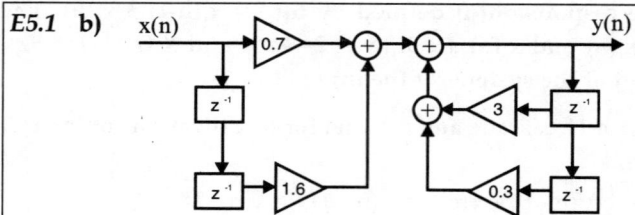

Fig E5.1b.1: *Block diagram.*

E5.2 a) $y(n) = \left[-2.775(0.1)^n + 3.375\left(\frac{1}{2}\right)^n\right]u(n)$ **b)** $y(n) = [0.47 - 0.02(-0.1)^n + 6.65(-2)^n]u(n)$

E5.3 a) $h(n) = \left(\frac{1}{7}\right)^{(n-3)}u(n-3)$ **b)** $h(n) = \left[4\left(\frac{1}{6}\right)^n - \frac{3}{2}\left(\frac{1}{9}\right)^n + \left(\frac{3}{2}\right)\left(\frac{1}{3}\right)^n\right]u(n)$

E5.4 a) $y(n) = \{1, 7, 18, \underset{\uparrow}{20}, -6, -16, 5, 3, -2\}$ **b)** $y(n) = \displaystyle\sum_{k=0}^{n} a^k$;   for $n = 0, 1, 2$

$$= \sum_{k=n-2}^{n} a^k \quad ; \quad \text{for } n > 2$$

E5.5 $y(n) = \{0.5, \underset{\uparrow}{-1}, 2, -2.5, 3.5, -1.5, 1, -0.5, -0.5, 1, -2\}$

E5.6 $H(e^{j\omega}) = 2\sin\frac{\omega}{2}e^{j\left(\frac{\pi-\omega}{2}\right)}$; $h(n) = \delta(n) - \delta(n-1)$

$|H(e^{j\omega})| = \left|2\sin\left(\frac{\omega}{2}\right)\right|$; $\angle H(e^{j\omega}) = -\left(\frac{\pi-\omega}{2}\right)$; for $\omega = -\pi$ to 0

$$= \frac{\pi-\omega}{2} \quad ; \quad \text{for } \omega = 0 \text{ to } \pi$$

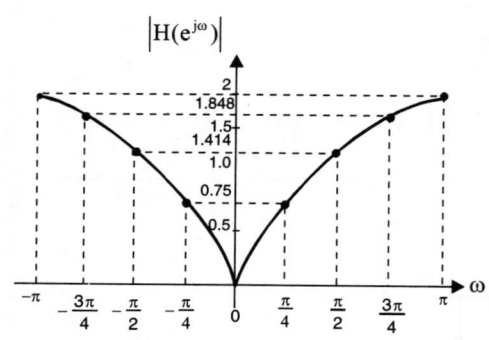

FigE5.6.1: *Magnitude spectrum.*

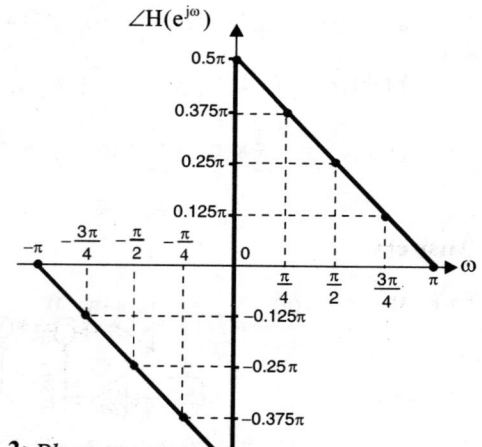

Fig E5.6.2: *Phase spectrum.*

E5.7 $H(e^{j\omega}) = 2\cos\left(\frac{\omega}{2}\right)e^{-j\frac{\omega}{2}}$; $|H(e^{j\omega})| = 2\cos\left(\frac{\omega}{2}\right)$

$\angle H(e^{j\omega}) = -\frac{\omega}{2}$

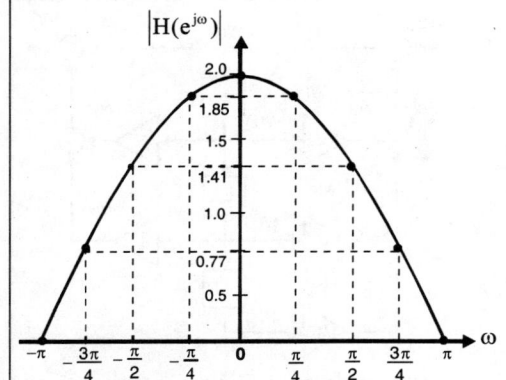

Fig E5.7.1: *Magnitude spectrum.*

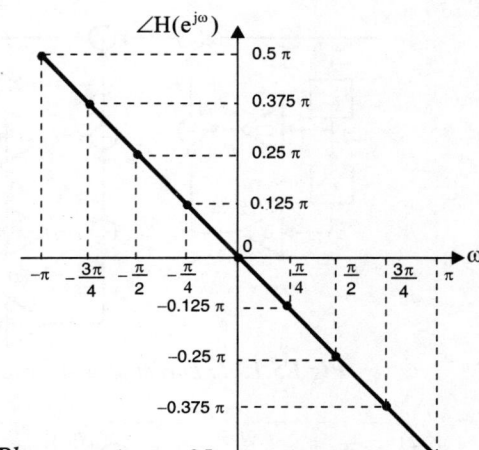

Fig E5.7.2: *Phase spectrum.*

E5.8 $y(n) = \{ -4, \ -6, \ -8, \ -8, \ -1, \ -7, \ -2 \}$
                              ↑

E5.9 $H(z) = \dfrac{1 - 0.6z^{-1}}{1 - 0.2z^{-1}}$; $H(e^{j\omega}) = \dfrac{1 - 0.6e^{-j\omega}}{1 - 0.2e^{-j\omega}}$

$h(n) = 0.2^n \, u(n) - 0.6 \, (0.2)^{n-1} \, u(n-1)$

$|H(e^{j\omega})| = \sqrt{\dfrac{1.36 - 1.2\cos\omega}{1.04 - 0.4\cos\omega}}$; $\angle H(e^{j\omega}) = \tan^{-1}\left(\dfrac{0.4\sin\omega}{1.12 - 0.8\cos\omega}\right)$

E5.10   **a)** $H(z) = \dfrac{z}{z^2 + 2z - 3}$; $h(n) = \frac{1}{4}[1 - (-3)^n]u(n)$

b) $H(z) = \dfrac{2z^2}{z^2 - \frac{7}{4}z + \frac{5}{8}}$; $h(n) = \frac{1}{3}\left[-4\left(\frac{1}{2}\right)^n + 10\left(\frac{5}{4}\right)^n\right]u(n)$

c) $H(z) = \dfrac{0.2z^2 - 5z}{z^2 - 0.6z + 0.08}$; $h(n) = [24.8(0.2)^n + 24.6(0.4)^n]u(n)$

d) $H(z) = \dfrac{1 + \frac{2}{3}z^{-1}}{1 - \frac{3}{2}z^{-1}}$; $h(n) = \left(\frac{3}{2}\right)^n u(n) + \frac{2}{3}\left(\frac{3}{2}\right)^{n-1} u(n-1)$

E5.11   **i) Stable system**

ROC : $0.5 < |z| < 3$; $h(n) = 2\left(\frac{1}{2}\right)^n u(n) - 4(3)^n u(-n-1)$

ii) Casual system

ROC : $|z| > 3$; $h(n) = 2\left(\frac{1}{2}\right)^n u(n) + 4(3)^n u(n)$

E5.12 $y(n) = \frac{1}{8}[1 - 49(-7)^n]u(n)$

E5.13	$y(n) = [\,0.056(0.5)^n - 0.444(-1)^n - 0.111\,(-0.25)^n\,]\,u(n)$
E5.14	$y(n) = \{0, 0, 2, -4, 10, -20\,\}$ $\uparrow$
E5.15	

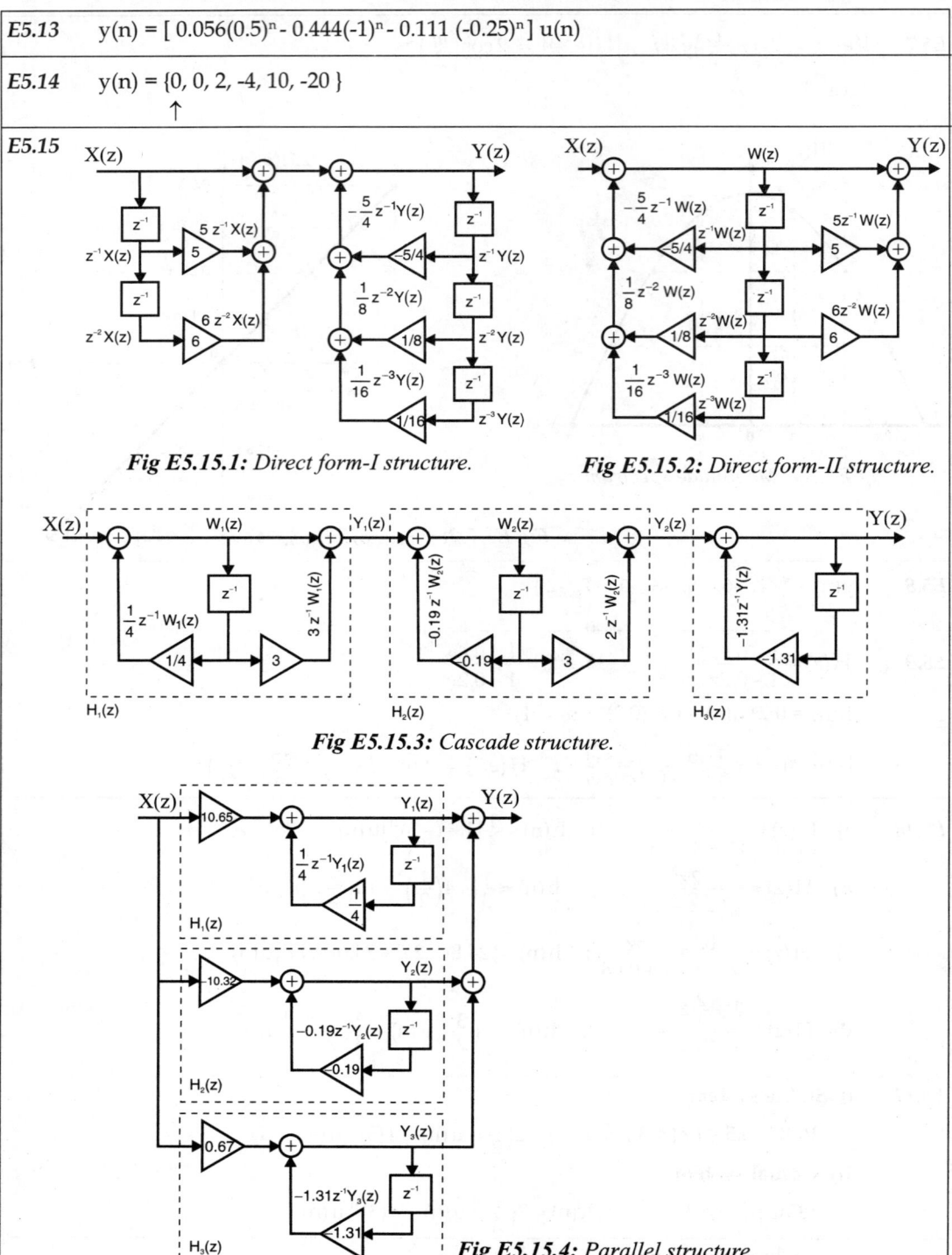

Fig E5.15.1: *Direct form-I structure.*   **Fig E5.15.2:** *Direct form-II structure.*

Fig E5.15.3: *Cascade structure.*

Fig E5.15.4: *Parallel structure.*

E5.16

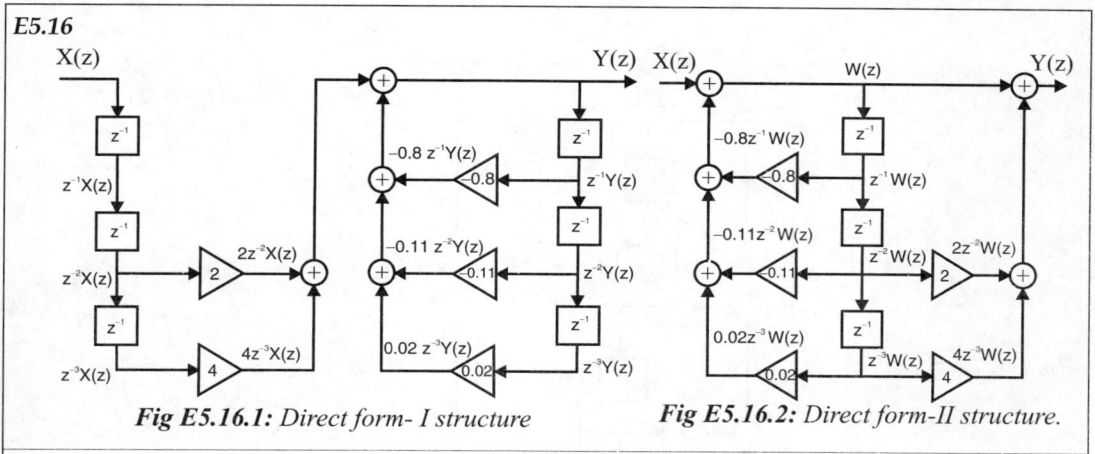

Fig E5.16.1: *Direct form- I structure*

Fig E5.16.2: *Direct form-II structure.*

E5.17

Fig E5.17.1: *Direct form-I structure.*

Fig E5.17.2: *Direct form-II structure.*

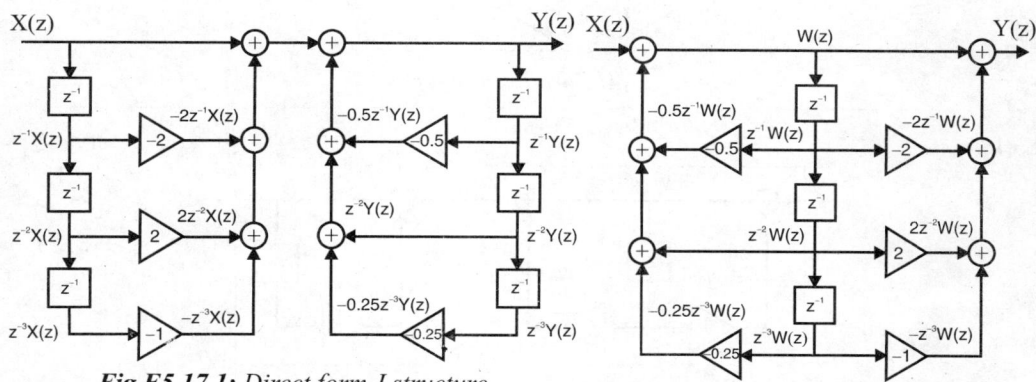

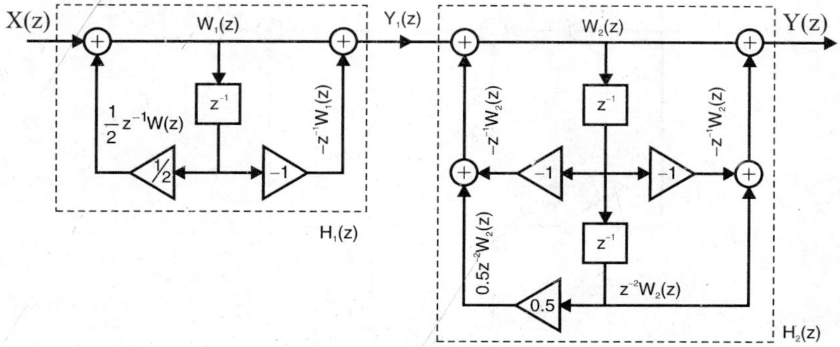

Fig E5.17.3: *Cascade structure.*

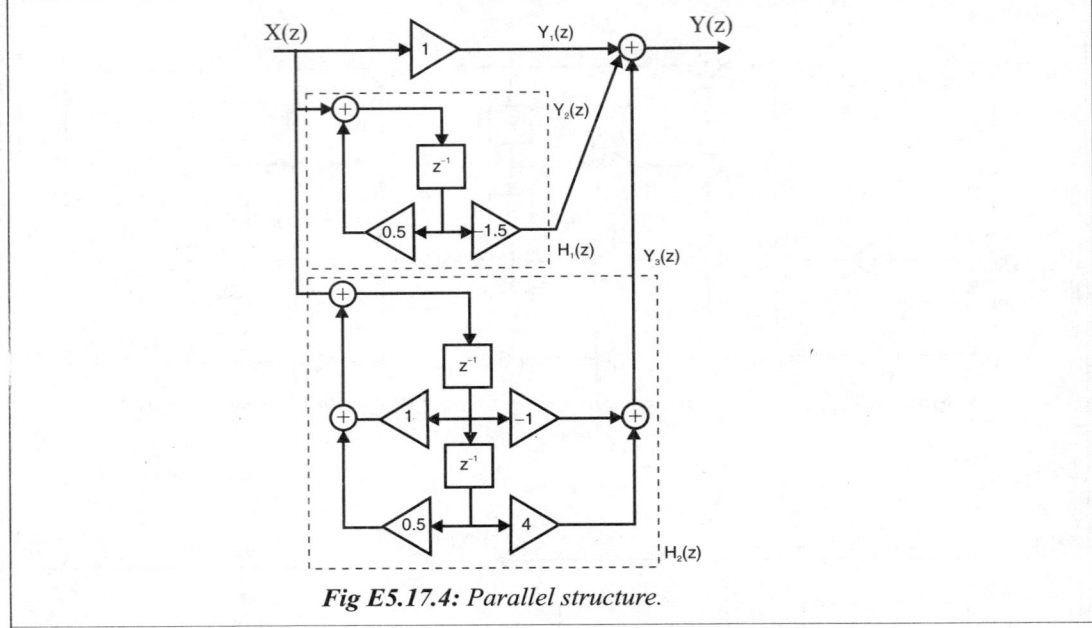

Fig E5.17.4: *Parallel structure.*

E5.18

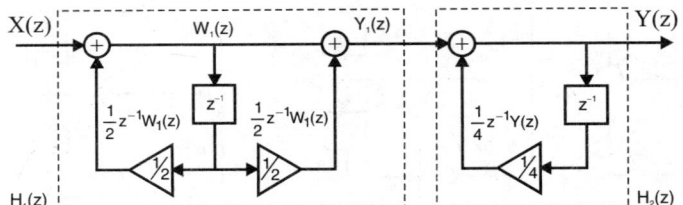

Fig E5.18.1: *Cascade structure.*

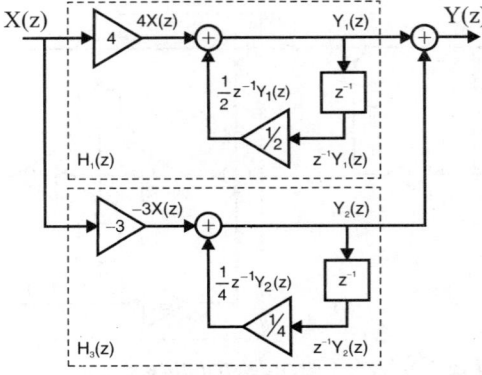

Fig E5.18.2: *Parallel structure.*

E5.19 a)

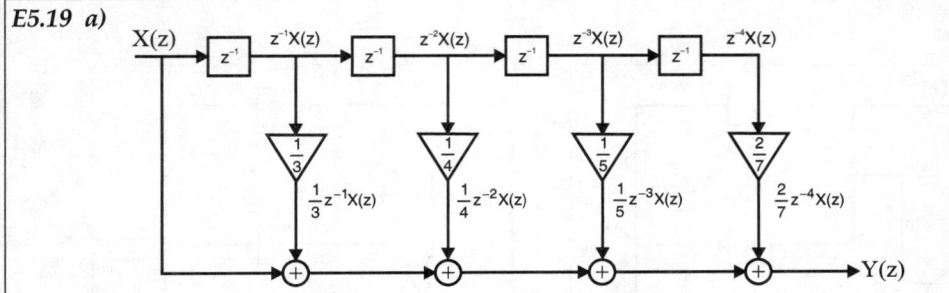

Fig E5.19.a: *Direct form structure.*

E5.19 b)

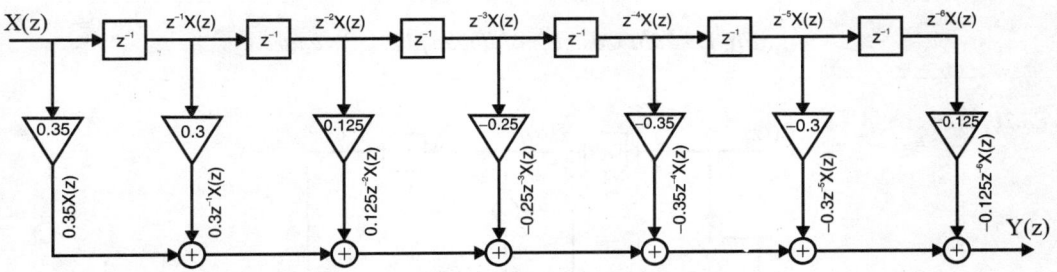

Fig E5.19.b: *Direct form structure.*

E5.20 a)

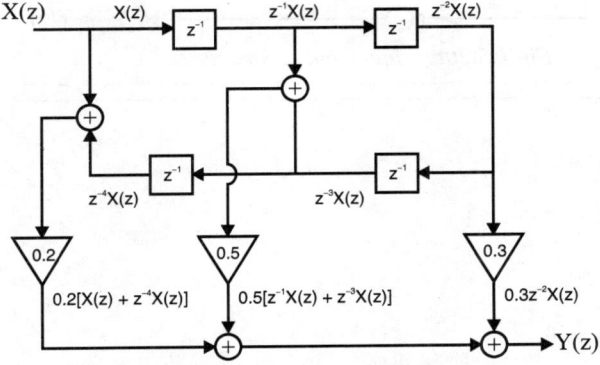

Fig E5.20a: *Linear phase structure.*

E5.20 b)

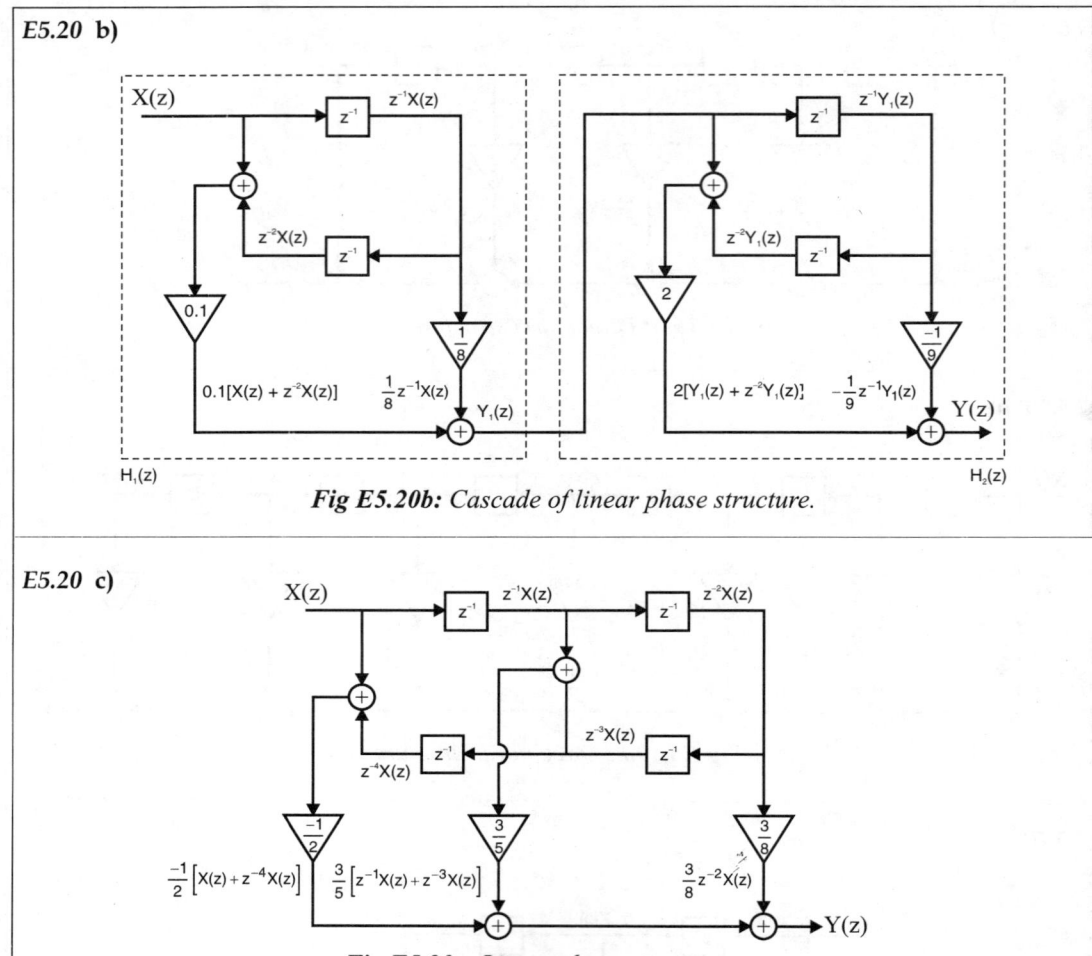

Fig E5.20b: *Cascade of linear phase structure.*

E5.20 c)

Fig E5.20c: *Linear phase structure.*

Important Mathematical Relations

A1.1 Trigonometric Identities

$$\tan \theta = \frac{\sin \theta}{\cos \theta}, \qquad \cot \theta = \frac{1}{\tan \theta}$$

$$\sec \theta = \frac{1}{\cos \theta}, \qquad \operatorname{cosec} \theta = \frac{1}{\sin \theta}$$

$$\sin^2 \theta + \cos^2 \theta = 1, \quad 1 + \tan^2 \theta = \sec^2 \theta$$

$$1 + \cot^2 \theta = \operatorname{cosec}^2 \theta$$

$$\sin(A \pm B) = \sin A \cos B \pm \cos A \sin B$$

$$\cos(A \pm B) = \cos A \cos B \pm \sin A \sin B$$

$$2 \sin A \sin B = \cos(A + B) - \cos(A - B)$$

$$2 \sin A \cos B = \sin(A + B) + \sin(A - B)$$

$$2 \cos A \cos B = \cos(A + B) + \cos(A - B)$$

$$\sin A + \sin B = 2 \sin \frac{A+B}{2} \cos \frac{A-B}{2}$$

$$\sin A - \sin B = 2 \cos \frac{A+B}{2} \sin \frac{A-B}{2}$$

$$\cos A + \cos B = 2 \cos \frac{A+B}{2} \cos \frac{A-B}{2}$$

$$\cos A - \cos B = -2 \sin \frac{A+B}{2} \sin \frac{A-B}{2}$$

$$\cos(\theta \pm 90°) = \pm \sin \theta$$

$$\sin(\theta \pm 90°) = \pm \cos \theta$$

$$\tan(\theta \pm 90°) = -\cot \theta$$

$$\cos(\theta \pm 180°) = -\cos \theta$$

$$\sin(\theta \pm 180°) = -\sin \theta$$

$$\tan(\theta \pm 180°) = \tan \theta$$

$$\sin 2\theta = 2 \sin \theta \cos \theta$$

$$\cos 2\theta = \cos^2 \theta - \sin^2 \theta$$

$$\cos^2 \theta = \frac{1 + 2 \cos \theta}{2}$$

$$\sin^2 \theta = \frac{1 - \cos 2\theta}{2}$$

$$\tan 2\theta = \frac{2 \tan \theta}{1 - \tan^2 \theta}$$

$$\sin \theta = \frac{e^{j\theta} - e^{-j\theta}}{2j}, \qquad \cos \theta = \frac{e^{j\theta} + e^{-j\theta}}{2}$$

$$e^{j\theta} = \cos \theta + j \sin \theta$$

A1.2 Complex variables

A complex number 'z' may be represented as,

$$z = x + jy = r\angle\theta = re^{j\theta} = r(\cos \theta + j \sin \theta)$$

Where, $x = \operatorname{Re}(z) = r\cos\theta$, $\quad y = \operatorname{Im}(z) = r \sin\theta$

$$r = |z| = \sqrt{x^2 + y^2}, \qquad \theta = \tan^{-1}\frac{y}{x}$$

$$j = \sqrt{-1}, \quad \frac{1}{j} = -j, \quad j^2 = -1$$

The conjugate of the complex number 'z' may be represented as,

$$z^* = x - jy = r\angle -\theta = re^{-j\theta} = r(\cos \theta - j \sin \theta)$$

Demovier's theorem : $\left(e^{j\theta}\right)^n = e^{jn\theta} = \cos n\theta + j \sin n\theta$

Let z_1 and z_2 be two complex number defined as $z_1 = x_1 + jy_1$ and $z_2 = x_2 + jy_2$.

Now $z_1 = z_2$ only if $x_1 = x_2$ and $y_1 = y_2$.

$$z_1 \pm z_2 = (x_1 + x_2) \pm j(y_1 + y_2)$$

$$z_1 z_2 = (x_1 x_2 - y_1 y_2) + j(x_1 y_2 + x_2 y_1) \text{ or } z_1 z_2 = r_1 r_2 \, e^{j(\theta_1 + \theta_2)} = r_1 r_2 \angle (\theta_1 + \theta_2)$$

$$\frac{z_1}{z_2} = \frac{(x_1 + jy_1)}{(x_2 + jy_2)} \times \frac{(x_2 - jy_2)}{(x_2 - jy_2)} = \frac{x_1 x_2 + y_1 y_2}{x_2^2 + y_2^2} + j\frac{x_2 y_1 - x_1 y_2}{x_2^2 + y_2^2}$$

or $$\frac{z_1}{z_2} = \frac{r_1}{r_2} e^{j(\theta_1 - \theta_2)} = \frac{r_1}{r_2} \angle (\theta_1 - \theta_2)$$

The following relations hold good for a complex number 'z'

$$\sqrt{z} = \sqrt{x + jy} = \sqrt{re^{j\theta}} = \sqrt{r} \, e^{j\frac{\theta}{2}} = \sqrt{r} \angle \frac{\theta}{2}$$

$$z^n = (x + jy)^n = r^n e^{jn\theta} = r^n \angle n\theta \quad \text{where, n is a integer.}$$

$$z^{1/n} = (x + jy)^{1/n} = r^{1/n} e^{j\theta/n} \, r^{1/n} \angle \left(\frac{\theta}{n} + \frac{2\pi k}{n}\right) \text{ for } k = 0, 1, 2, \ldots \ldots n-1$$

$$\ln z = \ln \left(re^{j\theta}\right) = \ln r + \ln e^{j\theta} = \ln r + j(\theta + 2k\pi) \quad \text{where k is an integer.}$$

A1.3 Hyperbolic Functions

$$\sinh x = \frac{e^x - e^{-x}}{2}, \qquad\qquad \cosh x = \frac{e^x + e^{-x}}{2}$$

$$\tanh x = \frac{\sinh x}{\cosh x}, \qquad\qquad \cosh x = \frac{1}{\tanh x}$$

$$\operatorname{cosech} x = \frac{1}{\sinh x}, \qquad\qquad \operatorname{sech} x = \frac{1}{\cosh x}$$

$$\sin jx = j\sinh x, \qquad\qquad \cos jx = \cosh x$$

$$\sinh jx = j\sin x, \qquad\qquad \cosh jx = \cos x$$

$$\sinh (x \pm y) = \sinh x \, \cosh y \pm \cosh x \, \sinh y$$

$$\cosh (x \pm y) = \cosh x \, \cosh y \pm \sinh x \, \sinh y$$

$$\sinh (x \pm y) = \sinh x \, \cos y \pm j\cosh x \, \sin y$$

$$\cosh (x \pm y) = \sinh x \, \cos y \pm j\cosh x \, \sin y$$

$$\tanh (x \pm y) = \frac{\sinh 2x}{\cosh 2x + \cos 2y} + j\frac{\sin 2y}{\cosh 2x + \cos 2y}$$

$$\cosh^2 x - \sinh^2 x = 1$$

$$\operatorname{sech}^2 x - \tanh^2 x = 1$$

$$\sin (x \pm y) = \sin x \, \cosh y \pm \cos x \, \sinh y$$

$$\cos (x \pm jy) = \cos x \, \cosh y \pm j\sin x \, \sinh y$$

A1.4 Derivatives

Let, $U = U(x)$, $V = V(x)$, and a = constant

$$\frac{d}{dx}(a\,U) = a\,\frac{dU}{dx}$$

$$\frac{d}{dx}(U\,V) = U\,\frac{dV}{dx} + V\,\frac{dU}{dx}$$

$$\frac{d}{dx}\left[\frac{U}{V}\right] = \frac{V\frac{dU}{dx} - U\frac{dV}{dx}}{V^2}$$

$$\frac{d}{dx}(a\,U^n) = n\,a\,U^{n-1}$$

$$\frac{d}{dx}\log_a U = \frac{\log_a e}{U}\frac{dU}{dx}$$

$$\frac{d}{dx}\ln U = \frac{1}{U}\frac{dU}{dx}$$

$$\frac{d}{dx}a^U = a^U\ln a\,\frac{dU}{dx}$$

$$\frac{d}{dx}e^U = e^U\,\frac{dU}{dx}$$

$$\frac{d}{dx}U^V = VU^{V-1}\frac{dU}{dx} + U^V\ln U\frac{dV}{dx}$$

$$\frac{d}{dx}\sin U = \cos U\,\frac{dU}{dx}$$

$$\frac{d}{dx}\cos U = -\sin U\,\frac{dU}{dx}$$

$$\frac{d}{dx}\tan U = \sec^2 U\,\frac{dU}{dx}$$

A1.5 Indefinite Integrals

Let, $U = U(x)$, $V = V(x)$, and a = constant

$$\int a\,dx = ax + C$$

$$\int U\,dV = U\,V - \int V\,dU$$

$$\int U^n \, dU = \frac{U^{n+1}}{n+1} + C, \quad n \neq -1$$

$$\int \frac{1}{U} \, dU = \ln U + C$$

$$\int a^U \, dU = \frac{a^U}{\ln a} + C, \quad a > 0 \text{ and } a \neq 1$$

$$\int e^U \, dU = e^U + C$$

$$\int e^{ax} \, dx = \frac{1}{a} e^{ax} + C$$

$$\int x e^{ax} \, dx = \frac{e^{ax}}{a^2} (ax - 1) + C$$

$$\int x^2 e^{ax} \, dx = \frac{e^{ax}}{a^3} (a^2 x^2 - 2ax + 2) + C$$

$$\int \ln x \, dx = x \ln x - x + C$$

$$\int \sin ax \, dx = -\frac{1}{a} \cos ax + C$$

$$\int \cos ax \, dx = \frac{1}{a} \sin ax + C$$

$$\int \tan ax \, dx = \frac{1}{a} \ln (\sec ax) + C = -\frac{1}{a}(\cos ax) + C$$

$$\int \sec ax \, dx = \frac{1}{a} (\sec ax + \tan ax) + C$$

MATLAB Commands and Functions

Operators and Special Characters	
+	Plus; addition operator.
−	Minus; subtraction operator.
*	Scalar and matrix multiplication operator.
.*	Array multiplication operator.
∧	Scalar and matrix exponentiation operator.
.∧	Array exponentiation operator.
\	Left-division operator.
/	Right-division operator.
.\	Array left-division operator.
./	Array right-division operator.
:	Colon; generates regularly spaced elements and represents an entire row/column.
( )	Parentheses; encloses function arguments and array indices; overrides precedence.
[ ]	Brackets; enclosures array elements.
.	Decimal point.
...	Ellipsis; line-continuation operator.
,	Comma; separates statements and elements in a row.
;	Semicolon; separates columns and suppresses display.
%	Percent sign; designates a comment and specifies formatting.
_	Quote sign and transpose operator.
._	Nonconjugated transpose operator.
=	Assignment (replacement) operator.

Logical and Relational Operators	
==	Relational operator : equal to.
~=	Relational operator : not equal to.
<	Relational operator : less than.
<=	Relational operator : less than or equal to.
>	Relational operator : greater than.
>=	Relational operator : greater than or equal to.
&	Logical operator : AND.
|	Logical operator : OR.
~	Logical operator : NOT.
xor	Logical operator : EXCLUSIVE OR.

Special Variables and Constants	
ans	Most recent answer.
eps	Accuracy of floating-point precision.
i,j	The imaginary unit ; $\sqrt{-1}$.
Inf	Infinity.
NaN	Undefined numerical result (not a number).
pi	The number π .

Commands for Managing a Session	
clc	Clears Command window.
clear	Removes variables from memory.
exist	Checks for existence of file or variable.
global	Declares variables to be global.
help	Searches for a help topic.
lookfor	Searches help entries for a keyword.
quit	Stops MATLAB.
who	Lists current variables.
whos	Lists current variables (long display).

Input/Output Commands	
disp	Displays contents of an array or string.
fscanf	Read formatted data from a file.
format	Controls screen-display format.
fprintf	Performs formatted writes to screen or file.
input	Displays prompts and waits for input.
;	Suppresses screen printing.

Format Codes for fprintf and fscanf	
%s	Format as a string.
%d	Format as an integer.
%f	Format as a floating point value.
%e	Format as a floating point value in scientific notation.
%g	Format in the most compact form : %f or %e.
\n	Insert a new line in the output string.
\t	Insert a tab in the output string.

Array Commands	
`cat`	Concatenates arrays.
`find`	Finds indices of nonzero elements.
`length`	Computes number of elements.
`linspace`	Creates regularly spaced vector.
`logspace`	Creates logarithmically spaced vector.
`max`	Returns largest element.
`min`	Returns smallest element.
`prod`	Product of each column.
`reshape`	Change size
`size`	Computes array size.
`sort`	Sorts each column.
`sum`	Sums each column.

Special Matrices	
`eye`	Creates an identity matrix.
`ones`	Creates an array of ones.
`zeros`	Creates an array of zeros.

Program Flow Control	
`break`	Terminates execution of a loop.
`case`	Provides alternate execution paths within switch structure.
`else`	Delineates alternate block of statements.
`elseif`	Conditionally executes statements.
`end`	Terminates for, while, and if statements.
`error`	Displays error messages.
`for`	Repeats statements a specific number of times
`if`	Executes statements conditionally.
`otherwise`	Default part of switch statement.
`return`	Return to the invoking function.
`switch`	Directs program execution by comparing point with case expressions.
`warning`	Display a warning message.
`while`	Repeats statements an indefinite number of times.

Basic xy Plotting Commands	
axis	Sets axis limits.
fplot	Intelligent plotting of functions.
grid	Displays gridlines.
plot	Generates xy plot.
print	Prints plot or saves plot to a file
title	Puts text at top of plot.
xlabel	Adds text label to x-axis.
ylabel	Adds text label to y-axis.

Plot Enhancement Commands	
axes	Creates axes objects.
close	Closes the current plot.
close	Closes all plots.
figure	Opens a new figure window.
gtext	Enables label placement by mouse.
hold	Freezes current plot.
legend	Legend placement by mouse.
refresh	Redraws current figure window.
set	Specifies properties of objects such as axes.
subplot	Creates plots in subwindows.
text	Places string in figure.

Specialized Plot Commands	
bar	Creates bar chart.
loglog	Creates log-log plot.
polar	Creates polar plot.
semilogx	Creates semilog plot (logarithmic abscissa).
semilogy	Creates semilog plot (logarithmic ordinate).
stairs	Creates stairs plot.
stem	Creates stem plot.

Convolution Functions	
conv(x,h)	Returns convolution of x and h.
deconv(y,x)	Returns deconvolution of y and x.

Logical Functions	
any	True if any elements are nonzero.
all	True if all elements are nonzero.
find	Finds indices of nonzero elements.
finite	True if elements are finite.
isnan	True if elements are undefined.
isinf	True if elements are infinite.
isempty	True if matrix is empty.
isreal	True if all elements are real.

Exponential and Logarithmic Functions	
exp(x)	Exponential; e^x.
log(x)	Natural logarithm; $\ln(x)$.
log10(x)	Common (base 10) logarithm; $\log(x) = \log_{10}(x)$.
sqrt(x)	Square root of x; $\sqrt{x}$.

Trigonometric Functions	
acos(x)	Inverse cosine; $\cos^{-1}(x)$.
acot(x)	Inverse cotangent; $\cot^{-1}(x)$.
acsc(x)	Inverse cosecant; $\mathrm{cosec}^{-1}(x)$.
asec(x)	Inverse secant; $\sec^{-1}(x)$.
asin(x)	Inverse sine; $\sin^{-1}(x)$.
atan(x)	Inverse tangent; $\tan^{-1}(x)$.
atan2(y,x)	Four-quadrant inverse tangent.
cos(x)	Cosine; $\cos(x)$.
cot(x)	Cotangent; $\cot(x)$.
csc(x)	Cosecant; $\mathrm{cosec}(x)$.
sec(x)	Secant; $\sec(x)$.
sin(x)	Sine; $\sin(x)$.
tan(x)	Tangent; $\tan(x)$.

Hyperbolic Functions	
`acosh(x)`	Inverse hyperbolic cosine; $\cosh^{-1}(x)$.
`acoth(x)`	Inverse hyperbolic cotangent; $\coth^{-1}(x)$.
`acsch(x)`	Inverse hyperbolic cosecant; $\operatorname{cosech}^{-1}(x)$.
`asech(x)`	Inverse hyperbolic secant; $\operatorname{sech}^{-1}(x)$.
`asinh(x)`	Inverse hyperbolic sine; $\sinh^{-1}(x)$.
`atanh(x)`	Inverse hyperbolic tangent; $\tanh^{-1}(x)$.
`cosh(x)`	Hyperbolic cosine; $\cosh(x)$.
`coth(x)`	Hyperbolic cotangent; $\cosh(x)/\sinh(x)$.
`csch(x)`	Hyperbolic cosecant; $1/\sinh(x)$.
`sech(x)`	Hyperbolic secant; $1/\cosh(x)$.
`sinh(x)`	Hyperbolic sine; $\sinh(x)$.
`tanh(x)`	Hyperbolic tangent; $\sinh(x)/\cosh(x)$.

Complex Functions	
`abs(x)`	Absolute value; $|x|$.
`angle(x)`	Angle of a complex number x.
`conj(x)`	Complex conjugate of x.
`imag(x)`	Imaginary part of a complex number x.
`real(x)`	Real part of a complex number x.

State Space Functions	
`ss2tf`	Computes transfer function from state model.
`tf2ss`	Computes state model from transfer function.

Transform Functions	
`fft`	Computes DFT via FFT.
`fourier`	Returns the Fourier transform.
`ifft`	Computes inverse DFT via FFT.
`ifourier`	Returns the inverse Fourier transform.
`ilaplace`	Returns the inverse Laplace transform.
`iztrans`	Returns the inverse $\mathcal{Z}$-transform.
`laplace`	Returns the Laplace transform.
`ztrans`	Returns the $\mathcal{Z}$-transform.

Summary of Various Standard Transform Pairs

Table A3.1: Standard Continous Time Fourier Transform Paris

$x(t)$	$X(j\Omega)$
$\delta(t)$	1
$\delta(t-t_0)$	$e^{-j\Omega t_0}$
A where, A is constant	$2\pi A\delta(\Omega)$
$u(t)$	$\pi\delta(\Omega) + \dfrac{1}{j\Omega}$
$\text{sgn}(t)$	$\dfrac{2}{j\Omega}$
$t\, u(t)$	$\dfrac{1}{(j\Omega)^2}$
$\dfrac{t^{m-1}}{(m-1)!}\, u(t)$ where, $m = 1,2,3, \ldots$	$\dfrac{1}{(j\Omega)^m}$
$t^m\, u(t)$ where, $m = 1,2,3, \ldots$	$\dfrac{m!}{(j\Omega)^{m+1}}$
$e^{-at}\, u(t)$	$\dfrac{1}{j\Omega + a}$
$t\, e^{-at}\, u(t)$	$\dfrac{1}{(j\Omega + a)^2}$
$Ae^{-a|t|}$	$\dfrac{2Aa}{a^2 + \Omega^2}$
$Ae^{j\Omega_0 t}$	$2\pi A\, \delta(\Omega - \Omega_0)$
$\sin \Omega_0 t$	$\dfrac{\pi}{j}\left[\delta(\Omega - \Omega_0) - \delta(\Omega + \Omega_0)\right]$
$\cos \Omega_0 t$	$\pi\left[\delta(\Omega - \Omega_0) + \delta(\Omega + \Omega_0)\right]$

Table A3.2: Standard Laplace Transform Pairs

Note: σ = *Real part of* s

x(t)	X(s)	ROC
$\delta(t)$	1	Entire s-plane
$u(t)$	$\dfrac{1}{s}$	$\sigma > 0$
$t\, u(t)$	$\dfrac{1}{s^2}$	$\sigma > 0$
$\dfrac{t^m - 1}{(m-1)!} u(t)$ where, $m = 1, 2, 3, ...$	$\dfrac{1}{s^m}$	$\sigma > 0$
$e^{-at}\, u(t)$	$\dfrac{1}{s+a}$	$\sigma > -a$
$-e^{-at}\, u(-t)$	$\dfrac{1}{s+a}$	$\sigma < -a$
$t^m\, u(t)$ where, $m = 1, 2, 3, ...$	$\dfrac{m!}{s^{m+1}}$	$\sigma > 0$
$t\, e^{-at}\, u(t)$	$\dfrac{1}{(s+a)^2}$	$\sigma > -a$
$\dfrac{1}{(m-1)!}\, t^{m-1}\, e^{-at}\, u(t)$ where, $m = 1, 2, 3, ...$	$\dfrac{1}{(s+a)^m}$	$\sigma > -a$
$t^m\, e^{-at}\, u(t)$ where, $m = 1, 2, 3, ...$	$\dfrac{m!}{(s+a)^{m+1}}$	$\sigma > -a$
$\sin\Omega_0 t\, u(t)$	$\dfrac{\Omega_0}{s^2 + \Omega_0^2}$	$\sigma > 0$
$\cos\Omega_0 t\, u(t)$	$\dfrac{s}{s^2 + \Omega_0^2}$	$\sigma > 0$
$\sinh \Omega_0 t\, u(t)$	$\dfrac{\Omega_0}{s^2 - \Omega_0^2}$	$\sigma > \Omega_0$
$\cosh \Omega_0 t\, u(t)$	$\dfrac{s}{s^2 - \Omega_0^2}$	$\sigma > \Omega_0$
$e^{-at}\, \sin\Omega_0 t\, u(t)$	$\dfrac{\Omega_0}{(s+a)^2 + \Omega_0^2}$	$\sigma > -a$
$e^{-at}\, \cos\Omega_0 t\, u(t)$	$\dfrac{s+a}{(s+a)^2 + \Omega_0^2}$	$\sigma > -a$

Table A3.3: Summary of Laplace and Fourier Tranform for Causal Signals

$x(t)$ for $t = 0$ to ∞	$X(s)$	$X(j\Omega)$ $\left[X(j\Omega) = X(s)\big\vert_{s=j\Omega} \right]$
$\delta(t)$	1	1
$u(t)$	$\dfrac{1}{s}$	$\dfrac{1}{j\Omega}$
$t\,u(t)$	$\dfrac{1}{s^2}$	$\dfrac{1}{(j\Omega)^2}$
$\dfrac{t^{n-1}}{(n-1)!}\,u(t)$ where, $n = 1,2,3, ...$	$\dfrac{1}{s^n}$	$\dfrac{1}{(j\Omega)^n}$
$t^n u(t)$ where, $n = 1,2,3, ...$	$\dfrac{n!}{s^{n+1}}$	$\dfrac{n!}{(j\Omega)^{n+1}}$
$e^{-at} u(t)$	$\dfrac{1}{s+a}$	$\dfrac{1}{j\Omega+a}$
$t\,e^{-at} u(t)$	$\dfrac{1}{(s+a)^2}$	$\dfrac{1}{(j\Omega+a)^2}$
$\sin\Omega_0 t\, u(t)$	$\dfrac{\Omega_0}{s^2+\Omega_0^2}$	$\dfrac{\Omega_0}{(j\Omega)^2+\Omega_0^2} = \dfrac{\Omega_0}{\Omega_0^2-\Omega^2}$
$\cos\Omega_0 t\, u(t)$	$\dfrac{s}{s^2+\Omega_0^2}$	$\dfrac{j\Omega}{(j\Omega)^2+\Omega_0^2} = \dfrac{j\Omega}{\Omega_0^2-\Omega^2}$
$\sinh\Omega_0 t\, u(t)$	$\dfrac{\Omega_0}{s^2-\Omega_0^2}$	$\dfrac{\Omega_0}{(j\Omega)^2-\Omega_0^2} = \dfrac{-\Omega}{\Omega^2+\Omega_0^2}$
$\cosh\Omega_0 t\, u(t)$	$\dfrac{s}{s^2-\Omega_0^2}$	$\dfrac{j\Omega}{(j\Omega)^2-\Omega_0^2} = \dfrac{-j\Omega}{\Omega^2+\Omega_0^2}$
$e^{-at}\sin\Omega_0 t\, u(t)$	$\dfrac{\Omega_0}{(s+a)^2+\Omega_0^2}$	$\dfrac{\Omega_0}{(j\Omega+a)^2+\Omega_0^2}$
$e^{-at}\cos\Omega_0 t\, u(t)$	$\dfrac{s+a}{(s+a)^2+\Omega_0^2}$	$\dfrac{j\Omega+a}{(j\Omega+a)^2+\Omega_0^2}$

Table A3.4: Standard Z-transform Pairs

x(t)	x(n)	X(z) With positive power of z	X(z) With negative power of z	ROC
	$\delta(n)$	1	1	Entire z-plane
	u(n) or 1	$\dfrac{z}{z-1}$	$\dfrac{1}{1-z^{-1}}$	$|z|>1$
	$a^n\,u(n)$	$\dfrac{z}{z-a}$	$\dfrac{1}{1-az^{-1}}$	$|z|>|a|$
	$n\,a^n\,u(n)$	$\dfrac{az}{(z-a)^2}$	$\dfrac{az^{-1}}{(1-az^{-1})^2}$	$|z|>|a|$
	$n^2\,a^n\,u(n)$	$\dfrac{az\,(z+a)}{(z-a)^3}$	$\dfrac{az^{-1}\,(1+az^{-1})}{(1-az^{-1})^3}$	$|z|>|a|$
	$-\,a^n\,u(-n-1)$	$\dfrac{z}{z-a}$	$\dfrac{1}{1-az^{-1}}$	$|z|<|a|$
	$-na^n\,u(-n-1)$	$\dfrac{az}{(z-a)^2}$	$\dfrac{az^{-1}}{(1-az^{-1})^2}$	$|z|<|a|$
$t\,u(t)$	$nT\,u(nT)$	$\dfrac{Tz}{(z-1)^2}$	$\dfrac{Tz^{-1}}{(1-z^{-1})^2}$	$|z|>1$
$t^2\,u(t)$	$(nT)^2\,u(nT)$	$\dfrac{T^2 z\,(z+1)}{(z-1)^3}$	$\dfrac{T^2 z^{-1}\,(1+z^{-1})}{(1-z^{-1})^3}$	$|z|>1$
$e^{-at}\,u(t)$	$e^{-anT}\,u(nT)$	$\dfrac{z}{z-e^{-aT}}$	$\dfrac{1}{1-e^{-aT}z^{-1}}$	$|z|>|e^{-aT}|$
$t\,e^{-at}\,u(t)$	$nT\,e^{-anT}\,u(nT)$	$\dfrac{zT\,e^{-aT}}{(z-e^{-aT})^2}$	$\dfrac{z^{-1}T\,e^{-aT}}{(1-e^{-aT}z^{-1})^2}$	$|z|>|e^{-aT}|$
$\sin\Omega_0 t\,u(t)$	$\sin\Omega_0 nT\,u(nT)$ $=\sin\omega n\,u(nT)$ where, $\omega=\Omega_0 T$	$\dfrac{z\sin\omega}{z^2-2z\cos\omega+1}$	$\dfrac{z^{-1}\sin\omega}{1-2z^{-1}\cos\omega+z^{-2}}$	$|z|>1$
$\cos\Omega_0 t\,u(t)$	$\cos\Omega_0 nT\,u(nT)$ $=\cos\omega n\,u(nT)$ where, $\omega=\Omega_0 T$	$\dfrac{z\,(z-\cos\omega)}{z^2-2z\cos\omega+1}$	$\dfrac{1-z^{-1}\cos\omega}{1-2z^{-1}\cos\omega+z^{-2}}$	$|z|>1$

Note : 1. *The signals multiplied by u(n) are causal signals (defined for $n \geq 0$).*
2. *The signals multiplied by u(-n-1) are anticausal signals (defined for $n \leq 0$).*

Table A3.5: Standard Discrete Time Fourier Transform Pairs

x(t)	x(n)	$X(e^{j\omega})$ with positive power of $e^{j\omega}$	$X(e^{j\omega})$ with negative power of $e^{j\omega}$
	$\delta(n)$	1	1
	$\delta(n-n_0)$	$\dfrac{1}{e^{j\omega n_0}}$	$e^{-j\omega n_0}$
	$u(n)$	$\dfrac{e^{j\omega}}{e^{j\omega}-1} + \displaystyle\sum_{m=-\infty}^{+\infty}\pi\delta(\omega-2\pi m)$	$\dfrac{1}{1-e^{-j\omega}} + \displaystyle\sum_{m=-\infty}^{+\infty}\pi\delta(\omega-2\pi m)$
	$a^n\, u(n)$	$\dfrac{e^{j\omega}}{e^{j\omega}-a}$	$\dfrac{1}{1-ae^{-j\omega}}$
	$n\, a^n\, u(n)$	$\dfrac{ae^{j\omega}}{\left(e^{j\omega}-a\right)^2}$	$\dfrac{ae^{-j\omega}}{\left(1-ae^{-j\omega}\right)^2}$
	$n^2\, a^n\, u(n)$	$\dfrac{ae^{j\omega}\left(e^{j\omega}+a\right)}{\left(e^{j\omega}-a\right)^3}$	$\dfrac{ae^{-j\omega}\left(1+ae^{-j\omega}\right)}{\left(1-ae^{-j\omega}\right)^3}$
$e^{-at}u(t)$	$e^{-anT}u(nT)$	$\dfrac{e^{j\omega}}{e^{j\omega}-e^{-aT}}$	$\dfrac{1}{1-e^{-j\omega}e^{-aT}}$
	1	$2\pi\displaystyle\sum_{m=-\infty}^{+\infty}\delta(\omega-2\pi m)$	
	$a^{|n|}$	$\dfrac{1-a^2}{1-2a\cos\omega+a^2}$	
	$\displaystyle\sum_{m=-\infty}^{+\infty}\delta(n-mN)$	$\dfrac{2\pi}{N}\displaystyle\sum_{m=-\infty}^{+\infty}\delta\!\left(\omega-\dfrac{2\pi m}{N}\right)$	
$e^{j\Omega_0 t}$	$e^{j\Omega_0 nt}=e^{j\omega_0 n}$ where, $\omega_0=\Omega_0 T$	$2\pi\displaystyle\sum_{m=-\infty}^{+\infty}\delta(\omega-\omega_0-2\pi m)$	
$\sin\Omega_0 t$	$\sin\Omega_0 nT$ $=\sin\omega_0 n$ where, $\omega_0=\Omega_0 T$	$\dfrac{\pi}{j}\displaystyle\sum_{m=-\infty}^{+\infty}\left[\delta(\omega-\omega_0-2\pi m)-\delta(\omega+\omega_0-2\pi m)\right]$	
$\cos\Omega_0 t$	$\cos\Omega_0 nT$ $=\cos\omega_0 n$ where, $\omega_0=\Omega_0 T$	$\pi\displaystyle\sum_{m=-\infty}^{+\infty}\left[\delta(\omega-\omega_0-2\pi m)+\delta(\omega+\omega_0-2\pi m)\right]$	

Table A3.6: Standard Common Z-transform and Fourier Transform Pairs

x(t)	x(n)	X(z)	$X(e^{j\omega})$
	$\delta(n)$	1	1
	$a^n\, u(n) \quad ; \quad \|a\| < 1$	$\dfrac{z}{z-a}$	$\dfrac{e^{j\omega}}{e^{j\omega}-a}$
	$n\, a^n\, u(n) \quad ; \quad \|a\| < 1$	$\dfrac{az}{(z-a)^2}$	$\dfrac{ae^{j\omega}}{(e^{j\omega}-a)^2}$
	$n^2\, a^n\, u(n) \quad ; \quad \|a\| < 1$	$\dfrac{az(z+a)}{(z-a)^3}$	$\dfrac{ae^{j\omega}(e^{j\omega}+a)}{(e^{j\omega}-a)^3}$
$e^{-at}\, u(t)$	$e^{-anT}u(nT) \quad ; \quad \|e^{-aT}\| < 1$	$\dfrac{z}{z-e^{-aT}}$	$\dfrac{e^{j\omega}}{e^{j\omega}-e^{-aT}}$
$te^{-at}\, u(t)$	$nTe^{-anT}u(nT) \quad ; \quad \|e^{-aT}\| < 1$	$\dfrac{zTe^{-aT}}{(z-e^{-aT})^2}$	$\dfrac{e^{j\omega}Te^{-aT}}{(e^{j\omega}-e^{-aT})^2}$

Summary of Properties of Various Transform

Table A4.1: Properties of Laplace Transform

Note: $\mathcal{L}\{x(t)\} = X(s); \quad \mathcal{L}\{x_1(t)\} = X_1(s); \quad \mathcal{L}\{x_2(t)\} = X_2(s)$

Property	Time Domain Signal	s-domain Signal
Amplitude scaling	$A\,x(t)$	$A\,X(s)$
Linearity	$a_1 x_1(t) \pm a_2 x_2(t)$	$a_1 X_1(s) \pm a_2 X_2(s)$
Time differentiation	$\dfrac{d}{dt}x(t)$	$s\,X(s) - x(0)$
	$\dfrac{d^m}{dt^m}x(t)$ where m = 1, 2, 3	$s^m X(s) - \displaystyle\sum_{K=1}^{m} s^{m-K}\,\dfrac{d^{(K-1)}\,x(t)}{dt^{K-1}}\bigg|_{t=0}$
Time integration	$\displaystyle\int x(t)\,dt$	$\dfrac{X(s)}{s} + \dfrac{\left[\int x(t)\,dt\right]\big|_{t=0}}{s}$
	$\displaystyle\int .....\int x(t)\,(dt)^m$ where m = 1, 2, 3	$\dfrac{X(s)}{s^m} + \displaystyle\sum_{K=1}^{m} \dfrac{1}{s^{m-K+1}}\left[\int .....\int x(t)\,(dt)^k\right]\bigg|_{t=0}$
Frequency shifting	$e^{\pm at}x(t)$	$X(s \mp a)$
Time shifting	$x(t \pm a)$	$e^{\pm as}\,X(s)$
Frequency differentiation	$t\,x(t)$	$-\dfrac{dX(s)}{ds}$
	$t^m x(t)$ where m = 1, 2, 3	$(-1)^m \dfrac{d^m}{ds^m}\,X(s)$
Frequency integration	$\dfrac{1}{t}\,x(t)$	$\displaystyle\int_s^{\infty} X(s)\,ds$
Time scaling	$x(at)$	$\dfrac{1}{|a|}\,X\!\left(\dfrac{s}{a}\right)$
Peri.odicity	$x(t + mT)$ where m = 1, 2, 3.... T = Period	$\dfrac{1}{1-e^{-sT}}\displaystyle\int_0^{T} x_1(t)\,e^{-st}\,dt$ where, $x_1(t)$ is one period of $x(t)$
Initial value theorem	$\underset{t \to 0}{\text{Lt}}\ x(t) = x(0)$	$\underset{s \to \infty}{\text{Lt}}\ s\,X(s)$
Final value theorem	$\underset{t \to \infty}{\text{Lt}}\ x(t) = x(\infty)$	$\underset{s \to 0}{\text{Lt}}\ s\,X(s)$
Convolution theorem	$x_1(t) * x_2(t)$ $= \displaystyle\int_{-\infty}^{+\infty} x_1(\lambda)\,x_2(t-\lambda)\,d\lambda$	$X_1(s)\,X_2(s)$

Table A4.2: Properties of Exponential Form of Fourier Series Coefficients

Note : c_n and d_n are exponential form of Fourier series coefficients of x(t) and y(t) respectively.

Property	Continuous Time Periodic Signal	Fourier Series Coefficient
Linearity	$A\,x(t) + B\,y(t)$	$A\,c_n + B\,d_n$
Time shifting	$x(t - t_0)$	$c_n e^{-jn\Omega_0 t_0}$
Frequency shifting	$e^{-jm\Omega_0 t}\,x(t)$	c_{n-m}
Conjugation	$x^*(t)$	c_{-n}^*
Time reversal	$x(-t)$	c_{-n}
Time scaling	$x(\alpha t)\;;\;\alpha > 0$ [x(t) is periodic with period T/α]	c_n (No change in Fourier coefficient)
Multiplication	$x(t)\,y(t)$	$\displaystyle\sum_{m=-\infty}^{+\infty} c_m\,d_{n-m}$
Differentiation	$\dfrac{d}{dt}x(t)$	$jn\Omega_0 c_n$
Integration	$\displaystyle\int_{-\infty}^{t} x(t)\,dt$ (Finite valued and periodic only if $a_0 = 0$)	$\dfrac{1}{jn\Omega_0}\,c_n$
Periodic convolution	$\displaystyle\int_T x(\tau)\,y(t-\tau)\,d\tau$	$T\,c_n\,d_n$
Symmetry of real signals	x(t) is real	$c_n = c_{-n}^*$ $|c_n| = |c_{-n}|\;;\;\angle c_n = -\angle c_{-n}$ $\mathrm{Re}\{c_n\} = \mathrm{Re}\{-c_n\}$ $\mathrm{Im}\{c_n\} = -\mathrm{Im}\{c_{-n}\}$
Real and even	x(t) is real and even	c_n are real and even
Real and odd	x(t) is real and odd	c_n are imaginary and odd
Parseval's relation	Average power, P of x(t) is defined as $P = \dfrac{1}{T}\displaystyle\int_T |x(t)|^2\,dt$	The average power, P in terms of Fourier series coefficients is $P = \displaystyle\sum_{n=-\infty}^{+\infty}|c_n|^2$
	$\dfrac{1}{T}\displaystyle\int_T |x(t)|^2\,dt = \sum_{n=-\infty}^{+\infty}|c_n|^2$	

Note: 1. The term $|c_n|^2$ respresents the power in n^{th} harmonic component of x(t). The total average power in a periodic signal is equal to the sum of power in all of its harmonics.

2. The term $|c_n|^2$ for n = 0, 1, 2, ... is the distribution of power as a function of frequency and so it is called **power density spectrum** or **power spectral density** of the periodic signal.

Table A4.3: Summary of Properties of Fourier Transform

Let, $\mathcal{F}\{x(t)\} = X(j\Omega)$; $\mathcal{F}\{x_1(t)\} = X_1(j\Omega)$; $\mathcal{F}\{x_2(t)\} = X_2(j\Omega)$

Property	Time Domain Signal	Frequency Domain Signal
Linearity	$a_1\,x_1(t) + a_2\,x_2(t)$	$a_1\,X_1(j\Omega) + a_2\,X_2(j\Omega)$
Time shifting	$x(t - t_0)$	$e^{-j\Omega t_0}\,X(j\Omega)$
Time scaling	$x(at)$	$\frac{1}{\|a\|}X\!\left(\frac{j\Omega}{a}\right)$
Time reversal	$x(-t)$	$X(-j\Omega)$
Conjugation	$x^*(t)$	$X^*(-j\Omega)$
Frequency shifting	$e^{j\Omega_0 t}\,x(t)$	$X(j(\Omega - \Omega_0))$
Time differentiation	$\dfrac{d}{dt}\,x(t)$	$j\Omega\,X(j\Omega)$
Time integration	$\displaystyle\int_{-\infty}^{t} x(\tau)\,d\tau$	$\dfrac{X(j\Omega)}{j\Omega} + \pi X(0)\,\delta(\Omega)$
Frequency differentiation	$t\,x(t)$	$j\dfrac{d}{d\Omega}\,X(j\Omega)$
Time convolution	$x_1(t) * x_2(t) = \displaystyle\int_{-\infty}^{+\infty} x_1(\tau)x_2(t-\tau)\,d\tau$	$X_1(j\Omega)\,X_2(j\Omega)$
Frequency convolution (or Multiplication)	$x_1(t)\,x_2(t)$	$\dfrac{1}{2\pi}\displaystyle\int_{\lambda=-\infty}^{\lambda=+\infty} X_1(j\lambda)\,X_2(j(\Omega-\lambda))\,d\lambda$
Symmetry of real signals	$x(t)$ is real	$X(j\Omega) = X^*(j\Omega)$ $\|X(j\Omega)\| = \|X(-j\Omega)\|; \angle X(j\Omega) = -\angle X(-j\Omega)$ $\mathrm{Re}\{X(j\Omega)\} = \mathrm{Re}\{X(-j\Omega)\}$ $\mathrm{Im}\{X(j\Omega)\} = -\mathrm{Im}\{X(-j\Omega)\}$
Real and even	$x(t)$ is real and even	$X(j\Omega)$ are real and even
Real and odd	$x(t)$ is real and odd	$X(j\Omega)$ are imaginary and odd
Duality	If $x_2(t) \equiv X_1(j\Omega)$ [i.e. $x_2(t)$ and $X_1(j\Omega)$ are similar functions] then $X_2(j\Omega) \equiv 2\pi x_1(-j\Omega)$ [i.e. $X_2(j\Omega)$ and $2\pi x_1(-j\Omega)$ are similar functions]	
Area under a frequency domain signal	$\displaystyle\int_{-\infty}^{+\infty} X(j\Omega)\,d\Omega = 2\pi\,x(0)$	
Area under a time domain signal	$\displaystyle\int_{-\infty}^{+\infty} x(t)\,dt = X(0)$	
Parseval's relation	Energy in time domain is, $E = \displaystyle\int_{-\infty}^{+\infty} \|x(t)\|^2\,dt$	Energy in frequency domain is, $E = \dfrac{1}{2\pi}\displaystyle\int_{-\infty}^{+\infty} \|X(j\Omega)\|^2\,d\Omega$
	$\displaystyle\int_{-\alpha}^{+\infty} \|x(t)\|^2\,dt = \dfrac{1}{2\pi}\displaystyle\int_{-\infty}^{+\infty} \|X(j\Omega)\|^2\,d\Omega$	

Table A4.4: Summary of Properties of $\mathcal{Z}$-Transform

Note : $X(z) = \mathcal{Z}\{x(n)\}$; $X_1(z) = \mathcal{Z}\{x_1(n)\}$; $X_2(z) = \mathcal{Z}\{x_2(n)\}$; $Y(z) = \mathcal{Z}\{y(n)\}$

Property		Discrete time signal	$\mathcal{Z}$-transform
Linearity		$a_1 x_1(n) + a_2 x_2(n)$	$a_1 X_1(z) + a_2 X_2(z)$
Shifting $(m \geq 0)$	$x(n)$; for $n \geq 0$	$x(n-m)$	$z^{-m} X(z) + \displaystyle\sum_{i=1}^{m} x(-i) z^{-(m-i)}$
		$x(n+m)$	$z^m X(z) - \displaystyle\sum_{i=0}^{m-1} x(i) z^{m-i}$
	$x(n)$; for all n	$x(n-m)$	$z^{-m} X(z)$
		$x(n+m)$	$z^m X(z)$
Multiplication by n^m (or differentiation in z-domain)		$n^m x(n)$	$\left(-z \dfrac{d}{dz}\right)^m X(z)$
Scaling in z-domain (or multiplication by a^n)		$a^n x(n)$	$X(a^{-1} z)$
Time reversal		$x(-n)$	$X(z^{-1})$
Conjugation		$x^*(n)$	$X^*(z^*)$
Convolution		$x_1(n) * x_2(n) = \displaystyle\sum_{m=-\infty}^{+\infty} x_1(m) x_2(n-m)$	$X_1(z) X_2(z)$
Corrrelation		$r_{xy}(m) = \displaystyle\sum_{m=-\infty}^{+\infty} x(n) y(n-m)$	$X(z) Y(z^{-1})$
Initial value		$x(0) = \underset{z \to \infty}{Lt} X(z)$	
Final value		$x(\infty) = \underset{z \to 1}{Lt} (1 - z^{-1}) X(z)$ $= \underset{z \to 1}{Lt} \left(\dfrac{z-1}{z}\right) X(z)$ if $X(z)$ is analytic for $\|z\| > 1$	
Complex convolution theorem		$x_1(n) x_2(n)$	$\dfrac{1}{2\pi j} \displaystyle\oint_C X_1(v) X_2\left(\dfrac{z}{v}\right) v^{-1} dv$
Parseval's relation		$\displaystyle\sum_{n=-\infty}^{+\infty} x_1(n) x_2^*(n) = \dfrac{1}{2\pi j} \oint_C X_1(z) X_2\left(\dfrac{1}{z^*}\right) z^{-1} dv$	

Table A4.5: Properties of Discrete Time Fourier Transform

Note : $X(e^{jw}) = \mathcal{F}\{x(n)\}$; $X_1(e^{jw}) = \mathcal{F}\{x_1(n)\}$; $X_2(e^{jw}) = \mathcal{F}\{x_2(n)\}$; $Y(e^{jw}) = F\{y(n)\}$

Property	Discrete time signal	Fourier transform
Linearity	$a_1 x_1(n) + a_2 x_2(n)$	$a_1 X_1(e^{j\omega}) + a_2 X_2(e^{j\omega})$
Periodicity	$x(n)$	$X(e^{j\omega + 2\pi m}) = X(e^{j\omega})$
Time shifting	$x(n-m)$	$e^{-j\omega m} X(e^{j\omega})$
Time reversal	$x(-n)$	$X(e^{-j\omega})$
Conjugation	$x^*(n)$	$X^*(e^{-j\omega})$
Frequency shifting	$e^{j\omega_0 n} x(n)$	$X(e^{j(\omega - \omega_0)})$
Multiplication	$x_1(n) \, x_2(n)$	$\dfrac{1}{2\pi} \displaystyle\int_{-\pi}^{+\pi} X_1(e^{j\lambda}) X_2(e^{j(\omega - \lambda)}) \, d\lambda$
Differentiation in frequency domain	$n \, x(n)$	$j\dfrac{d}{d\omega} X(e^{j\omega})$
Convolution	$x_1(n) * x_2(n) = \displaystyle\sum_{n=-\infty}^{+\infty} x_1(m) \, x_2(n-m)$	$X_1(e^{j\omega}) X_2(e^{j\omega})$
Correlation	$r_{xy}(m) = \displaystyle\sum_{m=-\infty}^{+\infty} x(n) y(n-m)$	$X(e^{j\omega}) Y(e^{-j\omega})$
Symmetry of real signals	$x(n)$ is real	$X(e^{j\omega}) = X^*(e^{-j\omega})$ $\text{Re}\{X(e^{j\omega})\} = \text{Re}\{X(e^{-j\omega})\}$ $\text{Im}\{X(e^{j\omega})\} = \text{Im}\{X(e^{-j\omega})\}$ $\|X(e^{j\omega})\| = \|X(e^{-j\omega})\|, \angle X(e^{-j\omega}) = -\angle X(e^{-j\omega})$
Symmetry of real and even signal	$x(n)$ is real and even	$X(e^{j\omega})$ is real and even
Symmetry of real and odd signal	$x(n)$ is real and odd	$X(e^{j\omega})$ is imaginary and odd
Parseval's relation	$\displaystyle\sum_{n=-\infty}^{+\infty} x_1(n) x_2^*(n)$	$\dfrac{1}{2\pi} \displaystyle\int_{-\pi}^{+\pi} X_1(e^{j\omega}) X_2^*(e^{j\omega}) \, d\omega$
Parseval's relation	Energy in time domain, $E = \displaystyle\sum_{n=-\infty}^{+\infty} |x(n)|^2$	Energy in frequency domain, $E = \dfrac{1}{2\pi} \displaystyle\int_{-\pi}^{+\pi} |X(e^{j\omega})|^2 \, d\omega$

Note : The term $|X(e^{j\omega})|^2$ represents the distribution of energy as a function of frequency and so it is called energy density spectrum or energy spectral density.

Table A4.6: Parsevals Relation in Various Transforms

Parseval's relation in continuous time fourier series	Average Power of P, x(t) is defined as, $$P = \frac{1}{T}\int_T \|x(t)\|^2\, dt$$	The average power, P in terms of Fourier series coefficients is, $$P = \sum_{n=-\infty}^{+\infty} \|c_n\|^2$$
	$$\frac{1}{T}\int_T \|x(t)\|^2\, dt = \sum_{n=-\infty}^{+\infty} \|c_n\|^2$$	
Parseval's relation in continuous time fourier transform	The energy in Time domain is $$E = \int_{n=-\infty}^{\infty} \|x(t)\|^2\, dt$$	The energy in frequency domain is $$E = \frac{1}{2\pi}\int_{-\infty}^{\infty} \|X(j\Omega)\|^2\, d\Omega$$
	$$\int_{-\infty}^{\infty} \|x(t)\|^2\, dt = \frac{1}{2\pi}\|X(j\Omega)\|^2\, d\Omega$$	
Parseval's relation in $\mathcal{Z}$-transform	$$\sum_{n=-\infty}^{+\infty} x_1(n)\, x_2^*(n) = \frac{1}{2\pi j}\oint X_1(z) X_2^*\left(\frac{1}{z^*}\right) z^{-1}\, dz$$	
Parseval's relation in discrete time fourier series	Average power P of x(n) is defined as $$P = \frac{1}{N}\sum_{n=0}^{N-1}\|x(n)\|^2$$	The Average power P in terms of Fourier series coefficient is, $$P = \sum_{k=0}^{N-1}\|c_k\|^2$$
	$$\frac{1}{N}\sum_{n=0}^{N-1}\|x(n)\|^2 = \sum_{k=0}^{N-1}\|c_k\|^2$$	
Parseval's relation in discrete time fourier transform	Energy E in time domain, $$E = \sum_{n=-\infty}^{+\infty}\|x(n)\|^2$$	Energy E in frequency domain, $$E = \frac{1}{2\pi}\int_{-\pi}^{\pi}\|X(e^{j\omega})\|^2\, d\omega$$
	$$\sum_{n=-\infty}^{+\infty}\|x(n)\|^2 = \frac{1}{2\pi}\int_{-\pi}^{\pi}\|X(e^{j\omega})\|^2\, d\omega$$	

B.E./B.Tech. DEGREE EXAMINATION, NOVEMBER/DECEMBER-2015
Third Semester
Electronics and Communication Engineering
EC 6303-SIGNAL AND SYSTEMS
(Regulation 2013)

Time : 3 hours   Maximum : 100 marks

Answer all questions

PART A - (10 × 2 = 20 Marks)

1. Find the value of the integral $\int_{-\infty}^{\infty} e^{-2t} \delta(t+2)\, dt$

    Chapter 1, SA - Q1.23 (iv)   *[Page No - 1.127]*

2. Give the relation between continous time unit impulse function δ(t). Step function u(t) and ramp function r(t).

    Chapter 1, Section 1.2.4 *[Page No - 1.22]*

3. State Dirichlets conditions.

    Chapter 2, Section 2.1.2 *[Page No - 2.2]*

4. Give the relation between Fourier transform and laplace transform.

    Chapter 2, Section 2.11 *[Page No - 2.128]*

5. What is u(t – 2) ∗ δ(t – 1)? Where ∗ represents convolution

    Chapter 3, SA - Q3.6   *[Page No - 3.112]*

6. Given the differential equation representation of a system $\dfrac{d^2}{dt^2}y(t) + 2\dfrac{d}{dt}y(t) - 3y(t) = 2x(t)$. Find the frequency response H(jΩ)

    Chapter 3, SA 3.15   *[Page No - 3.115]*

7. State the need for sampling.

    Chapter 4, SA- Q4.1   *[Page No - 4.91]*

8. Find the z - transform and its assosiated ROC for x[n] = { 1, –1, 2, 3, 4}

    Chapter 4, SA- Q4.12   *[Page No - 4.95]*   ↑

9. Distinguish between recursive and non- recursive systems.

    Chapter 1, Section 1.97 *[Page No - 1.107]*

10. Convolve the following signals, x[n] = { 1,3, 5} and h[n] = { 1, 4, –1}

    Chapter 5, SA- Q 5.5   *[Page No - 5.139]*

PART B -- (5 × 16 = 80 Marks)

11. (a) Given x[n] = {1,4,3,–1,2}. Plot the following signals   (16)

    (i) x[–n –1]   ↑   (ii) $x\left[-\dfrac{n}{2}\right]$

    (iii) x[–2n +1]   (iv) $x\left[-\dfrac{n}{2}+2\right]$

    Chapter 1, Example 1.6 *[Page No - 1.40]*

                              Or

    (b) Given the input- output relationship of a continous time system y(t) = t x(–t). Determine whether the system is causal, stable, linear and time invariant.

    Chapter 1, Example 1.25 *[Page No - 1.85]*

12. (a) State and Prove any four properties of Fourier transform.   (16)

    Chapter 2, Section 2.5.7 *[Page No - 2.43]*

Or

(b) Find the Laplace transform and its associated ROC for the signal x(t) = te$^{-2|t|}$.   (16)

Chapter 2, Example 2.25 (f) *[Page No - 2.88]*

13. (a) Convolve the following signals: (16)

x(t) = e^{-2t} u(t − 2)

h(t) = e^{-2t} u(t)

Chapter 3, Example 3.6(e) *[Page No - 3.23]*

Or

(b) The input - ouput of a causal LTI system are related by the differential equation   (16)

$$\frac{d^2}{dt^2} y(t) + 6 \frac{d}{dt} y(t) + 8y(t) = 2x(t).$$

(i) Find the impulse response h(t)

(ii) Find the response y(t) of the system if x(t) = u(t)

Hint: Use Fourier transform.

Chapter 3, Example 3.37 *[Page No - 3.105]*

14. (a) State and explain sampling theorem both in time and frequency domain with necessary
quantitative analysis and illustrations.   (16)

Chapter 4, Section 4.1 *[Page No - 4.1]*

Or

(b) State and Prove any two properties of DTFT and any two properties of Z - transform.

Chapter 4, Section 4.2.3 & 4.6 *[Page No - 4.9 & 4.48]*   (16)

15. (a) Convolve the following signals:

$$x[n] = \left(\frac{1}{2}\right)^{n-2} u[n-2]$$ (16)

h[n] = u[n + 2]

Chapter 5, Example 5.10 *[Page No - 5.33]*

Or

(b) Consider an LTI system with impulse response h[n] = α^n u(n) and input to the system is
x(n) = β^n u(n) with | α | & | β | < 1. Determine response y[n].   (16)

(i) When $\alpha = \beta$

(ii) When $\alpha \neq \beta$

Using DTFT

Chapter 5, Example 5.13 *[Page No - 5.49]*

B.E./B.Tech. DEGREE EXAMINATION, APRIL/MAY-2015
Third Semester
Electronics and Communication Engineering
EC6303 -SIGNAL AND SYSTEMS
(Regulation 2013)

Time : 3 hours   Maximum : 100 marks

Answer all questions

PART A - (10 × 2 = 20 Marks

1. Define a power signal.
    Chapter 1, Section - 1.4.4   *[Page No - 1.52]*

2. How the impulse response of a discrete time system is useful in determinig its stability & causality?
    Chapter 1, Section 1.9.4 & 1.9.5   *[Page No - 1.100 & 1.103]*

3. Find the Fourier coefficients of the signal $x(t) = 1 + \sin 2\omega t + 2\cos 2\omega t + \cos\left(3\omega t + \frac{\pi}{3}\right)$
    Chapter 2, SA - Q2.2.   *[Page No - 2.130]*

4. Draw the spectrum of a CT rectangular pulse.
    Chapter 2, Table 2.6   *[Page No - 2.62]*

5. Given x(t) = δ(t). Find X(s) and X(ω).
    Chapter 3, Section 2.6.1, SA Q2.23   *[Page No - 2.53 & 2.138]*

6. State the convolution integral
    Chapter 3, Section 3.4   *[Page No - 3.14]*

7. Determine the Nyquist sampling rate for x(t) = sin (200 πt) + 3sin ²(120 πt)
    Chapter 4, SA - Q 4.2   *[Page No - 4.91]*

8. List the methods used for finding the inverse z - transform.
    Chapter 4, Section 4.8   *[Page No - 4.67]*

9. Name the basic building blocks used in LTIDT system block diagram.
    Chapter 5, Table 5.1   *[Page No - 5.3]*

10. Write the nth order difference equation.
    Chapter 5, Section 5.10   *[Page No - 5.96]*

PART-B -- (5 ×16 = 80 marks)

11 (a) (i) Give an account for the classification of signals in detail   (10)
           Chapter 1, Section 1.4 & 1.5   *[Page No - 1.41 & 1.56]*

     (ii) Sketch the following signals.
        (1) $[u(t-2)+u(t-4)]$ (2) $(t-4)[u(t-2)-u(t-4)]$ (6)
        Chapter 1, Example1.4(a,b)   *[Page No - 1.26]*
                                      Or

  (b) (i) Check if $x(t) = 4\cos\left(3\pi t + \frac{\pi}{4}\right) + 2\cos(4\pi t)$ is periodic   (6)
        Chapter 1, Example 1.8(e)   *[Page No - 1.47]*

     (ii) For the system $y(n)=[\log x(n)]$, check for linearity, causality, time invariance and stability
        Chapter 1, Example 1.42   *[Page No - 1.114]*   (10)

12 (a) (i) Determine the fourier series representation for a periodic ramp signal with unit amplitude and a period T.   (10)
        Chapter 2, Example 2.8   *[Page No - 2.27]*

     (ii) Find the fourier transform of $x(t) = t\,e^{-at}\,u(t)$ (6)
        Chapter 2, Example 2.13 (c)   *[Page No - 2.66]*

Or

(b) (i)   If $x(t) \Leftrightarrow X(\omega)$. Then using time shifting property show the $x(t+T)+x(t-T) \Leftrightarrow 2X(\omega)\cos\omega T$

        Chapter 2, Example 2.14   **[Page No - 2.67]**   (6)

    (ii)   Find the inverse Laplace transform of $X(s) = \dfrac{8s+10}{(s+1)(s+2)}$ (10)

        Chapter 2, Example 2.43 (a)   **[Page No - 2.124]**

13. (a) (i)   Solve the differential equation $(D^2+3D+2)\,y(t) = D\,x(t)$ using the input $x(t) = 10\,e^{-2t}$ and with initial condition $y(0') = 2$ and $y(0') = 3$.   (10)

        Chapter 3, Example 3.14   **[Page No - 3.48]**

    (ii)   Draw the block diagram representation for $H(s) = \dfrac{4s+28}{s^2+6s+5}$ (6)

        Chapter 3, Example 3.32   **[Page No - 3.84]**

Or

(b) (i)   For a LTI system with $H(s) = \dfrac{s+5}{s^2+4s+3}$ find the differential equation. Find the system ouput $y(t)$ to the input $x(t) = e^{-2t}\,u(t)$.   (10)

        Chapter 3, Example 3.23   **[Page No - 3.63]**

    (ii)   Using graphical method convolve $x(t) = e^{-2t}\,u(t)$ with $h(t) = u(t+2)$.   (6)

        Chapter 3, Example 3.11(b)   **[Page No - 3.35]**

14 (a) (i)   A continous time sinusoid $\cos(2\pi ft+\theta)$ is sampled at a rate $f_0 = 1000$ Hz. Determine the resulting signal samples, if the input signal frequency f is 400 Hz, 600 Hz and 1000 Hz respectively.   (8)

        Chapter 4, Example 4.4   **[Page No - 4.5]**

    (ii)   Prove the following DTFT properties

        (1) $n\,x(n) \Leftrightarrow j\dfrac{dX(j\Omega)}{d\Omega}$ (2) $x(n)\,e^{j\Omega_0 n} \Leftrightarrow X(\Omega-\Omega_0)$ (8)

        Chapter 4, Section 4.2.3   **[Page No -4.11]**

Or

(b) (i)   Find the DTFT of $x(n) = \left(\dfrac{1}{2}\right)^{n-1} u(n-1)$ (5)

        Chapter 4, Example4.10   **[Page No - 4.24]**

    (ii)   Using suitable z-transform properties. Find $X(z)$ of $x(n) = (n-2)\left(\dfrac{1}{2}\right)^{n-2} u(n-2)$.   (6)

        Chapter 4, Example 4.26   **[Page No - 4.62]**

    (iii)   Find the z - transform of $x(n) = a^{|n|}$ $0 < a < 1$.   (5)

        Chapter 4, Example 4.21(e)   **[Page No - 4.45]**

15 (a) (i)   Determine the impulse response and step response

        $y(n)+y(n-1)-2y(n-2) = x(n-1)+2x(n-2)$ (10)

        Chapter 5, Example 5.41   **[Page No - 5.90]**

    (ii)   Find the convolution sum between $x(n) = \{1,4,3,2\}$ and $h(n) = \{1,3,2,1\}$ (6)

        Chapter 5, Example 5.4   **[Page No - 5.19]**

Or

(b) (i)   A causal system has $x(n) = \delta(n)+\dfrac{1}{4}\delta(n-1)-\dfrac{1}{8}\delta(n-2)$ and $y(n) = \delta(n)-\dfrac{3}{4}\delta(n-1)$.

        Find the impulse response and output if $x(n) = \left(\dfrac{1}{2}\right)^n u(n)$.   (12)

        Chapter 5, Example 5.42   **[Page No - 5.91]**

    (ii)   Compare recursive and nonrecursive systems.   (4)

        Chapter 5, Section 1.9.7   **[Page No - 1.107]**

B.E./B.Tech. DEGREE EXAMINATION, NOVEMBER/DECEMBER-2014
Third Semester
Electronics and Communication Engineering
EC 6303-SIGNAL AND SYSTEMS
(Regulation 2013)

Time : 3 hours   Maximum : 100 marks

Answer all questions
PART A - (10 × 2 = 20 Marks)

1.   State two properties of unit impulse function.
         Chapter 1, Section1.2.2   *[Page No - 1.11]*

2.   Draw the following signals:
      (a) u(t) – u(t – 10)
         Chapter 1, SA- Q1.16   *[Page No - 1.124]*
      (b) $(1/2)^n$ u(n – 1).
         Chapter 1, SA - Q1.18   *[Page No - 1.125]*

3.   State the condition for the convergence of Fourier series representation of continous time periodic signals.
         Chapter 2, Section 2.1.2   *[Page No - 2.2]*

4.   Find the ROC of the Laplace transform of x(t) = u(t).
         Chapter 2, Example 2.25(a) *[Page No - 2.86]*

5.   Draw the block diagram of the LTI system described by $\dfrac{dy(t)}{dt} + y(t) = 0.1x(t)$
         Chapter 3, SA - Q3.13   *[Page No - 3.114]*

6.   Find the y(n) = x(n –1) ∗ δ(n + 2).
         Chapter 5, SA - Q5.9 *[Page No - 5.140]*

7.   Find the DTFT of x(n) = δ(n) + δ(n – 1).
         Chapter 4, SA - Q 4.4   *[Page No - 4.92]*

8.   State and prove the time folding property of z- transform.
         Chapter 4, Section 4.6   *[Page No - 4.52]*

9.   Give the impulse response of a linear time invariant as h(n) = sin πn, check whether the system is stable or not.
         Chapter 5, SA - Q 5.17   *[Page No. - 5.143]*

10.   In terms of ROC, state the condition for an LTI discrete time system to be causal and stable.
         Chapter 5, Section 5.7.5   *[Page No -5.62]*

PART- B -- (5 ×16 = 80 marks)

11 (a)   Check whether the following signals are periodic/aperiodic signals.
          (i) x(t) = cos 2t + sin t/5.
             Chapter 1, Example 1.8 (d)   *[Page No - 1.47]*   (8)
          (ii) x(n) = 3 + cos π/2n + cos 2n.
             Chapter 1, Example 1.12 (f)   *[Page No - 1.59]*   (8)
                                            Or
     (b)   Check whether the following system is linear, causal, time invariant and stable.   (16)
          (i) y(n) = x(n) – x[n –1]
             Chapter 1, Example 1.39   *[Page No - 1.108]*

(ii) $y(t) = \dfrac{d}{dt} x(t)$.

Chapter 1, Example 1.26   *[Page No - 1.86]*

12 (a) Find the Fourier series coefficients of the following signal and plot the spectrum of the signal.   (16)

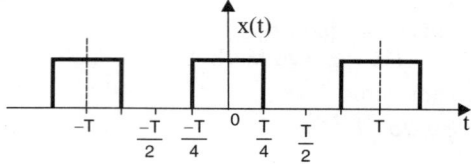

Chapter 2, Spectrum Section 2.4 & Example 2.1   *[Page No - 2.16 & 2.15]*

Or

(b) Find the spectrum of $x(t) = e^{-2|t|}$. Plot the spectrum of the signal.   (16)

Chapter 2, Example 2.24   *[Page No - 2.16]*

13 (a) Find the overall impulse response of the following system.   (16)

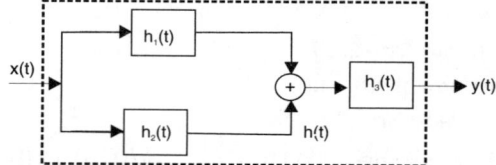

Here $h_1(t) = e^{-2t}u(t)$, $h_2(t) = \delta(t) - \delta(t-1)$ and $h_3(t) = \delta(t)$ and also find the output of the system for the input $x(t) = e^{-t}u(t)$ using convolution integral.

Chapter 3, Example 3.5   *[Page No. - 3.20]*

(b) An LTI system is represented by $\dfrac{d^2}{dt^2} y(t) + 4 \dfrac{d}{dt} y(t) + 4y(t) = x(t)$ with initial conditions

$y(\bar{0}) = 0$; $y(\bar{0}) = 1$, Find the output of the system, when the input is $x|t| = e^{-t} u(t)$ (16)

Chapter 3, Example 3.14   *[Page No - 3.48]*

14 (a) State and Prove sampling theorem for a band limited signal.   (16)

Chapter 4, Section 4.1   *[Page No - 4.1]*

Or

(b) Find inverse z - transform of $X(z) = \dfrac{z^{-1}}{1 - 0.25 z^{-1} - 0.375 z^{-2}}$ for   (16)

(i) ROC $|z| > 0.75$ (ii) ROC $|z| < 0.5$

Chapter 4, Example 4.36   *[Page No - 4.85]*

15 (a) Compute $y(n) = x(n) * h(n)$, where $x(n) = (1/2)^{n-2} u(n-2)$ and $h(n) = u(n-2)$ (16)

Chapter 5, Example 5.10   *[Page No - 5.33]*

Or

(b) LTI discrete time system $y(n) = 3/2 \, y(n-1) - 1/2 \, y(n-2) + x(n) + x(n-1)$ is given an input $x(n) = u(n)$ (16)

(i) Find the transfer function of the system.

(ii) Find the impulse response of the system.

Chapter 5, Example 5.27   *[Page No - 5.73]*

B.E./B.Tech. DEGREE EXAMINATION, MAY/JUNE-2014
Third Semester
Electronics and Communication Engineering
EC 2204/EC 35/EC 1202 A080290015/10144 EC 305 -SIGNAL AND SYSTEMS
(Regulation 2008/2010)

Time : 3 hours   Maximum : 100 marks

Answer all questions.

PART A - (10 × 2 = 20 Marks)

1.   Sketch the following signals
   (a)   x(t) = 2t for all t
         Chapter 1, SA - Q 1/15   *[Page No - 1.124]*
   (b)   x(n) = 2n – 3 for all n
         Chapter 1, SA - Q1.17   *[Page No - 1.125]*

2.   Given x[n] = [1, – 4,3,1,5,2]. Represents x[n] in terms of weighted shifted impulse functions
         Chapter 1, SA -Q1.26   *[Page No -1.128]*

3.   State the conditions for convergence of fourier series.
         Chapter 2, Section 2.1.2   *[Page No - 2.2]*

4.   State any two properties of ROC of Laplace transform X(s) of a signal x(t).
         Chapter 2, Section 2.8.3   *[Page No - 2.85]*

5.   State the necessary and sufficient condition for an LTI continous time system to be causal.
         Chapter 3, Section 1.7.4   *[Page No - 1.76]*

6.   Find the differential equation relating the input and output a CT system represented by
   $H(j\Omega) = \dfrac{4}{(j\Omega)^2 + 8j\Omega + 4}$
         Chapter 3, SA - Q 3.18   *[Page No - 3.116]*

7.   What is an anti-aliasing filter?
         Chapter 4, Section 4.32   *[Page No - 4.20]*

8.   State the multiplication property of DTFT.
         Chapter 4, Section 4.6   *[Page No - 4.55]*

9.   Find the overall impulse response h(n) = when two systems $h_1(n)$ = u(n) and
   $h_2(n)$ = δ(n)+2 δ(n –1) are series.
         Chapter 5, SA - Q 5.11   *[Page No - 5.143]*

10. Using z-transform, check whether the following system is stable $H(z) = \dfrac{z}{z - \frac{1}{2}} + \dfrac{2z}{z-3}$ for ROC $\frac{1}{2} < |z| < 3$
         Chapter 5, SA - Q 5.18   *[Page No - 5.143]*

PART - B -- (5 ×16 = 80 marks)

11. (a) (i)   A continous time signal is defined as, $x(t) = \frac{1}{6}(t+2)$, $-2 \le t \le 4$;
         Sketch the waveforms: = 0   ; otherwise
         1) x(t)   2) x(t+1)   3) x(2t)   4) x(t/2).   (8)
         Chapter 1, Example 1.3   *[Page No - 1.125]*

   (ii) Determine whether the discrete time sequence, $x[n] = \sin\left(\frac{3\pi}{7}n + \frac{\pi}{4}\right) + \cos\frac{\pi}{3}n$ is periodic or not
         Chapter 1, Example 1.12(g)   *[Page No - 1.60]*   (8)

Or

(b) Test the following system for linearity and stability.

     a) $y(t) = e^{x(t)}$ (8)

         Chapter 1, Example 1.18(e), 1.22(e)   *[Page No - 1.74 & 1.80]*

     b) $y(n) = x(n-1)$ (8)

         Chapter 1, Example 1.35 (d) , 1.41(b)   *[Page No - 1.104 & 1.112]*

12. (a) Find the Fourier series coefficients of the signal shown below: (16)

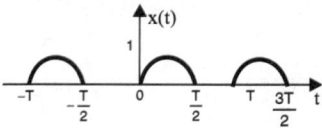

         Chapter 2, Example 2.4   *[Page No - 2.20]*

Or

(b) Find the inverse lapalce transform of $X(s) = \dfrac{1}{(s+5)(s-3)}$ if the ROC is,   (16)

     i) $-5 < Re\{s\} < 3$ ii) $Re\{s\} > 3$

         Chapter 2, Example 2.45   *[Page No - 2.127]*

13. (a) Using convolution integral determine the response of a CTLTI system y(t) given input
$x(t) = e^{-\alpha t} u(t)$ and impulse response $h(t) = e^{-\beta t}u(t)$, $|\alpha| < 1$, $|\beta| < 1$.   (16)

         Chapter 3, Example 3.8   *[Page No - 3.25]*

Or

(b) Find the frequency response of the system shown below: (16)

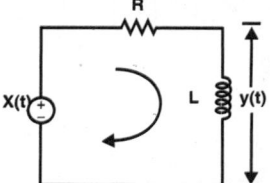

         Chapter 3, Example 3.39   *[Page No - 3.107]*

14 (a) Using convolution property of DTFT, find the inverse DTFT of $X(e^{j\omega}) = \dfrac{1}{(1-\alpha e^{-j\omega})^2}$, $|\alpha| < 1$

         Chapter 5, Example 5.7   *[Page No - 5.27]*   (16)

Or

(b) Find the inverse z- transform of $X(z) = \dfrac{z^2}{(z-0.5)(z-1)^2}$, $|z| > 1$.   (16)

         Chapter 4, Example 4.33   *[Page No - 4.83]*

15 (a) Find the convolution of sum of $x[n] = r[n]$ and $h[n] = u[n]$.   (16)

         Chapter 5, Example 5.11   *[Page No - 5.34]*

Or

(b) A causal LTI system is described by $y[n] - \dfrac{5}{6} y[n-1] + \dfrac{1}{6} y[n-2] = x[n]$ where x[n] is the
input to the system h[n] is the impulse response of the system. Find   (16)

     (i) System function H(z)
     (ii) Impulse response h(n)

         Chapter 5, Example 5.23   *[Page No - 5.69]*

B.E./B.Tech. DEGREE EXAMINATION, NOVEMBER/DECMBER-2013
Third Semester
Electronics and Communication Engineering
EC 2204/EC 35/EC 1202 A080290015/10144 EC 3056\ -SIGNAL AND SYSTEMS
(Regulation 2008/2010)

Time : 3hours   Maximum : 100 marks

Answer all questions.

PART A - (10 × 2 = 20 Marks)

1. Give the mathematical & graphical representation of continous time & discrete unit impuse function.
   Chapter 1, Section 1.2.1 & 1.3.4 *[Page No - 1.7 & 1.31]*

2. What are the conditions for a system to be LTI system?
   Chapter 1, Section 1.6 *[Page No - 1.67]*

3. State Dirichlet's conditions.
   Chapter 2, Section 2.1.2 *[Page No - 2.2]*

4. Give the equation for trignometric Fourier series.
   Chapter 2, Section 2.1.1 *[Page No - 2.1]*

5. What are the three elementary operations in block diagram representation of continous time system?
   Chapter 3, Section 3.2, Table- 3.1 *[Page No - 3.4]*

6. Check whether the causal system with transfer function $H(s) = \dfrac{1}{s-2}$ is stable.
   Chapter 3, SA- Q 3.10 *[Page No - 3.113]*

7. What is aliasing?
   Chapter 4, Section 4.1.1 *[Page No - 4.2]*

8. Definre unilateral and bilateral z transform.
   Chapter 4, Section 4.4.2 *[Page No - 4.32]*

9. Define convolution sum with its equation.
   Chapter 5, Section 5.4 *[Page No - 5.12]*

10. Check whether the system with system function $H(z) = \dfrac{1}{1-\frac{1}{2}z^{-1}} + \dfrac{1}{1-2z^{-1}}$ with ROC |z| < 1/2 is causal or stable.
    Chapter 5, SA - Q 5.16 *[Page No - 5.142]*

PART - B -- (5 ×16 = 80 marks)

11 (a) (i) Determine whether the signal x(t) = sin 20πt + sin 5πt is periodic and if it is periodic, find the fundamental period
   Chapter 1, Example 1.18(c) *[Page No - 1.46]* (5)

(ii) Define energy and power signals. Find whether the signal $x(n) = \left(\dfrac{1}{2}\right)^n u(n)$ is energy or power signal and calculate their energy or power.
   Chapter 1, Example 1.15(a) *[Page No -1.65]* (5)

(iii) Discuss various forms of real and complex exponential signals with graphical representation
   Chapter 1, Section 1.2.1 (7) *[Page No - 1.9]* (6)

Or

(b) Determine whether the discrete time system y(n) = x(n) cos (ωn) is (16)
   (i) Memoryless (ii) Stable (iii) Causal (iv) Time invariant
   Chapter 1, Example 1.40 *[Page No - 1.109]*

12 (a) (i) Find the exponential Fourier series of the waveform. (10)
   Chapter 2, Example 2.5 *[Page No - 2.22]*

   (ii) Find the Fourier transform of the signal x(t) = e^{-altl} (6)
   Chapter 2, Section 2.6.3 *[Page No - 2.54]*

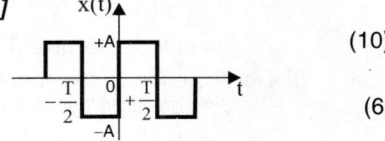

Or

(b) (i) Find the Laplace transform of the signal $f(t) = e^{-at} \sin \omega t$ (8)

         Chapter 2, Example 2.26(d)   *[Page No - 2.91]*

    (ii) Find the inverse Fourier transform of the rectangular spectrum given by $X(j\omega) = \begin{cases} 1, & -<\omega<W \\ 0, & |\omega|>W \end{cases}$

         Chapter 2, Example 2.17   *[Page No - 2.69]*

                                                             (8)

13 (a) (i) Define convolution integral and derive its equation.

         Chapter 3, Section 3.4.3   *[Page No- 3.16]*   (8)

    (ii) A Stable LTI system is charaterized by the differential equation. Find the frequency   (8)

         response and impulse response using Fourier transform $\dfrac{d^2 y(t)}{dt^2} + 4\dfrac{dy(t)}{dt} + 3y(t) = \dfrac{dx(t)}{dt} + 2x(t)$

         Chapter 3, Example 3.36   *[Page No- 3.104]*

Or

(b) (i) Draw direct form, cascade form and parellel form of a system with system function   (8)

         $H(s) = \dfrac{1}{(s+1)(s+2)}$

         Chapter 3, Example 3.35   *[Page No- 3.95]*

    (ii) Determine the state variable description   (8)

         corresponding to the block diagram given below

         Chapter 3, Section   *[Not in Regulation 2013 Syllabus]*

14 (a) (i) Determine the discrete time Fourier transform $x(n) = a^{|n|}, |a| < 1$.   (8)

         Chapter 4, Example 4.8   *[Page No - 4.22]*

    (ii) Find the z-transform and ROC of the sequence $x(n) = r^n \cos(n\theta) u(n)$.   (8)

         Chapter 4, Example 4.25   *[Page No - 4.61]*

Or

(b) (i) State and prove the following properties of z-transform   (8)

         (1) Linearity   (2) Time shifting   (3) Differentiation   (4) Correlation

         Chapter 4, Section 4.6   *[Page No - 4.48, 4.51& 4.53*   *]*

    (ii) Find the inverse z- transform of the function $X(z) = \dfrac{1+z^{-1}}{\left(1 - \dfrac{2}{3}z^{-1}\right)}$ ROC $|z| > \dfrac{2}{3}$.   (8)

         Chapter 4, Example 4.34   *[Page No - 4.83]*

15 (a) (i) Compute convolution sum of the following sequence $x(n) = \begin{cases} 1, 0 \le n \le 4 \\ 0, \text{Otherwise} \end{cases}$ & $h(n) = \begin{cases} \alpha^n, 0 \le n \le 6 \\ 0, \text{Otherwise} \end{cases}$

         Chapter 5, Example 5.5   *[Page No - 5.23]*   (10)

    (ii) Draw direct form I and direct form II implementation of the system described difference

         equation. $y(n) + \dfrac{1}{4}y(n-1) + \dfrac{1}{8}y(n-2) = x(n) + x(n-1)$.   (6)

         Chapter 5, Example 5.46   *[Page No- 5.101]*

Or

(b) (i) Determine the transfer function and the impulse response for the causal LTI system   (8)

         described by the difference equation using z transform.

         $y(n) - \dfrac{1}{4}y(n-1) - \dfrac{3}{8}y(n-2) = -x(n) + 2x(n-1)$.

         Chapter 5, Example 5.24   *[Page No - 5.70]*

    (ii) Develope the state variable description for   (8)

         the discrete time system given below.

         Chapter 5, Section   *[Not in Regulation 2013 Syllabus]*

B.E./B.Tech. DEGREE EXAMINATION, MAY/JUNE-2013
Third Semester
Electronics and Communication Engineering
EC 2204/EC 35/EC 1202 A/10144 EC 305/080290015 -SIGNAL AND SYSTEMS
(Regulation 2008/2010)

Time : 3 hours   Maximum : 100 marks

Answer all questions.
PART A - (10 × 2 = 20 Marks

1.   Check whether the discrtete time signal sin 3n is periodic.
     Chapter 1, SA - Q 1.5   *[Page No - 1.121]*

2.   Define a random signal.
     Chapter 1, Section1.4.1   *[Page No - 1.42]*

3.   State the time scaling property of Laplace transform.
     Chapter 2, Section 2.9   *[Page No - 2.105]*

4.   State the fourier transform property of a DC signal of amplitude 1?
     Chapter 2, Section 2.6.4   *[Page No - 2.55]*

5.   Define the convolution integral.
     Chapter 3, Section 3.4   *[Page No -3.14]*

6.   What is the condition for a LTI system to be stable?
     Chapter 3, Section 1.7.5   *[Page No - 1.79]*

7.   What is the z-transform of $\delta(n + k)$?
     Chapter 4, SA Q 4.13   *[Page No - 4.95]*

8.   What is aliasing?
     Chapter 4, Section 4.1.1   *[Page No - 4.2]*

9.   Is the discrete time system described by the difference equation $y(n) = x(-n)$ causal.
     Chapter 1, Example 1.34(d)   *[Page No - 1.102]*

10. If $X(\omega)$ is the DTFT of $x(n)$, what is the DTFT of $x^{*}(-n)$?
     Chapter 4, SA - Q 4.11   *[Page No - 4.94]*
PART-B -- (5 ×16 = 80 marks)

11 (a) (i)   Define an energy and power signal.   (4)
             Chapter 1, Section 1.4.4   *[Page No - 1.52]*

   (ii) Determine the whether the following signal are energy or power and calculate their
         energy or power.

         (1) $x(n) = \left(\frac{1}{2}\right)^n u(n)$ (2) $x(t) = rec\left(\frac{t}{T_0}\right)$ (3) $x(t) = \cos^2(\omega_0 t)$ (12)
         Chapter 1, Example 1.11 (e f), 1.15 (a) *[Page No - 1.55, 1.65]*
                                          Or
   (b) (i)   Define unit step, ramp pulse, impulse and exponential signals. Obtain the relationship
             between the unit step function and unit ramp function.   (10)
             Chapter 1, Section 1.2.1   *[Page No - 1.7, 1.9 & 1.22]*

   (ii) Find the fundamental period T of the signal $x(n) = \cos(n\pi/2) - \sin(n\pi/8) + 3\cos(n\pi/4) + \pi/2$
         Chapter 1, Example 1.13   *[Page No - 1.60]*

12 (a) (i)   Compute the Laplace transform of $x(t) = e^{-b|t|}$ for the cases of b < 0 and b > 0.   (10)
             Chapter 2, Example 2.28   *[Page No- 2.92]*

(ii) State and prove Parseval's theorem of Fourier transform. (6)

Chapter 2, Section 2.5.7 **[Page No - 2.49]**

Or

(b) (i) Determine the Fourier series representation of the half wave rectifier output shown in figure below. x(t) (8)

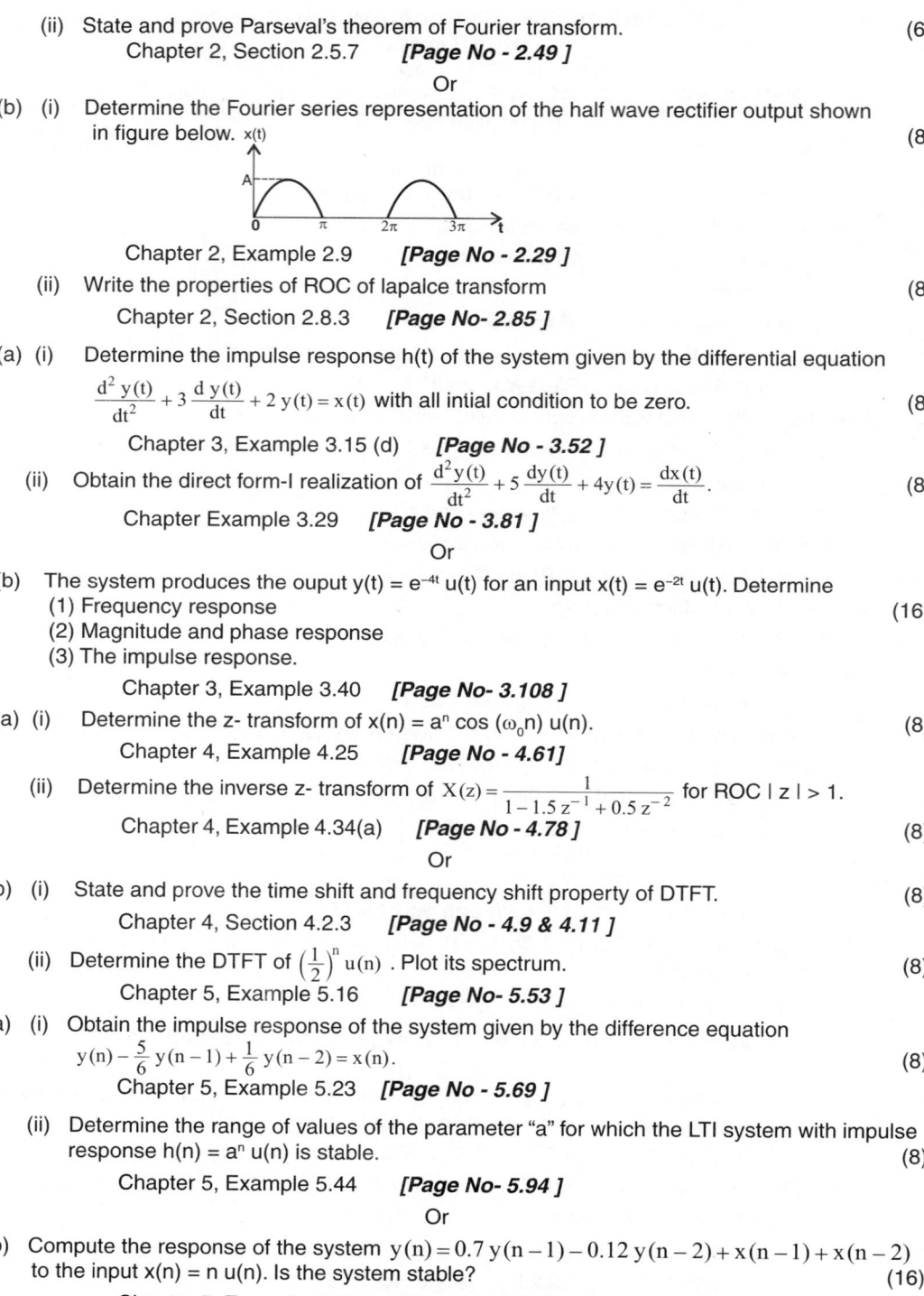

Chapter 2, Example 2.9 **[Page No - 2.29]**

(ii) Write the properties of ROC of lapalce transform (8)

Chapter 2, Section 2.8.3 **[Page No- 2.85]**

13 (a) (i) Determine the impulse response h(t) of the system given by the differential equation

$\dfrac{d^2 y(t)}{dt^2} + 3 \dfrac{d\,y(t)}{dt} + 2\,y(t) = x(t)$ with all intial condition to be zero. (8)

Chapter 3, Example 3.15 (d) **[Page No - 3.52]**

(ii) Obtain the direct form-I realization of $\dfrac{d^2 y(t)}{dt^2} + 5 \dfrac{dy(t)}{dt} + 4y(t) = \dfrac{dx(t)}{dt}$. (8)

Chapter Example 3.29 **[Page No - 3.81]**

Or

(b) The system produces the ouput y(t) = e^{-4t} u(t) for an input x(t) = e^{-2t} u(t). Determine
(1) Frequency response (16)
(2) Magnitude and phase response
(3) The impulse response.

Chapter 3, Example 3.40 **[Page No- 3.108]**

14 (a) (i) Determine the z- transform of x(n) = a^n cos (ω_0n) u(n). (8)

Chapter 4, Example 4.25 **[Page No - 4.61]**

(ii) Determine the inverse z- transform of $X(z) = \dfrac{1}{1 - 1.5\,z^{-1} + 0.5\,z^{-2}}$ for ROC I z I > 1.

Chapter 4, Example 4.34(a) **[Page No - 4.78]** (8)

Or

(b) (i) State and prove the time shift and frequency shift property of DTFT. (8)

Chapter 4, Section 4.2.3 **[Page No - 4.9 & 4.11]**

(ii) Determine the DTFT of $\left(\frac{1}{2}\right)^n$ u(n) . Plot its spectrum. (8)

Chapter 5, Example 5.16 **[Page No- 5.53]**

15 (a) (i) Obtain the impulse response of the system given by the difference equation

$y(n) - \dfrac{5}{6}\,y(n-1) + \dfrac{1}{6}\,y(n-2) = x(n)$. (8)

Chapter 5, Example 5.23 **[Page No - 5.69]**

(ii) Determine the range of values of the parameter "a" for which the LTI system with impulse response h(n) = a^n u(n) is stable. (8)

Chapter 5, Example 5.44 **[Page No- 5.94]**

Or

(b) Compute the response of the system $y(n) = 0.7\,y(n-1) - 0.12\,y(n-2) + x(n-1) + x(n-2)$ to the input x(n) = n u(n). Is the system stable? (16)

Chapter 5, Example 5.29 **[Page No 5.75]**

B.E./B.Tech. DEGREE EXAMINATION, NOVEMBER/DECEMBER-2012
Third Semester
Electronics and Communication Engineering
EC 2204/EC 35/EC 1202 A/10144 EC 305/080290015 -SIGNAL AND SYSTEMS
(Regulation 2008)

Time : 3 hours   Maximum : 100 marks

Answer all questions.

PART A - (10 × 2 = 20 Marks

1. . Determine whether the following signal is energy or power signal and calculate the energy or power
   $x(t) = e^{-2t} u(t)$
   Chapter 1, Example 1.11 (a)   *[Page No - 1.53]*

2. Check whether the following system is static (or) dynamic & also causal (or) non- causal : y(n)= x(2n)
   Chapter 1, SA - Q 1.10, Example 1.34 (c)   *[Page No - 1.123 & 1.102]*

3. Give synthesis and analysis equations of continous time Fourier transform
   Chapter 2, Section 2.5.1   *[Page No - 2.41]*

4. Define the region of convergence of the Laplace transform.
   Chapter 2, Section 2.8   *[Page No. - 2.80]*

5. List and draw the basic element for the block diagram representation of the continous time system
   Chapter 3, Section 3.2, Table 3.1   *[Page No - 3.4]*

6. Check the causality of the system with impulse response $h(t) = e^{-t} u(t)$.
   Chapter 3, SA - Q 3.11   *[Page No - 3.114]*

7. Define DTFT and inverse DTFT.
   Chapter 3, Section 4.2.1& 4.2.2   *[Page No - 4.6 & 4.7]*

8. State the convolutin property of the z-transform.
   Chapter 4, Section 4.6   *[Page No - 4.52]*

9. Convolve the following two sequences $x(n) = \{1,1,1,1\}$ and $h(n) = \{3,2\}$
   Chapter 5, SA - Q 5.4   *[Page No - 5.139]*

10. A causal LTI system has impulse response h(n) for which the z-transform is
    $H(z) = \dfrac{1+z^{-1}}{(1-0.5\,z^{-1})(1+0.25\,z^{-1})}$. Is the system stable? Explain.
    Chapter 5, SA - Q 5.19   *[Page No - 5.143]*

PART-B -- (5 ×16 = 80 marks)

11 (a) (i) How are the signals classified? Explain.   (8)
              Chapter 1, Section 1.4 & 1.5   *[Page No - 1.41 & 1.56]*

   (ii) Determine whether the following signal is periodic. If peridic determine the fundamental
        period. $x(t) = 3\cos t + 4\cos \dfrac{t}{2}$ (4)
              Chapter 1, Example 1.8(f)   *[Page No - 1.48]*

   (iii) Give the equation & draw the waveform of discrete time real & complex exponential signal
              Chapter 1, Section 1.3.4   *[Page No - 1.32 & 1.34]*   (4)

Or

(b) (i) Determine whether the following system is linear, time invariant, stable and invertible
        (i) $y(n) = x^2(n)$ (ii) $y(n) = x(-n)$ (10)
              Chapter 1, Example 1.44   *[Page No - 1.116]*

   (ii) Define LTI system, the properties of LTI system and explain.   (6)
              Chapter 1, Section 1.7   *[Page No - 1.68]*

12 (a) (i) State Dirchelet's conditions. Also its importance.   (4)
              Chapter 2, Section 2.1.2   *[Page No - 2.2]*

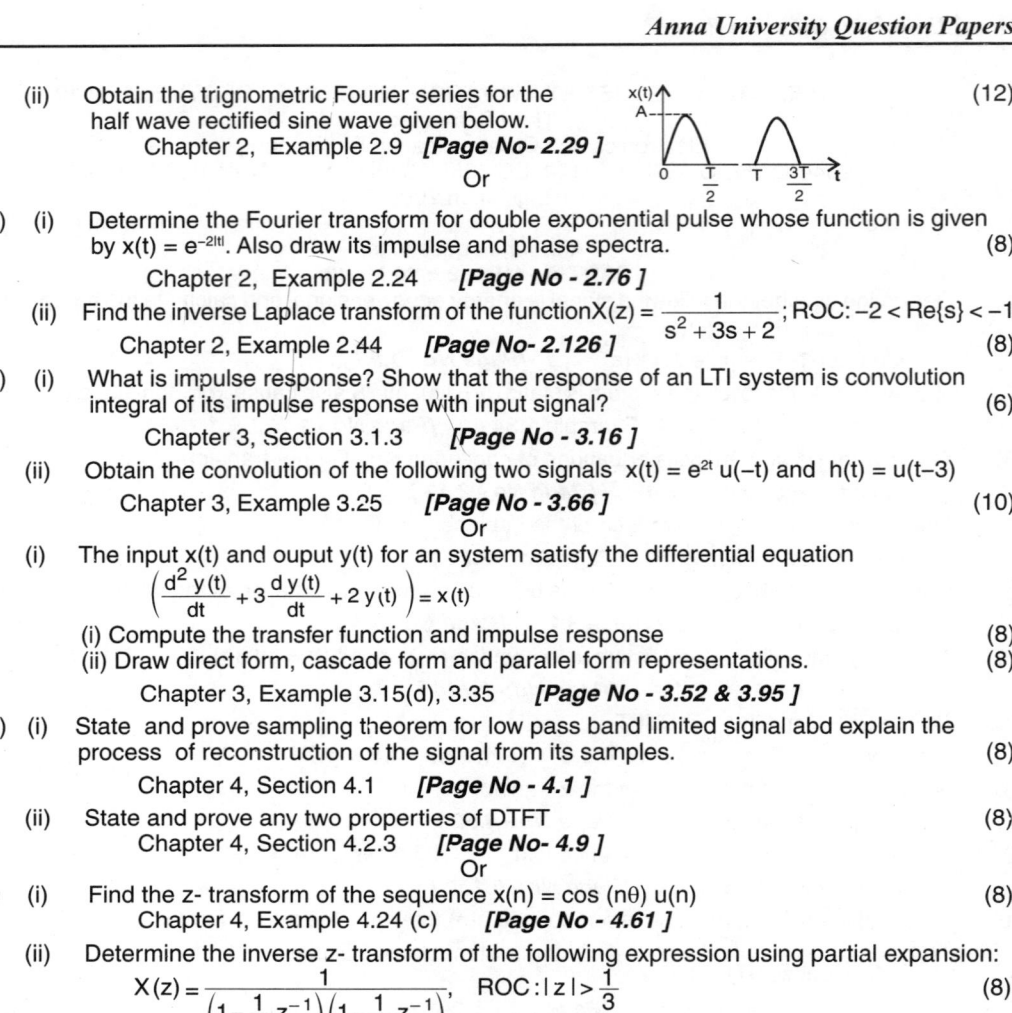

(ii)   Obtain the trignometric Fourier series for the   (12)
        half wave rectified sine wave given below.
                 Chapter 2, Example 2.9 *[Page No- 2.29]*
                               Or

(b) (i)   Determine the Fourier transform for double exponential pulse whose function is given
        by $x(t) = e^{-2|t|}$. Also draw its impulse and phase spectra.   (8)
            Chapter 2, Example 2.24   *[Page No - 2.76]*

(ii) Find the inverse Laplace transform of the function $X(z) = \dfrac{1}{s^2 + 3s + 2}$; ROC: $-2 < \text{Re}\{s\} < -1$

            Chapter 2, Example 2.44   *[Page No- 2.126]*   (8)

13 (a) (i)   What is impulse response? Show that the response of an LTI system is convolution
          integral of its impulse response with input signal?   (6)
            Chapter 3, Section 3.1.3   *[Page No - 3.16]*

(ii) Obtain the convolution of the following two signals $x(t) = e^{2t}\, u(-t)$ and $h(t) = u(t-3)$
            Chapter 3, Example 3.25   *[Page No - 3.66]*   (10)
                              Or

(b) (i)   The input $x(t)$ and ouput $y(t)$ for an system satisfy the differential equation

$$\left(\frac{d^2 y(t)}{dt} + 3 \frac{d\,y(t)}{dt} + 2\,y(t) \right) = x(t)$$

        (i) Compute the transfer function and impulse response   (8)
        (ii) Draw direct form, cascade form and parallel form representations.   (8)
          Chapter 3, Example 3.15(d), 3.35   *[Page No - 3.52 & 3.95]*

14 (a) (i)   State and prove sampling theorem for low pass band limited signal abd explain the
          process of reconstruction of the signal from its samples.   (8)
            Chapter 4, Section 4.1   *[Page No - 4.1]*

(ii) State and prove any two properties of DTFT   (8)
             Chapter 4, Section 4.2.3   *[Page No- 4.9]*
                            Or

(b) (i)   Find the z- transform of the sequence $x(n) = \cos(n\theta)\,u(n)$ (8)
         Chapter 4, Example 4.24 (c)   *[Page No - 4.61]*

(ii) Determine the inverse z- transform of the following expression using partial expansion:

$$X(z) = \frac{1}{\left(1 - \frac{1}{3} z^{-1}\right)\left(1 - \frac{1}{6} z^{-1}\right)}, \quad \text{ROC}: |z| > \frac{1}{3} \qquad (8)$$

            Chapter 4, Example 4.35   *[Page No- 4.84]*

15 (a) (i)   Find the system function and the impulse response $h(n)$ for a system described by the
          following input - output relationship. $y(n) = \frac{1}{3}\, y(n-1) + 3\,x(n)$.   (6)
             Chapter 5, Example 5.22   *[Page No- 5.69]*

(ii) A linear time - invariant system is characterized by the system function.Specify the ROC

      of $H(z)$, $H(z) = \dfrac{3 - 4 z^{-1}}{1 - 3.5\,z^{-1} + 1.5\,z^{-2}}$ and determine $h(n)$ for the following conditions:

      (1) The system is stabe   (2) The system is causal   (3) The system is anti- causal
          Chapter 5, Example 5.45   *[Page No - 5.94]*   (10)
                            Or

(b) (i)   Derive the necessary and sufficient condition for BIBO stability of an LSI system.
             Chapter 1, Section 1.7.5   *[Page No - 1.79]*   (8)

(ii) Draw the direct form, cascade form and parallel form block diagrams of the following

      system function $H(z) = \dfrac{1}{\left(1 + \frac{1}{2} z^{-1}\right)\left(1 - \frac{1}{4} z^{-1}\right)}$ (8)

          Chapter 5, Example 5.57   *[Page No- 5.126]*

B.E./B.Tech. DEGREE EXAMINATION, MAY/JUNE-2012
Third Semester
Electronics and Communication Engineering
EC 2204/EC 35/EC 1202 A/10144 EC 305/080290015 -SIGNAL AND SYSTEMS
(Regulation 2008)

Time : 3 hours   Maximum : 100 marks
Answer all questions.
PART A - (10 × 2 = 20 Marks

1. Verify whether the system described by the equation is linear and time invariant $y(t) = x(t^2)$.
    Chapter 1, Example 1.18(b) & 1.16(c)   *[Page No - 1.73 & 1.74]*

2. Find the fundamental period of the given signal, $x[n] = \sin\left(\dfrac{6\pi}{7}n + 1\right)$.
    Chapter 1, Example 1.12   *[Page No - 1.57]*

3. Define Nyquist Rate.
    Chapter 2, Section 4.1.1   *[Page No - 4.3]*

4. Determine the fouier series coefficients for the signal cos ωt.
    Chapter 2, SA - 2.1   *[Page No- 2.130]*

5. Determine the Laplace transform of the signal $\delta(t - 5)$ and $u(t - 5)$.
    Chapter 3, Example 2.27(a) & 2.37(a)   *[Page No- 2.92 & 2.114]*

6. Determine the convolution of the signals $x[n] = \{2,-1,3,2\}$ and $h(n) = \{1, -1,1,1\}$.
    Chapter 3, SA - Q 5.4 *[Page No. 5.139]*

7. Prove the time shifting property of discrete time Fourier transform.
    Chapter 4, Section 4.2.3   *[Page No- 4.9]*

8. State the final value theorem.
    Chapter 4, Section 4.6   *[Page No- 4.54]*

9. List the advantages of the state variable representation of a system.
    [Not in Regulation 2013 Syllabus]

10. Find the system function for the given difference equatin $y(n) = 0.5\, y(n -1) + x(n)$.
    Chapter 5, SA - Q 5.8   *[Page No- 5.140]*

PART-B-- (5 ×16 = 80 marks)

11 (a) Determine whether the systems described by the following input - output equation are
        linear, dynamic, causal and time variant.   (16)

   (i) $y_1(t) = x(t - 3) + (3 - t)$ (ii) $y_2(t) = \dfrac{dx(t)}{dt}$

   (iii) $y_1[n] = n\, x[n] + b\, x^2[n]$ (iii) Even $x[n - 1]$

        Chapter 1, Example 1.26,1.27 (a) & 1.41(a,b)   *[Page No - 1.86 , 1.87 & 1.111   ]*
                                        Or
    (b) A Discrete time system is given as $y(n) = y^2(n - 1) + x(n)$. A bounded input of $x(n) = 2\delta(n)$
        is applied to the system. Assume that the system is initially relaxed. Check whether system
        is stable or unstable.   (16)

        Chapter 1, Example 1.36   *[Page No - 1.104]*

12 (a) (i) Prove the scaling and time shifting properties of Laplace transform.   (8)
            Chapter 2, Section 2.9   *[Page No - 2.103 & 2.105]*
       (ii) Determine the Laplace transform of $x(t) = e^{-at}\cos \omega t\, u(t)$.   (8)
            Chapter 2, Example 2.26(e)   *[Page No - 2.91]*

Or

(b) (i) State and prove the Fourier transform of the following signal in terms of X(jω);

$x(t - t_0)$, $x(t)e^{j\omega t}$.   (8)

Chapter 2, Section 2.57   *[Page No - 2.44 & 2.45]*

(ii) Find the complex exponential Fourier series coefficients of the signal

$x(t) = \sin 3\pi t + 2\cos 4\pi t$.   (8)

Chapter 2, SA - Q 2.22   *[Page No- 2.137]*

13 (a) Compute and plot the convolution y(t) of the given signals

(i) $x(t) = u(t - 3) - u(t - 5)$, $h(t) = e^{-3t}u(t)$.   (8)

Chapter 3, Example 3.21(e)   *[Page No - 3.61]*

(ii) $x(t) = u(t)$, $h(t) = e^{-t}u(t)$ (8)

Chapter 3, Example 3.21 (b)   *[Page No -3.59]*

Or

(b) The LTI system is charaterised by the impulse response function given   (16)

$H(s) = \dfrac{1}{(s+10)}$ ROC : Re > 10. Determine the output of a system when it is excited by

the input $x(t) = -2e^{-2t}u(-t) - 3e^{-3t}u(t)$.

Chapter 3, Example 3.22   *[Page No - 3.62]*

14 (a) (i) Determine the Z - transform and sketch the pole zero plot with the ROC for each of the following signals.

(i) $x[n] = (0.5)^n u[n] - (1/3)^n u[n]$ (8)

(ii) $x[n] = (1/2)^n u[n] = (1/3)^n u[n-1]$.   (8)

Chapter 4, Example 4.22(a,b)   *[Page No- 4.46 & 4.47]*

Or

(b) (i) Find the inverse Z- transform of the $\dfrac{1}{(z^2 - 1.2z + 0.2)}$ (8)

Chapter 4, Section 4.3(d)   *[Page No - 4.80]*

(ii) Express the Fourier transform of the following signals in terms of $X(e^{j\omega})$ (8)

(1) $X_1[n] = X[1-n]$.

(2) $X_2[n] = (n-1)^2 x[n]$.

Chapter 4, Example 4.11(a,b)   *[Page No - 4.24]*

15 (a) (i) Find the impulse response of the difference equation.   (8)

$y(n) - 2y(n-2) + y(n-1) + 3y(n-3) = x(n) + 2x(n-1)$.

Chapter 5, Example 5.20   *[Page No - 5.67]*

(ii) Find the state variable matrices A, B,C amd D for the input -output relation given by

$y(n) = 6y(n-1) + 4y(n-2) + x(n) + 10x(n-1) + 12x(n-2)$ (8)

[Not in Regulation 2013 Syllabus]

Or

(b) (i) Draw the direct form-I block diagram representation for the system function   (8)

$H(z) = \dfrac{1 + 2z^{-1} - 20z^{-2} - 20z^{-3} - 5z^{-4} + 6z^{-6}}{1 + 0.5z^{-1} - 0.25z^{-2}}$

Chapter 5, Example 5.48   *[Page No - 5.104]*

(ii) Find the input x(n) which produces output y(n) = {3,8,14,8,3}, when passed through the system having h(n) = {1,2,3}   (8)

Chapter 5, Example 5.37   *[Page No - 5.86]*

B.E./B.Tech. DEGREE EXAMINATION, NOVEMBER/DECEMBER-2011
Third Semester
Electronics and Communication Engineering
EC 2204/EC 147303 -SIGNAL AND SYSTEMS
(Regulation 2008)

Time : 3hours   Maximum : 100 marks

Answer all questions.

PART A - (10 × 2 = 20 Marks

1. State the properties of LTI system.

    Chapter 1, Section 1.7   *[Page No - 1.68]*

2. Draw the function $\pi(2t+3)$ when $x(t) = \begin{cases} 1, & -\frac{1}{2} < t < \frac{1}{2} \\ 0, & \text{elsewhere.} \end{cases}$

    Chapter 1, SA - Q1.13   *[Page No - 1.123*   *]*

3. Draw the single sided spectrum for $x(t) = 7 + 10\cos\left(40\,\pi t + \frac{\pi}{2}\right)$

    Chapter 2, SA - Q2.7   *[Page No 2.132]*

4. What are the Laplace transform of $\delta(t)$ and u(t)?

    Chapter 2, SA - Q 2.23   *[Page No - 2.138]*

5. Given the system impulse response h(t), state the conditions for stability and causality.

    Chapter 3, Section1.74 & 1.75   *[Page No - 1.76 & 1.79]*

6. Write the equation for the complete response of a C.T. system in terms of state transition matrix.

    [Not in Regulation 2013 Syllabus]

7. What is the main condition to be satisfied to avoid aliasing?

    Chapter 4, Section 4.17   *[Page No - 4.2]*

8. Find the z- transform of $x(n) = a^n\,u(n)$, |a| < 1.

    Chapter 4, SA - Q4.14   *[Page No - 4.95]*

9. Find the convolution of the two sequences of x(n) = {1,1,1,1} and h(n) = {2,2}.

    Chapter 5, SA - Q5.4   *[Page No - 5.139]*

10. What is the overall impulse response h(n) when two system with impulse responses $h_1(n)$ and $h_2(n)$ are connected in parallel and in series?

    Chapter 5, SA - Q5.12   *[Page No - 5.141]*

PART-B (5 ×16 = 80 marks)

11 (a)   Find out whether the following systems are: (16)

        (1) Linear or non- linear   (2) Causal or non-causal

        (3) Fixed or time- variant   (4) Dynamic or instantaneous.

        (i) $y(n) = x(n) + \dfrac{1}{x(n-1)}$ (ii) $\dfrac{d^3 y(t)}{dt^3} + \dfrac{4\,d^2 y(t)}{dt^2} + \dfrac{5\,d y(t)}{dt} + 2\,y^2(t) = x(t).$

            Chapter 1, Example 1.41 (c), 1.27(b)   *[Page No - 1.88 & 1.113]*

                                Or

    (b)   Explain all classification of DT signals with examples for each category.   (16)

           Chapter 1, Section 1.5   *[Page No - 1.56]*

12 (a)   Find the trignometric Fourier series of the given signal. From the result calculate the
          coefficients of exponential Fourier series   (16)

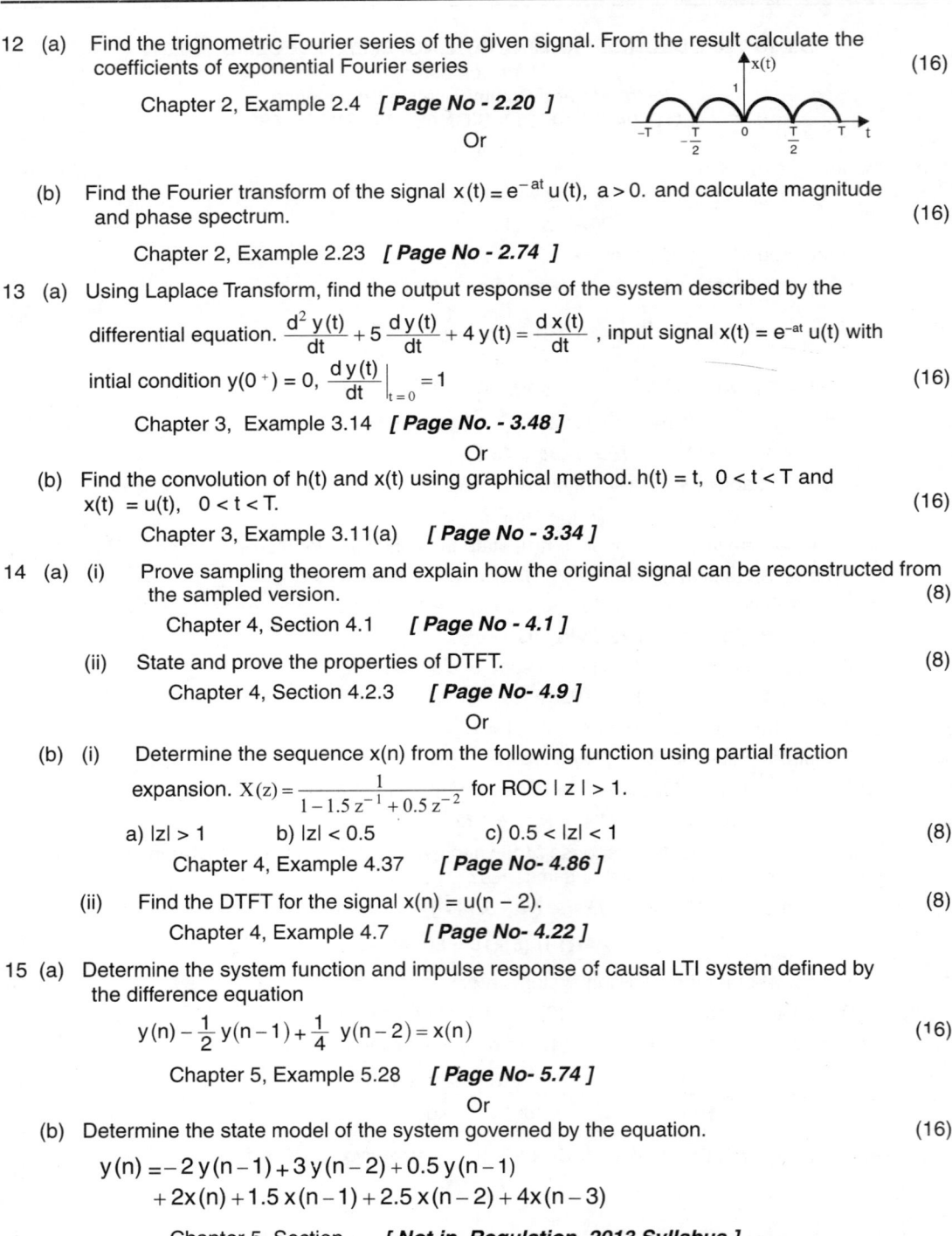

          Chapter 2, Example 2.4   *[Page No - 2.20]*

                                        Or

    (b)   Find the Fourier transform of the signal $x(t) = e^{-at} u(t)$, $a > 0$. and calculate magnitude
          and phase spectrum.   (16)

          Chapter 2, Example 2.23   *[Page No - 2.74]*

13 (a)   Using Laplace Transform, find the output response of the system described by the

          differential equation. $\dfrac{d^2 y(t)}{dt} + 5\dfrac{d y(t)}{dt} + 4 y(t) = \dfrac{d x(t)}{dt}$, input signal $x(t) = e^{-at} u(t)$ with

          intial condition $y(0^+) = 0$, $\dfrac{d y(t)}{dt}\Big|_{t=0} = 1$ (16)

          Chapter 3, Example 3.14   *[Page No. - 3.48]*

                                        Or

    (b)   Find the convolution of h(t) and x(t) using graphical method. h(t) = t, 0 < t < T and
          x(t) = u(t), 0 < t < T.   (16)

          Chapter 3, Example 3.11(a)   *[Page No - 3.34]*

14 (a) (i)   Prove sampling theorem and explain how the original signal can be reconstructed from
                the sampled version.   (8)

                Chapter 4, Section 4.1   *[Page No - 4.1]*

         (ii)   State and prove the properties of DTFT.   (8)

                Chapter 4, Section 4.2.3   *[Page No- 4.9]*

                                        Or

    (b) (i)   Determine the sequence x(n) from the following function using partial fraction

                expansion. $X(z) = \dfrac{1}{1 - 1.5 z^{-1} + 0.5 z^{-2}}$ for ROC | z | > 1.

                a) |z| > 1   b) |z| < 0.5   c) 0.5 < |z| < 1   (8)

                Chapter 4, Example 4.37   *[Page No- 4.86]*

         (ii)   Find the DTFT for the signal x(n) = u(n − 2).   (8)

                Chapter 4, Example 4.7   *[Page No- 4.22]*

15 (a)   Determine the system function and impulse response of causal LTI system defined by
          the difference equation

          $y(n) - \dfrac{1}{2} y(n-1) + \dfrac{1}{4} y(n-2) = x(n)$ (16)

          Chapter 5, Example 5.28   *[Page No- 5.74]*

                                        Or

    (b)   Determine the state model of the system governed by the equation.   (16)

          $y(n) = -2 y(n-1) + 3 y(n-2) + 0.5 y(n-1)$
          $\quad + 2x(n) + 1.5 x(n-1) + 2.5 x(n-2) + 4x(n-3)$

          Chapter 5, Section   *[Not in Regulation 2013 Syllabus]*

B.E./B.Tech. DEGREE EXAMINATION, MAY/JUNE-2011
Third Semester
Electronics and Communication Engineering
EC 2204-SIGNAL AND SYSTEMS
(Regulation 2008)

Time : 3 hours   Maximum : 100 marks

Answer all questions.

PART A - (10 × 2 = 20 Marks)

1. Define the step and impulse function in discrete signals.
   Chapter 1, Section 1.3.4   *[Page No - 1.31]*
2. Distingiush between deterministic and random signals.
   Chapter 1, Section 1.4.1   *[Page No - 1.42]*
3. Obtain the Fourier series coefficients for $x(n) = \sin \omega_0 n$.
   [Not in Regulation 2013 Syllabus]
4. State the initial and final value theorem of Laplace transforms.
   Chapter 2, Section 2.9   *[Page No - 2.106 & 2.107]*
5. What is meant by impulse response of any system ?
   Chapter 3, Section 3.1.1   *[Page No - 3.1]*
6. Define convolution is integral of continous time systems.
   Chapter 3, Section 3.4   *[Page No - 3.14]*
7. Define sampling theorem.
   Chapter 4, Section 4.1.1   *[Page No - 4.3]*
8. What is ROC in z - transforms?
   Chapter 4, Section 4.5   *[Page No - 4.35]*
9. Obtain the convolution of
   (a) $x(n) * \delta(n)$ (b) $\delta(n) * [h_1(n) + h_2(n)]$.
   Chapter 5, SA - Q5.6   *[Page No- 5.140]*
10. Write the difference equation for non recursive system.
    Chapter 1, Section 1.9.7   *[Page No- 1.107]*

PART-B (5 ×16 = 80 marks)

11. (a) (i) Check the following for linearity, time invariance, causality and stability:
    $y(n) = x(n) + n\,x(n + 1)$.   (8)
    Chapter 1, Example 1.43   *[Page No - 1.115]*

    (ii) Check whether the following are periodic:
    (1) $x(n) = \sin\left(\dfrac{6\pi}{7}n + 1\right)$ (2) $x(n) = e^{j\frac{3\pi}{5}\left(n + \frac{1}{2}\right)}$ (8)
    Chapter 1, Example 1.12 (a,h)   *[Page No - 1.57 & 1.60]*

    Or

    (b) (i) Describe the basic properties of system with examples.   (10)
    Chapter 1, Section 1.7 & 1.9   *[Page No - 1.68 & 1.90]*

    (ii) Sketch the following signals: where r(t) is a ramp signal.
    (1) $x(t) = r(t)$ (2) $x(t) = r(-t + 2)$ (3) $x(t) = -2\,r(t)$ (6)
    Chapter 1, Example 1.5   *[Page No - 1.27]*

12 (a) (i) Distinguish between Fourier series and Fourier transforms.   (4)
   Chapter 2, Section 2.5.6, Table 2.4   *[Page No- 2.43]*

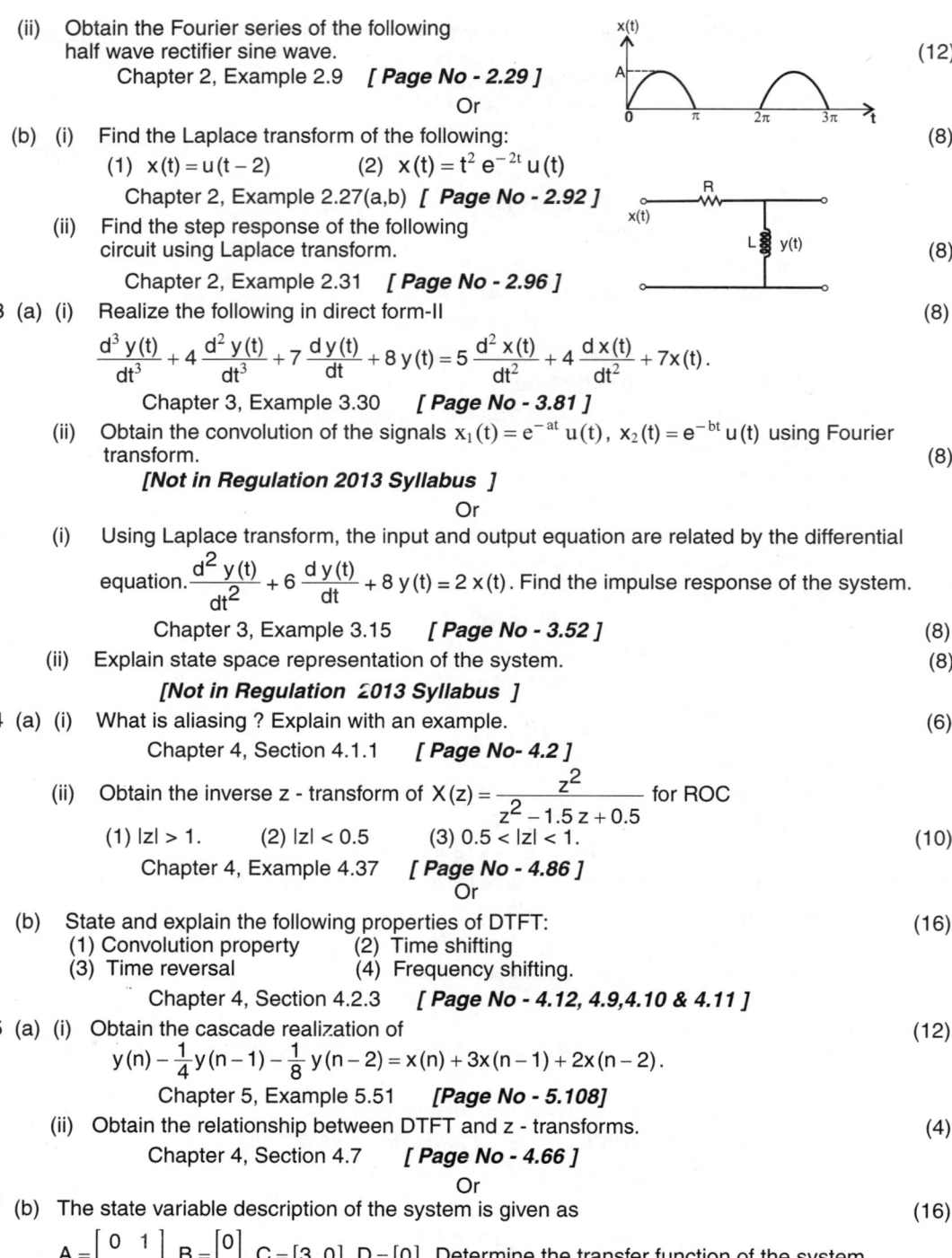

(ii) Obtain the Fourier series of the following half wave rectifier sine wave.   (12)

         Chapter 2, Example 2.9   *[Page No - 2.29]*

                          Or

(b) (i) Find the Laplace transform of the following: (8)

         (1) $x(t) = u(t-2)$ (2) $x(t) = t^2 e^{-2t} u(t)$

         Chapter 2, Example 2.27(a,b) *[Page No - 2.92]*

   (ii) Find the step response of the following circuit using Laplace transform.   (8)

         Chapter 2, Example 2.31   *[Page No - 2.96]*

13 (a) (i) Realize the following in direct form-II   (8)

$$\frac{d^3 y(t)}{dt^3} + 4\frac{d^2 y(t)}{dt^3} + 7\frac{d y(t)}{dt} + 8y(t) = 5\frac{d^2 x(t)}{dt^2} + 4\frac{d x(t)}{dt^2} + 7x(t).$$

         Chapter 3, Example 3.30   *[Page No - 3.81]*

   (ii) Obtain the convolution of the signals $x_1(t) = e^{-at} u(t)$, $x_2(t) = e^{-bt} u(t)$ using Fourier transform.   (8)

         [Not in Regulation 2013 Syllabus]

                          Or

   (i) Using Laplace transform, the input and output equation are related by the differential

equation. $\frac{d^2 y(t)}{dt^2} + 6\frac{d y(t)}{dt} + 8y(t) = 2x(t)$. Find the impulse response of the system.

         Chapter 3, Example 3.15   *[Page No - 3.52]*   (8)

   (ii) Explain state space representation of the system.   (8)

         [Not in Regulation 2013 Syllabus]

14 (a) (i) What is aliasing ? Explain with an example.   (6)

         Chapter 4, Section 4.1.1   *[Page No- 4.2]*

   (ii) Obtain the inverse z - transform of $X(z) = \dfrac{z^2}{z^2 - 1.5z + 0.5}$ for ROC

         (1) $|z| > 1$.   (2) $|z| < 0.5$ (3) $0.5 < |z| < 1$.   (10)

         Chapter 4, Example 4.37   *[Page No - 4.86]*

                          Or

(b) State and explain the following properties of DTFT: (16)

     (1) Convolution property   (2) Time shifting

     (3) Time reversal   (4) Frequency shifting.

         Chapter 4, Section 4.2.3   *[Page No - 4.12, 4.9,4.10 & 4.11]*

15 (a) (i) Obtain the cascade realization of   (12)

$$y(n) - \frac{1}{4}y(n-1) - \frac{1}{8}y(n-2) = x(n) + 3x(n-1) + 2x(n-2).$$

         Chapter 5, Example 5.51   *[Page No - 5.108]*

   (ii) Obtain the relationship between DTFT and z - transforms.   (4)

         Chapter 4, Section 4.7   *[Page No - 4.66]*

                          Or

(b) The state variable description of the system is given as   (16)

$A = \begin{bmatrix} 0 & 1 \\ -1 & 1 \end{bmatrix}$, $B = \begin{bmatrix} 0 \\ 2 \end{bmatrix}$, $C = [3 \ 0]$, $D = [0]$. Determine the transfer function of the system.

         [Not in Regulation 2013 Syllabus]

INDEX